World Geomorphological Landscapes

Series editor

Piotr Migoń, Wroclaw. Poland

More information about this series at http://www.springer.com/series/10852

Casey D. Allen
Editor

Landscapes and Landforms of the Lesser Antilles

Springer

Editor
Casey D. Allen
General Education
Western Governors University
Salt Lake City, UT
USA

ISSN 2213-2090 ISSN 2213-2104 (electronic)
World Geomorphological Landscapes
ISBN 978-3-319-55785-4 ISBN 978-3-319-55787-8 (eBook)
DOI 10.1007/978-3-319-55787-8

Library of Congress Control Number: 2017935564

Printed on acid-free paper

This Springer imprint is published by Springer Nature
The registered company is Springer International Publishing AG
The registered company address is: Gewerbestrasse 11, 6330 Cham, Switzerland

This Volume is dedicated to all peoples of the Lesser Antilles—past, present, and future. Your resilience and patriotism during even the most difficult times serves as an inspiration. May the Caribbean Sun continue to shine within you and your families.

Foreword

Lesser Antilles

The Lesser Antilles, which include the Leeward Islands, the Windward Islands, and the Leeward Antilles of the eastern Caribbean, were once described as "Edens of Delight," and they certainly comprise some of the most striking and intriguing landscapes on Earth. Used in 1925 as a model by William Morris Davis for his cycle of island development, they have never, until now, been the subject of a comprehensive geomorphological appreciation. They extend from the Virgin Islands in the north to Trinidad and Tobago in the south, and 1140 km from Aruba in the west to Barbados in the east. Consisting of some six hundred tropical islands, in total they only cover a relatively modest land area (less than half the size of Albania), but they exemplify major, horseshoe-shaped island arcs, with numerous volcanic features, including stratovolcanoes, granitic plutons, and also carbonate formations (including reefs) on which numerous karst features have developed. This well-structured volume provides an analysis of the tectonic history of the area, a discussion of its climatic conditions, and a survey of the human influence and transformation of its diverse landscapes. Above all, it discusses the main island clusters, drawing attention to their spectacular landforms, their land use histories, their heritage concerns, and the geomorphological hazards (including tsunamis, earthquakes, landslides, and hurricanes) with which their inhabitants have to contend. As befits an area of such splendid geomorphological phenomena, it is also beautifully illustrated with plates and maps and will become the first port of call for all those interested in how the landscapes of the islands have evolved. Moreover, it will stimulate further research. It is also a very worthy new volume in a series of books—The *World Geomorphological Landscapes*—that are doing much to draw attention to the magnificence and significance of our planet's major geomorphological regions.

Andrew Goudie
Emeritus Professor of Geography, University of Oxford, Oxford, UK
Past President of the International Association of Geomorphologists

Series Editor Preface

Landforms and landscapes vary enormously across the Earth, from high mountains to endless plains. At a smaller scale, Nature often surprises us creating shapes which look improbable. Many physical landscapes are so immensely beautiful that they received the highest possible recognition—they hold the status of World Heritage properties. Apart from often being immensely scenic, landscapes tell stories which not uncommonly can be traced back in time for tens of million years and include unique events. In addition, many landscapes owe their appearance and harmony not solely to the natural forces. Since centuries, or even millennia, they have been shaped by humans who modified hillslopes, river courses, and coastlines, and erected structures which often blend with the natural landforms to form inseparable entities.

These landscapes are studied by geomorphology—"the Science of Scenery"—a part of earth sciences that focuses on landforms, their assemblages, and surface and subsurface processes that molded them in the past and that change them today. Shapes of landforms and regularities of their spatial distribution, their origin, evolution, and ages are the subject of research. Geomorphology is also a science of considerable practical importance since many geomorphic processes occur so suddenly and unexpectedly, and with such a force, that they pose significant hazards to human populations and not uncommonly result in considerable damage or even casualties.

To show the importance of geomorphology in understanding the landscape, and to present the beauty and diversity of the geomorphological sceneries across the world, we have launched a new book series *World Geomorphological Landscapes*. It aims to be a scientific library of monographs that present and explain physical landscapes, focusing on both representative and uniquely spectacular examples. Each book will contain details on geomorphology of a particular country or a geographically coherent region. This volume introduces an area which is among the least familiar for world geomorphologists: The Lesser Antilles. While some islands occasionally headline the news—mainly when struck by natural disasters (Montserrat being the most recent example) and others provided key evidence of Quaternary sea-level change (Barbados)—the majority are largely under-researched and forgotten. And yet they are highly diverse geomorphologically, displaying an array of different landscapes ranging from volcanic and karstic to coastal and reef. Collectively, they tell the fascinating story of a (mostly) volcanic archipelago's geomorphic evolution, and this book undoubtedly helps in appreciating this understudied region's landscapes and landforms.

The *World Geomorphological Landscapes* series is produced under the scientific patronage of the International Association of Geomorphologists—a society that brings together geomorphologists from all around the world. The IAG was established in 1989 and is an independent scientific association affiliated with the International Geographical Union and the International Union of Geological Sciences. Among its main aims are to promote geomorphology and to foster dissemination of geomorphological knowledge. I believe that this lavishly illustrated series, which sticks to the scientific rigor, is the most appropriate means to fulfill these aims and to serve the geoscientific community. To this end, my great thanks go to the Editor, Dr. Casey D. Allen, for his initiative to produce the Lesser Antilles volume and his excellent coordination of work leading to a high degree of coherence between chapters. I am also most grateful to all individual contributors who agreed to add the task of writing chapters to their busy agendas and delivered high-quality final products.

Piotr Migoń

Acknowledgements

My most profound and generous thanks go to Kaelin Groom for her help with reviewing the early stages of the final manuscript, as well as her assistance with double-checking and editing the graphics (and a few other components) in this volume. Her keen eye and artistic abilities lay far beyond mine. She also graciously offered her cartographic prowess throughout the book, creating maps from scratch. I have the utmost respect for her as a colleague and greatly appreciate her companionship throughout the editing and writing process. Many thanks also go to the governments and people of each island in this volume, as several agencies and individuals helped to fill missing imagery gaps and confirm information. I would also like to sincerely acknowledge the chapter authors, especially those who submitted their chapters promptly and those who stepped-in at the last minute to help tidy-up and finish-off a few remaining chapters. I am grateful for your willingness to work on a sometimes extremely tight deadline. Finally, to Piotr Migoń, the Series Editor, for encouragement, support, and the tireless hours he puts into helping the *World Geomorphological Landscapes* series: thank you.

Washington, UT (USA) Casey D. Allen

Contents

Editor and Contributors

About the Editor

Casey D. Allen An award-winning Traditional Geographer *and* Educator recognized for research that spans the physical and social sciences as well as humanities, Dr. Allen maintains wide-ranging interests that coincide with his passion for fieldwork and peripatetics. For nearly 20 years, he has served in various capacities at several different universities (and other jobs), including a stint as instructor and coordinator of St. George's University's BS/MD program (Grenada), where he became enamored with the Caribbean. His current research focuses on validating the importance of experiential education through fieldwork—including his international field study programs, *Sustainability in the Caribbean* and *Geography by Rail*®—using the mediums of geomorphology and humanistic geography broadly speaking, and rock/cultural stone decay and landscape/sense of place more specifically. He also has expertise and keen interests in biological soil crusts, geo/digital humanities, and geoarchaeology. Check his website for updates (https://caseallen.com) and follow his eclectic tweets on Twitter: @caseallen.

Contributors

Jeanette C. Arkle is currently a Ph.D. geology candidate at the University of Cincinnati, Ohio, USA. She completed MS and BS geology degrees, and a BA geography degree from California State University, Fullerton. Her research focuses on crustal deformation, bedrock exhumation, and geomorphology of mountain belts. To investigate the evolution of landscapes and tectonic-surface process interactions, she uses low-temperature thermochronology, terrestrial cosmogenic nuclide dating, and optically stimulated luminescence dating, in tandem with GIS geomorphic analyses and field techniques. She has worked primarily in tectonically active mountain belts including southern Alaska, the northwest Himalaya, southern California and the southeast Caribbean.

Frederick Chambers is Associate Professor in the Department of Geography and Environmental Sciences at the University of Colorado Denver where he previously served as Department Chair and Director of the Master of Environmental Sciences program on two separate occasions. His current research interests are focused in four areas of concentration: (1) mineral weathering and micro-climatological interrelationships of basalt flows on the Big Island of Hawaii, (2) investigation of historical urban heat islands in both old mining towns and the "Rust Belt" of northeastern U.S., (3) continuing investigations into small glaciers in the western U.S. and their responses to climate change, (4) monitoring and assessing the glaciers in the Northern Patagonia Icefields. This research has been supported by the National Geographic Society, United States Geological Survey, National Park Service, and several internal grants from the University of Colorado Denver. Dr. Chambers is also an avid SCUBA diver with a burgeoning interest in the Caribbean.

Sean Chenoweth an Associate Professor of Geography at University of Louisiana at Monroe, USA, **Michael (Sean) Chenoweth** has been conducting research in the Caribbean since 1999. Much of Dr. Chenoweth's focus has been on karst geomorphology and human land use associated with these landscapes. Field work in the Jamaican Cockpit Country led him to seek out similar areas in the Caribbean for a cockpit karst correlation project he is currently pursuing. He also has a strong interest in geospatial technologies including GIS, remote sensing, GPS and applications involving unmanned aerial systems. Recently, he helped to create a non-profit organization dedicated to the preservation and enhancement of the Poverty Point World Heritage Site. Some of his hobbies are bicycling, photography, scuba diving and amateur radio.

Mick Day is an Emeritus Professor of Geography at the University of Wisconsin-Milwaukee, USA, specializing in karst landscapes, their geomorphology, land use and conservation. His regional focus is on the Caribbean, Central America and Southeast Asia.

Patti Day is a Lecturer in Geography at the University of Wisconsin-Milwaukee, USA, specializing in geographic information, its production, acquisition, curation and usage. She is particularly focused on legal issues pertaining to geographic information access, and has global interests extending beyond the Caribbean to North America, Europe and the Pacific.

Alison DeGraff Ollivierre is a geographer, certified geographic information systems (GIS) professional, and award-winning cartographer, who works for National Geographic Maps. **Alison DeGraff Ollivierre** has lived and worked in Saint Vincent and the Grenadines since 2011. Her initial research spanned both SVG and Grenada and focused on the facilitation of a participatory mapping project to develop a comprehensive local knowledge GIS database of important historical, cultural, and ecological heritage sites throughout the islands. Ollivierre additionally assisted with the development of a collaborative marine multi-use zoning plan for the Grenada Bank and is co-authoring and conducting research for an avian field guide for the transboundary Grenadines that highlights both scientific and local ecological knowledge. She holds a BA in Geography from Middlebury College (Vermont, USA) and an MSc in Geoinformatics from the University of the West Indies, St. Augustine (Trinidad and Tobago) where she completed her thesis research on the use of participatory mapping in Caribbean small island developing states to address climate change.

Susanna L. Diller completed her MS degree (Geography) at the University of New Mexico, Albuquerque, USA. She earned her BA degree (Geography) from the University of Colorado Denver, where her honors thesis was a GIS analysis of Twitter conversations about the boyband One Direction, which included extended field time in the UK. A traditional geographer with varied research interests that include geomorphology, cultural geography, sport geography, and GIS, her graduate studies focus on the social role of fountains in public space, particularly in semi-arid cities such as Albuquerque and Denver. Susanna currently works as part of the cyberinfrastructure team with New Mexico EPSCoR (Experimental Program to Stimulate Competitive Research), and has been increasing her regional expertise to include the Caribbean and Europe.

Russell Fielding earned his PhD in geography at Louisiana State University in 2010. He participated in the Canada-US Fulbright Program, spending a year researching and teaching with the Institute of Island Studies at the University of Prince Edward Island. As for research, Fielding is interested, broadly, in questions of subsistence, cultural tradition, and resource conservation. His current research projects focus on issues of food security with regard to the artisanal whaling operation in St. Vincent and local freshwater production through the incineration of municipal waste in St. Barthélemy.

W. Travis Garmon is a hydrogeochemist who completed a MS degree in Geology from the University of Arkansas, and BS degrees in both Geology and Geography from Western Kentucky University. His current research focuses on Mississippi Valley-Type ore deposits in KY and TN, with emphasis on determining the source region(s) of metals found within the deposits using bulk geochemistry and both stable and radiogenic isotopes. His previous research has included dye-tracing in karst aquifers in both south-central KY and south-eastern MN to determine aquifer dynamics and aid in tracking contaminant transport. He remains well-versed in tectonics, and has always held an interest in the Caribbean.

Kaelin M. Groom Fervently dedicated to fieldwork, discovery, and maintaining the delicate balance between historic preservation and educational experience/exposure, **Kaelin M. Groom** completed her PhD in Environmental Dynamics at the University of Arkansas (USA), where she also obtained an MA in Geography. She earned her BA, also in Geography, from the University of Colorado Denver. Her specialties include geomorphology, cartography, cultural resource management, and heritage tourism. With wide interests, Kaelin's current research includes analyzing cavernous rock decay (tafoni) formation and development, quantifying tangible impacts of tourism in culturally protected landscapes, and serving as an advocate for rapid and mixed methods field assessments in rock decay and heritage management. Professionally, she has worked as a research consultant with both domestic and international agencies such as the King Fahd Center for Middle East Studies, USAID's Middle East SCHEP team, US National Park Service, Grenada National Museum, and the Petra National Trust.

Ayumi Kuramae is finalizing her BSc thesis (Coastal Marine Management-Marine Biology) at the University of Applied Sciences Van-Hall Larenstein, Leeuwarden (Netherlands). She did an international study abroad program at Bangor University (School of Ocean Sciences) where she further specialized in marine ecosystems and processes, conservation, and exploitation. She has been collaborating on different tropical marine research projects—such as benthic mapping of Saba's coastal waters and distribution and population status of L. gigas in Anguilla—while also conducting volunteer work in the Dutch and British West Indies (Caribbean). She plans on continuing with her passion by expanding her knowledge and experience in (tropical) marine ecology by completing and MSc in marine biology, and her graduate studies focus on the evaluation of waste management and the possible impacts of solid waste and beach debris on St. Eustatius. Further research interests are coral reef ecology, ecosystem resilience, and regime shifts.

E. Arnold Modlin, Jr Ph.D. is the lone (but not lonely) geographer in the History and Interdisciplinary Studies Department at Norfolk State University in Norfolk, Virginia, USA. Arnold's research interests revolve around how people use landscapes to remember and forget painful pasts, particularly as they relate to issues such as slavery, colonialism, and civil rights. His main research focus over the last few years deals with how the memory of slavery is talked about contrarily in different places in the U.S. South and the Caribbean. Dr. Modlin believes that the struggle to remember difficult pasts connects linguistic, emotional, affective and sensorial expressions with geographies of racial, gendered, generational and economic difference and built and "natural" landscapes. Some of his work has been published in Tourism Studies: An International Journal, Southeastern Geographer, Journal of Heritage Tourism, and the edited volume Social Memory and Heritage Tourism Methodologies.

Elizabeth Nelson A Ph.D. candidate in geography at the University of South Carolina, **Elizabeth (Beth) Nelson** received her MA in geography from Arizona State University, and her baccalaureate (also Geography) from University of Nebraska Kearny. Her active research engages with human geography in the sociopolitical relationships involved in migratory patterns. Currently, her focus is specifically North African migrations to France and other colonial/postcolonial migratory conditions. During the course of her career, she has researched both physical and human geographic characteristics, many involving energy resources and human impacts on the environment. Ranging from hydrological and solar energy resources to political issues involving the extraction and use of fossil fuels, her varied research and careers have followed political engagements with various human-environment or human-political situations. Beth works hard to keep her passion, research topics, areas of expertise, and teaching in both physical and human geography up-to-date and relevant.

Lewis A. Owen received his PhD in geomorphology from the University of Leicester, U.K., in 1988. Before joining the University of Cincinnati in 2004, he held positions at the Hong Kong Baptist University, Royal Holloway—University of London, and the University of

California—Riverside. Professor Owen's research focuses on the Quaternary geology and geomorphology of tectonically active mountain belts and their forelands, particularly in the Himalayan–Tibetan orogen and the Cordilleras of North and South America. He has also undertaken research in other tectonically active regions, including the Caribbean, Red Sea margin in Yemen and the Atlas and Anti-Atlas Mountains of Morocco. In 2011, Professor Owen was awarded the Busk Medal from the Royal Geographic Society for his field research in Quaternary history and geomorphology in tectonically active areas.

Lauren Parkinson recently graduated cum laude from Kennesaw State University with an BA in Geography and a Minor in International Affairs. She was selected as the Department of Geography and Anthropology's 2015 Outstanding Student in Geography and the 2016 Outstanding Senior in Geography due to her involvement with Gamma Theta Upsilon, dedication to academic excellence, and her commitment to research. She will continue her education by studying natural resource management and sustainability in a graduate program in the near future.

Amy E. Potter an Assistant Professor of Geography in the Department of History at Armstrong State University in Savannah, Georgia, USA, **Amy E. Potter**, Ph.D., has research interests that center on the larger themes of cultural justice and the African Diaspora. She has conducted ethnographic fieldwork in the Caribbean and U.S. South. In the Caribbean, she examined how the island of Barbuda has transitioned, in part through the mechanism of migration, from an agriculture and grazing economy to that of tourism and the larger implications for the island's common property. Her current research, funded by the National Science Foundation, examines racialized southern heritage landscapes with a focus on plantation house museums. Some of her work has been published in the Journal of Cultural Geography, Historical Geography, Southeastern Geographer, Journal of Heritage Tourism, Island Studies Journal, and the edited volume Social Memory and Heritage Tourism Methodologies of which she is an editor.

Jennifer L. Rahn is an Associate Professor of Geography at Samford University in Birmingham, Alabama, USA. She first traveled to Saba in 1990 to learn how to scuba dive, and then spent two years as a dive master and tour guide on the island trails (before they were official trails). She has also spent significant time on Statia, and has explored 12 other Caribbean islands. Jennifer continued to visit Saba regularly and, in 2008, began teaching a study-abroad diving course in conjunction with Sea Saba Dive Center. Her coastal geomorphology research on Saba and Statia includes sub-aerial and sub-aqueous beach profile monitoring on sand and cobble beaches. On Saba specifically, her research also includes coral reef mapping and coral nursery implementation and monitoring, among other projects. She collaborates with and volunteers for the Saba Conservation Foundation and the Saba Marine Park whenever she is on island (about 3 months a year).

Richard Edward Arnold Robertson Originally from the island of St. Vincent, **Richard E. A. Robertson** is an experienced field researcher and academic who has been working in the Caribbean for over 25 years. He joined the staff at the Seismic Research Centre in Trinidad in 1993 where he now serves as Director. His research interests include volcano monitoring, hazard and risk assessment, crisis communications, risk perception and management, volcano geodesy and magma genesis. He has a keen interest in the dissemination of scientific information to vulnerable island communities and has published academic books, books chapters, and numerous refereed articles in his areas of expertise. He has worked on various projects including: the ongoing eruption of the Soufrière Hills Volcano on Montserrat, establishment of volcano monitoring networks, public education, and outreach programs throughout the Eastern Caribbean, and the lecturing and supervision of geoscience students. He is an advisor to disaster management organizations throughout the Eastern Caribbean.

Phillip P. Schmutz Ph.D. is an Assistant Professor in the Department of Earth and Environmental Sciences at the University of West Florida in Pensacola, Florida, USA. Phillip's research interests center on themes of environmental monitoring and modeling of coastal/beach/dune environments with an emphasis in beach hydrology. He has conducted fieldwork along the coast of Texas, Florida, North Carolina and Mexico as well as the island of Dominica. His doctoral research was funded by the National Science Foundation's Doctoral Dissertation Research Improvement (DDRI) award. He conducted his post-doctoral research at White Sands National Monument through a National Park Service grant. His work has been published in the Journal of Coastal Research, Aeolian Research, and Earth Surface Processes and Landforms.

Ryan Sincavage is currently a PhD candidate in Environmental Engineering at Vanderbilt University in Nashville, Tennessee, USA. He completed a B.S. degree in Earth Sciences from the Pennsylvania State University and a M.S. degree in Geology from the University of Colorado Boulder. His research focuses on sedimentology and stratigraphy of large fluvial systems and deltas, particularly the interactions of climate, tectonics, and autogenic processes on channel mobility, surface morphology, and preservation of sediments. Current active research focuses on the Holocene Brahmaputra River avulsion history of Sylhet Basin, a tectonically active sub-basin within the Ganges-Brahmaputra-Meghna Delta (GMBD) of Bangladesh, and the coupling of tectonic deformation and stratigraphic architecture in the Indo-Burman fold belt of eastern India. He primarily uses field observation of outcrops as well as laboratory analyses (grain size measurements and X-rayfluorescence) of sediments to quantitatively assess provenance and fluvial system behavior.

Vanessa Slinger-Friedman is an Associate Professor of Geography in the Department of Geography and Anthropology, and the Associate Director of Distance Education for the College of Humanities and Social Sciences at Kennesaw State University (USA). Originally from Trinidad, Dr. Slinger-Friedman obtained her MA in Latin American Studies and PhD in Geography from the University of Florida. Her work has included a World Bank sponsored study in Mexico and El Salvador of Vetiver grass technology for soil erosion control, the use of an agroforestry system for Amazonian urban resettlement in Acre, Brazil, and the use of ecotourism on Dominica, W.I. for economic development and nature preservation. Dr. Slinger-Friedman has a regional focus on Latin America, the Caribbean, and the SE United States, where she has researched ecotourism and the impact of Latino immigration respectively. Her other research interests include innovative pedagogy, online teaching, and intercultural competence related to study abroad.

Donald M. Thieme is a geomorphologist and soil scientist. Dr. Thieme studies the distributions and properties of sediments and soils to identify the direction and rate of landscape change. In addition to direct field examination, sampling, and laboratory analysis, he uses shallow geophysical methods for geological and archaeological research. His work includes studies of the effects of both climate change and human activities on soils. Dr. Thieme has only recently taken an interest in the landscapes of the Caribbean region, his previous field experience being confined to the continental United States and Mexico.

John C. Weber is a Professor of Geology at Grand Valley State University (USA) and also a regular summer field camp Instructor at the Yellowstone Bighorn Research Association facility for the University of Houston. In his research, he uses structural geology, GPS geodesy, thermochronology, and tectonic geomorphology to study plate and microplate motion and neotectonics in the SE Caribbean, northern Adriatic, and circum-Caucasus. He also interfaces with industry through teaching and consulting.

Tirzha Zabarauskas is a graduate student at University of Colorado Denver (USA) pursuing a Master of Arts in Applied Geography and Geospatial Science. She holds a BA in Geography with minors in geology and educational studies. She is a member of Phi Theta Kappa honor society and Philanthropic Education Organization. Tirzha spends her summers teaching STEM-focused camps to students in Boulder and Denver, Colorado. She is an avid photographer and traveler—having visited several Caribbean islands thus far in addition to other locales—and can be found doing either when not working. She lives with her husband and daughter in a suburb of Denver. Though she has burning interests in the Earth Sciences, her current research is focused on spatial literacy and education for primary school students.

Small Islands, Intriguing Landscapes 1

Casey D. Allen

Abstract

The Lesser Antilles comprises three main island groups: Leeward Islands, Windward Islands, and Leeward Antilles. Stretching from the Virgin Islands in the north to Trinidad and Tobago in the south and encompassing Aruba, Bonaire, and Curaçao to the east, the Lesser Antilles remain a geomorphologically and anthorpogenically diverse region. While this chapter introduces offers the reader a (very) brief overview of this fascinating and under-studied world region, its main focus rests in explaining this volume's structure and function, including notes regarding vernacular, historical accuracy, and the splendid cartography.

Keywords

Lesser Antilles • Leeward Antilles • Windward Islands • Leeward Islands

1.1 Introduction

While often neglected geomorphologically in both study and literature, the Lesser Antilles islands in the Caribbean contain outstanding landscapes and landforms: granitic islands in the north, active and highly explosive volcanoes in the center, mostly extinct volcanics in the south, and a few (sometimes large) sedimentary uplift blocks scattered throughout the archipelago. Spatially, the Lesser Antilles consist of a vast territorial swath spanning about 940 km from the British and US Virgin Islands in the north to Trinidad and Tobago in the south, and approximately 1140 km from Aruba in the west to Barbados in the east (over 1,000,000 km^2, Fig. 1.1). In that otherwise watery expanse, the Lesser Antilles claim no less than 600 individuals yet mostly uninhabited islands, forming—with the exceptions of outlying Barbados and Aruba, Bonaire, and Curaçao (the ABCs)—an arcing boundary between the Caribbean Sea and the Atlantic Ocean. This positioning on Caribbean Plate's edge creates a landscape of volcanic features: from plutons and stratovolcanoes to geothermal springs and lava flows, as well as hazards associated with regular heavy precipitation, fire, volcanically induced tsunamis, hurricanes, and, somewhat paradoxically, drought. Additionally, geologic uplift and subsequent rock decay continue to occur, exposing sedimentary landforms such as coralline shelves and carbonate reefs, as well as general karst landscape features.

The Lesser Antilles' human occupance also played a role in shaping the islands' landscapes, and, in many instances, these actions (individually, say a landholder, or as a group, say a country) have influenced landforms. As Amerindians began settling the Islands, migrating northward from mainland South America or southward from larger islands such as Puerto Rico and Hispaniola, they brought with them various cultural traits, including agricultural practices. Some left records in the form of rock engravings (petroglyphs) and additional tangible artifacts, others intermarried with European colonists or African slaves, and their progenies still inhabit the islands today. These peoples made use of the landforms, as evinced by Lesser Antillean archaeological sites being found near freshwater rivers, along bays with coral reefs, and in caves. Evidence for pre-Columbian populations also exists in the form of numerous "worked stone" (*cupules* and/or grind stones) on several islands,

C.D. Allen (✉)
General Education, Western Governors University,
Salt Lake City, UT, USA
e-mail: caseallen@gmail.com

C.D. Allen (ed.), *Landscapes and Landforms of the Lesser Antilles*,
World Geomorphological Landscapes, DOI 10.1007/978-3-319-55787-8_1

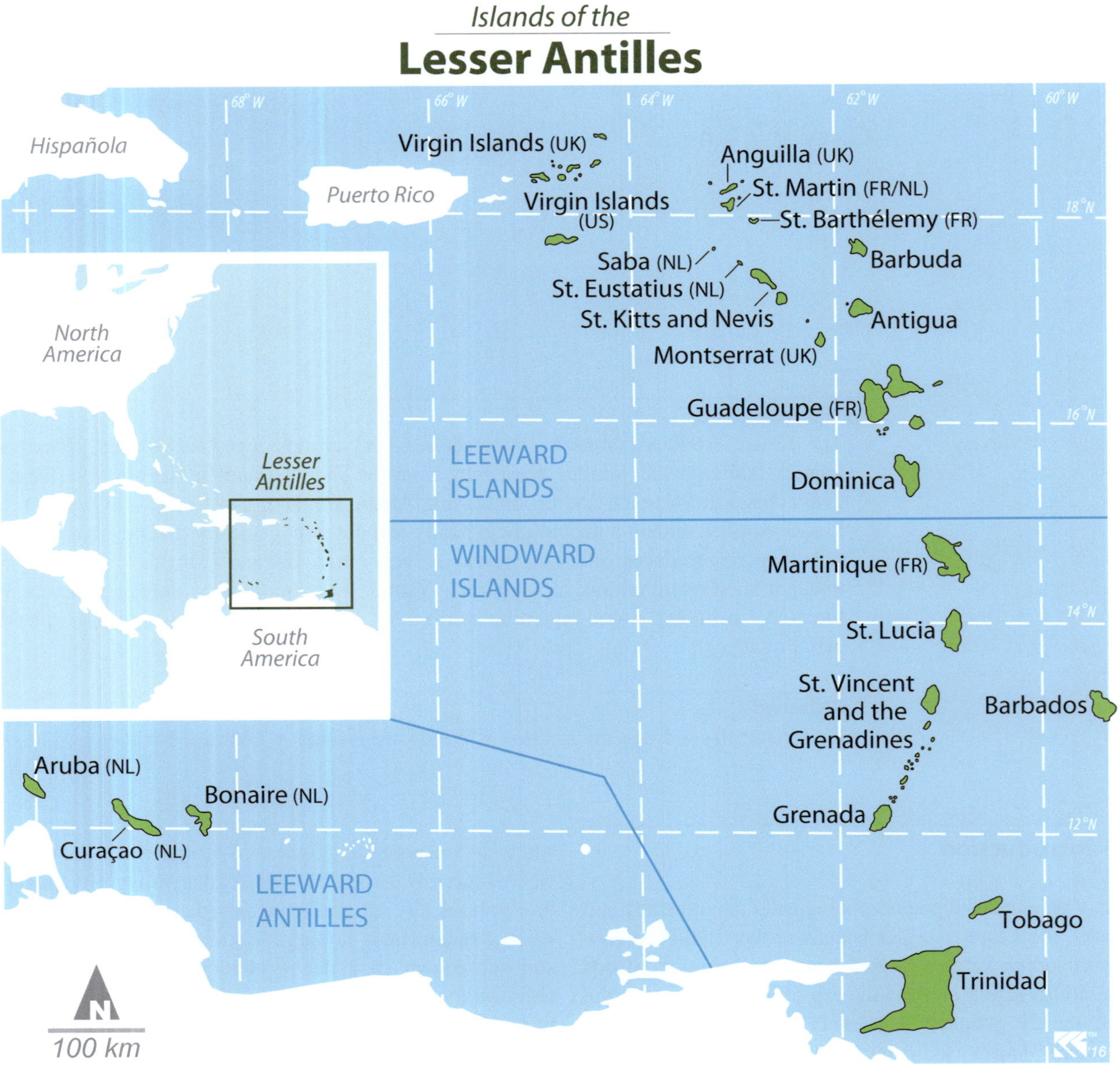

Fig. 1.1 General map of the Lesser Antilles identifying the three subregions—Leeward Islands, Windward Islands, and Leeward Antilles—as well as the primary islands/island groups that will be highlighted in this volume. Cartography by K.M. Groom

demonstrating that these peoples understood—if only subconsciously—at least the basics of how rock hardness and mineral grain size can interact to provide them with surfaces for grinding substances and creating tools and other implements (e.g., smooth, polished granitic rocks have been found at archaeological sites on islands where no granite is present).

After Columbus' "discovery" of the (then) New World, a different influence on Lesser Antillean landscapes and landforms began. Colonial settlements began appearing, slowly populating each island with a land tenure system that was sometimes sustainable, but mostly not. Plantations began to expand, and soon more labor was needed, sparking the Trans-Atlantic Slave Trade, where a majority of West African slaves ended up on plantations in the Caribbean. To this day, those slaves' *obeah* belief system survives on nearly every West Indian island in some form. Beyond human occupance, as landholdings expanded rapidly, island landscapes changed: Previously forested lands gave way to sugar, cotton, tobacco, and indigo fields, exposing precious (and often nutrient-rich) volcanic soils to the elements. Insect-infested lagoons were dredged, changing coastlines and depositional processes. Where encampments began, forts were built to protect interests, most using local stone and local limestone mortar, some of which still stand today.

Through the centuries after European "discovery," strong cash crop economies developed, necessitating the need for trade agreements and treaties, all while pirates and privateers continued to make some of the West Indies' small coves and ports infamous. As technology developed, so did economies and politics. Some islands sought (and subsequently won) independence from their colonial power—sometimes easily, but more often than not, with arduous struggle—while others remained part of the Commonwealth or colonial power. Toponymic remnants from these time periods can still be found throughout the Lesser Antilles. Still, it took until the latter part of the twentieth century (1960s–1980s) for islands to gain their independence, with some still struggling for their autonomy today (e.g., St. Kitts and Nevis) while other islands stay part of the original colonial power (e.g., Martinique, for example, where the official currency is the euro). Outside interferences, including non-colonial interests, have also influenced Lesser Antillean development. The infamous *Operation Urgent Fury* in 1983 led by American forces on the British Commonwealth Nation of Grenada, for example, highlighted perceived political turmoil of the region. While other industries occur—agricultural export, some fertilizer and salt production, and petroleum in the case of Trinidad and Tobago and the ABCs—the most common among the Lesser Antilles continues to be tourism, or at least the potential for it.

The Lesser Antilles of today represent a lively collection of economies, politics, cultures, and geomorphology. To showcase these vibrant characteristics, following this *Introduction*, the volume includes a general overview of regional tectonics and geology, landform evolution/geomorphology, and basic climate/climatic geomorphology. The tectonics and geomorphology discussions span from Eocene to the Quaternary (when island formation along the Lesser Antilles Arc began), while the climate component encompasses recent history to present. Then, this volume continues with separate and in-depth chapters on landscapes and landforms associated with individual islands (e.g., Dominica) or island sets (e.g., the Virgin Islands). Each chapter also contains two specific, yet geomorphologically applicable topics, especially for the Lesser Antilles: *Heritage and Tourism* and *Hazards*. A heritage and tourism component is included because, not unlike other regions of the world, each island has its own struggles with governmental regulations when it comes to tourism, and much of the tourism centers around geomorphic features and phenomena ("geotourism," e.g., beaches, reefs, rainforests, mountains, and even rock decay; see Dowling and Newsome 2006 for an overview), while also being (potentially) influenced by hazards. The Lesser Antilles are plagued not just by hurricanes and their associated risks, but also by volcanics, earthquakes, fire, landslides, and, interestingly, fire and drought. And each of these events—sometimes in tandem with others, sometimes solo —often plays a role in changing island geomorphology. Every chapter includes, at a minimum, the following topics:

- *Introduction*. A basic overview of the chapter's main points.
- *Setting*. Or sometimes being listed as "Geologic Setting," this section remains a short and concise overview of the island's or island set's formation 9 and sometimes geochronology), often including basic climate and other related information.
- *Landforms*. A discussion of major and minor landforms, including notable interior (e.g., mountain), coastal (e.g., bays, beaches), and offshore landforms (e.g., coral reefs), and sometimes geochronology.
- *Landscape*. Sometimes listed in specific chapters as "Environmental History and Landscape Change" or "Landscape and History", this section is meant to represent a bridge between geomorphology and culture. This often includes anthropogeomorphologic analyses or archeogeomorphologic discussions, for example, including both past and present peoples' histories and their link with the geomorphology.
- *Heritage and Tourism*. As a (the) main economic resource on each of the Lesser Antilles islands is usually tourism, this subsection represents a mostly concise review of impacts that heritage resources management and tourism have on the larger (geomorphic) landscape.
- *Hazards*. A review of past hazards and their geomorphic impacts, as well as evaluation of potential and future hazards, and what they could mean for the island's or island sets' (geomorphic) future.
- *Conclusion*. A concise review of the previous six topics, noting any potential future outlooks for the island or island set from a geomorphic point of view.

Finally, it should be noted that writing a chapter on a Caribbean island's landforms *and* landscapes remains a delicate endeavor. Some islands have adequate information to pen a chapter (though not necessarily in English), others less so. Indeed, the region remains rich with research opportunities, but with so much to do, authoring a chapter for an edited volume often takes a backseat to primary research agendas. Additionally, the author(s) must be part geomorphologist, part anthropologist, part political scientist, part historian, part hazard specialist—and more—and then be able to weave each together coherently. Chapter authors have striven for a balance between landforms and landscapes, although at times, one may seem to be heavier than the other. Still, when considering how little (current) research has been conducted throughout the Lesser Antilles compared to other locales, the fact that some semblance of balance can even exist remains, perhaps, a minor miracle. Sometimes finding a balance necessitates coauthors, while other times a person tackled "their" island solo, and this complex ability should not be ignored, nor should an author's claim that "their" island is the most beautiful, or most pristine, or most unspoiled, or has the best geomorphology in the Caribbean. These seemingly boastful comments are deemed allowable for this volume because the editor believes everyone should be afforded an opportunity to promote their passion for place. After being presented with evidence provided by these regional experts, the reader is invited to decide for themselves which island—which "… visually stunning …" and "… great landscape …" (Goudie 2002, 65)—they find most intriguing in this small, oft-understudied World Geomorphological Landscape.

1.2 A Few Notes About this Volume

1.2.1 Lesser Antilles Toponyms

Throughout the Lesser Antilles, colonialism shows itself most readily in place names (toponymy). The presence of local place names, usually never officially recorded or, at best, inaccurately recorded by the original scribe, further confuses Lesser Antillean toponymy. Often times, even with the best intentions, mistakes happened and, whether from misinterpretation of a local word, creolization, or faulty record keeping, the misspelling spread, sometimes making its way into vernacular. Even on islands that are still today an overseas territory of their once colonial power, toponymy can reveal much about an island's historical occupance. One common aspect rests in naming convention differences between European nations. The French, for example, tended to favor naming places based on physical features, and many of these survive throughout the Lesser Antilles: *Anse*, *La Soufriére*, *Morne*, and *Pitons*, for example. British colonists, on the other hand, favored naming places after famous leaders and people, giving the Lesser Antilles an abundance of George's, Mark's, and John's, as well as others. Similar linguistic influences can be found throughout the Caribbean for (former) Spanish, Dutch, and even Swedish colonies. That said, while some place names might be referred to differently on official records, this volume strives to offer the most recognized names associated with locations on each island, regardless of colonial influence, as these most often represent the local vernacular/convention.

1.2.2 Lesser Antilles Historical Accuracy

Most dating of pre-Columbian (i.e., pre-European contact) Lesser Antillean artifacts and occupational periods remain contextual. In the Lesser Antilles specifically, not all scientists agree on ages of rock art (petroglyphs) and other archaeological artifacts. Discrepancies arise because no precise technique has yet been devised to establish definitive dating in tropical regions, and most Amerindians left no (decipherable) written record. Each chapter in this volume, however, follows current research trends, with any deviation being specifically noted in that chapter. Similarly, the term *Amerindian* is used exclusively to refer to aboriginal/native peoples of the West Indies—those inhabitants prior to Columbus' "discovery"—as opposed to the (mostly) West African slave descendants that currently inhabit most islands, often referred to as *Afro-Caribbean* in the literature to denote the mixing between them and the original Amerindian population.

1.2.3 Lesser Antilles References

The reader may notice throughout this volume that some citations come from the internet (websites) and trade publications. Additionally, some chapters may continually reference only the same handful of peer-reviewed articles and/or books. While this is perhaps not normal academic practice, it remains warranted for this volume. It seems that in many cases throughout the Lesser Antilles, once a piece of (geomorphic) research has been completed, no one else continues or updates it, leaving many older references. Indeed, being a mostly under-researched region (the smaller islands in particular) means much of the data comes from either firsthand research experiences of the chapter author(s) themselves, a small cadre of researchers (sometimes several decades old and sometimes in a non-English language), and touristic paraphernalia—including Ministry of Tourism websites, guidebooks (online), and white papers. These types of references should not be seen as diminishing the quality of research presented in this volume, but instead should be taken as a call for more (field-based) academic research to be conducted in the region. Every effort has been made to verify the accuracy of Web-based references by the editor when peer-reviewed literature was unavailable.

1.2.4 Cartographers' Note

Maps in this volume are for informational and reference purposes only and are not prepared for legal, engineering, navigational, or surveying endeavors. The cartographic products herein have been created with the highest degree of accuracy possible, and the cartographer, editors, chapter authors, or publisher cannot be held responsible for any damages from omissions or misuse of maps contained in this volume. Depiction of boundaries is non-authoritative, and any misrepresentation of coastlines, features, cities, towns, etc., is not done with malicious intent, but most likely due to a paucity in reliable data or old information. For example, some islands may appear to be more topographically complex than their surrounding islands (e.g., Dominica), but this is not necessarily the case. These visual differences represent discrepancies among the precision of data available for each individual island/island set. The appearance of commercial establishments featured on these maps does not imply endorsement of them by either the cartographer, editors, chapter authors, or publisher, but simply reflects important locational features. All maps are copyrighted by the respective cartographer, and reproduction without legal consent may result in prosecution.

References

Dowling RK, Newsome D (2006) Geotourism. Routledge

Goudie AS (2002) Aesthetics and relevance in geomorphological outreach. Geomorphology 47:245–249

Geologic and Tectonic Background of the Lesser Antilles

2

W. Travis Garmon, Casey D. Allen and Kaelin M. Groom

Abstract

A case study of the geomorphology of the Lesser Antilles island arc reveals, in its entirety, the influence of numerous geological forces and events. Most notably, these include the products of plate tectonics, volcanism, and carbonate marine reef formation. North of Dominica the island arc splits into two separate chains. The easternmost archipelago of these chains is largely comprised of extinct volcanoes that have since become the core of carbonate reef growth. The westernmost archipelago of the island arc and the southern half of the overall Lesser Antilles are still active volcanic complexes formed due to partial melting of subducting oceanic crust. Orogenic uplift due to transform plate tectonics and thrust faulting is observable in the southern Leeward Antilles.

Keywords

Caribbean • Lesser Antilles • Island arc • Tectonics • Archipelago

2.1 Introduction

The variety of distinctive geologic formations, from sedimentary to volcanic, separate the Lesser Antilles island arc into three distinctive island groups: the Leeward Islands, the Windward Islands, and the Leeward Antilles Islands (Fig. 2.1). Encompassing the northern section of the arc, the major Leeward Islands (and island sets) tend to be smaller in size than their Windward counterparts and include—from northwest to southeast—the US and British Virgin Islands, Anguilla, Saint Martin, Saint Barthelemy, Saba, Saint Eustatius, Saint Kitts, Nevis, Barbuda, Antigua, Montserrat, and Guadeloupe. Depending on sources, Dominica can be classified as either a Leeward or Windward Island. This volume follows the more recent literature that treats Dominica as the northernmost Windward Island—those larger West Indian islands that contain the southern arm of the Lesser Antilles island arc—followed in a southerly direction by Martinique, Saint Lucia, Barbados, Saint Vincent and the Grenadines, Grenada, Tobago, and Trinidad. Though not usually associated with the Lesser Antilles specifically, the Leeward Antilles—Aruba, Curaçao, and Bonaire off the northern coast of Venezuela (often called the "ABC islands" or "ABCs")—remain spatially and geologically distinct from the rest of the Lesser Antilles. Still, they are included in this volume for ease of reference, because no other volume contains an overarching review of their landscapes and landforms.

W.T. Garmon
Independent Scholar, Burkesville, KY, USA
e-mail: WTgarmon@gmail.com

C.D. Allen (✉)
General Education, Western Governors University, Salt Lake City, UT, USA
e-mail: caseallen@gmail.com

K.M. Groom
Department of Geosciences and King Fahd Center for Middle East Studies, University of Arkansas, Fayetteville, AR, USA
e-mail: kmgroom@uark.edu

C.D. Allen (ed.), *Landscapes and Landforms of the Lesser Antilles*,
World Geomorphological Landscapes, DOI 10.1007/978-3-319-55787-8_2

Fig. 2.1 Tectonics of the Antilles Island Chain, illustrating plate boundaries and fault zones surrounding and within the Caribbean Plate. Subduction zones along the eastern and northeastern margins have led to volcanism, which produced the Windward Islands, the inner arc of the Leeward Islands, and the cores of the outer arc of the Leeward Islands. Transform boundaries and folding along the southern extent of the Caribbean Plate have uplifted the sedimentary units that comprise the Leeward Antilles. The directions of plate movement and prominent tectonic features in the region in relation to the Lesser Antilles shown in green. Cartography by K.M. Groom

2.2 Geologic History and Formation of the Lesser Antilles

2.2.1 Tectonics of the Atlantic Basin and Caribbean Sea

The formation of the Atlantic Basin began with the separation of the supercontinent Pangaea roughly 175 million years ago during the Jurassic. Rifting between the former continents of Laurasia (modern-day Europe, Asia, Greenland, and North America) and Gondwana (modern-day Africa, South America, Australia, Antarctica, and India), initiated the formation of the northern Atlantic Ocean. At the turn of the Cretaceous, roughly 25 million years later, Gondwana began to separate into the continents we recognize today, forming the southern Atlantic. These two different rifting events, each forming and continuing to widen the Atlantic Ocean, are centered on the submarine mountain chain known as the Mid-Atlantic Ridge—the divergent boundary between the African and American plates. As these plates move apart, decompression melting in the upper mantle produces magma flows which cool to form basalt and gabbro, leading to new oceanic crust. Subduction of the Caribbean Plate under the South American Plate began around 80 million years ago during the Late Cretaceous (Bouysse 1988; Mann 1999; Macdonald et al. 2000), and 40 million years ago, volcanism began because of that subduction (Fig. 2.2; see also Smith et al. 1980; Bouysse 1988; Kerr et al. 1996; Kerr et al. 2003).

Oceanic crust increases in both thickness and density with age. As the plate edges drift further from the divergent zone from which they were produced, they cool, and additional mantle material is accreted onto the bottom of the plate as formerly plastic mantle rocks become more rigid due to a drop in temperature. Furthermore, a steady "rain" of pelagic sediments within the ocean water column falls onto submerged crust, leading to the subaqueous formation of layers of marine shales, limestones, and sandstones. The longer an oceanic plate exists, the thicker the buildup of sedimentary rock, and the heavier the plate becomes. Ultimately, older oceanic plate will be sufficiently heavier and denser than bordering younger oceanic or continental plates, initiating subduction of the older plate. For these reasons, most oceanic crust is relatively young compared to continental crust, which does not subduct except in very rare circumstances. The eastern boundary of Caribbean Plate, also known as the Lesser Antilles subduction zone, is one of these rare instances where the largely continental South American Plate is subducting under the mostly oceanic Caribbean Plate, initiating volcanism and tectonic uplift.

Additional plate boundaries and interactions in the region include several transform boundaries, when two plates slide

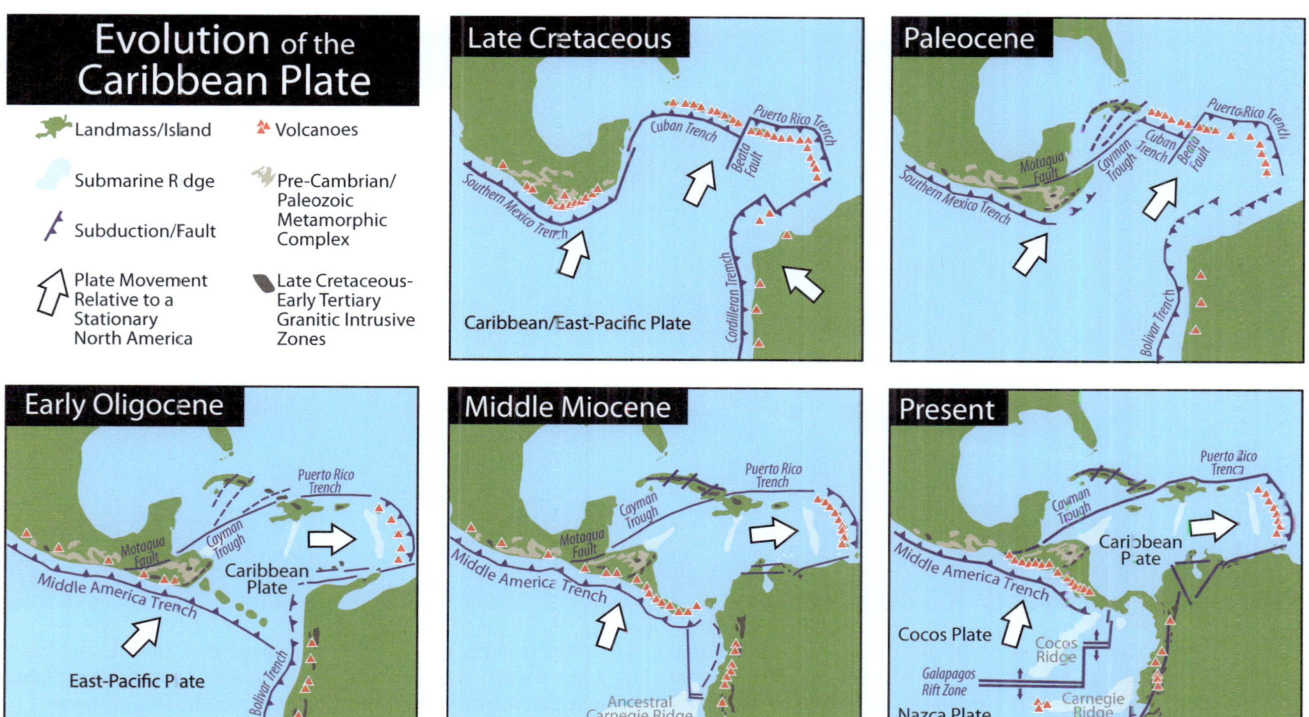

Fig. 2.2 Caribbean Plate Tectonic Evolution, including the Caribbean Plate's subduction under the South American Plate, beginning around 80 million years ago. Figure by K.M. Groom, modified from Malfait and Dinkleman 1972

past one another in opposite directions. Similar to (and often producing) strike-slip faults, these boundaries rarely produce orogenic events without additional deformational components. Brecciation and fracturing directly on the plate edge is a common feature at these locales. California's San Andreas Fault is a famous example of this sort of tectonic plate boundary, where the northward motion of the Pacific Plate relative to the North American Plate results in a massive right-lateral strike-slip fault. The relative motion of the adjoining plates defines left-lateral and right-lateral strike-slip faults. A similar mechanism to the San Andreas exists at multiple points throughout the Caribbean (Fig. 2.3; see also Edgar et al. 1971; Malfait and Dinkelman 1972). The southern margin of the Caribbean Plate expresses a right-lateral transform component relative to the South American Plate (the Caribbean Plate is moving eastward, relative to both the North and South American plates). In concert with subduction and transformational motion, slight compression along the southern margin has resulted in folding and thrust (steep reverse) faulting—yielding minor orogenic uplift to expose some of the Leeward Antilles, Barbados, Trinidad, and Tobago, along with many smaller islands (Fig. 2.1; see also Ave-Lallemant and Sisson 2005; Levander et al. 2006; Van der Lelij et al. 2007).

2.2.2 Lesser Antillean Volcanism

The Leeward Islands contain two distinct volcanic island arcs of different ages, resulting from a slight shift in plate interactions. The outer arc is the easternmost archipelago and, at 40 million years old, is the older of the two. This arc consists of extinct volcanic cores that have since decayed and developed marine reefs (Christman 1953; Malfait and Dinkelman 1972; Bouysse 1988; Bouysse et al. 1990; Marshall et al. 1997; Macdonald et al. 2000; Van der Lelij et al. 2007). The younger (20 million years) inner arc is still primarily active, marking the current site of the South American Plate and Caribbean Plate subduction zone (Malfait and Dinkelman 1972; Smith et al. 1980; Bouysse 1988; Bouysse et al. 1990; Macdonald et al. 2000; Robool and Smith 2004; García-Casco et al. 2011). Southward from Dominica, both the outer and inner arcs are superimposed, and many of the volcanoes are still active in the Windward Islands.

The geochemistry of volcanic eruptions (metal-rich effusive versus silica-rich explosive) determines the size and shape of the volcano itself. Effusive mafic eruptions result in shield volcanoes, which have a large surface and gentle slope due to the lower viscosity of mafic lavas. As the

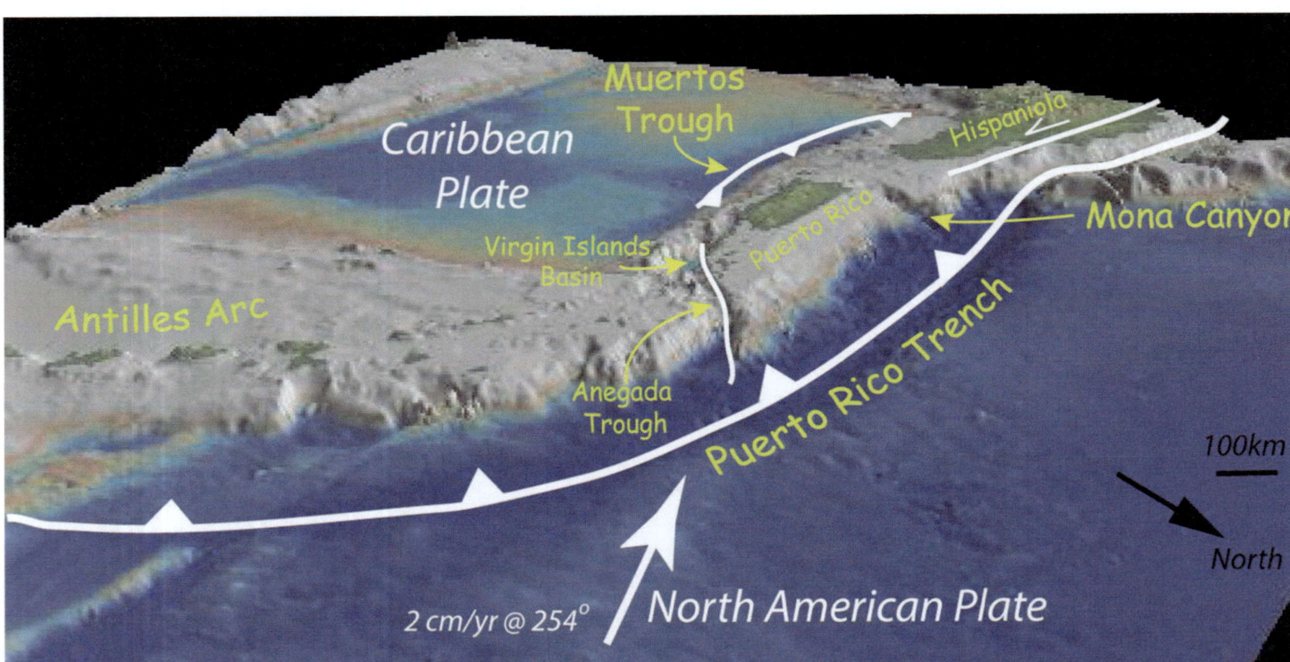

Fig. 2.3 A three-dimensional view of the Puerto Rico Trench (the Antilles Arc runs along its ridge). The northern boundary of the Caribbean Plate exhibits left-lateral strike-slip faulting along the North American Plate border to Puerto Rico, where the boundary transitions into a combination of transform and convergent (subduction) zones form the Puerto Rico Trench. Image courtesy of NOAA (http://oceanexplorer.noaa.gov/okeanos/explorations/ex1502/background/geology/welcome.html)

mafic lava is erupted, it can flow somewhat freely and thus spreads out in large, thin layers before cooling. Alternatively, high-viscosity felsic and intermediate lavas are more prone to the formation of stratovolcanoes, which have much steeper topography. Because of the felsic lava's high resistance to flow, the lava cools and solidifies before it has a chance to spread over a large geographical area. After multiple eruptions from the same vent, the igneous rocks form a steep mountain, and subsequent erupted material creeps down the sides of the volcano, forming igneous strata from each non-explosive event. Explosive events generate volcanic ash–molten rock and dust blasted into the atmosphere. Gravity soon overcomes the momentum of the erupted material, and the ash cloud collapses back to the surface of the Earth. When the collapse occurs quickly, it produces pyroclastic flows, which are effectively rivers of rapidly flowing superheated ash—an occurrence common in much of the Lesser Antilles (Sigurdsson et al. 1980). Occasionally, following a sudden purge of material within the magma chamber, the overlying volcanic complex can collapse in itself leaving a crater-like feature at the surface known as a caldera. Calderas are common on several islands throughout the Lesser Antilles, most notably on the isle of Dominica.

Nineteen active volcanic complexes are subaerially exposed throughout the Lesser Antilles, along with one active submarine volcano. The isle of Dominica is host to nine currently active volcanoes and was formed as multiple stratovolcanoes and calderas grew and overlapped one another. Volcanic complexes specific to Dominica include Morne Au Diable, Morne Diablotins, Morne Trois Pitons, Wotten Waven Caldera, Valley of Desolation, Watt Mountain, Morne Anglais, Grande Soufriere Hills, and Morne Plat Pays. Although all of the Dominican volcanoes are considered active and experience periodic minor eruptions, a substantial eruption has not been recorded on the island since Europeans first began exploring the Caribbean in the 1600s. Moving further south into the Windward Islands, active volcanoes are found on Martinique (Mt. Pelee), St. Lucia (Qualibou), St. Vincent (Soufriere), and Grenada (Mt. St. Catherine); however again, eruptions of these are rare occurrence, with geothermal activity (such as hot springs) being a main feature among them. An underwater active volcano named Kick 'em Jenny, north of the Grenada coast, has shown sign of rumbling as recent as August 2015.

Leeward Islands within the inner arc region are predominantly covered in Quaternary and Tertiary volcanic deposits. Most isles throughout the region are amalgamations of multiple stratovolcanoes. The compositions of most igneous rocks throughout the area range from mafic basalt to intermediate andesite (Kerr et al. 1996; Sinton et al. 1998; Macdonald et al. 2000; Robool and Smith 2004; García--Casco et al. 2011). Dominant minerals within the andesite regimes are biotite- and sodium-rich feldspars; with olivine, pyroxene, amphibole, and both sodium- and calcium-rich feldspars dominating the basaltic flows (Kerr et al. 1996; Robool and Smith 2004). The volcanic islands are littered with welded tuffs, pumice, tephra, and breccias, which are the product of pyroclastic flows typical of explosive eruptions (Kerr et al. 1996; Robool and Smith 2004). Basalt clasts (pieces of previously solidified basaltic lava flows preserved in a more recent flow that were not re-assimilated as melt) are noted within several of the non-explosive but still high-viscosity andesite eruptions, suggesting that intrusion of basaltic magma into andesitic magma chambers instigated several recent eruptions in the northern islands (Kerr et al. 1996; Robool and Smith 2004; Kerr and Tarney 2005; García-Casco et al. 2011). The andesitic chambers are proposed to have formed from the combined melt of oceanic basalt and gabbro with melted marine sediments during the subduction of the South American Plate (Malfait and Dinkelman 1972; Kerr et al. 1996; Robool and Smith 2004). Multiple dipping limestone beds of Oligocene age also outcrop on several of the inner arc islands, though were likely uplifted by a younger volcanic dome formation as opposed to a tectonic orogeny (Bouysse 1988; Mann 1999; Robool and Smith 2004; García-Casco et al. 2011).

The now-extinct outer arc of the Leeward Islands was first formed through Eocene volcanism and later modified by marine sedimentation from the Eocene to modern day. The limestones throughout the region require a subaqueous setting to be deposited, meaning the deposits are only formed in areas that were below sea level during their corresponding epoch of formation. Many forms of marine life precipitate an aragonite or calcite (both of which are polymorphs of calcium carbonate) shell throughout their life span. Upon their deaths, these shells fall to the bottom of the marine water column and accumulate on the seafloor, building layers of biogenic carbonate sediments which are later compacted and lithified to form limestone units. Coral and bryozoan reefs function in much the same manner—precipitating carbonate skeletons on existing structures. As these creatures die, they are covered in the carbonate frames of new reef-forming creatures, and the structure grows both outward and upward, provided it remains below sea level.

2.2.3 Lesser Antillean Unconformities

If a landscape transitions from subaqueous to subaerial, through uplift or sea-level change, marine carbonates can no longer be deposited. Once deposition of sediments ceases, a shift to an erosion-dominated environment often occurs, resulting in the removal of lithologic material. These periods of non-deposition and/or erosion are geologically represented by unconformities, where no record of natural events for a length of time is preserved in the stratigraphic column. Three primary types of unconformities exist, all of which are present in the Lesser Antilles. Nonconformities exist at the boundary between igneous and sedimentary rocks, such as those seen between the marine sediments and volcanic ash flows of the inner arc Leeward Islands. Disconformities describe gaps in rock deposition between relatively undeformed flat-lying sediments, observable on the sediment-dominated outer arc of the Leeward Islands and throughout the orogenically uplifted Leeward Antilles. Angular unconformities occur between flat-lying and inclined beds, in which a bed is uplifted/tilted, partially eroded, and then additional sediments are deposited in flat-lying strata above the remnants of the inclined beds, much like the volcanically inclined limestone beds on the isle of St. Eustatius.

Three regional unconformities are present in the Lesser Antilles. The first occurs between the Oligocene and Miocene volcanic and limestone deposits, the second at about 5.4 million years ago between Miocene and Pliocene volcanism, and the most recent at about 2.8 million years ago between the Pliocene and Quaternary volcanic ash flows. The earliest unconformity is most evident in the southern Windward Islands and the outer arc of the Leeward Islands, for example, the exposures on Antigua and Trinidad. The islands of Grenada and Tobago display examples of the Miocene–Pliocene unconformity in the Windward Islands, and the most recent Pliocene–Quaternary unconformity is observable in both the Windward Islands and the inner arc of the Leeward Islands at places such as Saba and Barbuda due to the active volcanism throughout both regions (Smith et al. 1980; Robool and Smith 2004). Fossil evidence suggests that these unconformities correspond with periods of relative sea-level drop in the region, which produced land bridges linking the island chain to the Greater Antilles to the north and to South America to the south (Marshall et al. 1997). While shifts in tectonic motion can cause changes in relative sea level, it is more likely the conditions necessary for these Lesser Antilles unconformities were the result of global climate change. Widespread evidence suggests periods when global sea levels have fallen in response to cooler global climates sequestering large volumes of water in ice at the poles as well as both alpine and continental glaciers (Marshall et al. 1997).

2.2.4 Leeward Antilles

Regionally and geomorphologically distinct, the Leeward Antilles—not to be confused with the Leeward Islands of the Lesser Antilles—are composed of the "ABCs" and several smaller islands owned by Venezuela and stretch along the southern extent of the Caribbean Plate just north of the South American coast. Unlike many of the Leeward and Windward Islands, the Leeward Antilles lack modern volcanism and consist of predominantly orogenically uplifted limestones (Levander et al. 2006; Van der Lelij et al. 2007). Transform faulting, reverse faulting, and minor subduction at the boundary with the South American Plate caused fold/thrust processes in portions of the South American Plate's continental shelf, bringing blocks of the seafloor upward to intersect the ocean's surface (Edgar et al. 1971; Levander et al. 2006; Van der Lelij et al. 2007; Viruete et al. 2008). Most of the islands within this region are merely exposed reefs and sandbars, although Bonaire is notable for being a reef uplifted due to volcanic activity (Van der Lelij et al. 2007). Exhumation of the Leeward Antilles was not a single geochronologically constrained event, as multiple uplifting periods brought the crustal blocks upward during the Cretaceous period, Paleocene, and Eocene epochs (Bouysse 1988; Van der Lelij et al. 2007).

2.3 A Brief Overview of Lesser Antillean Climate

Bounding the Caribbean Sea, and located roughly between latitudes of 10 and 19°N, the climate zone of the Lesser Antilles is easily identifiable as tropical. Although average rainfall varies from island to island, most precipitation throughout the region occurs during the latter half of the year —forming distinctive a "wet season." Year-round temperatures fluctuate very little, with approximate averages ranging from low of 22 °C to high of 29 °C (Chenoweth and Divine 2008). The summer and autumn hurricane season also brings substantial rainfall as hurricanes, tropical storms, and tropical depressions make landfall over the island arc. Though these data for the Lesser Antilles *as a region* are difficult to find, Jury et al. (2007) note that from the Virgin Islands to Barbados at least, a dominant midsummer to autumnal precipitation pattern generally occurs. Within the Lesser Antilles Arc, and including the ABCs, precipitation—and therefore climate—remained variable. For example, orographic uplift on the taller volcanic islands produces rainfall events as lifting unsaturated air parcels cool at the dry adiabatic lapse rate until saturation is achieved, inducing cloud formation and precipitation (Chenoweth and Divine 2008). Annual variations in subregional precipitation and storm frequency/intensity in the Lesser Antilles are also influenced by larger regional oscillations such as the North Atlantic Oscillation (NAO) and the El Niño Sothern Oscillation (ENSO), and even more so in the southern latitudes (Jury et al. 2007). For more specific information regarding climate in the Lesser Antilles, each subsequent chapter contains a basic climatological overview and, where significant, the climatic geomorphology is also discussed.

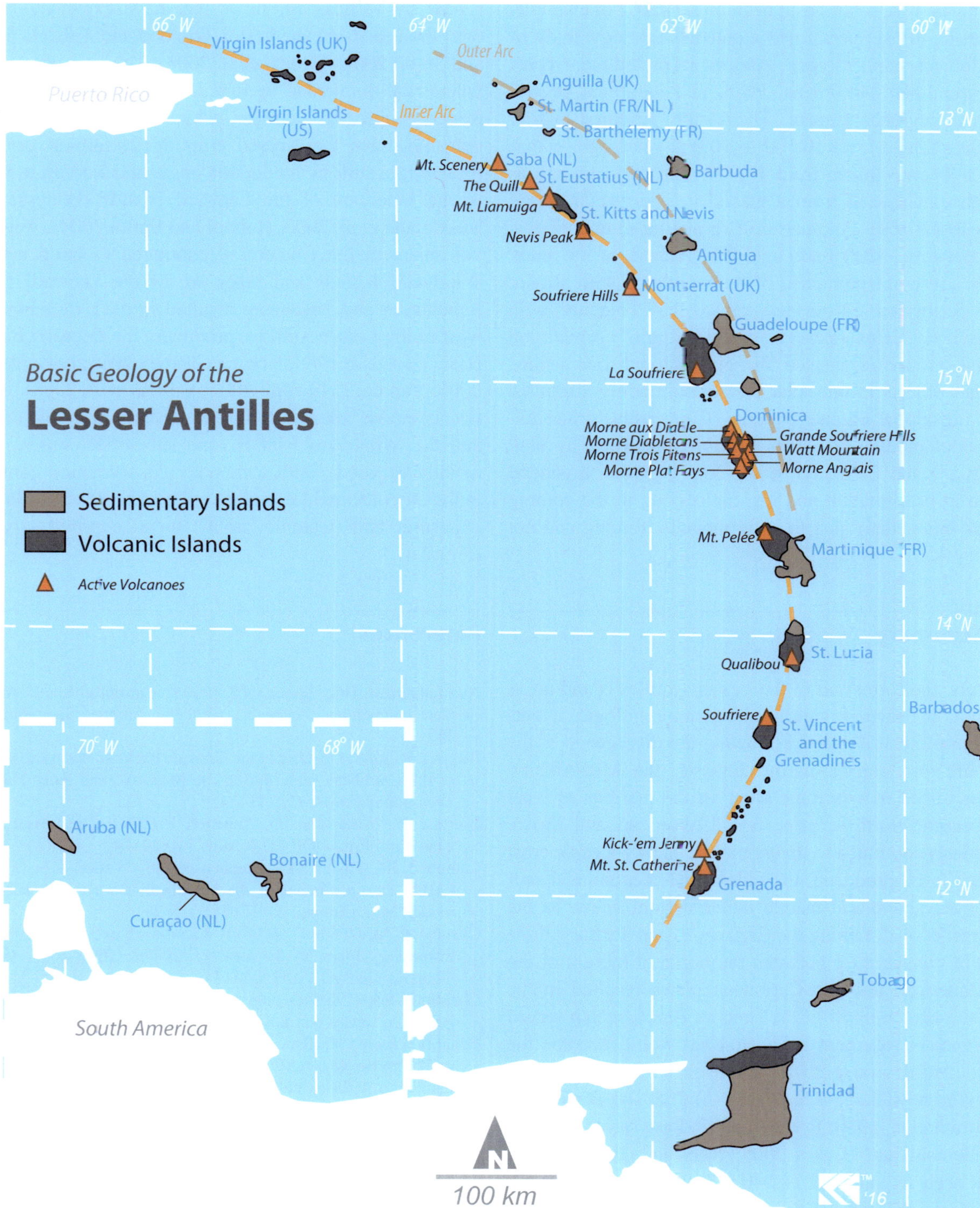

Fig. 2.4 Basic geology of the Lesser Antilles, illustrating dominant lithologies across the island chain. The outer Leeward Islands (outer arc) are the dominantly sedimentary islands of the northernmost arc, and the inner Leeward Islands (inner arc) are the dominantly igneous islands of the northernmost arc, converging at Dominica. Southward from Dominica, the Windward Islands have outcroppings of both igneous and sedimentary units. The Leeward Antilles, represented by the "ABC Islands," are composed of sedimentary units uplifted during orogenic events along the boundary between the Caribbean and South American plates. Cartography by K.M. Groom

Overall, Lesser Antillean climate may be somewhat consistent now, but other research suggests this may not always be the case—nor has it been in the past. For example, Hodell et al. (1991) performed a paleoclimatology study of the Caribbean using oxygen isotopes from sediment cores taken from Lake Miragoane, Haiti, and identified several changes in regional climate. Their study revealed a drier Caribbean climate from 10,500 to 10,000 years ago, with a shift to a wetter climate from 10,000 to 7000 years ago as the Earth transitioned from a glacial to interglacial period and sea levels rose. Another shift to a drier climatological regime occurred 3200 years ago. The authors of the study highlight the correspondence of their findings with similar studies performed across the globe that coincide with Milankovitch Cycles—periodic shifts in Earth's orbital and rotational properties caused by gravitational interactions with other objects in our solar system commonly attributed to affecting global climate patterns. Conversely, future climate model predictions, such as those presented by Hall et al. (2013) and Campbell et al. (2011), predict a general decrease in precipitation and regional drying in the coming centuries more likely due to anthropogenic forcing and not natural cycles.

2.4 Summary

Ultimately, the formation of the Lesser Antilles island arc is the result of several significant geomorphologic events spanning millions of years. Following the divergence of the supercontinent Pangaea in the Jurassic, the Mid-Atlantic Ridge produced new oceanic crust on the margins of what would become North America, South American, and Africa. As the plates continued to diverge and a "new" oceanic crust continued to be produced, the older crust accumulated sediments from above and mantle material from below as the mafic crust cooled. During the Cretaceous and Tertiary, three periods of convergence between the northern margin of the South American and the Caribbean plates resulted in the thrust-fault-driven uplift of the Bonaire Block, which served as the igneous basement cores of what would become the Leeward Antilles on both plates, even as carbonate reefs grew on the now shallow basement (Malfait and Dinkelman 1972; Smith et al. 1980; Bouysse 1988; Bouysse et al. 1990; Ave-Lallemant and Sisson 2005; Kerr and Tarney 2005; García-Casco et al. 2011). 40 million years ago, that increased weight and density resulted in the onset of subduction of the South American Plate beneath the Caribbean Plate. This process resulted in Tertiary-aged volcanism, which gave rise to the Windward Islands and the outer arc of the Leeward Islands, both of which are products of intermediate and mafic stratovolcanic activity. The subsequent exposed island stratigraphy reveals frequent intermixed explosive and effusive andesitic eruptions, which are represented by vertical ash flows that have expanded laterally onto the island-arc shelves of Trinidad and Tobago (Robool and Smith 2004; Viruete et al. 2008). The outer arc Leeward Islands are dominated by marine limestone reef deposits of Miocene age and younger, which are built upon extinct volcanic cores that ceased erupting 20 million years ago following a shift in subduction angle and leading to the rise of the inner arc of the Leeward Islands (Bouysse 1988; Macdonald et al. 2000; Robool and Smith 2004). Additional volcanism in the Quaternary continued to build upon the Windward Islands and inner arc of the Leeward Islands. Subduction and volcanism resulted in plate thinning to the west of the Lesser Antilles, producing the Grenada Back-arc Basin (Bouysse 1988; Kerr and Tarney 2005; Levander et al. 2006; Van der Lelij et al. 2007; Viruete et al. 2008). At least twenty active volcanic complexes remain along the inner arc and the Windward Islands (one submarine, nineteen subaerially exposed), with carbonate reef formation continuing in the outer arc and Leeward Antilles, except where subaerial exposure and tectonic uplift have exposed dry land (Fig. 2.4).

References

Ave-Lallemant HG, Sisson VB (2005) Caribbean-South American plate interactions, Venezuela. Geol Soc America Spec Pap 394:1–353

Bouysse P (1988) Opening of the Grenada back-arc basin and evolution of the Caribbean plate during the Mesozoic and early Paleogene. Tectonophysics 149(1):121–143

Bouysse P, Westercamp D, Andreieff P (1990) The Lesser Antilles island arc. Proc ODP Sci Results 110:29–44

Campbell JD, Taylor MA, Stephenson TS, Watson RA, Whyte FS (2011) Future climate of the Caribbean from a regional climate model. Int J Climatol 31(12):1866–1878

Chenoweth M, Divine D (2008) A document based 318-year record of tropical cyclones in the Lesser Antilles, 1690–2007. Geochem Geophys Geosyst 9:1525–2027

Christman RA (1953) Geology of St. Bartholomew, St. Martin, and Anguilla, Lesser Antilles. Geol Soc Am Bull 64(1):65–96

Edgar NT, Ewing JI, Hennion J (1971) Seismic refraction and reflection in Caribbean Sea. AAPG Bull 55:833–870

García-Casco A, Iturralde-Vinent MA, Proenza JA (2011) Subduction Zones of the Caribbean: the sedimentary, magmatic, metamorphic and ore-deposit records. Geol Acta: An Int Earth Sci J 9(3):217–224

Hall TC, Sealy AM, Stephenson TS, Kusunoki S, Taylor MA, Chen AA, Kitoh A (2013) Future climate of the Caribbean from a super-high-resolution atmospheric general circulation model. Theor Appl Climatol 113(1–2):271–287

Hodell DA, Curtis JH, Jones GA, Higuera-Gundy A, Brenner M, Binford MW, Dorsey KT (1991) Reconstruction of Caribbean climate change over the past 10,500 years. Nature 352:790–793

Jury M, Malmgren BA, Winter A (2007) Subregional precipitation climate of the Caribbean and relationships with ENSO and NAO. J Geophys Res: Atmo 112(D16)

Kerr AC, Tarney J (2005) Tectonic evolution of the Caribbean and northwestern South America; the case for accretion of two late cretaceous oceanic plateaus. Geology 33(4):269–272

Kerr AC, Tarney J, Marriner GF, Nivia A, Klaver GT, Saunders AS (1996) The geochemistry and tectonic setting of Late Cretaceous Caribbean and Colombian volcanism. J South American Earth Sci 9(1–2):111–120

Kerr AC, White RV, Thompson PME, Tarney J, Saunders AS (2003) No oceanic plateau; no Caribbean Plate? The seminal role of an oceanic plateau in Caribbean Plate evolution. AAPG Mem 79:126–158

Levander A, Schmitz M, Ave Lallemant HG, Zelt CA, Sawyer DS, Magnani MB, Mann P, Christeson G, Wright J, Pavlis G, Pindell J (2006) Evolution of the southern Caribbean plate boundary. EOS Trans 87(9):97–100

Macdonald R, Hawkesworth CJ, Heath E (2000) The Lesser Antilles volcanic chain: a study in arc magmatism. Earth Sci Rev 49(1–4):1–76

Malfait BT, Dinkelman MG (1972) Circum-Caribbean tectonic and igneous activity and the evolution of the Caribbean plate. Geol Soc Am Bull 83(2):251–271

Mann P (1999) Caribbean sedimentary basins; classification and tectonic setting from Jurassic to present. Sediment Basins World 4:3–31

Marshall LG, Sempere T, Butler RF (1997) Chronostratigraphy of mammal-bearing Paleocene of South America. J S Am Earth Sci 10 (1):63

Robool J, Smith AL (2004) Volcanology of Saba and St. Eustatius, northern Lesser Antilles. University of Puerto Rico, Department of Geology

Sigurdsson H, Sparks RSJ, Carey ST, Huang TC (1980) Volcanogenic sedimentation in the Lesser Antilles arc. J Geol 523–540

Sinton C, Duncan R, Storey M, Lewis J, Estrada J (1998) An oceanic flood basalt province within the Caribbean Plate. Earth Planet Sci Lett 155(3–4):221–235

Smith AL, Roobol MJ, Gunn BM (1980) The Lesser Antilles—a discussion of the island arc magmatism. Bull Volc 43(2):287–302

Van der Lelij R, Spikings R, Kounov A, Cosca M, Chew D (2007) Thermal and tectonic history of the Leeward Antilles: Aruba and Bonaire. University of Switzerland, Department of Mineralogy

Viruete JE, Joubert M, Urien P, Friedman R, Weis D, Ullrich T, Perez-Estaun A (2008) Caribbean island-arc rifting and back-arc basin development in the late Cretaceous: geochemical, isotopic and geochronological evidence from central Hispaniola. Lithos 104(1–4):378–404

3 The Virgin Islands

Kaelin M. Groom, Ryan Sincavage and Frederick Chambers

Abstract

The Virgin Islands, politically divided between the UK and USA, are a widespread island grouping at the northernmost end of the Lesser Antilles. Spanning nearly 135 km^2, the Virgin Islands consist of eight primary islands including (from north to south) Anegada, Virgin Gorda, Jost van Dyke, Tortola, St. Thomas, St. John, Water Island, and St. Croix. In addition to these main islands, several other smaller surrounding islands and cays fall under the jurisdiction of the Virgin Islands. Despite, or perhaps due to, broader spatial distribution, the Virgin Islands host a myriad of unique landscapes and landforms including volcanoes, valleys, reefs, and geologic anomalies.

Keywords

Virgin Islands • Tourism • Geology • Geomorphology

3.1 Introduction

Serving as the gateway into the Lesser Antilles, the Virgin Islands are an unexpectedly diverse island group—geologically, spatially, and politically (Fig. 3.1). Scattered along an active fault boundary, the islands vary from volcanic to sedimentary, with the occasional intrusive outcrop dotting the landscape. The islands range in size from well over 200 km^2 to just barely large enough to breach the ocean surface. Some are more than 65 km from the rest of the island chain. Their key position in the Caribbean has made them a desirable outpost throughout both the colonial and modern times for several larger Western powers including Spain, Denmark, Britain, and the USA. Currently, the Virgin Islands are divided into two distinctly governed nations: the British Virgin Islands (sometimes abbreviated to BVI) to the north/east and the US Virgin Islands (often "US Virgin Islands" or USVI) to the south and west. Having been under some form of British jurisdiction since the late 1600s, the British Virgin Islands include the main Islands of Tortola, Virgin Gorda, Anegada, and Jost van Dyke, along with over 30 smaller islands (Fig. 3.2). The US Virgin Islands include St. Croix, St. Thomas, St. John, and Water Island, plus a collection of smaller surrounding islands (Fig. 3.3). The US Virgin Islands were ruled by the Danish Monarchy during the colonial era and were not sold to the USA until 1917, after century-long negotiations (Dookhan 1974). Although the Virgin Islands only cover 500 km^2 in total landmass—roughly the size of Long Island, NY—they contain remarkably varied landforms, landscapes, tourism ventures, and hazards.

K.M. Groom (✉)
Department of Geosciences and King Fahd Center for Middle East Studies, University of Arkansas, Fayetteville, AR, USA
e-mail: kmgroom@uark.edu

R. Sincavage
Department of Earth and Environmental Sciences, Vanderbilt University, Nashville, TN, USA
e-mail: ryan.s.sincavage@vanderbilt.edu

F. Chambers
Department of Geography and Environmental Sciences, University of Colorado Denver, Denver, CO, USA
e-mail: frederick.chambers@ucdenver.edu

3.2 Setting

The Virgin Islands lie at the eastern end of the Greater Antilles and northern end of the Lesser Antilles, at the transition from active subduction of the Lesser Antilles Arc

C.D. Allen (ed.), *Landscapes and Landforms of the Lesser Antilles*,
World Geomorphological Landscapes, DOI 10.1007/978-3-319-55787-8_3

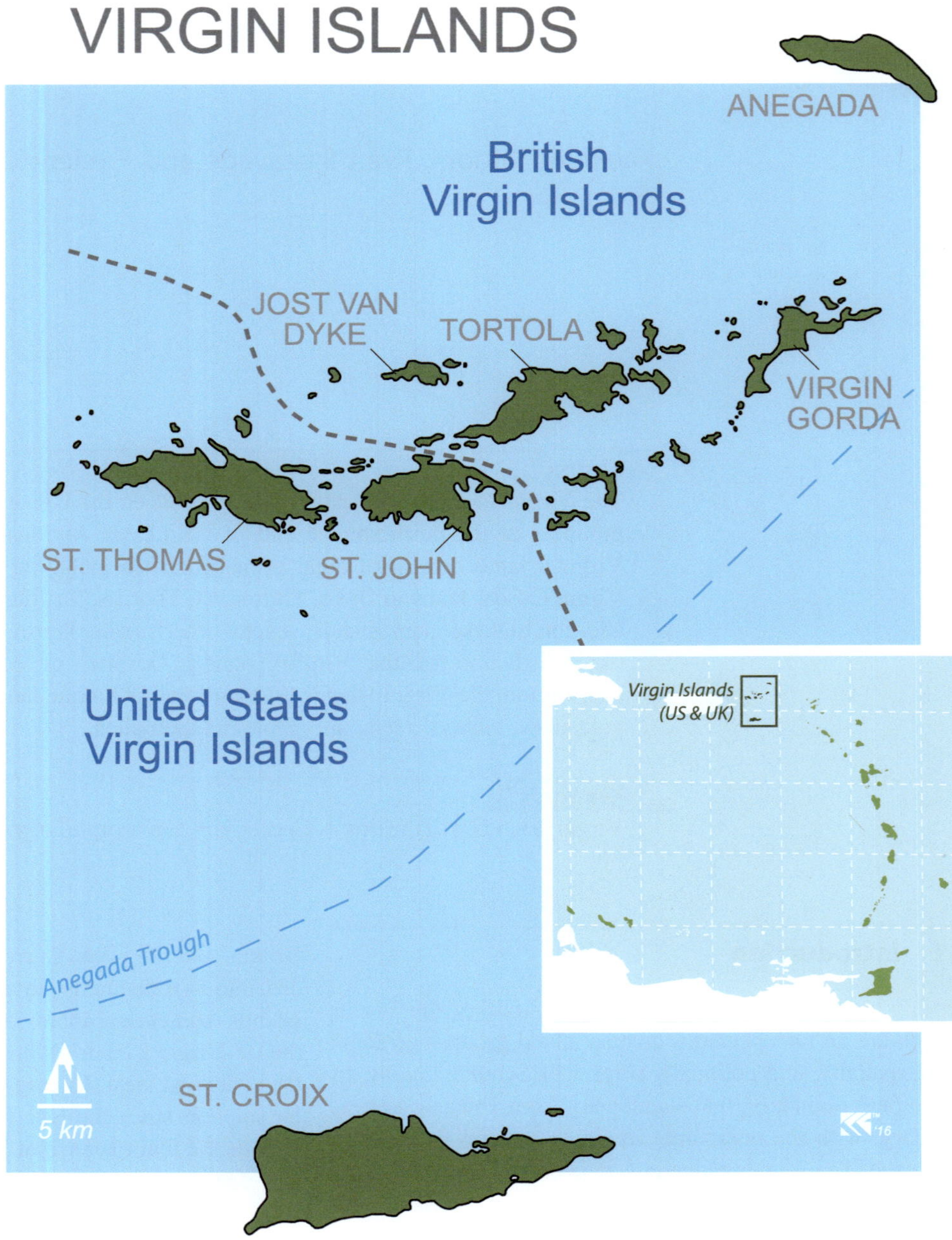

Fig. 3.1 Basic map of the Virgin Islands displaying spatial distribution, political affiliation, capital cities, the distance between St. Croix and Anegada from the rest of the Virgin Islands, as well as the Anegada Trough that separates St. Croix from the other islands. Cartography by K.M. Groom

and the strike-slip tectonic regime associated with the northern boundary of the Caribbean Plate (See Fig. 2.2). As described by Lewis and Draper (1990), the region can be divided into two east-northeast to west-southwest trending tectonic zones separated by the Anegada Trough, a ~50-km-wide extensional zone between the Caribbean Plate and the Puerto Rico-Virgin Islands microplate (Jany et al. 1990). The Northern Zone is essentially a continuation of Puerto Rico and contains the US Virgin Islands of St. Thomas and St. John and the British Virgin Islands of Tortola, Virgin Gorda, and Anegada. Structurally, the area is generally defined by beds dipping steeply to the north, likely associated with post-Eocene regional tectonics and not related to the intrusion of the Virgin Gorda batholith (Helsley 1971; Lewis and Draper 1990). The Southern Zone is geologically distinct from the Northern Zone, consisting of primarily sedimentary rocks (in contrast to the dominance of igneous rocks in the Northern Zone). It consists of a submarine ridge parallel to and south of the Anegada Trough, and the island of St. Croix, which offers the only exposures from which to describe the geology of this zone. Speed et al. (1979) described the structural complexities of St. Croix as having formed over three phases of deformation. In a general sense, the island consists of a graben structure at its center which hosted the accumulation of Tertiary sediments, surrounded on either side by horsts of Cretaceous-aged rocks.

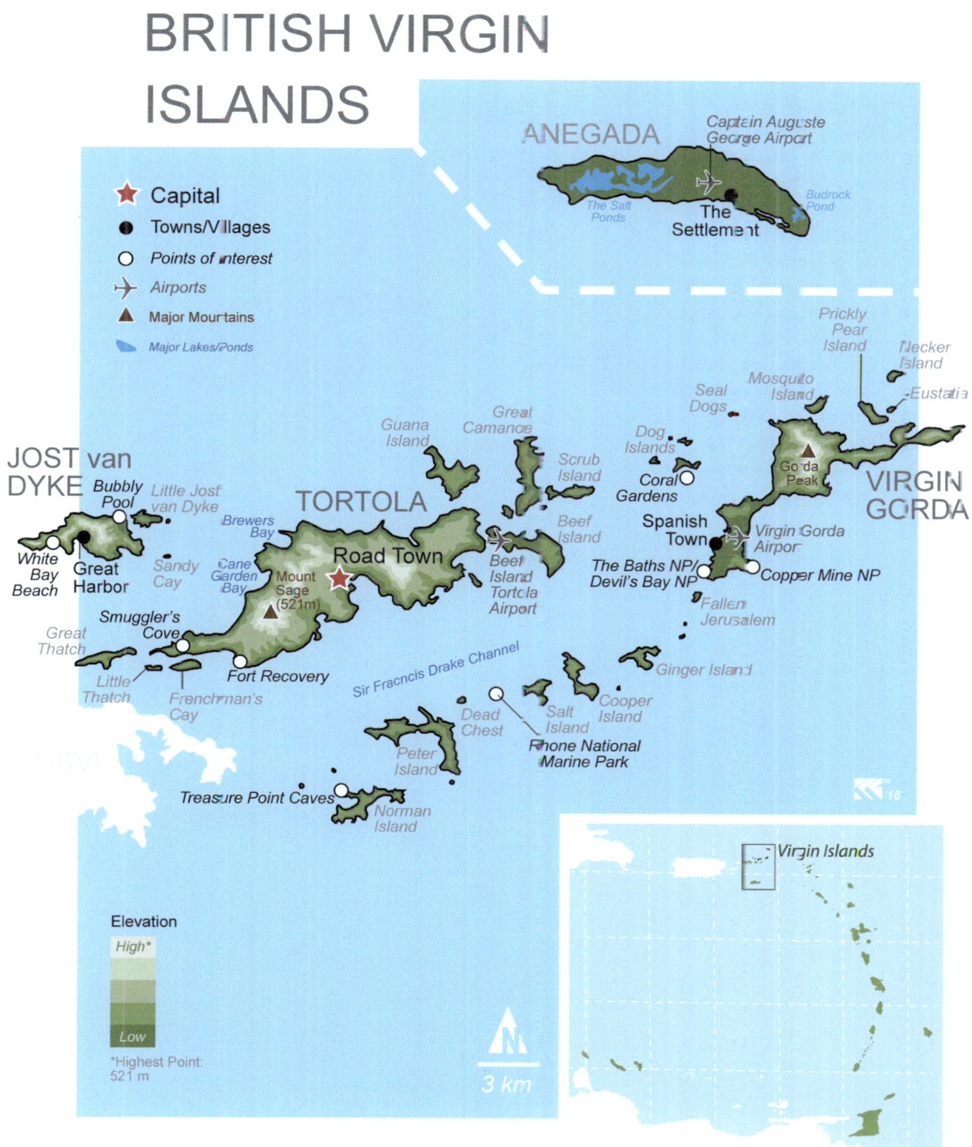

Fig. 3.2 General physiographic map of the primary British Virgin Islands and several of the larger surround islands showing political, physical, and cultural points of interest, as well as proximity to the USVI along the western border. Note the stark physiographic difference between Anegada and other main islands. Be aware that the actual distance between Anegada and the other BVI is larger than displayed on this map. The islands are shown closer to allow for more detailed representation of the islands' features. For a more accurate representation of distance, see Fig. 3.1. Cartography by K.M. Groom

The structure likely formed as the result of east-west extension in the Early Tertiary.

Despite the relatively small areal extent of the Virgin Islands, there are surprisingly diverse landforms along their interiors, coastlines, and offshore. Pioneering efforts at documenting the rich geomorphic variability of the US Virgin Islands were made by Meyerhoff (1926) and Cedarstrom (1941). As evidenced by the difference in tectonic regimes and geology of the Northern and Southern Zones of the Virgin Islands, there are similar differences in the physiography of the two regions. Islands in the Northern Zone are generally higher in elevation, with Mount Sage (521 m) on Tortola and Crown Mountain (473 m) on St. Thomas as the high points of the British and US Virgin Islands, respectively, whereas the high point of St. Croix (Mount Eagle) reaches an elevation of only 354 m. Much of the Northern Zone is contiguous with Puerto Rico as part of a submarine platform that was subaerially exposed during lowstands associated with Pleistocene glaciation (Rankin 2002). In both zones, the climate is considered tropical, but with a fair amount of variability across each individual island. Average precipitation values as high as 1000–1250 mm occur within the wettest areas of the islands,

Fig. 3.3 General physiographic map of the primary US Virgin Islands with major surround islands displaying political, physical, and cultural points of interest, as well as proximity to the BVI along the eastern border. Notice the extent of the Virgin Islands National Park on St. John and the wide plains of St. Croix. Be aware that the actual distance between St. Croix and the other USVI is larger than displayed on this map. The islands are shown closer to allow for more detailed representation of the islands' features. For a more accurate representation of distance, see Fig. 3.1. Cartography by K.M. Groom

which in general are found on the western (windward) sides. Vegetation varies with average precipitation, with desert-like scrub brush in the dry regions and dense tropical forests found in the more humid areas (Fig. 3.4).

3.3 Geochronology and Landforms

The diverse geology of the Virgin Islands is reflected in the resultant landforms and geomorphology. In general, the ages of the rock bodies in the Virgin Islands become younger in a north to south direction, with igneous rocks as old as Aptian–Albian (Early Cretaceous) exposed on the US Virgin Islands and Early Miocene sedimentary rocks found on St. Croix. Table 3.1 is a general stratigraphic summary of the Northern and Southern Zones of Virgin Islands, with nomenclature compiled from the work of numerous authors (Donnely 1966; Whetten 1966; Helsley 1971; Rankin 2002).

3.3.1 Geochronology

The oldest rocks in the Northern Zone are the Early Cretaceous Lameshur Volcanic-Intrusive Complex (as named by Rankin 2002), which consist of the Water Island Formation and the Careen Hill Intrusive Suite. This complex is dominated (~80%) by intrusive and extrusive igneous rocks of varying composition, particularly keratophyre (including lava flows, breccias, and dikes), with lesser amounts (~20%) of basalt and gabbro, and minor (1% or less)

Fig. 3.4 Examples of the diverse vegetation regimes found in the Virgin Islands. **a** Lush rainforest vegetation on Sage Mountain, Tortola. **b** Dry cactus vegetation, including *Stenocereus fimbriatus* and *Melocactus intortus*, on the Dead Chest National Park, BVI. **c** Caribbean dry forest surrounding a salt pond on Ginger Island, BVI. **d** Red mangroves, *Rhizophora mangle*, along the south east coast of Anegada. Photographs by Nancy Pascoe courtesy of the National Parks Trust of the Virgin Islands

amounts of radiolarian chert. The Lameshur Volcanic-Intrusive Complex is overlain by the Late Cretaceous Louisenhoj Formation, up to 4000 m of volcaniclastic materials, followed conformably by the Outer Brass Formation, a thin limestone unit containing planktonic foraminifera yielding Turonian to Santonian (Early- to mid-Late Cretaceous) ages. The Tutu Formation, conformably overlying the Outer Brass Formation, consists of the Picara Member, a conglomerate consisting of clasts of the underlying Louisenhoj Formation, a thin limestone lens known as the Congo Cay Member, and the overlying Mandal Member, consisting of thinner, finer-grained materials than the Picara Member. The Eocene Tortola and Necker Formations unconformably overly the Tutu Formation and represent the youngest stratified units in the Northern Zone. These units consist of more volcaniclastic debris, as well as occasional foraminifera-bearing limestone units that aided in age determination.

All the above stratified units are to some extent crosscut by mid- to late-Eocene intrusions. The two largest intrusive bodies are known as the Virgin Gorda batholith (located underneath the northeastern regions of the British Virgin Islands) and the Narrows batholith (located in the straits between St. John and Tortola). The composition of these plutonic bodies is predominantly tonalite, and there are excellent exposures of the Narrows batholith intruding the Mandal Member of the Tutu Formation along the north shore of Mary Point (Rankin 2002). Associated with these batholiths are many dikes, ranging from a few centimeters to 10s of meters in thickness. These features vary in composition from diabase to diorite, and contact metamorphism of the Eocene layers is not uncommon. While a detailed chronology of these bodies has not been undertaken, they have all been determined to be of approximately the same age, although it is likely that they did not all form synchronously.

Table 3.1 Basic stratigraphic table for the Virgin Islands based on various geologic studies and resources

Age	Group	Formation	Member	General lithology
Eocene	Virgin Island group			Mixture of volcaniclastic debris and limestone units
		Necker formation		
		Tortola formation		
Cretaceous	*Unconformity*			
		Tutu formation		
			Mandal member	Thinly bedded, fine-grained sandstone
			Congo bay member	Limestone lens
			Picara member	Conglomerate with clasts of Louisenhoj formation
		Outer brass formation		Thin limestone unit
		Louisenhoj formation		Volcaniclastic materials
	Unconformity			
	Lameshur volcanic intrusive complex			Complex mixture of lava flows, dikes, and chert
		Water Island formation		
		Careen hill intrusive suite		

3.3.2 Landforms

The interior landforms of Northern Zone are dominated by steep slopes covering much of each island's landmass, with 80% of the island of St. John consisting of slopes steeper than 30% (Anderson 1994). These ridges are typically incised by deep cuts (locally called ghauts) that often are focused on the near-vertical dikes prevalent throughout the region (Rankin 2002). Colluvium is common on most slopes, and virtually, all valleys contain alluvium of varying thickness. Large boulders that appear to be in place are common on the ridge tops. Their origins have been interpreted to be the result of subtropical rock decay—selective rock decay, differentiation of the rock mass into corestones and saprolite, and subsequent removal of loose saprolite. Many of the ghauts are also filled with boulders and exhibit a lack of vegetation, presumably the result of frequent flash flooding (Rankin 1984).

In contrast to the prevalence of igneous rocks in the Northern Zone of the Virgin Islands, most of the exposures in the Southern Zone consist of sedimentary rocks, and they are primarily found on the island of St. Croix. The Caledonia Formation, potentially as old as Cenomanian (Early to Late Cretaceous), consists of alternating bands of centimeter scale mudstones with abundant microfossils and graded sandstone and conglomerate beds with clasts of volcanic origin. This formation has been interpreted as distal turbidite deposits originating from a volcanic island (Speed et al. 1979). Complex interfingering relationships exist between the Caledonia Formation and the formations it grades into laterally and vertically, including the tuffaceous sandstones known as the Allandale Formation and the Judith Fancy Formation and the turbidites of the Cane Valley Formation. The youngest sedimentary rock bodies in the region are of Late Oligocene to Middle Miocene in age and are found in the central graben valley of St. Croix. The Jealousy Formation has limited outcrop exposures, but has been described as mainly mudstone with some interbedded conglomerates (Whetten 1966). Conformably overlying the Jealousy Formation is the Kingshill Formation, a flat-lying to gently folded fossil-rich limestone with abundant foraminifera and gastropods, indicative of a fossil coral reef that was later uplifted.

St. Croix's interior geomorphology can be subdivided into three zones: the Northside Range, the coastal plain, and the East End Range (Whetten 1966). The Northside Range is found on the northwestern side of the island and trends toward the east, gradually narrowing and merging with the coastal plain. Similarities exist between the topography of this range and the predominant topography of the Northern Zone, with deeply dissected ridges containing steep-walled, V-shaped valleys filled with alluvium. The highest elevations of the Northside Range create a rolling ridgeline that Meyerhoff (1926) correlated with similar surfaces in Puerto Rico and the northern Virgin Islands as a "peneplain," but this interpretation has generally been dismissed as unrealistic for such a small number of relatively low-discharge streams (Whetten 1966). The coastal plain consists of a relatively flat surface on the western side of the island that gradually grades into hillier terrain to the east as it converges with the East End Range. This range has similar characteristics and elevations to the Northside Range, and combining these two ranges represents the uplifted horsts surrounding the down-dropped graben of the coastal plain.

Both tectonic zones of the Virgin Islands contain similar coastal and offshore landforms. Due to the rugged topography and resistant underlying bedrock, sea cliffs are common features along most islands within the system. These rocky headlands serve as a barrier separating the thin beaches that surround the islands, often framing a protected cove or bay (Fig. 3.5). Fringing many of the islands are extensive coral reefs, many of which experienced massive losses in recent years due to increased sedimentation associated with island development, increases in macroalgae triggered by the death of coral species that fed on them, disease, and frequency and intensity of Atlantic Basin hurricanes. The health of the coral reefs, specifically as it relates to increased sediment flux, will be explored in more detail in the hazards section later in this chapter.

3.4 Landscape

When broadening the view of the Virgin Islands to include historical and cultural components, the imposing and lasting influence of Imperial Colonialism on the people and landscapes of the Virgin Islands becomes evident. The islands' proximity to historic trade routes between Europe and North

Fig. 3.5 Coastal landforms on St. Thomas and examples of modern utilization of those features. **a** The famous Magens Bay Beach (lower right of the image), St. Thomas, from Hull Bay Road. The bay is protected by large jutting headlands visible in the background and foreground. The island of Big Hans Lollick can be seen in the upper right behind the headlands. **b** Celebrity Cruise ship docked at Havensight Point, along with several yachts and sail boats (foreground and lower right side of the image) in a St. Thomas Harbor. While the naturally deep and protective harbor now hosts a myriad of pleasure/tourist ships, it was similarly attractive for trade and piracy centuries earlier. Photographs by B.J. Barnes

America and central location in the Caribbean gives them unique strategic importance. Dookhan (1974, p. 2) even goes as far as to say the Virgin Islands "occupy the keystone of the West Indies arch." This has led them to be initially coveted not only for their economic potential but also, and perhaps more importantly, for their military value. European "discovery" was by way of Columbus' voyages in 1493. The indigenous peoples of the islands were comprised of Ciboney, Caribs, and Arawaks—representative of most native populations in the Caribbean (Dookhan 1974)—and they left behind numerous rock carvings and petroglyphs, mainly on St. John and St. Croix (Haviser and Strecker 2006). The 1600s saw various European countries competing in attempts to colonize the region, focusing primarily on the efforts of Spain, England, France, and Denmark. Ultimately, by the mid-1700s, the Danes and British had prevailed in establishing lasting colonies on what are now recognized as the Virgin Islands. The British Virgin Islands became a cohesive entity following the British takeover of Tortola in 1672 and annexation of Anegada and Virgin Gorda in 1680. At about the same time, the Danish West India Guinea Company was establishing permanent settlements on St. Thomas and St. Johns. St. Croix was purchased from the French in 1733, thereby unifying the three islands as the Danish West Indies, sometimes referred to as the Danish Antilles, later falling under the direct rule of the Denmark–Norway Monarchy a few decades later (Dookhan 1974).

These political transitions were far from peaceful—violent conflicts between locals and foreigners, competing colonial forces, and rampant piracy—and this left a lasting impression on the landscapes of the Virgin Islands. Large stone fortresses were erected along the coasts of the primary islands, such as Fort Christian on St. Thomas and Fort Burt on Tortola. These not only created a protective component in the landscapes themselves, but also required intensive stone quarrying and extraction of natural resources on the islands—making the way for modern quarry businesses such as Aggregate Inc. on St. Croix. These forts were erected to deter political competition, but also to protect the main port cities from pirates and buccaneers. In fact, much of piracy's "Golden Age" (1690–1730) took place in the North Atlantic and Caribbean Sea (Hallwood and Miceli 2015). The unique features of the Virgin Islands that attracted colonial powers also made them prime targets for piracy, such as the natural deepwater harbors, central location, and continuous regional trade, as well as the numerous surrounding islands and coves that could serve as hideouts. Many of the smaller islands in the Virgin Islands still have feature names that reflect their dubious past: Dead Chest Island, Treasure Point Caves, Deadman's Bay, to name a few.

Once colonial powers were fully established by the end of the eighteenth century, there were two primary and complementary economies on the islands of the region: sugar and slave trade. Sugar plantations were a dominant economic venture during much of the colonial years, and slave trade from Africa was used both to supply labor for the cane fields and also as a commodity unto itself (Armstrong 2003). The central location of the Virgin Islands encouraged the development of both enterprises with notable success. Both the British and Danes employed the largely profitable, and equally popular, triangular trade model: European products were sold for slaves along West African coast, and the slaves were brought to the Caribbean to work plantations, which, in turn, supplied sugar and rum for trade back in Europe (Hall and Higman 1992). With the triangular trade in full effect, thousands of African slaves were transported to the Virgin Islands, replacing indentured employees, such as poor or criminal deportees from Europe, who had been working the fields to that point (Armstrong 2003).

The massive influx of slaves had such an impact on the islands that one scholar stated: "One of the most important aspects of the History of the British Virgin Islands is slavery" (Pickering 1987, p. 45). St. Croix, in particular, was especially profitable during this time due to its unique topography: The fertile coastal plain is constantly fed by water sourced from both northern and eastern highlands, making it ideal for intensive agriculture. Even the prevailing northeast trade winds powered the large windmills crushing sugarcane (Dookhan 1974). While sugar plantations were established on most colonized Caribbean Islands, very few were as productive as those on St. Croix (Macpherson 1973). Cotton farming was also attempted in the Danish West Indies, but it proved too taxing on the soils and the efforts were abandoned for far more lucrative sugarcane crops (Tyson 1992).

After several slave revolts and pressure from Europe, the British Empire abolished slavery in 1834, followed by Danish abolition in 1848, and the Virgin Islands began the long and difficult process of gaining self-governance. While most freed slaves stayed on the plantations to work for minimal compensation, a large portion moved to economic centers such as the port cites of Charlotte Amalie, St. Thomas and Christiansted and Frederiksted, St. Croix to work at the docks (Hall 1985). In fact, St. Thomas is said to have some of the best natural deepwater harbors in the Caribbean (CIA: Virgin Islands 2016). These natural harbors and strategic trade cites of the Virgin Islands allowed steady trade of goods throughout the region as an economic alternative, as sugarcane production gradually declined following the abolition of slavery, increased production of sugar beets in Europe and USA, and a series of devastating hurricanes (Dookhan 1974). Around the turn of the twentieth century, property values on the British Virgin Islands were so low that freed slaves could purchase land of their own—inspiring renewed national pride. Around the same time, the USA brokered a deal with Denmark to purchase the Danish West Indies for $25,000,000 in gold and renamed them the US

Virgin Islands (Dookhan 1974). The US valuation of the islands was primarily one of the military nature. World War I was winding down, and the area was viewed as having strategic importance to long-term US geopolitical interests in the region.

Today, the US Virgin Islands are an unincorporated organized territory of the US and the British Virgin Islands are a self-governing member of the Commonwealth of Nations, otherwise known as the British Commonwealth, after gaining autonomy in the 1960s. Despite the now revitalized economies and national identity, the lasting effects of colonialism, slavery, and plantation occupation can still be seen in the landscapes throughout the islands. Many of the lavish plantation houses and defensive forts erected during imperial rule remain reminders of the islands' turbulent and, to many, painful history. However, some of that history is now being repurposed as a resource to further economic development by means of heritage/historical tourism—such as the conversion of plantation homes into interpretive museums and/or memorials—one of the Virgin Islands' many tourism outlets.

3.5 Heritage and Tourism

As with many of the islands of the Lesser Antilles, tourism is the primary economic industry in the Virgin Islands (both for US and the UK). As per the CIA World Fact Book, tourism accounts for nearly 45% of the national income in the British Virgin Islands and almost 60% in the US Virgin Islands (CIA: Virgin Islands 2016). Despite being such small islands, they offer a wide variety of tourism resources and support several different specialties of the tourism sector: heritage, resort, relaxation/spa, adventure, aquatic, natural/ecotourism, and more. These interrelating branches of tourism often incorporate elements of the built landscape, such as Fort Burt on Tortola and the Annaberg Sugar Plantation on St. John, but also rely heavily on natural landforms and environments for context, advertising, and as attractions in and of themselves. In other words, certain elements of the physical landscapes facilitate tourism, such as the natural harbor of St. Thomas-Cruz Bay once promoting colonial trade now docking cruise ships, while others actively draw and attract tourism, such as Magens Bay Beach on St. Thomas, famous for its gentle waters protected by large peninsulas on either side of the half-mile stretch of sand. The natural beauty and diversity of the Virgin Islands provide unique tourism assets but also incorporate potentially fragile natural heritage, which is managed differently between the US and British governments.

On the British Virgin Islands, areas of significant value, whether historic and/or natural, marine and/or terrestrial, are designated as national parks and managed by the BVI National Parks Trust, established in 1961 under the Ministry of Natural Resources and Labour. The Trust currently manages over 20 different parks and protected areas that include geologic anomalies to the region, such as the Baths and the Bubbly Pool, as well as iconic and picturesque landscapes, such as Sage Mountain or Sandy Cay (Fig. 3.6). The Baths National Park, located adjacent to Devil's Bay National Park on the southern tip of Virgin Gorda, consists of scattered massive spheroidally decayed granite boulders creating stunning sea pools and secluded caves along the sandy shoreline. For something a bit more exciting, the Bubbly Pool—which used to be a "locals' secret" before becoming a national park—is a small alcove on the northern shore of Jost van Dyke surrounded by large boulders. When the tide changes, large swells from the Atlantic Ocean break over the protective boulders letting only sea foam and white waters rush into the pool, creating what some call a "natural Jacuzzi"—thus the name "Bubbly Pool" (Jost van Dyke 2013). Established in 1964 and covering 64 acres of central Tortola, Sage Mountain National Park was the first National Park in the British Virgin Islands and is the highest point in all the Virgin Islands. Not only does the park protect a thriving forest of white cedar and mahogany trees, but it also represents the beginning of nature conservation and natural resource management in the British Virgin Islands (Sage Mountain National Park 2016). Equally as iconic, the small and uninhabited island of Sandy Cay situated between Jost van Dyke and Tortola consists of nothing but soft sandy beaches, crystal clear water, a small salt pond at its center, and green-hued rocks from significant copper ore deposits in the area (Sandy Cay: BVI 2012). Sandy Cay is a relatively new addition to the National Park system being transferred to the trust by the Laurence Rockefeller Estate in 2008 and represents one of the most intact dry tropical forests in the world (Zaluski 2014).

Since the US Virgin Islands are an unincorporated territory of the USA, instead of an autonomous nation like the British Virgin Islands, the establishment and management of protected areas are slightly different. There are a few nationally protected areas governed by the US National Park Service (NPS), but nowhere near as many as the British Virgin Islands. The only full park is the Virgin Islands National Park that covers most of St. John. Additional properties include the alleged landing site of Christopher Columbus at the Salt River National Historic Park on St. Croix, Buck Island Reef National Monument off the north-east coast of St. Croix (not to be confused with the Buck Island National Wildlife Reserve just south of St. Thomas), and the three-mile-long Virgin Islands Coral Reef National Monument along the southern coast of St. John (Virgin Island: Parks 2016) (Fig. 3.7). While there are relatively few nationally protected assets, they are quite large. For example, Virgin Islands National Park is well over 7000 acres and Buck Island with its surrounding reefs total over 19,000 acres of protected land (Virgin Islands: Parks 2016). The US

Fig. 3.6 Examples of National Parks and geologic points of interest in the British Virgin Islands. **a** Aerial view of the uninhabited Sandy Cay Habitat Management Area off the coast of Jost van Dyke. Photograph by I. Bahadoor courtesy of the National Parks Trust of the Virgin Islands. **b** Dry vegetation along the steep cliffs at the Dead Chest National Park. Photograph by N. Pascoe courtesy of the National Parks Trust of the Virgin Islands. **c** One of the large covered pools at the Baths National Park on the southwest tip of Virgin Gorda. Photograph by S. Estes. **d** An external view of the monolithic granite outcrop and surrounding beaches of the Baths National Park. Photograph by N. Pascoe courtesy of the National Parks Trust of the Virgin Islands

Virgin Islands also have a collection of local governed "Territory Protected Areas"—though these are mostly marine zones—which are popular tourism destinations.

Beyond the terrestrial landscapes, both the US and British Virgin Islands boast an array of beaches, reefs, and other marine heritage/tourism assets. Some of the most famous beaches include Trunk Bay (part of the Virgin Islands National Park on St. John) and Sandy Point on St. Croix (which is the longest beach in the USVI.); nearly, all of Anegada's shoreline is one continuous beach. Offshore, the Virgin Islands have several well-established reef systems that are popular for snorkeling and scuba diving, much like those found in the Salt River Canyon, featuring a 300 m plummet to the sea floor, off St. Croix's north coast (St. Croix Diving 2016). Other popular dive sites showcase more anomalous geologic features, unique to the Virgin Islands, such as Blond Rock, an incomplete seamount between Salt and Dead Chest Islands rising from nearly 20 m at the sea floor to a mere 4.5 m from the water's surface (Bulter 2007), or the Tunnels of Thatch, a series of submerged volcanic tubes and trenches running under Thatch Cay to the northwest of St. Thomas. Recently, both the US and British Virgin Islands have been creating several artificial reefs, such as the two decommissioned WWII Navy Barges purposely submerged off the coast of St. Thomas and the sunken 360 Shorts Airplane in the Coral Gardens of Great Dog Island, to boost coral growth and revitalize marine habitats damaged by over-use, storms, or other forms of environmental stress.

3.6 Hazards

Being situated in an area of complex tectonics, in an open ocean basin, with a burgeoning population, it is not surprising that the Virgin Islands are subjected to a vast array of natural hazards. The steep slopes combined with tropical moisture make the threats of mass wasting events ever present. Seismic hazards associated with the numerous

Fig. 3.7 Landscapes and landforms on St. John, USVI. **a** Landscape view of Coral Harbor and the town of Coral Bay (*center*) from Centerline Road on St. John, looking to the south. The land on the *left side* of the photograph is the East End of St. John. Norman Island may be seen on the *horizon* on the *left*. **b** The picturesque John's Folly Bay and the Sabbat Channel between St. John and Leduck Island (just visible in the far *right* of the image). Fringing the Virgin Islands National Park, John's Folly (both the village and bay) on the southeastern coast is one of the few areas on St. John not managed by the US NPS. Tortola (*Left*) and Peter Island (*Right*) can be seen on the horizon in the background. **c** Looking northwest on a secluded beach with snorkeling between Long Bay and Pelican Rock on the East End of St. John. Photographs by C.M. Fielding

interacting plates and microplates are a constant threat. Tropical storms and hurricanes frequently carve a path of destruction through the region. It is not only these hazards in and of themselves that cause concerns for the inhabitants of the Virgin Islands, but also the side effects of these factors that can wreak havoc.

Using an extensive network of GPS data, Jansma et al. (2000) demonstrated that Puerto Rico and the northern islands of the Virgin Islands are actually part of a smaller microplate associated with the larger Caribbean Plate. As the North American plate moves west relative to the microplate associated with the Virgin Islands, a left-lateral strike-slip fault zone comparable to some of the largest and well-known systems of the world (e.g., the San Andreas in California) is created. While a study of historical earthquakes indicated that the release of energy in events within the Virgin Islands has been orders of magnitude lower than those within Puerto Rico (Doser et al. 2005), there is sedimentologic and stratigraphic evidence to suggest a major tsunami has hit the island of Anegada in the British Islands over the past several hundred years (Atwater et al. 2012). Despite the relatively low population of the Virgin Islands in comparison with the neighboring countries such as Puerto Rico, Cuba, and the Dominican Republic, because of the small size of the islands, the population density is high. Furthermore, the steep slopes of the islands are prone to mass wasting during energy release in a seismic event, and many of the houses are built with substandard construction materials and techniques. Seismic risk for the Virgin Islands is considered high due to these combinations of physical and human factors.

Another coupled human–environment hazard is the production and transport of sediment delivered to the marine ecosystems surrounding these islands. As development increases in the Virgin Islands, so does the density of roads (paved and unpaved) that dissect the islands. Regional studies (MacDonald et al. 2001; Ramos-Scharron and MacDonald 2007) have demonstrated that unpaved roads are the primary source of sediment for coastal waters, and that relatively minor precipitation events can generate disproportionately large volumes of sediment. Coral reef cover in the Caribbean has been in decline for at least the last 30 years, and increased sediment flux is considered a major factor in the health of corals adjacent to small islands (Gardner et al. 2003). As development and populations continue to increase in the Virgin Islands, it will become increasingly important to consider the significance of sediment production and transport in the design of infrastructure and control structures.

3.7 Conclusion

Boasting celebrated beaches, serene waters, emerald green mountains, and some of the best Scuba diving in the world (Bulter 2007), the Virgin Islands seem to have it all. Even elements of the islands' tumultuous past have been transformed by the local population, such as the conversion of the Whim Estate on St. Croix, a large sugar plantation, into a museum that not only commemorates the hardships of slavery but also celebrates their resilience and transition into free Crucians—people of St. Croix (Estate Whim Museum 2016). Indeed, even those quintessential deserted islands with nothing but beaches and a single palm tree that can only be reached by boat featured in countless travel magazines and tour brochures can be found in the Virgin Islands. An impressive myriad of unique geologic structures including soaring pinnacles, underwater lava tubes, colossal granite outcrops (i.e., the Baths, see Sect. 3.5), and towering volcanic peaks also sprinkle the islands. This is not to say the inhabitants of the islands are without their challenges, but the fact that all these features—alongside distinctive thriving cultures and histories—exist within a terrestrial area no larger than Long Island, NY, is a humbling reminder that there are more to landscapes than sheer size.

References

Anderson DM (1994) Analysis and modeling of erosion hazards and sediment delivery on St. John, U.S. Virgin Islands. Technical Report NPS/NRWRD/NRTR-94–34, US National Park Service Water Resources Division, Fort Collins, CO

Armstrong DV (2003) Creole transformation from slavery to freedom: historical archaeology of the east end community. University Press of Florida, St. John, Virgin Islands

Atwater BF, ten Brink US, Buckley M, Halley RS, Jaffe BE, Lopez-Venegas AM, Reinhardt EG, Tuttle MP, Watt S, Wei Y (2012) Geomorphic and stratigraphic evidence for an unusual tsunami or storm a few centuries ago at Anegada, British Virgin Islands. Nat Hazards 63:51–84

British Virgin Islands. (Updated 2016) CIA World Fact Book Online. URL: https://www.cia.gov/library/publications/the-world-factbook/geos/vi.html

Butler B (May 2 2007) Destination Top Ten: British Virgin Islands. *SCUBA diving* online magazine.URL: http://www.scubadiving.com/destination-top-10-british-virgin-islands

Cedarstrom DJ (1941) Notes on the physiography of St. Croix, Virgin Islands. Am J Sci 239(8):554–576

CIA: Virgin Islands. (Updated 2016) CIA world fact book online.URL: https://www.cia.gov/library/publications/the-world-factbook/geos/vq.html

Donnely TW (1966) Geology of St. Thomas and St. John, U.S. Virgin Islands. In: Hess HH (ed) Caribbean geological investigations: Geological Society of America Memoir 98, pp 85–176

Dookhan I (1974) A history of the Virgin Islands of the United States. Canoe Press, Kingston, Jamaica

Doser DI, Rodriguez CM, Flores C (2005) Historical earthquakes of the Puerto Rico-Virgin Islands region (1915–1963). Geol Soc Am Spec Pap 385:103–114

Estate Whim Museum (Accessed 12 Apr 2016) St. Croix Landmarks Society official website.URL: http://stcroixlandmarks.com/museums/estate-whim-museum

Gardner TA, Côté IM, Gill JA, Grant A, Watkinson AR (2003) Long term region-wide declines in Caribbean corals. Science 301:958–960

Hall NA (1985) Maritime maroons: "Grand Marronage" from the Danish West Indies. The William and Mary Quarterly: A Magazine of Early American History and, 476–498

Hall NA, Higman BW (1992) Slave society in the Danish West Indies: St. Thomas, St. John, and St. Croix. Aarhus Universitetsforlag

Hallwood CP, Miceli TJ (2015) Piracy in the golden age, 1690–1730: Lessons for Today. In: Maritime piracy and its control: an economic analysis. Palgrave Macmillan, US, pp 97–114

Haviser JB, Strecker M (2006) 'Zone 2: Caribbean Area and north-coastal South America' in rock art of Latin America and the Caribbean: Thematic study by The International Council on Monuments and Sites. Available http://www.icomos.org/studies/rock-latinamerica.htm

Helsley CE (1971) Summary of the geology of the British Virgin Islands. In: Mattson PH (ed) Transactions of the fifth Caribbean geological conference. St. Thomas, U.S. Virgin Islands 1968: New York, Queens College Press, pp 69–76

Jansma P, Mattioli G, Lopez A, DeMets C, Dixon T, Mann P, Calais E (2000) Neotectonics of Puerto Rico and the Virgin Islands, northeastern Caribbean from GPS geodesy. Tectonics 19:1021–1037

Jany I, Scanlon KM, Mauffret A (1990) Geological interpretation of combined Seabeam, Gloria and seismic data from Anegada Passage (Virgin Islands, North Caribbean). Mar Geophys Res 12:173–196

"Jost van Dyke: The "Barefoot" Island (Updated 2013) www.b-v-i.com URL: http://b-v-i.com/JostVanDyke/default.htm#Bubbly Pool

Lewis JF, Draper G (1990) Geology and tectonic evolution of the northern Caribbean margin. In Dengo G, Case JE (eds) The Caribbean region: Boulder, Colorado, Geological Society of America. The Geology of North America, v. H, pp 77–140

MacDonald LH, Sampson RW, Anderson DM (2001) Runoff and road erosion at the plot and road segment scales, St. John, US Virgin Islands. Earth Surf Proc Land 26:251–272

Macpherson J (1973) Caribbean lands: a geography of the West Indies, 3rd edn. Addison-Wesley Longman Ltd, Kingston, Jamaica

Meyerhoff HA (1926) Geology of the Virgin Islands, Culebra and Vieques: Physiography. N. Y.. Acad. Sci. Scientific Survey of Puerto Rico and the Virgin Islands 4:72–219
Pickering VW (1987) A concise history of the British Virgin Islands: from the Amerindians to 1986. Falcon Publications International
Ramos-Scharron CE, MacDonald LH (2007) Runoff and suspended sediment yields from an unpaved road segment, St John, US Virgin Islands. Hydrol Process 21:35–50
Rankin DW (1984) Geology of the U.S. Virgin Islands, a progress report. In: Gori PL, Hays WW (eds) A Workshop on "Earthquake hazards in the Virgin Islands region," Apr 9–10 1984, St. Thomas, Virgin Islands: U.S. Geological Survey Open-File Report 84–762, pp 83–96
Rankin DW (2002) Geology of St. John, U.S. Virgin Islands. USGS Professional Paper 1631
Sage Mountain National Park. (Updated 2016) www.bvitourism.com.URL: http://www.bvitourism.com/activity/sage-mountain-national-park
Sandy Cay, British Virgin Islands. (Updated 2012) www bvivacation. com.URL: http://www.bvivacation.com/sea-sand-sail/beaches-other-islands-cays-sandy-cay.php#.VwtVozZ97Ac
Speed RC, Gerhard LC, McKee EH (1979) Ages of deposition, deformation, and intrusions of Cretaceous rocks, eastern St Croix, Virgin Islands. Geol Soc Amer Bull 90(629):632
St. Croix Diving. Accessed 11 Apr 2016 www.visitusvi.com.URL: http://www.visitusvi.com/stcroix/diving
Tyson GF (1992) On the periphery of the peripheries: the cotton plantations of St. Croix, Danish West Indies, 1735–1815. J Caribb Hist 26(1):1
Virgin Islands: Parks. (Accessed 11 Apr 2016) National Park Service Official Website.URL: https://www.nps.gov/state/vi/index.htm
Whetten JT (1966) Geology of St Croix. U.S. Virgin Islands. In: Caribbean geological investigations 98. Geol. Soc. Amer. Mem., pp 177 239
Zaluski S (2014) Sandy cay: on the trail of "nature's" little secrets. Jost van Dykes Preservation Society Online Library. URL: http://jvdps.org/library-links-info/

Anguilla

4

Susanna L. Diller, Casey D. Allen, Ayumi Kuramae
and Donald M. Thieme

Abstract

Located east of the US Virgin Islands and 8 km north of Saint Martin, Anguilla is a low-lying Limestone Caribbee with no permanent standing water other than brackish salt ponds. With a volcanic historic long since passed, Anguilla has no mountainous features. Its limestone composition resulted in the formation of numerous caves, some of which are home to unique Amerindian religious sites. Anguilla's lack of soil coverage resulted in a relative lack of historic and modern plantation agriculture, and the island relied heavily on the export of salt and lobster until a tourism industry developed around the 33 public beaches that ring the island. Currently, tourism is the leading industry on Anguilla and has impacted both the environment and local awareness of environmental health.

Keywords

Anguilla • Beaches • Limestone • Erosion • Tourism

4.1 Introduction

Anguilla is one of the most northerly Leeward Islands in the Lesser Antilles, located east of the US Virgin Islands and 8 km north of Saint Martin/Sint Maarten (Fig. 4.1). Despite evidence of volcanic activity in the far past, Anguilla is a Limestone Caribbee—an island with limestone-dominant rock type. Because of this, Anguilla is a homogenously low-lying island with little relief (Bruce 2000), the highest point being Crocus Hill (65 m). Anguilla's water supply comes from rain and desalination plants, as the island has no permanent standing water other than salt ponds (Pan 2007). These salt ponds represent feeding grounds for several bird species and provide protection against flooding of living areas, businesses and roads (Anguilla 2008). Other than the salt ponds, the most dominant features on Anguilla are the caves, most notably Pitch Apple Hole, Cavanaugh Cave (alternately Cavannah or Kavanaugh), and Fountain Cavern, the latter of which is a national park that has been nominated for UNESCO recognition (Rogers 2010).

Anguilla is 26 km long by 5 km at the widest point, and this oblong shape led to its current name, which means "eel-shaped," while the original Amerindian name, *Malliouhana,* means "arrow-shaped" (Cambridge 2012b). Though some have suggested the first European to see Anguilla was Christopher Columbus, there is no evidence to that effect, and the first European contact was generally attributed to the French in 1565. After that, there was limited European exposure until the English, coming from St. Kitts, colonized the island in 1650. Despite a few French attempts at invasion, the British successfully retained control of the island

S.L. Diller (✉)
Department of Geography and Environmental Studies, University of New Mexico, Albuquerque, USA
e-mail: sldiller14@gmail.com

C.D. Allen
General Education, Western Governors University, Salt Lake City, UT, USA
e-mail: caseallen@gmail.com

A. Kuramae
University of Applied Sciences, Van-Hall Larenstein, Valdosta, Leeuwarden, Netherlands
e-mail: ayumikuramae@gmail.com

D.M. Thieme
Department of Physics, Astronomy, and Geosciences, Valdosta State University, Valdosta, GA, USA
e-mail: dmthieme@valdosta.edu

C.D. Allen (ed.), *Landscapes and Landforms of the Lesser Antilles*,
World Geomorphological Landscapes, DOI 10.1007/978-3-319-55787-8_4

Fig. 4.1 General physiographic map of Anguilla with surrounding islands and cays. Despite its fairly rocky topography, especially in the center of the island, the highest point of Anguilla is only 65 m. The Anguilla Channel to the south separates Anguilla from nearby St. Martin. The close proximity allows for a regular ferry service between St. Martin, Anguilla, Dog Island, and Scrub Island. Cartography by K. M. Groom

(Martin 1839; Staff 2015). In 1967, England oversaw the unified dependency of St. Kitts-Nevis-Anguilla, a move that most Anguillian opposed. The island revolted, and in 1971, they were returned to full British control. In 1980, Anguilla formally seceded from the unified dependency and has remained a British Overseas Territory since then (Staff 2015).

With a total land area of 90 km^2 and 61 km of coastline peppered with beautiful bays, Anguilla has come to be viewed by some as the best beach destination in the Caribbean (Tisdall 2010). Tourism remains Anguilla's number one industry, spurred on by more than 33 beaches, all of which are open to the public. The cays and islands found off the coast also serve as tourist attractions, with larger cays such as Scilly Cay and Prickly Pear Cay, and islands such as Scrub Island and Dog Island easily accessible from the main island. Along with the cays and islands, coral reefs ring the coast, most notably an extensive barrier reef to the north of the island. The reef systems are important for preventing beach erosion and helping mitigate damage from hurricanes, but coral health has reportedly declined in recent years. This has impacted offshore fisheries, a significant issue as fishing is one of Anguilla's main industries, employing 20% of the population (Hodge 2011). Anguilla has additionally developed a reputation as a tax haven, and offshore banking services are the second largest industry.

4.2 Setting

4.2.1 Geology

The island of Anguilla, along with St. Martin/Sint Maarten, St. Barthélemy, and the Sombrero and Dog islets, sits atop approximately 4500 km^2 of raised seafloor called the St. Barthélemy-Anguilla platform (Bouysse 1979, 1983, 1986). The platform slopes gradually to the southwest but has a much steeper eastern slope with submarine rock outcrops of early Tertiary age or older (Bouysse 1986).

The Lesser Antilles are divided into Limestone or Volcanic Caribbees, and Anguilla is one of the former (Quinn 2015: 415). On most of the Limestone Caribbees, limestone and other sedimentary material on the island's surface covers older volcanic arc remnants or metamorphic rocks. Topographically, Anguilla and the other Limestone Caribbees have less relief and fewer volcanic or tectonic landforms than the islands created by recent volcanic eruptions. Even among the Caribbees, Anguilla is particularly flat and low-lying, as are most of the small cays which occur on islets to the north and west of the main island (Hamylton 2013). On the main island, relief is greatest along the north shore of Anguilla, where the highest elevation occurs at the cliff overlooking Crocus Bay at 65 m above sea level (Bouysse 1986; Christman 1953; Mitchell (n. d.a); Stanley 2009; Smith 2013).

Geologically, Anguilla itself represents a textbook example of a Limestone Caribbee (Quinn 2015). Beneath the main sedimentary formations, the island consists of igneous basement rock dating to the middle Eocene, 56–41 million years ago (Ma). This is the remnant of a lava platform resulting from volcanic activity on nearby Saint Martin (Bruce 2000; Hamylton 2013). The basalt from the middle Eocene is most prominent on the northern end of the island and is visually distinct from the overlying limestone. In addition to the basalt, tuff and breccia from the middle to late Eocene (41–33 Ma) are also found around Anguilla and are visible along Crocus Bay on the north side of the island (Smith 2013; Wynne 2013). Above the basement rock, two layers of limestone form the body of Anguilla. The first is a blend of marly limestone and black shale originating in the Oligocene (33–23 Ma); the second is cavernous limestone laid down in the early Miocene (23–15 Ma). This layer comprises the visible cliffs on Anguilla and is riddled with fossils of a wide variety of fauna (Smith 2013).

Where limestone is visible at the surface or is overlain by only a thin veneer of sand, the topography is characterized by dolines and other karst features (Stanley 2009). Active collapse or subsidence has occurred even in some of the most developed areas such as The Valley, Anguilla's capital. Underground caverns occur in the limestones of Miocene and Oligocene ages, and these have been explored and utilized as sites for habitation as well as ritual use extending back many thousands of years (Crock 2008).

4.2.2 Climate

Anguilla is characterized as an "arid" island by Mitchell (n. d.a) with a hot and dry climate (Stanley 2009: 5) including a mean annual temperature of 28 °C (82.4 °F) and mean annual rainfall of 1017 mm (40 inches). At least 60% of Anguilla's land surface is exposed rock. Where the limestone has decayed to soil, it does successfully support crops. Because of the limited soil, the island's vegetation mainly consists of small trees and low scrub inland and sea grape and palm trees along the coasts. The limestone does not support water retention well, and though there are some wells on the island, most rainwater escapes back to the sea (Bruce 2000). The only wetlands are the salt ponds found throughout the island. These ponds are remnants of higher sea levels and are primarily brackish.

4.2.3 Reefs

Anguilla is noted for its spectacular and ecologically important coral reefs and beaches. In the shallow water off the north coast of Anguilla, 14,600 hectares of sublittoral substrate provides what should be ideal conditions for coral (Wynne 2013). The 17-km-long reef along the north coast is considered to be one of the most important reefs in the eastern Caribbean both because it helps protect against beach erosion and because it is host to many endangered and ecologically important marine life, such as spotted eagle rays, turtles, and the commercially fished queen conchs (Hodge 2011; Putney 1982). Living and fossil coral occur within the surf zone of the active shoreline on the north side of the island while "fringing" reefs a bit further offshore predominate to the south (Stanley 2009; Wynne 2013). Though coral conditions have been poorly monitored in Anguilla until more recently, experts from the Anguilla Department of Fisheries and Marine Resources find coral coverage and health have deteriorated since the 1990s. Most notably, coral reef coverage in the previously mentioned shallow water substrate declined from 14% in 1990 to 4% in 2009 (Wynne 2013).

4.3 Landforms

Because of Anguilla's low-relief limestone composition, it is without some of the dramatic mountainous features found on other Caribbean islands. What Anguilla lacks in the way of volcanic peaks, however, it makes up for with beautiful public beaches, an array of limestone caves, and various cays and islands offshore. The most practically important landforms on Anguilla are the salt ponds, the island's only permanent wetlands. There are 25 salt ponds on Anguilla, 22 of which are natural and formed thousands of years ago. The other three were created as part of more recent landscaping efforts. Along with serving a variety of ecosystems services such as flood mitigation during heavy rainfall, coastal protection during storm surges, and sediment trapping that catch contaminants coming off the mainland, these ponds provide a habitat for 130 species of birds, some 25% of which are endangered or threatened species (Anguilla 2008; Hodge 2011).

Despite the importance of the salt ponds, Anguilla is best known for its beautiful beaches. While the standard picture of an Anguillian beach is white sand stretching along the open coast, there are also cliffs in the north which enclose bays with smaller, more secluded beaches. Little Bay is one of the most famous enclosed beaches on Anguilla. It is surrounded to the west and east by cliffs and caves, and about 100 meters offshore from Little Bay there are substantial sea grass beds (Fig. 4.2). These mainly consist of *Syringodium* spp. which attracts marine turtles that come to feed on the sea grass (Hoggarth 2001). A two-mile-long white sand beach on the northeastern face of the island, Shoal Bay East (not be confused with Shoal Bay Beach), is widely considered the best beach in Anguilla (Jamason 2014; Staff 2011).

Anguilla's two national parks, Big Spring and Fountain Cavern, are both centered around limestone caves. Big Spring, located on the northeast end of the island, provided freshwater to the local residents until 1995, when contamination by Hurricane Luis rendered the water undrinkable. Fountain Cavern, on the same side of the island as Big Spring but further to the west near Shoal Bay, covers approximately six hectares and contains a limestone cave located 20 m underground (Anguillian 2015; Staff n.d.).

Aside from these notable examples, other limestone caves are found throughout the island. Cavanaugh Cave, located between North Hill and Katouche Bay, is the safest cave in Anguilla to explore. It was mined for phosphate through the nineteenth century and widely considered to be the site where fossils of *Amblyrhiza Inundata*, a large, extinct rodent were discovered (Green 1998). Despite the historic significance and enduring stability of Cavanaugh Cave, arguably the most famous cave in Anguilla is the archaeologically important Pitch Apple Hole, located in Katouche Bay on the north side of the island (Biknevicus et al. 1993). The cave gets its name for the Pitch Apple Tree whose roots extend far to the bottom of the cave, creating an open ceiling in the cave. Katouche Bay leads into one of Anguilla's only true valleys, Katouche Valley. Located between Crocus Hill and North Hill, Katouche Valley is the location of the only natural rain forest area on the island (Connor 2013). In addition to caves, there are other limestone formations found around the island. These include the natural arches such as the one found in Blolly Ham Bay. These are typically found on the west side of Anguilla and form due to erosion by wind, rain, and waves along the coast (Fig. 4.3).

Offshore, various islands and cays provide additional beach real estate and snorkeling opportunities. Cays are small islands that form on the surface of coral reef. When currents slow or shift, sediment builds up on the reef until it breaks the surface, gradually forming a stable surface (Editors 2016). There are many cays around Anguilla, with varying amounts of vegetation or habitable space. Scilly Cay is a tiny coral island located to the northeast of Anguilla. The cay has little vegetation with a few palm trees and a small sand beach. It is surrounded by waters with patch reefs, sea grass, and algal beds (Fig. 4.4). The Prickly Pear Cays, comprised of an east and west island, remain divided by a narrow channel on the Anguilla Bank. The West Cay is narrow and rocky, with no place to land a boat safely, while the East Cay is longer and easier to access (Blunt 1863). Prickly Pear Reef, which hosts an underwater canyon around the cays that ranges in depth from 12 to 21 m—including an underwater chimney formation, ledges, and caves—is part of why the cays have such a draw for snorkelers (Huber 2014).

Fig. 4.2 A beach enclosed by limestone cliffs and cave at the outer end of Little Bay, a smaller bay in the nook of Crocus Bay. Usually, the beach is exposed with calm seas, ground seas inundates the beach reaching the cliffs. The cliffs are dominated with dry scrubs (e.g., *Rondeletia anguillensis*) and cacti species. Photograph by A.K. Izioka

Of Anguilla's islands, one of the most prominent is Dog Island, an uninhabited island located 13 km NW off Anguilla. The island has a surface area of 207 ha with a rocky landscape formed by low cliffs that alternate with sandy beaches. It has a salt pond to the south and three small cays off the west and north coasts. Dog Island, like many of the islands of Anguilla, is classified as an Important Bird Area (Soanes et al. 2015, 2016), as it is home to three globally important bird colonies: Brown boobies, Masked boobies, and Frigate birds. The island's landscape structure and low relief provide the perfect nesting grounds for these birds (Soanes et al. 2015).

Another of Anguilla's relatively larger islands is Sombrero Island. Sombrero Island is important for many birds, just like Dog Island; however, it is also important for marine animals such as marine turtles (Kuramae Izioka nd). Sombrero Island is 1.6 by 6 km, with a total area of 38 ha. It got its name from its odd shape, which loosely resembles a sombrero hat from above. The vegetation on the island is very scarce, though Lazell (1964) noted some cactus species (*Opuntia antillana*) and Norton (1989) encountered more species such as *Sesuvium portulacastrum, Euphorbia serpens, E. mesembrianthemifolia, Heliotropium curassavacum, Portulaca oleracea, and Fimbristylis cymosa*. The island also hosts a lighthouse which passing ships often use as a landmark (Lazell 1964).

Sandy Island is off Anguilla's northwestern coast and is made up of fine sand and surrounded by shallow reefs. Forces of nature such as hurricanes, tropical storms, and severe wave action renew and reshape the island every year, occasionally stripping all vegetation from the island. This is part of why Sand Island has only limited vegetation. Except for some palm trees, all the vegetation is small growth plants (Fig. 4.5).

Fig. 4.3 Examples of coastal landforms on Anguilla. (**a**) Natural arch in the West-End of Anguilla. The arch is formed by erosion of currents, salt, and rain during ground seas. Vegetation of dry bush dominates this area, with some Aloe (*bottom left*) and cacti growing here and there. (**b**) A second natural arch at the same location with less vegetation growing on *top*. This arch is mainly formed by wave action and erosion exacerbating existing fractures in the sea cliff. Photographs by A.K. Izioka

Anguilla also hosts a few other small islands off its coast that should be noted here. Scrub Island has a unique mixture of coral and scrub. It has some rough coastal areas, particularly along the eastern side where the coastline primarily consists of low cliff formations. In the northern and southern parts of the island, the coast is made up of sandy beaches that slope into the sea. The beaches are important for marine turtles as breeding areas. Nesting activity has been recorded several times by Green Turtles and Hawksbills. The nearby Little Scrub Island is rough and rocky around the whole island. The waters surrounding Little Scrub are choppy and have impacted the island's inhospitable shoreline.

Seal Island is a small, narrow island between Dog Island and Anguilla. Perhaps more eel-shaped than Anguilla itself, Seal Island is sometimes hard to access because of the 8 km reef that runs between it and Anguilla. The reef, and its proximity to Anguilla, makes Seal Island a popular location for divers. The Flirt Rocks are two small islands, located about 6 km northwest of the Prickly Pear Cays. Both Great Flirt, which stands to a height of 6 m above sea level, and Little Flirt, which reaches 2.4 m above sea level, are rocky and unwelcoming (Blunt 1863). Anguillita is the southernmost feature of Anguilla. It is located off the southwestern coast and is primarily accessible via sea kayak. It is little known compared to features like Scrub Island, and most often frequented by divers who visit the coral reefs 5–20 m below the surface, and underwater caves, both of which host a variety of marine species (Hoggarth 2001; Wynne 2013).

4.4 Landscape and History

While Anguilla has so little relief that it went undetected by Europeans for much longer than some other Caribbean islands, its landscape has been shaped just as much by the features *not* found on the island as those that are. When Anguilla was first settled by the Arawak people, they were drawn to the island not for the verdant forests and lush farmland, but for the fertile fishing its coral reefs fostered, although the small quantities of soil on Anguilla were sufficient for the subsistence farming efforts the Arawak undertook. The two main caverns on the island, Fountain and Big Spring, which are a product of the island's limestone composition, became sacred sites for the Arawak (Rogers 2010).

Despite the importance of these caverns in the Arawak culture and the island's long Arawak and subsequent Carib history, there are virtually no remaining place-names on the island of Amerindian origin. One exception is Zemi Beach, which takes its name from the religious artifacts found nearby. Many of the current place names are slightly ironic. The Valley, for example, is the capital city on an island whose highest point is a mere 65 m above sea level. Others are descriptive: Shoal Bay or North Valley, and still others reference residents of the island: Cauls Pond or Hughes' Estate. Though many of these name origins are not known for certain, it is clear they are all from the period of European occupation. Interestingly, there are no names of African

Fig. 4.4 View of the southern end of Scilly Cay looking north from Anguilla. Scilly Cay is a coral island located on the East-End of Anguilla with low-lying vegetation. The northern side often experiences large waves during rough weather, but offshore coral reefs help protect against coastal erosion. The small island houses a large restaurant and enjoying a meal on Scilly Island has become one of Anguilla's most famous tourist attractions. Photograph by A.K. Izicka

origin or influence, even though the British did bring slaves to Anguilla (Mitchell 2010).

When the British first arrived on Anguilla, they realized that the soil would support crops of tobacco, corn, and cotton, the last of which had previously been cultivated by the Arawak and was already growing on the island (Mitchell n.d.b; Halstead n.d.). These first crops supported the initial British settlement, but were destroyed by a group of Caribs invading from nearby islands. The settlement itself was never eradicated, and the invasion was beaten back. The British subsequently attempted the same kind of plantation agriculture found throughout the Caribbean. Large numbers of slaves were imported from Africa, which became a problem when it became apparent that the soil conditions and arid climate were not conducive to large plantation operations on Anguilla. Though some sugarcane, rum, mahogany, cotton, and rum were all produced and exported from Anguilla, the plantations were small and low volume. This resulted in many of the slaves working on personal subsistence plots rather than the plantations, until the Emancipation Act passed in 1834 (Countries n.d.).

Today, the consensus is that only 13% of Anguilla's land cover is arable. The crops grown now are almost exclusively for domestic consumption, typically grown on plots of less than one-quarter hectare (Meditz 1987). The scrub plant cover on much of the island is more suited to grazing than agriculture; cattle, goats, sheep, and pigs are all raised for both domestic consumption and export, though these animals in turn have contributed to further deforestation (Mitchell n.d.b). All traditional agricultural practices undertaken on Anguilla are made possible by British investment in desalination plants, which help mitigate the absence of permanent freshwater, one of the island's most notable features. Anguilla's only permanent wetlands, the salt ponds, provided a steady export product until the 1970s, when a combination of overmining and significant hurricanes compromised the salt supply. Anguilla primarily exports through the Caribbean, with

Fig. 4.5 Sandy Island north of Road Bay, Anguilla. Similar to Scilly Cay, Sandy Island is composed of fine sand and coral and is protected by surrounding shallow reefs. There is no freshwater present on the island so vegetation is limited to a few palm trees and low-lying vegetation. The forested hills of Anguilla are visible in the background. Photograph by A.K. Izioka

virtually no exports going to America or Europe. When the salt supply dropped, Anguilla's export markets turned elsewhere (Meditz 1987).

The inability to develop significant land-based exports has pushed Anguilla to explore other avenues of income. These include offshore banking services, but the fishing and tourism industries have had a greater impact on the landscape. Along with other fish, lobster is one of Anguilla's primary exports, though sustained overfishing has now greatly depleted the local marine population. Anguilla never developed a tourist draw related to sport fishing, but visitors to the island find some of the most highly regarded seafood cuisine in the Caribbean. Tourists mostly visit the island for the beaches, which are an increasing concern for the government, as awareness of coastal erosion and reef health increase (Hoggarth 2001). Despite its low relief, the island still hosts points of interest. In addition to the caves, unique ecozones like Katouche Valley, and the offshore cays and islands, there are numerous sites around Anguilla where scuba and snorkeling are popular, including a fully underwater marine park and extensive coral reefs (Jamason 2014). Anguilla does not cater to cruise lines, and while some Caribbean cruises will stop at the island for a few hours, the island is largely protected from the massive population fluctuations that accompany cruise tourism (Swanson 2014). The increasing economic impact of tourism may lead to a revision of current policies regarding cruise ships, but for now Anguilla does well relying on the draw of some of the most beautiful beaches in the Caribbean (Fig. 4.6).

4.5 Heritage and Tourism

Heritage tourism has not yet been fully realized on Anguilla, but opportunity for expansion exists. There are 42 known Arawak sites on Anguilla, some of which have been dated back as far as 1500 CE, others as recently as 600 CE (Cambridge 2012b). The most significant heritage site on Anguilla, Fountain Cavern, displays a deep religious significance. Inside the cavern, the top of a five-meter-long stalagmite has been carved into what experts say is a representation of the *Jochahu*, the spirit of fertility. Fountain Cavern (and Big Spring) also contain numerous petroglyphs, and many Zemis, three-pointed stones thought to ward off evil, were found in the cave (Cambridge 2012a). Fountain Cavern used to be open to the public, and although it remains part of one of Anguilla's two national parks, the cave itself is now closed for

Fig. 4.6 The Four Seasons Resort and Residences Anguilla, located on the north side of the main island's far eastern edge. Notice the pristine white sand beaches lining the coast in both directions. The island mountains of St. Martin separated by the Anguilla channel are visible in the background. Image from the Four Seasons Web site (http://www.fourseasons.com/anguilla/)

preservation. It was nominated for UNESCO status, with overtures of ultimately turning it into a show cave, but has yet to receive that recognition (Rogers 2010).

In many ways, Anguilla's environmental conditions have served as a buffer against conflicts that plague other Caribbean islands. For example, though its location was in a politically desirable location for the purpose of securing safe trade routes, it lacked natural resources that would have made it a target of exploitation. Anguilla was attacked once in 1688 by some 300 Irish colonists who had been deported from the French-controlled St. Christopher. They caused significant damage and are today referred to as the "wild Irish," but these colonists ultimately settled peacefully and integrated into the local community (Mitchell, n.d.c). Two other significant points of early conflict were the attempted French invasions in 1745 and 1796, neither of which resulted in a lasting power exchange between France and England (Halstead, n.d.). Unlike some Caribbean islands, which saw multiple power exchanges between two or more countries throughout their formative years, Anguilla was relatively even-keeled politically.

The lack of natural resources also kept expansive plantation agriculture from taking hold, which prevented class stratification from developing as rigidly as it did on other islands. During the economic downturn following emancipation, many former slave owners left Anguilla, seeking better fortune elsewhere in the Caribbean or returning to Europe. This left a majority African population on the island, and today the minority white population is well integrated (Halstead n.d.). It has been argued that Anguilla's continued status as a British Dependency has limited its growth, but the only significant social conflict in Anguilla's history came from the population seeking direct British governance (Countries n.d.).

The absence of agricultural resources has not proven to be a deterrent to tourism, however, as tourism on Anguilla has historically been based on the appeal of its picturesque beaches. The island's more than 33 public beaches remain mostly undeveloped and nearly always uncrowded (Tourism 2016). Few other geomorphic points of interest that would serve as a tourist draw exist on the island, though there are many opportunities around the island for snorkeling, diving, or short boat trips to the surrounding cays and islands. Aside from a few low years, tourism in Anguilla seems to be on the rise. While potentially a good thing for locals, a growing concern surrounding tourism impacts continues, especially the need to conserve its primary tourist attraction: beaches. One primary concern about tourism rests in direct damage to reefs by snorkeling, diving, and/or boating. Increased tourism has also resulted in greater stress on fisheries, for both recreational fishing and restaurant demand (Hoggarth 2001).

On the other hand, tourism has contributed to growing community awareness surrounding the overall state of coral reef health, which has been degrading for a variety of reasons not directly related to tourism in the past decades, including coastal development and unregulated discharge. Hurricane Luis in 1995 first brought to the foreground concerns about beach erosion, but tourism has turned those concerns into more of a pressing issue. If the beaches were to degrade significantly, the resulting drop in tourism could devastate Anguilla's economy. While gradually adopting more proactive conservation policies for both heritage and the environment, little of the island is formally protected. Some of the outlying islands are designated as important bird areas, and the habitat value of these islands and the salt ponds on the mainland is increasingly recognized, but official protection remains haphazard at best (Anguilla 2008).

4.6 Hazards

Compared to other islands of the Lesser Antilles, Anguilla faces relatively few hazards. Because of the absence of volcanic activity and subsequent low relief on the island, there is no danger of either volcanic eruption or landslides. The latter element is doubly important, because Anguilla's lack of permanent water means that flood events resulting from periods of intense rainfall could be compounded if Anguilla had greater altitudinal variation. Currently, Anguilla's primary hazards center around hurricanes, earthquakes, tsunamis, and floods. Though flood events often coincide with hurricane activity, they are not restricted to these times, and can be significant and damaging from concentrated rainfall events. November 2011, for example, saw a 30-year storm drop 150 mm of rain in less than 12 hours. Following a period of consistent but seasonally normative rain (and fully saturated soil), increasing runoff led to flash floods that caused property damage across low-lying parts of the island (Admin 2011).

Anguilla has not experienced a recent tsunami (though it has been subject to storm surges directly related to hurricanes), but in December of 2011 it became the first non-US territory in the Caribbean to be certified as Tsunami Ready. This certification requires development of a hazard mitigation plan, emergency management plan, public notification system, and tsunami warning plan. Emergency managers on the island also conducted training scenarios and completed a public outreach program relating to the new warning system. Since Anguilla is considered at serious risk for earthquake activity, development of a tsunami response plan was a wise precaution. Earthquakes are frequent near Anguilla, though most occurrences are typically minor. Buildings should ideally be designed with earthquake preparedness in mind, but this can come at substantial initial cost and requires importing often-costly materials to the island. This has not been standard practice in the past, but the Government of Anguilla recently adopted a building code with stricter standards. It may take time for consistent implementation, but it demonstrates a commitment to building safety (Gibbs 1998).

These new building standards are as concerned with structural stability in a hurricane as an earthquake—a challenge since the components that make a building hurricane resistant are often opposite of the elements that are earthquake resistant (Gibbs 1998). Hurricanes are a more significant recurring hazard to Anguilla, however, and have far-reaching implications for the entire island. Hurricane Luis, a Category 4 storm in 1995, was one of the most significant hurricanes to make landfall in Anguilla's recent history. Luis caused significant reef damage, inundating at least one cay, and eroding measured beaches by 40%. While the beaches gained back an average 75% of the sand area over the next several years, the actual shoreline receded inland 8.7 m (Hoggarth 2001; Cambers 1996).

Hurricane Luis also served the unfortunate double function of increasing hurricane awareness in Anguilla. Since the last hurricane of comparable strength had been Hurricane Donna, a Category 5 in 1960, residents of Anguilla anticipated they had a few decades before another similar storm hit (Rappaport 1995). It took only four years before the next analogous storm hit: Hurricane Lenny, in 1999. Hitting the island after Hurricanes Georges (in 1998) and Jose (early in 1999), Hurricane Lenny made landfall as a Category 4 (Cambers 1999). Lenny caused flooding up to 4.3 m deep in The Valley, resulting in $65.8 million in damages (Pan 2007). Hurricane Earl, a Category 3 storm in 2010 took out the power on Anguilla and caused $4.3 million in damage, but no loss of life (Caribbean 2014). Most recently, Hurricane Gonzalo (in 2014), a Category 1 storm, caused flood damage in the amount of $500,000 (Anguillian 2014).

Anguilla's hurricane frequency is standard for Caribbean islands, but remains a cause for concern, particularly for an island whose beaches are the primary attraction. Beach monitoring stations installed in 1992 help keep track of erosion, and Anguilla has implemented beach stabilizing fences to some effect (Cambers 2007, 2009; UNESCO 2003). These measures help mitigate the damage caused by hurricanes, but are ultimately only slowing the changes that rising sea levels and storm impacts are inflicting on the coastline (IPCC 2007). More substantial measures may become necessary in future years, and hopefully the government will take a proactive stance on developing a plan for future beach management.

4.7 Conclusion

It would be easy to look at Anguilla, a flat body of land with little relief, and say it lacks the drama or variety of the surrounding islands. To do so would be to miss a special place with a range of features to offer residents and visitors alike. Anguilla is an amalgam of distinctive elements that set it apart from other island nations in the Lesser Antilles. The salt ponds that form its only permanent wetlands provided a unique export for an island that lacked other natural resources typically produced on islands in the area. Now, those same ponds play ecological roles by protecting the coast and providing habitat for an abundance of bird species. The cays and coral reefs that litter the coast around the island extend Anguilla's habitat areas, while offering exposure to a variety of marine life for snorkelers and scuba divers.

Tourists have been a blessing and a curse for Anguilla. While undoubtedly important, as tourism represents the number one industry on the island, increased tourism has negatively impacted reef health and overall environmental quality. Yet tourism has also raised local awareness of beach erosion and a range of reef health issues, leading to multiple avenues of governmental reform. Erosion management efforts are timely, as the area has entered a cyclical period of increased hurricane activity that will continue to alter the coastlines.

Some things have changed significantly on Anguilla since it was inhabited by the Arawak people who relied heavily on the sea for food, growing only light vegetable crops and cotton, and relying on Big Spring for freshwater. Now, desalination plants provide drinking water to the island. But there is still only minimal agriculture, mostly subsistence farming done in backyard plots. The limestone caverns that the Amerindians revered as sacred sites are still present, unique heritage sites that offer a glimpse into the long pre-European history of the island. And the beaches remain, pristine, quiet expanses of sand on an island that has not yet embraced the intense tourist experience of cruise ships.

References

Admin (2011) Freak rainstorm floods Anguilla. *The Anguillian*. Retrieved 25 May 2016 from http://theanguillian.com/2011/12/freak-rainstorm-floods-anguilla

Anguillian (2014) Gonzalo Pounds Anguilla—Tourism Plant Intact. *The Anguillian*. Retrieved 24 May 2016 from http://theanguillian.com/2014/10/gonzalo-pounds-anguilla-tourism-plant-intact

Anguillian (2015) Fountain National Park—An Anguillian Treasure Coming. *The Anguillian*. Retrieved 24 May 2016. from http://theanguillian.com/2015/08/fountain-national-park-an-anguillian-treasure-coming

Biknevicus AR, McFarlane DA, MacPhee RDE (1993) Body size in Amblyrhiza inundata (Rodentia, Caviomorpha), an Extinct Magafaunal Rodent from the Anguilla Bank, West Indies: Estimates and Implications. Am Mus Novitates 3079: 1–25

Blunt EM (1863) The American coast pilot: containing directions for the principal harbors, capes, and headlands, on the coast of north and part of South America… with the prevailing winds, setting of the currents, and the latitudes and longitudes of the principal harbors and capes; Together with tide tables and variation. Edmund and George W. Blunt, New York: NY

Bouysse P (1979) Caractères morphostructuraux et evolution géodynamique de l'arc insulaire des Petites Antilles. Bulletin du Bureau de Recherches Géologiques et Minières, series, IV, nos 3-4-1976: 85–210

Bouysse P (1986) Introduction to geology of the lesser antilles. In: Dengo G, Case JE (eds) The geology of North America, volume H: the Caribbean region. Geological Society of North America. Boulder, pp 141–143

Bouysse P, Guennoc P (1983) Données sur la structure de l'arc insulaire des Petites Antilles, entre Sainte-Lucie et Anguilla. Mar Geol 53:131–136

Bruce A (2000) Anguilla. Geol Today 113:112–113

Cambers G (1996) Hurricane impact on beaches in the eastern Caribbean Islands 1989–1995—Effects of the 1995 Hurricanes on the Islands from Anguilla to Dominica. Environment and Development in Coastal Regions and in Small Islands. Retrieved 24 May 2016 from http://www.unesco.org/csi/act/cosalc/hurl.htm

Cambers G (1999) Late hurricanes: a message for the region. Planning for coastal change in eastern Caribbean. College Station, Mayaguez: Puerto Rico

Cambers G (2007) Impact of climate change on the beaches of the Caribbean. In Paper prepared for commonwealth association of planners regional conference, 24–27 June 2007

Cambers G (2009) Caribbean beach changes and climate change adaptation. Aquat Ecosyst Health Manage 12(2):168–176

Cambridge N (2012a) July 22. Arawak Indians—an Anguillan history. History in an Hour. Retrieved 24 May 2016 from http://www.historyinanhour.com/2012/07/22/arawak-indians-an-anguillan-history/

Cambridge N (2012b) Discover the fountain cavern national park. Luxury Caribbean: Anguilla. Retrieved 21 May 2016 from http://www.luxury-caribbean-news.com/discover-the-fountain-cavern-national-park/

Caribbean Catastrophe Risk Insurance Facility (2014) CCRIF to make first payout on excess rainfall policy to Anguilla after hurricane Gonzalo rains. Grand Cayman, Cayman Islands

Christman RA (1953) Geology of St. Bartholomew, St. Martin, and Anguilla, Lesser Antilles. Geol Soc Am Bull 64:85–96

Connor A (2013) Katouche Bay: Three (or more) For the Price of One. Design Anguilla Magazine. Retrieved 25 May 2016 from http://designanguilla.com/katouche-bay/

Countries and Their Cultures. n.d. *Culture of Anguilla*. Retrieved 26 May 2016 from http://www.everyculture.com/A-Bo/Anguilla.html

Crock J (2008) Anguilla. In: Sanz N (ed) Rock art in the Caribbean. UNESCO, Paris, pp 264–268

Gibbs T (1998) Vulnerability assessment of selected buildings designated as shelters: Anguilla. USAID-OAS Caribbean Disaster Mitigation Project. Consulting Engineers Partnership Ltd

Green B (1998) Cavanaugh cave, an adventure. Anguilla News. Retrieved 24 May 2016 from http://news.ai/ref/cavanaugh.html

Hamylton S, Andrefouet S (2013) An appraisal of the extent and geomorphological diversity of the coral reefs of the United Kingdom dependent territories. In Sheppard CRC (ed) Coral reefs of the United Kingdom overseas territories. Springer, Dordrecht, pp 1–11

Hodge KVD (2011) Anguilla. UK Overseas Territories and Crown Dependencies: 2011 Biodiversity snapshot. Department of Environment: Government of Anguilla

Hoggarth D (2001) Management Plan for the Marine Parks of Anguilla. Prepared for: Organisation of Eastern Caribbean States Natural Resources Management Unit St Lucia. Department for International Development

Huber J, Huber J (2014) Best dives of Anguilla, Antigua, and Barbuda. Hunter Publishing, Inc. Edison: NJ, USA

IPCC (2007) Summary for policy makers. In: Climate CHANGE 2007: the physical science basis. Contribution of working group i to the fourth assessment report of the intergovernmental panel on climate change. Cambridge University Press, Cambridge, UK

Jamason R (2014) Best snorkeling in Anguilla. Snorkeling Online. Retrieved 24 May 2016 from http://www.snorkelingonline.com/pages/best-snorkeling-anguilla

Kuramae Izioka LA (nd) Baseline study of the benthic analysis of Sombrero Island. Department of Fisheries and Marine Resources

Lazell JD (1964) The reptiles of Sombrero, West Indies. Copeia 4:716–718

Martin RM (1839) Statistics of the colonies of the british empire: from the official records of the colonial office. W.H. Allen and Company, London, England

Meditz SW, Hanratty DM ed. (1987) Caribbean Islands: a country study—British Virgin Islands, Anguilla and Montserrat—Economy. Washington: GPO for the Library of Congress

Mitchell D (2010) Place Names of Anguilla. A Presentation to the Anguilla Archaeological and Historical Society. Retrieved 21 may 2016 from http://www.aahsanguilla.com/Selected%20Readings/PLACE_NAMES.pdf

Mitchell D (n.d.a.) Geology and Botany. Anguilla From the Archives Series. The Anguilla Archaeological and Historical Society. Retrieved 21 May 2016 from http://www.aahsanguilla.com/Selected%20Readings/Geology%20and%20Botany.pdf

Mitchell D (n.d.b.) Cotton and Salt. Anguilla From the Archives Series. The Anguilla Archaeological and Historical Society. Retrieved 21 May 2016 from http://www.aahsanguilla.com/Selected%20Readings/Cotton%20and%20Salt.pdf

Mitchell D (n.d.c.) The First Generation. Anguilla From the Archives Series. The Anguilla Archaeological and Historical Society. Retrieved 21 May 2016 from http://www.aahsanguilla.com/Selected%20Readings/4.%20First%20Generation.pdf

Norton RL (1989) First west Indian record of the black noddy and nesting of masked booby at Sombrero Island, Lesser Antilles. Colonial Waterbirds 12:120–122

Pan American Health Organization (2007) Anguilla. Health in the Americas 2007, Volume II—Countries. Retrieved 20 May 2016 from http://www1.paho.org/HIA/archivosvol2/paisesing/Anguilla%20English.pdf

Putney AD (1982) Survey of conservation priorities in the lesser antilles. Final Report. Caribbean Environment Technical Report. Caribbean Conservation Association

Quinn JA, Woodward SL (2015) Warth's landscape: an encyclopedia of the world's geographic features [2 Volumes]. ABC-CLIO, Santa Barbara: CA, USA

Rappaport E, Fernandez-Partagas J (1995) The deadliest atlantic tropical cyclones 1492–1996. National Hurricane Center, Miami

Rogers MF, Hodge KVD (2010) UK tentative list of potential sites for world heritage nomination: application form, fountain cave—Anguilla, British West Indies. UNESCO World Heritage Team, Department for Culture, Media, and Sport. Retrieved 26 May 2016 from http://www.worldheritagesite.org/countries/The%20Fountain%20Cavern%20(Anguilla).pdf

Staff (n.d.) Anguilla—Attractions. *iExplore*. Retrieved 26 May 2016 from http://www.iexplore.com/articles/travel-guides/caribbean/anguilla/attractions

Smith AL, Roobol MJ, Mattioli GS, Fryxell JE, Daly GE, Fernandez LA (2013) The volcanic geology of the mid-arc Island of dominica, lesser antilles: the surface expression of an Island-arc batholith. In: Special Paper 496, The Geological Society of America. Boulder: CO

Soanes LM, Bright JA, Bolton M, Millett J, Mukhida F, Green JA (2015) Foraging behaviour of Brown Boobies *Sula leucogaster* in Anguilla, Lesser Antilles: preliminary identification of at-sea distribution using a time-in-area approach. Bird Conservation International 25(1):87–96

Soanes LM, Bright JA, Carter D, Dias MP, Fleming T, Gumbs K, Hughes G, Mukhida F, Green JA (2016) Important foraging areas of seabirds from Anguilla, Caribbean: implications for marine spatial planning. Mar Policy 70:85–92

Staff (2011) Top 10 exotic beach destinations: shoal bay east, Anguilla. Coastal Living. Retrieved 26 May 2016 from http://www.coastalliving.com/travel/top-10/top-10-exotic-beach-destinations/shoal-bay-east-anguilla

Staff (2015) Anguilla. World Atlas. Retrieved 26 May 2016 from http://www.worldatlas.com/webimage/countrys/namerica/caribb/ai.htm

Stanley MG (2009) Geoconservation in the overseas Territories of the UK. Joint Nature Conservation Committee (JNCC), Petersborough, UK

Swanson D (2014) 5 Islands where cruise ships won't cramp your style. Miami Herald. Retrieved 21 May 2016 from http://www.miamiherald.com/living/travel/caribbean-travel/article3594587.html

The Anguilla National Trust and The Department of Environment (2008) Salt Pond Ecosystemsand the National Biodiversity Strategy and Action Plan. *Overseas Territories* and the National Biodiversity Strategy and Action Plan. *Overseas Territories Environment Programme through the Department of Environment: Government of Anguilla*

The Editors of Encyclopædia Britannica (2016) Cay. Encyclopædia Britannica. Retrieved 26 May 2016 from http://www.britannica.com/science/cay

The Halstead Trust (n.d.) Relics of Empire: Anguilla. Migration to, From & Within the British Isles. Retrieved 25 may 2016 from http://www.exodus2013.co.uk/relics-of-empire-anguilla/

Tisdall N (2010) Best caribbean beaches? Try Anguilla. Telegraph: Travel. Retrieved 20 May 2016 from http://www.telegraph.co.uk/

travel/sunandsea/7954070/Best-Caribbean-beaches-Try-Anguilla.html
Tourism (2016) Anguilla, British West Indies. Tourist Board, The Valley Anguilla: WI. The Government of Anguilla
UNESCO (2003) Anguilla (U.K. Territory). Environment and Development in Coastal Regions and in Small Islands—Caribbean—Beach Erosion, Anguilla. Retrieved from http://www.unesco.org/csi/act/other/anguilla.htm
Wynne SP (2013) Coral Reefs of Anguilla. Coral Reefs of the United Kingdom Overseas Territories. Coral Reefs of the World, Vol 4, pp 13–22. Springer, Berlin

Saint Martin/Sint Maarten and Saint Barthélemy

5

Russell Fielding

Abstract

St. Barthélemy and St. Martin are small islands among the northernmost of the outer, coralline Lesser Antilles arc and are separated from one another by a narrow channel. Both islands are characterized by undulating coastlines with cove-like white and golden sand beaches and impounded salt ponds, offshore islets, dry landscapes (though the higher peaks on St. Martin do support slightly more rain-dependent vegetation), and landforms combining ancient volcanics with more recent sedimentary geology. St. Barthélemy and the northern part of St. Martin are culturally and politically connected to France, while the southern part of St. Martin (known as Sint Maarten) is linked to the Netherlands. Both islands are hilly, and while St. Martin's larger size affords it some space for coastal plains, St. Barthélemy has significantly less flat land. Both islands are popular tourist destinations with St. Martin attracting more cruise ships and resort-based tourists and St. Barthélemy focusing on a smaller-scale, luxury tourism niche. Tropical cyclones occasionally threaten these islands, but the greatest hazards may be droughts and any disruption to the energy system since rainwater catchment and desalination provide the majority of the freshwater to both islands.

Keywords

Arid • Binational • Coralline • Mountainous • Tourism

A Toponymic Note

In this chapter, both islands will be referred to by shorthand names: Saint Barthélemy will be called "St. Barth" (read with a silent 'h'). This island is also frequently referred to in the literature as "St. Bart's" or "St. Barts," and less frequently in modern times as "St. Bartholomew," though this toponym is used in certain historical texts and maps. With regard to Saint Martin/Sint Maarten, following Benoît (2008) and other authors, distinct names for the two parts of the island, for the island as a whole, and during explicit discussion of the divided nature of the island will be used. The island of Saint Martin/Sint Maarten will be designated as simply "St. Martin." When discussing the distinct sides, the Dutch side will explicitly be referred to with the toponym "Sint Maarten," while the French side will be referred to as "Saint Martin", with no abbreviation of the word "Saint". When discussing the divided political geography of the island itself, both names will be used, separated by the virgule (i.e. "Saint Martin/Sint Maarten").

R. Fielding (✉)
Department of Earth and Environmental Systems,
University of the South,
Sewanee, TN 37383, USA
e-mail: russell.fielding@sewanee.edu

C.D. Allen (ed.), *Landscapes and Landforms of the Lesser Antilles*,
World Geomorphological Landscapes, DOI 10.1007/978-3-319-55787-8_5

5.1 Introduction

5.1.1 Orientation

Saint Martin/Sint Maarten (Fig. 5.1) and Saint Barthélemy (Fig. 5.2) are two small islands (87 and 21 km^2, respectively) located near the northern limit of the Lesser Antilles archipelago. Only Anguilla and the Virgin Islands reach further north. Politically, Saint Barthélemy is an overseas collectivity (*collectivité d'outre-mer*) of France, having achieved that status with independence from Guadeloupe in 2007. Saint Martin/Sint Maarten is a binational island, as its dual names indicate, divided by a roughly east–west border separating the French and Dutch sides of the rather triangular island. The larger (53 km^2), northern, French side is called Saint Martin, and the smaller (34 km^2), southern, Dutch side is called Sint Maarten. Saint Martin, like St. Barthélemy, is an overseas collectivity of France. Sint Maarten was a part of the Netherlands Antilles until that union's dissolution in 2010, after which Sint Maarten became an autonomous country within the Kingdom of the Netherlands.

The islands of St. Barth and St. Martin evoke landscapes and seascapes that come as close to the stereotypical image of "tropical paradise" as exist anywhere. Beaches of soft white or golden sand and warm, electric-blue water are backed by thickly vegetated mountains, alive with fruit trees, tropical birds, land snails, tortoises, and iguanas. Of course, as with other islands in the Caribbean, human activity, from subsistence agriculture to slave-based industries to international tourism, has altered the landscape on each island in its own way, both historically and today.

5.1.2 Previous Work

The most detailed, accurate, and influential early academic work on the landscapes and landforms of these two islands is the 1950 Ph.D. dissertation by Christman, later condensed and published as a journal article by the same author (1953). Christman's *Geologic Map of Saint Bartholomew* (Fig. 5.3) remains the standard some sixty-five years after its creation, although a 1983 map by Westercamp and Andreieff is more recent and technically superior. Christman's, however, remains popular owing perhaps to its sepia-tinted, monochromatic beauty, and English legend and accompanying text. The most accurate and functional colonial-era cartography of St. Martin, St. Barth, and several other islands in the region was done by Samuel Fahlberg in the eighteenth century (Reinhartz 2012).

In addition to the foundational work by Christman, notable earlier research into the geology, geography, and physical landforms of St. Martin and St. Barth includes Cleve's (1871) work, Spencer's (1901) report, and Vaughan's extensive work throughout the region, conducted throughout much of the first third of the twentieth century (e.g., 1918, 1926).

More recent work has focused upon specific questions of volcanism, tectonics, sedimentary geology, caverns, and biogeography. In addition to their geological interest, each island—but primarily St. Barth—has been the subject of anthropological and linguistic studies (Benoist 1964; Maher 2013), focusing upon its isolation, both as a function of its insularity and as exhibited within the island itself, owing both to topography and settlement patterns.

5.2 Setting

Previous literature on the Lesser Antilles region has failed to establish agreement regarding the island arc to which St. Barth and St. Martin belong. Fink and Fairbridge (1975) place both within the outer, "limestone" arc, while Cambers (2010) places St. Barth in the inner, "volcanic" arc and leaves St. Martin in the outer, "coralline" arc. Standing on the islands, evidence of their volcanic origins, especially in contrast to the nearby low and sandy Anguilla, remains apparent. These volcanic features, however, are not geologically recent and are in many places overlain by Miocene or later sedimentary rock. The islands in the inner, recent-volcanic arc, however, often show basaltic or pyroclastic geology at or near the surface. Many have erupted within historical time, including the most recent major eruption in the region—that of the Soufrière Hills volcano on Montserrat in 1995. By contrast, most of the rocks of volcanic origin on St. Martin and St. Barth date primarily from the Eocene to the Oligocene (Tomblin 1975). For this reason, this chapter places both islands firmly within the outer arc, while noting that the two arcs do indeed converge in the immediate vicinity, closing the gap that separates, for example, St. Vincent from Barbados or even Antigua from Barbuda. Only at Guadeloupe, where the two arcs are close enough to converge within one island, is the gap narrower.

The island of St. Martin lies between 18.0° and 18.1° North latitude and between 63.2° and 63.0° West longitude. Shaped roughly like an equilateral triangle leaning to the right, this island contains the point of highest elevation within the broad undersea plateau, called the Anguilla Bank, from which both islands rise. St. Martin is an island of hills, valleys, and impounded bays. The coastline is irregular in both morphology and topography, with sandy beaches, spits, and bars, alternating with rocky coasts, themselves comprised of limestones, marls, basalts, and porphyrites (Vroman 1968). The beaches consist of white to golden coralline sand, not the dark, volcanic sand of the geologically younger, volcanic islands to the south. The basement rocks on St. Martin are comprised of the ancient, silicified volcanic tuffs common to the other islands in the region (St. Barth and Anguilla especially), some of which remain exposed, forming the island's mountainous areas (Schwartz 2010). On St. Martin, these tuffs form the Pointe Blanche formation, named for the cape at the island's southernmost point, an area where the formation is most visible (Sypkens-Smit 1995). On St. Barth, rocks of the same age and provenance are referred to as the St. Bartholomew formation (Christman 1950).

Fig. 5.1 General physiographic map of St. Martin and surrounding islands. The island is separated into the French Saint Martin and the Dutch Sint Maarten. Place names and the descriptions of physiographic features such as rivers, lakes, and mountains reflect this political division (e.g., river vs. ravine). While the island officially has two separate place names, the more popular "St. Martin" is cartographically used to represent both nations. Cartography by K.M. Groom

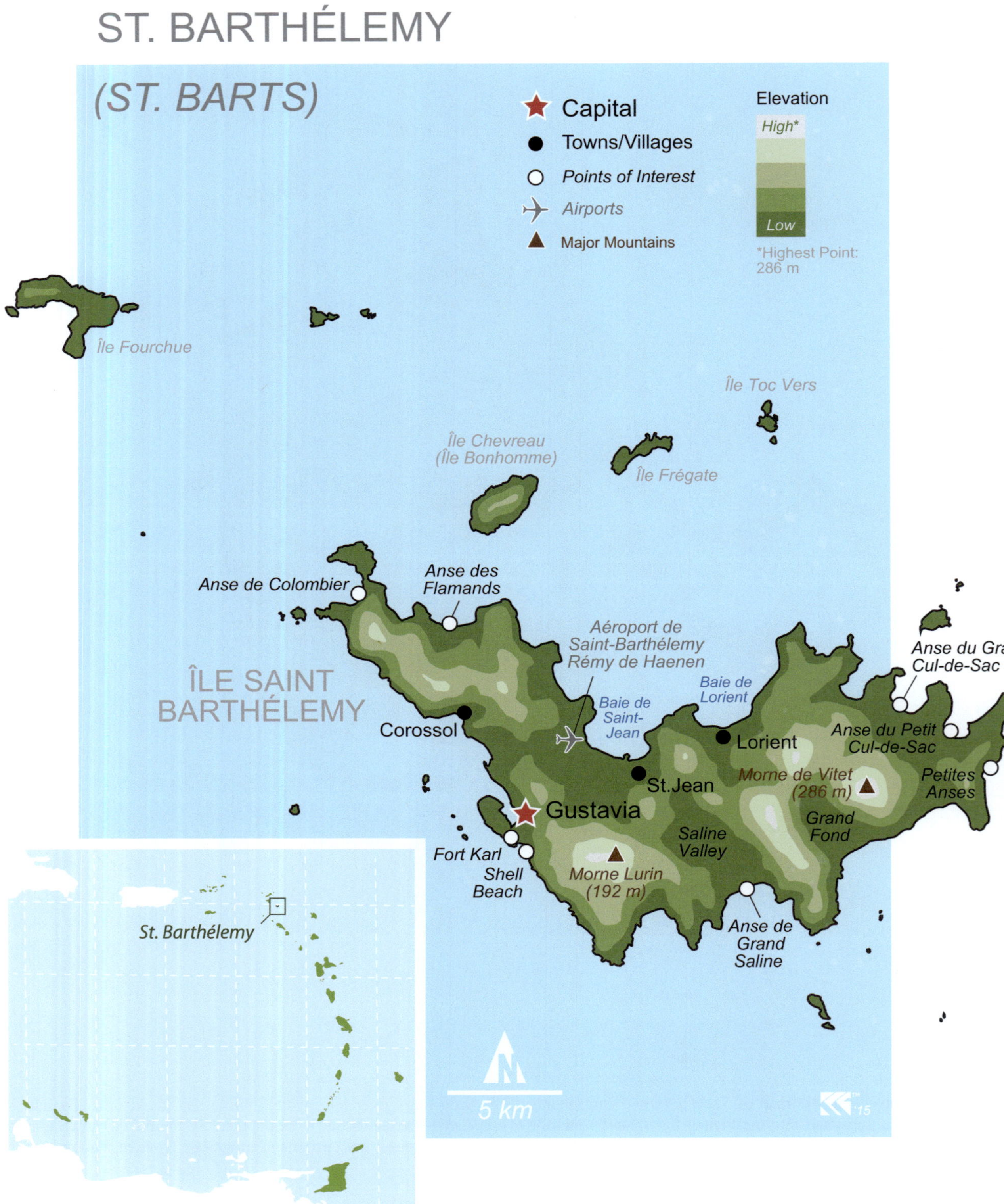

Fig. 5.2 General physiographic map of St. Barthélemy, colloquially known as St. Barts with surrounding islands. French place names and physical descriptions signify the islands political affiliation with France. Cartography by K.M. Groom

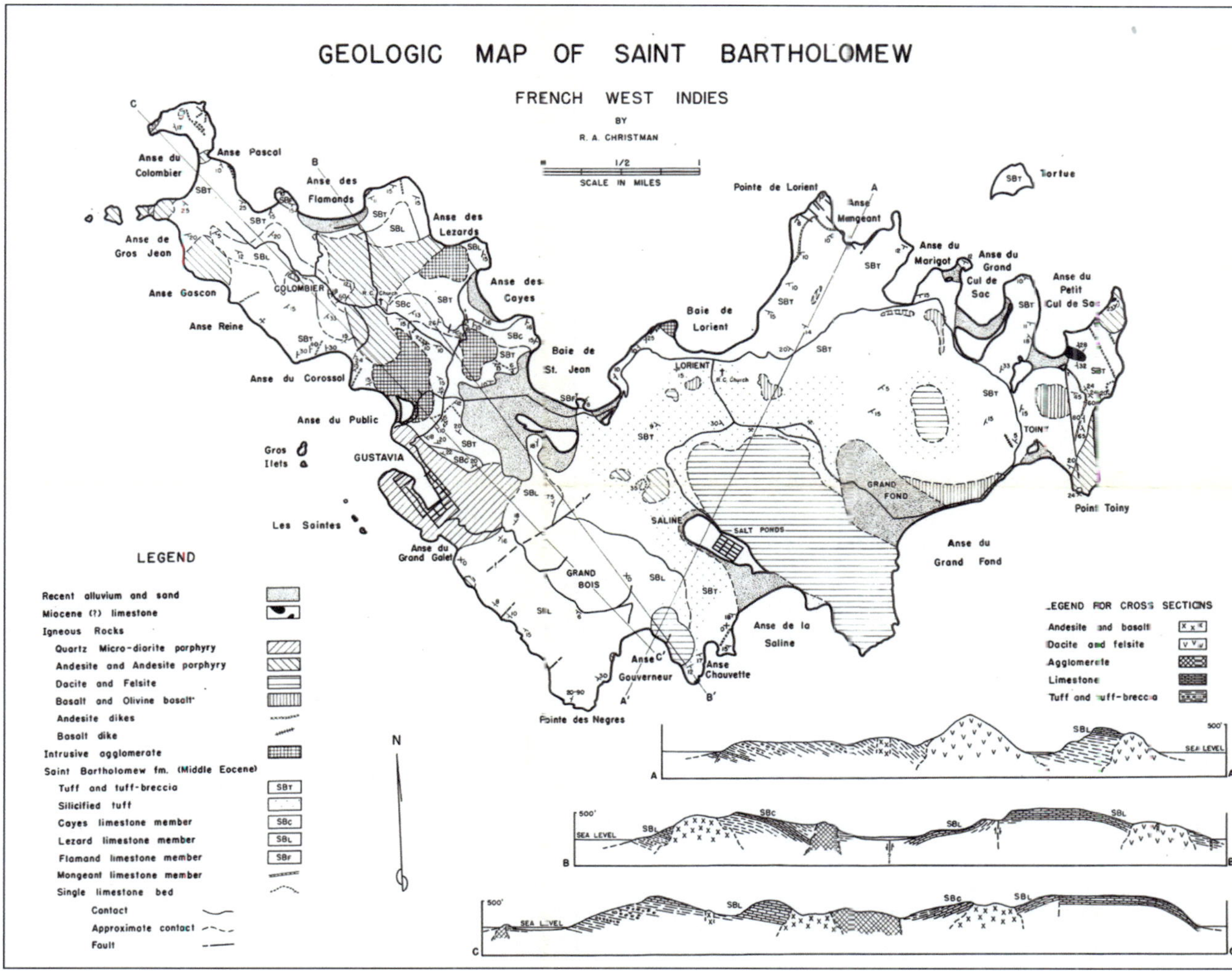

Fig. 5.3 "Geologic Map of Saint Bartholomew" by R.A. Christman, published in 1953. The detailed regional categorization, three cross sections running along the northwest–southeast and northeast–southwest diagonals, and pleasant aesthetic enhance the popularity of this map

St. Barth lies at 17.9° North latitude and between 62.9° and 62.8° West longitude. The island is shaped like a crescent opening north, or, if one is so inclined, like a smirking smile. The rolling coastline alternates between rocky shores and embayed coves, most of which feature sandy beaches at their landward extents. Like St. Martin, the structure of St. Barth is primarily composed of ancient volcanics overlain by Eocene limestones. Westercamp and Andreieff (1983) found evidence for three separate phases of volcanism, the first two occurring below sea level and coincident with the sedimentation processes that led to the limestone layers. According to these researchers, the volcanic center that produced the tuffs found on both St. Barth and St. Martin is thought to have been located near St. Barth's present north coast and offshore near the present islets of Île Chevreau (also called Île Bonhomme) and Île Frégate.

Both islands experience a climate marked by stable temperatures and seasonally variable precipitation. Average monthly temperatures range from 25 to 28 °C throughout the year. Diurnal temperature variation is more extreme than seasonal variation, ranging by about 5 °C from day to night. Monthly rainfall varies throughout the year, with a marked shift from rainy to dry seasons. On both islands, rainfall peaks at about 130 mm in October and is at a minimum, about 45 mm, in February. St. Martin's more mountainous interior does afford that island slightly more rainfall at higher elevations than St. Barth receives.

The soils of St. Barth are thin, rocky, dry, and relatively infertile. The native flora is largely xerophytic and agriculture has long been challenging. On St. Martin, owing to the higher elevations and the precipitation that this topography captures, soil layers are thicker and ephemeral streams leave deep, dry "ghauts" in some valleys, a feature mostly absent from the St. Barth landscape.

5.3 Landforms

5.3.1 General Landscapes

Rather than being dominated by a single volcano, as are several of the islands in the Lesser Antilles (e.g., Saba or Nevis), both St. Martin and St. Barth are characterized by numerous peaks, valleys, and—especially in the case of St. Martin—low coastal plains. Christman (1950) identified two separate mountain chains traversing St. Martin: the larger of which is roughly geographically central on the island, extending from Koolbaai Berg (Cole Bay Hill) in the south to Mount O'Reilly in the north (though perhaps a better origin is at Cay Bay Hill, slightly further south than Cole Bay Hill). This cordillera includes Pic Paradis, the island's highest point of elevation (424 m). Christman's second mountain range is a smaller, southeastern chain stretching from the island's southernmost point, Pointe Blanche, to Les Deux Frères, the two hills that stand behind the southern shore of the Baie de l'Embouchure, on the island's east coast.

Between these mountain chains lies the major valley on St. Martin, which extends from Philipsburg in the south to Baie Orientale in the northeast, then makes a westward traverse of the island, broken only by low hills, across to Grand Case, on the northwestern coast. This western extension of the valley is where the smaller, French side airport is located.

The other major valley on St. Martin begins at the Great Salt Pond and Little Bay on the island's south coast and trends northward. This valley is enclosed on three sides by ridges that converge at the international boundary to form the island's larger mountain chain. The valley, which is located entirely within the Dutch side, contains the villages of Cul-de-Sac and St. Peters.

The Terres Basses peninsula, as the name suggests (at least to Francophiles), is a low-lying area, punctuated by a single rise—Morne Rouge at 52 m. This peninsula, which Schwartz (2010, 303) describes as a "double tombolo," but which could also be seen as a single, large tombolo with a flooded interior, encloses Simpson Bay Lagoon (sometimes spelled *Simson*), the island's largest inland water body (Fig. 5.4). The two isthmuses that frame the bay are low, narrow, and predominantly sandy, though limestone terraces and cliffs may be found on the southern isthmus from Cupecoy to Maho Bay (Fig. 5.5) and on Terres Basses proper from Point Plum to Pointe du Bluff. The cliffs can reach heights of up to six meters, though most are much lower, and occasionally contain erosional features such as arches. The island's international airport is located on the southern spit.

St. Barth is a V-shaped island—others, beginning with Christman (1953, 67) have understandably likened it to a "boomerang"—with three main mountainous areas containing two major valleys and numerous small coastal plains. The westernmost mountainous area is the northwestern, leeward side of the island where peaks remain lower than 200 m but drop precipitously toward the beaches at Flamands, Colombier, and, to some extent, Corossol. The central mountains are found in the south of the island—the point of the V—and form a single complex constituting the foothills of Morne Lurin (192 m), this region's highest peak. The eastern mountains form a semicircle on the island's windward side, hemming in the Grand Fond Valley (Fig. 5.6), which faces the predominant wind and waves of the island's southeast coast. In the eastern portion of this semicircle of peaks is found Morne de Vitet, the summit of which, at 286 m, is St. Barth's highest point.

Between the central and eastern mountainous areas lies the Saline valley, the location of St. Barth's largest site for salt making, now defunct. Between the foothills of Morne Lurin and the mountains on the island's leeward side lies a saddle, the northern slope of which is the site for the island's airport. The airport itself has attained a certain degree of infamy as one of the more difficult landings, certainly in the Caribbean, perhaps in the world. The 650-m runway, short by almost any standard, slopes downhill[1] toward St. Jean Beach, where it terminates mere meters from the waterline (Fig. 5.7).

Around this mountainous skeleton, the island's deeply embayed coastline spreads like a fringe (Fig. 5.8). Eroded tongues of ancient lava flows, shaped by prevailing currents and capped in some places by cemented sediments, laid bare in others, circumscribe the island with an undulating ring of bays and headlands that trap coastal sediment, forming discrete beaches and imbuing a sense of place to each as its own entity.

The St. Barthélemy Channel separates St. Martin to the north and St. Barth to the south by a distance just less than 20 km at its narrowest point. Numerous islets, rocks, and reefs dot this channel, creating convenient anchorages or dangerous hazards, depending upon the attentiveness of the sailor. Two of these islets, Fourchue and Tintamarre, feature interesting geologies and human histories.

Fourchue is a small (roughly 0.5 km^2), uninhabited islet in the St. Barthélemy Channel belonging to St. Barth. This islet is rocky and denuded—the former quality an example

[1]The slope is downhill on the most common approach. Sometimes, though, owing to wind conditions, controllers opt to bring pilots in over the beach, in which case the runway slopes uphill from the perspective of those aboard the landing plane and terminates below a hilltop traffic roundabout on a busy stretch of road.

Fig. 5.4 Aerial view looking north toward Simpson Bay Lagoon, center, framed by its twin isthmuses. Mainland St. Martin is to the *right*; the Terres Basses peninsula is out of the frame to the *left*. The low island in the background is Anguilla. Photograph by D Fielding

Fig. 5.5 Limestone terraces at Cupecoy on St. Martin. Photograph by R. Fielding

of the region's volcanic history and the latter owing to its resident herd of goats—and is a popular destination for yachters and snorkelers. Historically, the islet has been associated with both pirates and slave traders and occupies a rather liminal space within the historical political geography of St. Barth. As islands often do, Fourchue became a sort of unofficial entrepôt for illicit commodities—including human cargo—that would otherwise have been forbidden (Maher 2013).

Île Tintamarre is a small (0.9 km^2), flat island located 3 km off St. Martin. Politically, it is a part of Saint Martin. Currently uninhabited, Tintamarre is a popular day-trip destination for tourists, who come for its beaches, reefs, and mud bath. The island was once home to a population of more than 100 people and even supported an international airline, the *Compagnie Aerienne Antillaise*, run by the late Rémy de Haenen, a celebrated resident and local official on St. Barth (Casius 2005).

Fig. 5.6 The Grand Fond valley on St. Barthélemy. Photograph by R. Fielding

Fig. 5.7 The runway at Aéroport St-Jean-Gustav III (SBH), recently renamed for Rémy de Haenen, on St. Barthélemy, as seen from an airplane making the most common approach. Baie de Saint-Jean is visible at the end of the runway. Image from Fielding (2014)

According to one mid-twentieth century historian, St. Barth "lacks almost all natural advantages except an excellent harbor" (Ekman 1964a, 17). Indeed, Gustavia does represent one of the best natural harbors in the region (Fig. 5.9), provided the vessel's draft is not too deep. This natural harbor is framed by *La Pointe*, the peninsula on the south, which adjoins the main island at a low saddle beneath the 34-m hill on which Fort Karl formerly stood. The saddle gives way to Shell Beach, one of the more unique beaches on St. Barth.

Shell Beach (*Anse de Grand Galet*) is not made up of pebbles as the French name would imply, but of bivalve and mollusk shells, deposited on the beach by waves after having been dredged from Gustavia harbor. This beach also features large (up to 4 m diameter) boulders of sedimentary rock. Some of this beachrock contains shells that retain their natural color and shape, an indication to one researcher of their young age (Vroman 1968).

The eastern coast of St. Barth, and to a lesser extent, Lorient Bay, exhibits an interesting example of coral

Fig. 5.8 Northeastern view of the undulating St. Barthélemy coastline, including the rocky Petite Anse, Anse des Flamands, and in the distance, Baie de Saint-Jean. Also visible are three offshore islets, from left: Île Chevreau (also called Île Bonhomme), Île Frégate, and Île Toc Vers. The sargassum raft barely shown at the bottom of the image was part of an island-wide presence during the winter of 2014–15 and continuing to the time of this writing (see Hazards section). Photograph by R. Fielding

pavement. The attached reef has colonized nearly the entirety of two bays—Anse de Grand Cul-de-Sac and Anse de Petit Cul-de-Sac, as well as the nearly beachless east-facing coast known as Petites Anses.

5.3.2 Salt Ponds, Baymouth Bars, and Piscines Naturelles

Both islands display several variations of inland saltwater features, both natural and anthropogenic. When circumnavigating St. Martin, one notices the abundance of baymouth bars and salt lagoons contained therein. Perhaps the most prominent are Simpson Bay Lagoon—contained within the isthmuses that connect the main island to the Terres Basses peninsula—and the Great Salt Pond, held back by the baymouth bar upon which the town of Philipsburg, the capital of Sint Maarten, is situated (Fig. 5.10).

On St. Barth, the large salt pond near the eponymous *Anse de Grand Saline* is no longer in operation but stands as a landscape of remembrance to one of the island's major historical industries—salt making (Fig. 5.11). The dryness of this island limited its agricultural potential (and challenges its present, seemingly-insatiable development for tourism) but made it an ideal site for evaporation ponds. The island government has recently begun efforts toward environmental restoration at the site of the former salt pond.

In at least two locations on St. Barth, ancient coastward lava flows have formed natural pools, replenished by breaking waves but otherwise separated from the sea. These sites are known as the Washing Machine, owing to the turbulence caused when large waves enter, and, simply, *Le Piscine Naturelle*, or the natural swimming pool (Fig. 5.12). Both are popular with swimmers, though the former is more accessible than the latter owing to a walking path that does not require the hiker to cross private property.

5.3.3 Caverns

Both islands feature caverns in places where the soft limestone is exposed and susceptible to erosion. Sypkens-Smit (1995) refers to, and includes photographic plates and schematic maps of, caverns at several locations along the Terres Basses peninsula of St. Martin. These caves have yielded important findings related to the paleofauna

Fig. 5.9 Aerial view, facing southeast, of Gustavia, St. Barth, showcasing the natural harbor. Photograph by D. Fielding

Fig. 5.10 Aerial view, looking north, of Philipsburg, Sint Maarten, situated between the Great Salt Pond (Groote Zoutpan) behind the town and Great Bay (Groote Baai) in the foreground. Photograph by D. Fielding

(McFarlane 2013) and human prehistory (Haviser 1991) of the region. Sadly, from the perspective of science, and inexcusably from the perspective of national heritage, one of the more artifact-rich caves on St. Martin was "definitively ruined some hours after the discovery by using explosives and pouring in concrete" by a hotel construction team (Dubelaar 1985: 171). On St. Barth, no artifacts or significant fossils have been reported, but Lenoble et al. (2012) have documented and charted numerous caverns and rock overhangs. These tend to be clustered in the south and east of the island with the greatest concentration among the foothills of Morne Lurin and near the coasts at their bases.

Fig. 5.11 The salt ponds at Saline on St. Barth. Saline Beach (Anse de Grande Saline) is visible between the mountains in the background. Photograph by R. Fielding

Fig. 5.12 The "Washing Machine" on St. Barth. Photograph by R. Fielding

5.4 Heritage and Tourism

5.4.1 History

The standard history of St. Martin is contained in the straightforwardly titled, *History of Sint Maarten and Saint Martin* (Hartog 1981). While dated, this text addresses much of the history of the island before mass tourism began to flourish there, though it does so from a decidedly Eurocentric perspective. Another valuable, locally flavored history is *Beyond the Tourist Trap: A Study of St. Maarten Culture* (Sypkens-Smit 1995), which is, as the name implies, primarily focused on the island's Dutch side.

The best locally produced (and certainly locally flavored) history of St. Barth is Bourdin's *Histoire de St. Barthélemy* (1978), published originally in French with side-by-side English translation. Maher's (2013) text on the linguistic geography of St. Barth provides a surprisingly thorough account of the history of the region, including not only St. Barth and St. Martin, but also St. Kitts, Guadeloupe, and islands further afield. Maher relies heavily upon the work of Per Tingbrand for information about the Swedish period of St. Barth's history.

5.4.2 Colonialism and Politics

Saint Martin/Sint Maarten is famously the "world's smallest binational island" (Schwartz 2010: 302). In response to frequent disputes between the French and Dutch settlements on the island, the 1648 Treaty of Concordia established the first border—altered several times later—between the two sides. While the existence of a border has led to the diverging political structures on the island, the actual border poses no barrier to travel, whatsoever. One would only stop to take a souvenir photograph—never to show a passport. At each point where the road crosses the border, a small marker stands with the French and Dutch flags on either side. The border, of course, does not separate regions of physical geographical difference, nor are the human geographies noticeably different on either side, except in a few minor ways. English is more prevalent on the Dutch side than on the French, but neither side is difficult for a monolingual Anglophone traveler.

St. Barth has been referred to as "history's shuttlecock" (Maher 2013: 29), owing to its back-and-forth transfer, first between the French and British and later between the French and Swedish colonial governments. This volley was not unique to St. Barth, of course, and could describe any number of islands throughout the Lesser Antilles. The near-century of Swedish colonial rule (1784–1878) gave St. Barth the name of its capital, Gustavia (previously known as *le carénage*, simply, "the place where boats are hauled out of the water"[2]), as well as an enduring cultural affinity for the Scandinavian country. St. Barth stands alone as a formerly Swedish Caribbean possession.

5.4.3 Tourism

St. Martin has embraced tourism, to the nearly full exclusion of other industries, since the 1956 economic development plan initiated by the Technical Economic Council of the Netherlands Antilles. According to Hartog (1981), it was this plan that rejected agriculture, livestock, fishing, mining, and industry as viable parts of St. Martin's economic future. During the next decade, the 1960s, several projects were started with the goal of developing the island's infrastructure for tourism. The specific focus was on the airport, water and electricity supplies, roads, and hotels. Today Sint Maarten—and to a lesser extent, Saint Martin—widely embraces tourism. The port in Philipsburg is equipped to handle multiple large cruise ships simultaneously. Princess Juliana International Airport is one of the busiest in the region and one of the few that can accommodate large, long-haul jets. The island hosts over one million visitors per year and offers resorts, casinos, tours, and watersports for activity-seeking tourists.

Tourism on St. Barth is of a different character than tourism on St. Martin. While the industry has its start at roughly the same time period, the mid-twentieth century, the current status is quite different. Maher (2013) traces the beginning of tourism there to the arrival of David Rockefeller in 1958. After Rockefeller bought property on the

[2]Technically speaking, the town is called Gustavia and the port was called *le carénage*. The earliest evidence of the distinct identities of the port and the town is from a March 24, 1788, report to Stockholm from Johan Norderling, a Swedish judiciary, who wrote, "The Carénage (one portion of the harbor of Gustavia) is above all else unceasingly beautiful; I never refrain from admiring it" (cited in Ekman 1964b: 8). There remains some confusion about the Le Carénage/Gustavia distinction within the popular press today. For example, one tourist book states that "the town of Carénage was renamed Gustavia in honor of the Swedish king" (Didcott and Didcott 1997: 7). The actual situation seems to be that there was no proper town near the port until Sweden took possession of St. Barth and built up the harbor area. Thus, the name *Gustavia* and the town itself emerged simultaneously at or near the site of the port, which had been—and for a while continued to be—known simply by its descriptor, *le carénage*. Maher's (2013) history supports this interpretation, as do certain earlier documents, for example, a periodical from 1800, which first refers to "Le Carenage, near which stands Gustavia, the sole town in the colony," and later states that "Le Carenage had no town belonging to it before the island was in possession of Sweden" (Anonymous 1800: 68). Today, the port is officially called the Port of Gustavia (*Port de Gustavia*) and still "looks as if it had been drawn by a child asked to sketch an ideal harbor" (Ekman 1964a: 224).

island and built a vacation home, other wealthy foreigners—first Americans, then Europeans—began to follow suit. By the end of the 1970s, the economy of St. Barth had been thoroughly transformed from the agrarian and maritime character of the past to one wholly focused upon tourism. Despite the single focus on one sector, tourism development has been strictly regulated on St. Barth. While many environmental activists, scholars, and journalists decry the saturation of the island's roadways with cars, its slips with yachts, and its beaches with villas and guesthouses, the trend in St. Barth remains one focused on luxury, rather than mass, tourism. Hotels remain small, both in physical size and number of guest rooms allowed in each. Prices remain high, compared to other Caribbean islands, and the port in Gustavia cannot accommodate large cruise ships (although ships do occasionally anchor offshore and transport tourists to St. Barth via tender). The wealthy clientele whose presence on St. Barth precipitated the island's tourism development have continued to constitute its target market even as the island is "discovered" by a more middle- to upper-middle-class tourist set. Still, according to Cousin and Chavin (2013: 188), the touristic appeal of St. Barth continues to be based upon its "insular exclusiveness, tropical exoticism, French refinement, and Euro-American cosmopolitanism."

5.5 Hazards

Both islands are vulnerable to hazards that might cut off their supply of imported resources. St. Barth, especially, is dependent upon imports for much of its subsistence. With no fresh surface water or groundwater, St. Barth has long been vulnerable to drought. Despite the waste-to-energy powered desalination plant, in operation since 2001, fresh water on St. Barth remains in short supply (Fielding 2014). Additionally, the fuel for the incinerator—municipal and industrial solid waste—is primarily obtained through imports in the form of consumer goods and packaging. Any hazard, such as a hurricane, that disrupts the importation of these goods would also threaten the fuel supply for the desalination plant.

5.5.1 Hurricanes

The location of St. Martin and St. Barth in the northern Lesser Antilles places the islands within the path of tropical cyclones. According to analysis by Caron (2011, 192), both islands have been "flooded" by five storm surge waves, produced by Category 4 or 5 hurricanes since 1851. The eye of a storm need not pass directly overhead to cause substantial damage. Hurricane Omar, as cited by Caron (2011), passed 100 km north of St. Barth—thus, at least 70 km north of St. Martin—and still resulted in major damages to both islands. Damage from hurricanes includes both the destruction of infrastructure and the erosion of coastlines.

5.5.2 Earthquakes and Tsunamis

While neither St. Martin nor St. Barth sits on a known fault line, the nearby seismic zones off Puerto Rico and further south, in the Windward Islands, are known to produce earthquakes that can cause tsunamis. Caribbean volcanoes are also rarely, but powerfully, tsunamigenic. Despite the multitude of potential causal events, such as earthquakes, terrestrial volcanic eruptions, submarine volcanic eruptions, and landslides, both in the region and further afield, tsunamis in the Caribbean are rare. One team of researchers analyzed the data and historical descriptions related to 91 extreme wave events that occurred throughout the Caribbean region between 1498 and 2000 and determined that between 27 and 36 of these wave events represented true tsunamis (Lander et al. 2002). Of these, only one is definitively known to have caused damage on St. Barth and St. Martin the November 18, 1867, tsunami caused by a 7.5-magnitude earthquake with an epicenter located between St. Croix and St. Thomas in the Danish West Indies—now the US Virgin Islands (Zahibo and Pelinovsky 2001). An earlier teletsunami, generated across the Atlantic in 1755 by an earthquake near Lisbon, Portugal, caused a 4.5-m wave on St. Martin. Though no specific effects were recorded from this tsunami, one author states that "lowlands on most other French islands were inundated," indicating the possibility of flooding and damage on both St. Barth and St. Martin (Lander et al. 2002, 64).

5.5.3 Nearshore Development

Both islands display frequent tourism-oriented development in close proximity to the shore. Perhaps this is best epitomized by Princess Juliana International Airport (SXM) and the famous Maho Beach on Sint Maarten. This west-facing, 320-m-long beach is situated directly across a two-lane street from the base of the airport's only runway. Because of SXM's role as a regional hub—one of the few airports in the Lesser Antilles that receives trans-Atlantic flights—large airliners regularly approach low over the beach, which is open to the public. The low-flying jumbo jets have become a major tourist attraction in their own right (Fig. 5.13). The spectacle notwithstanding, the proximity of the airport to the beach, serves to highlight the issue of coastal development and its exacerbation of both islands' vulnerability to a host of hazards including coastal erosion, storm surge from

Fig. 5.13 A KLM 747, arriving from Amsterdam, approaches Princess Juliana International Airport over Maho Beach for landing. Photograph by D. Fielding

hurricanes, tsunami, and perhaps most insidious—sea level rise due to climate change.

5.5.4 Sargassum

While not a hazard to life or property per se, the recent influx of sargassum seaweed (*Sargassum natans* and *S. fluitans*) has certainly been hazardous to both islands' tourism-based economies. This issue is by no means restricted to St. Martin and St. Barth. Rather, it has affected nearly all the islands and mainland beaches in the Caribbean basin. While some amount of sargassum has always been present in the waters and on the beaches of the Caribbean, 2011 saw an unprecedented increase in its quantity. This influx was repeated in 2014 and 2015, owing to the right combination of sea-surface temperatures, nutrient availability, and ocean currents bringing sargassum from its blooming sites in the Atlantic into the Caribbean (Doyle and Franks 2015). The sargassum problem is one of overabundance. A small amount of the plant is beneficial as habitat for marine fauna while at sea and performs services of beach nourishment and shoreline stabilization onshore (Doyle and Franks 2015). In larger quantities, however, the associated odor, insects, and unsightliness of sargassum on beaches, along with the risk of damage to fishing gear and boat motors, tend to outweigh the plant's natural benefits. While innovative uses for sargassum are being tested—including use as a fertilizer, animal feed, or dune stabilizer—most efforts in the Caribbean, including on St. Martin and St. Barth, focus on techniques for its speedy removal from beaches frequented by tourists. In St. Martin, this effort is primarily the responsibility of beachfront resorts, while in St. Barth the municipality manages the removal of sargassum from popular beaches.

5.6 Conclusion

St. Martin and St. Barth are exemplary forms of Lesser Antillean geography, both human and physical. While both lie within the outer, limestone arc, evidence of past volcanism remains on the surface or just below. The beaches, mountains, bays, and vegetation are typical of the region but perhaps somewhat more concentrated here, owing to the islands' small areas. Each island also reflects the history of European colonialism in the Caribbean—Saint Martin/Sint Maarten has its famous international border and St. Barth, as "history's shuttlecock" (Maher 2013, 29), has its history of European power shifts woven into its very fabric. While the climatic and tectonic processes that shaped the islands are still evident in the form of natural hazards, tourists continue to arrive—along with their attendant infrastructural growth and development—drawn by the densely packed variations in human and physical geography present on these small, lovely islands.

Acknowledgements Fieldwork that directly contributed to the completion of this chapter was supported by the Faculty Research Grants program at the University of the South. Other fieldwork, which helped acquaint the author with these islands through research on other topics, was funded by the University of Denver's Internationalization Grant program and the American Geographical Society's Bowman Expedition Fund. Finally, special thanks to my wife, Diane Cooper Fielding, who accompanies me in the field often, but most recently (and memorably) to the islands discussed in this chapter while eight months pregnant. Figures 5.4, 5.7, 5.9, 5.10, and 5.11 are hers.

References

Anonymous (1800) Account of the Swedish Island of St. Bartholomew, in the West Indies. Monthly Mag Am Rev 3:68–70

Benoist J (1964) Saint-Barthélemy: physical anthropology of an isolate. Am J Phys Anthropol 22(4):473–487

Benoît C (2008) Saint Martin's change of political status: inscribing borders and immigration laws onto geographical space. New West Indian Guide/Nieuwe West-Indische Gids 82(3–4):211–235

Bourdin G (1978) Histoire de St. Barthélemy. Porter Henry, New York

Cambers G (2010) Lesser Antilles. In: Bird ECF (ed) Encyclopedia of the world's coastal landforms. Springer, Dordrecht, pp 299–310

Caron V (2011) Contrasted textural and taphonomic properties of high-energy wave deposits cemented in beachrocks (St. Bartholomew Island, French West Indies). Sed Geol 237:189–208

Casius J (2005) Compagnie Aerienne Antillaise. Am Aviat Hist Soc J 50(1):55–66

Cleve PT (1871) On the geology of the North-Eastern West India islands. Norstedt and Söner, Stockholm

Christman RA (1950) Geology of Saint Bartholomew, Saint Martin, and Anguilla, West Indies. Ph.D. Dissertation. Princeton University

Christman RA (1953) Geology of St. Bartholomew, St. Martin, and Anguilla, Lesser Antilles. Bull Geol Soc Am 64:65–96

Cousin B, Chauvin S (2013) Islanders, Immigrants, and Millionaires: the Dynamics of Upper-Class Segregation in St. Barts, French West Indies. In: Hay I (ed) Geographies of the super-rich. Edward Elgar Publishing, Cheltenham, UK, pp 186–200

Didcott C, Didcott C (1997) St. Barth: French West Indies. Concepts, Waitsfield, Vermont

Doyle E, Franks J (2015) Sargassum fact sheet. Gulf and Caribbean Fisheries Institute, Marathon, Florida

Dubelaar CN (1985) Petroglyphs In The Lesser Antilles. Lat Am Indian Lit J 1-2:163–173

Ekman E (1964a) St. Barthélemy and the French Revolution. Caribbean Studies 3(4):17–29

Ekman E (1964b) A Swedish career in the tropics: Johan Norderling (1760–1828). Swed Pioneer Hist Rev 15(1):3–32

Fielding R (2014) "The good garbage": waste to water in the small island environment of St. Barthélemy. Focus Geogr 57(1):1–13

Fink LK Jr, Fairbridge RW (1975) Leeward Islands. In: Fairbridge RW (ed) Encyclopedia of world regional geography, part 1: Western Hemisphere (Including Antarctica and Australia). Dowden, Hutchinson, and Ross, Stroudsburg, PA, pp 339–341

Hartog J (1981) History of Sint Maarten and Saint Martin. The Sint Maarten Jaycees, Philipsburg, Sint Maarten

Haviser J (1991) Development of a prehistoric interaction sphere in the northern Lesser Antilles. New West Indian Guide/Nieuwe West-Indische Gids 65(3–4):129–151

Lander JF, Whiteside LS, Lockridge PA (2002) A brief history of tsunamis in the Caribbean sea. Sci Tsunami Hazards 20(1):57–94

Lenoble A, Queffelec A, Stouvenot C (2012) Grottes et abris de l'île de Saint Barthélemy. Spelunca 126:1–9

Maher J (2013) The survival of people and languages: schooners goats and cassava in St. Barthélemy, French West Indies. Brill, Leiden

McFarlane DA (2013) Limestone caves and the Quaternary record of terrestrial tetrapods on islands. J Geol Soc 170:535–538

Reinhartz D (2012) The Caribbean cartography of Samuel Fahlberg. In: Liebenberg E, Demhardt IJ (eds) History of cartography: lecture notes in geoinformation and cartography, vol 6. Springer-Verlag, Berlin, pp 21–34

Schwartz ML (2010) St. Martin. In: Bird ECF (ed) Encyclopedia of the world's coastal landforms. Springer, Dordrecht, pp 302–303

Spencer JWW (1901) On the geological and physical development of Anguilla, St. Martin, St. Bartholomew, and Sombrero. Q J Geol Soc London 57:520–533

Sypkens-Smit MP (1995) Beyond the tourist trap: a study of St. Maarten culture. Natuurwetenschappelijke Studiekring voor het Caraïbisch Gebied, Amsterdam

Tomblin JF (1975) The Lesser Antilles and Aves Ridge. In: Nairn AEM, Stehli FG (eds) The Gulf of Mexico and the Caribbean. Springer, New York, pp 467–500

Vaughan TW (1918) Geologic history of Central America and the West Indies during Cenozoic time. Bull Geol Soc Am 29:615–630

Vaughan TW (1926) Notes on the igneous rocks of the northeast West Indies and on the geology of the island of Anguilla. J Wash Acad Sci 16:345–358

Vroman M (1968) The marine algal vegetation of St. Martin, St. Eustatius and Saba (Netherlands Antilles). Stud Flora Curaçao Caribb Islands 2:1–120

Westercamp D, Andreieff P (1983) Carte Géologique de St. Barthélemy au 1/20 000e—Note Explicative. Bulletin de Recherche Géologique et Minière 5–38

Zahibo N, Pelinovsky EN (2001) Evaluation of tsunami risk in the Lesser Antilles. Nat Hazards Earth Syst Sci 1(4):221–231

6 Saba and St. Eustatius (Statia)

Jennifer L. Rahn

Abstract

Saba and St. Eustatius/Sint Eustatius (Statia) represent geologically young (less than a million years) Pleistocene island arc volcanoes that rise at the northernmost part of the Lesser Antilles. Both are small islands less than 21 km^2 in size with populations around 2000, and both have stratovolcanic features and many Pelean domes with adjacent pyroclastic flows. As a single volcano, Saba's steep slopes drop off precipitously into the sea, while St. Eustatius lies at the northern end of a shallow submarine bank that links to St. Kitts and Nevis to the southeast. Similar in geologic history and perhaps co-eruptive (though Statia is slightly older and has more erosional features and some uplift), the islands also have similar climates, winds, waves, and ocean current regimes, but lack often-found-in-the-Caribbean geomorphic features such as dunes, perennial streams, wetlands, karst, aquifers, and inland water bodies. Saba and Statia both have heritage sites, including a few (small) Amerindian archaeological sites, but most historical research centers on post-European history including maritime figures, agriculture, and slavery. Owing to their small sizes, neither has been developed very heavily, though Saba is famous for its scuba diving and Statia is known for its oil terminal. Both islands face similar environmental hazards that include hurricanes, earthquakes and volcanism, and drought.

Keywords

Pelean dome • Coral reef • Scuba • Volcano • Landform

6.1 Introduction

Saba and St. Eustatius/Sint Eustatius (Statia) (Figs. 6.1 and 6.2) are both located near the 17° north latitude and 63° west longitude lines. The smaller of the two, Saba (*say-bah*), hosts about 2000 full-time residents on it 12 km^2, while around 3500 people make the 21 km^2 St. Eustatius—*Statia* (*stay-shuh*) as locals call it—home. Separated by 27 km, Saba and Statia sit 45 km southwest (12 min by plane) and 61 km south (20 min by plane) of St. Martin/St. Maarten, respectively. Both islands are special municipalities of the Kingdom of the Netherlands, and the top of Saba's volcano, Mount Scenery, represents the highest point in the Dutch Kingdom (877 m). Saba has four villages, Zion's Hill or Hells Gate, Windwardside, St. Johns, and The Bottom, which serves the island's administrative capital. Statia does not have towns in the classic sense, only a settlement, Oranjestad, the capital. While most people from Saba call themselves Sabans, in the village of Hells Gate, they use Sabians. On Statia, locals refer to themselves as Statians. As of October 10, 2010 (10/10/10, colloquially know as ten-ten-ten), Saba, St. Eustatius, and Bonaire became special municipalities of the Netherlands and are also referred to as the "BES islands" or "Caribisch Nederland." Day-to-day local affairs are managed by an elected Island Council, presided over by a Lieutenant-Governor. The Dutch have jurisdiction, environmental say (both islands are regarded by the European Union as an overseas territory and subject to

J.L. Rahn (✉)
Department of Geography, Samford University, Alabama, USA
e-mail: jlrahn@gmail.com

C.D. Allen (ed.), *Landscapes and Landforms of the Lesser Antilles*,
World Geomorphological Landscapes, DOI 10.1007/978-3-319-55787-8_6

Fig. 6.1 General physiographic map of Saba and surrounding features, such as Diamond Rock and the Saba National Marine Park run by the Saba Conservation Foundation. The small mountainous island is only 13 km^2 (5 mi^2) but rises from sea level to 877 m at the peak of Mt. Scenery. Cartography by K.M. Groom

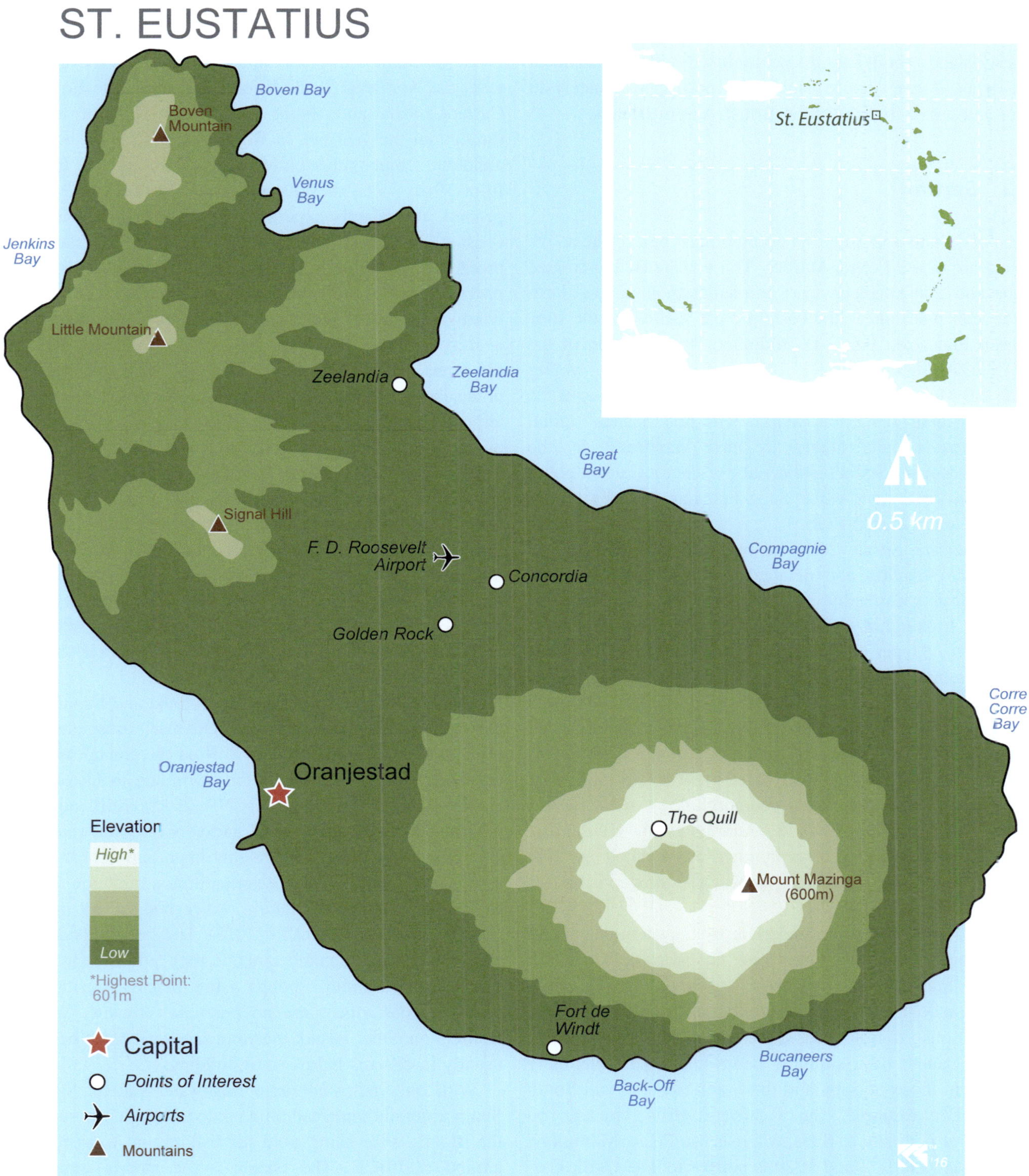

Fig. 6.2 General physiographic map of St. Eustatius. Slightly larger than its neighboring Saba, St. Eustatius is dominated by the large volcanic crater called "The Quill" to the south and a few smaller mountains in the north. Most of the island's population resides within the central lowlands. Cartography by K.M. Groom

EU regulations) and sponsor and subsidize the major projects on the island that affect its landscape and landforms. These islands should not to be confused with the island of Eustatia (30 acres in size) and Saba Rock (approximately an acre and a half in size) in the British Virgin Islands.

6.2 Setting

Saba and Statia represent the northernmost part of the active island arc of the Lesser Antilles. The volcanoes that created Saba and Statia formed from the subduction of the North American plate under the eastern-most section of the Caribbean Plate (see Fig. 2.1). Although the formation of the Lesser Antilles volcanic arc is geologically young (5.3–2.6 Ma), Saba and Statia are the youngest islands of the arc (less than a million years), with the islands' initial underwater activity likely to have happened in the mid-Pleistocene. Statia's subaerial presence occurred about 1 million years ago, while Saba first appeared above sea level around 500,000 BCE.

Roobol and Smith (2004) summarized Saba and Statia's structural evolution and detailed geology, petrology, and petrogenesis specifically, by drawing on previous works of Saba by Molengraff (1886), LaCroix (1890, 1893), Hovey (1905a, b), Sapper (1904), and Perret (1942), and of Statia by Westermann and Kiel (1961), Maclure (1817), Cleve (1871), Molengraff (1886, 1931), Sapper (1903, 1904), and Hardy and Rodrigues (1947). The Quaternary sea-level curve for the Caribbean region shows that sea level fluctuated from +6 to −24 m relative to present sea level, thus affecting the size, shape, and geomorphology of the islands through time (Emiliani 1978; Aubrey et al. 1988). The subsequent volcanic geomorphology was created by both constructive and destructive processes, now visible as landforms on both islands.

Saba is believed to be underlain by a system of parallel dikes emplaced into a fault zone. In contrast, The Quill volcano (Statia) is underlain by a magma chamber in which crystal fractionation occurs to produce a wide range of lava compositions. It has been successfully demonstrated that other volcanoes have similarly contrasted crustal plumbing systems to those of Saba and the Statia (Roobol and Smith 2004). For example, the Soufriere Hills volcano on Montserrat is similar to Saba, whereas Mt. Pelee, Martinique that erupted in 1902 and 1929 is similar to The Quill (Caribbean Volcanoes 2015). Geologists have shown that the island of St. Eustatius is gradually subsiding (DCBIODATA 2015) and there are local reports on Saba of uplift (Eddie Hassell personal communications).

6.2.1 Saba and Statia Climate

Saba and Statia's tropical climate is governed by their latitude, the Northeast Trade Winds, and the North Equatorial Current. Saba and Statia average of 12 h of daylight throughout the year (11 in December and 13 in July), and although temperatures extreems vary throughout the year from 19 to 36 C, they usually range from 25 to 28 C. In general, the islands experience sunny weather, with light constant Northeast Trade Winds that bring constant cool breezes and ample moisture (Statia Tourism 2015). The prevailing winds drive the dominant waves to hit the eastern, windward sides of the islands. Multidirectional wind gusts average 16 km/h (10 mph) and on Saba can reach 40 km/h (25 mph), which often causes all flights to be canceled. On both islands, precipitation varies from 980 to 1110 mm. Both temperatures and precipitation vary on the islands depending on the geographic location (higher precipitation on the windward sides, for example) and local elevation (e.g., cooler temperatures at higher elevations and orographic effect). Several events cause the islands to be relatively cool: The tops of the mountains have frequent cloud cover, winds blow up through the ravines bringing a cool breeze to the higher elevations, and the number of hours of direct sunshine is reduced by the shadow of surrounding mountains. In the case of The Bottom on Saba, this results in only about six hours of direct sunlight each day (DCBIODATA 2015).

While some precipitation falls on Saba and Statia during the hurricane season (June to October) associated with tropical storms, approximately 80% of the annual rainfall is concentrated in the rainy season beginning in October and lasting through December (DCNA 2015). Still, rainfall is inconsistent from year to year, and very dry ones can follow very wet years. Notable droughts happened in 1952/1953, 1982/1983, and 2015 where agricultural production dropped to a minimum, and drinking water had to be imported occasionally (DCBIODATA 2015). The steep hillsides, thin soil layers, and fractured igneous rock prevent the natural catchment of rainfall. Though a few brackish springs exist near the coast, there are no perennial streams, lakes, or aquifers on either island, meaning rainwater must be stored, usually in cisterns below most buildings.

With only rainwater as a freshwater source, Saba and Statia generally remained uninhabited, though it is suspected the islands were being used for fishing and foraging since around 500 BCE. The islands went mostly unused by European Colonialists until the seventeenth century and, though currently Dutch in nationality, the English language remains in strong use, demonstrating the persistence of early British Colonists. While Saba and Statia were a colony of the

Dutch West Indies by 1828, it was not until a couple of decades later that they would become Dutch dependencies (1845), and more than a century after that before they would become part of the Netherlands Antilles (1954). Throughout the twentieth century, Saba and Statia became and remained quiet islands, with tourism (primarily hiking and scuba diving) and archaeological studies beginning in the 1980s.

6.2.2 Saba Geologic History and Evolution

The deceptively simple stratovolcano-like appearance of Saba is misleading because Mt. Scenery, the highest peak on the island, is just one of about twenty Pelean-like domes. These domes, and their associated aprons of pyroclastic material, represent a combination of newer domes sitting eccentrically on top of older Pelean dome complexes (Roobol and Smith 2004). In terms of volcanology, Saba is a single volcanic complex, the subaerial part of which has dimensions that measure 50% of those defined by the 500 m submarine contour. Submarine morphology slopes uniformly away from the island in all directions (Fig. 6.3).

6.2.3 Statia Geologic History and Evolution

St Eustatius lies at the north end of an 80-km-long shallow submarine bank (the edges of which are 180 m below sea level themselves) on which the islands of St. Kitts and Nevis also lie. The geology of St. Eustatius can be divided into three very different landscapes (Fig. 6.4). The oldest geological unit on the island is the northern hills—the Northern Centers as Roobol and Smith (2004) called them—composed of five morphologically distinct older volcanoes, built of lava flows and Pelean domes, and pyroclastic aprons of block and ash deposits, each having varying degrees of erosion. The three youngest centers, although eroded, retain their original volcanic features, and lava flows and lithified pyroclastic deposits from these centers, well exposed in the steep sea cliffs around the northern part of the island, have been dated to an age of less than 1 million years. The southern slopes of the northernmost hill are orange-colored, marking the site of former fumarolic activity. When first formed, the Northern Centers appeared as an independent volcanic island with surrounding shallow water carbonate banks and patch reefs containing volcanic deposits.

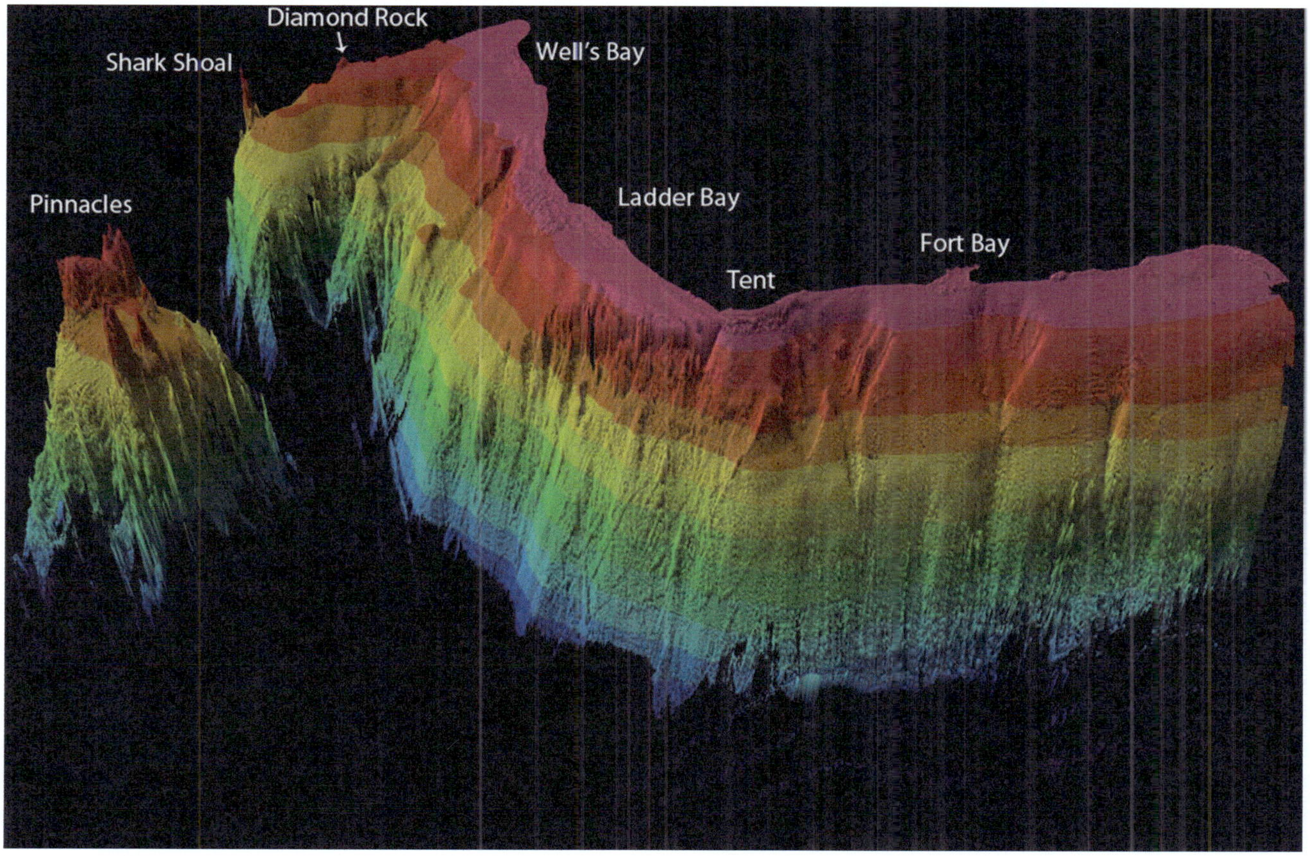

Fig. 6.3 Bathymetric map of Saba. Offshore Saba to the west 1.3 km, a single parasitic submarine Pelean dome (called The Pinnacles) rises from depths of 300 m to only 23 m below sea level. The platform's top, roughly the size of a football field, represents one of the most famous and spectacular dive sites around the island. Image courtesy of http://www.sabapark.org/marine_park/diving/

The Northern Centers may well be broadly contemporaneous (~500 ka) with the older stratigraphic division of Saba including the Diamond Rock–Torrens Point center. At this time, the islands of Saba and Statia were smaller and had different forms to the present-day islands (Roobol and Smith 2004).

Of intermediate age is Sugar Loaf-White Wall formation, a 1.2 km (0.7 mile) uptilted wall and ridge of shallow marine limestone on the extreme south end of the island embedded in the flanks of The Quill. Most likely, it is part of the limestone cap of the submarine platform uptilted by the emplacement of a dome (Roobol and Smith 2004), and a similar structure exists at Brimstone Hill on St. Kitts. Roobol and Smith (2004) characterize this formation as an unusual and interesting documentation of shallow-water phreatomagmatic activity over a period of several hundred thousand years.

The youngest geological unit is the single volcanic cone, The Quill, at the island's south. This morphologically youthful open-crater stratovolcano has flanks sweeping up to angles of 500 at the crater rim. The crater has a diameter of 800 m, and the eastern rim forms the highest point on the island at 600 m asl, though the crater floor rests at 278 m asl. The distal deposits of The Quill have largely been removed by erosion except where they form the nearly flat-lying central part of the island and terminate against the steep hills of the eroded Northern Centers. The Quill, apart from a few lava dome remnants exposed in the inner walls of the crater, is almost entirely a pyroclastic volcano, and the sea cliffs cut into the flank deposits reveal some of the best pyroclastic sections in the Lesser Antilles (Roobol and Smith 2004).

6.3 Landforms

Because they are both young Pleistocene island arc volcanoes, Saba and Statia have similar stratovolcanic features and many Pelean-type domes with adjacent unstable rock debris. Saba is a single volcano whose steep slopes drop off precipitously into the sea, while St. Eustatius is a more complex and older volcano resting at the northern end of an 80-km-long shallow (<180 m deep) submarine bank. Both islands have similar geologic histories and perhaps even co-eruptive, but Statia is older and has more erosional features, sedimentary structures (carbonate depositional sequences), and experienced some uplift. Common Caribbean island geomorphic features that are absent on Saba and Statia are perennial streams, wetlands, karst, aquifers, dunes, and inland water bodies, although there are a few brackish coastal wells. The islands' prime geomorphic agents include precipitation, rock decay, and gravity, each contributing to steeply eroded gullies ("guts") and coastal erosion, but anthropogenic activities, such as free-roaming goats and poor landuse/farming practices (particularly on Statia), also diminish vegetation growth and increase erosion.

6.3.1 Saba Interior Landforms

Saba's largest and most obvious landform, Mt. Scenery, lies approximately in the center of the island, dominating the geomorphic landscape (Fig. 6.5). Because of its classic volcanic shape, it is often assumed this is the main and only volcano of the island. Roobol and Smith (2004) explain, however, that the stratovolcano-like appearance of Saba results from a series of Pelean dome composites and their surrounding pyroclastic aprons that sit atop even older Pelean dome complexes. Because of its archetypal volcanic silhouette and steep coastal cliffs, the producers of the original 1933 King Kong movie used it as the backdrop of Skull Island (National Geographic Society 2007).

Another youthful volcanic feature, the only visible basaltic andesitic lava flow at the northeast corner of Saba, runs across the landscape, cradling the zigzag road (with 18 hairpin curves) up from the airport to Hell's Gate (aka Zion's Hill). The lava's high viscosity resulted in well-preserved lava levees that can clearly be seen as small, narrow, long hills on either side of the main flow (Roobol and Smith 2004). The airport lies at the terminal end of this flow, and the flat area of the airport is the flow covering a pre-existing wave cut platform. This is the largest area of flat elevation on the island, and the Juancho E. Yrausquin Airport (built in 1963) located here claims to have the shortest commercial runway in the world—a mere 400 m. It has also earned the superlative of being on the list of "scariest landings," as the approach is initially into a cliff face, coupled with a sharp bank-left to align with the runway.

6.3.2 Statia Interior Landforms

Statia's main geomorphic features are divided into three distinct landscapes composed of two volcanic areas (in the northwest and southeast) separated by a central, northward sloping plain (Fig. 6.6). In the northwest, The Quill, a symmetrical stratovolcano, also represents the second highest mountain in the Netherlands (600 m, see Fig. 6.7). Formed contemporaneously with Saba around 50 ka, The Quill contains a 2-km-wide crater. Its tropical rainforest vegetation-covered walls drop steeply, 322 m to the crater floor (at 273 m asl). The southeast, dominated by Mt. Mazinga, supports a small cloud forest on part of its crater rim, and the southern flank drops steeply to the sea where a wall and ridge of shallow marine limestone, known as the Sugar Loaf-White Wall (Fig. 6.8), runs east to west for

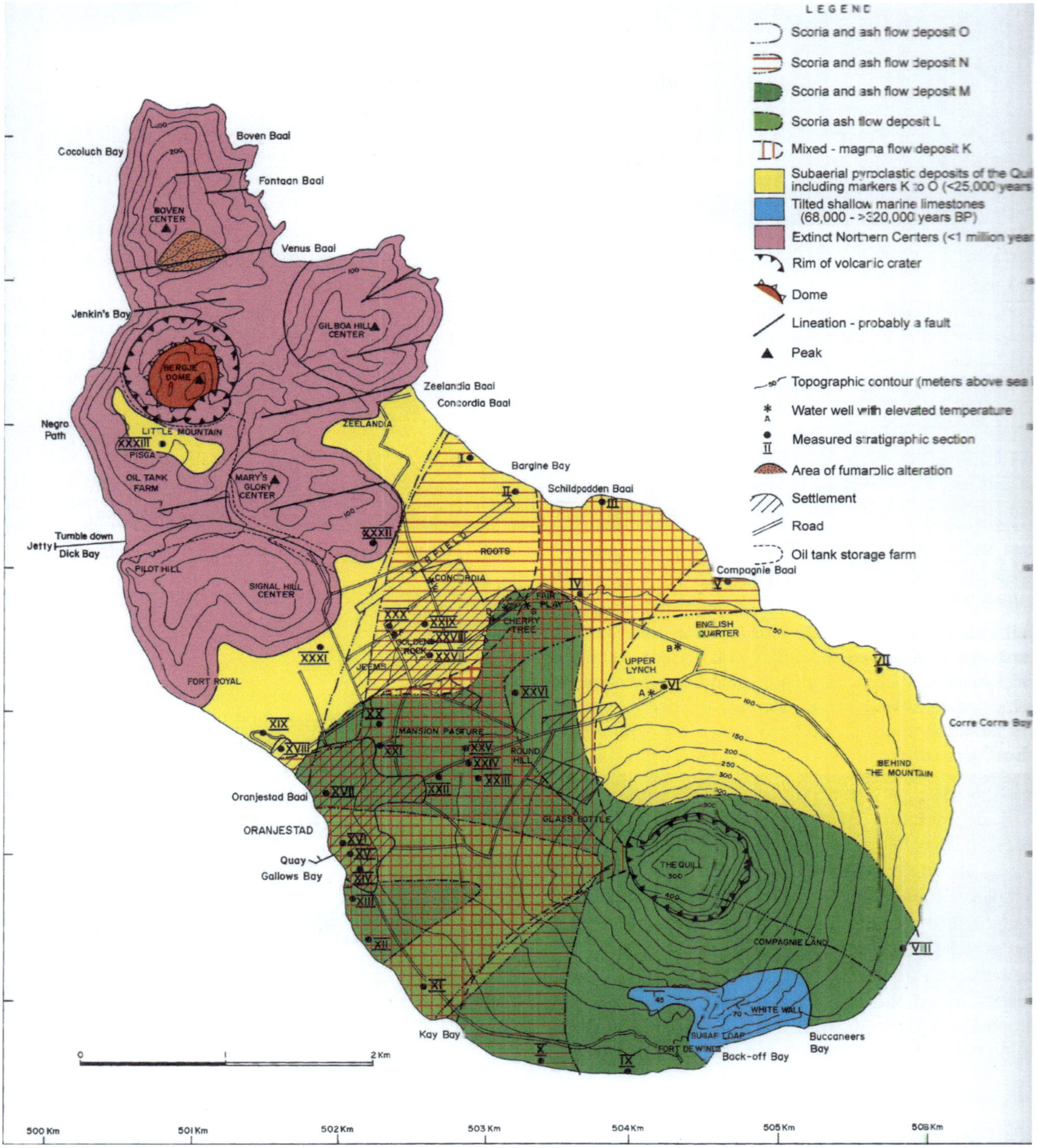

Fig. 6.4 Geologic map of Statia. The southeast end of the island is dominated by a 600 m high dormant volcano, The Quill. In the north end of the island, lower elevation hills have formed from an eroded, extinct volcanic complex (DCNA 2015). Between these two formations, in the central part of the island, is a low sloping 6 km^2 plain at 30–80 m above sea level composed of relatively fertile volcanic soils (DCBIODATA 2015). Map courtesy of http://caribbeanvolcanoes.com/st-eustatius-map/

about 1 km (Roobol and Smith 2004). Perhaps formed as part of the limestone cap of the submarine platform that was uptilted by dome growth, these marine sediments mixed with ash and pumice over time, eventually lithifying into strata visible from the "Around the Mountain" trail and Fort de Windt (Roobol and Smith 2004). On the western side of the

Fig. 6.5 Aerial photograph of Saba' south side with Mt. Scenery dome in the middle of the island. *Center left* in the *shade* is The Bottom. Surrounding The Bottom is the *horseshoe-shaped*, collapse scar of a volcanic lateral blast. Below The Bottom on the coast is Fort Bay, where the piers at the harbor are the only place boats can land on Saba. Photograph by L. Avery

crater, a pyroclastic flow fans out to the coast, and hosts the island's capital, Oranjestad (Roobol and Smith 2004). The northern slope of The Quill gradually drops down to the flat central plain, the Kultuurvlakte ("Culture Plain"), a 6 km^2 fertile plain containing volcanic soil, and where most of the roads, houses, and agriculture occur.

Statia's northern landscape—called the Boven Hills by locals and the Northern Centers by Roobol and Smith (2004) —consists of an extinct Quaternary (500 ka) volcanic complex of unevenly eroded hills. These hills initially comprised the entire island, and outcrops of the lava flows and pyroclastic deposits can be seen in adjacent sea cliffs. In the northern part, the Boven mountain's southern slopes are orange-colored, marking the site of former fumarolic activity (Roobol and Smith 2004). In 1979, on the westernmost hill of the Northern Centers, the Pisgah hilltop was flattened, and excavation began for the construction of an 11.3-million-barrel oil storage and transshipment facility/port. The rest of the Northern Center area is an uninhabited wilderness area with a drier weather pattern than The Quill producing savannah-like vegetation and fauna (DCNA 2015).

6.3.3 Coastal Processes and Landforms

Coastal features on both islands include a few black/brown volcanic sand beaches, more common cobble beaches, and coastlines made up of large rock, coral rubble, and cliffs. The length of coastline of Saba and Statia are 15 and 21 km, respectively. Water temperatures range from 25 to 29 C (77–85 °F), which support nearshore coral reefs. Prevailing Northeasterly Trade Winds drive waves to hit the Atlantic (Windward) sides of the islands, while the North Equatorial Current arrives from the southeast. Average wave heights range between 1 and 2 m, with strong North Atlantic winter storms generating swells up to 12 m along the north coasts. As with other small islands, waves tend to wrap around the islands, creating complex interference wave patterns on the southwest coasts.

6.3.4 Saba Coastal Landforms

Saba's outline shows a series of headlands separated by arcuate embayments (Fig. 6.9). These promontories consist of lava flows and domes, with bays (being) formed in the volcanoclastic host rocks themselves (Roobol and Smith 2004). The steep slopes on Saba continue underwater to approximately 500 m below sea level (Fig. 6.10). Saba's coast is roughly divided into thirds amongst sand and/or cobble beaches, rocky coast, and cliffs—especially steep cliffs because Saba has no shallow platform to prevent being constantly battered by wave action. There is only one permanent sandy beach at Cave of Rum Bay (Fig. 6.11), best accessible by boat.

Fig. 6.6 Aerial view of Statia's Culture Plain (Kultuurvlakte) and surrounding features. Below The Quill and its crater in the foreground, the Culture Plain (Kultuurvlakte) is visible *mid-photograph* with Oranjestad along the left coast. The Northern Centers, seen in the background, are the remains of an extinct quaternary (500,000 BP) volcanic complex of unevenly eroded hills. Photograph by J. Haviser

This beach is maintained by sediment from hillslope failure of the rapidly eroding guts below Mary's Point.

Wells Bay hosts a seasonal sandy beach that usually appears March through October and disappears in the winter months, exposing well-rounded cobbles. The Wells Bay sandy beach can also disappear for years at a time, when sand sediments are washed offshore below wave base (Rahn et al. 2015). The northern end of the Well's Bay beach transitions into the Torren's Point formation with sea stacks, arches, swim-through tunnels, and caves (Fig. 6.12), and an overhanging >90° cliff of pyroclastic material serves as a backdrop to the beach, the bottom of which illustrates a classic wave cut notch (Fig. 6.13).

Meanwhile, Cove Bay beach—nourished (2010–2016) by white sand dredged from St. Martin and backed by large boulders as revetment—provides an accessible, public sandy beach. Sea Turtles have recently (2013–2015) nested on Cave of Rum Bay and Cove Bay beaches. The cobble beaches on Saba have rounded volcanic clasts and large coral rubble fragments broken from the reef and deposited by hurricane waves and subsequently modified by less energetic waves. In total, nine cobble/pebble beaches surround Saba, with Tent Reef beach being the most accessible and most widely used recreationally by residents. Some recreational swimming occurs at Fort Bay, but this bay is the harbor on the island, is a busy working port, and is primarily used for fishing, cargo, and dive boats. On the northeast coast, below the airport, a hiking/bouldering trail winds between jagged rust-red lava rocks. This area, called the Tide Pools, technically contains wave splash-pools since the tides are so small (Kjerfve 1981). The Tide Pools showcase resplendent tafoni, as well as a few fulgurites (Fig. 6.14).

6.3.5 Statia Coastal Landforms

The majority of St. Eustatius is steep cliff coast, but because a shallow water platform ($\sim$10 m) surrounds the island,

Fig. 6.7 The Quill Crater as seen from above. Tropical rain forest covers the walls and the crater floor. A small cloud forest exists on part of the crater rim near Mount Mazinga. Photograph by J. Haviser

waves are milder than on Saba. The Atlantic coast has average wave heights 1–2 m while the Caribbean, leeward coast, waves average less than 30 cm (St Eustatius Management Plan 2007). Monitoring of cliff erosion between 2004 and 2006 showed erosion up to 2.0 m in one year, resulting in major landslides with chunks of rock over 1 m falling regularly (STENAPA 2007). Six sandy beaches occur around the island adjacent to the Culture Plain, though "sandy" can be a loose term, as most beaches remain a mix of sand, gravel, cobble, and coral rubble. The north coast of Statia contains an escarpment with a series of five, 300-m, closely spaced, undulating, cliffed-coast headlands (Fig. 6.15).

6.3.6 Nearshore and offshore Landforms of Saba

Saba hosts five distinct geomorphic forms of coral reef: coral-encrusted colluvial debris, spur and groove, antecedent bar-beach ridge veneer, sea stacks surrounded by reef, and offshore seamounts/pinnacles. Coral-encrusted colluvial debris from the adjacent steep slopes are the most common reef structures surrounding Saba and consist of hard and soft coral and sponges built on top of cobble- to boulder-sized rocks (van't Hof 1991). Spur and groove reef formations occur on the island's leeward side between the Hot Springs and Babylon dive sites. In the Tent Bay area, a shore-parallel reef and wall called Tent Reef was exposed as an offshore bar and beach ridge system at sea level around 8000 years ago. As this system submerged with sea-level rise, corals kept up with water levels and formed a shallow one to two-meter-thick veneer on top of the rocks and beach sediments. While diving this site, you can often see the well-rounded former volcanic beach cobbles beneath the coral layer. At the west end, this reef becomes a series of coral outcrops topping a vertical wall that drops 30–65 m (van't Hof 1991).

A 900 m^3 sea stack, Diamond Rock rises from Saba's northwestern sea floor. It is a favorite bird roost site and so-called because its silhouette is diamond shaped, and shines from a distance because of the white bird gvano's high reflectivity. The entire subaqueous diameter of Diamond Rock is covered with corals and other coral reef ecosystem inhabitants. Between Diamond Rock and Torrens Point, there are a series of smaller, drowned sea stacks also covered with

Fig. 6.8 Sugar loaf ridge viewed from Fort de Windt, southern St. Eustatius. It is an uptilted wall and ridge of shallow marine limestone on the extreme south end of the island imbedded in the flanks of The Quill. This feature formed from the shallow-water carbonate banks and patch reefs around 500 000 BP. Photograph by J. Haviser

corals. The most spectacular reef formations of Saba are the offshore "seamounts" about one-half km to the west. These "seamount" formations are probably parasitic domes formed by magma chambers connected to the main one below Saba (see Fig. 6.3).

6.3.7 Saba Bank

About 5 km southwest of Saba, separated from the island by a 700-m deep trough, lies a large (2200 km^2 or 849 mi^2) submarine carbonate bank, or atoll, called the Saba Bank. It is the third largest of its kind in the world, and spectacularly rich in marine biodiversity (DCNA 2015). The flat top of this atoll ranges in depth from 20 to 50 m, elevated about 1800 m above the surrounding sea floor. Various oil companies, government agencies, and academic institutions researched Saba Bank geology between 1970 and 1999, with approximately 4300 km of seismic data acquired over the Saba Bank area (Church and Allison 2004). Church and Allison (2004) summarize the Saba Bank geochronology as remnants of volcanic islands from the Cretaceous volcanic island arc with a thin Tertiary section overlying a thick pre-Eocene sequence and an eastern area with a thick Tertiary sedimentary section overlying Eocene volcanics and an even older (perhaps Cretaceous?) sequence. The upper layers indicate early Miocene fluvial-deltaic volcaniclastics and late Miocene–Pliocene carbonate deposition, capped with Quaternary coral reefs. The coral reefs are found primarily along the east and southeast edges of the Bank and contain a rich cover and diversity of reef-building corals. The Saba Bank hosts various reef types, from patch reefs to spur and groove type reefs with sandy channels. Each of these provides a hard substrate for coral and other animals to settle on, which in turn attracts fish and an abundance of other invertebrates (Saba Bank National Park 2012). The extensive reef development of the Saba Bank (estimates put it

Fig. 6.9 Northeast coast of Saba. The peaks of Mt. Scenery (*right*) and Booby Hill (*left*) rise above the rugged northeast coast of Saba. The airport on Flat Point (the only large flat area on Saba) and the hairpin road going up from there to Hells Gate are on Saba's only visible basaltic andesitic lava flow (*right foreground*). Well-preserved lava levees can be seen cradling the zigzag road. On the coast in the *central part* of the photograph are Spring Bay (*left center*) and Cove Bay (*right center*, adjacent to airport Flat Point area). On the coast below Flat Point are the Tide Pools. Photograph by J. Haviser

around 150 km^2) provides a rich fishing ground for Saban fishermen and hosts large populations of reef fish and associated spawning grounds.

6.3.8 Nearshore and offshore Landforms of Statia

The nearshore landforms on Statia's surrounding shallow platform also include coral reefs. The platform is the narrowest in the north and southeast. In the south, the steep slope of The Quill volcano continues into the sea. Here, corals enhance a vertical wall of the Drop Off, Grand Canyon, and The Cliffs dive sites. Statia dive shops lay claim to areas of coral-encrusted lava flows, boulders, and patch reefs, as well as more than 100 sunken trading ships, submerged anchors, and cannons. A number of artificial reefs have been created by the Marine Park since active management commenced in 1997. Some of these artificial reefs have been created for fishing purposes, and others for divers (Statia Park 2015).

6.3.9 Landscape

Little is known of Saba and Statia's Amerindian occupance, though archaeological excavations continue and most evidence points to basic hunting and gathering (and fishing) occurring around and on the islands as early as 500 BCE (SABARC 2015). Mostly uninhabited until the seventeenth century, Saba and Statia remained untouched until Colonial powers began expanding. Saba's rugged landscape and treacherous coastline that makes boat landings difficult, kept its historic population small (no greater than 3000). Saba had

Fig. 6.10 Image of the west coast of Saba showcasing Mt Scenery (the *highest point* in the photograph), Great, Paris and, Bunker hills to the *right*, and Diamond Rock about 425 m (1400 ft) offshore in the foreground. The seabed between Torrens Point and Diamond Rock is very rugged with irregular outcrops of dome remnants that house abundant coral reef communities, making the area a favored dive site. To the *right* and *left* of Torrens Point are the deeply incised and eroded ridges and guts that are successions of massive lithified, coarse block, and ash flow deposits. Photograph by C. Meijvogel

a small agricultural economy mainly for local consumption, but the people themselves were in demand, even today, for their navigation skills and seamanship (Haviser 2001).

Although Mt. Scenery dominates the island, Saba also hosts numerous hills giving the island an irregular topography, and the island's four villages sit on the gentler slopes between the peaks of hills. The village of The Bottom, for example, sits in the horseshoe-shaped, collapse scar of a volcanic lateral blast. As you descend into The Bottom, from the road there is an excellent (and much photographed) view of some of the newer volcanic dome hills that looks like, and is called, the whale (Fig. 6.16). The Whale's Tail represents an outstanding example of a well-preserved volcanic spine (or spike), usually created by dome growth and collapse. The Whale's Tail is also the teepee tent-like feature after which Tent Bay is named.

To transport goods, for many years, locals trekked through grueling elevation changes on twisting trails with perhaps a lone donkey to help. In the 1930s, Sabans asked the Dutch government to build a road on the island. Primarily experienced with engineering roads on flat ground the Dutch government considered it a foolhardy task due to the island's extreme topography. Eschewing the opinion of the Dutch engineers, Sabans decided it *was* possible to build a road on Saba. Today, the 14 km road that half-circumnavigates the island stands as a testament to the impressive local terrain knowledge and clever application of design and construction that allowed for slope stabilization on tropical mountain roads where heavy rain, steep grades, and weak soils and rocks combine to create slope instability. The outer side of the road usually has a steep drop off, and the inside has a steep cliff rising above it. The sole main road on Saba, winding-up from the airport on the northeast side, through the four villages, and terminating at the southwest coast in Fort Bay, traverses four ecological zones from dry scrub to dense tropical vegetation. The other half of the island is actually too steep for road construction and consists primarily of spectacular successions of block and ash flow deposits forming steep and precipitous slopes (Roobol and Smith 2004). These western slopes are so steep that there are no settlements in the area and, although there is a hiking trail there called the North Coast Trail, it is dangerous and requires a local guide.

Today, Saba and Statia rely on tourism in the form of world-class scuba diving and oil transshipment. In fact, because the rugged topography of Saba permits only small-area development, and most of that occurs above 240 m, few terrestrial human impacts have affected the reefs. On the other hand, potentially the most conspicuous coastal

Fig. 6.11 Cave of Rum Bay on the northeast side of Saba taken from the air with a drone. Cave of Rum Bay beach is about 150 m long 15 m wide, nestled between the headlands of Torrens Point to the south (*right* of photo) and Great Point to the north (not in photo). Photograph b C. Meijvogel

feature on Statia is its oil terminal, located in the northeast corner of the island on Little Mountain. Here, oil is offloaded from large tankers from other countries, modified, and then reloaded back on tankers that can distribute it to other places. Typically, tens of tankers are lined up offshore waiting for their turn for this exchange.

Even though they are part of the Netherlands, English is the primary language on both Saba and Statia—a holdover from Colonial times. But more than this, Saba and Statia remain quite different from other Caribbean islands (including the other Dutch Caribbean islands) in terms of heritage and culture. For example, Saba's architecture has evolved differently from the majority of other Caribbean islands, displaying mostly white buildings, with red roofs and green shutters that give the island a fairy tale-like ambience. Though small in size, the islands retain distinct physical and cultural characteristics, often enhanced by their heritage and tourism-focused economies centered on geotourism (see Dowling and Newsome 2006a, b for overview).

6.4 Heritage and Tourism

6.4.1 Heritage on Saba and Statia

The human history of both islands began with Amerindian peoples utilizing the land and sea for fishing, hunting, and foraging around 500 BCE. Columbus sailed by the islands on his second voyage in 1493, but both islands were deserted when Europeans visited in the early seventeenth century, perhaps due to an extended drought (Statia History 2015). European settlement on Saba and Statia occurred in 1640 and 1636, respectively. The British were the majority of the early settlers on both islands, and English is still the primary language spoken on Saba and Statia. During colonial times, the islands' populations and ownership varied. Statia's geographic location and its landscape with an accessible harbor and flat central plain allowed for a robust commercial and maritime economy, and up to 100 plantations. Statia's plantation economy consisted of various crops

Fig. 6.12 Oldest exposed rocks on Saba at Torrens Point on the northwest shore. These rocks include sea stacks, arches, caves, and swim-troughs that are remnants of a Pelean dome heavily hydraulically eroded. Photograph by J. Rahn

including tobacco, cotton, coffee, indigo, and its most important crop, sugar (Haviser 2001). Numbers for Statia's largest population range from perhaps 10,000 to 20,000, though this was a largely transient population in the mid-1700s (Statia Tourism 2015). Other commercially exploited natural resources over the centuries included sulfur deposits on Saba and sisal and pumice on Statia (Haviser 2001).

Throughout the seventeenth and eighteenth centuries, both islands were claimed successively by the Dutch, French, English (and Saba by the Spanish), with Statia changing hands 22 times. The islands first became a Dutch West Indies Colony in 1828, before being included as a Dutch dependency in 1845 along with other Caribbean Islands. It was not until the mid-1950s that the islands became part of the Netherlands Antilles (Britannica 2015). Since then, and up to present, Saba and Statia remain quiet islands that rely heavily on tourism (primarily hiking and scuba diving) for disposable income. archaeological studies on the islands are recent, beginning just a few decades ago in the 1980s. The Saba Archaeological Center (SABARC) seeks to integrate archaeological work with community-oriented projects and have succeeded in attracting and training youth in their preservation activities while promoting pride in local heritage. In February 2016, SABARC opened the Saba Heritage Center in the villages of Windwardside. The center endeavors include "living history" demonstrations, artifact exhibitions, technical laboratory facilities for scientific research, and a public facility for lectures on heritage (SABARC 2015). SABARC also has ongoing excavations at several Amerindian and colonial sites and often conducts weekend digs with local youths.

6.4.2 Coastal and Marine Conservation and Tourism on Saba and Statia

The national parks of Saba have been under the domain of the nonprofit Saba Conservation Foundation (SCF).

Fig. 6.13 Wells Bay beach: a seasonal sandy pocket-beach that usually disappears in wither months, replaces by small ($\sim$20 cm) volcanic cobbles. Locally called "wandering beach "or "disappearing beach," it is the only road-accessible, natural sandy beach on the island. At the bottom of the dramatic cliff is a wave cut notch (*center* of photograph) and the northern end of Well's Bay (*left* side of photograph) transitions into Torren's Point. The cliff face is unstable and beachgoers need to be aware of the danger of falling rock. Photograph by J. Rahn

Institutionalized forms of nature conservation on Saba date back to 1987, when the first protected area, the Saba Marine Park (SMP), was established, coinciding with formal incorporation of the SCF. The SMP circles the entire island from the high watermark to a depth of 60 m, including the seabed and overlying waters, and comprising a total area of approximately 1300 ha (see Fig. 6.1). A zoning plan divides the park for various recreational and commercial uses, including a system of permanent mooring buoys to facilitate diving and prevent damage to corals across 30 different dive sites. One of the few self-sustaining marine parks anywhere in the world, the SMP raises revenue through visitor fees (visitors pay US $3 per dive and $1 per night in hotels), souvenir sales, and donations (Saba Park 2015). In 2010, the Saba Bank National Park (SBNP) became an additional underwater-protected area run by The Saba Bank Management Unit (SBMU). An independently operating organization under the SCF umbrella, the SBMU is designed to safeguard the wealth of biodiversity of the entire 2680 km^2 (268,000 ha), flat-topped seamount of the Saba Bank. In particular, it regulates the prime fisheries to ensure their sustainability since Saba Bank remains an important economic resource for Saba, with the fisheries on the Bank contributing about 8% to the island's economy and providing full-time employment to 20 people and part-time employment for an additional 30 people (Dilrosun 2002). In 2012, Saba Bank became the world's thirteenth Particularly Sensitive Sea Area (PSSA), with designation from the International Maritime Organization (IMO). In the past, many freighters, tankers, and cruise ships passed over the Saba Bank and frequently anchored in the shallows, causing significant damage to the coral reefs and other bottom-dwelling ecosystems such as conch feeding grounds. After the IMO declared it an area to be avoided, oil tankers (and other large ships) were no longer allowed to pass over it, which reduced anchor damage on the fragile reef ecosystems (Saba Bank National Park 2012). Scuba diving-related tourism, which includes hotels and restaurants, comprises nearly one-third of Saba's economy. There are three dive shops on the island, and dive tourist numbers

Fig. 6.14 Distal end of Saba's only visible basaltic andesitic lava flow in the Flat Point area is known as the Tide Pools. These sharp, red volcanic rocks are a great place to climb and watch the northern-swell waves crash against Saba's shore. Hikers and climbers should be very aware of rouge waves in this environment, as several Sabans have been swept away. Old Booby Hill is visible in the background. Photograph by J. Rahn

hover around 8000 divers completing as many as 20,000 dives annually.

The national parks of St. Eustatius, which comprise The Quill/Boven Park, the Botanical Garden, and the Marine Park, remain under the control of the nonprofit NGO, St. Eustatius National Parks (STENAPA). The St. Eustatius National Marine Park was created in 1996 and extends around the entire island from the high waterline to 30-m depth contour. The park covers an area of more than 27 km^2 and protects a variety of habitats, including pristine coral, eighteenth century shipwrecks, and modern-day artificial reefs to promote fishing and dive tourism (Statia Park 2015). Within the park are two actively managed reserves in which no fishing or anchoring is permitted in hopes of conserving marine biodiversity, protect fish stocks, and promote sustainable tourism. Annually, around 11,000 visitors visit St. Eustatius, staying an average of eight nights. Figures from Statia Park (2010) determined that approximately 80% of those visitors engage in coral reef related activities.

Both the Saba and St. Eustatius Marine Parks work closely with the three local dive centers on each island to ensure that diving practices minimize impact on the reefs. Saba and Statia's marine environments are also a home, migratory stop over, or breeding site for four IUCN Red List Species, 10 CITES Appendix I species, and 98 Appendix II species (St. Eustatius Marine Park Management Plan 2007). The Yarari Marine Mammal Sanctuary is the fourth Caribbean marine mammal sanctuary and includes the Caribbean Netherlands' territorial waters of Saba, St. Eustatius, and Bonaire. The waters surrounding Saba and Bonaire became

Fig. 6.15 Aerial photograph of Statias' windward coast with The Quill volcano surrounded by clouds. Oranjestad, the island's single town and capitol, is visible to the *right* of the image. The *center* of the photograph is dominated by the Culture Plain, Kultuurvlakte, where most of the agriculture occurs. Historically, cultivation went high up the slopes of The Quill. Photograph by J. Haviser

Fig. 6.16 Landform known as "The Whale" on Saba. **a** View of The Whale (*center left*) from the road descending from St. John's to The Bottom. The much-photographed anthropomorphized volcanic domes of Paris Hill and Great Hill create the whale's body and the well-preserved volcanic spine serves as the Whale's Tail. **b** Close-up view of climbers on the Whale's Tail displaying the ridge's rough volcanic geology and lush vegetation. Photographs by J. Rahn

shark sanctuaries in 2015 that cover the islands' full exclusive economic zones. All commercial shark fishing is prohibited up to 200 nautical miles from each island.

From a landform perspective, Saba offers underwater explorers a unique treat, with several dive sites encompassing underwater pinnacles covered completely by coral that begin at 150 m below sea level and rise to 30 m below sea level. Deep sea pelagics frequently swim by and rare corals, such as black coral (*Antipatharia*), grow at deeper depths. This combination of marine ecosystems and (semi-stable) subaerial geology make Saba and Statia one of the more unique Caribbean locations in terms of underwater attractions.

6.4.3 Terrestrial Conservation and Tourism on Saba and Statia

In 2012, the Saba Conservation Foundation created a terrestrial park, the Saba National Land Park, a pie-shaped, 43-ha tract of land on the north coast of Saba, formerly owned by the Sulfur Mining Company. The area has important biological, geological, and historical values, as well as encompassing all vegetation zones present on Saba, an abandoned sulfur mine, hot springs, and serving as an important nesting area for Red-billed tropicbirds (*Phaethon aethereus*) and several other species of seabirds (Saba Park 2015). At the top of Mt. Scenery (877 m), an 8.6-ha plot is being considered for acquisition as an Elfin Forest Reserve. Reached by climbing 1064 Saban-built, rock-hewn steps from the village of Windwardside (Fig. 6.17), this area hosts montane cloud forest, also known as elfin forest, and consists of two primary tropical rainforest areas that represent climax vegetation. Saba's cloud forest is unusual because the Mountain Mahogany (*Freziera undulata*) is the "signature tree" and because the canopy is significantly higher than in other cloud forests (van't Hof 2010). Van't Hof also believes that this was a climax forest of approximately 500 years until four hurricanes from 1989 to 1999 devastated the area and only a few survived.

The SCF is also responsible for hiking trails (the Saba Trail Network), and they staff the Saba Trail and Information Centre in the village of Windwardside. Mapping and maintenance of the more than one-dozen hiking trails is done by park rangers and a significant amount of volunteer help. The SCF fulfills an important role as advisory body for the government and as management agency for protected areas. They are also committed to the belief that a stronger island economy will result from the sustainable use of Saba's rich and virtually unspoiled resources (Saba Park 2015).

Other nonprofits on Saba include Sea and Learn, which sponsors a month-long event each October. Sea and Learn on Saba is a nonprofit foundation that "brings together the local community, diverse nature experts, and visitors to understand the value of preserving, protecting, and sustaining the local natural and cultural resources and heritage on Saba, and worldwide; and to educate the potential safe-keepers in order to sustain viability and infinitum as an ecotourism destination" (Sea and Learn 2015). There are also two other heritage museums on Saba in Windwardside: the Harry L. Johnson Museum and the Dutch Museum. The former includes an original Saban cottage in the middle of a garden highlighting Saba's marine heritage (and located in a public park), while the latter houses collections of Dutch antiques from the 1500s to 1900s.

On St. Eustatius, STENAPA protects The Quill/Boven Park and the Botanical Garden. The Quill/Boven National Park was pronounced the first national park of the Netherlands Antilles in 1998. The park was created "to protect Statia's unique biodiversity and to ensure sustainable use by all stakeholders." This 5.4 km^2 park consists of two subsectors, The Quill and White Wall, and the Boven area covering five hills in the north of St Eustatius. Varying types of habitat are protected, ranging from elfin forest at the top of The Quill volcano to thorny woodland on the lower slopes. The park gives guided tours to visitors and maintains a network of 10 trails in The Quill sector (Statia Park 2015). The Miriam Schmidt Botanical Garden was also established in 1998 to display the island's flora (and fauna). This garden is adjacent to The Quill National Park boundary and extends for over 5-ha across the southern slopes of The Quill.

Established in 2004, the St. Eustatius Center for Archaeological Research (SECAR) provides a permanent archaeological presence on Statia focusing on basic heritage resource management issues working with the locals. SECAR is an active and vibrant research group involved in education, pioneering research techniques, and protecting underwater and terrestrial cultural heritage resources, including more than 100 sunken trading ships, submerged anchors, and cannons off the shores of Statia and terrestrial Amerindian and colonial excavations (SECAR 2015).

6.5 Hazards

6.5.1 Meteorological Hazards

Meteorological events (as well as their secondary and tertiary effects) represent the greatest hazard each island faces, and as world and local climates continue to change, smaller islands such as Saba and Statia feel the alterations first. The most frequent and costly natural hazard to Saba and Statia are the effects of tropical storms, hurricanes, and increasingly, drought. Thought there is little to no meteorological data on either island, because of their small populations, preparedness, and community spirt, there has been little loss

Fig. 6.17 Sign at the bottom of the Mt Scenery trail in Windwardside. At the top of the trail, there is a plaque stating, in Dutch, that Mt Scenery is the highest point in the Netherlands. Photograph by J. Rahn

of life due to tropical storms on Saba or Statia. Historical damage has included the removal of roofs, flooding, tree and other airborne debris, soil erosion, and small landslides. Power, telecommunications services, and food deliveries can be disrupted for a few days to weeks, depending on the extent of the storm. Flash flooding can occur on steep slopes, and roads in these areas can become ephemeral streams directing fast-flowing water to lower elevations. On Saba, this becomes a travel hazard since there is only one road, while the only low-lying flood-prone areas on Statia occur in the Culture Plain.

Wet and dry periods on Saba and Statia can be pronounced. In the last decade, drought has become a problem, with climate change seeming to affect the frequency and magnitude of these events. This is especially a problem since most water on Saba and Statia is provided by roof catchment collection and subsequently stored in cisterns. Local islanders are very water-conscious and aware of the amount of water they have for daily use. Measures like 2 minute showers, dish-washing by hand, small-capacity washing machines (or hand-washing clothes), and flushing toilets only when necessary are common island-wide. However, groups of foreigners (including the approximately 400 people associated with the Saba Medical School, the hundreds of people connected with the Statia oil terminal, and tourists) have little care for their water consumption, and this exacerbates the problem. In severe droughts, water has to be shipped to the islands. Additional problems occur because of the lack of a distribution system for water, where the current distribution system relies on privately owned trucks with 1000-gallon tanks to move water across the islands. Both islands have desalinization plants, but they have limited production volumes, and Statia's has not been operational for three years (Ryan Espersen personal communication).

High winds can also be a problem, preventing flights from landing on the small airstrips for days at a time. Usually occurring in winter and accompanied by large ocean swells, the winds can often preclude boats from running as well, exacerbating the problem further, as most food is shipped to the islands once per week.

Mass wasting can occur from high precipitation, ground shaking (earthquake or human- induced), and when gravity and time take their toll on the steep, thinly soiled, lightly vegetated slopes on both islands. On Saba, mass wasting occurs on both terrestrial and on steep submarine slopes. Most commonly, on both islands, it will occur along the coast where waves erode cliff bases.

6.5.2 Seismic and Volcanic Hazards

Historically, both islands have low-seismic dangers, and no damaging earthquakes or tsunamis have been recorded on Saba or Statia. The main risks would come from earthquakes in the Caribbean region possibly producing a tsunami (very rare: only two have been documented historically in the 1700's) or from local earthquake swarm precursor to a volcanic eruption. A seismic swarm occurred near Saba in 1992 (Ambeh and Lynch 1995), but the magnitudes were not large enough to cause any damage. In June of 2015, seismologists from the Dutch Meteorological Institute (KNMI) installed four seismometers each on Saba and Statia. The real-time feed can sometimes be viewed (http://www.knmidc.org/seismology/?NA_SEUT_—although some seismometers intermittently cease functioning and occasionally the Web site is down). Large earthquake induced mass movement events would hit Saba especially hard because it has only one road, and landslides could block access to the only two evacuation points at Fort Bay and the airport All of Saba's infrastructure, except the Fort Bay port and the airport—both of which lie near sea level—would not be at risk to tsunami waves. For example, the power plant used to be at 3 m above sea level in Fort Bay, but was moved to a 100 m elevation (along the road above Fort Bay) in February 2016. Statia would receive more damage from a tsunami, and possible effects there would include oil spills from the terminal, as well as some infrastructure located at low elevations near the west coast.

For both Saba and St. Eustatius, there are no historical records of eruptive activity. The most recent eruption on Saba and Statia were 1600 AD and 400 AD, respectively, as determined geologically (Hartog 1975, 1976). Volcanic hazard and risk assessments were made for Saba and Statia in 1977 (Fournier and Tomblin) and again in 1997 (Roobol, Smith, and Tomblin). In 2005, the Volcanic Hazard Atlas of the Lesser Antilles was published by the University of the West Indies, where Roobol, Smith, and Tomblin wrote the Saba and Statia chapters that contained a summary of the work the authors carried out on these islands over the past 25 years. Saba is clearly an active volcano as indicated by the presence of hot springs with temperatures around 72 and 55 °C and The Quill, even though it has no external manifestations, is still regarded as an active volcano (Roobol et al. 1997). Using geohazard interpretations reconstructed from the stratigraphic records of the entire pyroclastic history, Roobol and Smith (1989) note that the present period of volcanic inactivity on Saba and Statia does not indicate the extinction of these volcanoes, but rather periods of dormancy. It should be stated, however, that the overall volcanic hazard for these two are similar to other Lesser Antilles Caribbean volcanoes, and Roobol et al. (1997) recommended that the governments should have plans and designated funds to evacuate each island in 1–2 days. As of December 2016, the local governments have been working with the Dutch on these plans, but they have not been released to the public. Additionally, continuous seismic and monthly hot springs monitoring should be taking place. Ash fall from other nearby Caribbean islands might also be a hazard on Saba and Statia as, for example, Montserrat's 1995 eruption column that spread an ash across Saba and Statia.

6.5.3 Other Hazards

Though small islands in size, Saba and Statia still face hazard challenges beyond storms and drought. Neither Saba nor Statia are self-sufficient as far as water, food, or energy are concerned, and the delicate balance of subsistence could easily be affected by a variety of environmental problems and hazards including coral reef bleaching, alien species invasion, land removal, solid waste, and oil refinery-related hazards. The coral reefs ecosystems of Saba and Statia have had little subaerial human effects since populations are so small and many of the reefs are not near human habitations. The largest aquatic impact has been various Caribbean coral reef diseases, large scale herbivore removal with the demise of the spiny sea urchin (*Diadema antillarum*) in the 1980s, and higher than normal sea surface temperatures. Coral bleaching has occurred on these reefs. The worst event happened in 2005, but there have been several minor cases. Beach erosion on Statia (especially Zeelandia Beach) could affect sea turtle nesting.

Additionally, free-roaming goats on both islands are a large problem, causing soil degradation, loss of organic matter, reduced water retention, and overall erosion in semiarid rangelands (Rojer 1997). Although goats occur mostly below the 300-m contour level, in arid times they often graze above 600 meters. According to Romeijn (1987) and Rojet (1997), areas where goats have influenced the vegetation on the major part of the island and large parts of Saba and Statia remain permanently deforested, and in some areas, a complete absence of herbaceous and shrub layers in the forest is due to excessive goat grazing (Debrot 2010).

There have been unsuccessful efforts on both islands to cull, restrict, and better manage the roaming goats.

Removal and movement of large amounts of land for private commercial purposes continues to have effects on the ecological integrity of both islands. On Saba, 1.6 ha (19,000 yds^3) of Fort Hill was removed as part of a rock crushing operation, resulting in approximately 100,000 tons of rock (by estimate of a Saban to the author). Continuing expansion of the stone crushing facility has consequences for the geological formations, scenery, biodiversity in the sea, and dive tourism (Sybesma and Visser 1996; Rojer 1997). On Statia, the top of Signal Hill was leveled about 0.23 km to situate the oil terminal tanks. Signal Hill and Fort Hill were prime sea bird nesting sites. On Saba construction debris dumping (of up to 500 dump truck loads) has occurred on the coastline adjacent to Tent Reef dive sites, and that much sediment, were it to wash onto the nearby reefs, could decimate the fragile ecosystem. Efforts by this author and the SCF resulted in a $100,000 federal grant to remove the fill, but the project has still not occurred. To this author's knowledge, although required by The Netherlands (and the EU), environmental impact assessments do not occur prior to large projects on either island.

A large oil spill potentially affecting both islands could be a hazard, considering the 11-million-gallon Nustar Energy oil storage and distribution facility on Statia. On a busy day, ten large tankers come and go off-loading and on-loading fuel. Based on prevailing winds and waves, the most likely oil spill scenario has the oil heading directly to Saba, with limited localized effect on Statia. Though no major damage resulted from them, there have been several small oil spills between 2009 and 2015, and in response, the Dutch government has provided Saba and Statia with oil spill cleanup, training, and equipment. In 2010, the Saba government and SCF, in consultation with Dutch and EU scientists, decided to not use dispersant if there is a large oil spill because Saba's coastal resources are primary subaqueous (the coral reefs) and dispersant would submerge the oil and harm the reefs. Instead, the plan centers on allowing oil to hit their shores and float past the island, where perhaps dispersant would then be used at sea. On Statia, the oil terminal has affected ecotourism on the island, at least from a visual standpoint, since the main inhabited area hosts many oil storage tanks.

6.6 Conclusion

The landforms of Saba and St. Eustatius originate from subduction in the Lesser Antilles, resulting in Pleistocene island arc volcanics. Both islands have stratovolcanic landscapes and many Pelean-type domes with adjacent pyroclastic flows. Saba is a single volcano whose steep slopes drop off precipitously into the sea, while St Eustatius lies on a shallow submarine bank. They both have similar geologic histories and landforms and are perhaps co-eruptive, but Statia is slightly older, has more erosional features, a flat central plain, and experienced some uplift. Both islands have young, visually predominant volcanos that represent the highest (Saba) and second-highest (Statia) elevations in the Kingdom of the Netherlands. Their steep slopes are embeded with ephemeral, waterworn ravines locally called guts—some of their volcanic landforms being obscured by dense tropical vegetation—and most of their coastlines are rocky (abrasional) cliffs. While other Caribbean Islands contain karst, lakes, continually flowing rivers, and even abundant aeolian features, Saba and Statia lack these. Climatological events of tropical storms, hurricanes, heavy precipitation, and drought have been the most impactful hazards. Changing regional climate regimes will likely be the biggest environmental hazard and geomorphic agent in the future. Like other Lesser Antilles islands, seismic and volcanic activity also poses possible risks, including landslides, but these are rare events on Saba and Statia. Some of the best coral reefs in the Caribbean surround both islands and are important for tourism, especially on Saba where scuba-related tourism accounts for about a third of the economy. Saba and Statia both have influential NGOs that work with the government to protect their natural environments and cultural history, with each island hosting active Amerindian and colonial archaeological sites and nonprofit archaeological organizations.

Saba and Statia are atypical of other Caribbean islands because of their small sizes, rugged topographies, and lack of large resorts, development, and industry. They have maintained their relatively pristine environments because of low human impact. A healthy natural environment is important to the priorities and pride of the local peoples. Indeed, Saba is known as "The Unspoiled Queen" in the Caribbean, and inhabitants of Saba and Statia form close knit, small-town like societies allowing them to interconnect with their environment and with each other. While Sabans and Statians are community-oriented, environmentally aware, inquisitive, and work together to be resilient in the face of adversity, finding solutions to current and future problems will require continued solidarity and good relations between communities, governments, and the rest of the Caribbean.

Acknowledgements The author would like to thank the people of Saba, the Saba Conservation Foundation Saba Marine Park, Sea Saba Dive Center, SABARC, and Samford University for funding fieldwork on Saba.

References

Ambeh WB, Lynch L (1995) The earthquake sequence of June 1992 near Saba, West Indies. Tectonophysics 246(4):225–243

Aubrey DG, Emory KO, Uchupi E (1988) Changing coastal levels of South America and the Caribbean region from tide-gauge records. Tectonophysics 154:269–284

Britannica (2015) Available via http://www.britannica.com/place/Sint-Eustatius. Accessed 12 Jan 2016

Caribbean Volcanoes (2015) Available via http://caribbeanvolcanoes.com. Accessed 12 Jan 2016

Church RE, Allison KR (2004) The petroleum potential of the Saba Bank Area, Netherlands Antilles. Available via http://www.searchanddiscovery.com/pdfz/documents/2004/allison/images/allison.pdf.html. Accessed 12 Jan 2016

Cleve PT (1871) On the geology of the North-eastern West India Islands: Kongl. Svenska Vetenskaps-Akademiens Handlingar 9, v. 12

DCBIODATA (2015) Dutch Caribbean biodiversity explorer. Available via http://www.dcbiodata.net/explorer. Accessed 15 Feb 2016

DCNA (2015) The Dutch Caribbean nature alliance. Available via http://www.dcnanature.org/about-dcna. Accessed 15 Feb 2016

Debrot AO (2010) Available via https://www.wageningenur.nl/en/project/Harmful-invasive-alien-species-IAS-in-the-Caribbean-Netherlands.htm. Accessed 15 Feb 2016

Dilrosun F (2002) Spiny Lobster Fishery of the Saba Bank. 53rd Gulf and Caribbean Fisheries Institute. Available via http://www.dcbd.nl/sites/www.dcbd.nl/files/documents/Dilrosun%202010%20Spiny%20lobster%20fishing%20Saba%20Bank.pdf. Accessed 12 Jan 2016

Dowling R, Newsome D (2006a) Geotourism. Elsevier, Oxford

Dowling RK, Newsome D (2006b) Geotourism. Routledge, Chicago

Emiliani C (1978) The cause of the ice ages. Earth Planet Sci Lett 37: 349–352

Fournier EM, Tomblin JF (1977) Volcanic risk and monitoring on the island of Saba and St. Eustatius. UNESCO Restricted technical report RP/1975-76/2.161.4. Available via http://unesdoc.unesco.org/images/0002/000243/024338eo.pdf. Accessed 12 Jan 2016

Hardy F, Rodrigues G (1947) The agricultural soils of St. Kitts-Nevis with notes on Statia (Dutch): Studies in West Indian Soils 13, The Imperial College of Tropical Agriculture, Trinidad

Hartog J (1975) History of Saba: Van Guilder N.V., Saba, Netherlands Antilles

Hartog J (1976) History of St. Eustatius: De Wit Stores N.V., Aruba, Netherlands Antilles

Haviser JB (2001) Historical archaeology in the Netherlands Antilles and Aruba. In: Farnsworth P (ed) Island lives: historical archaeologies of the Caribbean, University of Alabama Press, pp 30–81

Hovey EO (1905a) Volcanoes of Martinique, Guadeloupe, and Saba: Report 8th International Geographic Congress, Washington, 1904, p 447–451

Hovey EO (1905b) Volcanoes of St. Vincent, St. Kitts, and Statia: report 8th international geographic Congress, Washington, 1904, p 452–454

Kjerfve, B (1981) Tides of the Caribbean Sea. J Geophys Res: 4243–4247

LaCroix A (1890) Sur la composition mineralogique des roches volcaniques de la Martinique et de l'isle Saba: Comptes Rendus hebdomadaires des séances de l'Academie des Sciences, Paris 111: 71–73

LaCroix A (1893), Les enclaves des roches volcaniques: Macon, Protat, p 48, 156–157

Maclure W (1817) Observations on the geology of the West India Islands, from Barbados to Santa Cruz, inclusive. J Acad Nat Sci Philadelphia 1: 134–149

Molengraff GAF (1886) De geologie van het eiland St. Eustatius. Eene bijdrage tot de kennis der Nederlandsche Kolonien: [Ac. thesis]: Leiden, University of Utrecht

Molengraaff GAF (1931) Saba, St. Eustatius (Statia) and St. Martin: Leidsche Geologische Mededeelingen 5: 715–729

National Geographic Society (2007) Available via http://www.nationalgeographic.com/adventure/travel/caribbean_saba_vacations.html. Accessed 15 Feb 2016

Perret FA (1942) Notes on the volcanism of the West Indies, Proceedings of the Eighth American Scientific Congress, Washington 1940, Geological Sciences, Washington, p 751–756

Rahn JL, Lannon H, Mossa J (2015) Diver depth-gauge profiling beyond wading depths: a new simple method for underwater surverying. J Coastal Res 31(2):505–511

Rojer, A (1997) Biological Inventory Saba. KNAP project 96–10. Carmabi Foundation. Available via http://www.sabapark.org/downloads/Biological%20Inventory%20Saba.PDF. Accessed 12 Jan 2016

Romeijn P (1987) Saba (N.A.), Bos en Nationale Parken. Vakgroep Bosteelt en Bosecologie

Roobol MJ, Smith AL (1989) Volcanic and associated hazards in the Lesser Antilles. In: Latter JH (ed) Volcanic hazards—assessment and monitoring. Springer, Berlin, pp 57–85

Roobol MJ, Smith AL (2004), Volcanology of Saba and St. Eustatius, Northern Lesser Antilles 2004, Royal Netherlands Academy of Arts and Sciences (Koninklijke Nederlandse Akademie Van Wetenschappen), Amsterdam, the Netherlands

Roobol MJ, Smith AL, Tomblin JF (1997) An assessment of the volcanic hazard on the islands of Saba and St. Eustatius in the northern Lesser Antilles: Special Publication of the Netherlands Geological Survey

Saba Bank National Park (2012) United Nations Environment Program Report. Available via http://www.car-spaw-rac.org/IMG/pdf/Report_Saba_Bank_National_Park-3.pdf. Accessed 12 Jan 2016

Saba Park (2015) Available via http://www.sabapark.org. Accessed 15 Feb 2016

SABARC (2015) Available via http://www.saba-news.com/sabarc-saba-archelogy-center/. Accessed 15 Feb 2016

Saba Tourist Bureau (2015) Available via http://www.sabatourism.com. Accessed 12 Jan 2016

Sapper K (1903) Ein Besuch von S. Eustatius und Saba. Centralblatt fur Mineralogie 1:314–318

Sapper K (1904) Die vulcanischen Kleinen Antillen und die Ausbruche der Jahre 1902 und 1903, Neues Jahrbuch fur Mineralogie 2: 1–70

Sea and Learn (2015) Available via http://www.seaandlearn.com/. Accessed 12 Jan 2016

SECAR (2015) Available via http://secar.org/en/st-eustatius-center-for-archaeological-research. Accessed 15 Feb 2016

Statia Park (2010) Available via http://www.statiapark.org/downloads/downloads/2010%20Statia%20National%20Marine%20Park%20Economic%20Valuation.pdf. Accessed 12 Jan 2016

Statia Park (2015) Available via http://www.statiapark.org/. Accessed 15 Feb 2016

Statia Tourism (2015) Available via http://www.statiatourism.com/essential_facts.htm%20-%20car. Accessed 15 Feb 2016

Statia History (2015) Available via http://www.steustatiushistory.org/StatiaHistoryandArchaeology.htm. Accessed 12 Jan 2016

St. Eustatius Marine Park Management Plan (2007) Available via http://www.spaw-palisting.org/uploads/files/4245e721fbcd354b8098 4c74aab0694b156085dd.pdf. Accessed 12 Jan 2016

STENAPA (2007) St Eustatius National Marine Park Management Plan 2007. Available via http://www.statiapark.org/downloads/downloads/St%20Eustatius%20National%20Marine%20Park%20Management%20Plan%202007.pdf. Accessed 15 Feb 2016

Sybesma J, Visser N (1996) "Een vuiltje aan de lucht" Milieugevolgen van de Stone crusher op Saba. Rapport VOMIL. Sectie Milieu en Natuur

van't Hof T (1991) Guide to the Saba marine park. Saba Conservation Foundation

van't Hof T (2010) Saba's unique cloud forest: and how it evolved during a series of major hurricanes. Self-published

Westermann JH, Kiel H (1961) The geology of Saba and St. Eustatius: Uitgaven Natuurwetenschappelijke Studiekring Voor Suriname en de Nederlandse Antillen', Utrecht, No. 24

St. Kitts and Nevis

7

Richard Edward Arnold Robertson

Abstract

The islands of St. Kitts and Nevis are part of a binary federation located within the northern Leeward Islands. Their landscapes are as much a result of their geological history as they are the manifestation of interactions of man and nature over time. From the central volcanic mountains draped with tropical rainforests, to the gently sloping pyroclastic fans that have provided rich soils for agriculture and land for settlement, to the coastline dominated by cliffs, to its variety of inshore and marine natural resources, the islands morphology is a key facet of its existence. The islands economy has always depended on the bounty of the land, first for agriculture and increasingly for a tourism product that is based largely on its natural environment.

Keywords

St. Kitts • Nevis • Volcanic • Environment

7.1 Introduction

The Federation of Saint Christopher (St. Kitts) and Nevis consists of three small islands: the inhabited islands of Saint Christopher (St. Kitts) and Nevis and the uninhabited Booby Island (Fig. 7.1). This Small Island Developing State is located in the northern section of the Leeward Islands and is part of the geo-political grouping called the Organization of Eastern Caribbean States.[1] Although there is only a small ocean channel (knows as the "Narrows") that physically separates the two islands, their separation is significant in terms of geo-political setting. Although provision is made under the binary "Federation" for the two islands to officially operate as one, a special relationship exists between them, which means that they function together and separately at the same time.

The most dominant influence on the landscape of both islands has been their volcanic history, which has resulted in a combination of volcanic mountains, rain forests, and unspoiled beaches. The fact that both islands have an adequate supply of groundwater and an abundance of well-drained fields on gently sloping land (<10°) suitable for agriculture is to a large extent also due to the underlying geology. Single, young volcanic centers dominate each island, comprising the core of central volcanic mountains, which are densely vegetated by rainforest. These central mountains remain surrounded by gently sloping land that falls away uniformly toward the coastline in all directions (Fig. 7.2a).

St. Kitts is 176 km^2 (68 mi^2) in area and roughly oval in shape with a narrow neck of land extending like a handle toward the southeast. It consists of a central mountain range dominated by Mt Liamuiga and surrounded by cultivated slopes, dissected by dry river valleys (ghauts) that drain into the sea. The southern sections of the central range enclose a

[1]This is an inter-governmental organization dedicated to economic harmonization and integration, protection of human and legal rights, and the encouragement of good governance between countries and dependencies in the Eastern Caribbean. It also performs the role of spreading responsibility and liability in the event of natural disaster, such as a hurricane.

R.E.A. Robertson (✉)
The UWI Seismic Research Centre, St. Augustine, Trinidad (West Indies)
e-mail: richard.robertson@sta.uwi.edu

C.D. Allen (ed.), *Landscapes and Landforms of the Lesser Antilles*,
World Geomorphological Landscapes, DOI 10.1007/978-3-319-55787-8_7

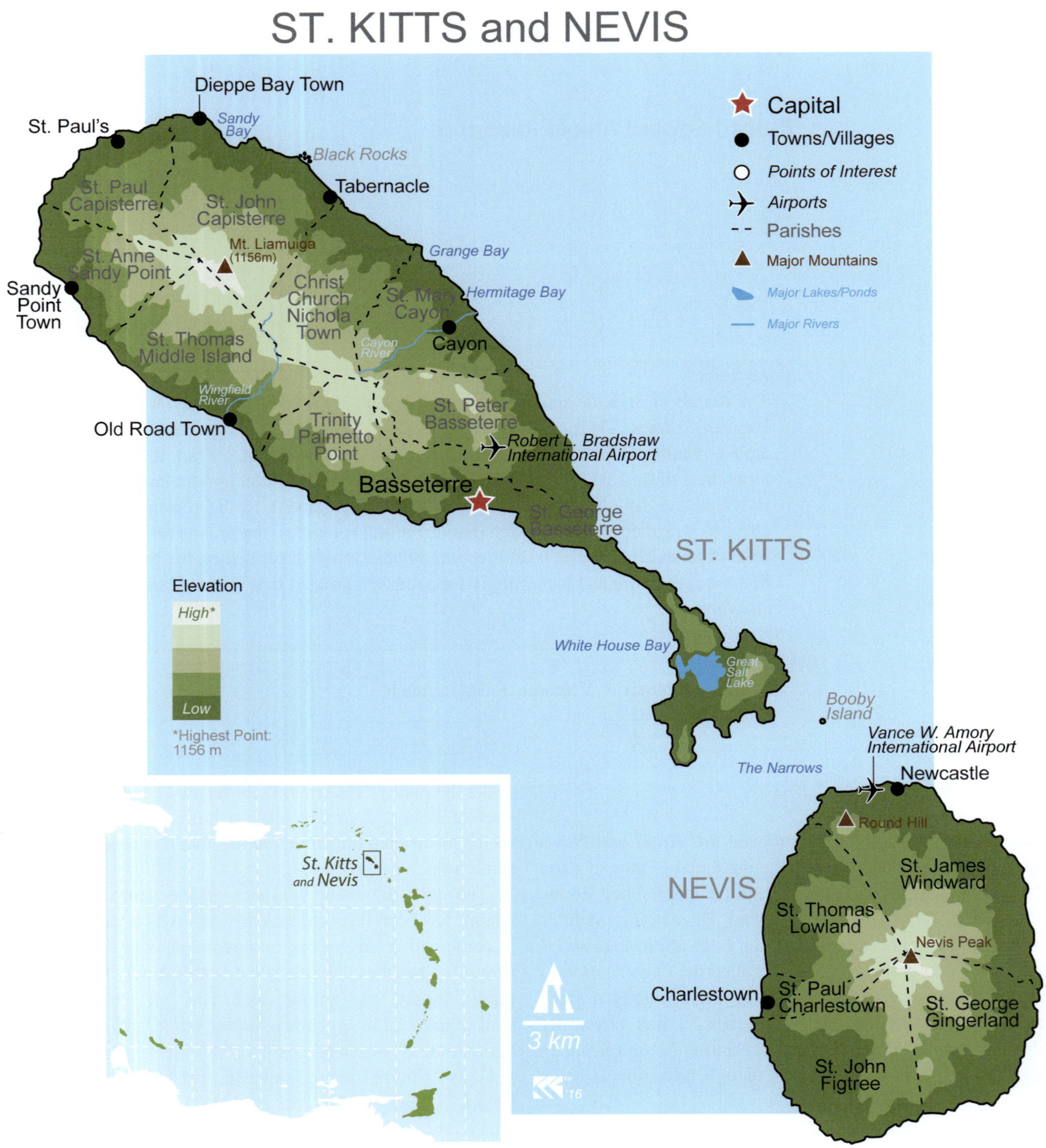

Fig. 7.1 General physiographic map of St. Kitts and Nevis highlighting the islands' mountainous topographies and political parish boundaries. The waterway separating the two main islands, the Narrows, contains the small island of Booby Island. Cartography by K.M. Groom

spacious fertile valley that opens into the capital city of Basseterre. Although golden sandy beaches are common in the Southeast Peninsula, most of the beaches on the island are of gray and brown volcanic sand. Manifestations of recent volcanic activity include active fumaroles in the crater of Mt Liamuiga (1156 m, where an ephemeral lake exists—see Fig. 7.2b) and past periods of intense earthquake activity associated with the volcano.

Fig. 7.2 Mountainous features of St. Kitts and Nevis. **a** Offshore view of St. Kitts showing its core of densely vegetated, central volcanic mountains (*left* to *right* are Brimstone Hill, Mt. Liamuiga, Middle Range, Southeast Range), surrounded by gentle cultivated slopes that drain to the sea. Basseterre, the capital of St. Kitts with its reclaimed waterfront area, is in the foreground. **b** The steep-walled summit crater of Mt. Liamuiga volcano is a 900 m (width) × 244 m (depth) feature that contains low temperature fumaroles, a small ephemeral lake visible at the bottom of the image and a dense cover of mainly cloud forest. Photographs courtesy of R. Robertson, The UWI Seismic Research Centre

At 93 km^2, Nevis is also dominated by a central peak (Nevis Peak, 985 m) and dissected by deep ghauts that run from the peak to the sea. No regular stream flow occurs in the ghauts except during heavy rains, and the island has no significant bays, inlets, or cays, but has long stretches of golden, sandy beaches around most of the island. Vegetation cover is extensive but not dense, and many wetlands occur along the leeward coast.

Understanding and appreciation of the value of the landscape on both islands have spawned the creation of various organizations, policies, and practices in an attempt to manage the island's heritage resources. Institutionalized concern is reflected in extensive government policies and legislation as well as in such organizations as the National Conservations Commission, the St. Christopher Heritage Society, and the Nevis Historical and Conservation Society. The creation of national parks as well as the organization of various beautification and environmental management projects represents specific actions that focus on preservation of heritage resources.

7.2 Setting

Volcanic in origin, St. Kitts (17° 15′ N, 62° 45′ W) and Nevis (3 km to the southeast, at 17° 10′ N, 62° 35′ W), is located in the northern part of the Lesser Antilles chain of islands in the Eastern Caribbean, upon a submarine ridge that extends from St. Eustatius to just south of Nevis, covering an area of 80 km × 16 km. The coastal shelf widens slightly between both islands, and a total area of 595 km^2 lies within the territorial waters of St. Kitts and Nevis (Island Resources Foundation 1991). The islands have a combined area of 269 km^2 and a population of approximately 46,000 persons (2010 census). With a Tropical Marine Climate that is strongly influenced by steady northeast trade winds and tropical oceanic cyclonic movements, the Federation experiences warm temperatures averaging about 28 °C, with small seasonal and diurnal variations. At high altitudes, temperatures can drop below 17 °C. Rainfall is mainly orographic, affected by a central mountain range that extends from Mt. Liamuiga (1156 m) to the Southeast Range (901 m). Precipitation increases in amount and frequency with altitude. The yearly average rainfall is 1620 mm, but it can range from ~1000 mm in coastal areas to ~3800 mm in the central mountains. Except for the Southeast Peninsula, which is very dry, rainfall is fairly well distributed throughout the island with a seasonal variation between the "wet" and "dry" season. The wet season extends from August to November, while the driest months are from January to April. Relative humidity is low in the dry season and high in the wet season with a mean value of 76%. The prevailing winds blow from the northeast (NE trade winds) with speeds ranging from 15 to 30 kph.

The volcanic landscape has resulted in beaches that consist of a variety of mostly volcanic materials ranging from black sand to cobbles, rocks, and boulders. In the Southeast Peninsula, south of Friar's Bay, beaches are comprised of marine sands consisting of coral and shell materials, and at Sandy Point, Dieppe Bay, and Conaree there is a mixture of land-based and marine sand (Island Resources Foundation 1991). In addition to the coral and volcanic beaches, the coastal areas of St. Kitts and Nevis abound in shore and marine natural resources including coral reefs, mangroves, saltwater lagoons, and sea grass beds.

The increasing importance of tourism to the economy of both islands makes these coastal areas a vital asset that needs to be protected from the effects of human activities and the ravages of storms. The coastal zone has been affected by developments along the shore, including farming, quarrying (sand mining for construction), sewage disposal, and land reclamation. Human activities have placed tremendous stress on the ecological balance of coral reefs, sea grass beds, and mangroves leading to a rapid decline of marine ecosystems. Elevated nutrient levels in the coastal zone due to runoff from urban and agricultural areas along ghauts and storm drains has led to imbalances to the marine ecosystems (Williams 2013). Wetland infilling for residential homes and hotels due to development pressure and natural hazards is causing the islands to lose its coastal mangrove habitats.

7.3 Landforms

7.3.1 St. Kitts Landforms

7.3.1.1 Major Features

St. Kitts is an elongated island oriented NW–SE, 36.8 km (23 mi) long and roughly oval shaped with a narrow neck of land extending from the southeastern end (Fig. 7.3). It generally rises from the coastline toward its mountainous interior. The island is suspended on an underwater ridge which has a northwest to southeast axis from which Nevis also arises. The physical landscape of St. Kitts has a rugged backbone characterized by three volcanic centers dominated by Mt. Liamuiga, which rises with a pronounced crater to 1156 m.

Southeast of Mt. Liamuiga the Central Range and Southeast Range continue the trend after which the land descends into the Basseterre Valley. The Central and

Fig. 7.3 Elongated shape of St. Kitts with its rugged backbone of volcanic centers that trend NW–SE is clearly seen on a largely cloudless day from Nevis Peak, Nevis. The contrasting morphology of the youthful Mt Liamuiga volcano (*upper left*) to the older dissected domes of the Salt Pond Peninsula (*mid-center*) is well illustrated in the photograph. The lush slopes of Nevis Peak occupy the foreground of the image. Photograph courtesy of R. Robertson, The UWI Seismic Research Centre

Southeast Range consist of a number of irregular peaks; the highest of which is Verchild's mountain at 975 m (3200 ft). The slopes in these ranges are generally steeper and shorter toward the leeward coast. The main mountainous interior blends downslope into a gradually sloping coastal plain that is interrupted by minor domes at Brimstone Hill, Ottley's Mountain, Sandy Point Hill, and Monkey Hill. Most of the flat and moderately sloping land occurs in these areas near the coast, which as a result, has been the areas used for urban and agricultural development.

7.3.1.2 Volcanic Centers

Although the core of St. Kitts is comprised of older (possibly Eocene) volcanic material, three younger volcanic centers encompass most of the island: the Southeast Range, the Middle Range, and Central Range dominated by Mt. Liamuiga (Martin-Kaye 1959). These are Pliocene to Pleistocene in age and become progressively younger toward the northwest (Hovey 1903, 1905; Sapper 1904; Davis 1924; Martin-Kaye 1959; Fels 1903; Earle 1925). The Salt Pond Peninsula is an old dissected landscape that consists of many low, rounded hills that comprise the topographic highs and reach a maximum height of 319 m, with flat, low-lying areas, and salt ponds comprising the topographic lows. Geologically, this area consists of lava domes and volcaniclastic deposits, which may have originally been pyroclastic flows and lahars from past eruptions. The oldest rocks exposed on the island (2.77 ± 0.3 Ma, Baker 1969) are found in this region (Fig. 7.4). Lang and Carroll (1966) suggested that the St. Anthony's Peak, the highest point on the Peninsula, might have been the site of a volcanic crater.

Younger Pleistocene volcanic centers (Southeast Range, Middle Range, and Mt. Liamuiga) that consist mainly of andesitic pyroclastic deposits are found along the island's central spine. They occur as ash, reworked sands and gravels, cobbles, and boulders. The Southeast Range rises up to 900 m and is a younger volcanic center made up of lava flows and volcaniclastic deposits. Due to poor exposures, little is known about its past eruptive history, but Baker (1985) obtained an age of 1 million years for a lava flow on its southern slopes. The Middle Range is heavily forested with few outcrops of hard rock and with difficult access to exposures of its geology. It has a summit lake and youthful appearance, but no age dates have been obtained and so it is difficult to make any conclusions about its eruptive history.

Located on the northwestern most mountain and being the youngest volcanic center on the island, Mt. Liamuiga represents a typical stratovolcano, composed of alternating layers of andesitic and basaltic lava flows, agglomerates, and pyroclastics (Fig. 7.5a). The last eruption is estimated to have been about 1620 years before present (Baker 1985). Its young age, associated geothermal and seismic activity, suggests that it is a potentially active volcano that is likely to erupt again in the future (Robertson 2005). There is a steep, summit crater ~900 m wide, and 244 m deep which contains a small ephemeral lake and has active low temperature (up to 100 °C) fumaroles along the crater walls (Robertson 2005). An old Carib legend suggests that Brimstone Hill, an andesite lava dome located on its lower flanks, grew out of the Mt. Liamuiga (Baker 1985), hinting that pre-Columbian peoples may have witnessed the growth of this dome.

7.3.1.3 Southeast Peninsula

The Southeast Peninsula is a cluster of seven older islands that have become linked by recent beach and saline marsh deposits. These were originally tombolos, which have widened into flat sedimentary plains and marshland linking the seven preexisting small islands/islets. The hills on the Peninsula are smoothly rounded with slightly convex peaks. The largest is St. Anthony's Peak at 319 m. Eight saline ponds that vary in size from 160 to 1.6 ha occur in the area.

The Peninsula consists of two distinct features (1) a narrow spit or bar that is slightly more than 0.5 km wide and about 4.5 km in length that extends in a southeasterly directly from Timothy Hill at Frigate Bay to Salt Pond Hill and (2) a larger, roughly triangular tombolo cluster surrounding the largest saline pond, 440 acre Great Salt Pond (Island Resources Foundation 1991).

7.3.1.4 Coastal Features

The coastline of St. Kitts consists largely of cliffs that range in height from 15 to 30 m. At Black Rocks, a basalt lava flow forms prominent distinctly black coastal cliffs (Fig. 7.5b). Narrow black sand beaches comprised of the eroded products from the volcanic hinterland occur at the base of most cliffs. In the northwest, the cliffs are lower, and some of the beaches are wider and comprised of yellow sand (Island Resources Foundation 1991). There are long stretches of fine yellow sand beaches extending from Conaree Beach on to the southeast of the island. Unfortunately, excessive sediment loads, biocide runoff, and other land-based sources of pollution continue to be the main causes of reef damage leading to coastal erosion across the islands (Island Resources Foundation 1991; Williams 2013).

7.3.2 Nevis Landforms

7.3.2.1 Major Features

Nevis is an oval-shaped island, 12.3 km long (7.64 mi) and 9.6 km (5.96 mi) wide at its widest part with its topography dominated by the central Nevis Peak. Although it is comprised of seven volcanic centers strung out SE–NW, the dominance of Nevis Peak on its topography is such that the peak overshadows all other topographic features, giving the island the appearance of a classic volcano-island. Windy Hill

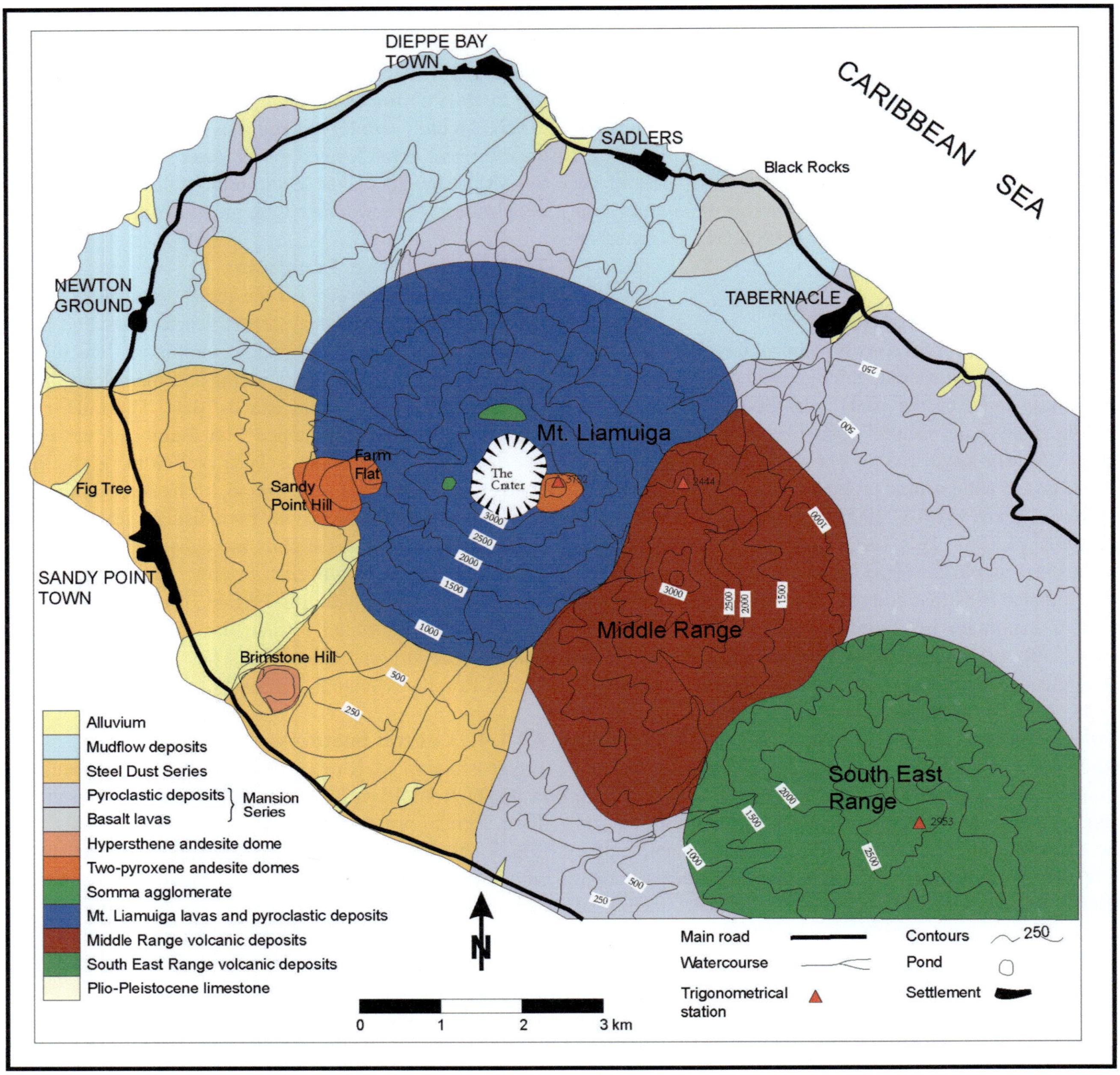

Fig. 7.4 Geological sketch map of the northwestern part of St. Kitts (adapted from Baker 1969) showing the main geological formations that make up the island. Note the different lava and pyroclastic flow events that cover much of the island's landscape. Cartography by Shahiba Ali from the Volcanic Hazard Atlas of the Lesser Antilles (see Robertson 2005)

(309 m) also called Round Hill in the northwest and Saddle Hill (381 m) in the southeast end of the island line up with Nevis Peak in the center to form a similar north-northwest to south-southeast trending trend as the more pronounced orientation of the mountains in St. Kitts.

7.3.2.2 Volcanic Centers

Seven volcanic centers have been identified on the island of Nevis: Hurricane Hill, Round Hill, Cades Bay, Saddles Hill, Red Cliff, Butlers Mountain, and Nevis Peak, all considered to be the remains of lava domes that were created by quiet, effusive eruptions (Hutton 1965; Hutton and Nockolds 1978; Fig. 7.6). The oldest rocks are found in the northwest of the island at Round Hill (3.43 Ma) and the youngest are associated with Nevis Peak (0.10–0.98 Ma; Hutton and Nockolds 1978).

Nevis Peak, rising to 985 m and representing the island's highest point, is comprised of a central cone with flank

Fig. 7.5 Examples of volcanic landforms on St. Kitts. **a** Mt. Liamuiga volcano, which rises up to 3792 ft. (1156 m), the youngest volcanic center on the island of St. Kitts and its highest mountain. Its lower flanks are occupied by farmland and small villages but at higher elevations tropical rainforest and cloud forest takes over. **b** Black Rocks (also called Black Stone), a notable rock formation that is an old lava flow from Mt. Liamuiga volcano, located on the northeastern coast of St. Kitts. This black basaltic rock was formed during the early evolution of Mt. Liamuiga volcano. Photographs courtesy of R. Robertson, The UWI Seismic Research Centre

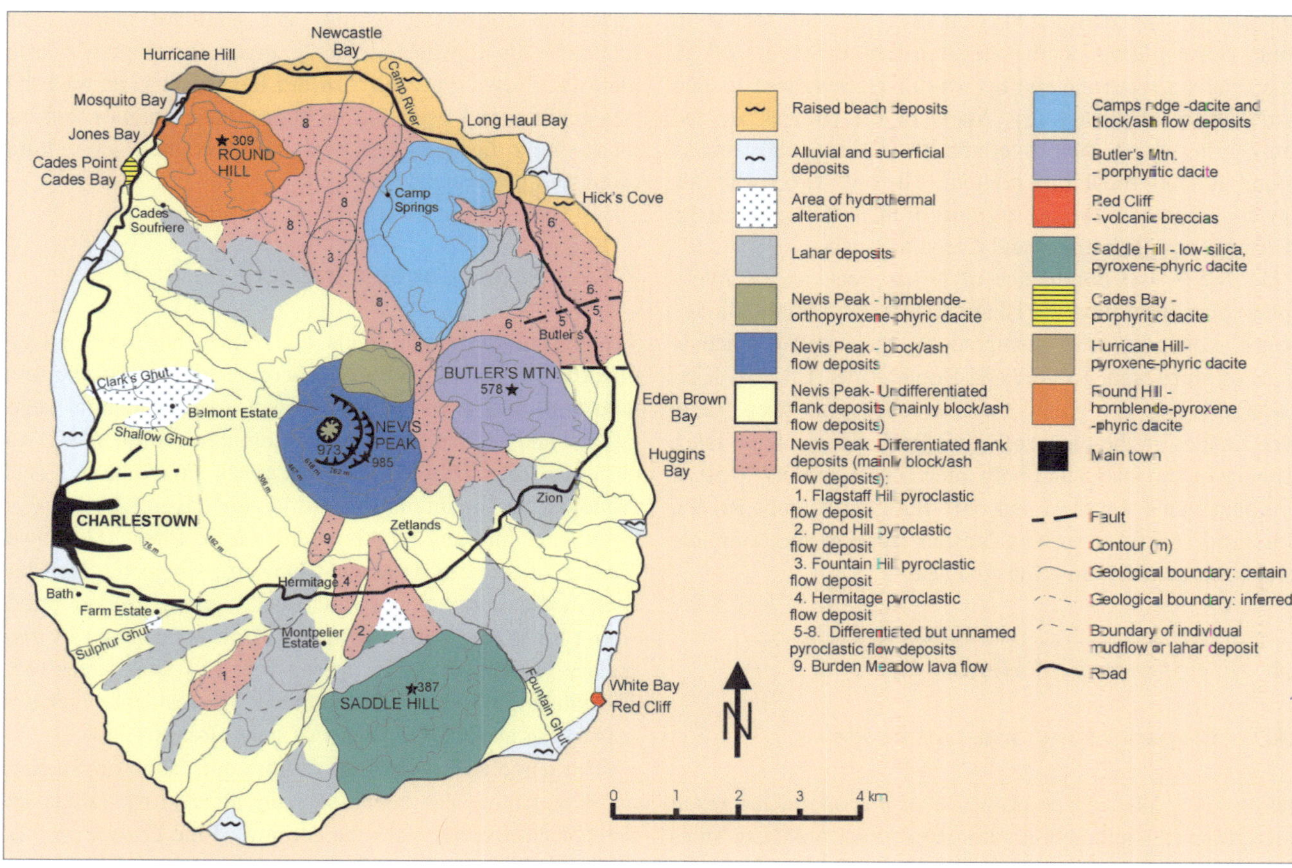

Fig. 7.6 Generalized geological map of Nevis Island (modified from Hutton and Nockolds 1978) showing all the major geologic features on the island. Cartography by Shahiba Ali from the Volcanic Hazard Atlas of the Lesser Antilles (see Simpson 2005)

deposits radiating outward toward the sea and two younger lava domes. Past eruptions have produced lava domes and associated volcaniclastic deposits that range in compositions from andesite to dacite (58–65% SiO_2; Hutton and Nockolds 1978).

Although there are large areas of hydrothermally altered rocks found throughout the island, only two localities exist with focused fumarolic activity: Cades Bay and Farm Estate. These are relatively low-temperature (53–100 °C) fumaroles characterized by warm, hydrothermally altered ground, small boiling pools, and weakly steaming vents (Robertson 2005). Geothermal activity on Nevis is also exhibited at a number of hot springs—the hottest being Bath Estate and Cades Bay beach. At these locations, the emission of hot gases under water provides a "champagne" effect.

7.3.2.3 Microclimates and Drainage

The presence of several peaks on Nevis results in a range of microclimates that vary greatly with height, location, and orientation. Possibly due to the lower elevation of its central mountain, the average rainfall is lower than on St. Kitts with considerable variation from year to year and month to month. Rain falling on the elevated mountain core of St. Kitts drains toward the sea by way of river valleys (called ghauts) that are mostly dry. Many of the streams are fed from springs and their overland flow disappears underground at about the 250 m contour. The only two rivers that flow consistently to the sea for most of the wet season are the relatively large Wingfield and Canyon Rivers.

On Nevis, the dominance of Nevis Peak as the major topographic feature on the island results in a radial drainage pattern, which is only disrupted by the smaller volcanic cones of Hurricane, Saddle, and Round Hills. Like St. Kitts drainage is via dry ghauts, which develop as deep, steep sided ravines on the flanks of Nevis Peak. There are no lakes or ponds on the island and only intermittent streams. Most of the ghauts are ephemeral and only Bath and Camps Rivers, which are fed from springs located close to the shoreline (<1.6 km), actually flow into the sea.

7.4 Landscape and History

7.4.1 Biogeography and Landscape

The biogeographical face of the country is as much a result of nature as it is interaction with people over time. Centuries of successful sugar cultivation have left an aesthetically pleasing, orderly, well-proportioned rural landscape on St. Kitts. In contrast, the long, less successful experience with sugar (until the 1950s) in Nevis and later free-grazing goats, sheep, cattle, and pigs has scarred and ravaged the landscape. Lack of this protective vegetation cover has resulted in greater erosive impact of wind and water erosion and sediment flow downslope that impact coastal areas and reefs (Island Resources Foundation 1991). The human-environment impact on Nevis differs significantly from that on St. Kitts.

Vegetation coverage on both islands comprises a variation of Rainforest, Dry Evergreen Forest, Dry Scrub Land, Palm Break, Elfin Woodland, and Disturbed lowlands that resulted from farming (Ministry of Sustainable Development 2014). The dominant influence on vegetation distribution on both islands is the effect of moisture-laden trade winds forced upwards by their prominent central peaks. Cooling of the moisture-laden air results in rain that falls more consistently on the upper slopes and leads to differential distribution between the windward and leeward side of the island. This has led to the development of classic island vegetation: Elfin Woodland and Palm Brake in the highest areas, giving way to rain forest on lower slopes, then seasonal and dry evergreen forests, and finally transitioning to littoral and mangrove along the coast. An upland forest belt occupies the steeper part of the central mountainous interior that comprises the northwest, central, and southeast ranges. This forest blends downslope into a gradually sloping coastal plain that flows gracefully toward the sea. It is an area that was traditionally occupied by sugarcane plantation but has increasingly become occupied by diversified agricultural crops and pastureland.

7.4.2 Anthropogenic Impacts

The vegetation on St. Kitts and Nevis has been greatly disturbed by human activity. Intensive land use in the lowland areas has removed all semblance of the natural vegetation, and agricultural crops dominate most areas not currently urbanized (Island Resources Foundation 1991). Although the mountains are still largely covered by forest, it is unlikely that any of this is still virgin forest. The lower reaches of forested areas (244–457 m), where the land has been under cultivation (with sugar, cotton, ground provisions) for a long period of time, are now occupied by secondary growth of dry evergreen forest of small trees (15–18 m in height) on abandoned farms. The resident vegetation consists of about 243 species of trees (Beard 1949).

The range and abundance of flora and fauna on St. Kitts and Nevis have also been adversely affected by human settlement particularly with the coming of the Europeans who replaced the indigenous Caribs and Arawaks. Modification, and in some cases elimination, of habitats and their associated wildlife followed the wholesale clearing of vegetation and changes to the landscape for estate development (sugarcane cultivation) and infrastructure development. Europeans civilization also introduced exotic animal species such as cats, dogs, rats, pigs, and mongoose that significantly affected native habitats and local populations of wildlife species such as ground-dwelling birds, beach-nesting sea turtles, agouti, and iguanas. The relatively inaccessible central mountainous backbone of St. Kitts (extending from Mt. Liamuiga to Olivees) and the island's largely remote Southeast Peninsula are areas where wildlife has been largely untouched and preserved. On Nevis, the combination of Nevis Peak and lower population densities has served to moderate the loss of wildlife.

Located along the island's coastal belt, the wetlands represent St. Kitts' most important habitat for migratory and marine avian species. The best example of a mangrove swamp on the island, Greatheeds Pond, resides on the Windward coast at Conaree, just below the Canada Hills. Other wetland habitats are found at Friar's Bay, Frigate Bay, and the pond system of the Southeast Peninsula. The island is among the few in the Eastern Caribbean to have large pond systems, and certainly unique in having one large pond (Salt Pond) that attracts a large number of shorebirds

7.4.3 Historical Impacts

When St. Kitts and Nevis was first sighted and named by Columbus in 1493, Carib Indians inhabited the island. They called the island "Liamuiga" (fertile land) and were the last of several Amerindian migrations from South America during the prehistoric period. These included the hunter-gathering people called the Siboney who arrived around 4500 BP, followed by the agriculturally based Arawaks who arrived around 2000 BP. Archaeological investigations by Goodwin (1979, 1980) indicate that Amerindian land use at this time was often high. Centuries of slash and burn agriculture destroyed much of the primary forest on both islands well before European arrival.

On Nevis, preceramic people occupied the island as far back as 4000 BP. The island was affected by similar migrations of indigenous peoples from South America, but its written history did not begin until the arrival of Columbus in 1493. The Caribs called the island "Oualie" (land of beautiful water), but the island's current name was derived from the Spanish "Nuestra Senora de las Nieves" (Our Lady of the Snows), so called due to the ubiquitous and continual cloud cover that encircles Nevis Peak (Fig. 7.7).

Columbus claimed both islands in the name of Spain, but since the Spanish focused more on the large islands in the region, they made little effort to colonize the two islands. This enabled the Caribs to remain unchallenged on the island until the early seventeenth century when land-hungry northern Europeans descended on the Eastern Caribbean. The first permanent European settlement on the island occurred in 1624 when Thomas Warner landed with a small party of British men (Island Resources Foundation 1991). The Caribs allowed him to establish a settlement near their village at Old Road. The French, led by Pierre Belain, arrived a few months after and were also welcomed by the Caribs. As was the practice at the time, French and British settlers began to enslave (and in some cases massacre) their indigenous hosts by 1626. Thereafter the island was divided between the two nations who maintained a peaceful co-existence since they had to guard against attacks by the Spanish. In 1628, the British expanded to Nevis when a group led by Anthony Hilton moved there and established a settlement named Jamestown.

In the early parts of the seventeenth century, St. Kitts was used as a base by the two colonial powers to expand their influence in the region with the French occupying Guadeloupe and Martinique, and the British expanding to Antigua and Montserrat. St. Kitts soon developed a reputation as one of the most fertile sugar colonies although intermittent colonial warfare between 1666 and 1708 prevented its emergence as a full-fledged sugar monoculture. During this period, the French and British attacked and seized each other's territory on St. Kitts and twice the French overran Nevis. The British eventually gained control, and in 1713 France ceded its portion of St. Kitts. The sugar industry expanded rapidly thereafter, and between 1715 and 1735, sugar monoculture supported by African slave labor replaced the preexisting diversified economy of the island. The island's population expanded significantly after this period as well, giving St. Kitts one of the highest population densities in the Caribbean by the middle of the eighteenth century (Island Resources Foundation 1991).

The main reason for colonization of the islands by the British was for sugar and Sea Island cotton export. Sugar has dominated the economic life of the two islands from colonial times until recently, when the focus has turned to tourism. All of this activity depends heavily on, and is affected by, the environment—all of which remains dominated by the volcanic landscape. Prior to the 1970s, most of St. Kitts' water needs were satisfied by surface springs and streams. Since then, there has been increasing focus on the island's groundwater resources as a source of supply to meet growing demand and ensure greater reliability. Still, on St. Kitts, springs, rather than groundwater, continue to provide the largest proportion of the total water supply. On Nevis, all water needs (domestic, agricultural, and industrial) are met by a combination of surface, rain, and groundwater sources, with groundwater providing the largest proportion of their public piped water supply.

Despite over 300 hundred years of continuous and intensive use for agriculture, St. Kitts' and Nevis' soils have generally stood up well. This is largely a result of the beneficial properties (chemical and physical qualities) resulting

Fig. 7.7 View of Nevis Peak from the ferry between St. Kitts and Nevis. A 3232 ft. (985 m) high stratovolcano, Nevis Peak is the highest point on the island of Nevis. Its central cone rises up behind the capital Charlestown and flank deposits radiate outward toward the sea in all directions. On days when the peak is completely free of clouds, there are remarkable views from its summit of the surrounding islands. Photograph courtesy of R. Robertson, The UWI Seismic Research Centre

from recent volcanism on the islands, since fresh volcanic materials can give rise to fertile soils up to 20 years after deposition (Hardy 1939). Rapid soil formation resulting from the abundance of fragmentary volcanic ejecta has also reduced the impact of soil erosion on cultivated slopes on both islands.

7.4.4 Geothermal Resources

The relatively young volcanic geology of St. Kitts and Nevis has provided it with significant untapped high-temperature geothermal resources with the potential for the development of geothermal power. Given that renewable energy resources (wind and solar) provide only 5.7% to their total installed power capacity of 56.4 MW, development of these resources has the potential to dramatically alter the energy balance of these islands (Alexander et al. 2015). Investigations undertaken by Energy and Climate Partnership for the Americas on a potential interconnection between St. Kitts and Nevis and Puerto Rico concluded that such an interconnection could simultaneously reduce oil and natural gas use on the islands and promote geothermal development (Alexander et al. 2015).

Investment and development plans have increased in recent years with exploratory drilling already completed in Nevis, and geothermal exploration exercises started in St. Kitts where consideration is being given to the generation of up to 20 MW for the initial phase. In both instances, however, plans for development have been challenged by political opponents and by lack of incentives for private sector participation (Alexander et al. 2015). On Nevis, the US Department of State and other partners in the USA are providing assistance.

St. Kitts and Nevis have separate utilities with exclusive rights to operate power systems on each island, and current developments demonstrate each island is utilizing different approaches to explore and develop their geothermal resources. Additionally, the high upfront capital needs, including the exploratory drilling, presents a major barrier to development of operational geothermal power. The relatively small power demand on both islands poses a challenge to profitability and points to the need for regional interconnection, collaboration, and trade in development of these resources.

7.5 Heritage and Tourism

The uniqueness of St. Kitts and Nevis, which has helped to shape the national identity and public image, is derived in part from the country's distinctive, dramatic, and spacious profile—all a direct result of its geologic past (Island Resources Foundation 1991). Each island is dominated by a single, volcanic edifice surrounded by a wide expanse of gently sloping fertile land that uniformly spreads toward the sea in all directions. Unlike other volcanic islands in the Lesser Antilles Island Arc, there is no complex interior landscape rising abruptly from the coastline. The hinterland of both islands is clearly visible from the coastline, often in one continuous parade of microhabitat variation and attitudinally conditioned biodiversity (Island Resources Foundation 1991).

Over the past two decades, the Federation has been transitioning from a way of life that evolved over four centuries of sugarcane cultivation to a political economy influenced by investments in the development of tourism, the challenges of international trade, and an uncertain environmental future (Williams 2013; Fig. 7.8). The sustainable management of land is a critical issue for the economic and social well-being of the country (Williams 2013). Partly in response to this, the Government of St. Kitts and Nevis (GOSKN) has established a number of laws, regulations, guidelines, and standards designed to govern the Federation's sustainable use and exploitation of the terrestrial and coastal resources. Each of these provides policies and guidelines for dealing with environmental issues with the overall intention of promoting sustainable development of the island.

Over the past few decades, a number of laws have also been passed to provide for the management, development, and protection of the natural environment. The legislation requires the preparation of Environmental Impact Assessments for projects that are likely to impact natural resources. Although there is no specific legislation that deals with soil conservation, there are legislative provisions that cover forestry and land tenancy—issues often related to soil conservation. One of the key issues is the existence of "landless" livestock farmers who graze animals in unrestricted areas.

St. Kitts and Nevis is signatory to five, and Party to 41 international environmental agreements, of which 18 are deemed to be the most important (Williams 2013). The obligations of the twin-island federation under a number of these international agreements have been recognized in national law. Despite the number of policies and laws that exists, Williams (2013) noted that outdated legislation and poor enforcement along with limited exposure of legal personnel to environmental issues and impacts are major weaknesses that minimize the effectiveness of the environmental regulations.

There are several designated national parks or protected areas on the islands. Brimstone Hill National Park Fortress (St. Kitts) and the Bath Hotel (Nevis) were two of the first, but recently the Central Forest Reserve National Park (CFRNP), the Royal Basseterre Valley Park (RBVP), St. Mary's Biosphere Reserve, and Frigate Bay Salt Pond have been declared protected. In addition, all land above the 300 m contour on Nevis is totally protected through administrative means and there are plans to establish a Nevis Peak Park to encompass this area.

Brimstone Hill Fortress National Park is a UNESCO World Heritage Site, and a key component of St. Kitts tourism product. Designed by British military engineers and constructed with skill, strength, and endurance of African slaves, it features great historical, cultural, and architectural significance. Although its primary focus is historical and cultural preservation, it also serves as an unofficial bird and monkey sanctuary. Alongside Brimstone Hill, the other early tourism draw for the Federation was the Bath Hotel, built in 1778 on a site overlooking Charlestown, the capital of Nevis, and adjacent to natural hot springs. Similar in function and style to its namesake in the UK, it first attracted many European travelers who would come for the hot springs.

The Central Forest Reserve National Park (CFRNP) is one of the largest protected areas on both islands encompassing 50 km^2 of cloud forest that provides habitat to a number of threatened birds and represents the last stand of undisturbed tropical rainforest on St. Kitts. The area is most important for its water resources, with an estimated 30% of St. Kitts' potable water supply originating from surface runoff within the Park. Heritage resources include scenic trails, as well as budge wood (*Myrtaceae*) for fish pots; bamboo, coconut, and calabash for traditional arts and crafts; and plants for traditional medicines.

The Basseterre Valley Park, at ~2 km^2 and immediately adjacent to the capital city Basseterre, houses an aquifer that provides 40% of the water supply for Basseterre. The water supply, however, is under threat from human-induced impacts such as fertilizer usage, sewage treatment and disposal, and storm water runoff along roadways. While still under development, plans have been put forth to provide a variety of development such as trails, parking lots, train station, and a large rainforest arboretum.

St. Mary's Biosphere Reserve and Frigate Bay Salt Pond are the most recently declared protected areas on the islands. Comprising cloud forests, mangroves, and coral reefs, St. Mary's Biosphere Reserve represents an important site in terms of biological diversity and is, in fact, one of the most

Fig. 7.8 Various tourism resources, both historic and geologic/geomorphologic, found on St. Kitts and Nevis. **a** Brimstone Hill Fortress National Park—a UNESCO World Heritage Site of historical, cultural, and architectural significance that is a prominent feature on the landscape of the island of St. Kitts. Its construction is a testimony to the ingenuity of British military engineers who designed it and to the skill, strength, and endurance of African slaves who built and maintained it. **b** The historic Church of Immaculate Conception located in Basseterre was built in or about 1856 and grew to become a beacon of Catholicism on St. Kitts during the ensuing years. Steeped in the history and traditions of Portuguese and Irish immigrants, the parish is one of two cathedrals in the Diocese of St. John's-Basseterre. **c** St. George's Anglican Church in Basseterre, one of the oldest surviving Anglican churches in the English-speaking Eastern Caribbean; a major tourist attraction and landmark on the island of St. Kitts. Its history, which consists of significant impacts by natural hazards (the 1842 earthquake, 1843 hurricane, and several others since 1989), along with human impact (the Great Fire of 1867), is symbolic of the two key influences on the evolution of the landscape on the island. This Church was originally built by the French as a Catholic Church but was subsequently burnt by English soldiers and rebuilt. **d** A green clock tower and water fountain that is known as Piccadilly Circus, or The Circus marks the main square of Basseterre, the Capital City of St. Kitts. This area was once a slave market for African slaves, who were one of the driving forces in the modification of the landscape on the island. It is a memorial to the former president of the General Legislative Council, Thomas Berkeley courtesy of R. Robertson, The UWI Seismic Research Centre

diverse natural communities on St. Kitts. Frigate Bay Salt Pond is located in an area that supports a number of bird species and is linked to the breeding of three such species. There are plans for its rehabilitation and generation of eco-tourism and educational experiences.

Two key NGOs remain actively involved in projects related to heritage conservation, environmental management, and protection in St. Kitts and Nevis: the *Saint Christopher National Trust* (SCNT) and the *Nevis Historical and Conservation Society* (NHCS). The SCNT evolved from the St. Christopher Heritage Society, which was incorporated as a private company in 1994. The Trust manages the National Museum in the Old Treasury Building in Basseterre and has as its key objective the protection, conservation, interpretation, and enhancement of the natural environment of St. Kitts, including its animals and plant life. The Trust provides a forum for the exchange of ideas, information, and knowledge on a range of issues related to conservation and heritage. The NHCS, established in 1980 to conserve the natural, cultural, and historic resources of Nevis, and its adjacent marine areas have worked to institute projects and policies designed not only to preserve Nevis' unique history and environment, but also to make that heritage accessible and intelligible to locals and visitors to the island.

7.6 Hazards

St. Kitts and Nevis remain exposed to a wide range of hazards, similar to other islands in the Lesser Antilles: hurricanes, floods, droughts, landslides, earthquakes, and volcanic activity. An estimated 39% of the population is considered to be at risk from such hazards (GFDRR 2010). The islands are among the top 60 countries exposed to risk of mortality and within the top 40 who would suffer significant economic risk from two or more hazards (GFDRR 2010). Risk to GDP is estimated to be 65% (GFDRR 2010). Destruction caused by natural disasters has had a negative impact on tourist arrivals and as such affects an important foreign exchange service sector.

The principal hazards affecting both islands are those related to hydro-meteorological events (high winds and rainfall due to tropical storms and hurricanes). The tendency toward urbanization with its resulting increased exposure to wind damage makes the islands increasingly vulnerable to wind storms. In addition, tourism development, the main trust for the island's economic development, is focused in the coastal zone, which is exposed to hurricane and storm surge impacts. The islands' topography limits the areas prone to the flooding which invariably results from prolonged rainfall and storm surge events, resulting in relatively few flood-prone areas. Those places prone to flooding, however, are usually located in low-lying coastal areas (e.g., coastal fishing villages) and along stream passages, and given that the general preference on St. Kitts to live in low-lying areas close to the sea, vulnerability to flooding and related storm surges is high. This is less so in Nevis where the settlement pattern is more dispersed, due to its legacy farming practices. Areas used for tourism and port facilities, however, are particularly exposed to storm surge events since these are also located near the coastline.

Hurricanes have had the most devastating effect on the islands with over 20 named storms passing within 60 nautical miles of the Federation between 1995 and 2004 (Carter 2010). Nine of these (three in particular were Category 3 or higher) made landfall on the territory. Hurricane damage to St. Kitts and Nevis was estimated at 85 and 140% of GDP in 1995 and 1998, respectively—significantly impacting the economy. In 1998, damages of Hurricane George were estimated to be US$445 million, including damage to 80% of the housing (GFDRR 2010). This hurricane significantly impacted not only natural resources closely linked to economic activities, but also funds required for rebuilding infrastructure and provision of social services. In 2008, Hurricane Omar, a Category 4 storm, passed 150 km east of the islands, but still caused significant damage from wind and storm surge.

Perhaps the greatest hazard potential lies in St. Kitts' and Nevis' volcanic origins. Mt. Liamuiga, the active volcanic center on St. Kitts, is considered likely to erupt in the future despite not currently showing any signs of increased activity (Robertson 2005). Based on its history, future events are likely to be explosive magmatic eruptions accompanied by the formation of a lava dome. The most devastating volcanic hazards (pyroclastic density currents and lahars) are gravity controlled, and strongly affected by St. Kitts' topography, relegating these major impacts to river valleys and associated outwash plains. This will most directly affect the northern parts of the island, but thick ash fall deposits could extend as far as the southern parts of the island, posing a threat to aviation especially (Robertson 2005).

Nevis Peak represents the active center on Nevis and the most likely location of future eruptions for that island. Although there have been no recent signs of increased activity, frequent shallow earthquake swarms and hydrothermal activity indicate that this center is still volcanically active, and an increase in activity could occur in the future (Simpson 2005). Based on past eruptions, the most likely future event would be a lava dome-forming eruption from the summit of Nevis Peak. Given the size of the island and the potential impact of hazardous volcanic activity, a major volcanic eruption would require the evacuation of the entire island (Simpson 2005). The Southern Peninsula of St. Kitts could also be affected by ash fall from an eruption of Nevis Peak.

7.7 Conclusion

The Federation of St. Kitts and Nevis is unique in the region in terms of the openness of its landscape and the bifurcated nature of its governance. Its volcanic geology has imprinted itself on all aspects of the landscape with densely vegetated central volcanic mountains that help seed the islands water supply, surrounded by gently sloping pyroclastic fans that have provided rich soils for agriculture and land for settlement. This picturesque landscape has been able to withstand decades of mono-crop agriculture and remains supportive of continued farming, settlement, and infrastructure.

The island has a rich history beginning with settlement by migratory Amerindian peoples originating in South America to eventual and subsequent conquests by the Spanish, French, and British colonial powers. After several decades of relative neglect, the islands rose in importance to at one time being the base for French and British colonial expansion in the region. With the onset of plantation agriculture built on African slave labor, the island was for many decades dominated by sugarcane cultivation, which left a lasting legacy on the landscape.

The interaction of people and the environment has also left its imprint on flora and fauna, to the point that very few untouched areas remain. With increasing realization of the importance of careful environmental management to help ensure sustainable development, there has been greater emphasis placed on implementation of policies and guidelines backed by laws to enforce a more careful management of the land. Like other Lesser Antillean countries, St. Kitts and Nevis continue to struggle with the challenges of Small Island Developing States, while seeking to develop to their fullest potential. The fundamental importance of its landscape and the impact of its geology on all facets of life on the island remain evident. Maintaining an appropriate balance between the demands of economic development while minimizing harmful impacts on the environment will continue to be a challenge well into the future.

References

Alexander O, Konold M, Auth K, Musolino E (2015) Philip Killeen (2015): Caribbean sustainable energy roadmap and strategy (C-SERMS): baseline report and assessment. Worldwatch Institute, Washington, DC

Baker PE (1969) The geological history of Mt. Misery volcano, St. Kitts, West Indies. Overseas Geol and Min Res 10(3):207–230

Baker PE (1985) Volcanic hazards on St. Kitts and Montserrat, West Indies. J Geol Soc Lond 142:279–295

Beard J (1949) The natural vegetation of the Windward and Leeward Islands. Clarendon Press, Oxford

Carter SS (2010) National Environmental Summary Federation of St. Kitts & Nevis. United Nations Environment Program, 27p

Davis WM (1924) The formation of the Lesser Antilles. Proc Nat Acad Sci Wash 10:205–211

Earle KW (1925) Reports on the geology of St. Kitts-Nevis, B.W.I. and of Anguilla, B.W.I. Crown Agents for Colonies, pp 1–50

Fels G (1903) Ein Anorthitauswurfling von der Insel S Christopher Z Kristallogr Miner 37:450–460

Lang D, Carroll D (1966) St. Kitts and Nevis soil and land use survey no. 16. Imp. Coll. Trop. Agri., St. Augustine, Trinidad

Martin-Kaye PHA (1959) Reports on the geology of the Leeward and British Virgin Islands: Castries. Voice Publication Company Ltd, St. Lucia, pp 1–117

Ministry of Sustainable Development (2014) St. Christopher and Nevis Natural Biodiversity, Strategy and Action Plan 2014–2020. Government of St. Kitts and Nevis, 100p

GFDRR (2010) Disaster risk management in Latin America and the Caribbean Region: GFDRR Country NotesSt. Kitts and NevisGlobal Facility for Disaster Risk Reduction, The World Bank, 10p

Goodwin R (1979) The prehistoric cultural ecology of St. Kitts, West Indies: a case study in island archaeology. Ph.D. dissertation, Arizona State Univ., Tempe, Arizona

Goodwin R (1980) Demographic change and crab-shell dichotomy. In: Proceedings of the eighth international congress for the study of pre-Columbian cultures of the Lesser Antilles, Tempe, Arizona, pp 50–65

Hardy F (1939) Soil erosion in St. Vincent, B.W.I. Trop Agric 16:58–65

Hovey EO (1903) Martinique and St. Vincent revisited. Am Mus J 3:41–45

Hovey EO (1905) Volcanoes of St. Vincent, St. Kitts and Statia. Rep. 8th International Geography Congress, Washington, pp 452–454

Hutton CO (1965) The mineralogy and petrology of Nevis, Leeward Islands, British West Indies. Fourth Caribbean Geological Conference, Trinidad, pp 383–388

Hutton CO, Nockolds SR (1978) The petrology of Nevis, Leeward Islands, West Indies. Overseas Geol Miner Res 52:1–31

Island Resources Foundation (1991) St. Kitts and Nevis Country Environmental Profile. Island Resources Foundation and Caribbean Conservation Association

Robertson REA (2005) St. Kitts. In: Lindsay JM, Robertson REA, Shepherd JB, Ali S (eds) Volcanic hazard atlas of the Lesser Antilles. Seismic Research Unit, The University of the West Indies, Trinidad and Tobago, W.I, pp 204–217

Sapper K (1904) Die Vulkanischen Kleinen Antillen und die Ausbruch der Jahre 1902 und 1903. N Jahrb Miner 2:1–70

Simpson K (2005) Nevis. In: Lindsay JM, Robertson REA, Shepherd JB, Ali S (eds) Volcanic hazard atlas of the Lesser Antilles. Seismic Research Unit, The University of the West Indies, Trinidad and Tobago, W.I, pp 170–180

Williams PI (2013) St. Kitts and Nevis Land Policy Issues Paper. Prepared for The Social and Sustainable Development Division (SSDD) of the Organization of Eastern Caribbean States (OECS), Morne Fortune, Castries, Saint Lucia, 130p

Antigua and Barbuda

8

Amy E. Potter, Sean Chenoweth and Mick Day

Abstract

The country of Antigua and Barbuda comprises two islands located in the Lesser Antilles of the Caribbean. While both islands are positioned on the Barbuda Bank, they each consist of a unique geologic setting: Antigua is composed of volcanic rock, clay, and limestone, while Barbuda is largely limestone with prominent cave features. Their unique attributes are not just geologic, as the two islands also have differing histories that developed in part due to environmental constraints—Antigua's enslaved African population grew sugarcane as a British colony, while enslaved Barbudans raised livestock as a leased entity of the British Codrington family because of shallow soils resulting from the island's karst topography. Today, Antigua is much more developed relative to Barbuda in terms of tourism and has a greater population density. Barbuda's population is small and development is limited as a result of Barbuda's unique land tenure of common property. Still, despite their differences, both islands face similar environmental hazards including drought, hurricanes, soil erosion, flooding, and the long-term risks posed by climate change.

Keywords

Antigua and Barbuda • Caribbean • Common property • Tourism • Karst

8.1 Introduction

Antigua and Barbuda have land areas of about 280 and 160 km^2, respectively, with coastlines some 153 and 107 km in length (Loveless 1960; UN 2002; Carr and Heyman 2009; GA&B 2015a; Figs. 8.1 and 8.2). Geologically, they are part of the Limestone Caribbees (Martin-Kaye 1969; Blume 1974)—the Cenozoic carbonate-rock-dominated outer band of the Lesser Antilles island arc. Barbuda is dominated by carbonate rocks, primarily limestones of Pliocene age. Antigua consists of limestones, volcanic rocks, and clays ranging in age from mid-through late Oligocene (Wadge 1994; Donovan et al. 2014).

Elevations range from sea level to 402 m at Mount Obama (formerly Boggy Peak) in Antigua and to 38 m in the Highlands of Barbuda (GA&B 2015a). On and within the Antiguan limestones have developed karst landscapes, produced essentially by carbonate dissolution, containing dolines/sinkholes, dry valley systems, residual hills, and calcareous mollisols and vertisols systems. A discontinuous escarpment marks the boundary between the limestones and the Antiguan clays, on which a rolling central plain has developed, and the volcanic rocks of the southwest persist as prominent, steep-sided peaks and valleys bordered by a narrow coastal plain (CCA 1991). Barbuda is almost entirely limestone, similarly karstified as Antigua, but with extensive

A.E. Potter (✉)
Department of History, Armstrong State University, Savannah, GA, USA
e-mail: amy.potter@armstrong.edu

S. Chenoweth
Department of Atmospheric Sciences, University of Louisiana at Monroe, Monroe, USA
e-mail: chenoweth@ulm.edu

M. Day
Department of Geography, University of Wisconsin-Milwaukee, Milwaukee, USA
e-mail: mickday@uwm.edu

C.D. Allen (ed.), *Landscapes and Landforms of the Lesser Antilles*, World Geomorphological Landscapes, DOI 10.1007/978-3-319-55787-8_8

Fig. 8.1 General physiographic map of Antigua with major surrounding islands. The highest point, Mt. Obama, was renamed from Boggy Peak in 2009. Cartography by K.M. Groom

overlying sands (Martin-Kaye 1959; CCA 1991; UNEP 2010).

The easterly trade winds help moderate the Lesser Antilles summer temperatures. Because of the prevailing northeasterly trade winds, the eastern sides and windward slopes of Caribbean islands, such as the Barbuda Highlands, receive more rainfall. Overall, the climate is dry, sub-humid with average temperatures ranging seasonally between 24 °C (75F) and 29 °C (84F) (CCA 1991; ABMS 2015), though Barbuda can be classified as a dry-winter tropical climate (Rudloff 1981). Antigua's annual rainfall from 1981 to 2000 averaged 1187.5 mm/46.75 in (Antigua and Barbuda Meteorological Services 2016). Annual rainfall is characterized by extreme inter-year variability, with a distinct dry season from January through April and an August-November wet season (Barragne-Bigot and Yearwood 1991; CCA 1991: UNCCD 2005). Soils reflect the underlying geologic substrates, being dominantly shallow calcareous mollisols and vertisols over the limestones, sands, and clays in the Central Plain of Antigua, and acidic clay loams on the volcanic rocks (Hill 1966; CCA 1991). The annual cycle of temperature exhibits a summer-winter variation with a small range. Low temperatures around 20 °C in February and high temperatures during September about 32 °C result in an annual range of 12 °C for Barbuda (National Ocean Service 2015). The annual rainfall range for Barbuda is 750–900 mm (Oliver 2005), and hurricane season is from June to November with a peak of activity in September, prior to the rainfall maximum in November for the Caribbean.

Fig. 8.2 General physiographic map of Barbuda. Since St. John's, Antigua, serves as the administrative capital for both islands, Barbuda has no capital city. Codrington Village is the population center for Barbuda. Cartography by K.M. Groom

Barbuda has a resident population of about 2000 and Antigua 90,000 with population densities of about 13/km^2 (32 per square mile) and 321/km^2 (833 per square mile), respectively (GA&B 2015a; CIA 2015). Annual population increase is currently about 1.3% (CIA 2015), but visitor numbers have increased steadily to about 800,000 annually (GA&B 2015b), effectively doubling the population during peak season. The population, about 24% of which is urban, is concentrated in northwestern Antigua, particularly around St. John's (22,000) (CIA 2015; GA&B 2015a).

Human impact on the landscape has been long term and profound, resulting in wholesale environmental changes (Watts 1987; Day 2010a, b). Drought and water supply are fundamental problems, as is accelerated erosion and, paradoxically, occasional flooding. Hurricanes, occasional earthquakes, and potentially volcanic ash fallout and tsunamis pose additional threats (Wadge 1994). Environmental issues are intensified by the combination of limited land area, high population density, and mass tourism.

8.2 Antigua

8.2.1 Setting

Representing an exposed portion of the extensive Barbuda Bank, the volcanic basis of Antigua is mid-Oligocene, with overlying sedimentary rocks of later Oligocene age (Martin-Kaye 1959; Donovan et al. 2014). The stratigraphic succession is the most complete in the Limestone Caribbees,

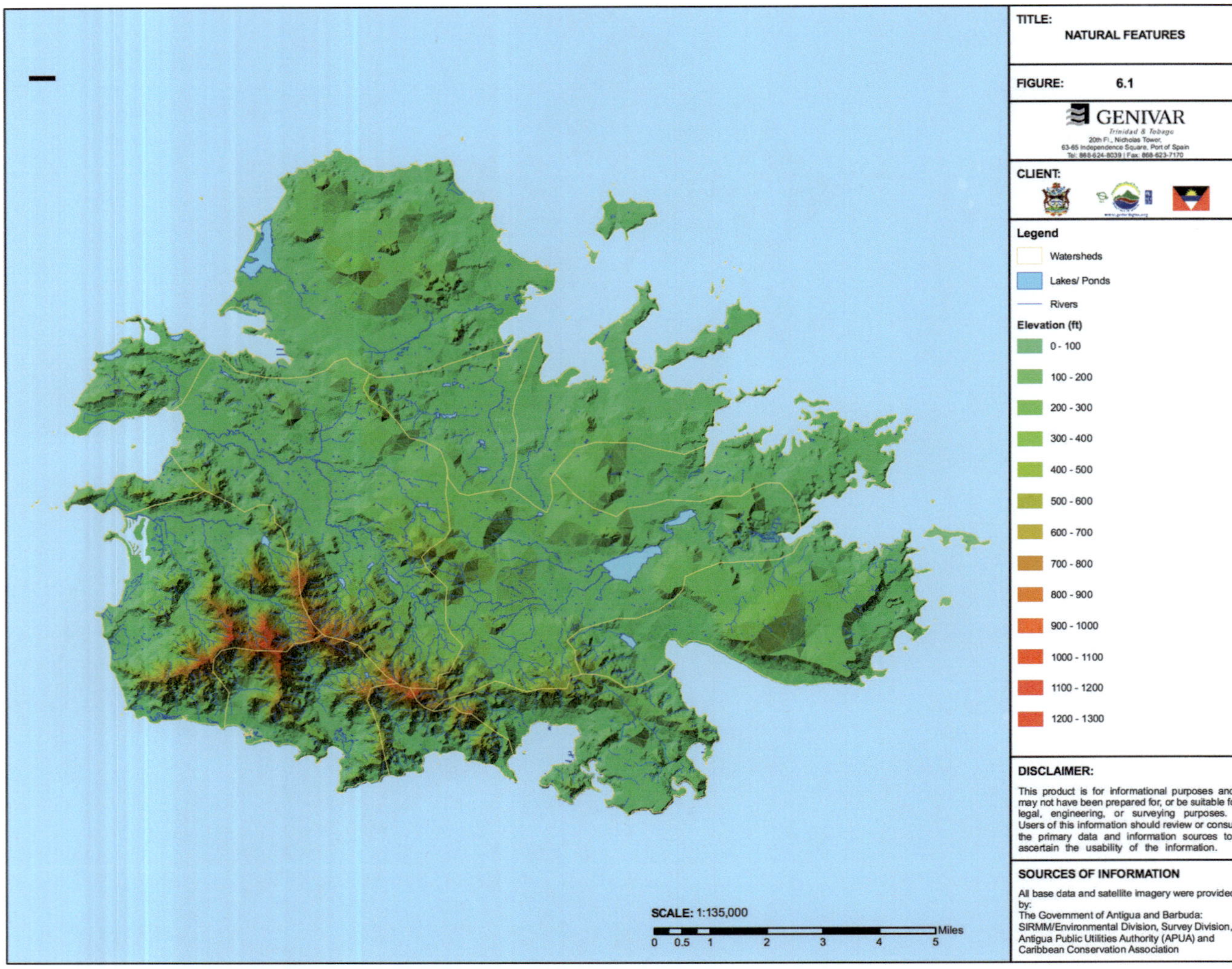

Fig. 8.3 Topographic map of Antigua, demonstrating a general dipping trend toward the northeast at an average of about 10, with the oldest outcrops in the southwest. The two main NW–SE-running valleys (between the island's high points) represent extensive strike-slip faulting. Figure courtesy of GENIVAR, Trinidad & Tobago (http://www.environmentdivision.info/wp-content/uploads/2012/01/NPDP-SIRMZP-2012.pdf)

demonstrating declining episodic volcanism and intermittent but increasing marine transgression throughout the Oligocene (Weiss 1994; Donovan et al. 2014) (Fig. 8.3).

The mid-Oligocene basal volcanics of southwestern Antigua are a suite of mostly calcalkaline andesites, although ranging from quartz basalts to dacites (Christman 1972; Weiss 1994; Tomblin 2005). Many are extensively metamorphosed, transported, decayed, and eroded, although volcanic cones and debris flows are still evident (Tomblin 2005). The rocks are dominantly blocky agglomerates, incorporating pyroclastic tuffs, lava flows, and intrusives. There are also some marine deposits, including carbonate lenses, and the total accumulation may exceed 2000 m in thickness (Martin-Kaye 1959).

The Central Plain Group is composed of perhaps a 1000 m thickness of agglomerates, mudstones, volcanic tuffs, and conglomerates, of both marine and terrestrial origin, with considerable local variability and local deformation. Terrestrial agglomerates and tuffs dominate, but there are also chert beds and limestone lenses, together with basaltic intrusions (Martin-Kaye 1959; Tomblin 2005).

The limestones and marls of northeastern Antigua belong to the Oligocene-aged Antigua Formation (Martin-Kaye 1959; Weiss 1994; Donovan et al. 2014). Older facies represent shallow-water coralline biostromes and patch reefs, but younger limestones are fine-grained, deeper-water deposits (Weiss 1994). The Antigua Formation attains a thickness of at least 550 m and dips generally at <15° toward the northeast (Martin-Kaye 1959). The limestones are highly variable, ranging from (bio)micrites through (bio) sparites, and many are highly fossiliferous. They also vary considerably in both purity and mechanical strength, reflecting rhythmic depositional sequences and variations in diagenesis (Weiss 1994; Day 2007).

8.2.2 Landforms

On and within the limestones, carbonate dissolution and associated processes have developed a karst landscape in Antigua dominated by dolines/sinkholes, dry valleys, residual hills, and springs (Day 1986, 2007). Surface water is rare here, except during particularly intense rains, when sinkholes may fill with localized ephemeral runoff, and normally dry valleys may channel dangerous floodwaters (Fig. 8.4a). Except during these events, drainage is predominantly subterranean, via rapid infiltration into, and percolation through, the thin soils, then transmission at various rates via pores, fissures, and conduits in the limestones to shallow, lenticular groundwater aquifers (Martin-Kaye 1956, 1959).

The most conspicuous landforms in the limestone area of Antigua are the discontinuous southwest-facing escarpment, which marks the boundary with the Central Plain, and the four northeast-trending dry and seasonally active allogenic valley systems that breach the escarpment and dissect the limestone upland into distinct blocks (Day 2002, 2007). Within the limestone blocks occur shorter autogenic dry valleys, generally shallow and without distinct stream channels, and mostly in agricultural use. These valley systems are significant geomorphologically in that they represent earlier phases in the karst development when secondary permeability was not sufficiently developed to permit total subterranean drainage, and runoff was via "normal" fluvial valleys (Smith 1975). As secondary permeability has increased through carbonate dissolution, drainage has increasingly been diverted underground, and the valleys have been essentially abandoned, only to be activated during intense rainfall, when the capacities of the karst plumbing system and its overlying soils are exceeded (Smith 1975; Day 2002). As such, in a sense, they are ephemeral or "occasional" drainage systems, and their persistence indicates that karstification has not yet progressed to the extent that surface drainage has been completely abandoned and the valleys obliterated by doline/sinkhole development, as appears to be the case in older karst landscapes elsewhere in the Caribbean (Day 2002).

Dolines/Sinkholes are often regarded as diagnostic of karst landscape (Ford and Williams 2007), and at least 45 occur in Antigua (Day 2007). In Antigua, dolines are most numerous in the central and southeast parts of the karst, although densities are generally low (Fig. 8.4b). Antiguan dolines, for the most part, are broad and shallow, with a tendency toward clustering. It has been suggested that smaller, individual dolines are near-circular, and probably dissolutional, but larger ones are asymmetric and probably represent constructional variability in the carbonate depositional environment (Weiss 1994; Day 2007).

The relationship between the autogenic dry valleys and the dolines/sinkholes is not entirely clear, although it is probable that formation of the former precedes that of the latter (Smith 1975; Day 2002). They may be competitive, in the sense that they have developed preferentially under different hydrological conditions, or they may reflect spatial variations in surface slope and/or carbonate lithology.

Many of the cliffed limestone slopes on Antigua have "ledge and notch" profiles, which reflect erosional exploitation of rhythmic carbonate depositional sequences (Weiss 1994). Depositional variations may also be manifest in the littoral karren (Lundberg 2004) and in the "…scattered, rather abrupt hills…" (Martin-Kaye 1959, p. 1) that also occur within the limestones (Multer et al. 1986; Weiss 1994; Day 2007).

There are few known caves in Antigua, since their development is restricted by the impurity of the limestones and the limited allogenic recharge (Martin-Kaye 1959; Gurnee 1961; Day 2002). Five types of carbonate island karst caves have been recognized in Barbados (Kambesis and Machel 2013), although these have not been identified

Fig. 8.4 Example Karst features on Antigua. **a** Typical dry valley in Antigua, typified by drought-resistant grasses and trees. Photograph by M. Day, 1975. **b** Typical small, circular doline/sinkhole manifested by the darker green area in the image's center. Photograph by M. Day, 1975

categorically in Antigua (Day 2009): (1) epigene caves formed by freshwater dissolution in the upland interior; (2) flank margin caves formed by mixing-zone dissolution along modern and older coastlines; (3) littoral caves formed by contemporary marine corrosion; (4) mechanical caves formed by bedrock mass movement; and (5) hybrid (polygenetic) caves, which constitute the majority, often involving modification of flank margin caves. There are natural bridges, which may be cave remnants, at several locations, including Soldier Point and Devils Bridge (Donovan et al. 2013).

Antigua exemplifies the composite type of the Carbonate Island Karst Model (Mylroie and Vacher 1999). The geologically young carbonate rocks have not been buried beyond the range of meteoric diagenesis, and they interact both with fresh and saline groundwater that is affected by glacio-eustatic and tectonic changes in sea level, forming immature (essentially Neogene) eogenetic karst.

The most prominent landforms in the volcanic terrain of southwestern Antigua are pronounced residual hills, often conical in overall morphology, dissected by steep-sided alluviated valleys displaying a radial drainage pattern moderated by the northeasterly dip (Martin-Kaye 1959). Steeper slopes are marked by various types of slope failure scars, particularly those resulting from debris flows.

The Central Plain is more varied than the name suggests, with rolling topography and relative relief in excess of 100 m culminating in the elongated central ridge of Belmont/Drews Hill and the southwestern escarpment overlooking the Bendals Valley (Martin-Kaye 1959). The overall morphology is that of a faulted monoclinal valley, transitioning southwestward to a monoclonal ridge (Martin-Kaye 1959).

Coastal landforms illustrate the multifaceted interaction between terrestrial and marine geomorphic processes, with restricted tidal platforms, cliffs with wave cut notches, littoral caves, stacks, arches, and extensive sand bars, spits, dunes, and beaches. The carbonate coastline in particular is deeply embayed and cliffed. Antigua's sand beaches are renowned (de Albuquerque and McElroy 1995), and there are marked energetic and morphological contrasts between beaches with a Caribbean aspect versus those subject to long-fetch Atlantic wave action.

Fringing, patch, and bank barrier reefs off the west, north, and east coasts represent contemporary analogues of the older features now exposed on the island itself, although the species composition differs (Multer et al. 1986).

8.2.3 Landscape

Human impact on the landscapes of Antigua and Barbuda has been long term and severe, as throughout the Caribbean (Watts 1987; Day 1993, 2010a, b). Amerindians perhaps arrived by 3500 BCE, and Antigua, in particular, has a rich legacy of pre-Columbian settlements, having been an important regional source of chert/flint for tool-making (Davis 1993). Native populations had profound influences on the natural vegetation (Harris 1965; Watts 1987). The island was colonized by the British in 1632 and was strongly fortified subsequently, resulting in a rich military heritage (CCA 1991; Museum of Antigua and Barbuda 2015).

The physical landscape of Antigua was devastated during the early colonial period by almost complete clearance of the natural vegetation and the institution of cotton, then sugar cane monoculture (Barker 1981; Watts 1987, 1988; Cooper and Bowen 2001). A few cave entrances, sinkholes, steeper-sided gullies and hillsides, and offshore islands were spared from clearance, and these represent important ecological remnants, although invasive species are highly competitive and increasingly challenging (FAO 2003; Gore-Francis 2013). Destruction of the natural vegetation undoubtedly changed the hydrologic regime and may have contributed significantly to historical aridification (Fig. 8.5; UNCCD 2005).

Antigua maintained its agricultural basis until the end of the twentieth century, shifting to livestock production and intensive production of produce for hotels and restaurants. Tourism is now ascendant, and much agricultural land has been converted to other uses (Barker 1981; CCA 1991; UNCCD 2005; Gore-Francis 2013).

8.2.4 Heritage and Tourism

Tourism is the leading economic activity in the country (CCA 1991; CIA 2015), and much of it focuses on the coastal zone, particularly beaches and reefs (Gore-Francis 2013). Annual tourist arrivals in 2014 totaled about 800,000 (Government of Antigua and Barbuda 2015b), and tourism accounts for 70% of GDP, 85% of foreign exchange earnings, and 40% of employment (Government of Antigua and Barbuda 2004; Gore-Francis 2013; CIA 2015). Mass tourism places major demands on the water supply system (CCA 1991) and on coastal zone resources (de Albuquerque and McElroy 1995).

Efforts have been made to encourage heritage tourism, agro-tourism, and ecotourism (Cooper and Bowen 2000; UN 2012), but these remain subsidiary to traditional tourism. Nelson's Dockyard, Shirley Heights, and other sites capitalize on colonial naval and other military heritage, with the former also being an important regional sailing center.

Antigua and Barbuda have only 11 formal protected areas, in which some 15% of the land area is at least theoretically conserved, primarily under the National Parks Act of 1984. Other areas, such as Santa Maria Hill, Corbison's

Fig. 8.5 Goat grazing near a denuded limestone slope in Antigua. This type of aridification has not been halted, as livestock and agricultural production have become somewhat more prevalent in recent decades. Photo courtesy of the Environment Division, Ministry of Public Works and Environment, Government of A&B

Point, and the Potworks and Collins reservoirs, are accorded variable levels of informal protection, and an integrated national protected areas system has been proposed, although not implemented, under the auspices of the National Biodiversity Strategy and Action Plan (GA&B 2001; Gore-Francis 2013). Antigua and Barbuda lag behind other Caribbean countries in protection of their karst landscape (Kueny and Day 1998; Day 2011), although important conservation efforts are being made on Great Bird Island and at other sites within the karst. The southwestern volcanic terrane also has considerable ecotourism potential, although this has not yet been fully realized.

8.2.5 Hazards and Environmental Issues

Drought and tropical storms, including hurricanes (with associated floods), are the major natural hazards affecting Antigua and Barbuda (GFDRR 2010), and projected anthropogenic climate change increases these threats, as elsewhere throughout the Caribbean karst (Farrell et al. 2007; Cambers 2009; Day and Chenoweth 2009a, b). Increased hurricane and drought frequencies have been experienced in recent years, with consequences including increased flooding, erosion, sedimentation, and decline in agricultural productivity (UNCCD 2005). Institutional, legislative and societal shortcomings, and questionable economic and development priorities exacerbate these problems (CCA 1991; Lorah 1996).

8.2.5.1 Drought

Drought (meteorological, agricultural, and hydrological) is a perennial, although increasing problem, particularly within the karstlands, where serious or severe drought historically occurs every 5–10 years (Martin-Kaye 1956; Lewis 1984, CCA 1991; Jackson 2001; OAS 2001; FAO 2004; A&BMS 2015). Severe drought produced by little rainfall and intense evapo-transpiration, has occurred 15 times since 1928, most recently in 2002/2003 and serious drought has occurred 16 times, notably in 1983–1984, and currently since July 2013 (A&BMS 2015).

8.2.5.2 Hurricanes and Flooding

Tropical cyclones, including storms and hurricanes threaten Antigua and Barbuda regularly, with 50 hurricanes and 59 tropical storms recorded since 1851 (A&BMS 2015).

Notable damage occurred during category 4 hurricanes: Hugo in 1989, Luis in 1995, and Omar in 2008 (A&BMS 2015). Localized flooding is a considerable risk in Antigua because of the combination of cyclonic rainfall intensity, shallow soils with limited permeability, dense drainage networks, over-grazing, and urbanization (Cooper 2001). Flooding in the karst is a particular danger, since sinkholes and dry valleys contain water only occasionally, and the potential risk of flooding is often underestimated.

8.2.5.3 Soil Erosion

Accelerated soil erosion and land degradation is a severe local problem, particularly in areas with large numbers of goats, and where trees and shrubs are cleared or burned on slopes to encourage pasture grasses (UNCCD 2005; UN 2012). The problem is further exacerbated because 60% of the land in Antigua is government-owned, giving farmers little incentive to invest in land improvement (Lemel et al. 1988; FAO 2003).

In the volcanic rock areas, hazards include accelerated soil erosion, gullying, landslides, and other mass movements, many of which are human-induced through over-grazing and other disturbance (UNCCD 2005). Impacts on human residents have so far been minimal, but slope failures represent a potentially significant hazard, although they are considered generally to be "not a pressing problem" (GFDRR 2010, p. 87).

8.2.5.4 Earthquakes

There is a moderate earthquake risk in Antigua and Barbuda, with particularly notable events in 1690, 1843, and 1974 (McCann et al. 1982; CCA 1991). Regional tectonism and volcanism elsewhere in the Caribbean also pose potential risks of tsunami impact and volcanic ashfall (CCA 1991: Donovan and Jackson 1994; GFDRR 2010).

8.2.5.5 Water Supply

Water supply difficulties are largely a function of the contemporary and historical rainfall regime and the islands' various landscapes. Surface drainage is usually of short duration, and water runs off quickly, particularly from impervious surfaces in urban areas, causing flooding and non-point pollution, and being effectively lost to human use through poor conservation practices (Barragne-Bigot and Yearwood 1991; Williams 1994; Mylroie and Carew 1997). Seawater intrusion into the shallow coastal aquifer is also an ongoing threat (Sandberg and Barnes 2004).

8.2.5.6 Degradation of Coastal Ecosystems

Degradation of coastal ecosystems is particularly acute around the St. John's urban core and in tourism areas along the coast, where wetlands are being degraded and infilled, coastal water quality is decreasing, coral reefs are in decline, and unapproved sand mining remains problematic (CCA 1991; de Albuquerque and McElroy 1995; Carr and Heyman 2009). Tourism and other infrastructure is infringing upon beaches, and beach erosion has become significant and endemic (Baldwin 2000). Such environmental impacts adversely affect residents and tourism alike (Lorah 1996; Carr and Heyman 2009).

8.3 Barbuda

8.3.1 Setting

Although separated by a mere 42 km of water, Antigua and Barbuda are quite unique in terms of their history and geography. This section captures some of the nuances that make the island of Barbuda altogether similar, yet also distinctly different, from its sister-island counterpart. The two islands are bathymetrically connected by the Barbuda Bank along the 200 m isobath. Earthquakes that occur in this region are associated with plate boundary subduction zone tectonics. The Leeward Islands of the Lesser Antilles (Limestone Caribbees) were formed during the Tertiary. Barbuda is composed of Pleistocene coral limestone, lying about 50 km east of the extinct volcanic chain arc axis but, rather than at its center, Barbuda lies uniquely on the northeast margin of its bank. The rocks are nearly flat-lying and limestone sequences are thin. Between each marine stage, depositional hiatuses occurred with karstic erosion creating an environment of offlapping carbonate units (Brasier and Donahue 1985).

The land area of Barbuda is approximately 150 km^2, with an east–west extent of about 13 km, and northwest to southeast axis of about 22 km. Codrington Lagoon is approximately 20 km^2 with an east–west extent of about 2.5 km and north–south range of 9 km. Barbuda has little relief except for the Highlands in the east rising to a height of only 38 m in the north, and gradually decreasing in elevation toward the south (Fig. 8.6). Still, four main geologic formations are found: Highlands, Beazer, Codrington, and Palmetto Point. These span the geologic time scale from Miocene to Holocene.

The Highlands Formation was deposited between the depths of 20 and 100 m during the mid-Miocene to early Pliocene. This unit forms the table-like inlier of the Highlands landscape, represented by lime wackestone—a cream-colored, fine-grained, and porcellaneous rock (Brasier and Donahue 1985). The Beazer Formation, comprised of a fine-grained, porcellaneous, hard, compact, creamy-white limestone, represents an older Pleistocene deposit, while the Codrington Formation—composed of well-preserved fossiliferous mollusk and coral sediments that have recrystallized to a lesser degree, was most likely deposited during a

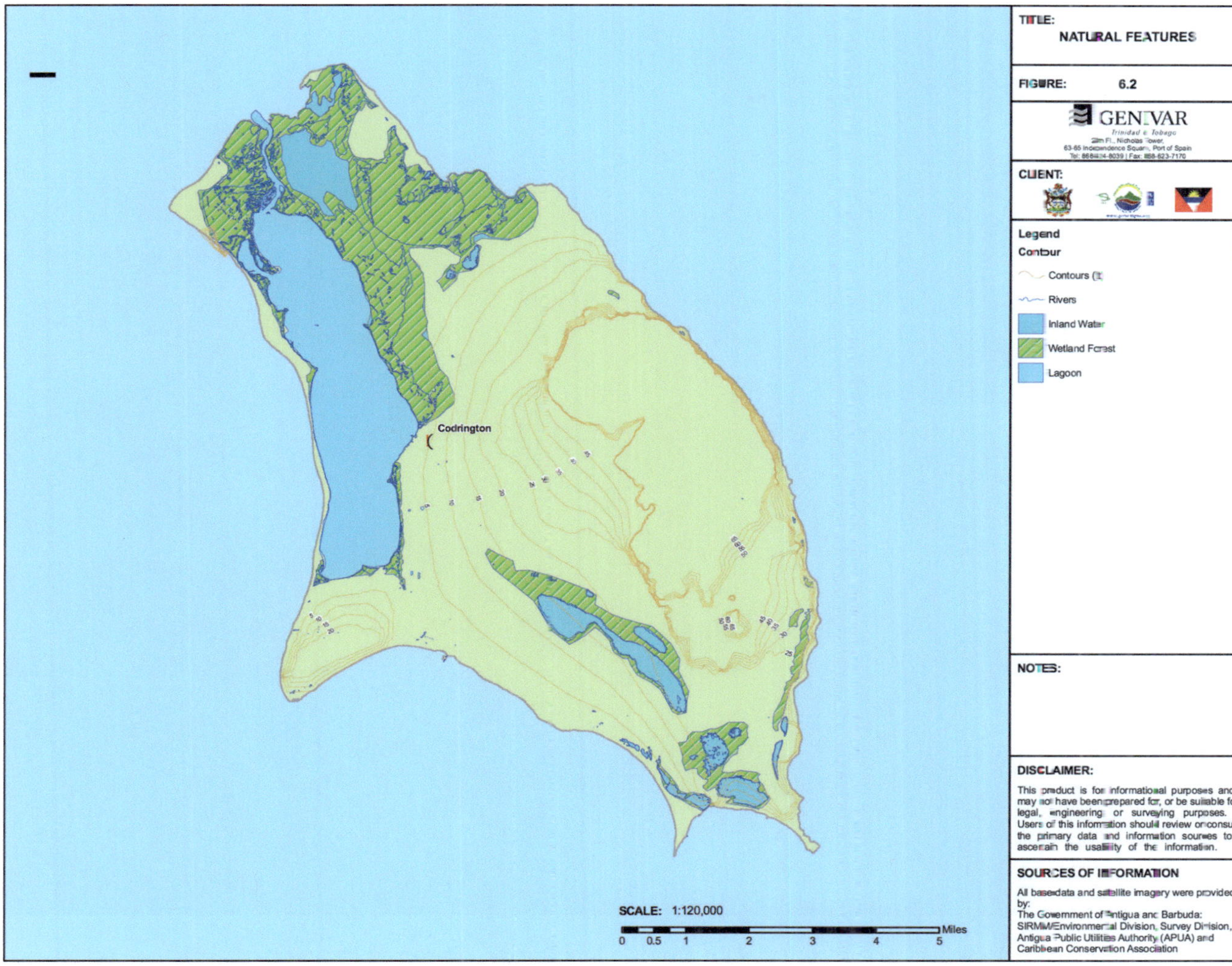

Fig. 8.6 Topographic map of Barbuda, displaying the lack of topographic variation. The mean and median elevations are both 26 m (with a mode of 7 m), with approximately 58% of the island being less than 10 m above sea level (USGS 2004). Figure courtesy of GENIVAR, Trinidad & Tobago (http://www.environmentdivision.info/wp-content/uploads/2012/01/NPDP-SIRMZP-2012.pdf)

high stand of the Last Interglacial in the late Pleistocene (Brasier and Donahue 1985; Brasier and Mather 1975).The Palmetto Point Formation representative of post-Pleistocene accumulations of sand, occurs at several locations on the island: Palmetto Point, Cedar Tree Point, Cocoa Point, dunes of the windward coast, and the sand bar forming the western enclosure of Codrington Lagoon (Brasier and Mather 1975).

8.3.2 Landforms

Sinkholes (dolines) are found in the Highlands and in an area called the Caves (local name for a sinkhole) south of Hog Point and east of Two Foot Bay (Fig. 8.7). The entrance surface trail to Darby Cave is a talus slope that leads to the bottom of the sinkhole similar to those found in the Jamaican Cockpit Country (Day and Chenoweth 2004). Even though Barbuda's karst landscape has no known surface streams, large trees and ferns cluster on the damp soil, reminiscent of a tropical rain forest (Harris 1965). There are sinkholes several times larger in surface area than the deepest (Darby Cave, see Fig. 8.8) located to the west-southwest and along the eastern Highlands near Pigeon Cliff (D.O.S. 1970, Sheet 1D). Care must be taken when walking around the eastern edge of the Highlands because many of the sinkholes are the collapsed roof of escarpment caves.

The Caves area is characterized by flat-lying outcropped limestone beach ridge facies of the Codrington Formation (Braiser and Mather 1975), with thin patches of soil and sinkhole clusters. Sinkholes are usually round or oval shaped in plan form and steep sided ranging in depth from 0.8 to 2.5 m. Length measurements range from 0.6 to 8.4 m and width dimensions from 0.5 to 5.6 m. Sinkholes with water present have a typical salinity range of 2.2–2.6 parts per

Fig. 8.7 Two Foot Bay is the site of caves and a view of the Atlantic side of Barbuda. (Photograph by Amy Potter, 2007)

thousand. There are several caves located in the interior of the Highlands: Garling Cave, Bryant Cave, Keyser Cave, and Dark Cave. Of these, only Dark Cave displays a sinkhole entrance that connects a series of large rooms, one of which contains freshwater pools with blind isopods (Nicholson 2007; Lawrence 2015).

There are a series of saline salt flashes in the southeastern portion of the island. Spring tides and storms intermittently flood these low-lying areas with sea water which evaporates to form thick layers of white and pink colored salt.

Barbuda is fringed by reefs primarily along its eastern and northern coasts. From Spanish Point in the southeast to Hog Point in the northeast, the windward coast is protected by a continuous reef structure absorbing the energy from the Atlantic Ocean waves. North of Hog Point lie Cobb and Goat Reefs which are important for local fishing due to ease of access from Codrington Lagoon through the Creek and the large area of calm, shallow water behind the reefs in this area. The leeward coast consists of patch reefs creating isolated navigational hazards. Along the south coast from Martello Tower to Spanish Well Point are groups of coral reef near the shore. South of Cocoa Point and Gravenor Bay are patch reefs and a large offshore assemblage called Palaster Reef.

The use of these landscape and landform formations including the island's coastal resources has figured notably in Barbuda's human occupation. The first evidence of a human settlement occurs around 4000 BC by the Siboney. A second more prominent occupation by the Saladoid peoples from what is today Venezuela dates to around 150 BC (Watters et al. 1992; Perdikaris et al. 2013). These early Barbudans left behind prominent shell middens along the island's southern coast. In their research exploring Barbuda's eastern coast caves, Perdikaris et al. (2013) documented the longitudinal importance of these karst features from the Archaic period to present as a place of shelter, ritual, and feasting. Their research uncovered evidence of Archaic lithics, as well as artifacts and faunal material that could have been produced by Obeah (local Caribbean folk religion) rituals.

Fig. 8.8 Though larger caves can be found on Barbuda, Darby Cave represents the deepest sinkhole on the island, measuring approximately 25 m deep and 120 m in diameter. Photograph by Amy Potter, 2010

8.3.3 Landscapes

The Marginal Plain includes all elevations less than 25 m which is approximately 80% of the island. Besides the sandy beaches, there are two soil series in the Marginal Plain: the Blackmere, a gray-brown loam, and the Codrington, a black clay (Charter 1937; Harris 1965). These two soil groups are poorly drained and alkaline, supporting mainly thorn-scrub evergreen woodland.

The Highlands are a limestone plateau dominating the eastern portion of Barbuda (Fig. 8.9). The north and eastern slopes are demarcated by an escarpment. The base of the escarpment initiates around the 8-m contour and steeply extends upward about 15 m. The escarpment is riddled with numerous caves of various dimensions facing the Atlantic Ocean. Intricate tropical rillenkarren and pankarren karst formations cover the outcropped limestone along the escarpment. Another dominant feature of the escarpment base are large boulders (>5 m) detached from the cliff face. The uplands are covered with dense woods, red soil, and rough rock outcrops. The Barbuda (terra rossa) soil series is reddish-brown, slightly alkaline, and well drained (Charter 1937; Harris 1965).

On the western side of the island lies Codrington Lagoon. It is shallow, hypersaline, and fringed by red mangrove trees, with a depth range of 1–4 m, and a salinity typically around 40 parts per thousand. The Lagoon's bottom is covered with sea grasses, algae, and a few scattered coral species. There is a narrow entrance to the lagoon from the ocean in the north locally known as "the Creek."

8.3.4 Heritage and Tourism

Barbudan heritage and identity are inextricably tied to a form of communal-land tenure that was formalized into law in 2008 (Antigua and Barbuda 2008) but developed over several hundred years of land use. For nearly two centuries

Fig. 8.9 The view looking out from Barbuda's Highlands region at the site of "Willy Bob," the ruins of the Codrington residence. The Highlands hosts the highest elevations on the island. Photograph by Amy Potter, 2007

(1680–1870), the British crown leased the island to the Codrington family of England who used the enslaved African population to grow food crops and raise livestock to provision sugar plantations (Betty's Hope) on neighboring Antigua. Unlike its Caribbean counterparts, the island was not suited to large-area agriculture due to its semiarid climate and karst topography (Harris 1965; Berleant-Schiller 1977). The Codrington family's and others' repeated attempts to establish plantations for cotton and other crops failed, in part due to the shallow soils, limited average annual rainfall, a long winter-dry season, frequent droughts, and a scarcity of surface water (Berleant-Schiller 1977). The enslaved and eventually free African-descended Barbudans overcame these perceived limitations developing an economy of subsistence agriculture, fishing, charcoal, and open-range livestock (Berleant-Schiller 1977; Lowenthal and Clarke 2007) centered on the communal-land tenure.

Over time, Barbuda's traditional economy of cattle herding and agriculture work began to wane as wage labor on the island expanded in the areas of sand mining, lobster export, and tourism (Potter and Sluyter 2010, 2012; Potter 2011, 2015a). In the 1960s, the first luxury hotel, Coco Point Lodge, provided construction jobs and then steady employment in the form of service jobs during the winter-tourist season.[1] Other luxury resorts on the south coast followed suit in the 1980s and 1990s, particularly the K-Club and The Beach House, providing additional seasonal employment.[2] Lighthouse Bay Resort, an endeavor by a Barbudan descendant, opened in 2008 across the lagoon, and smaller tourist accommodations have also sprung up on

[1]The resort employs approximately 80 Barbudans seasonally. This estimate is based on a conversation with an employee in December of 2013.

[2]Barbuda's tourism industry is wrought with failed initiatives; the K-Club has been vacant for years while The Beach House sits in disrepair. Perdikaris and Hejtmanek (2015) have described this as a boom/bust mentality.

Fig. 8.10 A major attraction in Barbuda is the Palmetto Point pink sand beach. The sand is a mix of coquina shell and coral fragments and is bottled and sold as a souvenir in Codrington Village. Photograph by Amy Potter, July 2010

North Beach. As economies rebound after the 2008 financial crisis, there has been a flurry of activity as the Barbuda Council entertains greater development of Barbuda's commons by the tourism sector, with the most recent approval of a $250 million investment by Robert DeNiro (Paradise Found) of the former K-Club Resort. These development initiatives are increasingly reliant on outside investment capital, which provide the funds for tourism infrastructure. The communal-land tenure has stunted Barbuda's tourism development relative to its Caribbean neighbors because of the limitations placed on foreign land ownership. Resorts that invest in Barbuda must lease land rather than buy outright.

In comparison with Antigua, Barbuda only receives a small percentage of the larger visitor total that travels to the twin-island country. Barbuda welcomed roughly 10,804 visitors in 2014, if you combine data including aircraft, ferry, and other watercraft. If you factor these numbers alongside the 249,316 total visitors counted in Tourist Stop Over statistics from the Caribbean Tourism Organization for the entire country (Antigua and Barbuda), roughly 4% of visitors in 2014 made their way to Barbuda.

Barbuda's stay-over tourism is small, attracting high-end guests.[3] In addition, day trips entice visitors via catamaran, ferry, and airplane from Antigua to experience the island's heritage sites (Amerindian petroglyphs in the caves at Two Foot Bay), the frigate bird sanctuary, and beaches. Two of the centerpieces of Barbudan tourism are the frigate bird colony as well as the island's pink sand beaches (Fig. 8.10). However, tourism initiatives often collide with the protection of these resources. A more recent tourism development at Cedar Tree Point cleared mangroves in the spring of 2014 in an area within the Codrington Lagoon National Park in close proximity to the Frigate Bird Colony (Butler 2013). Residents

[3]Room rates during off-peak season for Coco Point (rates are per room per night, based on double occupancy, and quoted in U.S. currency) range from $1050–1680 depending on the type of room. (See: http://cocopoint.com/guest-information/rates/, accessed July 2015).

Fig. 8.11 Sand mining has historically served as an important source of revenue for the Barbuda Council but has come with serious environmental and economic consequences as sand supplies dwindle through overuse. Photograph by Amy Potter, 2007

raised concern that the $1.9 million project did not properly consult the people of Barbuda and could ultimately disrupt the frigate bird habitat over the long term (Butler 2013).[4]

Tourism in Barbuda is not without its challenges. Until now, employment at resorts has largely been seasonal (December to April). The Barbuda Council was able to provide jobs to offset this seasonal employment in the past during the slow summer months because of revenues from sand mining; however, as this economic stream disappears, the Council has limited funds to keep this employment practice in place (Fig. 8.11).

Transportation infrastructure to the island is often fraught with problems. Several of the high-end resorts have their own landing strip outside of the main airport in Codrington, but sea-going vessels can experience mishaps, as in 2013, when an inoperable ferry impacted the number of tourists (2258 via ferry visitors in 2013 versus 8226 visitors via ferry in 2014). In addition, certain months of the year are particularly difficult for watercraft traveling the Caribbean Sea, where rough waters can impact the daily operations of the ferry to and from the island (particularly in the month of December). The daily flights from V.C. Bird International Airport (Antigua) to Codrington have not been without their headaches either (Observer Media 2014a; Webmaster in Local 2015). Flight cancellations and plane maintenance all impact island accessibility for not only tourists, but also Barbudans.

Barbuda is facing the usual pressures that come with tourism as a Small Island Developing State. The wages available to islanders are generally low, and there is fierce competition for jobs from non-Barbudans as new resorts necessitate skilled labor. Unique to Barbuda, however, are the negative impacts caused by tourism development in relation to the island's common property (loss of traditional resources and limited land access).

8.3.5 Hazards

Historically, Barbudans have shown tremendous resilience in coping with various hazards, be it drought, hurricanes (flooding), or earthquakes (Boger et al. 2014). For example, Hurricane Donna of 1960 and Luis of 1995 caused significant damage to homes within the village. Luis was one of the

[4]Despite expressed environmental concerns, the project proceeded.

major impetuses for the expansion of homes outside the traditional core of the village of Codrington. Hurricane Luis even breached a narrow strip of land across the lagoon that has since filled in.

Outside intense rainfall events wrought by tropical storms and hurricanes, Berleant–Schiller (1977, p. 87) documented cycles of drought and how these periods historically impacted past land use:

> Commercial cultivations disappeared in a dry spell, whereas subsistence cultivations contracted and stock-keeping expanded. Barbudans gave up attempting to control animals during drought because the crop production with which they might interfere is less important. During wet periods, when crop production is more successful, more human control is applied to animals.

A growing concern on the island not only centers on the cost of imported food, but also for those who continue to produce agriculture in line with the islanders' historic land uses suffering from increasing water salinity in their ground water sources (Boger et al. 2016). Boger et al. (2014) highlight how water quality is intimately tied to food production and food security. If Barbudans are going to produce food they have to contend with predicted decreases in annual rainfall and rising sea levels that will increase the salinity of the ground water for parts of the island.

Anthropogenic activities have also accelerated natural hazard risk. While Cambers (2009) notes these activities are certainly problematic in most Caribbean islands, Barbudans are particularly feeling the effects of sand mining, erosion, and overfishing.

Beginning in the 1970s, Red Jacket Mines, a company based out of Louisiana, began to mine in an area located near the southwest coast (Palmetto Point). Over time, the sand export revenue has served as an important source of employment for Barbudans and revenue (upwards of $5 million ECD) for the federal government and local governing body the Barbuda Council[5] (Potter and Sluyter 2012; Observer Media 2014b). While valuable to the Barbudan economy, mining has destroyed the ecological integrity of the natural sand dune system leaving the island vulnerable to hurricanes (de Albuquerque and McElroy 1995), typhoons, groundwater contamination (Campbell 2006), and climate change.

Beach erosion is also a concern. A 2010 *Observer* piece calls attention to threatened tourism developments following Hurricane Earl and Igor that severely eroded beaches in both Antigua and Barbuda (Observer News Observer News 2010a, b). Local environmentalists even question how development projects cause harm to the natural dune system.

Barbudans have traditionally relied on fishing in both the lagoon and surrounding ocean waters for subsistence and economy. Overfishing in the lagoon has put pressure on Barbudans to fish in waters further away from the island's shores, and there is currently a ban on fishing in the lagoon and surrounding waters.

[5]The Barbuda Council is the locally elected governing body established in 1976.

8.4 Conclusion

While the islands of Antigua and Barbuda comprise one country, we have sought to highlight their considerable differences in geology and historical development. Larger in land area and with greater elevations, Antigua is comprised of limestone, clay, and volcanic rock, while Barbuda is predominantly limestone, with caves on the Atlantic coast. Both islands exhibit little in the way of surface water and suffer from drought. Limited water supply is a constant concern, accelerated by erosion, and flooding during rainy times.

The long-term human impact is much more pronounced on the island of Antigua, with an agricultural legacy that incorporated enslaved African labor to grow sugarcane. Barbuda's karst topography posed certain difficulties to large-scale agriculture, with a colonial economy dominated by the use of enslaved labor to graze livestock. At present, Antigua is much more developed, particularly in terms of its tourism economy, which makes it especially susceptible to reverberations felt during global economic downturns. Barbuda's land tenure of common property has in some ways delayed this economic direction but limited revenues have shifted its leaders' attention toward an economy centered on tourism development of the commons. Barbuda's common property is a unique feature that has given islanders access to their lands when other Caribbean peoples are increasingly priced out of their island homes as they face competition and rising costs associated with tourism development. The question remains, however, whether Barbuda's unique land tenure will persist against the backdrop of this increasingly tourism-driven economy (Potter and Sluyter 2012; Potter 2015b).

Despite their differences, Antigua and Barbuda face many of the challenges common to Small Island Developing States (SIDS) especially in terms of climate change. These challenges include changing rainfall patterns, flooding, warming sea-surface temperatures, increased occurrence of natural disasters (such as hurricanes), and accelerated coastal erosion (Brown 2005). Environmental issues are then intensified by the combination of limited land area, high population density, and mass tourism, particularly for Antigua. These challenges aside, both islands showcase fantastic geomorphology, and a population that mostly understands effects of resource overuse. With enhanced programming and management driven by local leadership, Antigua and Barbuda could address these pressing issues associated with their apparent economic shortcomings.

Acknowledgements Mick Day acknowledges the invaluable assistance provided by the late Desmond Nicholson, and the financial support provided by the Department of Geography and the Center for Latin American and Caribbean Studies at the University of Wisconsin-Milwaukee. Amy Potter acknowledges the financial support of Armstrong State University and the collaborations with Andrew Sluyter, Sophia Perdikaris, Rebecca Boger, John Mussington and most importantly the people of Barbuda.

References

Antigua and Barbuda Meteorological Services (2015) Drought episodes experienced in Antigua (1928–2015). www.antiguamet.com

Antigua and Barbuda Meteorological Services (2016) Climate Variability and Change—Rainfall. www.antiguamet.com

Antigua and Barbuda (2008) The Barbuda Land Act 2007. Official Gaz 28(5): 1–18. Available at faolex.fao.org/docs/pdf/ant78070.pdf

Baldwin J (2000) Tourism development, wetland degradation and beach erosion in Antigua, West Indies. Tourism Geographies 2:193–218

Barker GH (1981) Antigua and Barbuda: an agricultural profile. Caribbean Agricultural and Research and Development Institute, St. Augustine

Barragne-Bigot P, Yearwood V (1991) Antigua and Barbuda. In: Falkland A (ed) Hydrology and water resources of small islands: a practical guide. UNESCO, Paris, pp 328–332

Berleant-Schiller R (1977) The social and economic role of cattle in Barbuda. Geogr Rev 67:299–309

Blume H (1974) The Caribbean islands. Longman, London

Brasier MD, Mather JD (1975) The stratigraphy of Barbuda, West Indies. Geol Mag 112(3): 271–282. Cambridge University Press

Brasier MD, Donahue J (1985) Barbuda—an emerging reef lagoon complex on the edge of the Lesser Antilles island arc. J Geol Soc Lond 142(6): 1101–1117

Boger R, Perdikaris S, Potter AE, Mussington J, Murphy R, Thomas L, Gore C, Finch D (2014) Water resouces and the historic wells of Barbuda: tradition, heritage and hope for a sustainable future. I Stud J

Boger R, Perdikaris S, Potter AE, Adams J (2016) Sustainable resilience in Barbuda: Learning from the past and developing strategies for the future. Int J Sustain 12(4):1–14

Brown I (2005) Impact of climate change on Caribbean agriculture. Jam Observer

Butler R (2013) Controversial $1.9 million hotel on the cards for Barbuda. The Antigua Observer. 10 Dec 2013. http://antiguaobserver.com/controversial-1-9-million-hotel-on-the-cards-for-barbuda/

Cambers G (2009) Caribbean beach changes and climatic change adaptation. Aquat Ecosyst Health Manage 12(2):168–176

Campbell P (2006) Barbuda sand mining proves difficult to manage. Antigua Sun, 24 October

Caribbean Conservation Association (1991) Antigua and Barbuda: country environmental profile. CCA, St. Michael, Barbados

Carr LM, Heyman WD (2009) Jamaica bound? Marine resources and management at a crossroads in Antigua and Barbuda. Geogr J 175 (1):17–38

Charter CF (1937) Soil survey of Antigua and Barbuda. Leeward Islands, Crown Agents, London

Christman RA (1972) Volcanic geology of southwestern Antigua, B.W. I. Geol Soc Am Mem 132:439–448

CIA (2015) World Factbook: Antigua and Barbuda. https://www.cia.gov/library/publications/the-world-factbook/geos/ac.html

Cooper B, Bowen V (2001) Integrating management of watersheds and coastal areas in small island developing states of the Caribbean. St. John's: Environment Division, Ministry of Tourism and Environment, Government of Antigua and Barbuda

Cooper V (2001) Inland Flood Hazard Mapping for Antigua and Barbuda: Summary Report. OAS/USAID. http://www.oas.org/pgdm/hazmap/flood/abfldsum.htm

Davis DO (1993) Archaic blade production on Antigua, West Indies. Am Antiq 58:688–697

Day MJ (1986) Karst in Antigua, West Indies. Proceedings of the 9th International Congress of Speleology (1): 218–220

Day MJ (1993) Human impacts on Caribbean and Central American karst. In: Karst Terrains: environmental changes and human impact. Catena Supplement 25. Catena Verlag: Cremlingen-Destedt, 109–125

Day MJ (2002) The role of valley systems in the evolution of tropical karstlands. In: Gabrovsek F, Zalozba ZRC (ed) Evolution of Karst: from Prekarst to Cessation, pp 235–241

Day MJ (2007) The karstlands of Antigua, their landuse and conservation. Geogr J 173(2):170–186

Day MJ (2009) Eastern Caribbean. In: Caves and Karst of the USA. In: AN Palmer, MV Palmer (ed) Nat Speleol Soc: 332, 346–347

Day MJ (2010a) Human interaction with Caribbean karst landscapes: past, present and future. Acta Carsologica 39(1):137–146

Day MJ (2010b) Challenges to sustainability of the Caribbean karst. Geologia Croatica 63(2):149–154

Day MJ (2011) Protection of karst landscapes in the developing world: lessons from Central America, the Caribbean and Southeast Asia. In: VanBeynen P (ed) Karst management. Springer, New York, pp 439–458

Day MJ, Chenoweth MS (2004) Encyclopedia of cave and karst science. Cockpit Country cone karst, Jamaica. pp 233–235. London: Fitzroy-Dearborn; Francis and Taylor Group

Day MJ, Chenoweth MS (2009a) Anthropogenic environmental change on the Caribbean karst: Chapter 5. In: Global change and Caribbean vulnerability: environment, economy and society at risk. University of West Indies Press, Jamaica

Day MJ, Chenoweth MS (2009b) Potential impacts of anthropogenic environmental change on the Caribbean karst. In: Barker B, Dodman B, McGregor D, Kingston D (eds) Global change and Caribbean vulnerability, UWI Press, pp 100–122

de Albuquerque K, McElroy JL (1995) Antigua and Barbuda: a legacy of environmental degradation, policy failure, and coastal decline. Washington DC: USAID Environmental and Natural Resources Policy and Training Project, Supplementary Paper 5

Donovan SK, Jackson TA (eds) (1994) Caribbean geology: an introduction. UWI Publishers' Association, Kingston

Donovan SK, Harper DAT, Jackson TA, Portrell RW (2013) A note on a coastal natural bridge in Antigua, West Indies. Cave Karst Sci 40 (3):105–108

Donovan SK, Jackson TA, Harper DAT, Portell RW, Renema W (2014) The upper Oligocene of Antigua: the volcanic to limestone transition in a limestone Caribbee. Geol Today 30(4):151–158

D.O.S. (1970) Directorate of overseas survey, British Government Ministry of Overseas. Series 257, Barbuda. Tolworth, Surrey, Great Britain

FAO (2003) National forestry action plan, Antigua and Barbuda. http://www.fao.org/docrep/X5681E/x5681e00.htm

FAO (2004) Irrigation in Antigua and Barbuda. http://www.fao.org/Regional/LAmerica/paises/h2o/antigua.htm

Farrell D, Nurse L, Moseley L (2007) Managing water resources in the face of climate change: a Caribbean perspective. University of the West Indies, Trinidad, p 11. http://64.28.139.231/conferences/salises/documents/Farrell%20D.pdf

Ford DC, Williams PW (2007) Karst hydrogeology and geomorphology. Wiley, Chichester, UK and Hoboken NJ

Gurnee RH (1961) The caves of Antigua and Barbuda. Int Speleol 1:14–18

Global Facility for Disaster Reduction and Recovery (2010) Disaster risk management in Latin America and the Caribbean region: Antigua and Barbados. GFDRR, Washington DC, pp 85–91

Gore-Francis J (2013) Antigua and Barbuda: small island developing states 2014 preparatory progress report. GO&B, St. Johns, p 47

Government of Antigua and Barbuda (2001) Biodiversity strategy and action plan for Antigua and Barbuda. http://biodiv.org/doc/world/ag/ag-nbsap-01-en.doc

Government of A&B (2004) Antigua and Barbuda: National Strategy to Strengthen Trade-Related Capacity. Government Printer, St Johns

Government of Antigua and Barbuda (2015a) About Antigua and Barbuda. http://www.ab.gov.ag

Government of Antigua and Barbuda (2015b) Tourist arrivals: year in review 2014. Research and Statistics Department, Ministry of Tourism, Economic Development, St. Johns

Harris DR (1965) Plants, animals, and man in the outer Leeward islands, West Indies, An ecological study of Antigua, Barbuda, and Anguilla. University of California Press

Hill ID (1966) Antigua. soil land-use survey of the British Caribbean. World Land Use Survey 19. Bude, Ebbingford, UK

Investment & Energy (2014) 16 p. http://members.antiguahotels.org/wp-content/uploads/2014/02/YearInReview2014-21.pdf

Jackson I (2001) Drought hazard assessment and mapping for Antigua and Barbuda. Organization of American States/USAID, St. John's, p 68

Kambesis PN, Machel HG (2013) Caves and karst of Barbados. In: Lace MJ, Mylroie JE (eds) Coastal Karst Landforms, Dordrecht: Springer, 227–244

Kueny JA, Day MJ (1998) An assessment of protected karst landscapes in the Caribbean. Caribb Geogr 9(2):87–100

Lawrence J (2015) Barbuda and Betty's Hope: The Codrington Connection. Sugarmill Tales, St. John's, Antigua

Lemel HW, Stanfield D, Deterville H (1988) A study of the land tenure structure of Antigua and Barbuda: public and private interests in land. U.W. Madison Land Tenure Center, Madison, WI

Lewis J (1984) Multi-hazard history of Antigua. Disasters 8:190–197

Lorah P (1996) An unsustainable path: tourism's vulnerability to environmental decline in Antigua. Caribb Geogr 6(1):28–39

Loveless AR (1960) The vegetation of Antigua, West Indies. J Ecol 48 (3):495–527

Lowenthal D, Clarke C (2007) Triumph of the commons: barbuda belongs to all Barbudans together. In: Besson J, Momsen J (eds), Caribbean land and development revisited. New York, Palgrave Macmillan, pp. 147–158

Lundberg J (2004) Coastal karst. In: Gunn J (ed) Encyclopedia of caves and karst science. Taylor and Francis, New York, pp 231–233

Martin-Kaye PHA (1956) The water resources of Antigua and Barbuda, B.W.I. La Penitence, British Guiana: British Guiana Lithographic Co

Martin-Kaye PHA (1959) Reports on the geology of the Leeward and British Virgin islands. Voice Publishing Co, Castries

Martin-Kaye PHA (1969) A summary of the geology of the Lesser Antilles. Overseas Geol Miner Res 10:172–206

McCann WR, Dewey JW, Murphy AJ, Harding ST (1982) A large normal-fault earthquake in the overriding wedge of the Lesser Antilles Subduction Zone: the earthquake of 8 October 1974. Bull Seismol Soc Am 72(6):2267–2283

Multer G, Weiss M, Nicholson D (1986) Antigua: Reefs, Rocks and Highroads of History. Leeward Islands Science Associates, St. John's

Museum of Antigua and Barbuda (2015) Historical sites. http://antiguahistory.net/Museum/Historical.htm

Mylroie JE, Carew JL (1997) Land use and carbonate island karst. In: Beck BF, Stephenson JB (eds) The engineering geology and hydrogeology of karst terranes. Balkema, Rotterdam, pp 3–12

Mylroie JE, Vacher HL (1999) A conceptual view of carbonate island karst. In: Palmer AN, Palmer MV, and Sasowsky ID (eds) Karst modelling. Karst Waters Institute, Charles Town, pp 48–57

National Ocean Service (2015) National Data Buoy Center; Station BARA9-9761115-Barbuda; Climate Summary Tables. http://www.ndbc.noaa.gov/station_history.php?station=bara9

Nicholson DV (2007) Heritage Treasures of Antigua and Barbuda. Museum of Antigua and Barbuda. Antigua, St. Johns, 114 p

OAS (2001) Drought risk Antigua. http://www.oas.org/pgdm/hazmap/drought/ant_map.pdf

Observer Media (2014a) Plans to expand Barbuda sand mining on hold. The Daily Observer. 29 Apr 2014. http://antiguaobserver.com/plans-to-expand-barbuda-sand-mining-on-hold/ Accessed 6 Feb 2015

Observer Media (2014b) Government unaware Barbuda without airlift. The daily observer 10 Dec 2014 http://antiguaobserver.com/government-unaware-barbuda-without-airlift/. Accessed 26 July 2015

Observer News (2010a) Barbudan hotelier denies claims of his property's demise. 28 Sept 2010. http://antiguaobserver.com/barbudan-hotelier-denies-claims-of-his-propertys-demise/. Accessed 23 Feb 2015

Observer News (2010b) Developer, environmentalist continue to clash over lighthouse bay. 29 Sept 2010. http://antiguaobserver.com/developer-environmentalist-continue-to-clash-over-lighthouse-bay/ Accessed 23 Feb 2015

Oliver JE (2005) The encyclopedia of world climatology. Springer, Netherlands. 854 p

Open File Report 97-470-K Map Showing Geology, Oil and Gas Fields, and Geologic Provinces of the Caribbean Region, which was compiled as a reference for the U.S. Geological Survey World Energy Project; URL: http://pubs.usgs.gov/of/1997/ofr-97-470/OF97-470K/graphic/data.html

Perdikaris S, Grouard S, Hambrecht G, Hicks M, Mebane-Cruz A, Persaud R (2013) The caves of Barbuda's Eastern Coast: long term occupation, ethnohistory and ritual. Caribb Connections 3(1) 1–9

Perdikaris S, Hejtmanek K (2015) The sea will rise, Barbuda will survive: environment and time consciousness. Proceedings from Rethinking Environmental Consciousness, Sundvall, Sweden, 6–8 Dec 2014 (in press)

Potter AE (2011) Transnational spaces and communal land tenure in a Caribbean place: "Barbuda is for Barbudans." PhD dissertation, Louisiana State University, Department of Geography and Anthropology

Potter AE (2015a) The commons as a tourist commodity: mapping memories and changing sense of place on the island of Barbuda. In Hanna S, Potter AE, Modlin EA, Carter P, Butler D (eds) Social memory and heritage tourism methodologies. Routledge, London

Potter AE (2015b) "Fighting for the rock at home and abroad": barbuda voice newspaper as a transnational space. Historical Geography 43:139–157

Potter AE, Sluyter A (2010) Renegotiating Barbuda's commons: recent changes in Barbudan open-range cattle herding. J Cult Geogr 27 (2):129–150

Potter AE, Sluyter A (2012) Barbuda: a Caribbean island in transition. Focus Geogr 55(4):140–145

Rudloff W (1981) World-climates, with tables of climatic data and practical suggestions. Wissenschaftliche Verlagsgesellschaft mbH Stuttgart

Sandberg SK, Barnes H (2004) Geophysical methods applied to saltwater intrusion in Antigua. In: Symposium on the application of geophysics to engineering and environmental problems. In: Allred B (ed) Wheat ridge, CO: environmental and engineering geophysical society, pp 936–943
Smith DI (1975) The problem of limestone dry valleys—implications of recent work in limestone hydrology. In: Peel RF, Chisholm M, Haggett P (eds) Processes in physical and human geography. Heinemann, London, pp 130–147
Tomblin J (2005) The geology of Antigua, Barbuda and Redonda. The Author, St. John's
UN (2002) Antigua and Barbuda: country profile. http://www.un.org/esa/agenda21/natlinfo/wssd/antigua.pdf
UN (2012) Stock Taking Report, Rio+20, Antigua and Barbuda. New York: UN, 50 p. http://www.uncsd2012.org/content/documents/518Antigua%20Report.pdf
UNCCD (2005) Draft National Action Plan for Antigua and Barbuda. New York: UNCCD, 61 p. http://www.unccd.int/actionprogrammes/lac/national/2005/antigua_and_barbuda-eng.pdf
UNEP (2010) National Environmental Summary: Antigua and Barbuda. UNEP, New York, 30 p. http://www.pnuma.org/publicaciones/FINAL%20NES%20Antigua%20Barbuda%20Nov%202010-%20edited.pdf
USGS (2004) Shuttle radar topography mission, 1 Arc Second Global scene n17_w062_1arc_v3.tif
Wadge G (1994) The Lesser Antilles. In: Donovan SK, Jackson TA (eds) Caribbean geology: an introduction. UWI Publishers' Association, Kingston, pp 167–177
Watters DR, Donahue J, Stuckenrath R (1992) Paleoshorelines and the Prehistory of Barbuda, West Indies. In: Paleoshorelines and Prehistory: An Investigation of Method. Lucille Lewis Johnson. CRC Press, London
Watts D (1987) The West Indies: patterns of development, culture and environmental change since 1492. Cambridge University Press, Cambridge
Watts D (1988) Development and renewable resource depletion in the Caribbean. J Biogeogr 15(1):119–126
Webmaster in Local (2015) Barbuda getting airways soon to ease travel issues. Caribbean Times Antigua and Barbuda. April 2 2015. http://www.caribbeantimes.ag/?p=1298. Accessed 25 July 2015
Weiss NP (1994) Oligocene limestones of Antigua, West Indies: Neptune succeeds Vulcan. Carib J Sci 30:1–29
Williams C (1994) Towards the sustainable management of freshwater resources, with particular reference to Antigua-Barbuda. Water Manage 1:9–47

Montserrat

9

Elizabeth Nelson

Abstract

Montserrat, a smaller Caribbean island of the Lesser Antilles island arc, has a considerable contribution to the regional geomorphology: the active Soufrière Hills Volcano. The volcano consists of a cluster of andesite domes with flanks of sediment deposited by pyroclastic lava flows. The Soufrière Hills Volcano eruptions can be traced back thousands of years, but more recently the volcano became active again in 1995 with the latest recorded eruption in 2013. The volcanic activity involves both eruptive and dome collapse processes creating ashfalls, debris avalanches, and pyroclastic deposits. Coastal advancement occurs through mudflows, floods, and wind. The Soufrière Hills volcanic eruptions are among the largest volcanic events to occur in the Lesser Antilles and have created substantial exclusion zones on the island for safety. The seismic activity on and near the island and the Soufrière Hills Volcano are monitored closely by the Montserrat Volcano Observatory.

Keywords

Volcano • Andesite domes • Pyroclastic flow • Soufrière hills • Exclusion zone

9.1 Introduction

The sleeping Soufrière Hills volcano awakened in 1995 with a catastrophic explosion. The island at the time was still recovering from 1989 Hurricane Hugo and now faced the decades-long process of living life with an active volcano. Since the initial eruption, the former capital city of Plymouth has been completely destroyed and buried by volcanic material. Two-thirds of the island's 1995 population—12,000—left the island, though some continue to trickle back, and today around 4500 live there (Donovan et al. 2013; UNICEF 2013). These persevering islanders live on a small portion of the land area in the north, while the southern part of the island consists of a large, uninhabitable exclusion zone, though much of the area is still accessible during the daytime under governmental permission.

Montserrat is an Overseas Territory of the United Kingdom of Great Britain and Northern Ireland (UK) in the Lesser Antilles (Fig. 9.1). Before the 1995 eruption, it had a thriving economy and was nearing independence from the UK. Now, Montserrat is dependent on aid from the UK and other foreign aid agencies (Skelton 2000). The capital city-now ghost town and exclusion zones created by volcanic devastation are in the place of a once-bustling Caribbean locale—now quiet, eerie, and despondent, mere remnants of a culture once achieved—taken over by a series of volcanic events, and finally left to survive and eke out an existence on the remaining inhabitable portion of the island. The proud island residents who have stayed behind include native Montserratians and migrant populations. Together, they make up a most unique culture in a landscape of at once lunar-like desolation and lush rainforest, spawning the island's moniker, the "Emerald Isle of the Caribbean" (Fig. 9.2).

There is a complex intertwining of physical and human geographies, of geomorphology and volcanology in Montserrat. Due to the relatively recent developments,

E. Nelson (✉)
Department of Geography, University of South Carolina, Columbia, USA
e-mail: bethtnelson@gmail.com

C.D. Allen (ed.), *Landscapes and Landforms of the Lesser Antilles*,
World Geomorphological Landscapes, DOI 10.1007/978-3-319-55787-8_9

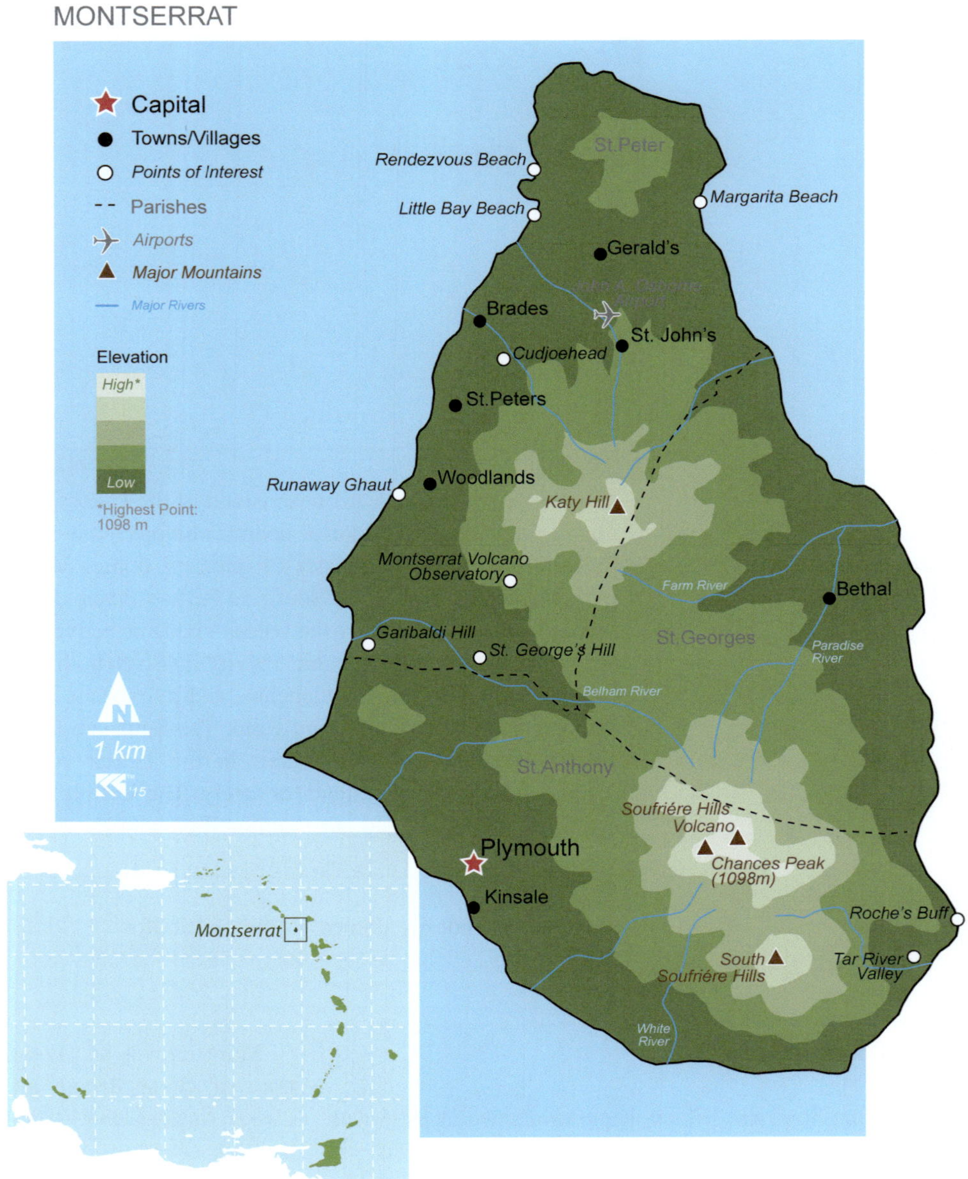

Fig. 9.1 General physiographic map of Montserrat. While Plymouth is still officially the Island's capital, it is no longer habitable—much like most of the island's southern half. Other cities in the southern region, such as Kindsale, are also abandoned. For more information on the extent of the established volcanic extrusion zones, see Fig. 9.3 and the corresponding Table 9.1. Cartography by K.M. Groom

literature and current assessments of various aspects of the island are no longer relevant. This chapter thus relies heavily on the post-eruption research of a small group of scientists, most heavily centered at the Montserrat Volcano Observatory (MVO). Additionally, information is included from scientists conducting ground-breaking research of an active volcano and simultaneously work with local, colonial, and international audiences. Before the 1995 eruption, geologic interest in Montserrat was limited to a small amount of literature, most notably by British scientists, geologist MacGregor (1938), and physicist Powell (1938), who together developed a foundational understanding of the geology of Montserrat. Post-eruption, many groups collected and published information on the island, including the University of the West Indies (UWI) (see *Volcanic Hazard Atlas of the Lesser Antilles*, UWI 2005), the Geological Society of London (2002), and the MVO.

This chapter begins with a geologic overview of the island, followed by an outline of its major landforms. With the island's physical features established, it then delves into the island's cultural landscape, drawing from an anthropogeomorphological approach in addition to traditional geographic landscape inquiries and how the general landscape has changed over time. This leads into an overview of the heritage and tourism aspects of the island before finally ending with a discussion regarding the hazards of the island, including past, present, and future outlooks for these hazards and their potential mitigation for inhabitants of Montserrat.

Fig. 9.2 A ground view of the Soufrière Hills Volcano on the Emerald Isle, with lush green land cover visible, and the volcanic peak is covered by clouds, as it is most days. Notice the extensive lava flows to both the *right* and *left* of peak. Photograph by J. Davenport

9.2 Setting

Montserrat (16°45′N, 62°10′W) is situated toward the northern end of the Lesser Antilles island arc (Le Friant et al. 2004). The humid tropical marine climate is moderated by the prevailing northeasterly trade winds. Because of the topography, annual and seasonal precipitation varies dramatically. The rainy season is typically between July and December (Weaver 1995, p. 597).

As a result of the subduction of the North American plate beneath the Caribbean plate, this island experiences the combined outcomes of tectonic plate activity of both strike-slip and subduction in addition to volcanic processes (Le Friant et al. 2004). The Lesser Antilles island arc includes the outer and inner arcs, both with characteristically different landforms and land-creation processes (Le Friant et al. 2004). The island lies at the western edge of the shallow platform between these arcs and has a shallow (less than 140 m) submarine shelf, with steep margins that fall off to the Grenada Basin in the west and to the Atlantic in the east (Fig. 9.3; see also Deplus et al. 2001). The sea floor around Montserrat has numerous fault scarps and coastal coves. On the east, south, and west coastal areas, hummocky topography has been identified as debris avalanche deposits due to major volcanic landslides (Deplus et al. 2001).

Montserrat is one of the smaller islands in the Antilles. The subaerial portion of Montserrat is 16.5 km long from north to south and 10 km wide from east to west (Le Friant et al. 2004; Hincks et al. 2005). Composed nearly exclusively of volcanic rock, there are four volcanic massifs that make up the 100 km^2 of total land area. The youngest massif, collectively called 'the Soufrières', consists of the Soufrière Hills and South Soufrière Hills (the active volcano), and these are sometimes referenced as one massif (Le Friant et al. 2004). However, for the purpose of clarity in this chapter, they are considered separate massifs. The other massifs, in order of youth, are the Centre Hills and Silver Hills. The erosional maturity of the massifs decreases progressively from north to south, indicating a southward migration of volcanism over time, and following the movement of the underlying tectonic plates (Harford et al. 2002). Migration in accordance with the tectonic plates occurs at a time averaged rate of ~6 km/Ma parallel to the trench and ~2 km/Ma perpendicular to and away from the

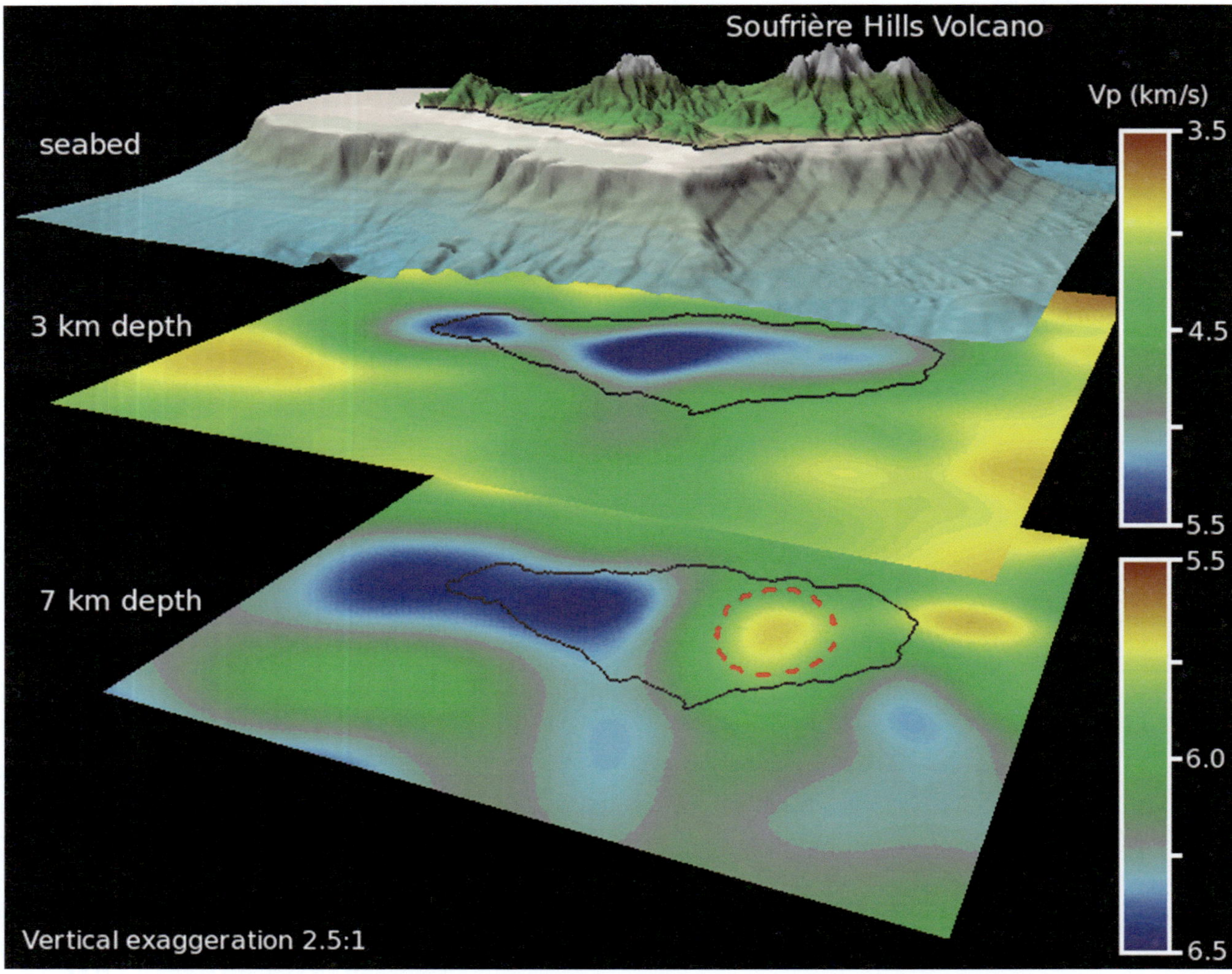

Fig. 9.3 Three-dimensional seismic velocity model of Montserrat at seabed, 3 and 7 km depths. (Animated model also available online from https://geoazur.oca.eu/spip.php?article1544). Model created by M. Paulatto in 2015, based on data from Paulatto et al. (2012)

trench, and the long axis of the island is aligned approximately north–south (Hincks et al. 2005).

Montserrat is mountainous with small areas of coastal lowlands and a densely vegetated interior of tropical rainforests. There are a limited number of beaches and no inland bodies of water. Rock exposures exist in road cuttings, coastal areas, quarries, and the steep inland cliffs (Hincks et al. 2005). After 20 years of volcanic activity, the fallout and subsequent landscape changes have created vast expanses on the flanks of the Soufrière complex, buried by pyroclastic flows, lahars, and ashfalls, particularly in the deep valleys. Local researchers call these "ghauts" (sometimes referred to as "guts" on non-French influenced islands) as clefts in the hills that commonly function as drainage paths for water and low-lying areas (Hincks et al. 2005). The infill of the major ghauts by pyroclastic deposits and deposition at the mouths of several ghauts has extended the coastline, adding new areas of land (Hincks et al. 2005).

Nearly all of this additional coastline lies in the exclusion zone, which covers two-thirds of the island and has access organized by level of danger, concluding with the absolute exclusion zone that is accessible only by helicopter and only with the express permission of the local police commissioner (Fig. 9.4). The hazard levels are jointly determined by the MVO and the Disaster Management Coordination Agency (DMCA) of Montserrat (Table 9.1).

Before the volcanic activity began, the Soufrières consisted of prehistoric lava domes, and the upper flanks were divided by ghauts that fed rivers on the outer flanks. The sustained volcanic activity, however, has significant impacts on the landscape of the Soufrière complex and, in many places, the preexisting vegetation has been completely

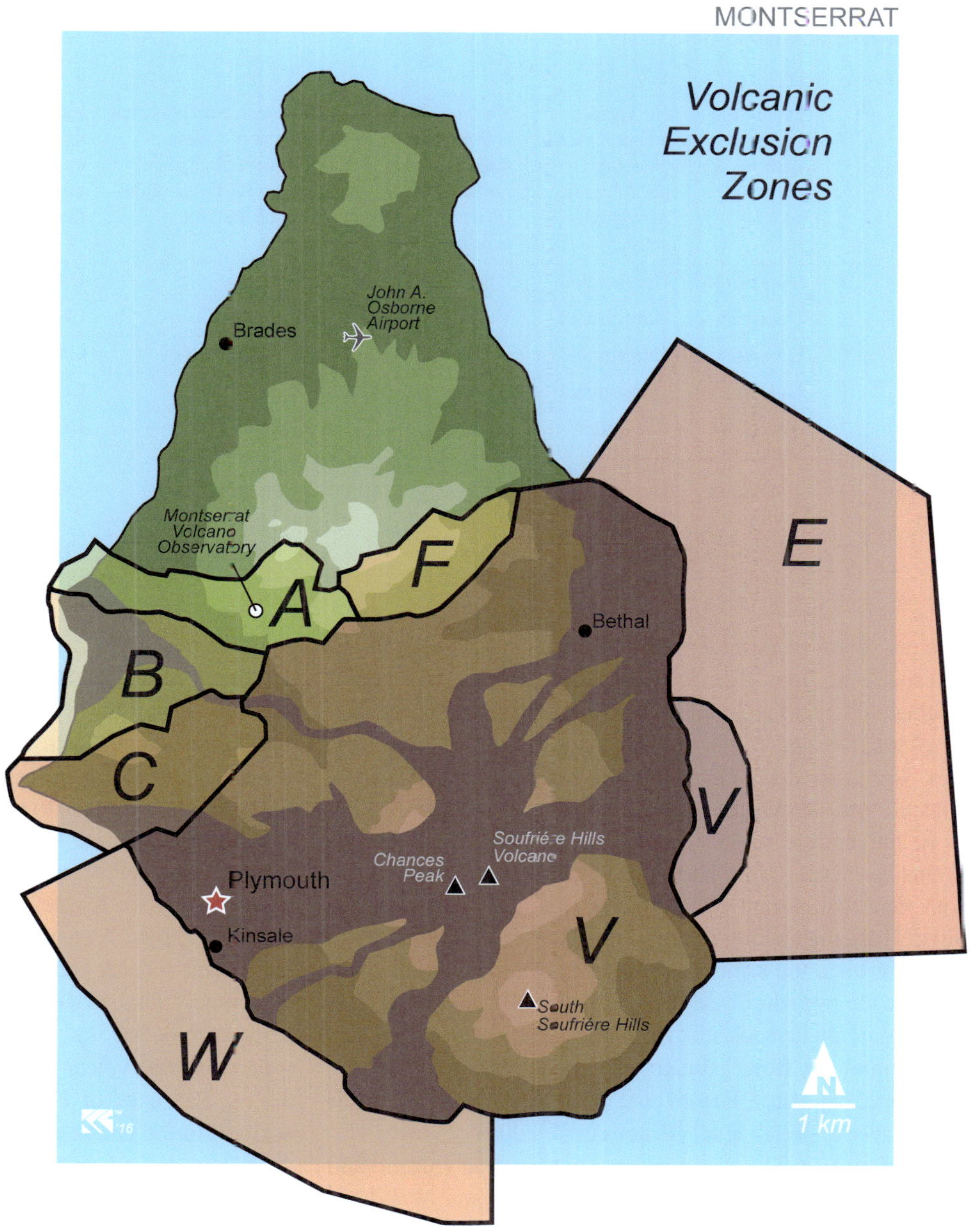

Fig. 9.4 Volcanic exclusion zones on Montserrat. Zones are divided into terrestrial and oceanic designations along with varying degrees of risk/hazard level. Some of the oceanic zones protrude several kilometers from the coast. Any sea vessel wishing to circumnavigate the island must travel around these zones. For further explanation of each zone, see Table 9.1. Cartography by K. M. Groom

destroyed by pyroclastic flows, ashfall, and acidic gas (Hincks et al. 2005).

9.3 Landforms

The island of Montserrat has undergone incredible landform changes during its volcanic activity. The volcano changed both terrestrial and marine ecosystems and, unlike many current marine habitats of the world where detrimental effects are mostly anthropogenic, it has been the pyroclastic flows of the Soufrière Hills Volcano that has most greatly affected Montserrat's marine environment (Myers 2013). The island has varying underwater topography containing, for the most part, untouched and unexplored reefs (Myers 2013). Above ground, the main subaerial landforms on Montserrat today are Silver Hills, Centre Hills, the Soufrière Hills Volcano, South Soufrière Hills, Roche's Bluff, Garibaldi Hill, and St. George's Hill. Prior to the initial eruption, the highest point on the island was Chances Peak (included in the Soufrière Hills volcanic complex) at 914 m above the sea level. Since April 22, 2003, however, the Chances Peak dome has measured 1098 m (Hincks et al. 2005).

Across the island, the Silver Hills volcanic massif was active between 2.6 and 1.2 million years ago (MacGregor 1938). It is composed of porphyritic hornblende andesitic

Table 9.1 Government-monitored exclusion zones. Time of access and personnel restrictions depend on the time of day and hazard levels defined by the Montserrat Volcano Observatory (MVO) and Disaster Management Coordination Agency (DMCA)

Hazard level		0	1	2	3	4	5
		More than one year with no measured activity	No surface activity for an extended period. Low measured activity	No surface activity that threatens the north or west. Low measured activity	Mild surface activity that threatens the west. Significant change in measured activity. High levels of activity	Lava extrusion that threatens the north or west. Large unstable dome to the north or west	Threat of large pyroclastic flows to the north or northwest. Threat of lateral blast or sector collapse
Inland zones	A	Unrestricted	Unrestricted	Unrestricted	Unrestricted	Unrestricted	Controlled access
	B	Unrestricted	Unrestricted	Unrestricted	Unrestricted	Controlled access	Controlled access
	C	Unrestricted	Unrestricted	Daytime Access	Controlled access	Controlled access	Essential workers
	F	Unrestricted	Unrestricted	Daytime Access	Daytime Access	Controlled access	Controlled access
	V	Access to some areas	Controlled access	Essential workers	Essential workers	Essential workers	Essential workers
Maritime zones	W	Unrestricted	Daytime transit	Daytime transit	Daytime transit	Essential workers	Essential workers
	E	Unrestricted	Daytime transit	Essential workers	Essential workers	Essential workers	Essential workers

lava, as observed in the eroded lava domes and sequence of volcaniclastic beds (Hincks et al. 2005). The highest point reaches 403 m above sea level. The submarine shelf is well developed around Silver Hills, and it is believed that this area is the eroded remains of a previously much larger volcanic massif (MacGregor 1938).

In the middle of the island, the Centre Hills area has extensive coastal cliffs and includes a mountainous interior with dense rainforest cover (Hincks et al. 2005). Like the Silver Hills, the mountains are made of porphyritic hornblende andesitic lava, the remnants of lava domes that collapsed to produce the pyroclastic deposits on the flanks. The material is predominantly pyroclastic flow deposit, but also includes pumice-and-ash flow, pumice fall, lahars, debris avalanche, and fluvial deposits (Harford et al. 2002). The highest point is 741 m above the sea level (Hincks et al. 2005).

The Soufrière Hills Volcano is a Peléen-type volcano and the only active volcanic massif on the island. Pre-1995 eruption, the massif contained five steep-sided andesitic lava domes, many of which are now covered by new material that has buried landforms during the volcanic activity. At the point of publishing, the volcano is still active, yet displays aprons of volcaniclastic deposits that flank the four remaining lava domes. The deposits are, according to Hincks et al. (2005:150), "…predominantly pyroclastic flow deposits related to dome collapse and lahar deposits, with subsidiary debris avalanche deposits, pumice flow deposits, fluvial deposits, stratified fallout tephras and paleosols." Existence of two-pyroxene andesite and a hornblende-hypersthene andesite indicates a transition from a period of basaltic volcanism (Harford et al. 2002). Additionally, three deposits of debris avalanches have been identified on the lower submarine flanks offshore from the Tar River valley, White River valley, and the once-capital city of Plymouth (Le Friant et al. 2004).

The smaller South Soufrière Hills are formed of lava flows and volcaniclastic beds of breccias, as well as some scoria-fall deposits, yet the lavas and pyroclastic flow deposits are of basaltic to basaltic–andesite composition. The area displays scars of an earlier dome collapse, and today its highest point is 756 m above sea level (Hincks et al. 2005).

A non-massif, Roche's Bluff, on the southeast side of the Soufrière Hills Volcano, is an important landform on Montserrat because the rocks of the bluff are "…cross cut by numerous faults, and it is interpreted as being an uplifted beach" (Hincks et al. 2005, p. 150). The bluff consists of steeply dipping submarine volcaniclastic rocks associated with thin limestones. The ages of both coral and foraminifera fossils, in addition to the coarse grained nature of the deposits, indicate deposition around one million years ago in an offshore shelf environment with an andesitic center. This is consistent with a link to the Centre Hills volcanic massif (Hincks et al. 2005).

Finally, the non-massif Garibaldi Hill and St. George's Hill are significant landforms, in that they have a distinctive geology on the flanks of the Soufrière Hills. Composed of andesitic lavas, lahars, pumice fall deposits, and laharic breccias in addition to pyroclastic flow deposits, this area has also been cut by numerous faults. This is consistent with the Centre Hills volcanic massif or an early product of the Soufrières (Hincks et al. 2005).

9.3.1 Soufrière Hills Volcanic Activity

The events of the Soufrière Hills Volcano during the timeframe of 1995–2013 were dramatic, near-constant, and often violent in experience. The Montserrat Volcano Observatory remains a world-renowned volcanic observatory, and the Soufrière Hills Volcano is one of the most researched, monitored, and actively managed volcanoes in the world. While a summary of events can be found in Table 9.2, the following subsections outline in greater detail the type of volcanic activity occurring on Montserrat, as well as the effects and consequences of those events, over the past ~20 years.

9.3.1.1 Modern Volcanic Activity: The Soufrière Hills Volcano 1995–1999

On July 18, 1995, the Soufrière Hills Volcano sprang to life with a burst of phreatic activity (Young et al. 1998). Small explosions were followed by ash and steam, and on August 21, 1995, larger phreatic explosions generated cold base surges and engulfed the capital city of Plymouth, prompting the first evacuation of the capital (Hincks et al. 2005).

Table 9.2 Timeline of activity on Montserrat. Data collected from MVO Web site (http://www.mvo.ms/), October 2016

Volcanic activity	Time period
Seismic activity	January 1992–July 1995
Phreatic explosions	July 1995–November 1995
Dome growth and collapse	November 1995–August 1997
Vulcanian explosions	August 1997
Dome growth	August 1997–September 1997
Vulcanian explosions	September 1997–October 1997
Dome growth	October 1997–December 1997
Major collapse	December 1997
Dome growth	January 1998–March 1998
Dome degradation and residual activity	March 1998–June 1999
Major Collapse	July 1999
Dome degradation and residual activity	August 1999–February 2000
Dome growth	November 1999–February 2000
Collapse	March 2000
Dome growth	March 2000–July 2001
Major collapse	July 2001
Dome growth	August 2001–July 2003
Major collapse	July 2003
Cessation of dome growth	July 2003–January 2009
Vulcanian explosions	November 2009–February 2010
Dome collapse	March 2013–April 2013

Phreatic explosions with steam and ash columns up to 3 km high lasted several months, and the largest of these ash clouds engulfed Plymouth in total darkness for up to 30 min (Young et al. 1998). A full evacuation of southern Montserrat was ordered in December 1995. As slopes destabilized, the first pyroclastic flows began in March 1996, increasing in magnitude and culminating in a massive dome collapse on September 17, 1996 (Young et al. 1998). This collapse triggered the first magmatic explosion of the eruption. It also removed an estimated 40% of the dome, depositing this material southward and extending the coastline where Plymouth once stood (Young et al. 1998; Hincks et al. 2005).

In October 1996, a second phase of dome growth began, triggering multiple smaller dome collapses. In December 1996, the first pumiceous pyroclastic flow accompanied by cyclic inflation and deflation of the dome caused periodic partial dome collapses into January 1997, sending pyroclastic flows down the Tar River valley (Young et al. 1998, p. 3391). On June 25, 1997, another massive dome collapse occurred to the east, sending an estimated 5 million cubic meters of material in a sustained pyroclastic flow down Mosquito Ghaut, killing 19 people in its wake (Hincks et al. 2005; Young et al. 1998). In August 1997, pyroclastic flows again travelled to the sea, through Fort Ghaut, and further destroyed and buried Plymouth, triggering Vulcanian explosions for more than a week. On September 21, 1997, another major dome collapsed with an estimated volume of 14 million cubic meters and a second round of 75 Vulcanian explosions occurred over the course of the following month. All of these events led up to December 26, 1997, when a failure of the upper flanks of Galway's Wall created a debris avalanche that triggered a volcanic blast that added 10 km^2 to the southwest of the island and covered the former airport and St. Patrick's village (Fig. 9.5; see also Hincks et al. 2005).

From March 1998 to November 1999, there was no measured dome growth, and the dome was continually degraded by small collapses, rockfalls, and pyroclastic flows. On July 03, 1998, another major dome collapse occurred, taking off about 25% of the dome and sending 20–30 million cubic meters of material down the eastern Tar River valley in pyroclastic flows, carrying ballistic blocks up to one kilometer (Hincks et al. 2005). After this event, heavy rains in September, November, and December created voluminous mudflows down all flanks of the volcano. Relentlessly, activity increased again in the spring of 1999 with a series of explosive ash-venting occurrences and another dome collapse on July 20, 1999 (Fig. 9.6).

9.3.1.2 Modern Volcanic Activity: The Soufrière Hills Volcano 1999–2005

Two hurricanes swept through the island in 1999: Hurricane Floyd in September and Hurricane Lenny in November,

Fig. 9.5 Examples of life ever changed by volcanic destruction. **a** An Exclusion Zone sign at the entrance of Zone C. All Exclusion Zones are readily identifiable by signs that indicate which zone you are in and gates that close when access is not permitted. Photograph by E. Nelson. **b** Buried village of St. Patrick's and site of former airport, Brambles Airport; the aviation tower is barely visible at the *right* of the photograph. Photograph by J. Davenport

triggering mudflows that were followed by small explosions from the volcano (Hincks et al. 2005). These were the first hurricanes to hit Montserrat since the devastating Hurricane Hugo in 1989.

On November 27, 1999, a new lava dome was measured in the base of the old dome crater and was the first sign of extrusive growth in 20 months. After months of growth, large pyroclastic flows from the new dome again poured down the Tar River valley in February 2000. Heavy rains in the spring of 2000, initiated another major collapse on 20 March that year, with a volume of 20 million cubic meters (Hincks et al. 2005). The growth of the dome continued moderately into 2001 with minor collapses, some pyroclastic flows, and rockfalls, and on May 18, 2001, a new lobe extrusion occurred to the south. In the following months, near-continuous rockfall activity occurred until, in July, torrential rainfall and mudflows down the Belham River valley were followed by the second collapse of the new dome (July 29, 2001). This near-continuous nine-hour event shed 45 million cubic meters of earth and sent pyroclastic flows down the Tar River valley, out to sea and into the air where particulate matter from Montserrat was recorded as far away as Puerto Rico and the Virgin Islands (Hincks et al. 2005). In November 2001, the direction of growth changed from east to west, but activity remained low for the rest of the year.

The following year 2002 was characterized by the impressive production of large spines. Lava spines are characterized as abrupt expulsions of volcanic material, most often shaped as a column (spine) or as a slab. Spines are forceful in creation yet are weak structures as landforms and erode very quickly. On February 25, 2002, for example, one 90-m spine appeared overnight at an altitude of 1080 m (Hincks et al. 2005, p. 156). In the spring of 2002, rockfall and seismic activity escalated but then declined and remained at low levels until July 21, 2002, when a new lobe began to form on the north face. By September, small collapse events generated pyroclastic flows that reached the east coast, and in October 2002, growth switched again to the northeast and prompted the evacuation of the margins of the Belham Valley on October 06, 2002.

By 2003, now in its eighth year of activity, the Soufrière Hills Volcano continued its rumblings with dome growth in the directions of north and northeast. In March of 2003, the dome reached its greatest altitude in recorded history, 1098 m, and by July had increased in volume to about 22 million cubic meters (Hincks et al. 2005). On July 12 and 13, 2003, the dome collapsed again, this time with an estimated collapse volume of 120 million cubic meters. This, coupled with heavy rainfall on 12 July, triggered mudflows in the Belham Valley. Low-volume pyroclastic flows now travelled down the Tar River, increasing in frequency and intensity (Hincks et al. 2005, p. 160). The larger flows reached the sea and produced phreatic explosions, alongside an estimated 1.2 million tons of ash falling over the populated areas of the island. The pyroclastic flows slowed on July 13, 2003, but three volcanic explosions occurred after the collapse and produced an ash cloud that rose 15 km into the sky with gravel and large rock fragments falling over the populated areas of the island —"raining gravel" again on the island (Hincks et al. 2005).

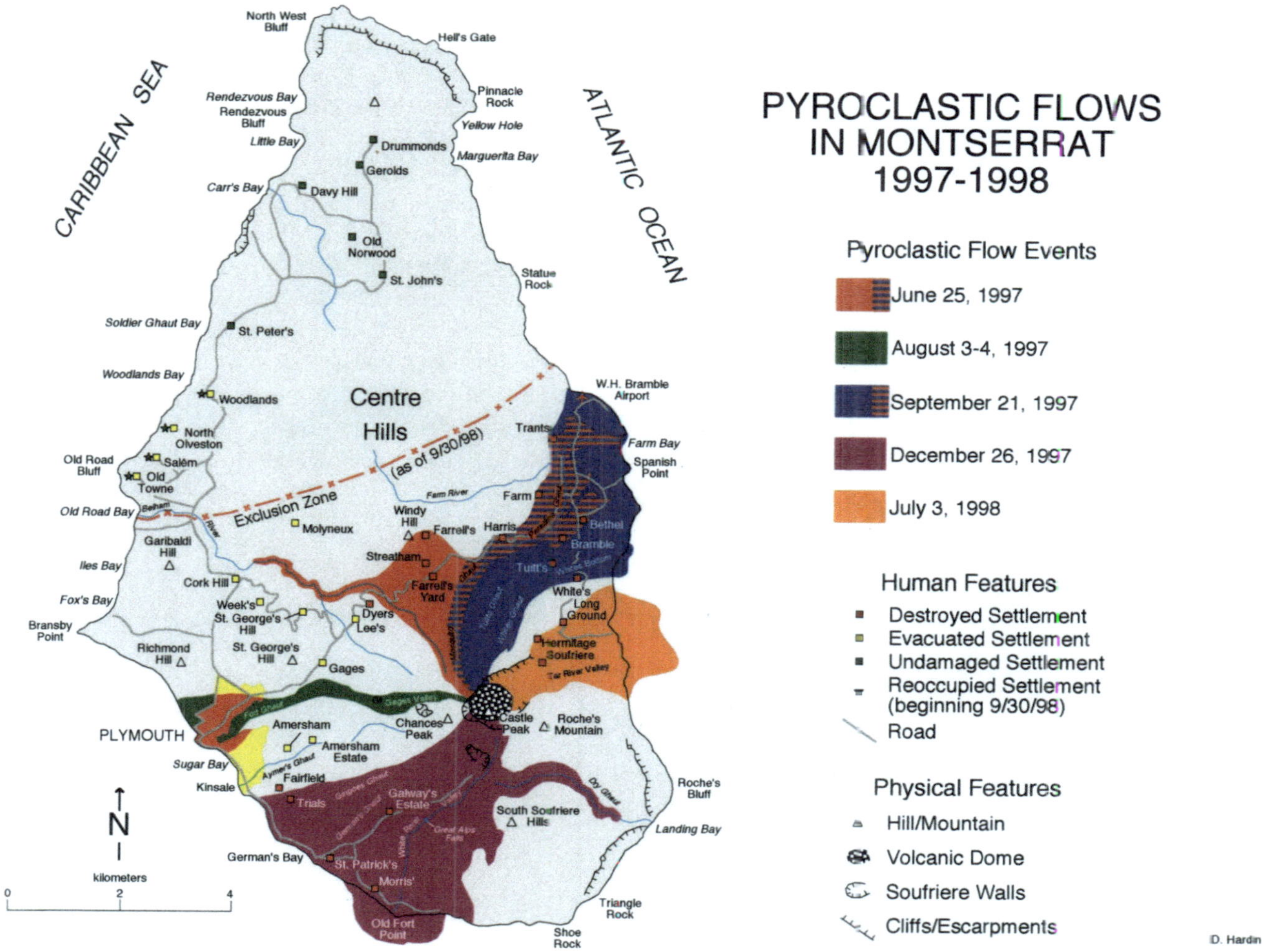

Fig. 9.6 Map of areas of pyroclastic flows on the island of Montserrat during the years 1997 and 1998. From Hardin, retrieved from: http://www.hoeckmann.de/karten/amerika-nord/montserrat/index-en.htm

For the rest of 2003 and 2004, only tremors and minor ash venting occurred until March 03, 2004, when another explosion and dome collapse occurred. This time, it sent pyroclastic flows down the Tar River and into the sea, and ash clouds reached nearly 7 km into the air (Hincks et al. 2005, p. 160). The spring of 2004 brought moderate-to-low tremors, but on May 21, 2004, intense rain triggered massive mudflows in the Belham Valley with waves reaching up to 2 m high (Hincks et al. 2005). In August of 2005, an eruption created a 1.8-km plume into the sky (MVO 2015).

9.3.1.3 Modern Volcanic Activity: The Soufrière Hills Volcano 2005–2013

On December 24, 2006, pyroclastic flows became visible again, and on January 08, 2007, an evacuation order was issued for the Lower Belham Valley (MVO 2015). On July 28, 2008, pyroclastic flows again reached Plymouth, caused by a dome collapse on the lava dome's eastern side, creating an ash cloud 12 km tall (Chardot et al. 2010). During 2009 and 2010, extrusion of the lava dome triggered the most recent large event at the Soufrière Hills Volcano, and included the several rapidly developed spines and lobes that extruded from the volcano. On February 05, 2010, pyroclastic flows rapidly filled surrounding valleys on multiple sides of the volcano, and on February 11, 2010, a partial dome collapse again sent large ash clouds into the sky (MVO 2015; Stinton et al. 2014).

Volcanic activity began again in November 2009 and continued through February 2010 with ash venting and an explosion that created pyroclastic flows in several directions down the sides of the volcano (MVO 2015). In 2013, the MVO reported a pyroclastic flow that travelled down the Tar River Valley in March. The deposits indicated a failure of a large slab that had been moving away from the dome (MVO

2015). Heavy rainfall occurred after this event triggering lahars in the surrounding valleys (MVO 2015). Figure 9.7 displays some visual evidence of these events.

9.3.2 Rivers and Beaches

The rivers of Montserrat have been affected by the volcanic activity, and some have been filled by pyroclastic debris, many are conduits for lahars that also affect the sedimentation of the river basin. Montserrat has 11 rivers (including tributaries) that include Farm River, Lee River, Paradise River (altered by pyroclastic flows), Tar River valley (destroyed by pyroclastic flows), Hot River, White River (destroyed by pyroclastic flows), Belham River, Dyer's River, Nantes River, Collins River, and Lawyers River.

The beaches and bays on Montserrat vary in size, location, and material makeup. The island has both black sand and white sand beaches. The bays of Montserrat include Bunkum Bay, Carr's Bay, Foxes Bay, Isle Bay Beach (Old Road Bay), Lime Kiln Bay, Little Bay, and Marguerita Bay (Montserrat Tourist Board 2012). Many of these beaches are inaccessible post-eruption, as they are located in the exclusion zones, for example, Foxes Bay has only recently been declared safe, located in Zone B (see zone reference chart). However, the volcano also caused coastal advancement that extended the Isle Bay Beach (Old Road Bay). Little Bay is the only bay deep enough to accommodate for small watercraft, and there are no bays on the island capable of accommodating large cruise ships. The only white sand beach on the island is Rendezvous Beach; this beach is located in an uninhabited area of the island and is only

Fig. 9.7 An aerial view of the Soufrière Hills Volcano looking north. Volcanic processes that altered the physical landscape of the island of Montserrat—such as pyroclastic flows, coastal advancement, previous dome collapses, and an ash plume—are all visible in this image, including the deltaic pattern, some lava flows created at their termini, and the partially buried remnants of the small village of Bethel (*center-right*). Photograph by J.-M. Bardintzeff, 2010. Retrieved from http://44.svt.free.fr/jpg/volcanologue.htm

accessible by boat or hike. Woodlands beach is the most notable black sand beach, with the material makeup of volcanic sand, one of the few in the Lesser Antilles region.

9.4 Cultural Landscape

The ongoing eruption has had a significant impact on both the physical and cultural landscapes of the island. Across the island, vegetation and much of the tropical rainforest has been destroyed or damaged by pyroclastic flows, ashfall, and acidic gas. Low-lying areas and ghauts have been filled in by pyroclastic flows, lahars, and ashfalls; additionally, these flows have created coastal expansion (Hincks et al. 2005).

This physical standpoint provides an entry point to frame an understanding of Montserrat's physical and social landscape interplay. It offers a manifestation of "…complex and emergent relationships that can be used as a platform" to understand multiple, converging viewpoints and understandings (Wilcock et al. 2013, p. 573). Ethnogeomorphology, for this study, is used as "…a geomorphic lens to discuss 'relationality'—or the way in which processes emerge in nondeterministic and non-causative ways" (Wilcock et al. 2013, p. 574). Landscape analysis builds on the delineation of landforms, processes, evolutionary adjustments, and the associated applications of these understandings (Brierley et al. 2013). Expanding on this foundation, within human geography, relations at a landscape scale form connections between place and home and strongly influence land use, following historical settlement patterns, social, and cultural interactions (Wilcock et al. 2013).

The ethnogeomorphology of Montserrat, because of the volcanic history and subsequent destruction of the former capital, economy, and way of life for Montserratians, requires concepts of connectivity and space–time relationality to frame the understanding. This perspective is grounded on the literature and on the experience of Montserrat pre- and post-eruption, and engagement with the community of Montserratians to examine knowledge production and concerns for environmental negotiation and decision-making.

9.4.1 Background

Montserrat, one of Britain's oldest colonies, was a quiet and relatively prosperous island in the Caribbean before the volcanic eruption, with good levels of employment, and high standards of education and health care. Montserrat held one of the highest standards of living in the Caribbean (Skelton 2000). The capital city of Plymouth was a flourishing town with a mature infrastructure of municipal functions including schools, a medical university, medical resources, heavy tourist traffic, and other successful commercial and industrial activities. It was a mainstay on the stopover itinerary of major cruise lines, had a large number of second homes for expats, and ran multiple hotels ranging in luxury with ocean views that were competitive and easily accessible (Weaver 1995).

The first modern disaster to strike was Hurricane Hugo, a category four hurricane which passed over the island in 1989 and destroyed 90% of the island's structures (Harden and Pulsipher 1992). Millions of dollars in aid poured into the recovery of the island with a major focus on rebuilding the port city of Plymouth. Yet only a few years later, the island, and specifically the city of Plymouth, would again come under attack by a natural disaster.

In July 1995, the long dormant the Soufrière Hills Volcano became active, triggering a chain reaction of events for the residents to evacuate and make decisions and life changes, in most cases permanently (Skelton 2000). The volcanic event brought immediate worldwide attention to the island, and while thousands of the resident population out-migrated, hundreds of scientists in-migrated to study the volcano, to advise on evacuation, and to quickly address the changing status of an active volcano and evacuation plan (Donovan et al. 2013).

The areas now enclosed in the exclusion zones to the east, south, and central areas include the former capital city of Plymouth (Fig. 9.8), the original airport, and many villages that were once agricultural villages, although most residents had paid work in Plymouth (Skelton 1996). The north of the island, currently the only inhabited area, is much dryer and less fertile. It was designated as a safe zone from the volcanic activity although ash is regularly blown over the region. The areas now included in the exclusion zone made up 74% of the households in Montserrat pre-eruption. These residents had to relocate to makeshift accommodations in shelters, churches, and schools, or with friends and family in the north, on other Caribbean islands, or the UK (Skelton 2000).

Scientists from the UK and around the world descended upon Montserrat after the initial eruption, participated in the monitoring of the volcano, the decision-making processes about the evacuation, and other safety concerns. This mix of local and foreign participants in a time of crisis became a dysfunctional process, spawning infighting and resentment on both sides. The typically risk-averse western scientists and UK officials, feeling responsibility on the part of the colonial holding, were quick to issue large evacuation orders, where local and non-western decision-makers were more contentious of the effects of false orders of evacuations and the emotional impact they had on local residents (Donovan et al. 2013). Where foreign visitors had the readily available option of leaving the island and going home at any time, those actually being evacuated had to make quick decisions and experience the excruciating trauma of a rapid departure from their beloved homes.

Fig. 9.8 Desolated capital city of Plymouth, covered by pyroclastic flow. Collapsing houses, in stark contrast to the barren ashfall and the lush greenery surrounding the flow path, now dominate the landscape. Photograph by J. Davenport

9.4.2 The Landscape of Past-Present

A visit to the exclusion zone and seeing a street of abandoned houses leaves an outsider with a sense of voyeurism. The sadness and loss, while intangible, descend heavily, almost like the pyroclastic flow that caused the abandonment. The feeling is intense, unforgiving, and overwhelming. Entering the derelict homes is indicative of a forceful, hurried, and frantic departure. Swimming suits still hang to dry over shower curtain rods, magazines from 1995 lay open to the last page of reading, closets still full of clothing, drawers left open with items from around the house, a collection of random items that make up a life, scattered about in the rush to evacuate, captured in a moment in time, when the residents had to leave. Many homes have been inundated with mudflows, destroyed by the hot gases of the pyroclastic flows, yet as with most disastrous events, oddities in the destruction remain. Where the chemical reactions of the pyroclastic material melted safes and swallowed metal desks and furniture, cloth curtains and paper documents remained. When a pyroclastic flow-delivered ballistic did not ignite a home, a boulder may have cracked the side wall and stayed there forever captured in this moment of time and location. The abandoned homes hold indications of the unknown and sporadic evacuations. Some storefronts have actual signs of their "temporarily relocated" businesses (these became permanent relocations). A mass of debris collected outside one home, it was the stopping point of a flow of trees and other debris. In another home, someone took off a pair of glasses, set them on the counter and just left (Fig. 9.9).

This arrival of the past, of the thriving city of Plymouth, arrives in the present in a preserved and respected way—a type of arrested decay. Arrested decay is a concept by Western scholars to preserve "ghost towns" or other abandoned areas in a way that stabilizes the existing infrastructure and allows organic decay as the buildings slowly give way to weathering and decaying processes, where the general appearance and layout of the streets and houses are retained but not repaired (DeLyser 1999). Arrested decay also addresses the particulate matter that accumulates in preserved or abandoned sites, mandating that neither the artifacts or the dust, or in this case the ash, that covers the entire area of abandoned homes is removed (DeLyser 1999). The ash-covered neighborhood had air of authenticity, many

Fig. 9.9 Select images exemplifying the urgent and sudden evacuation of Plymouth during the 1995 eruption. **a** A swimming suit hanging over the shower to dry in 1995. This is a very common sight in the abandoned houses. **b** Closest *left* with clothing and items still hanging on the door in an abandoned house. **c** A pair of glasses, set on the counter and *left* behind when the evacuation was ordered. **d** East Caribbean coins minted in 1994, just before the volcanic explosions began, lay untouched in an abandoned house. All photographs by E. Nelson

homes had footprints, also preserved in the ash, and some homes had no footprints (DeLyser 1999).

It was the homes without footprints that offered a visitor a true glimpse, again a voyeuristic glance, into the island lifestyle of the former residents. There are no written signs or organized tours of the areas surrounding the exclusion zones, and arrested decay is not an enforced policy in this particular neighborhood, a relatively unregulated area. This lack of written interpretation leaves the visitor with a sense of authenticity through the power of visual appearance to create meaning (Strazdes 2013). Authenticity engages with the historical and cultural landscapes and often leaves questions of an "authentic heritage," yet this landscape in Montserrat leaves no thought to interpretation, like many of the features of the island, the abrupt and violent nature of this town's abandonment, and the 20 years of ash, dust, and volcanic material that have collected in a very convincing way that leaves no questions of authenticity.

9.5 Heritage and Tourism

The island was first inhabited by Arawak people, and subsequently by the Carib, or Kalinago Amerindians. The Caribs called the island Alliouagana, meaning "Land of the Prickly Bush" (Fergus 1975). In 1493, Christopher Columbus claimed the island for Spain on his second voyage to the new world. He named the island "Santa Maria de

Montserrat" after the Virgin of Montserrat in Catalonia, Spain (Minahan 2009). In 1632, Irish-Catholics from nearby Saint Kitts colonized the island, fleeing anti-Catholic violence and indentured servitude. These Irish immigrants were followed by Irish-Catholics from Virginia, also fleeing religious persecution (Skelton 2000). Montserrat's nickname, "The Emerald Isle of the Caribbean," comes from its Irish heritage and interpreted resemblance to the other Emerald Isle, Ireland (Fergus 1984).

Despite the Irish colonists, it was Britain and France who fought for possession of the island for most of the 18th century. The island has hallmarks and regions named for the historical battles that raged on its soil. Runaway Ghaut was the site of a bloody colonial battle between the British and the French, which as its name implies, ended with one side retreating. In 1783, the battles ended, and Montserrat was declared a British possession (CIA 2015).

Like other Caribbean islands, Montserrat had a thriving slave trade linked to plantations. The first African slaves arrived in the 1660s (CIA 2015). While Montserrat has an Irish heritage as well, including the Irish indentured servants, the modern population is overwhelmingly of African descent, albeit with Irish surnames, creating an interesting confluence of history (Messenger 1994). The slave trade gradually increased from approximately 1000 slaves in 1678–7000 in 1810, when the enslaved population outnumbered the white settlers (Fergus 1984). These slaves became free in 1834 after slavery was abolished in British Overseas Territories, following the passage of the Slavery Abolition Act of 1833 in the UK (Fergus 1984).

Numerous slave rebellions occurred during the time of slavery, most notably one led by a slave named Cudjoe. Local legend claims that Cudjoe plotted a rebellion, escaped, and was caught by white slave owners. They beheaded him and hung his head from a silk cotton tree as a warning to other slaves. The silk cotton tree still stands today in a village aptly named Cudjoe Head. In honor of this rebellion and the island's strong African heritage, the island holds a Cudjoe Head celebration every summer (Montserrat Tourist Board 2011).

Another celebrated slave rebellion took place on St. Patrick's Day in 1768, when slaves rebelled, but failed to achieve freedom (Fergus 1984). This controversial depiction of the holiday is portrayed in tourist literature:

> Montserrat is the only country in the world apart from Ireland that observes St. Patrick's Day with a holiday. The weeklong celebration highlights Montserrat's local culture and Irish heritage that culminates with St. Patrick's Day on March 17 (Montserrat Tourist Board 2011).

While the festival is full of typical St. Patrick's Day revelry, complete with shamrocks and leprechauns, the locals consider the festival a solemn honor to the slavery history of the island.

The development of the tourism industry in Montserrat has, for the last 20 years, taken a back seat to the redevelopment of the island's infrastructure and, quite plainly, to the restarting of life as a new island. The volcanic eruption altered the island forever, destroying touristic activities and the previous capital city. Nonetheless, it has also created touristic activities in the exploration of the abandoned city, and has attracted hundreds of social, physical, and environmental scientists. The island holds a story of survival through its volcanic past and will undoubtedly continue to thrive and grow while maintaining a sense of seclusion that is particularly appealing to many tourists. Montserrat is truly a gem, the Emerald Isle, an exciting and different locale in the Caribbean, albeit a potentially dangerous gem. Looming in the background, rarely visible, even on sunny days, is the Soufrière Hills Volcano. Unseen, but always felt, a constant reminder of danger, of the unknown, of fear. The hazard of the volcano has become a way of life for Montserratians, from explosion to pyroclastic flow, from evacuation to revival.

9.6 Hazards

Montserrat is a volcanic island, and the active Soufrière Hills Volcano represents its most relevant hazard. But volcanism involves additional hazards of earthquakes (seismicity), volcanic blasts and explosions, dome collapses, pyroclastic flows, tephra falls, ballistics, and mudflow (Hincks et al. 2005). Additionally, the island remains geographically located in an area with a high incidence of hurricanes, making Montserrat an exceptionally likely place for catastrophes.

Since 1900, seven hurricanes have affected Montserrat. Of those, Hurricane Hugo in 1989 was the most devastating, and it destroyed a majority of the island (Weaver 1995, p. 597). Previous to Hugo, an entire generation had been living a disaster-free existence, since the last destructive storm of any kind had occurred in the late 1920s, and the volcano was still dormant (Harden and Pulsipher 1992). However, after 12 h of unrelenting wind and rain, "Every structure on the island suffered damage and 90% of the 12,000 inhabitants were left homeless, at least temporarily. All roads were rendered impassable by fallen trees and downed power lines; and the schools, churches and hospitals were badly damaged or destroyed" (Harden and Pulsipher 1992, p. 9).

Yet, the descriptions of Hurricane Hugo, and local references to it, were all pale in comparison with the volcanic activity. As terrible as Hurricane Hugo was, the difference between a hurricane and volcanic fallout is access. After the hurricane, inhabitants could rebuild. After the volcanic eruption, inhabitants were evacuated permanently.

An introduction to the volcanology of Montserrat, and the various hazards within the larger hazard of a volcanic eruption, must be given for understanding the Soufrière Hills Volcano. A major factor in determining the level of hazard across the island is the dome formation and eruption, the rate and direction of those processes, and the associated topography of the projected hazardous area (Hincks et al. 2005). For this section then, after a brief explanation of these phenomena, a quick review of the island's history of monitoring is offered before ending with an in-depth report of the volcanic events leading up to today.

9.6.1 Volcanic Activity

Dome collapse is caused by two mechanisms: first, an endogenous collapse involving internal forces within the volcano that destabilizes the dome by causing a surge of lava that always occurs in the direction of the dome growth (Watts et al. 2002); and the second mechanism is specifically rainfall-induced, creating an environment for a weakening of the dome and subsequent collapse (Matthews et al. 2002). Both of these mechanisms occur on Montserrat.

Pyroclastic flows typically follow a dome collapse and are high-speed, gravity-driven flows of debris, hot particles and gas (Hincks et al. 2005, p. 161). As gravity-based flows, they tend to follow the topography of the landscape, yet more powerful surges are capable of exiting a typical route and covering an unpredictable area and amount of land, much like the waters of a flooded river (Fig. 9.10).

Tephra fallout represents the most widespread volcanic hazard, caused from massive plumes of ash and rock created by phreatic explosions, pyroclastic flows, and explosions. The material enters the atmosphere and, depending on density and weight, falls back down to the earth. As Montserrat has predominately easterly low-level wind, much of the ash from tephra fallout is blown to the west, away from the currently inhabited areas. Yet occasionally, less common wind directions will cause significant ashfall and rock-fragment fall in the populated area—"raining gravel"—as the locals say (Bonadonna et al. 2002).

Ballistics are large blocks of rock, ejected from the volcano at speeds and temperatures high enough to cause significant damage within a 2.5 km radius of the dome. Also known as hot ejecta and volcanic bombs, the ballistics can cause fires when they come into contact with flammable material, like a wooden house (Hincks et al. 2005).

Volcanic mudflows, also called lahars, are caused by intense rainfall that mobilizes the loose deposits on the flanks of the volcano. These can move debris of all sizes and, because of their viscous nature, can completely bury cars, roads, and buildings, hampering access to critical services (Hincks et al. 2005, p. 161). On the island, near the mouth of the Belham River, deposits have been measured up to 3 m deep, and mud carried down Fort Ghaut continues to bury the city of Plymouth (Hincks et al. 2005).

9.6.2 Volcanic Monitoring on Montserrat

The most significant hazards involved with a volcano, because they underlie everything else, are the eruptions themselves. Volcanic monitoring began in Montserrat in 1936 by two geologists, C.F. Powell and A.G. MacGregor, sent by the Royal Society of London. These geologists produced the first detailed description of the geology of Montserrat (MacGregor 1938). Powell set up a seismograph network that was in operation until 1951 (Hincks et al. 2005; Powell 1938). The Seismic Research Unit of UWI (SRU) installed four seismograph stations in 1966 and piloted studies of ground-deformation measurement. Subsequent installations occurred between 1956 and 1986 (Hincks et al. 2005). The seismograph network was destroyed by Hurricane Hugo in 1989 and was restored in 1992 and has been greatly enhanced since the eruptions began in 1995.

After the initial eruption of the Soufrière Hills Volcano the Montserrat Volcano Observatory (MVO) was established to provide constant, intensive monitoring of the volcano. During the early stages of the crisis, through 1999, the MVO brought in scientists from the British Geological Survey (BGS), universities from the UK and the USA, and the SRU team, all working with local technical teams. Beginning in 1998, the scientific contributions of the MVO have been managed by the BGS and in 2003, the MVO moved into a purpose-built building observatory with a full view of the volcano (Hincks et al. 2005). Volcanic reporting continues daily with reports from the MVO Web site and on the MVO Facebook page that reaches across informational boundaries to reach the widest audience possible, both scientist and non-scientist alike, with up-to-date information about the volcano (MVO 2015)

The MVO conducts visual, deformation, gas, and environmental monitoring. These techniques involve aerial- and ground-based photography, as well as topographic surveying. They use leading technological advancements of photography to model the dome and dome growth (MVO 2015). Using GPS measurements and Electronic Distance Measurements (EDM), the MVO monitors the deformation of the flanks of the volcano, within a variable of a few millimeters (Hincks et al. 2005).

Earthquakes are common on Montserrat. Between 1933 and 1937, 3290 earthquakes were reported (Young et al. 1998). After these years of heightened activity, only 200 earthquakes were reported in 1937 and 1938 (Shepherd et al. 2002). This pattern of seismicity repeated itself nearly 30 years later in 1966 with heightened activity but it declined severely, and by

Fig. 9.10 Extended coastline from a pyroclastic flow as it enters the ocean creating new landmass for the island. Notice the fluvial-like features in both the resulting erosional and depositional formations. Photograph by S. O'Meara 2004

November 1967, "it was no longer thought necessary to continue intensive monitoring" (Hincks et al. 2005, p. 153; Shepherd et al. 2002). Yet in 1992, after decades of inactivity, seismic activity increased sharply, and from the beginning of 1992 to July 1995, the average number of daily earthquakes was higher than at any time since the 1938 observations. The phase of eruptions between 1995 and 1999 coincided with shallow, low-frequency earthquakes originating in a small area, often occurring in swarms and lasting for hours (Hincks et al. 2005).

9.6.2.1 Future Outlook

The Soufrière Hills Volcano has been relatively quiet for the past few years. Currently, MVO scientists continue to monitor the changing conditions. Hazard and risk assessments are carried out at six-month intervals by a panel of MVO personnel and scientific advisors since December 1997 (MVO 2015). The panel's reports are shared with the Governor of Montserrat and provide a basis for decisions concerning the exclusion zones and emergency procedures.

As the island is vulnerable to multiple disasters, the DMCA and the MVO work together to create safe and reliable forecasts of disasters and prepare themselves and the population of Montserrat accordingly. With a state-of-the-art warning system in place, and in collaboration with the other Caribbean islands' warning systems, Montserrat stands ready for whatever disaster may befall them in the future. Yet they also patiently await a few decades of peacefulness. Perhaps years of quiescence and economic growth are in order after the past decades of struggle.

9.7 Conclusion

The convergence and urgency of the geomorphologic landscape of Montserrat and its effects on the residents of the island offer a story of survival. At a time when other islands and nations were struggling with different disasters and different political quandaries, Montserratians lived a moment-to-moment existence in reverence to the volcano. As the volcano erupted, towns were evacuated, scientists flew in from all over the world, and the island experienced an influx of people and knowledge while simultaneously experiencing widespread out-migration—first to neighboring islands and, in many cases, on to the colonial homeland of the UK. Before Hurricane Hugo struck in 1989, the island was on the verge of independence from the UK, yet after the hurricane and more permanently devastating volcanic events of the Soufrière Hills Volcano, they are today, completely dependent on aid from the UK and other foreign agencies.

This vacillation from quiet, beautiful, and peaceful Montserrat, to violent, desperate, and traumatized Montserrat was no easy transition. Montserratians carry their past with them and, while they are still a proud and strong people thankful for their precious resources, they carry with them a fear and respect for the unrelenting volcano as well. Many locals speak of nature as a force not to be questioned, for they well know that nature will do what nature will do and humans have very little to do with the processes.

Those processes of land accretion and erosion, as outlined in this chapter, have created a Montserrat of beautiful and breathtaking mountains, hills, and the volcano. Additionally, beautiful beaches, nearly untouched by the ravages of mass tourism, make the island spectacular in the Caribbean. It is these very processes, however, that destroyed their former capital city of Plymouth by pyroclastic flows following multiple dome collapses of the volcano, triggering land deformation on an enormous scale that has forever altered Montserrat's landscape. This, coupled with the peculiar engagement with a legendary music scene and desired destination for affluent resident tourists, makes Montserratians rather extraordinary, especially for such a small island. They are an everyday folk, yet are outstanding in their hospitality, their resilience, and their ability to adapt, embrace, and carry on. Through a violent history of colonization, slave trade, and modern natural disasters, this island is a place of wonderment for anyone daring enough to discover it. It takes a certain kind of person to travel to Montserrat, but an even more special kind of person to *be* Montserratian.

References

Bonadonna C, Mayberry G, Calder E, Sparks R, Choux C, Jackson P … Young S (2002) Tephra fallout in the eruption of Soufrière Hills Volcano, Montserrat. In: The Eruption of the Soufrière Hills Volcano, Montserrat, from 1995 to 1999, vol 21. Geological Society of London, London, pp 483–516

Brierley G, Fryirs K, Cullum C, Tadaki M, Huang HQ, Blue B (2013) Reading the landscape: integrating the theory and practice of geomorphology to develop place-based understandings of river systems. Prog Phys Geogr 37(5):601–621. doi:10.1177/0309133313490007

Chardot L, Voight B, Foroozan R, Sacks S, Linde A, Stewart R, Widiwijayanti C (2010) Explosion dynamics from strainmeter and microbarometer observations, Soufrière Hills Volcano, Montserrat: 2008–2009: SHV EXPLOSIONS 2008–2009. Geophys Res Lett 37 (19). doi:10.1029/2010GL044661

CIA (2015) The World Factbook. Retrieved 15 Aug 2015, from https://www.cia.gov/library/publications/the-world-factbook/geos/mh.html

DeLyser D (1999) Authenticity on the ground: engaging the past in a California ghost town. Ann Assoc Am Geogr 89(4):602–632

Deplus C, Le Friant A, Boudon G, Komorowski J-C, Villemant B, Harford C, Ceminée J-L (2001) Submarine evidence for large-scale debris avalanches in the Lesser Antilles Arc. Earth Planet Sci Lett 192(2):145–157. doi:10.1016/S0012-821X(01)00444-7

Donovan AR, Bravo M, Oppenheimer C (2013) Co-production of an institution: Montserrat Volcano Observatory and social dependence on science. Sci Public Policy 40(2):171–186. doi:10.1093/scipol/scs078

Fergus HA (1975) History of Alliouagana: a short history of Montserrat. University Centre

Fergus HA (1984) Montserrat: emerald Isle of the Caribbean. 2nd edn. Macmillan Caribbean, London

Harden C, Pulsipher L (1992) Come a nasty gale: the response to Hurricane Hugo on the island of Montserrat, West Indies. Focus 42 (2):9

Harford C, Pringle M, Sparks R, Young S (2002) The volcanic evolution of Montserrat using 40Ar/39Ar geochronology. In: The eruptions of Soufrière Hills Volcano, Montserrat, from 1995–1999, pp 93–113

Hincks T, Sparks S, Dunkley P, Cole P (2005) Montserrat. In: Volcanic hazard atlas of the Lesser Antilles. Seismic Research Center, Trinidad and Tobago, pp 147–169

Le Friant A, Harford CL, Deplus C, Boudon G, Sparks RSJ, Herd RA, Komorowski JC (2004) Geomorphological evolution of Montserrat (West Indies): importance of flank collapse and erosional processes. J Geol Soc 161(1):147–160

MacGregor AG (1938) The Royal Society expedition to Montserrat, B. W.I. The volcanic history and petrology of Montserrat with observations on Mt. Pelée in Martinique. Philos Trans R Society, London, pp 1–90

Matthews A, Barclay J, Carn S, Thompson G, Alexander J, Herd R, Williams C (2002) Rainfall-induced volcanic activity on Montserrat. Geophys Res Lett 29(13), art no-1644

Messenger JC (1994) St. Patrick's Day in the other emerald isle. Eire-Ireland 29(1):12–23

Minahan J (2009) The complete guide to national symbols and emblems, vol 2. Greenwood Press

Montserrat Tourist Board (2011) Guide to Montserrat. Montserrat Tourist Board

Montserrat Tourist Board (2012) Rediscover Montserrat. Montserrat Tourist Board

MVO (2015) Montserrat Volcano Observatory. Retrieved 17 Aug 2015, from http://www.mvo.ms/

Myers A (2013) Coral Reefs of Montserrat. In: Sheppard CRC (ed) Coral Reefs of the United Kingdom Overseas Territories. Springer, Netherlands, pp 89–96. Retrieved from http://link.springer.com/chapter/10.1007/978-94-007-5965-7_8

Paulatto M, Annen C, Henstock TJ, Kiddle E Munshull TA, Sparks RSJ, Voight B (2012) Magma chamber properties from integrated seismic tomography and thermal modeling at Montserrat. Geochem Geophys Geosys 13(1)

Powell CF (1938) The Royal Society expedition to Montserrat, B.W.I. Final Report. Philos Trans R Society, London, pp 1–34

Shepherd J, Robertson R, Lynch L, Latchman J (2002) Montserrat: scientific investigations and political reactions in the period preceding the 1995 eruption. Paper presented at UWI conference, UWI, Montserrat

Skelton T (1996) "Cultures of land" in the Caribbean: a contribution to the debate on development and culture. Eur J Dev 8(2):71–92

Skelton (2000) Political uncertainties and natural disasters: Montserratian identity and colonial status. Interventions 2(1):103–117. doi:10.1080/136980100360823

Stinton A, Cole PD, Odbert HM, Christopher T, Avard G, Bernstein M (2014) Dome growth and valley fill during Phase 5 (8 October 2009–11 February 2010) at the Soufrière Hills Volcano, Montserrat. In: The Eruption of the Soufrière Hills Volcano, Montserrat, from 2000 to 2010. Geographical Society, London, pp 113–131

Strazdes D (2013) The display of ruins: lessons from the ghost town of Bodie. Change Over Time 3(2):222–243

UNICEF (2013) Migration profiles: Montserrat. UNICEF. Retrieved from http://esa.un.org/MigGMGProfiles/indicators/files/Montserrat.pdf

UWI (2005) Volcanic hazard atlas of the lesser antilles, 1st edn. Seismic Research Centre, Trinidad and Tobago

Watts R, Herd R, Sparks R, Young S (2002) Growth patterns and emplacement of the andesite lava dome at the Soufrière Hills Volcano, Montserrat. In: The Eruption of the Soufrière Hills Volcano, Montserrat, from 1995 to 1999, vol 21. Geological Society of London, London, pp 115–152

Weaver D (1995) Alternative tourism in Montserrat. Tour Manag 16 (8):593–604

Wilcock D, Brierley G, Howitt R (2013) Ethnogeomorphology. Prog Phys Geogr 37(5):573–600. doi:10.1177/0309133313483164

Young S, Sparks RS, Aspinall W, Lynch L, Miller A, Robertson R, Shepherd J (1998) Overview of the eruption of Soufriere Hills volcano, Montserrat, 18 July 1995 to December 1997. Geophys Res Lett 25(18):3389–3392

Guadeloupe

10

E. Arnold Modlin Jr and Casey D. Allen

Abstract

Comprising a small set of several islands located at the southern end of the Leeward Islands in the Lesser Antilles, Guadeloupe is a French Overseas Department. The two larger islands are separated by less than 40 m in some sections giving Guadeloupe the geologic distinction of being part of both the Volcanic Caribbees (geologically younger) and the Limestone Caribbees (geologically older). Basse-Terre (the western main island) consists of a series of volcanically active andesitic massifs, while Grande-Terre is of lower terrain and a volcanic bedrock foundation overlain with carbonate sedimentary deposits. While various Amerindian groups visited and started settling in the islands between 4000 and 6000 BP, it was unknown to Europeans until Columbus' second voyage (November 1493). Once French colonists began settling the island after 1635, the landscape was dramatically transformed first by fighting with the Carib people for five years, and subsequently by bringing African slaves to cultivate sugar. Even today, Guadeloupeans represent a resilient people struggling with social and economic issues such as the high cost of living and high unemployment rates.

Keywords

Volcano • Basse-Terre • Grande-Terre • Marie-Galante • Soufriere • Vichy • Slavery

10.1 Introduction

With a land area of approximately 1628 km^2 and 581 km of coastline, as of 2015, the department of Guadeloupe had a total population of 400,132 people (INSEE 2016; Pruett and Cimino 2000). The island of Basse-Terre, Guadeloupe, sits 42 km north of Dominica and 56 km southeast of Montserrat, while the other major island Grande-Terre lies 61 km south of Antigua. A narrow channel 40–150 m wide and about five km long (*Rivière Salée*, or Salt River) separates these two largest islands of Guadeloupe. Sometimes Guadeloupe is referred to as *l'Île Papillon* (Butterfly Island), because Basse-Terre and Grande-Terre form a butterfly-like shape (Fig. 10.1). As an Overseas Department of France (also known as an Overseas Region since 2003), Guadeloupe is part of the European Union and uses the Euro as currency, although it is not a part of the Schengen area (US Department of State 2016). Citizens of Guadeloupe remain citizens of France, and the islands' inhabitants are represented in the French national government by officials elected locally on the island.

Geologically, Guadeloupe remains quite complex (Fig. 10.2). The islands have three basic geochemical compositions. Grande-Terre, Marie-Galante, and the Îles de la Petite Terre are part of the Limestone Caribbees, islands that represent an arc of volcanic activity that ceased millions of years ago, and are overlain by carbonate deposits. To the east is La Désirade which has a volcanic base that is older

E.A. Modlin Jr (✉)
History and Interdisciplinary Studies Department,
Norfolk State University, Norfolk, VA, USA
e-mail: arnoldmodlin@gmail.com

C.D. Allen
General Education, Western Governors University,
Salt Lake City, UT, USA
e-mail: caseallen@gmail.com

C.D. Allen (ed.), *Landscapes and Landforms of the Lesser Antilles*,
World Geomorphological Landscapes, DOI 10.1007/978-3-319-55787-8_10

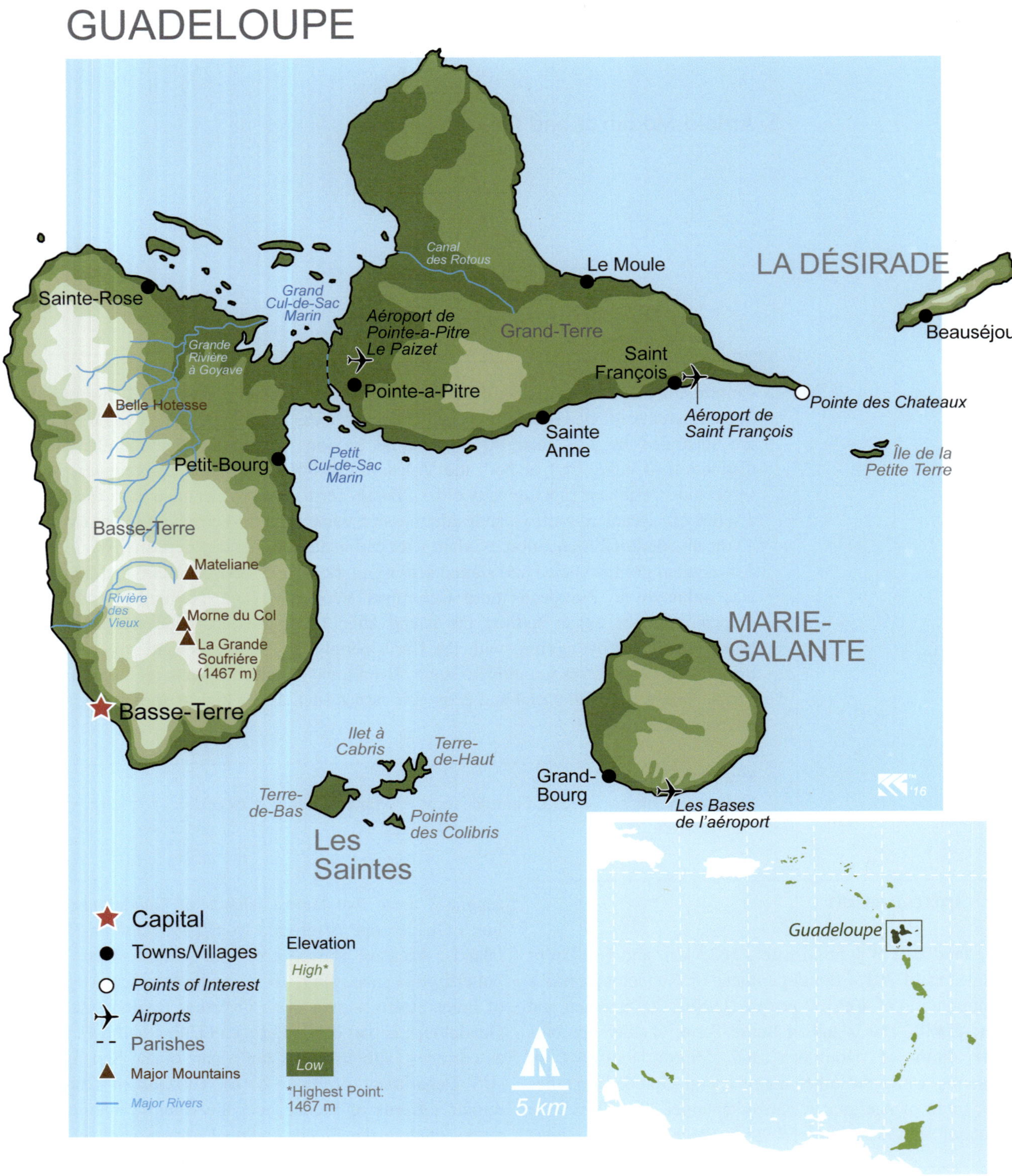

Fig. 10.1 General physiographic map of Guadeloupe and numerous surrounding islands. Many of the small neighboring islands are populated, such as Marie-Galante to the southeast. The bow-shaped main island of Guadeloupe is noticeably more mountainous in the western Basse-Terre region than the eastern Grande-Terre region. Cartography by K.M. Groom

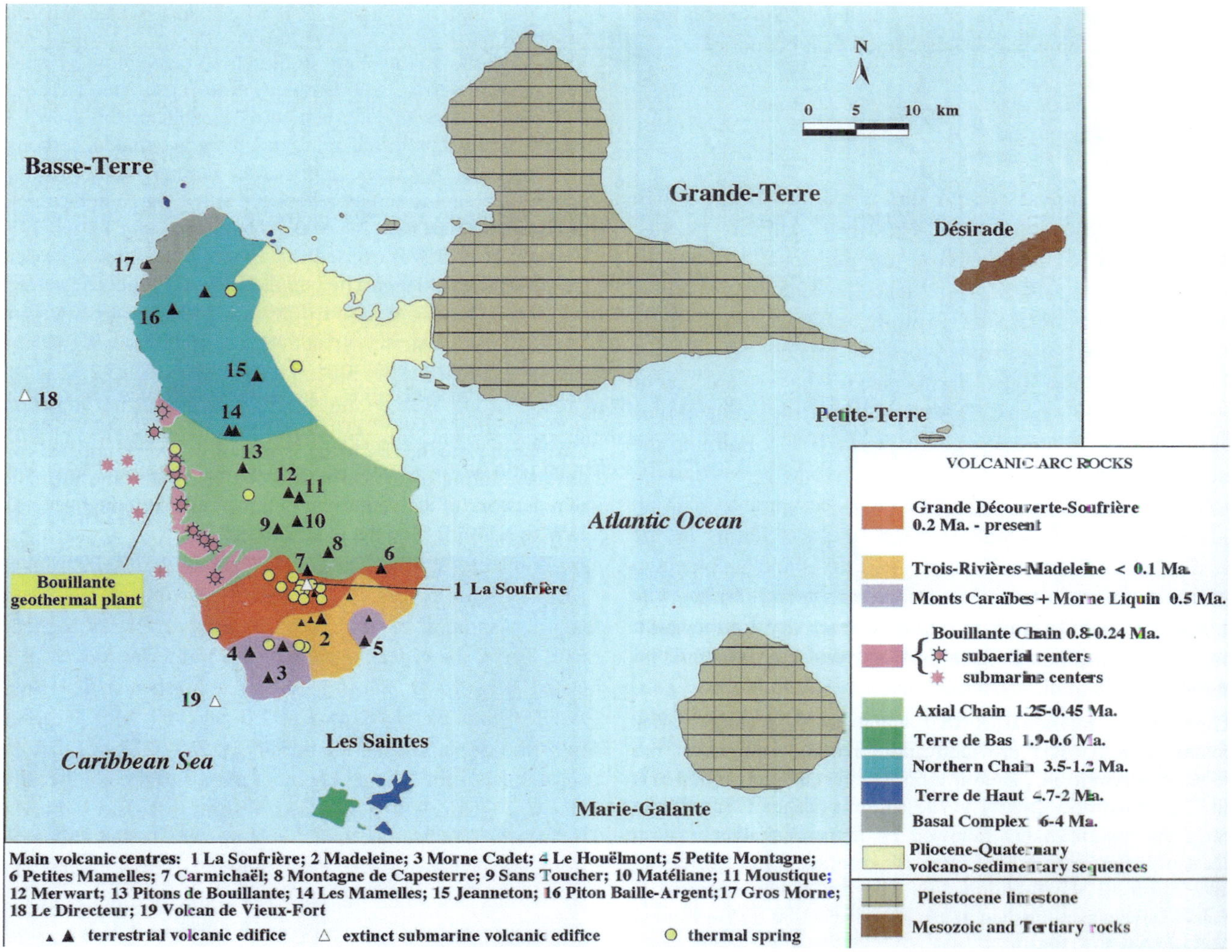

Fig. 10.2 Geological map of Guadeloupe displaying its distinctness across the major islands. Map courtesy of http://caribbeanvolcanoes.com/guadeloupe-map/. See also Bouysee et al. (1990)

and mineralogically different from the rest of the Limestone Caribbees. To their west, Basse-Terre and Les Saintes islands are part of the Volcanic Caribbees, which are the result of more recent volcanic activity and have a higher profile than the Guadeloupean Limestone Caribbees (Bouysee et al. 1990). La Grande Soufrière, a 1467-m-high volcano on Basse-Terre, although quiet at the time of this writing, represents the highest point in the Lesser Antilles and still considered an active volcano.

In addition to natural forces that shaped the landscape, anthropological forces transformed the islands that make up Guadeloupe. The first migratory waves of Amerindian groups passed through, and in some cases stayed, prior to 2000 B.C.E. (Allaire 1997). Columbus discovered and landed on the islands of Marie-Galante and Basse-Terre in November 1493. Caribs continued to inhabit on the islands until at least 1640 when the French largely displaced them. In 1635, the French settled the islands with the intention of raising tobacco. Sugar and slavery would shape the landscape from the 1660s to the abolition of slavery in 1848, when indentured servants would then take many of the formerly enslaved people's places in the sugarcane fields. Many Guadeloupeans would fight for the French in World War I and World War II (Jennings 2001). During the World War II years of 1940–1943, Vichy France maintained control of the island. In 1946, Guadeloupe became an Overseas Department of France. Today, tourism on the island takes a variety of forms, from cruise day-stops to sun-sea-sand visitors, ecotourism, and active tourism, although participating in one of these does not necessary preclude one from participating in other activities. As important as tourism is to the economy, other industries also contribute substantially to the local economy, and yet many Guadeloupeans see a need to further develop and diversify their local economy.

10.2 Setting

Guadeloupe comprises over a dozen islands—six of which are continually inhabited. The two largest islands, Basse-Terre and Grande-Terre, remain separated by the narrow Salt River. Basse-Terre hosts more mountains and is larger than its counterparts, even though the name literally means "lowland"—so-named by the French because the settlement of Basse-Terre was located on the leeward side of the mountainous island, which blocked the trade winds coming from the northeast (Rogoziński 1999). Forming the other half of the Caribbean Butterfly, Grande-Terre—though not as large as Basse-Terre—is large in comparison with all other islands under Guadeloupe's purview, including *Îles de la Petite Terre*, or Petite Terre—a two-island archipelago with a combined total area of approximately 1.7 km^2. The larger, southern island is Terre-de-Bas, and the smaller island to the northeast, Terre-de-Haut. While uninhabited by people, the surrounding water serves as a natural reserve for a number of Lesser Antilles Iguanas as well as sea turtle nesting areas (Réserves Naturelles 2016).

To the northeast of the Petite Terre Islands, and due east of southern Grande-Terre, is the inhabited island, La Désirade. The Petite Terre Islands are administratively cared for by the local government (commune) of La Désirade. La Désirade is located 8 km from Grande-Terre. Long and narrow (11 km by 2 km, on average), the island of La Désirade is the oldest part of Guadeloupe with research on the west end of the island and just offshore, indicating rock formation ages of over 140 Ma (Mattinson et al. 1980). La Désirade's major settlement is Beauséjour on the east end of the island. It is connected to Saint-Fraçois on Grande-Terre by ferry.

The third largest island of the archipelago of Guadeloupe is Marie-Galante, located about 26 km south of Grande-Terre and 30 km southeast of Basse-Terre. Ferries connect the island to Point-á-Pitre and Saint-Fraçois on Grande-Terre. The landscape was once dominated by sugar plantations, but the decline of sugarcane production on the island reduced job opportunities, and the population has dropped substantially to ~11,000 in 2012 (INSEE 2015b). One of the Limestone Caribbees, the island is made of limestone deposited, in some cases up to 200 m, on top of volcanic rock (Collina-Girard 2002; Fig. 10.3x).

To the west of Marie-Galante and south of Basse-Terre is *Îles des Saintes* ("Islands of the Saintes," often shorted to "Les Saintes"). Only two of the islands of this archipelago are inhabited by people. Resulting from volcanic activity, some of these islands are so small and located so close to each other that only a few feet of shallow water separate them from another island in the group. Finally, another often overlooked island lies between the northern sections of Basse-Terre and Grande-Terre; Ile à Fajou is an island located in the Grand Cul-de-Sac Marin.

10.3 Landforms

Despite the main islands' closeness—as little as 40 m apart in some sections—Basse-Terre (the western main island) and Grande-Terre are geologically quite different. Basse-Terre and Les Saintes islands are identified with the new or recent arc of volcanic activity aligned with the Volcanic Caribbees. Grande-Terre, Marie-Galante, and the Petite Islands are covered with calcium carbonate and considered part of the Limestone Caribbees, an older arc of volcanic activity, though the entire archipelago is moving at approximately 20 mm a year (DeMets et al. 2000).

Basse-Terre consists of a series of volcanic andesitic massifs (Samper et al. 2007). Generally speaking, on Basse-Terre, the older the massif, the further northwest it is located (Fig. 10.3). Starting from the northwest, the Basil Complex was active between 2.0 and 6.0 Ma. Heading south, the Septentrional Chain, active 3.5–1.15 Ma, covers over a third of the island. The next massif to the southeast is the Axial Chain which was active from ~1.0 to 0.44 Ma. The Bouillante Chain to the west of the Axial Chain hugs the Caribbean coast and was active between 1.2 and 0.9 Ma. The Monts Caraïbes massif is at the southern tip of Basse-Terre and was active between 600 and 400 Ka. North of Monts Caraïbes and south of the Bouillante and Axial Chains is the Composite Volcano of La Grande Découverte (CVGD), a still-active massif of La Grande Soufrière volcano. It has been active from 8.5 Ka to present. Other parts of the CVGD massif have been active for the last 200,000 years (Feuillet et al. 2002, Samper et al. 2007). Traveling from the northwest edge of Basse-Terre toward the south-central part of the island, Basse-Terre's profile increases in height until the highest point, La Grande Soufrière. The northeastern corner of the island is its lowest part, on average (Fig. 10.4).

Basse-Terre's mighty La Soufrière dominates the landscape (Fig. 10.5). The resultant steep slopes created by numerous eruptions and subsequent erosion events has led to steep, almost feral reliefs in some areas. Surrounding La Soufrière is Guadeloupe National Park, which hosts numerous natural features, including geothermal vents, a crater lake (Lac Flammarion, Fig. 10.6), and majestic waterfalls such as those along the Carbet River (Fig. 10.7). Although these types of features remain typical of many Lesser Antillean volcanic islands, what sets Guadeloupe apart is the ability to capitalize on them in terms of

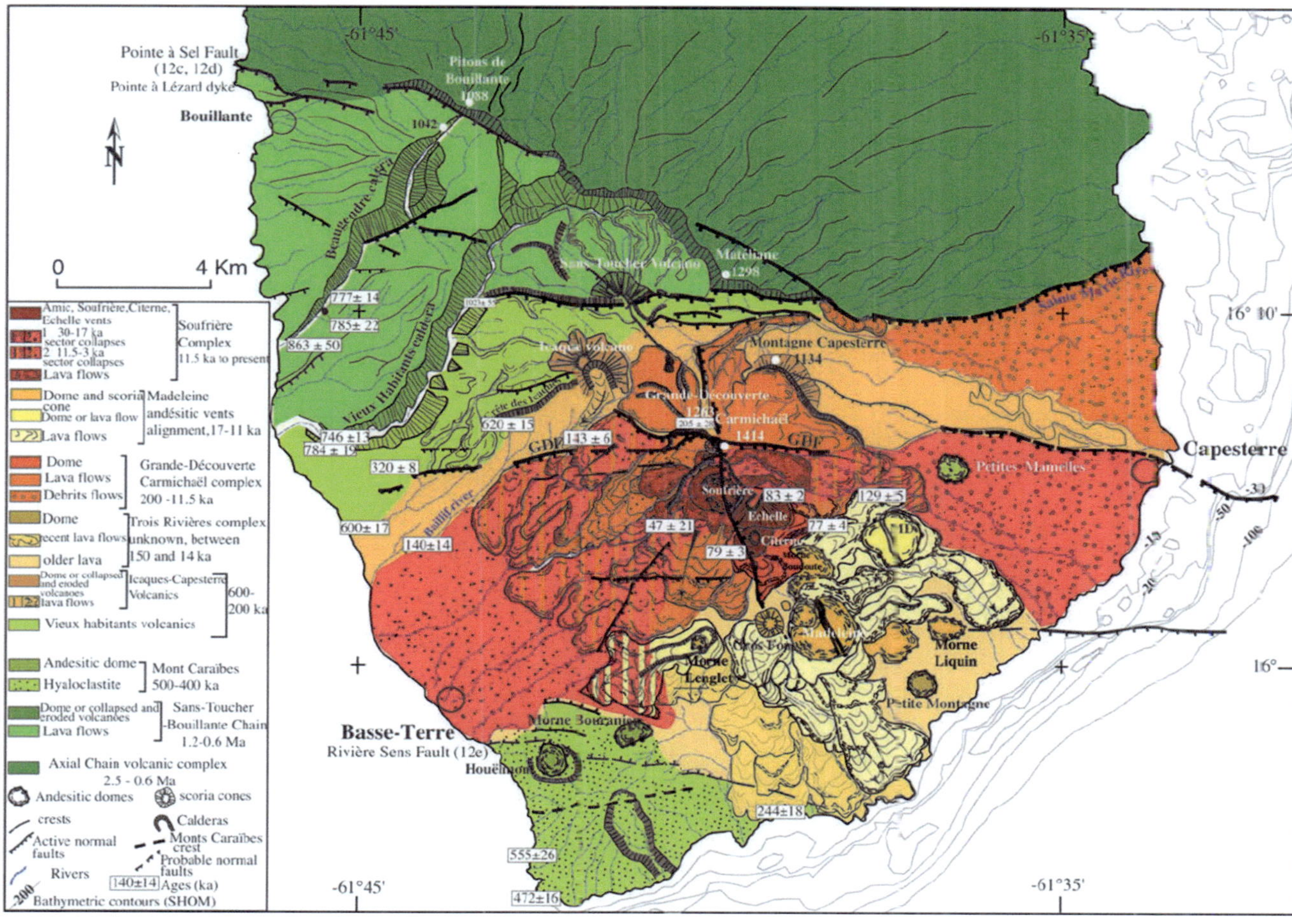

Fig. 10.3 Geological map of the southern portion of Basse-Terre surrounding La Soufrière. Note the general age trends (southeast to northwest, white boxes). Figure from Feuillet et al. (2002), available online at https://www.researchgate.net/publication/228472964_Arc_parallel_extension_localization_of_volcanic_complexes_in_Guadeloupe_Lesser_Antilles, and compiled from various sources

economics (e.g., tourism), protection (national parks and reserves), and energy production (a functioning geothermal power plant; see Jaud and Lamethe 1985).

While not flat, Grande-Terre consists of lower terrain than most of Basse-Terre (Fig. 10.8). The foundational bedrock of the island is volcanic, with the last 5.0 Ma being an active period of carbonate depositing (Cornée et al. 2012). Being at the south end of the eastern arc, Grande-Terre—and Marie-Galante—became inactive more recently than the northern Limestone Caribbees such as Sombrero, Anguilla, and Barbuda (Robertson 2009). On Grande-Terre, Cornée et al. (2012) identified four periods of sedimentary buildup—four unique sedimentary sequences—each of which is separated by a period of subaerial erosion.

Grande-Terre represents one of the major areas of Caribbean karst lands in the Lesser Antilles. On the islands of Guadeloupe, most caves associated with the karst landforms are on or near the coast (Lenoble et al. 2015). Among karst type caves, only one wet cave appears in interior Marie-Galante while two wet caves and one dry cave appear in interior Grande-Terre. The majority of karst caves are on the eastern side of Grande-Terre, and southeastern Marie-Galante (Fig. 10.9), while over a dozen dry caves are on the southeastern two-thirds of La Desirade (Lenoble et al. 2015).

La Désirade on the other hand represents a geologically unique island within the Guadeloupean archipelago (Fig. 10.10). Current understanding about the movement of the Caribbean Plate recognizes that it shifted from the area now associated with the Pacific region (Corsini 2011). Research on rock formations on the east end of La Désirade and just off of its eastern coast dates some of the igneous rocks in that area at over 140 Ma (Mattinson et al. 1980; Mattinson et al. 2008; Corsini 2011). That is, even though the island is overlain by carbonate deposits—a characteristic shared with Grande-Terre and Marie-Galante, two of the Limestone Caribbees—La Désirade reveals its own complex geology and geologic history.

The Grand Cul-de-Sac Marin, a shallow water area with coral reefs, mangroves and seagrass beds, rests between the northern sections of Basse-Terre and Grande-Terre. This

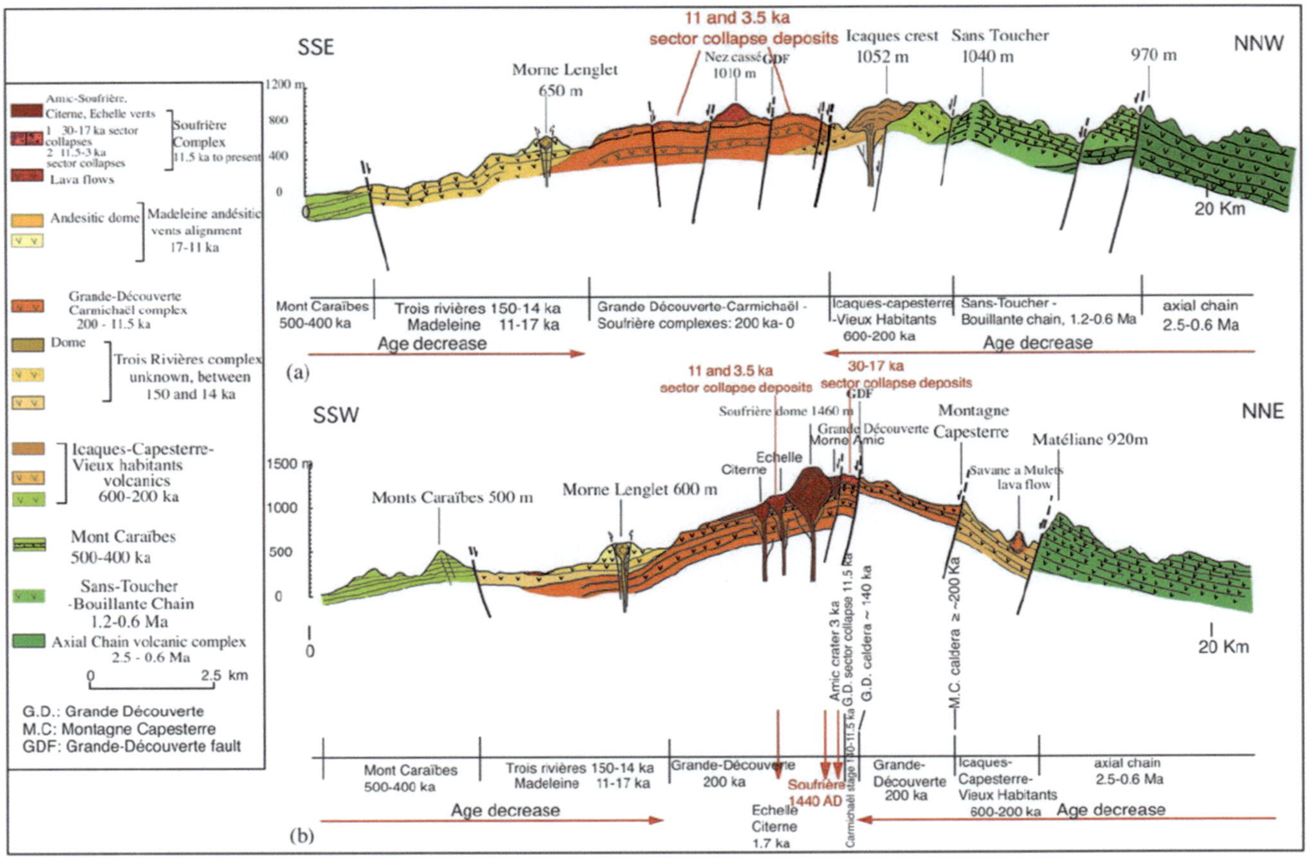

Fig. 10.4 Geological profile of Basse-Terre's southern portion, demonstrating the island's volcanic complexity. This profile runs south–southeast to north–northwest (**a**) and south–southwest to north–northeast (**b**), bisecting approximately 1.2 km south of La Soufrière. Figure from Feuillet et al. (2002), and available online at https://www.researchgate.net/publication/228472964_Arc_parallel_extension_localization_of_volcanic_complexes_in_Guadeloupe_Lesser_Antilles

Fig. 10.5 Views of La Soufrière Volcano. **a** Basse-Terre as seen from Les Saintes. La Soufrière, just *right* (east) of center is the highest point. The other peak to the *left* (west) is the Monts Caraïbes massif. Photograph taken by Wikiuser Friman on February 26, 2008. Uploaded to Wikimedia Commons and shared under a Creative Commons License on March 15, 2008. **b** On top of volcano La Soufrière at island Basse-Terre, Guadeloupe, looking toward La Découverte in the direction of Gouffre Tarissan, South Crater. Photograph taken by Wikiuser Miebner. Uploaded to Wikimedia Commons and shared under a Creative Commons License on June 19, 2011. Required tag for both images: http://de.wikipedia.org/wiki/Benutzer:Miebner

Fig. 10.6 Lake Flammarion in Guadeloupe National Park, a crater lake filled by rainfall, approximately 1.2 km southeast of La Soufrière. Though a volcanic island, Basse-Terre has few such lakes, most likely due to its rugged topography. Image courtesy of Guadeloupe National Park: http://www.guadeloupe-parcnational.fr/fr/galeries-photo/paysages-de-la-guadeloupe

area contains mangroves known to exist for at least two millennia (Lallier-Vergés et al. 2008). Within this broader ecological zone are numerous smaller zones with species of flora and fauna that are sensitive to changes in salinity. Sea level rise will be a challenge for this area causing changes in the chemistry across this bay and wetlands. Because of the environmental sensitivity of this area, further research is needed to understand how this region in general will react to sea level rise and climate change, as well as how incremental changes in one smaller area will affect neighboring areas.

10.4 Landscapes

Like all of islands in the Lesser Antilles, there was a sense of both crossroads and destination for many of the early peoples to Guadeloupe. Early native people migrated to and through the territory prior to 2000 BCE (Higman 2011), but archaeologists have not found much evidence of these peoples' existence in Guadeloupe prior to 300 BCE. When Columbus met the Caribs in Guadeloupe, they lived in small communities of 20–30 houses situated around a central square and buried their dead in ways that reveal cosmic beliefs and the existence of an afterlife and respect for ancestors (Rodríguez 1997). By the time of Columbus' second voyage, Caribs lived on parts of Guadeloupe–the double island had fresh water, and the Caribs were most likely in regular contact with other Caribs and native people elsewhere in the immediate and larger Caribbean world.

The French colonized Guadeloupe in 1635, not only encroaching on land first settled by native peoples, but also taking food from them. The ensuing violent struggle for the islands of Guadeloupe costs many French and Amerindian lives, culminating with the confinement of Caribs to the mountainous section of Basse-Terre in 1641, and eventually

Fig. 10.7 The third waterfall along the Carbet River, which begins on the lower slopes of La Soufrière, and flows in a southerly direction. A popular tourist attraction in Guadeloupe National Park, this series of three falls (125, 110, and 20 m high, respectively) traverses the rugged rainforest and volcanic terrain, decreasing in height as the river makes its way to the sea. Image courtesy of Guadeloupe National Park: http://www.guadeloupe-parcnational.fr/fr/galeries-photo/paysages-de-la-guadeloupe

allowed or forced to migrate to Dominica and/or St. Vincent (Pons 2013; Watts 1987). Meanwhile the French started growing their tobacco-based colony, switching to a sugar-dominated colony a few decades later (Dunn 2012). The few thousand Europeans were soon joined by tens of thousands of enslaved Africans whose work made the island one of the French Sugar Islands (Curtain 1969). Clearing the land was a key requirement of sugar cultivation, and even more pressure was put on the original forests as wood was initially used as fuel for sugar refining—a practice the French quickly discouraged before denudating the landscape permanently (Watts 1987).

Following a false abolition period from 1794 to 1802, a drop in sugar production occurred alongside many slaves entering the field of piracy (Rodigneaux 2015). Still, after this period, sugar planters resurrected their business with the import of thousands more indentured servants (Northrup 2000). By 1900, the indentured system came to an end, and fields began to fallow. By World War II, fallow land was seen not as resting and regenerating land for future exploitation, but as unpatriotic. The Vichy representatives on the island pushed for extracting food from everywhere on the island, even while many Guadeloupeans suffered from hunger (Jennings 2001).

Fig. 10.8 A view along Grande-Terre's coast from Anse-Betrand cliff in the far north, demonstrating its lower relief in comparison with Basse-Terre. Image courtesy of Guadeloupe National Park, available online: http://www.guadeloupe-parcnational.fr/fr/galeries-photo/paysages-de-la-guadeloupe

Since then, the Department of Guadeloupe has made major changes to the landscape. The international airport, Pointe-à-Pitre Le Raizet, was designed to handle a much higher volume of air traffic than it does, in hopes of expanding tourism options. Pristine beaches and a range of hotels dot many of the islands, and modern, paved roads are often crowded with vehicles. Agriculture still represents a significant portion of the economy and continues to shape the landscape. Offshore, large parts of the water between northern Basse-Terre and Grande-Terre are preserved as the Grand Cul-de-Sac Marine Nature Reserve, while additional land in central Basse-Terre has been set aside as a national park. Many of the past landscape moments are still reflected in areas as relict or historic landscapes and often serve as touchstones for memory and heritage on the island.

10.5 Heritage and Tourism

While heritage and tourism are often connected, heritage goes beyond history by making personal and emotional links to the past, considered unique to the social group. But they can become tools for tourism promoters to market to outsiders. This commodification of the past (and heritage sites) over time can become contentious, where some locals feel that their heritage is being cheapened, while visitors might not understand the full value and deep meaning of these past moments, events, and locations to the local population (Compare Nuryanti 1996 for some of the issues associated with the relationship between tourism and heritage).

Fig. 10.9 Limestone features in Saint-Louis de Marie-Galante, Guadeloupe: Gueule Grand Gouffre, a geomorphological complex including a littoral chasm (from a roof collapse) and a natural arch (bottom-center of image). Sometimes called "Trou a Diable," this coastal sinkhole is a popular tourist destination. The trail/road leading to the viewpoint is visible along the northern rim of the chasm. Photograph taken by Wikiuser: Thesoulfulwriter. Uploaded to Wikimedia Commons and shared through a Creative Commons License on January 5, 2009

10.5.1 History and Heritage

For Guadeloupe, initial Columbian contact occurred from 3 to 10 November 1493, on Columbus' second voyage. He did not leave settlers to colonize the island, but over the next century, Europeans stopped for water from the islands after crossing the Atlantic and for occasional slave raiding of the Caribs (Delpuech 2011). In 1635, the French created a settlement with 550 settlers led by Charles Lienard and Jean Duplessis Ossonville (Pendery 2010). Sick, hungry, and viewing the natives' lives as of less value than their own, some of the French followed Governor Liénarde's advice and started raiding the Caribs gardens and reserved food stores (Boucher 1992). The French and the Caribs fought over the island for four years, with the French finally prevailing by the addition of French soldiers from St. Christopher in 1640 (Boucher 1992).

By the beginning of the eighteenth century, Guadeloupe made the transition from an uncertain business interest of a few French citizens to a colony of the French King (Higman 2011). It also transitioned from a society with slaves to a slave society (Curtain 1969). During the mid-eighteenth century and into the early nineteenth century, Guadeloupe was captured by the British from the French a number of times, and also spent 15 months in Swedish hands—though largely in name only as British governors remained as administrators (Tingbrand 2002). In 1794, the French government decreed that slavery was abolished in all colonies, only to have that decree reversed in 1802 by Napoleon—who was married to a Martinican planter's daughter (Bélénus 2015). When the news reached Guadeloupe, the once-enslaved population did not allow themselves to so easily be subjugated again. Napoleon sent troops to Guadeloupe who killed thousands of free Blacks who refused to be re-enslaved. Instead, they chose to fight until their deaths (Bélénus 2015). Still slavery was violently reinstituted on the island and would not be abolished until 1848.

Many formerly enslaved people started subsistence farming, building their own homes on land that they worked for their own needs (Fig. 10.11a). Planters continued to use coersive labor, importing indentured servants from West Africa, Vietnam, China, and India (Northrup 2000). By

Fig. 10.10 View of La Désirade (Background, left of center) from the east coast of Grande-Terre, Guadeloupe. The small island, only 2 km by 11 km, is dominated by its distinctive central plateau and unique geology. The white sand beach and sea stump in the foreground are indicative of the picturesque coastal landforms found on many of Guadeloupe's islands. Photograph taken by Wikiuser: Jayen466 on August 22, 2007. Uploaded to Wikimedia Commons and shared through the photographer releasing it to the public domain on January 12, 2009

1900, the indentured servant system was almost over, though Guadeloupe was still a colony of the French with an economy based on exporting agricultural crops such as sugar, pineapples, and bananas (Fig. 10.11b). During the first half of the twentieth century, agriculturally the economy was reconfiguring of toward Guadeloupean food self-sufficiency by leaving no field fallow (Jennings 2001).

While Guadeloupe was declared an Overseas Department shortly after World War II—like Martinique, French Guiana, and Réunion—its citizens' relationship with Paris unfolded in different ways than the other Overseas Departments. In some ways, Guadeloupe and Martinique shared important moments in history because of their close geographic relationship and because Guadeloupe was administered through Martinique for much of its colonial history. Recently, Guadeloupeans have been more vocal about their economic difficulties, and on January 20, 2009, forty-eight organizations came together under the name Lyannaj Kont Pwofitasyon (LKP) and listed 165 demands that included a 200-euro per month increase in pay for those working at minimum wage (Bonilla 2009). Though not seen as a colony in the eyes of the French government, the insular economy of Guadeloupe is unlike the economy of Continental France. Because of the cost of importing goods, taxation, and limited competition in certain industries, items in Guadeloupe can cost 20–170% more than in Metropole France (Bonilla 2009). Bonilla's (2009) report on the strike reveals not only just how workers in Guadeloupe got important concessions, but also exposes the struggle between those running small and large businesses on the island. Large employers refused

Fig. 10.11 Historic postcards displaying the lives of freed slaves in Guadeloupe. **a** Title: Guadeloupe—Cabane d'Indiens. Postcard was created prior to February 1907. Images like these show how some Guadeloupeans were forced to live around the turn of the twentieth century. This image should be considered with a critical eye, as the assumptions that images like this encourage the viewer to make tell more about the viewer than the lives of those pictured in the image (Modlin, 2015). This is an unused postcard from the author's personal collection. **b** Title: Compagnie des Antilles.—Propríetaire de la marque Rhum Chauvet. (English: "Company of the Antilles—Owner of the Rhum Chavet brand") Bottom text: 54—Récolte d'ananas à la Guadeloupe (English: "Pineapple harvest in Guadeloupe"). Image shows workers harvesting pineapples. Photograph taken prior to December 1915. This is an unused postcard from the author's personal collection

to negotiate with the strikers, but the UCEG, "which represented smaller, local Guadeloupean business owners" stepped forward and used the chance to challenge the dominance of businesses owned by the white elites (Bonilla 2009:10). On March 4, 2009 a final agreement was signed for Guadeloupe.[1]

[1]Other Overseas Departments followed Guadeloupe's lead. Martinique signed an agreement on March 11, 2009. Réunion had labor unrest shortly afterward and a general strike also occurred in France starting March 19, 2009 (Bonilla 2009).

10.5.2 Tourism

For many of the islands in this volume, tourism remains a staple in their economies. This often means either attracting more tourists, or getting them to stay longer. As a commodity—what the tourism experience is—there are few ways to make attractions fit a solid cost-benefit model. In the Caribbean, many islands compete for the same set of tourists. Guadeloupe, like other Lesser Antillean islands, benefits in a number of ways from longer-staying tourists: More money is spent, a greater respect for the environmental impact can be engendered in tourists, and some of these tourists will move away from the high-volume tourism areas, spreading out the possible economic benefits of their visits. In places such as Marie-Galante, Les Saintes, and La Désirade, longer stopovers are a necessity, as just getting to those islands can be an ordeal requiring arriving by air to Point-á-Pitre, clearing the airport, getting transportation to the appropriate ferry dock, and then continuing on to the destination hotel on one of these islands. Finding a balance of locals' and tourists' needs and wants will probably be one of those issues constantly under negotiation.

Still, Wong (2015) makes a strong argument that Caribbean tourism can repeat colonial exploitation, as many of the jobs created are low-wage jobs employed by multinational hospitality companies that pressure islands to keep minimum wages low to boost their own profits. However, tourism is now such a significant part of many Caribbean economies that major changes in its role locally need to be thought through with solid plans about what will replace it. This is true of Guadeloupe. Over 100,000 cruise tourists visited Guadeloupe each year between 2009 and 2011, and over 150,000 cruise tourists each year in 2012 and 2013, but in most instances, these tourists stay for less than 24 h (Table 10.1). However, "stopover tourism" in Guadeloupe, while showing some reaction to European financial issues, is interesting, especially when compared to other places in the Caribbean. For example, French tourists, who make up between 70 and 80% of the tourists to Guadeloupe tend to take longer vacations than tourists from other countries, and this potentially affects the number of days on average that a stopover tourist spends in the Overseas Department. According to UNWTO (2015), however, stopover tourists on average spend between 13 and 14 days in Guadeloupe. Yet, it is at this point that a qualification has to be made about these data. Five years of data are not enough to project the direction of trends.[2]

Cruise ship tourism, while bringing in hundreds or even thousands of tourists at once, does not necessarily contribute large amounts of income per passenger to the local economy. Part of this is because tourists have the choice to eat on the ship or the island, to wander the island on their own, or participate in prepackaged shore excursions arranged by and paid for through the cruise company. Additionally, many tourists return home via airlines from the cruise's final destination with weight limits on what they can carry home in their bags, thus limiting what they purchase at each stop. Further, ports of call compete with each other, and cruise companies can decide to stop visiting a port for reasons ranging from simply trying a new destination or concern over port infrastructure to perceptions of tourists about the value or safety of certain places. Generally, but not always, negative events such as unrest or local violence can cause a drop in the number of ships that visit a port one to two years later.

Tourism also remains susceptible to macroeconomic events, such as political upheavals or financial crises, and these undoubtedly affect the number of cruise passengers and stopover visitors to Guadeloupe. The twentieth century found Guadeloupeans still in the midst of international political, ideological, and economic forces, leading to challenges that revolve mainly around tourism at the human–environment interface. With a population of 400,000—and another 400,000 or more stopover visitors and dozens of cruise ships stopping annually—for example, plenty of challenges remain, such dealing with waste disposal and fresh water consumption. As one of the larger, more spread-out territories of the Lesser Antilles, Guadeloupe also encourages an increase in car ownership and rentals, leading to potential pollution problems.

10.6 Hazards

Like other Lesser Antillean islands, Guadeloupe is at risk of a number of hazards including hurricanes and other tropical storms, volcanic eruptions, earthquakes, tsunamis, and longer-term hydrologic and climatic changes such as sea level rise. The following subsections focus on how each of these has, and potentially will, affect the Guadeloupean islands.

10.6.1 Tropical Storms and Hurricanes

Hurricanes and tropical storms have struck or brushed the Guadeloupean islands dozens of times since the first European settlement. Just since 1851, over three dozen tropical storms or hurricanes have grazed or crossed part of the group of islands that make up Guadeloupe[3] (NOAA 2016). Some of these storms have caused significant damage and loss of life. The "Great Hurricane" of 1928 claimed 1200 Guadeloupean lives, and a hurricane in 1776 killed over 6000

[2]Despite searching through numerous sources, the only reliable data of tourism numbers for Guadeloupe are from the UNWTO. These data were limited to the years between 2009 and 2013, and the detailed report that the UNWTO produced for Guadeloupe contains significant gaps.

[3]Thirty-seven tropical storms or hurricanes have come within 70 km of Pointe-à-Pitre. This radius includes all of the territory of the department without including neighboring islands. As such some larger storms have had effects on the island such as storm surge, wind, or excessive rainfall events. As an example, Hurricane David in 1979 passed south of Dominica but was close enough to the island to caused major damage in Les Saintes, Marie-Galante, and Basse-Terre with crop damage estimated at over $100 million (NOAA 2016; Lawrence 1979).

Table 10.1 Estimated number of tourists (Stopover) by year. Data from United Nations World Tourism Organization (UNWTO 2015). Data for all years of cruise passenger arrivals and Overnight Stopover tourists for 2011 and 2012 rounded to nearest thousand by UNWTO

Year	Total overnight stopover tourists	Total cruise passenger arrivals
2009	346,507	111,000
2010	392,282	105,000
2011	317,000	102,000
2012	325,000	162,000
2013	487,416	158,000

Table 10.2 Significant Earthquake recorded events in Guadeloupe since 1690 with a Mercalli Intensity of IV or higher (Most Data from Dorel 1981: 694)

Date earthquake was felt in Guadeloupe	Mercalli intensity
April 5, 1690	IX
January 11, 1839	VI
February 8, 1843	IX
May 16, 1851	VII
April 29, 1897	IX
December 25, 1969	VI
October 8, 1974	VII
November 21, 2004	VIII*

Understanding the differences in Mercalli Intensity, the rating of VIII applies to the island of Terre-de-Bass, Les Saintes. It would be lower on parts of Basse-Terre. However, the magnitude was still M_w 6.3
*Intensity of VIII is based on magnitude and reports of damage in Les Saintes on November 21, 2004. This intensity would be lower in Basse-Terre, the capital and even lower in Point-á-Pitre

(Lawrence 1979; NHC 2005). Tropical storm activity does not have to directly "hit" the islands of Guadeloupe for there to be a substantial impact. For examples, Hurricane David destroyed the banana crop on the island, although the eye of the storm passed south of Dominica and west of Guadeloupe causing substantial damage to crops on the islands.

The introduction of more reliable, satellite-based tropical storm tracking and warning systems have substantially reduced the likelihood of extreme human casualty numbers, allowing people time to prepare and occupy more durable structures which are not likely to be destroyed by high winds or flooded by storm surge and high waves. Yet, this does not eliminate the potential damage to agriculture, nor do such warning systems reduce the likelihood or volume of landslides. For example, Tropical Storm Helena in October 1963, caused over 250 landslides of various sizes along the steepest slopes of the interior of Basse-Terre, mostly centered around the watersheds of Beaugendre, Vieux Habitants, Capesterre, and the upper parts of Bras-David and Goyave (Allemand et al. 2014). Allemand et al. (2014) extrapolated that this one extreme meteorological event caused a denudation rate of roughly 1.6 mm averaged over the few days, spread across the affected watersheds—the same amount of erosion that would take four to twelve years on average without extreme meteorological events in these watersheds (see also Le Bivic et al. 2014).

10.6.2 Volcanic, Seismic, and Tsunamic Activity

While each Guadeloupean island consists of base volcanic rock, today only La Soufrière on Basse-Terre is considered active. The volcano has erupted eight times since French settlement in 1635 (Global Volcanism Program 2013), most recently in between July 8, 1976 and March 1, 1977. This eruption period led to the evacuation of the Guadeloupean capital, and second largest city, Basse-Terre, as well as surrounding areas that ultimately displaced over 72,000 people for a few months (Fiske 1979). Even with better monitoring and advances in understanding of volcanos in general—and La Soufrière in particular—there are still risks, such as people settling a mere 2.5 km from of the summit.

Lying along and near many faults, each of the islands also remains susceptible to damage from earthquakes, recorded as far back as April 5, 1690 (Table 10.2). A relatively recent earthquake in Guadeloupe occurred on November 21, 2004. With the epicenter located approximately 15 km southeast of the Les Saintes archipelago, it killed a five-year-old girl in Trois-Rivières in southern Basse-Terre Island and injured one other person on Basse-Terre and ten people on the Les Saintes' island of Terre-de-Bas (Nikolkina et al. 2010; Zahibo et al. 2005). Numerous buildings were severely damaged and destroyed on Terre-de-Bass and in Trois-Rivières, and the damaging effects were noted in

Portsmouth, Dominica (25 km from the epicenter), and Pointe-à-Pitre (65 miles from the epicenter). The most intense seismic events do not demonstrate a regular frequency, partly because the study of earthquakes in Guadeloupe is based on fairly short mechanically captured seismic histories and fairly recent historical records.

Connected to earthquakes and volcanic activity, Guadeloupe has experienced a number of tsunamis in its recorded history. Rogers et al. (2013) describe a number of factors that influence the likelihood and height of tsunamis. In researching for a predictive model of the tsunami that could have been felt after the February 08, 1843 megathrust event, the authors noted that while the earthquake was quite destructive on parts of Guadeloupe and recorded in historical records, they did not find anything in historical records about an associated tsunami. More recent events like the 2004 Les Saintes earthquake and three pyroclastic flows from nearby Montserrat reaching the sea (1997, 2003, and 2006), however, propagated one- to two-meter-tall tsunamis. Researchers indicate that distance from seafloor to sea surface, reefs, approaching direction of a particular tsunami, and even the locational relationship of some of the smaller islands of the Guadeloupean archipelago have significant influence amplifying or mitigated the damaging effects of a major tsunami (Rogers et al. 2013). This does not mean that tsunamis are insignificant events, as even the November 21, 2004 Mw 6.3 earthquake caused an 80-cm drop and 70-cm rise, which created a significant amount of water force striking the coast (Zahibo et al. 2005).

10.6.3 Hydrologic and Climatic Hazards

Research indicates that sea level rise and climate change are no longer processes predicted for the future. Evidence indicates that for decades, ocean and sea levels have risen and that this will continue to do so for at least the rest of this century. The rising levels of the major saltwater bodies of the Earth, like the Caribbean Sea and Atlantic Ocean, are due, at least in part, to anthropological forces such as major deforestation—often through burning—and the use of carbon-based fuels, leading to climate change on a global scale.[4] Understanding the full effects of climate change in a particular area requires a deep understanding of the global, regional, and local atmospheric, geologic, and hydrological processes.

Sea level rise (SLR) cannot be ignored on Guadeloupe and its many islands, but its effects will not be evenly distributed. Like some other Caribbean islands, Guadeloupe has a large population with a significant tourism industry that is, in part, based on its sea resources—beaches, food, and hospitality venues, as well as coral reef diving and mangrove habitats. Scott et al. (2012) predict an SLR range of 18–59 cm between 1900 and 2100, noting SLR does not follow a global pattern, but that the Caribbean in general will experience it more dramatically (Compare Tamisiea and Mitrovica 2011). While trying to understand all of the processes associated with sea level rise and attempting to predict it, these numbers—as small as they seem—will affect mangroves and coral reefs causing damage in some areas that will remove or degrade some of the coastal protections from which the islands of Guadeloupe have benefited. The damages of SLR can include saltwater intrusion, beach erosion, and even the endangerment of seaside buildings and infrastructure (Scott et al. 2012).

Just as the effects of SLR will vary from location to location, so will the effects of climate change, especially in terms of Guadeloupe's karst landscape. As a dissolutional landscape, karst regions remain at the mercy of precipitation. As precipitation increases, so does erosion in karst landscapes, beginning slowly, but eventually giving way to collapse (e.g., Fig. 10.9). Day (2010) notes that not only will climate change affect Caribbean karst landscapes, but the eroding landscape will, eventually, also affect biological systems as well as social and economic systems. Thus, climate change will have an impact on the broader set of processes of environmental change. Scientists expect biodiversity to decrease (May 2010). The acceleration of karst damage will affect karst hydrology with the potential for more seasonal flooding and prolonged drought (May 2010). The cause of the more frequent dry season drought will be due to a reduction in soil moisture budgets, with soil having more difficulty holding moisture for longer periods of time. Collapsing stone deposits and the denuding topsoil due to increased flash flooding—events that start and end quickly, yet are devastating when they occur—can also open up points of quick infiltration of water into the local water tables. Rapid infiltration of water into subterranean water conduits makes these karst water sources vulnerable to pollutants, including biological contaminants such as microorganisms.

10.7 Conclusion

Geologically, Guadeloupe remains representative of the Volcanic Caribbees and the Limestone Caribbees, as well as an older, Late Jurassic igneous formation on the eastern end of the easternmost island La Désirade. Traveling from east to west, the island group as a whole becomes more mountainous with the active volcano, La Soufrière at 1467 m,

[4]The use of the concept "climate change" is intentional. It recognizes that in some areas temperature might actually decrease as well as major changes in other climate factors such as rainfall, air quality, and power and direction of ocean currents.

serving as the highest point not only in Guadeloupe, but also in the entire Lesser Antilles. The main island of Basse-Terre with its mainly volcanic landscape sits juxtaposed next to Grande-Terre that, while also relatively flat, contains older volcanics overlain by more recent, thick lithologies of limestone. The other dozen or so inhabited islands that comprise Guadeloupe remain just as varied geologically as the two main (and much larger-in-size) islands: some igneous, others limestones, and varying in age from Late Jurassic to recent volcanic eruptions.

Humans visited Guadeloupe as early as 2000 BCE, and numerous Amerindian migrations occurred before Columbus sighted Guadeloupe in November 1493. In the late 1630s, hundreds of French Colonists arrived with the intention to exclusively exploit Guadeloupe, a goal they achieved barely a decade later. Tobacco farms gave way to sugar plantations in the late seventeenth century, and hundreds of thousands of enslaved Africans were imported to transform the landscape. Even after emancipation in 1848, tens of thousands of additional indentured servants were brought in over the next fifty years (Northrup 2000).

Tourism and agriculture represent important industries in Guadeloupe today, but there is still a need to further diversify the economy. While hundreds of thousands of tourists visit Guadeloupe's intriguing landscapes each year, it hosts a multibillion dollar agricultural sector. Yet, tourism and agriculture have not led to strong employment on Guadeloupe's islands. Even many of the residents who do work, find it difficult to live because the cost of living is so high, even though, along with Martinique, Guadeloupe retains the highest income for minimum wage workers in the Caribbean. These economic difficulties will remain a challenge facing Guadeloupe into the foreseeable future. Even among the young adults who doubtlessly feel the pull to places where they might have better economic opportunities, some return after getting their college degrees in Metropole France with the determination to help Guadeloupe meet its challenges in the future.

Though susceptible to common Caribbean island hazards (e.g., hurricanes, volcanic eruptions, tsunamis, and sea level rise), Guadeloupe remains a geomorphologically varied island. Visitors can encounter any number of landforms and resultant processes by taking a short walk from any of the main towns. Guadeloupe's complex geology yields many types of distinct volcanic, karst, and coastal landforms: from abundant geothermal springs, steep-relief slopes, and tall waterfalls to dolines, sea arches, and crater lakes. Indeed, when it comes to breadth of landform type, Guadeloupe may be one of the most varied in the Caribbean. Hopefully, its natural beauty continues to inspire researchers in the coming decades, to help more fully understand its geomorphologic diversity.

References

Allaire L (1997) The lesser antilles before columbus. In: Wilson SM (ed) The indigenous people of the Caribbean. University of Florida Press, Gainesville, pp 20–28

Allemand P, Delacourt C, Lajeunesse E, Devauchelle O, Beauducel F (2014) Erosive effects of the storm Helena (1963) on Basse Terre Island (Guadeloupe—lesser Antilles Arc). Geomorphology 206 (2014):79–86

Arbell M (2002) The Jewish Nation of the Caribbean: the Spanish-Portuguese Jewish Settlements in the Guianas. Gefen Publishing House, Jerusalem

Bélénus R (2015) How slavery became reestablished in guadeloupe and the French colonies. In: L'Étang T (ed) Memorial ACTe: exploring slavery and the African slave trade in the caribbean and around the world (trans: Beaver S, Whittaker J, Ricaut J). Caribbean Centre for the Expressions and Memory of African Slave Trade & Slavery, Pointe-à-Pitre, Guadeloupe, pp 280–285

Bonilla Y (2009) Guadeloupe on strike: a new political chapter in the French Antilles. NACLA report on the Americas, May/June 2009, pp 6–10

Boucher P (1992) Cannibal encounters: Europeans and Island Caribs. Johns Hopkins University Press, Baltimore, pp 1492–1763

Boudon G, Komorowski J-C, Villemant B, Semet MP (2008) A new scenario for the last magmatic eruption of La Soufrière of Guadeloupe (Lesser Antilles) in 1530 A.D. Evidence from stratigraphy radiocarbon dating and magmatic evolution of erupted products. J Volcanol Geoth Res 178(3):474–490

Bouysse P, Westercamp D, Andreieff P (1990) The lesser antilles Island Arc. Proc ODP Sci Results 110(1):29–44

Cérol MJ (1992) What history tells us about the development of creole in guadeloupe. New WestIndian Guide/Niéuwe West-Indische Gids 66(1&2):61–76

Collina-Girard J (2002) Underwater mapping of late quaternary submerged shorelines in the western mediterranean sea and the caribbean sea. Quatern Int 92:63–72

Cornée J-J, Léticée J-L, Münch P, Quillévérés F, Lebrun J-F, Moissette P, Braga J-C, M-D M, de Min L, Oudet J, Randrianasolo A (2012) Sedimentology, palaeoenvironments and biostratigraphy of the Pliocene-Pleistocene carbonate platform of Grande-Terre (Guadeloupe, Lesser Antilles forearc). Sedimentology 59:1426–1451

Corsini M, Lardeaux JM, Verati C, Voitus E, Balagne (2011) Discovery of lower cretaceous synmetamorphic thrust tectonics in French lesser Antilles (La Désirade Island, Guadeloupe): implications for caribbean geodynamics. tectonics 30(TC4005):1–15

Cosgrove D (1985) Prospect, perspective and the evolution of the landscape idea. Trans Inst Br Geogr 10(1):45–62

Creswell T (2003) Landscape and the obliteration of practice. In: Anderson K, Domash M, Pile S, Thrift N (eds) Handbook of cultural geography. pp 269–281

Curtain PD (1969) The Atlantic slave trade: a census. University of Wisconsin Press, Madison

Delpuech A (2011) Archéologie historique en Guadeloupe. Les nouvelles de l'archéologie 108/109 Web address: http://nda.revues.org/156

DeMets C, Jansma, PE, Mattioli GS, Dixon TH, Farina F, Bilham R, Calais E, Mann P (2000) GPS geodetic constraints on Caribbean-North America plate motion. Geophys Res Lett 437–440

Dorel J (1981) Seismicity and seismic gap in the Lesser Antilles arc and earthquake hazard in Guadeloupe. Geophys J Int 67(3):679–695

Dunn RS (2012) Sugar and slaves: the rise of the planter class in the english West Indies. University of North Carolina Press, Chapel Hill, NC, pp 1625–1713

Feuillet N, Manighetti I, Tapponnier P, Jacques E (2002) Arc parallel extension and localization of volcanic complexes in Guadeloupe, Lesser Antilles. J Geophys Res: Solid Earth 107(B12)

Fiske RS (1979) Volcano hazards—lessons learned in the Eastern Caribbean. In: Whitmore FC, Williams ME (eds) Resources for the twenty-first century—proceedings of the international centennial symposium of the U.S. Geological Survey, Held at Reston, Virginia, 14–19 Oct 1979, pp 256–260

Frédéric R (2015) Plantation Society. In: L'Étang T (ed) Memorial ACTe: Exploring Slavery and the African Slave Trade in the Caribbean and around the World (trans: Beaver S, Whittaker J, Ricaut J). Caribbean Centre for the Expressions and Memory of African Slave Trade & Slavery, Pointe-à-Pitre, Guadeloupe, pp 164–165

Global Volcanism Program (2013) Soufrière Guadeloupe (360060) in volcanoes of the world, v. 4.4.3. In: Venzke E (ed) Smithsonian institution. Accessed 16 Jun 2016, Web Address: http://volcano.si.edu/volcano.cfm?vn=360060

Heath E (2011) Creating rural citizens in guadeloupe in the early third French Republic. Slavery Abolition 32(2):289–307

Heiken G, Crowe B, McGetchin T, West F, Eichelberger J, Bartram D, Peterson R, Wohletz K (1980) Phreatic eruption clouds: the activity of soufrière de gaudeloupe, F.W.I., August–October 1976. Bull Volcanol 43(2):383–395

Higman BW (2011) A concise history of the Caribbean. Cambridge University Press, Cambridge

Institut national de la statisique et des etudes économiques (INSEE) (2015a) Enquête emploi en continu en Guadeloupe : Stabilité du chômage en 2015. Accessed 25 May 2016. http://www.insee.fr/fr/themes/document.asp?reg_id=26&ref_id=24110

Institut national de la statisique et des etudes économiques (INSEE) (2015b) EPCI de La CC de Marie-Galante (249710047). Accessed on: 19 June 2016. Web address: http://www.insee.fr/fr/themes/comparateur.asp?codgeo=epci-249710047

Institut national de la statisique et des etudes économiques (INSEE) (2016) Population de 1968 à 2015: comparaisons régionales et départementales. Accessed 25 May 2016. http://www.insee.fr/fr/themes/tableau.asp?reg_id=99&ref_id=TCRD_004#col_1=8

Jaud P, Lamethe D (1985) The bouillante geothermal power-plant. Guadeloupe. Geothermics 14(2–3):197–205

Jennings Eric T (2001) Vichy in the Tropics: Pétain's National Revolution in Madagascar, Guadeloupe and Indochina 1940–1944. Standford University Press, Stanford, CA

Lallier-Vergés E, Marchand C, Disnar J-R, Lottier Nathalie (2008) Origin and diagenesis of lignin and carbohydrates in mangrove sediments of Guadeloupe (French West Indies): evidence for a two-step evolution of organic deposits. Chem Geol 255:388–398

Lawrence M (1979) Hurricane david preliminary report, page 3. National Hurricane Center. Accessed June 1 2016. Web Address: http://www.nhc.noaa.gov/archive/storm_wallets/atlantic/atl1979-prelim/david/prelim03.gif

Le Bivic R, Allemand P, Delacourt C, Quiquerez A (2014) Erosive effects of the storms HELENA (1963) and HUGO (1989) on Basse-Terre island (Guadeloupe-Lesser Antilles Arc). Geophysical Research Abstracts, vol 16, EGU2014–11541

Lenoble A, Queffelec A, Bonnissent D, Stouvenot C (2015) Rock art taphonomy in Lesser Antilles: study of wall weathering and engravings preservation in two preColumbian caves on Marie-Galante Island. Proceedings of the 25th International Congress for Caribbean Archaeology. pp 634–657

Mattinson JM, Fink LK Jr, Hopson CA (1980) Geochronologic and isotopic study of the La Désirade island basement complex: Jurassic oceanic crust in the Lesser Antilles? Contrib Miner Petrol 71:237–245

Mattinson JM, Pessagno EA, Jr, Montgomery H, Hopson CA (2008) Late Jurassic age of oceanic basement at La Désirade Island, Lesser Antilles arc. Geol Soc Am Special Papers 438:175–190

May M (2010) Challenges to sustainability in the Caribbean karst. Geol Croat 62(2):149–154

National Hurricane Center (NHC) (2005) The deadliest atlantic tropical cyclones, 1492–1996. Web Address: http://www.nhc.noaa.gov/pastdeadlyapp1.shtml. Accessed on: 15 June 2016

National Oceanic and Atmospheric Administration (NOAA). (2016) Historical hurricane tracks. Accessed 15 June 2016. https://coast.noaa.gov/hurricanes/

Nikolkina I, Zahibo N, Pelinovsky E (2010) Tsunami in Guadeloupe (Caribbean Sea). The Open Oceanogr J 4:44–49

Northrup D (2000) Indentured Indians in the French Antilles. Les immigrants indiens engagés aux Antilles françaises. Revue française d'histoire d'outre-mer. 87(325):245–271

Nuryanti W (1996) Heritage and postmodern tourism. Ann Tourism Res 23(2):249–260

Pendery SR (2010) A survey of French Fortifications in the New World 1530–1650. First Forts: Essays on the Archaeology of Proto-colonial Fortifications. Leiden: Brill, pp 41–63

Pons FM (2013) History of the caribbean. Markus Wiener Publishers, Princeton, NJ

Pruett L, Cimino J (2000) Unpublished data and figures calculated by Pruett and Cimino using "World Vector Shoreline, United States Defense Mapping Agency 1989"

Réserves Naturelles de France (2016) Îles de la Petite-Terre. Web address: http://www.reserves-naturelles.org/iles-de-la-petite-terre. Accessed 19 June 2016

Reuters. (2007) Hurricane destroys Martinique, Guadeloupe bananas. Accessed 24 May 2016. http://www.reuters.com/article/us-storm-dean-bananas-idUSL1861011420070818

Robertson REA (2009) Antilles geology. Encyclopedia of islands. University of California Press, pp 29–35

Rodigneaux M (2015) The Pirates of Guadeloupe Et the Quasi-War with the United States. In: L'Étang T (ed) Memorial ACTe: Exploring Slavery and the African Slave Trade in the Caribbean and around the World (trans: Beaver S, Whittaker J, Ricaut J). Caribbean Centre for the Expressions and Memory of African Slave Trade & Slavery, Pointe-à-Pitre, Guadeloupe, p 270

Rodríguez M (1997) Religious beliefs of the saladoid people. In: Wilson SM (ed) The indigenous people of the caribbean. University of Florida Press, Gainesville, FL, pp 80–87

Rogers J, Dudon B, Zahibo N (2013) Tsunami hazard assessment of Guadeloupe Island (F.W.I.) related to a megathrust rupture on the Lesser Antilles subduction interface. Nat Hazards Earth Syst Sci 13:1169–1183

Rogoziński J (1999) A brief history of the caribbean: from the arawak and carib to the present. Plume, New York

Samper A, Quidelleur P, Mollex D (2007) Timing of effusive volcanism and collapse events within an oceanic arc island: Basse-Terre, Guadeloupe archipelago (Lesser Antilles Arc). Earth Planet Sc Lett 175–191

Scott D, Simpson MC, Sim Ryan (2012) The vulnerability of Caribbean coastal tourism to scenarios of climate change related sea level rise. J Sustain Tourism 20(6):883–898

Tamisiea ME, Mitrovica JX (2011) The moving boundaries of sea level change: understanding the origins of geographic variability. Oceanography 24(2) 24–39

Tingbrand Per (2002) A swedish interlude in the caribbean. Forum Navale 57:64–92

United Nations World Tourism Organization UNWTO (2015) Guadeloupe: country-specific: arrivals of non-resident tourists at national borders, by country of residence 2009–2013 (04.2015)

U.S. Department of State: Bureau of Consular Affairs (2016) U.S. Passports and International Travel: France. Last Modified: February 29 2016. Last Accessed 31 May 2016. https://travel.state.gov/content/passports/en/country/france.html

U.S. Geological Survey (USGS) (2013) The severity of an earthquake. Accessed on 16 June 2016. Web address: http://pubs.usgs.gov/gip/

earthq4/severitygip.html. Page Last Modified: 11 January 2013, 12:51:44 PM

U.S. Geological Survey (USGS) (2016) Magnitude/ Intensity Comparison. Accessed on 16 June 2016. Web address: http://earthquake.usgs.gov/learn/topics/mag_vs_int.php. Page Last Modified: 07 April 2016 17:22:31 UTC

Watts D (1987) The West Indies patterns of development, culture, and environmental change since 1492. Cambridge University Press, Cambridge, UK

Wong A (2015) Caribbean island tourism: pathways to continued colonial servitude. Études caribéennes 31–32. Accessed 5 April 2017. http://etudescaribeennes.revues.org/7524

Zahibo N, Pelinovsky E, Okat E, Yalçiner A, Kharif C, Talipova T, Kozelkov A (2005) The earthquake and tsunami of november 21 2004 at Les Saintes, Guadeloupe. Lesser Antilles. Sci Tsunami Hazards 23(1):25–39

Dominica

11

Vanessa Slinger-Friedman

Abstract

Located along the eastern boundary of the Caribbean Plate, Dominica is one of the most volcanically active countries in the Caribbean. Its volcanic nature results in a rugged mountainous landscape that is lush and fertile and creates the backdrop for a rich biodiversity of plants and animals. Dominica, with its mountain peaks, subaerial volcanoes, crater lake, coastal dike, and numerous rivers, has been shaped over millions of years by volcanic and hydrological processes, which have built up and worn down the island, respectively. Dominica's physical geography played a role in helping the native Carib Amerindians resist European colonization, and today contributes to Dominica's economy by supplying the landscape for the agriculture, fishing, and tourism that takes place on the island. Dominica's economy and parts of its culture are a reflection of its landscapes and landforms. Conversely, the influence of various cultures and economic activities on Dominica is reflected in impacts on the geomorphic landscape.

Keywords

Dominica • Volcanic • Biodiversity • Caribs • Ecotourism

11.1 Introduction

Dominica, the largest and northernmost Windward Island in the Caribbean, is located in the central Lesser Antilles (Fig. 11.1). The volcanic nature of this island means that it is replete with high, rocky coasts and a mountainous spine, peaked by the highest mountain in the eastern Caribbean, Morne Diablotins (1447 m). At least 60% of the island is covered in slopes of 30° or more. There are clear signs of geologically recent volcanic activity on Dominica, including hot springs, the warm bubbling waters of Champagne Beach, and the famous flooded fumarole known as the "Boiling Lake" in the Valley of Desolation (Caribbean Geology 1994).

While its 92 miles of coastline provides Dominica with domestic fishing grounds and some offshore reefs, the volcanically based, fertile mountainous interior of Dominica is covered by dense rainforest, which incorporates 60% of the island (Fig. 11.2). This forest and some of its inhabitants are protected by three national parks. The southernmost park, centered on Morne Trois Pitons, a UNESCO World Heritage Site, is considered to be one of the richest biodiversity spots in the Caribbean. The combination of steep terrain and high annual rainfall across the island provides the conditions for numerous rivers, streams, and waterfalls.

When Christopher Columbus tried to describe Dominica's appearance to the King and Queen of Spain, he supposedly threw down a piece of crumpled parchment to indicate its many mountains and rugged nature (Fig. 3.3; "Dominica" New World Encyclopedia 2013). The Caribs, the indigenous people residing on the island at the time of Columbus's arrival, had

V. Slinger-Friedman (✉)
Kennesaw State University, MD 2201, SO Bldg. 22, Rm. 4042, GA 30144-5591 Kennesaw, GA, USA
e-mail: vslinger@kennesaw.edu

C.D. Allen (ed.), *Landscapes and Landforms of the Lesser Antilles*,
World Geomorphological Landscapes, DOI 10.1007/978-3-319-55787-8_11

Fig. 11.1 General physiographic map of Dominica. Note the complex topography and various volcanic peaks positioned along a central mountain ridge that runs the length of the island. Cartography by K.M. Groom

Fig. 11.2 Hills of Marigot in the northeast of Dominica as seen from the air. These densely vegetated hills represent the rugged and forested landscape that dominates Dominica. Photograph by V. Slinger-Friedman

named the land "Wai'tukubuli" meaning "tall is her body" (Honychurch 1991). By the 1700s, control of Dominica changed hands from the French to the British, and the Caribs on the island were the only remaining indigenous group living in the Caribbean. Today, they live as a community on a 3700 acres (1500 hectare) mountainside reserve on the northeast of Dominica (Slinger 2000).

Dominica's economy has been dependent on agriculture, particularly bananas that thrive in the fertile volcanic soils. More recently, cruise ship tourism and ecotourism are the main drivers of Dominica's economy. Dominica's rugged landscape and volcanic black sand beaches deterred the creation of a traditional three "S"—sun, sea, and sand—mass tourism on the island. Instead, in the 1990s, the Dominican government began to focus on its varied and distinctive landscape, tropical vegetation, and unique cultural groups to develop Dominica as an ecotourism destination. Both agriculture and tourism have been adversely affected by the hazards of hurricanes.

11.2 Setting

11.2.1 Dominica's Interior

Dominica, and nearby Martinique and Guadeloupe, all have areas greater than 750 km^2 and lie at the center of the Lesser Antilles western, and geologically younger, Island Arc (Steiner 2003). Dominica was created starting with volcanic eruptions along the eastern margin of the Caribbean Plate. With the exception of some sedimentary rocks and corals along its west coast, Dominica is predominantly made up of andesite and dacitic volcanic rocks (DeGraff 1998). Rocks from three different periods or arcs—Miocene, Pliocene, and Quaternary—compose Dominica (Smith et al. 2013). Now deeply decayed, the oldest Miocene rocks, dating between 6.92 and 5.22 million years ago (Ma), form the basal portion of the island and are found along the eastern Atlantic coast (McCarthy et al. 2005). A huge stratovolcano, named the Cochrane-Mahaut Centre, was formed out of lava and ash during the Pliocene. This landform has eroded over time and now exists as remnants in the form of the mountain peaks of the south central part of the island (including Morne Cabrits Marons, Morne Cola Anglais, Deux Saisons, Morne Boyer, Morne Couronne, and Morne Negres Marrons). Further to the center and north, there are the highly dissected remnants of other Pliocene (2.83 Ma) volcanoes, including Morne Espagnol and Pelean, which are associated with Pliocene activity at Morne Diablotins. These areas of Pliocene rocks do not experience seismic or geothermal activity and are considered unlikely to become volcanically active again.

Older Pleistocene rocks and volcanoes are located in the north of Dominica, some built upon a Pliocene rock base. Examples of these areas include the peninsula of Morne aux Diables and the largest volcano on the island, Morne Diablotins. Seismic activity in these areas indicates that there is the possibility of future eruptions. However, about 1 Ma ago during the younger Pleistocene period, the center of volcanism switched to the south of Dominica with six major active volcanoes, including Morne Trois Pitons, Wotten Waven, Watt Mountain, Grand Soufrière Hills, Morne Anglais, and the Plat Pays volcanic complex. The youngest dated volcanic deposits on Dominica are on the large dome of Morne Patates, which is associated with the active Plat Pays Volcano (Roobol, J, and Smith, n.d.). Also active in the south is the Valley of Desolation, a major geothermal area. Steam explosions, or phreatic activity, last took place here in 1997. Dominica has nine active subaerial volcanoes, each with its own radial drainage system; however, it has not experienced a major magmatic eruption since the eruption of Morne Patates about 500 years ago (Steiner 2015). For this reason, Dominica has well-preserved tropical rainforest. Currently, the volcanoes in the south of Dominica are linked with geothermal and/or seismic activity and are considered likely to erupt again (Roobol and Smith, n.d.). This causes some concern since Roseau, the country's south-based capital, lies in the potential path of pyroclastic flows from the Wotten Waven caldera on its eastern side (Fig. 11.3).

11.2.2 Dominica's Coastal Setting

Dominica's coastal shelf is narrow, measuring less than 1 km wide in parts. Its narrow, stony, or black sand beaches exist mainly in embayments, separated by lengths of steep cliffs. Embayments on the south and west of the island generally have straight sections (Fig. 11.4). Comparatively, embayments on the north and east sides are more indented (Cambers 1996). There is an area of coral sand and volcanic sand beaches on the north coast between Hampstead and Woodford Hill. A narrow coastal shelf and the impact of fluvial sediment outfall from its many rivers limit Dominica's environment for reef development (Steiner 2003). The distribution of the island's reefs has been significantly influenced by the northeast trade winds and the impact of storm-induced waves. The eastern, windward side of Dominica lacks a wide island shelf and protective structures such as barrier reefs, and also experiences turbulent waters and the full force of Atlantic waves (Fig. 11.5). As a result, there is a contrast between the sessile communities on the windward and leeward (western) side of the island. The eastern side of Dominica has only a few coral reefs that can be found in north-facing sub-bays that give protection from the Atlantic's waves. The larger reefs exist on the calmer, southwestern section of the leeward side of Dominica, in areas with wide coast shelves and in some of the bays along the north coast (Steiner 2015).

Overall, the island is characterized by high-relief terrestrial topography such as domes, peaks, and ridges along with a narrow sublittoral zone. The country's central spine is a northwest–southeast axis of steep volcanic slopes and deep gorges with an elevation that varies between 300 and 1400 m. The narrow coastal plain is studded with sea cliffs, and typically has narrow volcanic stone or black sand beaches with limited development of coral reefs.

11.3 Landforms

11.3.1 Terrain Features

Many of Dominica's landforms relate to its volcanic nature. The interior of Dominica features rugged mountains and ridges, many of which are volcanic, the highest of which is Morne Diablotins (1447 m). Dominica has nine volcanoes, compared to the other volcanic islands of the Lesser Antilles that have only one each. Morne Trois Pitons (1387 m) is the second highest mountain on the island and a dormant volcano (Fig. 11.6). It is shaped by three peaks at its top and located in the south within the Morne Trois Pitons National Park, a UNESCO World Heritage Site (whc.unesco.org). The park is the site of verdant tropical rainforests, 50 fumaroles, many hot springs, a "Boiling Lake," and five volcanoes. The Morne Trois Pitons National Park was inscribed as a World Heritage Site based on the World Heritage List's Natural Criteria (viii) and (x):

> Criterion (viii): Be outstanding examples representing major stages of earth's history, including the record of life, significant on-going geological processes in the development of landforms, or significant geomorphic or physiographic features (IUCN 2008, p. 13)

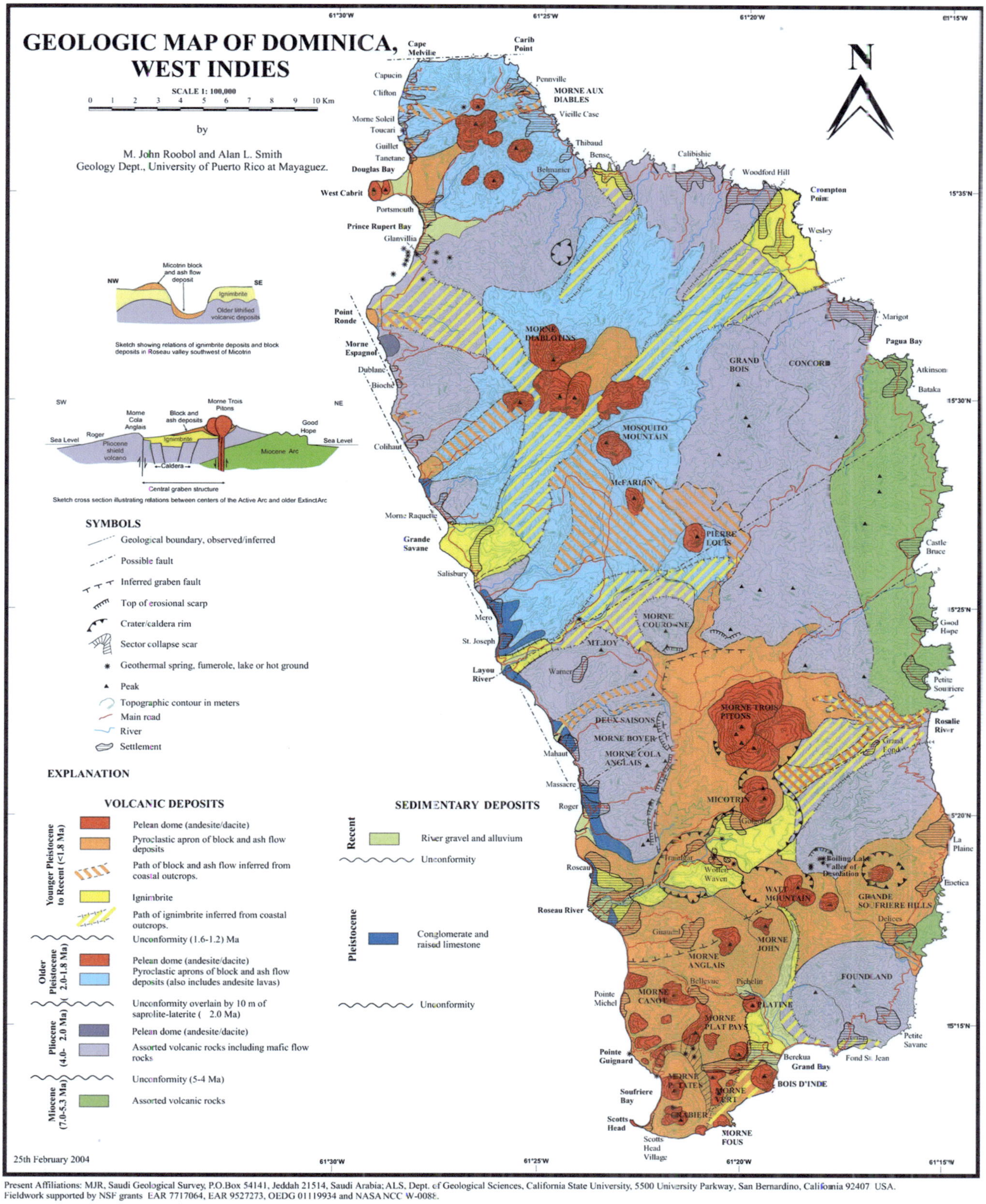

Fig. 11.3 Digital elevation model (DEM) of Dominica displaying its rugged topography. Image courtesy of http://caribbeanvolcanoes.com/green-volcanoes/, retrieved January 2017

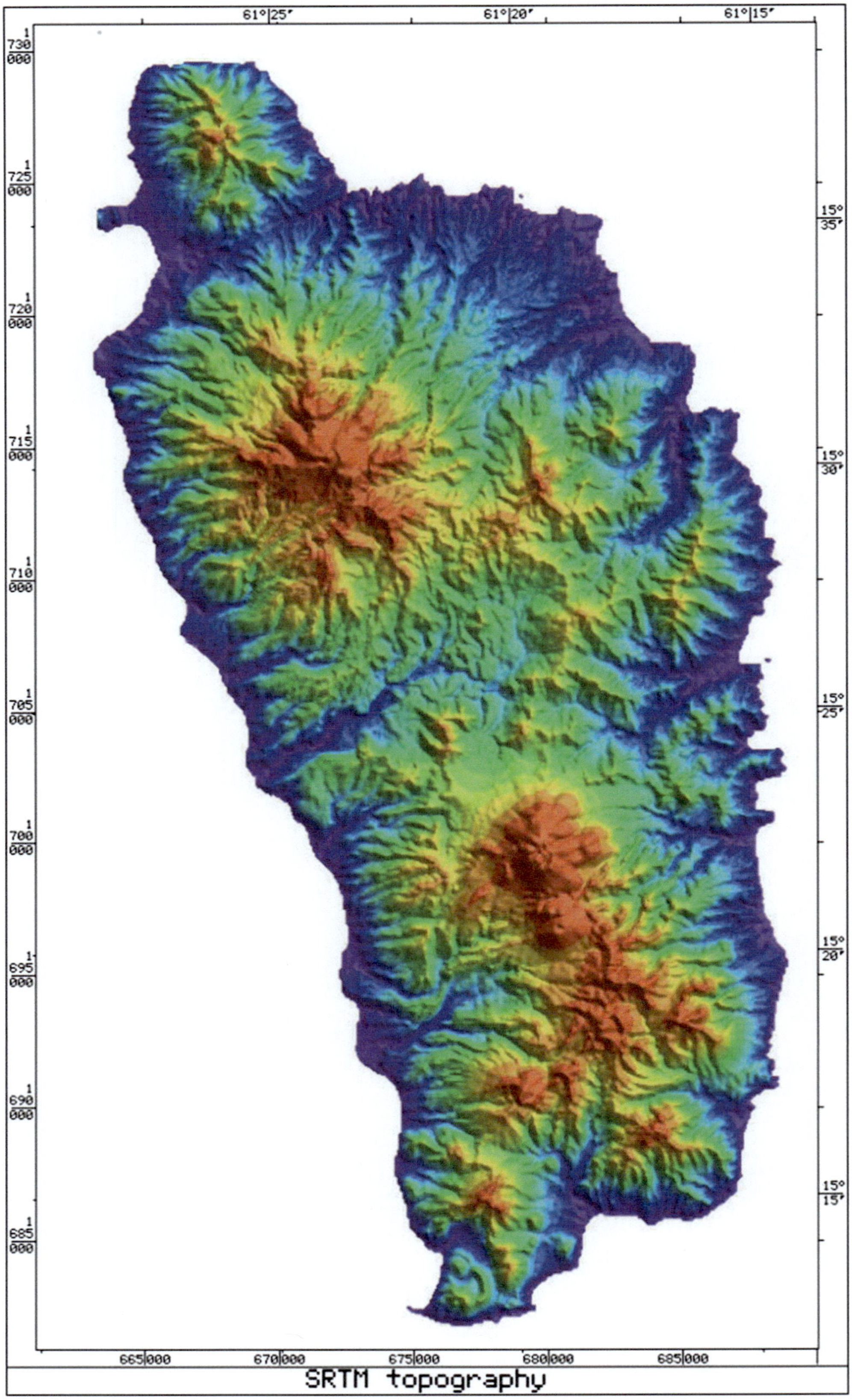

Fig. 11.4 Geologic map with accompanying profile of Dominica, demonstrating its complex geology and volcanics. Image courtesy of Caribbean Handbook on Risk Information Management, available at http://www.charim.net/sites/default/files/handbook/usecase/8/83%20dominica%20geological%20map.jpg (authorship embedded in map), retrieved January 2017

Fig. 11.5 Cove near the hamlet of Salibia on Dominica's east coast showing the island's narrow, steep-sided coast with exposed volcanic rocks, with a small rocky beach in the lower left corner of the image and the formation of sea caves along the cliffs in the background. Photograph by V. Slinger-Friedman

> Criterion (x): contain the most important and significant natural habitats for in situ conservation of biological diversity, including those containing threatened species of outstanding universal value from the point of view of science or conservation. (IUCN 2008, p. 15)

Found within the Morne Trois Pitons National Park at an altitude of 750 m, Boiling Lake is the world's second largest high-temperature volcanic crater lake after New Zealand's Frying Pan Lake, at approximately 63 m across (Fournier et al. 2009; Fig. 11.7a). The lake is an area of high hydrothermal activity generally considered to be very stable while maintaining a constant temperature (180–197 °F or 80–90 °C), occasionally punctuated by irregular periods of rapid draining and refilling. During these times, it can drain within 24 h and fill back up just as quickly (Fournier et al. 2009). The lake is a flooded fumarole, with a vent in the earth's surface that leads directly down to volcanic magma. Filled by rainwater and a perennial stream that feeds into it, Boiling Lake's exit flows into the Riviere Blanche (White River), so named because of its milky white color brought on by the Lake's high sulfur content. The White River is also the source for Victoria Falls, one of the many waterfalls in Dominica.

Along the trail to the Boiling Lake lies the Valley of Desolation (Fig. 11.7b). This active volcanic area has steam vents, fumaroles, hot springs, and boiling mud pools (Holahan 2009). There are almost always fumarolic clouds in the air, smelling of sulfur (Fontaine 2007). The hike to the Boiling Lake starts at Titou Gorge, a deep pool warmed by a hot spring. It then passes through an area of tropical rainforest before transitioning into the Valley of Desolation,

Fig. 11.6 Morne Trois Pitons …Photograph by J. Hains, available online at http://static.panoramio.com/photos/original/42592079.jpg, from http://www.worldtravelserver.com/travel/en/dominica/rosalie/gallery_rosalie/photo_42592079-Morne+Trois+Pitons+-+Dominica.html. Retrieved January 2017

whose harsh environment only supports limited vegetation in the form of mosses, lichens, and grasses. Hot springs arising from submerged fumaroles, and similar to the ones in the Valley of Desolation, can be also found in several other parts of Dominica, including those at Wotten Waven in the south, and in the dormant crater of Morne Aux Diables in the northern tip of the island.

11.3.2 Hydrological Features

Dominica has a humid tropical marine climate. Average yearly rainfall varies from 500 cm (196.9 in) on the windward coast, 900 cm (354.3 in) in the interior, and 180 cm (70.9 in) on the leeward coast. With 83 significant waterways, a total of 365 rills and brooks, and many freshwater lakes, Dominica is resplendent with water resources. Trafalgar Falls (Fig. 11.8) consists of twin waterfalls that have cold-water pools downstream and hot ponds to one side that are filled by springs from underground, and Middleham Falls (Fig. 11.9)—one of the Dominica's tallest—sits in a pristine setting in the northwest of Morne Trois Pitons National Park (Ralston 2014). Boeri Lake, a circular shape crater lake, is the highest lake in Dominica at 869 m. Both Boeri Lake and Freshwater Lake are located in the Morne Trois Pitons National Park, with Freshwater Lake in the lip of a volcanic crater (Ralston 2014). They are separated by Morne Micotrin (Macaque), which was formed in the crater of the old volcano. Dominica's widest river, the Indian River, originates from Morne Diablotins and exits out into the Caribbean Sea at Portsmouth in the north (Fig. 11.10a). Flowing through a coastal wetland and mangroves, and lined with buttressed Bwa Mang trees whose roots reach out up to 6 m, this brackish river is home to barracudas, crayfish, and the green heron (Love and Gabbett 2012).

Fig. 11.7 Examples of geothermal/volcanic activity on Dominica. **a** Boiling Lake, a volcanic crater lake formed by the flooding of a fumarole, is a World Heritage Site and a popular tourist destination. The lake is located near Watt Mountain in the southern range. **b** Sulfuric river valley and steam vents in the Valley of Desolation, an active volcanic field with numerous vents and hot springs just south of Watt Mountain. Photographs by V. Slinger-Friedman

11.3.3 Coastal and Marine Features

The volcanic activity on Dominica extends to the coast and impacts some of the coastal and marine landforms. Dominica's narrow coastal zone is composed of steep rocky shorelines and volcanic black sand beaches such as Londonderry Beach and Castle Bruce Beach on Dominica's eastern coast. At Wavine Cyrique, water from a river plunges almost 46 m into the tidal surges of the Atlantic (Ralston 2014). Similarly, Crayfish (or Isulukati) River has a waterfall that cascades from rock pools into the ocean (Fig. 11.10b). In the northeast, an ancient volcanic rock formation known as L'escalier Tete Chien forms a natural staircase to the sea. This volcanic dike occurred when a sheet of rock formed in a fracture in the preexisting rock. The dike rock is impermeable and becomes more prominent as the surrounding rock forms erode over time. Off the southwest end of Dominica is Soufriere Bay, an extinct volcanic crater whose walls drop to an uncharted depth as a lava chute (Scott's Head Soufriere Marine Reserve 2003–2004). Dominica's Champagne Reef, located in this area, is so named due to the warm gases rising up from thermal vents on the ocean floor, which give the appearance of bubbles in a glass of champagne. It is a shallow submarine hydrothermal area (McCarthy et al. 2005). Beautiful coral reefs are located here, as are numerous underwater pinnacles, such as Scotts Head Pinnacle and Swiss cheese, as well as walls, such as Cachacrou and La Sorciere. East coast reefs consist of patch and fringing reefs characterized by low live cover and less diversity than west coast or leeward reefs (Steiner 2003).

11.4 Landscape

11.4.1 Impact of the Indigenous Caribs on Dominica's Landscape

Past and present cultures on Dominica have relied on the surrounding landscapes and landforms for their survival, livelihoods, and pleasure. As a result, they have had to modify and accommodate the natural environment in numerous ways. Dominica's Amerindians, including the Caribs and groups that existed before them, have been an integral part of the environment and natural cycles, because they relied on the environment for hunting and gathering and subsistence agriculture. Volcanic peaks took on particular significance as they were viewed as the life-giving source of all the natural resources on the islands. Volcanoes were represented in the conch shell, stone, and clay in the form of a religious object known as a zemi. Zemis represent the spirit that gives fertility to and brings rain to the land (Honychurch, n.d.).

The Caribs connected to their landscape in other ways. The Indian River is so named because of the known settlement of Caribs on the upper banks of that river at Prince Rupert's Bay in Portsmouth. This location provided for all of their needs with a 'sheltered bay, fresh water, reefs and fishing banks, land for cultivation and abundant forests bearing … wood, thatch, bark, fruit and herbs…" (Honychurch, n.d.). The Caribs' fierce nature and Dominica's mountainous terrain helped them resist colonization for a

Fig. 11.8 Trafalgar Falls as seen from the trail. A set of falls, locally named "mommy" and "Daddy," the falls themselves remain difficult to reach. The resultant, nearby step pools—both hot and cold—however, are readily visited by tourists. Image courtesy of Hans Hillewaert, shared under Creative Commons, original available at https://commons.wikimedia.org/wiki/File:Rainforest_at_Trafalgar_Falls_(Dominica).jpg

longer period than on any other island in the Caribbean. When the French and British eventually colonized the island, the Caribs retreated to the very rugged, unoccupied northeast of Dominica. In general, the inaccessible coast and mountainous terrain provided a safe haven for the Caribs and a launching point to make raids against Europeans. The 3700-acre (15 km^2) area in the northeast was granted to the Caribs by the British Crown in 1903 and is today known as the Carib Territory or Reserve (Fig. 11.11; Slinger 2000). It is home to a small community of approximately 3000 Carib descendants who mainly make a living from agriculture and tourism (Fig. 11.12). Agriculture, in the form of bananas, coconuts, cassava, yam, sweet potatoes, and fruits, flourishes in the fertile volcanic soils of the region. Tourism in the Carib Territory stems from the tourists' desire to experience their indigenous culture, and see other major landform-related attractions including Crayfish or Isulukati River and L'escalier Tete Chien. The Caribs' intense use of their commonly owned land has created some concerns by having the worst soil erosion in Dominica. It has also experienced stream loss through deforestation (Dominica Export-Import Trade and Business Directory 2013).

Fig. 11.9 Middleham Falls, located in the northwest of Morne Trois Pitons National Park. At almost 100 m, these Falls represent some of the tallest on Dominica, but remain lesser touristed owing to their hour-plus hike through sometimes steep and rugged rainforest. Image courtesy of Hans Hillewaert, shared under Creative Commons, original available at https://commons.wikimedia.org/wiki/File:Middleham_Falls,_Dominica.JPG, retrieved January 2017

11.4.2 Colonial Arrivals and Their Impact on Dominica's Landscape

The European colonists, African slaves, and both of their descendants have also used the landscape for varying purposes. Wood was the predominant building material in colonial Dominica; however, colonial fortifications (such as the buildings on the Cabrits Garrison in Portsmouth), stone houses, and some foundations for wooden houses were constructed using volcanic stone found scattered in the hillsides. Coral limestone, collected on the nearby reefs, was used to produce a powder that was mixed with aggregate, water, and molasses and then used as a mortar to bind walls (Steiner 2015). This was practiced until the 1950s when cement became more available and cost-effective.

Most of the water sources for agriculture originate in Dominica's central mountain peaks. Dominica's lack of flat land has played an important role in colonial and modern agriculture. The high-producing plantations found on other Caribbean islands could not be as easily accommodated on Dominica. Early estates focused on coffee, sugar, limes, and cocoa, and later bananas, which became the major export crop. Dominica's mountainous terrain has always imposed physical limitations and transportation difficulties. To some

Fig. 11.10 Examples of Dominica's many rivers and fluvial landforms. **a** River guide and boats aside the Indian River, Dominica's widest river. Its course through picturesque coastal wetlands and mangroves, along with the fact that scenes from *The Pirates of the Caribbean: Dead Man's Chest* were filmed here, makes the Indian River a popular tourism attraction. **b** Cascading waterfall of Crayfish River (also known as Isulukati River) flowing into the Atlantic Ocean through a series of rocky, volcanic outcrops. Photographs by V. Slinger-Friedman

Fig. 11.11 Part of "Carib Territory," in the northeast of Dominica. Notice the small stretch of sandy beach in this image, a rarity on Dominica. Image courtesy of Hans Hillewaert, shared under Creative Commons, original available at https://commons.wikimedia.org/wiki/File:Carib_Territory_(Dominica).jpg. Retrieved January 2017

Fig. 11.12 A Carib woman selling *larouma* reed baskets to tourists. Tourists come to the Carib Territory to experience the indigenous culture and visit the landform-related attractions in the area. Photograph by V. Slinger-Friedman

extent, development in Dominica has gone hand in hand with agriculture. For instance, roads were constructed for the purpose of opening more land up to development and for transporting products to market (Nelson 2010). Agriculture has brought benefits in terms of jobs and livelihoods, but it has also had adverse effects on the environment. Much of the littoral woodland, on the exposed eastern or windward slopes of the island, has been cleared for agriculture, predominantly for monoculture banana plantations. Another issue related to agriculture and the environment is illustrated in a study by Bradshaw et al. (2010), which evaluated the relative environmental impact of countries and ranked

Dominica as one of the worst countries in proportional fertilizer use.

11.4.3 Other Contemporary Activities and Their Impact on Dominica's Landscape

The human-related impacts on Dominica's landscape and geology have also come from the actions of foreigners who have sought or have been given concessions to exploit Dominica's geological resources. In 1995, Broken Hill Proprietary (BHP), an Australian mining company came close to obtaining the rights to open a major copper mining operation from the government, potentially compromising 10% of Dominica's land surface. Activists worked to stop this from happening, resulting in BHP pulling out of Dominica (Atherton Martin 2015). A Canadian company, Dom-Can Timbers, using "best Canadian practices," including skidding logs to spar a tree, a practice acceptable in the temperate forests of Canada, destroyed areas of Dominica's tropical forests by taking down every tree and sapling in the affected area and causing gully erosion and soil compaction in the skid path. Despite these events, Dominica's harsh, rugged landscape is the main reason why the country does not face greater issues with deforestation, since most trees are located in mountainous regions that are difficult to reach.

While Dominica's water bodies are not navigable, many streams, waterfalls, and lakes are sources of hydroelectric power. The first form of hydropower on the island was derived from water wheels used for processing sugarcane. Electricity generation from hydropower sources started in 1951. There are currently three hydropower plants with a total capacity of 6.4 MW, providing 40% of the country's electricity. (UNIDO and ICSHP 2013). Modifications to the Freshwater Lake to accommodate the development of a giant man-made reservoir to provide for the hydroelectric stations caused the lake to be excavated, streams redirected, and the water level rose by over 6 m ("Dominica Export-Import Trade and Business Directory" 2013). The potential of geothermal energy is currently being explored through the drilling of test wells. Environmentalists have expressed apprehension over the impact of deep drilling associated with geothermal power on the water table and the release of noxious gases.

In the 1980s, non-extractive uses of the coral reefs, include SCUBA and snorkeling, emerged mainly in the southeast. However, Dominica's limited coral and reef structures have been impacted by both natural hazards and a growing population living predominantly along the shores and relying intensively on the marine resources (Steiner 2003). Fluvial sediment fallouts have influenced reef development, especially during the nineteenth century when agriculture-related deforestation was at its peak. Furthermore, pollutants from industry and household wastes often reach the coast because they are not treated and disposed of properly. Impacts of this pollution include suffocation of marine organisms such as corals and sponges, coral bleaching in Dominica's few coral growth areas in the Soufriere Pinnacle, and polluted beaches.

Fishing is a major local industry in Dominica, with fish protein comprising 13% of the Dominican diet, compared to 7% in the rest of the Caribbean and 6% in the rest of the world (Raymond, n.d.). The country's geology, including the rapid drop-off of the west coast (100 fathoms per half mile), allows for different types of fisheries, including fringing reef, shelf slope, and open sea. Unlike many other Caribbean islands, Dominica does not have sea grass bed and barrier reef fisheries. Reef fish are considered to be most intensively fished, to the point of being overexploited. While some Dominican fishermen have adopted modern fishing technologies, the vast majority of them harvest inshore using seine nets (Schwarzlose 2002). The north coast fringing reefs are the site of marine algae collection to make sea moss, a traditional beverage.

11.5 Heritage and Tourism

Tourism in Dominica is centered on its geomorphic landscapes. Whereas most other Caribbean islands have capitalized on the traditional "sun, sea, and sand" type of tourism, Dominica's mountainous terrain, rocky shores, and black sand beaches deterred its development in that direction. Instead, the Dominican government has focused, in policy and mostly in practice, on developing an ecotourism industry. Ecotourism involves visitors enjoying, learning about, and helping to sustain the natural environment and culture of a location. The varied geomorphic landforms of Dominica—hot springs, waterfalls, boiling fumaroles, mountain peaks, lake, rivers, wetlands, mangroves, dykes, and reefs, along with their associated vegetation and wildlife—fit well with this type of tourism. Concurrent with ecotourism, the government has also promoted cruise ship tourism, though this form of tourism comes with concerns over the impacts a large number of visitors at ecologically sensitive or fragile nature sites could have on the island's overall health.

In some respects, the Dominican government has implemented policies and practices to manage the island's heritage resources. These are reflected in land and marine parks that promote protection of Dominica's landforms, rainforest, and reefs. By establishing the Morne Trois Pitons National Park in 1975, the Morne Diablotin National Park in 2000, and the

Soufriere/Scott's Head Marine Research in 2001, one-fifth of the island has been allocated to national parks and protected areas (Patterson and Rodriguez 2003). Additionally, legislation, such as the Forestry and Wildlife Act, the Fisheries Act, and Beach Control Act, has been instituted to protect the island's natural environment. Research indicates that Dominicans are aware of the importance of their heritage resources to both the ecotourism industry and their livelihoods. This view has led to many personal conservation actions by individuals in the tourism industry that are considered to be beneficial for the island's landforms, including community and individual beach and river cleanups, placing boulders to protect against erosion, growing and using organic produce, using Green Globe recommended cleaning products (such as vinegar to clean furniture), educating locals and tourists on conservation, preventing tourists from taking coral, building good sewage systems due to proximity to the sea, using solar heating systems for energy, and composting (Slinger-Friedman 2009).

Still, government, corporations, and individuals have at times made decisions and taken actions related to the tourism industry that have adverse or, at best, unknown effects on the country's landforms. Some areas of concern involve the following:

- The lack of impact assessment at certain tourist sites that are receiving increasing numbers of tourists (Watty 2008).
- The creation of hydroelectric power developments at Trafalgar Falls, Titou Gorge, and Freshwater Lake that have increased noise pollution in these tourist areas and diminished water flow to Trafalgar Falls, a key tourist attraction (Evans and Williams 1997; Gayle 1997).
- Consideration of a cable car system to take large numbers of tourists to the Morne Trois Pitons National Park. This cable car project was stopped in part due to public outcry and to the threat by UNESCO to strip the park of its designation as a World Heritage Site (UNESCO 1998), and the plans of a previous government to build an international airport that would displace farming communities, destroy mountains, and cut down forest (Fontaine 2003; Watty 2008).
- Government sanctioned quarrying near the Emerald Pool Waterfall has contributed to increasing amounts of sediment to its waters.

Some authors have noted the Dominican government's ambivalence over promoting nature-based tourism versus mass tourism or some other types of development. Actions such as developing hydropower facilities in potentially sensitive areas in the Morne Trois Pitons Park and trying to attract an oil refinery to the north of the island support this concern (Weaver 2004; Klak 2007). Researchers in this area suggest that this wavering demonstrates the desperate actions of a fiscally impoverished government seeking economic development by any means possible (Weaver 2004; Klak 2007).

11.6 Hazards

Dominica is exposed to a wide range of natural hazards. The most common and historically most significant are tropical storms and hurricanes (Benson et al. 2001). Other potential but lesser hazards include droughts, storm surges, floods, bush fires, and tsunamis. The coastal location of most of its population and infrastructure reflects its rocky and mountainous physical topography and means that Dominica's population is especially vulnerable to strong winds and high seas, although the island has a high potential for mass wasting events, such as landslides (Fig. 11.13).

11.6.1 Hurricanes

Dominica experiences two seasons, a wet period between June and October and a dry period between February and May. Temperatures range from 85 °F in the northern winter months and 90 °F in the summer. The island lies on the hurricane path that affects this region most between September and November. While Dominica has fortunately not experienced many hurricanes, some of the ones that have landed on the island have caused extensive damage. This includes Category 4 Hurricane David in 1979, Hurricane Hugo in 1989, Hurricanes Luis and Marilyn in 1995, Hurricane Lenny in 1999, and Hurricane Dean in 2007. Hurricanes Luis and Marilyn caused $66 million in damage to infrastructure and agriculture (Gayle 1997) and Hurricane David left 80% of the island's population homeless. More recently in 2015, Tropical storm Erica with its associated flooding and landslides caused 20 deaths, and an additional 50 people went missing. The coastal town of Petite Savanne in the southeast had to be evacuated because of potential landslides.

Hurricane David left visible impacts on the geomorphology in the south. These included the narrowing of the 300-m-long isthmus joining Scotts Head Point to the mainland by 15 m (Cambers 1996), and the replacement of sand with stones on a beach on the Caribbean side of the isthmus. After a beach monitoring program was established in Dominica in 1987, more measurable impacts of the hurricanes were derived. Following Hurricanes Gabriella and Hugo in 1989, on average, all beaches narrowed by 4.4 m and the land edge retreated inland 1.7 m. While the beaches recovered 93% of their profile area and width values, they were an average of 3 m narrower than before the hurricanes (Cambers 1996).

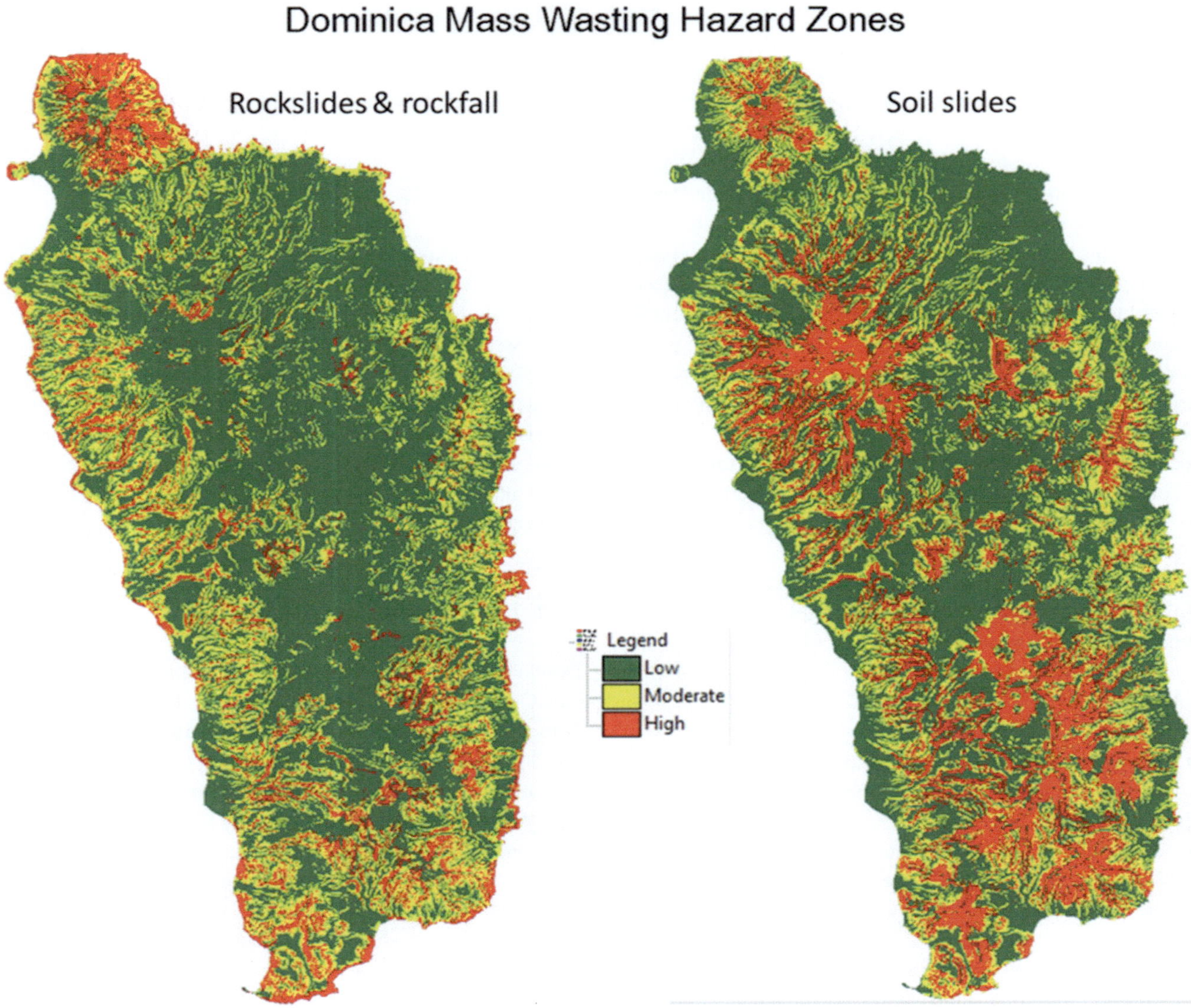

Fig. 11.13 Map of susceptible mass wasting regions in Dominica. Note that high rockslide and rock fall events tend to occur along Dominica's rugged, rocky coast, while landslides ("soil slides") areas remain confined to the island's steeper interior regions and along steep valleys. Map courtesy of Caribbean Handbook on Risk Information Management, available at http://www.charim.net/use/83, retrieved January 2017

11.6.2 Landslides

In Dominica, landslides are typically found to be debris flows, with volcanic bedrock, high rainfall, flooding often associated with hurricanes and tropical storms, and the steep slopes contributing to significant landslide activity (James and DeGraff 2012). The Good Hope landslide in 1987 is an example of a landslide induced from a prolonged period of rain during the wet season (DeGraff 1998). Deforestation of watersheds and construction on unstable soil aggravate the risk of landslides, particularly when it occurs on mountainous topography such as that found in Dominica. The Layou River Valley landslide in 1997 involved the formation and failure of two landslide dams. The first landslide dam was over 15 m high and it created a solid plug in the river before it was breached, sending 300 million gallons of water down the valley. The breach in the second landslide dam seven days later flooded the Layou River Valley with an estimated 350 million gallons of water. Significant deposition of sediments occurred along the river, reducing its carrying capacity and impacting its channel. In association with these two landslide dams, another landslide dam formed on the Matthieu River tributary of the Layou River. Water behind the dam built up into a lake—officially called Matthieu Lake, but dubbed "Miracle Lake" by locals (Fig. 11.14)—that attained a depth of over 43 m and persisted for almost

Fig. 11.14 "Miracle Lake," as the locals called it, remained for 13 years, eight months, and three days before collapsing in 2011 sending huge amounts of water and sediment rushing down the steep-sided valley to the ocean, widening and perhaps replenishing beach sand (James 2014)

14 years before finally breaching in 2011 (James and DeGraff 2012). Geological impacts of this breach include the emptying of the Matthieu Lake, severe stream bank erosion downstream at the junction of the Layou and Matthieu rivers, and aggradation of the valley floor. Additionally, the large volumes of fresh pumice sand from the collapse of the Matthieu dam that were deposited into the Layou River estuary widened the Layou Beach, making it the widest sandy stretch on the island in 2012. It is believed, but not yet confirmed, that Mero Beach and Salisbury Beach have also been renourished with sand from the collapsed dam (James 2014).

11.6.3 Tectonic Activity

Dominica is volcanic in nature and geologically young. Sulfur springs have the potential to cause landslides, boiling pools, and phreatic (steam) eruptions. The last volcanic eruption in Dominica was roughly 500 years ago. Although there has been only one volcanic event in Dominica's recorded history, there have been eight seismic crises in the past 50 years, and the island has been on a volcanic alert since 1998, highlighting its susceptibility to future volcanic activity and the related risk of earthquakes. Earthquakes large enough to be recorded seismically occur once or twice a year. However, there have been locally felt swarms of small earthquakes from time to time over the past two hundred years (Abraham n.d.; Seismic Research Center 2015; Lindsay et al. 2005). Major earthquakes and aftershocks have occurred in 2004 (6.3 Richter scale) and in 2014 (6.7 Richter scale). The earthquake in 2004 caused landslides in areas such as Portsmouth that had already been affected by heavy rains. Phreatic explosions in the Valley of Desolation in 1880 and 1997, recent earthquakes, and the presence of one of the highest concentrations of potentially active volcanoes in the world indicate that future eruptions are highly likely and the potential for change in the island's geomorphic future is strong.

11.7 Conclusion

Travel logs of several historic visitors to Dominica aptly describe the landforms and landscapes that they encountered. Trollope, a British traveler on an unnamed ship coming to the Caribbean to conduct, "affairs of the state," described Dominica as by far the most picturesque of all these islands and expressed the desire "to be off and rambling among those green mountains …" (Trollope 1867). An American traveler, De Forest Day (1899), echoed Trollope's sentiments about Dominica when she noted that all the elements of beauty which make the other islands such a feast to the eyes could be found a hundredfold in Dominica. Visitor's travel logs also recounted natural beauties in Dominica such as the Freshwater Lake, Boiling Lake, and forests, "the like of which we poor commonplace mortals had never before dreamed" (De Forest Day 1899).

These travelers' descriptions capture the variety and majesty of Dominica's landforms. They also depict a pure and untouched landscape, coastline, and waters. This is the vision that Dominica's Division of Tourism seeks to present and capitalize on as they promote Dominica as the "Nature Island of the Caribbean." This image is perhaps slightly removed from the reality due to the effects that nature and humans have had in modifying its pristine landscapes that early travelers encountered. Dominica's regular volcanic activity and frequent landslides often change its topography after each hazard event. While rainforest vegetation covers about 66% of Dominica's land surface, human-induced impacts on Dominica's physical landscape have occurred as people embedded their ecological footprint on the island—lightly at first, with the Caribs use of the landscapes to carve out a largely subsistence living, and then more heavily in the form of hydroelectric power plants, increased population pressure for tourism or habitation, timber extraction, and agriculture.

Despite these modifications, it has been suggested that Dominica would be one of the only islands that Columbus would recognize if he came back to the region today (Doyle 2005; Daniells et al. 2008). Its landscape has impeded Dominica from large-scale development in the form of traditional 3S tourism or any other major industry. Conversely, its outstanding natural features have allowed it to remain as intact as it is today, providing a unique backdrop for its nature-based ecotourism product. It is an island that engages its visitors to interact with its geological landscape rather than lay passively on pool decks. Perhaps the most strikingly unique feature of Dominica is the community of the Caribs who were protected to some extent by its rugged, tropical landscape and whose descendants still inhabit the island today.

References

Abraham W (n.d.) Dominica's Seismicity. Public Seismic Network Inc. Web 17Aug 2015. http://www.dpsninc.org/index.php/seismicity

Atherton Martin (2015) The goldman environmental Prize. Web 14 Aug 2015. http://www.goldmanprize.org/recipient/atherton-martin/

Benson C, Clay E, Michael FV, Robertson AW (2001) Dominica: natural disasters and economics development in a small Island State. Working paper. ODI annual report

Bradshaw C, Giam Xingli JA, Sodhi NS (2010) Evaluating the relative environmental impact of countries. Plos ONE 5.5: 1–16. Academic search complete. web. 20 March 2015

Cambers G (1996) Hurricane impact on beaches in the eastern caribbean Islands 1989–1995. Environment and development in coastal regions and in small Islands. COSALC coast and beach stability in the Lesser Antilles. Web. 6 Aug 2015. http://www.unesco.org/csi/act/cosalc/hur8.htm

Caribbean Geology (1994) Caribbean Islands Handbook. Ed. Sarah Cameron and Ben Box. 6th ed. Chicago: Passport. 37–38. Print

Daniells EA, Ackley JW, Carter RE, Muelleman PJ, Rudman SM, Turk PA, Vélez Espinet NJ, White LA, Wyszynski NN (2008) An annotated checklist of the amphibians and reptiles of dominica, West Indies. Iguana: Conservation, Natural History, and Husbandry of Reptiles. 15(3):130–141

De Forest Day D (1899) The cruise of the Scythian in the West Indies. F. Tennyson Neely Publishing, London

DeGraff J (1998) Natural hazards and disasters: Landslides in Dominica. International Landslide Research Group (ILRG) Newsletter. University of the West Indies (UWI), 12(1). http://www.mona.uwi.edu/uds/Land_Dominica.html

"Dominica" New World Encyclopedia. (2013). 15 July 2015. http://www.newworldencyclopedia.org/entry/Dominica

"Dominica Export-Import Trade and Business Directory: Strategic Information and Contacts" (2013). USA International Business Publications (Ed). Washington DC, USA. Volume 1

Doyle C (2005) The only island that columbus would recognize. Caribbean Compass, Nov 2005. http://www.caribbeancompass.com/dominicadesti.htm

Evans PGH, Williams D (1997) Development and management of nature sites—integrating conservation with ecotourism in dominica, Project Report No. 4, Brussels, European Community Project No. B7-5040-24. Ecosystems Ltd

Fontaine T (2003) 'No International Airport for Dominica', TheDominican.net, No. 42, May 7. www.thedominican.net/articles/airport.htm

Fontaine T (2007) The history and pictures of dominica's volcanoes—Morne Watt. The History and Pictures of Dominica's Volcanoes-Morne Watt. Web. 8 Feb. 2015. http://www.thedominican.net/articles/watt.htm

Fournier N, Witham F, Moreau-Fournier M, Bardou L (2009) Boiling Lake of Dominica, West Indies: High-temperature Volcanic Crater Lake Dynamics. J Geophys Res 114

Gayle DJ (1997) Ecotourism: Fad or Future? Hemisphere: a magazine of the Americas, 8(1):20–23

Holahan D (2009) "Rugged Dominica worth the ruined sneakers." Boston.com. The New York times, 18 Jan. 2009. Web. 4 Feb. 2015. http://www.boston.com/travel/getaways/caribbean/articles/2009/01/18/rugged_dominica_worth_the_ruined_sneakers/?page=full

Honychurch L (1991) Dominica: Isle of adventure. Macmillan, Basingstoke

Honychurch L (n.d.) Aspects of Carib/Kalinago culture. Web 17 July 2015. http://www.lennoxhonychurch.com/article.cfm?id=389

IUCN International Union for Conservation of Nature. (2008) Outstanding universal value: standards for natural world heritage. A compendium on standards for inscriptions of natural properties on the world heritage list. IUCN, Gland, Switzerland. Web Nov 30 2016 https://portals.iucn.org/library/efiles/documents/2008-036.pdf

James A (2014) So, what's with Matthieu or Miracle Lake? The Sun. May 11 2014. Web Aug 27 2015 http://sundominica.com/articles/so-whats-with-matthieu-or-miracle-lake-1238/

James A, DeGraff J (2012) The Draining of matthieu landslide-dam lake, Dominicam West Indies. Landslides. 9(4):529–537

Klak T (2007) Sustainable ecotourism development in central america and the caribbean: review of debates and conceptual reformulation. Geography Compass 1(5):1037–1057

Lindsay JM, Smith Alan L, John Roobol M, Stasiuk Mark V (2005) Dominica. In: Lindsay JM, Robertson REA, Shepherd J, Ali S (eds) Volcanic hazard atlas of the Lesser Antilles. University of the West Indies, Seismic Research Center

Love B, Gabbett M (2012) DOMINICA Photo Gallery 4- Cabrits National Park & Indian River. Green Global Travel. 25 Web. 26 Apr. 2015. http://greenglobaltravel.com/2012/07/25/dominica-photo-gallery-4-cabrits-national-park-indian-river/

McCarthy KT, Pichler T, Price RE (2005) Geochemistry of Champagne Hot Spring shallow hydrothermal vent field and associated elements, Dominica, Lesser Antilles. Chemical Geology 224 (2005) 55–68. Elsevier Publishers

Nelson V (2010) Dominica's lime industry: agriculture and identity in the early twentieth century. National identities 12.3: 219–236. Academic Search Complete. Web. 20 Feb. 2015

Patterson T, Rodriguez L (2003) The political ecology of tourism in the commonwealth of dominica. In: Gössling S (ed) Tourism and development in tropical Islands: political ecology perspectives. Edward Elgar, Cheltenham, pp 60–87

Ralston J (2014) "Who Needs a Beach Who You've Got This?" Waterfalls, rivers and rain forest on Dominica. New York times. http://www.nytimes.com/2014/11/09/travel/waterfalls-rivers-and-rain-forest-on-dominica.html?_r=0

Raymond T (n.d.). "Wise coastal management: pollution issues in Dominica." UNESCO.ORG. United Nations Educational, Scientific and Cultural Organization. Web. 18 Mar. 2015

Roobol J, Smith A (n.d.) "Geology of Dominica." *Caribbean Volcanoes*. Web. 9 Feb. 2015. <http://www.caribbeanvolcanoes.com/dominica/geology.htm>

Schwarzlose J (2002) An Analysis of the fisheries of Dominica. TAMU Study Abroad. Springfield, Dominica. Web 29 Aug 2015. http://dominica.tamu.edu/student%20projects/Dominica%20Projects%20pdf%20copy/Schwarzlose_John.pdf

Scott's Head Soufriere Marine Reserve (2003–2004). Web 23 Aug 2015. http://www.avirtualdominica.com/ssmr/

Seismic Research Center (2015) The University of West Indies Seismic Research Center. Web 29 Aug 2015. http://www.uwiseismic.com/

Slinger V (2000) Ecotourism in the last indigenous caribbean community, Ann Tourism Res 27:520–523

Slinger-Friedman V (2009) Ecotourism in dominica: studying the potential for economic development, environmental protection and cultural conservation. I Stud J 4:3–24

Smith AL, Roobol MJ, Mattioli GS, Fryxell JE, Daly GE, Fernandez LA (2013) The volcanic geology of the mid-arc island of dominica, lesser antilles. Geol Soc Am Spec Pap 496:1–2

Steiner SCC (2003) Stony corals and reefs of Dominica. Atoll research bulletin. Issued by the National Museum of Natural History, Smithsonian Institution, Washington, D.C., U.S.A

Steiner SCC (2015) Coral reefs of Dominica. Annalen des Naturhistorischen Museums in Wien, B. pps. 47–119. http://www.itme.org/reports/ITME_RReports33_2015_Steiner_Coral%20Reefs_Dominica_web.pdf

Trollope A (1867) The west indies and the spanish main, 6th edn Chapman and Hall, London

UNESCO (1998) Convention concerning the protection of the world cultural and natural heritage, Bureau of the World Heritage Committee, 22nd Session, Paris, France. http://whc.unesco.org/archive/repbur98a.htm

UNESCO (n.d.) "Morne Trois Pitons National Park." United Nations Educational, Scientific and Cultural Organization. Web. 24 Apr. 2015. http://whc.unesco.org/en/list/814

UNIDO and ICSHP (2013) United Nations Industrial Development Organization and International Center on Small Hydro Power. World Small Hydropower Development Report 2013: Dominica. Web 14 Aug 2015. http://www.smallhydroworld.org/fileadmin/user_upload/pdf/Americas_Caribbean/WSHFDR_2013_Dominica.pdf

Watty WRF (2008) Feature Address W. R. Franklin Watty, Chairman of diaspora affairs, dominica academy of arts and sciences at the annual awards gala dominica hotels and tourism association. http://da-academy.org/Watty_Address_DHTA.pdf

Weaver DB (2004) Manifestations of ecotourism in the Caribbean. In: Duval DT (ed) Tourism in the Caribbean: Trends. Routledge, Development, Prospects, New York, pp 172–186

12 Martinique

E. Arnold Modlin Jr and Casey D. Allen

Abstract

Martinique is an island located at the northern end of the Windward Islands in the Lesser Antilles. One of the larger islands of the Lesser Antilles, Martinique, is unique because of its location near the convergence of two volcanic arcs on the eastern edge of the Caribbean Plate. The two volcanic arcs cross Martinique on close to parallel paths that run south to northwest. The arc associated with the Volcanic Caribbees is a few miles west of the arc connected to the Limestone Caribbees. Martinique's location has influenced not only its geologic past, but also its anthropological past and present. The shaping of the history of the island is a series of interactions between the geology of the island including specific geological events, its climate including specific weather events, and the social interactions of people native to the Americas, Europeans, Africans, South Asians, and East Asians. Approximately, 380,000 people call Martinique home. Still, the island is a tourism destination primarily for Europeans staying for a few days and cruise boat passengers staying for a few hours. However, tourism officials in Martinique are trying to diversify the origins of tourists vacationing in the islands. Located in a portion of the Caribbean susceptible to hurricanes, with a volcano that is still active, together with labor issues, Martinique faces challenges as it continues into the twenty-first century.

Keywords

Volcano • Mount pelée • Tsunami • Saint-Pierre • 1902 • Négritude • Tourism

12.1 Introduction

Martinique is located 33 km north of St. Lucia, 430 km north of the South American coast, and 42 km (26 miles) south of Dominica (Fig. 12.1). It has a land area of approximately 1128 km^2 with 350 km of coastline (Bocquené and Franco 2005). As an overseas department of France (also known as an overseas region since 2003), Martinique is part of the European Union and uses the Euro as currency, although it is not a part of the Schengen Area (US Department of State 2016). Citizens of Martinique are citizens of France, and the island's inhabitants are represented in the French national government by officials elected locally on the island. Approximately, 378,243 people lived on the island as of 2015 (INSEE 2016).

At its base, Martinique was formed by a series of volcanic activities. Due to land movement and a series of volcanic eruptions that stretch over at least the recent 8–10 million years, the island was created. The oldest southeast and east-central part of Martinique formed over 8–10 Ma. Mount Jacob, the largest volcano in the Lesser Antilles, was active from 5.2 to 1.5 Ma. Trois Îlets Peninsula area (southwest) was volcanically active from 1 to 3 Ma. *Pitons du Carbet* (north of Fort-de-France) were active between 800 and

E.A. Modlin Jr (✉)
History and Interdisciplinary Studies Department, Norfolk State University, Norfolk, VA, USA
e-mail: arnoldmodlin@gmail.com

C.D. Allen
General Education, Western Governors University, Salt Lake City, UT, USA
e-mail: caseallen@gmail.com

C.D. Allen (ed.), *Landscapes and Landforms of the Lesser Antilles*, World Geomorphological Landscapes, DOI 10.1007/978-3-319-55787-8_12

Fig. 12.1 General physiographic map of Martinique and surrounding islands displaying key features and points of interest highlighted in this chapter. Note the mountainous northern region with lower plains in the island's center. The majority of small, uninhabited islands are along the western coast. Cartography by K.M. Groom

300 Ka. Mount Pelée has been active for the last 10,000 years.

Though the Lesser Antilles are often thought of as a backwater to other places in the world, Martinique and Martiniquais have often taken a primary place in history—though not always in a good way. Archaeological evidence indicates that prehistoric people settled the island around 100 BCE, though Martinique might have been settled much earlier than that as it was part of the pathway of prehistoric people migrating north from South America. Although the Spanish "discovered" the islands in 1502, they considered it one of the *islas inútiles* (meaning "useless islands"). Over a century later, the French eventually settled it. In time, it was transformed into a sugar island based on the labor of enslaved Africans and their descendants. After emancipation of enslaved people, the planter class turned to imported indentured workers from Africa and Asia. Eventually, even this coercive labor system faded out. Yet, the unique experience of ones who were the descendants of those forcefully migrated to the island was expressed in a number of ways. *Négritudes* and *Créolité* are just two theoretical and activist stances enunciated and developed by Martiniquais in response to French colonialism of the island and its people.

Today, Martinique and Martiniquais face a number of challenges and hazards. High unemployment and low employment participation rates in Martinique exacerbate the effects of other hazardous events and processes. These other events include volcanic eruptions and seismic activity, tsunamis, tropical storm activity, overuse of natural resources, and the challenges of climate change and sea level rise.

12.1.1 Setting

Martinique's location created a dramatic setting for the island's geologic and anthropologic histories. Located due north of St. Lucia and southwest of Dominica, Martinique is oriented from southeast toward northwest. South of the islands midpoint on the west (Caribbean) side of the island is the present-day Department capital of Fort-de-France which lies on the north side of the Fort-de-France Bay (sometimes known as Flamands Bay[1]).

To the south of the bay is a peninsula created by volcanic activity. Located on the northern (bay) side of the peninsula is Les Trois Îlets, birthplace of Josephine, the daughter of a planter and the first wife of Napoleon Bonaparte. Most of the southern third of Martinique, including the peninsula, are in the sub-prefecture of Le Marin, home to a quarter of the population of Martinique. Continuing counterclockwise around the island, the next sub-prefecture is Le Trinité, which is oriented northeast toward the Atlantic Ocean. The Caravelle Peninsula stretches east into the Atlantic Ocean approximately 15 km. About a fifth of the total population of Martinique lives in the sub-prefecture of Le Trinité. The final sub-prefecture is in the northwestern corner of Martinique. It is the sub-prefecture of Saint-Pierre, which is further divided into cantons including Saint-Pierre, Le Morne-Rouge and Le Prêcheur. Parts of the sub-prefecture, particularly Saint-Pierre and Le Morne-Rouge, were the sites of extensive damage and loss of life due to the eruption of Mount Pelée in 1902.

Historically, the 08 May 1902 eruption of Mount Pelèe (see Sect. 12.4.2 for a detailed account), which killed tens of thousands of people, served as a turning point for research on volcanos in the Caribbean and elsewhere, with scientists like the geographer Heilprin (1903) using Martinique as a research laboratory for studying geology, geochemistry, and geophysics of volcanos. Indeed, this focus on volcanism in northern Martinique has led to less focus on the southeastern part of the island that shares some characteristics in common with Grande-Terre (part of Guadeloupe), and other Limestone Caribbees.

12.2 Landforms

To say that Caribbean geology is complex is an understatement. This is true too with Martinique. The older, eastern arc nearer to the eastern edge of the Caribbean Plate is a curvature of islands known as the Limestone Caribbees. The recent (new) arc, to the west of the older arc, is a track of islands and portions of islands that are the result of recent volcanic activity and are known as the Volcanic Caribbees (Fig. 12.2; see Bouysse et al. 1990; Germa et al. 2010). Chronologically, and generally speaking, the oldest rocks on Martinique—a basal complex and series from the Late Oligocene—reside in the island's southern half, with rocks from the Mid- to Late Miocene comprising most of the southeastern and middle eastern sides of the island, and Pleistocene-age rocks making up most of the island's north half. The historical analysis of geology and geomorphology has, for Martinique, made huge advances over the last century in some ways, while still being focused on only certain questions such as volcanism in Martinique's northern and central areas, paying less attention to the southeastern section of the island. Even though Martinique is often classified on its whole as one of the Volcanic Caribbees, parts of the eastern and southern sections of the island share more in common with the Limestone Caribbees (Germa 2010). Since 1902, the geologic focus in Martinique has primarily been on the two-part question of what exactly is Martinique's geologic past as it relates to volcanism and how can the

[1]Not to be confused with the Anse des Flamands in St. Barthélemy.

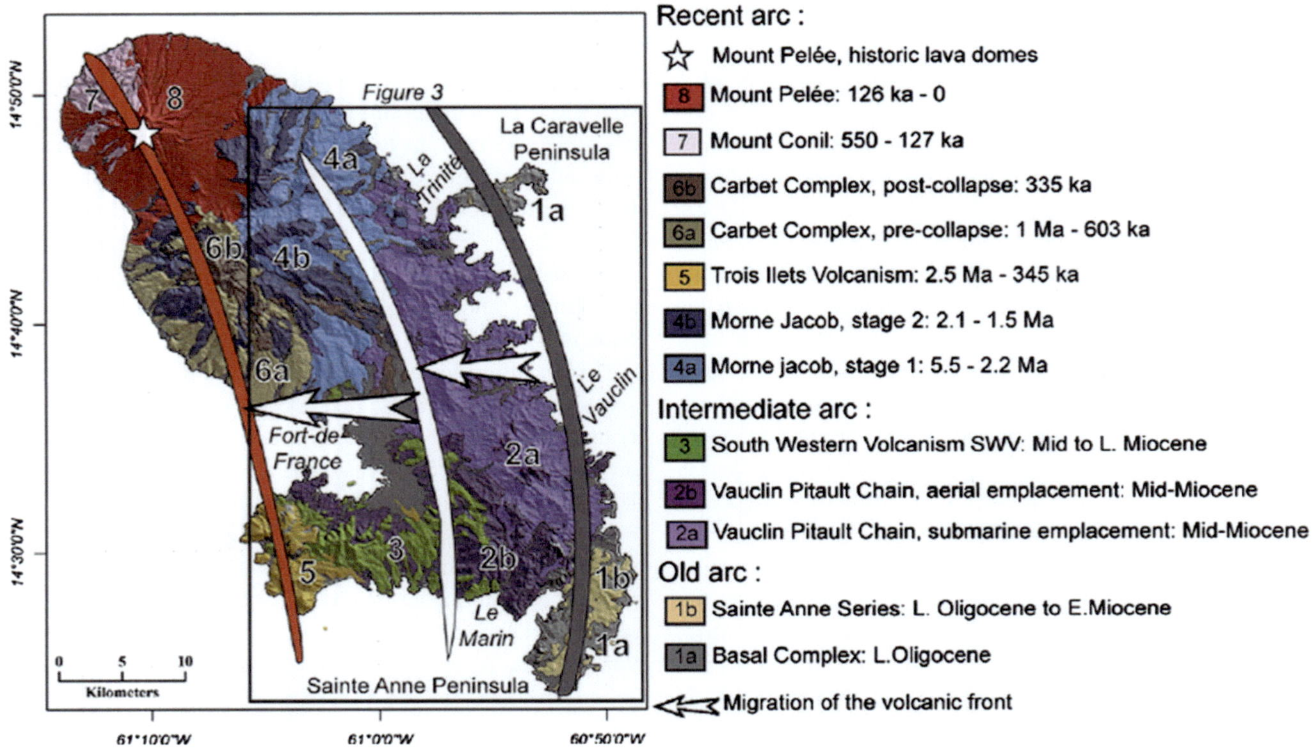

Fig. 12.2 Basic geologic map of Martinique displaying its distinct volcanic units, from Germa et al. (2011b: 124), who note "1: old arc (*1a* Basal Complex, *1b* Saint Anne Series); *2* Vauclin—Pitault Chain (*2a* submarine emplacement, *2b* aerial emplacement); *3* Southwestern volcanism; *4* Morne Jacob shield volcano (*4a* stage 1 tholeiitic, *4b* stage 2 calc-alkaline); *5* Trois Ilets Volcanism; *6* Carbet complex (*6a* pre-collapse stage, *6b* post-collapse stage); *7* Mount Conil; *8* Mount Pelée." *White arrows* show the westward displacement of the volcanic front. The *white star* indicates Mount Pelée volcano. *Gray*, *white*, and *orange parallel lines* (from east to west) indicate the location of the old, intermediate, and recent arcs, respectively. Ages for the recent arc complexes (4, 5, 6, 7 and 8) are from Germa et al. (2010, 2011a). Epochs of activity for old and intermediate arcs (Oligocene and Miocene) are from Westercamp et al. (1989). Available online at: https://www.researchgate.net/figure/235828104_fig2_Fig-2-Simpli-fi-ed-geological-map-of-Martinique-Island-and-its-eight-volcanic-units

human disaster that occurred in May 1902 be prevented from occurring again.

Out of this research, our understanding of Martinique has moved from anecdotal and superstitious to a better understanding of volcanic mechanisms and geochemistry on the island. For example, in 1902, National Geographic reprinted part of Lafcadio Hearn's book from (1890), *Two Years in the French West Indies* which made the statement,

> "No description could give the reader a just idea of what Martinique is, configuratively, so well as the simple statement that although less than fifty miles in extreme length, and less than twenty in average breadth, there are upward of *four hundred mountains* in this little island, or of what at least might be termed mountains elsewhere…Ninety-one only of the principal mountains have been named" (Hearn 1902 italics original).

The northern, central, and western parts of the island are mountainous, at times with steep inclines. However, four hundred mountains is an overestimation of the number of mountains on Martinique—although that number could include the many hill crests nowadays considered to be part of larger mountains (Fig. 12.3).

12.2.1 Volcanics

It is understandable why Martinique's volcanic geology and geomorphology get so much attention, as the island's genesis and growth are the result of the activity of dozens of volcanos. Mount Jacob, southeast of Mount Pelée, is the largest volcano in the Lesser Antilles covering an area of 350 km^2 (Germa 2010). Based on their research, Germa et al. (2010) note that the volcano was continuously active through most of the period of 5.2–1.5 Ma. *Pitons du Carbet,* a set of twelves volcanic domes in Martinique located south-southeast of Mount Pelée and southwest of Mount Jacob, were active from 770 to 331 Ka (Germa et al.

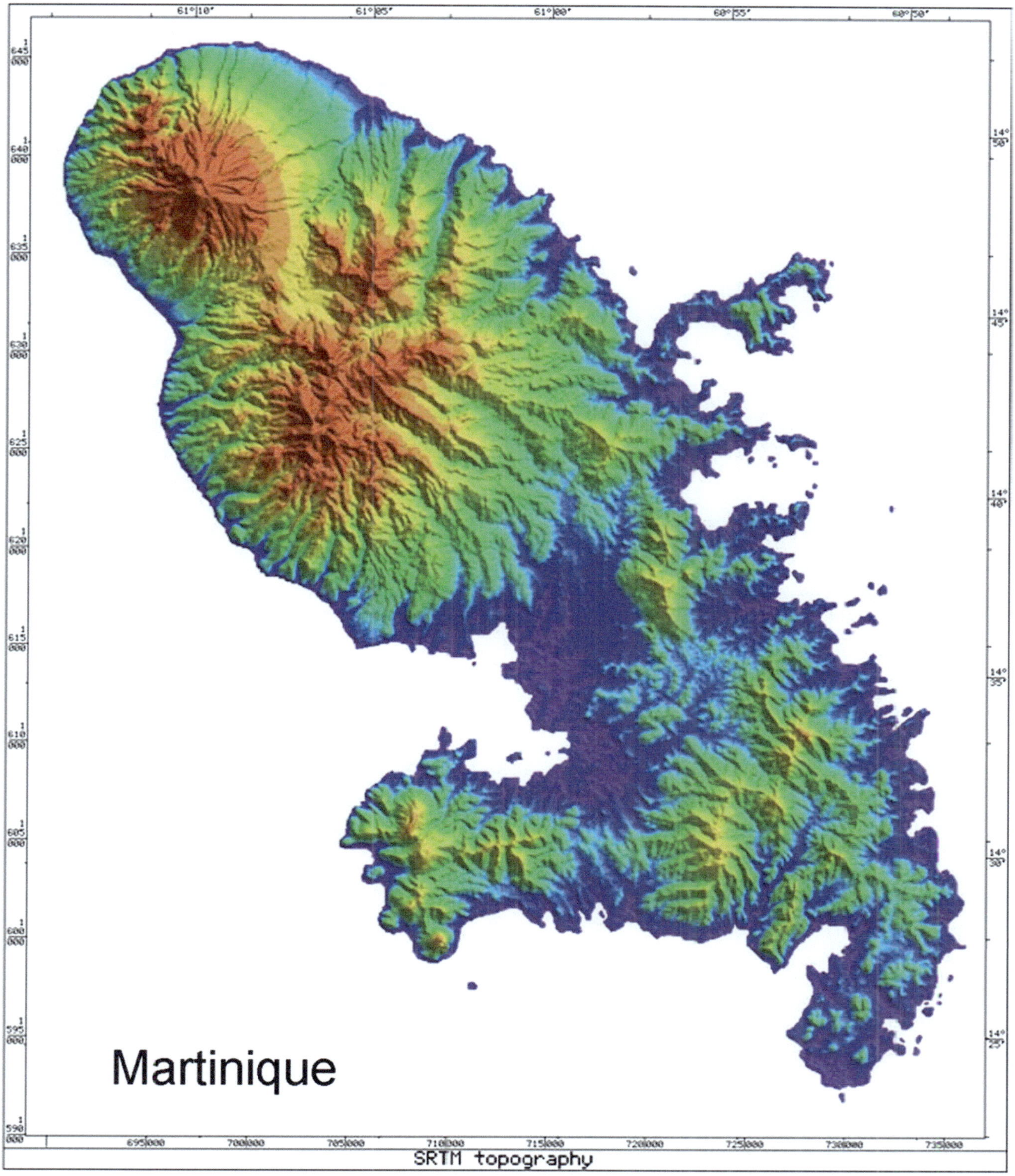

Fig. 12.3 Digital elevation model (DEM) of Martinique. *Red* denotes high elevation regions (in the island's northern half, where Mt. Pelée is located), with the elevation decreasing according to color (i.e., *red*, *orange*, *yellow*, *green*, *teal* (light *bluish-green*), *blue*, and *purple*). Note the topographic variance between the north (recent geology) and south (older geology) portions of the island. *Image source* http://caribbeanvolcanoes.com/martinique-radar-map/

Fig. 12.4 Coastal capital city of Fort-de-France as seen from Trois Îlets Peninsula across Flamands Bay. Parts of the Pitons du Carbet are visible beyond the city. Photograph by E.A. Modlin

2011a). Mount Conil to the northwest of Mount Pelée is much older, since Pelée's activity covers much of it (Germa 2011a). The Trois Îlets Peninsula on Martinique's southwest coast (Fig. 12.4) contains a series of less-studied volcanos with the western tip of the peninsula aligned on the recent arc south of the Carbet Complex, Mount Pelée, and Mount Conil (Germa 2011a). Data on the Trois Îlets volcanism date from the late 1980s and early 1990s, with some dating being challenged by results from newer methods. Still, it is generally assumed the range of activity for this area is fairly recent at over 3–1 Ma ago (Germa 2011a; Germa 2010). Extreme southern, eastern, and southeastern portions of Martinique including the La Caravelle Peninsula and Saint-Anne Peninsula are associated with the old (Eastern) Arc. Geologically, that means they are connected to the Limestone Caribbees' volcanic activity in these areas. Some researchers put the last activity in south and central parts of Martinique at almost 8–10 Ma (Germa et al. 2010).

While much of northern Martinique[2] was created due to the effects of multiple prehistoric volcanos, many of which are now extinct, Mount Pelée continues to be active over the last 10,000 years (Global Volcanism Program 2013a, b). Research has indicated that shortly prior to the arrival of French colonists in 1635, there was an eruption.[3] Eruptions recorded after 1635 include the eruptions of 1792, 1851–1852, 1902–1905, and 1929–1932 (Global Volcanism Program 2013a, b).

12.2.2 Coastal Forms

Martinique remains situated between the Atlantic Ocean and Caribbean Sea as all islands in the West Indies are, resulting in rugged, wind-swept eastern coasts (windward side) and calmer, gentler western coasts (leeward side), resulting in similar coastal landforms to other West Indian islands: the development of various colored beaches, coastal reef development, and mangrove ecosystems. In 2008, Schleupner estimated that about 79 km of Martinique's coastline was lined with mangrove forest, which on the island as a whole covered approximately 1850 ha (4571.4 acres), or around 6% of the total island surface. Over half of these mangroves are found in the Fort-de-France Bay with smaller mangrove forests found on portions of most of the bays along the east-central, southeastern, and southern coast.

[2]The Limestone covered portion of the island also rests on even earlier volcanic work.

[3]"Pelée" in French can mean "peeled" or "bare" which could be an indication that Mount Pelée was cleared of vegetation on top because of the recent eruption when they French arrived in 1635.

These mangroves serve as nurseries for a variety of fish, are rich in mollusks and crabs, and are home to a variety of birds, reptiles, and mammals (Schleupner 2008).

Martinique claims around three dozen beaches, most of which are sandy. Generally speaking, young volcanics in the north result in varying sizes of black sand beaches—both long stretches and small, hidden coves. The island's rugged eastern shore, facing the Atlantic Ocean and constantly battered by the northeast trade winds, hosts several beaches, though only a few remain viable options for water activities. Surrounding the island's southern and southwestern (and quite older, geologically) portions are renowned white sand beaches with calmer waters of the Caribbean Sea and beautiful palm-lined stretches such as *Plage de Salines* (Fig. 12.5). Though southern beaches often attract many tourists, so-called deserted beaches can also be found around the island (Fig. 12.6).

Off coast, the coral reefs surrounding parts of Martinique are most extensive on the eastern and southern coasts. Heavily damaged coral reefs are also present in the Fort-de-France Bay, while no reefs are present on the north and west coasts of Martinique (Schleupner 2008). Where reefs are present, they tend to slow down erosion and protect the shoreline, although major meteorological events such as swells from tropical storms can still cause damage.

12.3 Landscapes

12.3.1 Historical Landscapes

For thousands of years, Martinique's landscapes have been shaped by not just geomorphic forces, but also human forces. Prehistorically, migrants from the South American

Fig. 12.5 Martinique's longest stretch of white sand beach, *Plage de Salines*. Nourished by the erosion of some of the island's oldest volcanics, this remains a favorite beach for tourists. Images courtesy of Nicolas Bouthors, available from https://commons.wikimedia.org/wiki/File:Martinique_Beach_(Salines).jpg

Fig. 12.6 Hidden beach on Guadeloupe's far northwestern coast, *Anse des Galets* contains a mixture of *black* and *white* sand deposits, giving it a *gray-ish-brown* color. Beaches like this are often created mixing currents that transport sediments from both older and younger volcanics. Image courtesy of Jean and Nathalie, available from https://commons.wikimedia.org/wiki/File:Martinique_-_Anse_des_Galets.jpg

regions around the Orinoco River started to move up the Lesser Antilles prior to 2000 BCE (Higman 2011). Much of these nomadic peoples' material culture seems to have been too fragile to show up in the archaeological evidence collected from sites in Martinique thus far. Still, at least 100 Amerindian sites have been identified on Martinique, and evidence suggests the first permanent inhabitants arrived around the first century BC, though their interaction with the landscape remains an ongoing study (Bérard et al. 2014).

Direct European contact with the island did not occur until June 1502, on Columbus' fourth voyage.[4] Largely, the Spanish ignored the island as a whole, with two exceptions. First, they were responsible for the possible introduction of some of the invasive species such as pigs and goats (Borroto-Páez and Woods 2012). Second, in 1512, Martinique was declared one of the *islas inútiles* which allowed for the Spanish to capture Amerindians from the islands and remove them elsewhere in the Spanish colonies to work the gold and silver mines. In 1635, under the aegis of the French stock company *Compagnie des Îles de l'Amérique*, Pierre Bélain, Sieur d'Esnambuc settled in Saint-Pierre with approximately 100 colonists and claimed the entire island for France and the King. Despite Bélain's death within a few months, the colony continued to grow. Initially, the colonists focused on raising tobacco, even bringing the tobacco plants with them when they first arrived (Singler 1996). In Martinique and elsewhere, tobacco did not tend to create the concentrated wealth that sugar later would (de Gurbert 2014), partly because tobacco exhausted soil quickly and

[4]Columbus was told that the island existed on his first voyage and that it was occupied only by women who wore gold or copper. This idea excited Columbus and rivaled European interests with the Seven Cities and El Dorado (Columbus 1987: 172, 174).

Fig. 12.7 Two-room cabin built in the style and with the materials often used by rural Black Martinicans both before and after emancipation. While this structure is not representative of all of the housing types occupied by formerly enslaved people, it does demonstrate a dominate house type found up on steep mountains. Because of the building's size, the yard around the home was seen as an extension to the family's living space. Photograph by E.A. Modlin

constantly required new sites after a few years to produce meaningful yields (Berlin and Morgan 1993). While tobacco was not new to the Caribbean, the way that Europeans chose to grow it changed the landscape, with each farmer needing to clear a few acres every three to four years, while letting recently used land lay fallow. Though less documented, cotton was also grown during this period (Zamor 2015), and both would be grown for profit, shaping the landscape by clearing fields appropriate for each crop's production.

Sugar and slavery served as significant factors of the history of Martinique during the late seventeenth to early nineteenth centuries. The transition from tobacco to sugar occurred later on Martinique than in the Portuguese and Anglophone colonies of the Western Hemisphere, but still transformed the landscape by consolidating land away from small landowners to larger plantations. While sugar dominated the landscape, some planters also experimented with coffee and cacao (chocolate), further altering the landscape (Ruf and Schroth 2004; Pendergrast 2010). Cacao and coffee were grown in areas where there was at least some tree coverage, but this usually involved cutting down some of the tall trees in the canopy so filtered sunlight could reach the forest floor. Additionally, cacao was produced in Martinique on unfertilized land, and once the soil would no longer support its production, the cultivator would shift to new land (Ruf and Schroth 2004).

In Martinique, slavery was finally abolished in 1848, but this did not necessarily mean that life significantly improved economically for most Black Martiniquais. The ideal agricultural land was controlled by European planters and their offspring, including some mixed-race descendants. Meanwhile, many Black Martiniquais moved to some of the steep mountain slopes and lived in small two-room homes—primarily for sleeping and storage—with much of life happening outside of the cabin, including subsistence farming[5] (Fig. 12.7)

[5]Much of this information comes from a tour of La Savane des Esclaves, a museum located near Trois Îlets in December 2015. The museum was built between 2000 and 2004 and opened by Gilbert Larose in December 2004 (For additional information see La Savane des Esclaves 2016)

Other landscape changing events and processes have shaped Martinique and landscapes on parts of the island. For example, the end of slavery did not mean the end of coercive labor in Martinique or the dramatic reshaping of the rural, agricultural landscape away from the dominance of sugar (see Sect. 12.5.1 for more specific details). The eruption of Mount Pelée on 08 and 20 May 1902, reshaped the physical appearance on part of the northwest section of the island (see Sect. 12.4.2 for specifics). Finally, tourism has influenced the landscapes of Martinique with the preservation of certain areas and the distribution of tourism southward away from Saint-Pierre to Fort-de-France and Trois Îlets.

12.3.2 Mount Pelée and Saint-Pierre

While Mount Pelée (Fig. 12.8) erupted twice after the arrival of the French colonists between 1635 and 1852, the 1902 eruptions of Mount Pelée stand out as the worst volcanic natural disaster of the twentieth century. Understanding the loss of life can be difficult from our temporal standing over a century later, but, quite simply, the natural disaster was made worse by decisions made by some individuals locally —particularly the island's governor. The 1851–1852 eruption event was evidently a steam or phreatic eruption or set of eruptions, which fifty years later would have been remembered by the older individuals who lived in the area as relatively minor.

Even as it became apparent that the eruption event occurring was different than the one fifty years earlier, an upcoming election in May 1902 influenced some elected officials to minimize the potential danger just a few days prior to the election. As Mount Pelée had started minor activity a few years earlier, the feeling among some was that a few days more was not that risky. This was despite the fact that ash had fallen for a few days before the 08 May 1902 eruption. Research since the eruption suggests that hot mud and possibly lava flowed from the opening on the southwest side of Mount Pelée during the earlier eruptions. Depending on location, ash deposits ranged from a few centimeters up to a meter—though this depth of deposits might have been the result of air movement from the volcanic eruption on 08

Fig. 12.8 Mount Pelée. This view, looking almost directly north from a beach near the town of Le Carbet on the western coast showcases the volcano's steep sides and enormous presence on the island. To demonstrate its size, the summit lies roughly 15 km from the point this image was taken. Image courtesy of anonymous, available at https://commons.wikimedia.org/wiki/File:La_Pel%C3%A9e_vue_du_Carbet.jpg

Fig. 12.9 Mount Pelée (cloud covered, to the *right* of center) with Saint-Pierre in the foreground. The volcanic eruption of 08 May 1902 came from this (southwest) side of the volcano, reaching this part of Saint-Pierre—approximately 8 km from the volcano's summit—in less than two minutes. The top of the volcano is usually covered by cloud due to moisture condensing as it rises over the mountain. Photograph by E.A. Modlin

May (Heilprin 1903). Finally, news that Soufrière, on Saint Vincent erupted on 06 May 1902, led some to believe that pressure under the surface of the region was relieved, lending additional weight to the false sense of security.

The delay in evacuation proved deadly. On May 08, 1902, two days before the election, the volcano erupted at approximately 7:50 am (Fig. 12.9). The pyroclastic surge killed 30,000 people in just a few minutes. Because of the unexpectedness of the explosion, the simultaneous death of the island's governor, who with his wife was in Saint-Pierre, and the extreme heat emanating from the ruins, days passed before rescuers and others could enter the city proper. That day in May (08 May) was not when the then-current eruption period ended. Less than two weeks later, it was followed by an eruption that killed more people participating in relief work, searching for survivors, and surveying the damage (Heilprin 1903).

The May 1902 event preceded an eruption on August 30, 1902, which killed about 800 people in the Morne-Rouge area and about 300 in neighboring communities (Scarth 2002). The 30 August 1902 event was followed by another geological oddity. The eruptions that occurred from September 1902–1904 resulted in another noticeable, if temporary, topographical change of the island: A tall viscous spine rose from the earth that eventually peaked at over 300 meters before collapsing (Fig. 12.10; Westercamp and Traineau 1983). The last eruption of Mount Pelée was in 1929 (Global Volcanism Program 2013a, b). Today, the volcano is monitored, so hopefully future volcanic events will not lead to human loss of life, though real estate property built in the area could still make future eruptions economically costly.

As one of the most impressive cities in the French Caribbean prior to 08 May 1902, Saint-Pierre doubtlessly had tourists visit the city, but its sudden destruction started a new era for tourism in the area. Over the next few years, tourists visited by ship and toured the ruins. Scarth (2002) notes that this type of tourism started in January 1903 (Fig. 12.11a). While some tourists were "lucky enough" to have their pictures taken at the site, others relied on the dozens of different postcards that were created—some showing supposed survivors of the eruption, others demonstrating the damage, and still others showing people touring the site. One postcard sender, using racist terms, revealed that some locals had developed different ways to be paid entertainers for visitors to the site (see Fig. 12.11b).

12.4 Heritage and Tourism

12.4.1 Heritage

Martinique might have seemed minor and useless to the Spanish, yet it provided the French who maintained control over the island for most of its European colonial history, a part of their foothold in the struggle to claim and colonize islands in the Caribbean. Martinique's transition to the sugar industry, although later than the Portuguese and British, was

Fig. 12.10 SAINT-PIERRE (Martinique)—Title: "Le Cône du Mont Pelé. S'élevant à 250 m. de hauteur á la suite des éruptions volcaniques." The spine or "Needle of Pelée," approximately 250 m in this image, reached as high at 300 m by some sources. Though not mailed, the format of this postcard is from "Divided Back Period" usually published between February 28, 1907 and December 31, 1915. This postcard is from E.A. Modlin's personal collection

no less profound for Black laborers and their descendants. Although the majority of its history since 1634 has been as a part of the French empire, the British controlled the island three times, most significantly during the period when the French temporarily abolished slavery in its colonies and then reinstituted it in 1802[6] (Bélénus 2015; Sainton 2015). Freedom in 1848 for Black Martiniquais left the planter class looking for other laborers. The planter class turned to indentured servants who came from Africa, India, China, Vietnam, and Portugal. While more attention has been given to such groups in the British Empire, those imported into the French islands worked as hard and in many ways continue to be another set of outcasts in French society today (Northrup 2000). Mortality among some of the indentured workers was high enough that upward of one-third did not live to complete their five-year contract (Northrup 2000).

The island's residents lived—and continue to live—in a highly stratified society racially. Racism and its responses—including race-based resistance—continued well into the twentieth century. Out of resistance to colonialism and racism, the poet, author and future mayor of Fort-de-France, Aimé Césaire collaborated with other individuals under French colonial rule and developed the concept of *Négritude*. The idea *Négritude* appropriated the racially derogatory word for black people and used it to develop a strong sense of self-worth, which opposed the idea of assimilation, while still being politically involved, even if it meant forceful resistance (Njami 2015). In response to *Négritude*, *Créolité* emerged with significant contributors to the emerging style of discourse including the Martiniquais Patrick Chamoiseau, Raphaël Confiant, and Jean Bernabé—among others. This literary and philosophical set of thoughts builds on Edouard Glissant's *Antillianité* concepts which stressed a full idea of what "Caribbean-ness" is, including input from those of African descent, but also those of indigenous, European, East Indian, and Chinese descent. This created a unique set of worldviews deeply influenced by the synthesis communities that they created over the generations in the Caribbean. A significant mark of *Créolité* is that their works usually are initially published in both French and Creole.

12.4.2 Tourism

Heritage—the process of taking ownership of the past and making it your own—often ties tightly to tourism, particularly in places where the past is used to market a place as different. Martinique's tourism plans have not until recently seemed to heavily market heritage tourism. In many ways, the focus has been on "sun, sea, and sand" tourists. This focus served it well prior to 2000. According to the Caribbean Tourism Organization, from 2003 to 2014, tourist stopovers varied between 443,202 and 503,107, a difference of 12% (Table 12.1). Issues that affected these numbers are varied. First, there is the issue that Martinique competes with many other "sea, sun, and sand" destinations both in the Caribbean and elsewhere (Dehoorne and Augier 2011). Additionally, during the twenty-first century, North

[6]Thus, theoretically, Martiniquais did not go through the trauma of being freed and then re-enslaved as occurred in Guadeloupe, though René Bélénus (2015: 280) points out that in Guadeloupe this period could be seem as "false abolition" (see the chapter on Guadeloupe for more information on this). The main reason that local whites surrendered the island to Britain was over the issue of maintaining slavery, which Britain indicated it would do if it gained control of the island (Bélénus 2015).

(a)
MARTINIQUE
Leboullanger, éditeur à Fort-de-France (Martinique) - 2021
RUINES DE ST-PIERRE
Dear Jim, Passed near this to day & had a good view.
Terrible ruins. Best love,
Father
30/9/03

(b)
MARTINIQUE
Dear Edith: Rather a change from the St Pierre you saw. There were 1/2 dozen Martinique boats & boys to dive for our pennies. On shore a few darkey derelicts selling relics of the destruction
With love
Leboullanger, éditeur à Fort-de-France (Martinique) - 2226
Phot. Cochet
VUE GÉNÉRALE DE ST-PIERRE DÉTRUIT - QUARTIER DU MOUILLAGE

◀ **Fig. 12.11** Examples of historic postcards highlighting the devastating effects of the 1900s eruption of Mount Pelée. **a** Title: "Ruins of St.-Pierre." A father wrote in part to his son in Toronto, "Passed near this to day & had a good view. Terrible ruins." The postcard is dated September 30, 1903. This postcard is from the "Undivided Back Period" published prior to February 28, 1907. For more information on the value of tourists' comment on postcards and general guidelines to dating postcards, please see Modlin (2014). **b** Title: "Vue Générale de St-Pierre Détruit—Quartier du Mouillage." Tourist's text says, in part, "Rather a change from the St. Pierre you saw. There were ½ dozen Martinique boats and boys to dive for our pennies. On shore a few darkey [sic] derelicts selling relics of the destruction." The card was canceled (mailed) on March 25, 1906. Despite the racist term, the author discusses some of the economic activities that are geared toward tourists visiting the site. This postcard is a very early example of a postcard from the "Divided Back Period" usually published between February 28, 1907 and December 31, 1915. Both postcards from E.A. Modlin's personal collection

Table 12.1 Number of all tourists who arrived by plane or cruise boat (Caribbean Tourism Organization 2015)

Year	Total overnight stopover tourists	Total cruise passenger arrivals
2003	445,424	268,542
2004	470,891	159,416
2005	484,127	93,064
2006	502,053	95,812
2007	503,107	71,683
2008	481,226	87,079
2009	443,202	69,749
2010	476,492	74,634
2011	496,538	41,142
2012	487,359	93,515
2013	489,706	103,770
2014	489,561	177,786

Cruise numbers were particularly volatile as Martinique competes against other Caribbean ports of call. Additionally, the very low numbers of arrivals by plane in 2009 and cruise ship in 2011 might be related to strikes on Martinique in 2009 (Chrisafis 2009; BBC 2009). Because of advanced planning and commitments, it usually takes a couple of years for Cruise companies to respond to such events. This pattern was noticeable in Curaçao in cruise tourists visit in 1971 after unrest in 1969 (see Chap. 18 in *Landscapes and Landforms of the Lesser Antilles*). While not reaching the high of 2007, stopover growth held steady around 487,000 to 490,000 from 2011 to 2014. Cruise passenger arrivals swung more dramatically in the years between 2003 and 2017. 2003, 2004, and 2005 saw dramatic drop-off of cruise tourists year-over-year, a trajectory that continued on average until the "bottom" was hit at the low of 2011, which was only about 15.3% of the number of such passengers in 2003 when 268,542 cruise passengers arrived in Martinique. Since 2011, the number of cruise passengers has risen to 177,786 in 2014

American tourist number declined, while French tourist numbers increased. For English-speaking tourists, there is also the issue that English is not as widely spoken as in some other parts of the Caribbean or even in Metropole France. Yet, the Martinique Tourism Authority (2015a) is trying to attract more English speakers by having English-speaking street guides for cruise passengers visiting the island.

The argument that by linking tourism and environmental concern, Martinique could draw a niche tourism clientele that looks for sustainable development—often meaning minimal development of some areas of the island—has been taken up most notably by Dehoorne and Augier (2011). They advocate for refocusing from mass tourism to ecotourism, where fewer tourists would participate in tourism activities and each tourist would have a smaller impact on the environment. While Dehoorne and Augier (2011) are quick to point out that Martinique has a great variety of flora and fauna, as well as geographic and geologic diversity, they do not clearly show how an increase in ecotourism will fill the financial gap of a drop in mass tourism. That said, those advocating for more socially and environmentally responsible touring like ecotourism recognize that in some ways Martinique's pursuit of more and more tourists is having an effect on the environment and will be ecologically unsustainable in the future.

As with other Caribbean islands, tourism remains a complex social topic. Hundreds of thousands of tourists visiting the island yearly can bring billions of dollars into the economy both directly and indirectly. Yet, this income is not evenly distributed to all Martiniquais. Ownership of the property and businesses that get much of this income are, in most cases, in the hands of a small population of descendants of the European colonial settlers (about 1% of the population). Meanwhile, the island struggles with high unemployment rates—19% as of late 2014, although that is down from 25% a few years earlier (Morisme and Greliche 2007; INSEE 2015).

12.5 Hazards

Martinique, like other Lesser Antillean islands, remains at risk of a number of hazards including hurricanes and tropical storms, flooding and landslides, earthquakes, volcanos, tsunamis, mosquito-borne illnesses, diseases contracted by contaminated water, and, in lower areas, even susceptibility to sea level rise.

12.5.1 Tropical Storms and Hurricanes

Hurricanes and tropical storms have struck or brushed the islands dozens of times since the first European settlement. Since 1864, 32 tropical storms or hurricanes have brushed or crossed part of the island (NOAA 2016), some causing significant damage and loss of life. The hurricanes of 1780 killed 9000 in Saint-Pierre due to storm surge and 4000 French sailors off of the coast, while a storm in 1891 killed 700 individuals (Pérez, n.d.). Other major storm-related events include Tropical Storm Dorothy in 1970 (killed 44 people), Hurricane David in 1979 (destroyed 500 homes, 80 boats, and injured at least 20 people), and Hurricane Dean in 2007 that heavily damaged Martinique's banana plantations and sugar farms affecting the employment of thousands of Martiniquais (Perrusset and Bouguen 1970; Reuters 2007).

12.5.2 Seismic and Tsunamic Activity

Heavy rainfall from named tropical storms can cause and has caused flooding and landslides in Martinique. Flooding can also occur for a number of other reasons, such as extreme meteorological events—like the 05 May 2009 rains that caused landslides and killed four people (Paravisini-Gebert 2009). In addition to meteorological events, the steep grades of the hills and mountains on the island are susceptible to landslides during seismic events from earthquakes and volcanic activity, even if the seismic event happens off-island.

Low-lying villages, towns, and cities that line the coast, including the capital city of Fort-de-French, are also vulnerable to tsunamis with 28 recorded tsunami events reaching Martinique between 1501 and 2004 (Fig. 12.5). Most of these events were local or regional (within the Caribbean area), although a few originated outside of the region, most significantly the 1755 Lisbon earthquake-initiated tsunami which traveled across the Atlantic Ocean and was observable in the Caribbean seven to nine hours after the earthquake (Accary and Roger 2010). Volcanic activity and the associated seismic activity near Mount Pelée caused a number of tsunamis that affected the island's northwest portion. Away from the areas immediately adjacent to Mount Pelèe, other parts of the islands, such as La Trinité on the Atlantic coast and Fort-de-France, have sustained major damage by one or more tsunamis over the years. For example, Fort Royal (renamed to Fort-de-France) experienced a tsunami in 1839 (11 January) that killed 400 people and destroyed 400 homes (Accary and Roger 2010).

12.5.3 Pollution and Sea Level Rise

Coral reef communities and mangroves in Martinique remain susceptible to development, pollution, and sea-adjacent development. Martinique lost about a third of the mangroves on the island between 1972 and 1992, primarily to development (Gabrie et al. 2004; as cited in Schleupner 2008). This is concerning and shows that while attitudes toward mangroves have largely shifted from viewing them as "unhealthy and inhospitable" (Avau et al. 2011), they are still not receiving the protection they should. Meanwhile, the coral reefs, particularly those in the Fort-de-France Bay, are also struggling to deal with chemical runoff and pollution that both harm coral and encourage an increase in algae at the cost of coral (Littler et al. 1993).

sea level rise (SLR) cannot be ignored on Martinique. Like other Caribbean islands, Martinique has a significant tourism industry that is, in part, based on its sea resources—seaside beaches, food, and hospitality venues Scott et al. (2012) predict a SLR range of 18–59 cm between 1900 and 2100, noting SLR does not follow a global pattern, but that the Caribbean in general will experience it more dramatically (Compare Tamisiea and Mitrovica 2011). While trying to understand all of the processes associated with sea level rise and attempting to predict them, these numbers—as small as they seem—will affect mangroves and coral reefs causing damage in some areas that will remove or degrade some of the coastal protections. The damages of SLR can include saltwater intrusion, beach erosion, and even the endangerment of seaside buildings and infrastructure (Scott et al. 2012). Among the possible effects of sea level rise are the destruction of coral reefs and the "disappearance or redistribution of wetlands and lowlands" (Schleupner 2008). While some mangroves have room to migrate landward in response to sea level rise, others on the island do not have such available space (Soares 2009. Compare Schleupner 2005)

12.6 Conclusion

By many measures, Martinique is a unique island. Geologically, it should be considered part Volcanic Caribbees *and* part Limestone Caribbees. The heavily populated southern part of the island is lower in elevation, but still hilly. The middle and northern portions of Martinique are mountainous, reflecting more recent volcanic activity of over a dozen volcanos. One of these volcanos—Mount Pelée—is still active and, currently, is the second tallest mountain in the Lesser Antilles.

As an island located in the hurricane belt with a population approaching 400,000, a number of hazards and challenges face the people of Martinique. Biological and environmental zones along the coasts—both below the surface and adjacent to it—are facing stress due to direct human activity such as development and climate change caused in part by human activity worldwide. Estimates place sea level rise in parts of the Caribbean at 20–50 cm by the middle of the century (Schleupner 2005). Coral and mangroves in certain areas might be able to keep up such sea level rise, but other places perhaps not.

Yet, not all challenges are environmentally centered. Socially, the past is very much with the citizens of Martinique, with slavery and indentured servitude leaving their legacy. Many of the descendants of freed slaves and completed-contract indentured servants still struggle to rise from the bottom, economically. With high unemployment rates and low labor participation rates, many young Martiniquais are looking to other locations for careers—places such as Metropole France and North American cities.

These potential negatives aside, however, there are some very positive things happening in Martinique. The government is investing in infrastructure, and over the last decade, employment has been on the rise. Some inclusive philosophical work is coming from Martiniquais writers and thinkers, such as *Créolité*. While not ignoring the troubling history of the Caribbean, this movement is attempting to open a dialogue about past heritage—acknowledging the variety of people who are part of Caribbean societies. Instead of being top-down focused, the use of Creole—as a language and as identity—attempts to refocus the conversation in a way that gives those often marginalized a voice in shaping their identity. Like most other Lesser Antillean nations, Martinique has room for improvement in various sectors, but it is working to take steps in a direction that will potentially mitigate those shortcomings.

References

Accary F, Roger J (2010) Tsunami catalogue and vulnerability of martinique (Lesser Antilles, France). Int J Tsunami Soc 29(3):148–170

Avau J, Cunha-Lignon M, De Myttenaere B, Godart M-F, Dahdouh-Guebas F (2011) The commercial images promoting Caribbean mangroves to tourists: case studies in Jamaica, Guadeloupe and Martinique. J Coast Res, Special Issue 64:1277–1281

BBC (2009) France faces unrest in Caribbean. BBC News. Published February 12 2009. Accessed May 30 2016. Web address: http://news.bbc.co.uk/2/hi/europe/7885683.stm

Bélénus R (2015) How slavery became reestablished in Guadeloupe and the French colonies. In: Thierry L'Étang (ed) Memorial ACTe: exploring slavery and the African slave trade in the Caribbean and around the world. Translated by Simon Beaver, Jason Whittaker and Jérémie Ricaut. Pointe-à-Pitre, Guadeloupe: Caribbean Centre for the Expressions and Memory of African Slave Trade & Slavery. pp 280–285

Bérard B, Espersen R, White C (2014) Martinique. Encyclopedia of Caribbean archaeology. University Press of Florida, Gainesville, pp 132–134

Berlin I, Morgan PD (1993) Introduction: labor and the shaping of slave life in the Americas. In: Berlin I, Morgan PD (eds) Cultivation and culture: labor and the shaping of slave life in the Americas. University of Virginia Press, Charlottesville, pp 1–45

Bocquené G, Franco A (2005) Pesticide contamination of the coastline of Martinique. Mar Pollut Bull 51(5–7):612–619

Borroto-Páez R, Woods CA (2012) Status and impact of introduced Mammal in the West Indies. In: Borroto-Paez R, Woods CA, Sergile FE (eds) Terrestrial Mammals of the West Indies. Florida Museum of Natural History and Wacahoota Press, Gainesville, pp 241–257

Bouysse P, Westercamp D, Andreieff P (1990) The Lesser Antilles island arc. Proc ODP Sci Results 110(1):29–44

Caribbean Tourism Organization (2015) Latest Statistic Tables 2003-July 2015. Web Address: http://www.onecaribbean.org/statistics/2003-july-2015-tourism-stats/. Accessed: June 10, 2016

Chrisafis A (2009) France faces revolt over poverty on its Caribbean islands. The Guardian. Published online on February. 11 2009. Accessed May 30 2016. http://www.theguardian.com/world/2009/feb/12/france-revolts-guadeloupe-martinique

Columbus C (1987) The log of Christopher Columbus. Fuson, Rober H. (translator). Translation based on Las Casas' abstract of the Log with additions from his Histoia and Hernando Columbus's Historie [of the Columbus family]. New York: McGraw

Curtain PD (1969) The Atlantic slave trade: a census. University of Wisconsin Press, Madison

de Gurbert, GP (2014) Tobacco: the commodification of the Caribbean and the origins of globalization. In Sansavior E, Scholar R (eds) Caribbean globalizations, 1492 to the Present Day, pp 155–172

Dehoorne O, Augier D (2011) Toward a new tourism policy in the French West Indies: The end of mass tourism resorts and a new policy for sustainable tourism and ecotourism. Études Caribéennes. Issue 19. Accessed June 6 2016. Web Address: http://etudescaribeennes.revues.org/5262

Gabrie C, Bouchon Y, Bouchon C (2004) State of the environment at La Martinique. Les récifs dans les DOM-TOM. Initiative française pour les récifs coralliens. IFRECOR. Marseille

Germa A, Quidelleur X, Labanieh S, Lahitte P, Chauvel C (2010) The eruptive history of Morne Jacob volcano (Martinique Island, French West Indies): geochronology, geomorphology and geochemistry of the earliest volcanism in the recent Lesser Antilles arc. J Volcanol Geoth Res 198:297–310

Germa A, Quidelleur X, Lahitte P, Labanieh S, Chauvel C (2011a) The KeAr CassignoleGillot technique applied to western Martinique lavas: a record of Lesser Antilles arc activity from 2 Ma to Mount Pelée volcanism. Quat Geochronol 6:341–355

Germa A, Quidelleur X, Labanieh S, Chauvel C, Lahitte P (2011b) The volcanic evolution of Martinique Island: Insights from K-Ar dating into the Lesser Antilles arc migration since the Oligocene. J Volcanol Geoth Res 208(2011):122–135

Global Volcanism Program (2013a) Smithsonian Institute: National Museum of Natural History. "Pelée". Accessed May 24 2016. http://volcano.si.edu/volcano.cfm?vn=360120

Global Volcanism Program (2013b) Pelee (360120) in Volcanoes of the World, v. 4.4.3. Venzke, E (ed). Smithsonian Institution. Downloaded 13 Jun 2016 (http://volcano.si.edu/volcano.cfm?vn=360120)

Hearn L (1890) Two years in the French West Indies. Harper & Brothers, New York

Hearn L (1902) Lafcadio hearn on the island and people of martinique. Natl Geogr 8(6):214–216

Heilprin A (1903) Mont Pelée and the tragedy of martinique: a study of the great catastrophes of 1902, with observations and experiences in the field. J. B. Lippincott Company, Philadelphia

Higman BW (2011) A concise history of the Caribbean. Cambridge University Press, Cambridge

Institut national de la statisique et des etudes économiques (INSEE) (2015) Enquête emploi en continu en Martinique: Stabilité du chômage en 2015. Accessed May 30 2016. http://www.insee.fr/fr/themes/document.asp?reg_id=23&ref_id=24111

Institut national de la statisique et des etudes économiques (INSEE) (2016) Population de 1968 à 2015: comparaisons régionales et départementales. Accessed May 24 2016. http://www.insee.fr/fr/themes/tableau.asp?reg_id=99&ref_id=TCRD_004#col_1=8

La Savane des Esclaves (2016) La Savane des Esclaves. Accessed: June 6 2016. Web address: http://www.lasavanedesesclaves.fr/

Littler MM, Littler DS, Lapointe BE (1993) Modification of tropical reef community structure due to cultural eutrophication: the southwest coast of martinique. Proc Seventh Int Coral Reefs Symp, Guam 1992(1):335–343

Martinique Tourism Authority (2015) Good outlook for Martinique's cruise traffic. Web Address: http://s34127.gridserver.com/pdf/martinique-cruise-updates.pdf. Accessed June 13 2016

Martinique Tourism Authority (2015a) New initiatives in Martinique improve product delivery and passenger satisfaction. Web Address: http://s34127.gridserver.com/pdf/martinique-cruise-updates.pdf. Accessed June 13 2016

Modlin EA (2014) A market or "A relic of barbarism"? Towards a more inclusive analysis of social memory on postcards. In: Hanna SP, Potter AE, Modlin EA, Jr, Carter P, Butler DL (eds) Social memory and heritage tourism methodologies. Routledge, New York, pp 170–188

Morisme, Éric and Agnès Greliche (2007) Les comptes économiques de la Martinique en 2006. Institut national de la statisique et des etudes économiques (INSEE)

National Oceanic and Atmospheric Administration (NOAA) (2016) Historical hurricane tracks. Accessed May 24 2016. https://coast.noaa.gov/hurricanes/

Njami S (2015) Mutation of black identity. In: L'Étang T (ed) Memorial ACTe: exploring slavery and the African slave trade in the Caribbean and around the world. Translated by Simon Beaver, Jason Whittaker and Jérémie Ricaut. Pointe-à-Pitre, Guadeloupe: Caribbean Centre for the Expressions and Memory of African Slave Trade & Slavery. pp 344–349

Northrup D (2000) Indentured Indians in the French Antilles. Les immigrants indiens engagés aux Antilles françaises. Revue française d'histoire d'outre-mer 87(325):245–271

Paravisini-Gebert L (2009) 4 dead as floods and landslides hit Martinique. Repeating islands. Accessed May 24 2016. https://repeatingislands.com/2009/05/06/4-dead-as-floods-and-landslides-hit-martinique/

Pendergranst M (2010) Uncommon grounds: the history of coffee and how it transformed our world (Revised edition). Basic Books. New York

Pérez O (n.d.) Notes on the tropical cyclones of Puerto Rico, 1508—1970. Atlantic Oceanographic and Meteorological Laboratory. Accessed May 24 2016. http://www.aoml.noaa.gov/hrd/data_sub/perez_11_20.pdf

Perrusset M, Bouguen P (1970) La Tempête Tropicale Dorothy—Page 3. Direction de la Meteorologie Nationale: Météorologique du Groupe Antilles-Guyane. Accessed May 24 2016. http://www.nhc.noaa.gov/archive/storm_wallets/cdmp/dvd0032-jpg/1970/atlantic/dorothy/postevent/servicemeterologique.04.jpg

Reuters (2007) Hurricane destroys Martinique, Guadeloupe bananas. Accessed May 24 2016. http://www.reuters.com/article/us-storm-dean-bananas-idUSL1861011420070818

Ruf F, Schroth G (2004) Chocolate forests and monocultures: a historical review of cocoa growing and its conflicting role in tropical deforestation and forest conservation. In Schroth G, da Fonseca GA B, Harvey CA, Gascon C, Vasconcelos HL, A-M N Izac (eds) Agroforestry and biodiversity conservation in tropical landscapes. Island Press, Washington, pp 107–134

Sainton J-P (2015) The revolution, the first abolition of 1793–1794 and its impact on the Caribbean (1794–1804). In: Thierry L'Étang (ed) Memorial ACTe: exploring slavery and the African slave trade in the Caribbean and around the world. Translated by Simon Beaver, Jason Whittaker and Jérémie Ricaut. Pointe-à-Pitre Guadeloupe: Caribbean Centre for the Expressions and Memory of African Slave Trade & Slavery. pp. 274–279

Scarth A (2002) La catastrophe: the eruption of Mount Pelée, the worst volcanic eruption of the twentieth century. Oxford University Press, Oxford

Schleupner C (2005) Spatial Analysis As Tool for Sensitivity Assessment of Sea Level Rise Impacts on Martinique. FNU-71. Accessed: June 1 2015. Web address: http://epub.sub.uni-hamburg.de/epub/volltexte/2012/16427/pdf/Spatial_Analysis_as_Tool_for_Sensitivity_Assessment_FNU_71.pdf

Schleupner C (2008) Evaluation of coastal squeeze and its consequences for the Caribbean island Martinique. Ocean Coast Manag 51(5):383–390

Scott D, Simpson MC, Sim R (2012) The vulnerability of Caribbean coastal tourism to scenarios of climate change related sea level rise. J Sustain Tourism 20(6):883–898

Singler JV (1996) The demographics of creole genesis in the Caribbean: a comparison of Martinique and Haiti. In: Arends J (ed) The early stages of creolization. John Benjamins Publishing Company, Amsterdam, pp 203–232

Soares MLG (2009) A conceptual model for the responses of mangrove forests to sea level rise. J Coast Res, Special Issue 56:267–271

Tamisiea ME, Mitrovica JX (2011) The moving boundaries of sea level change: understanding the origins of geographic variability. Oceanography 24(2):24–39

U.S. Department of State: Bureau of Consular Affairs (2016) U.S. Passports and International Travel: France. Last Modified: February 29 2016. Last Accessed: May 31 2016. https://travel.state.gov/content/passports/en/country/france.html

Westercamp D, Traineau H (1983) The past 5000 years of volcanic activity at Mt. Pelee Martinique (F.W.I.): Implications for assessment of volcanic hazards. *Explosive Volcanism* 17(1-4):159–185

Westercamp D, Andreieff P, Bouysse P, Cottez S, Battistini R (1989) Martinique; carte géologique à 1/50 000. In: BRGM (ed) 246 pp

Zamor H (2015) The history of Martiniquan rum. Int J Humanit Cult Stud 2(2):557–568

St. Lucia

13

Vanessa Slinger-Friedman, Susanna Diller and Lauren Parkinson

Abstract

Volcanic processes have crafted an undeniably beautiful geological landscape on the island of St. Lucia. The young island is rugged and mountainous with narrow valleys. St. Lucia's topology reflects its volcanic origin, including the Barre de l'Isle volcanic axial range, the dramatic twin Pitons, the Qualibou Caldera, and Sulphur Springs, one of the hottest and most active geothermal fields in the Lesser Antilles. Many of these outstanding geomorphic features are why the Pitons Management Area in the southwest of St. Lucia was inscribed as a UNESCO World Heritage Site. The second largest of the Windward Islands, St. Lucia, has been home to Amerindian populations, pirates, whalers, runaway slaves, and European colonists. Today, the island's geomorphic features continue to play an important role in providing the base for agriculture and for attracting tourists from around the world. Both of these activities, while important for the island's economy, are susceptible to the hazards of hurricanes, floods, and landslides experienced on St. Lucia; and in turn, both these activities and the natural hazards have impacts on the island's geological landscapes.

Keywords

St. Lucia • Volcanic • Agriculture • 3S tourism • Alternative tourism

13.1 Introduction

With its contrasting volcanic, mountainous peaks and its flat white sand beaches, and well-formed harbors, St. Lucia is a strikingly beautiful island. Perhaps its most prominent topographic features are the volcanic axial range, known as the Barre de l'Isle, extending down the length of the island and the famous steep-sided seaside Pitons. The mountains rise to the south peaking at Mount Gimie (950 m). The main attraction for most visitors to St. Lucia are the rainforest-clad twin mountain peaks of the Pitons, along with white sand beaches, natural harbors, coral reefs, and sulphur springs.

St. Lucia is composed of its mainland and the Maria Islands in the southeast (Fig. 13.1). The island was once inhabited by indigenous Arawaks (200–400 AD) and Caribs (800 AD) and vigorously fought over by English and French colonists during the seventeenth and eighteenth centuries. Over time, the island's geological landscapes played several roles, including the provider of a subsistence living for the Amerindian groups, a pirate hideout, a whaling station, a refuge for runaway slaves, the backdrop for extensive plantations, and the setting for a tourism haven. For each of these activities, the island has experienced a corresponding adjustments of its geomorphic landscape.

St. Lucia's mainstay agriculture and tourism industries are vulnerable to hazards such as hurricanes, floods, and landslides due to its location in the hurricane belt and its geological rugged landscape. Likewise, the geological

V. Slinger-Friedman (✉) · L. Parkinson
Department of Geography and Anthropology, Kennesaw State University, Kennesaw, GA, USA
e-mail: vslinger@kennesaw.edu

L. Parkinson
e-mail: laurenhparkinson@gmail.com

S. Diller
Department of Geography and Environmental Studies, University of New Mexico, Albuquerque, USA
e-mail: sldiller14@gmail.com

C.D. Allen (ed.), *Landscapes and Landforms of the Lesser Antilles*, World Geomorphological Landscapes, DOI 10.1007/978-3-319-55787-8_13

Fig. 13.1 General physiographic map of St. Lucia and smaller surrounding islands, such as the Maria Islands off the southeast coast. Copious rivers and valleys radiate from the central mountains and the various peaks that dot the coast, such as Petit Piton and Gros Piton on the southwest coast. Cartography by K.M. Groom

landscape is impacted by the economic activities taking place on its surface. Over time, the focus of tourism has shifted to include the promotion of its natural geological landscapes in a form of alternative tourism. The concern is that this form of tourism might be just as destructive as the traditional mass tourism product that has been so popular in St. Lucia due to the ecologically sensitive settings of the most visited locations. Only time will tell whether environmental policies put into place by the government will reduce the potentially harmful impacts of its humans on its landscape.

13.2 Setting

13.2.1 St. Lucia's Interior

St. Lucia, part of the of the Windward Island chain of the Lesser Antilles, is 620 km^2 and has 158 km of coastline (Nations n.d.). Like the majority of the Windward Islands, St. Lucia is volcanic in origin, and most of the geology reflects that history (Fig. 13.2). St. Lucia lies on the eastern margin of the Caribbean tectonic plate, and the island was formed in a series of volcanic processes that took place between 20 thousand and 19 million years ago. It is believed that there were three primary periods of island building for St. Lucia (Island Profiles n.d.a). The first took place between 9 and 19 million years ago, and formed the area between present-day Castries and the northern tip of the island. The second took place 5–6 million years ago, and formed a distinct landmass that has become the southern portion of the island. Between 1 and 2 million years ago, there was a significant period of volcanic activity in the present-day Mount Gimie region, which connected the two prior island formations and built up the center of St. Lucia (The Island and Nature n.d.). The Pitons were formed in the most recent period of volcanic activity, which started sometime in the last 1 million years in the Soufriere Bay region of St. Lucia (Island Profiles n.d.b). In the south, the young rock makes it easier to identify the various basaltic andesite centers in this area, which include Mount Gomier, Morne Caillandre, Moule a Chique, Beauséjour, St. Urbain, Mt. Tourney, and Savannes (Island Profiles n.d.a).

There is alignment between Morne Caillandre, Beauséjour, and Mt. Tourney, indicating that they are part of the same structural distribution (Lindsay et al. 2002). A large depression begins north of Soufriere and runs south of Mt. Tabac to the east, almost reaching Mt. Gimie before it loops unevenly to the south, just below Gros Piton. Known as the Qualibou Depression (Fig. 13.3), this area contains basalt rocks dated from 5–6 million years ago, while the Gros and Petit Pitons within the depression are as young as 290 thousand years old and give a more precise date range for the age of the region (Lindsay et al. 2002). Originally, the depression was thought to be one large caldera, but Wohletz et al. (1986) determined that only a portion of the area was actually a caldera. The initial depression resulted from subsequent landslides or other forms of structural collapse (The Island and Nature n.d.; Lindsay et al. 2002). There are debris avalanche deposits off the coast of St. Lucia near Soufriere that are potentially related to the Qualibou Depression (About 2010). Based on the assessment of sediment layers on top of the debris, the collapse is estimated to have happened approximately 300 thousand years ago. Though volcanic flows near Mount Gimie cut off at the margins of the depression, indicating that volcanic activity was taking place in the region before the depression formed, the actual Qualibou Caldera developed after the depression (Lindsay et al. 2002). The only limestone on the island is in this area, north of Soufriere on the eastern coast (Earle 1924).

In what is referred to as the Central Series, the rocks have been dated within six age ranges running from 10.4 million years old (Miocene) in lava flows west of Dennery, to 2.8 million years old (Pliocene) in lava flows near Derriere Dos. This region of St. Lucia is heavily forested and has the most significant relief on the island, with the second and third highest peaks, the Gros and Petit Pitons, that rises to 770 and 743 m, respectively. The Piton Mitan Ridge, which is part of a 2909 hectare area designated as a UNESCO World Heritage Site and known as the Pitons Management Area, links these two volcanic spires (UNESCO n.d.).

North of the Rousseau Valley, the older volcanic features are significantly eroded, which limit the ability to identify specific volcanic centers (The Island and Nature n.d.). This volcanic region is known as the Northern Series, and is comprised of basalt and andesite lava flows and pyroclastic deposits (Government 2001). Based on comparison with similar rocks in Martinique, Newman (1965) suggested that these rocks are from the Eocene period, but nothing on St. Lucia has been definitively dated to that time period. The oldest basalt in this region of St. Lucia has been dated to 18 million years old, in the Miocene (Lindsay et al. 2002; Smith et al. 2013). The youngest centers in the Northern Series, dating from the Late Miocene to the Early Pliocene, are Mt. Pimard, Vigie, and Mt. Monier, which have an area of weak geothermal activity on its south flank in Ravine Raisinard (Smith et al. 2013).

13.2.2 St. Lucia's Coastal Setting

St. Lucia's coastal shelf has an area of 522 km^2, and is narrow along the western coast (0.1 km) and wider around the rest of the island, reaching a width of 5 km in places (Mohammed and Lindop 2015). Reefs are found along 44% of St. Lucia's coastline (Maps 2008), with the highest

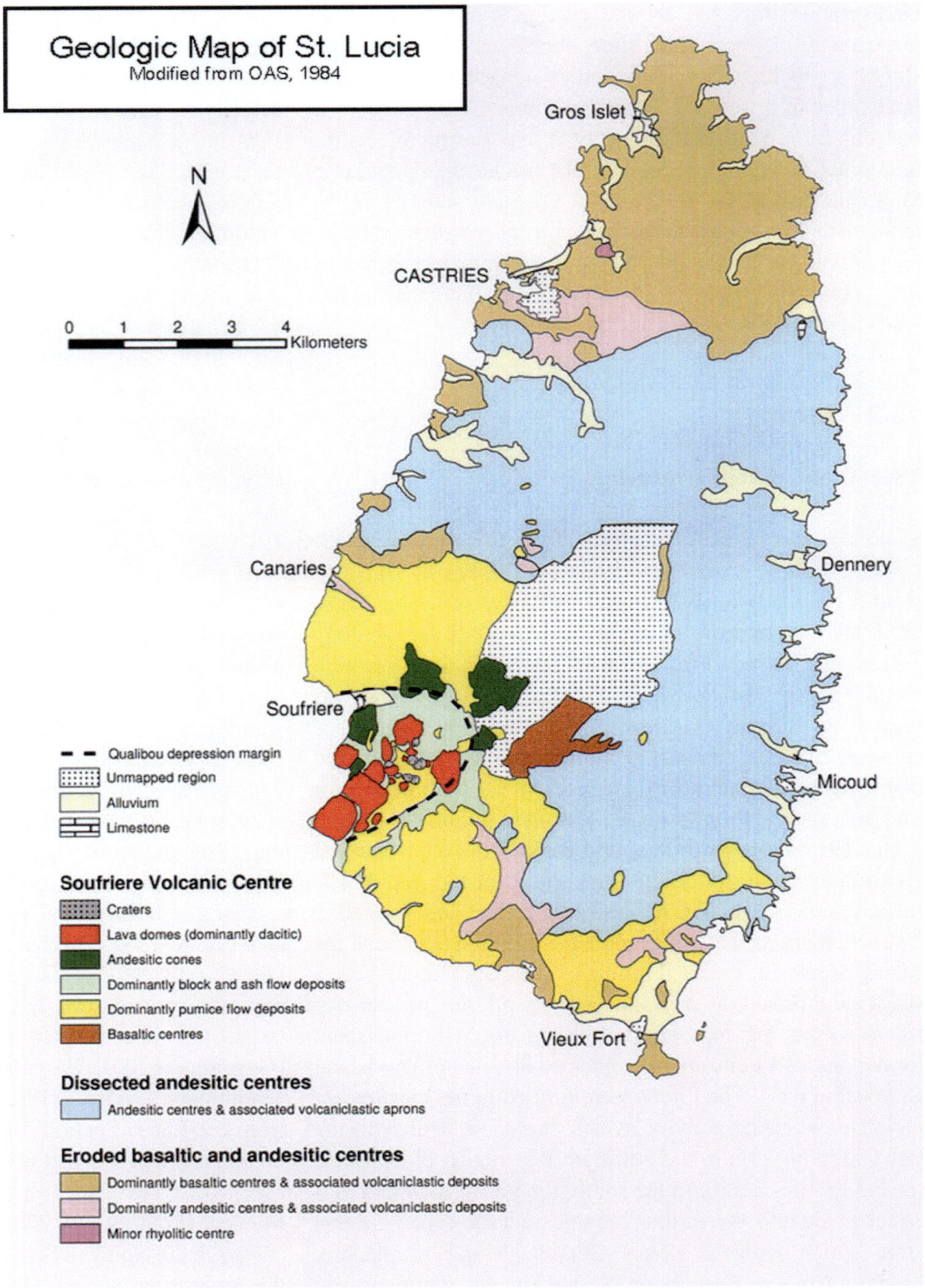

Fig. 13.2 Geologic map of St. Lucia displaying its varied volcanics. Generally, older rock lies to the north while younger rock can be found in the south. The region around Soufriere (especially the *red-colored* areas) represents perhaps the most recent volcanics on the island. Image courtesy of http://uwiseismic.com/General.aspx?id=72

concentration of reefs found on the east and south coasts. The reefs on the east coast are mostly patch reefs, isolated concentrations of coral growth, or fragmented fringe reef systems (St. Lucia n.d.). This relatively high degree of uneven reef distribution on the windward side of the island is likely due to the exposure to the forceful waters of the Atlantic. On the sheltered leeward side of the island, the narrow fringing reef grows on submerged volcanic rock, and it is the most vibrant and dynamic reef ecosystem on St. Lucia (Mohammed and Lindop 2015).

Fig. 13.3 The Qualibou Depression on the southern end of the island. The steep cliffs of the half-moon depression encompass the small volcanic peak of Petit Piton and the town of Soufrière. The formation and origin of this depression have been debated throughout recent history. Photograph by A. DeGraff Ollivierre

13.3 Landforms

13.3.1 Terrain Features

After Dominica, St. Lucia is the second largest of the Windward Islands. It is composed of the mainland and two smaller islands, the Maria Islands, off of its southeastern coast. Pigeon Island is located on the northwestern coast of St. Lucia, and in 1972, it was connected to the mainland by a human-made dirt causeway (Pigeon Island National Landmark). St. Lucia is a volcanically formed, mountainous island, with a north-south running backbone ridge called the *Barre de l'Isle* (Fig. 13.4). Heading south along this spine, the mountains generally increase in size, peaking at 950 m with Mount Gimie (St. Lucia 2008). Located on the western coast of the island southwest of Mount Gimie, the Gros Piton (770 m) and the Petit Piton (743 m) rise up dramatically from the sea and are visible from most parts of the island (Fig. 13.5). These mountains are so emblematic of St. Lucia that they are even featured on the country's flag, one triangle inside the other (Morales 2011). The two steep-sided spires are plugs, the resistant sections of an eroded volcanic dome, and are linked by the Piton Mitan Ridge. They sit along the edges of the Soufriere Volcanic Center, which encompasses the surrounding areas, including the town of Soufriere (Pitons Management Area Saint Lucia).

At the center of the Qualibou Caldera, a volcanic valley, and situated over a magma chamber two kilometers below the surface lies Sulphur Springs (Geology n.d.); it is marketed as the Caribbean's only "drive-in volcano" ("Sulphur Springs Park"). While not technically a volcano, it is one of the hottest and most active geothermal sites in the Lesser Antilles. The region is part of a UNESCO World Heritage Site, the Pitons Management Area (UNESCO n.d.), which also includes the Pitons. This area was inscribed as a World Heritage Site having met criteria (vii) and (viii):

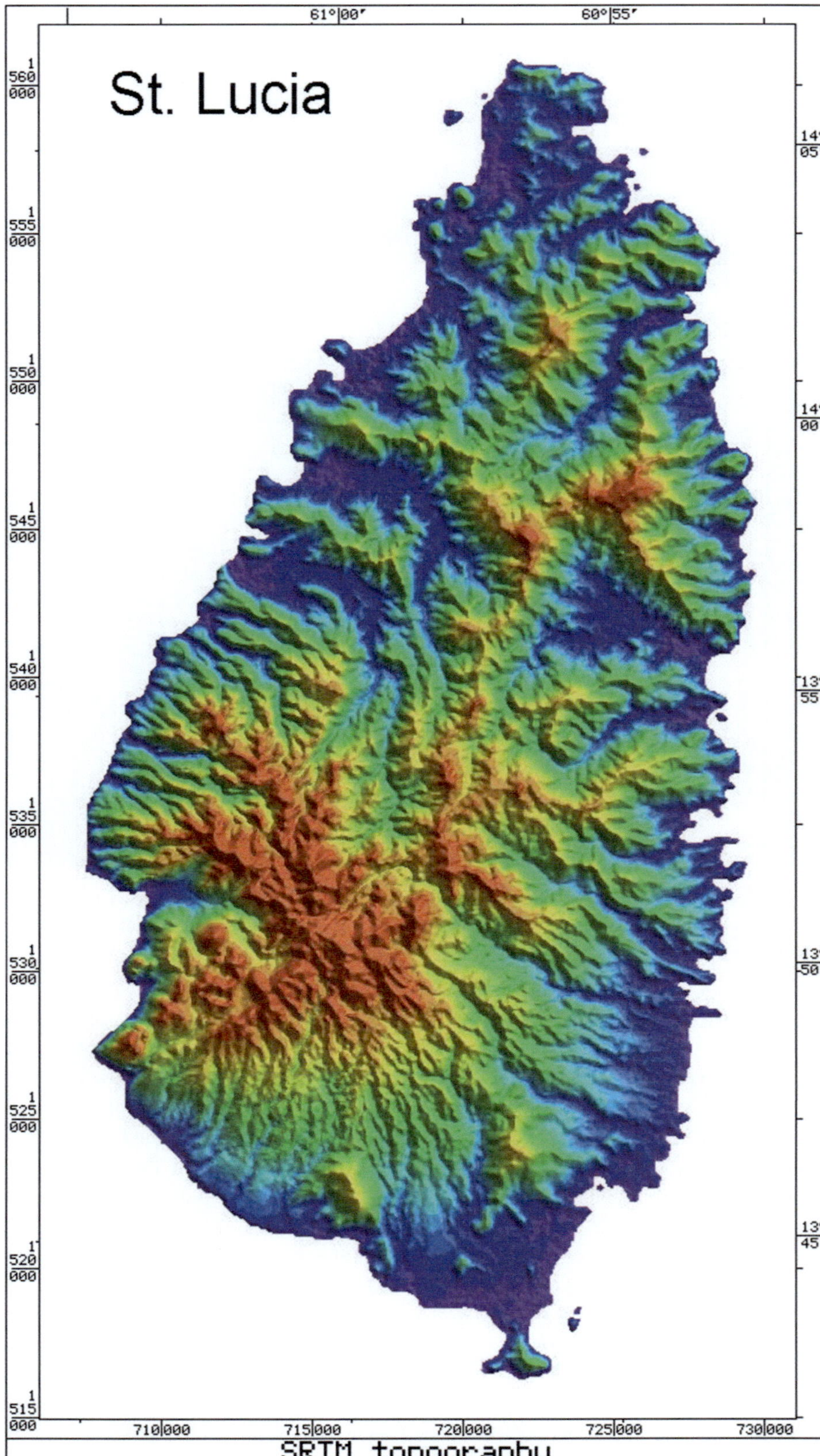

Fig. 13.4 Digital Elevation Model (DEM) showing St. Lucia's mountainous regions (*rust-colored* areas) and more gentle topography hugging the small coastal regions (*blue-colored* and *purple-colored* areas). The *Barre de l'Isle*, a generally north-south trending ridge, can be seen along the *center* of the image, with the highest elevations found along the ridge's southernmost points

Fig. 13.5 Perhaps the most iconic features on the island of St. Lucia are the volcanic mountainous peaks of Gros Piton and Petit Piton near the town of Soufrière. Both peaks are included in the 29 km^2 Pitons Management Area, which was declared an UNESCO World Heritage Site in 2004. Photograph by A. Potter

Criterion (vii): The Pitons Management Area derives its primary visual impact and aesthetic qualities from the Pitons, two adjacent forest-clad volcanic lava domes rising abruptly from the sea to heights greater than 700 m. The Pitons predominate over the St. Lucian landscape, being visible from virtually every part of the island and providing a distinctive landmark for seafarers. The combination of the Pitons against the backdrop of green tropical vegetation and a varying topography in the marine foreground gives the area its superlative beauty.

Criterion (viii): The Pitons Management Area contains the greater part of a collapsed stratovolcano contained within the volcanic system, known to geologists as the Soufriere Volcanic Center. Prominent within the volcanic landscape are two eroded remnants of lava domes, Gros Piton and Petit Piton. "The Pitons occur with a variety of other volcanic features including cumulo-domes, explosion craters, pyroclastic deposits (pumice and ash), and lava flows. Collectively, these fully illustrate the volcanic history of an andesitic composite volcano associated with crustal plate subduction" (http://whc.unesco.org/en/list/1161).

Morne Soufriere and Rabot Ridge border the Qualibou Caldera to the east and west, respectively. The slopes of these mountains that face Sulphur Springs are cut by numerous faults, which allow rainwater to penetrate deep into the earth, become heated with hot rocks, and return via pressure to the surface. As a result, the site includes a host of geothermal and volcanic landforms, such as hot springs, sulfur-spewing fumaroles reaching temperatures of 172 °C, pools of boiling water, and *soufrieres*, which are volcanic vents which emanate steam and chemicals, such as hydrogen sulfide (St. Lucia 1994). This area is also the location of two mineral pools attaining temperatures up to 90 °C (Hot Water n.d.). Other areas of geothermal activity exist outside of the Qualibou Caldera including the warm springs at Jalousie and Choiseul, underwater vents at Black Bay and offshore between Anse Mamin and Soufriere Bay, and areas of "cold Soufrière" (cold fumarolic activity which are acidic, roughly 28 °C, and associated with areas of highly altered rock) near Bois Demanje and in De Mailly Village (Lindsay et al. 2002) (Fig. 13.6).

13.3.2 Hydrological Features

St. Lucia has a tropical climate with two distinct seasons: a wet season from June to September and a dry season from February to May. The northern and southern coasts average 1500–1750 mm of rain a year, while the interior of the island nearly triples that number at about 3990 mm a year. There are nine major rivers on St. Lucia, including Cul de Sac, Cannelles, Dennery, Fond, Piaye, Doree, Canaries, Marquis, and Roseau. The Roseau River, on the western side south of Castries, is the longest river on the island (Nations n.d.). All of the rivers on the island run from the central highlands out to the coast and are evenly distributed around the island. The mountainous landscape combined with the high yearly rainfall makes St. Lucia an ideal environment for waterfalls. The Enbas Saut Waterfall Trail leads hikers through the thickly forested slopes of Mount Gimie, ending at a waterfall that is slowly eroding away the volcanic rock (Hastings 2015). With a drop of a little over 9 m, the Piton Waterfalls is in the south of the island and located near the Petit Pitons. On the northeast of the island is the Latille Waterfall, with a drop of about 7 m. Sault Falls (also known as Dennery Falls or Errand Falls), on the eastern side of the island, has a longer drop of roughly 17 m ("Top 5 Waterfalls in Saint Lucia"). Diamond Waterfalls, in the south, is set within Diamond Botanical Gardens. At 17 m high, the falls are notable not only for their height, but also for the spectacular coloration of the surrounding rock. The water comes

Fig. 13.6 An example of geothermal activity on St. Lucia, expansive Sulphur Springs Park near Soufrière volcano is prided as being the Caribbean's only drive-in volcano. The viewing platform with visitor in the *upper left* of the image provides scale. Photograph by A. DeGraff Ollivierre

down from the sulphur springs, and mixes with rainwater, carrying sulfur, copper sulfate, magnesium, iron, manganese, and calcium to the brink, where it colors the rock underneath (Fig. 13.7). The multicolored rock makes the water appear to change color from moment to moment (Caribbean n.d). The heated water from the springs blends with rainwater to create warm pools below ("Top 5 Waterfalls in Saint Lucia" n.d.).

13.3.3 Coastal and Marine Features

The influence of volcanic processes extends down to St. Lucia's coasts where black sand beaches can be found alongside the white sand beaches. Along the coast are natural harbors with white sand beaches. These include Marigot Bay, Rodney Bay with Reduit Beach, and Pigeon Island Park with Pigeon Point Beach (Porter and Prince 2007). There are 90 km^2 of coral reefs in total around St. Lucia. The most extensive reefs are located on eastern and southern coasts, but the west coast boasts outstanding reefs that grow out of the sides of volcanic rock ("Reefs in St. Lucia"). At the base of the Pitons are the Piton Wall and Superman's Flight, the latter having been used in the movie Superman III ("Diving" 2015). Both provide habitat for corals, fish, and other ocean life. Anse La Raye Wall is a steep underwater feature with fire coral in the shallow areas and a large variety of sponges and soft coral in deeper waters ("St. Lucia Dive Sites" n.d.). Finally, St. Lucia's southeast tidal shores support Mankòtè Mangrove, which cover 40 ha (99 acres).

13.4 Landscape

Prior to European discovery of St. Lucia in the sixteenth century, the island was inhabited by the Arawak people as far back as 200–400 CE (Things to do n.d.). The Arawak

Fig. 13.7 Diamond Waterfalls, a geothermal feature in the Diamond Botanical Gardens, displays discoloration and mineral deposits from both rainwater and naturally occurring chemicals in the underlying rock. The unique features of the waterfall and the pleasant setting make it a popular tourist destination. Photograph by A. Potter

were farmers and fishermen, and both utilized the natural resources of St. Lucia and maintained a spiritual connection with the land (St. Lucia n.d.). Gros Piton, called Yokahu, was worshipped as the god of fire, thunder, and rain. Petit Piton, Atabeyra, was the god of fertility and food. The Arawak also believed that the spirit of their gods was housed in the boiling water of the Sulphur Springs (The Soufriere Foundation—History of Soufriere 2010a, b, c, d). By contrast, the warlike Caribs, who came after them in 800 CE (All 2014), believed the Sulphur Springs was a god, and threw human sacrifices into the water during periods of increased hydrothermal activity. This practice gave name to the depression in which Sulphur Springs is located, as Qualibou means "at the place of death" (The Soufriere Foundation—History of Soufriere 2010a, b, c, d). The original Carib name for St. Lucia was Iouanalao, meaning, "land of iguana." This name took various forms over time, before settling on Hewanorra, now the name of St. Lucia's airport.

Credit for discovering St. Lucia goes to Juan de Cosa, who first labeled St. Lucia as El Falcon on a map dated to 1499. By 1502, the island was marked on a globe and labeled as Santa Lucia (All n.d.). The first documented European to actually set foot on the island was the French pirate François le Clerc, known at the time as Jambe de Bois, meaning "wooden leg" (Past and Present n.d.). He lived on Pigeon Island off the northwest coast of St. Lucia and attacked Spanish ships as they passed (Things, n.d.). Pigeon Island also played an important role during the French and British tug-of-war over territories in the West Indies, during which time St. Lucia became known as the "Helen of the West" (About n.d.). The British built a naval base on Pigeon Island in 1778, a move that was crucial to their success in subsequent battles with the French. Along with its history with piracy and its role as a strategic piece in British military control of St. Lucia, Pigeon Island served as a whaling station from 1909 to 1952, when legislation controlling whaling was enacted. Pigeon Island National Park was founded in 1992 as the culmination of an ongoing preservation project, both for cultural and natural heritage values (Sites n.d.). Where mainland St. Lucia meets the artificial bridge connecting to Pigeon Island is the town of Gros Islet. Once a small fishing village, Gros Islet has become one of the more popular places on the island today (Fig. 13.8).

Fig. 13.8 Aerial view of Pigeon Island, which has played an integral role as a naval base and whaling station. The now-ruined Fort Rodney atop the outcrop in the *lower left* of the image was an important British military outpost in the late 1700s, as the view from the *top* apparently included the French Navel Base on neighboring Martinique. Photograph by A. DeGraff Ollivierre

The colonial period saw many of the most dramatic changes to St. Lucia's landscape. Two Frenchmen built the first plantation on the island in 1765, and by the time the British finally took control in 1814, there were 50 operational plantations (All n.d.). Throughout this time, slaves were brought to the island from Africa. Although Britain outlawed slavery in 1807, this legislation did not reach St. Lucia until 1834, and slavery was not truly abolished until 1838. Emancipation did not correspond to a shift in power, however, and white landowners still held control over the island's resources (All, n.d.). This meant that the livelihood for many of the inhabitants consisted of intensive fishing, woodcutting, or charcoal making (Heritage n.d.). Leading up to World War II, charcoal was one of St. Lucia's top exports, but in the 1980s, concerns about rapid deforestation led to a significant reduction in this industry. Deforestation for fuel wood is still an issue today due to spotty enforcement of regulations (UNESCO 2010).

During the Second World War, St. Lucia became home to a US Air Force base on 1200 acres, which had been allocated by the British Empire. With its construction, several natural land features needed to be altered or removed, including streams and mangroves, in order to make way for the infrastructure (Burns and Novelli 2011). This meant that the installation had a significant impact on the environment, but after the base was dismantled and the land handed back, St. Lucia was left with roads, sewage lines, electricity, an airport, and a harbor. Despite the intrusion from the USA, this infrastructure enabled the future development of the agriculture and tourism industries.

Bananas became St. Lucia's primary agricultural export after large-scale sugarcane production diminished with the dissolution of slavery. Today, bananas are still St. Lucia's leading export (Exports 2013), though only 5% of national GDP comes from agriculture, while 80% comes from tourism and services (Lashley 2013; United 2009). Even though St. Lucia has been transitioning from an agriculturally based economy to a tourism and services-based economy since the early 1990s, agriculture is still a prominent feature on the island, and the significantly reduced agricultural industry is still actively shaping the landscape. The Roseau River feeds the banana plantations in the Roseau Valley, and it is a significant source of drinking water for the island (All n.d.). The Roseau Valley plantations are the most extensive agricultural area on the island, and the success of hillside agriculture in Roseau inspired similar work in Cul de Sac Valley and Mabouya Valley (Department 1986). In eastern St. Lucia, Mabouya Valley has been used heavily for agriculture, despite steep sides susceptible to landslides and a previously swampy valley floor that needed to be drained before it could be used. Modern erosion control techniques have led to hillside agricultural developments, including bananas, coconuts, and mangoes. Mabouya Valley is home to a nature and heritage park called Fond D'Or, where a colonial sugar mill was built over a former Amerindian settlement. This park provides an interesting cross section of the way landscape has been used through time (Heritage Sites n.d.).

Mangrove use on St. Lucia for fuel by French and British colonial powers began at Mankòtè Mangrove in 1760. Currently, there are only 18 mangrove sites remaining in St. Lucia, all of them are now Marine Reserves due to government recognition of their ecosystem value. The mangroves serve as the main source of nutrients for the island's natural fish nursery. However, enforcement of these reserves is highly variable, and the general public opinion still undervalues the mangroves (Wilkie 2005).

Other than for existing tourism purposes, the geothermal features on St. Lucia have not been tapped for use. While long explored through surveys and the drilling of shallow wells, the contemporary development of the geothermal potential of Sulphur Springs has been limited by lack of finances and the high cost of geothermal exploration (Geothermal energy in St Lucia 1989). These are examples of the ways that the striking landscapes in St. Lucia have been shaped by centuries of human impact, agricultural production, and the gradual increase of tourism. These examples serve to illustrate the ways that mixed development now informs the island's overall appearance, with some landscapes secluded and protected by the government, and others under endless fields of agricultural production or impacted by tourist pressure.

13.5 Heritage and Tourism

With the emergence of mass tourism in the Caribbean region in the mid-1960s, St. Lucia became a popular destination and progressed along the lines of large area foreign-owned development (Fig. 13.9). This type of tourism has been characterized by the standard 3S (sun, sea, and sand) tourism product and cruise ship tourism (Fig. 13.10). As the island has transitioned from agriculture to tourism, with a noticeable increase in visitors and rapid hotel growth, there has been a concurrent rise of resource use conflicts and planning challenges as land, labor, and capital are competed over (Fig. 13.11; McElroy 2003). Tourism and its related infrastructure have also brought threats to the environment in St. Lucia. Inadequate sewage treatment facilities have historically led to degraded shore marine water quality and coral reef death. Government action to establish sewage and wastewater treatment plants in the most heavily populated tourist areas like Rodney Bay area has attempted to reduce these impacts. Other tourism-related impacts include coral damage from divers and boat anchors, and sand compaction from heavily trafficked beaches (Nagle n.d.).

The construction of coastal tourism facilities and roads in ecologically sensitive areas has caused stress in marine environments. The building of tourism infrastructure has resulted in the destruction of mangroves and beaches. In addition, sand mining, dredging, and sewage dumping caused the environmental degredation of a lagoon. Wetland areas function to create the ideal conditions for coral reef development; hence, their removal impacts the formation and health of coral reefs and the flora and fauna that inhabit them. A mangrove was destroyed in order to build the Rodney Bay Marina. The construction of the Pigeon Island causeway, which involved sand mining for construction and coastal erosion from coastal engineering works, altered the coastal water flows, caused siltation of coral reefs, and brought about the virtual wiping out of the nearshore fishing industry in Gros Islet (Nagle n.d.).

Since the 1990s, there has been a growing awareness in St. Lucia of the potential and need for promoting an alternative tourism based on its nature and culture. As a result, several localized ecotourism opportunities have been developed in St. Lucia and these have been instrumental in the island being labeled as a "Nature Heritage Tourism" destination (Weaver 2001). The tourism industry in St. Lucia is fueled by the island's beautiful natural landscapes, dominated by tropical rainforest. The island's government recognizes the tourist appeal of these areas, and 13% of the island is reserved forest (Arce 2009). St. Lucia's National Rainforest, covering 7690 hectares, is in a protected area in the southeast of the island, which also encompasses the Pitons (Natural Attractions n.d.). While the mountainous twin Pitons provided a safe haven and base for plotting rebellion for runaway slaves in the eighteenth century, today, tourists hike the Gros and Petit Pitons. Petit Piton involves a technical climb to reach the summit, and safety equipment must be brought in (Trend 2015). There is no hiking allowed in any protected rainforest without paying an entrance fee to

Fig. 13.9 Aerial view of Gros Islet, the town at the connection between Pigeon Island and the mainland, which is a popular tourism destination in and of itself with beach resorts, inland port, and coral reefs. Pigeon Island is to the *left* of the image. Photograph by A. DeGraff Ollivierre

the St. Lucia Forestry Department, and trails are typically guided (Ministry of Agriculture, Food Production, Fisheries, Cooperatives & Rural Development n.d.).

The Pitons Management Area (PMA) is a multiple use protected area of 2909 hectares. Located near the town of Soufriere, the PMA consists of both marine and terrestrial environments, encompassing the Pitons and the Sulphur Springs. This site was given World Heritage Status in 2004 and is St. Lucia's most visited attraction with 200,000 visitors annually (Nicholas et al. 2009). Reefs protect 44% of St. Lucia's coastline (Maps and Data 2008), and there are 20 marine protected areas (MPAs) in the country, many of which have significant coral reef content (St. Lucia n.d.). The MPA contained in the Piton Management Area has up to 60% coral reef coverage (UNESCO n.d.). Many of the other MPAs have experienced highly variable management practices, with differing success in protection and maintenance. The most well maintained of St. Lucia's MPAs is the Soufriere Marine Management Area (SMMA), an 11-km stretch along the southwestern coast of St. Lucia (Renard 2001).

The Soufriere Marine Management Area (SMMA) protects Soufriere Bay as well as the surrounding waters. The SMMA is divided into five different zones dependent on the activity allowed ("Pitons Management Area Saint Lucia"). Unfortunately, this type of care toward its natural landscapes has not always been apparent, as seen in 1969 when the wetlands surrounding the town of Gros Islet, once a small fishing village, were drained to reduce the available breeding ground for sand flies, which were irritating to the tourists. The land was thus reclaimed but with unforeseen consequences, such as stronger currents that began to erode some of the nearby beaches (Nagle 1999).

The northern city of Gros Islet and nearby Rodney Bay have become major scenic backdrops for the island's tourism marketing efforts. In 1970, a development plan was executed

Fig. 13.10 An example of tourism impacts and activity on St. Lucia, with cruise ships docked at Castries. The majority of tourists that visit the island do so as part of Caribbean cruises and spend most of their time in Castries, although various tours are offered to other parts of the island. Photograph by A. Potter

by the Caribbean Development Corporation to make use of the well-situated area (McGregor 1998). To the south of Gros Islet is Rodney Bay, named after an English admiral ("Pigeon Island National Landmark"). Shortly, after draining the wetlands near Gros Islet, an artificial lagoon was installed in Rodney Bay. This bay, with expansive Reduit Beach, was utilized heavily by pirates in the 1700s and 1800s, and its notoriety has added to its tourist appeal (Things to do n.d.). Rodney Bay's popularity has limited development in Marigot Bay to the south. Surrounded on three sides by steep, heavily forested hills, Marigot Bay is a secluded inlet. This bay, considered one of the most picturesque in the Caribbean, was featured in the 1967 "Dr. Doolittle" film. Long popular with yachters, it has been built out in recent years to accommodate increased tourist demand (Visions of St. Lucia Tourist Guide n.d.). This increased use of coastal resources is one of the most common landscape conflicts in St. Lucia today. Traditional fishing practices are slowly being edged out by increasing use of coastal regions by yachts, tourists diving to coral reefs, and coastal development that leads to erosion, and pollution of the water (Morales 2011).

13.6 Hazards

St. Lucia is exposed to a number of natural hazards, such as tropical storms, hurricanes, heavy rainfall, winds, occasional droughts, landslides, earthquakes, and eruptions. Weather-related hazards, like hurricanes, are the most severe climatic disturbance in St. Lucia. In 1980, Hurricane Allen destroyed the agricultural sector and killed nine people (Meditz and Hanratty 1987). The hurricane seasons of the 1970s and 1980s were relatively tame in comparison with the 1990s, when seven devastating hurricanes or storms hit the Caribbean. The inhabited and more developed coastal areas between Gros Islet and Castries, where 60% of the population is living, were greatly impacted. This part of St. Lucia is particularly vulnerable because it is composed of low-lying, flood-prone land. Loose regulations and outright

Fig. 13.11 Local man collecting mineral-rich volcanic mud by bucket to ensure an ample supply at the Sulphur Springs Volcano Mud Baths near Soufrière for tourists. While the therapeutic mud accumulates naturally in the springs, increased tourist demand has made it necessary to manually retrieve the substance from areas upstream. Photograph by A. Potter

illegal development along the coast have worsened these problems. Between 1970 and 1999, St. Lucia suffered economic losses equally 272% of the country's GDP due to hurricane and hurricane-related damage, such as flooding or storm surges. In terms of geological impacts, Hurricane Lenny, an unusual west to east traveling hurricane occurring in 1999, washed away beaches from the white-sanded western coast (Mycoo 2011).

Flood and coastal erosion threats in St. Lucia come mainly from storm surge and coastal wave action. Low-lying coastal areas, such as Dennery and the area of Anse La Raye, are especially susceptible (The World Bank and GFDRR 2010). Coral reef habitat has also been shown to suffer from the effects of storms and sedimentation (Hawkins et al. 2006). St. Lucia, along with five other Caribbean nations, was listed as one of the world's top 40 climate hot spots. Models that attempt to predict the effects of climate change on St. Lucia estimate drier seasons, more heat waves and droughts, increased rainfall when it does occur, and rising seas levels, which could particularly affect St. Lucia's predominant coastal development. In addition, climate change can also cause the sea temperature to change, leading to more intense storms (The World Bank and GFDRR 2010).

With its steep slopes and patterns of heavy rainfall, St. Lucia is particularly prone to landslides. Tropical storm Debbie was responsible for extensive landslides in 1994. A recent initiative in collaboration with the World Bank has been hailed as a success story in the Caribbean. When Hurricane Tomas hit the island in 2010, hillside communities in St. Lucia that had implemented the initiative were not affected by landslides. St. Lucia's vulnerability to adverse natural events has been reduced, through improvements such as rehabilitation bridges, coastal protection works, and small mitigation works in various communities across the island. They have significantly increased the percentage of population with access to improved infrastructure, from 30% to an estimated 80% (The World Bank 2013).

St. Lucia's location on the eastern margin of the Caribbean Plate means that it is commonly subjected to lowest magnitude, low-risk earthquakes (The World Bank and GFDRR 2010). The Qualibou Caldera, with its Soufrière Volcanic Center, is considered to be the only volcanically active center on the island. The last recorded eruption was in 1766, and the most recent activity was a swarm of minor volcanic earthquakes in 2000. Some of these earthquakes were located beneath older basaltic centers previously considered to be "dead," such as Mount Gomier (Lindsay et al. 2002). The phreatic (steam) eruption in 1766 deposited a thin layer of fine ash-like material of mud, old altered rock, and mineral fragments over the area. Within the Soufrière Volcanic Center, St. Lucia's Sulphur Springs is an active geothermal area with numerous hot springs, bubbling mud pools, and fumaroles. Underground acidic water below Sulphur Springs causes rock alteration, resulting in soft and unstable soil that is prone to landslides, especially triggered by earthquakes. An example of such a landslide was reported on the Terre Blanche flank of Sulphur Springs in 1990. Over time, any migration of geothermal activity at Sulphur Springs will potentially increase the chances for landslides in areas with steep slopes (Lindsay et al. 2002). Landslides in geothermal areas in St. Lucia, in turn, can affect the internal material of a system, causing a hydrothermal eruption.

Many mountains in southern St. Lucia are thought to be the work of one volcano with many vents. This volcano, Soufrière Volcanic Centre (SVC), is the only volcano on the island and it is unknown when it may erupt again. Many of the other Lesser Antilles islands have a single, potentially active volcano with a history of activity. However, St. Lucia does not have one clear volcanic vent from which future eruptions are expected, making it challenging to predict when and where the next magma eruption will take place (Lindsay et al. 2002). Prior eruption history for the island suggests that the Soufriere Volcanic Center is the best potential location for future eruption on St. Lucia. The most likely scenario for future activity is a short-lived (a few hours or days) phreatic or hydrothermal eruption (not containing any magma) from the Sulphur Springs area with effects contained to the surrounding area (University of the West Indies 2009–2011).

13.7 Conclusion

St. Lucia's dramatic landscapes have provided for its inhabitants over the island's history. Today, this landscape provides all the ingredients for a successful tourism industry. Tourism in St. Lucia is evolving from the traditional sun, sea, and sand to include alternative tourism attractions such as land- and sea-based nature tourism and cultural heritage tourism. While this brings in more tourists and a greater potential stream of capital for development, it also means that formerly relatively untouched geological landscapes are being exposed to environmental degradation. A concern would be that if the natural landscape and ecological systems suffer from too much degradation, this could negatively impact the island's tourist capacity, and thus reduce the desired development impact from tourism. Directed government policies and actions can and do play a role in reducing the harmful effects of agriculture and tourism on the environment. There has been a significant decrease in environmental impacts despite the rise in human footprint because of the revenue generated from tourism and the St. Lucian government being proactive (Sharmon 2005). Will it be enough, however, as St. Lucia tries to develop its economy using a strategy that promotes tourism to its most ecologically sensitive geological sites?

References

About Diamond Falls Botanical Gardens (n.d.) Diamond Falls Botanical Gardens & Mineral Baths, Saint Lucia. Web 16 Oct 2015. http://diamondstlucia.com/about-diamond-falls/

About Soufriere (2010) Geology. The Soufriere Foundation. Web 18 Oct 2015.http://soufrierefoundation.org/about-soufriere/geology

All About St Lucia. (n.d.) Resources and usage. All About Saint Lucia. Web 18 Oct 2015. http://allaboutstlucia.com/resources-usage/

Arce JP (2009) Protected areas of St. Lucia. Eoearth. Web 12 Oct 2015. http://www.eoearth.org/view/article/155412/

Atlapedia Online (n.d.) Latimer Clarke Corporation Pty Ltd. Web 26 Oct 2015. http://www.atlapedia.com/online/countries/stlucia.htm

Burns P, Novelli M (2011) Sustainable tourism and National Park Development in St. Lucia. In: Tourism and social identities: global frameworks and local realities. Elsevier, New York City

Caribbean Diamond Falls (n.d.) Diamond falls (Soufriere, St. Lucia). World of Waterfalls. Web 18 Oct 2015. http://www.world-of-waterfalls.com/caribbean-diamond-falls.html

Diamond Falls Mineral Baths Spa (n.d.) Diamond falls. Botanical Gardens & Mineral Baths, Saint Lucia. Web 16 Oct 2015. http://diamondstlucia.com/diamond-falls-attraction/diamond-falls-mineral-baths-spa/

Diving (2015) Scuba St. Lucia. Web 29 Oct 2015. http://www.scubastlucia.com/diving.html

Earle KW (1924) Geological survey of the windward and leeward islands: the geology of St. Lucia. S.I.:s.n.

Geothermal energy in St Lucia (1989) Minerals and Energy—Raw Materials Report. Vol 6, Iss. 4

Government of St. Lucia (2001) Integrating the management of watersheds and coastal areas in St. Lucia. Issued by: Water Resources Management Unit, Ministry of Agriculture, Forestry, and Fisheries

Hastings K (2015) 14 Top-rated tourist attractions in St Lucia. PlanetWare Inc. Web 6 Jan 2016. http://www.planetware.com/tourist-attractions/st-lucia-stl.htm

Hawkins JP, Roberts CM, Dytham C, Schelten C, Nugues MM (2006) Effects of habitat characteristics and sedimentation on performance of marine reserves in St. Lucia. Bio Conserv 127 (4):487–499

Heritage Sites (n.d.) Fond d'Or nature and heritage park. Heritage Tours, Saint Lucia. Web. 16 Oct 2015 http://www.heritagetoursstlucia.org/heritage_site/fond-dor-nature-heritage-park/

Island Profiles (n.d.-a) St. Lucia—Geology. Seismic Research Center—University of the West Indies. Web 16 Oct 2015. http://www.uwiseismic.com/General.aspx?id=72

Island Profiles (n.d.-b) St. Lucia—Volcanism. Seismic Research Center—University of the West Indies. Web 16 Oct 2015. http://www.uwiseismic.com/General.aspx?id=72

Lashley J, Moore W (2013) Private sector assessment of Saint Lucia. Issuedby: Inter-American Development Bank.

Lindsay J, David J, Shepherd J, Ephraim J (2002) Scientific supplement to the volcanic hazard assessment for Saint Lucia, Lesser Antilles. Issued by: Seismic Research Unit, The University of the West Indies, St. Augustine, Trinidad and Tobago

Maps and Data (2008) Shoreline protected by coral reefs—St. Lucia. World Resources Institute. Web 14 Oct 2015. http://www.wri.org/resources/maps/shoreline-protected-coral-reefs-st-lucia

Maria Islands Nature Reserve. Saint Lucia National Trust. Saint Lucia National Trust (2011) Web 26 Oc. 2015. http://www.slunatrust.org/sites/maria-island-nature-reserve/

Mc Elroy JL (2003) Tourism development in small islands across the world. Geografiska Annaler. Series B Human Geography. Column 85, Issue 4, pp 231–242

McGregor DFM, Barker D, Evans S (1998) Resource sustainability and Caribbean development. Press, University of the West Indies, Kingston

Meditz SW, Hanratty DM (eds) (1987) Caribbean islands: a country study. Washington: GPO for the Library of Congress. http://countrystudies.us/caribbean-islands/

Ministry of Agriculture, Food Production, Fisheries, Cooperatives and Rural Development (n.d.) Forestry. Malff. Web 18 Oct 2015

Mohammed E, Lindop A (2015) Working Paper#2015—53: St. Lucia: Reconstructed Fisheries Catches 1950–2010. Issued by: Fisheries Centre, The University of British Columbia

Morales O (2011) Saint Lucia and the effect of climate change in its sustainability. The Mandala Projects. American University. Web 26 Oct 2015

Mycoo M (2011) Natural hazard risk reduction: Making St. Lucia safe in an era of increased hurricanes and associated events. Natural Hazards Review 12.1 pp 37–45. Environment Complete. Web 28 Oct 2015

Nagle G (1999) The impact of tourism in coastal areas. Focus on geography: tourism, leisure and recreation. Oxford UP, Oxford, pp 32–38 (Print)

Nagel G (n.d.) Geo Factsheet: The environmental impact of tourism in the Caribbean. No. 204. www.curriculum-press.co.uk

Nations Encyclopedia (n.d.) Saint Lucia. Encyclopedia of the nations. Web 10 Oct 2015. http://www.nationsencyclopedia.com/geography/Morocco-to-Slovakia/Saint-Lucia.html

Natural Attractions (n.d.) Natural Attractions of St. Lucia. St. Lucia, Simply Beautiful. Web 18 Oct 2015. http://www.geographia.com/st-lucia/lceco01.htm

Newman WR (1965) A report on general and economic geological studies, St. Lucia, West Indies. Issued by: The United Nations

Nicholas L, Thapa B, Pennington-Gray L (2009) Public sector perspectives and policy implications for the pitons management area world heritage site, St. Lucia. Int J Sustain Dev Word Ecol 16 (3):205–216

O'Keefe T (n.d.) St. Lucia Maria islands nature reserve. Guide To Caribbean Vacations. Web 26 Oct 2015. http://www.guidetocaribbeanvacations.com/st_lucia/MariaIslands.htm

Past and Present (n.d.) History of St. Lucia. Caribya!. Web 29 Oct 2015. http://caribya.com/st.lucia/history/

Pigeon Island National Landmark. Saint Lucia National Trust. Saint Lucia National Trust. Web 27 Oct 2015. http://slunatrust.org/sites/pigeon-island-national-landmark/

Pitons Management Area Saint Lucia. United Nations Environmental Programme, 1 May 2011. Web 26 Oct 2015. http://www.thesalmons.org/lynn/wh-wcmc/Saint%20Lucia%20-%20Pitons.pdf

Porter D, Prince D (2007) St. Lucia. Caribbean for Dummies. 4th ed. Hoboken: Wiley: pp 376–406. Print

Reefs in St. Lucia. Climate and Reefs.org. Australia Caribbean Coral Reef Collaboration. Web 29 Oct 2015. http://climateandreefs.org/st-lucia/

Renard Y (2001) Case of the Soufriere Marine Management Area (SMMA), St. Lucia. Issued by: Caribbean Natural Resources Institute

Saint Lucia Reduces Vulnerability. http://www.worldbank.org/en/results/2013/09/06/saint-lucia-vulnerability-natural-disasters. The World Bank, 13 Sept 2013 Web 7 Oct 2015

Sharmon J (2005) Sustainable Tourism in St. Lucia. IISD Home IISD. Accessed 11 Oct 2015

Smith AL, Roobol MJ, Mattioli GS, Fryxell JE, Daly GE, Fernandez LA (2013) The volcanic geology of the mid-arc Island of Dominica, Lesser Antilles: the surface expression of an island-arc batholith. The Geological Society of America, Boulder, CO

St. Lucia (1994) Caribbean islands handbook. Ed. Sarah Cameron and Ben Box. 6th ed. Chicago: Passport. pp 612–626. Print

St. Lucia (2008) South America, Central America and the Caribbean. Ed. Jacqueline West. 16th edn. Routledge: London. pp 809–818. Print

St. Lucia Dive Sites. Dive Fair Helen. Dive Fair Helen. Web 29 Oct 2015

St. Lucia (n.d.) Reefs in St Lucia. Australia Caribbean Coral Reef Collaboration. Web. 13 Oct 2015. http://climateandreefs.org/st-lucia/

The Island and Nature (n.d.) "Geology." All About St Lucia. Web 14 Oct 2015. http://allaboutstlucia.com/geology/

The Soufriere Foundation (2010) Sulphur Springs Park. Soufriere Regional Development Foundation. Web 26 Oct 2015. http://soufrierefoundation.org/discover/attractions/sulphur_springs_park/

The Soufriere Foundation (2010) Geology. Soufriere Regional Development Foundation. Web. 16 Oct. 2015. http://soufrierefoundation.org/sulphur-springs/geology

The Soufriere Foundation (2010) Hot Water Baths. Soufriere Regional Development Foundation. Web 16 Oct 2015. http://soufrierefoundation.org/sulphur-springs/hot-water-baths

The Soufriere Foundation (2010) History of Soufriere. Web 6 Jan 2016. http://soufrierefoundation.org/about-soufriere/history

The World Bank (2013) Saint Lucia: Leaders in reducing landslide risk. Web 5 Jan 2016. http://www.worldbank.org/en/news/feature/2013/06/11/saint-lucia-oecs-leader-reducing-landslide-risk

The World Bank and Global Facility For Disaster Reduction and Recovery (GFDRR) (2010) Disaster Risk Management in Latin America and the Caribbean Region. Web 6 Jan 2016. http://www.gfdrr.org/sites/gfdrr.org/files/DRM_LAC_CountryPrograms.pdf

Things to do (n.d.). Rodney Bay. St. Lucia Now. Web 16 Oct 2015. http://stlucianow.com/rodney-bay-village/

Top 5 Waterfalls in Saint Lucia. Real St. Lucia Tours. Real St. Lucia Tours. Web 29 Oct 2015. http://realstluciatours.com/St_2.html

Trend N (2015) Climbing in St Lucia: the best view in the Caribbean. Telegraph.co.uk. Web 18 Oct 2015. http://www.telegraph.co.uk/travel/destinations/centralamericaandcaribbean/saintlucia/11351100/Climbing-in-St-Lucia-the-best-view-in-the-Caribbean.html

UNESCO (n.d.) Pitons Management Area. World Heritage List. Web 13 Oct 2015. http://whc.unesco.org/en/list/1161

United Nations Development Programme (2009) National issues paper on climate change and the water sector in Saint Lucia. UNDP Environment and Energy Group, Issued by

University of the West Indies (2009–2011) Island Profiles: St. Lucia—Volcanism The University of the West Indies Seismic Research Center. Web 27 Oct 2015. http://www.uwiseismic.com/General.aspx?id=74

Visions of St. Lucia Tourist Guide (n.d.) The Sights of St Lucia. Official Publication of the St Lucia Hotel & Tourism Association. Web 13 Oct 2015

Weaver DB (2001) Mass tourism and alternative tourism in the Caribbean. In: Harrison D (ed) Tourism and the less developed world: issues and case studies. CABI Publishing

Welcome to My St Lucia Beaches. My St Lucia. My St Lucia. Web 30 Oct 2015. http://mystlucia.org/info_htm/beaches.htm

Wilkie ML (2005) Global forest resources assessment 2005, thematic study on Mangroves, Saint Lucia country profile. Issued by: United Nations Food and Agriculture Organization

Wohletz K et al (1986) The Qualibou Caldera. St. Lucia, West Indies J Volcanol Geoth Res 27(1-2):77–115

Barbados

14

Mick Day and Patti Day

Abstract

Barbados is unique in the Lesser Antilles, in that it is an exposed portion of an accretionary prism, with no volcanic basis. Pleistocene carbonates cover some 85% of the island, and terraced karst terrain, with dry valleys and dolines/sinkholes dominates the landscape. The Scotland District is composed of siliclastics, which have been disturbed by fluvial erosion and mass movements. Distinct wet and dry seasons result in paradoxical flooding and drought, both accentuated by human interference, particularly colonial agricultural clearance and twentieth-century mass tourism, and risks are likely to increase because of projected anthropogenic climate change. Maintaining adequate water supplies is a perennial problem, and occasional flooding represents another major hazard. Mass movements afflict the Scotland District, and there are under-appreciated karst collapse and tsunami risks. Conservation-based land use management and planning is poorly developed, yet vital to residents and a sustainable tourism-based economy.

Keywords

Accretionary sediments • Limestones • Reefs • Karst • Drought • Water supply • Floods • Human impacts

14.1 Introduction

Barbados, the most easterly of the Caribbean islands, has a land area of about 432 km^2 and a coastline some 92 km in length (Carter and Singh 2010; Government of Barbados 2010; Fig. 14.1). Barbados is unique within the Lesser Antilles in that it has no volcanic basis, being composed entirely of sedimentary rocks, with the exception of minor volcanic ash beds (Humphrey 1997; Speed 2012). Geologically, the island is dominated by carbonate rocks (primarily limestones) of Pleistocene age (2.6 Ma–11.7 ka), which were formed predominantly as fringing coral reefs and have now been uplifted tectonically as a series of terraces that occupy about 85% of the island (Day 1983; Humphrey 1997; Donovan 2005 Government of Barbados 2010; Speed 2012; Kambesis and Machel 2013). Separated from the overlying limestones by a thin sheet of Miocene chalks and marls, the remainder of the land area is underlain by older sandstones, siltstones, and clays of the Tertiary Scotland Series (Patel 1995; Humphrey 1997; Government of Barbados 2009, 2010; Speed 2012).

Elevations range from sea level to 340 m at Mount Hillaby (Donovan 2005). Numerous terraces have been documented at elevations ranging from 3 m to 192 m (Humphrey 1997), with particularly conspicuous examples at about 30 m (First High Cliff, 125 ka) and 170 m (Second High Cliff, 460 ka) serving as litho- and chronostratigraphic markers separating the Coral Rock Formation into Lower, Middle, and Upper members (Fig. 14.2; Humphrey 1997). On and within the limestones has developed a karst landscape, produced essentially by carbonate dissolution, containing numerous dolines (sinkholes), extensive dry valley

M. Day (✉) · P. Day
Department of Geography, University of Wisconsin, Milwaukee, USA
e-mail: mickday@uwm.edu

P. Day
e-mail: p8d@uwm.edu

C.D. Allen (ed.), *Landscapes and Landforms of the Lesser Antilles*, World Geomorphological Landscapes, DOI 10.1007/978-3-319-55787-8_14

Fig. 14.1 General physiographic map of Barbados. Low-lying plains dominate the southern region with single perennial river, Constitution River, flowing from the central highlands. Cartography by K.M. Groom

Fig. 14.2 The Second High Cliff of Barbados. Note the limestone outcrops on the cliff and the small doline (*light-colored area*) at the vegetation break in the middle distance, just *left of center*. Photo by M. Day

systems, springs, and cave systems (Kambesis and Machel 2013). The poorly consolidated sedimentary rocks of the Scotland District are extensively eroded by gullying and slope failures (Patel 1995; Government of Barbados 2010; Speed 2012).

The climate is dry sub-humid with temperatures between 20 and 30 C. Mean annual rainfall ranges with elevation between 1254 and 1650 mm, with evapotranspiration generally exceeding precipitation, and with a distinct dry season from January through May and an August–December wet season (Government of Barbados 2009, 2010; Carter and Singh 2010). Soils on the limestones are mostly mollisols and vertisols, with complex components including detrital carbonates, Antillean volcanic ash, and Saharan dust (Muhs 2001). Soils in the Scotland District are immature and strongly controlled by parent materials, being dominated by sands and clays (Patel 1995).

Barbados represents perhaps "…the wealthiest and most developed country in the Eastern Caribbean," with a resident population nearing 300,000 (CIA 2015). The population density at about 638/km^2 is the highest in the Caribbean and among the densest 20 in the World (CIA 2015). Annual population increase has been effectively limited to about 0.3% (Government of Barbados 2010; CIA 2015) but visitor numbers have increased steadily to about 500,000 annually (Caribbean Tourism Association 2011), effectively doubling the population impact. The resident population, of which about 40% is urban, remains concentrated along the west, south, and southeast coasts particularly around Bridgetown in St. Michael parish and also in St. Philip, Christ Church, St. James, and St. Peter (Government of Barbados 2010).

Human impact on the landscape has been long term and profound, resulting in wholesale environmental changes. Drought and water supply are fundamental problems, as well as accelerated erosion and, paradoxically, occasional flooding. Environmental issues are intensified by the combination of limited land area, the predominance of karst landscape, (ironically) relative economic wealth, and mass tourism.

14.2 Setting

Barbados is the only emergent portion of the Barbados Ridge Accretionary Prism, "…an elongate, arcuate accretionary complex…east of the Lesser Antilles magmatic island arc and the Tobago Trough forearc basin," formed by subduction of the South American Plate under the Caribbean Plate (Humphrey 1997: 382; Donovan 2005). An Eocene marine mixed sedimentary basement is overlain by Miocene chalks, marls, and radiolarites, which in turn are capped by up to about 100 m of Pleistocene limestones (Fig. 14.3; Speed 2012; Limonta et al. 2014).

The island is dominated areally by Pleistocene limestones of the Coral Rock Formation, or "Coral Cap," most of which originated over the past 600 ka during eustatic high sea level stands as a series of discontinuous coral reefs fringing the Tertiary island core (Bender et al. 1979; Poole and Barker 1983). Subsequent and ongoing tectonic movements during eustatic lowstands have elevated the reefs, forming a series of terraces. Each of these essentially represents a separate reef complex (Mesolella et al. 1969), with terrace risers consisting of rear-zone, reef-crest, and fore-reef lithologies,

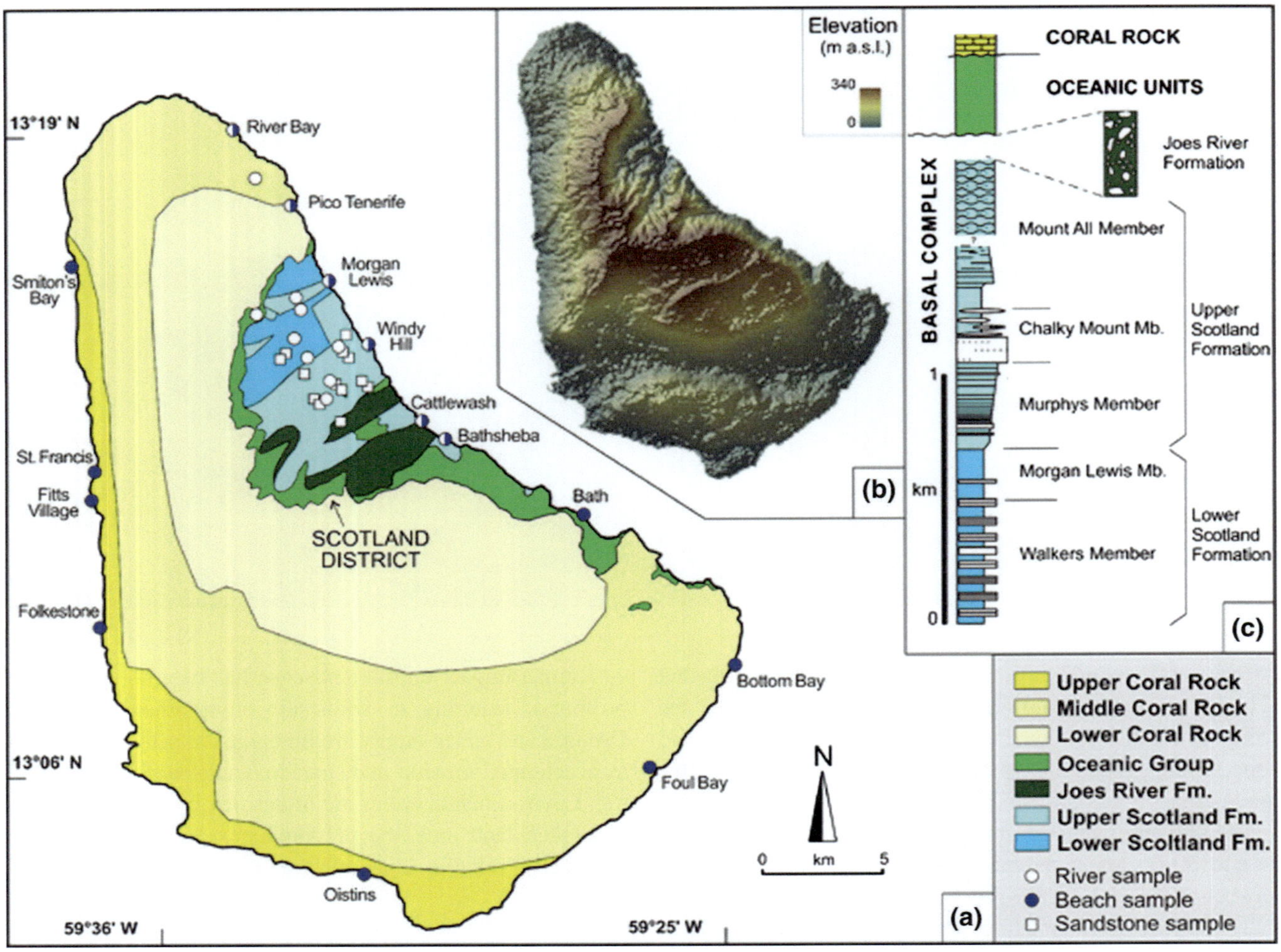

Fig. 14.3 Geologic map of Barbados (**a**) by (Limonta et al. 2014), with accompanying topographic relief image (**b**), and stratigraphic column (**c**), with sample points denoted by small circles. Available from https://www.researchgate.net/profile/Eduardo_Garzanti/publication/268528051/figure/fig1/AS:272681375367224@1442023714925/Figure-3-Geological-sketch-map-a-topography-b-and-stratigraphy-c-of-Barbados.png

and landings composed of backreef deposits (Humphrey 1997), although at least locally the situation is more complex (Schellmann and Radtke 2004a, b; Speed and Cheng 2004). Highstand reef deposition probably occurred over periods of 10–15 ky^{-1} (Humphrey 1997) and uplift rates have been estimated at between 0.12–0.22 m ky^{-1} (Jones 2009) and 0.3–0.4 m ky^{-1} (Speed 1994).

The older and higher limestones in the center of the island are relatively well-lithified biomicrites, but most of the later Pleistocene carbonates are poorly consolidated, although interactions with meteoric and mixing-zone waters have resulted in a wide range of diagenetic modification (Humphrey 1997). The Coral Rock Formation, although variable, averages about 70 m in thickness and overlies Oligocene–Miocene chalks and marls that act as a regional aquiclude (Humphrey 1997).

The sedimentary basement beneath the limestones consists of four major units of structurally complex allochthonous tertiary marine rocks exposed only in the Scotland District (northeast of the island), where the overlying limestone has been eroded (Speed 1994, 2012; Donovan 2005). The Scotland Formation, an Eocene complex of terrigenous turbidites and hemipelagic/pelagic radiolarites, is overlain by Miocene prism-cover rocks deposited in the Woodbourne Trough, and in thrust contact with these are nappes of calcareous (hemi) pelagic rocks interbedded with volcanogenic ashes, constituting the Oceanic Formation. Intruding into these three units are "…tectonic diapirs consisting of a mélange of organic mud matrix" (Humphrey 1997: 383).

14.3 Landforms

Speed (2012) recognized four geomorphic zones reflecting the geological structure and topography: the Central Highlands, the Terraced Flank and the Windward Slope on the limestones, and the Scotland District. On and within the limestones, carbonate dissolution and associated processes have developed a karst landscape dominated by dolines (sinkholes), dry valleys, caves, and springs (Fermor 1972; Day 1983, 2009; Kambesis and Machel 2013). Surface water is rare here, except during particularly intense rains, when sinkholes might fill with localized ephemeral runoff, and normally dry valleys may channel dangerous floodwaters. Except during these events, drainage is predominantly subterranean, via rapid infiltration into and percolation through the thin soils, then transmission at various rates via pores, fissures, and conduits through the limestones to the groundwater aquifer and/or to terrestrial and coastal springs.

The most conspicuous karst landforms are the extensive dry valley systems (Fig. 14.4), which in many aspects represent the Caribbean type-example (Fermor 1972; Day 2002). Ranging from shallow to steep-sided, with or without distinct stream channels, and wooded or farmed, these valley systems are significant geomorphologically, because they represent earlier phases in the karst development when secondary permeability was not sufficiently developed to permit total subterranean drainage, and runoff occurred via "normal" fluvial valleys (Smith 1975). As secondary permeability has increased through carbonate dissolution, drainage has increasingly been diverted underground, and the valleys have been essentially abandoned, only to be activated during intense rainfall, when the capacities of the karst plumbing system and its overlying soils are exceeded (Smith 1975; Day 2002). As such, in a sense, they represent ephemeral drainage systems, and their persistence indicates that karstification has not yet progressed to the extent that surface drainage has been completely abandoned and the valleys obliterated by doline development, as appears to be the case in older karst landscapes elsewhere in the Caribbean (Day 2002).

The precise development of the Barbados dry valleys remains contentious among researchers, with both surface drainage and speleogenesis implicated (Kambesis and Machel 2013). Many valleys are clearly surface erosional features, but others resemble collapsed caves. Still others appear to have polygenetic origins (Speed 2012), with Kambesis and Machel (2013: 10) advocating a complex scenario involving five stages: (1) initial surface erosion, (2) flooding during sea-level rise, and the development of flank margin caves in the valley walls, (3) renewed surface erosion and vadose speleothem formation in the valley flank caves following sea-level fall, (4) partial marine erosion associated with further sea-level rise, and (5) "…final exposure…through tectonic uplift, resulting in wide gullies with opened flank margin caves that are lined with speleothems."

Dolines are often regarded as diagnostic of karst landscape (Ford and Williams 2007), and more than 2800 occur in Barbados (Kambesis and Machel 2013), with initiation still occurring through small collapses, both within and beyond existing depressions (Day 1983; Fig. 14.5). Dolines are generally shallow and near-circular, with a tendency

Fig. 14.4 Coastal dry valley in Christ Church parish with a dry channel in the valley axis. Most of rivers depicted radiating from the highlands on maps of Barbados are actually dry valleys and ephemeral streams such as this one. Photo by M. Day

Fig. 14.5 Localized collapse within a doline/sinkhole in St. Thomas parish with exposed limestone and gullying on the doline flanks. Photo by M. Day

toward clustering, and sampling has suggested systematic variations in their density, with increasing densities up to about 150 m elevations, and lower densities at higher elevations (Day 1993). Rainfall often collects in larger interfluvial dolines, while smaller shafts within the valleys transmit water rapidly into the karst aquifer (Jones and Banner 2003).

The relationship between dry valleys and dolines is not entirely clear, although it is probable that formation of the former precedes that of the latter (Smith 1975; Day 2002). Fermor (1972) suggested that the features were competitive, in the sense that each developed preferentially under different hydrological conditions. It has also been suggested that dolines typify flatter areas, whereas valleys occur on more steeply sloping surfaces, although the nature of this relationship is not explicit, and there may be distinct doline subpopulations in valley and interfluve locations (Day 1983; Jones et al. 2000, 2003).

There are over 100 known caves in Barbados (Kambesis and Machel 2013). Harrison's Cave, in the Highlands, is the best-known on the island (Hobbs 1994) and is developed for tourism, as is Animal Flower Cave, on the north coast. Kambesis and Machel (2013) recognize five types of caves: (1) *epigene*, formed by freshwater dissolution in the upland interior, (2) *flank margins*, formed by mixing-zone dissolution along modern and older coastlines, (3) *littoral*, formed by contemporary marine corrasion, (4) *mechanical*, formed by bedrock mass movement, and (5) *hybrid* (polygenetic), which constitute the majority on Barbados, often involving modification of flank margin caves.

Depending on interpretation, Barbados exemplifies either the composite or the carbonate-cover type of the carbonate island karst model (Mylroie and Vacher 1999; Mylroie and Carew 2000; Mylroie 2004). The geologically young carbonate rocks have not been buried beyond the range of meteoric diagenesis, and they interact both with fresh and saline groundwater that is affected by glacio-eustatic and tectonic changes in sea level, forming immature (essentially Neogene) eogenetic karst (Mylroie 2004). Significantly, however, the geological configuration is such that the older non-carbonate rocks of the Scotland District do not contribute significant allogenic discharge to the karst landscape, precluding development of a composite speleogenetic situation, and resulting in cave development taking place above and at the carbonate–non-carbonate contact. Kambesis and Machel (2013) ascribe Barbados to the composite category, suggesting that the epigene karst receives both autogenic and allogenic recharge, as well as hypogene karst being decoupled from the surficial hydrology.

The most prominent landform in the Scotland District is an unstable east-facing escarpment, below which detached limestone blocks are evident (Fig. 14.6). Below the escarpment are steep gullies, landslides, debris flows, slumps, and other mass movement features, generally correlated with geology and slope angle (Patel 1995; Hearn et al. 2001; Hodgson et al. 2002). There is evidence of slab or toppling failure from the escarpment, and headward gully erosion close to the escarpment has captured headwater drainage from limestone valleys, "…forming preferred locations of infiltration" locally (Hodgson et al. 2002: 282). Much of the accelerated erosion is human-induced, caused by removal of the pre-European forest cover and subsequent land use practices (Carson and Tam 1977; Patel 1995). Beaches are siliclastic, locally sandy or pebbly, and highly dynamic, with

Fig. 14.6 Large detached limestone block forming a coastal stack in St Joseph parish with a person for scale. The block has developed a wave undercut notch. Photo by M. Day

localized coastal sand dunes (Patel 1995). Coastal carbonate landforms illustrate the multifaceted interaction between terrestrial and marine geomorphic processes, with abrasional terraces, restricted tidal platforms (Fig. 14.7), littoral karren, cliffs with wave-cut notches, littoral caves indicative of saline-freshwater mixing, sea stacks, sea arches, carbonate clastic beaches, and localized sand dunes (Fig. 14.8; Lundberg 2004; Speed 2012).

The coastline of the Scotland District is particularly dynamic, featuring rapid delivery of sediments through fluvial erosion and slope processes, and effective removal by long-fetch Atlantic wave action. Cliffed erosional shorelines are characteristic of the windward northern and eastern coasts, contrasting with the sandy low-energy coasts to the leeward. Fringing reefs, particularly off the leeward coast represent contemporary analogues of the older features now exposed on the island itself, although the species composition differs (James et al. 1977).

14.4 Landscape

As with most other Caribbean islands, human impact on the landscape of Barbados has been long-term and severe (Watts 1987, 1988; Day 1993, 2010a, 2010b). Although Amerindians could have perhaps arrived as early as 2600 BCE (Fitzpatrick 2011), later dates of 350–650 BCE are more commonly accepted (Drewett 1993). The native population abandoned the island in the 1500s, following Spanish slave-raiding, when it was named by Portuguese. Barbados was claimed subsequently by the British in 1625 and gained independence in 1966 (Beckles 2006).

The physical landscape was devastated during the early colonial period by almost complete clearing of natural vegetation and the institution of sugar cane monoculture (Randall 1970; Watts 1987, 1988; Richardson 1997). A few steeper-sided gullies were spared, and these represent important ecological remnants, although invasive species are highly competitive (Waugh 2009). Destruction of the natural vegetation undoubtedly changed the hydrologic regime and may have contributed significantly to historical aridification (Watts 1987, 1988).

The nature of maronnage (resistance by escaped slaves) during the early colonial period was constrained by the limited area of the island and the subdued topography, which provided scant permanent isolation and little defensibility, but escaped slaves used caves and more remote wooded valleys as hiding places (Kambesis and Machel 2013) and took advantage of both the physical and the built landscape to create at least temporary refuge (Handler 1997).

Environmental issues that have been assessed as being of national priority include the inadequacy and unreliability of freshwater supplies, the degradation of coastal ecosystems, waste management challenges, the limited land area and inefficiency of land use, and natural hazards, including the challenges posed by climate change (Carter and Singh 2010). The overall environmental situation in Barbados mirrors that throughout the Caribbean, but it is magnified by the combination of limited land area, the predominance of karst landscape, (ironically) relative economic wealth and mass tourism.

14.4.1 Hydrologic Implications

Water supply difficulties are largely a function of the contemporary and historical rainfall regime and the island's predominantly karst landscape. Surface drainage is usually of short duration, and water runs off quickly, particularly from impervious surfaces in urban areas, causing flooding

Fig. 14.7 A view overlooking the Atlantic Ocean from Barbados' northern coast, near Animal Flower Cave. Barbados is resplendent with such coastal karstscapes, such as these abrasional terraces and restricted tidal platforms. Photo by T. Zabarauskas

Fig. 14.8 Looking east along Barbados' northern coast, near Animal Flower Cave. Note the wave-cut platforms and notches, sea caves, and small sea arch (just *right of center*). A small platform/sea stump can also be seen just offshore (and to the *left*) of the far background, just below the horizon. Other coastal karstscapes exhibit similar features. Photo by T. Zabarauskas

and non-point pollution, and being effectively lost to human use. Land use in Barbados has changed considerably since the 1960s, with urban land increasing from 21.2 to 37.6%, and arable land declining from 57.7 to 46.2% (UNDSEC 2000; UNFAO 2004a).

In non-urban limestone areas, water usually drains underground rapidly, either diffusely via pores and fissures or discretely via sinkholes, drainage ("suck") wells and valley bases, to replenish a shallow and volumetrically restricted lenticular or wedge-shaped karst groundwater aquifer that varies in thickness seasonally from less than 5 m to nearly 20 m (Humphrey 1997; Jones et al. 2000). This overlies deeper saline water, separated by a mixing zone of similarly variable thickness (Humphrey 1997). Above the elevation of the groundwater aquifer, upper surface (about 20 m) water movement is vadose and occurs via a spectrum of voids in the limestones ranging from individual pores, through dissolutionally enlarged fissures to large conduits (caves). The contact with the underlying relatively impermeable Tertiary rocks acts as an aquiclude (Humphrey 1997). Freshwater aquifer recharge occurs almost entirely in the wet season and represents 15–30% of precipitation (Jones et al. 2000, 2003). This represents the major source of potable water, which is extracted via 21 municipal wells (with horizontal basal adits designed to reduce drawdown), and the Three-Houses and Porey's springs. There are also in excess of 100 private agricultural irrigation wells, and abstractions currently approach potential ground water yield estimates of 59.0 million m^3 per year in an average year and approximately 45 million m^3 per year in a 1 in 15 year drought (United Nations 1998; Government of Barbados 2010). Further restrictions result from limitations of the water distribution system, and available supply is below the 1000 m^3 per capita threshold at which a country is classified as "water scarce" (UNFAO 2004b), although three seawater desalination plants now supplement groundwater supplies, a wastewater recycling plan is now in operation (Johnson and Mwansa 2007), and there is an increasing awareness of the need for integrated water resources management (Farrell et al. 2007).

Through land use ordinances dating to 1963, so-called water protection zones were designed to provide partial safeguards against potential groundwater loss and bacterial contamination, and overall groundwater quality is good (Carter and Singh 2010). "Development" is most restricted in zone 1, but is unrestricted in zone 5. This zoning provides no effective protection against chemical pollution, and increasing nitrate and pesticide levels resulting from agriculture activities are of concern, as are incidents of accidental and negligent industrial and domestic pollution (Chilton et al. 1990; Carter and Singh 2010). Seawater intrusion into the shallow coastal aquifer is also an ongoing threat (Humphrey 1997).

14.4.2 Resource Conservation Efforts

Several agencies share responsibility for land use and resource management on Barbados, including soil conservation, environmental protection, sustainable agriculture, and planning, broadly under the auspices of the Natural Heritage Department (NHD), which was established in 2005. Rapid soil erosion is widely recognized as a problem within the Scotland District, but affects the karstscape as well. About 140,000 m^3 of limestone is quarried annually, mostly for road construction and maintenance (Government of Barbados 2009), and quarry expansion is a complex and contentious issue (Dey and Ramcharan 2008).

Solid waste management and disposal represent long-term and increasing problems, driven particularly by rising volumes of residential and tourist consumer waste. The capacity of the main sanitary landfill at Mangrove Pond, St. Thomas is limited, and illegal dumping is particularly problematic in dolines, valleys, and quarries, where it poses a serious potential for groundwater contamination and human health impacts. Hazardous waste represents a special problem, and recycling efforts seek to divert recoverable materials (Carter and Singh 2010).

Government agencies, international and national NGOs, and other groups have developed multiple and overlapping policies and programs to address environmental issues. Relevant government policies addressing water supplies, sustainable resource development, waste management, increasing land use efficiency, and improving disaster management include the National Physical Development Plan (1998), National Strategic Plan 2006–2025, Barbados Sustainable Development Policy (2004), and the Barbados Medium Term Strategy 2010–2014. Government retains its stronghold over physical development planning, effectively restricting the participation of the public and NGOs (Pugh and Potter 2002), and there remains a need for better coordination and effective implementation of plans and policies to achieve environmental protection and sustainable development (Mycoo 2006; Carter and Singh 2010).

All these environmental conditions notwithstanding, Barbados maintained its strong agricultural basis until the end of the twentieth century, following the ascension of tourism and the conversion of agricultural land to other uses (UNDSEC 2000; UNFAO 2004a). With annual tourist arrivals totaling some 500,000 (Caribbean Tourism Association 2011), the Barbados karstscape is among the most intensely visited globally (Day and Hall 2012), and this large, but temporary population influx places major demands on the water-supply system and the coastal environment (Burke 2002; Charara et al. 2011; Cashman et al. 2012).

14.5 Heritage and Tourism

The preponderance of Barbados tourism is focused on beaches, coastal hotels, relaxation, and hedonism (Dharmaratne and Brathwaite 1998), with only a minor component involving ecotourism and attention to local heritage (Mycoo 2006). Sustainability is highly questionable, both from environmental and economic perspectives, and projected climate change poses exacerbated challenges (Mycoo 2006, 2014; Cashman et al. 2012). The "…transformation of rural areas into up-scale leisure amenity landscapes…" (Bunce 2008: 969) creates conflicts between offshore elites and local interests and constrains implementation of sustainable environmental and development agendas (Fig. 14.9; Bunce 2008).

Barbados has only four IUCN-category protected areas, the Folkestone Marine Reserve (II), Harrison's Cave (III), Welchman Hall Gully (IV), and Turner's Hall Wood Nature Reserve (Ia). Other areas, such as Carlisle Bay, Jack-in-the-Box Gully, the Farley Hill Barbados Wildlife Reserve, and the Graeme Hall (Swamp) Nature Sanctuary (a threatened Ramsar wetland), are also accorded variable levels of informal protection, and 13 protected areas, including the 7000+ ha Barbados National Park are recognized within a prospective national protected areas system (Government of Barbados 2010).

Within the karstscape, Harrison's Cave has been a significant tourist destination since its commercial opening in 1981 (Hobbs 1994) and is touted as Barbados' "number one attraction," offering a variety of tours including those by electric tram and "eco-adventure" tours (Harrison's Cave 2015). Nearby Welchman Hall Gully has been developed as a

Fig. 14.9 One of several real estate lots for sale, looking south over the Atlantic Ocean from Inch Marlow near the island's most southern point. Notice the relatively flat relief, creating naturally (mostly) flat building plots. Photo by T. Zabarauskas

botanical garden by the Barbados National Trust, with a focus on restoration of native flora, and Farley Hill National Park features the ruins of a colonial mansion. By comparison, Barbados lags behind other Caribbean countries in protection of its karst landscape (Kueny and Day 1998; Day 2011).

Despite its relative inaccessibility from the airport and major tourism areas of the south and west coasts, the Scotland District has considerable ecotourism potential (Hearn et al. 2001). Despite the overall sensitivity of the steep slopes and poorly consolidated bedrock, the area has attractive and unspoiled high-energy beaches, with considerable potential for watersports, camping and hiking, and other nature-based activities (Patel 1995).

Degradation of coastal ecosystems is particularly acute in the highly populated southwest urban corridor and in tourism areas along the west coast, where infrastructure is infringing upon beaches, wetlands are being degraded and infilled, coastal water quality is decreasing, and coral reefs are in decline (Leitch and Harbor 1999; Lewis 2002). Such environmental impacts adversely affect residents and tourism alike (Carter and Singh 2010; Cashman et al. 2012). A Coastal Zone Management Unit was established in 1996, and a proposed Marine Management Authority potentially augurs well for improved integrated coastal zone management and this will also most likely affect the tourism industry.

14.6 Hazards and Environmental Issues

Drought and hurricanes (with associated floods) are the major natural hazards affecting Barbados, and projected anthropogenic climate change increases these threats, as elsewhere throughout the Caribbean karst (Day and Chenoweth 2009). Increased hurricane and drought frequencies have been experienced in recent years, with consequences including increased flooding, erosion, sedimentation, and decline in agricultural productivity (Carter and Singh 2010). Institutional, legislative, and societal shortcomings as well as questionable economic and development priorities exacerbate these problems (Pugh and Potter 2002; Bunce 2008; Carter and Singh 2010). Projected climate change has major implications both for the natural environment and human activities such as tourism (Cashman et al. 2012), with considerable differences of opinion about appropriate responses (Belle and Bramwell 2005).

Drought represents a perennial problem in Barbados, and human activities have only exacerbated it (Watts 1987; Richardson 1997). Seasonal water deficiencies, particularly for agricultural needs, are to be anticipated and can be prepared for, but longer-term rainfall insufficiencies are less predictable and, in tandem with inadequate storage and transmission capacities and unrestrained usage, pose serious societal problems (Peters 2015). Extended drought cycles with periodicities of about 50 years have been recognized (Burton 1995), and rainfall variability has been attributed to both the El Nino Southern Oscillation (ENSO) (Laing 2004) and the North Atlantic Oscillation (NAO) (Charlery et al. 2006). Barbados has experienced numerous periods of more-than-seasonal drought, most recently in 2009–2010 (Peters 2015), with subsequent environmental and human impacts.

The majority of Caribbean hurricanes pass north of Barbados (Mah and Stearn 1986), so their impact is generally limited to moderate winds and heavy rainfall, which may still be damaging. Still, eleven hurricanes, three of which caused severe damage, were recorded between 1651 and 1850 on Barbados, with the 1780 hurricane killing over 4000 people. Subsequent "hits" were recorded in 1895, 1955 (Hurricane

Fig. 14.10 Comparison of coastal landforms between windward and leeward coasts. A high-energy coast in St Lucy parish displays steep, eroded cliffs and limited tidal platforms (**a**), while a low-energy beach in St James parish, beckons the visitor to relax (**b**). The boat in (**b**) sits on a low beach berm, and the breaking waves indicate an offshore reef. Photos by M. Day

Janet), 1979 (Hurricane David), and 1980 (Hurricane Allen) (Mah and Stearn 1986). The most recent to hit the island, Hurricane Thomas in 2010, caused only minor damage.

Flooding in the karstscape is a particular danger, since dolines and dry valleys contain water only occasionally, and the potential risk of flooding is often underestimated. Extensive networks of check dams and drainage ("suck") wells were installed to ameliorate flooding and aid groundwater replenishment during the colonial period, but these have now largely fallen into disrepair. The damaging flood potential in the karstscape, particularly where urbanization has occurred, is demonstrated by the 1995 flooding in the vicinity of Holetown and Weston, in St James parish (Leitch and Harbor 1999).

Surface collapse over karst cavities is generally a minor and localized hazard, although it may be accentuated by increasing groundwater withdrawal (Humphrey 1997). The most frequent result is agricultural inconvenience, but structural damage occurs occasionally, and a surface collapse over a cave in the Brittons Hill area of Bridgetown in 2007 resulted in five deaths (Government of Barbados 2009; Kambesis and Machel 2013). Welchman Hall Gully is also reported to be a collapsed feature (Donovan 2005).

In the Scotland District, hazards include accelerated soil erosion, gullying, landslides, and other mass movements, many of which are human-induced (Carson and Tam 1977) and some of which have caused structural damage and abandonment (Hearn et al. 2001; Hodgson et al. 2002). The Boscobel Landslip of October 1901 displaced some 10 million cubic meters of material, and mass movements of similar magnitude continue to represent a potentially significant hazard (Cruden et al. 2014).

Although tectonism and volcanism do not pose immediate local threats to Barbados, there is a risk of tsunami impact, resulting from earthquake activity elsewhere in the Caribbean (Scheffers and Kelletat 2006). Recognizing that earthquake and tsunami risk in the Caribbean has long been underrated, the Caribbean Tsunami Information Center was established in Barbados in 2013 and remains the only tsunami-warning mechanism for many Lesser Antillean islands.

14.7 Conclusion

Barbados is unique geologically within the Lesser Antilles, but its karst landscape—dolines, dry valleys, and caves—serves as a model for karst elsewhere in the limestone Caribbees. The raised reef terraces are Caribbean type examples, with considerable paleoenvironmental significance and parallels with contemporary reefs and coastlines. Landforms in the Scotland District provide a marked contrast, exemplifying considerable dynamism in the form of extensive slope instability and a readily erodible coastal fabric.

Visitors often regard Barbados as a Caribbean paradise, with warm weather, fine beaches, and interesting coastlines (Fig. 14.10), a verdant landscape, a vibrant, thriving population, and a sound economy, but the reality is somewhat different. The climate poses extreme risks of drought and flood, and the geology compounds these through the prevalence of karst landscape, with its inherent hydrological complexities, and the presence of poorly consolidated rocks in the Scotland District that are easily eroded and subject to rapid mass movements. Coupled with a history of landscape devastation, poorly developed conservation practices, unsustainable mass tourism, and projected climate change, Barbados faces a difficult future, in which rational land use planning needs to give full weight to the underlying geological and geomorphological frameworks.

Acknowledgements We acknowledge a preliminary (2006) contribution to this review by Leah Schultz, and the financial support provided by the Department of Geography and the Center for Latin American and Caribbean Studies at the University of Wisconsin-Milwaukee.

References

Beckles HMcD (2006) A history of Barbados: from Amerindian settlement to Caribbean single market. Cambridge University Press, New York

Belle N, Bramwell B (2005) Climate change and small island tourism: policy maker and industry perspectives in Barbados. J Travel Res 44:32–41

Bender ML, Fairbanks RG, Taylor FW, Matthews RK, Mesolella KJ (1979) Uranium-series dating of the Pleistocene reef tracts of Barbados, West Indies. Geol Soc Am Bull 90:577–594

Bunce M (2008) The 'leisuring' of rural landscapes in Barbados: new spatialities and the implications for sustainability of small island states. Geoforum 39(2):969–979

Burke RI (2002) Environment and tourism: examining the relationship between tourism and the environment in Barbados and St. Lucia. Caribbean Policy Development Center. http://www.sia-acp.org/acp/download/acp_eu_sia_tourismenvironment_case_study.pdf

Burton S (1995) Long-term fluctuations in rainfall over Barbados. Caribbean Meteorological Institute Technical Note 30

Caribbean Tourism Association (2011) Available via http://www.onecaribbean.org/content/files/Strep1AnguillaToBonaire2010.pdf. Accessed 15 Sept 2015

Carson MA, Tam SW (1977) The land conservation conundrum of eastern Barbados. Ann Assoc Am Geogr 67(2):185–203

Carter SS, Singh A (2010) National environmental summary: Barbados 2010. UNEP Regional Office for Latin America and the Caribbean, 56 p. Available via http://www.pnuma.org/publicaciones/FINAL%20Barbados%20NES%20Nov%202010-%20edited.pdf. Accessed 18 Sept 2015

Cashman A, Cumberbatch J, Moore W (2012) The effects of climate change on tourism in small states: evidence from the Barbados case. Tourism Rev 67(3):17–29

Charara N, Cashman A, Bonnell R, Gehr R (2011) Water use efficiency in the hotel sector of Barbados. J Sustain Tourism 19(2):231–245

Charlery J, Nurse L, Whitehall K (2006) Exploring the relationship between the North Atlantic oscillation and rainfall patterns in Barbados. Int J Climatol 26:819–827

Chilton PJ, Vlugman AA, Foster SSD (1990) Ground-water pollution risk assessment for public water supply sources in Barbados. In: Krishna HJ, Quinones-Aponte V, Gomez-Gomez F, Morris GL, Middleburg VA (eds) Tropical hydrology and caribbean water resources. American Water Resources Association, pp 279–289

CIA (2015) The World Factbook: Barbados. https://www.cia.gov/library/publications/the-world- factbook/geos/bb.html

Cruden D, Machel HG, Knox J, Goddard R (2014) The 'Boscobel Landslip' of October 1st 1901—the largest historic landslide in Barbados, West Indies. Landslides 11(4):673–684

Day MJ (1983) Doline morphology and development in Barbados. Ann Assoc Am Geogr 73(2):206–219

Day MJ (1993) Human impacts on Caribbean and Central American karst. In: Karst Terrains: environmental changes and human impact. Catena Supplement 25. Catena Verlag, Cremlingen-Destedt, pp 109–125

Day MJ (2002) The role of valley systems in the evolution of tropical karstlands. In: F Gabrovsek (ed) Evolution of karst: from prekarst to cessation, Zalozba ZRC, pp 235–241

Day MJ (2009) Eastern Caribbean. In: Palmer AN, Palmer MV (eds) Caves and karst of the USA, Nat Speleol Soc: 332, 346–347

Day MJ (2010a) Human interaction with Caribbean karst landscapes: past, present and future. Acta Carsologica 39(1):137–146

Day MJ (2010b) Challenges to sustainability of the Caribbean karst. Geologia Croatica 63(2):149–154

Day MJ (2011) Protection of karst landscapes in the developing world: lessons from Central America, the Caribbean and Southeast Asia. In: VanBeynen P (ed) Karst management. Springer, New York, pp 439–458

Day MJ, Chenoweth MS (2009) Potential impacts of anthropogenic environmental change on the Caribbean karst. In: Barker B, Dodman D, McGregor D (eds) Global change and Caribbean vulnerability. UWI Press, Kingston, pp 100–122

Day MJ, Hall AB (2012) On tourism in tropical karst. Cave Karst Sci 39(3):119–122

Dey PK, Ramcharan EK (2008) Analytic hierarchy process helps select site for limestone quarry expansion in Barbados. J Environ Manage 88(4):1384–1395

Dharmaratne GS, Brathwaite AE (1998) Economic valuation of the coastline for tourism in Barbados. J Travel Res 37(2):128–144

Donovan SK (2005) The geology of Barbados: a field guide. Caribb J Earth Sci 38:21–33

Drewett P (1993) Excavations at Heywoods, Barbados, and the economic basis of the Suazoid period in the Lesser Antilles. J Barbados Mus Hist Soc 38:113–137

Farrell D, Nurse L, Moseley L (2007) Managing water resources in the face of climate change: a Caribbean perspective. University of the West Indies, Trinidad, 11 p. Available via http://64.28.139.231/conferences/salises/documents/Farrell%20D.pdf. Accessed 15 Sept 2015

Fermor J (1972) The dry valleys of Barbados: a critical review of their pattern and origin. Trans Inst Br Geogr 57:153–165

Fitzpatrick SM (2011) Verification of an archaic age occupation on Barbados, southern Lesser Antilles. Radiocarbon 53(4):595–604

Ford DC, Williams PW (2007) Karst hydrogeology and geomorphology. Wiley, Chichester

Government of Barbados (2009) National Report to the United Nations Commission for Sustainable Development (UNCSD) Cycle 18/19 (2009/2010), 52 p. Available at http://www.un.org/esa/dsd/dsd_aofw_ni/ni_pdfs/NationalReports/barbados/Full_text.pdf. Accessed 15 Sept 2015

Government of Barbados (2010) Barbados National Assessment Report, 128 p. Available via http://www.sids2014.org/content/documents/18Barbados-MSI-NAR2010.pdf. Accessed 15 Sept 2015

Handler JS (1997) Escaping Slavery in a Caribbean Plantation Society: Marronage in Barbados, 1650s–1830s. New West Indian Guide 71 (3/4):183–225

Harrison's Cave (2015) http://www.harrisonscave.com/index.php

Hearn GJ, Hodgson IF, Woddy S (2001) GIS-based landslide hazard mapping in the Scotland District, Barbados. In: Griffiths JS (ed) Land Surface Evaluation for Engineering Practice, ed. Geological Society of London, Engineering Geology Special Publications, vol 18, 151–157

Hobbs HH (ed) (1994) A study of environmental factors in Harrison's cave, Barbados, West Indies. National Speleological Society, Huntsville

Hodgson IF, Hearn GJ, Lucas J (2002) Landslide hazard assessment for land management and development planning, Scotland District, Barbados. In: McInnes RG, Jakeways J (eds) Instability: planning and management. Thomas Telford, London, pp 281–290

Humphrey JD (1997) Geology and hydrogeology of Barbados. In: Vacher HL, Quinn TM (eds) Geology and hydrogeology of carbonate islands. Elsevier, Amsterdam, pp 381–406
James NP, Stearn CW, Harrison RS (1977) Field guidebook to modern and pleistocene reef carbonates, Barbados. W.I. Third International Coral Reef Symposium, Miami, p 30
Johnson JA, Mwansa BJ (2007) Development of water reuse regulations and the implementation of water augmentation strategies—a case study of the Barbados water authority. Proc Water Environ Fed Conf 2007:4843–4859
Jones IC, Banner JL, Humphrey JD (2000) Estimating recharge in a tropical karst aquifer. Water Resour Res 36(5):1289–1299
Jones IC, Banner JL (2003) Estimating recharge thresholds in tropical karst island aquifers: Barbados, Puerto Rico and Guam. J Hydrol 278:131–143
Jones RW (2009) Stratigraphy, palaeoenvironmental interpretation and uplift history of Barbados based on foraminiferal and other palaeontological evidence. J Micropalaeontol 28:37–44
Kambesis PN, Machel HG (2013) Caves and karst of Barbados. In: Lace MJ, Mylroie JE (eds) Coastal karst landforms. Springer, Dordrecht, pp 227–244
Kueny JA, Day MJ (1998) An assessment of protected karst landscapes in the Caribbean. Caribb Geogr 9(2):87–100
Laing AG (2004) Cases of heavy precipitation and flash floods in the Caribbean during El Nino winters. J Hydrometeorol 5:577–594
Leitch C, Harbor J (1999) Impacts of land-use change on freshwater runoff into the near- coastal zone, Holetown watershed, Barbados: Comparisons of long-term to single-storm effects. J Soil Water Conserv 54(3):584–592
Lewis JB (2002) Evidence from aerial photography of structural loss of coral reefs at Barbados, West Indies. Coral Reefs 21:49–56
Limonta M, Garzanti E, Resentini A, Andò S, Boni M, Bechstädt T (2014) Multicyclic sediment transfer along and across convergent plate boundaries (Barbados, Lesser Antilles). Basin Res 2014:1–18
Lundberg J (2004) Coastal karst. In: Gunn J (ed) Encyclopedia of caves and karst science. Fitzroy Dearborn, New York, pp 231–233
Mah AJ, Stearn CW (1986) The effect of Hurricane Allen on the Bellairs fringing reef, Barbados. Coral Reefs 4:169–176
Mesolella KJ, Matthews RK, Broecker WS, Thurber DL (1969) The astronomical theory of climate change: Barbados data. J Geol 77:250–274
Muhs DR (2001) Evolution of soils on Quaternary reef terraces of Barbados, West Indies. Quat Res 56(1):66–78
Mycoo M (2006) Sustainable tourism using regulations, market mechanisms and green certification: a case study of Barbados. J Sustain Tourism 14(5):489–511
Mycoo M (2014) Sustainable tourism, climate change and sea level rise adaptation policies in Barbados. Nat Resour Forum 38 (1):47–57
Mylroie JE (2004) Speleogenesis: coastal and oceanic settings. In: Gunn J (ed) Encyclopedia of caves and karst science. Fitzroy Dearborn, New York, pp 674–677
Mylroie JE, Carew JL (2000) Speleogenesis in coastal and oceanic settings. In: Klimchouk AB, Ford DC, Palmer AN, Dreybrodt W, Huntsville AL (eds) Speleogenesis: evolution of karst aquifers, National Speleological Society, pp 226–233
Mylroie JE, Vacher HL (1999) A conceptual view of carbonate island karst. In: Palmer AN, Palmer MV, Sasowsky ID (eds) Karst modelling, Charles Town, Karst Waters Institute, WV, 48–57
Patel F (1995) Coastal development and geomorphological processes: Scotland district, Barbados. In: Barker D, McGregor DFM (eds) Environment and development in the Caribbean: geographical perspectives. UWI Press, Kingston, pp 209–232
Peters EJ (2015) The 2009/2010 Caribbean drought: a case study. Disasters 39(4):738–751
Poole EG, Barker LH (1983) The geology of Barbados. 1:50,000 sheet. Directorate of overseas surveys, U.K. and Government of Barbados
Pugh J, Potter RB (2002) Rolling back the state and physical development planning: the case of Barbados. Singap J Trop Geogr 21(2):183–199
Randall RE (1970) Vegetation and the environment on the Barbados coast. J Ecol 58(1):155–172
Richardson BC (1997) Economy and environment in the Caribbean: Barbados and the Windwards in the late 1800s. University Press of Florida, Gainesville FL
Scheffers A, Kelletat D (2006) New evidence and datings of Holocene paleo-tsunami events in the Caribbean (Barbados, St. Martin and Anguilla). In: Mercado-Irizarry A, Liu P (eds) Caribbean Tsunami Hazard, Singapore: World Scientific Publishing, pp 178–202
Schellmann G, Radtke U (2004a) A revised morpho- and chronostratigraphy of the late and middle pleistocene coral reef terraces on Southern Barbados (West Indies). Earth-Sci Rev 64(3–4):157–187
Schellmann G, Radtke U (2004b) The marine quaternary of Barbados Kolner Geographische Arbeiten 81. Koln: Geographisches Institute der Universitat zu Koln
Smith DI (1975) The problem of limestone dry valleys—implications of recent work in limestone hydrology. In: Peel RF, Chisholm M, Haggett P (eds) Processes in physical and human geography. Heinemann, London, pp 130–147
Speed RC (1994) Barbados and the Lesser Antilles Forearc. In: Donovan SK, Jackson TA (eds) Caribbean geology: an introduction. UWI Publishers, Kingston, pp 179–192
Speed RC (2012) Geology and geomorphology of Barbados. Geological Society of America Special Paper 491. Boulder CO: Geological Society of America, 63 p
Speed RC, Cheng H (2004) Evolution of marine terraces and sea level in the last interglacial, Cave Hill, Barbados. Bull Geol Soc Am 116 (1–2):219–232
United Nations (1998) Natural resource aspects of sustainable development in Barbados. Available via http://www.un.org/esa/agenda21/natlinfo/countr/barbados/natur.htm. Accessed 15 Sept 2015
United Nations Food and Agriculture Organisation (2004) Country profile: Barbados. Available at http://www.fao.org/es/ess/yearbook/vol_1_2/pdf/barbados/pdf. Accessed 15 Sept 2015
United Nations Food and Agriculture Organisation (2004b) Water Resources in Barbados. Available at http://www.fao.org/ag/agl/aglw/aquastat/countries/barbados/index.stm. Accessed 15 Sept 2015
United Nations Development System for the Eastern Caribbean (2000) Sub-regional common assessment of Barbados and OECS. Available at http://www undg.org/documents/1399_Barbados_CCA_-_Barbados_2000.pdf. Accessed 15 Sept 2015
Watts D (1987) The West Indies: patterns of development, culture and environmental change since 1492. Cambridge University Press, Cambridge
Watts D (1988) Development and renewable resource depletion in the Caribbean. J Biogeogr 15(1):119–126
Waugh J (2009) Trade and invasive species in the Caribbean: a universe of risk. IUCN, Gland, Switzerland

Saint Vincent and the Grenadines

15

Russell Fielding and Alison DeGraff Ollivierre

Abstract

Saint Vincent and the Grenadines is a southern Caribbean archipelago consisting of one major island, seven smaller, inhabited islands in the Grenadines, and numerous other uninhabited islets and cays. Geologically and culturally, St. Vincent differs from the Grenadines. St. Vincent is a large island, forested and mountainous, heavily dependent upon agriculture, and entirely the product of volcanic activity. St. Vincent's stratovolcano, La Soufrière (1178 m), has erupted several times since European settlement, most recently in 1979. The Grenadines consist of small, tourism- and fishery-dependent islands, which are more arid, and geologically much older and more complex. The entire archipelago is subject to a suite of natural hazards, owing to its geography, with St. Vincent particularly susceptible to volcanic hazards.

Keywords

Saint Vincent • Grenadines • Volcanic • Coralline • Archipelago

15.1 Introduction

15.1.1 Orientation

St. Vincent—lushly forested on the flanks of its volcanic peak, La Soufrière—together with the Grenadines, is billed in some tourism literature as "the Caribbean you're looking for" (SVG Tourism Authority 2009). The main island's interior landscape is steep, rugged, and largely undeveloped. The coastline is urbanized in places—most notably the area surrounding the capital of Kingstown—but far from completely developed. In fact, St. Vincent's coastal road does not even travel the entire island's circumference, such is the lack of development in the hinterlands.

R. Fielding (✉)
Department of Earth and Environmental Systems, University of the South, Sewanee, TN, USA
e-mail: russell.fielding@sewanee.edu

A. DeGraff Ollivierre
Tombolo Maps & Design, Denver, CO, USA
e-mail: aly.ollivierre@gmail.com

C.D. Allen (ed.), *Landscapes and Landforms of the Lesser Antilles*, World Geomorphological Landscapes, DOI 10.1007/978-3-319-55787-8_15

St. Vincent is a relatively large island (344 km^2) situated between 13° 23′ and 13° 07′ North latitude (Fig. 15.1), while the Grenadines are a chain of smaller islands (the inhabited of which range from 0.4 to 16 km^2), islets, rocks, and reefs dotting the shallow sea between the larger islands of St. Vincent in the north and Grenada in the south (Fig. 15.2). The international boundary between St. Vincent and the Grenadines and Grenada passes through the archipelago between the (relatively) larger inhabited Grenadine Islands of Union Island and Carriacou. According to local knowledge, the northernmost tip of Carriacou, at Gun Point, is also technically part of St. Vincent and the Grenadines, due to the political boundary originally being drawn along a line of latitude. Apart from this artifact land border, the border falls between the two small islands Petit St. Vincent (PSV) and Petite Martinique (PM), which lie less than a kilometer apart. This chapter will limit its discussion of the Grenadines to only those islands politically linked with St. Vincent, as Chap. 16 will cover Grenada and its Grenadines.

At the northern end of the archipelago, the island of St Vincent serves as the economic, cultural, and political hub of the nation. Physically and culturally, the Grenadine islands

Fig. 15.1 General physiographic map of St. Vincent depicting radiating rivers and central mountain range, as well as St. Vincent's location just north of the Grenadines (see inset map). Cartography by K.M. Groom and A. Ollivierre

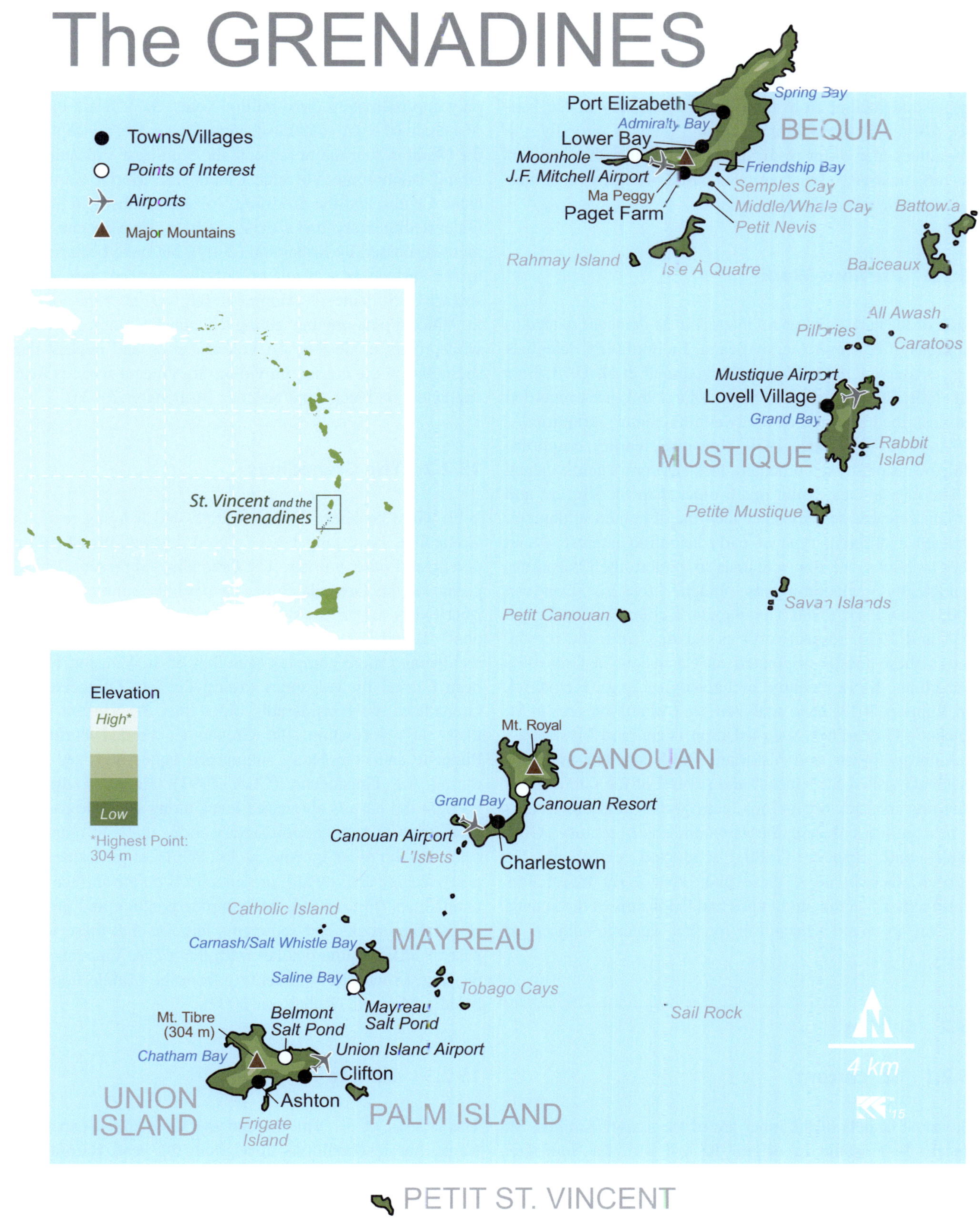

Fig. 15.2 General physiographic map of the Grenadines. In order to maintain accurate size and distance between islands, topographic representation is not as detailed as other maps in this volume. Please note: The names of the inhabited islands are shown in all caps whereas names of uninhabited islands and cays are lowercase. Cartography by K.M. Groom and A. Ollivierre

differ starkly from St. Vincent. In terms of landscape, the Grenadines represent more of the stereotypical Caribbean island scenery, white sand beaches, crystalline blue water, and bright coral reefs, while St. Vincent's shoreline rises steeper and rockier from the ocean, with many of the beaches marked by dark-colored, volcanic sand. Culturally, the Grenadines also differ from St. Vincent in that they support less urbanization, with economies based primarily on fishing and tourism, rather than agriculture.

15.1.2 Previous Work

Much of the early geological interest in St. Vincent related to eruptions of its iconic stratovolcano, La Soufrière. Scientists (e.g., Anderson 1903; Flett 1908; Aspinall et al. 1973; Barr and Heffter 1982; Fiske and Sigurdsson 1982) have recorded and described all three of the twentieth-century eruptions—1902, 1971–1972, and 1979—however, earlier eruptions, while occasionally documented, lack rigorous scientific data.

The Grenadines, being much older than St. Vincent and with no volcanic activity since the time of human settlement, attracted a different type of early scientific interest. As in other locations, the first scientists to explore the Grenadines geologically—paleontologists—sought fossils. Donovan (2003) cites Earle's and Trenchmann's collections (taken in 1924 and 1934, respectively) as the first from the region, though these fossils originated on Carriacou, in Grenada's Grenadines. More recently, archaeologists (e.g., Fitzpatrick and Kappers 2013) have analyzed pre-Columbian sites in St. Vincent's Grenadines, such asUnion Island and Mustique.

Anthropologists and historians (both amateur and professional) visited St. Vincent and several of the Grenadines, searching for rock engravings, petroglyphs, and other signs of past cultures during the late nineteenth century. Ober (1880) and Brinton (1889) conducted early, notable research. Occasionally, these and other early researchers would remark on the landscapes and landforms of the islands as a way to create a backdrop for their specific subjects of study.

15.2 Setting

15.2.1 St. Vincent

Like most islands of the inner arc of the Lesser Antilles, St. Vincent is volcanic in origin. Its volcano, La Soufrière, remains one of the most active in the region, having erupted both frequently and recently. The last major eruption occurred in 1979, the latest of three during the past century. Magma produced by the subduction of the South American Plate under the Caribbean Plate reaches the subsea surface at many points along the island arc, forming seamounts; islands occur when these seamounts extend above sea level. The majority of the island's structure, therefore, is basaltic and pyroclastic in composition. Relatively geologically young, at only approximately three million years, St. Vincent consists of a chain of stratovolcanic centers. Geologists conceptualize the island in four major regions: the Southeast Volcanics, the Grand Bonhomme Volcanic Center, the Morne Garu Volcanic Center, and the Soufrière Volcano (Robertson 2003b). Older sedimentary rocks exist, such as the Miocene specimens discussed by Robertson (2009), but have been brought to the surface as a result of uplifting and are rare on the surface of St. Vincent. Ample rainfall nourishes the forest in St. Vincent's mountains, giving the island a "rugged beauty," while rivers erode the soft volcanic rock and deposit dark sediment on the coasts, providing St. Vincent with its many characteristic black sand beaches (Robertson 2003a).

15.2.2 The Grenadines

With white beaches, shallow waters, and fringing reefs, the Grenadines have long been favored destinations of sailors, divers, and other tourists. The Grenadines represent the high points of the Grenada Bank, "a partially submerged ridge [that exists as] an extension of the South American Continent" (Bland 1871, p. 60), making the geologic history of the archipelago more complex than that of St. Vincent. Having been formed by volcanoes during the late Oligocene, the Grenadines are considerably older than St. Vincent. Since their early formation, the Grenadines eroded during the Pliocene and completely submerged under high sea levels during the Pleistocene (CCA 1991). Regional uplifting brought the islands above sea level, along with the coralline limestones and other sedimentary rocks that—in many places—top the older igneous rocks. While ample evidence of past volcanic activity still presents itself on the surface in the Grenadines (Fig. 15.3a), the complex geology and presence of coralline rocks and sediments indicate that these islands share a history quite distinct from that of St. Vincent, where the oldest rocks are sedimentary from the Middle Eocene to Middle Miocene (Robertson 2009).

15.2.3 Climate

The islands of St. Vincent and the Grenadines experience warm, humid conditions throughout the year. Rainfall and temperature vary seasonally, but the variation in the former is significantly more pronounced. During the dry season (December–May), the islands receive 76–90 mm of precipitation per month. The wet season (June–November) brings 190–215 mm of rain per month. These ranges represent the

Fig. 15.3 Examples of volcanic landforms in St. Vincent and the Grenadines. **a** An exposed dike on Petit Nevis (foreground), a Grenadine island just south of Bequia. Photograph by A. Ollivierre. **b** A view inside the volcanic crater atop La Soufrière on St. Vincent. Visible in the center is the magma dome; to the right is evidence of a recent landslide. Photograph by P. Cole

means of observed measurements taken between 1970 and 1999 (McSweeney et al. 2010a).

These figures represent countrywide averages, so significant variation exists, both among and within the islands. Those with higher elevations see more rainfall, which is concentrated specifically in those high-elevation mountains. The peak of La Soufrière and surrounding mountainous areas, the wettest parts of the country, receive up to 580 mm of rain per month, while the low valleys and windward coastal plains can be quite dry, experiencing much lower monthly rainfall totals (CCA 1991). Across the archipelago, precipitation generally decreases from north to south, owing to the climatological influence of La Soufrière (and the other high mountains on St. Vincent) and the relatively low elevations and small islands further south in the Grenadines. Climatological models predict a decrease in overall annual precipitation, based on observed trends since 1960 (McSweeney et al. 2010a, b).

With regard to temperature, elevation again drives variation—in this case outweighing both seasonality and diurnal fluctuation (CCA 1991). Between 1970 and 1999, observed mean temperatures for the entire archipelago ranged from 25.6 to 27.3 °C. Climatologists have observed a trend of increasing temperatures since 1960 and predict an annual increase in the average temperature between 0.6 and 2.3 °C by 2060 (McSweeney et al. 2010a, b). Models predict that both warming and decreasing precipitation will be more pronounced and more rapid in the Grenadines than in St. Vincent.

15.3 Landscapes and Landforms

With numerous inhabited and uninhabited islands to the south of the main island, St. Vincent and the Grenadines, as an archipelago, spans a north–south range of nearly 100 km. To accommodate its vastness, this section begins in the north at the island of St. Vincent and moves southward through each major Grenadine island and several minor islets, cays and rocks. This section ends at the small island called Petit St. Vincent (PSV)—the southernmost in the country—after expounding on the main landscape feature(s) associated with each island or islet.

15.3.1 Leeward and Windward Sides of St. Vincent Island

The dichotomy of a windward (eastern) and leeward (western) side common to many islands in the Lesser Antilles is present on St. Vincent, but seems less pronounced here than on other islands. Barbados sits 159 km to the east of St. Vincent, making St. Vincent the only Lesser Antillean Island with a similarly sized, equilatitudinal island on its windward side—the direction from which the dominant winds blow. This means that, even though Barbados has minimal relief, its orographic processes still affect St. Vincent's eastern coast (Watts 1987). While St. Vincent's windward side is dry compared to the island's lusher leeward

Fig. 15.4 Comparison of vegetation and landscapes between the leeward and windward coasts of St. Vincent. **a** A forested landscape on St. Vincent's leeward side near Vermont. **b** An arid landscape on the windward side of St. Vincent near Brighton. Photographs by R. Fielding

side, its climate is perhaps not as dry as it would be without the presence of Barbados (Fig. 15.4).

Coastal morphology also differs from one side of St. Vincent to the other. The windward side features rocky coasts—many with steep cliffs—and a few coastal plains, with the steepest finding their origin in the ridges leading down from the island's major mountain peaks. The leeward coast undulates between rocky headlands and sandy beaches. The beaches are composed of volcanic sediment transported by rivers and longshore currents, which leaves them dark in color. The one white sand beach on the leeward side can be found at a luxury resort in Buccament Bay. The developers nourished the beach artificially with imported sand. Lighter sand—more gray than white—is found on St. Vincent's southeastern side, near Indian Bay and Villa Beach. Here, currents nourish the beach from offshore, rather than sediments from terrestrial rocks of volcanic origin, eroded and carried coastward by fluvial transport.

[1]Maps and written records ascribe a range of elevations to La Soufrière. We rely upon the work of Robertson (2005), whom we consider authoritative, for our figure, but acknowledge that the easily remembered measurement of 1234 m appears most often on maps and in print. A recent amateur expedition (Gilbertson and Gilbertson, Personal communication) took GPS readings from various promontories around the crater rim and settled on a maximum of 1236 m. Part of the discrepancy may result from imprecise survey methods or from measurements being taken from various locations around the crater rim and not necessarily the highest point. Actual change in the height of the mountain, owing to volcanic or tectonic activity, cannot be thoroughly dismissed.

15.3.2 La Soufrière

The highest point, and most recognizable landform, on St. Vincent is the summit of La Soufrière (1178 m).[1] Viewed from its base, the mountain appears steep, dark, lush, and impenetrable, with clouds often obscuring the summit. On the rare clear day, La Soufrière reveals a spectacular steep-walled crater with a pyroclastic volcanic dome, dark and smoldering, rising from within (Fig. 15.3b).

The entire northern quarter of St. Vincent, exclusive of some immediately coastal areas, consists of pyroclastic deposits and lava flows from the Soufrière volcano. Upon approaching the crater along the foot paths that rise from either coast (meeting in the middle at the peak), the vegetation shifts from rain forest to cloud forest as the terrain gets steeper, higher, and more geologically fractured. Scarps and rifts appear—some of which carry ephemeral streams—as indicators of this mountain's dynamic nature. Researchers have recently discovered debris avalanche deposits from scarp collapses on La Soufrière offshore during high-resolution bathymetry data-collection cruises. This has led to a new understanding of the evolution of the volcano—and subsequently the entire island—and the magnitude of the eruptions during its formation (Le Friant et al. 2009).

15.3.3 Other Mountains on St. Vincent

The massive presence of La Soufrière dominates St. Vincent's topography. Local residents often refer to it simply as "the volcano." Geographically though, "the volcano" and its flanks only cover the northern quarter of the island's area.

Two major rivers, the Wallibou (not to be confused with the Wallilabou) and the Rabacca, flowing west and east, respectively, have their sources on the southern slopes of La Soufrière, separated by a high, narrow, north–south running ridge. These rivers and the gorges they carve serve to separate La Soufrière from the rest of the island. The southern region comprises a complex of mountains and hills, all volcanic in origin, but none as recent nor as starkly impressive as La Soufrière. Moving south from La Soufrière, the first peaks encountered are the Morne Garu mountains, dominated by Richmond Peak (1077 m) and Brisbane Peak (932 m). Further south, the Cumberland and Colonarie rivers again nearly bisect the island as they flow down from the northern flanks of the Bonhomme Peaks: Grand Bonhomme (970 m) and Petit Bonhomme (756 m). This mountainous area also serves as the source for the Yambou River, which carves the Mesopotamia Valley, one of St. Vincent's deepest, lushest, and most verdant regions. South of "Mespo" (as locals call it), Mt. St. Andrew (735 m) provides a steep backdrop to the capital city, Kingstown, located in front of a natural harbor on the island's southwest coast.

The major mountain peaks of St. Vincent stand in a roughly north-south alignment, but the landscape obscures this neat arrangement. Numerous smaller peaks, sharp ridgelines, and deep valleys—all the result of erosion and dissection of the southern, pre-Soufrière volcanic centers—radiate outward from the mountains, lending St. Vincent a sense of topographic randomness with sharp rises, steep descents, dark valleys, and a few coastal plains. St. Vincent is an island for mountain people—a place where choosing between volcanic topography and coastal morphology is unnecessary.

15.3.4 Geothermal Areas

Found only in the area inside the crater of La Soufrière and on the southern flanks of the volcano, geothermal activity includes fumaroles at the base of the dome within the crater and warm springs (37 °C) along the upper Wallibou River (Robertson 2005). Owing to the risk of future eruptions, experts see these sites' potential for geothermal energy development as unlikely (CCA 1991).

15.3.5 Airport Development

At the time of writing, the largest and most landscape-altering development project in St. Vincent and the Grenadines was the construction of the Argyle International Airport on St. Vincent's windward side. The new airport has finally reached the completion of its politically charged, oft-delayed, decade-long planning and construction process. Replacing the current E.T. Joshua Airport, Argyle, will handle larger jets, potentially opening St. Vincent to more international tourism. The airport development provoked controversy among several groups in St. Vincent and the Grenadines, with concerns ranging from safety (e.g., strong crosswinds common on the windward side of the island) to opposition to the major landscape changes that the airport construction required. For example, one tourism official stated that "to build this airport, engineers had to move three mountains, fill two valleys, create embankments, and span a river" (Myers 2015). Additionally, the construction of the runway uncovered significant archaeological findings, including the postholes from at least eleven structures, stone tools, pottery, and the buried remains of several humans. A hastily organized, yet remarkably effective rescue excavation determined that the site had been a Carib settlement from the late sixteenth to early seventeenth century (Hofman and Hoogland 2012). The most important finding from the site involved the commingling of native and colonial artifacts, suggesting a degree of cultural interactions not previously documented.

15.3.6 Salt Ponds: Recreational and Productive

St. Vincent features at least two so-called salt ponds, formed by lava flows that reached the sea and created small pools, sealed off from the ocean during low tide, but capable of being replenished at high tide. Both are on the windward side—a small one at Brighton and a larger one at Owia—and are popular as swimming areas for locals and visitors alike (Fig. 15.5a). By contrast, water bodies called salt ponds in the Grenadines rarely get used recreationally. Instead, locals (primarily on Union Island and Mayreau) harvest salt, a culturally important heritage that dates back to the 1700s when colonial settlers produced salt for domestic use and export (Fig. 15.5b). These salt ponds and lagoons additionally provide essential habitat for resident and migratory birds, with many being recognized as BirdLife International important bird areas (IBAs) (Culzac-Wilson 2008).

15.3.7 The Grenadines

The largest of St. Vincent's Grenadines is Bequia (16 km^2), an S-shaped island surrounded by a number of smaller offshore cays. The Bequia Channel—8.3 km at its narrowest point—separates this northernmost Grenadine island from St. Vincent. Bequia is an island of steep, rolling hills,

Fig. 15.5 Comparison of salt ponds on St. Vincent versus those found in the Grenadines. **a** The Owia Salt Pond, on St. Vincent's northeastern coast, with a swimmer for scale. The relative clarity of the water and picturesque setting make it a popular natural attraction. Photograph by R. Fielding. **b** An aerial view of the salt pond on the southern peninsula of Mayreau in the Grenadines. The murky water and small salt mining retaining pond visible in the center of the image discourages swimmers. Photograph by A. Ollivierre

Fig. 15.6 Conglomerate beach rock with basaltic clasts on Bequia, located on a headland between Lower Bay and Princess Margaret Beach. Photograph by R. Fielding

vegetated coves, and small, sandy coastal plains. Volcanic in origin, as are all of the Grenadines, Bequia has a complex geologic history. The beach sand ranges from white to golden, evidence of the island's surficial coralline nature, although basaltic clasts occur in some of the beach rock, underscoring its volcanic origin (Fig. 15.6).

Perhaps Bequia's most renowned natural feature is Admiralty Bay, a large, deep bay on the island's protected leeward side. At the northern end of this bay sits Port Elizabeth, Bequia's principal settlement, while other notable features on the island include Spring Bay on the windward side and Friendship Bay on the south coast. An interesting housing development called Moonhole is found at the southwestern-most extent of the island. Developed around a natural arch by the same name at the tip of Bequia, it was first developed in the 1960s using local materials including olivine-rich basaltic stone and whalebone to create unique holiday homes. While the structure inside the arch itself has

Fig. 15.7 Aerial view of the J.F. Mitchell Airport on the southern coast of Bequia. Photograph by A. Ollivierre

fallen into disrepair and is considered unsafe to visit (following a rockfall that preceded its abandonment), the rest of the site remains an eclectic part of the island's history and current settlement pattern.

Built on reclaimed land along the south coast near the community of Paget Farm, the creation of Bequia's airport (Fig. 15.7) resulted in the destruction of a beach, a barrier reef, and shallow lagoon habitats. This subsequently created further problems for the community due to an inadequate and malfunctioning drainage system that has resulted in severe problems with stagnant waters and the accumulation of marine debris (Jaja 2013).

Several islets off the south coast of Bequia also warrant attention. The first is Semples Cay, an islet at the mouth of Friendship Bay. To some, Bequia earns a certain amount of fame—notoriety to others—as the only Caribbean community with an aboriginal subsistence whaling quota from the International Whaling Commission (Adams 1971; Ward 1995). Based around Friendship Bay, on the island's south coast, Bequia whalers pursue humpback whales (*Megaptera novaeangliae*) in an area extending approximately 19 km off the windward side of Bequia and encompassing the unprotected waters around the island of Mustique's marine conservation buffer, along with several smaller cays (see chart in Adams 1971, pp. 64–65). Captured whales are brought to Semples Cay for processing. This is the only remaining whale processing station out of at least seven that once stood on Bequia and the nearby islets. East of Bequia lie the uninhabited islets of Battowia and Baliceaux. These arid islets lack natural freshwater sources and find use primarily as grazing land for livestock. The history of Baliceaux, however, belies its pastoral present. During the early colonial era, the British used this islet as a natural prison—a holding place for native Caribs exiled from St. Vincent and later deported to Central America (see *Heritage*).

Continuing south, the next inhabited island is Mustique. This private island is owned by the Mustique Company and has been developed almost exclusively as an island of holiday homes and rental villas, although a small community of local Mustiquans remains in the town of Lovell Village. What once was a derelict sugar island has been transformed into an impeccably manicured resort, the island's eponymous mosquitoes (the name evolved as a corruption of *moustique*, the French word for *mosquito*) controlled through the draining of some of the wetlands in which they formerly bred.

The channel between Mustique and the next inhabited island to the south, Canouan, is dotted with several small cays and islets important for seabird nesting and roosting, but otherwise represents the longest stretch of open water within the Grenadines archipelago. Another luxury resort company owns two-thirds of Canouan. This resort achieved local infamy for its golf club whose runoff of herbicides and fungicides led to the degradation of Canouan's extensive coral reef (Scott and Horrocks 1993). Despite this, the main industry on the island remains tourism—particularly

Fig. 15.8 Aerial view, facing south, of the Carnash tombolo on Mayreau. Photograph by A. Ollivierre

associated with the resort—while traditional fisheries, though weakened along with the health of the reef, maintain a smaller contribution to the local economy.

South of Canouan is the small island of Mayreau, home to not more than 300 people. Mayreau's landforms include a classic tombolo (Fig. 15.8) at its northern end where Mt. Carbuit is joined to the main island by a narrow strand, the western side of which forms the back beach of Carnash (or Salt Whistle) Bay, a favorite anchorage for yachts. The northern point of the island features an example of coral pavement (Fig. 15.9), exposed during low tide. At Mayreau's south end, a wide isthmus featuring a large salt pond in its center connects Lande-Ici Hill to the rest of the island. To the west of this isthmus, on the leeward side of the island, is Saline Bay, home to Mayreau's main port—by which this island maintains its sole connection (as it has no airport) to the rest of the archipelago.

Due east of Mayreau stand the uninhabited, but noteworthy, Tobago Cays. This tightly clustered group of four islets, Petit Rameau, Petit Bateau, Jamesby, and Baradal, plus a fifth outlier—Petit Tabac—rise above the shallow sand, grass, and coral-covered seafloor just meters below the surface. Narrow channels of only 100–200 m separate the four main cays, while Petit Tabac lies west of the reef from which the other cays sit, about a kilometer from the rest of the group. The Grenadines contain the most extensive coral reef system in the Eastern Caribbean, with each island supporting fringing, patch, or barrier reefs, and a variety of ecologically important seagrass and mangrove habitats. The Tobago Cays Marine Park (TCMP) (originally established in 1997, but ineffective until its relaunching by the government in 2006) protects the two largest of these reefs, Horseshoe Reef and World's End Reef (Fig. 15.10a), that surround the Tobago Cays in the southern Grenadines (Mahon et al. 2004; DeGraff and Baldwin 2013). The decision to include the entire neighboring, fisheries-dependent island of Mayreau in this protected area, however, continues to provoke controversy in the community (Hoggarth 2007).

South of Mayreau sits Union Island, home to the highest point in the Grenadine archipelago, Mount Tibre[2] (304 m), and the largest mangrove forest in the country (25 ha). One of the most controversial development projects in the Grenadines is the failed Ashton Marina project (Price and Price 1998), which had intended to build a 300-berth marina and replace the existing lagoon and its surrounding reefs with condominiums and a golf course. When the Italian development firm abandoned construction due to bankruptcy in 1995, the marina berths remained, causing a stagnation and eutrophication of the water in the lagoon, which critically reduced the marine habitats and marine life (Sorenson 2007).

One somewhat unique geologic feature within the Caribbean, an outcrop of columnar basalt, may be found on Frigate Island, just south of—and artificially connected to—Union Island (Fig. 15.10b). This feature was exposed during mining operations to produce rocks and fill for a marina

[2]This mountain has several similar alternate names. *Taboi* is seen most often on official maps. *Tibre* is the most commonly used local name. Other maps and charts label the peak as *Toboi* or *Tabor*. The confusion is perhaps a result of variations in pronunciation or a cartographic transcription error propagated to subsequent maps.

Fig. 15.9 Coral pavement, exposed and eroded on the north coast of Mayreau. Photograph by A. Ollivierre

Fig. 15.10 Examples of unique landforms in the Grenadines, both terrestrial and aquatic. **a** Horseshoe Reef (*foreground*) and World's End Reef (*background*) with Petit Tabac, one of the Tobago Cays, in the center. **b** Columnar basalt on Frigate Island, a small cay off of Ashton Bay, Union Island, that was artificially joined with Union in 1994–95 in a failed attempt to build a marina in Ashton Lagoon (Price and Price 1998). Photographs by A. Ollivierre

project (Price and Price 1998). When lava cools, it contracts, both vertically and horizontally. Horizontal contractions can lead to jointing—or cracking—of the hardening rock. In basalts, these joints often result in a hexagonal shape, although irregular cooling can sometimes produce columns with less regular shapes and varying numbers of sides. Another example of columnar basalt exists in the Cumberland valley on St. Vincent. Early visitors to the island (e.g., Davy 1854; Sharp 1890) described and photographed the geology of this site, indicating that its exposure likely resulted from natural surface movement processes.

East and southeast of Union Island stand the small private resort islands of Palm Island and Petit St. Vincent. Both have received numerous international tourism awards and recognitions and are known for their luxury accommodations. Palm Island, known as Prune Island before its sale to resort developers, served as a leper colony for stricken inhabitants of Union Island from the late 1700s into the early 1800s.

15.4 Heritage and Tourism

15.4.1 Heritage

Evidence of human occupation in St. Vincent and the Grenadines exists from as early as 2000 BCE. Hairouna (the indigenous name for St. Vincent) remained a stronghold for native inhabitants well into the colonial era. While other islands were being depopulated through disease, genocide, and forced relocations, the native population on Hairouna continued to thrive. Only on Dominica was the maintenance of Carib culture and population greater. Although Columbus sighted the island in 1498—on January 22, the feast day of Saint Vincent—Europeans did not begin to settle on St. Vincent until the late seventeenth century (Young 1993). The French came first, basing their operation at Barrouallie and establishing farms, primarily on the leeward side. Today, the names of many villages and geographic features, especially on the island's west coast, retain their French roots.

In 1635, the arrival of Africans on Bequia, survivors of the wreck of a slave ship [or possibly two shipwrecks (Matthei and Smith 2008)], augmented the Carib population in St. Vincent and the Grenadines. As depicted in an art installation (Fig. 15.11a) at the museum in Fort Charlotte (located on a hill above Kingstown), the survivors eventually came to St. Vincent, where the local Caribs welcomed them. Over time, the Africans assimilated into the Carib culture, and their descendants became known as "Black Caribs." This mixed-ancestry population gained a reputation for their strong defense of their land against colonizers (Kirby and Martin 1972). Despite having declared St. Vincent, along with Dominica, a refuge for the native Carib people (also called "Yellow Caribs") displaced by colonization on other islands, the British became interested in settling on St. Vincent in 1763. Soon thereafter, a three-way contest for control of the island began with the British, French, and Carib peoples in near-continuous conflict.

In 1796, the British defeated the Caribs. The victors captured over 4000 native prisoners and placed them temporarily on Baliceaux, a previously uninhabited Grenadine islet between Bequia and Mustique. Over half of those

Fig. 15.11 Remnants of St. Vincent and the Grenadines' native heritage. **a** A 1972 painting by William Linzee Prescott, on display at Fort Charlotte, St. Vincent. The caption (*not pictured*) reads, "A few Black Caribs who had escaped Abercrombie's dragnet were eventually settled at Greiggs, shown here. Their descendants still display that fierce independence that so characterized the BLACK CARIBS." Photograph by R. Fielding. **b** Work stone in Spring Bay, Bequia. Such stones would have been used to grind flour or pastes from seeds, meat, plants, etc. Photograph by A. Ollivierre

captured died during their three-month interment on Baliceaux; the rest were deported to Roatán, an island off the Caribbean coast of Honduras. From Roatán, the deportees—the Garifuna as they came to be known—crossed to the Honduran mainland and spread along the Caribbean coast of Central America where their descendants live today (Taylor 2012). St. Vincent and the Grenadines maintains a strong connection to its Carib heritage despite only a small percentage (2–4%) of the population claiming direct Carib ancestry. In March each year, Carib descendants from St. Vincent, Dominica, and throughout Central America make a pilgrimage to Baliceaux, the site where their ancestors were held and where many died. Pilgrims are often overcome by emotion as they experience first hand the dry, shadeless, inhospitable landscape to which their ancestors were banished.

15.4.2 Petroglyphs and Stone Artifacts

According to one early Caribbean anthropologist, "throughout the West Indian archipelago there is nothing of greater archaeological importance than the St. Vincent petroglyphs" (Huckerby 1914, p. 239). Researchers have found and recorded twelve separate petroglyph panels—each with numerous motifs—in St. Vincent, all of which are either found along the coast or the island's river valleys (with the exception of the area around La Soufrière). For many years, a petroglyph also survived on the northern side of Canouan in the Grenadines; however, this artifact disappeared during the construction of the resort. Petroglyphs in St. Vincent and the Grenadines share a similar motif style with engravings in Grenada, leading researchers to hypothesize that inhabitants of both main islands maintained contact, or perhaps familial relations, with one another (Haviser and Strecker 2006).

Work stones, also known as *polissoirs* (polishers, grinders) or cupules, once served as mortars for sharpening tools and for grinding dried or fibrous food sources and may still be found in the river valleys and beaches of St. Vincent and on Bequia (Martin 2006; Bradford 2001). Haviser and Strecker (2006, p. 79), perhaps somewhat hyperbolically, claim that work stones have been found on "practically every beach" on St. Vincent. Despite this apparent exaggeration, Kirby (1969), a Vincentian prehistorian, cataloged over fifty work stones in St. Vincent alone, with the vast majority found on the island's beaches (Fig. 15.11b).

15.4.3 Early to Modern Agro-Processing

During the plantation era, sugarcane was a major crop, both for local consumption and international export as raw or processed sugar, molasses, and rum. Many signs of this economy remain visible on the landscape. Remnants of cotton mills and lime kilns for producing cement also dot the landscape, though many have been left to ruin and overgrow with weeds and "cashi" (cacti). Overcultivation of the Grenadines' hilly landscapes with cotton, sugarcane, and subsistence crops such as cassava, pigeon peas, and corn led to erosion and gullying, as well as soil nutrient loss—a problem further propagated in the sharecropping land tenancy system after emancipation. As cash agriculture declined in the seasonally dry Grenadines, hardy livestock such as goats, sheep, and fowl became the leading exports, alongside the ever-persistent fisheries industry (Adams 1978). In St. Vincent, sugarcane cultivation dominated much of the coast, though the crop did not fare as well in the foothills of the wet forested mountains. The cotton industry followed sugar and led to similar, if not more severe, erosion problems. Experimentation with other crops—especially arrowroot—ensued, with varying degrees of success until, finally, bananas emerged as the largest export for the island, booming in the 1980s as a result of preferential pricing arrangements in Europe, and then declining as prices dropped amidst foreign market shifts to Latin American bananas (Grossman 1998). Today, bananas still comprise the majority of St. Vincent's official agricultural production, with illicit marijuana cultivation supplementing legal agriculture to an unquantified, but large, degree (INCSR 2006).

15.4.4 Tourism

Many, perhaps most, tourists entering the country are destined for the Grenadines. Sailors, divers, snorkelers, and those who simply want to lie on a pristine beach are drawn to this archipelago of small islands with white sand beaches and extensive reef systems in brilliant turquoise waters. The Grenadines also host a number of exclusive resorts and private villas, the latter primarily on Mustique, though the number of villas on Bequia is growing. St. Vincent continues to work hard at attracting adventure tourists, and recently established its first all-inclusive luxury resort. Despite these developments, agriculture in St. Vincent still competes with tourism as a major part of the country's GDP—the two industries contributing 7.5 and 12.6%, respectively—a rare economic situation in this region (World Bank 2013).

The tourism industry in St. Vincent boasts a long history of environmental protection for recreational purposes. For example, the St. Vincent Botanical Gardens are, by some count, the oldest extant botanical gardens in the Western Hemisphere, dating to 1765 (Anonymous 1922; Howard 1954). The gardens feature endemic species from around the Caribbean, as well as other tropical locations. One of the most famous plants in the gardens is an individual breadfruit tree (*Artocarpus altilis*), said to have been brought to St. Vincent from Tahiti personally by Captain Bligh on his first voyage after the infamous *Bounty* mutiny (Howard 1954). Breadfruit has a long and complicated history in the Caribbean, being a beloved, yet humble, local delicacy as well as a botanical link to the region's history of slavery and the perceived need, on the part of the planters, for local, abundant, high-caloric food to fuel the workforce. Another long-standing example of environmental protection in St. Vincent is the Kings Hill Forest Reserve, with protection of its 55 acres of dry forest commencing in 1791 (CCA [Caribbean Conservation Association] 1991).

15.5 Hazards

As with the rest of the Eastern Caribbean, the islands of St. Vincent and the Grenadines remain particularly vulnerable to natural hazards (Anderson et al. 2011). That heightened vulnerability stays strongly tied to the demographic, economic, and geopolitical situation in the Lesser Antilles which favors relatively high population densities on islands or island groups of small geographic size with relatively low GDP (Anderson et al. 2011). The result is that a single hazardous event can have a disproportionate effect on a small island's population, infrastructure, and GDP. Indeed, in the case of tropical cyclones, the storm systems themselves are usually many times larger than the individual islands over which they track.

Researchers have catalogued St. Vincent's hazards numerous times (e.g., Boruff and Cutter 2007; Robertson 2009). Among these, Boruff and Cutter (2007) compiled records for natural hazards in St. Vincent, considering historical sources spanning the entire twentieth century. From these data, we see that fire occurs most frequently, although the mix of qualitative and quantitative data in the tally makes for difficult direct comparison among specific hazards. Perhaps the most striking lesson from Boruff and Cutter's (2007) list is the sheer variety of hazards that have occurred and are likely to continue, if not intensify, as a result of global climate change. As hazards intimately connect to geomorphology and landscape anywhere, especially in St. Vincent and the Grenadines, our discussion of past, current, and potential hazards here will necessarily be among the most in-depth of this chapter.

15.5.1 Fire

A scholar considering Lesser Antillean natural hazard vulnerabilities would likely not think of fire first. Caribbean fires do not make headlines in North America or Europe as readily as volcanic eruptions or major hurricanes. This lack of newsworthiness may originate from the fact that most fires in the Caribbean are human-caused—intentionally or unintentionally—though, as Richardson (2004, p. 196) concludes in his study of fire in the Caribbean, they are "never under complete control." In St. Vincent and the Grenadines, the process of making charcoal has had disastrous ecological effects—both from over-zealous, yet controlled, burning, and from fires spreading out of control and burning unintendedly large swaths of forest. The colonial government passed the first regulations on the production of charcoal in St. Vincent in 1839 and amended the laws several times throughout the nineteenth century (Richardson 2004). The procedure to legally produce charcoal was so complicated, expensive, and laden with bureaucracy, however, that many would-be law-abiding charcoal makers opted to practice their craft illicitly. This often took place in the hinterlands, away from the watchful eyes of colonial administrators based in Kingstown, an early analogue to today's illicit, yet normally uncontested, remote marijuana farms. The essentially unregulated production of charcoal through felling and burning trees in situ resulted in such dramatic deforestation that the once-forested slopes of Mt. St. Andrew had become, in the words of one late nineteenth-century writer, reduced to "complete barrenness" (Hooper 1886, pp. 9–10 [cited in Richardson 2004, p. 89]).

Even the smaller islands of the Grenadines were susceptible to the hazards wrought by human-caused fires. Richardson (2004, p. 90) notes that, despite their remoteness with respect to the seat of colonial administration, the Grenadines "supplied slooploads of charcoal to the Bridgetown [Barbados] market, to the detriment of their own insular habitats." Considering the small land area of the Grenadines, the general lack of precipitation and surface water, and the resultant meager forest cover, the disproportionate impact of even a single "sloopload" to the ecological health of a Grenadine forest community becomes apparent. Today, there is virtually no first growth forest remaining.

15.5.2 Volcanic Eruptions

While most hazards to which St. Vincent is vulnerable occur more frequently than volcanic eruptions, the ever-present sight of La Soufrière, combined with the high number of casualties and large economic losses due to its regular—yet infrequent—eruptions, may serve to make perceptions of volcanic risk more widespread than that of other hazards (Robertson 1995; Scarlett 2014). For example, according to Boruff and Cutter (2007), between 1955 and 2004, over 1500 human lives on St. Vincent were lost due to the gases, debris flows, and ejecta associated with the eruptions. La

Soufrière's eruptions have claimed a death toll greater by an order of magnitude than the second leading cause of natural hazards death: windstorms (presumably as part of tropical cyclones). The subtitle to a contemporary account of the 1902 eruption, reading boldly, "DEATH! DEATH! EVERYWHERE!" does not seem as sensationalistic when the sheer number of victims is considered (Garesché 1902, p. 5, capitalization in original). Still, volcanic eruptions are among the least frequent of all natural hazards experienced on St. Vincent and do not occur at all in the Grenadines. So, while the volcano looms over everyday Vincentian life and wreaks enormous havoc when it erupts, volcanic hazards occur perhaps once in a generation, not nearly as frequent as other, less disastrous hazards.

Eruptions have not only caused human deaths and injuries, but have reshaped the island landscape as well. The changes to the physical landscape are straightforward: nearly all the surface rock on St. Vincent is basaltic or pyroclastic in origin. In terms of human geography and the built landscape, Richardson (2004, p. 17) presciently states that

> The rapid transformation of St. Vincent from a sugarcane island in the 1880s and 1890s to an island of smallholders in the early twentieth century probably was brought about by an 1898 hurricane followed by a volcanic eruption in 1902, twin disasters that drove away most of the island's planter class.

Twenty-five volcanic eruptions occurred in the Eastern Caribbean between 1780 and 2006 (Lara 2006), the deadliest being Mt. Pelée on Martinique in 1902, which killed over 28,000 people. During that event, La Soufrière also erupted, though the casualties were considerably fewer, owing to the sparse population in the north of the island. These simultaneous eruptions of neighboring volcanoes demonstrate the long-known fact that "there are sub-marine communications between the burning mountains or volcanoes in each of [the islands]" (Anderson 1785, p. 30). La Soufrière last erupted in 1979, claiming no lives but forcing the evacuation of over 15,000 people, due to the increased populations on the north of the island by then. The typical "St. Vincent style" eruption (Robertson 2005, p. 247) is highly explosive, preceded by frequent, strong earthquakes, and involves the emission of toxic gases, lahars, and the ejection of large volumes of new material (Boruff and Cutter 2007).

Robertson (2005), the preeminent geologist studying La Soufrière, expects the volcano to continue to erupt at least once per century. The next eruption could be explosive, like the 1902 eruption, or effusive, like that of 1979. In the event of an eruption, the northern quarter of the island faces the greatest risk from pyroclastic flows, volcanic projectiles, and lahars. This would include the settlement of Georgetown, the largest on the windward side. Ashfall could affect more than half the island, reaching as far south as Barrouallie on the leeward side and the Bonhomme Peaks in the island's center. Even the so-called Safe Zone—essentially the southern third of the island—would not escape the effects of a major eruption, as evacuees from the northern area would increase the burden on housing and services in southern communities such as Layou, Mesopotamia, and, most predominantly, Kingstown.

15.5.3 Earthquakes and Tsunamis

St. Vincent and the Grenadines' location near the Caribbean and the North and South American tectonic plate boundaries and the presence of an active volcano on St. Vincent (and several others in the region) leads to common tectonic and volcanic earthquakes. Although earthquakes occur often, they are not usually strong or destructive—a trend that Robertson (2009) attributes to the relatively slow plate convergence at the subduction zone (~2 mm/year). The Seismic Research Centre at the University of the West Indies (UWI) maintains five seismic stations and eight GPS stations, and monitors changes in ground angle at several dry tilt sites on St. Vincent (UWI 2011).

Tomblin (1981) points out the added vulnerability of reclaimed or filled land in waterfront development—a common construction strategy in St. Vincent and other Caribbean islands—to liquefaction events during earthquakes. The inherently low elevation of reclaimed land also increases its vulnerability to storm surge, sea-level rise, and tsunamis. For example, Tomblin (1981) cites the infamous 1692 earthquake and subsequent tsunami that partially and permanently submerged the settlement of Port Royal, Jamaica. Port Royal had not been built upon reclaimed land, but on similarly soft, accreted sediment on a coastal spit. This historical event, well known throughout the Caribbean and beyond, should serve as a warning to nearshore developers around the region

Despite the occurrence of earthquakes in St. Vincent, the zone between Grenada and St. Lucia—where St. Vincent and the Grenadines comprise the majority of the area—is the least seismically active in all the Lesser Antilles. By way of explanation, Robertson (2009, p. 32) mentions two competing hypotheses, attributing the less-than-average number of earthquakes either to the smooth descent of the subducting slab, in which case the scarcity of earthquakes may be seen as a reliable, ongoing trend, or, more ominously, "to the accumulation of strain energy, which is yet to reach its limit," a scenario that would indicate the future occurrence of a major, possibly catastrophic earthquake, resulting in the release of long accumulated energy.

Earthquakes and volcanoes in the region are rarely, but powerfully, tsunamigenic. One notable tsunami-like event involved a "tidal" run-up on the coastal plain near

Georgetown on St. Vincent's windward side, associated with the 1902 eruption of La Soufrière, which researchers have retrospectively estimated at just over three kilometers (Lander et al. 2002). Despite the multitude of potential causal events, such as earthquakes, terrestrial volcanic eruptions, submarine volcanic eruptions, and landslides, both in the region and further afield (earthquakes as far away as the Iberian Peninsula have resulted in destructive tsunamis to Caribbean shores), tsunamis in the Caribbean are uncommon. One team of researchers analyzed data related to 91 extreme wave events that occurred throughout the Caribbean region between 1498 and 2000 and determined that only about one-third represented true tsunamis (Lander et al. 2002).

In 1867, a strong earthquake centered in the Virgin Islands caused shaking as far away as Guadeloupe, generating a tsunami that affected many Caribbean islands farther from its epicenter. Owing to the large area affected, this earthquake and its associated tsunami represents one of the best-documented natural hazard events from the nineteenth century, even if some of the descriptions of local effects and observations are likely to have been exaggerated. While no record exists indicating that the earthquake was felt in St. Vincent and the Grenadines, effects of the tsunami were recorded for both St. Vincent and Bequia, the latter experiencing three large, yet smooth waves of about two meters' height (Zahibo and Pelinovsky 2001).

Another set of scholars (Le Friant et al. 2009) found evidence suggesting at least one major flank-collapse on La Soufrière during the past 50,000 years. The team also located a debris avalanche associated with this event, which deposited 9–10 km^3 of material offshore. If a similar collapse were to happen in the future, the combined results of the volcanic eruption, mass wasting, and the tsunami that would likely result would be catastrophic, posing a hazard to all islands in the region.

15.5.4 Flooding and Mass Wasting

Floods and landslides also threaten St. Vincent. The most recent natural hazard to have occurred on a national scale is the flood of December 24–25, 2013, when numerous rivers swollen by heavy rains overflowed their banks. According to a government report on the disaster (Locke et al. 2014), during one three-hour period on the evening of December 24, 278 mm of rain fell in the northeastern portion of the island—a hundred-year event. The government declared over a dozen specific disaster areas where the most intense flooding and erosion had occurred. The transport sector suffered largest economic losses with many bridges and sections of road washed out. Nine people were confirmed dead. At the time of writing, recovery—assisted by the World Bank and other international organizations—has made good progress but is still ongoing with some of the planned long-term actions expected to take up to 5 years to complete. While the flooding of December 2013 was a highly unusual event, damaging floods do occur, on average, about four times per century (Boruff and Cutter 2007).

Landslides and other examples of mass wasting occur more frequently. DeGraff et al. (1989) found debris flows to be the most common form of mass wasting in a three-island study area comprised of St. Vincent, St. Lucia, and Dominica. A common scenario for mass wasting in St. Vincent is a slope of greater than 35° failing at a depth of less than two meters as a result of heavy precipitation, often produced by a tropical cyclone (DeGraff et al. 1989). Considering that half the island's surface area is sloped at least 30°, and only 20% of the island has slopes less than 20°, the frequency of mass wasting events in St. Vincent is certainly not surprising (CCA 1991). Road-cutting and other forms of anthropogenic alteration of the landscape, including unplanned housing on steep, unreinforced slopes, exacerbate the landslide risk (Anderson et al. 2011). The alternating layers of ash, basaltic rock, and cemented pyroclastic material underlying the island's steep slopes allow for differential decay rates that often weaken lower layers and destabilize entire areas. Rock falls also occur on St. Vincent, usually where bedrock scarps occur on mountain slopes or where sea cliffs and rocks show well-defined jointing, or as a result of undercutting by wave action (DeGraff et al. 1989). Still, the most common trigger of mass wasting in St. Vincent is precipitation, although tectonic and volcanic activities also cause mass wasting, particularly in the form of mudflows, or lahars, associated with eruptions of La Soufrière (DeGraff et al. 1989; Robertson 2005).

Landslides and flooding are rarely experienced in the Grenadines. Very little surface water exists on most islands except for salt ponds, ephemeral pools and streams following periodic rain events, and built ponds used to water livestock. The slopes and elevations also remain generally lower in the Grenadines than on St. Vincent, so while minimal flooding in low-lying coastal areas may occur, it does not cause the same devastating effects as floods do on St. Vincent.

15.5.5 Hurricanes

Located at the Hurricane Belt's edge, strong hurricanes usually pass slightly to the north of St. Vincent and the Grenadines. Notable exceptions to this trend occur regularly, however, and many hurricanes and lesser cyclones have caused death, injury, or major damage on these islands including, most recently, Hurricanes Matthew in 2016, Tomas in 2010, Ivan in 2004, and Lenny in 1999.

15.5.6 Sand Mining

The removal of sand from beaches and mixing with concrete for construction purposes is a common, albeit illegal, practice throughout the country. Sand mining has a negative impact on coastal zones, minimizing the beach's effectiveness as a form of coastal protection, and affecting beach recreation activities—an essential draw for tourism in St. Vincent and the Grenadines (Christie et al. 2015). The Caribbean Conservation Association (CCA 1991, *xxi*) describes how "virtually all the beaches… accessible by road have been mined to some extent, but those closest to Kingstown and other population centers have been the most affected." Sand mining can result in erosion along the coast and require placement of seawalls, gabions, or groins to protect coastal properties and developments. Port Elizabeth, in Bequia, serves as a salient example of these attempts at mitigation. In addition, vehicles used to access the sand destroy fragile beach vegetation.

15.6 Conclusion

The archipelagic nation of St. Vincent and the Grenadines typifies Eastern Caribbean landscapes, but also stands out in several key ways. With examples of both large (by Lesser Antilles standards) and small islands; wet and dry environments; volcanic and sedimentary geology; and French, British, and Carib cultural heritage, this island nation presents something of a microcosm of the wider region. The typical Lesser Antillean landscapes that are lacking in St. Vincent—coralline beaches, offshore reefs—are abundant in the Grenadines, and the volcanic landforms either lacking or obscured by more recent geological activity in the Grenadines are found in abundance on the surface of St. Vincent. Still, there remain important ways in which this island nation is unique within the region. Foremost among these is the risk presented by the active volcano, La Soufrière, one of the most active in the Lesser Antilles. The economic focus on agriculture also stands out among the islands of the region and increases the vulnerability of St. Vincent and the Grenadines to certain hazards, especially climate-driven changes in precipitation patterns, which would affect agricultural islands more drastically than those exclusively focused on tourism.

Acknowledgements The authors would like to thank Vincent Reid and Richard Robertson for sharing their local scholarly expertise in the development of this chapter. Thanks also to Paul Cole of Plymouth University for allowing us to use his stunning photograph of the volcano on a rare cloudless day in Fig. 3 and to Matthew and Eric Gilbertson, who shared their hard-won and not previously published GPS measurement of La Soufrière's summit elevation.

References

Adams JE (1971) Historical geography of whaling in Bequia Island, West Indies. Caribb Stud 11:55–74

Adams JE (1978) Union island, West Indies: an historical and geographic sketch. Caribb Stud 18:5–45

Anderson J (1785) An account of Morne Garou, a mountain in the island of St. Vincent with a description of the volcano on its summit. Philos Trans R Soc 125:16–32

Anderson MG, Holcombe E, Blake JR, Ghesquire F, Holm-Nielsen N, Fisseha T (2011) Reducing landslide risk in communities: evidence from the Eastern Caribbean. Appl Geogr 31:590–599

Anderson T (1903) Recent volcanic eruptions in the West Indies. Geogr J 21(3):265–281

Anonymous (1922) Oldest American botanical gardens. Am Botanist 28(4):170

Aspinall WP, Sigurdsson H, Shepherd JB (1973) Eruption of Soufrière volcano on St. Vincent Island, 1971–1972. Science 181(4095) 117–124

Barr S, Heffter JL (1982) Meteorological analysis of the eruption of Soufrière in April 1979. Science 216(4550) 1109–1111

Bland T (1871) Notes relating to the physical geography and geology of, and the distribution of terrestrial mollusca in certain of the West India Islands. Proc Am Philos Soc 12(86):56–63

Boruff BJ, Cutter SL (2007) The environmental vulnerability of Caribbean island nations. Geogr Rev 97(1):24–45

Bradford MAC (2001) Caribbean perspectives on settlement patterns: the Windward Island study. Anthropology Department. Iowa City, Iowa, USA, University of Iowa. Doctor of Philosophy

Brinton DG (1889) On a petroglyph from the Island of St. Vincent, W. I. Proc Acad Nat Sci Philadelphia 41:417–420

CCA [Caribbean Conservation Association] (1991) St. Vincent and the Grenadines: country environmental profile. CCA, St. Michael, Barbados

Christie M, Remoundou K, Siwicka E, Wainwright W (2015) Valuing marine and coastal ecosystem service benefits: case study of St. Vincent and the Grenadines' proposed marine protected areas. Ecosyst Serv 11:115–127

Culzac-Wilson L (2008) St. Vincent & the Grenadines. In: Wege DC, Anadón-Irizarry V, Vincenty M (eds) Important bird areas in the Caribbean. BirdLife International, Cambridge, pp 295–308

Davy J (1854 [2012]) The West Indies, before and since slave emancipation. Routledge, New York

DeGraff A, Baldwin K (2013) Participatory mapping of heritage sites in the Grenadine Islands. CERMES Technical Report 65. Cave Hill, Barbados: Centre for Resource Management and Environmental Studies, The University of the West Indies

DeGraff JV, Bryce R, Jibson RW, Mora S, Rogers CT (1989) Landslides: their extent and economic significance in the Caribbean. In: Brabb EE, Harrod BL (eds) Landslides: extent and economic significance. Balkema, Rotterdam

Donovan SK (2003) Charles Taylor Trechmann and the development of Caribbean geology between the wars. Proc Geol Assoc 114: 345–354

Fiske RS, Sigurdsson H (1982) Soufriere volcano, St. Vincent: observations of its 1979 eruption from the ground, aircraft, and satellites. Science 216(4550):1105–1126

Fitzpatrick SM, Kappers M (2013) Preliminary investigation of pre-Columbian sites on the islands of Mustique and Union in the Grenadines, West Indies. Caribb J Sci 47(2–3):260–272

Flett JS (1908) Petrological notes on the products of the eruptions of May, 1902, at the Soufrière in St. Vincent. Philos Trans R Soc Lond Ser A 208:305–331

Garesché WA (1902) Complete story of the Martinique and St. Vincent horrors. L.G. Stahl, Chicago

Gilbertson E, Gilbertson M (2015) Personal communication with authors. See also: http://web.mit.edu/egilbert/Public/CountryHighPoints/stvincent.html

Grossman L (1998) The political ecology of bananas. University of North Carolina Press, Chapel Hill

Haviser JB, Strecker M (2006) Zone 2: Caribbean area and north-coastal South America. Rock art of Latin America and the Caribbean: thematic study. The International Council on Monuments and Sites, Paris, pp 43–83

Hofman CL, Hoogland MLP (2012) Caribbean encounters: rescue excavations at the early colonial Island Carib site of Argyle, St. Vincent. Analecta Praehistorica Leidensia 43(44):63–76

Hoggarth D (2007) Tobago Cays Marine Park 2007–2009 management plan final draft. Organisation of Eastern Caribbean States (OECS) and Environment and Sustainable Development Unit (ESDU)

Hooper EDM (1886) Report upon the forests of St. Vincent. Waterlow and Sons, London

Howard RA (1954) A history of the botanic garden of St. Vincent, British West Indies. Geogr Rev 44(3):381–393

Huckerby T (1914) Petroglyphs of St. Vincent, British West Indies. Am Anthropol 16(2):238–244

INCSR [International Narcotics Control Strategy Report] (2006) The Caribbean. US Department of State, Washington, DC

Jaja J (2013) Reclaiming Paget Farm: a participatory film and community restoration initiative in the Caribbean. [Video] Accessed online at http://reclaimingpagetfarm.com/promo-video-2/

Kirby IE (1969) Pre-Columbian monuments in stone. St. Vincent Archaeological and Historical Society, Kingstown

Kirby IE, Martin CI (1972) The rise and fall of the Black Caribs of St. Vincent. St. Vincent Archaeological and Historical Society, Kingstown

Lander JF, Whiteside LS, Lockridge PA (2002) A brief history of tsunamis in the Caribbean Sea. Sci Tsunami Haz 20(1):57–94

Lara OD (2006) Space and history in the Caribbean. Markus Wiener Publications, Princeton

Le Friant A, Boudon G, Arnulf A, Robertson REA (2009) Debris avalanche deposits offshore St. Vincent (West Indies): impact of flank-collapse events on the morphological evolution of the island. J Volcanol Geoth Res 179:1–10

Locke J, Rodrigues M, Pandey B, Meier G, Charles K (2014) Rapid damage and loss assessment (DaLA): December 24–25, 2013 floods—a report by the Government of Saint Vincent and the Grenadines. World Bank Group, Washington, DC

Mahon R, Almerigi S, McConney P, Ryan C, Whyte B (2004) Coastal resources and livelihoods in the Grenadine islands: facilitating change in self-organising systems. Proc Gulf Caribb Fish Inst 55:56–67

Martin K (2006) Appendix VIII: ICOMOS Caribbean rock art information request form for St. Vincent and the Grenadines. In: Rock art of Latin America and the Caribbean: thematic study. The International Council on Monuments and Sites, Paris, pp. 79–82

Matthei LM, Smith DA (2008) Flexible ethnic identity, adaptation, survival, resistance: the Garifuna in the world-system. Soc Identities 14(2):215–232

McSweeney C, New M, Lizcano G (2010a) UNDP climate change country profiles: St Vincent and the Grenadines. Accessed online at http://country-profiles.geog.ox.ac.uk

McSweeney C, New M, Lizcano G, Lu X (2010b). The UNDP climate change country profiles improving the accessibility of observed and projected climate information for studies of climate change in developing countries. Bull Am Meteorol Soc 91:157–166

Myers GN (2015) Airport key to St. Vincent's evolution. Travel Weekly. June 18. Accessed online at http://www.travelweekly.com/Caribbean-Travel/Airport-key-to-St-Vincents-evolution

Ober FA (1880) Camps in the Caribbees: the adventures of a naturalist in the Lesser Antilles. David Douglas, Edinburgh

Price WS, Price PG (1998) Paradise lost: a postmortem of Ashton Marina project ecological impact on Ashton Lagoon. Union Island Association for Ecological Preservation, Clifton, St. Vincent and the Grenadines

Richardson B (2004) Igniting the Caribbean's past: fire in British West Indian history. University of North Carolina Press, Chapel Hill

Robertson REA (2009) Antilles, geology. In: Gillespie RG, Clague DA (eds) Encyclopaedia of islands. University of California Press, Oakland, pp 29–35

Robertson REA (2005) St. Vincent. In: Lindsay JM, Robertson REA, Shepherd JB, Ali S (eds) Volcanic hazard atlas of the Lesser Antilles. St. Augustine, Trinidad & Tobago: Seismic Research Unit, University of the West Indies. pp 240–261

Robertson REA (2003a) Making use of geology—the relevance of geology and geological information to the development process in St Vincent and the Grenadines. Paper presented at the St. Vincent and the Grenadines Country Conference, Kingstown, St. Vincent. May 22–24

Robertson REA (2003b) The volcanic geology of the pre-Soufriere rocks of St. Vincent, West Indies. Ph.D. thesis. University of the West Indies

Robertson REA (1995) An assessment of the risk from future eruptions of the Soufriere volcano of St. Vincent, West Indies. Nat Hazards 11(2):163–191

Scarlett JP (2014) Volcanic risk perceptions of La Soufrière, St. Vincent, East Caribbean. M.Sc. thesis. Lancaster University

Scott NM, Horrocks JA (1993) WIDECAST sea turtle recovery action plan for St. Vincent and the Grenadines. Kingston, Jamaica: UNEP Caribbean Environment Programme

Sharp B (1890) An account of the Vincelonian volcano. Proc Acad Nat Sci Philadelphia 42:289–295

Sorenson L (2007) Participatory planning workshop for the restoration of Ashton Lagoon: workshop proceedings and final report. Union Island, St. Vincent and the Grenadines: The Society for the Conservation and Study of Caribbean Birds and the Sustainable Grenadines Project

SVG Tourism Authority (2009) Discover St. Vincent and the Grenadines. Accessed online at http://discoversvg.com/

Taylor C (2012) The Black Carib wars: freedom, survival and the making of the Garifuna. Signal Books, Oxford

Tomblin J (1981) Earthquakes, volcanoes and hurricanes: a review of natural hazards and vulnerability in the West Indies. Ambio 10(6):340–344

UWI [University of the West Indies Seismic Research Centre] (2011) St. Vincent—monitoring. Accessed online at http://www.uwiseismic.com/General.aspx?id=70

Ward N (1995) Blows, mon, blows! A history of Bequia whaling. Gecko Productions, Woods Hole

Watts D (1987) The West Indies: patterns of development, culture, and environmental change since 1492. Cambridge University Press, Cambridge

World Bank (2013) Databank. Accessed online at http://data.worldbank.org/country/st-vincent-and-the-grenadines

Young VH (1993) Becoming West Indian: culture, self, and nation in St. Vincent. Smithsonian Institution Press, Washington, DC

Zahibo N, Pelinovsky EN (2001) Evaluation of tsunami risk in the Lesser Antilles. Nat Hazards Earth Syst Sci 1(4):221–231

Grenada: the Spice Isle

16

Casey D. Allen, Susanna L. Diller and Tirzha Zabarauskas

Abstract

Located on the southern end of the Lesser Antilles' Windward Island chain, the tri-island nation of Grenada is comprised of primarily volcanic islands. Grenada proper retains a mountainous landscape with a variety of volcanic features and fertile soil that supports the island's agricultural economy and helps to promote tourism. The nation's other two islands, Carriacou and Petite Martinique, both have mountainous interiors as well (at least for their size), though significantly less population. Grenada's strategic position made it a desirable territory for both French and English colonial powers, but its landscape gave the Amerindian people an edge in resisting colonization. Though there is little Amerindian cultural presence on the island now, their legacy can be seen around the island in petroglyphs, a few archaeological sites, and with place names like *Morne de Sateur's* (Leaper's Cliff). Today, the landscape supports both the primary industry, agriculture, and the largest portion of the economy, tourism. But these endeavors remain constrained by the islands' landforms and left to the whims of Nature which plays a grand role in changing the landscape every few decades. As more on-Island studies are completed, general awareness, mitigation, and management of precious resources such as soil, sand, and tourism, will undoubtedly help Grenada to continue providing for its inhabitants.

Keywords

Grenada • Volcanic • Hurricane • Petroglyph • Ecotourism • Beach • Rainforest

Note

Throughout this chapter, the word Grenada is used to denote both the main island and the tri-island nation itself. When discussing Carriacou or Petite Martinique, those names are used specifically. Every effort has been made so the distinction between Grenada-the-Tri-Island-Nation and Grenada-the-Main-Island is clear throughout this chapter.

C.D. Allen (✉)
General Education, Western Governors University,
Salt Lake City, UT, USA
e-mail: caseallen@gmail.com

T. Zabarauskas
Department of Geography and Environmental Sciences,
University of Colorado Denver, Denver, USA
e-mail: tirzha.zabarauskas@ucdenver.edu

S.L. Diller
Department of Geography and Environmental Studies,
University of New Mexico, Albuquerque, USA
e-mail: SLdiller14@gmail.com

16.1 Introduction

Though it officially administers over a dozen islands, The Nation of Grenada is often referred to as a "tri-island nation", as only three main islands are permanently inhabited (Fig. 16.1). The main island of Grenada (*gruh-nay-duh*, with an area of 306 km^2) straddles the twelfth degree north latitude line, with a population varying between 95,000 and 110,000. Carriacou (*carry-uh-coo*, with an area of 34 km^2, see Fig. 16.2), 37 km north of the main island, hosts a population of ~4500, and its northern tip, Gun Point, is unique because it technically represents a land border with St. Vincent and the Grenadines. When the Grenadine islands were split between St. Vincent and Grenada for the purpose of administrative efficiency, the boundary accidentally

C.D. Allen (ed.), *Landscapes and Landforms of the Lesser Antilles*,
World Geomorphological Landscapes, DOI 10.1007/978-3-319-55787-8_16

Fig. 16.1 General physiographic map of Grenada, Carriacou, and Petite Martinique, along with smaller adjacent islands. Various uninhabited islands surround the main island of Grenada, as well as in between Grenada and Carriacou. Cartography by K.M. Groom

transected the northern-most portion of Carriacou. Though Gun Point is officially part of St. Vincent and the Grenadines, it is left to Carriacou's jurisdiction (though the "point" itself and the surrounding land are actually privately owned). Petite Martinique (*peh-tee-mar-tin-eek*, with an area of ~3 km^2, see Fig. 16.3) another ~4 km east of Carriacou boasts a population of ~1000. Situated between St. Vincent and the Grenadines to the north, and Trinidad and Tobago to the south then, the Tri-Island Nation of Grenada lies approximately 160 km north of Venezuela and represents the southernmost Windward Island of the Lesser Antilles, resting near the Lesser Antilles arc where the North and (the extreme northeast portion of the South) American plate subduct west under the Caribbean Plate.

A volcanic island with a mountainous interior, Grenada's main island rises to its highest point at Mt. St. Catherine (840 m) in the central-northern half of the island (Fig. 16.4). Evidence of volcanic activity includes the crater lakes of Lake Antoine, Levera Pond, Grand Etang Lake, the twin calderas at Port Louis and St. George's harbor, and the geothermally active River Sallee "Boiling Springs." Numerous rivers radiate from the center of the island, and Grenada's mountainous terrain allows for several dramatic waterfalls, including those at Concord, Mt. Carmel, Seven

Fig. 16.2 Carriacou as seen from the south. The island is not as small as the visitor might believe—the faint mountains in the background are still the same island—and is very hilly, especially for its size. Even though Carriacou represents the largest Grenadine island (area-wise), it is still part of the tri-island nation of Grenada. Photo by C.D. Allen, 2016

Fig. 16.3 Island of Petite Martinique as seen from the north side of Carriacou (*foreground*). While Petite Martinique (large island, *center right*) is part of the tri-island nation of Grenada, Petit St. Vincent (*left*), Fota (small rock island *mid-center*), and Petit Dominique (*mid-center-right*) are not. Photo by C.D. Allen, 2016

Sisters, and Annandale. For their sizes, both Carriacou and Petite Martinique are mountainous as well, with high points of 291 and 230 m, respectively.

The main island of Grenada has 121 km of coastline and, like other Caribbean Islands, a number of beautiful beaches, including Grand Anse Beach—the island's most popular, and one of its most picturesque. There are a number of dive sites around the island, including shipwrecks and a protected underwater sculpture park (Molinière Sculpture Park) that can be viewed by snorkelers or SCUBA divers. Carriacou is known locally as the "Land of The Reefs" and its most notable dive site is Magic Garden, which features bubbling geothermal vents (Crask 2012). In addition to supporting the tourist industry, Grenada's largest portion of its economy, these reefs are also home to fish—Grenada's third-largest economic sector. Inland, agricultural production (Grenada's second largest economic sector) has a foothold in the islands' rich and productive volcanic soil.

Grenada's original name, given by the Amerindian inhabitants, was *Camerhogne*, the meaning of which is unknown. Its first European name, given by Christopher Columbus, was *Concepción*. Later, Grenada's mountainous green terrain led to its renaming, *Granada*, by Spanish sailors (Martin 2007). During the subsequent French takeover, it was changed to *Grenade*, and then, under British control, Grenada (*gruh-NAY-duh*). As plantations flourished and the island was found to support various cash crops, different spices were cultivated, eventually leading to the island's *Isle of Spice* moniker. Today, the Amerindian influence on Grenada has largely been lost—except for the rock art (petroglyphs) found around the island, and small

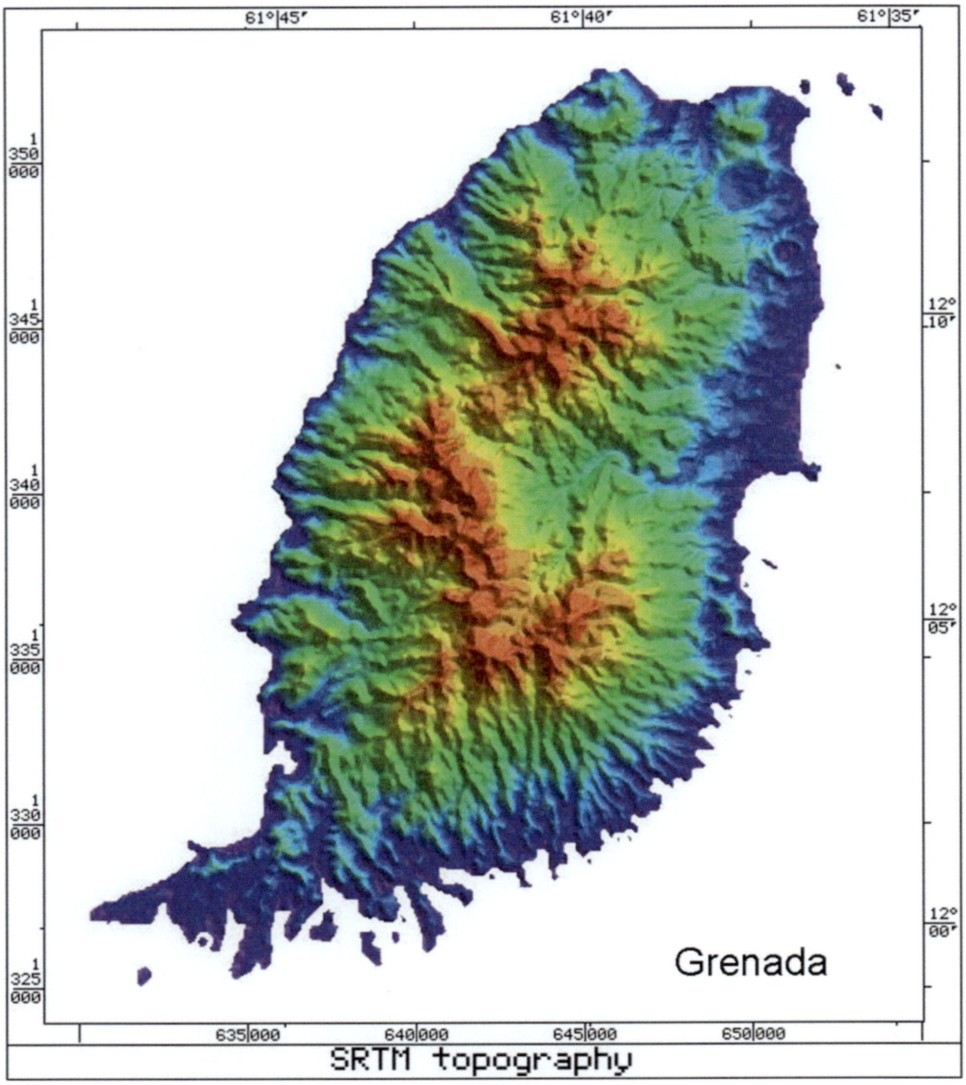

Fig. 16.4 Digital Elevation Model (DEM) of Grenada, with highest points represented by the *orange*, rust-colored region in the island's interior. Although rather steep in the interior, notice the gentle slopes on the islands east (windward) and southern sides (*purple* and *dark blue-colored regions*). Carriacou and Petite Martinique share a similar topography as Grenada, with a mountainous interior and scattered, flatter areas of sediment accumulation leading to beach creation. DEM courtesy of http://caribbeanvolcanoes.com/radar-grenada/. Retrieved Oct 2016

archaeological sites scattered predominantly on the Island's Eastern coast—though some locals claim to be of Carib or other Amerindian ancestry.

The tri-island nation of Grenada has a humid tropical marine climate, characterized by an average temperature range of 24–30 °C. Little variation in length of day, temperature, or relative humidity occurs due to its near-equatorial latitude. The dry season generally lasts from January to May, and the wet season from June to December. During the wet season, residents on Carriacou and Petite Martinique, which have no rivers or streams, collect rainwater for drinking (AQUASTAT 2012). As with other Caribbean Island Nations, Grenada can be dramatically affected by natural events. Although it lies at the cusp of the Caribbean Hurricane Belt, 50-year events continue to wreak havoc. In recent history, Hurricanes Emily (2005), Ivan (2004), and Janet (1955) devastated the Island, and some areas are still recovering from the effects of Emily and Ivan more than a decade later. While large volcanic events on the islands have ceased, geothermal activity and an offshore underwater volcano, Kick 'em Jenny—which erupted most recently in 2015 and continues to rumble and cause the occasional earthquake—demonstrate continuing potential for volcanic impacts.

16.2 Geologic Setting

Geologically speaking, Grenada is largely composed of andesite domes, basalt flows, and the (sedimentary) Tufton Hall Formation (Fig. 16.5). While all three main islands share similar geologic compositions, the largest island

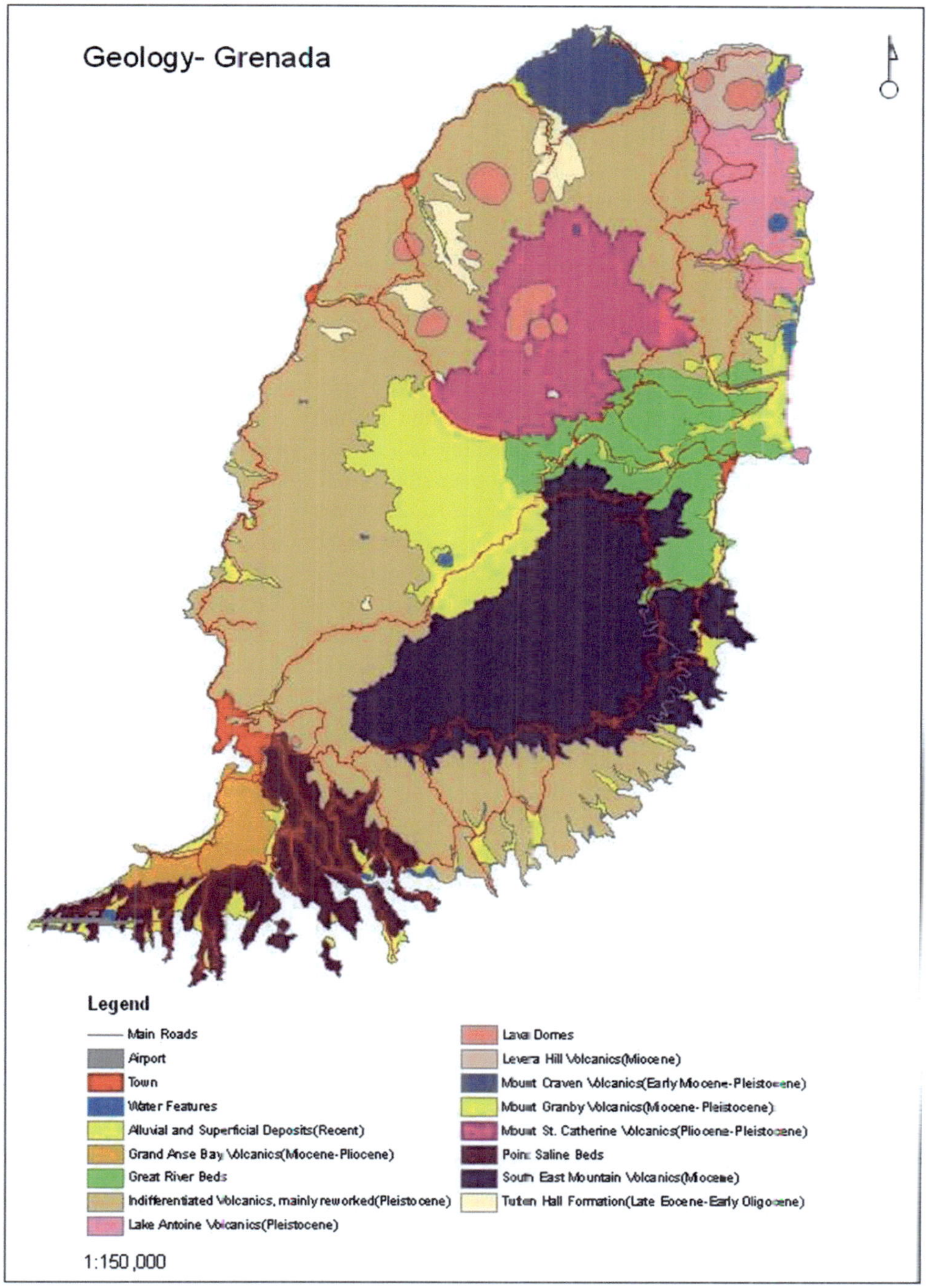

Fig. 16.5 Geologic map of Grenada. Even though Grenada was formed volcanically like other West Indian islands, it retains a geologic diversity such as lava (andesite) domes (salmon-colored ovals in the north of the map), Miocene volcanics, and Late Eocence–Early Oligocene sedimentary uplift regions (*light tan-colored* blotches in the northwest of the map). Map courtesy of Working Together for Grenada, an initiative sponsored by The Caribbean Natural Resources Institute (CANARI), available online here: https://sites.google.com/site/grenadabioconserv/home. Retrieved Oct 2016

features the most prominent geology. Northern Grenada holds the oldest volcanic rocks on the islands, with andesite domes of Mt. Alexander, Mt. William, and Mt. Rodney having been dated to 21 mya (Arculus 1976). Levera Hill and Levera Island (also known as Sugar Loaf) are examples of andesite intrusion through the Tufton Hall Formation, and Grand Etang Lake and Lake Antoine are well-preserved examples of explosion craters dating back to the Pleistocene (Arculus 1976). Volcanic activity in Grenada has mostly ceased, with one exception being the active submarine volcano Kick 'em Jenny, approximately 8 km north of Grenada and 20 km south of Carriacou. Additionally, a few hot springs remain on the island indicating continued geothermal activity. Generally, Grenada's coasts exhibit less relief than the interior geomorphology, having been subjected to mass movement episodes. Mudslides, landslides, lahars, and fluvial deposits have largely covered Grenada's southern regions, forcing a gentler topography. Similarly, eastern

Grenada is composed of volcanic deposits and sediment. While past and present watersheds and mass wasting events have helped shape the southern and eastern coasts into less rugged terrain, the western coast still displays the steeper landscape typical of Windward Islands due to asymmetric volcanic eruptions and windward–leeward differences (Grenada Forestry Department 1988).

16.3 Landforms

Analyzing Grenada's geomorphologic past represents a challenging exercise for even the most well-versed geomorphologist. Like other tropical islands, vegetation is dense and rock decay has worked its magic creating (sometimes thick) soils. Grenada also hosts great topographic relief, making stratigraphic recreation problematic. Additionally, between the Eocene (50 mya) and Miocene (~ 25 mya), the Lesser Antilles volcanic arc underwent extreme geologic stresses, twisting, buckling, and warping. Though these activities occurred mostly underwater, the result was a fresh reorganization of geology, driven by still-active volcanic processes (Arculus 1976). Today, Grenada's current volcanically dominant landscape remains underlain by this Lower and Upper Miocene basement sedimentary rock, known as the Tufton Hall Formation, which formed in the then-shallow sea and consisted of mainly mud, sand, and silt. Between Tufton Hall stratum, volcanics have been found (Arculus 1976), leading geologists to believe that volcanic activity was occurring here even during the Ologicene (~ 37 to 25 mya). Uplifted, folded, and faulted in many locations during this ancient timeframe, remnants of the Tufton Hall Formation can still be seen on the surface, primarily in the northwest of the main island and near Annandale Falls.

16.3.1 Grenada's Volcanism and Tectonics

While some studies focused on geochemistry tectonics are available (Saunders et al. 1985; Urzua et al. 2015; White 2015; Speed et al. 1993; Tomblin 1974, 1975; Weeks et al. 1971), gaining precise geologic understanding of Grenada is still difficult due to a host of factors: changing volcanic centers, rock type/age mixing, unknown rock sources, and inter-event erosion, to name a few. Arculus (1976) wrote perhaps the most definitive work, though he also acknowledges that much data is missing. According to Arculus (1976), Grenada's greatest volcanism occurred during the Pliocene (5–2 mya) and into the Pleistocene (2 mya–10,000 ya). Pyroclastic events, basaltic and andesitic lava, and intense volcanic eruptions associated with the Mt. Granby and Mt. Qua Qua regions happened during this timeframe. The geologic record displays these sometimes-furious events in the transition from basaltic and andesitic lavas, exhibiting the eruptions' cyclical character. Southern Grenada showcases the lava flows, tuff deposits, and lahars associated with this timeframe, though the source evidence for these events is probably buried (Arculus 1976; White 2015). These events, however, most likely resulted in the comparatively mild relief of the island's southern portion (Arculus 1976; White 2015).

Mt. Alexander features one of only a few examples of exposed limestone in Grenada, providing strong evidence for geologic uplift after the Pleistocene, as the area has limestone soils not present in other parts of the island nation. Iron and aluminum oxides dominate soils in the interior of the island, giving the latosols distinct red and yellow-red colors. These nutrient-poor soils require close management as they are fragile and easily disrupted (Grenada Forest Department 1988). Arculus (1976) speculated that perhaps Grenada's most recent volcanic massif is Mt. St. Catherine, which probably had its genesis during the Pliocene, but lasted throughout the Pleistocene. The dacitic and andesitic lava flows located northwest of Mt. St. Catherine's summit lay on top of extrusive basaltic lavas, while to the summit's west, some of the highest volume pyroclastic flows can be found (Arculus 1976).

Grenada's high volcanic activity ended violently, with massive explosions around the island, as evinced by the large calderas of Lake Antoine, Grand Etang, and the St. George's area. St. George's harbor and its neighboring Port Louis represents one of the most unique landforms on Grenada, being one of a few examples worldwide of a double caldera (Fig. 16.6; Hearn 1890). The inner harbor was once swampy, and blocked from access by baymouth bars, but these were originally removed by the Colonial French to improve marine access, and dredging continues to this day (Martin 2007, 2013). Arculus (1976) notes the Lake Antoine caldera (in the island's north) as being Grenada's best example of a true tufaceous ring, while the (rare) double caldera of St. George's harbor and Port Louis give Grenada its two main natural harbors and a large source of gravel (scoria) used in construction. Perhaps the most obvious caldera on the island, Grand Etang, represents the youngest volcanic structure, forming from three closely spaced events around 12,000 years ago (Arculus 1976). Recent history records scant volcanic activity on Grenada, except for its geothermal activity in the form of hot springs. While most of these are found around the Mt. St Catherine area, a few

Fig. 16.6 This image (created from three separate photographs "stitched" together), taken from Fort George, displays a rare double caldera—making the ports of St. George's Harbor (bay on the *left*) and Port Louis (bay on the *right*) possible. St. George's Harbor, locally known as the Carenage, was naturally deep and used since Colonial times for commerce and visitors alike. It also hosted the island's first cruise ship terminal until the early twenty-first century, when it was moved to the Esplanade, located on the other side of the hill. Commercial vessels and ferries still depart from St. George's Harbor. Port Louis, originally a swamp dredged by the Colonial French, is now the main marina and harbor for boating tourists to moor. Photos by C. D. Allen, 2016, stitching of panorama by K.M. Groom, 2017

others are known near River Sallee and Peggy's Whim Springs—the former being more sulfurous, the latter not so much, and both of which are more lukewarm than hot.

16.3.2 Underwater Volcanism

Approximately 8 km north of Grenada, however, between the main island and Carriacou, lies an active submarine volcano, Kick 'em Jenny. Roughly 180 m below sea level, it has erupted at least a dozen times since record keeping began nearly a century ago. Thought to be the only active submarine volcano in the Eastern Caribbean, Kick 'em Jenny is 1300 m high and 300 m wide at the summit (Global Volcanism Program 2005; Gilser 2006; Seismic Research Center 2016). Although no eruption has broken the surface since 1939, there is still cause for concern, as an eruption in 2001 made the ocean surface bubble, prompting more monitoring and studies (Gilser 2006). In July 2015, the University of West Indies Seismic Research Center noted highly unusual activity (extreme bubbling, regular earthquakes), leading them to issue a warning about its impending eruption in the near future.

16.3.3 Geomorphological Features

Numerous examples of Grenada's geomorphological past remain evident around the island, and for the casual observer, many can be seen readily along most roadsides. Driving the main road around the Island, for example, affords opportunities to view classic lahars, prominent geologic contacts (Fig. 16.7), and pyroclastic flows, as well as coastal features such as tombolos, sea stumps (Fig. 16.8), and long stretches of beach (Fig. 16.9). Moving through Grenada's interior, baked soils (Fig. 16.10), laterite, and (small) feral relief is visible from a vehicle. The keen observer will also note occasional buried soils, pillow basalts (Fig. 16.11), sea caves (Fig. 16.12), sea arches (Fig. 16.8), and the aforementioned andesite cones and calderas (Fig. 16.13).

On the northeast side of the island, Lake Antoine is a mere 6.5 m above sea level, with a maximum depth of 30 m. The lake was formed 12,000–15,000 years ago during the final stages of volcanic activity in the area and is approximately 16 acres in size, second only to Grand Etang Lake (UNDESA). Lake Antoine is a national landmark in Grenada located near the River Sallee Boiling Springs, south of Levera National Park. The six springs reach varying temperatures, but never more than of 35 °C, and the largest measures 5 m wide and 2 m deep. Though undeveloped, these springs are designated as a protected natural landmark and have a religious significance for some local inhabitants, as evinced by their use by a local church for baptisms. There are other sulfur springs, formed by the combination of rainwater and volcanic hydrogen sulfide, and scattered around the northern end of the island. Though also referred to as boiling springs, most are tepid, but indicate continued geothermal activity.

From Levera, on the north end of Grenada, the confluence of the Atlantic Ocean and Caribbean Sea is visible from the shore. The Atlantic is frequently choppy with a great deal of turbidity and suspended sediment that renders the water a kind of deep blue opaqueness. Lower turbidity and suspended sediment loads mean the waters of the Caribbean

Fig. 16.7 Geologic contact between volcanic tuff (*bottom*, *lighter-colored* layer) and lahar deposits (*top*, rocky layer), common in the southeastern areas of the island. This contact can be seen along the La Sagesse road in St. David's parish. Photo by C.D. Allen. May 2013

Fig. 16.8 Common coastal features along Grenada's rugged, northern horn coasts: a small sea stump (*background*, in sea) and sea arch (*foreground*, *center*) just off the north coast outside of the town of Sauteurs. Photo by C.D. Allen, 2009

have greater visibility than the Atlantic (Crask 2012). The striking difference between the east and west side of the island is easily visible on a map (Fig. 16.1). The island's windward side is typified by the consistent Northeast Trade Winds that stunt and twist vegetation as well as create rough sea currents, steep cliffs, and sea caves (Fig. 16.14). Much of the plant growth is scrub, including cacti in such profusion that the French dubbed an area on the south end of the island

Fig. 16.9 A view from Duquesne bay, looking northeast. Visible in the far background, Ronde Island lies ~9 km off Grenada's north coast, and the underwater volcano, Kick 'em Jenny lies approximately 5 km west of Ronde Island. Photo by C.D. Allen, 2012

L'Anse aux Epines ("Bay of Spines"). The leeward side is typified by calm waters and sheltered bays, allowing for the formation of coral reefs and uniform beaches with well-sorted, well-rounded sand grains (Fig. 16.15). White sand beaches result from (mostly) crushed shells, coral, and other sediments. Several beaches also exhibit a combination of the two types with the slightly larger black sand grains settling on top of the white sand (Destination Guide nd), while others, such as Levera Beach, host white sand adjacent to a lava flow, due to extreme currents in the region. Black sand beaches on Grenada are the result of (mostly) decaying basalt flows being eroded over thousands of years and re-deposited on the shore.

Each of the islands—Grenada, Carriacou, and Petite Martinique—were shaped by their volcanic pasts, and some of these striking landforms have been recognized and protected by the government. Officially, the tri-island nation contains four national parks: Grand Etang, Mt. Saint Catherine, Levera, and High North (on Carriacou) (Huber et al. 2012). There are also 31 nationally protected natural sites, both inland and marine, ranging from Concord Falls Natural Landmark to Mt. Hartman Dove Preserve (Arce 2008; Huber et al. 2012). Additionally, two historic sites in the city of St. George's are on the tentative list for UNESCO designation. Mt. St. Catherine National Park boasts Grenada's tallest point (840 m), as well as three other points above 610 m: Mt. Granby, Mt. Lebanon, and Mt. Sinai. Grand Etang National Park hosts some of the most dramatic features on the island, including Grand Etang Lake—a 36-acre freshwater lake in a caldera 550 m above sea level with an average depth of 6 m (Fig. 16.16). Grand Etang Lake provides the majority of freshwater for the island, and numerous streams feed into and run off the lake (Fritz et al. 2011). Seven Sisters Falls, considered one of the most beautiful waterfalls on the island, is also located within Grand Etang National Park.

16.4 Landscape

16.4.1 Historical Occupation

Prior to the influx of European influence in the 1700s, Grenada had not seen widespread landscape transformation. The Arawak and, subsequently, Carib peoples left a relatively minor agricultural and developmental footprint on the island, and there is no evidence of significant deforestation pre-European contact. The island was first noted by Christopher Columbus in 1498, and he named it *Concepción*. Grenada's mountainous green terrain led to its renaming to Granada by Spanish sailors, but despite its visual familiarity, the Spaniards were unable to establish a colony on Grenada due to violent resistance from the Carib Amerindian. The French were the first European power to

Fig. 16.10 Geologic contact between Pleistocene lava flow and paleosol. Here, as hot lava flowed over the then-established soil, it "baked" and oxidized the soil, turning the soil to a *rusty orange* color. This feature can be found along River Road which runs in a generally east–west direction, just south of the National Stadium. Photo by C.D. Allen, 2009

establish a settlement, in the south, by bartering to buy property from the Caribs. They quickly overstepped their welcome by attempting to expand their control to the whole island, beginning a period of sustained conflict with the Caribs. This conflict culminated in a battle during which the remaining Caribs are said to have jumped to their deaths rather than be taken as slaves. Although little historical evidence exists to support this claim (it is more likely they jumped into the ocean and swam to safety to escape the French forces), a monument to their plight exists on a cliff in the town of Sauteurs, which takes its name from the French, *Le Morne de Sauteurs* ("Leaper's Hill")—the location where the mass suicide supposedly occurred (History & Culture nd). Even though no physical evidence of the "leap" exists, "The site attracts visitors who are probably awed by the self-sacrifice but the outcome nonetheless evokes sadness and anger" (Martin 2007, p. 140).

16.4.2 The Rise of the "Spice Isle"

During the subsequent French takeover, Grenada's name was changed to *Grenade*, and finally, once under British control, formalized to Grenada (*GRUH-nay-duh*). During

Fig. 16.11 Exposed pillow basalt on the coast of Fort Juedy Point. Photo by C.D. Allen, 2009

Fig. 16.12 Sea cave at Fort Juedy Point with considerable-sized swells entering the cave, various terraces in the foreground, and people in the background for scale. Photo by C. D. Allen, 2009

the French–British tug of war over the island, what is now St. Georges was settled, Fort George was built, and indigo, cocoa, and coffee were brought to Grenada. Fort George, originally built as Fort Royale by the French, is an example of colonists responding to the natural landscape. Built on a volcanic spire, the fort was constructed back-to-front, to defend against attacks from the interior (Kilgore and Moore 2011). When the British took stable control of Grenada in 1783 as part of the Treaty of Versailles, they brought sugarcane and slave labor to the

Fig. 16.13 Iconic andesite cone of Sugar Loaf off the coast of Levera Beach on the northern tip of Grenada. Photo by C.D. Allen, 2011

Fig. 16.14 This image, taken from Galby Bay on Grenada's eastern coast, typifies the island's windward side coastal features, including low-lying, windswept vegetation (on the hillside, *right*), wave cut notches and rock fall debris (*right*), rough headlands (*center-left*), and a sea arch (*left-center*). At the headland's tip, a couple stands overlooking the Atlantic Ocean for scale. Photo by C.D. Allen, 2004

island Sugarcane, and the production of rum became the dominant form of economic production on the island. A water-intensive crop, sugarcane also requires a substantial amount of cleared land. While the first is not an issue for Grenada (average annual precipitation <1000 mm for the Nation), the latter has had lingering impacts on the island (AQUASTAT 2012).

In the mid-1800s, as the Far East spice markets were being decimated by pests, Grenada began experimenting with the growth of nutmeg (*Myristica fragrans*), and through

Fig. 16.15 Looking northward along the famous Grand Anse Beach on Grenada's southern west coast. The capital, St. Georges, is visible in the *upper left corner* of the image. Photos by C.D. Allen, 2009

trial and error discovered it to be a viable cash crop. Other spices, such as clove, cinnamon, allspice, and the Tonka bean (similar to vanilla), soon followed. Coffee was attempted, but never took hold, and eventually was replaced by cocoa. In the early twenty-first century, Grenada claims between one-fourth and one-third of the world nutmeg market (depending on the year), second only to Indonesia. The Island's other spice exports including cocoa are nominal on the world market, but function as a significant component of the Nation's economy. Cocoa, in the form of chocolate and its associated products, has become a boutique/niche crop for the island, with two main companies vying for market shares. Though it has yet to break into the world market at large, Grenada's chocolate has won several international awards over the years. Still, for such a small island to hold a disproportionate world market share of such an important spice (nutmeg), is remarkable, and Grenada prides itself on being *The Isle of Spice*.

16.4.3 Carriacou's Landscape

Carriacou, the second largest Grenadian island, stretches 1.6 km between Pegus Point in the south and Gun Point in the north. Just south of Gun Point lies Petite Carenage Bay, home to the Kido Ecological Research Station. Petite Carenage Beach, Mangrove Wetlands, Sparrow Bay, and Anse La Roche together comprise the Nation's largest nesting area for Hawksbill and Leatherback sea turtles, and the Petite Carenage Mangrove Wetlands represent the most developed mangrove ecosystem, while also serving as a bird sanctuary. Carriacou also hosts three marine protected areas: Sandy Island (off its west coast), Petite Dominique Protected Area (between the northeast coast of Carriacou and southwest of Petite Martinique), and Gun Point Marine Protected Area, which encompasses the northern tip of Carriacou (Caribbean Conservation Association 2003).

Carriacou rivals Grenada for snorkeling and dive sites, including Magic Garden, off Mabouya Island to the west of Carriacou, a dive site known for beautiful reefs and underwater volcanic activity (Nagel nd). Secluded Anse La Roche is considered the most scenic beach on Carriacou, and a popular snorkeling spot. Hillsborough Beach, however, with its white sand running the entire length of its namesake village (and Carriacou's largest populated village), proximity to amenities, and nearly transparent water makes it one of the more popular beaches on the island for snorkeling (Fig. 16.17; Attractions n.d.). Paradise Beach (also called L'Esterre), however, is located between Tyrrel Bay and Hillsborough and is the most popular local beach, spanning the length of the village of L'Esterre (~4 km). Mt. Carre (288 m, also called Chapeau Carre), the second highest point on Carriacou, is clearly visible from Tyrrel Bay (Crask

Fig. 16.16 Various features and attractions found in Grand Etang National Park. **a** Panorama of Grand Etang Lake in the center of the park. Steep slopes of the caldera surround the lake and thick reeds grow around the lakeshore. Photo by C.D. Allen, 2010. **b** One of the many lush hiking trails that transect the park, offering visitors the opportunity to immerse themselves Grenada's wilderness. Photo by K.M. Groom, 2012. **c** A closer view of Grand Etang Lake from the shore. Note the artificial retaining wall in the foreground intended to limit sedimentation and bank erosion into the island's main fresh water supply. Photo by K. M. Groom, 2012. **d** One of highlights of visiting the Grand Etang National Park is the possibility of feeding the wild Mona monkeys that inhabit the rainforest. Not native to the Caribbean, the Mona monkey population can be genetically traced back to a single pregnant female that was unintentionally transported to the island aboard in a slave ship from Africa. Photo by K.M. Groom, 2012

Fig. 16.17 Town of Hillsborough, Carriacou, as seen from the Princess Royal Hospital (Carriacou's only hospital and one of its main tourist vistas). Visible in this image are the island's lone airstrip (*upper center*), the main docking area/port authority (*center*), Hillsborough and Paradise beaches (*light-colored* sands running "up" from dock), and Sandy Island (the boomerang-shaped small island in the bay)—a popular day-trip snorkeling location. Photo by C.D. Allen, 2016

2012). High North Nature Reserve and National Park contains Mt. North (291 m), the highest point on Carriacou, (Natural, nd). The park has the least human impact of any area on Carriacou, and the ecology ranges from dry thorn scrub deciduous and evergreen forests to mangrove ecosystems (Huber et al. 2012).

16.4.4 Petite Martinique's Landscape

Like Grenada and Carriacou, Petite Martinique is also volcanic in origin. While significantly smaller than the other two islands, its interior is still mountainous, but unlike Carriacou and Grenada, it has no natural watersheds. There are three beaches on Petite Martinique, but only the one on the western shore, Main Beach, is open to the public. One of the other beaches is owned by a fuel company and serves as a way-lay station for their ships, and the third is restricted to government access (Home nd).

16.4.5 Expansion and Infamy

Housing has dramatically altered Grenada's landscape in recent decades. Since the late 1980s, Grenada has seen a steady influx of expatriates. While some of these people settled in the tourist hubs of Grand Anse and St. George's, others began searching for more secluded areas such as Westerhall and Fort Jeudy, and expanding into L'Anse Aux Epines. In the late twentieth and early twenty-first centuries, Grenada began to see a return-migration of locals who had made their fortune overseas and returned with disposable income. These factors have resulted in large swaths of land being deforested, hillsides being dug into, and even high-end resorts being built on smaller islands such as Hog and Calivigny Islands. Many of these homes are used only a few weeks each year by their owners and offered for rent during the year's remainder. Development has not been as apparent on Carriacou and Petite Martinique, though property is continually offered at reasonable prices.

One of Grenada's largest employers is St. George's University. Founded in 1976 as an American-run offshore medical school, it has since gained an international reputation for its programs, and now offers a full suite of university courses, including undergraduate studies. SGU had humble beginnings with a small campus located on Grand Anse beach, but later expanded to True Blue (and Prickly Bay) on the island's southwest peninsula, where the majority of its buildings are housed and classes are taught today. Since the 1990s, the campus has continued to expand, presently covering over 42 acres and hosting more than 50 buildings and nearly 6500 students in residence. The University is perhaps most remembered as being the focus of *Operation Urgent Fury*. In 1983, at the request of several Eastern Caribbean States, Ronald Reagan used the evacuation of mostly American students as a pretense for intervening in a bloody military coup, led by his former second-in-command, Bernard Coard, that ended with the execution of Prime Minister Maurice Bishop and several of his cabinet members. Memorials of the 1983 tragedy can be found around the island, and Grenada's Thanksgiving Day celebrated each year on 25 October—when the intervention began—serves as remembrance of that tumultuous time.

16.5 Heritage and Tourism

The impacts of tourism and management of heritage are often overlooked, especially in connection with geomorphologic value (Goudie 2002). Grenada depends heavily on tourism to help sustain its economy and provide the means to improve infrastructure, but management of its limited resources has often been misguided and detrimental, requiring improved practices that seldom come to fruition. While resort complexes still do not resemble the mass tourist locales of other islands (a consequence of Bishop's 1979 "New Tourism"), the tourist sector in Grenada continues to expand, and there is concern among some parties that large corporations will begin to put pressure on governmental officials to allow taller buildings. Currently, Grenada's landscape is attractive to tourists—not only for its beautiful beaches and splendid ocean activities, but also for its inland waterfalls and cultural heritage.

16.5.1 Beaches

Beaches represent a significant tourist draw for Grenada. Although the main island claims no fewer than 40 beaches, a three km stretch of white sand, Grand Anse beach, represents the island's most visited. Nearby Grand Anse are several other beautiful beaches, deserving of their names: Morne Rouge, Magazine, Parc a Beouf, and Pink Gin. More secluded and pristine beaches easily accessible on a short drive include La Sagesse, Bathway, Levera, and Sauteurs. With a little more effort, deserted and unspoiled beaches can be found, though sometimes they are named after their respective Bay or town/village, such as Grand Mal, Flamingo, Duquesne, Antoine, La Tante, Bacolet, and Granby to name several (Fig. 16.18).

As Grenada's Grand Anse beach becomes more crowded, tourists are looking for more unique beaches to spend their time. Two very different beaches offering a more undeveloped feeling are La Sagesse and Levera. La Sagesse, a "Nature Center" with nice lodging and a locally sourced restaurant, is nestled in a secluded bay with nearly a full km of gentle sloping beach. Though La Sagesse proper is shaded with palm trees and a couple cabanas, several hikes around the Nature Center are well marked and have outstanding views of the surrounding landscape, including wave cut platforms, small headlands, and ancient lahar deposits, as well as an abundance of tropical flora. The pristine setting

Fig. 16.18 Examples of Grenada's diverse beaches, in *both* color and surrounding landscapes. **a** Black sand beach at Black Bay as seen from off shore. The large basalt boulders on the *right-hand side* of the photograph exemplify the volcanic rocks that serve as parent material for the beach's black sand. Photo by C. D. Allen, 2013. **b** Mixed sands at Levera Beach. Golden colored sand with scattered patches of black sand occupy the center foreground of the image with rough basalt terrace covered with red-colored lithified scoria/lava flow in the background. Photo by C.D. Allen, 2012. **c** Tranquil white sands of the secluded La Luna Beach. The shallow waters display ample coral reefs (possible source for the white sand), and the layered volcanic lava flows that compose much of the headlands along the western coast are visible in the background. Photo by C.D. Allen, 2004

lures even the most hectic visitor to "lime"—a term Grenadians use to mean relaxing and being with friends and family.

At the other end of the island, Levera Beach is about as far away from the main tourist hub as a person can get. The beach also serves as a nesting area for the endangered leatherback turtle, adding to the area's lure. Though seemingly untouched, clear cutting in the early 2000s for a resort golf course increased water and sediment runoff in the area, and with increased interest from other private stakeholders for resorts on Grenada's northern shores, the probability of increased coastal erosion remains high. If the coastline is altered too drastically, it could greatly reduce the available area for nesting sea turtles (Maison et al. 2010). Likewise, the removal of ecologically important mangroves in the area increases the speed of erosion and destroys habitat for multiple species of land and water fauna (Moore et al. 2015). While currently stalled, should development continue in this region, it would be imperative the areas be carefully managed with insight from locals, authorities, environmental groups, and the resorts themselves.

16.5.2 Ocean-Based Tourism

An increasing threat to ecosystem stability and maintenance of natural resources are snorkelers and SCUBA divers visiting Grenada's beautiful coral reefs. While snorkeling and diving excursions represent a very important mode of revenue for the island, they also represent ways in which further degradation can occur. To help partially mitigate visitor impact to this fragile ecosystem, an underwater attraction—the world's first underwater sculpture gallery—was created in 2006 just off Grenada's west coast in Molinère Bay.

The artist, Jason deCaires Taylor, has since become renowned for his work and several subsequent underwater sculpture parks of his creation can now be visited at various locations around the world. The statues in Molinère Underwater Sculpture Park provide an artificial reef for wildlife and plants after much of the area's natural coral reefs were damaged and destroyed following hurricanes Ivan and Emily (Gocova 2013). Whether or not this unique attraction helps offset the damage caused by the hurricanes requires further study, as the benefits may be offset by the increased rate of visitation from divers to this unique area.

16.5.3 Waterfalls

Waterfalls represent another frequently visited geotourism attraction on Grenada (Fig. 16.19). Most popular among these are Annandale (Fig. 16.19a) and Concord Falls (Fig. 16.19b), though Royal Mt. Carmel Falls (Fig. 16.19c) is also an easy and rewarding hike. A fourth waterfall discovered just this century, St. Margaret Falls—locally known as Seven Sisters—sits near Grand Etang National Park but requires a guide and some scrambling. Annandale Falls has relatively easy road access and a short, self-guided, paved walking path from the small parking area to the falls. The road to Concord Falls, on the other hand, remains less developed—though recent improvements have been made to sections affected by landslides—and is accompanied by several souvenir stands and a snack bar. At Concord, easy views of the falls can be found on the built decks and stairs leading to the base, and visitors can descend all the way to the bottom for a different view, or for a dip in the water at the falls' base. With increased tourism, however, comes increased degradation, unless policies to the contrary are enacted. Concord's landscape was first altered by the addition of cement stairs and small shops overlooking its pristine location. To help ease Concord's congestion from multiple simultaneous tour groups, a large parking area above lower Concord Falls was completed in 2015. Though the project removed/rearranged a few tons of topsoil and altered both forest cover/ecology and tributary hydrology in the short term, the site recovered quickly and is now better equipped to handle large tourist groups, with streamflow turbidity and discharge back at normal levels, and surrounding vegetation replanted with thriving native flora barely a year later (Fig. 16.9b).

Containing two sets of falls, including the island's tallest, Mt. Carmel Falls' "entrance" begins at one of the main road's many hairpin turns on Grenada's eastern side. The short walk wanders the hiker through private, cultivated agricultural area including bananas, oranges, star fruit, nutmeg trees, and more. Visitors here are rewarded with cool, fresh water, and multiple pools for swimming.

Fig. 16.19 Three of Grenada's most famous waterfalls. **a** Annandale Falls from the nature walk and gardens leading to the falls. A half-kilometer to the northwest of the falls, an exposure of the Tufton Hall can be seen. The Grenadian flag in the *upper left* and viewing platform in the *lower right* of the image provide scale. Photo by C.D. Allen, 2015. **b** Concord Falls from the steps leading down to the river. From this pristine vantage point, the large parking lot, built above the fall in 2014, is no longer visible. Photo by C.D. Allen, 2016. **c** Mount Royal Carmel Falls, the tallest waterfall on Grenada, during the dry season. In the wet season, the falls span the entire knick point. Note the people for scale. Just downstream is a much lesser-graded waterfall with slick algae covered rocks, making a natural waterslide popular with tourists and locals alike. Photo by K.M. Groom, 2014

16.5.4 Amerindian Heritage and Tourism

Rock art (petroglyphs) on the island interests some tourists and also provides an important link to Grenada's Amerindian Heritage. Commonly referred to as "Carib Stones" and often found alongside cupules (grind/work stones), these engravings are named for the most recent Amerindian inhabitants (the Carib Indians), although they were more likely created by the Island Arawak/Taino people (Martin 2007, 2013; Thomas 2010; Dubelaar 1995; Marquet 2009). While several petroglyph sites exist on the island, management efforts have been haphazard at best, with best intentions often doing more damage than good to these priceless cultural heritage resources.

For example, after Hurricane Ivan (2004), funds were procured to help mitigate decay of the motifs by erecting a small cement wall around a large petroglyph panel at Duquesne Bay. While a good idea in theory, the enclosure has had the opposite effect of enhancing decay because it now traps sea water from storm surges as well as effluence from surrounding houses (Allen and Groom 2013a, b), though recent attempts have been made to remedy this problem. Similarly, a motif-abundant site near Mt. Rich previously in overall good condition was "cleaned" in mid-2015. When left to the elements, rocks naturally develop a patina (rock coating). These coatings often develop over hundreds and sometimes thousands of years depending on the environment, and usually serve to protect the rock from decaying. In an apparent effort to make the Mt. Rich petroglyphs more visible to tourists, local "caretakers" inadvertently scrubbed the rock coating off the stones, potentially leaving the motifs even more susceptible to rapid decay (Fig. 16.20). Though long-term outcome is still being studied, this act—done with the best of intentions to help a struggling economy but lacking appropriate technique and education—has possibly weakened the host rock and enhanced decay processes, possibly erasing a priceless heritage resource quicker than normal conditions. In 2016, a reassessment was conducted of the host rock geologic stability, and it was found to be recovering better than expected. To compound management issues, Grenada's petroglyph sites have no official protection and remain highly under studied. Still, monitoring efforts continue keeping a watchful eye of these priceless cultural heritage resources.

On Grenada's east coast lies Pearls Airport—once Grenada's main airport until Operation Urgent Fury in 1983—an archaeologically rich site of Arawak and Carib artifacts. While a few active digs took place before and after Pearls Airport ceased operations, official digs have not really occurred after 1990. The area holds high volumes of pottery sherds and figurines (Keegan 1991), but a large strip of the area remains covered by a runway that is now only used for occasional drag races by locals. While archaeological artifact sale is officially illegal in the Caribbean, some entrepreneurial locals dig their own pits looking for well-preserved pieces of pottery to sell to unsuspecting tourists that visit the airport for primarily historic reasons: Pearls contains a

Fig. 16.20 Comparison of the Mt. Rich petroglyphs before and after being "cleaned". **a** Large boulder containing the majority of Mt. Rich's petroglyphs, surrounded by thick vegetation. Dark rock varnish covers most of the rock's surface. Photo by C.D. Allen, May 2014. **b** The same boulder from a similar vantage point a year later. Much of the protective rock coating has been removed (compare coloration of the main boulder in the background with the smaller boulder in the foreground, which was not cleaned) leaving the petroglyphs more visible but completely exposed to potentially more intense decay processes. Large portions of the surrounding vegetation has also been cleared and stacked for burning. Photo by C.D. Allen, May 2015

derelict USSR (Soviet) and Cuban aircraft left behind after *Operation Urgent Fury*. While a seemingly harmless endeavor to dig up artifacts, doing so severely impacts site integrity and reduces the number of artifacts available for study. Currently, there is one official dig in the area at La Poterie, north of Pearl's, but no plans to preserve what remains there or at Pearl's. Still, that locals are becoming aware of these cultural heritage sites as potential tourist attractions, speaks to the growing economic importance tourism is having on the island, and means more educational outreach must be conducted if protection/conservation/management is desired.

16.5.5 Tourism on Carriacou and Petite Martinique

Carriacou and Petite Martinique receive far fewer tourists and house a much smaller permanent population. As tourism to Grenada increases, however, it remains like that visitation to these smaller islands will also increase. Ferries provide daily back-and-forth access between Grenada and Carriacou (and Carriacou to Petite Martinique), and during tourist season, the ferry company increases the number of return trips. This trend shows no sign of waning and may further alter the seascape between, as well as on the islands themselves. More construction will be required to accommodate visitors for overnight stays on Carriacou and Petite Martinique, which may lead to increased erosion of coastline, forest, and stone. As tourism remains Grenada's primary economic sector, it is important that the nation continues to develop that industry and strikes a balance between expansion and careful management of resources and habitats.

16.6 Hazards

Like other Lesser Antillean islands, Grenada has been, and will continue to be, susceptible to a myriad of hazards. At 12° north of the equator, the island nation lies on the hurricane belt's edge, but still impacted by the Inter Tropical Convergence Zone's (ITCZ) varying oscillations north and south over the equator throughout the year. The ITCZ is characterized by a band of clouds that circle the globe and produces often erratic weather patterns alternating between the static and the violent. Grenada's location leaves it vulnerable to tropical depressions, tropical storms, and the occasional high-category hurricanes. Additionally, the island nation is at risk of experiencing tsunamis and volcanic activity.

16.6.1 Hurricanes

Hurricanes and tropical storms are frequent in the Caribbean Sea, concentrated largely in what is known as Hurricane Alley. These storms begin formation at or near the equator off the west coast of Africa over warm ocean currents, strengthening as they move west and north. Grenada sits at the border of Hurricane Alley and, as a result, avoids many of the hurricanes that are directed north toward the Gulf of Mexico and east coast of the United States by the Coriolis Effect. However, several hurricanes have hit Grenada in recent history with devastating results. As a developing nation, Grenada is ill-equipped to recuperate from most high-category storms, and each major occurrence leaves the nation more susceptible to detrimental impacts of the next.

The most catastrophic event in recent history was Hurricane Ivan in 2004. The hurricane initially formed as a tropical depression, strengthening to a tropical storm on 03 September 3. On September 05, 2004, approximately 1850 km east of Grenada, the storm strengthened and was classified a category 3 hurricane on the Saffir-Simpson hurricane wind scale. The rapid strengthening of the storm at its latitude in the Atlantic Basin is a rare occurrence and, as the hurricane continued west, it weakened briefly and appeared to begin a northward path, but then shifted toward the Grenada with increasing ferocity. Nearly caught off guard, safety and mitigation processes were not in place when the newly strengthened now-category 3 (and building) Hurricane Ivan hit the nation on September 07, 2004. Originally believing Ivan had been downgraded and posed a less significant threat, residents were unprepared for its ferocity. Damage was severe and far-reaching, affecting every part of the tri-island nation, killing 39 people, and causing destruction to more than 80% of the island, resulting in more than $815 million USD damages—more than twice Grenada's GDP—prompting locals to rename the storm "Ivan the Terrible" (World Bank 2005; Ministry of Finance 2005). Barely a year later, in 2005, Hurricane Emily reached the northern end of Grenada as well as Petite Martinique and Carriacou, causing further damage and an additional death. Previous twentieth Century hurricanes had little effect on Grenada, other than Hurricane Janet (1955), which killed over 100 people and caused widespread damage to the Nation (Associated Press 1955).

But Hurricane Ivan affected more than infrastructure, as watersheds were effectively rearranged by the extensive damage which affected 90% of the vegetation, some of which is still recovering over a decade later (World Bank 2005). Additionally, 60% of cocoa trees, 60% of nutmeg trees, and nearly the entire banana crop were lost, as well as more than 50% of Grand Etang's rainforest canopy (World Bank 2005). The loss of nutmeg crops is particularly devastating as Grenada is responsible for more than one-third of the world's nutmeg supply, and nutmeg trees take over a decade to produce usable fruits. In conversation with a still-recovering nutmeg co-op outside of Gouyave, it was found that one plantation went from employing more than 100 people in 2004 to employing five after Ivan the Terrible (Allen 2012).

While Hurricane Ivan was a dry storm, the widespread demolishing of land cover led to extreme flash flooding during Hurricane Emily the following year. Nutrient-rich soils were moved or stripped away completely during this time, altering possible future crop yields. While damage from Hurricane Ivan was more moderate on Petite Martinique and Carriacou, these islands sustained heavier damages during Hurricane Emily. The potential for future damage and alteration to Grenada's islands from hurricanes is great. While large storms are more common north of the small island nation, increased erratic weather patterns as a result of climate change may change that, even though locals note a roughly 50-year cycle between major hurricane events. Still, future storms will increase erosive processes as a result of altered vegetation cover and damage to existing watersheds.

16.6.2 Mass Wasting

Deforested areas are far more susceptible to landslides and other erosion events, serious impact from flooding, and loss of top soil crucial for agricultural production. On a mountainous island such as Grenada, this can have serious consequences for development. Landslides can damage homes and agricultural land, and the potential for mass wasting events heightens concerns about infrastructure during hurricanes (Howell 1991). Of particular concern involving deforestation are precious mangrove forests. Mangroves not only provide a critical coastal habitat for a variety of species, but also serve to strengthen the coastline and prevent serious erosion from occurring during storms and storm surges. Beach sand also remains important for enhancing coastline strength but is threatened by sand mining efforts. The two resources—sand and mangroves—help buffer the coastline from wave impact, absorbing energy and limiting damage inland. Much of the western coast, which faces the powerful Atlantic Ocean, has been significantly deforested. This subjects the main road to erosion from waves at far greater frequency than would be the case if natural protections remained in place (Howell 1991).

16.6.3 Sand Mining

The Grenadian government recognized the negative impacts of sand mining on the environment, and in 2009, the practice was banned. Unfortunately, the resource limitations on such a small island meant that the need for sand in construction-pitted development interests against environmental concerns. Predictably, development won out. In 2013, Grenada reversed the ban on sand mining for two beaches on the island. The growth of the tourism sector and, to a lesser extent, business and housing has taken precedence over potentially irreversible environmental impacts. This move ignores the fact that sand mining puts Grenada's number-one tourist draw, its beaches, in crisis (Local News 2013).

16.6.4 Volcanic Hazards

Volcanic and seismic activity also poses a threat to Grenada and surrounding islets. Several extinct volcanic remnants exist on Grenada, but Kick 'em Jenny, Grenada's submarine volcano continues to cause concern. With increased activity of Kick 'em Jenny (its most recent eruption, as of publication, was July 2015; Seismic Research Center 2016), remote sensing and observation have been employed to monitor it. A safety perimeter has also been established to prevent divers and snorkelers from entering the area within 1.5 km of the cone (5 km during times of high activity), and boats must also steer clear of the area. An eruption, however, can alter the buoyancy of local sea water, posing a threat to passing ships (Gisler et al. 2006).

Eruptions on record have mostly been minor with little area disturbance, but a large eruption has the potential to breach the surface and release ash. It is not likely that Kick 'em Jenny will cease activity in the near future as its location on the Lesser Antilles arc acts to sustain conditions conducive to continued small eruptions (Seismic Research Center 2016). Submarine volcanoes have the potential to alter the landscape through continued eruption and increase in seamount size. Those that continually increase in size can breach the ocean's surface and form new islands. Once a volcanic island is formed, future eruptions would then impact nearby nations in perhaps severe ways through volcanic activities. Kick 'em Jenny's location near the equator could have high consequences for future global climate should the seamount breach the surface, as ash clouds block incoming solar radiation, and near the equator where solar radiation is the most constant, such affects are a prospect. Currently, however, research suggests that the potential for tsunami or catastrophic eruption events related to Kick 'em

Jenny is low, and it is most likely that impacts from future eruptions will remain minimal (Boruff and Cutter 2007; Seismic Research Center 2016).

16.6.5 Climate Change

Climate change will also be a factor in the hazards Grenada faces in the decades to come. Current models suggest a continued upward trend in average temperatures in the Caribbean, which is likely to cause longer dry seasons, erratic weather patterns, higher ocean temperatures, and an increased danger of drought (Cashman et al. 2010). Rising ocean temperatures have been attributed to fishery collapse in nearby Venezuela, and Grenada is already experiencing reduced populations of bait fish (Rustad 2012). The islands depend on fish resources, and a collapse could be calamitous. In addition to fish resources, the islands depend heavily on agriculture, which will be strained as temperatures rise and weather patterns alter. As tourism increases, further strain is placed on an already taxed water infrastructure that fluctuates with wet and dry cycles, as well as seasonal tourism. Additionally, sea level rise could dramatically reshape the islands, reducing available land for residents, visitors, and agriculture. According to a brief released in partnership between the Government of Grenada and the German Agency for International Cooperation (2014), a rise in sea level of 20 cm would reduce Grenada's beach areas by at least 40%. Warming waters will also contribute to the expansion of lionfish and other invasive species in the area.

16.6.6 Coral Reefs

The nation is also seeing degradation of coral reefs, perhaps due to climate change. Grenada has a substantial insular shelf, with an area of 3100 km^2. Because insular shelves tend to be shallower, and therefore receive more sunlight, and they tend to host the most expansive and diverse reef ecosystems (Howell 1991). There are 160 km^2 of coral reefs around Grenada, most extensively on the north, east, and south sides of the island, where the insular shelf is widest. The reefs off the north coast are considered to be the best and healthiest, while reefs to the south increasingly show reef degradation. This is likely correlated to rapid tourist development in the south raising levels of sewage, agrochemical pollution, and sedimentation from development processes (Bouchon 2004).

The combination of climate change and hazards—past, present, and future—in Grenada tends to increase difficulty for maintaining future goals. Landslides, floods, volcanic eruptions, tsunamis, and storms will remain a constant issue for Grenada. It is inevitable that Grenada will continue to be physically shaped by these events, but continuing to strive for proper management and mitigation will put the nation on the path to lower vulnerability and higher resilience.

16.7 Conclusion

Grenada, like its other Lesser Antillean cousins, offers stunning scenery full of a natural beauty and cultural richness. Its volcanic past has shaped it into an attractive landscape with accompanying striking landforms, and this variety is reminiscent of other Leeward Islands. From the north, where visitors to Levera Beach can ogle at andesite cone volcanics and lava flows while watching their step to avoid disturbing turtle nesting sites, to the interior's rainforest, where dramatic waterfalls such as Seven Sisters, Concord, Annandale, and Mt. Carmel invigorate the hiker—and from placid Grand Etang Lake surrounded by lush vegetation, to the southern coast and postcard-worthy Grand Anse Beach—Grenada offers something different at seemingly every turn. Though L'anse aux Epines and La Sagesse are less than half an hour's drive from each other on the southeast coast, L'anse aux Epines is a windswept area dominated by cactus, while La Sagesse remains a sheltered cove with soft sand and a vast beach next to a substantial mangrove. While there are no longer pure-blood Amerindians on the island, their presence is still felt in cultural sites such as Pearls Airport and La Poterie, their petroglyphs carved into basalt, and in place names. These important parts of Grenada's heritage are at risk of being lost, but interest and education could reverse the current path of loss and mismanagement. Cultural remnants from more recent developments face the same struggles as the Amerindian Heritage sites. There are abandoned plantations with unused water wheels and other holdovers from Grenada's colonial past, as well as numerous cultural sites—some that could offer meaningful insight into the history of the island, like Leaper's Hill, and some that offer incredible island views, such as Fort Frederick.

For the geomorphologist, Grenada offers an array of viewing and research choices: volcanic extrusions and sedimentary inclusions, fragile forests, endangered mangroves, delicate coral reefs, pristine beaches, beautiful bays, safe ports, and rugged cliffs. The Spice Isle also offers a variety of sites that may be of interest to cultural geomorphologists, such as Amerindian sites, historical plantations, sugar mills, rum distilleries, spice plantations, and festivals. From a biogeomorphologic perspective, Grenada also hosts an impressive number of flora and fauna, given its small areal extent, including several endemic species. And for those interested in climate change, hazards, and climatic geomorphology, the island nation offers a superb case study.

All these characteristics aside, for the past several decades, Grenada's natural resources have been taxed. Soil erosion, changing hydrologic regimes, stream and drinking water sedimentation, and even habitat loss represent serious concerns for the small island nation. Similar quandaries exist anthropogeomorphologically as well, where a lack of oversight regarding cultural resources continues to negatively impact important archaeological and historical sites. At the same time, however, to bolster the country's main economic component (tourism), better developed natural and cultural attractions are necessary. With increased catering to tourism, such as the completion of a Sandals resort in the southern part of the island, Grenada is facing a new set of challenges. Development is taking a toll on the landscape, and further challenges about tourist site suitability are continually raised because of the likelihood of landslides and increased concerns related to climate change events such as increased hurricanes, rising sea levels, and loss of beach material. Still, if Grenada can find a balance between the challenges of a growing tourist industry with protecting its natural beauty, The Spice Isle will continue to provide a wealth of experiences for residents and visitors alike far into the future. Indeed, as with most Caribbean tourist-driven economies, geomorphological wonders will play a pivotal role in shaping visitors' experiences.

References

Allen CD (2012) Personal interview with Dougaldston Estate, Gouyave, Grenada. 26 May 2012

Allen CD, Groom KM (2013a) A geologic assessment of Grenada's Carib stones. Int Newsl Rock Art 65:19–24

Allen CD, Groom KM (2013b) Evaluation of Grenada's "Carib Stones" via the rock art stability index. Appl Geogr 42:165–175

AQUASTAT (2012) Country overview: Grenada. Issued by: Food and Agriculture Organization of the United Nations

Arce JP (2008) Protected areas of Grenada. In: Encyclopedia of Earth. Available via http://www.eoearth.org/view/article/155396/. Accessed 19 Oct 2015

Arculus RJ (1976) Geology and geochemistry of the alkali basalt-andesite association of Grenada, Lesser Antilles island arc. Geol Soc Am Bull 87:612–624

Associated Press (1955) Janet leaves 108 dead in West Indies. Sarasota Herald-Tribune (Miami, Florida, 23 Sept)

Attractions (nd) Beaches on Carriacou. In: Caribya!. Available via http://caribya.com/carriacou/beaches/. Accessed 19 Oct 2015

Bishop M (1979). The new tourism. Lecture presented at Regional Conference on the Socio-cultural and Environmental Impacts of Tourism in Caribbean Countries. Holiday Inn. St. George's

Boruff BJ, Cutter SL (2007) The environmental vulnerability of Caribbean island nations. Geogr Rev 97(1):24

Bouchon C et al. (2004). Status of Coral Reefs in the French Caribbean Islands and Other Islands of the Eastern Antilles. In: Wilkinson C (ed), Status of Coral Reefs of the World: 2004. pp 493–509. Australian Institute of Marine Science, Townsville, Queensland, Australia

Cashman A, Nurse L, John C (2010) Climate change in the Caribbean: the water management implications. J Environ Dev 19(1):42–67

Caribbean Conservation Association (2003) Marine Protected Area (MPA) planning for Carriacou and Petite Martinique. Sustainable Integrated Development and Biodiversity Conservation in the Grenadine Islands, Coastal and Marine Management Programme, Caribbean Conservation Association, Barbados, Version 1, 43 p

Crask P (2012) Grenada: Carriacou—Petite Martinique. The Globe Pequot Press Inc, Guilford

Destination Guide (nd) Grenada's beaches. In: Grenada explorer. Available via http://www.grenadaexplorer.com/Eco-Tourism.htm. Accessed 20 Oct 2015

Dubelaar CN (1995) The petroglyphs of the Lesser Antilles, the Virgin Islands and Trinidad. Amsterdam: Foundation for Scientific Research in the Caribbean Region: Publication 135

Fritz SC, Björck S, Rigsby CA, Baker PA, Calder-Church A, Conley DJ (2011) Caribbean hydrological variability during the Holocene as reconstructed from crater lakes on the island of Grenada. J Quat Sci 26(8):829–838

Gilser G (2006) Two-dimensional simulations of explosive eruptions of Kick-em Jenny and other submarine volcanoes. Sci Tsunami Hazards 25(1):34–41

Gisler G, Weaver R, Mader C, Gittings, M (2006) Two-dimensional simulations of explosive eruptions of Kick-em Jenny and other submarine volcanos. Carib Tsunami Hazard 138

Global Volcanism Program (2005) Report on Kick 'em Jenny (Grenada). In: Wunderman R (ed) Bulletin of the Global Volcanism Network, 30:2. Smithsonian Institution

Gocova A (2013) Artlantis: Jason deCaires Taylor's underwater sculptures protect coral reefs and usher in a new era of tourism. Altern J 39(3):34

Grenada Forestry Department (1988) Plan and policy for a system of national parks and protected areas

Goudie AS (2002) Aesthetics and relevance in geomorphological outreach. Geomorphology 47:245–249

Government of Grenada, German Agency for International Cooperation (2014) Protecting our coastline: coastal zone management in Grenada. Climate Change Adaptation Strategies Programme

Hearn L (1890) Two years in the French West Indies. Harper & Brothers Publishers, New York

History & Culture (nd) Grenada: The Spice of the Caribbean. In: Geographia. Available via http://www.geographia.com/grenada/gdhis01.htm. Accessed 20 Oct 2015

Home (nd) Welcome to Petite Martinique. In: Petite Martinique. Avialable via http://www.petitemartinique.com/. Accessed 18 Oct 2015

Howell C (1991) Grenada: country environmental profile. Issued by: The Caribbean Conservation Association and The Government of Grenada

Huber R, Vincent G, MacFarland C, Meganck R (2012) Draft: plans and policy for a system of national parks and protected areas. Issued By: The Government of Grenada and The Organization of American States

Keegan W (1991) Miscellaneous Project Report Number 47. Florida Museum of Natural History

Kilgore BC, Moore A (2011) Adventure Guide to Grenada, St. Vincent and the Grenadines. Montreal, Quebec: Hunter Publishing, Inc

Local News (2013) Sand mining returning to Grenada. In: The New Today. Available via http://thenewtoday.gd/local-news/2013/04/28/sand-mining-returning-grenada/. Accessed 17 Oct 2015

Maison K, King R, Lloyd C, Eckert S (2010) Leatherback nest distribution and beach erosion pattern at Levera Beach, Grenada, West Indies. Mar Turt Newsl 127(1):9–12

Marquet SJ (2009) Contextual analysis of the Lesser Antillean Windward islands petroglyphs: methods and results. In: Hayward MH, Atkinson L-G, Cinquino MA (eds) Rock art of the Caribbean, University of Alabama Press, pp 147–160

Martin JA (2007) A to Z of Grenada Heritage. Macmillan Education, Oxford

Martin JA (2013) Island Caribs and French Settlers in Grenada: 1498–1763. Grenada National Museum Press

Ministry of Finance, Government of Grenada (February 22, 2005) IMF statement on Grenada

Moore G, Gilmer B, Schill S (2015) Distribution of mangrove habitats of Grenada and the Grenadines. J Coastal Res 31(1):155–162

Nagel C (nd) Scuba Diving in Carriacou, Caribbean. In: Dive Site Directory. Available via http://www.divesitedirectory.co.uk/dive_site_caribbean_carriacou_reef_magic_garden.html. Accessed 17 October 2015

Rustad H (2012) Climate change linked to Caribbean fishery collapse. Geographical 84(12):14

Seismic Research Center (2016) The University of West Indies Seimic Research Center. Available via http://www.uwiseismic.com/. Accessed 10 June 2016

Saunders JB, Bernoulli D, Martin-Kaye PHA (1985) Late Eocene deep-water clastics in Grenada, West Indies. Eclogae Geol Helv 78:469–485

Speed RC, Smith-Horowitz P, Perch-Nielsen K, Saunders J, Sanfilippo A (1993) Southern Lesser Antilles Arc Platform: pre-late Miocene stratigraphy, structure, and tectonic evolution. Geol Soc Am Spec Pap 277:1–98

Thomas H (2010) Petroglyphs of Grenada and a recently discovered petroglyph in St. Vincent, vol 1. BiblioBazaar

Tomblin JF (1974) Lesser Antilles. Geol Soc Lond Spec Publ 4:663–670

Tomblin JF (1975) The Lesser Antilles and Aves Ridge. In The Gulf of Mexico and the Caribbean. Springer, pp 467–500

UNDESA (2012) Climate change adaptation in Grenada: water resources, coastal ecosystems, and renewable energy. Issued by: Division for Sustainable Development of the United Nations Department of Economic and Social Affairs

Urzua L, Benavente O, Lovelock B, Brookes A, Ussher G (2015) Grenada geothermal surface exploration. In: Proceedings 37th New Zealand geothermal workshop, p 20

Weeks LA, Lattimore R, Harbison R, Bassinger B, Merrill G (1971) Structural relations among Lesser Antilles, Venezuela, and Trinidad-Tobago. AAPG Bull 55:1741–1752

White W (2015) A tale of two magma series: geochronology and geochemistry of volcanism on Grenada, Lesser Antilles. 2015 AGU Fall Meeting

World Bank, Latin America and the Caribbean Hazard Risk Management Unit (2005) Grenada: a nation rebuilding. An assessment of reconstruction and economic recovery one year after Hurricane Ivan. Washington, DC

Trinidad and Tobago

17

Jeanette C. Arkle, Lewis A. Owen and John C. Weber

Abstract

The nation of Trinidad and Tobago comprises the two southeastern-most islands in the Caribbean. Despite being neighbors and part of the same nation, the two islands have markedly different geology, landforms and landscapes. The geologic framework of these islands is the result of protracted and complex tectonism associated with Mesozoic rifting of Pangaea, circum-Caribbean subduction, and currently, Caribbean-South American transform plate motion. Tectonism, climatic shifts, and associated sea-level fluctuations throughout the Quaternary have shaped the islands' wetlands, drainage systems, modern and ancient reefs, marine terraces, and morphology of their shallow continental shelf. The islands proximity to mainland South America resulted in unique and diverse biota of continental origin and also had a marked influence on the migration and settlement of the earliest people in the West Indies. The natural landscapes, as well as society, and policies, of Trinidad and Tobago have been affected by agriculture, mining, urban development, and the hydrocarbon industry. Continued natural and human-induced environmental change, particularly related to sea-level rise, will lead to dramatic changes of the islands' landscapes.

Keywords

Landscape evolution • Tectonics • Tropical erosion • Quaternary geology • Caribbean

17.1 Introduction

Trinidad and Tobago emerge in the southeastern-most part of the Caribbean as a pair of islands separated by <30 km on NE corner of the South American continental shelf (Fig. 17.1). Despite being neighbors and part of the same nation, the two islands have markedly different landscapes. This is a consequence of their unrelated origins in the early Mesozoic circum-Caribbean and their different and protracted geologic and tectonic histories associated with interactions between the Caribbean and South American plates. Trinidad and Tobago's diverse geologic frameworks provide the context for the islands' varied landscapes, landforms, and geomorphic evolution, which have played a critical role in the nation's socioeconomic development, particularly as the leading producer of oil and gas in the Caribbean.

The Quaternary landscape evolution of Trinidad and Tobago is a product of their positions along the southeast corner of the Caribbean plate, their shallow emergence on the South American continental shelf, and, as a result, an impressive combination of geomorphic processes. Differential vertical tectonism has uplifted coral reefs, created marine terraces, and also caused extensive subsidence (e.g., Snoke et al. 2001; Weber 2005). The Central Range transform fault

J.C. Arkle (✉) · L.A. Owen
Department of Geology, University of Cincinnati, Cincinnati, OH, USA
e-mail: arklejc@mail.uc.edu

L.A. Owen
e-mail: lewis.owen@uc.edu

J.C. Weber
Department of Geology, Grand Valley State University, Allendale, MI, USA
e-mail: weberj@gvsu.edu

C.D. Allen (ed.), *Landscapes and Landforms of the Lesser Antilles*, World Geomorphological Landscapes, DOI 10.1007/978-3-319-55787-8_17

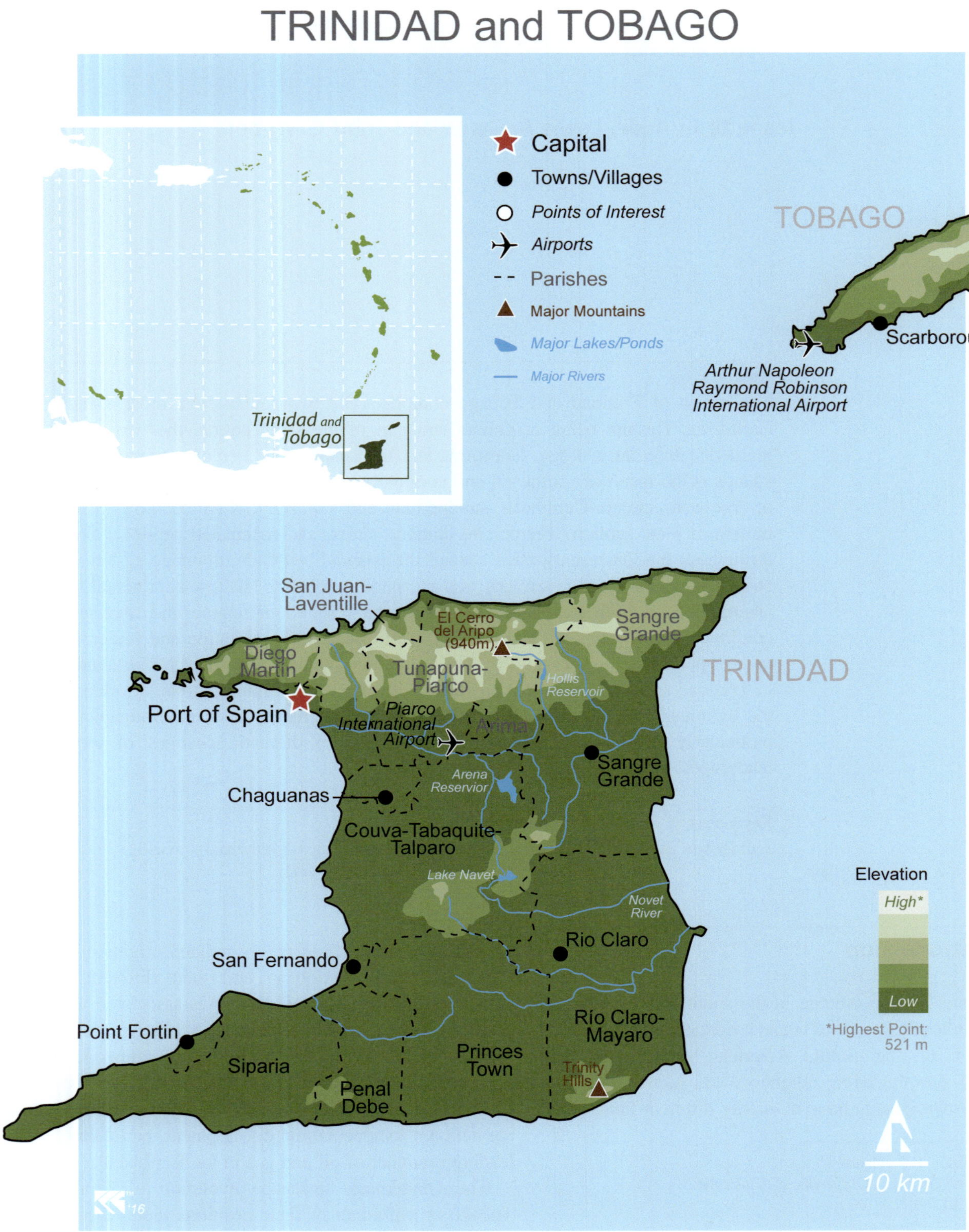

Fig. 17.1 General physiographic map of Trinidad and Tobago. Note Trinidad's low-lying plains in the south and highlands along its northern coast, as well as Tobago's more mountainous topography. Cartography by K.M. Groom

in Trinidad, which defines the Caribbean-South American plate boundary, has generated a striking topographic scar across the entire width of central Trinidad and other tectonic landforms including shutter ridges, linear drainages, and offset ridges (Prentice et al. 2001, 2010; Weber et al. 2001a, 2011).

Sea-level fluctuations, associated with climatic shifts, have exposed portions of the shallow continental shelf and resulted in the development of an extensive offshore network of paleochannels (Soto et al. 2007, 2011). These base-level changes have caused adjustments to fluvial systems and the redistribution of sedimentation and erosion throughout the islands' interior landscapes (Ramcharan 2004; Arkle et al. 2015). Periodically connected with mainland South America during sea-level low stands, Trinidad and Tobago are home to unique flora and fauna of continental origin (Kenney 2008). The severe tropical climate has caused intense weathering, ubiquitous landsliding, and flooding (ODPM 2014). Strong, sediment-rich ocean currents dominate coastal sedimentation and erosion and influence coral reef development (Darsan et al. 2012). Collectively, these dynamic geomorphic processes have sculpted a spectacular variety of landscapes and landforms throughout Trinidad and Tobago.

Since the initial arrival of humans in Trinidad at ~8 ka, the landscape has been modified by agriculture, mining, biotic resource extraction, and urban development (Harris 1973; Boomert 2000; Pagán-Jiménez et al. 2015; Siegel et al. 2015). In the twentieth–twenty-first centuries, the hydrocarbon industry and an emerging tourism sector have further transformed the landscape. Continued natural and human-induced environmental changes will result in more dramatic landscape changes. Sea-level rise associated with human-induced climate change, mass wasting, and frequent seismicity and large magnitude earthquakes are some of the greatest threats to Trinidad and Tobago (Nurse et al. 1998; ODPM 2014). The future of the islands' landscapes and their management remains in a critical balance between employment, tourism, ecological protection, and sociopolitical factors (Dillman 2015).

17.2 Setting

17.2.1 Current Setting

Trinidad and Tobago are located in the SE corner of the tectonically and geologically complex Caribbean-South American plate boundary. Currently, the plate boundary is a transform margin that strikes eastward, south of Venezuela's northern Cordillera de la Costa, steps to the right in the Gulf of Paria, and then strikes ENE through central Trinidad. The current relative Caribbean-South American plate motion is ~20 mm/yr to the east, directed ~N86°E in Trinidad (Weber et al. 2001a, 2011). The majority of the dextral shear strain, ~12–15 mm/yr, is accommodated on the Central Range fault that strikes ~N72°E through Trinidad and continues offshore where it essentially merges with faults and structures of the Barbados accretionary prism (Pérez et al. 2001; Weber et al. 2001a, 2011; DeMets et al. 1994; see Fig. 17.2). Given this configuration, the southern half of Trinidad is moving westward with the South American plate, while land north of the Central Range fault, including northern Trinidad and Tobago, travels with the east-directed Caribbean plate.

Trinidad is structurally characterized by an alternating series of roughly ENE striking mountains and basins (Fig. 17.3). The Northern Range, Trinidad's tallest mountains, that rise to 940 m above mean sea level (amsl), exposes low-grade metamorphic rocks that were derived from Mesozoic passive-margin sedimentary rocks (Fig. 17.4). Late Cretaceous to Pleistocene sedimentary rocks underlie central and southern Trinidad (Erlich et al. 1993). These sedimentary rocks are folded and faulted to form two sets of low-relief hills, the Central and Southern ranges, which are bounded to the north by the Northern and Southern basins, respectively.

Tobago is dominantly composed of Mesozoic oceanic-arc and forearc rocks that rise to 572 m amsl in the mountains called the Main Ridge (Snoke et al. 2001; Fig. 17.5). These rocks are structurally sunken and covered by a Pleistocene coralline raised reef in the lowlands of southern Tobago (Fig. 17.6). Tobago is characterized by a northern highland and southern lowland that are associated with a series of horst and graben blocks formed by Pliocene-Recent regional extension.

Trinidad and Tobago have a tropical maritime climate with wet (June–December) and dry (January–May) seasons. Mean annual temperatures range from ~21 to 31 °C and the humidity is high, ranging from ~50 to 100% and averaging >80% (WRA 2001). Mean annual rainfall is ~2000 mm with extreme events delivering over ~5000 mm in areas of the Northern Range in Trinidad and in the Main Ridge of Tobago (WRA 2001; Hijmans et al. 2005). The islands experience relatively constant and strong easterly trade winds, and sporadic wet season tropical storms (WRA 2001). Although located outside of the main Atlantic hurricane belt, hurricanes occasionally track across the islands. The most infamous hurricane was Flora in 1963, which was devastating for Tobago (ODPM 2014).

The Guiana Current is the principal ocean current that impacts Trinidad and Tobago's coastal ecology and morphology (Van Andel 1967). The Guiana Current flows northward along South America and then bifurcates to the west through the Columbus Channel along southern Trinidad, and to the north along the coastlines of eastern Trinidad

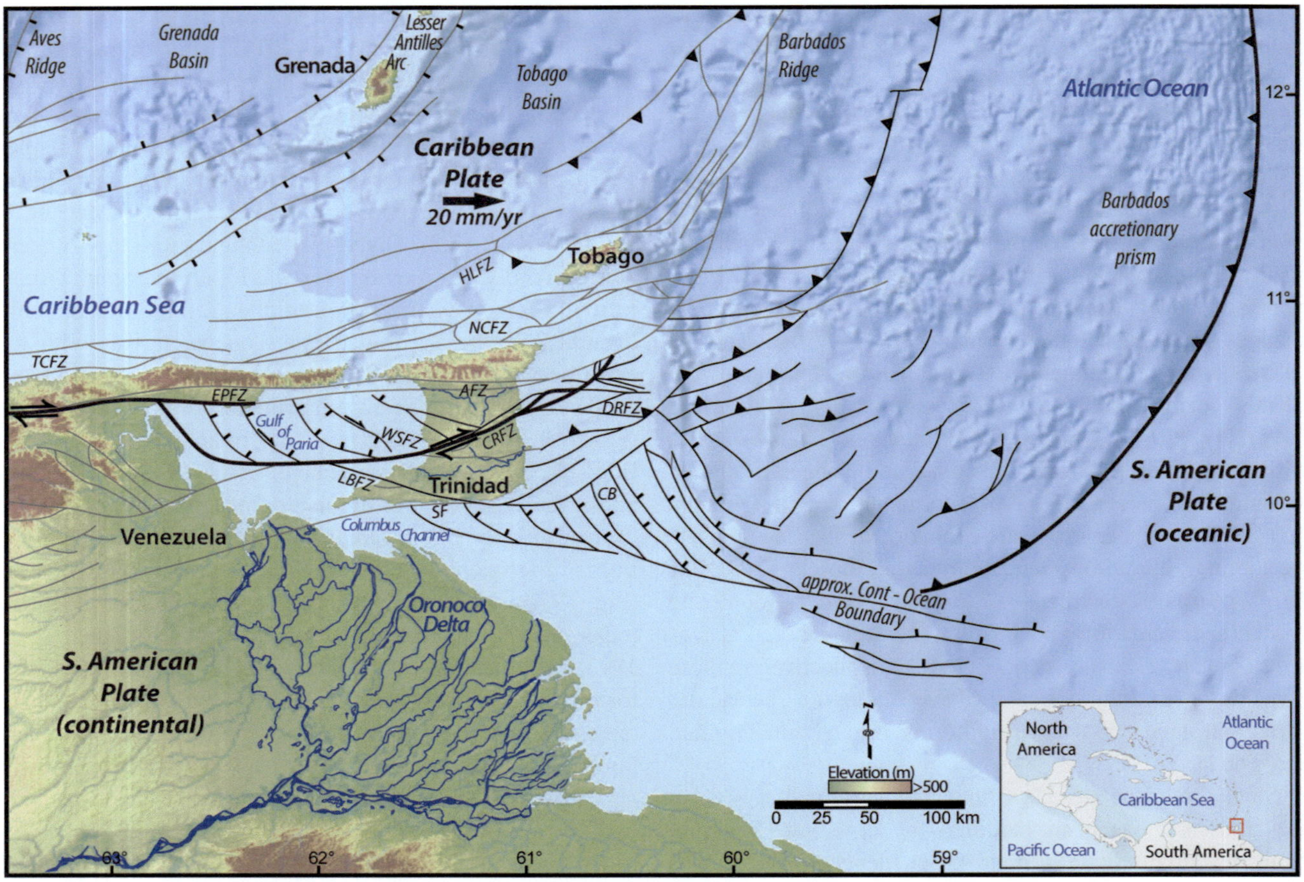

Fig. 17.2 Overview of the southeast Caribbean showing topography (0.5 km DEM from the US Geological Survey), bathymetry (from ESRI base maps and contributors therein), and major (*bold black lines*) and minor faults (*thin gray lines*) that show relative motion where known as thrust (teeth), normal (*bar*), and strike-slip (*arrows*) (after Soto et al. 2007, 2011; Prentice et al. 2010; Garciacaro et al. 2011 and references therein). Relative Caribbean-South American plate motion vector and velocity from Weber et al. (2001a, 2011), Pérez et al. (2001), and DeMets et al. (1994). *TCFZ* Tortuga Coche fault zone; *EPFZ* El Pilar fault zone; *AFZ* Arima fault zone; *CRFZ* Central Range fault zone; *WSFZ* Warm Springs fault zone; *LBFZ* Los Bajos fault; *SF* Soldado fault; *DRFZ* Darien Ridge fault zone; *CB* Columbus Basin; *NCFZ* North Coast fault zone; *STFZ* Southern Tobago fault zone; and *HLFZ* Hinge Line fault zone. Cartography by J.C. Arkle

and SE Tobago (Van Andel 1967). This current carries vast quantities of suspended sediment and plant detritus from the Orinoco and Amazon River deltas, which critically impacts coastal sedimentation and erosion, and, in large part, inhibits reef formation, particularly around Trinidad (Van Andel 1967; Laydoo 1991; Lapointe et al. 2010).

17.2.2 Geologic Setting and History

Trinidad and Tobago's geologic setting is complex and both islands have different geologic histories, structures, and rocks. Understanding the geological history is critical because it defines the context and "initial conditions" for the development of Trinidad and Tobago's landscapes and landforms.

Tobago's bedrock framework consists of three main lithologic belts that strike approximately eastward across the island and include: (i) a Cretaceous metavolcanic province, (ii) a Cretaceous intrusive igneous province, and (iii) a Plio-Pleistocene sedimentary sequence (Snoke et al. 2001; see Fig. 17.5). The Mesozoic oceanic forearc and arc rocks of Tobago formed far west of their current position and were rotated clockwise, possibly >90°, as they were translated eastward during oblique collision along the leading edge of the Caribbean plate (e.g., Erlich and Barrett 1990; Snoke et al. 2001). The intrusive igneous suites were emplaced at upper crustal levels during the late Cretaceous and experienced a protracted period of slow exhumation, through at least the Eocene (Cerveny and Snoke 1993), and then a pulse of increased exhumation during the late Neogene (Arkle et al. 2014). The metavolcanic and plutonic belts are bounded by the Central Tobago fault system (see Fig. 17.5). This normal-sense fault system and ductile shear zone strikes NE with down-to-the-south displacement and are interpreted to be related to extension that accommodated exhumation of the deep Tobago oceanic-arc rocks (Cerveny and

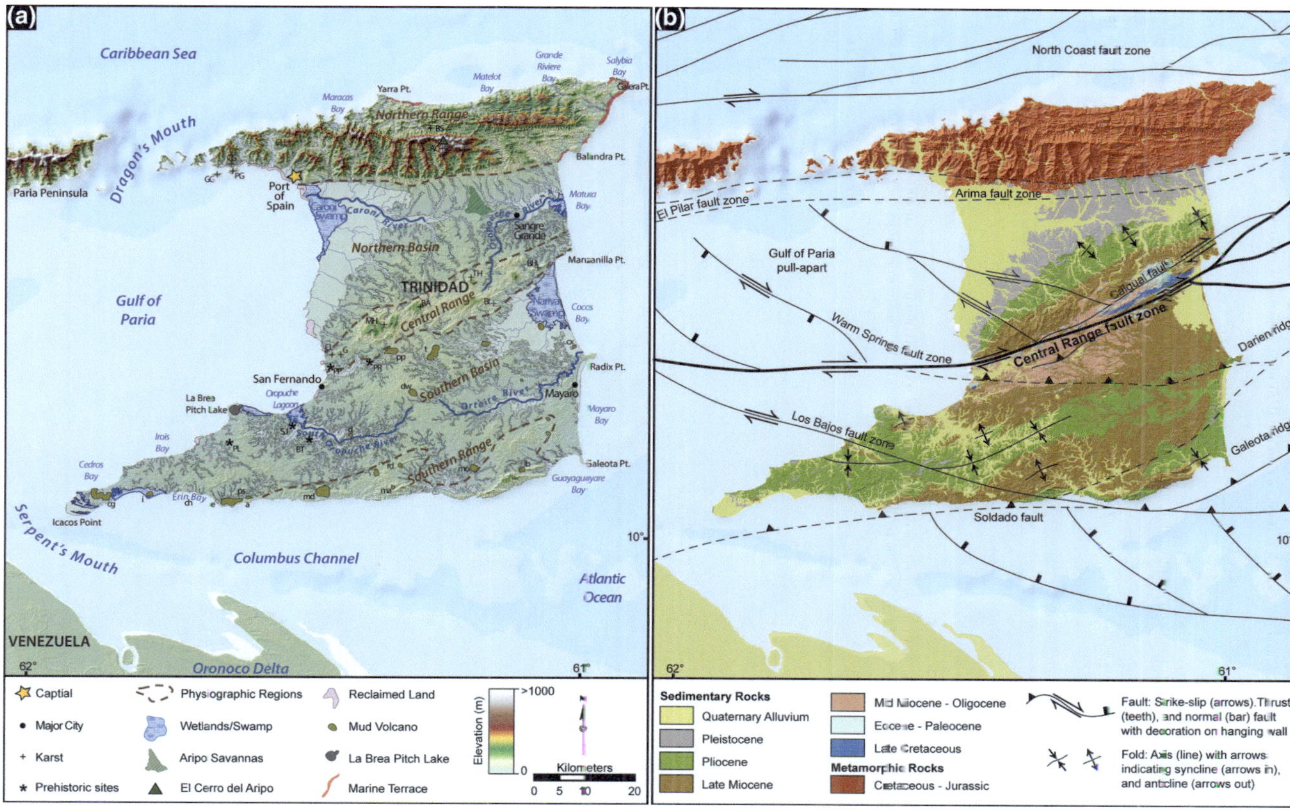

Fig. 17.3 Maps of Trinidad showing **a** the topography (30 m DEM from USGS Earth Explorer), the major highland (Northern, Central, and Southern Range) and lowland (Northern and Southern Basin) physiographic regions (*brown dashed lines*), trunk rivers (*bold blue lines*), and minor streams (*thin blue lines*), and **b** the geology. Approximate location of wetlands, swamps, lagoons, and reclaimed land are from the geologic map of de Verteuil et al. (2006). Prehistoric sites from Harris (1973): *PP* Pointe-a-Pierre; *PR* Poonah Road; *BT* Banwari Trace; *PL* Parrylands; *SJ* St. Johns. Mud volcanoes from Deville and Guerlais (2009) and Deville et al. (2003): *cg* Columbus group; *i* Islote; *ch* Chatham; *e* Erin; *ps* Palo Seco; *a* Anglais Point; *md* Morne Diablo; *ma* Marac; *rd* Rock Dome; *mo* Moruga; *lb* Lagon Bouffe; *d* Digity; *pp* Piparo; *dw* Devil's Woodyard; *t* Tabaquite; *cx* Cascadoux. Select karst outcrops from Erlich et al. (1993) and Day and Chenoweth (2004): *CL* Concord Limestone; *G* Gasparillo Limestone; *GL* Guaracara Limestone; *TL* Tabaquite Limestone; *TH* Tamana Hill; *BL* Biche Limestone; *BH* Brigand Hill; Hills; *C* Chaguaramas; *GG* Gaspar Grande; *PG* Point Gourde; *C* Cameron; *AC* Aripo Cave; *BS* Brasso Seco; **b** Simplified geologic map of Trinidad modified from de Verteuil et al. (2006), Soto et al. (2007, 2011), and Prentice et al. (2010). Cartography by J.C. Arkle

Snoke 1993; Snoke et al. 2001). The Central Tobago fault system is offset by a series of younger oblique-slip faults that strike NNW (Snoke et al. 2001).

The deep meta-arc rocks of Tobago's Main Ridge are structurally sunken and covered by the Rockly Bay Formation and a coralline platform in the south (see Fig. 17.5). The clastic rocks of the Rockly Bay Formation were deposited during the middle Pliocene as a graben-fill sequence and are also likely a transgressive unit associated with concordant sea-level rise (Donovan 1989). The coralline limestone platform of southern Tobago formed during the Pleistocene and is well exposed in parts of the Tobago lowlands (Donovan and Jackson 2010). The Southern Tobago fault system and additional extensional faults cut and down-drop this limestone platform (Snoke et al. 2001; Donovan and Jackson 2010). Active extension continues today in south Tobago. The 1982 earthquake swarm and the 6.7 M_w 1997 Tobago earthquake are testaments to active extension (Weber et al. 2015a).

Trinidad's Northern Range is composed of metasedimentary rocks with Jurassic-Cretaceous protolith ages. These rocks are a continuation of the rocks exposed to the west in Venezuela's Paria Peninsula and Coasta de Cordillera. The Northern Range metasedimentary rocks were probably initially deposited along a north-facing South American passive margin during the Mesozoic (Algar and Pindell 1993). Miocene oblique collision of the Caribbean plate with the South American passive-margin, deformed, metamorphosed, and exhumed Northern Range bedrock and was synchronous with folding and thrusting in south Trinidad (Algar and Pindell 1993; Weber et al. 2001b; Cruz et al. 2007). The Caribbean-South American plate boundary

Fig. 17.4 View of the Northern Range, Trinidad from the north coast. Photograph by J.C. Arkle

stepped southward and changed to the current phase of transform plate motion at ~ 10 Ma (Pindell et al. 1998). The transition in plate motion from oblique collision to transform motion probably also marked the onset of low-magnitude, upper-plate collapse and extension that dominate the youngest tectonic features in and around Tobago (Snoke et al. 2001; Weber et al. 2015a).

In central and south Trinidad, the bedrock is sedimentary. The oldest exposed units are composed of Late Cretaceous passive-margin organic-rich marlstone (Erlich et al. 1993). The Paleogene-Early Neogene section was deposited in a distal foreland basin in front of the advancing Caribbean plate. Late Pliocene to Pleistocene sedimentary rocks in Trinidad were mostly deposited in the paleo-Orinoco delta system (Erlich et al. 1993). On the north flank of the Central Range, the Late Neogene-Pleistocene rocks were deposited after intense folding and thrusting of pre-Middle Miocene strata. This succession of rocks was then tilted to the north. In south Trinidad, folding and faulting, probably related to bends and steps in the transform system, continued into the Pleistocene.

Paleoclimate records from the Cariaco Basin in Venezuela show climatic shifts, most broadly, from a cold and dry period at the end of late Pleistocene into the Holocene (~ 11.5 ka) that transitions to a protracted warm and wet period until the middle Holocene (10.5–5.4 ka), and then a gradual shift to cooler and arid conditions throughout the late Holocene (~ 5.4–2.8 ka; Peterson et al. 2000; Haug et al. 2001). Quaternary temperature records for tropical South America are limited and regionally variable, but most studies indicate that temperatures were ~ 5 °C cooler than present in low-latitude areas of South America during global glacial stages (van der Hammen and Hooghiemstra 2000; Baker and Fritz 2015). Climatic transitions into cooler and arid periods likely caused rapid shifts between arid grassland and wet forest and shifts montane vegetation downslope (van der Hammen and Hooghiemstra 2000; Hughen et al. 2004).

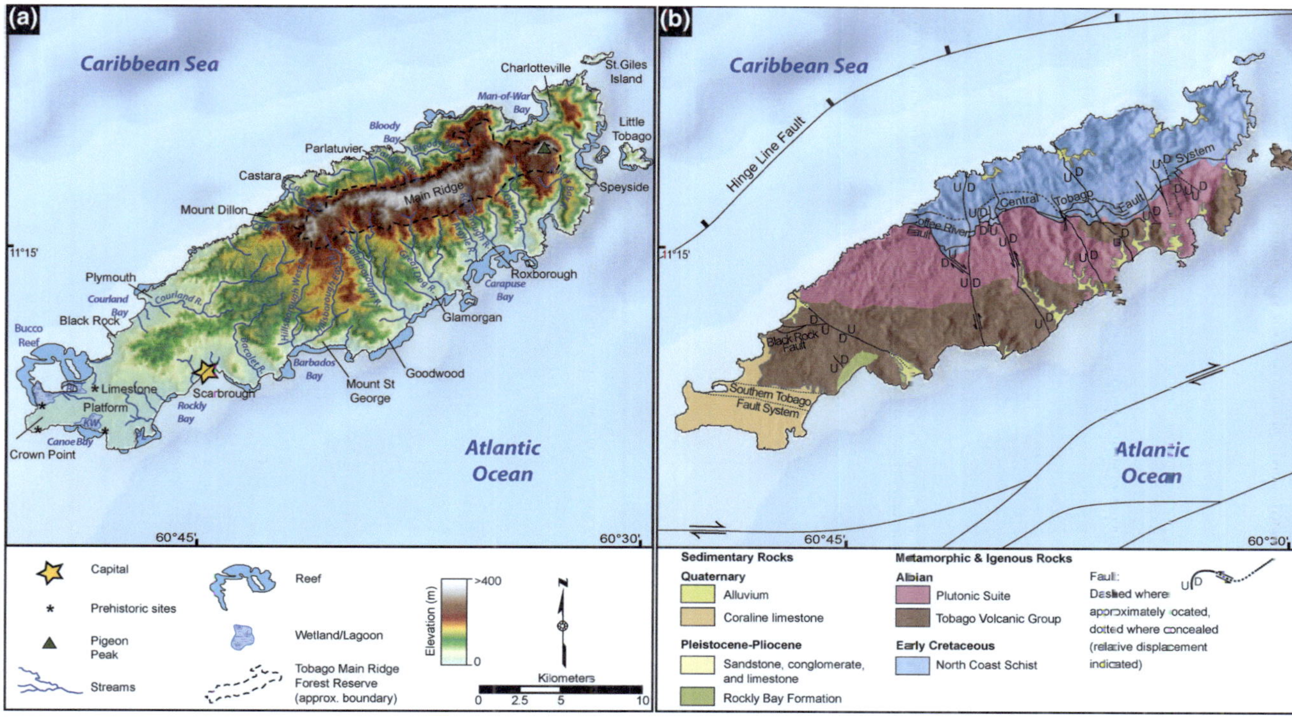

Fig. 17.5 Maps of Tobago showing **a** the topography (30 m DEM from USGS Earth Explorer), the Main Ridge Mountain highlands, the coralline limestone lowland, and major streams (*blue lines*), and **b** geology simplified from Snoke et al. (2001) and offshore faults from Garciacaro et al. (2011 and references therein). Reefs are from Hassanali (2013) and prehistoric sites from Steadman and Stokes (2002). Cartography by J.C. Arkle

Differential tectonic uplift in the region makes it extremely difficult to determine the absolute magnitudes of local sea-level fluctuations and the timing of when the islands were connected to mainland South America. However, if consistent with eustatic sea-level changes, the Caribbean may have been ~120 m lower than at present during the global last glacial maximum (e.g., Muhs et al. 2011, 2012). Trinidad, positioned aside the shallow (<30 m depth) Gulf of Paria, was likely connected to South America as recently as the Holocene, and Tobago was likely connected with the mainland during the Pleistocene (e.g., Boomert 2000).

17.3 Landforms

17.3.1 Tectonic Landforms

The large-scale (>10 km^2) landforms of Trinidad and Tobago reflect their complex tectonic and paleoclimatic history. The framework of Tobago's large-scale geomorphology is largely controlled by horst and graben structures. The island emerges as a structural high, as part of a series of regional extensional horst and graben blocks, positioned on the slightly stretched and extended southern edge of the Caribbean plate. The southern Tobago lowland lies over a down-dropped half-graben south of the South Tobago fault system, which also brings up the elevated horst block of the Main Ridge to the north. The Pleistocene limestone platform that forms the southern lowland is one of Tobago's most prominent sunken tectonic landforms. The seismically defined Southern Tobago fault system strikes eastwards across the platform, but is mostly buried and generally lacks surface expression (Snoke et al. 2001). Sea-level fluctuations and tectonic motion in southern Tobago have also generated numerous modern fringing reefs, which are more abundant and developed on the leeward side of the island, protected from severe weather tracks and sediment-laid ocean currents.

Past oblique-collision and active dextral Caribbean-South American plate deformation has created a suite of ENE striking bedrock structures that form the fundamental framework of Trinidad's large-scale geomorphology. This is highlighted by the occurrence of the alternating series of small topographic highs and low basins on the island (Fig. 17.1).

The geomorphology of Trinidad's Northern Range is dominantly controlled by neotectonic subsidence in the west and surface uplift in the east, in combination with a steep east-to-west gradient in rainfall (Weber 2005; Ritter and Weber 2007; Arkle et al. 2015). Eastern and western "blocks" of the Northern Range "split" at Yarra Point and are defined by an *en echelon* dextral break in the main

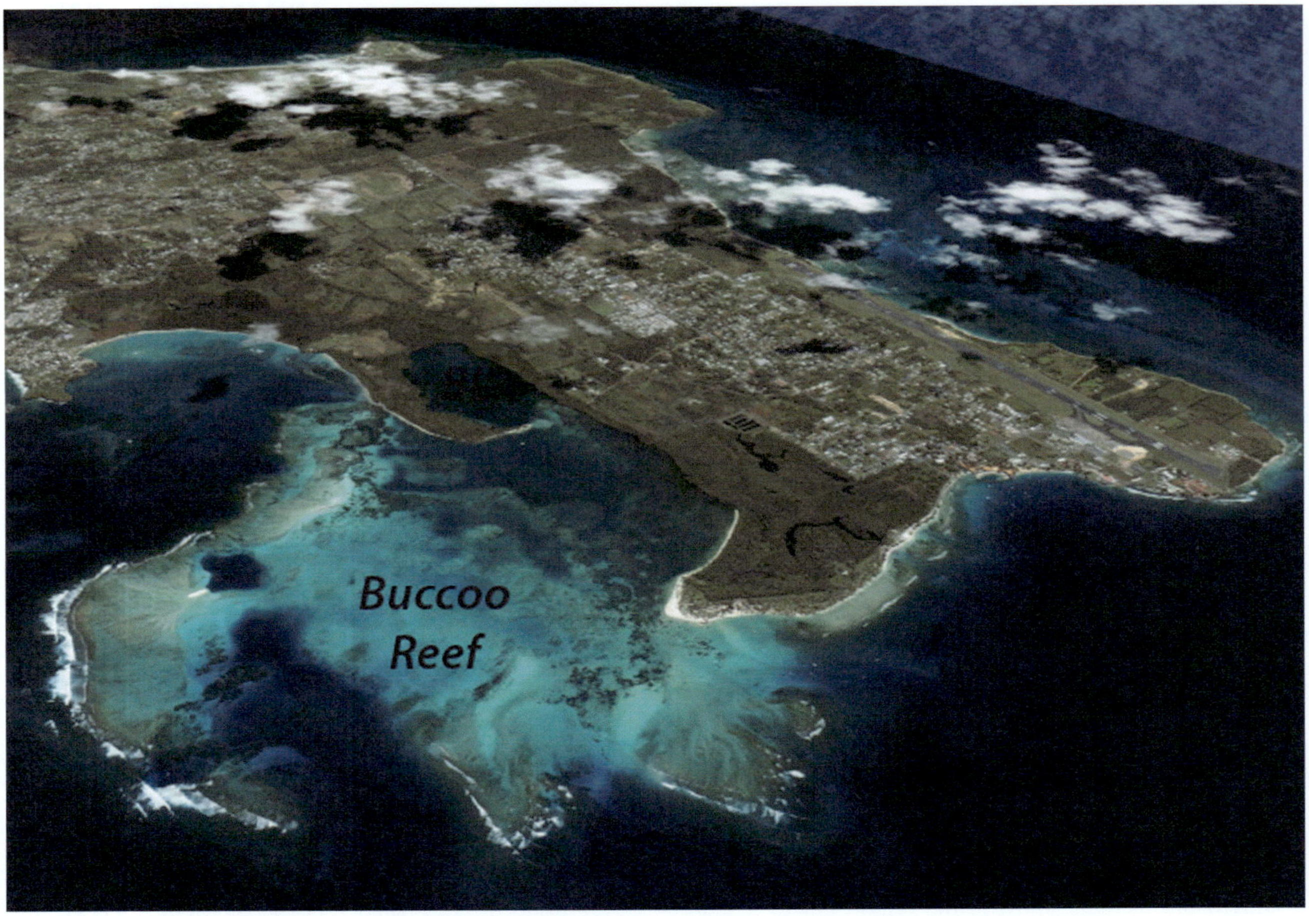

Fig. 17.6 Oblique areal view to the southeast of the coralline limestone raised reef and the relatively flat, low-lying topography of Tobago's southern lowlands. The foreground shows Buccoo Reef and the Bon Accord Lagoon (BL), Tobago's largest modern reef system and wetland. Photograph from Google Earth ca. December, 2004

ridgeline (N–S drainage divide; see Fig. 17.3). Geomorphic transitions include a relatively straight coastline punctuated by marine terraces in the east, to a highly scalloped and embayed coastline in the west, drowned island topography in the Gulf of Paria, and westward decreases of mountain front sinuosity and catchment outlet elevations. Alluvial fans bound the eastern mountain front, yet the correlative alluvial units in the west have subsided to depths over 100 m and are concealed by the Caroni Swamp (Weber 2005; Ritter and Weber 2007; Arkle et al. 2015). Quaternary erosion rates for catchments that drain the Northern Range are ~ 8 times greater in the east compared to those of the western catchments, reflecting both tectonic eastside-up tilting and high precipitation in the east (Arkle et al. 2015).

The Arima fault-line scarp bounds the southern flank of the Northern Range and defines a sharp transition from Trinidad's high-relief topography of the Northern Range to the Northern Basin (Algar and Pindell 1993; Weber et al. 2001b). Fault gouge is present along the Arima fault trace, yet there is no geomorphic or geodetic evidence of recent motion along this fault (Algar and Pindell 1993; Weber et al. 2001b). This suggests that it is probably an inactive exhumed fault with differential erosion of the relatively harder schist and quartzite to the north versus the softer slate to the south. The fault-line scarp strikes eastward and is exposed from the western offshore islands near Gaspar Grande, along the foothills of Port of Spain, and near the Village of Arima (Algar and Pindell 1993).

Southern and central Trinidad are characterized by a well-defined ENE structural and topographic grain, which is also parallel to the active Central Range fault zone. The contact between the weak shale-rich rocks of the Nariva Belt and more resistant Pliocene-Pleistocene Orinoco sand-rich deltaic rocks of the Southern Basin also forms a major eastward-trending escarpment (Fig. 17.7). This escarpment traverses the entire width of the island and from Point Radix in the east, continues offshore as the Darien Ridge, and is likely a fault-bounded contractile ridge. The Southern Range consists of a series of east striking *en echelon* ridges such as the Trinity and Cat's

Fig. 17.7 Low-relief hummocky hills of the Southern Basin, Trinidad. Photograph by J.C. Arkle

Hills. Collectively these features, both ancient and active, control the distribution and nature of Quaternary sedimentation, fluvial systems, coastline geometry, and endogenic landforms (e.g., mud volcanoes and asphalt seeps) in central and southern Trinidad.

The Central Range fault, the modern transform plate boundary, creates a major topographic lineament that strikes ~072° across the entire width of central Trinidad (Prentice et al. 2001, 2010; Weber et al. 2001a; see Fig. 17.3). Linear drainages and saddles and troughs are topographically aligned along the Central Range fault. Small, steep-sloped hills mark a series of shutter ridges that block and pond streams. Linear scarps striking NNE along the fault trace appear as subtle low-relief rolling bulges. Deflected ridges and streams that drain the hills across the fault trace have right-lateral offsets (Weber et al. 2001a; Crosby et al. 2009; Prentice et al. 2010) These landforms are subtle in the Central Range hills, probably due to the intense weathering and erosion of the soft shale-rich bedrock rock in Trinidad's tropical climate, as well as substantial modifications to the landscape due to agriculture, historically mostly sugarcane cultivation (Crosby et al. 2009; Prentice et al. 2010).

Soto et al. (2007, 2011) used 3D seismic reflection data to map the submarine continuation of the Central Range fault eastwards for another 60 km into the offshore. They also map several other faults and structures, and a complex network of paleofluvial systems that are developed in the apron of young, shallow continental shelf sediments. Offsets along these paleofluvial channels provides long-term slip rate estimates of ~17–19 mm/yr along the Central Range fault offshore (Soto el al. 2007, 2011), which is consistent with the onshore geodetic estimates (Pérez et al. 2001; Weber et al. 2001a, 2011).

17.3.2 Mud Volcanoes and Related Landforms

Trinidad's most unique and transient landforms are generated by asphalt seeps and mud volcanoes, which are concentrated in southern Trinidad (Fig. 17.8). Related active hydrocarbon features are regionally widespread and extend over several hundred kilometers, from the Barbados accretionary wedge into Venezuela (e.g., Higgins and Saunders 1974; Babb and Mann 1999; Pindell et al. 1998). Active petroleum generation occurs offshore and onshore in southern Trinidad, resulting in oil and gas accumulations in complex faulted and folded structures (Babb and Mann 1999; Pindell et al. 1998). Onshore Trinidad, most mud volcanism is located in clay- and sand-rich folded sedimentary rocks. It is driven in part by active oil and gas generation at depth, with local temperature and pressure changes likely controlling the cyclic activity and eruptive style (Dia et al. 1999; Deville and Guerlais 2009).

Eruption styles of mud volcanoes in southern Trinidad range from slow, continuous expulsion, to punctuated, effusive catastrophic eruption, which creates an assortment of landforms (e.g., Higgins and Saunders 1974; Deville and Guerlais 2009). Fields of mud (tassiks) surround small, steep-sloped (>25°) cones and domes composed of stacked mudflows that reach over 150 m in height and span from a few cm up to 2 km in diameter (Higgins and Saunders 1974; Dia et al. 1999; Deville et al. 2003; Deville and Guerlais 2009). The surfaces of tassik fields solidify and desiccate into forms that include concentric and hexagonal shapes; some are capped by polygenic breccia with clasts that include quartz and sulfide pebbles (Higgins and Saunders 1974; Dia et al. 1999; Deville et al. 2003).

Mud pools and lakes are dynamic and range from a few cm up to several hundred meters in diameter. Lagon Bouffe, for example, expels mud continuously from several vents and forms large pools (>100 m wide) of fluid and viscous mud (Dia et al. 1999; Deville and Guerlais 2009; see Fig. 17.3). Mud pools can generate very smooth-textured "stemmed glass" surfaces, and convection processes in larger pools are associated with displaced mud flanked by ribbed, ring-shaped rolls (Higgins and Saunders 1974; Dia et al. 1999; Deville et al. 2003; Deville and Guerlais 2009). Some of these landforms are short-lived. For example, in 1964 the Chatham mud island located in Erin Bay along the Columbus Channel formed an $\sim$8 m high by an $\sim$0.1-km^2 island in just a few days. However, the mud island only lasted $\sim$8 months above sea level as it became compacted and was quickly eroded to below sea level (Higgins and Saunders 1967).

Pitch Lake, a popular tourist destination in Trinidad ($\sim$20,000 annual visitors), is the largest natural active asphalt deposit in the world, spanning an area of >0.5 km^2 with an estimated depth of >75 m (UNESCO 2015; see Fig. 17.8). It is located on Trinidad's southwest peninsula, $\sim$1 km from the coast, near the village of La Brea, and is elevated to $\sim$85 m amsl, and is on the crest of an anticline of Neogene sedimentary rocks (Kugler 1959). Pitch Lake expels a mix of heavy oil, bitumen (asphalt), and water. In aerial view, this asphalt lake forms hexagonal surface textures, concentric ribbed and ringed rolls, and small mounds composed of brecciated rock. Convection cells generate meter-sized bulges and features analogous with spreading ridges, smoothed fissures, and mini-subduction zones.

Pitch Lake is recorded in Sir Walter Raleigh's ship's log as a source of pitch that was used for patching his damaged ships. Amerindian pottery and the partial remains of a prehistoric fauna are preserved in Pitch Lake (UNESCO 2015). Several asphalt streams shown on Kugler's (1959) geological map of Trinidad indicate that Pitch Lake used to flow freely into the Gulf of Paria. However, mining of the pitch has lowered the level of the "lake" and it no longer drains externally, and only minor traces of asphalt streams can be seen in the topography today.

17.3.3 Drainage Networks and Fluvial Landforms

Drainage networks and streams developed in Trinidad and Tobago are relatively small, and many are ephemeral. Flash flooding is common, especially during the wet season (July–December), and occurs as overbank channel floods and as sheetwash flow. Landslides, slumping, and severe erosion also occur frequently during wet season floods and form serious natural hazards (ODMP 2014).

In Tobago, surface runoff is limited mainly to ephemeral streams in the lowlands, as surface water quickly infiltrates this porous limestone platform (Lapointe et al. 2010). Streams that drain the leeward (NW) flank of the Main Ridge are short, with steep gradients compared to streams on the windward (SE) flank that have developed longer, more entrenched and comprehensive watersheds. The majority of the streams in the Main Ridge, including the Coffee, Hillsborough, Goldsborough, and Queens rivers, are developed along oblique-slip faults that strike NNW. The headwater regions of streams typically have bedrock channels that transition downstream, to boulder covered, and then to coarse-to-fine-grained alluvial channels. The resistant

Fig. 17.8 Selected special features in Trinidad. **a** Karren dissolution developed on Cretaceous-Jurassic metalimestone bedrock exposed in a corridor of blocks (~10 m height) near the Aripo Cave in the Northern Range. Photograph by J.C. Arkle. **b** Mangrove wetlands in Caroni Swamp, the largest brackish water ecosystem in the country and habitat for the national bird, the scarlet ibis (Eudocimus ruber). Photograph by J.C. Arkle. **c** The Banwari skeleton, collected from a burial site in San Francique, southeast Trinidad, is estimated at 6000–8000 cal yr BP and is the oldest known archaeological site in the Caribbean. Photograph courtesy of the Department of Life Sciences, The University of the West Indies, St. Augustine. **d** Pitch Lake, the largest active asphalt deposit in the world, is located on Trinidad's southwest peninsula in the village of La Brea. A tour guide pulls viscous asphalt into the air with a stick to show tourists. Photograph by Dr. B. Crowley, University of Cincinnati

bedrock units form numerous stepped and hanging waterfalls and plunge pools, such as Argyle (Tobago's highest at ~55 m) and Twin Rivers (~30 m high) waterfalls (see Fig. 17.9), which commonly occur at major lithologic transitions.

Trinidad has four low-gradient truck rivers in the lowlands that generally flow along the axis of four major drainage basins; in the Southern Basin, the Ortoire and South Oropuche rivers, and the Caroni and North Oropuche rivers in the Northern Basin. These main truck rivers have broad, alluvial channels and small tightly spaced meanders. Low-gradient alluvial tributary streams drain the low-relief hills in central and southern Trinidad and dissect the soft shale-rich bedrock. Streams draining the north- and south-oriented catchments of the Northern Range are relatively evenly spaced and have numerous waterfalls and plunge pools. Stream channels on the east side of the mountains are V-shaped, and some have steep (>45°), narrow (<1 m) channels. Streams in western catchments have relatively flat, broad (>1 km) valley floors that cover the basement with over 100 m of sediment (Weber 2005; Ritter and Weber 2007; Arkle et al. 2015). Small river terraces (<20 m long) are present along several streams that drain the south flank of the Northern Range (Kugler 1959).

Holocene sea-level rise caused the islands' rivers to aggrade, and created and modified many of the coastal wetlands (Ramcharan 2004). During sea-level low stands, streams flowed through subaerial portions of the Gulf of

Fig. 17.9 Waterfall (~5 m high) in Castara River, northwest coast of Tobago, developed along NNW striking structures in the North Coast Schist. Photograph by J.C. Arkle

Paria. Trinidad's eastern coastline was ~80 km east of the present shoreline during Pleistocene glacial stages (Warne et al. 2002). The morphology of the eastern offshore is characterized by a gradual eastward sloping continental shelf composed of crystalline basement with a cover of Cretaceous to Quaternary strata. The Quaternary sediment on the slope is incised by a deep and complex network of paleofluvial channels, with depths to channel bases of >100 m below modern sea level (Soto et al. 2007, 2011). East-directed paleofluvial systems there align with some of the modern drainages in eastern Trinidad (Soto et al. 2007, 2011).

The offshore paleofluvial systems may have been important conduits, in terms of the geobiological history of the modern fluvial networks in the Northern Range. Northern Range streams are habitat for the freshwater guppy (*Poecilia reticulata*), which, from an evolutionary perspective, is one of the most well-studied animal species on Earth (e.g., Alexander et al. 2006; Willing et al. 2010). Highly differentiated guppy populations in the Northern Range are concentrated on the north flank of mountains, and on the south flank in the Caroni River watershed and in the Oropuche River watershed (Alexander et al. 2006). The west-flowing Caroni and east-flowing Oropuche Rivers are separated by a major but subtle drainage divide that overlies the Guatapajaro Anticline in the Northern Basin (see Fig. 17.3). Phylogenetic and morphological analyses indicate that the Caroni network guppies are closely related to Venezuelan populations to the west of Trinidad, whereas their Oropuche neighbors share features with populations to the south of Trinidad and in eastern Venezuela and Guyana (Alexander et al. 2006; Willing et al. 2010). Weber et al. (2015b) postulated that, in addition to distal fluvial connections during sea-level low stands, range-wide eastside-up tilting during the Quaternary may have contributed to partial and progressive replacement of guppy stocks in the Northern Range streams. One possibility is that as northern Trinidad was tilted, the Guatapajaro divide may have acted like a rolling hinge, where streams were captured and added to either the Caroni network, with eastern hinge roll, or the Oropuche network, with western hinge roll, which would have then separated the two stocks of guppies (Weber et al. 2015b).

17.3.4 Alluvial Fans

Alluvial fans are not well developed on Tobago, and alluvial deposits are concentrated mainly along streams draining the Main Ridge, such as the Courland River. In contrast, impressive alluvial fan complexes are present in northern Trinidad, particularly along the southern mountain front of the Northern Range (see Fig. 17.3). The alluvial fan deposits are only exposed along the eastern mountain front and have been mapped recently as the Valencia gravels (e.g., de Verteuil et al. 2006). The Valencia gravels were initially mapped as the Pleistocene Cedros Formation on Kugler's (1959) geologic map, which is a type section in southern Trinidad of sediment that was interpreted to have been deposited by the Orinoco River. However, the vein-quartz-pebble gravels and clean quartz sands clearly indicate that the Valencia alluvium was derived from the Northern Range. In the western Northern Basin, range front alluvial fans that correlate with those in the east have subsided to depths over 100 m and are covered by the Caroni Swamp and swamp sediment. These deposits are well known as "the Northern gravels" and are a very important source of groundwater in and around Port of Spain (Weber et al. 2011).

The northern alluvial fans are highly dissected by modern streams and tributaries that flow southward out of the Northern Range. These alluvial fans consist of multiple lobes with abandoned elevated terrace surfaces. Alluvial fan deposits are broad relative to the contributing upstream catchments (see Fig. 17.3). South-flowing streams from the Northern Range deeply entrench the alluvial fans, and modern deposition is focused along the alluvial fan toes. Fan-head entrenchment likely results from a combination of dynamic climatic and tectonic interactions including a reduction of the magnitude and rate of Northern Range erosion during Holocene warming, sea-level rise, and neotectonic uplift of the eastern Northern Range (Weber 2005; Arkle et al. 2015).

17.3.5 Wetlands

Trinidad and Tobago contain a variety of coastal and inland wetlands including swamps, bogs, marshes, and lagoons (see Figs. 17.3 and 17.5). In Trinidad, the majority of wetlands occur along the coastlines, such as the Los Blanquizales Lagoon and Rousillac Swamp, whereas other small wetlands have developed along streams, such as the North and South Oropuche lagoons. The majority of wetlands in Tobago are developed along coastal outlets of streams that drain the Main Ridge such as in Bloody Bay and Lucy Vale in Speyside and are protected by ephemeral barrier beaches. Wetlands located on Tobago's southern limestone lowlands include the Kilgwyn and Little Rockly Bay wetlands, and the Petit Trou Lagoon.

Bon Accord Lagoon is the largest wetland in Tobago covering an area of ~ 1 km^2 and is bound by the Buccoo Reef along the southwest coast (Laydoo 1991; see Figs. 17.5 and 17.6). Biologically, the lagoon supports sea grass, dominantly turtle grass (*Thalassia testudinum*), and is fringed by several species of mangrove (Kenny 1976). Freshwater exchange occurs through small perennial streams and by groundwater transport through the porous limestone (Lapointe et al. 2010). Inflowing water, however, carries pollutants from the developed lowlands, including those from several sewage treatment plants, which put the water quality and ecological health of the lagoon and Buccoo Reef at risk (Lapointe et al. 2010).

The Caroni Swamp in Trinidad is the largest brackish water ecosystem in the country, covering an area of >55 km^2, and is characterized biologically as an herbaceous and mangrove forest wetland (Gibbes et al. 2009; see Fig. 17.8). The Caroni Swamp is a coastal wetland located in the western Northern Basin and is fed by the Caroni River and several streams that drain from the Northern and Central ranges. Tidal exchange occurs over the low-lying topography along the western coast, which is associated with tectonic extension and subsidence in the Gulf of Paria. Local communities harvest fish, crabs, and oysters, as well as other fauna and flora, and the swamp is a popular ecotourism destination due, in part, to its proximal location to Port of Spain. The areas around and within the swamp boundaries have been intensely modified and developed into agricultural and urban land (Gibbes et al. 2009). In addition, land along the northern edge of the swamp has been altered and polluted by Trinidad's large Beetham Landfill.

The Nariva Swamp in Trinidad is the country's largest freshwater wetland, covering an area of >62 km^2 (Gibbes et al. 2009). The swamp is located along the east coast and is bound to the north by the Central Range and to the south by the Darrien Ridge escarpment. The Nariva Swamp and the majority (90%) of coastline along this stretch is, on average, <2 m amsl (Ramcharan 2004). Radiocarbon ages and fossil pollen show a varied vegetation history and that the wetland has existed since at least 6 ka (Ramcharan 2004). The Nariva River drains the swamp, which flows in a shore-parallel depression, bound by the Manzanilla barrier beach and subsidiary sandbars. The Manzanilla barrier beach protects the wetland from high-energy Atlantic waves and assists in regulating salinity levels in the swamp (Gibbes et al. 2009).

Freshwater lakes are limited to artificially constructed reservoirs, such as the Hollis, Navet, and Arena reservoirs in Trinidad, and the Hillsborough Reservoir in Tobago (WASA 2015). The nation's wetlands, both natural and constructed, serve as important ecological habitats for biota including the national bird, the scarlet ibis (*Eudocimus ruber* common to the Caroni Swamp), and the West Indian manatee (*Trichechus manatus* present in the Nariva Swamp), among many others (WASA 2015). Communities around these wetland areas are dependent on their resources.

17.3.6 Coastlines and Coastal Landforms

17.3.6.1 Shorelines and Coastlines

The coastal regions of Trinidad and Tobago vary from the high-energy, windward coasts bordering the Atlantic Ocean to the relatively protected leeward coasts bordering the Caribbean Sea and Gulf of Paria.

Tobago's windward SE coast, bordering the Atlantic, consists of numerous large bays and beaches. In the SW, the coastline along the limestone platform consists of elevated wave-cut cliffs that contain small sea caves and are punctuated by beaches composed of coarse-grained carbonate sand. The central and northern portions of the SE coast consist of large arcuate bays where the resistant oceanic-arc bedrock crops out along the shorelines. Numerous sea arches and sea stacks have formed in the rocky headlands, as well as fringing reefs (see Figs. 17.5 and 17.10).

Although wave and wind action are less intense along Tobago's leeward NW coast, the bathymetry and onshore topography steepen to the NE. Thus, the north and central stretch of this coastline is rocky with steep sea cliffs and hillslopes and consists of intermittent beaches formed in small embayments. Some barrier beaches, baymouth bars, and lagoons occur along stream outlets that drain the Main Ridge, such as those at Bloody Bay. To the south, the continental shelf is shallow and supports numerous, large fringing reefs such as the Buccoo Reef (Fig. 17.6). This segment of coastline consists of long beaches and bays with prominent headlands including Crown Point, Pigeon Point, Mt. Irvine, Black Rock, and Plymouth (from south to north).

Trinidad's Northern Range rises from the Caribbean Sea to form a rugged northern coastline, which is shaped by its bedrock geology and high-energy wave and wind action (see Fig. 17.4). Yarra Point marks a significant transition in coastal morphology, from a highly embayed, drowned coastline in the west, to a semi-linear, elevated coastline in the east; this transition is controlled by Quaternary eastside-up tilting of northern Trinidad (Weber 2005; Arkle et al. 2015). To the east of Yarra Point, steep sea cliffs, marine terraces, rocky headlands, and occasional sea stacks characterize the coastline. Subsidence to the west of Yarra Point, has led to the capture of watersheds by the sea and has left drainage divides and their flanks elevated at or near sea level, where small beaches bound by semi-circular bays have developed, such as Maracas Bay (Weber 2005). Sandbars, beach berms, and barrier beaches are present to both the west and east of Yarra Point. Back-bay areas that contain localized brackish tidal channels, isolated wetlands, and lagoons, including the Maracas Swamp and Madamas Lagoon (see Figs. 17.3 and 17.11), are critical for protection of the delicate wetland biota in the northern bays.

The eastern Atlantic coastline of Trinidad has the longest and straightest beaches on the island, namely Matura, Manzanilla, and Mayaro beaches. Manzanilla Beach is an impressive ~20-km-long barrier beach composed of fine sand and silt across which small ephemeral streams drain (Fig. 17.3). The beaches, sandbars, and shoreline along the east coast are maintained by sediment brought by the Guiana Current (Darsan et al. 2013). However, human-induced erosion from agriculture and recreation has caused severe coastal erosion. Manzanilla Beach may also likely be undergoing severe and rapid retreat due to sea-level rise (e.g., Bedell et al. 1997).

Trinidad's southern coast consists of sea cliffs, beaches, and small wetlands. This coastline is bordered in the east by the Trinity Hills of the Southern Range that rise to 325 m amsl, by low coastal plains in the west, and by the Columbus Channel to the south. The Columbus Channel is a shallow (maximum depth of ~55 m) and narrow waterway that separates Trinidad from mainland South America—by only ~14 km at Icacos Point (see Fig. 17.2). The west-directed Guiana Current flows through the Columbus Channel and has created narrow asymmetric beaches that contain abundant silt and nutrients transported from the Orinoco and Amazon rivers (Darsan et al. 2012). Beach erosion and hillslope failures are severe along the southern coastline due to strong the currents and weakly consolidated sedimentary bedrock (Darsan et al. 2012; Darsan and Alexis 2014).

Trinidad's west coast borders the Gulf of Paria, a large extensional pull-apart basin where submerged and buried normal faults connect the El Pilar and Central Range transform faults (Flinch et al. 1999; Babb and Mann 1999). Tectonic extension results in surface subsidence along the west coast. This creates drowned islands and rocky embayments in the north, mudflats and mangroves along the central-west coast, and low-lying beaches and cliffs in the south (Darsan et al. 2012). The Gulf of Paria is a major depocenter of Orinoco River-derived sediment, with very shallow water depths that, aside from the deep inlets and outlets, reaches to only ~30 m below mean sea level (Van Andel and Sachs 1964). Although the Gulf of Paria is a major depocenter, coastal erosion is a severe problem for

Fig. 17.10 Rocky headlands along the northeast Atlantic coast of Tobago south of Speyside, showing a sea arch and sea stacks. Photograph by J.C. Arkle

many of the shorelines along this stretch of coast (Darsan et al. 2012; Darsan and Alexis 2014).

17.3.6.2 Marine Terraces

Marine terraces are present in both Trinidad and Tobago. The Pleistocene coralline raised reef of southern Tobago's lowlands is exposed over an area of >27 km^2 and has an average thickness of 12 m (Wadge and Hudson 1986; see Fig. 17.6). The dominant fauna are benthic mollusks and hermatypic corals, which are nicely preserved in many areas (Trechmann 1934; Maxwell 1948). This fossil reef formed over the oceanic-arc and forearc bedrock of Tobago in a structurally low, down-dropped half-graben, likely during the last interglacial (Wadge and Hudson 1986; Snoke et al. 2001; Donovan and Jackson 2010). Subsequent uplift and tilting has exposed the reef platform to heights of up to ~10 m amsl along the SSW coast and up to ~30 m amsl in the NE (Donovan and Jackson 2010). The sharp unconformity between the underlying Mesozoic bedrock and the overlying coral limestone is apparent in coastal outcrops at Back Bay to the north of Plymouth (Donovan and Jackson 2010).

In Trinidad, marine terraces occur both as single platforms and as multiple terrace flights that punctuate the emergent NE coastline. Marine terraces are intermittently exposed between Yarra and Galera points and then southward to near Balandra Point at elevations from a few up to ~45 m amsl (Kugler 1959; Barr 1963). Most of the marine terraces consist of an eroded metamorphic strath surface with an overlying cover of unconsolidated quartz-rich, sand- to gravel-sized sediment that generally lacks any carbonate marine fossil debris. The terraces are all mapped as Pleistocene (Kugler 1959; Barr 1963), but recent OSL dating on quartz-rich sands and gravels indicates some may be as young as the late Holocene (Arkle et al. 2015). At least four flights of marine terraces are present and well preserved near

Fig. 17.11 Barrier beach berm, and lagoon of the Madamas River outlet along the north coast of Trinidad. Photograph by J.C. Arkle

the villages of Blanchisseuse and Yarra, but dense vegetation and tropical weathering have degraded and perhaps concealed the continuation of well-defined terrace flights along this coastline. In contrast, at some localities such as Tompire Bay, along the NE coast, a marine terrace is nearly continuously exposed for over 0.5 km at elevation ~20 m amsl (see Fig. 17.12).

17.3.6.3 Reefs

Reefs are abundant around Tobago, but are limited in Trinidad due to the high sediment load derived from the Orinoco and Amazon Rivers and, where present, tend to be dominated by sediment-tolerant species (Bouchon et al. 2008; Kenny 2008). Trinidad's only extensive coral reef is located along the NE coast and forms a flat, broad reef platform that extends ~100 m offshore from Salybia Beach. Reef growth in Salybia Bay is enhanced by its connection with the Caribbean Sea and its protected location (Kenny 2008). Salybia Beach is Trinidad's only white sand beach, composed of broken branching coral and other primary reef-dwelling carbonate fragments.

Tobago is flanked by numerous active patch and fringing reefs that stretch for >90 km around its coastline (nearly 50% of Tobago's coastline) (Laydoo 1991; Kenny 2008; see Fig. 17.5). Tobago's NW coastline, along the Caribbean Sea, is largely protected from the NE weather tracks and high-energy waves and has a higher diversity of coral species than on the Atlantic side of the island (Bouchon et al. 2008). Reefs on the Caribbean coasts support coral-gorgonian communities, whereas macro-algae and sponges dominate on the Atlantic coasts (Bouchon et al. 2008). Fringing reefs also occur along many of Tobago's satellite islands and are particularly abundant along Sisters Rocks, Little Tobago, and Saint Giles Islands.

Buccoo Reef is the largest modern reef system and covers an area of >7 km^2 extending seaward from between Booby

Fig. 17.12 Marine terrace, a metamorphic strath surface with an overlying cover of unconsolidated sand-gravel raised about 30 m amsl, located at Tompire Beach along the northeast coast of Trinidad. Photographs by J.C. Arkle

Point and Pigeon Point in SW Tobago (see Fig. 17.6; Laydoo 1991; Kenny 2008). Freshwater streams enter the reef at Buccoo Bay and in the Bon Accord Lagoon. Buccoo Reef is arcuate and consists of numerous emergent reef flats, which are structurally continuous but fragmented and covered by sediment in places (Laydoo 1991; Kenny 2008). Buccoo Reef Marine Park was established in 1973 as a designated protected marine area. The reef attracts tens of thousands of visitors per year, and this together with urban development and industrial activities in the southern lowlands has caused severe damage, erosion, and alteration to the reef (Laydoo 1991; Hassanali 2013).

17.3.7 Karstlands and Karst Landforms

Trinidad and Tobago have an impressive variety of karst landforms and landscapes, although these are fairly limited in area (Day and Chenoweth 2004). According to Day and Chenoweth (2004), limestone outcrops represent only $\sim 1.6\%$ of Trinidad's and $\sim 8.1\%$ of Tobago's land areas, with most karst located in Trinidad's Northern and Central ranges and in Tobago's southwestern lowlands.

Karst is present only in the coralline limestone platform of the southern lowlands in Tobago (see Figs. 17.5 and 17.13). Rugged and pitted dissolution surfaces are present where the limestone is devoid of vegetation and soil, particularly along the southern coastline. Dry valley systems, ≤ 16 m wide, ≤ 1 km long, and ≤ 2 m deep, are present on the relatively flat carbonate platform. Dolines also occur across the platform surface and are usually ~ 5 m in diameter and <2 m deep (Day and Chenoweth 2004). Several small caves are formed in the sea cliffs along the south coast including Robinson Crusoe Cave (Eshelman and Grady 1990). Narrow openings generally impede access to these caves, but the subterranean cave system is likely fairly

Fig. 17.13 Karst and collapse features developed in the Late Pleistocene coralline limestone near the Robinson Crusoe cave along the south coast of Tobago. Photograph by J.C. Arkle

extensive with complex networks of tunnels and subcaverns (Eshelman and Grady 1990; Day and Chenoweth 2004).

Karst features are present throughout central Trinidad and are developed mainly in the Miocene limestones, particularly in the Tamana Formation, which crops out in the Central Range (Kugler 1959; Erlich et al. 1993; Day and Chenoweth 2004; see Fig. 17.3). The limestone bedrock is exposed as resistant topographic highs, distinct from their softer bounding sedimentary rocks, in a series of northeast trending hills, e.g., the Tamana, Biche, Bassin, and Brigand hills. Ephemeral streams, springs, and dolines are present in the Tamana and Brigand hills (Day and Chenoweth 2004). An impressive cave system on the north flank of Tamana Hill consists of the Tamana Main (>130 m long) and Tamana Dry ($\sim$50 m long) caves (Darlington 1993). The cave consists of several large walk-in chambers, throats, and chimneys, and narrow crawl shafts. The Tamana Main chamber is floored by a small northward-flowing stream that drains beyond the generally accessible passage into the "Far Deep" (Darlington 1993). A variety of depositional and dissolution speleothems occur in this wet and humid cave, including fluted stalactites and flowstone. The Tamana cave system also harbors a significant ecosystem, including $\sim$11 species of bats with a population that can exceed 1.5 million (Goodwin and Greenhall 1961).

The most extensive karst and cave systems in Trinidad are developed within the metacarbonate bedrock (Cretaceous to Jurassic protolith ages; Cenozoic metamorphic ages) of the Northern Range and in the western offshore islands (Shaw 1993; Day and Chenoweth 2004; see Fig. 17.8). Development of karren is abundant on uncovered surfaces and dolines, and associated ephemeral springs are common. The largest known cave system in Trinidad, the Aripo Main Cave, is >860 m long, and is located south of El Cerro del Aripo peak at an altitude of $\sim$800 m amsl (Shaw 1993; see Fig. 17.3). The cave walls and ceiling are lined with intricate and colorful flowstone and speleothems that are generally bulky and include coned stalagmites, large, rounded stalactites, and ribbed columns.

Small vegas, flat-floored intermountain valleys, similar to poljes in karst nomenclature (e.g., Ford and Williams 2007) are scattered throughout the Northern Range. These intermountain valleys are consistently located where metacarbonate layers have been mapped and identified in hillslopes and stream channels (see Fig. 17.3). Some of these vega regions were developed into prolific cocoa plantations during colonial times because of their rich soils and originally tall, dense forest cover and cool, damp microclimates. Segments of metacarbonate stream channels also form small gorges, some of which are lined with tufa, and many of which form several lithologically controlled knickpoints or waterfalls with associated plunge pools such as the 20-m-high Sobo Falls near Brasso Seco.

Day and Chenoweth (2004) identify an area of >6 km^2 of polygonal cockpit karst in Cameron, a block of hills at ~350 m located between Maraval and Petit valleys (see Fig. 17.3). They describe cockpit depressions that reach diameters of 100 m and are generally 15–20 m deep, bound by moderately steep hillslopes (~25°) and serrated ridges. Day and Chenoweth (2004) describe similar features on Paramin Hill, a region famous for the cultivation of culinary herbs, which is located between Maraval village and Maracas Bay. The Cameron karstscape is drained by a star-shaped centripetal drainage system formed along the hillslopes and is perhaps internally drained through concealed cave systems (Day and Chenoweth 2004).

The western offshore islands are composed of the Cretaceous Laventille limestone that contains abundant karst and caves, which are situated both below (drown caves) and above sea level, including the popular Gasparee Cave (Shaw 1993). Caves in northern Trinidad are of great ecologic importance as they harbor unique fauna such as the oilbird (*Steatornia caripensis*; Day and Mueller 2004), blind catfish (*Caecorhamdia urichi*; Shaw 1993), and numerous species of bats (Goodwin and Greenhall 1961).

17.4 Landscape

17.4.1 First Human Interactions with the Landscapes

Trinidad and Tobago have a protracted history of settlement and the longest documented human interactions with landscapes in the West Indies (e.g., Boomert 2000; Siegel et al. 2015). Trinidad has the oldest documented archaeological sites in the West Indies, Banwari Trace, and St. John, which date the earliest people (Amerindians) on the island to ~8 ka (Harris 1973; Boomert 2000; Pagán-Jiménez et al. 2015; see Fig. 17.8). (Note: archaeological ages reported have been calibrated to calendar years.) Archaeological evidence from Tobago indicates earliest occupation of Amerindians before ~5.2 ka, which currently postdates the occupation ages on other West Indies islands farther to the north such as Grenada (~5.6 ka; Siegel et al. 2015).

Trinidad is considered the "first step" on the path of initial human migration and diffusion out of mainland South America into the West Indies (Boomert 2000, 2002; Wilson 2007). This view is based on the Banwari Trace and Trinidad's proximity to the South American mainland, their intermittent connections during sea-level low stands, as well as navigational ease, and resource availability. Although the exact timing and location of a land bridge(s) between mainland South America to Trinidad and Tobago is not known, Trinidad's last mainland connection was likely during the early Holocene and Tobago's most recent connection may have been during the Pleistocene (Boomert 2000). The Gulf of Paria likely did not exist, and Trinidad was connected with the mainland during the global last glacial maximum, while Tobago was likely connected to Trinidad along the NNE coast (Boomert 2000).

Waterways were likely critical for the Amerindians (Boomert 2000, 2009). Boomert (2009) suggests that coastal routes were generally more favored than overland travel routes and interisland migration and trade routes were likely assisted along stream networks, lagoons, and wetlands. Thick tropical vegetation and the Northern Range and Cordillera de la Costa likely formed significant barriers to early human interisland migration and interaction (Boomert 2009; see Fig. 17.2). Early people in central and southern Trinidad interacted most intensely with those in bordering regions of mainland South America, whereas early people that settled along Trinidad's north coast were aligned more closely with other maritime occupants along the Venezuelan coast, Tobago, and the southern islands of the Lesser Antilles (Boomert 2000, 2002, 2009). Throughout early occupation, waterways facilitated Trinidad's development into an important center of cultural exchange and communication (Boomert 2000, 2013; Wilson 2007).

Trinidad and Tobago offered varied geologic and biotic natural resources for Amerindians. Stone artifacts recovered from Amerindian sites on the two islands include intricate ornamental artifacts carved from the islands' different lithologies, and the hard bedrock from Tobago was particularly valuable for cutting tools (Boomert 2000, 2009, 2013). Trinidad and Tobago's intermittent connections and proximity to the mainland also influenced their unique terrestrial biota, which are primarily of continental origin, similar to that of nearby South America (Kenny 2008). The species richness was likely advantageous for the initial occupants providing opportunistic foraging, collecting, fishing, and hunting, including the widespread availability of large game (Steadman and Stokes 2002; Boomert 2013; Siegel et al. 2015).

Paleoenvironmental evidence from Nariva Swamp indicates that the Amerindians used fire to manage and maintain habitats (Siegel et al. 2015). Pagán-Jiménez et al. (2015) identified numerous cultivar and wild taxa from starch grains on stone artifacts collected at the St. John Archaeology Site (see Fig. 17.3). These authors suggest that Trinidad's early occupants ($\sim$7.7 ka) perhaps transported domesticated and wild plant species from mainland South America. Collectively, these findings offer a significant insight into human interactions with the landscape, as highlighted by Siegal et al.'s (2015) submission that the earliest-known occupants of Trinidad were modifying, creating, and managing the landscape roughly 8 ka.

17.4.2 Landscapes Transformed: Colonization to the Twenty-first Century

In 1498, on his third voyage, Christopher Columbus sighted the Trinity Hills in the Southern Range, landed on Trinidad, and claimed the island as a Spanish colony. This "ownership" lasted until 1797 when Trinidad was claimed as a British colony. Tobago, a western spelling of the Amerindian word for tobacco, was fought over by the Spanish, British, French, and Dutch because of its strategic position and fertile land, until 1889 when both Trinidad and Tobago became a British territory. Upon the arrival of Europeans, the Amerindian populations of Trinidad and Tobago were very quickly reduced. The natural vegetation and landscapes of Trinidad and Tobago were transformed by agriculture and plantation-style farming throughout the sixteenth–nineteenth centuries. Trinidad and Tobago developed slave and indentured servant populations during this period. Large areas of Trinidad and Tobago's landscapes were cultivated for tobacco, cacao, and sugarcane (Newson 1976).

Development and proliferation of the country's hydrocarbon industry that quickly followed the drilling of Trinidad's first oil well in 1857 brought significant and irreversible changes to Trinidad and Tobago's landscapes (Higgins 1996). Oil extracted from Trinidad constituted $\sim$40% of the British Empire's total oil production by 1930 (Dillman 2015). In 1962, Trinidad and Tobago gained independence from the British Empire and then became a republic in 1976. The two-island nation is now the leading producer of oil and gas in the Caribbean, which comprised $\sim$45% of their GDP in 2012 (Ministry of Energy and Energy Industries 2015). Although, the agricultural economy has declined during the twentieth–twenty-first centuries, cultivation of Trinidad and Tobago's lands including dredging and modification of wetlands, river channels, and other ecologically sensitive areas has caused significant landscape changes. Extensive limestone quarrying has significantly modified some of the karst terrains such as in the Laventille and Montserrate hills. The diverse landscapes of Trinidad and Tobago have been a great asset and have helped directed the trajectory of people there for roughly 8000 years.

17.5 Heritage and Tourism

17.5.1 Heritage Tourism

Trinidad and Tobago have a diverse and rich cultural blend of people, locally called a "callaloo," that includes Amerindian, Spanish, French, English, African, Indian, Chinese, Syrian, and Lebanese influences. Heritage tourism is an emerging focus for both islands. Tourism contributes only $\sim$4% annually to the economic output (TDC 2015) and employs only $\sim$10.4% of the labor force. Of all Caribbean overnight visitor arrivals in 2011, Trinidad and Tobago had $\sim$400,000 visitors representing only $\sim$2% of Caribbean tourism (TDC 2015). The Government of Trinidad and Tobago recognize the potential and positive economic impacts that tourism could have and has an active policy of encouraging the growth of the tourism trade (Wall and Ali 1977; TDC 2015). With the hydrocarbon industry's dominance in Trinidad, tourism on the island is promoted for its heritage activities, including major annual events such as Carnival, whereas Tobago is better positioned to become a typical "sun, sea, and sand" island of leisure like many of the other Caribbean islands (TDC 2015).

Trinidad and Tobago's heritage sites and historic landmarks include the karat-roofed leepay homes of indentured servants, sugar and cocoa sheds, the barrack houses of slaves, urban chattel houses, the great houses of the Port of Spain Savannah, various estates (plantations), and historic churches, mosques, and temples (Jordan 2013). These historic landmarks, however, are overshadowed by Trinidad's annual Carnival that is celebrated before Ash Wednesday, typically in February or March. The Carnival is responsible for the largest influx of tourists, over $\sim$40,000 annually, which come from around the world (TDC 2015). The Carnival dates to the eighteenth century and showcases the island's heritage through traditional music, dance, and food, and two major street parties—J'Ouvert ("I arise"), a "dirty" mud and cocoa butter nighttime masquerade, which is followed by the daytime "pretty" masquerade. While this event takes over nearly the whole of Port of Spain, it (particularly the "pretty" masquerade) is centered on the Queen's Park Savannah, which is the largest (>1 km^2) public space in Trinidad, and the oldest recreation ground (165 years old) in the West Indies (Jordan 2013). The large number of visitors during this roughly weeklong celebration has lead to tremendous pollution problems in the city and surrounding areas, and it is taxing on the country's natural resources. The need for infrastructure to support such an event, as well as the rapidly growing urban population, has lead to significant changes of the landscape, particularly in the form of hasty development and expansion, particularly in and around Port of Spain (Wall and Ali 1977).

17.5.2 Ecotourism

The natural landscapes of Trinidad and Tobago have the potential for becoming a major destination for ecotourism, particularly because it is one of the most unique and species-rich nations in the region (Kenny 2008; TDC 2015). Notably, the Trinidad and Tobago National Commission was successful in having three sites added to UNESCO's World Heritage Tentative Listing in 2011. These included the Banwari Trace Archaeological Site (cultural), La Brea Pitch Lake (natural), and the Tobago Main Ridge Forest Reserve (mixed) (Jordan 2013; UNESCO 2015).

Trinidad and Tobago's rainforests, habitat to unique biota of continental origin, are popular ecotourism destinations (UNESCO 2015). The Main Ridge Forest Reserve on Tobago encompasses nearly 40 km^2 of mountainous rainforest and is the oldest (since 1776) piece of protected land and forest reserve in the Western Hemisphere (UNESCO 2015). Trinidad's Northern Range includes significant biopreservation sites such as the Asa Wright Nature Reserve (www.asawright.org) that contains over 2200 flowering plant species, 97 native mammals, 400 birds, 55 reptiles, 25 amphibians, and 617 butterflies. Impacts of tourism on Trinidad and Tobago' rainforests have not been quantified. However, undisturbed rainforest in Trinidad's Northern Range has been reduced to isolated patches due to piecemeal degradation related to urbanization, agriculture, mining, and timber production (Day and Chenoweth 2004).

Snorkeling and scuba diving are popular tourist activities, concentrated mainly in Tobago along the southwest coast and around its northeastern satellite islands. The Buccoo Reef Marine Park was designated as a marine protected area in 1973; it has also been a Ramsar Site (see http://www.ramsar.org/) since 2005, and the Buccoo Reef Trust and the University of the West Indies started monitoring at Buccoo Reef Marine Park in 2007 (Laydoo 1991). However, local communities are very dependent on tourism, and thus, land on the nearby limestone platform has been developed for tourist accommodations and corals have been harvested for tourist souvenirs (Wall and Ali 1977; Hassanali 2013). Other threats to Buccoo Reef include sewage pollution, climate change and associated mass-bleaching events, and various coral diseases (Bouchon et al. 2008; Laydoo 1991; Lapointe et al. 2010).

Sandy beaches along the north and east coasts of Trinidad are home to the second largest nesting assemblage of the leatherback sea turtle (*Dermochelys coriacea*) in the Atlantic (e.g., Mycoo and Gobin 2013). Roughly 3000 turtles nest annually on the beach at Grande Riviere along the northeast coast (Mycoo and Gobin 2013). These delicate nesting sites are impacted by both natural and human influences including coastal development, illegal harvesting, climate change, coastal erosion, and thick accumulations of seaweed (*Sargassum*) on the beaches (Mycoo and Gobin 2013; Weber et al. 2015c). The village of Grande Riviere serves as an excellent model for how regions can transition from exploitation to protected ecotourism destinations.

Many other areas of ecological importance and popular ecotourism destinations are present throughout the interior of Trinidad, including the Aripo savannah, and the Caroni and Nariva swamps (see Fig. 17.3). The Aripo savannah is a unique edaphic landscape in the Northern Basin, designated an Environmentally Sensitive Area in 2007, that covers roughly 18 km^2. The area contains several localized dry savannahs with the dominant grass (*Paspalum pulchellum*) developed on clay-rich sandy soil (Richardson 1963). The Caroni and Nariva swamps were designated as protected areas in 1936 and 1968, respectively, and are both major tourist attractions. Traditional activities such as castnet fishing, small-scale rice farming, and the harvest of many of the marine and brackish species, such as oysters, are permitted in the swamps. These activities, as well as ecotourism, have in large part, driven controversial land cover shits in the Caroni and Nariva swamps (e.g., Gibbes et al. 2009). Perceptions of Trinidad and Tobago's landscapes, and thus management of them, remain perched in critical balance between employment, tourism, and conservation, among other environmental and social-political factors (Dillman 2015).

17.6 Hazards

Trinidad and Tobago have experienced several destructive natural disasters, which have been recorded since the seventeenth century. The largest historic earthquake to shake Trinidad was in 1766 and is estimated to have had a magnitude of ~8. Earthquake damage was so severe that Trinidad's people petitioned the King of Spain for resettlement to another Caribbean island. The largest recorded earthquake (M_w 6.7) occurred in 1997, ~5 km offshore south of Tobago along an unnamed fault and caused over US$25 million in damages, injured several people, and left many people homeless in Tobago (ODPM 2014; Weber et al. 2015a).

Even though Trinidad and Tobago are positioned outside of the main Atlantic hurricane belt, the islands have been damaged by several hurricanes. In 1963, Tobago was devastated by one of the deadliest Atlantic hurricanes (Category

3) in recorded history, Hurricane Flora, with winds >160 km/h. The hurricane destroyed or damaged 6250 of Tobago's 7500 houses and at least 18 lives were lost. Property and crop damage on the island amounted to US $30 million (ODPM 2014). The islands regularly experience large tropical storms often associated with distant hurricanes, particularly during the wet season, and annually cause tremendous damage to infrastructure and occasional loss of life (ODPM 2014).

Fires and droughts are among the most common natural hazards in Trinidad and Tobago (ODPM 2014). They are typically associated with the dry season, and year-round high winds. Floods and mass movements are also common. Mass movements range from landslides on steep hillslopes like those in the Northern Range to slow, but destructive creep, especially in the expanding clay soils in central and southern Trinidad. Liquefaction of unconsolidated sediment and long-term subsidence in western Trinidad collectively result in erosion and have caused dramatic changes to the landscape.

Sea-level rise, frequent seismicity, and large magnitude earthquakes pose some of the greatest threats to the islands' (Nurse et al. 1998; ODPM 2014). Low-laying coastal areas in Trinidad and Tobago, like those in other small-island nations of the Caribbean, are extremely vulnerable to sea-level rise as a consequence of human-induced global climate change (e.g., Nurse et al. 1998). This is particularly confounded in Trinidad and Tobago because the islands' major population centers and infrastructure are located along its coastlines. Trinidad's capital, Port of Spain, is built in part on reclaimed land and located along the western coast, which is tectonically subsiding. Sea-level rise and climate change are also linked to potentially devastating hazards such as the loss of agricultural land and delicate ecosystems, coastal erosion, increased pollution, coral reef destruction, changes in aquifer volume and quality, and many other complex direct and indirect environmental responses to global climate change (e.g., Nurse et al. 1998). Efforts to minimize, respond to, and recover from natural hazards are a major concern of local government, community organizations, as well as other West Indian nations and non-governmental organizations abroad (ODPM 2014).

17.7 Conclusions

Trinidad and Tobago's distinct geologic histories and their shallow emergence along the southeast corner of the Caribbean plate throughout the Quaternary provide the context for the two islands' varied landscapes, landforms, and geomorphic evolution. Dynamic geomorphic processes have sculpted a spectacular variety of landscapes and landforms in Trinidad and Tobago. Tectonism, climatic shifts, and associated sea-level fluctuations have shaped the islands' wetlands, modern and ancient reefs, marine terraces, alluvial fan development, and shallow continental shelf. Significant changes in local and regional base level have caused adjustments in fluvial systems, and the redistribution of sedimentation and erosion throughout the islands' interior landscapes (e.g., Ramcharan 2004; Weber 2005; Soto et al. 2007; Arkle et al. 2015). The severe tropical climate causes ubiquitous landsliding and flooding and contributes to intense weathering and to the development of a variety of karst and cave systems (e.g., Day and Chenoweth 2004; ODPM 2014). Strong, sediment-rich ocean currents dominate coastal sedimentation and erosion and influence coral reef development (Darsan et al. 2012).

Trinidad and Tobago's geologic and geomorphic histories have also influenced their biota, the migration and settlement of its people, colonization, and economy. Intermittently connected to mainland South America throughout the Quaternary, the two islands were populated with rich and diverse continental flora and fauna (Kenny 2008). Trinidad played a critical role as a "first step" for initial human migration out of mainland South America and then as a central hub for trade as early Amerindians dispersed into the West Indies (Boomert 2000, 2002; Wilson 2007). Geologic structures, both ancient and active, control the distribution and nature of active hydrocarbon systems, and the most transient landforms on Trinidad, mud volcanoes and asphalt seeps. As the leading hydrocarbon producer in the Caribbean, tourism is an important and emerging economic sector for Trinidad and Tobago, with its diverse culture heritage, unique biodiversity, and varied landforms.

Natural and human-induced environmental changes continue to cause dramatic modifications of the islands' landscapes. Potential natural hazards include frequent and large magnitude earthquakes, climate change and associated sea-level rise, tropical storms and associated flooding, mass wasting, wind damage, and dry-season fires. Perceptions of the landscape and its management require a delicate balance between employment and tourism, and ecological protection (Dillman 2015).

Acknowledgements JCA and LAO thank the University of Cincinnati (UC) for funding fieldwork in Trinidad and Tobago. JCA thanks UC for funding her doctoral work on the geomorphology of Trinidad and Tobago. JCW thanks all of the generous, warm, friendly, and helpful Trinidadians and Tobagonians that he has interacted with while working and learning in the field.

References

Alexander HJ, Taylor JS, Wu SS-T, Breden F (2006) Parallel evolution and vicariance in the guppy (*Poecilia reticulata*) over multiple spatial and temporal scales. Evolution 60(11):2353–2369

Algar S, Pindell J (1993) Structure and deformation history of the Northern Range of Trinidad and adjacent areas. Tectonics 12 (4):814–829

Arkle JC, Weber J, Enkelmann E, Owen LA (2014) Exhumation in the Southeast Caribbean plate corner. In: Abs., Thermo 2014—the 14th international thermochronology conference, Chamonix, France, 8–12 Sept 2014

Arkle JC, Owen LA, Weber J, Moonan M, Enkelmann E (2015) Late Neogene-recent evolution of the Northern Range, Trinidad. In: Abs., Caribbean geological conference, Port of Spain, Trinidad, 17–21 May 2015

Babb S, Mann P (1999) Structural and sedimentary development of a Neogene transpressional plate boundary between the Caribbean and south America plates in Trinidad and the Gulf of Paria. In: Mann P (ed) Sedimentary Basins of the World, Caribbean Basins. Elsevier, Amsterdam, pp 495–557

Baker PA, Fritz SC (2015) Nature and causes of quaternary climate variation of tropical South America. Quat Sci Rev 124:31–47

Barr KW (1963) The geology of the Toco District, Trinidad, W.I., overseas geological surveys. Her Majesty's Stationary Office, London

Boomert A (2000) Trinidad, Tobago, and the Lower Orinoco interaction sphere: an archaeological/ethnohistorical study. Cairi Publications, Alkmaar, The Netherlands

Boomert A (2002) Amerindian-European encounters on and around Tobago (1498–ca. 1810). Antropológica 97(98):71–207

Boomert A (2009) Between the Mainland and the islands: the Amerindian cultural geography of Trinidad. Bull Peabody Museum Nat Hist 50(1):63–73

Boomert A (2013) Gateway to the mainland. The Oxford Handbook of Caribbean Archaeology, p 141

Bouchon C, Portillo P, Bouchon-Navaro Y, Louis M, Hoetjes P, De Meyer K, Macrae D, Armstrong H, Datadin V, Harding S, Mallela J, Parkinson R, van Bochove J, Wynne S, Lirman D, Herlan J, Baker A, Collado L, Nimrod S, Mitchell J, Morrall C, Isaac C (2008) Chapter 19. Status of coral reefs of the Lesser Antilles: The French West Indies, The Netherlands Antilles, Anguilla, Grenada, Trinidad and Tobago. In: Wilkinson C (ed) Status of coral reefs of the world 2008. Global Coral Reef Monitoring and Reef and Rainforest Research Centre, Townsville, Australia, pp 265–279

Cerveny PF, Snoke AW (1993) Thermochronologic data from Tobago, West Indies: constraints on the cooling and accretion history of Mesozoic oceanic-arc rocks in the southern Caribbean. Tectonics 12 (2):433–440

Crosby CJ, Prentice CS, Weber JC, Ragona D (2009) Logs of paleoseismic excavations across the Central Range fault, Trinidad. US Geological Survey Open-File Report, 2331–1258

Cruz L, Fayon A, Teyssier C, Weber JC (2007) Exhumation and deformation processes in transpressional orogens: the Venezuelan Paria Península, SE Caribbean-South American plate boundary. Geol Soc Am Spec Pap 434:149–165

Darlington J (1993) Recent work on the caves of Trinidad and Tobago. Acta Carsologica 22:77–87

Darsan J, Alexis C (2014) The impact of makeshift sandbag groynes on coastal geomorphology: a case study at Columbus Bay, Trinidad. Environ Nat Resour Res 4(1):94

Darsan J, Ramnath S, Alexis C (2012) Status of beaches and bays in Trinidad 2004–2008. Technical Report, Institute of Marine Affairs, Hilltop Lane, Chaguaramas, p 214

Darsan J, Asmath H, Jehu A (2013) Flood-risk mapping for storm surge and tsunami at Cocos Bay (Manzanilla) Trinidad. J Coast Conserv 17(3):679–689

Day MJ, Chenoweth MS (2004) The karstlands of Trinidad and Tobago, their land use and conservation. Geogr J 170(3):256–266

Day MJ, Mueller W (2004) Aves (birds). In: Gunn J (ed) The encyclopedia of caves and karst science. Taylor and Francis, New York, pp 130–131

de Verteuil L, Ramlal B, Weber J (2006) Trinidad geological GIS, module 1- surface geology and geography. Latinum, Ltd., Port-of-Spain, Trinidad

DeMets C, Gordon RG, Argus DF, Stein S (1994) Effect of recent revisions to the geomagnetic reversal time scale on estimates of current plate motions. Geophys Res Lett 21(20):2191–2194

Deville E, Guerlais SH (2009) Cyclic activity of mud volcanoes: evidences from Trinidad (SE Caribbean). Mar Pet Geol 26(9):1681–1691

Deville E, Battani A, Griboulard R, Guerlais S, Herbin JP, Houzay JP, Muller C, Prinzhofer A (2003) The origin and processes of mud volcanism: new insights from Trinidad. Geol Soc Lond Spec Publ 216(1):475–490

Dia AN, Castrec-Rouelle M, Boulègue J, Comeau P (1999) Trinidad mud volcanoes: where do the expelled fluids come from? Geochim Cosmochim Acta 63(7–8):1023–1038

Dillman J (2015) Colonizing paradise: landscape and empire in the British West Indies. University of Alabama Press

Donovan SK (1989) Palaeoecology and significance of barnacles in the mid-Pliocene Balanus Bed of Tobago, West Indies. Geol J 24 (4):239–250

Donovan SK, Jackson TA (2010) Classic localities explained 6: Tobago. Geol Today 26(6):233–239

Erlich RN, Barreit SF (1990) Cenozoic plate tectonic history of the northern Venezuela-Trinidad area. Tectonics 9:161–184

Erlich RN, Farfan PF, Hallock P (1993) Biostratigraphy, depositional environments, and diagenesis of the Tamana Formation, Trinidad: a tectonic marker horizon. Sedimentology 40(4):743–768

Eshelman RE, Grady F (1990) The caves of Crown point, Tobago, West Indies. Nat Speleol Soc Bull 52(1):16–20

Flinch JF, Rambaran V, Ali W, Lisa VD, Hernández G, Rodrigues K, Sams R (1999) Chapter 17, Structure of the Gulf of paria pull-apart basin (Eastern Venezuela-Trinidad). In: Mann P (ed) Sedimentary basins of the World, vol 4. Elsevier, pp 477–494

Ford D, Williams P (2007) Karst hydrogeology and geomorphology Chichester. Wiley, 562 p

Garciacaro E, Mann P, Escalona A (2011) Regional structure and tectonic history of the obliquely colliding Columbus foreland basin, offshore Trinidad and Venezuela. Mar Pet Geol 28(1):126–148

Gibbes C, Southworth J, Keys E (2009) Wetland conservation: change and fragmentation in Trinidad's protected areas Geoforum 40 (1):91–104

Goodwin GG, Greenhall AM (1961) A review of the bats of Trinidad and Tobago: descriptions, rabies infection, and ecology 122:191–301

Harris P (1973) Preliminary report on Banwari trace, a preceramic site in Trinidad. In: Proceedings of the fourth international congress for the study of Pre-Columbian cultures of the Lesser Antilles, St. Lucia Archaeological and Historical Society, Castries, pp 115–125

Hassanali K (2013) Towards sustainable tourism: the need to integrate conservation and development using the Buccoo Reef Marine Park, Tobago, West Indies. Nat Resour Forum 37(2):90–102

Haug GH, Hughen KA, Sigman DM, Peterson LC, Röhl U (2001) Southward migration of the intertropical convergence zone through the holocene. Science 293(5533):1304–1308

Higgins GE (1996) A history of Trinidad Oil, Trinidad Express Newspapers

Higgins GE, Saunders JB (1967) Report on the 1964 Chatham mud island, Erin Bay, Trinidad, West Indies. Am Assoc Petrol Geol Bull 51(1):55–64

Higgins GE, Saunders JB (1974) Mud Volcanoes. Their nature and origin. Verhandlungen Naturforschenden Gesselschaft in Basel 84:101–152

Hijmans RJ, Cameron SE, Parra JL, Jones PG, Jarvis A (2005) Very high resolution interpolated climate surfaces for global land areas. Int J Climatol 25(15):1965–1978

Hughen KA, Eglinton TI, Xu L, Makou M (2004) Abrupt tropical vegetation response to rapid climate changes. Science 304 (5679):1955–1959

Jordan L-A (2013) Managing built heritage for tourism in Trinidad and Tobago: challenges and opportunities. J Heritage Tourism 8(1):49–62

Kenny JS (1976) A preliminary study of the Buccoo Reef/Bon Accord complex with special reference to development and management. Department of Biological Sciences, University of the West Indies, St. Augustine

Kenny JS (2008) The biological diversity of Trinidad and Tobago: a naturalist's notes. Prospect Press/MEP, Port of Spain, 265 p

Kugler HG (1959) Geological map and sections of Trinidad. Port of Spain, Petroleum Association of Trinidad, scale 1:100,000. 1 sheet

Lapointe BE, Langton R, Bedford BJ, Potts AC, Day O, Hu C (2010) Land-based nutrient enrichment of the Buccoo Reef complex and fringing coral reefs of Tobago, West Indies. Marine Pollut Bull 60 (3):334–343

Laydoo RS (1991) A guide to the coral reefs of Tobago. Institute of Marine Affairs and the Asa Wright Nature Centre, Republic of Trinidad and Tobago

Maxwell JC (1948) Geology of Tobago, British West Indies. Geol Soc Am Bull 59(8):801–854

Ministry of Energy and Energy Industries (2015) Government of the Republic of Trinidad and Tobago, http://www.energy.gov.tt/our-business/oil-and-gas-industry/. Accessed 20 July 2015

Muhs DR, Simmons KR, Schumann RR, Halley RB (2011) Sea-level history of the past two interglacial periods: new evidence from U-series dating of reef corals from south Florida. Quat Sci Rev 30:570–590

Muhs DR, Pandolfi JM, Simmons KR, Schumann RR (2012) Sea-level history of past interglacial periods from uranium-series dating of corals, Curaçao, Leeward Antilles Islands. Quat Res 78(2):157–169

Mycoo M, Gobin J (2013) Coastal management, climate change adaptation and sustainability in small coastal communities: leatherback turtles and beach loss. Sustain Sci 8(3):441–453

Newson LA (1976) Aboriginal and Spanish Colonial Trinidad: a study in cultural contact. Academic Press, London, New York

Nurse LA, McLean RF, Suarez AG (1998) Small island states. In: The regional impacts of climate change: an assessment of vulnerability. In: Watson RT, Zinyowera MC, Moss RH (eds) A Special report of IPCC working group II. Cambridge University Press, Cambridge, United Kingdom and New York, NY, USA, pp 331–354

ODPM (Office of Disaster Preparedness and Management) (2014) Preliminary vulnerability assessment of Trinidad and Tobago. Prepared by Ms. Rishma Maharaj attached to the Mitigation Planning and Research Unit of the Office of Disaster Preparedness and Management, Tacarigua, Trinidad

Pagán-Jiménez JR, Rodríguez-Ramos R, Reid BA, van den Bel M, Hofman CL (2015) Early dispersals of maize and other food plants into the Southern Caribbean and Northeastern South America. Quat Sci Rev 123:231–246

Pérez OJ, Bilham R, Bendick R, Velandia JR, Hernández N, Moncayo C, Hoyer M, Kozuch M (2001) Velocity field across the Southern Caribbean plate boundary and estimates of Caribbean/South-American plate motion using GPS geodesy 1994–2000. Geophys Res Lett 28(15):2987–2990

Peterson LC, Haug GH, Hughen KA, Röhl U (2000) Rapid changes in the hydrologic cycle of the tropical atlantic during the last glacial. Science 290:1947–1951

Pindell JL, Higgs R, Dewey J (1998) Cenozoic palinspastic reconstruction, paleogeographic evolution, and hydrocarbon setting of the northern margin of South America. In: Pindell JL, Drake C (eds) Paleogeographic evolution and non-glacial eustasy, northern South America: SEPM (Society for Sedimentary Geology) Special Publication 58, pp 45–85

Prentice CS, Weber J, Crosby CJ (2001) Paleoseismic and geomorphic evidence for quaternary fault slip on the central range fault, South American-Caribbean Plate Boundary, Trinidad. EOS Trans Am Geophys Union 82:F928

Prentice CS, Weber JC, Crosby CJ, Ragona D (2010) Prehistoric earthquakes on the Caribbean-South American plate boundary, Central Range fault, Trinidad. Geology 38(8):675–678

Bedell M, Kurtis B, Kurtis P, Chicago Production C, Wttw and Public Media, V. 1997, Is Trinidad drowning?: [Chicago, Ill.], Public Media Video

Ramcharan EK (2004) Mid-to-late Holocene sea level influence on coastal wetland development in Trinidad. Quat Int 120(1):145–151

Richardson WD (1963) Observations on the vegetation and ecology of the Aripo Savannas, Trinidad. J Ecol 51(2):295–313

Ritter J, Weber J (2007) Geomorphology and quaternary geology of the Northern Range, Trinidad and Paria Peninsula, Venezuela: recording quaternary subsidence and uplift associated with a pull-apart basin. In: Proceedings, geological society of Trinidad and Tobago, fourth geological conference

Shaw T (1993) The history of cave studies in Trinidad, Jamaica, The Bahamas, and some other Caribbean islands. Acta Carsologica 22:15–76

Siegel PE, Jones JG, Pearsall DM, Dunning NP, Farrell P, Duncan NA, Curtis JH, Singh SK (2015) Paleoenvironmental evidence for first human colonization of the eastern Caribbean. Quat Sci Rev 129:275–295

Snoke AW, Rowe DW, Yule JD, Wadge G (2001) Petrologic and structural history of Tobago, West Indies: a fragment of the accreted Mesozoic oceanic arc of the southern Caribbean. Geol Soc Am Spec Pap 354:1–54

Soto DM, Mann P, Escalona A, Wood LJ (2007) Late Holocene strike-slip offset of a subsurface channel interpreted from three-dimensional seismic data, eastern offshore Trinidad. Geology 35(9):859–862

Soto D, Mann P, Escalona A (2011) Miocene-to-recent structure and basinal architecture along the central range strike-slip fault zone, eastern offshore Trinidad. Mar Pet Geol 28(1):212–234

Steadman D, Stokes A (2002) Changing exploitation of terrestrial vertebrates during the past 3000 years on Tobago, West Indies. Hum Ecol 30(3):339–367

TDC (Tourism Development Company Limited) (2015) Implementation of the Ministry of Tourism of the Government of the Republic of Trinidad and Tobago, http://www.tdc.co.tt/index.htm. Accessed July 20 2015

Trechmann CT (1934) Tertiary and quaternary beds of Tobago, West Indies. Geol Magazine 71(11):481–493

UNESCO (United Nations Educational, Scientific and Cultural Organization) (2015) World Heritage Centre, http://whc.unesco.org. Accessed 10 Aug 2015

Van Andel TH (1967) The Orinoco Delta. J Sediment Petrol 37:297–310

Van Andel TH, Sachs PL (1964) Sedimentation in the Gulf of Paria during the Holocene transgression: a subsurface acoustic refraction study. J Mar Res 22:30–50

van der Hammen T, Hooghiemstra H (2000) Neogene and quaternary history of vegetation, climate, and plant diversity in Amazonia. Quatern Sci Rev 19(8):725–742

Wadge G, Hudson D (1986) Neotectonics of southern Tobago. In: Rodrigues K (ed) Transactions of the 1st geologic conference of the geologic society of Trinidad and Tobago, Port of Spain, Trinidad, 10–12 July 1985, pp 7–20

Wall G, Ali IM (1977) The impact of tourism in Trinidad and Tobago. Ann Tourism Res 5:43–49

Warne A, Guevara E, Aslan A (2002) Late quaternary evolution of the Orinoco delta, Venezuela. J Coastal Res 18:225–253

WASA (Water and Sewerage Authority) (2015) http://www.wasa.gov.tt . Accessed 10 Aug 2015

Weber JC (2005) Neotectonics in the Trinidad and Tobago, West Indies segment of the Caribbean-South American plate boundary. Occas Pap Geolog Inst Hungary 204:21–29

Weber JC, Dixon TH, DeMets C, Ambeh WB, Jansma P, Mattioli G, Saleh J, Sella G, Bilham R, Pérez O (2001a) GPS estimate of relative motion between the Caribbean and South American plates, and geologic implications for Trinidad and Venezuela. Geology 29(1):75–78

Weber JC, Ferrill DA, Roden-Tice MK (2001b) Calcite and quartz microstructural geothermometry of low-grade metasedimentary rocks, Northern Range, Trinidad. J Struct Geol 23(1):93–112

Weber JC, Saleh J, Balkaransingh S, Dixon T, Ambeh W, Leong T, Rodriguez A, Miller K (2011) Triangulation-to-GPS and GPS-to-GPS geodesy in Trinidad, West Indies: Neotectonics, seismic risk, and geologic implications. Marine Pet Geol 28(1):200–211

Weber JC, Geirsson H, Latchman JL, Shaw K, La Femina P, Wdowinski S, Higgins M, Churches C, Norabuena E (2015a) Tectonic inversion in the Caribbean-South American plate boundary: GPS geodesy, seismology, and tectonics of the M_w 6.7 22 April 1997 Tobago earthquake: Tectonics, 1181–1194

Weber J, Arkle JC, Noriega N (2015b) Northern range, Trinidad: the guppy geomorphology connection. In: Abs., Caribbean geological conference, Port of Spain, Trinidad, 17–21 May 2015

Weber JC, William N, Arkle JC (2015c) A tale of two beaches: Tompire Bay, NE Trinidad. In: Coastal care. beach of the month, www.coastalcare.org

Willing E-M, Bentzen P, Van Oosterhout C, Hoffmann M, Cable J, Breden F, Weigel D, Dreyer C (2010) Genome-wide single nucleotide polymorphisms reveal population history and adaptive divergence in wild guppies. Mol Ecol 19(5):968–984

Wilson SM (2007) The archaeology of the Caribbean. Cambridge University Press, New York

WRA (Trinidad and Tobago Water Resources Agency) (2001) Integrating the management of watersheds and coastal areas in Trinidad and Tobago. Prepared by the Water Resource Agency for the Ministry of the Environment, Complex Independent Square, Port of Spain

Aruba, Bonaire, and Curaçao

18

Phillip P. Schmutz, Amy E. Potter and E. Arnold Modlin Jr

Abstract

Aruba, Bonaire, and Curaçao consist of five islands that make up the Leeward Islands of the Dutch Antilles off of the northern coast of South America, often referred to as the ABCs. Because of location, the islands share similar characteristics of climate, geology, geomorphology, and history, though each bears a variation of these characteristics. These largely limestone islands are part of the southern Caribbean dry zone. Historically, they were important centers of livestock grazing, aloe vera production, and salt export. All three islands in the twentieth century have been impacted by the growth of oil export in Venezuela. Curaçao, once the tourism leaders among the Dutch Caribbean Islands, is trying to catch up with Aruba in stopover tourists, while Bonaire has smaller but strong "active" tourism for its size. While their climate in the past was considered less than desirable for large-area agricultural pursuits, presently their dry, warm weather largely outside the hurricane belt is an attractive asset to the millions of tourists who visit them annually.

Keywords

Tourism • Dutch • Slavery • Salt • Oil

18.1 Introduction

Aruba, Bonaire, and Curaçao consist of five islands off of the northern coast of South America and comprise the Leeward Islands of the Dutch Antilles.[1] Although nicknamed the ABCs, when actually approaching the islands from west to east, the physical order of the islands would be Aruba, Curaçao, and then Bonaire. The islands range from 29 km off of the Venezuelan coast of South America in the case of Aruba, to 92 km off of the South American coast for Bonaire. All three territories are connected to the Netherlands, though in slightly different ways. Aruba and Curaçao are constituent countries of the Kingdom of the Netherlands, while Bonaire is a special municipality within the state of the Netherlands.

Aruba (Fig. 18.1) has a land area of approximately 179 km^2. Two islands comprise Curaçao (Fig. 18.2): the main island of Curaçao (444 km^2) and the uninhabited "little" or Klein Curacao southeast of the main island, which is 1.7 km^2 in size. Bonaire (Fig. 18.3) comprises the larger, inhabited island of Bonaire (287 km^2) and the smaller uninhabited island of Klein Bonaire (6 km^2) off its west coast. Elevations for all three islands range from sea level to a height of 165 m (Mt. Yamanola) on Aruba, 240 m (Brandaris Hill) on Bonaire, and 372 m (Mt. Christoffel) on

[1]The application of Leeward here is not to be confused with its use in the greater Lesser Antilles Caribbean region.

P.P. Schmutz
Department of Earth and Environmental Sciences, University of West Florida, Pensacola, FL, USA
e-mail: pschmutz@uwf.edu

A.E. Potter (✉)
Department of History, Armstrong State University, Savannah, GA, USA
e-mail: amy.potter@armstrong.edu

E. Arnold Modlin Jr
History and Interdisciplinary Studies Department, Norfolk State University, Norfolk, VA, USA
e-mail: arnoldmodlin@gmail.com

C.D. Allen (ed.), *Landscapes and Landforms of the Lesser Antilles*, World Geomorphological Landscapes, DOI 10.1007/978-3-319-55787-8_18

Fig. 18.1 General physiographic map of Aruba. Notice the long sandbars along the island's southern coast and the Arikok National Park covering much of the eastern end. Cartography by K.M. Groom

Fig. 18.2 General physiographic map of Curaçao. The slender island has many inland bays and waterways. Cartography by K.M. Groom

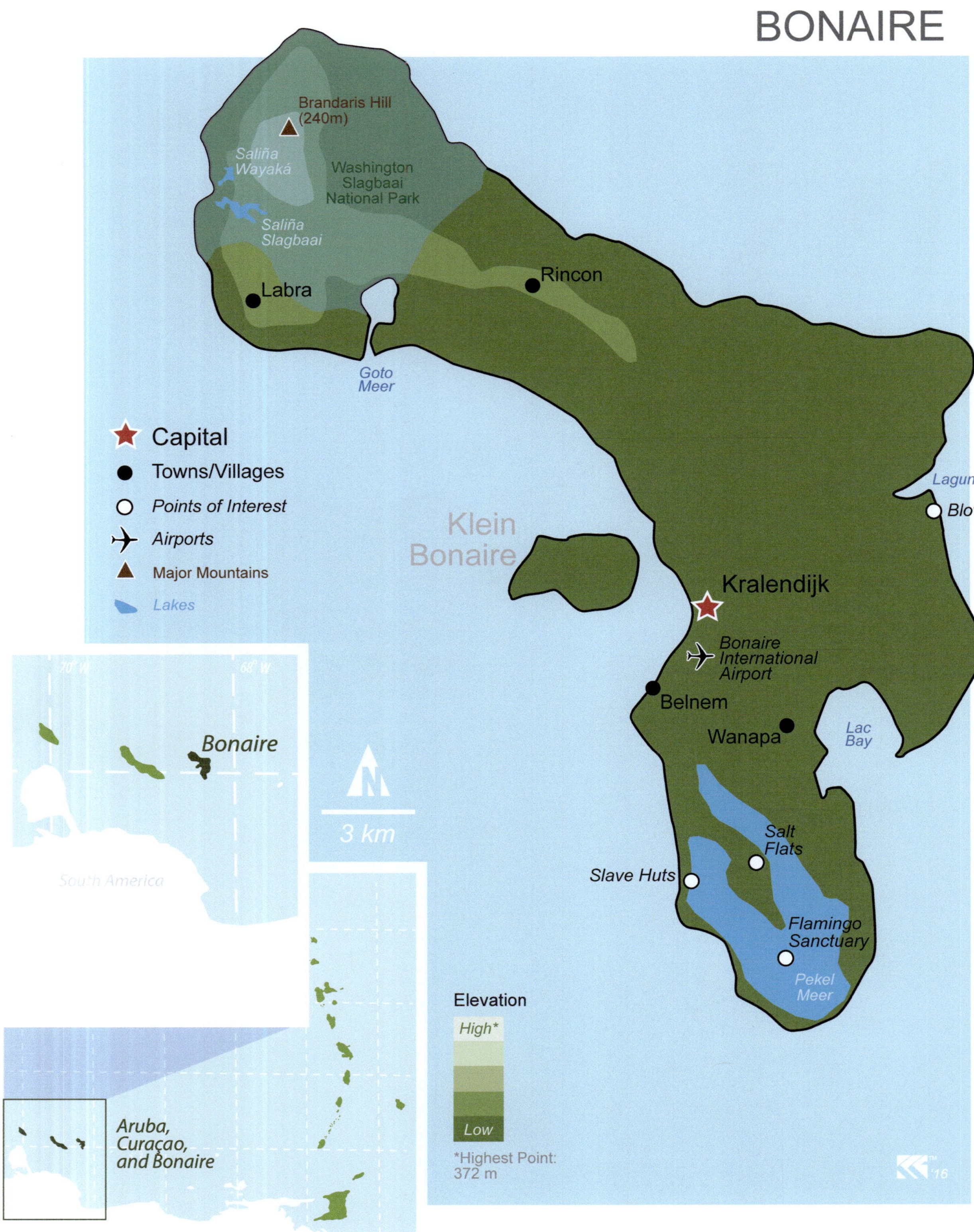

Fig. 18.3 General physiographic map of Bonaire. The southern end of the island is dominated by large salt flats and inland bays. There are saline lakes in the northwest but not nearly as large. The northern pools are also included in the Washington Slagbaai National Park. Cartography by K.M. Groom

Fig. 18.4 a Christoffel Mountain, the highest point on Curaçao at 372 m, with a small salt pan visible in the foreground. Photograph by Amy Potter, May 2012

Curaçao (Fig. 18.4). Both Klein Bonaire and Klein Curaçao are low islands just a few meters above high tide. Geologically, the islands are similarly composed of the Late Cretaceous and Paleogene volcanic and sedimentary rocks overlain by multiple sequences of Quaternary limestone terraces.

Located in the southern Caribbean dry zone, characterized by a tropical steppe, semiarid hot climate, the ABCs' average temperatures range seasonally between 26 and 29 °C (Department Meteorologio Aruba 2010; Meteorological Department of Curaçao 2016). Annual rainfall is characterized by extreme interyear variability, with a distinct dry season from January to September and an October to December wet season. Total annual rainfall varies between the islands: Aruba averages 470 mm, Curaçao averages 550 mm, and Bonaire averages 490 mm. The northern and eastern coasts for all three islands are exposed to consistent easterly trade winds and experience high wave energy, whereas the southern and western coasts are characterized by low-energy leeward conditions.

Curaçao has the largest population, with 158,986 people as of January 2016 (Central Bureau of Statistics Curaçao 2016). Aruba's population was 101,484 according to the 2010 census (Central Bureau of Statistics 2010), while Bonaire's population stood at 18,413 on January 1, 2014 (Centraal Bureau voor de Statistiek 2015). This puts the population density at 358 people per km^2 in Curaçao, 551 people per km^2 in Aruba, and 64 people per km^2 in Bonaire. Each of the islands has shown growth since their last censuses in either 2010 (Aruba) or 2011 (Curaçao and Bonaire).

18.2 Geologic Setting

Aruba, Bonaire, and Curaçao lie along a 200 km segment of the east–west-trending Leeward Antilles ridge within the Caribbean-South America plate boundary zone (Hippolyte and Mann 2011; Zapata et al. 2014). The Leeward Antilles ridge formed as part of the Cretaceous Great Arc of the Caribbean, as the arc was thrust against the northern margin of the South American continent during the westward advance of the Caribbean plate from the Pacific Ocean (Beardsley and Ave-Lallemant 2007; Gorney et al. 2007; Van der Lelij et al. 2010; Neill et al. 2011). This collision and the subsequent uplift of the continental margin, more than 5 km of vertical displacement, in the Late Cretaceous and Paleogene caused the folding, faulting, and metamorphism of the allochthonous igneous rocks that constitute the geologic core of the Leeward Antilles ridge and ABCs (Beets 1972; Beets et al. 1984; White et al. 1999; Beardsley and Ave-Lallemant 2007; Gorney et al. 2007). The ABCs were emergent throughout the Paleogene, upon which significant denudation of the volcanics occurred producing, in places, thick layers of clastic sedimentary rocks (Westermann 1932; De Buisonje 1974).

The geologic core of all three islands is similarly composed of the Late Cretaceous and Paleogene volcanic and sedimentary rocks; however, the lithology and sequence phases differ slightly between the islands. Aruba contains a basement sequence of stratified basalt flows, volcano-clastic deposits, and epiclastic strata, which are subsequently intruded by a large tonalitic batholith (Fig. 18.5; Westermann 1932; De Buisonje 1974; White et al. 1999; Van der Lelij et al. 2010). The greater part of the tonalitic batholith has been removed through denudation; however, segments of the batholith formation are still evident by the large boulder structures located in the central plain of Aruba (Fig. 18.6). On Curaçao, the geology is divided into three stratigraphic sequences: the Curaçao Lava Formation (tholeiitic basalts), the Knip Group (silica-rich rocks and clastic sediments), and the Midden-Curaçao Formation (conglomerate, sandstone, and shale) (Fig. 18.7; Beets 1972; De Buisonje 1974; Fouke et al. 1996; Kerr et al. 1996; White et al. 1999). On Bonaire, the basement sequence (Washikemba Formation) is composed of basaltic flows and pyroclastic deposits interbedded with pelagic chert and limestones (Fig. 18.8). This formation is overlain by three sequences of sedimentary strata: the Maastrichtian Rincon Formation (limestone clasts interbedded in turbidite conglomerates), the Soebi Blanco Formation (conglomerates, sandstones, and shales of fluviatile origin), and limestones of the Seroe Montagne Formation (Pijpers 1933; Beets 1972; Beets et al. 1977).

Sea-level fluctuations combined with a slow tectonic rise of the Leeward Antilles ridge throughout the Neogene and Quaternary produced numerous limestone terraces that unconformably overlay the Cretaceous and Paleogene volcanic and sedimentary rocks (Alexander 1961; De Buisonje 1974). On Curaçao and Bonaire, the limestone formations occur along the coast rim and within portions of the interior of the islands; however, on Aruba, the limestone terraces occur extensively along the leeward side of the island and, in small, isolated patches along the windward coastal rim (De Buisonjé 1974). The oldest limestone terrace is the Miocene formed Seroe Domi Formation, which occurs almost exclusively along the leeward coast of all three islands (De Buisonje 1974; Fouke et al. 1996). Fouke et al. (1996) established that the Seroe Domi Formation experienced three major episodes of limestone diagenesis and dolomitization taking place following successive Mio-Plio-Pleistocene depositional and subaerial exposures. The Seroe Domi Formation is in turn overlying Quaternary limestone terraces composed of lagoon and fringing reef coral–algal grainstones (Alexander 1961; Bandoian and Murray 1974; De Buisonje 1974; Fouke et al. 1996). The terraces are preserved as a set of four marine accumulation sequences, from oldest to youngest: Highest (150–90 m), Higher (80–50 m), Middle (15–45 m), and Lower (<10 m). The Highest terrace is found only on a narrow portion of Curaçao, while the Higher terrace is located along the windward sides of all three islands. The Middle and Lower terraces developed during a transgressive period of sea-level fall and are present along both the windward and leeward sides of all three islands. Both the Middle and Lower terraces show the most complete development and can be divided into three biozones [barrier reef and two lagoonal zones] (De Buisonje 1974; Fouke et al. 1996). Due to the level of reef development, the Middle and Lower terraces are separated by steep cliffs, which have extensive karst (e.g., cave and fissure) structures (Fig. 18.9). Eolianites deposits, up to 15 m thick in places, can be found atop the Higher, Middle and Lower terraces on the windward sides of all three islands (De Buisonjé 1974).

18.3 Landforms

Even though the ABCs were primarily formed by plate collisions that resulted in igneous rocks being formed and subsequently exposed, much of their surficial landforms remain sedimentary. This occurred during the Upper Eocene when the volcanics were overlain by sedimentary events such as reef terraces. Today, remnants of the volcanics can be seen on the surface in such formations as the Ayo and Casibari Rock Formations (Fig. 18.6), and the Hooiberg (haystack) on Aruba (Fig. 18.10), and the Brandaris on Bonaire (the island's highest point, at 241 m; Fig. 18.11). Karstic features are visible on the ABC's coasts as well, such as the limestone features at Shete Boka (Seven Inlets) National Park on Curacao's northern coast (Fig. 18.12). Though relatively flat in relief compared to most of their Lesser Antillean island cousins, the ABCs still boast significant geomorphologic areas, the most prominent types of which are featured in this section.

18.3.1 Coastal

The ABCs include a variety of sandy beach environments, varying from small sheltered coves to extensive stretches of

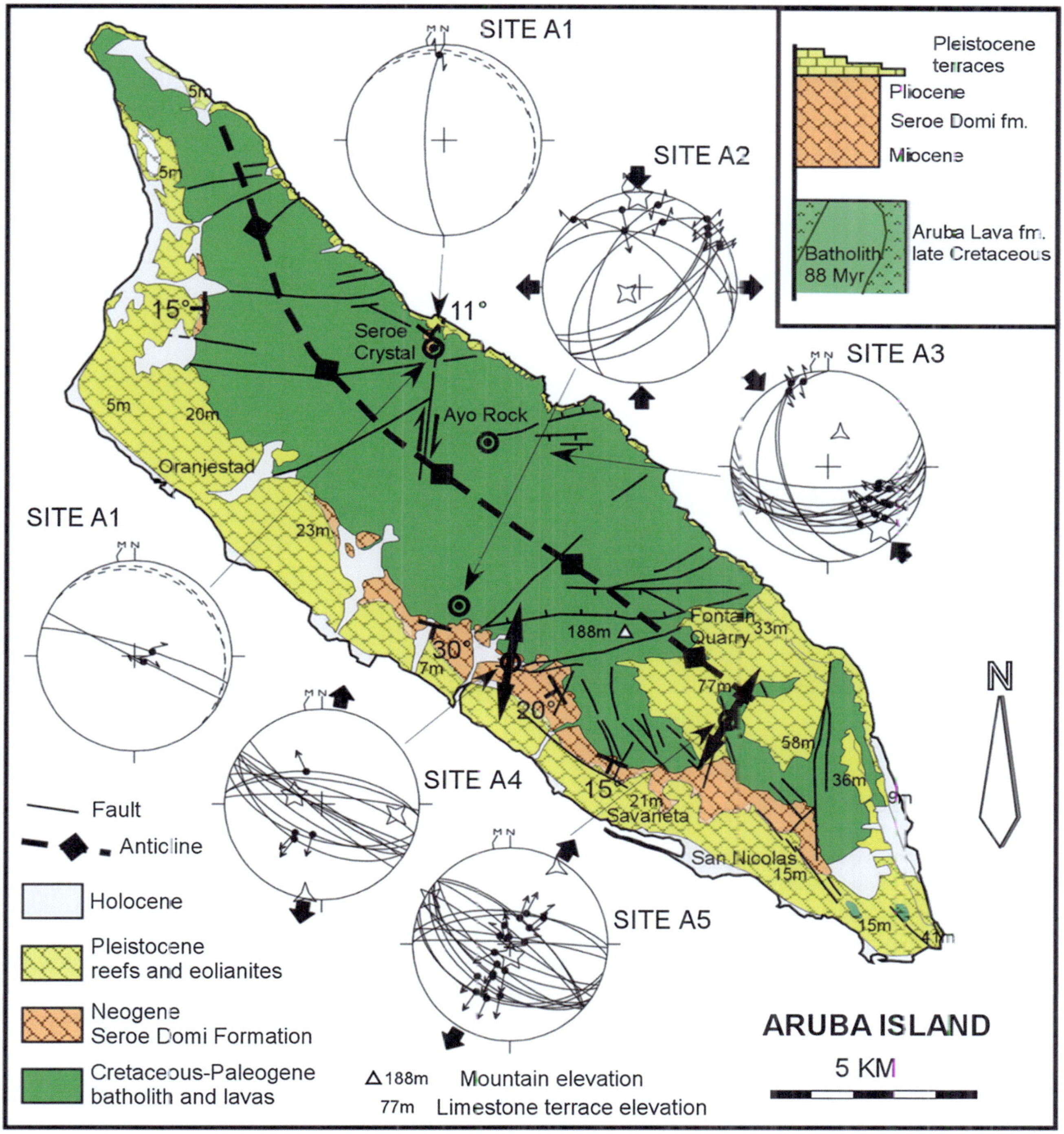

Fig. 18.5 Geologic map of Aruba displaying relative rock ages, faulting and folding, and a basic stratigraphic column. Rock ages generally decrease from east to west, with some scattered Holocene geology. Map from Hippolyte and Mann (2011, p. 263), available online for download at https://www.researchgate.net/publication/222888734_Neogene-Quaternary_tectonic_evolution_of_the_Leeward_Antilles_islands_Aruba_Bonaire_Curacao_from_fault_kinematic_analysis. Retrieved January 2017

white sand. Beaches along the windward coast tend to be rocky and experience high wave energy, whereas the western leeward side of the islands have more stretches of sandy shorelines. The most prevalent beach type on Curaçao and Bonaire is the sheltered pocket coves bordered by rocky headlands, whereas the extensive stretches of white sandy beaches are located in the central leeward coast of Aruba (e.g., Eagle, Palm, and Manchebo beaches) and throughout

Fig. 18.6 Ayo Rock Formation found in central Aruba. It is one of a number of boulder fields scattered throughout the island formed from the erosion of the tonalitic batholith. Photograph by Phillip Schmutz, May 2012

the uninhabited islands of Klein Curaçao and Klein Bonaire (Halcrow 2010).[2]

In addition to the sandy beach environments across the ABCs, the coastal landscape displays extensive coarse sediments and boulders. Recent studies have attributed this coarse debris along the coastal zone to the occurrence of three paleotsunamis that struck the islands around 450–500 BCE, 1500 BCE, and 3500 BCE (Scheffers 2002, 2004; Scheffers et al. 2005). Prior to these studies, tsunami impacts were unknown for the ABCs, and the debris formations were exclusively attributed to hurricane generated waves (e.g., De Buisonjé 1974). The extensive nature of tsunami deposits across the islands are illustrative of powerful systems that deposited several million tons of debris and boulders as far inland as 400 m and up to 12 m above mean sea level. The debris deposits are predominantly located along the windward coastlines of the islands with the most extensive geomorphic evidence on Bonaire and a significant reduction of deposits on Curaçao and Aruba due to a shadowing effect. Scheffer (2004) found three distinct geomorphic assemblages of tsunami debris: rubble ridges, rampart formations, and boulder fields with each formation exhibiting a distinct morphology and geographic distribution that the author related to coastal landform configuration. Although three different types of coarse debris deposits are documented, this does not mean that each of these deposits represents a single tsunami event. Radiocarbon dating from samples around the islands found that the different debris assemblages were deposited contemporaneously (Scheffer 2004; Scheffer et al. 2005).

[2]Detailed studies on the sandy beach environment of the ABCs are severely lacking, yet the many travel websites (e.g., aruba.com, Curaçao.com, and tourismbonaire.com) provide photographs and descriptions of the most popular beaches across the islands.

18.3.2 Estuarine

A number of inland pocket bays dot the coastal landscape of Bonaire and Curaçao, which result from the drowning of paleo-“rooien” (small rivers) valleys eroded into both the interior igneous rocks and limestone terrace rim. At least two such bays formerly existed on Aruba (e.g., Druif and Spaansch Lagoen), but the interior portions have been filled with alluvium (Alexander 1961). As a consequence of the

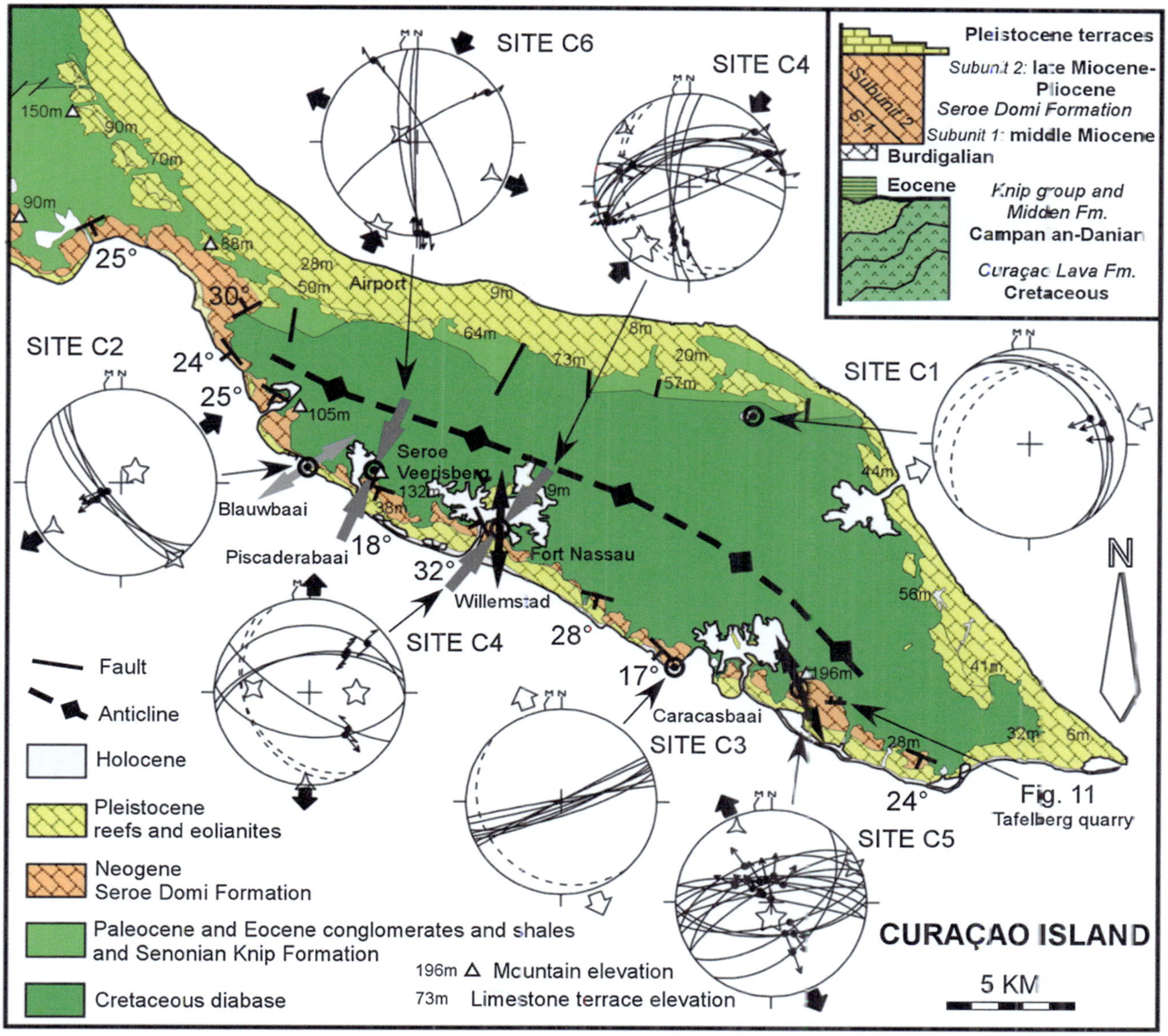

Fig. 18.7 Geologic map of Curaçao's middle and southern regions showing relative rock ages, faulting and folding, and a basic stratigraphic column. Note the juxtaposition of Cretaceous rock (*dark green*) with Pleistocene sedimentary. Compare with geology of Aruba (Fig. 18.5) Map from Hippolyte and Mann (2011, 265), available online for download at https://www.researchgate.net/publication/222888734_Neogene-Quaternary_tectonic_evolution_of_the_Leeward_Antilles_islands_Aruba_Bonaire_Curacao_from_fault_kinematic_analysis. Retrieved January 2017

poor water retention of the topsoil on the islands, rainfall quickly runs off into these bays creating permanent or temporarily flooded "saliñas" (hypersaline lakes) (Roos 1971). While the saliñas are hypersaline, they are far from abiotic, supporting significant flora and fauna populations, and serving as a key stopover habitat for many migratory birds. Because of their important ecosystem function, three of Bonaire's saliñas (Saliña Gotomeer, Saliña Slagbaai, and Pekelmeer) are internationally protected as RAMSAR sites by the Convention on Wetlands of International Importance.

In addition to the saliñas, a few of the inland bays function as intertidal estuary systems which contain the majority of the seagrass and mangrove communities on the islands. Lac Bay, located on the southeastern windward shore of Bonaire, supports the island's only significant mangrove and seagrass communities (Nagelkerkena et al. 2000). Lac Bay is the largest inland bay in the ABCs with an area of 7.5 km^2 and maximum water depth of 4.5 m. The ecological presence of mangroves within the bay is relatively rare as there is no significant riverine input into the system,

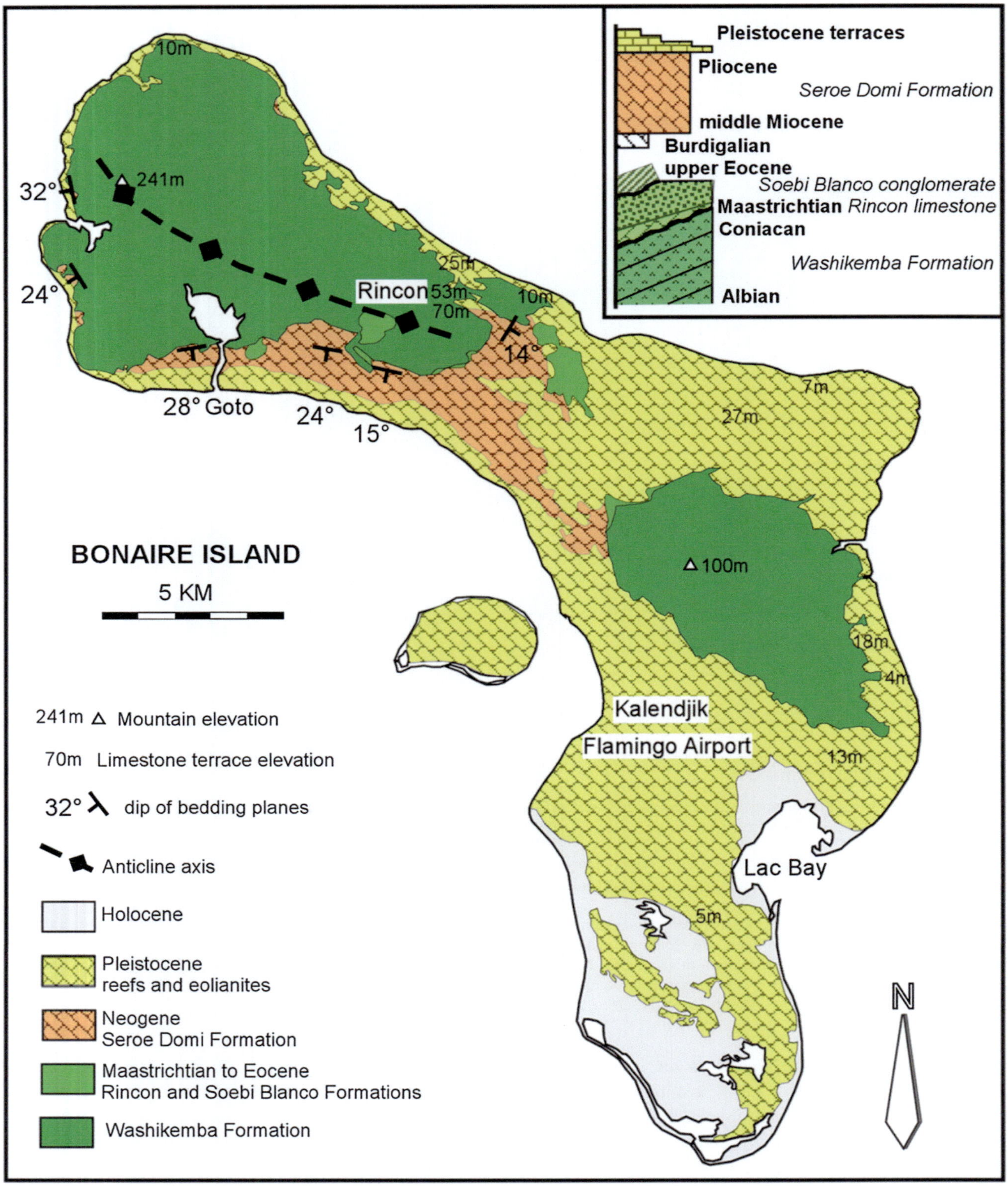

Fig. 18.8 Geologic Map of Bonaire with relative rock ages, faulting and folding, and a basic stratigraphic column. Notice the Cretaceous formations (*dark green color*) surrounded by Pleistocene deposits (*yellow*), and only one major folding event. Compare with geology of Aruba (Fig. 18.5). Map from Hippolyte and Mann (2011, 266), available online for download at https://www.researchgate.net/publication/222888734_Neogene-Quaternary_tectonic_evolution_of_the_Leeward_Antilles_islands_Aruba_Bonaire_Curacao_from_fault_kinematic_analysis. Retrieved January 2017

Fig. 18.9 Karst landform structures of the Middle and Lower limestone terraces on the windward coast of Curaçao. The harsh dry climate is reflected in the resilient vegetation, such as the large cactus in the foreground. Photograph by Phillip Schmutz, May 2012

making them particularly vulnerable to dieback during episodes of hypersaline conditions (van Moorsel and Meijer 1993 as cited in De Meyer 1998). The bay is internationally protected as a RAMSAR site by the Convention on Wetlands of International Importance. On Curaçao, the largest concentrations of seagrasses and mangroves can be found in Spaanse Water, located on the southeastern leeward coast of the island. Spaanse Water is the second largest inland bay on the island with an area of 3.2 km^2 and a mean depth of 5 m. It has been identified as a priority area for conservation and is part of the Curaçao Underwater Park (Debrot and de Freitas 1991). Debrot and de Freitas (1991) found that increased development in the bay has led to a dramatic decrease in species habitat to less than half of the spatial coverage of a century ago.

18.3.3 Dune

Dune fields are sparse on the islands, yet do exist along the windward coasts due to the presence of stable trade wind conditions. The largest dune fields occur on Aruba, with isolated small patches of sand dunes located at a few beaches on Bonaire and Curaçao. Research into the formation and evolution of the dunes are rare, with the work of Westermann (1932) providing the most extensive background. Westermann established that the dunes are of Holocene age consisting of limestone and quartzite sands, depending upon location. On Aruba, the largest dune field is located in the isolated northern tip of the island on the Hudishibana Plateau (Fig. 18.13). The dune field known locally as the California dunes is approximately one kilometer in length, with an area of 0.32 km^2. The dune field exhibits a quasi-parabolic shape and contains numerous vegetated nebkhas. Smaller dune fields occur within Arikok National Park on the southeastern portion of the island at Boca Prins and Dos Playa.

18.3.4 Karst

The limestone terraces along the coastal fringe of all three islands are highly permeable, and subterraneous solution of the limestone has produced numerous cave structures. The caves primarily exhibit a longitudinal shape, running perpendicular to the Quaternary coastlines, and can be found in several places along the cliff-face of the Middle, Lower, and

Fig. 18.10 View from Hooiberg (*Haystack*), Aruba. A volcanic outcrop… Photograph by D. Stanley, available online via https://commons.wikimedia.org/wiki/File:Summit_of_Hooiberg,_Aruba_(4901403229).jpg. Retrieved November 2016

Sereo Domi Formation terraces (Zonneveld et al. 1977). German geologist Karl Martin, professor of geology and director of the museum in Leiden, was the first to study the development of caves on the ABC islands in the late 1880s (Hummelinck 1979). Martin distinguishes two sets of cave formations: (1) caves that developed as a result of wave action, and (2) caves that owe their existence to openings in the cliff of the reef terrace and became enlarged via increased groundwater flow during at least two periods of more humid climatic conditions. Research by Zonneveld et al. (1977) and Hummelinck (1979) support evidence of Martin for the second set of cave formations pointing to the existence of extensive flowstone and horizontally layered dripstone structures, indicative of a more humid climate, overlain by stalactites and stalagmites, and indicative of less humid, drier climates. At present, most caves on the islands are "dead" caves as speleothem activity has completely stopped and are badly weathered (Hummelinck 1979). In addition to the cave structures, small aquifers occur below the limestone terraces across the islands, held within the layers of sedimentary sands and clays of igneous origin (De Buisonjé 1974). The aquifers consist of brackish groundwater reserves as the sources of water input primarily include sea spray and hypersaline rainwater, and the overall lack of fresh surface water (van Sambeek et al. 2000).

18.3.5 Coral Reef

The coral reefs across the ABCs have been described extensively by Roos (1964, 1971), Bak (1975, 1977), and van Duyl (1985), who also produced detailed maps of the shallow coral reef communities along the leeward shore of Bonaire. Extreme differences in wave conditions cause the reef communities on the windward and leeward coasts of the islands to fundamentally differ in terms of reef structure and species abundance (Bak 1977; Vermeij 2012). Along the windward shore of all three islands, coral development is sparse with an abundance of crustose coralline algae found in water depths up to 12 m and the presence of dense stands

Fig. 18.11 Looking over Bonaire's Gotomeer, a saltwater lagoon on the island's northern end that hosts the Caribbean's only pink flamingo nesting sites, to Brandaris, the island's highest point. Like the lower elevation peaks surrounding it, Brandaris is a volcanic outcrop in the Washikemba Formation that straddles the northernmost section of the island's anticlinal axis. Photograph by Bgabel, available online via https://commons.wikimedia.org/wiki/File:BONAIRE-goto-meer-1.jpg. Retrieved November 2016

of *Sargassum platycarpum* (brown macroalgae) extending to depths of 40 m. The reef structure along the western leeward coasts of the ABCs, however, differs slightly between the three islands. Bonaire and Curaçao support near continuous fringing coral reefs, with a total surface of 26 km^2 (10 mi^2) on Bonaire and 20 km^2 (7.75 mi^2) on Curaçao. The islands share a general pattern in reef formation, which consists of a submarine terrace extending seaward up to 250 m gradually sloping to a drop-off at 7–12 m depth. Beyond this, the reef slopes steeply at a 20–50° angle to depths of 50–55 m where it flattens out into a shelf. A second drop-off occurs beyond this to depths of 70–80 m, ending in a sandy plain at 80–90 m (van Duyl 1985). Bak (1977) found that coral cover and diversity increase with depth across the submarine terrace with the greatest level of cover and diversity occurring near the drop-off. Beyond the drop-off, coral cover and diversity decreases rapidly as light decreases and the influence of sedimentation is high. Research on the coral reefs of Aruba is significantly less prevalent than on Bonaire and Curaçao with the work of Roos (1971) and Bak (1975) providing the most extensive, yet minimal, analysis. The limited research on Aruba's reefs occurs in part because coral growth is limited to shallow depths due to a sandy bottom at depths of 20–30 m, which obstructs the formation of coral reefs (Roos 1971; Bak 1975). There are, however, locations along the north- and southwestern portions of the island (e.g., Arashi and Baby beaches) that support extensive coral development as the sandy flat bottom changes into a shingle bottom about 1 km offshore (Roos 1971).

18.4 Landscapes

The historic and present-day residents of Aruba, Curaçao, and Bonaire have modified, utilized, and exploited their landscape in a variety of ways.[3] All three ABCs experienced waves of migration from South America, particularly from

[3]Due to the overlapping nature of heritage and its implications for landscape interactions, we have chosen to detail this information in Sect. 18.5 *Heritage and Tourism*.

Fig. 18.12 The north coast of Curaçao, host to *Shete Boka National Park* (Seven Inlets), boasts spectacular views of limestone seacoasts, and resultant coastal features like sea alcoves/caves (shown here) as well as blowholes and natural bridges. Image courtesy of J. and R. Klotz, available via https://commons.wikimedia.org/wiki/File:SHETE_BOKA_NATIONAL_PARK,_CURACAO.jpg. Retrieved November 2016

what is now Venezuela. These Amerindian peoples utilized the marine resources surrounding them for food. They also took advantage of the islands' limestone features for tools and sought shelter in the caves. The *Kueba di Hato* caves on Curaçao as well as Christoffel National Park contain petroglyphs that are at least 1500 years old (Fig. 18.14). These same caves were also a place of refuge for escaped enslaved persons from Africa taken to Curaçao during the height of the transatlantic slave trade (Fig. 18.15). Bonaire's caves also contain petroglyphs, but were unlikely used as dwellings because there is no evidence of middens or shell heaps (Klomp 1986).

The Spanish introduced livestock including "goats, sheep, dogs, donkeys, cows, and pigs" on Aruba and Bonaire[4] (Abel 2000; Dresscher 2009, p. 4) and cattle on Curaçao (Rupert 2012). As these animals were not indigenous to the islands and "[T]he presence of these small and mobile grazers and browsers meant selection for thorn forest vegetation (acacia, mesquite, many cactus species, etc.) that could defend itself from foraging animals, and that today dominates the landscape" (Abel 2000, p. 20). Additional animals were introduced as well, including dogs, cats, horses, black rats (Curaçao), brown rats and house mice, one species of agouti (Curaçao), Whitetail deer (Curaçao), and other mammals (Borroto-Páez, and Woods 2012; Watts 1987).

Large-area agriculture, as is common in other islands of the Caribbean, did not occur on the ABCs under European colonizers because of climate.[5] The Dutch, however, worked with the dry climatic constraints of the islands establishing export industries centered on aloe vera and salt production through the use of enslaved African and indigenous labor and ingenuity. The physical landscape of Bonaire is

[4]The Spanish also introduced cattle and horses to Bonaire in 1527 (Hartog 1978).

[5]This is not to imply that there were not productive plots of land on the islands. For example, Bonaire had agricultural parcels called *kunuku* where sorghum and maize were cultivated (Klomp 1986).

Fig. 18.13 Dune field on the northern tip of Aruba, known locally as the California dunes for the offshore wreck of the famous ship "California." Photograph by Phillip Schmutz, May 2012

dominated still today by salt production located at Pekelmeer on the southern portion of the island, and the cultural landscape of Aruba bears the mark of a gold rush that persisted throughout the nineteenth century.

The discovery of oil in the Lake Maracaibo region of Venezuela in the twentieth century affected the landscape of each of the ABCs, although in different ways. Landscape modification included the construction of refineries and the dredging of harbors to allow for transhipment of oil (Fig. 18.16). Oil pollution from spills within the region has been documented on both Curaçao and Bonaire's beaches (Debrot et al. 1995, 2013).

Beach erosion is another consequence of human modifications stemming from development. In 1950, for example, Aruba deepened its harbor entrance, which had the unintended consequence of cutting off the natural sediment flow that nourished the island's beaches (Kohsiek et al. 1987). Hotels constructed during the 1960s and 1970s faced beach retreat by the late 1980s (Kohsiek et al. 1987:52).

18.5 Heritage and Tourism

The ABCs share a similar heritage that includes the influence of Amerindians from mainland South America, European Spaniards, Dutch, and enslaved Africans. All three islands have felt the impacts of Venezuelan oil in some capacity and the turn to tourism in the latter part of the twentieth century. While they are indeed similar in some respects, this section will focus independently on each of the three islands drawing out the nuances of their heritage particularly how each island developed into a tourist destination in the region. We will begin with Curaçao, as it was historically the dominant of the three.

18.5.1 Curaçao

The early indigenous inhabitants for all the ABCs were from mainland South America. About 30 sites investigated

Fig. 18.14 The entrance gate to the Hato caves, a popular tourist attraction located near the airport on Curaçao. The caves were an important hiding place for escaped enslaved persons and also contain Amerindian petroglyphs that are 1500 years old. Photograph by Amy Potter, May 2012

on Curaçao are from the Archaic Period (2540 BC–1840 CE), and these hunter-gatherer groups that ate turtle and crabs and used the limestone and shale to make tools lightly populated the island with one major permanent settlement (Saunders 2005). Half of the other sites belong to a later group from the Ceramic period (450 CE to the arrival of Europeans), dominated by the Dabajuroid culture diaspora from Venezuela near Lake Maracaibo (Saunders 2005).

Among Europeans, the first "discoverers" of Curaçao was the trio of Alonso de Ojeda, Amerigo Vespucci, and Juan de la Coas in May 1499, though a population has yet to be established for this time period (Rupert 2012).[6] In 1513, Viceroy Diego Colón, son of Christopher Columbus, declared Curaçao and its neighbors (Aruba and Bonaire) "useless islands" because they did not contain precious metals that he perceived to be marketable.[7] In doing this, Colón opened the door for the deportation from these islands' of all native people, most of whom ended up in Hispaniola to work at the gold mines (Rupert 2012; Watts 1987; Sauer 1966). The Spanish largely ignored the ABCs until 1526, when Juan de Ampíes was granted the right to repopulate the islands and brought 25 Native Americans to Curaçao (Rupert 2012, p. 21).

In 1633, when the Dutch were initially expelled from St. Martin, they turned their attention to Curaçao with its salt pans and attractive harbor. A year later in 1634, the Spanish governor of Curaçao surrendered and retreated. In 1636, the Dutch conquered Bonaire and Aruba, both of which were considered dependencies of Curaçao. The Dutch presence in the Caribbean was driven by a need for salt to supply the expanding herring industry, as well as cheese and butter (Römer 1984). The West India Company encouraged the settlement of colonists (the most notable being the Sephardic Jews). The Dutch were also interested in acquiring Curaçao

[6]Linda Rupert estimates that the population of Caquetios on Curaçao was 2000, while David Watts (1987, p. 109) and Carl Sauer (1966, 194) put the population much higher at 100,000, a number that today seems quite high. Rupert does not indicate whether this number was pre- or post-European contact, which could have had significant impact on the island's population due to the introduction of European diseases.

[7]As noted later in this section, gold was not discovered in Aruba until the nineteenth century.

Fig. 18.15 Inside Hato caves, Curaçao. Caves such as these host Amerindian carvings (petroglyphs) that have been dated to around 500 CE. Photograph by J. and R. Klotz, available via https://commons.wikimedia.org/wiki/File:HATO_CAVES_,_CURACAO.jpg. Retrieved November 2016

for its close proximity to Brazil as well as the West India company's involvement in the slave trade (Dresscher 2009). From 1634 to 1675, the island served as a naval base centered at the port of Willemstad. In 1675, the Dutch opened the islands to free trade, transforming the once small naval landholding by the Dutch into a major maritime commercial center and the hub of the Dutch Atlantic system (Rupert 2012). Between 1674 and 1689, Curaçao served as a slave depot for enslaved person arriving from Africa and "was the destination for 59% of all Dutch slave exports" with the majority then being re-exported to South America (Rupert 2012, p. 77).[8] Outside the port of Willemstad, merchants constructed large rural estates known as *landhuis* as a symbol of their growing wealth. Their primary function was relaxation rather than significant agricultural production with the main house emulating the large Caribbean sugar plantations but with a Dutch architectural style (Rupert 2012, 135). In 1863, slavery was abolished.

The twentieth century was marked by two dominant industries: oil refining and tourism. Shell Oil was established in Curaçao in 1918, while tourism had its beginnings in the early 1900s, though it took until the 1950s before tourism became a significant industry. Today, tourism accounts for 18.5% of the island's' economy[9] up from 10.5% in 2004 (Halcrow 2010; Croes et al. 2015). In addition, tourism spending has increased over the last 15 years by an annual average of 4% (Croes et al. 2015). A significant increase in cruise passengers occurred in 2013 as a result of increased port of call, and perhaps, because that same year, Curaçao was ranked for the first time in *Cruise Insight* magazine's "World's Top 50 Ports."

Still, oil refining poses challenges to tourism. As the Curaçao Tourism Bureau (2005, p. 30) notes, "The refinery (along with other industry) dominates the harbor area and is a source of air and water pollutants in the harbor and areas immediately west of its location. In that sense, the refinery limits the land available for tourism development and can be

[8]The Dutch, beginning in 1600, contributed to roughly 5% (half a million persons) of the transhipment of enslaved persons to the New World (Emmers 2006).

[9]2013 estimates.

Fig. 18.16 Baby Beach on the southern part of Aruba with the then Valero Refinery in the background. The installment of the refinery has heavily impacted the tourism industry in the area as it is often cited on traveler blogs and review Web sites, such as *TripAdvisor*®, as an eyesore decreasing the natural appeal of the beach. Photograph by Amy Potter, May 2012

an eyesore for visitors." In addition, larger socioeconomic issues facing the island could ultimately hamper the growth of tourism in the coming years: a weak economy, debt repayment, the uncertain future of the financial sector related to offshore financing legislation, 15% unemployment (youth unemployment is 35%), poverty incidence exceeding 15%, and "brain drain of educated Curaçaoans elsewhere" (Croes et al. 2015, p. 45). While tourism does relieve some unemployment, providing 16,503 jobs, there is still a disconnect in terms of the beneficiaries of the industry.

While coral reefs and beaches attract tourists, Curaçao also boasts a repertoire of notable architecture and historic sites that both the government and private institutions have sought to restore. Willemstad, the island's capital, has around 750 historic buildings and has been designated a UNESCO World Heritage Site (Fig. 18.17). Outside Willemstad, there are 55 examples of a *landhuis* (Gravette 2000). In terms of connecting to the island's African heritage, there are several sites of significance: *Landhuis* Knip's[10] Museo Tula, the Tula monument to the Curaçao slave revolt of 1795,[11] Museum Kurá Hulanda, an anthropological museum with a variety of exhibits which focus on civil rights and the transatlantic slave trade (Fig. 18.18), and the Kas di Pal'i Maishi Slave House and Museum (Curaçao Museum) (Fig 18.19).[12]

18.5.2 Aruba

In Aruba, the first Amerindians to arrive as early as 2500 BCE were fishers, hunters, and gatherers. Just before 1000 CE, a second wave of migrants, the Caquetios from South America, arrived and settled parts of the island and used ceramics and were agriculturalists (Rupert 2012). Waves of influence can be felt from Venezuela between 100 BCE and 1000 CE. Several sites excavated on the island reveal the influence of the people of the Dabajuroid ceramic tradition,

[10]Also called Kenepa.

[11]This was the largest slave revolt in Curaçao's history.

[12]This house is an example of house construction by descendants of enslaved persons living on the island during the seventeenth to early twentieth century.

Fig. 18.17 View of the colorful gabled townhouses in Willemstad from Santa Ana Bay. In the foreground is a boat from Venezuela on its way to the floating market. Photograph by Amy Potter, May 2012

which include Tanki Flip (950–1250 CE) and Santa Cruz (occupation dates to historic times). At Tanki Flip, "…investigations have found evidence of perhaps 100 inhabitants living in large oval-shaped and small structures, and surrounded by Dabajuroid pottery, hearths or kilns, stone tools, animal bones, and thousands of shell fragments" many of which were made of Queen Conch (Gravette 2000, p. 17). The mainland connection to Aruba continued through the Spanish and Dutch period, and until 1750, Aruba was an Amerindian reservation with close ties to mainland South America—except for a small group of residents from the West India Trading Company. During this time, the island was largely maintained as a horse ranch (Römer 1984).

While other important activities centered on the agricultural production of aloe vera cultivation,[13] salt production, and phosphate mining, gold was discovered in the nineteenth century. During the height of production (between 1824 and 1899, formally closing in 1913), 1,350,000 kg of gold was extracted (Gravette 2000). G.B. Bosch, a visiting preacher from Curaçao, recalled in 1836 that people left their occupations for more lucrative prospects in gold mining (Dresscher 2009, p. 15).

The 1918 discovery of oil in the Lake Maracaibo region had tremendous impacts on Aruba. Rather than opening a refinery in Venezuela—due to the turbulent political situation there—the oil companies situated their refinery on the eastern tip of Aruba because of its geography and climate.[14] The refineries would do well, with their heyday lasting from 1928 to 1948.

Tourism took on a more prominent role with decisive government intervention after the oil refinery, the Lago Oil & Transport Company, Ltd.[15]—which contributed to about 25% of the island's GDP and employed roughly 30–40% of the population—closed in 1985 (Vanegas and Croes 2003; Ridderstaat 2007). As the refinery shut down, Aruba also underwent a political shift, establishing a separate state in the Kingdom of the Netherlands. Aruba's gradual shift to tourism, however, took several decades. In 1947, Aruba

[13]At one time, Aruba, Bonaire, and Curaçao provided more than half of all aloe production in the world.

[14]They initially considered Curaçao, but the island's ports were already in use.

[15]At the time, Aruba was the site of the largest oil refinery in the world (Gravette 2000).

Fig. 18.18 The yard of Museum Kurá Hulanda, where enslaved Africans were auctioned off to destinations throughout South America. Photograph by Amy Potter, May 2012

established the Aruban Tourism Commission to explore the prospects of developing tourism for the island, leading to the establishment of the Aruba Tourism Bureau (now the Aruba Tourism Authority) in 1953 (Vanegas and Croes 2003), though it was not until 1959 that the island constructed its first 100-room hotel (Ridderstaat et al. 2014). Aruba intensified its efforts to attract tourists through advertising, infrastructure development (hotels and shopping malls, airport), and the expansion of air-seat capacity as a way to offset its economic loss of the oil refinery (Croes and Vanegas 2005; Ridderstaat et al. 2014).

Over the last several decades, international tourist arrivals increased from 129,000 tourists in 1975 to 721,000 in 2000 (Croes and Vanegas 2005). In 2014, the island surpassed one million stopover arrivals (Caribbean Tourism Organization 2016), the number of hotel rooms tripled (Ridderstaat et al. 2014), and cruise ship arrivals increased nearly tenfold in the same period (Ridderstaat et al. 2014; Caribbean Tourism Organization 2016).

Aruba has one of the highest living standards in the Caribbean, but over 85% of their economy depends directly or indirectly on tourism, making it vulnerable to global economic downturns. Many of the hotels are foreign-owned leading to capital leakage (Dresscher 2009; IMF Country Reports 2015; Caribbean Journal 2015). In addition, investment in infrastructure is ongoing and extensive, with Aruba currently undergoing a billion-dollar island revitalization and beautification project for hotels, roads, airport, water, energy, and sanitation (Caribbean Journal 2015).[16]

18.5.3 Bonaire

Bonaire's indigenous history shows evidence of a variant of the Dabajuroid cultural tradition (1000–1400 CE) that also can be felt on Curaçao and Aruba. Archaeologists have uncovered crude pottery figures, jars, cassava griddles, and documented rock paintings, which were part of spiritual life. Other tools used to prepare maize indicate these early inhabitants had a mixed diet that would have also included marine resources of fish, shellfish, and crabs. Most of the sites are located inland because these early inhabitants lived

[16]The government recently renovated the Reina Beatrix International Airport (*Caribbean Journal* 2015).

Fig. 18.19 Salt ponds on Bonaire. Photograph by A. Sieben, available via https://commons.wikimedia.org/wiki/File:Bonaire_Salzgewinnung.jpg. Retrieved November 2016

near their fields as freshwater was only available further inland (Saunders 2003; Klomp 1986).

From 1636 to 1868, Bonaire was exploited as a single large plantation,[17] initially by the West India Company and then by the government of the Netherlands. Bonaire never developed large-scale plantation agriculture. Instead, the colonial era was dominated by stockbreeding and the export of aloe divi-divi pods, brazilwood, and salt extracted in the southern region of the island. The island's dry climate facilitates high rates of evaporation that leads to salinization of inland lakes (the most notable of which is Pekelmeer) (Hartog 1957; Klomp 1986). Enslaved Africans were initially transported to the island to work in the salt industry.[18] In 1868, after emancipation, the Dutch government sold large tracts of land on Bonaire charting a new era of privately owned landholdings on the island. This time period was extremely difficult for those who were formerly enslaved, as they suffered low wages, continuous debt (similar to sharecropping in the USA), and reduced access to grazing lands (Abel 2000, p. 24; Klomp 1986).

With the establishment of Shell Oil in Curaçao in 1918 and Aruba's ESSO's LAGO in 1927, those living on Bonaire could seek out wage labor and work in the refineries or on the containers that transported oil. The Bonairean

[17]The term plantation here is not to be confused with plantations in other parts of the Caribbean. Ank Klomp describes Bonaire's plantations as similar to the Latin American hacienda. "One might picture a huge ranch on which a number of economic activities were pursued along with extensive stock raising" (p. 15).

[18]A report from 1806 documents 945 inhabitants on island: 72 white, 284 Indian, 225 freemen, 92 privately owned slaves, and 272 government-owned slaves (Hartog 1957, p. 108).

Petroleum Corporation, a fuel oil storage and transhipment terminal, was established in 1975.

In terms of tourism, the first hotel on the island opened in 1952. It was a former structure that served as an internment camp for 500 German nationals during World War II (Parker 1999). During the 1960s and 1970s, tourism on Bonaire was relatively small due to the island's limited infrastructure: four hotels (two of which had 57 rooms each) and the island's very short airport runway (Klomp 1986). In November of 1967, Bonaire received its first cruise boat visit by Norwegian American Line's *Bergensfjord* with 425 guests (Bongers 2013). In 1976, the airport added a new terminal, and in 1980, the single runway was extended. In the last several decades, Bonaire has experienced a significant increase in tourism but not on the scale of its island neighbors (Caribbean Tourism Organization 2016).

Most of the present-day tourism on Bonaire consists of "active tourism" with a particular emphasis on its marine resources (SCUBA diving). Other activities include kitesurfing, windsurfing, snorkeling, boating, and sailing. Land activities center on ecotourism, which include hiking, biking, horseback riding, caving, and birdwatching. Bonaire boasts one of the few flamingo breeding sites in the southern Caribbean with approximately 3000 breeding pairs located in the Pekelmeer flamingo sanctuary on the southern end of the island (Abel 2000; Vdauskaite 2015; Dutch Caribbean Nature Alliance 2016). Bonaire's reefs are some of the best preserved within the region and include the 2700 hectares of the National Marine Park of Bonaire, established in 1979[19] (Dutch Caribbean Nature Alliance 2016). The park includes the waters surrounding the island to a high-water mark depth of 200 feet (91 m).

In addition to Bonaire's coral reefs, the island has unique architecture. The oldest standing buildings—its slave huts—are located along the salt ponds on the southeastern side of the island. These free-standing stone structures with thatched roofs are a testament to the island's enslaved Africans and were built around 1820 (Gravette 2000). Like Curaçao, there are also planter's residences dating to the twentieth century, in addition to an ancient aloe oven at Karpata.

[19]"The early days of unregulated marine tourism on Bonaire impacted the reefs heavily with spearfishing, with anchor damage to corals and with siltation caused by onshore development. As these problems became evident, a coalition of local leaders together with The World Wildlife Fund succeeded in convincing Holland and the Antillean Government to intervene. The result was the creation of the Bonaire Marine Park (BMP) in 1979" (Parker 1999, p. 243).

18.6 Hazards

Climate change will pose a serious threat not only for the island's residents but also for the tourism industry (Debrot and Rob Bugter 2010). Scientists predict there will be increases in air and sea surface temperatures, coastal and beach erosion via sea-level rise, and an increase in the intensity and number of tropical cyclone systems (Ministry of Health and Nature 2014).

The islands formally fall outside the hurricane belt and are rarely impacted directly by tropical cyclones. On occasion, however, tropical cyclones do cause destruction through coastal erosion as refracted storm-wave swells alter, in the short run, sand movement in the littoral zone eastward (Kohsiek et al. 1987). On average, once every four years, a tropical cyclone (tropical storms up to category 5 hurricanes) occurs within a radius of 150 km, with a total of 38 tropical cyclones during the time period from 1850 to 2014 (Meteorological Department of Curaçao 2015). The most recent tropical cyclone to cause significant damage to the islands was Hurricane Lenny in 1999, resulting in considerable beach erosion and damage to the coral reefs across the islands as well as damage to historic structures at the Washington Slagbaai Park on Bonaire.

For islands such as Bonaire and Curaçao that attract tourists to their coral reefs, climate change will have a dramatic impact on these systems. Strelich (2015) analyzed data between 1982 and 2012 and observed an overall warming trend in sea surface temperatures over the past 30 years in the Caribbean, with the most significant increase occurring just north of South America. The warming ocean surface temperatures are impacting coral reef communities prompting coral bleaching events by altering ocean chemistry. It is estimated that as much as 80% of the living coral reef across the Caribbean are already gone (Economic Commission for Latin America and the Caribbean 2011, pp. 16–17) and the ABCs have not fared much better as episodes of massive coral bleaching have occurred over the past few decades (Vermeij 2012; Steneck et al. 2015). Additionally, Bak and Nieuwland (1995) found that since the 1970s, the health of the coral colonies has decreased across the shallow reef complex as a result of coastal development, sewage discharge, and artificial beach construction. A more recent report by The Ministry of Health and Nature in 2014 found that Curaçao's reefs are experiencing both human- and natural-related stresses because of overfishing, development, sewage discharge, chemical pollution, and artificial beach construction. Nevertheless, Bonaire and Curaçao support some of the healthiest coral reefs in the Caribbean, containing by far the greatest level of species diversity (Miloslavich et al. 2010; Vermeij 2012; Steneck et al. 2015).

Aside from the threat of climate change, there are a number of other hazards that will pose serious risks to the island's inhabitants and tourism infrastructure. Fresh groundwater resources are limited, and a large variety of natural and anthropogenic processes are influencing the groundwater quality. Sea spray, hypersaline rainwater, and lack of fresh surface water all contribute to the brackish nature of the groundwater resources, whereas anthropogenic effects introduce high concentrations of potassium, sulfate, and nitrate as a result of groundwater mixing with polluted wastewater (van Sambeek et al. 2000). The Department of Agriculture and Fisheries established a monitoring network on the islands in the late 1970s because of poor groundwater quality, though only the monitoring network on Curaçao is still operational (van Sambeek et al. 2000). These issues have prompted the production of drinking water through saltwater distillation (van Sambeek et al. 2000).

Although no written or oral sources detailing a tsunami impact exist since the occupation of the islands by the Spanish in 1527, there is historic evidence of tsunamis on the ABCs dating to only 500 years BCE (Scheffers 2002, 2004; Scheffers et al. 2005). The possibility of a new event remains plausible, and based on the location of transported debris from previous tsunami events, the height of onshore waves could have exceeded 12 m.

Beach pollution from oil spills and marine debris has been documented in a number of studies. For example, a study detailing tar pollution on Curaçao shows that over a 14-month period, January through May were the greatest months for contamination as a result of occasional oil spills. There was also evidence showing impact from more significant spills on the central-southwest coast (Debrot et al. 1995). The same authors (Debrot et al. 1999) also documented marine debris based on a study of 10 pocket beaches on Curaçao. They found higher amounts of debris on windward beaches, which included predominantly plastic and wood materials. Leeward-side beaches experienced debris from local recreational activity, while windward beaches were more likely to experience debris from the region. A study focused specifically on Bonaire found tar pollution to be less than that of Curaçao but still present (Debrot et al. 2013). In addition to oil spills, the oil refineries discharge smoke and noxious gases. While this pollution has generally been reduced, it still impacts air and water quality for areas on Curaçao, which include Willemstad, Parasasa, Piscaderabaai, and St. Michiel and serves as a visual eyesore for the tourism industry (Halcrow 2010) (Fig. 18.16).

18.7 Conclusion

Comprising the Leeward Dutch Antilles, Aruba, Bonaire, and Curaçao share a similar geology and climate. The geologic structure of the ABCs underlies the episodic evolution of the Leeward Antilles ridge since the Late Cretaceous. Although the lithology and sequence phases differ slightly between the islands, the geologic core of all three islands is composed of Late Cretaceous and Paleogene volcanic and sedimentary rocks overlain with multiple limestone terraces formed as a result of sea-level fluctuations and the tectonic rise of the Leeward Antilles ridge during the Neogene and Quaternary. Climatically, the ABCs exhibit a tropical semiarid climate with an average temperature around 28 °C.

Indigenous peoples were present on all three islands well before Europeans, and the landscape bears their imprint. Direct links to the Greater Antilles, particularly Hispaniola, started with the Spanish just after 1500 CE. Shortly afterward, the Spaniards declared the islands "useless" and attempted to depopulate them. A few enslaved Amerindians were reintroduced with their Spanish masters in the sixteenth century, but more direct links with the Lesser Antilles and Africa seem to date to the Dutch period, which began in 1634. Because of the Dutch's deep connection with the slave trade for over two centuries, these islands have an outsized role in historic slavery. The cultural imprint of the African diaspora (particularly in Curaçao and Bonaire) is deep, long-lasting, and meaningful. After the abolition of slavery, the islands became economically marginal for the Kingdom of the Netherlands. The discovery of oil in Venezuela early in the twentieth century impacted all three islands.

Tourism, which developed later in the twentieth century, is now important to the economy of all three islands. While historically Dutch activity centered on Curaçao, today Aruba dominates in the realm of tourism (cruise passenger arrivals and tourist stopover arrivals). The very characteristics that in the past gave the islands the moniker "useless" are now seen as desirable for this economic mainstay. With increasing visitors to the three ABC's, leaders will have to balance economic growth while also safeguarding resources (heritage, reefs, etc.). Bonaire's coral reef protection is some of the most impressive efforts seen globally. In addition, given Curacaos' unemployment rate, sustainable tourism development must also work to the benefit of island residents. Leaders must do all of this while also grappling with the inevitable adverse impacts of climate change.

References

Abel T (2000) Ecosystems, sociocultural systems and ecological economics for understanding development: the case of ecotourism on the island of Bonaire, N.A. Ph.D. Dissertation. University of Florida

Alexander CS (1961) The marine terraces of Aruba, Bonaire and Curaçao, Netherlands Antilles. Ann Assoc American Geogr 51:102–123

Bak RPM (1975) Ecological aspects of the distribution of reef corals in the Netherlands Antilles. Bijdragen tot de Dierkunde 45(2):181–190

Bak RPM (1977) Coral reefs and their zonation in the Netherlands Antilles. Stud Geol 4:3–16

Bak RPM, Nieuwland G (1995) Twenty years of change in coral communities over deep reef slopes along leeward coasts in the Netherlands Antilles. Bull Mar Sci 56(2):609–619

Bandoian CA, Murray RC (1974) Pliocene-Pleistocene carbonate rocks of Bonaire, Netherlands Antilles. Geol. Soc. Amer. Bull. 85:1243–1252

Beardsley AG, Ave-Lallemant HG (2007) Oblique collision and accretion of the Netherlands Leeward Antilles to South America. Tectonics 26(16)

Beets DJ (1972) Lithology and Stratigraphy of the Cretaceous and Danian succession of Curaçao. Ph.D thesis, p 153

Beets DJ, Klaver GT, Mac Gillavry HJ (1977) Geology of the cretaceous and early Tertiary of Bonaire, guide to the field excursions, in transactions, 8th Caribbean Geological Conference: Amsterdam, Gemeentelijke Universiteit Amsterdam (GUA) papers of geology. 10:18–28

Beets DJ, Maresch WV, Klaver A, Montana R, Bocchio R, Beunk FF, Moen HP (1984) Magmatic rock series and high-pressure metamorphism as constraints on the tectonic history of the southern Caribbean, in the Caribbean-South American plate boundary and regional tectonics. Geol Soc Am, Bull, p 162

Bongers E (2013) tourism History: Tourism History Part 13—Cruise Tourism: Part 1: Welcome the *Bergensfjord* and Sinterklaas. The Bonaire Reporter 20(16):9

Borroto-Páez R, Woods CA (2012) Status and Impact of Introduced Mammal in the West Indies. In: Borroto-Paez R, Woods CA, Sergile FE (eds) Terrestrial mammals of the West Indies. Florida Museum of Natural History and Wacahoota Press, Gainesville, Florida, pp 241–257

Caribbean Journal (2015) Why aruba tourism is booming—and what's next? http://caribjournal.com/2015/09/13/why-aruba-tourism-is-booming-and-whats-next/#

Caribbean Tourism Organization (2016) 2003—July 2015 Tourism stats. http://www.onecaribbean.org/statistics/2003-july-2015-tourism-stats/ Accessed May 10 2016

Centraal Bureau voor de Statistiek (2015) Caribisch Nederland; bevolkingsontwikkeling, geboorte, sterfte, migratie: 12 augustus 2015. Accessed May 23 2016. http://statline.cbs.nl/StatWeb/publication/?DM=SLNL&PA=80539ned&D1=0–1,9-10&D2=a&D3=a&HDR=T&STB=G1,G2&CHARTTYPE=1&VW=T

Central Bureau of Statistics (2010) Fifth population and housing census: September 29 2010, Aruba. Accessed May 23 2016. http://cbs.aw/wp/index.php/2012/07/06/census-2010/

Central Bureau of Statistics Curaçao (2016) Age Distribution Curaçao, January 1st. Accessed May 23 2016. www.cbs.cw/document.php?m=1&fileid=2228&f=b6064bb81f3308c7638186fdeed719d6&attachment=1&c=157

Croes R, Semrad KS, Rivera MA (2015) Curaçao: tourism master plan 2015–2020. http://www.Curacao.com/media/uploads/2015/10/07/CUR_TMP_Oct-05-15_Final.pdf

Croes RR, Vanegas M Sr (2005) An econometric study of tourist arrivals in Aruba and its implications. Tour Manag 26:879–890

Curaçao Tourism Bureau (2005) Master plan for tourism development. update 2005–2009. Willemstad: CTB. http://s3.amazonaws.com/zanran_storage/www.ctb.an/ContentPages/1209019495.pdf

Curaçao Tourism Bureau (2010) Cruise passengers and cruise calls 1950–2009. Last accessed June 4 2016. http://www.Curacao.com/media/uploads/2014/08/30/historic_overview_cruise_1950.pdf

De Meyer K (1998) Bonaire, Netherlands Antilles. In: UNESCO, CARICOMP—Caribbean coral reef, seagrass and mangrove sites. Coastal region and small island papers 3, UNESCO, Paris, xiv + 347 pp

Debrot AO, de Freitas JA (1991) Wilderness areas of exceptional conservation value in Curaçao, Netherlands Antilles. Neth Comm Int Nat Prot Meded 26:1–25

Debrot AO, Bugter R (2010) http://www.dcbd.nl/sites/www.dcbd.nl/files/documents/HD3229ClimateChangeEffects.pdf

Debrot AO, Bradshaw JE, Tiel AB (1995) Tar Contamination on Beaches in Curaçao, Netherlands Antilles. Mar Pollut Bulletin 30 (2):689–693

Debrot AO, Tiel Aubrey B, Bradshaw JE (1999) Beach Debris in Curaçao. Mar Pollut Bull 38(9):795–801

Debrot AO, van Rijn J, Bron PS, de León R (2013) A baseline assessment of beach debris and tar contamination in Bonaire, Southeastern Caribbean. Mar Pollut Bull 71(2013):325–329

De Buisonje PH (1974) Neogene and quaternary geology of Aruba, Curaçao and Bonaire, PhD thesis, Universiteit van Amsterdam, The Netherlands

Departmento Meteorologico Aruba (2010) Climate report. Climate averages 1981–2010. URL: http://www.meteo.aw/files/Download/climatnormals19812010.pdf

Dresscher E (2009) Aruba, an island navigating a globalizing world: a brief history. J Nat Archaeol Mus Aruba 1:1–33

Economic Commission for Latin America and the Caribbean (2011) An assessment of the economic impact of climate change on tourism sector in Aruba. http://repositorio.cepal.org/bitstream/handle/11362/38616/LCCARL303_en.pdf?sequence=1

Dutch Caribbean Nature Alliance (2016) http://www.dcnanature.org/bonaire-national-marine-park/

Emmer PC (2006) The dutch slave trade, 1500–1850. Berghahn Books, Oxford

Fouke BW, Beets CJ, Meyers WJ, Hanson GN, Melillo AJ (1996) 87Sr/86Sr chronostratigraphy and dolomitization history of the Seroe Domi formation, Curaçao (Netherlands Antilles). Facies 35:293–320

Gorney D, Escalona A, Mann P, Magnani MB (2007) Chronology of Cenozoic tectonic events in western Venezuela and the Leeward Antilles based on integration of offshore seismic reflection data and on-land geology. AAPG Bull 91:653–684

Gravette A (2000) Architectural Heritage of the Caribbean: An A–Z of Historic Buildings. Ian Randle Publishers, Kingston

Halcrow (2010) Strategic tourism master plan for the island of curaçao 2010–2014. http://static.bearingpointcaribbean.com/wp-content/uploads/2015/01/Strategic-Tourism-Master-Plan-Report-compressed.pdf

Hartog J (1957) Bonaire. Van Indianen tot toeristen. Aruba, Gebroeders de Wit

Hippolyte J-C, Mann P (2011) Neogene-Quaternary tectonic evolution of the Leeward Antilles islands (Aruba, Bonaire, Curacao) from fault kinematic analysis. Mar Pet Geol 28:259–277

Hummelinck PW (1979) Caves of the Netherland Antilles. Utrecht: Foundations for Scientific Research in Surinam and the Netherland Antilles (in Dutch and English). 176 pgs

Kerr AC, Tarney J, Marriner GF, Nivia A, Klaver G, Saunders AD (1996) The geochemistry and tectonic setting of late Cretaceous Caribbean and Colombian. J South Am Earth Sci 9:111–120

Klomp, Ank (1986) Politics on Bonaire: An Anthropological Study (trans: Dirk H, van der Elst Ph.D, Van Gorcum) Wolfeboro, New Hampshire

Kohsiek LHM, Hulsbergen CH, Terwindt JHJ (1987) Beach erosion along the west coast of Aruba, Netherlands Antilles. J Coastal Res 3 (1):37–53

Meteorological Department of Curaçao (2015) Climate reports. Hurricanes and tropical storms in the dutch caribbean. URL: http://www.meteo.cw/Data_www/pdf/pub/HurricanesTropicalStorms_DC.pdf

Meteorological Department of Curaçao (2016) URL: http://www.meteo.cw/climate.php?Lang=Eng&St=TNCC&Sws=R11

Miloslavich P, Dı´az JM, Klein E, Alvarado JJ, Dı´az C, et al (2010) Marine biodiversity in the Caribbean: regional estimates and distribution patterns. PLoSONE 5(8): e11916. doi:10.1371/journal.pone.0011916

Ministry of Health and Nature (2014) National report on Curaçao. http://www.tt.undp.org/content/dam/trinidad_tobago/docs/Democratic Governance/Publications/National_Report_C_2014_July.pdf

Nagelkerkena I, van der Velde G, Gorissen MW, Meijer GJ, Van't Hof T, den Hartog C (2000) Importance of mangroves, seagrass beds and the shallow coral reef as a nursery for important coral reef fishes, using a visual census technique. Estuar Coast Shelf Sci 51 (1): 31–44

Neill I, Kerr AC, Hastie A, Stanek KP, Millar IL (2011) Origin of the Aves Ridge and Dutch-Venezuelan Antilles: interaction of the Cretaceous 'Great Arc' and Caribbean-Colombian Oceanic Plateau? J Geol Soc Lond 168:333–347

Parker S (1999) Collaboration on tourism policy making: environmental and commercial sustainability on bonaire, NA. J Sustain Tourism 7(3–4):240–259

Pijpers PJ (1933) Geology and paleontology of bonaire. Geogr Geol Meded. Physiogr Geol Reeks. 8:103

IMF Country Reports (2015) Kingdom of the Netherlands–Aruba: 2015 Article IV Consulation Discussions–Staff Report; and Press Release. 15/116

Ridderstaat J (2007) The lago story: the compelling story of an oil company on the island of Aruba. Editorial Charuba

Ridderstaat J, Croes R, Nijkamp P (2014) Tourism and Long-run Economic Growth in Aruba. Int J Tourism Res 16:472–487

Robert SS, Suzanne NA, de León R, Rasher DB (2015) Executive summary: status and trends of bonaire's reefs in 2015: slow but steady signs of resilience. http://www.bmp.org/pdfs/Status%20of%20Bonaire's%20Coral%20Reefs%202015-FINAL.pdf

Römer RA (1984) The dutch colonial history in the Netherland Antilles. De Halve Maen 57(11–13):28

Roos PJ (1964) The distribution of reef corals in Curaçao. In: Studies on the Fauna: of Curaçao and Other Caribbean Islands 20:1–51. Utrecht, The Netherlands

Roos PJ (1971) The shallow water stony corals of the Netherlands Antilles. Studies of Fauna: Curaçao and Other Caribbean Islands, 37. Utrecht, The Netherlands, pp 1–108

Rupert Linda M (2012) Creolization and contraband: curaçao in the early modern atlantic world. University of Georgia Press, Athens, GA

Saunders NJ (2005) The peoples of the Caribbean: an encyclopedia of tradition and culture

Scheffers A (2002) Paleotsunami evidences from boulder deposits on Aruba, Curaçao and Bonaire. Sci Tsunami Hazards 20(1):26–37

Scheffers A (2004) Tsunami imprints on the Leeward Netherlands Antilles (Aruba, Curaçao, Bonaire) and their relation to other coastal problems. Quatern Int 120:163–172

Scheffers A, Scheffers S, Kelletat D (2005) Paleo-tsunami relics on the southern and central antillean island arc. J Coastal Res 21(2):263–273

Strelich L (2015) Sea surface temperatures on the rise in the Caribbean, Eos, 96. doi:10.1029/2015EO039535. Published on 13 November 2015

Van der Lelij R, Spikings R, Kerr AC, Kounov A, Cosca M, Chew D, Villagomez D (2010) Thermochronology and tectonics of the leeward antilles: evolution of the southern caribbean plate boundary zone. Tectonics 29:1–30

van Duyl FC (1985) Atlas of the living reefs of Curaçao and Bonaire (Netherlands Antilles). 117, Utrecht, The Netherlands, 37 pp

van Sambeek MHG, Eggenkamp HGM, Vissers MJM (2000) The groundwater quality of Aruba, Bonaire and Curaçao: a hydrogeochemical study. Geol en Mijnbouw/ Neth J Geosci 79(4):459–466

Vanegas M Sr, Croes RR (2003) Growth development and tourism in a small economy: evidence from Aruba. Int J Tourism Res 5:315–330

Vermeij MJA (2012) The current state of Curaçao's reefs. Carmabi report 38 pp

Vidauskaite R. (2015) Destination branding through wedding tourism: The Case of the Caribbean. A master's thesis University of Ljubljana. Accessed May 8 2016. http://www.cek.ef.uni-lj.si/magister/vidauskaite1917-B.pdf

Westermann JH (1932) The geology of Aruba. Ph.D thesis, University of Utrecht

White RV, Tarney J, Kerr AC, Saunders AD, Kempton PD, Pringle MS, Klaver A (1999) Modification of an oceanic plateau, Aruba, Dutch Caribbean: implications for the generation of continental crust. Lithos 46:43–68

Zapata S, Cardona A, Montes C, Valencia V, Vervoort J, Reiners P (2014) Provenance of the Eocene Soebi Blanco formation, Bonaire, Leeward Antilles: correlations with post-Eocene tectonic evolution of northern South America. J S Am Earth Sci 52:179–193

Zonneveld JIS, de Buisonje PH, Herweijer JP (1977) Geomorphology and denudation processes, in guide to the field excursions on Curaçao, Bonaire and Aruba, Netherlands Antilles (8th Caribbean Geological Conference, Curaçao 9–24 July 1977). GUA Papers of Geology, 10:56–68

Index

C.D. Allen (ed.), *Landscapes and Landforms of the Lesser Antilles*, World Geomorphological Landscapes, DOI 10.1007/978-3-319-55787-8

MIX
Papier aus verantwortungsvollen Quellen
Paper from responsible sources
FSC® C105338

If you have any concerns about our products,
you can contact us on
ProductSafety@springernature.com

In case Publisher is established outside the EU,
the EU authorized representative is:
Springer Nature Customer Service Center GmbH
Europaplatz 3, 69115 Heidelberg, Germany

Printed by Libri Plureos GmbH
in Hamburg, Germany

Light Cone 2015

Giovanni Salmè · Barbara Pasquini
Editors

Light Cone 2015

Originally published in *Few-Body Systems,*
Volume 57, Issues 6–8, 2016

Springer

Editors
Giovanni Salmè
INFN Sezione di Roma
Rome
Italy

Barbara Pasquini
Department of Physics
University of Pavia
Pavia
Italy

ISBN 978-3-319-50698-2

Library of Congress Control Number: 2016959541

Printed on acid-free paper

This Springer imprint is published by Springer Nature
The registered company is Springer International Publishing AG
The registered company address is: Gewerbestrasse 11, 6330 Cham, Switzerland

Contents

DOI 10.1007/978-3-319-50699-9_1
Reprinted from *Few-Body System* Journal, DOI 10.1007/s00601-016-1103-3

EDITORIAL

Barbara Pasquini · Giovanni Salmè

LIGHT-CONE 2015: Theoretical and experimental challenges onto the Light-Front

Received: 1 April 2016 / Accepted: 4 April 2016 / Published online: 25 April 2016

The 2015 LIGHT-CONE meeting was hosted at the National Laboratories of the Istituto Nazionale di Fisica Nucleare (INFN) in Frascati, Italy, on September 21–26 (http://www.roma1.infn.it/conference/light_cone_2015/). The local organizing committee was composed by B. Pasquini (Co-Chair, Pavia U. and INFN Pavia) and G. Salmè (Co-Chair, INFN Roma), M. Battaglieri (INFN Genova), D. Binosi (ECT*, Trento), M. Boglione (Torino U. and INFN Torino), U. D'Alesio (Cagliari U. and INFN Cagliari), M.P. Lombardo (LNF, INFN Frascati), M. Mirazita (LNF, INFN Frascati), S. Scopetta (Perugia U. and INFN Perugia), with Mrs. D. Pierluigi (LNF, INFN Frascati) providing administrative support.

This series of meetings is held under the auspices of the International Light Cone Advisory Committee (ILCAC), Inc. (http://www.ilcacinc.org).

The Workshop was supported in part by generous contributions from INFN, and the Jefferson Science Association (JSA). This year there was a remarkable number of applications for the McCartor Fellowships, that allow promising young scientists to attend and present their research at the annual meeting. In particular, thanks to the support from JSA, ILCAC awarded the McCartor Fellowships to eight young physicists.

LIGHT-CONE 2015 had 79 registered participants, and was also attended by several researchers working at the LNF. The program of the workshop consisted in a total of 65 talks and 6 posters. Each participant was invited to vote for the two best posters. The presenters of the two most voted posters (L. Bellantuono and C. Mondal) have been awarded the possibility to contribute with a peer-reviewed paper published in this issue. Emphasis on the interface between theory, phenomenology and experiments in hadron physics was a distinct feature of LC2015. The program included several experimental talks from Jefferson Lab (H. Avakian, M. Battaglieri, R. De Vita, S. Malace, S. Pisano, P. Rossi), COMPASS at CERN (F. Bradamante) and HERMES (L. Pappalardo). Furthermore, we had an overview talk about the future perspectives at an electron ion collider (A. Deshpande). These talks have stimulated alive discussions on both recent results and future perspectives of hadron physics at experimental facilities worldwide. The workshop covered also many topics in hadron phenomenology related to the deep inelastic processes under study at several facilities, including nucleon generalized parton distributions (D. Mueller, H. Moutarde) and Compton form factors (B. Bakker), transverse momentum dependent distributions (M. Anselmino, A. Bacchetta, M. Echevarria, L. Mantovani, A. Signori), and double parton distributions (W. Broniowski, J. Gaunt, M. Rinaldi). Other interesting topics of discussion were: the flavor asymmetries in the nucleon sea (C.-R. Ji), the spin and orbital angular momentum structure of

B. Pasquini (✉)
University of Pavia, Pavia, Italy
E-mail: Barbara.Pasquini@pv.infn.it

B. Pasquini
INFN Pavia, Pavia, Italy

G. Salmè
INFN Roma, Roma, Italy

the nucleon (M. Burkardt, C. Lorcé, S. Pisano, O. Teryaev, J. Zhang), hadron reactions and spectroscopy studies (S. Glazek, M. Bochicchio, A. Szczepaniak). Furthermore, results from QCD lattice studies covering many of the previous different topics were also presented (C. Alexandrou, J. Papavassiliou, S. Ryan, G. Bali). Several formal theoretical topics in QCD and QED, were illustrated, like some recent advancements in light-front field theory (J.R. Hiller, L. Martinovic, D. Kulshreshtha), the Faddeev approach for hadrons (G. Eichmann), the Minkowski-space description of composite systems (J. Carbonell, W. De Paula, T. Frederico, V. Karmanov, M. Viviani), light-front models (H. Choi, T. Pena), and issues involving bound states and zero-modes (W. Polyzou). Recent progress in light-front holography and AdS/QCD (Guy de Teramond, S. Cotogno, V. Lyubovitskij and A. Vega), as well as computational developments in many-body dynamics (J. Vary) were also highlighted. The closing presentation was given by S. Brodsky.

A half-day excursion was made to Villa Adriana in Tivoli and was followed by the Workshop dinner. During the dinner, the Chair of LC2015 and ILCAC presented awards to the 8 McCartor Fellowship recipients: Arkadiusz Trawinski (Warsaw U.), Xiaonu Xiong (Pavia U.), Kelly Chiu (SLAC), Frederik Van der Veken (CERN), Matteo Rinaldi (Perugia U.), Henry Lamm (Arizona State U.), Sabrina Cotogno (Nikhef and Vrije U., Amsterdam), and Andrea Signori (Nikhef and Vrije U., Amsterdam).

The reader can find more details about the mentioned presentations and many other interesting contributions-all peer-reviewed-in this special issue (LC2015 Proceedings). The detailed program and all the presentations can be found at the workshop website http://www.roma1.infn.it/conference/light_cone_2015/sc_program.html.

It is a pleasure to express our deep appreciation to all the participants, local organizers, scientific advisory committee members, the staff of LNF in Frascati and supporters who made this LC2015 meeting possible.

DOI 10.1007/978-3-319-50699-9_2
Reprinted from *Few-Body System* Journal, DOI 10.1007/s00601-016-1072-6

Mauro Anselmino

TMD Phenomenology

Received: 21 February 2016 / Accepted: 29 February 2016 / Published online: 18 March 2016

Abstract The transverse momentum dependent partonic distributions and fragmentation functions should reveal new properties of the three-dimensional structure of nucleons and of the quark hadronization process. Many experimental data are now available and much progress has been made in their phenomenological interpretation. A short summary of the situation is presented.

1 Introduction

The exploration of the structure of the nucleon—the most fundamental piece of matter in the visible Universe—is one of the most challenging open problems in physics. Up to 10–15 years ago our knowledge was essentially inspired by a one-dimensional model of the nucleon as a set of co-linearly moving quarks and gluons (partons). Data were obtained in high energy inclusive processes, mainly Deep Inelastic Scattering of leptons off nucleons (DIS, $\ell N \to \ell X$) or Drell–Yan processes (D–Y, $N N \to \ell^+\ell^- X$), interpreted in the QCD factorisation scheme according to which the cross sections are written as a convolution of Partonic Distribution Functions (PDFs) and Fragmentation Functions (FFs) with elementary hard interactions among constituents. The FFs can be measured in Semi-Inclusive Deep Inelastic Scattering (SIDIS, $\ell N \to \ell h X$) or e^+e^- annihilation ($e^+e^- \to h_1 h_2 X$).

More recently, a new phase has started in the study of the nucleon, aiming at resolving its full three-dimensional structure both in momentum and in co-ordinate space. Questions about the orbital motion of quarks, its correlation with the nucleon spin, the spatial distribution of valence and sea quarks, the average intrinsic momentum of unpolarised partons, have been posed and first answers are emerging. We consider here the nucleon structure in momentum space. This new information is encoded in the transverse momentum dependent parton distribution functions (TMD-PDFs) and fragmentation functions (TMD-FFs). At leading-twist there are 8 nucleon TMD-PDFs, which describe the distribution of partons, carrying a 3-momentum $\boldsymbol{q} = x\boldsymbol{P} + \boldsymbol{k}_\perp$ and spin $\boldsymbol{s}_q$, inside a nucleon with 3-momentum $\boldsymbol{P}$ and spin $\boldsymbol{S}$. Similarly, there are two TMD-FFs describing the fragmentation of a parton with 3-momentum $\boldsymbol{q}$ and spin $\boldsymbol{s}_q$ into a spinless or unpolarised hadron h with 3-momentum $\boldsymbol{P}_h = z\boldsymbol{q} + \boldsymbol{p}_\perp$. We refer collectively to TMD-PDFs and TMD-FFs as TMDs.

The parallel information on the three-dimensional nucleon structure in co-ordinate space can be obtained through the Generalised Parton Distributions (GPDs), which can be measured in exclusive reactions such as the lepto-production of a photon or of a meson or in photo-production of a lepton pair. Both the GPDs and the TMD-PDFs are particular limits of a vast class of functions, the so-called Wigner functions (or Generalised TMDs, GTMDs), which are the quantum mechanical version of the classical phase-space distributions. The really ultimate theoretical goal is that of reconstructing the nucleon Wigner functions; attempts to do that can be done, at the moment, by modelling the light-front nucleon wave functions.

M. Anselmino (✉)
Dipartimento di Fisica, Torino University and INFN, Via P. Giuria 1, 10125 Torino, Italy
E-mail: mauro.anselmino@to.infn.it

2 How Do We Access TMDs?

TMDs, like their integrated versions (PDFs and FFs), describe fundamental properties of the nucleons or of the hadronisation process; as such, they cannot be obtained from a basic theory, but must be extracted from experiments. What can be obtained from a basic theory, QCD in this case, is the Q^2 evolution of the TMDs, that is their dependence on the energy scale at which we are exploring the nucleons. Eventually, the TMD-PDFs could be built from some model of the nucleon wave function.

The usual approach adopted to extract TMDs from data is the QCD factorisation scheme, successfully used in the collinear case. For some processes—namely SIDIS, D-Y and e^+e^- annihilations—the factorisation has been proven to be valid, in selected kinematical regions, also with unintegrated PDFs and FFs [1], thus expressing measurable cross sections in terms of TMDs and elementary interactions. Schematically, one has, respectively for SIDIS, D-Y and e^+e^- processes:

$$d\sigma^{\rm SIDIS} = \sum_{q,\boldsymbol{s}_q,\boldsymbol{s}'_q} f_{q/N,S}(x,\boldsymbol{k}_\perp,\boldsymbol{s}_q;\,Q^2)\otimes d\hat\sigma^{\ell+q,s_q\to\ell+q,s'_q}\otimes D_{h/q,s'_q}(z,\boldsymbol{p}_\perp;\,Q^2) \tag{1}$$

$$d\sigma^{\rm D-Y} = \sum_{q,\boldsymbol{s}_q,\boldsymbol{s}_{\bar q}} f_{q/p,S_p}(x_1,\boldsymbol{k}_{\perp 1},\boldsymbol{s}_q;\,Q^2)\otimes f_{\bar q/N,S_N}(x_2,\boldsymbol{k}_{\perp 2},\boldsymbol{s}_{\bar q};\,Q^2)\otimes d\hat\sigma^{q,s_q+\bar q,s_{\bar q}\to\ell^++\ell^-} \tag{2}$$

$$d\sigma^{e^+e^-} = \sum_{q,\boldsymbol{s}_q,\boldsymbol{s}_{\bar q}} d\hat\sigma^{e^++e^-\to q,s_q+\bar q,s_{\bar q}}\otimes D_{h_1/q,s_q}(z_1,\boldsymbol{p}_{\perp 1},\boldsymbol{s}_q;\,Q^2)\otimes D_{h_2/\bar q,s_{\bar q}}(z_2,\boldsymbol{p}_{\perp 2},\boldsymbol{s}_{\bar q};\,Q^2)\,. \tag{3}$$

When fully expressed, these cross sections can be written as the sum of several pieces, each of which has a typical dependence on azimuthal angles, which is induced by the spins. Each piece contains a different convolution of TMDs thus allowing the separation of the 8 leading twist TMD-PDFs and the two leading twist TMD-FFs.

3 Experimental Results and Data

Since the introduction of the TMDs and the recognition of their role in the interpretation of spin dependent cross sections, many new experiments have confirmed their non negligible importance. The bulk of data come from SIDIS processes (HERMES, COMPASS and JLab Collaborations) for the TMD-PDFs, from e^+e^- annihilation (Belle, BaBar and BESIII Collaborations) for the TMD-FFs and from D-Y experiments regarding, so far, the unpolarised TMD-PDFs.

SIDIS azimuthal asymmetries [2,3] have definitely confirmed the non vanishing of the Sivers TMD, describing the spin dependent distribution of unpolarised partons inside a transversely polarised proton. These asymmetries originate from the convolution of the Sivers distribution with the unpolarised TMD-FFs, in polarised SIDIS. The FFs are known and it is relatively straightforward to extract, from data, information on the unknown Sivers functions. This has been done by several groups [4–6], assuming a simple shape of the Sivers TMD and of the unpolarised TMD-FF. Their dependence on the relevant variables is factorised: in $\boldsymbol{k}_\perp$ and (x, Q^2) for the Sivers TMD and in $\boldsymbol{p}_\perp$ and (z, Q^2) for the TMD-FFs. The dependence on the transverse motions is assumed to be Gaussian, with flavour and energy independent widths, $\langle k_\perp^2\rangle$ and $\langle p_\perp^2\rangle$.

Similarly, the Collins effect, describing the azimuthal dependence (around the quark direction) of unpolarised hadrons created in the fragmentation of a transversely polarised quark, has been observed in SIDIS [7–9] and $e^+e^- \to h_1\,h_2\,X$ [10–15] processes. In SIDIS the Collins TMD couples with the unknown transversity distribution, while azimuthal correlations in the production of two hadrons in opposite jets in e^+e^- annihilation are related to the product of two Collins functions. Thus, direct information [16] on the Collins function can be obtained from e^+e^- data and combined information [17,18] on the transversity distribution and the Collins function from SIDIS data. Independent information on the transversity distribution can be obtained by coupling it, in SIDIS or $p\,p$ processes, to a di-hadron FF [19,20].

Apart from spin dependent effects, which cannot be understood without TMDs, even for unpolarised processes the idealisation of a high energy nucleon as a collection of (almost) free partons, all moving co-linearly, is a clear oversimplification. It might work for very high energy particular processes, like $p\,p$ large P_T inclusive cross sections, but it is not adequate to describe, for example, the azimuthal distribution of hadrons produced in SIDIS processes, as observed in the $\gamma^* - N$ c.m. frame [21,22]. Indeed, the study of unpolarised SIDIS cross sections, and their dependence on the transverse momentum $\boldsymbol{P}_T$ of the final hadron,

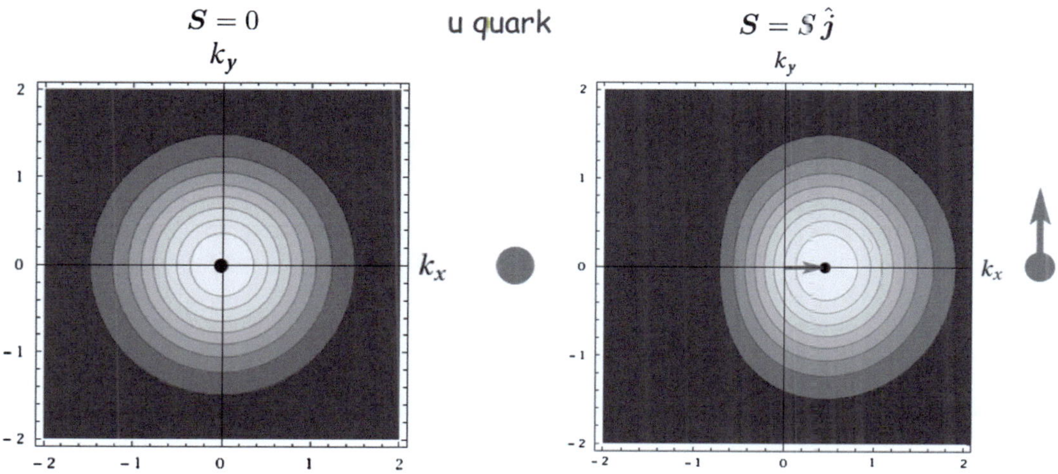

Fig. 1 Transverse momentum distribution, in units of 100 MeV, of u quarks inside an unpolarised proton (*left plot*) and a proton polarised along the $\hat{y}$ direction (*right plot*). *Brighter areas* are more densely populated. Courtesy of A. Prokudin

yields information on the average intrinsic motion of quarks in a nucleon and on the transverse distribution of hadrons produced in the fragmentation of a quark [23,24].

4 TMDs and the Nucleon Structure

The 8 leading twist TMDs have a partonic interpretation; their full knowledge should allow, at least in models, a reconstruction of the partonic nucleon structure in momentum space. In a complementary effort, the GPDs should give information on the spatial parton distributions. Indeed, the three-dimensional imaging of the nucleon is a major goal in the study of TMDs and GPDs [25].

So far, the study of the unpolarised TMDs [23,24] has confirmed the adequacy of a gaussian $k_\perp$ distribution in explaining HERMES [21] and COMPASS [22] multiplicities; it is still not clear, considering the available data, whether or not one should introduce a flavour dependence in the gaussian width. Similarly, the unpolarised TMD-FFs and the Collins functions which fit well the data [17] have a simple gaussian and flavour blind $p_\perp$ dependence.

The Sivers TMD is expected to have a special meaning and to be related to fundamental intrinsic properties of the nucleon. In fact, it describes the momentum distribution of unpolarised quarks and gluons inside a polarised proton with spin $\boldsymbol{S}$; then, in order to build a scalar, parity invariant quantity, $\boldsymbol{S}$ must couple to the only other available pseudo-vector, that is the parton orbital angular momentum, $\boldsymbol{L}_q$ or $\boldsymbol{L}_g$. There should be a way to link the Sivers function to the orbital motion of partons, which has been done so far only in models [6].

The presence of the Sivers effect results in an asymmetric $\boldsymbol{k}_\perp$ distribution of quarks inside a polarised proton; this has been nicely visualised in Fig. 1 [26], describing the difference, in the $\boldsymbol{k}_\perp$ distribution of u quarks, between an unpolarised and a polarised proton.

Another peculiar feature of the Sivers distribution is that—according to a QCD understanding of its origin from quark and gluon interactions—it should have opposite signs in SIDIS and D-Y processes, which remains to be tested. The Collins function, instead, is expected to be universal, describing a fundamental property in the hadronisation process of a transversely polarised quark.

5 TMDs and QCD Evolution

The QCD properties of the TMDs, in particular their QCD evolution, have been studied and, very recently, a sound and consistent factorisation scheme for TMDs has been developed [1,27,28] and the very first applications have been made [5,29,30]. The consistent and correct implementation of the TMD evolution in the analysis of the experimental data would give the study of the three-dimensional nucleon structure the same level of scientific soundness as the study of the integrated PDFs and FFs. However, TMD evolution is more difficult to implement, compared with the usual collinear evolution; it not only contains perturbative QCD contributions, but also non perturbative ones, which have to be parameterised and extracted from data, typically D-Y data. Then, different implementations of the TMD evolution can be found in the literature and even different evolution schemes [31,32].

So far, most data have been interpreted and analysed within the QCD factorisation scheme, with no, or limited, TMD evolution. Typically, the $k_\perp (p_\perp)$ dependence of the TMDs is factorised and only the collinear $x(z)$ part is evolved according to the usual DGLAP rules. Such a simplified approach is safe enough for SIDIS data, as they spread over a rather limited Q^2 range of few GeV^2. It might be more problematic for e^+e^- Belle and BaBar data, gathered at the larger value $s \simeq 100\ \text{GeV}^2$. However, also in this case, the simple approach might be reasonably safe for fitting azimuthal asymmetries as the TMD evolution strongly affects the cross sections, but much less their ratios.

In any case, it is crucially important to implement and test the TMD evolution scheme, for its own theoretical value and the future applications to higher energy data. A first hint of TMD evolution has been found by comparing COMPASS and HERMES Sivers asymmetry data, which slightly differ in Q^2 [5,29]. A first extraction of the combined transversity distribution and Collins function, from e^+e^- and SIDIS data has been performed in Ref. [30], with results not very different from those obtained with only collinear evolution.

6 Role of TMDs in Other Processes

Although TMD effects have been mainly investigated in SIDIS and e^+e^- processes, they might play important roles in several other processes. Interesting studies have pointed out that linearly polarised gluons inside unpolarised protons—a gluon TMD-PDF—affect the transverse momentum distribution of Higgs bosons produced via gluon-gluon fusion in $p\,p$ collisions [33]. They modify the distribution produced by unpolarised gluons in a characteristic way that could be used to determine whether the Higgs boson is a scalar or a pseudo-scalar particle. Similarly, the same gluon TMD-PDF originates azimuthal asymmetries in heavy quark pair production in unpolarised electron-proton and proton-proton collisions [34].

Data on the transverse momentum (q_T) distribution of Z bosons produced at LHC [35] could not be explained, at small q_T values, by the usual collinear PDF factorisation and need TMD-PDFs.

A special attention deserve the large transverse single spin asymmetries (SSAs, denoted by A_N) observed since a long time in the single inclusive production of a large P_T hadron in the interaction of polarised protons off unpolarised ones [36] (Fig. 2). For such processes, where only one large scale is present (P_T), the TMD factorisation has not been proven. However, if one assumes it to hold, the SSAs could be generated mainly by the Sivers TMD [37] and, to a lesser extent, by the Collins TMD-FF [38].

A different approach, known as twist-3 approach, is based on the collinear QCD factorisation scheme and involves as basic quantities higher-twist quark-gluon-quark correlations in the nucleon as well as in the hadronisation process. These correlators are related to moments of the TMDs, thus establishing a possible relation between the two approaches. However, the twist-3 correlator related to the $\boldsymbol{k}_\perp$ moment of the Sivers

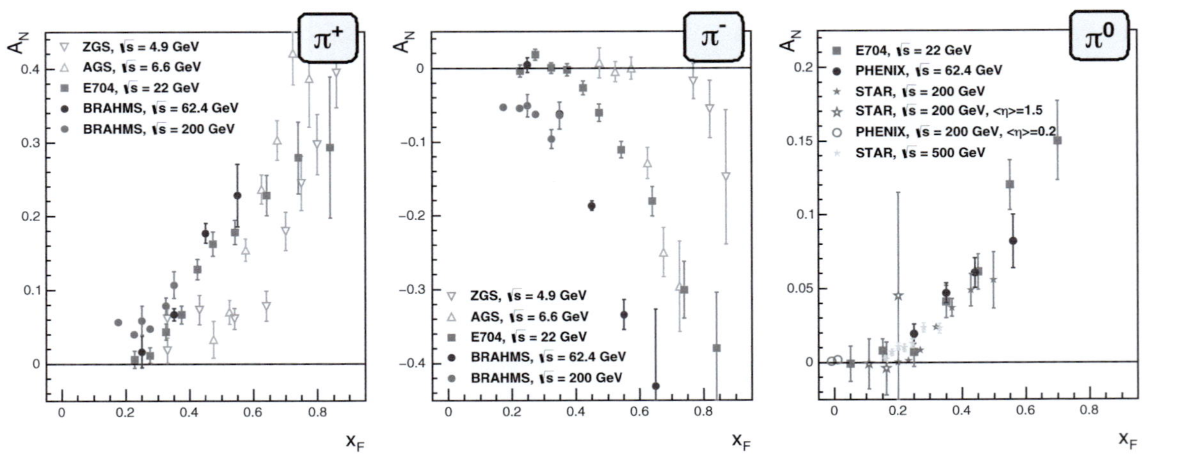

Fig. 2 A summary of the transverse single spin asymmetries $A_N = (d\sigma^\uparrow - d\sigma^\downarrow)/(d\sigma^\uparrow + d\sigma^\downarrow)$ measured in different experiments at different energies, from Ref. [36]

function, which should be dominant in the generation of the SSAs, gives sizeable values of A_N, but with the wrong sign [39]. A new kind of correlator in the fragmentation process seems to be more promising [40].

7 Future Experiments and Conclusions

The exploration of the three-dimensional partonic structure of nucleons is a fundamental and challenging task. In momentum space, this information is encoded in the transverse momentum dependent partonic distributions, which, in some physical observables, couple to transverse momentum dependent fragmentation functions. Among these TMDs, the Sivers and the Collins effects are well established, leading to transverse spin asymmetries which have been observed by several independent experiments. Gluon TMDs deserve a special attention; they might play a role at LHC.

A major issue in the modelling of the 3-D structure of hadrons is the orbital motion of quarks and, possibly, of gluons. Information on the total angular momentum of partons could be obtained from GPDs. Remaining in the TMD sector, the Sivers function, which depends on the proton spin, must be related to another pseudo-vector, namely the unpolarised parton orbital motion. Studies in this direction should be encouraged.

Several studies of TMD factorisation and TMD evolution have been recently completed; the full implementation, in the data analysis, of the TMD evolution is in progress. The most important issue of the sign change of the Sivers functions between D-Y and SIDIS processes has to be experimentally resolved.

Combined data from SIDIS, Drell–Yan, e^+e^- and $p\,p$ processes, with theoretical interpretation and phenomenological modelling, should lead to a true 3-D imaging of the proton. Great expectations are linked to new facilities, either going to run soon and being planned. The 12 GeV upgraded JLab results will certainly much expand our knowledge of TMDs and GPDs; RHIC and COMPASS polarised D-Y measurements are expected to measure very soon the sign of the Sivers function; above all, the dedicated Electron Ion Collider, being planned in the USA, will start a fundamental new phase in the exploration of the nucleon 3-D structure.

Acknowledgments Partial support from the Progetto di Ricerca di Ateneo/CSP" (codice TO-Call3-2012-0103) is gratefully acknowledged.

References

1. Collins, J.C.: Foundations of Perturbative QCD. Cambridge University Press, Cambridge (2011)
2. Airapetian, A., et al., (HERMES Collaboration): Observation of the Naive-T-odd sivers effect in deep-inelastic scattering. Phys. Rev. Lett. **103**, 152002 (2009)
3. Adolph, C., et al., (COMPASS Collaboration): Experimental investigation of transverse spin asymmetries in $\mu - p$ SIDIS processes: Sivers asymmetries. Phys. Lett. **B717**, 383 (2012)
4. Anselmino, M., Boglione, M., D'Alesio, U., Kotzinian, A., Melis, S., Murgia, F., Prokudin, A., Turk, C.: Sivers effect for pion and kaon production in semi-inclusive deep inelastic scattering. Eur. Phys. J. A **39**, 89 (2009)
5. Anselmino, M., Boglione, M., Melis, S.: A strategy towards the extraction of the Sivers function with TMD evolution. Phys. Rev. D **86**, 014028 (2012)
6. Bacchetta, A., Radici, M.: Constraining quark angular momentum through semi-inclusive measurements. Phys. Rev. Lett. **107**, 212001 (2011)
7. Airapetian, A., et al., (HERMES Collaboration): Effects of transversity in deep-inelastic scattering by polarized protons. Phys. Lett. **B693**, 11 (2010)
8. Adolph, C., et al., (COMPASS Collaboration): Collins and Sivers asymmetries in muon-production of pions and kaons off transversely polarised protons. Phys. Lett. **B744**, 250 (2015)
9. Adolph, C., et al., (COMPASS Collaboration): Experimental investigation of transverse spin asymmetries in $\mu - p$ SIDIS processes: Collins asymmetries. Phys. Lett. **B717**, 376 (2012)
10. Abe, K., et al., (Belle Collaboration): Measurement of azimuthal asymmetries in inclusive production of hadron pairs in e^+e^- annihilation at Belle. Phys. Rev. Lett. **96**, 232002 (2006)
11. Seidl, R., et al., (Belle Collaboration): Measurement of azimuthal asymmetries in inclusive production of hadron pairs in e^+e^- annihilation at $\sqrt{s} = 10.58$ GeV. Phys. Rev. **D78**, 032011 (2008)
12. Seidl, R., et al., (Belle Collaboration): Measurement of azimuthal asymmetries in inclusive production of hadron pairs in e^+e^- annihilation at $\sqrt{s} = 10.58$ GeV. Phys. Rev. **D86**, 039905(E) (2008)
13. Lees, J., et al., (BaBar Collaboration): Measurement of Collins asymmetries in inclusive production of charged pion pairs in e^+e^- annihilation at BaBaR. Phys. Rev. **D90**, 052003 (2014)
14. Aubert, B., et al., (BaBar Collaboration): Collins asymmetries in inclusive charged $K\,K$ and $K\,\pi$ pairs produced in e^+e^- annihilation. Phys. Rev. **D92**, 111101 (2015)
15. Ablikim, M., et al., (BESIII Collaboration): Measurement of azimuthal asymmetries in inclusive charged dipion production in e^+e^- annihilations at $\sqrt{s} = 3.65$ GeV. Phys. Rev. Lett. **116**, 042001 (2016)

16. Anselmino, M., Boglione, M., DAlesio, U., Gonzalez Hernandez, J.O., Melis, S., Murgia, F., Prokudin, A.: Extracting the kaon Collins function from e^+e^- hadron pair production data. Phys. Rev. D **93**, 034025 (2016)
17. Anselmino, M., Boglione, M., D'Alesio, U., Gonzalez Hernandez, J.O., Melis, S., Murgia, F., Prokudin, A.: Collins functions for pions from SIDIS and new e^+e^- data: a first glance at their transverse momentum dependence. Phys. Rev. D **92**, 114023 (2015)
18. Anselmino, M., Boglione, M., D'Alesio, U., Melis, S., Murgia, F., Prokudin, A.: Simultaneous extraction of transversity and Collins functions from new SIDIS and e^+e^- data. Phys. Rev. D **87**, 094019 (2013)
19. Radici, M., Courtoy, A., Bacchetta, A., Guagnelli, M.: Improved extraction of valence transversity distributions from inclusive dihadron production. JHEP **1505**, 123 (2015)
20. Bacchetta, A., Radici, M.: Dihadron interference fragmentation functions in proton-proton collisions. Phys. Rev. D **70**, 094032 (2004)
21. Airapetian, A., et al., (HERMES Collaboration): Multiplicities of charged pions and kaons from semi-inclusive deep-inelastic scattering by the proton and the deuteron. Phys. Rev. **D87**, 074029 (2013)
22. Adolph, C., et al., (COMPASS Collaboration): Hadron Transverse Momentum Distributions in Muon Deep Inelastic Scattering at 160 GeV/*c*. Eur. Phys. J. **C73**, 2531 (2013)
23. Anselmino, M., Boglione, M., Gonzalez, J.O., Melis, S., Prokudin, A.: Unpolarised transverse momentum dependent distribution and fragmentation functions from SIDIS multiplicities. JHEP **1404**, 005 (2014)
24. Signori, A., Bacchetta, A., Radici, M., Schnell, G.: Investigations into the flavour dependence of partonic transverse momentum. JHEP **1311**, 194 (2013)
25. Burkardt, M., Pasquini, B.: Modelling the nucleon structure. arXiv:1510.02567 [hep-ph], to appear in EPJA (2016)
26. Prokudin, A.: Private communication
27. Aybat, S.M., Rogers, T.C.: TMD parton distribution and fragmentation functions with QCD evolution. Phys. Rev. D **83**, 114042 (2011)
28. Aybat, S.M, Collins, J.C., Qiu, J.-W., Rogers, T.C.: The QCD evolution of the Sivers function. Phys. Rev. D **85**, 034043 (2012)
29. Aybat, S.M., Prokudin, A., Rogers, T.C.: Calculation of TMD evolution for transverse single spin asymmetry measurements. Phys. Rev. Lett. **108**, 242003 (2012)
30. Kang, Z.-B., Prokudin, A., Sun, P., Yuan, F.: Extraction of quark transversity distribution and collins fragmentation functions with QCD evolution. Phys. Rev. D **93**, 014009 (2016)
31. Echevarria, M.G., Idilbi, A., Schfer, A., Scimemi, I.: Model-independent evolution of transverse momentum dependent distribution functions (TMDs) at NNLL. Phys. J. C **73**, 2636 (2013)
32. Echevarria, M.G., Idilbi, A., Scimemi, I.: Soft and collinear factorisation and transverse momentum dependent parton distribution functions. Phys. Lett. B **726**, 795 (2013)
33. Boer, D., den Dunnen, W.J., Pisano, C., Schlegel, M., Vogelsang, W.: Linearly polarised gluons and the Higgs transverse momentum distribution. Phys. Rev. Lett. **108**, 032002 (2012)
34. Pisano, C., Boer, D., Brodsky, S.J., Buffing, M.G.A., Mulders, P.J.: Linear polarization of gluons and photons in unpolarised collider experiments. JHEP **1310**, 024 (2013)
35. Chatrchyan, S., et al., (CMS Collaboration): Measurement of the rapidity and transverse momentum distributions of Z Bosons in pp collisions at $\sqrt{s} = 7\,\text{TeV}$. Phys. Rev. **D85**, 032002 (2012)
36. Aschenauer, E.C., D'Alesio, U., Murgia, F.: TMDs and SSAs in hadronic interactions. arXiv:1512.05379 [hep-ph], to appear in EPJA (2015)
37. Anselmino, M., Boglione, M., D'Alesio, U., Melis, S., Murgia, F., Prokudin, A.: Sivers effect and the single spin asymmetry A_N in $p^\uparrow p \to h X$ processes. Phys. Rev. D **88**, 054023 (2013)
38. Anselmino, M., Boglione, M., D'Alesio, U., Leader, E., Melis, S., Murgia, F., Prokudin, A.: On the role of Collins effect in the single spin asymmetry A_N in $p^\uparrow p \to h X$ processes. Phys. Rev. D **8**, 6074032 (2012)
39. Kang, Z.-B., Qiu, J.-W., Vogelsang, W., Yuan, F.: An observation concerning the process dependence of the sivers functions. Phys. Rev. D **83**, 094001 (2011)
40. Kanazawa, K., Koike, Y., Metz, A., Pitonyak, D.: Towards an explanation of transverse single-spin asymmetries in proton-proton collisions: the role of fragmentation in collinear factorization. Phys. Rev. D **89**, 111501 (2014)

DOI 10.1007/978-3-319-50699-9_3
Reprinted from *Few-Body System* Journal, DOI 10.1007/s00601-016-1043-y

Cédric Lorcé · Keh-Fei Liu

Quark and Gluon Orbital Angular Momentum: Where Are We?

Received: 12 January 2016 / Accepted: 12 January 2016 / Published online: 8 February 2016

Abstract The orbital angular momentum of quarks and gluons contributes significantly to the proton spin budget and attracted a lot of attention in the recent years, both theoretically and experimentally. We summarize the various definitions of parton orbital angular momentum together with their relations with parton distributions functions. In particular, we highlight current theoretical puzzles and give some prospects.

1 Introduction

One of the major challenges in hadron physics is to unravel how the spin of the proton arises from the spin and orbital motion of its constituents. It appears that the quark and gluon spin contributions account for about 60 % of the proton spin budget [1–3], implying that 40 % should be accounted by the quark and gluon orbital angular momentum (OAM). This is a large fraction which reflects the relativistic nature of the quark–gluon bound state. The quark and gluon OAM, being a correlation between position and momentum, is more difficult to access experimentally than the spin. It depends also on how the total angular momentum (AM) is divided into separate quark and gluon contributions, which is intrinsically ambiguous due to quark–gluon couplings.

It has long been thought that only the kinetic (or mechanical) decomposition of the proton spin makes sense because its quark and gluon contributions can be extracted from experimental data and computed on the lattice. Recent theoretical and experimental progress have, however, shown that the canonical decomposition can also be accessed experimentally and computed on the lattice, though in a more complicated way. Kinetic and canonical decompositions appear to be complementary with their own advantages and disadvantages. For more detailed discussions, see the recent reviews [4,5].

We present here a short summary of the theoretical status of quark and gluon OAM. First we discuss the most general gauge-invariant decomposition of the proton spin and show how it is related to the other ones proposed in the literature. We also comment on its physical significance and stress the importance of the role played by the experimental configuration and the theoretical framework in deciding which explicit form to use. Then we summarize the relations between measurable parton distributions and the different contributions to the proton spin. Having established relations at the integrated level, we finally discuss the status of those proposed at the density level. More details and discussions including recent important lattice developments can be found in Ref. [6]

C. Lorcé (✉)
Centre de Physique Théorique, École polytechnique, CNRS, Université Paris-Saclay, 91128 Palaiseau, France
E-mail: cedric.lorce@polytechnique.fr

K.-F. Liu
Department of Physics and Astronomy, University of Kentucky, Lexington, KY 40506, USA

2 Gauge-Invariant Decomposition of Angular Momentum

The total AM operator in QCD can be decomposed as

$$\mathbf{J} = \mathbf{S}^q + \mathbf{S}^G + \mathbf{L}^q_{\text{kin}} + \mathbf{L}^G_{\text{can}} + \mathbf{L}_{\text{pot}}, \tag{1}$$

where the quark spin, gluon spin, quark kinetic OAM, gluon canonical OAM and potential OAM contributions are respectively given by

$$\begin{aligned}
\mathbf{S}^q &= \int \mathrm{d}^3r\, \psi^\dagger \tfrac{1}{2}\boldsymbol{\Sigma}\psi, & \mathbf{S}^G &= \int \mathrm{d}^3r\, \mathbf{E}^a \times \mathbf{A}^a_{\text{phys}}, \\
\mathbf{L}^q_{\text{kin}} &= \int \mathrm{d}^3r\, \psi^\dagger\,(\mathbf{r}\times i\mathbf{D})\,\psi, & \mathbf{L}^G_{\text{can}} &= -\int \mathrm{d}^3r\, E^{aj}\left(\mathbf{r}\times \boldsymbol{\mathcal{D}}^{ab}_{\text{pure}}\right) A^{bj}_{\text{phys}}, \\
\mathbf{L}_{\text{pot}} &= -\int \mathrm{d}^3r\, \rho^a\, \mathbf{r}\times \mathbf{A}^a_{\text{phys}}
\end{aligned} \tag{2}$$

We followed Chen et al. [7,8] and decomposed the gauge potential into a pure-gauge field and a "physical" field [9,10]

$$A^\mu = A^\mu_{\text{pure}} + A^\mu_{\text{phys}} \tag{3}$$

such that $F^{\mu\nu}_{\text{pure}} = \partial^\mu A^\nu_{\text{pure}} - \partial^\nu A^\mu_{\text{pure}} - ig[A^\mu_{\text{pure}}, A^\nu_{\text{pure}}] = 0$. Under a gauge transformation, A^μ_{pure} and A^μ_{phys} transform as

$$A^\mu_{\text{pure}} \mapsto U\left[A^\mu_{\text{pure}} + \tfrac{i}{g}\partial^\mu\right]U^{-1}, \qquad A^\mu_{\text{phys}} \mapsto U A^\mu_{\text{phys}} U^{-1}. \tag{4}$$

The pure-gauge covariant derivatives are defined like the ordinary covariant derivatives with the gauge potential A^μ replaced by the pure-gauge field A^μ_{pure}. This ensures explicit gauge invariance of each contribution in Eq. (2).

Due to the QCD equations of motion $\rho^a = g\psi^\dagger t^a \psi = \boldsymbol{\mathcal{D}}^{ab}\cdot\mathbf{E}^b$, the potential OAM can be interpreted as either a quark or a gluon contribution. The quark canonical OAM appearing in the (gauge-invariant) canonical decomposition is obtained by combining quark kinetic and potential OAM

$$\mathbf{L}^q_{\text{can}} = \mathbf{L}^q_{\text{kin}} + \mathbf{L}_{\text{pot}} = \int \mathrm{d}^3r\, \psi^\dagger\left(\mathbf{r}\times i\mathbf{D}_{\text{pure}}\right)\psi. \tag{5}$$

The gluon kinetic OAM appearing in the (gauge-invariant) kinetic decomposition is obtained by combining gluon canonical and potential OAM

$$\mathbf{L}^G_{\text{kin}} = \mathbf{L}^G_{\text{can}} + \mathbf{L}_{\text{pot}} = \int \mathrm{d}^3r\, \mathbf{r}\times\left[\left(\mathbf{A}^a_{\text{phys}}\times \boldsymbol{\mathfrak{D}}^{ab}_{\text{pure}}\right)\times \mathbf{E}^b\right], \tag{6}$$

where $\mathfrak{D}^\mu_{\text{pure}} = \frac{1}{2}(\mathcal{D}^\mu + \mathcal{D}^\mu_{\text{pure}})$ [11].

The decomposition of the gauge potential (3) essential for ensuring gauge invariance is not unique since

$$A^\mu_{\text{pure}} \mapsto A^\mu_{\text{pure}} + \partial^\mu C \qquad A^\mu_{\text{phys}} \mapsto A^\mu_{\text{phys}} - \partial^\mu C, \tag{7}$$

referred to as a Stueckelberg transformation [12,13], leaves the fundamental Lagrangian invariant. In practice, this is actually not an issue since the experimental conditions combined with the theoretical framework usually provide a natural decomposition. In experiments probing the internal structure of the proton, the off-shell probe indeed provides a natural direction [14] along which one can unambuously define spin and OAM contributions [13,15,16].

3 Accessing Angular Momentum with Parton Distributions

Ji [17] derived a remarkable relation between the total kinetic AM of quarks and gluons, and twist-2 generalized parton distributions (GPDs)

$$\left\langle J^{q,G}_{\text{kin}}\right\rangle = \tfrac{1}{2}\int \mathrm{d}x\, x\left[H^{q,G}(x,0,0)+E^{q,G}(x,0,0)\right]. \tag{8}$$

The kinetic OAM of quarks and gluons can then be obtained by subtracting the corresponding spin contributions,

$$\left\langle L^{q,G}_{\text{kin}}\right\rangle = \left\langle J^{q,G}_{\text{kin}}\right\rangle - \left\langle S^{q,G}\right\rangle. \tag{9}$$

which are given in the $\overline{MS}$ scheme by the first Mellin moment of the quark and gluon helicity distributions $\langle S^q\rangle = \frac{1}{2}\int \mathrm{d}x\,\Delta q(x)$, $\langle S^G\rangle = \int \mathrm{d}x\,\Delta g(x)$. Alternatively, it has been shown by Penttinen, Polyakov, Shuvaev and Strikman (PPSS) [18] that the quark kinetic OAM can also be directly expressed in terms of a two-parton twist-3 GPD [19–22]

$$\langle L^q_{\text{kin}}\rangle = -\int \mathrm{d}x\, x\, G^q_2(x,0,0). \tag{10}$$

Similarly, it is thought that transverse-momentum distributions (TMDs) could also give quantitative information about OAM. Quark model calculations motivated the following relations [23–26]

$$\begin{aligned}\langle L^q_{\text{can}}\rangle &= \int \mathrm{d}x\,\mathrm{d}^2k_\perp\left[h^q_1\left(x,\mathbf{k}^2_\perp\right) - g^q_{1L}\left(x,\mathbf{k}^2_\perp\right)\right],\\ &= -\int \mathrm{d}x\,\mathrm{d}^2k_\perp\,\frac{\mathbf{k}^2_\perp}{2M^2}\,h^{\perp q}_{1T}\left(x,\mathbf{k}^2_\perp\right),\end{aligned} \tag{11}$$

but they turned out to be valid only under some restricted condition, and not in general in QCD [27–29]. Alternatively, Burkardt [30,31] suggested that a chromodynamic lensing mechanism could relate the quark Sivers TMD $f^{\perp q}_{1T}(x,\mathbf{k}^2_\perp)$ and the quark GPD $E^q(x,\xi,t)$. Although this mechnaism can hardly be put on a firm theoretical ground, a variation of it by Bacchetta and Radici [32] has led to a new estimate of $\langle J^q_{\text{kin}}\rangle$, in good agreement with most common GPD extractions.

The most natural expression for the OAM is that of a phase-space integral [33,34]

$$\left\langle L^{q,G}_{\mathcal{W}}\right\rangle = \int \mathrm{d}x\,\mathrm{d}^2k_\perp\,\mathrm{d}^2b_\perp\,(\mathbf{b}_\perp\times\mathbf{k}_\perp)_z\,\rho^{q,G}_{++}(x,\mathbf{k}_\perp,\mathbf{b}_\perp;\mathcal{W}), \tag{12}$$

where the Wilson line $\mathcal{W}$ ensures gauge invariance. In a semi-classical interpretation, the Wigner distribution $\rho^{q,G}_{++}$ gives the (quasi-)probability for finding, at the transverse position $\mathbf{b}_\perp$ a quark or a gluon with momentum $(xP^+,\mathbf{k}_\perp)$ inside a longitudinally polarized ($\Lambda'=\Lambda=+$) proton. This Wigner distribution is related *via* Fourier transform to the generalized transverse-momentum dependent distributions (GTMDs) [27,35,36], leading to the simple relation [33,37,38]

$$\left\langle L^{q,G}_{\mathcal{W}}\right\rangle = -\int \mathrm{d}x\,\mathrm{d}^2k_\perp\,\frac{\mathbf{k}^2_\perp}{M^2}\,F^{q,G}_{14}(x,0,\mathbf{k}_\perp,\mathbf{0}_\perp;\mathcal{W}). \tag{13}$$

For a staple-like Wilson line $\mathcal{W}_\sqsupset$ [14], Eq. (13) gives the canonical OAM $\langle L^{q,G}_{\text{can}}\rangle = \langle L^{q,G}_{\mathcal{W}_\sqsupset}\rangle$ irrespective of whether the staple is future- or past-pointing [21,22,37,39,40]. For a straight Wilson line $\mathcal{W}_|$, it gives the kinetic version of the quark OAM $\langle L^q_{\text{kin}}\rangle = \langle L^q_{\mathcal{W}_|}\rangle$ [41] and the Ji–Xiong–Yuan (JXY) [39] definition of the gauge-invariant gluon OAM $\langle L^G_{\text{JXY}}\rangle = \langle L^G_{\mathcal{W}_|}\rangle$, where $\mathbf{L}^G_{\text{JXY}} = -\int \mathrm{d}^3r\,E^{aj}(\mathbf{r}\times\mathcal{D}^{ab})A^{bj}_{\text{phys}}$. Unfortunately, it is not known so far how to extract them from actual experiments, except possibly at small x [35]. Moreover, the GTMD F_{14} does not reduce to any GPD or TMD, and so cannot be directly constrained. However, in the last few years recent developments [42–44] opened the interesting possibility of computing GTMDs and OAM directly on the lattice. Moreover, phenomenological models, constrained by experimental data, also provide indirect access to GTMDs and hence OAM.

4 Angular Momentum at the Density Level

Angular momentum can also be defined at the density level and, therefore, be mapped in both position and momentum spaces. Many definitions, differing by superpotential terms of the form $\partial_\alpha X^{[\alpha\mu]\cdots}$, where square brackets stand for antisymmetrization of indices, have been proposed in the literature, creating a somewhat confusing situation. It is essential to keep track of these superpotential terms, since they affect the interpretation of the density.

Hoodbhoy et al. [45] defined higher moments of the kinetic AM densities and concluded that

$$\langle L^q_{\text{kin}}\rangle(x) = \tfrac{x}{2}\left[H^q(x,0,0) + E^q(x,0,0)\right] - \tfrac{1}{2}\Delta q(x). \tag{14}$$

However, this definition is somewhat *ad hoc* since a complicated tower $\Delta L^{+\cdots+}$ has been added to the natural simple one

$$\tilde{L}^{q,+\cdots+}_{\text{kin}} = \frac{1}{n+1}\sum_{j=0}^{n}\int \mathrm{d}^3r\,\overline{\psi}\gamma^+\left(iD^+\right)^j(\mathbf{r}_\perp\times i\mathbf{D}_\perp)_z\left(iD^+\right)^{n-j}\psi \tag{15}$$

in order to have it evolve as a leading-twist operator. Ji et al. [39,41,46] argued that the integrand of the Ji relation (8) $\frac{x}{2}\left[H^q(x,0,0)+E^q(x,0,0)\right]$ could naturally be identified with an angular momentum decomposition of the transverse component of the quark angular momentum in a transversely polarized target. However, a careful inspection revealed several caveats leading to the conclusion that this interpretation is not justified [47–49].

It is actually not so surprising that the integrand of the Ji relation (8) cannot be simply interpreted as the density of parton AM. Indeed, the covariant derivative in the definition of kinetic OAM necessarily implies the contribution of higher-twist parton distributions. Hägler et al. [20] proposed another density of OAM which now involves both twist-2 and twist-3 GPDs

$$\langle L^q_{\text{kin}}\rangle(x) = x\left[H^q(x,0,0) + E^q(x,0,0) + G^q_2(x,0,0) - 2G^q_4(x,0,0)\right] - \Delta q(x) \tag{16}$$

but it has been obtained within the Wandzura–Wilczek approximation, where the distinction between canonical and kinetic OAM disappears. In a detailed discussion of twist-3 GPDs, Hatta and Yoshida [21] stressed that the density of kinetic OAM is ambiguous because it involves two longitudinal momentum fractions x_1 and x_2, contrary to canonical OAM. This is related to the fact that contrary to ordinary derivatives, covariant derivatives do not commute and, therefore, do not admit a unique non-local generalization [40].

Clearly, canonical OAM is more amenable to a description at the density level. Following the work of Jaffe and Manohar [50], a natural density of quark and gluon canonical OAM in the light-front gauge has been defined by Harindranath and Kundu [51] and later improved by Bashinsky and Jaffe [15] who provided an explicit expression invariant under residual gauge transformations. Recently, it has been shown that the density of OAM can directly be defined in a gauge-invariant way in phase-space [33,34,40]

$$\left\langle L^{q,G}_{\mathcal{W}}\right\rangle(x,\mathbf{k}_\perp,\mathbf{b}_\perp) = (\mathbf{b}_\perp\times\mathbf{k}_\perp)_z\,\rho^{q,G}_{++}(x,\mathbf{k}_\perp,\mathbf{b}_\perp;\mathcal{W}). \tag{17}$$

Contrary to the integrated version, the density depends on whether the staple points toward the future $\mathcal{W}_\sqsupset$ or the past $\mathcal{W}_\sqsubset$ [6,52]. Indeed, using the constraints imposed by parity and time-reversal symmetries, the Wigner distribution of unpolarized quarks and gluons inside a longitudinally polarized proton can be decomposed into four contributions [52]

$$\rho^{q,G}_{++} = \rho^{q,G}_1 + (\mathbf{b}_\perp\cdot\mathbf{k}_\perp)\,\rho^{q,G}_2 + (\mathbf{b}_\perp\times\mathbf{k}_\perp)_z\,\rho^{q,G}_3 + (\mathbf{b}_\perp\times\mathbf{k}_\perp)_z\,(\mathbf{b}_\perp\cdot\mathbf{k}_\perp)\,\rho^{q,G}_4, \tag{18}$$

where $\rho^{q,G}_i \equiv \rho^{q,G}_i(x,\mathbf{k}^2_\perp,(\mathbf{b}_\perp\cdot\mathbf{k}_\perp)^2,\mathbf{b}^2_\perp;\mathcal{W})$. The functions ρ_1 and ρ_2 are related via Fourier transform to the real and imaginary parts of the GTMD F_{11}, and similarly for ρ_3 and ρ_4 with F_{14}. The coefficient $(\mathbf{b}_\perp\cdot\mathbf{k}_\perp)$ implies that ρ_2 and ρ_4 are naive T-odd, i.e. they change sign under $\mathcal{W}_\sqsupset \leftrightarrow \mathcal{W}_\sqsubset$. Integrating over $\mathbf{b}_\perp$ or $\mathbf{k}_\perp$, one is left with just ρ_3, i.e. the real part of F_{14}. On the other hand, a straight Wilson line $\mathcal{W}_|$ leads to the density of JXY quark and gluon OAM [39,41] provided that $\mathbf{k}_\perp$ is integrated over [40].

The Ji relation (8) has also been discussed in position space. Since the information about the spatial distribution of quarks and gluons is encoded in the t dependence of GPDs [53,54], Polyakov [55,56] suggested that the Ji relation generalized to $t \neq 0$ should provide information about the spatial distribution of kinetic OAM

$$\left\langle J^{q,G}_{\text{kin}}\right\rangle(t) = \frac{1}{2}\int \mathrm{d}x\, x\left[H^{q,G}(x,0,t) + E^{q,G}(x,0,t)\right]. \tag{19}$$

Interestingly, Adhikari and Burkardt [57] observed within the scalar diquark model with Pauli–Villars regularization that the quark kinetic and canonical OAM spatial densities do not coincide $\langle L^q_{\text{kin}}\rangle(\mathbf{b}_\perp) \neq \langle L^q_{\text{can}}\rangle(\mathbf{b}_\perp)$, contrary to their integrated counterparts $\langle L^q_{\text{kin}}\rangle = \langle L^q_{\text{can}}\rangle$. It should however be noted that these spatial densities have been defined as the Fourier transform of the t-dependent distribution. This is however not justified in all cases. For example, the total AM, is given by the Fourier transform of the combination $\langle J^q_{\text{kin}}\rangle(t) + t\,\frac{\partial}{\partial t}\langle J^q_{\text{kin}}\rangle(t)$ [4], not just $\langle J^q_{\text{kin}}\rangle(t)$. Note also that the spatial density of quark kinetic OAM has been defined as $\langle L^q_{\text{kin}}\rangle(\mathbf{b}_\perp) = \langle J^q_{\text{kin}}\rangle(\mathbf{b}_\perp) - \langle S^q\rangle(\mathbf{b}_\perp)$, but a superpotential ruins this relation at the density level [4]. Further investigations are therefore needed.

5 Summary

We briefly reviewed the recent developments about kinetic and canonical decompositions of the proton spin. We discussed in particular the issue of gauge invariance and the link with measurable parton distributions. We also critically commented the various definitions of quark and gluon angular momentum densities.

References

1. Aidala, C.A., Bass, S.D., Hasch, D., Mallot, G.K.: The spin structure of the nucleon. Rev. Mod. Phys. **85**, 655 (2013)
2. de Florian, D., Sassot, R., Stratmann, M., Vogelsang, W.: Evidence for polarization of gluons in the proton. Phys. Rev. Lett. **113**(1), 012001 (2014)
3. Nocera, E.R., NNPDF Collaboration, et al.: A first unbiased global determination of polarized PDFs and their uncertainties. Nucl. Phys. B **887**, 276 (2014)
4. Leader, E., Lorcé, C.: The angular momentum controversy: what's it all about and does it matter?. Phys. Rep. **541**, 163 (2014)
5. Wakamatsu, M.: Is gauge-invariant complete decomposition of the nucleon spin possible?. Int. J. Mod. Phys. A **29**, 1430012 (2014)
6. Liu, K.F., Lorcé, C.: The parton orbital angular momentum: status and prospects. arXiv:1508.00911 [hep-ph] (2015)
7. Chen, X.-S., Lu, X.-F., Sun, W.-M., Wang, F., Goldman, T.: Spin and orbital angular momentum in gauge theories: nucleon spin structure and multipole radiation revisited. Phys. Rev. Lett. **100**, 232002 (2008)
8. Chen, X.-S., Sun, W.-M., Lu, X.-F., Wang, F., Goldman, T.: Do gluons carry half of the nucleon momentum?. Phys. Rev. Lett. **103**, 062001 (2009)
9. Lorcé, C.: Gauge-covariant canonical formalism revisited with application to the proton spin decomposition. Phys. Rev. D **88**, 044037 (2013)
10. Lorcé, C.: Gauge symmetry and background independence: should the proton spin decomposition be path independent?. Nucl. Phys. A **925**, 1 (2014)
11. Lorcé, C.: The proton spin decomposition: path dependence and gauge symmetry. Int. J. Mod. Phys. Conf. Ser. **25**, 1460030 (2014)
12. Stoilov, M.N.: Hamiltonian analysis of the proposal by Chen et. al., PRL 100 (2008) 232002. Bulg. J. Phys. **41**, 251 (2014)
13. Lorcé, C.: Geometrical approach to the proton spin decomposition. Phys. Rev. D **87**, 034031 (2013)
14. Collins, J.: Foundations of perturbative QCD. Cambridge Univeristy Press, Cambridge (2011)
15. Bashinsky, S., Jaffe, R.L.: Quark and gluon orbital angular momentum and spin in hard processes. Nucl. Phys. B **536**, 303 (1998)
16. Wakamatsu, M.: On the two remaning issues in the gauge-invariant decomposition problem of the nucleon spin. Eur. Phys. J. A **51**(5), 52 (2015)
17. Ji, X.-D.: Gauge-invariant decomposition of nucleon spin. Phys. Rev. Lett. **78**, 610 (1997)
18. Penttinen, M., Polyakov, M.V., Shuvaev, A.G., Strikman, M.: DVCS amplitude in the parton model. Phys. Lett. B **491**, 96 (2000)
19. Kiptily, D.V., Polyakov, M.V.: Genuine twist three contributions to the generalized parton distributions from instantons. Eur. Phys. J. C **37**, 105 (2004)
20. Hägler, P., Mukherjee, A., Schäfer, A.: Quark orbital angular momentum in the Wandzura–Wilczek approximation. Phys. Lett. B **582**, 55 (2004)
21. Hatta, Y., Yoshida, S.: Twist analysis of the nucleon spin in QCD. JHEP **1210**, 080 (2012)
22. Lorcé, C.: The light-front gauge-invariant energy-momentum tensor. JHEP **1508**, 045 (2015)
23. Ma, B.Q., Schmidt, I.: The quark orbital angular momentum in a light cone representation. Phys. Rev. D **58**, 096008 (1998)
24. She, J., Zhu, J., Ma, B.Q.: Pretzelosity $h_{1T}^{\perp}$ and quark orbital angular momentum. Phys. Rev. D **79**, 054008 (2009)
25. Avakian, H., Efremov, A.V., Schweitzer, P., Yuan, F.: The transverse momentum dependent distribution functions in the bag model. Phys. Rev. D **81**, 074035 (2010)

26. Efremov, A.V., Schweitzer, P., Teryaev, O.V., Zavada, P.: Images of quark intrinsic motion in covariant parton model. PoS DIS **2010**, 253 (2010)
27. Lorcé, C., Pasquini, B., Vanderhaeghen, M.: Unified framework for generalized and transverse-momentum dependent parton distributions within a 3Q light-cone picture of the nucleon. JHEP **1105**, 041 (2011)
28. Lorcé, C., Pasquini, B.: On the origin of model relations among transverse-momentum dependent parton distributions. Phys. Rev. D **84**, 034039 (2011)
29. Lorcé, C., Pasquini, B.: Pretzelosity TMD and quark orbital angular momentum. Phys. Lett. B **710**, 486 (2012)
30. Burkardt, M.: Impact parameter dependent parton distributions and transverse single spin asymmetries. Phys. Rev. D **66**, 114005 (2002)
31. Burkardt, M.: Chromodynamic lensing and transverse single spin asymmetries. Nucl. Phys. A **735**, 185 (2004)
32. Bacchetta, A., Radici, M.: Constraining quark angular momentum through semi-inclusive measurements. Phys. Rev. Lett. **107**, 212001 (2011)
33. Lorcé, C., Pasquini, B.: Quark Wigner distributions and orbital angular momentum. Phys. Rev. D **84**, 014015 (2011)
34. Lorcé, C., Pasquini, B., Xiong, X., Yuan, F.: The quark orbital angular momentum from Wigner distributions and light-cone wave functions. Phys. Rev. D **85**, 114006 (2012)
35. Meissner, S., Metz, A., Schlegel, M.: Generalized parton correlation functions for a spin-1/2 hadron. JHEP **0908**, 056 (2009)
36. Lorcé, C., Pasquini, B.: Structure analysis of the generalized correlator of quark and gluon for a spin-1/2 target. JHEP **1309**, 138 (2013)
37. Hatta, Y.: Notes on the orbital angular momentum of quarks in the nucleon. Phys. Lett. B **708**, 186 (2012)
38. Kanazawa, K., Lorcé, C., Metz, A., Pasquini, B., Schlegel, M.: Twist-2 generalized TMDs and the spin/orbital structure of the nucleon. Phys. Rev. D **90**, 014028 (2014)
39. Ji, X., Xiong, X., Yuan, F.: Probing parton orbital angular momentum in longitudinally polarized nucleon. Phys. Rev. D **88**(1), 014041 (2013)
40. Lorcé, C.: Wilson lines and orbital angular momentum. Phys. Lett. B **719**, 185 (2013)
41. Ji, X., Xiong, X., Yuan, F.: Proton spin structure from measurable parton distributions. Phys. Rev. Lett. **109**, 152005 (2012)
42. Ji, X., Zhang, J.H., Zhao, Y.: Justifying the naive partonic sum rule for proton spin. Phys. Lett. B **743**, 180 (2015)
43. Sufian, R.S., Glatzmaier, M.J., Yang, Y.B., Liu, K.F., Sun, M.: Glue spin S_G in the longitudinally polarized nucleon. PoS Lattice **2014**, 166 (2015)
44. Zhao, Y., Liu, K.F., Yang, Y.: arXiv:1506.08832 [hep-ph]
45. Hoodbhoy, P., Ji, X.D., Lu, W.: Quark orbital-angular-momentum distribution in the nucleon. Phys. Rev. D **59**, 014013 (1999)
46. Ji, X., Xiong, X., Yuan, F.: Reply to arXiv:1211.3957 and arXiv:1211.4731 by Leader et al. and arXiv:1212.0761 by Harindranath et al. Phys. Rev. Lett. **111**, 039103 (2013)
47. Leader, E., Lorcé, C.: Comment on 'Proton spin structure from measurable parton distributions' by Ji, Xiong and Yuan (PRL109, 152005 (2012)). Phys. Rev. Lett. **111**, 039101 (2013)
48. Harindranath, A., Kundu, R., Mukherjee, A., Ratabole, R.: Comment on Phys. 109, 152005 (2012). Rev. Lett. Phys. Rev. Lett. **111**, 039102 (2013)
49. Harindranath, A., Kundu, R., Mukherjee, A.: On transverse spin sum rules. Phys. Lett. B **728**, 63 (2014)
50. Jaffe, R.L., Manohar, A.: The G(1) problem: fact and fantasy on the spin of the proton. Nucl. Phys. B **337**, 509 (1990)
51. Harindranath, A., Kundu, R.: On orbital angular momentum in deep inelastic scattering. Phys. Rev. D **59**, 116013 (1999)
52. Lorcé, C., Pasquini, B.: Multipole decomposition of the nucleon transverse phase space. arXiv:1512.06744 [hep-ph] (2015)
53. Burkardt, M.: Impact parameter dependent parton distributions and off forward parton distributions for zeta —> 0. Phys. Rev. D **62**, 071503 (2000) [Erratum-ibid. D **66**, 119903]
54. Burkardt, M.: Impact parameter space interpretation for generalized parton distributions. Int. J. Mod. Phys. A **18**, 173 (2003)
55. Polyakov, M.V.: Generalized parton distributions and strong forces inside nucleons and nuclei. Phys. Lett. B **555**, 57 (2003)
56. Goeke, K., Grabis, J., Ossmann, J., Polyakov, M.V., Schweitzer, P., Silva, A., Urbano, D.: Nucleon form-factors of the energy momentum tensor in the chiral quark-soliton model. Phys. Rev. D **75**, 094021 (2007)
57. Adhikari, L., Burkardt, M.: Distribution of angular momentum in the transverse plane. Nucl. Phys. Proc. Suppl. **251**(252), 105 (2014)

DOI 10.1007/978-3-319-50699-9_4
Reprinted from *Few-Body System* Journal, DOI 10.1007/s00601-016-1064-6

Matthias Burkardt

Quark Orbital Angular Momentum

Received: 13 January 2016 / Accepted: 16 February 2016 / Published online: 10 March 2016

Abstract Generalized parton distributions provide information on the distribution of quarks in impact parameter space. For transversely polarized nucleons, these impact parameter distributions are transversely distorted and this deviation from axial symmetry leads on average to a net transverse force from the spectators on the active quark in a DIS experiment. This force when acting along the whole trajectory of the active quark leads to transverse single-spin asymmetries. For a longitudinally polarized nucleon target, the transverse force implies a torque acting on the quark orbital angular momentum (OAM). The resulting change in OAM as the quark leaves the target equals the difference between the Jaffe–Manohar and Ji OAMs.

1 Angular Momentum Decompositions

Since the famous EMC experiments revealed that only a small fraction of the nucleon spin is due to quark spins [1], there has been a great interest in 'solving the spin puzzle', i.e. in decomposing the nucleon spin into contributions from quark/gluon spin and orbital degrees of freedom. In this effort, the Ji decomposition [2]

$$\frac{1}{2} = \frac{1}{2}\sum_q \Delta q + \sum_q L_q^z + J_g^z \tag{1}$$

appears to be very useful: through GPDs, not only the quark spin contributions Δq but also the quark total angular momenta $J_q \equiv \frac{1}{2}\Delta q + L_q^z$ (and by subtracting the spin piece also the the quark orbital angular momenta L_q^z) entering this decomposition can be accessed experimentally. In the Ji decomposition (1) the quark OAM is defined as the expectation value

$$L_q^z = \int d^3r \langle PS|q^\dagger \left(\mathbf{r} \times \frac{1}{i}\mathbf{D}\right)^z q|PS\rangle / \langle PS|PS\rangle \tag{2}$$

in a nucleon state polarized in the $+\hat{z}$ direction. Here $\mathbf{D} = \boldsymbol{\partial} - ig\mathbf{A}$ is the gauge-covariant derivative. The main advantages of this decomposition are that each term can be expressed as the expectation value of a manifestly gauge invariant local operator and that the quark total angular momentum $J^q = \frac{1}{2}\Delta q + L^q$ can be related to GPDs [2] and is thus accessible in deeply virtual Compton scattering and deeply virtual meson production and can also be calculated in lattice gauge theory.

This work was supported by the DOE under Grant Number DE-FG03-95ER40965.

M. Burkardt (✉)
Department of Physics, New Mexico State University, Las Cruces, NM, 88003, USA
E-mail: burkardt@nmsu.edu
Tel.: +1-575-6461928
Fax: +1-575-6461934

Jaffe and Manohar have proposed an alternative decomposition of the nucleon spin, which does have a partonic interpretation, and in which also two terms, $\frac{1}{2}\Delta q$ and ΔG, are experimentally accessible [3]

$$\frac{1}{2} = \frac{1}{2}\sum_q \Delta q + \sum_q \mathcal{L}^q + \Delta G + \mathcal{L}^g. \tag{3}$$

In this decomposition the quark OAM is defined as

$$\mathcal{L}^q \equiv \int d^3r \langle PS|q_+^\dagger \left(\mathbf{r}\times\frac{1}{i}\boldsymbol{\partial}\right)^z q_+|PS\rangle / \langle PS|PS\rangle, \tag{4}$$

where light-cone gauge $A^+ = 0$ is implied. Although Eq. (4) is not manifestly gauge invariant as written, gauge invariant extensions can be defined [4–6]. Indeed, manifestly gauge invariant definitions for each of the terms in Eq. (3) exist, which, with the exception of Δq, involve matrix elements of nonlocal operators. In light-cone gauge those nonlocal operators reduce to a local operator, such as Eq. (4).

2 TMDs and OAM from Wigner Distributions

Wigner distributions can be defined as off forward matrix elements of non-local correlation functions [7–10] with $P^+ = P^{+\prime}$, $P_\perp = -P'_\perp = \frac{q_\perp}{2}$

$$W^{\mathcal{U}}(x,\mathbf{b}_\perp,\mathbf{k}_\perp) \equiv \int \frac{d^2\mathbf{q}_\perp}{(2\pi)^2}\int \frac{d^2\xi_\perp d\xi^-}{(2\pi)^3} e^{-i\mathbf{q}_\perp\cdot\mathbf{b}_\perp} e^{i(xP^+\xi^- - \mathbf{k}_\perp\cdot\boldsymbol{\xi}_\perp)}\langle P'S'|\bar{q}\left(-\frac{\xi}{2}\right)\Gamma\mathcal{U}_{-\frac{\xi}{2},\frac{\xi}{2}}q\left(\frac{\xi}{2}\right)|PS\rangle. \tag{5}$$

Throughout this paper, we will chose $\mathbf{S} = \mathbf{S}' = \hat{\mathbf{z}}$. Furthermore, we will focus on the 'good' component by selecting $\Gamma = \gamma^+$. To ensure manifest gauge invariance, a Wilson line gauge link $\mathcal{U}_{-\frac{\xi}{2},\frac{\xi}{2}}$ connecting the quark field operators at position $-\frac{\xi}{2}$ and $\frac{\xi}{2}$ is included. The issue of choice of path for the Wilson line will be addressed below.

In terms of Wigner distributions, TMDs and OAM can be defined as [4]

$$\begin{aligned} f(x,\mathbf{k}_\perp) &= \int d^2\mathbf{b}_\perp W^{\mathcal{U}}(x,\mathbf{b}_\perp,\mathbf{k}_\perp) \\ L_{\mathcal{U}} &= \int dx\, d^2\mathbf{b}_\perp d^2\mathbf{k}_\perp\, (\mathbf{b}_\perp\times\mathbf{k}_\perp)^z\, W^{\mathcal{U}}(x,\mathbf{b}_\perp,\mathbf{k}_\perp). \end{aligned} \tag{6}$$

No issues with Heisenberg's uncertainty principle arise here: only perpendicular combinations of position $\mathbf{b}_\perp$ and momentum $\mathbf{k}_\perp$ are needed simultaneously in order to evaluate the integral for $L_{\mathcal{U}}$.

A straight line connecting $-\frac{\xi}{2}$ and $\frac{\xi}{2}$ for the Wilson line in $\mathcal{U}_{-\frac{\xi}{2},\frac{\xi}{2}}$ results in [8]

$$L^q_{straight} = L^q_{Ji}. \tag{7}$$

However, depending on the context, other choices for the path in the Wilson link $\mathcal{U}$ should be made. Indeed for TMDs probed in SIDIS the path should be taken to be a straight line to $x^- = \infty$ along (or, for regularization purposes, very close to) the light-cone. This particular choice ensures proper inclusion of the FSI experienced by the struck quark as it leaves the nucleon along a nearly light-like trajectory in the Bjorken limit. However, a Wilson line to $\xi^- = \infty$, for fixed $\boldsymbol{\xi}_\perp$ is not yet sufficient to render Wigner distributions manifestly gauge invariant, but a link at $\xi^- = \infty$ must be included to ensure manifest gauge invariance. While the latter may be unimportant in some gauges, it is crucial in light-cone gauge for the description of TMDs relevant for SIDIS [11,12].

Let $\mathcal{U}^{+LC}_{-\frac{\xi}{2},\frac{\xi}{2}}$ be the Wilson path ordered exponential obtained by first taking a Wilson line from $\left(-\frac{\xi^-}{2},-\frac{\xi_\perp}{2}\right)$ to $\left(\infty,-\frac{\xi_\perp}{2}\right)$, then to $\left(\infty,\frac{\xi_\perp}{2}\right)$, and then to $\left(\frac{\xi^-}{2},\frac{\xi_\perp}{2}\right)$, with each segment being a straight line (Fig. 1) [5,6].

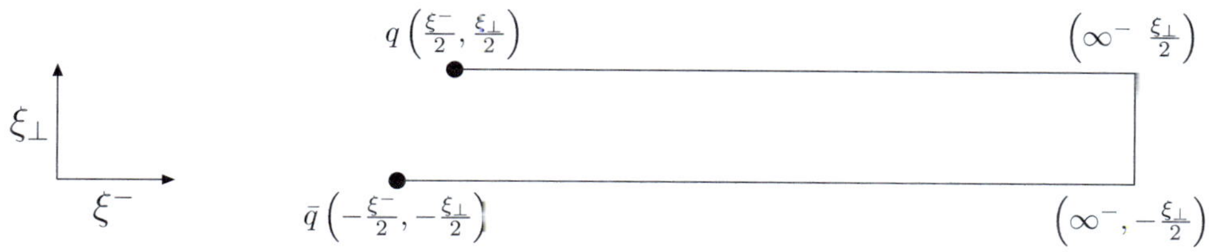

Fig. 1 Illustration of the path for the Wilson line gauge link $\mathcal{U}^{+LC}_{-\frac{\xi}{2},\frac{\xi}{2}}$ entering W^{+LC}

The shape of the segment at ∞ is irrelevant as the gauge field is pure gauge there, but it is still necessary to include a connection at ∞ and for simplicity we pick a straight line. Likewise, with a similar 'staple' to $-\infty$ we define the Wilson path ordered exponential $\mathcal{U}^{-LC}_{-\frac{\xi}{2},\frac{\xi}{2}}$, and using those light-like gauge links[1], we define

$$W^{\pm LC}(x,\mathbf{b}_\perp,\mathbf{k}_\perp) \equiv \int \frac{d^2\mathbf{q}_\perp}{(2\pi)^2}\int \frac{d^2\xi_\perp d\xi^-}{(2\pi)^3} e^{-i\mathbf{q}_\perp\cdot\mathbf{b}_\perp} e^{i(xP^+\xi^- - \mathbf{k}_\perp\cdot\boldsymbol{\xi}_\perp)}\langle P'P'|\bar{q}(0)\Gamma\mathcal{U}^{\pm LC}_{-\frac{\xi}{2},\frac{\xi}{2}}q(\xi)|PS\rangle. \quad (8)$$

This definition for W^{+LC} the same as that in Ref. [5,6] and similar to that of W_{LC} in Ref. [8] (the link segment at $\xi^- = \infty$ was not included in the definition of W_{LC}).

In light-cone gauge $A^+ = 0$, only the segment at $\xi^- = \pm\infty$ contributes and the OAM looks similar to the local manifestly gauge invariant expression, except

$$\mathbf{r}\times\mathbf{A}(\mathbf{r}) \longrightarrow \mathbf{r}\times\mathbf{A}(r^- = \pm\infty,\mathbf{r}_\perp). \quad (9)$$

From PT invariance one finds that $\mathcal{L}^q_+ = \mathcal{L}^q_-$ [5,6]. In the Bashinsky–Jaffe definition of OAM $\mathcal{L}^q_{BJ}$ [16], the vector potential in the gauge covariant derivative is replaced by

$$\frac{\int_{-\infty}^{\infty} dx^- A_\perp(r^-,\mathbf{r}_\perp)}{\int_{-\infty}^{\infty} dx^-} = \frac{1}{2}\left[A_\perp(r^- = \infty,\mathbf{r}_\perp) + A_\perp(r^- = -\infty,\mathbf{r}_\perp)\right], \quad (10)$$

and is thus equivalent to the light-cone-staple definition

$$\mathcal{L}^q_{BJ} = \frac{1}{2}\left(\mathcal{L}^q_+ + \mathcal{L}^q_-\right) = \mathcal{L}^q_+ = \mathcal{L}^q_-. \quad (11)$$

Imposing $A^+ = 0$ does not completely fix the gauge as one can still make r^--independent gauge transformations. If one fixes this residual gauge invariant by imposing anti-symmetric boundary conditions $A_\perp(r^- = -\infty,\mathbf{r}_\perp) = -A_\perp(r^- = \infty,\mathbf{r}_\perp)$ the vector potential at $r^- = \pm\infty$ cancels out in $\mathcal{L}^q_+ + \mathcal{L}^q_-$ and therefore, with the understanding of anti-symmetric boundary conditions at $r^- = \pm\infty$ the Jaffe–Manohar OAM becomes also identical to $\mathcal{L}^q_\pm$. This observation is crucial for understanding the difference between the Ji vs. Jaffe–Manohar OAM, which in light-cone gauge[2] involves only the replacement $A^i_\perp(\mathbf{r}) \longrightarrow A^i_\perp(r^- = \pm\infty,\mathbf{r}_\perp)$. Using

$$A^i_\perp(r^- = \infty,\mathbf{r}_\perp) - A^i_\perp(r^-,\mathbf{r}_\perp) = \int_{r^-}^{\infty} dz^-\partial_- A^i_\perp(z^-,\mathbf{r}_\perp) = \int_{r^-}^{\infty} dz^- G^{+i}(z^-,\mathbf{r}_\perp) \quad (12)$$

where $G^{+\perp} = \partial_- A^\perp$ is the gluon field strength tensor in $A^+ = 0$ gauge. Note that

$$-\sqrt{2}gG^{+y} \equiv -gG^{0y} - gG^{zy} = g\left(E^y - B^x\right) = g\left(\mathbf{E} - \mathbf{v}\times\mathbf{B}\right)^y \quad (13)$$

yields the $\hat{y}$ component of the color Lorentz force acting on a particle that moves with the velocity of light in the $-\hat{z}$ direction ($\mathbf{v} = (0,0,-1)$) — which is the direction of the momentum transfer in DIS [17,18]. Thus the difference between the Jaffe–Manohar and Ji[3] OAMs

$$\mathcal{L}^q - L^q = -g\int d^3x\,\langle P,S|\,\bar{q}(\mathbf{x})\gamma^+\left[\mathbf{x}\times\int_{x^-}^{\infty} dr^- F^{+\perp}(r^-,\mathbf{x}_\perp)\right]^z q(\mathbf{x})\,|P,S\rangle\,/\langle PS|PS\rangle \quad (14)$$

has the semiclassical interpretation of the change in OAM due to the torque from the FSI as the quark leaves the target: [20] while L^q represents the local and manifestly gauge invariant OAM of the quark *before* it has been struck by the γ^*, $\mathcal{L}^q$ represents the gauge invariant OAM *after* it has left the nucleon and moved to $r^- = \infty$.

[1] Subtleties in regularizing/renormalizing such objects are addressed in Ref. [13–15].
[2] As L^q involves a manifestly gauge invariant local operator, it can be evaluated in any gauge.
[3] Here we replaced $\gamma^0 \to \gamma^+$ in L^q as discussed in Ref. [19].

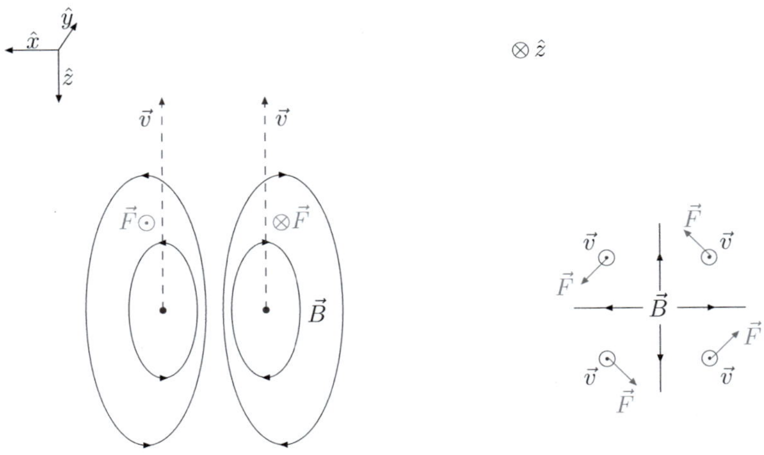

Fig. 2 Illustration of the torque acting on a positron moving in the $-\hat{z}$ direction through a magnetic dipole field caused by the magnetic moment of an electron polarized in the $+\hat{z}$ direction. **a** Side view; **b** top view. In this example the $\hat{z}$ component of the torque is negative as the positron leaves the bound state

3 Intuitive Picture for the Torque from Final State Interactions

In order to estimate the effect from the final state interactions on the quark OAM we first consider the effect on a positron moving through the magnetic dipole field of an electron, which is polarized in the $+\hat{z}$ direction. This should be the most simple analogy to a proton polarized in the $+\hat{z}$ direction because more quarks are polarized in the same direction as the nucleon spin and the color-electric force between the active quark and the spectators is attractive.

As illustrated in Fig. 2 the magnetic FSI leads to a negative torque. Since the example was chosen such that the signs of polarization and forces are the same as in the nucleon this implies that the color-magnetic torque acting on quarks ejected from the proton are negative as well.

The dipole example also illustrates why the torque vanishes immediately after the absorption of the virtual photon: when comparing the $\perp$ magnetix field (for fixed $\perp$ position) at positions at the 'front' and 'back' side of the nucleon are equal and opposite, and therefore the torque at the original position of the active quark averages to zero

$$\langle P, S|\bar{q}(0)\left[xF^{+y}(0) - yF^{+x}(0)\right]q(0)|P, S\rangle = 0. \tag{15}$$

Note that this cancellation is a consequence of averaging over all possible initial positions of the active quark. For a specific initial position the initial torque will in general be nonzero. While we used here an intuitive picture to motivate this result, it can also be shown to be a rigorous consequence of PT invariance.

The same example also illustrates the nonzero net effect from the final state interaction torque: although there is some partial cancelation from the 'front' and 'back' of the nucleon, after integrating along the trajectory of the ejected quark, the torque from the 'back' impacts more quarks than that from the 'front' resulting in a nonzero ensemble average.

4 Angular Momentum Decomposition for an Electron

In a recent paper [21] it was pointed out that for an e^- in QED, to $\mathcal{O}(\alpha)$, $\mathcal{L}^e = L^e$. This result corrects the previous calculation in [19], which was incomplete: for the massive Pauli–Villars photons, a wave function component describing longitudinally polarized photons needs to be included [22], which was omitted in [19]. Even though the wave function component describing a state with a longitudinal massive photons goes to zero as the Pauli–Villars mass goes to infinity, that component contributes a finite amount to the Ji OAM as shown in [21].

However, this result does not impact the fact that in general one still expects $\mathcal{L} \neq L$. This should be clear from the example in the previous section. It only implies that the electron to $\mathcal{O}(\alpha)$ is not a good example to illustrate that in general $\mathcal{L} \neq L$. Furthermore, the physical interpretation of the difference $\mathcal{L} - L$ being caused by the torque due to the FSI is not impacted by this result either.

References

1. Ashman, J. et al.: (EMC), A measurement of the spin asymmetry and determination of the structure function g_1 in deep inelastic muon-proton scattering. Phys. Lett. B **206**, 364 (1988)
2. Ji, X.: Gauge-invariant decomposition of nucleon spin. Phys. Rev. Lett. **78**, 610 (1997) [arXiv:hep-th/9603249]
3. Jaffe, R.L., Manohar, A.: The g_1 problem: fact and fantasy on the spin of the proton. Nucl. Phys. B **337**, 509 (1990)
4. Lorcé, C., Pasquini, B.: Quark Wigner distributions and orbital angular momentum. Phys. Rev. D **84**, 014015 (2011) [arXiv:1106.0139]
5. Hatta, Y.: Gluon polarization in the nucleon demystified. Phys. Rev. D **84**, 041701 (2011) [arXiv:1101.5989]
6. Hatta, Y.: Notes on the orbital angular momentum of quarks in the nucleon. Phys. Lett. B **708**, 186 (2012) [arXiv:1111.3547]
7. Belitsky, A.V., Ji, X., Yuan, F.: Quark imaging in the proton via quantum phase space distributions. Phys. Rev. D, **69**, 074014 (2004) [arXiv:hep-th/0307383]
8. Ji, X., Xiong, X., Yuan, F.: Proton spin structure from measurable parton distributions. Phys. Rev. Lett. **109**, 152005 (2012) [arXiv:1202.2843]
9. Metz, A., Meissner, S., Schlegel, M.: Nontrivial relations between GPDs and TMDs. Mod. Phys. Lett. A **24**, 2973 (2009)
10. Burkardt, M., Pasquini, B.: Modelling the nucleon structure. Eur. Phys. J. A (to appear) [arXiv:1510.02567]
11. Belitsky, A.V., Ji, X., Yuan, F.: Final state interactions and gauge invariant parton distributions. Nucl. Phys. B **656**, 165 (2003) [arXiv:hep-th/0307383]
12. Boer, D., Mulders, P.J., Pijlman, F.: Universality of T odd effects in single spin and azimuthal asymmetries. Nucl. Phys. B **667**, 201 (2003) [arXiv:hep-th/0303034]
13. Collins, J.C.: What exactly is a parton density?. Acta Phys. Polon. B **34**, 3103 (2003) [arXiv:hep-th/0304122]
14. Aybat, S.M. et al.: The QCD evolution of the sivers function. Phys. Rev. D **85**, 034043 (2012) [arXiv:1110.6428[hep-th]]
15. Cherednikov, I.: Cusped light-like Wilson loops in gauge theories. Phys. Part. Nucl. **44**, 250 (2013) [arXiv:1212.4345]
16. Bashinsky, S., Jaffe, R.L.: Quark and gluon orbital angular momentum and spin in hard processes. Nucl. Phys. B **536**, 303 (1998) [arXiv:hep-th/9804397]
17. Qiu, J., Sterman, G.: Single transverse spin asymmetries. Phys. Rev. Lett. **67**, 2264 (1991)
18. Burkardt, M.: Transverse force on quarks in DIS. Phys. Rev. D **88**, 114502 (2013) [arXiv:0810.3589]
19. Burkardt, M., Hikmat, B.C.: Angular momentum decomposition for an electron. Phys. Rev. D **79**, 071501 (2009) [arXiv:0812.1605]
20. Burkardt, M.: Parton orbital angular momentum and final state interactions. Phys. Rev. D **88**, 1 (2013) [arXiv:1205.2916]
21. Ji, X., et al.: Spin decomposition of electron in QED. [arXiv:1511.08817]
22. Burkardt, M., Langnau, A.: A Hamiltonian formulation of QED in (2+1) dimensions on the light cone. Phys. Rev. D **44**, 1187 (1991)

DOI 10.1007/978-3-319-50699-9_5
Reprinted from *Few-Body System* Journal, DOI 10.1007/s00601-016-1059-3

D. Müller

Covariance Constraints for Light Front Wave Functions

Received: 28 January 2016 / Accepted: 11 February 2016 / Published online: 5 March 2016

Abstract Light front wave functions (LFWFs) are often utilized to model parton distributions and form factors where their transverse and longitudinal momenta are tied to each other in some manner that is often guided by convenience. On the other hand, the cross talk of transverse and longitudinal momenta is governed by Poincaré symmetry and thus popular LFWF models are often not usable to model more intricate quantities such as generalized parton distributions. In this contribution a closer look to this issue is given and it is shown how to overcome the issue for two-body LFWFs.

1 Introduction

In quantum chromodynamics (QCD) inspired phenomenology it is popular to discuss the partonic picture of observables, such as form factors, structure functions of inclusive processes, spectrography, and many other quantities as well as partonic quantities, defined as matrix elements of two-body operators, such as distribution amplitude, parton distribution functions, generalized parton distributions (GPDs), transverse momentum dependent parton distributions, and phase space functions in terms of light front wave functions (LFWFs) [1–3]. Unfortunately, LFWFs could so far not be determined from the QCD dynamics. Since in a Hamilton approach the underlying Poincaré symmetry is not explicitly manifested, it arises for model builders the problem how in the LFWFs the longitudinal and transverse momenta are tied to each other. If one considers translation invariant parton distributions or hadronic distribution amplitudes, this issue does not imply *visible* inconsistencies, however, it becomes crucial if one deals with non-forward quantities such as generalized parton distribution or phase space functions.

In particular, for the phenomenology of deeply virtual Compton scattering [4–6] and deeply virtual meson production [7,8] the Poincaré covariance of GPDs is crucial. In the double distribution (DD) representation [4,9] and the double partial wave expansion, see, e.g., [10] and references therein, this property is manifestly implemented, however, if one likes to model GPDs in terms of LFWFs this might be considered as a theoretical challenge. In the LFWF overlap representation [11,12] the outer GPD region in the momentum fraction x and skewness η plane, i.e., $|\eta| \leq |x| \leq 1$, is obtained from the parton number conserved overlap while the central GPD region i.e., $|x| \leq |\eta|$, arises from the parton number changing LFWF overlap. The GPD covariance condition ensures that the GPD moments

$$\int_{-1}^{1} dx\, x^n H(x, \eta, t) = \sum_{\substack{m=0 \\ m \text{ even}}}^{n+1} H_{nm}(t)\eta^m, \tag{1}$$

D. Müller (✉)
Theoretical Physics Division, Rudjer Bošković Institute, 10000 Zagreb, Croatia
E-mail: dieter.mueller@tp2.rub.de; dieter.mueller@irb.hr

where t is the momentum transfer square, are even polynomials in η of order n (even n) or $n+1$ (odd n). Moreover, covariance also ensures that the amplitudes, given as convolution of partonic amplitudes and GPDs, satisfy fixed-t dispersion relations [13], e.g., at leading order accuracy they are written as

$$\mathcal{H}^{\sigma}(\xi,\vartheta,t) \stackrel{\mathrm{LO}}{=} \int_0^1 dx \frac{(1+\sigma)x+(1-\sigma)\xi}{\xi^2-x^2} H^{\sigma}(x,\eta=\vartheta x,t) + \frac{1+\sigma}{2}\mathcal{D}(\vartheta,t)\,, \tag{2}$$

where ξ is a Bjorken-like scaling variable, $\vartheta=\eta/\xi$ is the asymmetry parameter, $\sigma=\pm 1$ is the signature factor, and $\mathcal{D}(\vartheta,t)$ is a subtraction constant, called D-term form factor. However, for LFWFs that are ambiguously modeled as functions of longitudinal and transverse momenta it is very likely that the covariance property of GPD and so the analyticity of the amplitudes is violated. Also in many applications only the outer GPD region could be calculated from the parton conserved LFWF overlap and it is sometimes believed that the (unknown) non-conserved parton number LFWF overlap contribution will render GPDs that are covariant.

On the other hand, the Poincaré covariance property ensures that the outer region, i.e., the partonic s-channel view, and the central one, i.e., the partonic t-channel view, are dual to each other. More precisely, the central region can be uniquely mapped to the outer one [14] (this map was first applied for evolution kernels in [15]), while the inverse map is only unique in the charge odd sector, however, in the charge even sector a so-called D-term contribution [16], which entirely lives in the central region and vanishes on the cross-over line $|x|=|\eta|$, might appear. One might fix this contribution by some external requirements, however, no proof on general ground is known [17]. Thus, one might be able to consistently model the GPDs in terms of parton number conserved overlap of LFWFs, where duality is used to restore the full GPD. The condition, which is left, is on the functional form of LFWFs, i.e., how longitudinal and transverse momenta are tied to each other. The advantage of such a framework is besides a direct interpretation of experimental measurements in terms LFWFs also that positivity constraints, which are in their most general formulation are given impact parameter space [18], should be manifestly implemented in the LFWF overlap representation.

2 GPD Duality

The GPD support properties, given here within the restriction $|x|\le 1$, can be derived from the DD representation and might be written in terms of quark (q) and anti-quark ($\overline{q}$) building blocks

$$\begin{aligned} H(x,\eta,t) = &-\theta(x\le -|\eta|)\left[\mathrm{H}^{\overline{q}}(-x,\eta,t)+\mathrm{H}^{\overline{q}}(-x,-\eta,t)\right] + \theta(x\ge|\eta|)\left[\mathrm{H}^{q}(x,\eta,t)+\mathrm{H}^{q}(x,-\eta,t)\right] \\ &+\theta(|x|\le|\eta|)\left[-\mathrm{H}^{\overline{q}}(-x,\eta,t)+\mathrm{H}^{q}(x,\eta,t)\right], \end{aligned} \tag{3}$$

where the building block for the (anti-)quark GPD has the integral representation

$$\mathrm{H}^{i}(x,\eta,t) = \frac{1}{\eta}\int_0^{\frac{x+\eta}{1+\eta}} dy\, h(y,(x-y)/\eta,t)\,, \quad i\in\{u,\bar{u},d,\bar{d},\ldots\}. \tag{4}$$

Note that the DD $h(y,z,t)$, which has the symmetry property $h(y,-z,t)=h(y,z,t)$, might be undefined in the outer region for $y<\frac{x-\eta}{1-\eta}$, however, the GPD in this region is defined,

$$H(x\ge\eta,\eta,t) = \mathrm{H}^{i}(x,\eta,t)+\mathrm{H}^{i}(x,-\eta,t) = \frac{1}{\eta}\int_{\frac{x-\eta}{1-\eta}}^{\frac{x+\eta}{1+\eta}} dy\, h(y,(x-y)/\eta,t). \tag{5}$$

Formulae (3) and (4) ensure that the x-moments of the GPD are polynomials (1) and that amplitudes are calculable from (2). Utilizing the dispersive framework and the operator product expansion for doubly deep virtual Compton scattering in the Euclidean region [19,20] yields for $|\vartheta|\le 1$ the equality

$$\mathrm{PV}\int_{-1}^{1} dx \frac{1}{\xi-x} H^{\sigma}(x,\eta,t) = \mathrm{PV}\int_0^1 dx \frac{(1+\sigma)x+(1-\sigma)\xi}{\xi^2-x^2} H^{\sigma}(x,\eta=\vartheta x,t) + \frac{1+\sigma}{2}\mathcal{D}(\vartheta,t). \tag{6}$$

As annulled in the Introduction, we are interested to construct the full GPD from the outer region and so far various analytic procedures that are related to each other were suggested:

1. The truncated Taylor expansion of the moments $\int_\eta^1 dx\, x^n H(x,\eta,t)$, where a possible non-analytic part must be removed first, yields even polynomials in η of order n or $n+1$, see Ref. [14],

$$H_n^{\rm trun}(\eta,t|\sigma) = \sum_{m=0}^{n+1} \frac{1}{m!}\frac{d^m}{d\eta^m}\left\{\int_\eta^1 dx\, x^n\,[H(x,\eta,t)-\sigma H(-x,\eta,t)] - \text{non-analytic part}\right\}\Big|_{\eta=0}. \quad (7)$$

The analytic continuation of odd n ($\sigma=+1$) and even n ($\sigma=-1$) polynomials and an inverse Mellin transform allows to find the GPDs with definite signature $H^\sigma = H(x,\eta,t)-\sigma H(-x,\eta,t)$,

$$H^\sigma(x\ge 0,\eta,t) = \frac{1}{2\pi i}\int_{c-i\infty}^{c+i\infty} dx\, x^{-n-1} H_n^{\rm trun}(\eta,t|\sigma) + \frac{1+\sigma}{2}\theta(0\le x\le|\eta|)2\,\delta D(x/|\eta|,t). \quad (8)$$

These GPDs $H^\pm$ have the support $0\le x\le 1$, their continuation to negative x is done by antisymmetrization or symmetrization in x. For a signature even (or charge even) GPD it is allowed to add an additional $\delta D(z,t)=\delta D(-z,t)$ with $\delta D(\pm1,t)=0$ contribution to the intrinsic D-term.
2. The GPD on the r.h.s. of (6), i.e., $H^\sigma(x,\eta=\vartheta x)$, is only scanned in the outer GPD region and the l.h.s. can be inverted by a Hilbert transform or under certain assumptions by an alternative integral transform over the region $\xi\in[-1,1]$ [20], providing the full GPD for definite signature

$$H^\sigma(x,\eta,t) = \frac{\rm PV}{\pi^2}\int_{-\infty}^{\infty} d\xi \frac{\Re\mathrm{e}\mathcal{H}^\sigma(\xi,\eta/\xi,t)}{\xi-x} \ \text{or}\ H^\sigma(x,\eta,t) = \frac{\rm PV}{\pi^2}\int_{-1}^{1} d\xi \frac{\sqrt{1-x^2}}{\sqrt{1-\xi^2}}\frac{\Re\mathrm{e}\mathcal{H}^\sigma(\xi,\eta/\xi,t)}{\xi-x}, \quad (9)$$

where $\mathcal{H}^\sigma$, calculated from the dispersion integral (2), must be continued into the region $|\vartheta|\ge 1$.
3. Expressing the GPD in the outer region by the DD-integral (5) allows to read off the DD itself, i.e., the complete GPD can be restored [21,22], for an example see next section.

To exemplify the first two methods let us consider the photon GPD, obtained from a one-loop calculation [23], which reads up to a normalization factor in the outer region as following

$$H_1(x\ge\eta,\eta) = 1 - \frac{2x(1-x)}{1-\eta^2}. \quad (10)$$

The truncated moments (7) can be straightforwardly calculated and are even polynomials in η of order n (even n) or $n+1$ (odd n). With $\sigma=-(-1)^n$ they can be written in terms of a rational function

$$H_{1,n}^{\rm trun}(\eta|\sigma) = \frac{2-[1+\sigma]\eta^{1+n}}{2(n+1)} - \frac{2-[1+\sigma]\eta^{3+n}-[1-\sigma]\eta^{2+n}}{(2+n)(3+n)\left(1-\eta^2\right)}. \quad (11)$$

Apart from the highest possible power η^{n+1} for odd n these polynomials coincide with those of the originally calculated GPD, see Eq. (23) in [24]. If the D-term related η^{n+1} coefficient is known, the GPD can uniquely restored in the central region. In our example the addenda $\delta D_n = \frac{\eta^{n+1}}{n+2}$ should be added and the inverse Mellin transform (8) yields for $\sigma=+1$ the original GPD, written here as

$$\begin{aligned} H_1(x\ge 0,\eta) &= \frac{1}{2\pi i}\int_{c-i\infty}^{c+i\infty} dx\, x^{-n-1}\left[H_{1,n}^{\rm trun}(\eta|\sigma=+1)+\delta D_n\right],\quad \delta D_n = \frac{\eta^{n+1}}{n+2} \\ &= \theta(0\le x\le\eta)\frac{x(1-\eta)}{\eta(1+\eta)} + \theta(\eta\le x\le 1)\left[1-\frac{2(1-x)x}{1-\eta^2}\right] \quad \text{for}\quad \eta\ge 0. \end{aligned} \quad (12)$$

To employ the second method, we first calculate the $\sigma=+1$ amplitude from the dispersion relation (2), where the D-term form factor, considered as known, is evaluated from the D-term (26) of [24],

$$\begin{aligned} \mathcal{H}_1(\xi,\vartheta) &= \frac{2\vartheta\ln\frac{1+\vartheta}{1-\vartheta}+2\ln\left(1-\vartheta^2\right)}{\vartheta^2\left(1-\vartheta^2\xi^2\right)} + \frac{\left[1+(2-\vartheta^2)\xi^2\right]\ln\frac{\xi^2}{\xi^2-1}-2\xi\ln\frac{1+\xi}{\xi-1}}{1-\vartheta^2\xi^2} + \mathcal{D}(\vartheta), \\ \mathcal{D}(\vartheta) &= \int_{-\vartheta}^{\vartheta} dx \frac{2x\,D(x,\vartheta)}{1-x^2} = -\frac{\vartheta\ln\frac{1+\vartheta}{1-\vartheta}+\left(2-\vartheta^2\right)\ln\left(1-\vartheta^2\right)}{\vartheta^2}. \end{aligned} \quad (13)$$

Performing the Hilbert transform (9) yields the GPD (12). Note that the GPD can be alternatively calculated from the imaginary part of the amplitude, i.e., $H_1(x,\eta) = \Im\mathrm{m}\mathcal{H}_1(x - i\epsilon, \eta/[x - i\epsilon])/\pi$.

Let us emphasize that an approximate restoration of polynomiality yields incorrect results for the real part of amplitudes if they are calculated from the GPD convolution formula. That this effect can be large in the presence of Regge-like behavior has been exemplified in [25].

3 Functional Form of Two-Body LFWFs

The restoration of the full GPD from the parton number conserved LFWF overlap, i.e., from the outer GPD region, is only possible if the LFWFs respect the underlying Poincaré symmetry. There are various suggestions in the literature how covariance might be implemented in LFWFs [22,26,27]. The GPD covariance property has two aspects:

– restoration of t-dependence from the two-dimensional transverse momentum vector $\mathbf{\Delta}_\perp = (\Delta_x, \Delta_y)$
– ensuring that the resulting GPDs satisfy the polynomiality condition.

In Ref. [22] both of these problems were solved for effective two-body LFWFs. The restoration of t dependence can be ensured by a $(\mathbf{k}_\perp^2 - X(1-X)M^2)/(1-X)$ dependence and the GPD polynomiality constraints provide a further restriction on the functional form of LFWFs, which might be conveniently written as a Laplace transform

$$\phi(X, \mathbf{k}_\perp) = \int_0^\infty d\alpha\, \varphi(\alpha) \exp\left\{-\alpha \frac{\mathbf{k}_\perp^2 + (1-X)m^2 + X\lambda^2 - X(1-X)M^2}{(1-X)M^2}\right\}, \tag{14}$$

where X and m is the longitudinal momentum fraction and the mass of the struck quark, λ is the spectator quark mass, M is the hadron mass, $\mathbf{k}_\perp$ is the transverse momentum, and $\varphi(\alpha)$ might be somehow interpreted as a reduced wave function. The transverse and longitudinal momenta are tied to each other by the off-shell propagator [26]

$$M^2 - k_1^- - k_2^- = M^2 - \frac{\mathbf{k}_\perp^2 + m^2}{X} - \frac{\mathbf{k}_\perp^2 + \lambda^2}{1-X},$$

which, however, in (14) is additionally scaled by the momentum fraction X. The $\mathbf{k}_\perp$-unintegrated LFWF overlap or unintegrated GPD in the outer region is defined in such models as

$$\mathbf{H}(x \geq \eta, \eta, \mathbf{\Delta}_\perp, \mathbf{k}_\perp) = \frac{1}{1-x}\phi^*\left(\frac{x-\eta}{1-\eta}, \mathbf{k}_\perp - \frac{1-x}{1-\eta}\mathbf{\Delta}_\perp\right)\phi\left(\frac{x+\eta}{1+\eta}, \mathbf{k}_\perp\right), \tag{15}$$

where for simplicity we do not discuss here the quark and hadron spin content. For $x \geq \eta$ this GPD can be equivalently written in terms of an unintegrated DD representation [22],

$$\mathbf{H}(x, \eta, \mathbf{\Delta}_\perp, \mathbf{k}_\perp) = \int_0^1 dy \int_{-1+y}^{1-y} dz\, \delta(x - y - z\eta)\, \mathbf{h}\big(y, z, t, \overline{\mathbf{k}}_\perp\big), \tag{16}$$

which allows us to extend the support into the central region $-\eta \leq x \leq \eta$. This unintegrated DD can be expressed after some algebra in terms of the LFWF (14),

$$\mathbf{h}(y, z, t, \overline{\mathbf{k}}_\perp) = \frac{1}{2}\int_0^\infty dA\, A\, \varphi^*\left(A\frac{1-y+z}{2}\right)\varphi\left(A\frac{1-y-z}{2}\right)$$
$$\times \exp\left\{-A\left[(1-y)\frac{m^2}{M^2} + y\frac{\lambda^2}{M^2} - y(1-y) - \left[(1-y)^2 - z^2\right]\frac{t}{4M^2} + \frac{\overline{\mathbf{k}}_\perp^2}{M^2}\right]\right\}, \tag{17}$$

depending on $\overline{\mathbf{k}}_\perp = \mathbf{k}_\perp - (1-y+z)\mathbf{\Delta}_\perp/2$, the set $\{m, \lambda, M\}$ of mass parameters, and the reduced LFWF $\varphi(\alpha)$. The integration over the transverse degrees of freedom can trivially be performed and we recover from (16) the common DD-representation for a 'quark' GPD,

$$H(x, \eta, t) = \int_0^1 dy \int_{-1+y}^{1-y} dz\, \delta(x - y - z\eta)\, h(y, z, t), \quad h(y, z, t) = \iint d^2\overline{\mathbf{k}}_\perp\, \mathbf{h}(y, z, t, \overline{\mathbf{k}}_\perp). \tag{18}$$

which manifestly implements the GPD covariance properties. Two comments are in order.

- GPDs build with two-body LFWF models possess a cross-talk among t- and η-dependence.
- Other functional forms of two-body LFWFs than (14) might be not capable to deliver a GPD that respects covariance and, consequently, analyticity, i.e., the equality (6), can not be satisfied.

Let us exemplify these points, where for simplicity the mass parameters $M/2 = \lambda = m$ are equated and $\varphi(\alpha) = \delta(\alpha - \alpha_0)$ is taken in the ansatz (14). Plugging the resulting exponential LFWF

$$\phi(x, \mathbf{k}_\perp) = \exp\left\{-\alpha_0 \frac{\mathbf{k}_\perp^2 + m^2 - 4X(1-X)m^2}{4(1-X)m^2}\right\} \tag{19}$$

in the overlap representation (15) and performing the $\mathbf{k}_\perp$-integration yields a GPD in the other region,

$$H(x \geq \eta, \eta, t) \propto \exp\left\{-\frac{\alpha_0}{2}\left[\frac{(1-2x)^2}{1-x} - \frac{(1-x)t}{4m^2}\right]\right\}, \tag{20}$$

that is independent on η. Thus the amplitude (2) is ϑ-independent and the Hilbert transform (9) tells us that the extension into the central region is done by replacing the constraint $x \geq \eta$ by $x \geq 0$. Of course, this result can be also obtained from the DD representation (17) and (18).

Contrarily, the construction of a consistent GPD might be impossible if we take an arbitrary functional form. For instance, picking up one of the popular choices for modeling meson LFWFs [28],

$$\phi(x, \mathbf{k}_\perp) = \exp\left\{-\alpha_0 \frac{\mathbf{k}_\perp^2 + m^2 - 4X(1-X)m^2}{4(1-X)Xm^2}\right\}, \tag{21}$$

we immediately get from the overlap formula (15) the GPD in the outer region

$$H(x \geq \eta, \eta, t) \propto \frac{x^2-\eta^2}{x\left(1+\eta^2\right)-2\eta^2} \exp\left\{-\frac{\alpha_0(1-\eta^2)/2}{x(1+\eta^2)-2\eta^2}\left[\frac{x^2\left(1+\eta^2-2x\right)^2}{(1-x)\left(1-\eta^2\right)\left(x^2-\eta^2\right)} - \frac{(1-x)t}{4m^2}\right]\right\}. \tag{22}$$

Utilizing the first method of the preceding section, one realizes that the coefficients in the truncated Taylor expansion (7) suffer from essential singularities, except for the constant term. Thus, removing these singularities yields to a η-independent GPD. Also the second method fails, namely, the "spectral function" takes, e.g., for $t = 0$, the form

$$H(x, \eta = \vartheta x, t = 0) \propto \frac{x\left(1-\vartheta^2\right)}{1-x(2-x)\vartheta^2} \exp\left\{-\frac{\alpha_0}{2}\frac{\left[1-x(2-x\vartheta^2)\right]^2}{(1-x)x\left(1-\vartheta^2\right)\left[1-x(2-x)\vartheta^2\right]}\right\}. \tag{23}$$

and has for $0 \leq x \leq 1$ essential singularities at $\vartheta^2 = 1$ and $\vartheta^2 = 1/x(2-x)$. Consequently, the resulting amplitude (2) does not exist for $\vartheta^2 > 1$ and its limiting value at $\vartheta^2 = 1$ depends on the direction. It is not hard to realize that more general exponential LFWF ansätze are also not usable to build GPDs.

4 Conclusions

We recalled three methods that allow to restore the full GPD from the knowledge of the GPD in the outer region. All these methods are based on extension procedures. So far they can be only utilized if the GPD in the outer region is known analytically. The extension of GPDs is tied to the analytic properties of amplitudes which besides the dispersion relation representation with respect to the Bjorken-like scaling variable ξ must also possess certain properties with respect to the photon asymmetry parameter ϑ, namely, analyticity inside the unit circle and extendable on the segment $\vartheta \in [1, \infty]$ (the value on $[-\infty, -1]$ follows from symmetry requirements). The investigation of the remaining mathematical problems is important if one likes to have a numerical GPD extension procedure, which can then be conveniently employed in GPD phenomenology. Without relying on an external principle, the freedom which is left in the charge even sector is the so-called D-term, entirely living in the central GPD region.

From the partonic point of view it is tempting to interpret experimental measurements of deeply virtual Compton scattering and deeply virtual meson production in terms of (effective) LFWFs. Such a framework also offers the possibility to implement positivity constraints in the phenomenological analysis, however, here it is requested that the resulting amplitudes possess the correct analytic properties. It is expected that this is ensured if the LFWFs satisfy the Poincaré covariance constraints and so one can utilize GPD duality to restore (apart from the D-term) the full GPD in the central region, too. It was exemplified here within exponential LFWFs, used for mesons, that if such constraints are neglected the LFWFs might not be utilized to find the full GPD. Consequently, if one requires that theoretical constraints should be consistently implemented or one simply aims for a more universal description of hadronic phenomena, such popular ansätze are simply excluded.

Examples of effective two-body LFWFs overlap modeling were given in Ref. [22]. Thereby, a (t-independent) Regge behavior was implemented by means of a convolution with a spectator mass spectral density. Such models are generically in agreement with experimental and phenomenological findings, however, various problems remain to be solved, such as mimicking a t-dependent Regge behavior that ensures positivity by construction and flexibility of the LFWF parametrization, before on can develop a phenomenological LFWF framework that is suited for the description of experimental data.

Acknowledgments For discussions I am indebted to S. Brodsky, D. Chakrabarti, C. Lorce, B. Pasquini, M. Polyakov, K. Semenov-Tian-Shansky, G.F. de Teramond, O. Teryaev. This work has been supported by the Croatian Ministry of Science, Education and Sport (MSES) under the NEWFELPRO Grant Agreement No. 54.

References

1. Drell, S.D., Yan, T.M.: Connection of elastic electromagnetic nucleon form-factors at large q^2 and deep inelastic structure functions near threshold. Phys. Rev. Lett. **24**, 181 (1970)
2. Brodsky, S.J., Pauli, H.C., Pinsky, S.S.: Quantum chromodynamics and other field theories on the light cone. Phys. Rep. **301**, 299 (1998)
3. Boffi, S., Pasquini, B.: Generalized parton distributions and the structure of the nucleon. Riv. Nuovo Cim. **30**, 387 (2007)
4. Müller, D., Robaschik, D., Geyer, B., Dittes, F.M., Hořejši, J.: Wave functions, evolution equations and evolution kernels from light-ray operators of QCD. Fortschr. Phys. **42**, 101 (1994)
5. Radyushkin, A.V.: Scaling limit of deeply-virtual Compton scattering. Phys. Lett. B **380**, 417 (1996)
6. Ji, X.: Deeply-virtual Compton scattering. Phys. Rev. D **55**, 7114 (1997)
7. Radyushkin, A.V.: Asymmetric gluon distribution and hard diffractive electroproduction. Phys. Lett. B **385**, 333 (1996)
8. Collins, J., Frankfurt, L., Strikman, M.: Factorization for hard exclusive electroproduction of mesons in QCD. Phys. Rev. D **56**, 2982 (1997)
9. Radyushkin, A.V.: Non-forward parton distributions. Phys. Rev. D **56**, 5524 (1997)
10. Müller, D., Polyakov, M.V., Semenov-Tian-Shansky, K.M.: Dual parametrization of generalized parton distributions in two equivalent representations. JHEP **03**, 052 (2015)
11. Brodsky, S.J., Diehl, M., Hwang, D.S.: Light-cone wavefunction representation of deeply virtual Compton scattering. Nucl. Phys. B **596**, 99 (2001)
12. Diehl, M., Feldmann, T., Jakob, R., Kroll, P.: The overlap representation of skewed quark and gluon distributions. Nucl. Phys. B **596**, 33 (2001)
13. Teryaev, O.V.: Analytic properties of hard exclusive amplitudes. hep-ph/0510031 (2005)
14. Müller, D., Schäfer, A.: Complex conformal spin partial wave expansion of generalized parton distributions and distribution amplitudes. Nucl. Phys. B **739**, 1 (2006)
15. Dittes, F.M., Müller, D., Robaschik, D., Geyer, B., Hořejši, J.: The Altarelli–Parisi kernel as asymptotic limit of an extended Brodsky–Lepage kernel. Phys. Lett. B **209**, 325 (1988)
16. Polyakov, M.V., Weiss, C.: Skewed and double distributions in pion and nucleon. Phys. Rev. D **60**, 114017 (1999)
17. Müller, D., Semenov-Tian-Shansky, K.M.: $J = 0$ fixed pole and D-term form factor in deeply virtual Compton scattering. Phys. Rev. D **92**, 074025 (2015)
18. Pobylitsa, P.: Positivity bounds on generalized parton distributions in impact parameter representation. Phys. Rev. D **66**, 094002 (2002)
19. Kumericki, K., Müller, D., Passek-Kumericki, K.: Towards a fitting procedure for deeply virtual Compton scattering at next-to-leading order and beyond. Nucl. Phys. B **794**, 244 (2008)
20. Kumerički, K., Müller, D., Passek-Kumerički, K.: Sum rules and dualities for generalized parton distributions: Is there a holographic principle?. Eur. Phys. J. C **58**, 193 (2008)
21. Hwang, D.S., Müller, D.: Implication of the overlap representation for modelling generalized parton distributions. Phys. Lett. B **660**, 350 (2008)
22. Müller, D., Hwang, D.S.: The concept of phenomenological light-front wave functions—Regge improved diquark model predictions. arXiv:1407.1655 (2014)
23. Friot, S., Pire, B., Szymanowski, L.: Deeply virtual Compton scattering on a photon and generalized parton distributions in the photon. Phys. Lett. B **645**, 153 (2007)
24. Müller, D.: Generalized Parton distributions—visions, basics, and realities. Few Body Syst. **55**, 317 (2014)
25. Diehl, M., Ivanov, D.Y.: Dispersion representations for hard exclusive processes. Eur. Phys. J. C **52**, 919 (2007)
26. Brodsky, S.J., Huang, T., Lepage, G.P.: The Hadronic Wave Function in Quantum Chromodynamics. SLAC-PUB-2540 (1980)

27. Brodsky, S.J., Hiller, J.R., Hwang, D.S., Karmanov, V.A.: The covariant structure of light front wave functions and the behavior of hadronic form-factors. Phys. Rev. D **69**, 076001 (2004)
28. Huang, T., Zhong, T., Wu, X.G.: Determination of the pion distribution amplitude. Phys. Rev. D **88**, 034013 (2013)

DOI 10.1007/978-3-319-50699-9_6
Reprinted from *Few-Body System* Journal, DOI 10.1007/s00601-016-1094-0

Markus Diehl · Jonathan R. Gaunt · Daniel Ostermeier · Peter Plößl · Andreas Schäfer

Cancellation of Glauber Gluon Exchange in the Double Drell–Yan Process

Received: 30 January 2016 / Accepted: 16 March 2016 / Published online: 8 April 2016

Abstract For any factorisation proof, a crucial step is a demonstration of the cancellation of so-called Glauber gluons. We summarise a recent paper in which we demonstrated this cancellation for double Drell–Yan production (the double parton scattering process in which a pair of electroweak gauge bosons is produced), both for the integrated cross section and for the cross section differential in the boson transverse momenta.

1 Introduction

In order to make predictions at the LHC one relies on factorisation formulae that separate the short distance/high-scale dynamics of interest from the low-scale nonperturbative physics. The short distance physics is encoded in perturbatively-computable partonic cross sections, whilst the infra-red physics is encoded in parton distribution-type objects (and possibly also soft functions). In single parton scattering, factorisation has been rigorously proven for Drell–Yan production (or, more generally, colour-singlet production), for the total cross section and cross section differential in the p_T of the colour singlet system [8,12–14]. These factorisation formulae are correct at leading power—i.e. up to corrections of order Λ^2_{QCD}/Q^2, with Q the hard scale of the process. For the analogous process in double parton scattering, namely double Drell–Yan production, a factorisation formula was written down long ago for the total cross section [21,22] based on the analysis of the lowest-order Feynman diagrams. A factorisation formula for the cross section differential in the p_Ts of the colour singlet systems has also been written down using the same method [15]. However, serious attempts to rigorously justify these formulae have only begun in recent years [7,15,16,20].

M. Diehl
DESY, 22603 Hamburg, Germany
E-mail: markus.diehl@desy.de

J. R. Gaunt (✉)
Nikhef and VU Amsterdam, Science Park 105, Amsterdam 1098 XG, The Netherlands
E-mail: jgaunt@nikhef.nl

D. Ostermeier · P. Plößl · A. Schäfer
Institut für Theoretische Physik, Universitat Regensburg, 93040 Regensburg, Germany
E-mail: Daniel.Ostermeier@physik.uni-regensburg.de

P. Plößl
E-mail: peter.ploessl@physik.uni-regensburg.de

A. Schäfer
E-mail: Andreas.Schaefer@physik.uni-regensburg.de

One approach to proving factorisation formulae at leading power (and potentially beyond) is the so-called Collins–Soper–Sterman (CSS) approach, which is the one that was employed in [8,12–14] (there exists an alternative approach based on effective field theories—see e.g. [1–6,23] —but we will not discuss this here). Let us give a very brief review of this method. The first goal is to identify leading infra-red contributions in Feynman graphs contributing to the process of interest—contributions from small regions around the points at which certain lines go on shell, which despite having a smaller phase space are leading due to propagator denominators blowing up. These infra-red contributions are the ones that will have to be absorbed in the PDF-type objects. To be more precise, one needs to identify regions around pinch singularities—these are points where propagator denominators pinch the contour of the Feynman integral [strictly speaking the singularities only appear in the limit in which parton masses and small transverse momenta are set to zero]. If poles all converge on the Feynman contour from one side, one can perform a contour deformation to avoid them (in that case the contribution originally in the vicinity of the poles is subsumed into some other region). The identification of pinch singular points is facilitated by the Coleman–Norton theorem [11], which states that pinch singularities in Feynman graphs correspond to classically allowed processes, and the pinch singular surfaces for single and double Drell–Yan are known.

The pinch singularity analysis tells us nothing about the strength of the singularities, and so must be complemented with a power counting procedure to identify the leading power regions around the singularities. Doing this, one identifies several types of loop momentum scaling that can give rise to a leading power contribution. These are *collinear* (momentum close to some beam/jet direction), *(central) soft* (all momentum components small and of the same order), *hard* (all momentum components large and of order Q) and *Glauber*. The kinematics of Glauber particles is that of those mediating forward or small-angle scattering—i.e. a Glauber momentum r satisfies $|r^+r^-| \ll \mathbf{r}^2$, where $r^{\pm} = (r^0 \pm r^3)/\sqrt{2}$, $\mathbf{r} = (r^1, r^2)$.

Initially, there are many longitudinally-polarised gluon connections between parton lines in the collinear region and the union of the soft and Glauber regions, as well as between lines in the collinear region and the hard region. The next step of the factorisation procedure is to apply approximations appropriate to these momentum regions, followed by Ward identities in the sum over graphs to strip away these multiple attachments, yielding the separate functions that appear in the factorisation formula. Unfortunately, this procedure does not work for the Glauber modes, since the approximations needed to apply the Ward identities may not be made in this case. Thus, to achieve factorisation one must demonstrate that the contribution from the Glauber region cancels for the given observable. Note that by 'cancels' here, we mean that there is no remaining 'distinct' (pinched) Glauber contribution—there can (and will) be remaining contributions associated with the Glauber kinematic region, but these can be absorbed into the collinear or soft functions (this was re-emphasised recently in the effective field theory context in [23]). The cancellation of Glauber exchanges was achieved in single Drell–Yan (total cross section and cross section differential in colour singlet p_T) by CSS, and recently in double Drell–Yan (total cross section and cross section differential in colour singlet p_Ts) by us in [16]—here we summarise this latter work. In this short proceedings contribution we mainly limit ourselves to a discussion of the cancellation of Glauber modes at the one-gluon level in Sect. 2—this is useful to illustrate why the cancellation works for double Drell–Yan as it does for single Drell–Yan. We very briefly mention how the all-order proof works in Sect. 3.

2 Glauber Cancellation for One-Gluon Exchange

In order to show the cancellation of Glauber modes at the one-gluon level, we adopt a model in which all partons (except the exchanged Glauber gluon) are scalar (represented by a solid line), and the hadrons are also scalars (represented by a dashed line). These partons may be predominantly collinear to the lower hadron travelling in the plus direction (we colour plus-collinear lines in red), or collinear to the upper hadron travelling in the minus direction (we colour minus-collinear lines in blue). Since the Glauber cancellation argument below depends only on the analyticity properties of the Feynman integrands, which in turn is determined only by the propagator denominators, this argument applies beyond the model, also to diagrams in QCD.

The lowest-order diagram contributing to the double Drell–Yan amplitude (which one simply squares to get the lowest order contribution to the cross section) is given in Fig. 1, along with the nonzero virtual one-gluon corrections to this (these are combined with the tree-level amplitude to get the virtual one-gluon corrections to the cross section). The real one-gluon corrections to the tree-level amplitude squared are zero by colour considerations, and anyway can have no contribution associated with the Glauber kinematic region due to the on-shell constraint for the gluon. Let us consider the virtual corrections in turn, starting with the 'double box'

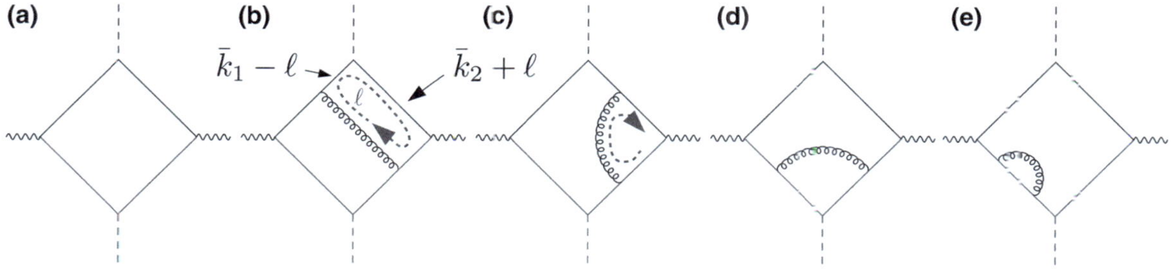

Fig. 1 Lowest order diagram contributing to double Drell–Yan in the model described in the text (**a**), and the nonzero virtual one-gluon corrections to this (**b–e**)

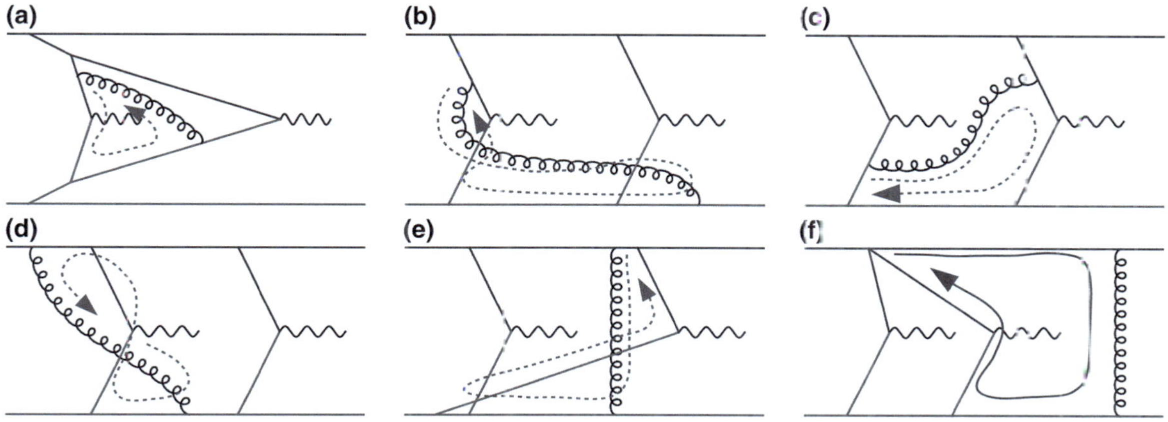

Fig. 2 Example graphs for the double Drell–Yan amplitude within the model described in the text. The incoming lines on the left of each graph could represent the proton, or partons emerging from the proton

graph in Fig. 1b. In this graph, the gluon momentum ℓ runs 'against' and 'with' a large minus momentum—this results in ℓ^+ being trapped at small values, as one can see by looking at the propagator denominators associated with these minus-collinear lines:

$$2\ell^+\bar{k}_2^- + \cdots + i\varepsilon \qquad -2\ell^+\bar{k}_1^- + \cdots + i\varepsilon \tag{1}$$

The terms indicated by '$\cdots$' are products of small components only. This is a general principle - to trap the plus/minus component of a soft momentum at small values, the momentum must run both 'against' and 'with' a collinear line (or multiple lines) with large minus/plus momentum. Now, since ℓ in Fig. 1b only runs against a large plus momentum, ℓ^- is not trapped, and we can deform ℓ^- until ℓ is out of the Glauber region (and is in the collinear region). So there is no distinct Glauber contribution associated with Fig. 1b.

In the gauge boson vertex correction graph, Fig. 1c, neither ℓ^+ nor ℓ^- is trapped, so we can deform the contour out of the Glauber region there too. This leaves us only with the hadron vertex correction and parton self-energy graphs in Figs. 1d, e, which are topologically factorised already, and thus can present no problem. Thus, for the one-gluon corrections to the tree-level graph, there is no problematic pinched Glauber contribution in the first place.

This kind of consideration can be extended to arbitrarily complex one-gluon diagrams in the model, some of which are sketched in Fig. 2. For many types of diagram, a routing can be found for the gluon momentum ℓ such that ℓ^+ and/or ℓ^- is not trapped at small values. This is the case for diagrams (a–e) in Fig. 2—for each of these diagrams a routing for ℓ that leaves one light-cone component untrapped is given by the green dashed arrow. The only graphs for which both ℓ^+ and ℓ^- are always pinched are the ones in which the gluon attaches to spectator partons that either go directly into the final state or only split into partons that go into the final state, both in the upper and lower parts of the graph. An example of this type of graph is given in Fig 2f (with a routing that gives ℓ^+, ℓ^- trapped denoted by the solid purple arrow). However, these graphs are of essentially the same nature as the Glauber-pinched graphs that appear for single Drell–Yan (except that the latter have the double Drell–Yan production subgraph, as on the left of Fi. 2f, replaced by a simpler single Drell–Yan production). For these graphs one can use the same unitarity-based argument as was used by CSS for single

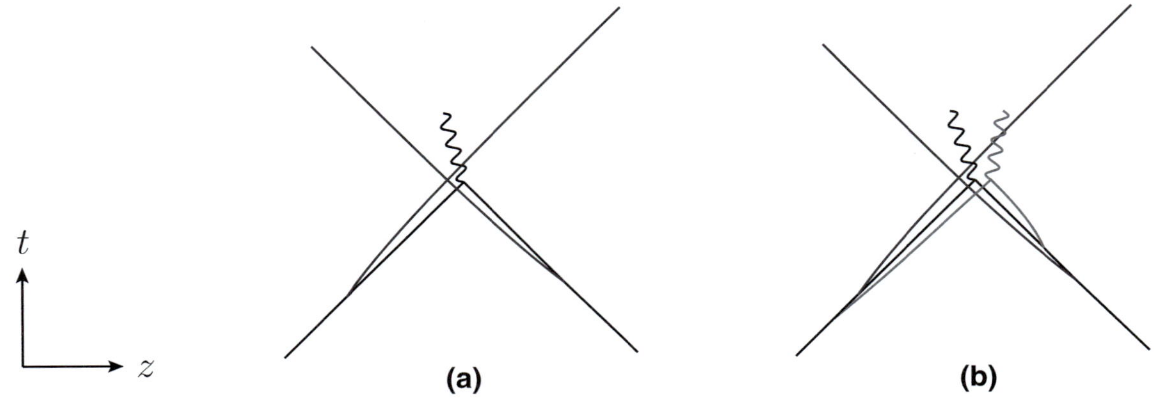

Fig. 3 Spacetime structure of the pinch surfaces for single Drell–Yan (*left*) and double Drell–Yan (*right*). The parton lines move along light-like paths and have been drawn with a slight curvature only for clarity. Figure taken from [15]

Drell–Yan (reviewed in e.g. [17]), to cancel the Glauber contribution after the sum over possible final-state cuts. For this argument to work it is essential that the observable be insensitive to the position of the cut in the 'final state' spectator-spectator system.

3 Glauber Gluon Cancellation at All orders

In order to demonstrate the cancellation of Glauber exchanges in double Drell–Yan production to all orders in QCD perturbation theory, we made use of the same techniques as were used in [12, 14] to show the all-order cancellation in single Drell–Yan—in particular, we used the light-front ordered version of QCD perturbation theory (LCPT) (see e.g. [9, 10, 18, 19, 24] for an introduction). We do not discuss the details of this proof here, referring the interested reader to [16]. The general idea is the same as the one-loop proof discussed above, however. Also in the all-order LCPT picture, one sees that from the point of view of the Glauber gluons, single and double scattering look rather similar, and that the troublesome 'final state' poles obstructing the deformation out of the Glauber region cancel after the sum over cuts. The argument is based on unitarity, and it can be cast into a form in which it is essentially the same for single and double Drell–Yan.

One can get some insight into why the Glauber cancellation proceeds for double Drell–Yan as it does for single Drell–Yan by looking at the pinch surfaces for the two processes in space–time, given in figure 11 of [15] and reproduced here as Fig. 3. In the diagrams are drawn the collinear lines (black and red solid lines) as well as the hard vertices producing the colour-singlet particles. One observes that the hard vertices occur at the same point in spacetime for double Drell–Yan, and that the locations of the collinear lines and hard vertices are the same for single and double Drell–Yan. Thus, from the perspective of soft long-range gluons the two processes look essentially the same, and the Glauber cancellation that works for single Drell–Yan should also work for double Drell–Yan, as we indeed found.

Acknowledgments We acknowledge support by BMBF (Grants 05P12WRFTE and 05P15WRCAA). J.G. acknowledges financial support from the European Community under the Ideas program QWORK (contract 320389).

References

1. Bauer, C.W., Stewart, I.W.: Invariant operators in collinear effective theory. Phys. Lett. B **516**, 134–142 (2001). doi:10.1016/S0370-2693(01)00902-9. arXiv:hep-th/0107001
2. Bauer, C.W., Fleming, S., Luke, M.E.: Summing Sudakov logarithms in $B \to X_s \gamma$ in effective field theory. Phys. Rev. D **63**, 014,006 (2000). doi:10.1103/PhysRevD.63.014006. arXiv:hep-th/0005275
3. Bauer, C.W., Fleming, S., Pirjol, D., Stewart, I.W.: An effective field theory for collinear and soft gluons: heavy to light decays. Phys. Rev. D **63**, 114,020 (2001). doi:10.1103/PhysRevD.63.114020. arXiv:hep-th/0011336

4. Bauer, C.W., Fleming, S., Pirjol, D., Rothstein, I.Z., Stewart, I.W.: Hard scattering factorization from effective field theory. Phys. Rev. D **66**, 014,017 (2002a). doi:10.1103/PhysRevD.66.014017. arXiv:hep-th/0202088
5. Bauer, C.W., Pirjol, D., Stewart, I.W.: Soft collinear factorization in effective field theory. Phys. Rev. D **65**, 054,022 (2002b). doi:10.1103/PhysRevD.65.054022. arXiv:hep-th/0109045
6. Beneke, M., Chapovsky, A., Diehl, M., Feldmann, T.: Soft collinear effective theory and heavy to light currents beyond leading power. Nuclear Phys. B **643**, 431–476 (2002). doi:10.1016/S0550-3213(02)00687-9. arXiv:hep-th/0206152
7. Blok, B., Dokshitser, Yu., Frankfurt, L., Strikman, M.: pQCD physics of multiparton interactions. Eur. Phys. J C **72**, 1963 (2012). arXiv:1106.5533 [hep-th]
8. Bodwin, G.T.: Factorization of the Drell–Yan cross-section in perturbation theory. Phys. Rev. D **31**, 2616 (1985), doi:10.1103/PhysRevD.34.3932, doi:10.1103/PhysRevD.31.2616, [Erratum: Phys. Rev.D34,3932(1986)]
9. Brodsky, S.J., Lepage, G.P.: Exclusive processes in quantum chromodynamics. In: Mueller, A.H. (ed.) Perturbative Quantum Chromodynamics. World Scientific, Sigapore (1989)
10. Chang, S.J., Ma, S.K.: Feynman rules and quantum electrodynamics at infinite momentum. Phys. Rev **180**, 1506–1513 (1969). doi:10.1103/PhysRev.180.1506
11. Coleman, S., Norton, R.E.: Singularities in the physical region. Singul. Phys. Reg. **38**, 438–442 (1965). doi:10.1007/BF02750472
12. Collins, J.: Foundations of Perturbative QCD. Cambridge University Press, Cambridge (2013)
13. Collins, J.C., Soper, D.E., Sterman, G.F.: Factorization for short distance hadron–Hadron Scattering. Nuclear Phys. B **261**, 104 (1985). doi:10.1016/0550-3213(85)90565-6
14. Collins, J.C., Soper, D.E., Sterman, G.F.: Soft gluons and factorization. Nuclear Phys. B **308**, 833 (1988). doi:10.1016/0550-3213(88)90130-7
15. Diehl, M., Ostermeier, D., Schafer, A.: Elements of a theory for multiparton interactions in QCD. JHEP **03**, 089 (2012). doi:10.1007/JHEP03(2012)089. arXiv:1111.0910 [hep-th]
16. Diehl, M., Gaunt, J.R., Ostermeier, D., Plößl, P., Schäfer, A.: Cancellation of Glauber gluon exchange in the double Drell–Yan process. JHEP **01**, 076 (2016). doi:10.1007/JHEP01(2016)076. arXiv:1510.08696 [hep-th]
17. Gaunt, J.R.: Glauber gluons and multiple parton interactions. JHEP **07**, 110 (2014). doi:10.1007/JHEP07(2014)110. arXiv:1405.2080 [hep-th]
18. Kochegov, Y.V., Levin, E.: Quantum Chromodynamics at High Energy. Cambridge University Press, Cambridge (2012)
19. Kogut, J.B., Soper, D.E.: Quantum electrodynamics in the infinite momentum frame. Phys. Rev. D **1**, 2901–2913 (1970). doi:10.1103/PhysRevD.1.2901
20. Manohar, A.V., Waalewijn, W.J.: A QCD analysis of double parton scattering: color correlations, interference effects and evolution. Phys. Rev. D **85**, 114,009 (2012). doi:10.1103/PhysRevD.85.114009. arXiv:1202.3794 [hep-th]
21. Mekhfi, M.: Multiparton processes: an application to double Drell–Yan. Phys. Rev. D **32**, 2371 (1985). doi:10.1103/PhysRevD.32.2371
22. Paver, N., Treleani, D.: Multi-quark scattering and large p_T jet production in hadronic collisions. Nuovo cim A **70**, 215 (1982). doi:10.1007/BF02814035
23. Rothstein, I.Z., Stewart, I.W.: An effective field theory for forward scattering and factorization violation (2016). arXiv:1601.04695 [hep-th]
24. Yan, T.M.: Quantum field theories in the infinite momentum frame. 4. Scattering matrix of vector and dirac fields and perturbation theory. Phys. Rev. D **7**, 1780–1800 (1973). doi:10.1103/PhysRevD.7.1780

DOI 10.1007/978-3-319-50699-9_7
Reprinted from *Few-Body System* Journal, DOI 10.1007/s00601-016-1087-z

Wojciech Broniowski · **Enrique Ruiz Arriola** · **Krzysztof Golec-Biernat**

Generalized Valon Model for Double Parton Distributions

Received: 31 January 2016 / Accepted: 9 March 2016 / Published online: 28 March 2016

Abstract We show how the double parton distributions may be obtained consistently from the many-body light-cone wave functions. We illustrate the method on the example of the pion with two Fock components. The procedure, by construction, satisfies the Gaunt–Stirling sum rules. The resulting single parton distributions of valence quarks and gluons are consistent with a phenomenological parametrization at a low scale.

Multiparton distributions were considered already in pre-QCD times [1] and in the early days of QCD [2–4]. Recently, the interest in these objects has been renewed (see, e.g., [5–11] and references therein), with expectations that the double parton distributions (dPDF's) of the nucleon may be accessible from the double parton scattering contribution in certain exclusive production channels at the LHC [12,13]. Whereas dPDF's are well defined objects, their explicit construction, or parametrization that can be used phenomenologically, is difficult to achieve in a way where all the formal constraints are satisfied. In particular, the frequently used product ansatz or its modifications is inconsistent with the basic requirements of the Gaunt–Stirling (GS) sum rules [14,15].

On the model side, there are only a few explicit calculations in the literature: the studies in the MIT bag model [16] or in the constituent quark model [17] implement relativity approximately, hence the support of the distributions extends outside of the physical region. These problems were mended in Refs. [18,19]. On the other hand, a simple *valon* model [20,21] applied to the nucleon in [22] leads to dPDF's satisfying all formal constraints and to reasonable single parton distributions (sPDF's) for the valence sector. In this talk we pursue this idea, extending the valon model to include higher Fock-state components, which allows us to study the gluon and sea content.

Talk presented by WB at the Light Cone 2015 Conference, Frascati, Italy, 21–25 September 2015. Supported by Polish National Science Center (Grant DEC-2015/17/B/ST2/01838), by Spanish DGI (Grant FIS2014-59386-P), and by Junta de Andalucía (Grant FQM225).

W. Broniowski · K. Golec-Biernat
Institute of Nuclear Physics, Polish Academy of Sciences, 31-342 Kraków, Poland

W. Broniowski (✉)
Institute of Physics, Jan Kochanowski University, 25-406 Kielce, Poland
E-mail: Wojciech.Broniowski@ifj.edu.pl

E. Ruiz Arriola
Departamento de Física Atómica, Molecular y Nuclear, Instituto Carlos I de Física Teórica y Computacional, Universidad de Granada, 18071 Granada, Spain

K. Golec-Biernat
Faculty of Mathematics and Natural Sciences, University of Rzeszów, 35-959 Rzeszow, Poland

For practical purposes, it would be highly desirable to have a simple working parametrization for dPDF's in analogy to parameterizations of sPDF's. The ultimate goal of our approach is to systematically construct dPDF's which on one side satisfy all formal constraints, and on the other side reproduce the known sPDF's.

In phenomenological applications one assumes for simplicity the transverse-longitudinal decoupling in the dPDF's, $\Gamma_{ij}(x_1, x_2, k_T) = D_{ij}(x_1, x_2) f(k_T)$, moreover, one frequently applies the ansatz $D_{ij}(x_1, x_2) = D_i(x_1) D_j(x_2) \theta(1 - x_1 - x_2)(1 - x_1 - x_2)^n/(1 - x_1)^{n_1}/(1 - x_2)^{n_2}$, where $D_i(x)$ are the sPDF's. However, this form violates the GS sum rules [14]:

$$\int_0^{1-y} dx\, D_{i_{\rm val} j}(x, y) = \int_0^{1-y} dx\, [D_{ij}(x, y) - D_{\bar{i} j}(x, y)] = (N_{i_{\rm val}} - \delta_{ij} + \delta_{\bar{i} j}) D_j(y), \qquad (1)$$

$$\sum_i \int_0^{1-y} dx\, x D_{ij}(x, y) = (1 - y) D_j(y), \quad \text{(and similarly for the second parton).}$$

This violation is formally a serious problem, as sum rules (1) follow from very basic field-theoretic features, namely the Fock-space decomposition of the hadron wave function and conservation laws. An attempt of a construction of the ansatz made in Ref. [23] met problems with the parton exchange symmetry and positivity. For the gluon sector of the proton, a formally successful ansatz has been obtained with the help of the Mellin moments [24].

The first issue we wish to point out in this talk is the non-uniqueness of the sPDF constraints. In other words, the sPDF's do not fix the dPDF's unambiguously. Suppose we have found a form of dPDF's which satisfies the sum rules (1). A sample function illustrating the situation follows from the valon model for the nucleon [22] with a single valence Fock component $|p\rangle = |uud\rangle$. It is presented in the left panel of Fig. 1. Projections lead to $D_u(x) = 2D_d(x) = 40x(1 - x)^3$, where these valence sPDF's satisfy the GS sum rules, as can be explicitly seen:

$$D_u(y) = \int dx\, D_{du}(x, y) = \int dx\, D_{uu}(x, y), \quad 2D_d(y) = \int dx\, D_{ud}(x, y), \qquad (2)$$

$$(1 - y) D_u(y) = \int dx\, x[D_{du}(x, y) + D_{uu}(x, y)], \quad (1 - y) D_d(y) = \int dx\, x D_{ud}(x, y).$$

To show that the solution for dPDF's leading to specified sPDF's is not unique, we may perturb the dPDF's by adding or subtracting strength in specific points, as shown in the right panel of Fig. 1. Then, by our specific choice of adding delta functions of strength ϵ and subtracting -2ϵ (or their multiplicities) at coordinates which are in proportion 2 : 5 : 8 does not contribute to the right-hand sides of the sum rule. One may smear this

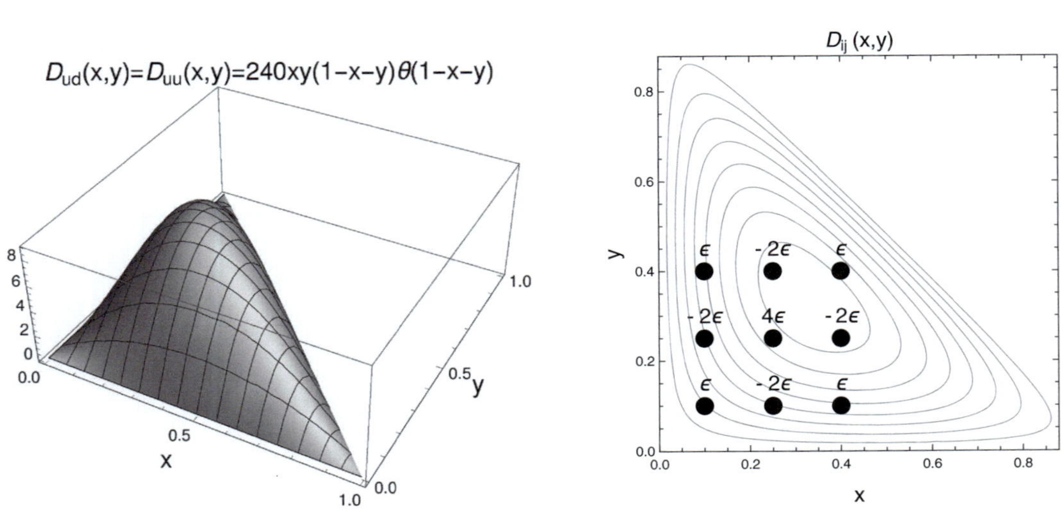

Fig. 1 *Left* example of the valence dPDF of the proton which satisfies the GS sum rules. *Right* schematic illustration of the non-uniqueness of the sPDF constraints

construction by using a distribution of points. From a general point of view, the non-uniqueness is obvious: one-particle distributions do not fix the two-particle distribution, which may contain correlations [25]. This shows that the "bottom-up" attempts to guess dPDF's with just the constraints from sPDF's are arbitrary.

As a remedy, we propose the "top-down" method which starts from the multiparticle light-cone wave function, extending the analysis of [22]. The approach guarantees the fulfillment of the formal requirement, in particular the GS sum rules (1) [15]. The Fock expansion of a hadron state in partonic constituents is assumed to have the form

$$|h\rangle = \sum_N \sum_{f_1 \ldots f_N} \int dx_1 \ldots dx_N \, \delta(1 - \sum_{k=1}^{N} x_k) \Psi_N(x_1 \ldots x_n; f_1 \ldots f_N) \, |x_1 \ldots x_N; f_1 \ldots f_N\rangle. \tag{3}$$

with f_i denoting the parton type. One should then model the N-parton Fock components Ψ_N's, obeying the constraints from the known sPDF's. Having this, one may compute the double distributions, $D_{f_1 f_2}(x_1, x_2)$, and also the higher-particle distributions if needed.

We emphasize that the presented analysis based on Eq. (3) is effectively one-dimensional, with the transverse degrees of freedom integrated out. Non perturbative transverse lattice calculations suggest that such a dimensional reduction is triggered at lower renormalization scales set by the lattice spacing, $Q \sim 1/a_\perp$. Then, the transverse degrees of freedom are effectively frozen when the transverse momenta of the quarks are smaller than the lattice spacing [26,27] (see also the discussion in Ref. [22]).

We make the simplifying assumption that the only correlations in the wave function come from the longitudinal momentum conservation: $1 = x_1 + x_2 + \cdots + x_n$ (the *generalized valon model*). Then

$$|\psi_N(x_1 \ldots x_n; f_1 \ldots f_N|^2 = A^2 \phi_{f_1}(x_1) \ldots \phi_{f_N}(x_N), \tag{4}$$

where $\phi_{f_1}(x) = |\psi_{f_i}(x)|^2$ and $\psi_{f_i}(x)$ is the one-body wave function of parton f_i, in principle computable in a dynamical model. Let the asymptotics of the single-parton functions be

$$\phi_{f_i}(x) \sim x^{\alpha_{f_i}-1} \quad \text{at } x \to 0, \qquad \phi_{f_i}(x) \sim (1-x)^{\beta_{f_i}} \quad \text{at } x \to 1, \tag{5}$$

where for integrability $\alpha_{f_i} > 0$ and $\beta_{f_i} > -1$. For $N = 2$ with parton types f and f' the resulting asymptotics for sPDF's at $x \to 0$ and $x \to 1$ is $D_f(x) \sim x^{\alpha_f + \beta_{f'} - 1}$ and $D_f(x) \sim (1-x)^{\beta_f + \alpha_{f'} - 1}$, whereas dPDF's assume the singular form $D_{ff'}(x, y) \sim \phi_f(x)\phi_{f'}(y)\delta(1 - x - y)$. The emergence of the singular part follows from the presence of just two partons. It would be washed out by the QCD evolution to higher scales [22], which redistributes the strength to higher Fock components. For $N \geq 3$

$$D_f(x) \sim x^{\alpha_f - 1} \quad \text{at } x \to 0, \qquad D_f(x) \sim (1-x)^{\beta_f + \alpha_{f_{(1}} + \cdots + \alpha_{f_{N-1)'}} - 1} \quad \text{at } x \to 1, \tag{6}$$

where $(\ldots)'$ indicates that index f is omitted in the sequence. We note that the $x \to 1$ behavior is sensitive to the low-x behavior of the other components, as in this limit the kinematics "pushes" them towards 0. The above formulas are useful in modeling, as they allow for matching of the asymptotic behavior of the single-particle functions ϕ_{f_i} to the phenomenologically known asymptotics of sPDF's.

Before we apply the extended valon model to a specific case, let us make some remarks on the choice of the scale and the QCD evolution. With an increasing scale Q_0, more and more partons at low x are generated, hence more and more Fock components are needed in the decomposition (3). For practical reasons of not dealing with too many components, it is then favorable to use the parameterizations for sPDF's at a lowest possible Q_0, such as, e.g., the GRV$_\pi$ [28] parametrization for the pion which uses Q_0 as low as 500 MeV. The tempting evolution to lower scales cannot be carried out too far down, as negative distributions are generated [29,30], as well as non-perturbative domain is entered. We also remark that the constituent quark models which do not have gluons or quarks at the *quark model scale* [31,32] do not generate sufficiently many gluons and sea quarks at higher experimental scales.

Explicitly, the GRV$_\pi$ [28] parametrization for the valence, gluon and sea sPDF's reads

$$\begin{aligned} xV(x) &= 0.52\left(0.38\sqrt{x} + 1\right)(1-x)^{0.37} x^{0.50}, \\ xg(x) &= \left(0.34\sqrt{x} + 0.68\right)(1-x)^{0.39} x^{0.48}, \quad xq_{\text{sea}}(x) = 0, \quad (Q_0 = 500\,\text{MeV}) \end{aligned} \tag{7}$$

In the following we will use the momentum fractions as constraints:

$$2\int dx\, xV(x) = 0.58, \qquad \int dx\, xg(x) = 0.42. \tag{8}$$

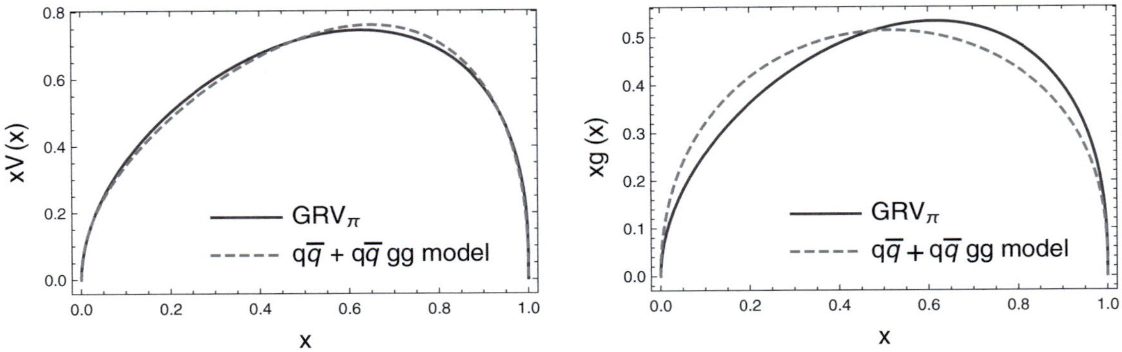

Fig. 2 The valence quark (*left*) and gluon (*right*) sPDF's of the pion from the two-component valon model (9, 12) compared to the the GRV$_\pi$ parametrization [28]

The average number of gluons inferred from Eq. (7) is $\int dx\, g(x) = 1.46$, hence a component with at least two gluons is necessary in the wave function.

We use the simple ansatz with just two Fock components:

$$|\pi^+\rangle = A|\bar{u}d\rangle + B|\bar{u}dgg\rangle. \tag{9}$$

Of course, this is a simplification, as we could have components with a single gluon and more than two gluons, as well as quark sea contributions. The corresponding sPDF's are:

$$D_{\bar{u}}(x) = A^2|\Psi_{\bar{u}d}(x, 1-x)|^2 + B^2 \int dx_3 dx_4 |\Psi_{\bar{u}dgg}(x, 1-x-x_3-x_4, x_3, x_4)|^2 = D_d(x), \tag{10}$$

$$D_g(x) = B^2 \int dx_1 dx_2 \left(|\Psi_{\bar{u}dgg}(x_1, x_2, x, 1-x-x_1-x_2)|^2 + |\Psi_{\bar{u}dgg}(x_1, x_2, 1-x-x_1-x_2, x)|^2\right).$$

The conditions

$$A^2 + B^2 = 1, \tag{11}$$
$$\int dx\, x[D_{\bar{u}}(x) + D_d(x)] = 0.58 \quad \text{or} \quad \int dx\, x D_g(x) = 0.42$$

provide constraints for parameters. We use the generalized valon ansatz

$$|\Psi_{\bar{u}d}(x_1, x_2)|^2 \sim f(x_1; a, b) f(x_2; a, b), \tag{12}$$
$$|\Psi_{\bar{u}dgg}(x_1, x_2, x_3, x_4)|^2 \sim f(x_1; \alpha_q, \beta_q) f(x_2; \alpha_q, \beta_q) f(x_3; \alpha_g, \beta_g) f(x_4; \alpha_g, \beta_g)$$

with $f(x; \alpha, \beta) = x^{\alpha-1}(1-x)^\beta$. The choice $a + b = 0.5$, $\alpha_q = 0.5$, $\beta_q = -0.09$, $\alpha_g = 0.48$, $\beta_g = -0.09$ is consistent with the asymptotic limit (6). Application of (11) yields $A^2 = 0.15$ and $B^2 = 0.85$, which means a strong dominance of the component with gluons over the $q\bar{q}$ component in the considered model.

The resulting sPDF's are shown in Fig. 2. We note a quite good (taking into account the simplicity of the model) agreement with the GRV$_\pi$ parametrization, especially for the valence quarks. We can now tackle with our main task, namely, the determination of dPDF's. As already mentioned, the $N = 2$ component generates a singular part $D_{\bar{u}d} \sim f(x; a, b) f(y; a, b) \delta(1-x-y) = D^{\text{sing}}_{\bar{u}d}(x, y)\delta(1-x-y)$. The $N = 4$ component leads, upon integration over x_3 and x_4, to broadly distributed functions. The result is shown in Fig. 3, proving that the proposed construction is practical.

An analogous construction for the case of the nucleon would be more challenging, as it requires at least three components in the Fock-space decomposition:

$$|p\rangle = A_{uud}|uud\rangle + \cdots + A_{uud\bar{q}q}|uudq\bar{q}\rangle + \cdots + A_{uudgg}|uudgg\rangle \tag{13}$$

This study, which would generalize the results for the valon model $|p\rangle = |uud\rangle$ considered in [22], is left for the future.

In conclusion, here are our main points:

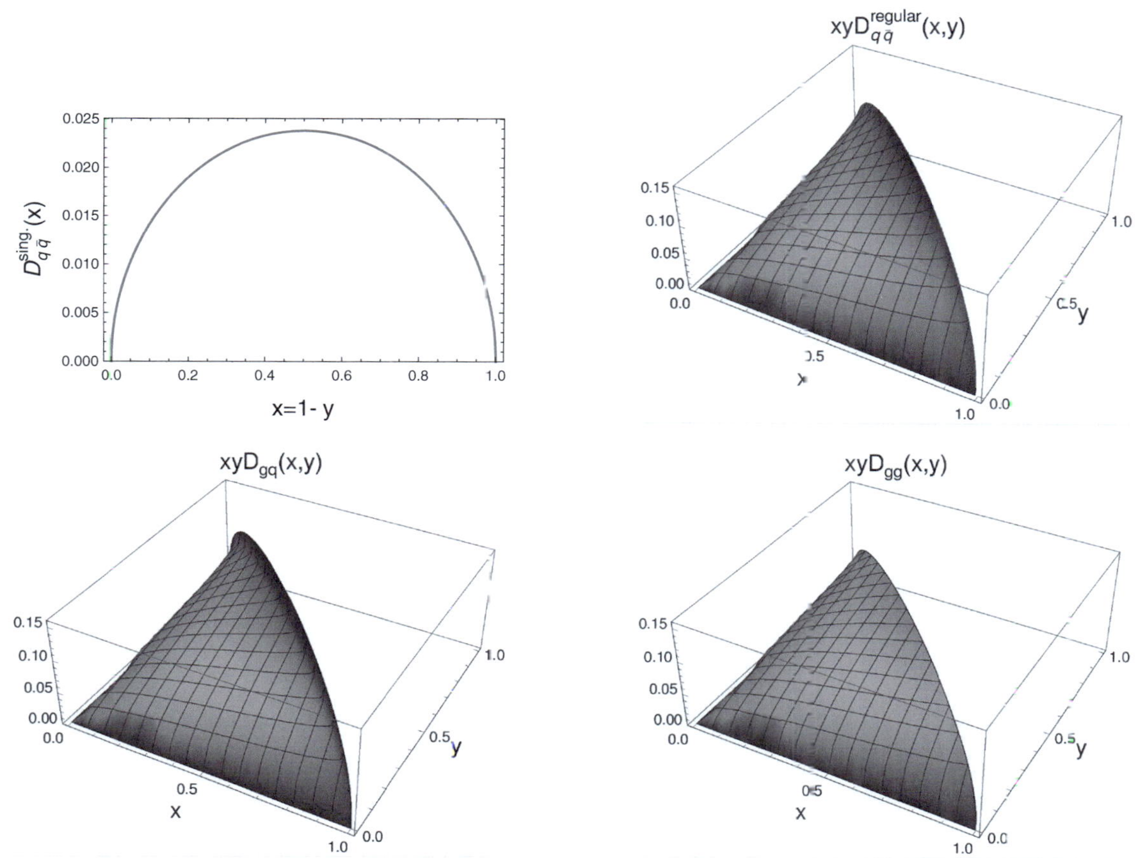

Fig. 3 dPDF's of the pion from the the two-component valon model (9, 12)

- The top-down strategy of constructing multi-parton distributions, which guarantees the formal features, is practical when the number of the Fock components is not too large. This is the case of dynamics at the lowest possible scale.
- The approach requires modeling of the light-cone wave functions, hence has physical input.
- Phenomenological sPDF's are used as constraints, but sPDF's alone cannot uniquely fix dPDF's.
- Many Fock components are needed to accurately reproduce the popular parameterizations of sPDF's, even at relatively low scales. The QCD evolution takes care of the generation of the higher Fock components at higher momentum scales.
- The valon model offers a simple ansatz at the initial low-energy scale that grasps the essential features with just the longitudinal momentum conservation, and the transverse degrees of freedom separated out from the dynamics.
- We note that the QCD evolution washes out the correlations in dPDF's at low x_1 and x_2, justifying the approximate validity of the product ansatz in that limit. However, outside of that region the correlations are substantial.

References

1. Kuti, J., Weisskopf, V.F.: Inelastic lepton–nucleon scattering and lepton pair production in the relativistic quark parton model. Phys. Rev. D **4**, 3418 (1971)

2. Konishi, K., Ukawa, A., Veneziano, G.: Jet calculus: a simple algorithm for resolving QCD Jets. Nucl. Phys. B **157**, 45 (1979)
3. Kirschner, R.: Generalized Lipatov–Altarelli–Parisi equations and jet calculus rules. Phys. Lett. B **84**, 266 (1979)
4. Shelest, V., Snigirev, A., Zinovev, G.: The multiparton distribution equations in QCD. Phys. Lett. B **113**, 325 (1982)
5. Bartalini, P., Berger, E., Blok, B., Calucci, R. Corke, et al.: (2011). arXiv:1111.0469
6. Snigirev, A.: Double-parton distributions in QCD. Phys. Atom. Nucl. **74**, 158 (2011)
7. Diehl, M., Ostermeier, D., Schafer, A.: Elements of a theory for multiparton interactions in QCD. JHEP **03**, 089 (2012). arXiv:1111.0910
8. Blok, B., Dokshitser, Yu., Frankfurt, L., Strikman, M.: pQCD physics of multiparton interactions. Eur. Phys. J. C **72**, 1963 (2012). arXiv:1106.5533
9. Blok, B., Dokshitzer, Yu., Frankfurt, L., Strikman, M.: Perturbative QCD correlations in multi-parton collisions. Eur. Phys. J. C **74**, 2926 (2014). arXiv:1306.3763
10. Kasemets, T., Mulders, P.J.: Constraining double parton correlations and interferences. Phys. Rev. D **91**, 014015 (2015). arXiv:1411.0726
11. Astalos, R. et al.: (2015). arXiv:1506.05829
12. d'Enterria, D., Snigirev, A.M.: Same-sign WW production in proton-nucleus collisions at the LHC as a signal for double parton scattering. Phys. Lett. B **718**, 1395 (2013). arXiv:1211.0197
13. Luszczak, M., Maciula, R., Szczurek, A.: Production of two $c\bar{c}$ pairs in double-parton scattering. Phys. Rev. D **85**, 094034 (2012). arXiv:1111.3255
14. Gaunt, J.R., Stirling, W.J.: Double parton distributions incorporating perturbative QCD evolution and momentum and quark number sum rules. JHEP **005**, 1003 (2010). arXiv:0910.4347
15. Gaunt, J.R.: *Double Parton Scattering in Proton–Proton Collisions*, Ph.D. thesis, Trinity College, University of Cambridge (July 2012)
16. Chang, H.-M., Manohar, A.V., Waalewijn, W.J.: Double parton correlations in the bag model. Phys. Rev. D **87**, 034009 (2013). arXiv:1211.3132
17. Rinaldi, M., Scopetta, S., Vento, V.: Double parton correlations in constituent quark models. Phys. Rev. D **87**, 114021 (2013). arXiv:1302.6462
18. Rinaldi, M., Scopetta, S., Traini, M., Vento, V.: Double parton correlations and constituent quark models: a Light Front approach to the valence sector. JHEP **12**, 028 (2014). arXiv:1409.1500
19. Rinaldi, M., Scopetta, S., Traini, M., Vento, V.: Double parton scattering: a study of the effective cross section within a Light-Front quark model. Phys. Lett. B **752**, 40 (2016). arXiv:1506.05742
20. Hwa, R.C., Zahir, M.S.: Parton and valon distributions in the nucleon. Phys. Rev. D **23**, 2539 (1981)
21. Hwa, R.C., Yang, C.: Parton distributions in the valon model. Phys. Rev. C **66**, 025204 (2002). arXiv:hep-ph/0202140
22. Broniowski, W., Arriola, E.Ruiz: Valence double parton distributions of the nucleon in a simple model. Few Body Syst. **55**, 381 (2014). arXiv:1310.8419
23. Golec-Biernat, K., Lewandowska, E.: How to impose initial conditions for QCD evolution of double parton distributions? Phys. Rev. D **90**(1), 014032 (2014). arXiv:1402.4079
24. Golec-Biernat, K., Lewandowska, E., Serino, M., Snyder, Z., Stasto, A.M.: Constraining the double gluon distribution by the single gluon distribution. Phys. Lett. **B750**, 559 (2015). arXiv:1507.08583
25. Snigirev, A.M., Snigireva, N.A., Zinovjev, G.M.: Perturbative and nonperturbative correlations in double parton distributions. Phys. Rev. D **90**(1), 014015 (2014). arXiv:1403.6947
26. Burkardt, M., Dalley, S.: The relativistic bound state problem in QCD: transverse lattice methods. Prog. Part. Nucl. Phys. **48**, 317 (2002). arXiv:hep-ph/0112007
27. Dalley, S., van de Sande, B.: Transverse lattice calculation of the pion light-cone wavefunctions. Phys. Rev. D **67**, 114507 (2003). arXiv:hep-ph/0212086
28. Gluck, M., Reya, E., Vogt, A.: Pionic parton distributions. Z. Phys. C **53**, 651 (1992)
29. Smith, C.H.L., Wolfram, S.: Positivity constraints on quark and gluon distributions in QCD. Nucl. Phys. **B138**, 333 (1978)
30. Ruiz, E.: Arriola, NLO evolution for large scale distances, positivity constraints and the low-energy model of the nucleon. Nucl. Phys. A **641**, 461 (1998)
31. Davidson, R., Ruiz Arriola, E.: Structure functions of pseudoscalar mesons in the SU(3) NJL model. Phys. Lett. B **348**, 163 (1995)
32. Broniowski, W., Ruiz Arriola, E., Golec-Biernat, K.: Generalized parton distributions of the pion in chiral quark models and their QCD evolution. Phys. Rev. **D77**, 034023 (2008). arXiv:0712.1012

DOI 10.1007/978-3-319-50699-9_8
Reprinted from *Few-Body System* Journal, DOI 10.1007/s00601-016-1048-6

Guy F. de Téramond

Hadron Physics from Superconformal Quantum Mechanics and Its Light-Front Holographic Embedding

Received: 25 January 2016 / Accepted: 30 January 2016 / Published online: 23 February 2016

Abstract The complex nonperturbative color-confining dynamics of QCD is well captured in a semiclassical effective theory based on superconformal quantum mechanics and its extension to the light-front. I describe here how this new approach to hadron physics incorporates confinement, the appearance of nearly massless pseudoscalar particles, and Regge spectroscopy consistent with experiment. It also gives remarkable connections between the meson and baryon spectrum across the light and heavy-light hadron spectrum. I also briefly discuss how higher spin states are consistently described in this framework by the holographic embedding of the superconformal theory in a higher dimensional semiclassical gravity theory.

1 Introduction

Quantum chromodynamics (QCD) describes the interactions of quark and gluons with remarkable success in the high energy regime where asymptotic freedom allows accurate computations using perturbation theory. However, in the low energy domain, where hadrons are the relevant degrees of freedom, QCD is nonperturbative and understanding the mechanism of confinement is still an unsolved problem. In fact, no analytic solution to the problem of confinement in QCD has been found yet in the physical realm [1]. According to the Kinoshita–Lee–Nauenberg theorem, perturbative QCD cannot describe confinement at any order in perturbation theory, at least in a simple way [2,3]. Nonperturbative methods such as lattice field theory, effective field theories, Schwinger–Dyson equations or the gauge/gravity correspondence, are thus needed to study confinement dynamics [1,4].

The problem of confinement is vastly complex. The QCD Lagrangian in the limit of massless quarks has no scale; still confinement and a mass gap should emerge from the quantum theory built upon the classical QCD conformal theory. The increase of the QCD coupling in the infrared regime implies that an infinite number of quark and gluons are dynamically intertwined, and a description of the strong dynamics in terms of the fundamental fields appearing in the QCD Lagrangian becomes analytically intractable. Indeed, the dynamical problem may become undecidable: In QCD this means that it is not possible to know whether the system is truly gapless, or whether increasing the lattice size would reveal it to be gapped [5].

Holographic methods provide new tools for the study of strongly correlated quantum systems. The AdS/CFT correspondence between gravity on a five-dimensional anti-de Sitter (AdS) space and conformal field theories (CFT) in physical space-time [6] is an explicit realization of the holographic principle. The gauge/gravity correspondence leads to a semiclassical gravity approximation for a strongly-coupled non-abelian quantum field theory, thus providing new physical insights into its nonperturbative dynamics.

The holographic mapping to light-front (LF) physics is particularly useful since quantization in the front form of dynamics [7] provides a relativistic and frame independent Hamiltonian framework to describe bound-state hadronic systems [8]. In fact, since the structure of the LF vacuum is particularly simple, it is possible to

G. F. de Téramond (✉)
Universidad de Costa Rica, San José, Costa Rica
E-mail: gdt@asterix.crnet.cr

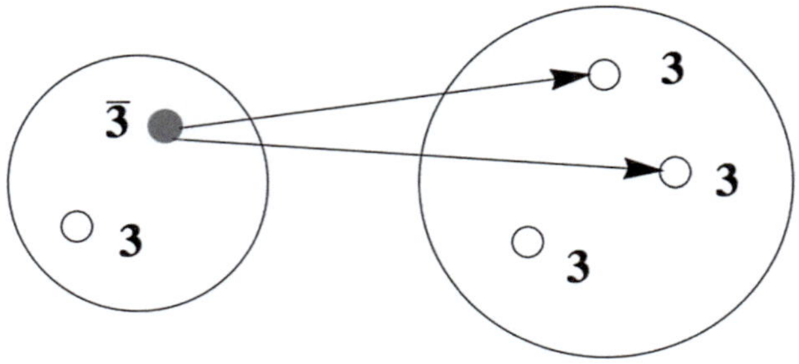

Fig. 1 In QCD hadronic dynamical supersymmetry is rooted in the dynamics of color $SU(3)$ where $\bar{\mathbf{3}} \sim \mathbf{3} \times \mathbf{3}$

retain a probabilistic quantum-mechanical interpretation in terms of hadronic wave functions. One can show that the eigenvalue equations in AdS space have a precise mapping to the relativistic semiclassical bound-state equations in the light front [9]. This connection also gives an exact relation between the holographic variable z of AdS space and the invariant impact separation light-front variable ζ.[1]

The holographic mapping implies that the non-trivial geometry of AdS space encodes all the kinematical aspects, whereas the modification of the action in AdS space—described in terms of a dilaton profile $\varphi(z)$, includes the confinement dynamics and determines the effective LF potential $U(\zeta)$ in the light-front bound-state equations [11]. The embedding of LF wave equations in AdS leads to an extension of the LF potential U to arbitrary spin from conformal symmetry breaking in the AdS action [11,12].[2] The light-front effective potential U acts on the valence state and incorporates and infinite number of higher Fock states and retarded interactions [14], but its derivation from QCD represents an unsurmountable task.

Since the light-front semiclassical approximation to the QCD Hamiltonian equation leads to a one-dimensional quantum field theory (QFT), it is natural to ask if there is a compelling connection of conformal quantum mechanics (QM) [15] with hadronic physics. Indeed, Brodsky, Dosch and I have shown recently that one can introduce a gap and a confinement scale $\sqrt{\lambda}$ by extending conformal QM to the light front [16]. The precise mapping of conformal QM to the light-front bound-state equations uniquely determines the form of the LF confinement potential U—and thus the dilaton profile φ, while keeping the conformal invariance of the action preserved [15]. This formalism uniquely determines the confinement potential U to have the form of a harmonic oscillator in the light-front transverse plane.[3] The resulting model is very successful in describing a wealth of hadronic data [18].

More recently Dosch, Brodsky and I have shown how the extremely complex nonperturbative QCD dynamics is well captured in a semiclassical effective theory based on superconformal quantum mechanics [19,20] and its extension to light-front physics [21–23]. This new approach to hadron physics incorporates two basic aspects one expects from QCD, but whose understanding has remained elusive: confinement and the appearance of nearly massless pseudoscalar particles. Furthermore, this framework gives remarkable connections between the light meson and nucleon spectra as well as predictions across the full heavy-light hadron spectra, where heavy quark masses break the conformal invariance, but the underlying supersymmetry still holds. It is important to recall that in the context of hadronic physics the supersymmetric relations are not a consequence of supersymmetric QCD at the level of fundamental fields but an emergent dynamical supersymmetry from color $SU(3)_C$. This relies on the fact that in $SU(3)_C$ a diquark can be in the same color representation as an antiquark, namely a $\bar{\mathbf{3}} \sim \mathbf{3} \times \mathbf{3}$ [24] as illustrated in Fig. 1. Hadronic supersymmetry was introduced by Miyazawa [25].

2 Superconformal Quantum Mechanics and Light-Front Bound-State Equations

We construct a superconformal algebraic structure in one dimension by introducing the fermionic operators Q, $Q^\dagger$, S and $S^\dagger$ [20]

[1] Light-front holographic methods were introduced by matching the electromagnetic form factor in AdS space with the corresponding light-front expression in physical space-time [10]. The invariant distance squared for a two-particle bound state is $\zeta^2 = x(1-x)b_\perp^2$, where x is the constituent momentum fraction, and $b_\perp$ is the transverse separation between the constituents.

[2] The potential U is related to the dilaton profile by [11,13] $U(\zeta, J) = \frac{1}{2}\varphi''(\zeta) + \frac{1}{4}\varphi'(\zeta)^2 + \frac{2J-3}{2\zeta}\varphi'(\zeta)$, where J is the total angular momentum of the hadron.

[3] For light-quark masses the harmonic oscillator potential in the light front is equivalent to the usual linear confining potential at large separation distances in the instant form of dynamics [17].

$$\tfrac{1}{2}\{Q, Q^\dagger\} = H, \quad \tfrac{1}{2}\{S, S^\dagger\} = K,$$
$$\{Q, S^\dagger\} = f - B + 2iD, \quad \{Q^\dagger, S\} = f - B - 2iD, \tag{1}$$

where Q is represented by $Q = \psi^\dagger(-d/dx + f/x)$ and $Q^\dagger = \psi(d/dx + f/x)$, with f a dimensionless quantity. The generator S is given by $S = \psi^\dagger x$ and $S^\dagger = \psi x$, while $B = \frac{1}{2}[\psi^\dagger, \psi]$ is a bosonic operator with $\{\psi, \psi^\dagger\} = 1$. The three generators of translation, dilatation and the special conformal transformation H, D and K

$$H = \frac{1}{2}\left(-\frac{d^2}{dx^2} + \frac{f^2 + 2Bf}{x^2}\right), \quad D = \frac{i}{4}\left(\frac{d}{dx}x + x\frac{d}{dx}\right), \quad K = \tfrac{1}{2}x^2, \tag{2}$$

satisfy the conformal algebra:

$$[H, D] = iH, \qquad [H, K] = 2iD, \qquad [K, D] = -iK. \tag{3}$$

Following Fubini and Rabinovici [20] we define a new fermionic operator R, a linear combination of the generators Q and S,

$$R_\lambda = Q + \lambda S, \qquad R_\lambda^\dagger = Q^\dagger + \lambda S^\dagger, \tag{4}$$

which generates a new Hamiltonian G

$$G = \{Q_\lambda, Q_\lambda^\dagger\}, \tag{5}$$

where by construction

$$\{R_\lambda, R_\lambda\} = \{R_\lambda^\dagger, R_\lambda^\dagger\} = 0 \qquad \text{and} \qquad [R_\lambda, G] = [R_\lambda^\dagger, G] = 0, \tag{6}$$

a set of relations which closes a graded algebra $sl(1/1)$, identical to Witten's supersymmetric quantum mechanics [26]. Indeed, since the Hamiltonian G commutes with R_λ; it thus follows that the states $|\phi\rangle$ and $R^\dagger|\phi\rangle$ have identical eigenvalues: If $|\phi_E\rangle$ is an eigenstate of G with $E \neq 0$, $G\,|\phi_E\rangle = E\,|\phi_E\rangle$, then $G\,R_\lambda^\dagger\,|\phi_E\rangle = R_\lambda^\dagger\,G\,|\phi_E\rangle = E\,R_\lambda^\dagger|\phi_E\rangle$, and thus $R_\lambda^\dagger\,|\phi_E\rangle$ is also an eigenstate of G with the same eigenvalue. Since the dimensions of the generators Q and S are different, a scale λ is introduced in the Hamiltonian in analogy with the previous treatment by de Alfaro et al. [15].

In a Pauli matrix representation Eq. (5) is given by $G = 2H + 2\lambda^2 K + 2\lambda\,(f - \sigma_3)$, leading to the eigenvalue equations [22]

$$\left(-\frac{d^2}{dx^2} + \lambda^2 x^2 + 2\lambda f - \lambda + \frac{4(f + \frac{1}{2})^2 - 1}{4x^2}\right)\phi_1 = E\,\phi_1, \tag{7}$$

$$\left(-\frac{d^2}{dx^2} + \lambda^2 x^2 + 2\lambda f + \lambda + \frac{4(f - \frac{1}{2})^2 - 1}{4x^2}\right)\phi_2 = E\,\phi_2, \tag{8}$$

with identical eigenvalues for $f \geq \frac{1}{2}$ and $\lambda > 0$: $E_n = 4\lambda\,(n + f + 1/2)$.

We now compare Eqs. (7) and (8) with the leading twist LF holographic equations which describe the pseudoscalar and nucleon orbital and radial excitation spectrum [18]

$$\left(-\frac{d^2}{d\zeta^2} + \frac{4L_M^2 - 1}{4\zeta^2} + \lambda_M^2\,\zeta^2 + 2\lambda_M(L_M - 1)\right)\phi_{Meson} = M^2\,\phi_{Meson}, \tag{9}$$

$$\left(-\frac{d^2}{d\zeta^2} + \frac{4L_N^2 - 1}{4\zeta^2} + \lambda_N^2\,\zeta^2 + 2\lambda_N(L_N + 1)\right)\phi_{Nucleon} = M^2\,\phi_{Nucleon}, \tag{10}$$

with eigenvalues

$$M_M^2 = 4\lambda_M\,(n + L_M)\,, \tag{11}$$
$$M_N^2 = 4\lambda_N\,(n + L_N + 1)\,. \tag{12}$$

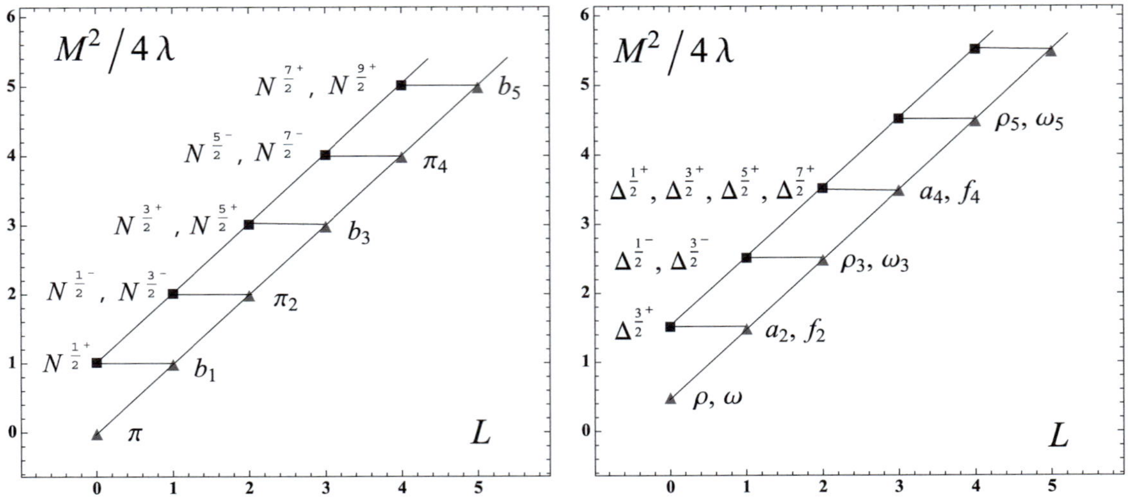

Fig. 2 Meson–baryon superconformal connection. The predicted values of M^2 in units of 4λ for pseudoscalar mesons (*red triangles*) and nucleons (*blue squares*) are plotted versus the orbital angular momentum L. The operator R_λ acts horizontally to the right and transforms a baryon with orbital L into a meson with orbital $L+1$. $R_\lambda^\dagger$ acts on the opposite direction and annihilates the π meson which has no nucleon partner (*left*). The ρ and ω have no baryonic partner, since it would imply a negative value of L (*right*). The hadron quantum number assignment is taken from Ref. [18] (color figure online)

Since R is a fermion operator which relates the upper and lower components in the multiplet $|\phi\rangle = \begin{pmatrix} \phi_1 \\ \phi_2 \end{pmatrix}$,

$$R_\lambda = \begin{pmatrix} 0 & -\frac{d}{dx} + \frac{f}{x} + \lambda\, x \\ 0 & 0 \end{pmatrix}, \qquad R_\lambda^\dagger = \begin{pmatrix} 0 & 0 \\ \frac{d}{dx} + \frac{f}{x} + \lambda\, x & 0 \end{pmatrix}, \tag{13}$$

it is tempting to relate the meson and nucleon bound-state wave functions by the supercharges within the superconformal algebra. Identifying the upper component equation (7) with the meson LF bound-state equation (9) and the lower component equation (8) with the nucleon equation (10) we find, upon the substitution $x \to \zeta$ and $E \to M^2$ the conditions $\lambda = \lambda_M = \lambda_N$, and $f = L_N + \frac{1}{2} = L_M - \frac{1}{2}$. Thus the remarkable relation $L_M = L_N + 1$ between the orbital angular momentum of the meson and nucleon superpartners [22]. This result is in agreement with the light-front holographic QCD spectral formulas Eqs. (11) and (12) for pseudoscalar mesons and nucleons respectively. The superconformal approach has, however, the advantage that mesons and nucleons are treated on the same footing and additional constant terms in the confinement potential are determined from the onset.[4]

The fermion operator $R^\dagger$ transforms a meson state $|M, L\rangle$ with orbital angular momentum L into a nucleon state $|N, L-1\rangle$ with orbital $L-1$: The superpartners have the same parity. The operator $R^\dagger$ also annihilates the lowest possible mass state, the pion, which has no nucleon superpartner [22]:

$$R^\dagger |M, L\rangle = |B, L-1\rangle, \qquad R^\dagger |M, L=0\rangle = 0. \tag{14}$$

Thus the special role of the pion in the supersymmetric approach to hadronic physics as a unique state of zero energy [see Fig. 2 (left)].

3 Baryon–Meson Dynamical Supersymmetry

For light hadronic systems the essential confinement dynamics is encoded in the supercharge generator R_λ which connects meson and nucleon bound-state wavefunctions. We expect a rather similar situation for the baryon superpartner trajectory of the ρ and ω vector mesons, the Δ baryons, the essential difference being an overall shift of the spectrum due to the spin interaction [see Fig. 2 (right)]. This is indeed the case experimentally, where the superconformal predictions are reproduced to a very good degree of accuracy [22]. A

[4] In Ref. [21] superconformal quantum mechanics was used to describe the nucleon spectrum. In this case, the supercharges relate the positive and negative chirality components of the nucleon wave functions.

crucial prediction observed in Fig. 2 is the fact that the lowest meson state in each family with $L = 0$ has no superpartner. Physically, this corresponds to the fact that the partner baryon would have $L = -1$. Indeed, the lowest meson state is annihilated by the supercharge $R_\lambda^\dagger$ and there is no baryon partner.[5]

It is remarkable that the underlying dynamical supersymmetry holds across the heavy-light hadron spectrum [23], even if conformal symmetry is badly broken by heavy quark masses. In this case the superpotential is no longer constrained by conformal symmetry and it is basically unknown. However, the supersymmetric predictions for mesons and baryons with charm and beauty are also observed to a surprising level of accuracy [23]. In particular, the crucial prediction for the lowest state $L = 0$ meson, namely that it has no supersymmetric baryon partner—as it would correspond to an $L = -1$ state, is confirmed in the observed heavy-light hadron spectrum [23].

4 Concluding Remarks

The classical QCD Lagrangian in the limit of zero-quark mass has no scale. The origin of a mass scale and confinement remains essentially a fundamental unsolved problem due to the vast complexity of the strong dynamics of quarks and gluons in the low energy domain [1,4]. Indeed the problem is of such complexity that it may become undecidable [5].

The connection of light-front dynamics, classical gravity in a higher-dimensional space and superconformal quantum mechanics brings new insights to our understanding of confinement dynamics and QCD. In particular, an indication of the emergence of a mass scale in a conformal invariant theory can be drawn from the mechanism proposed by de Alfaro et al. [15] in the context of one-dimensional QFT. Mathematically, this remarkable result follows from the isomorphism of the group $SO(2,1)$ of conformal QM with the one-dimensional conformal group $Conf\left(R^1\right)$. Since the generators of $Conf\left(R^1\right)$, H, D and K, have different dimensions, their relation with the generators of $SO(2,1)$ requires the introduction of a scale $\sqrt{\lambda}$. The scale $\sqrt{\lambda}$ plays the role of the confinement scale, which is directly related to a physical observable, such as a hadron mass. In this context, scales corresponding to scheme dependent perturbative computations at short distances, such as the QCD renormalization scale Λ_s, are determined by the nonperturbative scale $\sqrt{\lambda}$ [28]. In superconformal QM the Hamiltonian G (5) is the superposition of generators with different dimensions which also requires the introduction of a scale $\sqrt{\lambda}$ [20].

The mapping of the superconformal Hamiltonian to the light-front relativistic bound-state equations uniquely determines the confinement potential U, including constant terms in the potential. It also determines uniquely the dilaton profile φ in the embedding space. The holographic connection leads to the extension of the LF potential U to arbitrary spin [11]. This embedding is required to describe consistently all the light mesons and baryons and their superconformal connections [27]. For heavy quark masses conformal symmetry is broken but the supersymmetric predictions are observed all across the heavy-light spectrum [23], giving strong evidence of an underlying dynamical supersymmetry in hadron physics.

There is another important connection which we have not described here: the connection between the one-dimensional conformal group $Conf\left(R^1\right)$ and the group of isometries of AdS_2 in two dimensions, $SO(2,1)$, which underlies the AdS_2/CFT_1 correspondence [29]. AdS_2 becomes relevant to the emergence of its one-dimensional dual QFT in the infrared through holographic renormalization in the bulk [30]. Indeed the holographic flow of boundary theories to AdS_2 geometry in the infrared [31] may be relevant to the approach of hadron physics discussed in this article, since its dual one-dimensional quantum field theory is realized as conformal quantum mechanics.

Acknowledgments The results presented here are based on collaborations with Stanley J. Brodsky, Alexandre Deur, Hans Guenter Dosch and Cédric Lorcé. I want to thank the organizers of Light Cone 2015 for their hospitality at Frascati.

References

1. Brambilla, N. et al.: QCD and strongly coupled gauge theories: challenges and perspectives. Eur. Phys. J. C **74**, 2981 (2014)
2. Kinoshita, T.: Mass singularities of Feynman amplitudes. J. Math. Phys. **3**, 650 (1962)

[5] A few questions left open in Ref. [22], notably the nature of the spin interaction and the correct twist assignment of the wave functions in the $\rho - \Delta$ connections, have been addressed recently by enforcing superconformal symmetry in the holographic embedding of the corresponding light-front bound-state equations [27].

3. Lee, T.D., Nauenberg, M.: Degenerate systems and mass singularities. Phys. Rev. **133**, B1549 (1964)
4. Greensite, J.: An introduction to the confinement problem. Lect. Notes Phys. **821**, 1 (2011)
5. Cubitt, T.S., Perez-Garcia, D., Wolf, M.M.: Undecidability of the spectral gap. Nature **528**, 207 (2015); arXiv:1502.04135 [quant-ph]; arXiv:1502.04573 [quant-ph]
6. Maldacena, J.M.: The large-N limit of superconformal field theories and supergravity. Int. J. Theor. Phys. **38**, 1113 (1999)
7. Dirac, P.A.M.: Forms of relativistic dynamics. Rev. Mod. Phys. **21**, 392 (1949)
8. Brodsky, S.J., Pauli, H.C., Pinsky, S.S.: Quantum chromodynamics and other field theories on the light cone. Phys. Rep. **301**, 299 (1998)
9. de Teramond, G.F., Brodsky, S.J.: Light-front holography: a first approximation to QCD. Phys. Rev. Lett. **102**, 081601 (2009)
10. Brodsky, S.J., de Teramond, G.F.: Hadronic spectra and light-front wave functions in holographic QCD. Phys. Rev. Lett. **96**, 201601 (2006)
11. de Teramond, G.F., Dosch, H.G., Brodsky, S.J.: Kinematical and dynamical aspects of higher-spin bound-state equations in holographic QCD. Phys. Rev. D **87**, 075005 (2013)
12. Gutsche, T., Lyubovitskij, V.E., Schmidt, I., Vega, A.: Dilaton in a soft-wall holographic approach to mesons and baryons. Phys. Rev. D **85**, 076003 (2012)
13. de Teramond, G.F., Brodsky, S.J.: Gauge/gravity duality and hadron physics at the light front. AIP Conf. Proc. **1296**, 128 (2010)
14. Pauli, H.C.: On confinement in a light cone Hamiltonian for QCD. Eur. Phys. J. C **7**, 289 (1999)
15. de Alfaro, V., Fubini, S., Furlan, G.: Conformal invariance in quantum mechanics. Nuovo Cim. A **34**, 569 (1976)
16. Brodsky, S.J., de Teramond, G.F., Dosch, H.G.: Threefold complementary approach to holographic QCD. Phys. Lett. B **729**, 3 (2014)
17. Trawiński, A.P., Glazek, S.D., Brodsky, S.J., de Teramond, G.F., Dosch, H.G.: Effective confining potentials for QCD. Phys. Rev. D **90**, 074017 (2014)
18. Brodsky, S.J., de Teramond, G.F., Dosch, H.G., Erlich, J.: Light-front holographic QCD and emerging confinement. Phys. Rep. **584**, 1 (2015)
19. Akulov, V.P., Pashnev, A.I.: Quantum superconformal model in (1,2) space. Theor. Math. Phys. **56**, 862 (1983)
20. Fubini, S., Rabinovici, E.: Superconformal quantum mechanics. Nucl. Phys. B **245**, 17 (1984)
21. de Teramond, G.F., Dosch, H.G., Brodsky, S.J.: Baryon spectrum from superconformal quantum mechanics and its light-front holographic embedding. Phys. Rev. D **91**, 045040 (2015)
22. Dosch, H.G., de Teramond, G.F., Brodsky, S.J.: Superconformal baryon-meson symmetry and light-front holographic QCD. Phys. Rev. D **91**, 085016 (2015)
23. Dosch, H.G., de Teramond, G.F., Brodsky, S.J.: Supersymmetry across the light and heavy-light hadronic spectrum. Phys. Rev. D **92**, 074010 (2015)
24. Catto, S., Gursey, F.: Algebraic treatment of effective supersymmetry. Nuovo Cim. A **86**, 201 (1985)
25. Miyazawa, H.: Baryon number changing currents. Prog. Theor. Phys. **36**, 1266 (1966)
26. Witten, E.: Dynamical breaking of supersymmetry. Nucl. Phys. B **188**, 513 (1981)
27. Brodsky, S.J., de Teramond, G.F., Dosch, H.G., Lorce, C.: (in preparation)
28. Deur, A., Brodsky, S.J., de Teramond, G.F.: Connecting the hadron mass scale to the fundamental mass scale of quantum chromodynamics. Phys. Lett. B **750**, 528 (2015); On the interface between perturbative and nonperturbative QCD, arXiv:1601.06568 [hep-ph]
29. Chamon, C., Jackiw, R., Pi, S.-Y., Santos, L.: Conformal quantum mechanics as the CFT_1 dual to AdS_2. Phys. Lett. B **701**, 503 (2011)
30. Faulkner, T., Liu, H., McGreevy, J., Vegh, D.: Emergent quantum criticality, Fermi surfaces, and AdS_2. Phys. Rev. D **83**, 125002 (2011)
31. Heemskerk, I., Polchinski, J.: Holographic and Wilsonian renormalization groups. JHEP **1106**, 31 (2011)

DOI 10.1007/978-3-319-50699-9_9
Reprinted from *Few-Body System* Journal, DOI 10.1007/s00601-016-1065-5

Joannis Papavassiliou

Hadron Phenomenology from First-Principle QCD Studies

Received: 14 January 2016 / Accepted: 21 February 2016 / Published online: 18 March 2016

Abstract The form of the kernel that controls the dynamics of the Bethe–Salpeter equations is essential for obtaining quantitatively accurate predictions for the observable properties of hadrons. In the present work we briefly review the basic physical concepts and field-theoretic techniques employed in a first-principle derivation of a universal (process-independent) component of this kernel. This "top-down" approach combines nonperturbative ingredients obtained from lattice simulations and Dyson–Schwinger equations, and furnishes a renormalization-group invariant quark-gluon interaction strength, which is in excellent agreement with the corresponding quantity obtained from a systematic "bottom-up" treatment, where bound-state data are fitted within a well-defined truncation scheme.

1 Introduction

The spectrum and various physical properties of the mesons are traditionally obtained in the continuum by means of special integral equations, known as Bethe–Salpeter equations (BSEs) [1–9], whose general form is captured by the diagram in Fig. 1. In this particular eigenvalue equation, Γ denotes the so-called Bethe–Salpeter amplitude, and $\mathcal{K}$ the fully-amputated quark-antiquark scattering kernel. The details of the solutions obtained from BSEs depend crucially on the precise form of $\mathcal{K}$, and the nonperturbative information included in it. In fact, it is well-known that any self-consistent analysis based on BSEs must be intimately connected with nonperturbative phenomena such as chiral symmetry breaking, and quark and gluon mass generation, which are described by the dynamical equations obeyed by the Green's functions of the theory, namely the Dyson–Schwinger equations (DSEs) [10,11]. In this presentation we briefly review recent work that aims at a first-principle derivation of a special component of the BSE kernel [12], and compare the results obtained with phenomenologically successful "bottom-up" versions of the same quantity.

2 Definitions and Basic Ingredients

In the Landau gauge the gluon propagator is given by

$$i\Delta_{\mu\nu}(p) = -i\left[g_{\mu\nu} - p_\mu p_\nu/p^2\right]\Delta(p^2), \tag{1}$$

while the ghost propagator, $D(p^2)$, and its dressing function, $F(p^2)$, are related by $D(p^2) = F(p^2)/p^2$.

J. Papavassiliou (✉)
Department of Theoretical Physics and IFIC, University of Valencia and CSIC, 46100 Valencia, Spain
E-mail: Joannis.Papavassiliou@uv.es

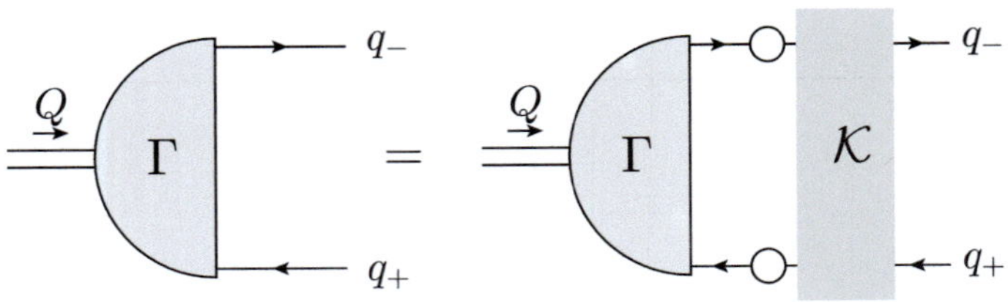

Fig. 1 The meson BSE and the kernel $\mathcal{K}$

In addition, consider a special two-point function, denoted by $\Lambda_{\mu\nu}(p)$, defined as

$$\begin{aligned}\Lambda_{\mu\nu}(p) &= g_{\mu\nu} - ig^2 C_A \int_k H^{(0)}_{\mu\rho} D(k+p) \Delta^{\rho\sigma}(k)\, H_{\sigma\nu}(-k-p,k,p), \\ &= g_{\mu\nu}[1+G(p^2)] + \frac{p_\mu p_\nu}{p^2} L(p^2);\end{aligned} \tag{2}$$

where C_A is the Casimir eigenvalue of the adjoint representation, and $\int_k \equiv \mu^{2\varepsilon}(2\pi)^{-d} \int d^d k$, with $d = 4 - \epsilon$ the dimension of space-time. The quantity $H_{\mu\nu}$ corresponds to the well-known ghost–gluon kernel that enters in the Slavnov–Taylor identity satisfied by the full three-gluon vertex [13,14]. In addition, $H_{\mu\nu}$ is related to the full gluon–ghost vertex, Γ_μ, whose tensorial structure is given by

$$-\Gamma_\mu = B_1 p_\mu + B_2 k_\mu, \tag{3}$$

where $B_i = B_i(-k-p,k,p)$, with k representing the momentum of the gluon and p the one of the anti-ghost. Specifically,

$$p^\nu H_{\mu\nu}(-k-p,k,p) = -i\Gamma_\mu(-k-p,k,p). \tag{4}$$

At tree-level, $H^{(0)}_{\mu\nu} = ig_{\mu\nu}$, and $\Gamma^{(0)}_\mu = -p_\mu$.

It turns out that the $1+G(p^2)$ and $L(p^2)$ defined in Eq. (2) are related to $F(p^2)$ by an exact relation (valid in Landau gauge only) [15]

$$F^{-1}(p^2) = 1 + G(p^2) + L(p^2). \tag{5}$$

3 The Universal and Renormalization-Group Invariant Part of the BS Kernel

Let us consider the kernel appearing in a typical Bethe–Salpeter equation (BSE), shown in Fig. 1, which is contained in the gray box. To make contact with earlier works, we will divide it by a factor of 4π, and will denote it by $\mathcal{K}$. The kernel $\mathcal{K}$ receives a "universal" (process-independent) contribution, whose origin is the pure gauge sector of the theory; in that sense, this contribution constitutes the common ingredient of any such kernel, regardless of the nature of the particles between which it is embedded.

The systematic diagrammatic identification of the precise pieces that constitute this particular quantity may be carried out following the procedure known in the literature as pinch technique (PT) [16]. In general, the upshot of this construction is the rearrangement of a physical amplitude into sub-amplitudes with very special properties; in particular, one obtains vertices that satisfy QED-like Ward identities and a gluon propagator that captures all the RG logarithms of the theory, see Fig. 2. In fact, it turns out that these latter quantities coincide precisely with the corresponding vertices and gluon propagator defined in the Background Field Method (BFM) [17]. This particular identification persists both perturbatively, to all orders, as well as nonperturbative, at the level of the corresponding DSEs.

In the case of the gluon propagator, the standard $\Delta(p^2)$ defined in Eq. (1), and the scalar cofactor of the PT-BFM gluon propagator, denoted by $\widehat{\Delta}(p^2)$, are related by the exact relation [18]

$$\Delta(p^2) = \widehat{\Delta}(p^2)[1 + G(p^2)]^2. \tag{6}$$

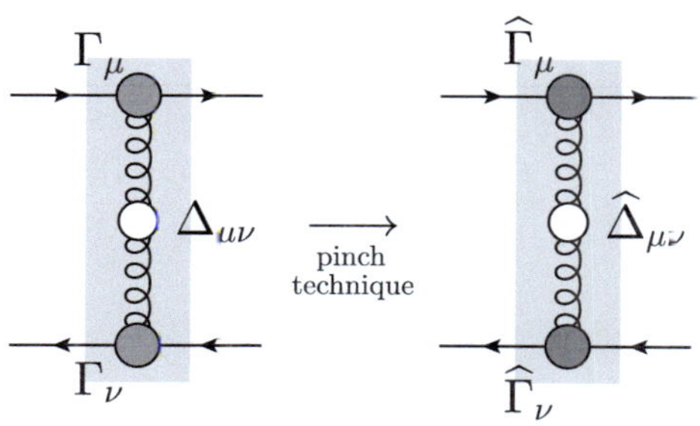

Fig. 2 The one-gluon exchange kernel before and after the pinch-technique rearrangement

At the one-loop level, and keeping only UV logarithms, one has [19]

$$1 + G(p^2) = 1 + \left(\frac{9}{4}\right)\frac{\alpha_s C_A}{12\pi}\ln\left(\frac{p^2}{\mu^2}\right); \quad \Delta^{-1}(p^2) = p^2\left[1 + \left(\frac{13}{2}\right)\frac{\alpha_s C_A}{12\pi}\ln\left(\frac{p^2}{\mu^2}\right)\right], \tag{7}$$

and thus

$$\widehat{\Delta}^{-1}(p^2) = p^2\left[1 + b\alpha_s\,\ln\left(\frac{p^2}{\mu^2}\right)\right], \tag{8}$$

where $b = 11C_A/12\pi$ is the first coefficient of the Yang–Mills β function, as it should [17].

Similarly, the PT-BFM quark–gluon vertex $\widehat{\Gamma}^a_\mu = \frac{\lambda^a}{2}\widehat{\Gamma}_\mu$, which satisfies the QED-like Ward identity

$$q^\mu \widehat{\Gamma}_\mu(q, p_2, -p_1) = S^{-1}(p_1) - S^{-1}(p_2), \tag{9}$$

is related to the conventional Γ_μ by the BQI

$$[1 + G(q^2)]\Gamma_\mu(q, p_2, -p_1) = \widehat{\Gamma}_\mu(q, p_2, -p_1) + S^{-1}(p_1)Q_\mu(q, p_2, -p_1) + \overline{Q}_\mu(-q, p_1, -p_2)S^{-1}(p_2), \tag{10}$$

where $S^{-1}(p)$ is the inverse of the full quark propagator, with $S^{-1}(p) = A(p^2)\not{p} - B(p^2)$, and the quantities Q_μ and $\overline{Q}_\mu$ are auxiliary three-point functions containing composite vertices. The important point for what follows is that the last two terms on the rhs of Eq. (10) vanish when the external quarks are on shell; otherwise, they cancel against other (process-dependent) contributions.

To see how these considerations apply to the case at hand, let us express $\mathcal{K}$ in terms of the basic field-theoretic quantities that comprise it, namely (suppressing all spinor indices)

$$\mathcal{K}(p, q_+, -q_-) = \alpha_s \Gamma_\mu(p, q_+, -p - q_+)\left(\frac{\lambda^a}{2}\right)\Delta^{\mu\nu}(p)\left(\frac{\lambda^a}{2}\right)\Gamma_\nu(-p, -q_-, p + q_-), \tag{11}$$

where $\alpha_s = g^2/4\pi$, and g is the gauge coupling.

As a consequence of Eqs. (10) and (6), Eq. (11) may be cast into the equivalent form

$$\mathcal{K}(p, q_+, -q_-) = \alpha_s \widehat{\Gamma}_\mu(p, q_+, -p - q_+)\left(\frac{\lambda^a}{2}\right)\widehat{\Delta}^{\mu\nu}(p)\left(\frac{\lambda^a}{2}\right)\widehat{\Gamma}_\nu(-p, -q_-, p + q_-). \tag{12}$$

In what follows we will focus only on the the part of $\widehat{\Gamma}_\mu$ that is proportional to γ_μ, namely $\widehat{\Gamma}_\mu = \gamma_\mu \widehat{\Gamma}_1 + \cdots$, so that the corresponding contribution to the $\mathcal{K}$ of Eq. (12), to be denoted by $\mathcal{C}$, is given by

$$\mathcal{C}(p, q_+, -q_-) = \underbrace{\alpha_s \widehat{\Delta}(p^2)}_{\text{universal}}\underbrace{[\widehat{\Gamma}_1(p, q_+, -p - q_+)(\gamma_\mu)\left(\frac{\lambda^a}{2}\right)P_{\mu\nu}(p)\left(\frac{\lambda^a}{2}\right)(\gamma_\nu)\widehat{\Gamma}_1(-p, -q_-, p + q_-)]}_{\text{process-dependent}}, \tag{13}$$

Due to the special Ward identities satisfied by the PT-BFM Green's functions [e.g., Eq. (9)], the (dimensionful) universal combination [20]

$$\widehat{d}(p^2) = \alpha_s \widehat{\Delta}(p^2) = \frac{\alpha_s \Delta(p^2)}{\left[1 + G(p^2)\right]^2}, \tag{14}$$

introduced in Eq. (13), is also renormalization-group invariant (RGI). Indeed, since $g(\mu^2) = Z_g^{-1}(\mu^2) g_0$ and $\widehat{\Delta}(p^2, \mu^2) = \widehat{Z}_A^{-1}(\mu^2)\widehat{\Delta}_0(p^2)$, where the "0" subscript indicates bare quantities, and the QED-like relation $Z_g = \widehat{Z}_A^{-1/2}$ is valid, we have that

$$\widehat{d}_0(p^2) = \alpha_{s0}\widehat{\Delta}_0(p^2) = \alpha_s(\mu^2)\widehat{\Delta}(p^2, \mu^2) = \widehat{d}(p^2) \tag{15}$$

maintains the same form before and after renormalization, i.e., it forms a RGI (μ-independent) quantity.

4 Nonperturbative Evaluation of $\widehat{d}(p^2)$

In this section we use a combination of lattice results and dynamical equations to determine the nonperturbative form of the fundamental quantity $\widehat{d}(p^2)$ given in Eq. (14). To that end, we will use for $\Delta(p^2)$ the lattice data from [21], whilst for the quantities $1 + G(p^2)$ and $L(p^2)$ we will use the following set of (renormalized) equations

$$\begin{aligned} 1 + G(p^2) &= Z_c + \frac{g^2 C_A}{d-1}\int_k \left[(d-2) + \frac{(k\cdot p)^2}{k^2 p^2}\right] B_1(-k,0,k)\Delta(k)D(k+p), \\ L(p^2) &= \frac{g^2 C_A}{d-1}\int_k \left[1 - \frac{(k\cdot p)^2}{k^2 p^2}\right] B_1(-k,0,k)\Delta(k)D(k+p), \end{aligned} \tag{16}$$

where the renormalization constant Z_c is determined from the MOM condition $F(\mu^2) = 1$. The assumptions and approximations employed in the derivation of the above set of equations have been explained in detail elsewhere [15,20]. In addition, the form factor B_1 that enters in the three equations of (16) has been computed from its own DSE, in the limit of vanishing ghost momentum [22]. Note that even though only $G(p^2)$ enters into the definition of Eq. (14), from Eq. (16) we will determine $G(p^2)$, $L(p^2)$, and $F(p^2)$; this in turn, will allow us to test numerically the validity of Eq. (5), which constitutes a nontrivial check of the entire numerical procedure.

The results of the numerical solution of Eq. (16) are shown in the right panel of Fig. 3. It is clear that the solutions obtained for $1 + G$ and L satisfy the fundamental relation (16) at a high level of accuracy; this provides a non-trivial test for the integration routines used in solving the ghost DSE. Note also that even though $L(p^2)$ vanishes at the origin, it has a non-vanishing support in the region of physical interest (see the top right panel of Fig. 3).

The quantity $\widehat{d}(p^2)$ is RGI, as is evident from the bottom left panel of Fig. 3. There, $\widehat{d}(p^2)$ is shown when evaluated at three different renormalization points: $\mu = 4.3$, 3.0 and 2.5 GeV, for which the corresponding coupling reads $\alpha_s(\mu^2) = 0.22$, 0.30 and 0.36 respectively. As can be seen in the bottom right panel, these values for the strong couplings are in very good agreement with those obtained from detailed calculations of the gauge coupling, $\alpha_{\rm MOM}$, in the momentum subtraction (MOM) scheme, for values of $\Lambda_{\rm QCD}$ between 250 and 320 MeV [23].

5 Comparison Between "Top-Down" and "Bottom-Up" Approaches

In order to make contact with the relevant literature on BSEs, we note that the quantity $\mathcal{C}$ appearing in Eq. (13) has been traditionally cast in the form

$$\mathcal{C}(p, q_+, -q_-) = \frac{\mathcal{I}(p^2)}{p^2} \otimes (\text{process-dependent}), \tag{17}$$

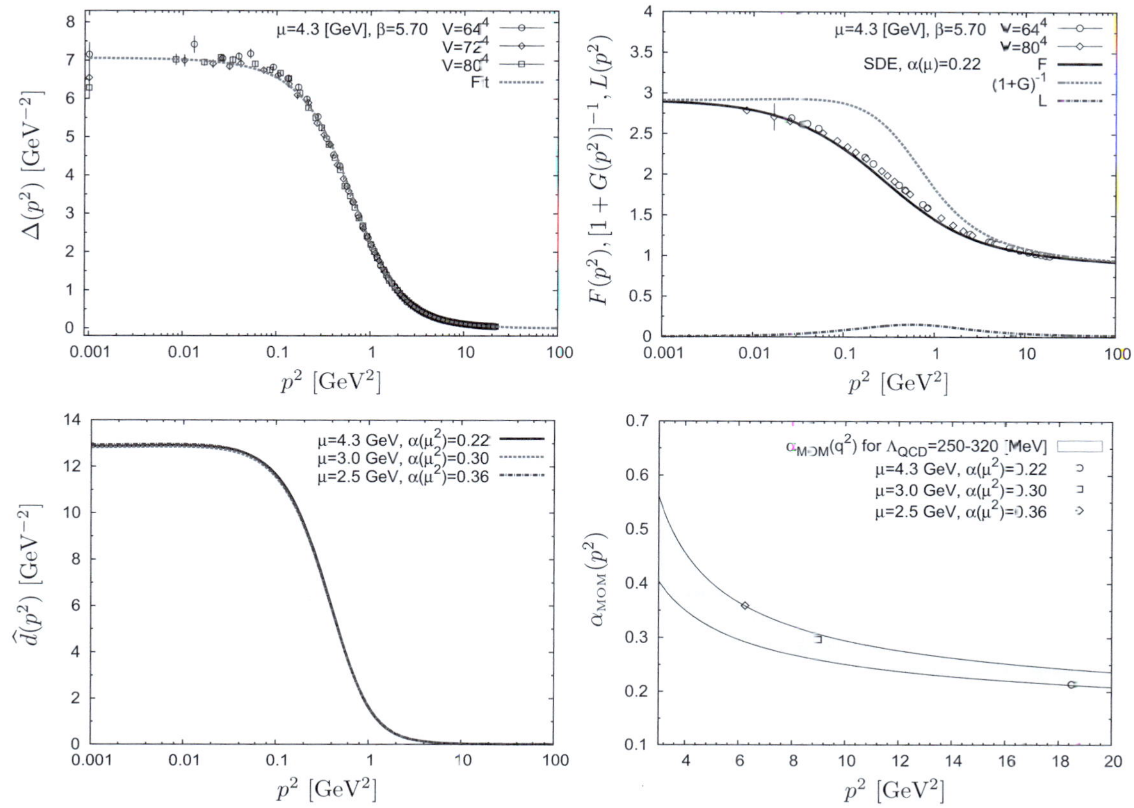

Fig. 3 The Landau gauge gluon propagator (*top left panel*) used for the numerical evaluation of the DSEs in (16) (*top right panel*) for the case $\mu = 4.3\,\text{GeV}$ (lattice data are from of [21]). The RGI quantity $\widehat{d}(p^2)$ evaluated at $\mu = 4.3$, 3.0 and 2.5 GeV (*bottom left panel*). The resulting curves are then rescaled by $\alpha_s(\mu^2) = 0.22$, 0.30 and 0.36, respectively, in compliance with values obtained for α_{MOM} when Λ_{QCD} varies between 250 and 320 MeV (*shaded area* on the *bottom right panel*) [23] (color figure online)

where $\mathcal{I}(p^2)$ is a dimensionless quantity, which may be interpreted as the effective interaction strength of the quark-gluon system. Evidently, in the PT-BFM case one must extract an analogous quantity from the $\widehat{d}(p^2)$ of Eq. (14) through multiplication by p^2, namely

$$\mathcal{I}(p^2) = p^2\widehat{d}(p^2). \tag{18}$$

At this point, the $\mathcal{I}(p^2)$ obtained from Eq. (18) can be compared with the corresponding quantities defined in the bottom-up framework. As reviewed elsewhere (see, e.g., [24]), successful explanations and predictions of numerous hadron observables can be obtained by choosing

$$\mathcal{I}(p^2) = p^2\mathcal{G}(p^2); \quad \mathcal{G}(p^2) = \frac{8\pi^2}{\omega^4}D\mathrm{e}^{-p^2/\omega^2} + \frac{8\pi^2\gamma_m\left(1 - \mathrm{e}^{-p^2/4m_t^2}\right)}{k^2\ln\left[\tau + (1 - p^2/\Lambda_{\text{QCD}}^2)^2\right]}, \tag{19}$$

where $\gamma_m = 12/(33 - 2N_f)$ [typically, $N_f = 4$], $\Lambda_{\text{QCD}} = 0.57\,\text{GeV}$ (in the MOM scheme); $\tau = \mathrm{e}^2 - 1$, $m_t = 0.5\,\text{GeV}$. Note that D and ω are not independent, but must be related by $D\omega = (\varsigma_{\text{G}})^3 = \text{const}$ and $\omega \in [0.4, 0.6]\,\text{GeV}$; then, one can reproduce a large body of observable properties of ground-state vector- and isospin-nonzero pesudoscalar mesons, as well as various properties of the nucleon and Δ resonance [25]. The parameter ς_{G} is fixed by requiring that the correct value of the pion decay constant f_π is reproduced; its precise value depends on the vertex Ansatz employed.

In the case of the rainbow-ladder (RL) truncation [26], corresponding to $\Gamma^\nu \sim \gamma_\nu$, one has $\varsigma_{\text{RL}} = 0.87\,\text{GeV}$. Instead, the improved truncation scheme of [25], which incorporates dynamical chiral symmetry breaking (DB) effects by using a more sophisticated representation for Γ^ν, gives $\varsigma_{\text{DB}} = 0.55\,\text{GeV}$.

The results of the top-down approach of [12] (denoted by DSE), and the two bottom-up approaches (RL and DB) summarized above, are shown in Fig. 4. It is clear that while the RL interaction strength is far larger

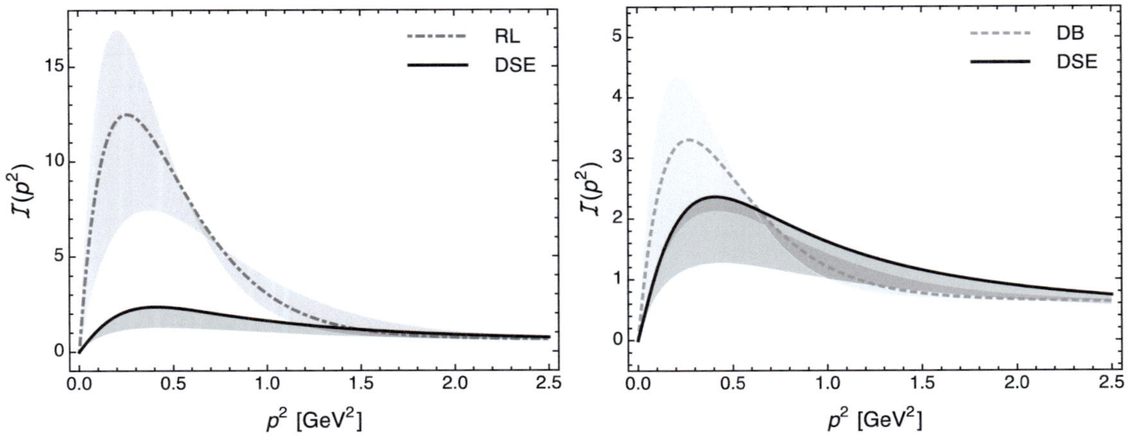

Fig. 4 Comparison of the interaction strength $\mathcal{I}$ evaluated in the RL, IRL and DSE schemes. The *shaded areas* represent phenomenologically acceptable ranges

than that of the DSE, the comparison between the DB and DSE results is very favorable. Evidently, the DB approach captures corrections that are not included in the RL, and is therefore much closer to the DSE result.

6 Conclusions

In this work we have reviewed recent developments towards a first-principle derivation of the BSE kernel from the dynamical equations obtained from the QCD Lagrangian. This particular effort, in turn, bridges to a large extent the gap between nonperturbative continuum QCD studies and bound-state phenomenology.

Acknowledgments This research is supported by the Spanish MEYC under Grants FPA2011-23596, FPA2014-53631-C2-1-P, and SEV-2014-0398, and the Generalitat Valenciana under Grant PrometeoII/2014/066. I would like to thank the organizers for their kind invitation and their warm hospitality during the workshop.

References

1. Jain, P., Munczek, H.J.: Q anti-q bound states in the Bethe-Salpeter formalism. Phys. Rev. D **48**, 5403 (1993)
2. Munczek, H.J.: Dynamical chiral symmetry breaking, Goldstone's theorem and the consistency of the Schwinger-Dyson and Bethe-Salpeter equations. Phys. Rev. D **52**, 4736 (1995)
3. Bender, A., Roberts, C.D., Von Smekal, L.: Goldstone theorem and diquark confinement beyond rainbow ladder approximation. Phys. Lett. B **380**, 7 (1996)
4. Maris, P., Roberts, C.D., Tandy, P.C.: Pion mass and decay constant. Phys. Lett. B **420**, 267 (1998)
5. Bender, A., Detmold, W., Roberts, C.D., Thomas, A.W.: Bethe-Salpeter equation and a nonperturbative quark gluon vertex. Phys. Rev. C **65**, 065203 (2002)
6. Chang, L., Roberts, C.D.: Sketching the Bethe-Salpeter kernel. Phys. Rev. Lett. **103**, 081601 (2009)
7. Fischer, C.S., Williams, R.: Probing the gluon self-interaction in light mesons. Phys. Rev. Lett. **103**, 122001 (2009)
8. Eichmann, G.: Hadron Properties from QCD Bound States. arXiv:0909.0703 [hep-ph]
9. Eichmann, G., Cloet, I.C., Alkofer, R., Krassnigg, A., Roberts, C.D.: Toward unifying the description of meson and baryon properties. Phys. Rev. C **79**, 012202 (2009)
10. Roberts, C.D., Williams, A.G.: Dyson-Schwinger equations and their application to hadronic physics. Prog. Part. Nucl. Phys. **33**, 477 (1994)
11. Maris, P., Roberts, C.D.: Dyson-Schwinger equations: a tool for hadron physics. Int. J. Mod. Phys. E **12**, 297 (2003)
12. Binosi, D., Chang, L., Papavassiliou, J., Roberts, C.D.: Bridging a gap between continuum-QCD and ab initio predictions of hadron observables. Phys. Lett. B **742**, 183 (2015)
13. Ball, J.S., Chiu, T.W.: Analytic properties of the vertex function in gauge theories. 1. Phys. Rev. D **22**, 2550 (1980)
14. Ball, J.S., Chiu, T.W.: Analytic properties of the vertex function in gauge theories. 2. Phys. Rev. D **23**, 3085 (1981)
15. Aguilar, A.C., Binosi, D., Papavassiliou, J.: Indirect determination of the Kugo-Ojima function from lattice data. JHEP **0911**, 066 (2009)
16. Binosi, D., Papavassiliou, J.: Pinch technique: theory and applications. Phys. Rep. **479**, 1 (2009)
17. Abbott, L.F.: The background field method beyond one loop. Nucl. Phys. B **185**, 189 (1981)
18. Binosi, D., Papavassiliou, J.: Pinch technique and the Batalin-Vilkovisky formalism. Phys. Rev. D **66**, 025024 (2002)
19. Aguilar, A.C., Binosi, D., Papavassiliou, J.: Gluon and ghost propagators in the Landau gauge: deriving lattice results from Schwinger-Dyson equations. Phys. Rev. D **78**, 025010 (2008)

20. Aguilar, A.C., Binosi, D., Papavassiliou, J., Rodriguez-Quintero, J.: Non-perturbative comparison of QCD effective charges. Phys. Rev. D **80**, 085018 (2009)
21. Bogolubsky, I.L., Ilgenfritz, E.M., Muller-Preussker, M., Sternbeck, A.: Lattice gluodynamics computation of Landau gauge Green's functions in the deep infrared. Phys. Lett. B **676**, 69 (2009)
22. Aguilar, A.C., Ibanez, D., Papavassiliou, J.: Ghost propagator and ghost-gluon vertex from Schwinger-Dyson equations. Phys. Rev. D **87**(11), 114020 (2013)
23. Boucaud, P., De Soto, F., Leroy, J.P., Le Yaouanc, A., Micheli, J., Pene, O., Rodriguez, J.-Q.: Ghost-gluon running coupling, power corrections and the determination of Lambda (MS-bar). Phys. Rev. D **79**, 014508 (2009)
24. Cloet, I.C., Roberts, C.D.: Explanation and prediction of observables using continuum strong QCD. Prog. Part. Nucl. Phys. **77**, 1 (2014)
25. Qin, S.X., Chang, L., Liu, Y.X., Roberts, C.D., Wilson, D.J.: Interaction model for the gap equation. Phys. Rev. C **84**, 042202 (2011)
26. Maris, P., Tandy, P.C.: Bethe-Salpeter study of vector meson masses and decay constants. Phys. Rev. C **60**, 055214 (1999)

DOI 10.1007/978-3-319-50699-9_10
Reprinted from *Few-Body System* Journal, DOI 10.1007/s00601-016-1091-3

Stanisław D. Głazek · Patryk Kubiczek

Proton Structure in High-Energy High-Multiplicity p–p Collisions

Received: 25 January 2016 / Accepted: 15 March 2016 / Published online: 5 April 2016

Abstract A few-body proton image, expected to be derivable from QCD in the renormalization group procedure for effective particles, is used within the Monte Carlo Glauber model to calculate the anisotropy coefficients in the initial collision-state of matter in high-energy high-multiplicity proton–proton interaction events. We estimate the ridge-like correlations in the final hadronic state by assuming their proportionality to the initial collision-state anisotropy. In our estimates, some distinct few-body proton structures appear capable of accounting for the magnitude of p–p ridge effect, with potentially discernible differences in dependence on multiplicity.

1 Introduction

Protons resist precise theoretical description of their internal dynamics in the Minkowski space–time for a long time by now. The simplest such picture, which is provided by the constituent quark model used to classify hadrons, is not precisely derived from QCD. The theory itself uses the Euclidean-space techniques that do not easily yield any real space–time image. In these circumstances, it is of interest to note that high-energy high-multiplicity proton–proton (*pp*) collisions may shed new light on the issue of proton structure. Namely, the numerous products in such collisions exhibit collective behavior that appears dependent on the initial state of colliding proton matter and the latter depends on the proton structure. Thus, the correlations among products in high-energy high-multiplicity *pp* collisions may report on the proton structure.

In particular, the CMS [1] and ATLAS [2] collaborations reported the collective flow in *pp* collisions that resembles the one observed in heavy-ion collisions [3–5]. Several authors discussed such flows in *pp* collisions [6–18] and we follow the insights of Ref. [6] in order to estimate the extent to which the high-energy high-multiplicity *pp* events are sensitive to the model of proton structure. Following the approach of Ref. [6] and the parameter choice such as in Ref. [15], used here, means making a strong assumption that the parton medium produced in the overlap region of *pp* collision at the LHC has similar hydrodynamical properties as that in heavy ion collisions at RHIC. The individual proton structures we consider are motivated by the general features of the renormalization group procedure for effective particles (RGPEP) in quantum field theory [19,20].

We find that, the effective picture of a quark and diquark with a gluon flux between them produces a different dependence of eccentricity and triangularity on multiplicity than the three-quark picture with a star-like junction

S. D. Głazek (✉)
Institute of Theoretical Physics, Faculty of Physics, University of Warsaw, Pasteura 5, 02-093 Warsaw, Poland
E-mail: stglazek@fuw.edu.pl

P. Kubiczek
Faculty of Physics, Astronomy and Applied Computer Science, Jagiellonian University, Łojasiewicza 11, 30-348 Kraków, Poland
E-mail: patryk.kubiczek@gmail.com

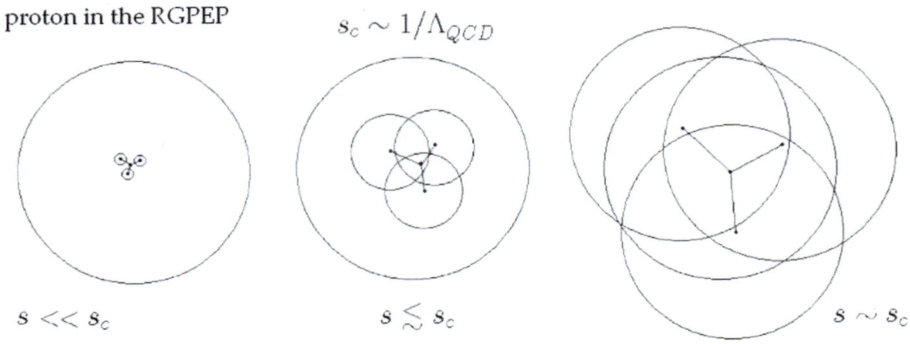

Fig. 1 Proton structure described using effective quarks of size s that is very much smaller than, smaller and comparable with the constituent quark size s_c, with the single large circle indicating the volume available for effective gluons irrespective of their corresponding size, see Ref. [19]

made of gluons does. According to this finding, the recent data for high-energy high-multiplicity events suggest a significant star-like gluon junction component in the proton structure. Our analysis also indicates a need for assessing the adequacy of the linear relationship used by us between asymmetries in the initial collision state, such as the eccentricity or triangularity, and the final state correlations in high-multiplicity events, such as the elliptic flow.

2 Proton Structure in *pp* Collisions

As mentioned in Sect. 1, the ridge-effect in *pp* scattering can be described using the hydrodynamic evolution of the asymmetric state of matter that results from one proton's quark and gluon distribution suddenly colliding with another's. The asymmetric state is meant to evolve according to laws of hydrodynamics until it eventually turns into the detected particles that emerge through hadronization in the final state, in which they exhibit the angular correlation over a long-range in rapidity, called the ridge. The final state ridge-like correlations, such as the elliptic flow, are thus related to the initial stage of *pp* collision whose nature depends on the proton structure. One of the key issues is thus how to describe the proton structure using QCD in the Minkowski space–time.

Conceptually, we approach this issue using the RGPEP [19,20], which is a candidate for providing the mathematical tools for describing protons as bound states of effective quarks and gluons of specific size s. The size parameter s plays the role of an arbitrary renormalization-group scale that can be adjusted to the physical process one wants to accurately describe in simplest possible terms. This condition means choosing the right variables for grasping the essence of physics most economically from the computational point of view. The scale dependence of the proton structure expected in the RGPEP is illustrated in Fig. 1 and the corresponding examples of the color structure are shown in Fig. 2. The expectation is based on the scale-dependent features of effective Hamiltonians, which imply the possibility that a relativistic bound-state eigenvalue problem can be equivalently written in terms of a few-body problem for sizable effective quarks and gluons instead of an infinite combination of bare point-like quarks and gluons of canonical QCD. Hence, the RGPEP provides the scheme in which the distribution of matter in proton can be imagined in terms of wave functions, or probability distributions for the effective quarks and gluons of size s.

The few-body picture of protons in QCD suggested by the RGPEP allows us to preliminarily model the proton structure in terms of shapes illustrated by two typical examples in Fig. 3 [21], knowing that such models can in future be verified in theory. We ask if the ridge effect can phenomenologically distinguish between the effective configurations.

We consider three types of configurations. The proton quark–diquark configuration, denoted by **I** and shown on the left-hand side of Fig. 3, is motivated by Refs. [17,18]. It is a superposition of a few Gaussians that represent a quark, a diquark and gluons forming a tube in between. The three-quark configuration, denoted by **Y** and shown on the right-hand side of Fig. 3, is motivated by Ref. [19]. It is a superposition of Gaussians that represent three quarks and additional gluons forming the Y-shaped junction. The shape of **Y** configuration is kept fixed. In addition, we consider the Gaussian fluctuating three-quark configuration, denoted by **G–f**, which

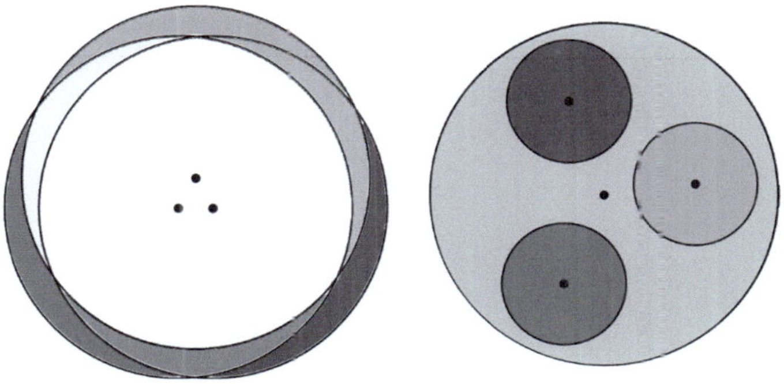

Fig. 2 *Color* structure of effective quarks for two values of the RGPEP scale parameter $s \sim s_c$ and $s < s_c$ in Fig. 1. Pions are meant to couple to nucleons in the constituent quark picture only at the nucleon boundaries, where *color* is not neutralized, and for smaller values of s the quarks form more localized objects with the gray area indicating the volume available for effective gluons of a similar size to the quarks (drawing from Ref. [22]) (color figure online)

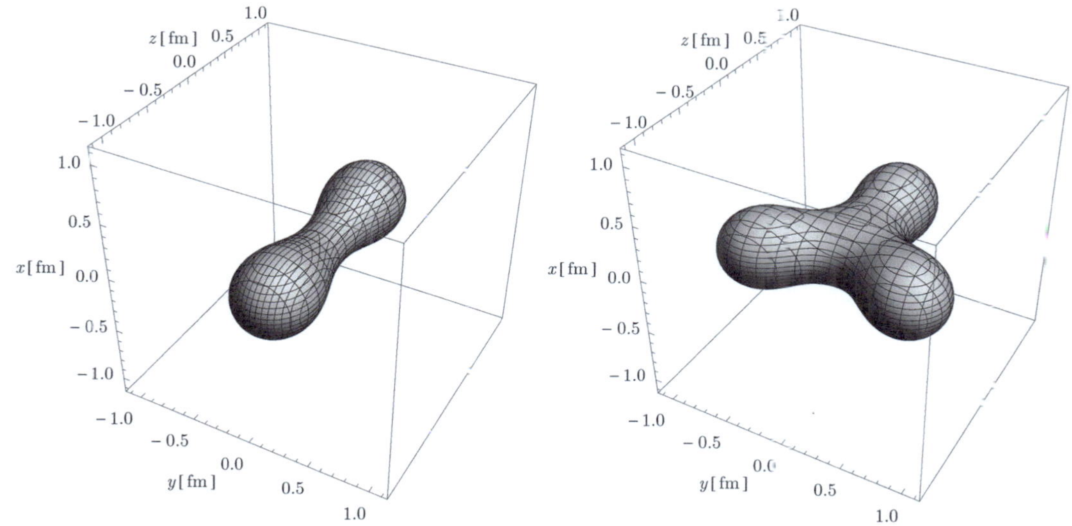

Fig. 3 Effective constituent configurations of typical size s in proton: on the *left* is the quark–diquark configuration labeled in the text as **I** and on the *right* is the three-quark configuration with a star-junction built from gluons labeled in the text by **Y**

is the same as the **Y** configuration but with the shape parameters generated according to Gaussian probability distributions. Details of all configurations we consider are available in Refs. [21,22].

3 Asymmetries in the Initial Stage of a *pp* Collision and the Final State Correlations

Following Ref. [6], we adapt a simple Glauber model, widely used for modelling high energy nuclear collisions [23], to describe the density of binary partonic collisions in scattering of two systems A and B,

$$n_{\text{coll}}(x, y; b, \mathbf{\Sigma}_A, \mathbf{\Sigma}_B) = \sigma_{gg} \int_{-\infty}^{\infty} dz\, \rho\left(x - \frac{b}{2}, y, z; \mathbf{\Sigma}_A\right) \int_{-\infty}^{\infty} dz'\, \rho\left(x + \frac{b}{2}, y, z'; \mathbf{\Sigma}_B\right), \tag{1}$$

which is a function of the coordinates x and y in the plane transverse to the colliding beams, the impact parameter b and the varying parameters $\mathbf{\Sigma}$ that identify the proton structure and its orientation in space. The coefficient σ_{gg} denotes a parton–parton scattering cross-section, in our estimates on the order of 4 mb, and ρ denotes the three-dimensional parton distribution described in Sect. 2.

Eccentricity ϵ_2 and triangularity ϵ_3 in the initial stage of pp collision are calculated using the formula [24]

$$\epsilon_n = \frac{\sqrt{\{s^n \cos(n\phi)\}^2 + \{s^n \sin(n\phi)\}^2}}{\{s^n\}}, \tag{2}$$

in which the curly brackets denote the expectation value

$$\{f(x,y)\} = \frac{\int dx\,dy\, f(x,y)\, n_{\rm coll}(x,y;b,\mathbf{\Sigma}_A,\mathbf{\Sigma}_B)}{\int dx\,dy\, n_{\rm coll}(x,y;b,\mathbf{\Sigma}_A,\mathbf{\Sigma}_B)}, \tag{3}$$

and coordinates are parameterized as $x = s\cos\phi$, $y = s\sin\phi$. The number of collisions in an event is

$$N_{\rm coll}(b,\mathbf{\Sigma}_A,\mathbf{\Sigma}_B) = \int dx\,dy\, n_{\rm coll}(x,y;b,\mathbf{\Sigma}_A,\mathbf{\Sigma}_B) \tag{4}$$

and the cross-section density in the impact parameter plane is

$$\sigma(b,\mathbf{\Sigma}_A,\mathbf{\Sigma}_B) = 1 - \left[1 - \frac{N_{\rm coll}(b,\mathbf{\Sigma}_A,\mathbf{\Sigma}_B)}{N_g^2}\right]^{N_g^2}. \tag{5}$$

The total pp cross-section is thus

$$\sigma_{pp} = \int_0^\infty 2\pi b\, db \int P(\mathbf{\Sigma}_A)\, d\mathbf{\Sigma}_A \int P(\mathbf{\Sigma}_B)\, d\mathbf{\Sigma}_B\, \sigma(b,\mathbf{\Sigma}_A,\mathbf{\Sigma}_B), \tag{6}$$

where $P(\mathbf{\Sigma})$ is the probability density for proton configuration $\mathbf{\Sigma}$. For any quantity Q, its expectation value in many collisions is

$$\langle Q\rangle = \frac{1}{\sigma_{pp}} \int_0^\infty 2\pi b\, db \int P(\mathbf{\Sigma}_A)\, d\mathbf{\Sigma}_A \int P(\mathbf{\Sigma}_B)\, d\mathbf{\Sigma}_B\, \sigma(b,\mathbf{\Sigma}_A,\mathbf{\Sigma}_B)\, Q(b,\mathbf{\Sigma}_A,\mathbf{\Sigma}_B)\,. \tag{7}$$

We used randomly oriented proton configurations in the Monte Carlo generation of about $3\cdot 10^5$ events for each proton model and estimated the averaged eccentricity ϵ_2 and triangularity ϵ_3 in the resulting samples. In our estimates, $\sigma_{gg} \sim 4.3\,$mb [15] and $\sigma_{pp} \sim 60\,$mb [25] required the number of scattering partons $N_g = 9 \pm 2$ to obtain agreement with data, assuming that the multiplicity $N = \alpha N_{\rm coll}$ and reproducing the value $\langle N\rangle = 30$ [26] for charged particles by choosing $\alpha = 8 \pm 3$. Our minimum bias results for eccentricity and triangularity are shown in Fig. 4

It is visible in Fig. 4 that the initial stage of pp collision is characterized by different multiplicity dependence of the asymmetries for different proton structures. In collisions of quark–diquark (**II**) and Gaussian-fluctuating (**G–f**) structures, the asymmetries decrease with multiplicity above about $N = 100$, while in the collisions of tripod three-quark configurations (**YY**) the initial stage asymmetries persist or even increase above $N = 100$.

In order to relate the eccentricity and triangularity to data, we note that the observable multiplicity distributions in transverse momentum p_T and pseudorapidity η are conventionally written as

$$\frac{d^3N}{d^2p_T d\eta} = \left\{1 + 2\sum_{n=1}^{\infty} v_n(p_T,\eta)\, \cos\left[n(\phi - \Phi_{RP})\right]\right\} \frac{d^2N}{2\pi p_T dp_T d\eta}, \tag{8}$$

where the reaction plane angle Φ_{RP} for colliding spherically symmetric distributions is illustrated in Fig. 5. In the actual events the angles Φ_{RP} are determined from the particle distributions.

Assuming that the averaged minimal bias elliptic flow parameter v_2 is proportional to the minimal bias eccentricity parameter ϵ_2 with coefficient order 0.3 [15], we obtain $v_2 \sim 0.11$, 0.14 and 0.09 for the proton models **II**, **YY** and **G–f**, respectively, while data indicates v_2 in the range 0.04–0.08 [5].

More generally, taking into account that the initial-stage asymmetry parameters ϵ_n in Eq. (2) and the final-state coefficients v_n in Eq. (8) are small, and assuming that a set of averaged coefficients v_n with different n depends approximately linearly on the set of averaged parameters ϵ_n, one can infer the dependence of v_n on

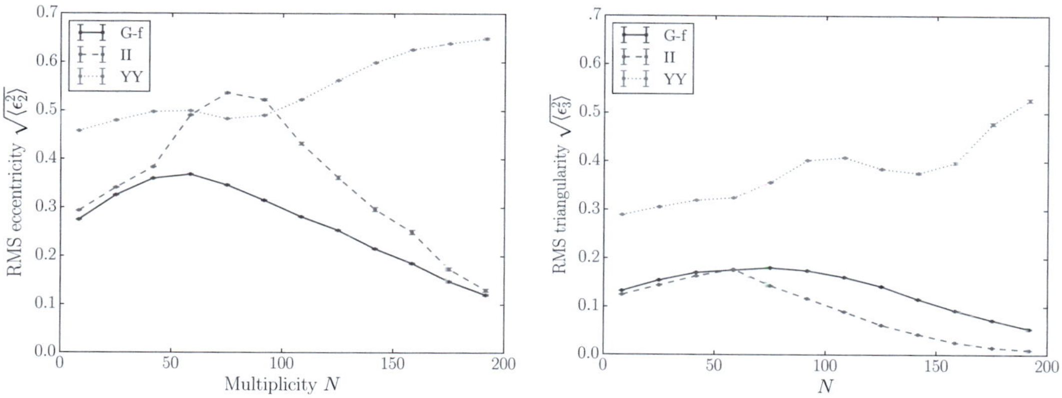

Fig. 4 Average eccentricity and triangularity obtained using Eq. (7) for three different types of proton structure in *pp* collisions. Green curves labeled **II** correspond to collisions of protons in configuration **I** in Fig. 3, and the red curves labeled **YY** correspond to collisions in configurations **Y** in Fig. 3. The *blue lines* labeled **G–f** correspond to the Gaussian fluctuating three-quark configuration in which the **Y**-type proton configurations appear with different shape parameters according to a Gaussian probability distribution [21]. Note the distinct multiplicity dependence in the case of **YY** configurations (color figure online)

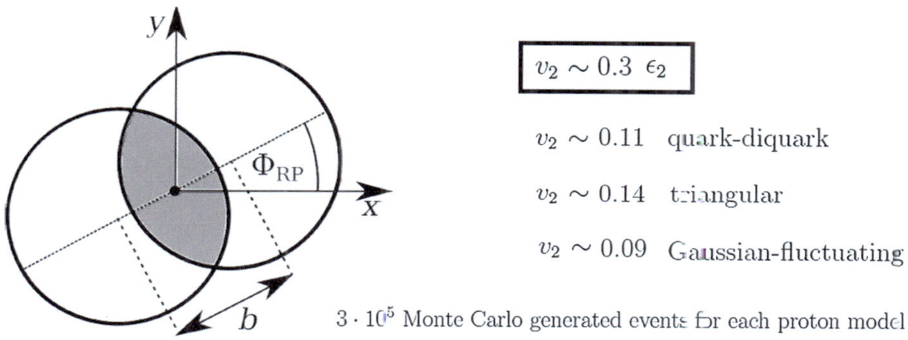

Fig. 5 View of the initial stage of *pp* collision along the beam for illustration of Eq. (8). The average value of the elliptic flow coefficient $v_2(p_T, \eta)$, denoted by v_2, is assumed to be proportional to the eccentricity $\sqrt{\langle\epsilon_2^2\rangle}$ obtained in Eq. (7) with a coefficient on the order of 0.3 [15]. The magnitudes of v_2 resulting from different proton models are displayed with quark–diquark = **II**, triangular = **YY**

multiplicity N from the dependence of ϵ_n on N. Accordingly, Fig. 4 suggests that the averaged elliptic flow v_2 and higher correlations in high-energy high-multiplicity *pp* events may indicate which configurations of effective quarks and gluons in the proton structure are most likely to occur. Namely, only the **YY** configurations lead to ϵ_2 and ϵ_3 that do not fall off for N exceeding about 100. Actually, recent data [2] from ATLAS Collaboration for *pp* collisions with $\sqrt{s} \sim 13$ TeV show that v_2 is stable for large N. According to our analysis, this finding favors the configuration **YY**.

It should be observed that the long-range, near-side angular correlations in *pp* collisions at LHC energies can also be studied in terms of multiparton interactions [27]. Interactions of four partons to four partons and beyond allow for linking the ridge effect to the models [28] and theory [29] of double parton distributions and their light-front analysis [30]. A unified approach is hence greatly desired.

4 Conclusion

Simple model estimates suggest that the correlations among final-state particles in high-energy high-multiplicity *pp* collisions are sensitive to the proton structure. It is not excluded that future sophisticated calculations will identify some spatial features of protons through precise interpretation of experimental data on these correlations. The multidimensional linear relationship between initial-state asymmetries and final-state flow parameters that we assume in our estimates requires verification, e.g., in the hydrodynamic model. If confirmed, it would provide an efficient way for studying the proton structure through correlations in high-multiplicity pp collisions using the universal matrices determined by the nature of assumed underlying collision dynamics. Interpreting the most recent LHC data on multiplicity dependence of the elliptic flow coefficient v_2 using our simple estimates, suggests that protons may often occur in the configuration of three effective quarks connected by a Y-shaped gluon junction.

References

1. CMS Collaboration: Observation of long-range, near-side angular correlations in proton-proton collisions at the LHC. J. High Energy Phys. **1009**, 091 (2010)
2. ATLAS Collaboration: Observation of long-range elliptic anisotropies in $\sqrt{s} = 13$ and 2.76 TeV pp collisions with the ATLAS detector (2015). arXiv:1509.04776 [hep-ex]
3. Białkowska, H.: The ridge effect from pp to Pb-Pb (and back). Acta Phys. Pol. B **43**, 705 (2012)
4. Li, W.: Observation of a Ridge correlation structure in high multiplicity proton-proton collisions: a brief review. Mod. Phys. Lett. A **27**, 1230018 (2012)
5. Bożek, P.: Elliptic flow in proton–proton collisions at sqrt(s) = 7 TeV. Eur. Phys. J. C **71**, 1530 (2011)
6. d'Enterria, D., et al.: Estimates of hadron azimuthal anisotropy from multiparton interactions in proton-proton collisions at sqrt(s) = 14 TeV. Eur. Phys. J. C **66**, 173 (2010)
7. Casalderrey-Solana, J., Wiedemann, U.A.: Eccentricity fluctuations make flow measurable in high multiplicity p–p collisions. Phys. Rev. Lett. **104**, 102301 (2010)
8. Prasad, S.K., et al.: Elliptic flow (v_2) in pp collisions at energies available at the CERN large Hadron Collider: hydrodynamical approach. Phys. Rev. C **82**, 024909 (2010)
9. Bożek, P.: Observation of the collective flow in proton-proton collisions. Acta Phys. Pol. B **41**, 837 (2010)
10. Ortona, G., et al.: Elliptic flow in high multiplicity proton-proton collisions at sqrt(s) = 14 TeV as a signature of deconfinement and quantum energy density fluctuations (2009). arXiv:0911.5158 [hep-ph]
11. Pierog, T., et al.: Collective flow in (anti)proton-proton collision at Tevatron and LHC (2010). arXiv:1005.4526 [hep-ph]
12. Bautista, I., et al.: Particle production azimuthal asymmetries in a clustering of color sources model. J. Phys. G **37**, 015103 (2010)
13. Avsar, E., et al.: Eccentricity and elliptic flow in proton-proton collisions from parton evolution. Phys. Lett. B **702**, 394 (2011)
14. Deng, W.T., et al.: Elliptic and triangular flow and their correlation in ultrarelativistic high multiplicity proton proton collisions at 14 TeV. Phys. Lett. B **711**, 301 (2012)
15. Drescher, H.J., et al.: The centrality dependence of elliptic flow, the hydrodynamic limit, and the viscosity of hot QCD. Phys. Rev. C **76**, 024905 (2007)
16. Alver, B., Roland, G.: Collision geometry fluctuations and triangular flow in heavy-ion collisions. Phys. Rev. C **81**, 054905 (2010)
17. Bjorken, J.D., et al.: Possible multiparticle ridge-like correlations in very high multiplicity proton–proton collisions. Phys. Lett. B **726**, 344 (2013)
18. Bjorken, J.D.: Double diffraction at zero impact parameter. Int. J. Mod. Phys. A **29**, 1446006 (2014)
19. Głazek, S.D.: Hypothesis of quark binding by condensation of Gluons in Hadrons. Few-Body Syst. **52**, 367 (2012)
20. Gómez-Rocha, M., Głazek, S.D.: Asymptotic freedom in the front-form Hamiltonian for quantum chromodynamics of gluons. Phys. Rev. D **92**, 065005 (2015)
21. Kubiczek, P., Głazek, S.D.: Manifestation of proton structure in the initial-state anisotropies in high-energy proton-proton collisions. Lith. J. Phys. **55**, 155 (2015)
22. Kubiczek, P.: Geometrical model of azimuthal correlations in high-multiplicity proton-proton collisions. Bachelor thesis, University of Warsaw, September (2014)
23. Miller, M.L., et al.: Glauber modelling in high energy nuclear collisions. Ann. Rev. Nucl. Part Sci. **57**, 205 (2007)
24. Blaizot, J.P., et al.: Correlations in the Monte Carlo Glauber model. Phys. Rev. C **90**, 034906 (2014)
25. CMS Collaboration: Measurement of the inelastic protonproton cross section at sqrt(s) = 7 TeV. Phys. Lett. B **722**, 5 (2013)
26. CMS Collaboration: Charged particle multiplicities in pp interactions at sqrt(s) = 0.9, 2.36 and 7 TeV. J. High Energy Phys. **1101**, 079 (2011)
27. Alderweireldt, S., Van Mechelen, P.: Obtaining the CMS ridge effect with multiparton interactions (2012). arXiv:1203.2048 [hep-ph], http://inspirehep.net/record/1093441
28. Broniowski, W., Ariola, E.R.: Valon model for double parton distributions. Contribution to LC2015 (2015)
29. Gaunt, J.: Theory of double parton scattering. Contribution to LC2015 (2015)
30. Rinaldi, M.: Double parton scattering and 3D proton structure: a light-front analysis. contribution to LC2015 (2015)

DOI 10.1007/978-3-319-50699-9_11
Reprinted from *Few-Body System* Journal, DOI 10.1007/s00601-016-1058-4

Matteo Rinaldi · Sergio Scopetta · Marco Traini · Vicente Vento

Double Parton Scattering and 3D Proton Structure: A Light-Front Analysis

Received: 15 January 2016 / Accepted: 11 February 2016 / Published online: 29 February 2016

Abstract Double parton scattering, occurring in high energy hadron–hadron collisions, e.g. at the LHC, is usually investigated through model dependent analyses of the so called effective cross section σ_{eff}. We present a dynamic approach to this fundamental quantity making use of a Light-Front model treatment. Within such a framework σ_{eff} is initially evaluated at low energy scale using the model and then, through QCD evolution, at high energy scale to reach the experimental conditions. Our numerical outcomes are consistent with the present experimental analyses of data in kinematical region we investigate. An important result of the present work is the x_i dependence of σ_{eff}, a feature directly connected to double parton correlations and which could unveil new information on the three dimensional structure of the proton.

1 Introduction

In hadron–hadron collisions, possible signals of Multi Parton Interactions (MPI), due to the interactions of at least two partons of each hadron, have been investigated for a long time [1]. Recently, thanks to the possibilities offered by the Large Hadron Collider (LHC), RUN 2, a lot of attention on this subject has been paid [2–4]. In the present talk, we focus our scrutiny on the simplest MPI process, double parton scattering (DPS). The cross section for this process, assuming factorization, depends on non-perturbative quantities, the so called double parton distribution functions (dPDFs), which describe the joint probability of finding two partons at a given transverse distance with given longitudinal momentum fractions. dPDFs are therefore related to the three-dimensional (3D) nucleon structure [5] and are non-perturbative quantities. Tractable model calculations

M. Rinaldi (✉) · S. Scopetta
Dipartimento di Fisica e Geologia, Università degli Studi di Perugia, Via A. Pascoli, 06123 Perugia, Italy
E-mail: matteo.rinaldi@pg.infn.it

M. Rinaldi · S. Scopetta
INFN Section of Perugia, Perugia, Italy

M. Traini
Dipartimento di Fisica, Università degli Studi di Trento, Via Sommarive 14, 38123 Povo Trento, Italy

M. Traini
INFN - TIFPA, Trento, Italy

V. Vento
Departament de Fisica Teòrica, Universitat de València, 46100 Burjassot, València, Spain

V. Vento
Institut de Fisica Corpuscular, Consejo Superior de Investigaciones Científicas, 46100 Burjassot, València, Spain

of the dPDFs have been proposed in the last few years (e.g. Refs. [6–8]). In particular, in Refs. [8,9], dPDFs have been evaluated by means of a Light-Front (LF) approach, which allows one to achieve a Poincaré covariant treatment and then to reproduce the essential dPDFs sum rules. Let us stress that even if there are no clear data from which dPDFs can be extracted, a signature of DPS has been observed and measured in several experiments [10–15]. In particular the DPS cross section is estimated through the so called "effective cross section", σ_{eff} and evaluated in model dependent ways. Despite large error bars σ_{eff} is found to be constant w.r.t. the center-of-mass energy of the collision and the longitudinal momentum fractions carried by the interacting partons. In order to calculate σ_{eff} in our dynamical model, a useful expression has been derived in Ref. [16]. Thanks to that study, σ_{eff} can be directly evaluated from the calculations of dPDFs. In particular, in [16] and here, use has been made of the results discussed in Ref. [8], in which dPDFs have been estimated by means of a LF model. In the next sections the main results of the calculations of σ_{eff} will be properly shown and described.

2 The Effective Cross Section

In this section, starting from the phenomenological definition of σ_{eff}, a direct relation between this quantity and standard parton distribution functions (PDFs) and dPDFs will be shown. σ_{eff}, is usually defined through a "pocket formula" when a DPS process produces two final states A and B [3]:

$$\sigma_{eff} = \frac{m}{2} \frac{\sigma_A^{pp'} \sigma_B^{pp'}}{\sigma_{double}^{pp}}. \tag{1}$$

Here, $m = 1$ if A and B are identical and $m = 2$ if they are different. In the numerator, $\sigma_{A(B)}^{pp'}$ is the differential cross section for the inclusive process $pp' \to A(B) + X$ and reads as follows:

$$\sigma_{A(B)}^{pp'}(x_{1(2)}, x'_{1(2)}, \mu_{1(2)}) = \sum_{i,k} F_i^p(x_{1(2)}, \mu_{1(2)}) F_k^{p'}(x'_{1(2)}, \mu_{1(2)})\, \hat{\sigma}_{ik}^{A(B)} \left(x_{1(2)}, x'_{1(2)}, \mu_{1(2)}\right), \tag{2}$$

where here $F_{i(k)}^p$ is the standard single PDF, the indices read $i, k = \{q, \bar{q}, g\}$ and $\mu_{1(2)}$ is the factorization scale for the process $A(B)$. On the other hand side, the denominator of Eq. (1) is the differential double parton scattering cross section:

$$\sigma_{double}^{pp}(x_1, x'_1, x_2, x'_2, \mu_1, \mu_2) = \frac{m}{2} \sum_{i,j,k,l} \int D_{ij}(x_1, x_2; \mathbf{k}_\perp, \mu_1, \mu_2)\, \hat{\sigma}_{ik}^A(x_1, x'_1, \mu_1)$$
$$\times D_{kl}(x'_1, x'_2; -\mathbf{k}_\perp, \mu_1, \mu_2) \hat{\sigma}_{jl}^B(x_2, x'_2, \mu_2) \frac{d\mathbf{k}_\perp}{(2\pi)^2}. \tag{3}$$

In Eq. (3), $D_{ij}(x_1, x_2; \mathbf{k}_\perp)$, often called "double generalized parton distributions" (${}_2GPDs$ [17,18]), depends on $\mathbf{k}_\perp$, the transverse momentum imbalance of the partons 1 and 2. ${}_2GPDs$ are the Fourier transform of the dPDFs $D_{ij}(x_1, x_2; \mathbf{r}_\perp)$, which depend on $\mathbf{r}_\perp$, the transverse separation, in coordinate space, between two partons. $D_{ij}(x_1, x_2; \mathbf{r}_\perp)$ represents the probability of finding parton pairs i, j with longitudinal momentum fractions x_1, x_2 and transverse distance $\mathbf{r}_\perp$. This is the non-perturbative input describing soft-physics. Let us point out the physical content of σ_{eff}. In Eq. (1), if the occurrence of the B were not biased by that of the A, instead of the ratio σ_B/σ_{eff} one would read σ_B/σ_{inel}, representing the probability to have B once A has taken place assuming rare hard multiple collisions. The difference between σ_{eff} and σ_{inel} measures therefore correlations between the interacting partons in the colliding proton.

Now a suitable expression of σ_{eff} for microscopic calculations, in particular highlighting the x_i-dependence of σ_{eff}, will be derived. First of all, let us assume that heavy quark flavors do not contribute in the process and the that elementary cross section, appearing in Eqs. (2) and (3), is basically (see Ref. [16] for details) $\hat{\sigma}_{ij}(x, x') = C_{ij}\bar{\sigma}(x, x')$, where $\bar{\sigma}(x, x')$ is a universal function, and C_{ij} are color factors, with $i, j = q, \bar{q}, g$, which stay in the ratio $C_{gg}{:}C_{qg}{:}C_{qq} = 1{:}(4/9){:}(4/9)^2$.

At this point, inserting Eqs. (2)–(3) in Eq. (1) together with these assumptions, one finds the following formal expression of σ_{eff}:

$$\sigma_{eff}(x_1, x'_1, x_2, x'_2) = \frac{\sum_{i,k,j,l} F_i(x_1) F_k(x'_1) F_j(x_2) F_l(x'_2) C_{ik} C_{jl}}{\sum_{i,j,k,l} C_{ik} C_{jl} \int D_{ij}(x_1, x_2; \mathbf{k}_\perp) D_{kl}(x'_1, x'_2; -\mathbf{k}_\perp) \frac{d\mathbf{k}_\perp}{(2\pi)^2}}. \tag{4}$$

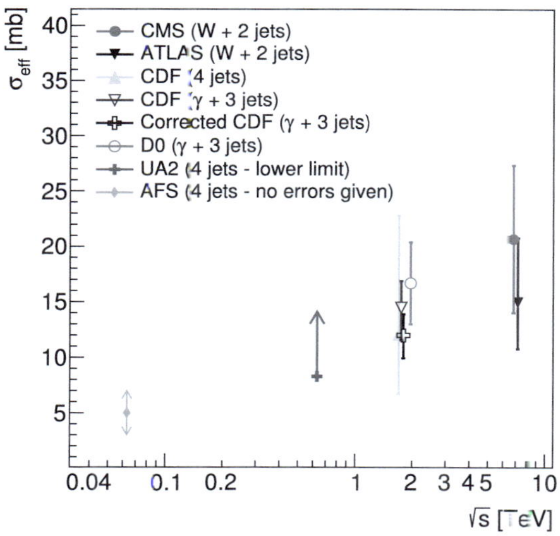

Fig. 1 Centre-of-mass energy dependence of σ_{eff} measured by different experiments using different processes [10–15]. The figure is taken from [15]

As one can see, in the above equation, assuming a complete factorized ansatz for dPDFs in terms of standard PDFs, e.g. $D_{ij}(x_1, x_2; \mathbf{k}_\perp) \propto F_i(x_1)F_j(x_2)T(\mathbf{k}_\perp)$, one obtains a constant value of σ_{eff} w.r.t. to x_i. This scenario is realized neglecting double parton correlations. This is the kind of condition often guessed for the extraction of σ_{eff} from data since there is no experimental information on dPDFs. The problems related to the uncorrelated ansatz are discussed in a number of papers (see, e.g. Ref. [2,3,19]). In particular, in the valence region we are interested here, this assumption is not supported by model calculations [6–8], due to the presence of double parton correlations. Before describing our results, the present experimental scenario is shown in Fig. 1, where σ_{eff}, measured by different experiments [10–15], which analyzed the latter quantity with different final states at different values of the center-of-mass energy, $\sqrt{s}$, has been plotted. As one can see, a constant value of σ_{eff} could be consistent with the data within the large experimental errorbars, which could hide the fundamental information on the possible x_i-dependence. Let us stress that such important dependence could open new ways to access the 3D nucleon structure [5].

Nowadays, fundamental information on the 3D structure of the nucleon, related to the transverse position of partons, is usually investigated thorough the study and the measurements of hard-exclusive processes such as deeply virtual Compton scattering (DVCS), which allows to access the Generalized Parton Distributions (GPDs) (see Ref. [20] for recent developments). Further information could be obtained via the investigation of dPDFs in its full x_i dependence which can be extracted from σ_{eff} in DPS. Let us remark that the information encoded in dPDFs are anyhow complementary to those provided by GPDs in impact parameter space. In fact, the latter quantities are one-body densities describing the probability of finding a parton with longitudinal momentum fraction x at a given transverse distance from the center of the target, while, dPDFs are two-body distributions related to the *relative* distance between two partons with given x_1 and x_2. In other words, the measurements of DPS could access information on the average transverse distance of two fast or slow partons: a very interesting dynamical feature, not included in GPDs.

3 Calculation of the Effective Cross Section Within a LF Approach

dPDFs can not be easily evaluated within QCD while they can be estimated at a low energy scale, $Q_0 \sim \Lambda_{QCD}$, using models, as extensively done for the PDFs (e.g. Ref. [21]). The results of these analyses should then be evolved using perturbative QCD (pQCD) in order to match data taken at an energy scale $Q > Q_0$. dPDF evolution is discussed e.g. in Refs. [2,3,22,23,28]. The analysis we are presenting makes use of the Poincaré covariant Light-Front model approach, allowing for a correct starting point for a precise pQCD evolution. Thanks to this feature, our model calculations can be relevant for the analysis of high-energy data.

The model, chosen in order to grasp the most relevant features of dPDFs, is the one presented and described in Ref. [24], being already used for the studies of other different distributions (see, e.g. Refs. [25,26] and references therein). The explicit calculation of $D_{ij}(x_1, x_2; \mathbf{k}_\perp, \mu)$ has been described in details in Refs [8,9].

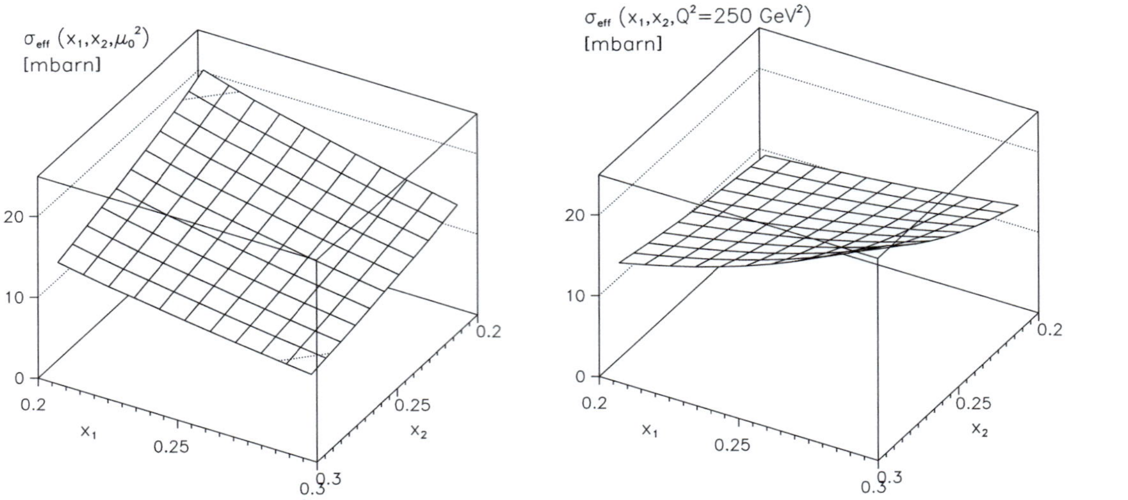

Fig. 2 $\sigma_{eff}(x_1, x_2, Q^2)$ for the values of x_1, x_2 measured in Ref. [10]. *Left panel* hadronic scale; *right panel* $Q^2 = 250\,\text{GeV}^2$

In particular, let us point out that strong correlation effects are present at the scale of the model and still sizable, in the valence region, at the experimental scale, i.e. after pQCD evolution. At low values of x, presently studied at the LHC, correlations become less relevant, although their effects are still important for the spin-dependent contributions to unpolarized proton scattering. In order to evaluate σ_{eff}, the single and the double PDFs, appearing in Eq. (4) have been calculated imposing a unique factorization scale $\mu_1 = \mu_2 = \mu_0$ (as in, e.g. Refs. [2,4,27]), where μ_0 is the hadronic scale, where only valence quarks are present. Here and in Ref. [16] we focus our analysis on the kinematics of the old AFS data [10], i.e. $x_1 \simeq x_1'$, $x_2 \simeq x_2'$ and $0.2 \le x_{1,2} \le 0.3$ and the average energy scale turns out to be $Q^2 \simeq 250\,\text{GeV}^2$. The results of the calculations are shown in Fig. 2, at the scale of the model, $\mu_0^2 \simeq 0.1\,\text{GeV}^2$, and after non-singlet evolution to Q^2 (details on the fixing of the hadronic scale and on the calculation of the QCD evolution can be found in Ref. [8]). As one can see, a strong $x_{1,2}$ dependence has been found in this narrow kinematical range at both the experimental and the model scales. One should notice that the three old experimental extractions of σ_{eff} [10–12], which include the valence region (cf. Fig. 1), lie in the obtained range of values of σ_{eff} shown in Fig. 2. Let us remark two important issues on the pQCD evolution of dPDFs. In our approach, there is only evolution on $x_{1,2}$, being the one on $k_\perp$ still an open challenge and, furthermore, since we are interested, for the moment being, on the valence region, only the homogenous part of the evolution equations has been taken into account in our calculation. Moreover, in Ref. [16], the average value of σ_{eff}, $\bar{\sigma}_{eff} \sim 10.9\,\text{mb}$, is in good agreement with the experimental data reported in Fig. 1.

4 Conclusions

In the present work, a suitable expression of σ_{eff} for microscopic calculations, in terms of dPDFs and standard PDFs, has been derived. In order to perform the calculation of σ_{eff}, use has been made of a relativistic Poincaré covariant quark model. In particular, our investigation predicts an x_i dependence of σ_{eff}, whose values are consistent with data including the valence region. Moreover, such dependence, found when a non singlet evolution of the valence distributions is performed, as well as when perturbatively generated gluons are included into the scheme, is a feature not easily read in the available data. The measurement of σ_{eff} in restricted x_i ranges would lead, therefore, to a first signature of double parton correlations in the proton, a novel and interesting aspect of the 3D structure of the nucleon.

Acknowledgments This work was supported partially through the McCartor Fund Fellowship program under the auspices of International Light-Cone Advisory Committee (ILCAC).

References

1. Paver, N., Treleani, D.: Multi-quark scattering and large P(t) jet production in hadronic collisions. Nuovo Cim. A **70**, 215 (1982)
2. Gaunt, J.R., Stirling, W.J.: Double parton distributions incorporating perturbative QCD evolution and momentum and quark number sum rules. J. High Energy Phys. **03**, 005 (2010)
3. Diehl, M., Ostermeier, D., Schafer, A.: Elements of a theory for multiparton interactions in QCD. J. High Energy Phys. **03**, 089 (2012)
4. Manohar, A.V., Waalewijn, W.J.: A QCD analysis of double parton scattering: color correlations, interference effects and evolution. Phys. Rev. D **85**, 114009 (2012)
5. Calucci, G., Treleani, D.: Proton structure in transverse space and the effective cross-section. Phys. Rev. D **60**, 054023 (1999)
6. Chang, H.-M., Manohar, A.V., Waalewijn, W.J.: Double parton correlations in the bag model. Phys. Rev. D **87**, 034009 (2013)
7. Rinaldi, M., Scopetta, S., Vento, V.: Double parton correlations in constituent quark models. Phys. Rev. D **87**(11), 114021 (2013)
8. Rinaldi, M., Scopetta, S., Traini, M., Vento, V.: Double parton correlations and constituent quark models: a Light Front approach to the valence sector. J. High Energy Phys. **12**, 028 (2014)
9. Rinaldi, M., Scopetta, S., Traini, M., Vento, V.: Double parton distributions in Light-Front constituent quark models. Few Body Syst. **56**(6-9), 515 (2015)
10. Akesson, T., Axial Field Spectrometer Collaboration, et al.: Double parton scattering in p–p collisions AT S**(1/2) = 63-GeV. Z. Phys. C **34**, 163 (1987)
11. Alitti, J., UA2 Collaboration, et al.: A study of multi-jet events at the CERN anti-p–p collider and a search for double parton scattering. Phys. Lett. B **268**, 145 (1991)
12. Abe, F., CDF Collaboration, et al.: Double parton scattering in $\bar{p}p$ collisions at $\sqrt{s} = 1.8$ TeV. Phys. Rev. D **56**, 3811 (1997)
13. Abazov, V.M., D0 Collaboration, et al.: Double parton interactions in γ+3 jet events in pp^{-} bar collisions $\sqrt{s} = 1.96$ TeV. Phys. Rev. D **81**, 052012 (2010)
14. Aad, G., ATLAS Collaboration, et al.: Measurement of hard double-parton interactions in $W(\rightarrow l\nu)+$ 2 jet events at $\sqrt{s} = 7$ TeV with the ATLAS detector. New J. Phys. **15**, 033038 (2013)
15. Chatrchyan, S., CMS Collaboration, et al.: Study of double parton scattering using W + 2-jet events in proton–proton collisions at $\sqrt{s} = 7$ TeV. J. High Energy Phys. **03**, 032 (2014)
16. Rinaldi, M., Scopetta, S., Traini, M., Vento, V.: Double parton scattering: a study of the effective cross section within a Light-Front quark model. Phys. Lett. B **752**, 40 (2016)
17. Blok, B., Dokshitzer, Y., Frankfurt, L., Strikman, M.: pQCD physics of multiparton interactions. Eur. Phys. J. C **72**, 1963 (2012)
18. Blok, B., Dokshitzer, Y., Frankfurt, L., Strikman, M.: Perturbative QCD correlations in multi-parton collisions. Eur. Phys. J. C **74**, 2926 (2014)
19. Echevarria, M.G., Kasemets, T., Mulders, P.J., Pisano, C.: Polarization effects in double open-charm production at LHCb. J. High Energy Phys. **04**, 034 (2015)
20. Guidal, M., Moutarde, H., Vanderhaeghen, M.: Generalized parton distributions in the valence region from deeply virtual compton scattering. Rep. Prog. Phys. **76**, 066202 (2013)
21. Traini, M., Mair, A., Zambarda, A., Vento, V.: Constituent quarks and parton distributions. Nucl. Phys. A **614**, 472 (1997)
22. Shelest, V.P., Snigirev, A.M., Zinovev, G.M.: The multiparton distribution equations in QCD. Phys. Lett. B **113**, 325 (1982)
23. Broniowski, W., Ruiz Arriola, E.: Valence double parton distributions of the nucleon in a simple model. Few Body Syst. **55**, 381 (2014)
24. Faccioli, P., Traini, M., Vento, V.: Polarized parton distributions and light front dynamics. Nucl. Phys. A **656**, 400 (1999)
25. Boffi, S., Pasquini, B., Traini, M.: Linking generalized parton distributions to constituent quark models. Nucl. Phys. B **649**, 243 (2003)
26. Traini, M.: NNLO nucleon parton distributions from a light-cone quark model dressed with its virtual meson cloud. Phys. Rev. D **89**, 034021 (2014)
27. Diehl, M., Kasemets, T., Keane, S.: Correlations in double parton distributions: effects of evolution. J. High Energy Phys. **1405**, 118 (2014)
28. Gaunt, J.R., Maciula, R., Szczurek, A.: Conventional versus single-ladder-splitting contributions to double parton scattering production of two quarkonia, two Higgs bosons and $c\bar{c}c\bar{c}$. Phys. Rev. D **90**(5), 054017 (2014)

DOI 10.1007/978-3-319-50699-9_12
Reprinted from *Few-Body System* Journal, DOI 10.1007/s00601-016-1074-4

CrossMark

D. Chakrabarti · C. Mondal · A. Mukherjee

On Gravitational Form Factors and Transverse Spin Sum Rule

Received: 15 February 2016 / Accepted: 1 March 2016 / Published online: 18 March 2016

Abstract Using the light front wave functions of the scalar quark–diquark model for nucleon predicted by the soft-wall AdS/QCD, we calculate the flavor dependent gravitational form factors. We evaluate the matrix element of Pauli–Lubanski operator in this model and show that the intrinsic spin sum rule involves the higher twist form factor $\bar{C}$. The longitudinal momentum densities in the transverse impact parameter space are also discussed for both unpolarized and transversely polarized nucleons.

1 Introduction

Spin structure of proton is one of the most intriguing and fundamental problems of hadron physics. Compared to longitudinal spin or helicity, understanding the transverse spin and transverse angular momentum of the proton is a much more involved problem. The longitudinal angular momentum operator is kinematical on the light-front, whereas the transverse angular momentum and rotation operators are dynamical. This implies that the the partonic structure of the transverse spin is different from that of the longitudinal spin structure of the proton. Recently, several sum rules have been proposed in the literature about transverse spin. A new transverse polarization sum rule was proposed in [1–3] which was interpreted at the partonic level. In [1,2] the matrix elements of the transverse component of the Pauli–Lubanski operator for a transversely polarized state are analyzed and related to the gravitational form factors (GFF) $A(0)$ and $B(0)$. It was pointed out [4–6]that the above result is frame dependent and the relation not only involves $A(0)$ and $B(0)$ but also the higher twist term $\bar{C}(0)$, and the contribution from $\bar{C}(0)$ is not suppressed. Using a phenomenological light front scalar quark–diquark model [7], we evaluate the GFFs for a transversely polarized proton from the energy momentum tensor and verify the sum rule for the transverse spin in a light front quark-diquark model.

The Fourier transform of the gravitational form factor in the impact parameter space has interesting interpretations [8,9]. The Fourier transform of the form factor $A(Q^2)$ gives the longitudinal momentum density (p^+ density) in the transverse impact parameter space. We find that for unpolarized nucleon the momentum density is axially symmetric whereas for a transversely polarized nucleon, the deviation from the axially symmetric distribution is found to be dipolar in nature.

D. Chakrabarti (✉) · C. Mondal
Department of Physics, Indian Institute of Technology Kanpur, Kanpur 208016, India
E-mail: dipankar@iitk.ac.in

C. Mondal
E-mail: mchandan@iitk.ac.in

A. Mukherjee
Department of Physics, Indian Institute of Technology Bombay, Powai, Mumbai 400076, India
E-mail: asmita@phy.iitb.ac.in

2 GFFs in the Light-Front Quark–Diquark Model for the Nucleon

We consider a light-front scalar quark–diquark model [7] for the nucleons where the two particle wave functions are modeled from the soft-wall AdS/QCD solution [10]. The light front wave function (LFWF) $\psi^{\lambda_N}_{\lambda_q q}(x, \mathbf{k}_\perp)$ (where nucleon helicity $\lambda_N = \pm$ and $\lambda_q = \pm$ indicate the spin up or down state of quark q) in the model are

$$\psi^{+}_{+q}(x, \mathbf{k}_\perp) = \varphi^{(1)}_q(x, \mathbf{k}_\perp), \quad \psi^{+}_{-q}(x, \mathbf{k}_\perp) = -\frac{k^1 + ik^2}{xM_n}\varphi^{(2)}_q(x, \mathbf{k}_\perp)$$

$$\psi^{-}_{+q}(x, \mathbf{k}_\perp) = \frac{k^1 - ik^2}{xM_n}\varphi^{(2)}_q(x, \mathbf{k}_\perp), \quad \psi^{-}_{-q}(x, \mathbf{k}_\perp) = \varphi^{(1)}_q(x, \mathbf{k}_\perp), \tag{1}$$

$$\varphi^{(i)}_q(x, \mathbf{k}_\perp) = N^{(i)}_q \frac{4\pi}{\kappa}\sqrt{\frac{\log(1/x)}{1-x}}\, x^{a^{(i)}_q}(1-x)^{b^{(i)}_q} \exp\left[-\frac{\mathbf{k}^2_\perp}{2\kappa^2}\frac{\log(1/x)}{(1-x)^2}\right]. \tag{2}$$

The normalization constants $N^{(i)}_q$ are fixed by the conditions that $F^u_1(0) = 2$, $F^u_2(0) = \kappa_u = 1.673$ and $F^d_1(0) = 1$, $F^d_2(0) = \kappa_d = -2.033$. The AdS/QCD parameter $\kappa = 406$ MeV [11,12] as determined by fitting the nucleon form factor in soft-wall AdS/QCD. The other parameters are fixed by fitting to the electromagnetic properties for the proton and neutron (electromagnetic form factors, magnetic moments, charge radii). The parameters used in [13] do not have proper normalization, here we have used the parameters with proper normalization [14]. The main conclusion of the paper does not depend on the values of the parameters.

The gravitational form factors(GFFs) which are related to the matrix elements of the the stress tensor ($T^{\mu\nu}$) play an important role in hadronic physics. For a spin 1/2 composite system, the matrix elements of $T^{\mu\nu}$ involve four gravitational FFs [1,6]

$$\langle P', S'|T^{\mu\nu}_i(0)|P, S\rangle = \bar{U}(P', S')\Big[-B_i(q^2)\frac{\bar{P}^\mu\bar{P}^\nu}{M_n} + (A_i(q^2) + B_i(q^2))\frac{1}{2}(\gamma^\mu\bar{P}^\nu + \gamma^\nu\bar{P}^\mu)$$
$$+ C_i(q^2)\frac{q^\mu q^\nu - q^2 g^{\mu\nu}}{M_n} + \bar{C}_i(q^2)M_n g^{\mu\nu}\Big]U(P, S), \tag{3}$$

where $\bar{P} = (P + P')/2$ and $q = P' - P$ and $A(q^2)$, $B(q^2)$, $C(q^2)$ and $\bar{C}(q^2)$ are the GFFs. According to Ji's sum rule [15,16], $2\langle J_q\rangle = A_q(0) + B_q(0)$. Using the two particle Fock states for $J^z = +\frac{1}{2}$ and $J^z = -\frac{1}{2}$ and the light-front wave functions given in Eq. (1), we evaluate the GFFs $A(q^2)$ and $B(q^2)$ depending on different flavors (struck quark) in the scalar diquark model. In Fig. 1 we show total $A(q^2)$ and $B(q^2)$ depending on different struck quarks and the contributions of quark and the diquark to the total spin-nonflip and spin-flip GFFs. At zero momentum transfer, $A(0) = A^q(0) + A^b(0) = 1$ and $B(0) = B^q(0) + B^b(0) = 0$ as expected.

3 Energy-Momentum Tensor and Pauli–Lubanski Operator

For transverse spin evaluation, consider a transversely polarized state (polarized along +ve x direction)

$$|P, S^{(1)}\rangle = \frac{1}{\sqrt{2}}\Big(|\Psi^{\uparrow}_{2p}(P^+, P^\perp)\rangle + |\Psi^{\downarrow}_{2p}(P^+, P^\perp)\rangle\Big), \tag{4}$$

where $\Psi^{\uparrow}_{2p}(\Psi^{\downarrow}_{2p})$ represents the two-particle Fock state corresponding to $J^z = +\frac{1}{2}(J_z = -\frac{1}{2})$. For the transversely polarized state we calculate the matrix elements of $T^{\mu\nu}$ for $\Psi^{\uparrow}_{2p}$ going to $\Psi^{\downarrow}_{2p}$ and $\Psi^{\downarrow}_{2p}$ going to $\Psi^{\uparrow}_{2p}$. For zero skewness, the matrix elements of $T^{\mu\nu}$ up to $\mathcal{O}(q)$ are [13]:

$$\langle P', S^{(1)}|T^{++}|P, S^{(1)}\rangle = B(Q^2)\frac{(P^+)^2}{M_n}(iq^2_\perp), \quad \langle P', S^{(1)}|T^{+1}|P, S^{(1)}\rangle = 0, \tag{5}$$

$$\langle P', S^{(1)}|T^{+-}|P, S^{(1)}\rangle = -M_n\big[A(Q^2) + 2\bar{C}(Q^2)\big](iq^2_\perp), \quad \langle P', S^{(1)}|T^{+2}|P, S^{(1)}\rangle = 0. \tag{6}$$

For nonzero skewness, the matrix element of T^{+2} has a term proportional to q^+ [6]. For zero skewness, the results of this quark–diquark model are consistent with [6]. In Fig. 2 we show the form factor $\bar{C}(Q^2)$ and the quark and the diquark contributions are shown in Fig. 2b, c respectively.

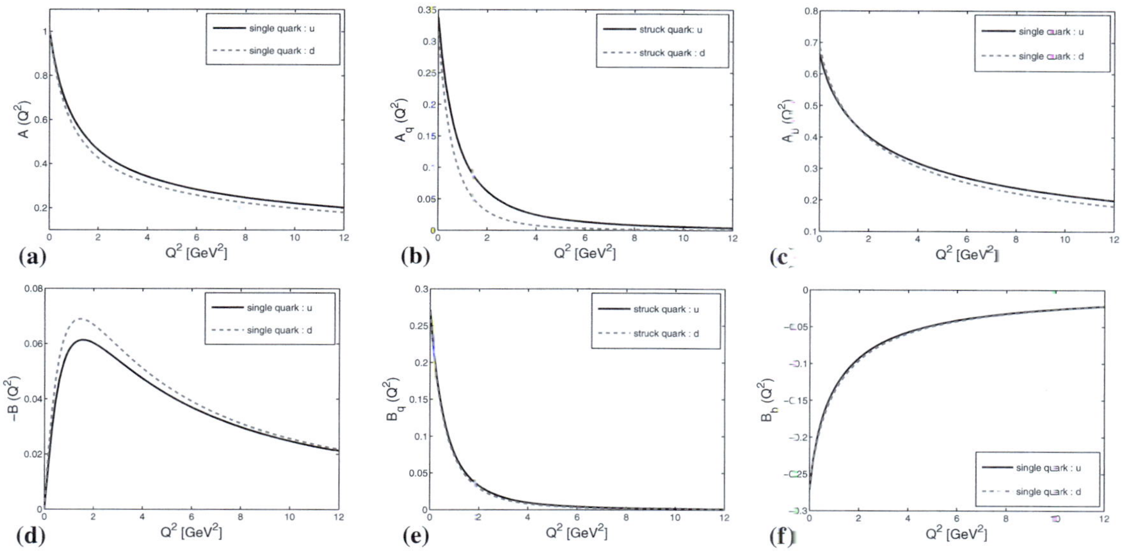

Fig. 1 Gravitational form factors and contributions of quark and diquark (when the single quark is u or d) to the nucleon form factors; upper panel for $A(Q^2)$ and lower panel for $B(Q^2)$

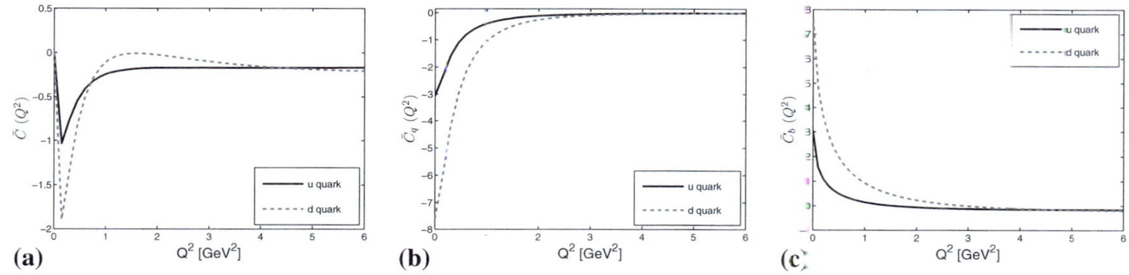

Fig. 2 (Color online) Plots of gravitational form factor $\bar{C}(Q^2)$. The contributions of quark and diquark (when the single quark is either u or d) to the nucleon are shown in **b** and **c**

The Pauli–Lubanski operator is defined as [6] $W_i^1 = \frac{1}{2}F_i^2 P^+ + \tilde{K}_i^3 P^2 - \frac{1}{2}\tilde{E}_i^2 P^-$, where F^j and E^j are the light front transverse rotation and transverse boost operators, K^3 is the longitudinal boost operator. The matrix elements of the operators F_i^2, $\tilde{K}_i^3$ and $\tilde{E}_i^2$ in a transversely polarized state are given by

$$\langle PS^{(1)}|F_i^2|PS^{(1)}\rangle = i(2\pi)^3\delta^3(0)\Big[\frac{\partial}{\partial\Delta_-}\langle P'S^{(1)}|T_i^{+2}(0)|PS^{(1)}\rangle - \frac{\partial}{\partial\Delta_2^{\perp}}\langle P'S^{(1)}|T_i^{+-}(0)|PS^{(1)}\rangle\Big]_{q=0}, \quad (7)$$

$$\langle PS^{(1)}|\tilde{K}_i^3|PS^{(1)}\rangle = -\frac{i}{2}(2\pi)^3\delta^3(0)\Big[\frac{\partial}{\partial\Delta_-}\langle P'S^{(1)}|T_i^{++}(0)|PS^{(1)}\rangle\Big]_{q=0}, \quad (8)$$

$$\langle PS^{(1)}|\tilde{E}_i^2|PS^{(1)}\rangle = -i(2\pi)^3\delta^3(0)\Big[\frac{\partial}{\partial\Delta_2^{\perp}}\langle P'S^{(1)}|T_i^{++}(0)|PS^{(1)}\rangle\Big]_{q=0}, \quad (9)$$

where $\Delta = P' - P$. Using the results of the individual matrix element in Eqs. (7), (8) and (9), the matrix element of the total Pauli–Lubanski operator W^1 evaluated in this quark–diquark model as

$$\frac{\langle PS^{(1)}|W_i^1|PS^{(1)}\rangle}{\langle PS^{(1)}|PS^{(1)}\rangle} = \frac{\langle PS^{(1)}|W_i^1|PS^{(1)}\rangle}{(2\pi)^3 2P^+\delta^3(0)} = \frac{M_n}{2}\Big[A_i(0) + B_i(0) + \bar{C}_i(0)\Big]. \quad (10)$$

The non-vanishing contribution of $\bar{C}_i$ to the matrix element of Pauli–Lubanski operator W_i^1 has been reported previously in [6,17,18].

For a massive particle like nucleon, the intrinsic spin operators can be related with the Pauli–Lubanski operators through the following relation $M_n\mathcal{J}^i = W^i - P^i\mathcal{J}^3 = W^i - P^i(J^3 + \frac{1}{P^+}(E^1P^2 - E^2P^1))$, where

J^3 is the helicity operator. The matrix element of the intrinsic spin operator in a transversely polarized state in the quark–diquark model is given by [13]

$$\frac{\langle PS^{(1)}|\mathcal{J}_i^1|PS^{(1)}\rangle}{\langle PS^{(1)}|PS^{(1)}\rangle} = \frac{1}{M_n}\frac{\langle PS^{(1)}|W_i^1|PS^{(1)}\rangle}{\langle PS^{(1)}|PS^{(1)}\rangle} = \frac{1}{2}\Big[A_i(0) + B_i(0) + \bar{C}_i(0)\Big]. \tag{11}$$

Though $\bar{C}_i(0)$ is a higher twist term, it is not suppressed in the angular momentum sum rule and not zero as can be seen from Fig. 2.

4 Longitudinal Momentum Density in Transverse Plane

For a unpolarized nucleon the longitudinal momentum density in the transverse plane [9] is defined as

$$\rho(b) = \int \frac{d^2q_\perp}{(2\pi)^2} A(Q^2) e^{iq_\perp.b_\perp} = \int_0^\infty \frac{dQ}{2\pi} Q J_0(Qb) A(Q^2), \tag{12}$$

where $b = |b_\perp|$ represents the impact parameter. Under isospin symmetry, the momentum density is same for both proton and neutron. Due to polarization, the density gets modified by the spin-flip form factor $B(Q^2)$. For transversely polarized nucleon, the momentum density is given by [9]

$$\rho_T(b) = \rho(b) + \sin(\phi_b - \phi_s) \int_0^\infty \frac{dQ}{2\pi} \frac{Q^2}{2M_n} J_1(bQ) B(Q^2), \tag{13}$$

where M_n is the mass of nucleon and $S_\perp = (\cos\phi_s \hat{x} + \sin\phi_s \hat{y})$ and $b_\perp = b(\cos\phi_b \hat{x} + \sin\phi_b \hat{y})$. Without loss of generality, the polarization of the nucleon is chosen along x-axis ie., $\phi_s = 0$. The second term in Eq. (13), provides the deviation from circular symmetry of the unpolarized density.

Results for the momentum density $\rho(b)$ for the active quark u for both unpolarized and the transversely polarized nucleon are shown in Fig. 3. Similar distributions are also observed for d-quark. The unpolarized densities are axially symmetric and have the peak at the center of the nucleon ($b = 0$). For the nucleon polarized along x-direction, the peak of densities gets shifted towards positive y-direction and the densities no

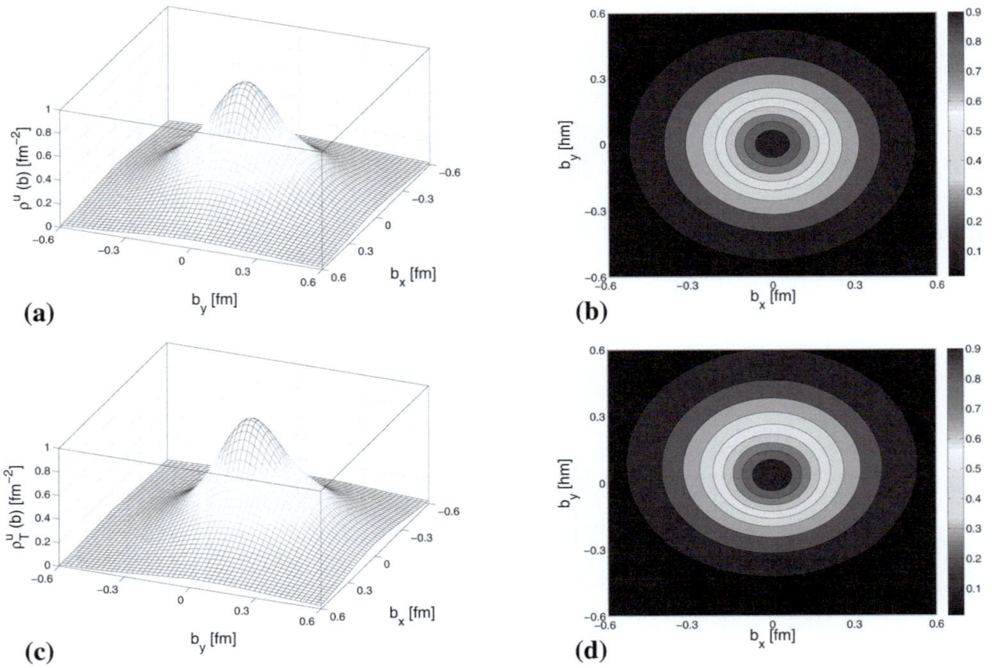

Fig. 3 p^+ densities for the active u quark in the transverse plane, upper panel for unpolarized nucleon, lower panel for nucleon polarized along x-direction. **b** and **d** are the top view of **a** and **c** respectively

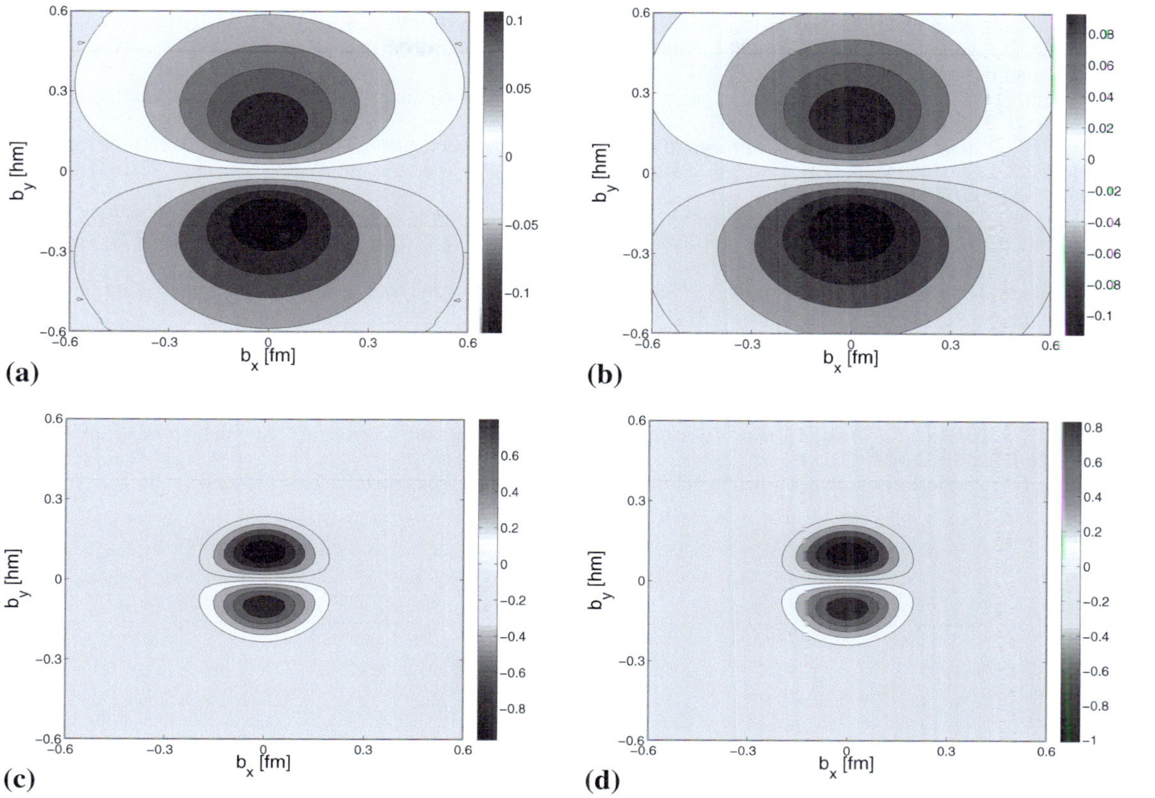

Fig. 4 The momentum density asymmetry $(\rho_T(b) - \rho(b))$ in the transverse plane for a nucleon polarized in x-direction, **a** for u quark **b** for d quark **c** for nucleon when the active quark is u and **d** for nucleon when the active quark is d

longer have the symmetry. The total nucleon p^+ density (fermionic plus bosonic) for the transversely polarized nucleon shows very little difference with respect to the symmetric density of unpolarized nucleon. However, removing the axially symmetric part of the density from $\rho_T(b)$ if we look at $(\rho_T(b) - \rho(b))$, one can find that the total angular-dependent part of the density displays a dipole pattern (see Fig. 4c, d). The angular-dependent part of the densities for active quarks u and d are shown in Fig. 4a, b respectively.

5 Summary

In a light-front scalar diquark model of nucleon, we have demonstrated that the sum rules for the intrinsic spin for a transversely polarized proton involve the form factors A_q, B_q and $\bar{C}_q$ in agreement with the claim in [6]. We have also evaluated the longitudinal momentum density (p^+ density) in the transverse plane for both unpolarized and polarized nucleon. For transversely polarized nucleon, the asymmetries in the distributions for an individual quark is quite large, but when the contributions from the quark and the bosonic diquark are considered, the overall asymmetries in the nucleon become small but are shown to be dipolar in nature.

References

1. Ji, X., Xiong, X., Yuan, F.: Transverse polarization of the nucleon in parton picture. Phys. Lett. B **717**, 214 (2012)
2. Ji, X., Xiong, X., Yuan, F.: Proton spin structure from measurable parton distributions. Phys. Rev. Lett. **109**, 15200 (2012)
3. Ji, X., Xiong, X., Yuan, F.: Reply to arXiv:1211.3957 and arXiv:1211.4731 by Leader et al. and arXiv:1212.0761 by Harindranath et al. Phys. Rev. Lett. **111**, 039103 (2013)
4. Leader, E., Lorce, C.: Comment on 'proton spin structure from measurable parton distributions' by Ji, Xiong and Yuan (PRL109, 152005 (2012)). Phys. Rev. Lett. **111**, 039101 (2013)
5. Harindranath, A., Kundu, R., Mukherjee, A., Ratabole, R.: Comment on Phys. Rev. Lett. 109, 152005 (2012). Phys. Rev. Lett. **111**, 039102 (2013)

6. Harindranath, A., Kundu, R., Mukherjee, A.: On transverse spin sum rules. Phys. Lett. B **728**, 63 (2014)
7. Gutsche, T., Lyubovitskij, V.E., Schmidt, I., Vega, A.: Light-front quark model consistent with Drell-Yan-West duality and quark counting rules. Phys. Rev. D **89**, 054033 (2014)
8. Selyugin, O.V., Teryaev, O.V.: Generalized parton distributions and description of Electromagnetic and graviton form factors of nucleon. Phys. Rev. D **79**, 033003 (2009)
9. Abidin, Z., Carlson, C.E.: Hadronic Momentum densities in the transverse plane. Phys. Rev. D **78**, 071502 (2008)
10. Brodsky, S.J., de Teramond, G.F.: Hadronic Form Factor Models and Spectroscopy Within the Gauge/Gravity Correspondence. arXiv:1203.4125 [hep-ph]
11. Chakrabarti, D., Mondal, C.: Generalized parton distributions for the proton in AdS/QCD. Phys. Rev. D **88**, 073006 (2013)
12. Chakrabarti, D., Mondal, C.: Nucleon and flavor form factors in a light front quark model in AdS/QCD. Eur. Phys. J. C **73**, 2671 (2013)
13. Chakrabarti, D., Mondal, C., Mukherjee, A.: Gravitational form factors and transverse spin sum rule in a light front quark-diquark model in AdS/QCD. Phys. Rev. D **91**, 114026 (2015)
14. Chakrabarti, D., Mondal, C.: Chiral-odd generalized parton distributions for proton in a light-front quark-diquark model. Phys. Rev. D **92**, 074012 (2015)
15. Ji, X.D.: Gauge-invariant decomposition of nucleon Spin. Phys. Rev. Lett. **78**, 610 (1997)
16. Ji, X.D.: Lorentz symmetry and the internal structure of the nucleon. Phys. Rev. D **58**, 056003 (1998)
17. Hatta, Y., Tanaka, K., Yoshida, S.: Twist-three relations of gluonic correlators for the transversely polarized nucleon. JHEP **1302**, 003 (2013)
18. Leader, E.: A critical assessment of the angular momentum sum rules. Phys. Lett. B **720**, 120 (2013)

DOI 10.1007/978-3-319-50699-9_13
Reprinted from *Few-Body System* Journal, DOI 10.1007/s00601-016-1090-4

Alessandro Bacchetta · Sabrina Cotogno · Barbara Pasquini

Internal Structure of the Pion Inspired by the AdS/QCD Correspondence

Received: 4 February 2016 / Accepted: 13 March 2016 / Published online: 12 April 2016

Abstract We present a study of the pion structure in the context of the AdS/QCD soft-wall model. This approach provides the light-front wave function of the pion in terms of a set of parameters that we fit to available experimental information on the electromagnetic form factor and parton distribution of the pion. We discuss the corresponding predictions for the unpolarized transverse momentum dependent parton distribution of the pion.

1 Introduction

AdS/QCD correspondence with soft-wall holographic models has been employed in a number of recent works to get new insights on hadronic physics [1–4]. Remarkably, light-front holographic QCD methods (see [5] and references therein for a complete review on the topic) have provided expressions for the light-front wave function (LFWF) of a meson valence Fock state [6,7]. This allows one to obtain direct information about many hadronic observables, which can be expressed in terms of overlaps of LFWFs. The outline of this work is as follows: In Sect. 2 we consider the expressions for the pion LFWF derived in [6,7]. We introduce the quark masses in a Lorentz invariant way and derive analytical expressions for the pion valence Parton Distribution Function (PDF) and electromagnetic Form Factor. We determine the free parameters of the LFWFs by fitting simultaneously the available experimental information on form factor and valence PDF. In Sect. 3, we discuss the corresponding predictions for the pion unpolarized valence Transverse Momentum Dependent parton distribution function (TMD). In the final section, we summarize the results.

2 The Pion LFWF

The LFWFs in the soft-wall model of the AdS/QCD correspondence were originally derived in two different matching procedures [6,7].

Presented by Sabrina Cotogno at Light Cone 2015, 21-25 September, 2015, INFN Frascati National Laboratories.

S. Cotogno (✉)
Vrije Universiteit and Nikhef, Science Park 105, Amsterdam, The Netherlands
E-mail: scotogno@nikhef.nl

A. Bacchetta · B. Pasquini
Dipartimento di Fisica, Università degli Studi di Pavia, and INFN, Sezione di Pavia, 27100 Pavia, Italy
E-mail: alessandro.bacchetta@unipv.it

B. Pasquini
E-mail: barbara.pasquini@unipv.it

Table 1 Results from the fit of the pure- and effective-valence LFWFs in different quark mass scenarios

LFWF	m (GeV)	κ (GeV)	Q_0 (GeV)	$\chi^2_{\text{d.o.f.}}$ (FF)	$\chi^2_{\text{d.o.f.}}$ (PDF)	$\chi^2_{\text{d.o.f.}}$ (Tot)
$\psi^V_{q\bar{q}/\pi}$	0.005 (fixed)	0.397 ± 0.003	0.500 ± 0.003	1.96	1.19	3.15
	0.200 (fixed)	0.351 ± 0.003	0.491 ± 0.003	9.10	2.66	11.76
	0.0500 ± 0.00004	0.371 ± 0.002	0.498 ± 0.002	1.84	0.41	2.25
$\psi^E_{q\bar{q}/\pi}$	0.005 (fixed)	0.261 ± 0.002	0.498 ± 0.003	4.25	1.19	5.44
	0.200 (fixed)	0.322 ± 0.002	0.630 ± 0.008	11.5	1.43	12.97
	0. (fixed)	0.262 ± 0.002	0.498 ± 0.003	4.17	1.21	5.38

From the matching procedure carried out in [6] the valence quark LFWF for the pion follows immediately and reads:

$$\psi^V_{q\bar{q}/\pi}(x,\, \boldsymbol{k}_\perp) \sim \frac{1}{\kappa\sqrt{(1-x)\,x}} e^{-\frac{1}{2}\frac{\boldsymbol{k}_\perp^2}{\kappa^2 x(1-x)}}. \tag{1}$$

The superscript (V) indicates that we are considering the LFWF for the "pure-valence" state for the pion including only the valence Fock state. The quark masses are included by completing the invariant mass of the system in the argument of the exponential as [8]:

$$M^2 = \sum_i \frac{m_i^2 + \boldsymbol{k}_{\perp i}^2}{x_i} = \frac{m^2 + \boldsymbol{k}_\perp^2}{x(1-x)}, \tag{2}$$

where $m = m_1 = m_2$ and, from momentum conservation, $\boldsymbol{k}_\perp = \boldsymbol{k}_{\perp 1} = -\boldsymbol{k}_{\perp 2}$ and $x = x_1 = 1 - x_2$. As a result, the expression in Eq. (1) becomes (see also [9,10])

$$\psi^V_{q\bar{q}/\pi}(x,\, \boldsymbol{k}_\perp) = A\frac{4\pi}{\kappa\sqrt{(1-x)\,x}} e^{-\frac{1}{2\kappa^2}\left(\frac{m^2}{x}+\frac{m^2}{1-x}+\frac{\boldsymbol{k}_\perp^2}{x(1-x)}\right)}. \tag{3}$$

where A is a normalization constant fixed by the conditions

$$\int_0^1 dx f(x;\, Q_0) = F(Q^2 = 0) = 1, \tag{4}$$

Using the LFWF overlap representation formulae [11], the PDF and the form factor read

$$f(x;\, Q_0) = A^2 e^{\left(-\frac{m^2}{\kappa^2 x(1-x)}\right)}; \quad F(Q^2) = A^2 \int_0^1 dx e^{\left(-\frac{m^2}{\kappa^2 x(1-x)} - \frac{Q^2(1-x)}{4\kappa^2 x}\right)}. \tag{5}$$

In Eq. (3) κ is a free parameter corresponding to the physical mass scale which breaks the conformal symmetry. We fix this parameter along with the quark mass m using the experimental data on the electromagnetic form factor of the pion [12–15] and the parametrization of the PDF in [16]. The latter comparison has to be done after evolving the PDF from the scale of the model Q_0 to the scale $Q = 5\,\text{GeV}$ of the parametrization. We do so by using the Hoppet evolution code [17] inside the fitting procedure, leaving the initial scale Q_0 of the model as an additional free parameter of the fit. For parameter m, we consider three scenarios: either we fix it to a value of 5 MeV (current quark mass), or to a value of 200 MeV (constituent quark mass), or we leave it as a free parameter in the fit.

Our results for the set of parameters in the different quark-mass scenarios are listed in the upper part of Table 1. The value of κ used in Refs. [6,18] is compatible with our results, while a larger value of $\kappa = 0.54\,\text{GeV}$ is needed to describe the hadronic masses and the Regge trajectories [10,19–21].

An alternative expression for the LFWF has been derived in [7], considering the matching with a dressed current. In this case, one obtains a LFWF which incorporates effects due to Fock states beyond the valence sector, and therefore represents an "effective" two-parton state of the pion. Introducing the quark-mass dependence as outlined above, the effective LFWF reads

$$\psi^E_{q\bar{q}/\pi}(x,\, \boldsymbol{k}_\perp) = 4\pi A\frac{\sqrt{\log\left(\frac{1}{x}\right)}}{\kappa\,(1-x)} e^{-\frac{\boldsymbol{k}_\perp^2+m^2}{2\kappa^2}\frac{\log\left(\frac{1}{x}\right)}{(1-x)^2}}. \tag{6}$$

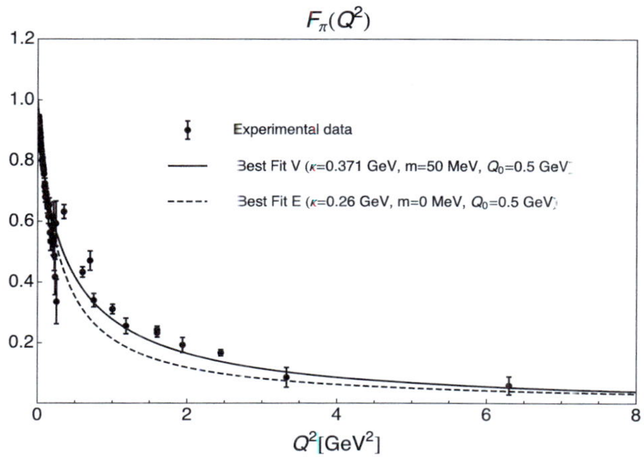

Fig. 1 Results for the electromagnetic form factor from the pure-valence LFWF (*solid curve*) and the effective LFWF (*dashed curve*) with the best-fit parameters. The experimental data are from Refs. [12–15]

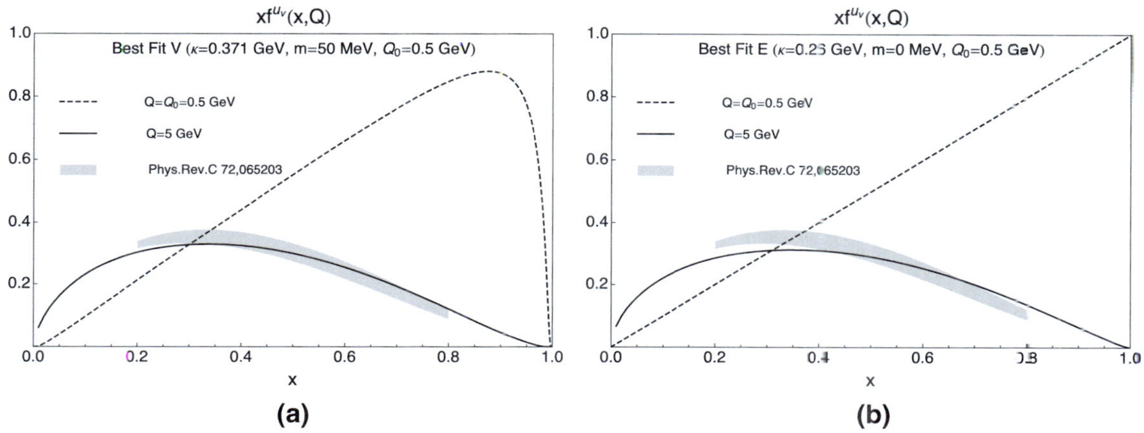

Fig. 2 Results for the quark PDF of the pion from the pure-valence LFWF (**a**) and the effective LFWF (**b**) with the best-fit parameters. *Dashed curves* at the initial scale of the model. *Solid curves* LO evolved to $Q = 5\,\text{GeV}$. *Dashed band* LO parametrization at $Q = 5\,\text{GeV}$ from Ref. [16]

The explicit expression for the form factor and PDF from the effective LFWF will be shown elsewhere [22]. Following the same fitting procedure as for the valence LFWF, with the quark mass fixed to three different values ($m = 0$, 5, and 200 MeV), we obtain the parameter values quoted in the lower part of Table 1.

In Fig. 1 we show the results for the pion form factor obtained from the pure-valence (solid curve) and effective (dashed curve) LFWFs, with the best-fit parameter set, i.e. the values with the best total $\chi^2_{\text{d.o.f}}$ in Table 1. The corresponding results for the PDF are shown in Fig. 2a, b, respectively. The dashed curves show the results at the hadronic scale, and the solid curves are obtained after leading-order (LO) evolution to $Q = 5\,\text{GeV}$. The shaded band corresponds to the results from the parametrization at $Q = 5\,\text{GeV}$ of Ref. [16] (shaded band). The results from the pure-valence LFWF are in good agreement with the available experimental information, while a worse comparison, especially for the form factor, is obtained in the case of the effective LFWF. Our results can be compared with those obtained in, e.g., Refs. [9,21,23,24].

3 TMD Analysis

The unpolarized TMD $f_1(x, \boldsymbol{k}_\perp^2)$ can be obtained from the following overlap of LFWF [25]

$$f_1\left(x, \boldsymbol{k}_\perp^2; Q_0^2\right) = \frac{1}{16\pi^3} |\psi_{q\bar{q}/\pi}(x,\ \boldsymbol{k}_\perp)|^2. \tag{7}$$

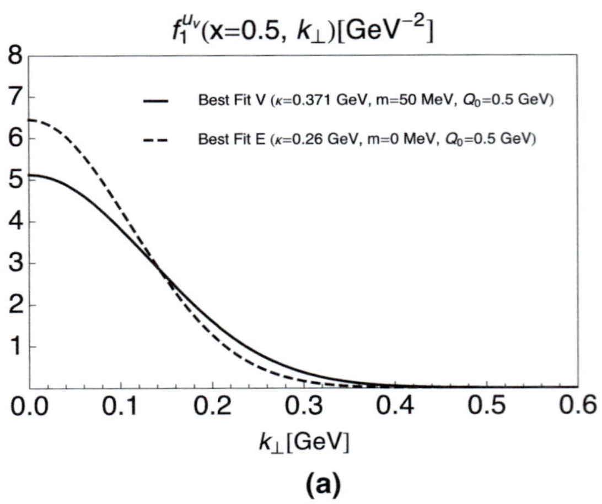

Fig. 3 **a** The unpolarized quark TMD of the pion as function of $k_\perp$ at fixed $x = 0.5$. **b** $\langle \boldsymbol{k}_\perp^2(x)\rangle$ as function of x. *Solid curves* results from the pure-valence LFWF. *Dashed curves* results from the effective LFWF

Using the expressions in Eqs. (3) and (6), one finds that the TMD in both models is a gaussian distribution in $k_\perp = |\boldsymbol{k}_\perp|$ with an x-dependent mean square transverse momenta given by

$$\langle \boldsymbol{k}_\perp^2(x)\rangle^V = \kappa^2 x(x-1), \quad \langle \boldsymbol{k}_\perp^2(x)\rangle^E = \frac{\kappa^2(x-1)^2}{\log(x)}. \tag{8}$$

In the pure-valence case (V), $\langle \boldsymbol{k}_\perp^2(x)\rangle$ is symmetric under exchange of $x \to 1-x$, with a maximum at $x = 0.5$. There is no such a symmetry in the effective case (E) and the maximum is $x = 0.28$ (Fig. 3a). Integration over x produces the values $\langle \boldsymbol{k}_\perp^2\rangle^V = 0.023\,\text{GeV}^2$ and $\langle \boldsymbol{k}_\perp^2\rangle^E = 0.020\,\text{GeV}^2$ at the scale of the model (Fig. 3b).

To compute the pion TMD at some higher scale, closer to experimental data, the effect of TMD evolution has to be included. The TMD is dramatically broadened, particularly at low x. These effects will be explicitly discussed in an upcoming work [22].

4 Conclusions

We presented a study of the pion LFWF obtained from AdS/QCD correspondence with a soft-wall holographic model. We consider two different variants of the LFWF, which we denote with "pure valence" and "effective" ones, already proposed in the literature. We obtained the expressions of the pion valence parton distribution function, the pion form factor and the pion unpolarized TMD. The free parameters of the model are the quark mass m, the scale of the conformal symmetry breaking κ, the initial scale of the model Q_0. We fixed these parameters by reproducing the pion form factor data and the parametrization of the pion valence PDF. The best fits were obtained with values of Q_0 around 0.50 GeV and values of κ below 400 MeV. The best description of the data was obtained with the pure-valence version, with $m = 0.050$ GeV. We studied the pion unpolarized TMD, computing in particular the mean square transverse momentum of the distribution. Comparison with experimental data will be possible only after the application of TMD evolution, which dramatically broadens the TMD when going from the model scale to the experimental one.

Acknowledgments This work was partially supported through the McCartor Fund Fellowship program under the auspices of International Light-Cone Advisory Committee (ILCAC), and by the European Research Council (ERC) under the programs QWORK (contract No. 320389) and 3DSPIN (contract No. 647981).

References

1. Erlich, J., Katz, E., Son, D.T., Stephanov, M.A.: QCD and a holographic model of hadrons. Phys. Rev. Lett. **95**, 261602 (2005)
2. Brodsky, S.J., de Teramond, G.F.: Hadronic spectra and light-front wavefunctions in holographic QCD. Phys. Rev. Lett. **96**, 201601 (2006)
3. de Teramond, G.F., Brodsky, S.J.: Light-front holography: a first approximation to QCD. Phys. Rev. Lett. **102**, 081601 (2009)
4. Karch, A., Katz, E., Son, D.T., Stephanov, M.A.: Linear confinement and AdS/QCD. Phys. Rev. D **74**, 015005 (2006)
5. Brodsky, S.J., de Teramond, G.F., Dosch, H.G., Erlich, J.: Light-front holographic QCD and emerging confinement. Phys. Rep. **584**, 1 (2015)
6. Brodsky, S.J., de Teramond, G.F.: Light-front dynamics and AdS/QCD correspondence: the pion form factor in the space- and time-like regions. Phys. Rev. D **77**, 056007 (2008)
7. Brodsky, S.J., Cao, F.G., de Teramond, G.F.: Meson transition form factors in light-front holographic QCD. Phys. Rev. D **84**, 075012 (2011)
8. Brodsky, S.J., de Teramond, G.F.: AdS/CFT and light-front QCD. Subnucl. Ser. **45**, 139 (2009)
9. Vega, A., Schmidt, I., Branz, T., Gutsche, T., Lyubovitskij, V.E.: Meson wave function from holographic models. Phys. Rev. D **80**, 055014 (2009)
10. Branz, T., Gutsche, T., Lyubovitskij, V.E., Schmidt, I., Vega, A.: Light and heavy mesons in a soft-wall holographic approach. Phys. Rev. D **82**, 074022 (2010)
11. Brodsky, S.J., Pauli, H.C., Pinsky, S.S.: Quantum chromodynamics and other field theories on the light cone. Phys. Rep. **301**, 299 (1998)
12. Amendolia, S.R., et al.: A measurement of the space-like pion electromagnetic form-factor. Nucl. Phys. B **277**, 168 (1986)
13. Brauel, P., et al.: Electroproduction of π^+n, π^-p and $K^+\Lambda$, $K^+\Sigma^0$ final states above the resonance region. Z. Phys. C **3**, 101 (1979)
14. Volmer, J., et al.: Measurement of the charged pion electromagnetic form-factor. Phys. Rev. Lett. **86**, 1713 (2001)
15. Bebek, C.J., et al.: Electroproduction of single pions at low epsilon and a measurement of the pion form-factor up to $q^2 =$ 10-GeV^2. Phys. Rev. D **17**, 1693 (1978)
16. Wijesooriya, K., Reimer, P.E., Holt, R.J.: The pion parton distribution function in the valence region. Phys. Rev. C **72**, 065203 (2005)
17. Salam, G.P., Rojo, J.: A higher order perturbative parton evolution toolkit (HOPPET). Comput. Phys. Commun. **180**, 120 (2009)
18. Vega, A., Schmidt, I.: Hadrons in AdS/QCD correspondence. Phys. Rev. D **79**, 055003 (2009)
19. Forkel, H., Beyer, M., Frederico, T.: Linear square-mass trajectories of radially and orbitally excited hadrons in holographic QCD. JHEP **0707**, 077 (2007)
20. Colangelo, P., De Fazio, F., Giannuzzi, F., Jugeau, F., Nicotri, S.: Light scalar mesons in the soft-wall model of AdS/QCD. Phys. Rev. D **78**, 055009 (2008)
21. Swarnkar, R., Chakrabarti, D.: Meson structure in light-front holographic QCD. Phys. Rev. D **92**(7), 074023 (2015)
22. Bacchetta, A., Cotogno, S., Pasquini, B.: Transverse structure of the pion from AdS/QCD correspondence
23. Gutsche, T., Lyubovitskij, V.E., Schmidt, I., Vega, A.: Light-front quark model consistent with Drell–Yan–West duality and quark counting rules. Phys. Rev. D **89**(5), 054033 (2014)
24. Gutsche, T., Lyubovitskij, V.E., Schmidt, I., Vega, A.: Pion light-front wave function, parton distribution and the electromagnetic form factor. J. Phys. G **42**(9), 095005 (2015)
25. Pasquini, B., Schweitzer, P.: Pion transverse momentum dependent parton distributions in a light-front constituent approach, and the Boer–Mulders effect in the pion-induced Drell–Yan process. Phys. Rev. D **90**(1), 014050 (2014)

DOI 10.1007/978-3-319-50699-9_14
Reprinted from *Few-Body System* Journal, DOI 10.1007/s00601-016-1056-6

Arkadiusz P. Trawiński

The Pion Renormalized Light-Cone Wave Function

Received: 23 January 2016 / Accepted: 9 February 2016 / Published online: 2 March 2016

Abstract An approximate light-cone wave function for the pion effective quark-antiquark Fock sector corresponding to a small value of the renormalization group parameter is presented. The approximate wave function is motivated by the LF-holography and the quadratic confinement potential in the front form of Hamiltonian dynamics, which is in harmony with the linear confining potential in the instant form. The pion radius, decay constant and form-factor are also presented.

1 Introduction

In this article, it is presented how to describe hadrons involved in high-energy interaction that are solutions of QCD eigenvalue equation in terms of their constituents. Hereafter, let us consider a meson as an example of a hadron, which can be written as

$$|\text{Meson}\rangle = \psi_{q\bar{q}}(\lambda)\,|q\bar{q};\lambda\rangle + \psi_{q\bar{q}g}(\lambda)\,|q\bar{q}\,g;\lambda\rangle + \psi_{q\bar{q}gg}(\lambda)\,|q\bar{q}\,gg;\lambda\rangle + \psi_{q\bar{q}q\bar{q}}(\lambda)\,|q\bar{q}\,q\bar{q};\lambda\rangle \\ +\psi_{q\bar{q}q\bar{q}g}(\lambda)\,|q\bar{q}\,q\bar{q}\,g;\lambda\rangle + \dots \tag{1}$$

The constituents' movements inside the hadron are encoded in a set of wave functions, $\psi_{q\bar{q}}(\lambda)$, $\psi_{q\bar{q}g}(\lambda)$, $\psi_{q\bar{q}gg}(\lambda)$, ... Each wave function carries information about a number of constituents, their types, electric charges, effective sizes, etc. The effective size of a constituent plays a key role for the presented description and its relation to the parameter λ which is discussed below. This size is not a physical observable, though it is a theoretical concept helpful for understanding of different properties of hadrons, such as mass, a radius, and form-factors. The calculation of these hadron properties are our ultimate goal.

In quantum field theory, which constitutes a mathematical language of high-energy physics constituents are quanta of a field dependent on space-time coordinates. The choice of the initial conditions in space-time leads to different formulations of the theory [1], out of which the two most known are the instant form and the front form of dynamics.

In the front form (FF) formulation of the theory, which we will focus on, constituents' wave functions are solutions of the Schrödinger-like eigenvalue equation

$$H_{\text{QCD}}|\text{Meson}\rangle = \big[K + V\big]|\text{Meson}\rangle = M^2\,|\text{Meson}\rangle\,, \tag{2}$$

Presented by APT at the Light Cone 2015 Conference, Frascati, Italy, September 21–25, 2015.
Work in collaboration with S. D. Głazek.

A. P. Trawiński (✉)
Faculty of Physics, Institute of Theoretical Physics, University of Warsaw, Warsaw, Poland
E-mail: Arkadiusz.Trawinski@fuw.edu.pl

Fig. 1 The figure represents three Hamiltonians in the RGPEP. From left, at $\lambda = \infty$, at typical value of λ and at $\lambda = \lambda_c$. The intensity of the cell color in each Hamiltonian represents the limit of the energy transfer between sectors. The higher and further to the left the cell is placed, the coupling to higher sectors represents. Hamiltonian at the confining scale, λ_c, is almost diagonal, which means that the lower sectors are decoupled from higher ones

where K is a kinetic part of the full QCD Hamiltonian H_{QCD} and V is its interaction part (see [2,3]). The correctly formulated quantum field theory has to be equipped with the renormalization group procedure, since otherwise is afflicted by singularities problems, which was pointed out in numerous papers (*e.g.* in Ref. [4]).

The RGPEP (the renormalization group procedure for effective particles) [5] allow to change Hamiltonian from one λ value to another. In the result diagonalizes Hamiltonian by a rotation parameterized by λ, see Fig. 1. In this procedure the renormalization group parameter limits the energy transfer in the interaction among quanta. An interaction term written for certain λ effectively allows changes of energy that are not greater than λ.

The aforementioned size of quanta is closely related to the renormalization group parameter of the field. If the size of quanta is small, the renormalization group parameter takes a large value and, if the size of a quanta is big, the renormalization group parameter takes a small value. If the renormalization group parameter is small enough and the change of number of quanta in the interaction is not possible, then confinement phenomenon occurs and the value of parameter for which this happens is called a confinement scale and we denote it by λ_c.

It is important to know that within the RGPEP approach the Hamiltonian is expressed in terms of creation and annihilation operators that depend on λ. These operators can be written using any λ value, but the total Hamiltonian stays unchanged. Thus, the choice of λ is arbitrary and depends only on us; we select λ to describe physics in a simplest way possible. It is easier to solve an eigenvalue equation for a small λ than for a large one.

2 The Ehrenfest Approach

In the previous section, we mentioned that it is more complicated to solve an eigenvalue equation for large λ values than for small, although the solutions for large λ are still preferred to describe high-energy processes. The Ehrenfest approach [6,7] works around this difficulty and allows one to calculate some special quantities in high-energy process. The example of such a quantity is mass of the hadron and its electromagnetic form-factor. For any λ, the Ehrenfest approach yields the equation that is similar to the equation for a small λ and, therefore, easier to solve. In the case of mass and form-factor equations, the same effective wave function is obtained regardless of the initial choice of the renormalization group parameter. This approach is similar to the Ehrenfest averaging in quantum mechanics (hence the name of the approach). The approach is based on averaging over constituents that are not involved in the process.

The averaging of the Eq. (2) leads to the equation

$$M^2 = \left\langle \int_{x_i, k_i^\perp} \psi_n^\dagger(\lambda) \left[\frac{(k_i^\perp)^2 + (m_i)^2}{x_i} + \frac{(k_i^\perp)^2 + (\mathcal{M}_i)^2}{1 - x_i} \right] \psi_n(\lambda) \right\rangle + \left\langle \text{connected interactions} \right\rangle, \quad (3)$$

where $\frac{(k_i^\perp)^2 + (m_i)^2}{x_i} + \frac{(k_i^\perp)^2 + (\mathcal{M}_i)^2}{1 - x_i}$ is a result of K acting on n-th component of the hadron described by the wave-function $\psi_n(\lambda)$. The quantities $m_i \equiv m_i(\lambda)$, x_i and $k_i^\perp$ are mass, longitudinal momenta fraction and transverse momenta of i-th constituent in the n-th sector respectively. The symbol $\left\langle \dots \right\rangle$ denotes averaging over i and integration over spectators, which gives

$$\left\langle \int_{x_i, k_i^\perp} (m_i)^2 \, |\psi_n(\lambda)|^2 \right\rangle = m_{\text{active}}^2 \,, \tag{4}$$

$$\left\langle \int_{x_i, k_i^\perp} (\mathcal{M}_i)^2 \, |\psi_n(\lambda)|^2 \right\rangle = m_{\text{core}}^2 \,, \tag{5}$$

where $\mathcal{M}_i$ is the invariant mass of the spectators (for a fixed n-th sector, spectators are all constituents except i-th). In the same manner, we obtain

$$K_{\text{eff}} = \frac{(k^\perp)^2 + m_{\text{active}}^2}{x} + \frac{(k^\perp)^2 + m_{\text{core}}^2}{1-x} \,, \tag{6}$$

$$U_{\text{eff}} = \left\langle \text{connected interactions} \right\rangle , \tag{7}$$

where x and $k^\perp$ are longitudinal momenta fraction and transverse momenta of an active constituent, which is effective quantum obtained by averaging over all selected constituents. As a result, we obtain the Ehrenfest equation for the hadron's mass,

$$\left[K_{\text{eff}} + U_{\text{eff}} \right] \psi(k^\perp, x) = M^2 \, \psi(k^\perp, x) \,, \tag{8}$$

in term of a new function $\psi(k^\perp, x)$ called the Ehrenfest function.

When we look at Eqs. (8) and (2) for $\lambda = \lambda_c$ in lowest sector, we arrive with two equations

- Ehrenfest equation: $$\left[\frac{(k^\perp)^2 + m_{\text{active}}^2}{x} + \frac{(k^\perp)^2 + m_{\text{core}}^2}{1-x} + U_{\text{eff}} \right] \psi(x, k^\perp) = M^2 \, \psi(x, k^\perp) \,, \tag{9}$$

- QCD equation for λ_c: $$\left[\frac{(k^\perp)^2 + m_q(\lambda_c)^2}{x} + \frac{(k^\perp)^2 + m_{\bar{q}}(\lambda_c)^2}{1-x} + V_{\text{QCD}}(\lambda_c) \right] \psi_{q\bar{q}}(x, k^\perp; \lambda_c) \approx M^2 \, \psi_{q\bar{q}}(x, k^\perp; \lambda_c) \,, \tag{10}$$

which lead us to formulation of a conjecture that $V_{\text{QCD}}(\lambda_c) \approx U_{\text{eff}}$ and $\psi_{q\bar{q}}(x, k^\perp; \lambda_c) \approx \psi(x, k^\perp)$.

The Ehrenfest averaging approach can be repeated [6] staring from the Drell-Yan formula on the form-factor [8]; then an effective formula for the form-factor is obtained

$$F(Q^2) = \int \frac{d^2k^\perp dx}{16\pi^3 x(1-x)} \, \psi\left(x, k^\perp + (1-x)q^\perp\right) \psi\left(x, k^\perp\right) \,, \tag{11}$$

where $Q^2 = (q^\perp)^2$. The same as in the case of the eigenvalue equation, Eq. (11) approximates the formula on the form-factor at λ_c but, in addition, it enables one to use it for energy transfered $Q^2 > \lambda_c^2$. Thus, instead of calculating the form-factor in process of scale $\lambda > \lambda_c$ and dealing with all difficulties of solving Eq. (2) for that λ, one can calculate $\psi_{q\bar{q}}(x, k^\perp; \lambda_c)$ which approximates $\psi(x, k^\perp)$ and use it to calculate the form-factor given in Eq. (11).

3 Pion as an Example of a Hadron

Prior to exemplifying the usage of the Ehrenfest approach, we first have to determine the effective potential $U_{\text{eff}} \approx V_{\text{QCD}}(\lambda_c)$. In the literature, one finds many suggestions of effective potentials. However, the light-front (LF) holography [9] approach uses exactly the same effective equation as we obtained in the Ehrenfest approach. So, taking the LF-holography potential, we come to agreement with all successful predictions that have been made in this scheme. This includes: hadron spectroscopy [10], hadron form-factors [11], the QCD running coupling at small Q^2 [12], ρ electroproduction [13], *etc.*

The original LF-holography potential is a harmonic potential $U_{\rm eff} = -\frac{1}{4}\varkappa^4 r_\perp^2$, where $r_\perp$ is a canonical conjugated variable to $p^\perp = \frac{1}{2}k^\perp/\sqrt{x(1-x)}$ and $\varkappa = 550\,\text{MeV}$. Note, that $U_{\rm eff}$ does not depend on λ. However, in this original formulation $m_{\rm active}$ and $m_{\rm core}$ are set 0. It is natural to generalize this approach to the case of massive quarks. It was suggested in Ref. [14], that for $m_{\rm active} = m_{\rm core} = m$, the original potential $U_{\rm eff} \sim \varkappa^4 r_\perp^2$ should be replaced by

$$U_{\rm eff} = -\frac{1}{4}\varkappa^4 \left(r_\perp^2 + r_3^2\right)\,, \tag{12}$$

where r_3 is a canonical conjugated variable to p^3. The definition of p^3 comes from equating the FF invariant mass square of two equally massive particles, see Eq. (6), with $(2m)^2 + (2\vec{p})^2$, where $\vec{p} = (p^\perp, p^3)$. Then, one obtains $p^3 = m\left(x - \frac{1}{2}\right)/\sqrt{x(1-x)}$. This approach differs form the standard one [15], when one equates the invariant mass square of two equally massive particles with $(2m)^2 + (2\vec{k})^2$, where $\vec{k} = (k^\perp, k^3)$ and $k^3 = (x - \frac{1}{2})\sqrt{\frac{m^2+(k^\perp)^2}{x(1-x)}}$.

The potential in Eq. (12) is the 3-dimensional harmonic oscillator. Thus, excitations in the transverse plane are paired with excitations in the 3rd-direction, and 3-dimensional rotational symmetry is restored, which overcomes the drawback of the original formulation. It has been shown that the quadratic potential, Eq. (12) in the FF, agrees with a linear potential σr ($r = \sqrt{r_\perp^2 + r_3^2}$) in the IF ($\sqrt{\sigma} = \varkappa/\sqrt{2}$) [14].

Solving Eq. (8) for pion as an example of a hadron, with $U_{\rm eff}$ given by Eq. (12) we arrive at

$$\psi_{q\bar{q}/\pi}(x, k^\perp; \lambda_c) \approx \psi(x, k^\perp) \propto \exp\left[-\frac{m_c^2 + (k^\perp)^2}{2\,x(1-x)\,\varkappa^2}\right], \tag{13}$$

where $m_c = m(\lambda_c)$. Up to this point, the spin properties of hadron constituents have not been taken into account. This property influences pion wave-functions but even approximate formula of $U_{\rm eff}$, that includes spin dependence of constituents in the FF, is not known. Only because of that, we can only suggest that the Ehrenfest pion wave function may read (see also Refs. [17,18])

$$\psi(x, k^\perp) = \sqrt{\mathcal{N}}\,\bar{u}\gamma^5 v\,\exp\left[-\frac{m_c^2 + (k^\perp)^2}{2\,x(1-x)\,\varkappa^2}\right], \quad \text{where} \quad \bar{u}\gamma^5 v = \begin{cases} \pm\dfrac{m_c}{\sqrt{x(1-x)}} & \text{for } {(\uparrow\downarrow) \atop (\downarrow\uparrow)} \\ \dfrac{k_1 \pm i k_2}{\sqrt{x(1-x)}} & \text{for } {(\uparrow\uparrow) \atop (\downarrow\downarrow)} \end{cases}, \tag{14}$$

when the spin dependence is taken into account. Then the form-factor, see Eq. (11), reads

$$F_\pi(Q^2) = 3\mathcal{N}\int \frac{d^2k^\perp dx}{16\pi^3 x(1-x)}\,2\left[\frac{m_c^2 + (k^\perp)^2}{x(1-x)} - \frac{1-x}{x}\frac{Q^2}{4}\right]\exp\left[-\frac{m_c^2 + (k^\perp)^2}{x(1-x)\varkappa^2} - \frac{1-x}{x}\frac{Q^2}{4\varkappa^2}\right], \tag{15}$$

where factor 3 comes from summing over colours and factor 2 comes from summing over spins. The condition $F(0) = 1$ fixes the value of $\mathcal{N}$. Figure 2 shows good agreement between calculated and experimentally obtained data [16] for $m_c = 300$ MeV and $\varkappa = 550$ MeV. These values of m_c and $\varkappa$ have been chosen in order to stay in agreement with experimentally measured pion radius and pion decay constant, too. The pion radius reads

$$r_\pi^2 = -6\,\frac{d}{dQ^2}F(Q^2)\bigg|_{Q^2=0}\,, \tag{16}$$

and for $m_c = 300$ MeV and $\varkappa = 550$ MeV one gets $r_\pi^2 = 10.86\ \text{GeV}^{-2} = 0.43\ \text{fm}^2$, what can be compared with the PDG [16] value $\langle r_\pi^2\rangle = 0.45(1)\ \text{fm}^2$.

The pion decay constant f_π is defined from the pion-to-vacuum matrix element of the weak current, $\langle 0|\,\bar{\Psi}_u(0)\,\gamma^+\,\Psi_d(0)\,|\pi^-\rangle = i f_\pi\,P_\pi^+$. Thus, we have

$$f_\pi = 3\sqrt{\mathcal{N}}\int \frac{d^2k^\perp dx}{16\pi^3 x(1-x)}\,4m_c\,\exp\left[-\frac{m^2 + (k^\perp)^2}{2x(1-x)\varkappa^2}\right], \tag{17}$$

where factor 3 comes from summing over colours. One gets $f_\pi = 134$ MeV for $m_c = 300$ MeV and $\varkappa = 550$ MeV, what can be compared with the PDG [16] value $\langle f_\pi\rangle = 130.4(2)$ MeV.

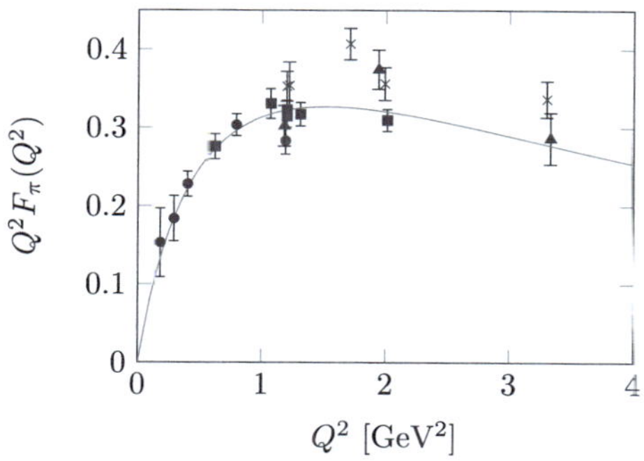

Fig. 2 (color online). Blue points show experimentally determined pion form-factor, see PDG [16]. The *solid red line* represents the pion form-factor given by Eq. (15) for $m_c = 300$ MeV and $\varkappa = 550$ MeV muliplied by $Q^2 = (q^\perp)^2$

4 Conclusion

We have justified the observation that the effective wave-function obtained in the Ehrenfest approach approximates the lowest Fock sector meson wave-function for the renormalization group parameter equal to the confinement scale. In result, it was possible to present an example of the pion wave function. The quantities such as pion radius, decay constant and form-factor were calculated using this wave function and they were found to stay in good agreement with the experiment.

Acknowledgments This work was supported partially through the McCartor Fund Fellowship program under the auspices of International Light-Cone Advisory Committee (ILCAC).

References

1. Dirac, P.A.M.: Forms of relativistic dynamics. Rev. Mod. Phys. **21**, 392 (1949)
2. Lepage, G.P., Brodsky, S.J., Huang, T., Mackenzie, P.B.: Hadronic Wave Functions In Qcd, FERMILAB-CONF-81-114-T
3. Brodsky, S.J., Pauli H.C., Pinsky, S.S.: Quantum chromodynamics and other field theories on the light cone. Phys. Rept. **301**, 299 (1998). arXiv:hep-ph/9705477
4. Dirac, P.A.M.: Quantum electrodynamics without dead wood. Phys. Rev. **139**, B684 (1965)
5. Głazek, S.D.: Reinterpretation of gluon condensate in dynamics of hadronic constituents. Acta Phys. Polon. B **42**, 1933 (2011). arXiv:1106.6100 [hep-th]
6. Głazek, S.D., Trawiński, A.P.: Model of the AdS/QFT duality. Phys. Rev. D **88**(10), 105025 (2013). arXiv:1307.2059 [hep-ph]
7. Trawiński, A.P., Głazek, S.D.: Example of a model for AdS/QFT duality. Few Body Syst. **55**, 463 (2014)
8. Drell, S.D., Yan, T.M.: Connection of elastic electromagnetic nucleon form-factors at large Q**2 and deep inelastic structure functions near threshold. Phys. Rev. Lett. **24**, 181 (1970)
9. de Téramond, G.F., Brodsky, S.J.: Hadronic spectrum of a holographic dual of QCD. Phys. Rev. Lett. **94**, 201601 (2005). arXiv:hep-th/0501022
10. de Téramond, G.F., Brodsky, S.J.: Light-front holography: a first approximation to QCD. Phys. Rev. Lett. **102**, 081601 (2009). arXiv:0809.4899 [hep-ph]
11. Brodsky, S.J., de Téramond, G.F.: Light-front dynamics and AdS/QCD correspondence: the pion form factor in the space- and time-like regions. Phys. Rev. D **77**, 056007 (2008). arXiv:0707.3859 [hep-ph]
12. Brodsky, S.J., de Téramond, G.F., Deur, A.: Nonperturbative QCD coupling and its β-function from light-front holography. Phys. Rev. D **81**, 096010 (2010). arXiv:1002.3948 [hep-ph]
13. Forshaw, J.R., Sandapen, R.: An AdS/QCD holographic wavefunction for the rho meson and diffractive rho meson electroproduction. Phys. Rev. Lett. **109**, 081601 (2012). arXiv:1203.6088 [hep-ph]
14. Trawiński, A.P., Głazek, S.D., Brodsky, S.J., de Téramond, G.F., Dosch, H.G.: Effective confining potentials for QCD. Phys. Rev. D **90**(7), 074017 (2014). arXiv:1403.5651 [hep-ph]
15. Bakker, B.L.G., Kondratyuk, L.A., Terentev, M.V.: On the formulation Of two-body and three-body relativistic equations employing light front dynamics. Nucl. Phys. B **158**, 497 (1979)
16. Olive, K.A. et al.: [Particle data group collaboration] Review of particle physics. Chin. Phys. C **38**, 090001 (2014)
17. Choi, H.M., Ji, C.R.: Mixing angles and electromagnetic properties of ground state pseudoscalar and vector meson nonets in the light cone quark model. Phys. Rev. D **59**, 074015 (1999). arXiv:hep-ph/9711450
18. Choi, H.M., Ji, C.R.: Distribution amplitudes and decay constants for (pi, K, rho, K*) mesons in light-front quark model. Phys. Rev. D **75**, 034019 (2007). arXiv:hep-ph/0701177

DOI 10.1007/978-3-319-50699-9_15
Reprinted from *Few-Body System* Journal, DOI 10.1007/s00601-016-1100-6

CrossMark

Marco Bochicchio

Glueball and Meson Spectrum in Large-N QCD

Received: 27 January 2016 / Accepted: 30 March 2016 / Published online: 20 April 2016

Abstract In the proceedings of HADRON 2015 we outlined a proposal for a string solution of large-N QCD, that is a candidate to satisfy fundamental properties such as the asymptotic freedom of the QCD S-matrix. Here we examine in more detail the implications for the large-N glueball and meson spectrum, that we compare with Particle Data Group (2015) and lattice gauge-theory computations at large N.

1 Introduction

It is generally believed that a solution of large-N QCD [1,2] may be a string theory. But the old Veneziano approach to a string theory of the strong interactions [3], and the modern one, based on Gauge/Gravity duality [4], encounter fundamental difficulties, both phenomenologically, in order to reproduce a spectrum of glueballs and mesons experimentally acceptable, and theoretically, in order to reproduce the asymptotic freedom of QCD on the string side [5,6]. In [7] we outlined a proposal for a string solution of large-N QCD belonging to a new class of twistorial topological strings [8–11] on (non-commutative) twistor space ($TTST$) [12], that is singled out by the requirement of asymptotic freedom for the S-matrix [7,13]. For a short introduction to the relevant topological theory on the field-theory side see [14]. Our proposal leads to a glueball and meson spectrum at large N compatible both with the experiments, i.e. the Particle Data Group (PDG) (2015), and with lattice gauge-theory computations at large N, as discussed in detail in the next Section.

2 Glueball and Meson Spectrum in the $TTST$

We summarize as follows the large-N glueball and meson spectrum implied by the $TTST$ [7]. $m_k^{(s)2} = (k + \frac{s}{2})\Lambda^2_{QCD}$, s even, $k = 1, 2, \ldots$ for glueballs in the Open-String sector. $m_k^{(s)2} = 2(k + \frac{s}{2})\Lambda^2_{QCD}$, s odd, $k = 1, 2, \cdots$ for glueballs in the Closed-String sector. $m_n^{(s)2} - m^2_{PGB} = \frac{1}{2}(n + s)\Lambda^2_{QCD}$, $s = 0, 1, \cdots$, $n = 0, 1, \cdots$ for mesons, $m_n^{(s)2} - m^2_{PGB} = \frac{1}{2}(n + s - \frac{1}{2})\Lambda^2_{QCD}$ $s = 1, \ldots, n = 0, 1, \cdots$ but for the case $s = 0$, for which $n = 1, \ldots$, for mesons. s is the spin, k, n are internal quantum numbers, parity and charge conjugation are unspecified. For massless quarks, i.e. in the chiral limit of QCD, the (string) spectrum has no free parameter but Λ_{QCD}, that in the scheme which we employ coincides by definition with the mass of the lightest glueball, that turns out to be of spin zero according to the preceding Equations, phenomenologically identified with $f_0(1505)$. Thus in our scheme $\Lambda_{QCD} = m_{lightestglueball} = 1505$ MeV, and this fixes once for all the QCD scale. For massive quarks, in the limit of unbroken vector flavor symmetry, the other free parameter is the common quark mass, that on the string side has as counterpart the mass of the Pseudo-Goldstone Boson

M. Bochicchio (✉)
INFN Roma1, Physics Department Sapienza University, Piazzale Aldo Moro 2, 00185 Rome, Italy
E-mail: marco.bochicchio@roma1.infn.it

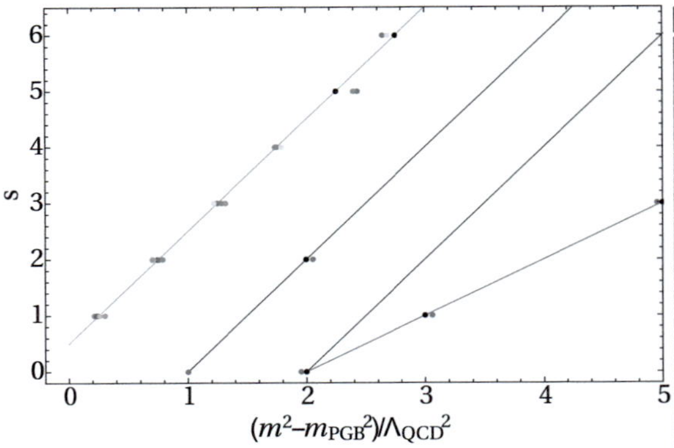

	J^{PC}	m' (MeV)	m_{PGB} (MeV)	$(m'/\Lambda_{QCD})^2$	Th.
$\rho(770)$	1^{--}	775	135	0.257	0.25
$a_2(1320)$	2^{++}	1318	135	0.759	0.75
$\rho_3(1690)$	3^{--}	1689	135	1.251	1.25
$a_4(2040)$	4^{++}	1996	135	1.751	1.75
$\rho_5(2350)$	5^{--}	2350	135	2.430	2.25
$a_6(2450)$	6^{++}	2450	135	2.642	2.75
$K^*(892)$	1^-	892	498	0.242	0.25
$K_2^*(1430)$	2^+	1426	498	0.788	0.75
$K_3^*(1780)$	3^-	1776	498	1.283	1.25
$K_4^*(2045)$	4^+	2045	498	1.737	1.75
$K_5^*(2380)$	5^-	2382	498	2.396	2.25
$\omega(782)$	1^{--}	783	135	0.262	0.25
$f_2(1270)$	2^{++}	1275	135	0.710	0.75
$\omega_3(1670)$	3^{--}	1667	135	1.219	1.25
$f_4(2050)$	4^{++}	2018	135	1.790	1.75
$f_6(2510)$	6^{++}	2469	135	2.683	2.75
$\varphi(1020)$	1^{--}	1019	674	0.258	0.25
$f_2(1430)$	2^{++}	1430	674	0.702	0.75
$\varphi_3(1850)$	3^{--}	1854	674	1.317	1.25
$D^*(2007)^0$	1^-	2007	1865	0.243	0.25
$D^*(2010)^+$	1^-	2010	1870	0.241	0.25
D_s^{*+}	1^-	2112	1969	0.259	0.25
J/ψ	1^{--}	3097	2984	0.304	0.25
B^*	1^-	5325	5280	0.214	0.25
B_s^*	1^-	5415	5367	0.231	0.25
$\Upsilon(1S)$	1^{--}	9460	9398	0.519	0.25

Fig. 1 Leading Regge trajectories in the *TTST* compared with PDG (2015). In the tables $m'^2 = m^2 - m_{PGB}^2$

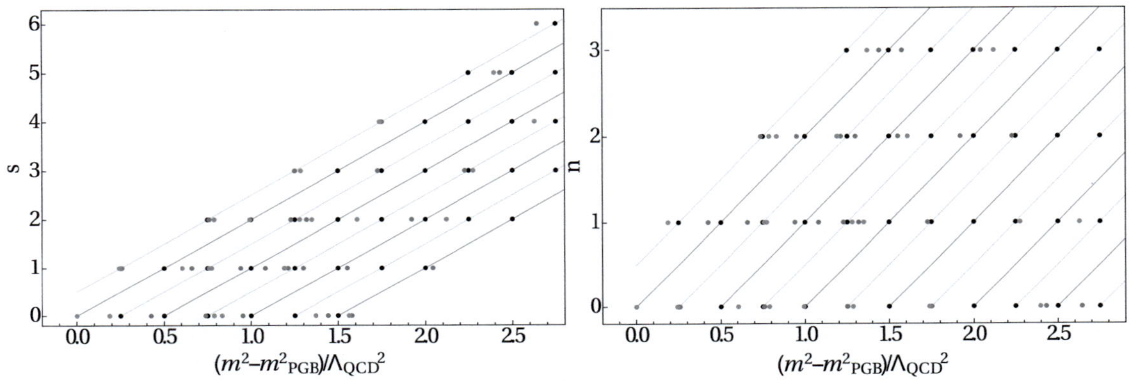

	J^{PC}	m (MeV)	m_{PGB} (MeV)	$(m'/\Lambda_{QCD})^2$	Th.
π^0	0^{-+}	135	135	0.000	0
$\rho(770)^0$	1^{--}	775	135	0.257	0.25
$a_0(980)$	0^{++}	990	135	0.425	0.5
$b_1(1235)$	1^{+-}	1230	135	0.659	0.5
$a_1(1260)$	1^{++}	1230	135	0.660	0.5
$\pi(1300)$	0^{-+}	1300	135	0.738	0.75
$a_2(1320)$	2^{++}	1318	135	0.759	0.75
$\rho(1450)$	1^{--}	1465	135	0.940	1
$a_0(1450)$	0^{++}	1474	135	0.951	1
$\rho(1570)$	1^{--}	1570	135	1.080	1
$\pi_1(1600)$	1^{-+}	1662	135	1.211	1.25
$a_1(1640)$	1^{++}	1647	135	1.190	1.25
$\pi_2(1670)$	2^{-+}	1672	135	1.226	1.25
$\rho_3(1690)$	3^{--}	1689	135	1.251	1.25
$\rho(1700)$	1^{--}	1720	135	1.298	1.25
$a_2(1700)$	2^{++}	1732	135	1.316	1.25
$\pi(1800)$	0^{-+}	1812	135	1.442	1.5
$\rho(1900)$	1^{--}	1880	135	1.552	1.5
$\pi(1880)$	0^{-+}	1895	135	1.577	1.5
$\rho_3(1990)$	3^{--}	1982	135	1.726	1.75
$a_4(2040)$	4^{++}	1996	135	1.751	1.75
$\pi_2(2100)$	2^{-+}	2090	135	1.920	2
$\rho(2150)$	1^{--}	2155	135	2.042	2
$\rho_3(2250)$	3^{--}	2250	135	2.227	2.25
$\rho_5(2350)$	5^{--}	2350	135	2.430	2.25
$a_6(2450)$	6^{++}	2450	135	2.642	2.75

	J^P	m (MeV)	m_{PGB} (MeV)	$(m'/\Lambda_{QCD})^2$	Th.
K^+	0^-	494	494	0.000	0
κ	0^+	819	494	0.189	0.25
$K^*(892)$	1^-	892	494	0.243	0.25
$K_1(1270)$	1^+	1272	494	0.607	0.50
$K_1(1400)$	1^+	1403	494	0.761	0.75
$K^*(1410)$	1^-	1414	494	0.775	0.75
$K_0^*(1430)$	0^+	1425	494	0.789	0.75
$K_2^*(1430)$	2^+	1426	494	0.790	0.75
$K(1460)$	0^-	1460	494	0.833	0.75
$K_2(1580)$	2^-	1580	494	0.995	1
$K^*(1680)$	1^-	1717	494	1.194	1.25
$K_2(1770)$	2^-	1773	494	1.280	1.25
$K_3^*(1780)$	3^-	1776	494	1.285	1.25
$K_2(1820)$	2^-	1816	494	1.348	1.25
$K(1830)$	0^-	1830	494	1.371	1.25
$K_0^*(1950)$	0^+	1945	494	1.563	1.5
$K_2^*(1980)$	2^+	1973	494	1.611	1.5
$K_4^*(2045)$	4^+	2045	494	1.739	1.75
$K_2(2250)$	2^-	2247	494	2.122	2.25
$K_3(2320)$	3^+	2324	494	2.277	2.25
$K_5^*(2380)$	5^-	2382	494	2.397	2.5
$K_4(2500)$	4^-	2490	494	2.630	2.75

Fig. 2 Isovector and strange Regge trajectories in the *TTST* compared with PDG (2015)

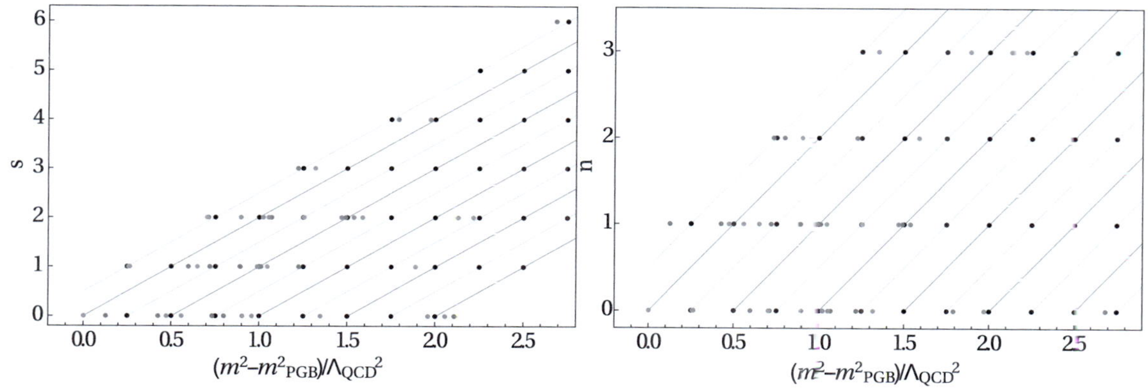

	J^{PC}	m (MeV)	m_{PGB} (MeV)	$(m'/\Lambda_{QCD})^2$	Th.
η	0^{-+}	548	548	0.000	0
$\omega(782)$	1^{--}	783	135	0.262	0.25
$\eta'(958)$	0^{-+}	958	958	0.000	0
σ	0^{++}	550	135	0.126	0.25
$f_0(980)$	0^{++}	990	135	0.425	0.5
$\varphi(1020)$	1^{--}	1019	674	0.258	0.25
$h_1(1170)$	1^{+-}	1170	135	0.596	0.5
$f_2(1270)$	2^{++}	1275	135	0.710	0.75
$f_1(1285)$	1^{++}	1282	135	0.717	0.75
$\eta(1295)$	0^{-+}	1294	135	0.731	0.75
$f_0(1370)$	0^{++}	1350	135	0.797	0.75
$h_1(1380)$	1^{+-}	1386	674	0.647	0.75
$\eta(1405)$	0^{-+}	1409	958	0.471	0.5
$\omega(1420)$	1^{--}	1425	135	0.888	1
$f_1(1420)$	1^{++}	1426	135	0.890	1
$f_2(1430)$	2^{++}	1430	674	0.702	0.75
$\eta(1475)$	0^{-+}	1476	958	0.557	0.5
$f_1(1510)$	1^{++}	1518	135	1.009	1
$f'_2(1525)$	2^{++}	1525	548	0.894	1
$f_2(1565)$	2^{++}	1562	135	1.069	1
$h_1(1595)$	1^{--}	1594	548	0.989	1
$\eta_2(1645)$	2^{++}	1617	548	1.022	1

	J^{PC}	m (MeV)	m_{PGB} (MeV)	$(m'/\Lambda_{QCD})^2$	Th.
$f_2(1640)$	2^{++}	1639	548	1.053	1
$\omega(1650)$	1^{--}	1670	135	1.223	1.25
$\omega_3(1670)$	3^{--}	1667	135	1.219	1.25
$\varphi(1680)$	1^{--}	1680	674	1.045	1
$f_0(1710)$	0^{++}	1720	958	0.901	1
$\eta(1760)$	0^{-+}	1756	958	0.956	1
$f_2(1810)$	2^{++}	1815	674	1.254	1.25
$\eta_2(1870)$	2^{-+}	1842	135	1.490	1.5
$\varphi_3(1850)$	3^{--}	1854	674	1.317	1.25
$f_2(1910)$	2^{++}	1903	548	1.466	1.5
$f_2(1950)$	2^{++}	1944	548	1.536	1.5
$f_0(2020)$	0^{++}	1992	958	1.347	1.25
$f_2(2010)$	2^{++}	2011	674	1.585	1.75
$f_4(2050)$	4^{++}	2018	135	1.790	1.75
$\varphi(2170)$	1^{--}	2175	674	1.888	1.75
$f_0(2200)$	0^{++}	2189	135	2.107	2
$\eta(2225)$	0^{-+}	2226	548	2.055	2
$f_2(2300)$	2^{++}	2297	674	2.129	2.25
$f_0(2330)$	0^{++}	2314	958	1.959	2
$f_4(2300)$	4^{++}	2320	958	1.971	2
$f_2(2340)$	2^{++}	2339	674	2.215	2.25
$f_6(2510)$	6^{++}	2469	135	2.683	2.75

Fig. 3 Isoscalar Regge trajectories in the *TTST* compared with PDG (2015)

m_{PGB}, to be fixed experimentally. Now, although also the vector flavor symmetry is broken in *QCD* because of the differences in the quark masses, as a first approximation we can apply our Equations for the spectrum to each flavor separately, employing the measured value of the mass of each *PGB*. For example, in the *TTST* we find at large N the following relation for the ρ mass as a function of the π mass: $\frac{m_\rho^2-m_\pi^2}{m^2_{lightestglueball}} = \frac{1}{4}$. We can readily see from the first line in the table of Fig. 1 that this relation, somehow surprisingly, is satisfied in the real word to an accuracy better than 3 %, despite the large difference of scales involved in the left-hand side and the fact that this Equation is supposed to hold exactly only in the large-N limit. In particular the agreement implies remarkable cancellations of the $\frac{1}{N}$ corrections between the mass of the *PGB* and of the corresponding vector meson, that extend to the whole leading Regge trajectory for each flavor but for the Υ, as displayed in Fig. 1. Indeed, the plot in Fig. 1 summarizes the universality of the intercepts (modulo the shift by m^2_{PGB} in the meson sector) and of the slopes (with ratios 1:1:$\frac{1}{2}$) of the leading Regge trajectories, in the meson (green line) and even/odd glueball (blue/red line) sectors respectively, implied by the *TTFT* versus the experiments. The states implied by the *TTST* are the black points. We do not discuss here their degeneracies.The colored points close to the green line are the actual experimental (i.e. *QCD* with $N = 3$) leading meson Regge trajectories for any flavor but for the Υ, labelled by the corresponding vector meson, red $= \rho$, green $= K^*$, cyan $= \omega$, orange $= \varphi$, purple $= D^*$, yellow $= D_s^*$, grey $= B^*$, brown $= B_s^*$, pink $= J/\Psi$ from the PDG (2015). For the φ trajectory $m^2_{PGB} = \frac{3}{4}m^2_\eta + \frac{1}{4}m^2_{\eta'}$, due to $\eta + \eta'$ mixing, that, being subleading in $\frac{1}{N}$, is taken into account only phenomenologically. The purple points close to the blue lines are the glueball candidates (actually degenerate in mass and mixed with mesons), $f_0(1505)$, $f_0(2100)$, $f_2(2150)$, compatible with results of the *BES* collaboration and with the *TTST*. The black points on the red line are odd-spin glueballs not experimentally observed yet, predicted by the *TTST* at large N with spin-1 and spin-3, and mass close to 2607 and 3365 MeV respectively, but in the reach of forthcoming experiments such as *BESIII*, *TOTEM + CMS*, *PANDA*, *J − PARK*, *LHCb*, and perhaps *JLab*. The relation for the ρ mass from the

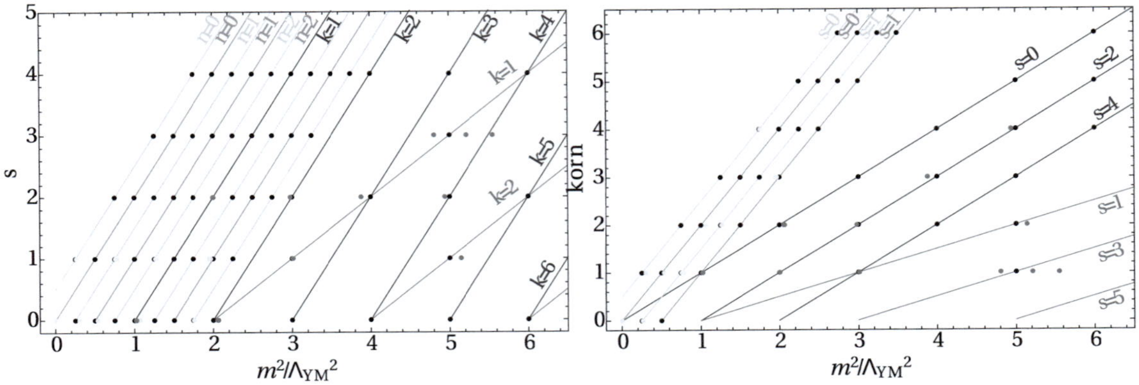

	m/Λ_{YM}	Th.	$(m/\Lambda_{YM})^2$	Th.
0^{++}	1.01	$\sqrt{1}=1$	1.02	1
0^{++*}	1.43	$\sqrt{2}=1.41$	2.06	2
0^{-+}	1.44	$\sqrt{2}=1.41$	2.06	2
2^{++}	1.42	$\sqrt{2}=1.41$	2.00	2
2^{-+}	1.73	$\sqrt{3}=1.73$	2.98	3
1^{+-}	1.73	$\sqrt{3}=1.73$	3.01	3
2^{++*}	1.97	$\sqrt{4}=2$	3.88	4
1^{--}	2.27	$\sqrt{5}=2.24$	5.14	5
2^{--}	2.22	$\sqrt{5}=2.24$	4.94	5
3^{++}	2.19	$\sqrt{5}=2.24$	4.80	5
3^{--}	2.28	$\sqrt{5}=2.24$	5.21	5
3^{+-}	2.36	$\sqrt{5}=2.24$	5.55	5
$?^{+-}$	2.40	?	5.74	?

	m/Λ_{YM}	Th.	$(m/\Lambda_{YM})^2$	Th.
0^{-+}	0	0	0	0
1^{--}	0.513	$\sqrt{\frac{1}{4}}=0.500$	0.264	$\frac{1}{4}=0.25$
0^{++}	0.551	$\sqrt{\frac{1}{4}}=0.500$	0.304	$\frac{1}{4}=0.25$
0^{++}	0.731	$\sqrt{\frac{1}{2}}=0.707$	0.534	$\frac{1}{2}=0.5$
1^{++}	0.871	$\sqrt{\frac{3}{4}}=0.866$	0.758	$\frac{3}{4}=0.75$
1^{+-}	0.883	$\sqrt{\frac{3}{4}}=0.866$	0.780	$\frac{3}{4}=0.75$
0^{-+*}	1.032	$\sqrt{1}=1$	1.066	1
1^{--*}	1.125	$\sqrt{\frac{5}{4}}=1.118$	1.266	$\frac{5}{4}=1.25$
0^{++*}	1.326	$\sqrt{\frac{7}{4}}=1.323$	1.758	$\frac{7}{4}=1.75$
1^{++*}	1.396	$\sqrt{2}=1.414$	1.949	2
1^{+-*}	1.403	$\sqrt{2}=1.414$	1.968	2

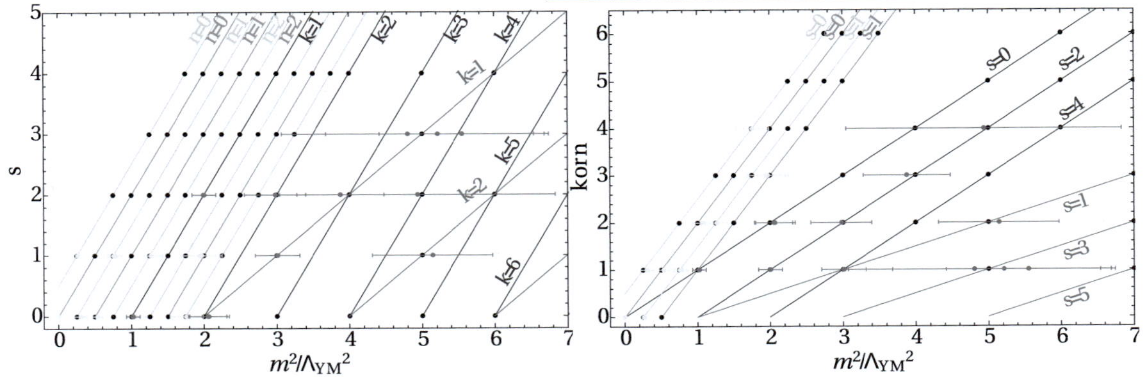

Fig. 4 Regge trajectories in the *TTST* compared with lattice gauge-theory computations at large N. In the tables $\Lambda_{YM} = m_{lightestglueball}$

TTST, $m_\rho^2 = \frac{1}{4} m_{lightestglueball}^2 + m_\pi^2$, reproduces also unquenched $SU(3)$ lattice gauge-theory computation by Jozef Dudek at HADRON 2015 [15] to an accuracy better than 1 %. In fact, at HADRON 2015 we played the following game with Dudek. He had fixed the *QCD* scale in his lattice computation by the experimental mass of the Ω baryon, and he had chosen unphysical values of the pion mass $m_\pi = 391, 236$ MeV because numerically it is difficult to reach the physical value, since it is close to the chiral limit. Then we gave the corresponding predictions of the *TTST* by our one-line formula, $m_\rho = 848{,}789$ MeV, that we compared with the values of the ρ mass, $m_\rho = 854, 783$ MeV, that Dudek had previously computed by a long run of a supercomputer, finding the aforementioned agreement to an accuracy better than 1 %. An interesting consequence of our Equations for the meson spectrum is that the meson spectra for different flavors in the real world should at least approximatively superimpose, between themselves and to the spectrum of the *TTST* in the chiral limit, once the shift by m_{PGB}^2 is taken into account, as shown in Figs. 2 and 3. In Fig. 2 the unflavored $I = 1$ meson Regge trajectories from the PDG (2015), labelled by the corresponding Pseudo-Goldstone Boson, are in red=π, and the strange-meson Regge trajectories in green=K^+. In Fig. 3 the isoscalar unflavored meson Regge trajectories from the PDG (2015), labelled by the corresponding Pseudo-Goldston Boson, are in red $= \pi$, green $= \eta$, orange $= \eta + \eta'$, cyan $= \eta'$. The *TTST* reproduces the spectrum of the PDG (2015) to a surprising accuracy of a few per cent or better for the leading Regge trajectory of any flavor (but for the Υ) and for the low-lying candidate glueballs, and with a reasonable accuracy the whole meson spectrum associated to the light quarks. Finally, it compares favorably with large-N lattice gauge-theory computations for mesons and glueballs in the massless limit as well. Indeed, in Fig. 4 the red and yellow points represent respectively glueballs and mesons actually found in quenched lattice computations involving $SU(8)$ glueballs [16] and $SU(17)$ mesons [17] for massless quarks. The mass of the ρ meson and the existence of a scalar state approximately degenerate in mass with the ρ meson in the chiral limit, reported in Fig. 4, are an updated computation of the collaboration [17] not published yet[1] but anticipated in the comments and plot at the end of [18]. Unfortunately, especially in the glueball sector the lattice computations are still affected by large statistical errors, as depicted in Fig. 4. Neverthless, if we assume the *TTST*, the statistical errors of the lattice computation are probably overestimated especially in the glueball sector, since the central values of the masses squared in the lattice computation match the *TTST* to an unreasonable accuracy, i.e. with an unreasonably small P value, given the reported the variance [16] of the masses in the glueball sector. This may be due to fluctuations of the string tension, instead of fluctuations of the ratios masses squared/string tension, in the sample of gauge configurations.

Acknowledgments We would like to thank Alessandro Pilloni for working out the wonderful colorful plots.

References

1. 't Hooft, G.: A planar diagram theory for strong interactions. Nucl. Phys. B **72**, 461 (1974)
2. Veneziano, G.: Some aspects of a unified approach to gauge, dual and Gribov theories. Nucl. Phys. B **117**, 519 (1976)
3. Veneziano, G.: Construction of a crossing-symmetric, Regge-behaved amplitude for linearly rising trajectories. Nuovo Cimento A **57**, 190 (1968)
4. Aharony, O., Gubser, S.S., Maldacena, J., Ooguri, H., Oz, Y.: Large N field theories, string theory and gravity. Phys. Rep. **323**, 183 (2000). arXiv:hep-th/9905111
5. Bochicchio, M., Muscinelli, S.P.: Ultraviolet asymptotics of glueball propagators. JHEP **08**, 064 (2013). arXiv:1304.6409 [hep-th]
6. Bochicchio, M.: Glueball and meson propagators of any spin in large-N QCD. Nucl. Phys. B **875**, 621 (2013). arXiv:1305.0273 [hep-th]
7. Bochicchio, M.: An asymptotic solution of large-N QCD, for the glueball and meson spectrum and the collinear S-matrix. In: Proceedings of the XVIth Conference on Hadron Spectroscopy (HADRON), AIP Conference Series (2015)
8. Nair, V.P.: A current algebra for some gauge theory amplitudes. Phys. Lett. B **214**, 215 (1988)
9. Witten, E.: Perturbative gauge theory as a string theory in twistor space. Commun. Math. Phys. **252** 189 (2004). arXiv:hep-th/0312171
10. Neitzke, A., Vafa, C.: N = 2 strings and the twistorial Calabi-Yau (2004, unpublished). arXiv:hep-th/0402128
11. Boels, R., Mason, L., Skinner, D.: Supersymmetric gauge theories in twistor space. JHEP **0702**, 014 (2007). arXiv:hep-th/0604040
12. Bochicchio, M.: The Yang-Mills string as the A-model on the twistor space of the complex two-dimensional projective space with fluxes and Wilson loops: the beta function (2008). arXiv:0811.2547 [hep-th]
13. Bochicchio, M.: An asymptotic solution of large-N QCD. In: Proceedings of QCD@work, EPJ Web of Conferences, vol. 80, p. 00010 (2014). arXiv:1409.5144 [hep-th]
14. Bochicchio, M.: Yang-Mills mass gap at large-N, non-commutative YM theory, topological quantum field theory and hyperfiniteness. IJMPD **24**(06) (2015). arXiv:1202.4476 [hep-th]
15. Dudek, J.: Hadron scattering and resonances in QCD. In: Proceedings of the XVIth Conference on Hadron Spectroscopy (HADRON), AIP Conference Series (2015)
16. Meyer, H.B.: Glueball Regge trajectories (2005). arXiv:hep-lat/0508002
17. Bali, G.S., Bursa, F., Castagnini, L., Collins, S., Del Debbio, L., Lucini, B., Panero, M.: Mesons in large-N QCD. JHEP **06**, 071 (2013). arXiv:1304.4437 [hep-th]
18. Bali, G.S., Castagnini, L., Lucini, B., Panero, M.: Large-N mesons. PoS (LATTICE 2013), vol. 100 (2013). arXiv:1311.7559

[1] Courtesy of B. Lucini.

DOI 10.1007/978-3-319-50699-9_16
Reprinted from *Few-Body System* Journal, DOI 10.1007/s00601-016-1080-6

Jorge Segovia

Elastic and Transition Form Factors in DSEs

Received: 7 January 2016 / Accepted: 3 March 2016 / Published online: 15 March 2016

Abstract A symmetry preserving framework for the study of continuum quantum chromodynamics (QCD) is obtained from a truncated solution of the QCD equations of motion or QCD's Dyson–Schwinger equations (DSEs). A nonperturbative solution of the DSEs enables the study of, e.g., hadrons as composites of dressed-quarks and dressed-gluons, the phenomena of confinement and dynamical chiral symmetry breaking, and therefrom an articulation of any connection between them. It is within this context that we present a unified study of Nucleon, Delta and Roper elastic and transition form factors, and compare predictions made using a framework built upon a Faddeev equation kernel and interaction vertices that possess QCD-like momentum dependence with results obtained using a symmetry-preserving treatment of a vector ⊗ vector contact-interaction.

1 Introduction

Nonperturbative QCD poses significant challenges. Primary amongst them is a need to chart the behaviour of QCD's running coupling and masses into the domain of infrared momenta. Contemporary theory is incapable of solving this problem alone but a collaboration with experiment holds a promise for progress. This effort can benefit substantially by exposing the structure of nucleon excited states and measuring the associated transition form factors at large momentum transfers [1]. Large momenta are needed in order to pierce the meson-cloud that, often to a significant extent, screens the dressed-quark core of all baryons [2,3]; and it is via the Q^2 evolution of form factors that one gains access to the running of QCD's coupling and masses from the infrared into the ultraviolet [4,5].

A unified QCD-based description of elastic and transition form factors involving the nucleon and its resonances has acquired additional significance owing to substantial progress in the extraction of transition electrocouplings, g_{vNN^*}, from meson electroproduction data, obtained primarily with the CLAS detector at the Thomas Jefferson National Accelerator Facility (JLab). The electrocouplings of all low-lying N^* states with mass less-than 1.6 GeV have been determined via independent analyses of $\pi^+ n$, $\pi^0 p$ and $\pi^+\pi^- p$ exclusive channels [6,7]; and preliminary results for the g_{vNN^*} electrocouplings of most high-lying N^* states with masses below 1.8 GeV have also been obtained from CLAS meson electroproduction data [1,8].

Many new insights have been revealed in a series of recent articles [9–13] focused on the calculation of the Nucleon, Delta and Roper elastic and transition form factors using a widely-used leading-order (rainbow-ladder) truncation of QCD's Dyson–Schwinger equations and comparing results between a QCD-based frame-

J. Segovia (✉)
Instituto Universitario de Física Fundamental y Matemáticas (IUFFyM), Universidad de Salamanca,
Plaza de la Merced 1-4, 37008 Salamanca, Spain
E-mail: segonza@usal.es
Tel.: +34-923-294434
Fax: +34-923-294584

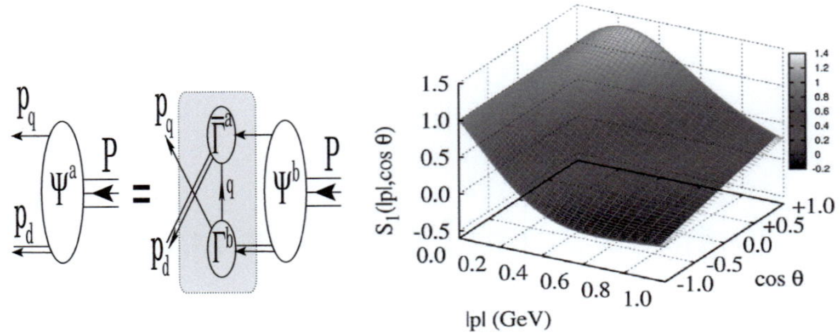

Fig. 1 *Left panel* Poincaré covariant Faddeev equation. Ψ is the Faddeev amplitude for a baryon of total momentum $P = p_q + p_d$, where $p_{q,d}$ are, respectively, the momenta of the quark and diquark within the bound-state. The *shaded area* demarcates the Faddeev equation kernel: *single line*, dressed-quark propagator; Γ, diquark correlation amplitude; and *double line*, diquark propagator. *Right panel* Dominant piece in the nucleon's eight-component Poincaré-covariant Faddeev amplitude: $S_1(|p|, \cos\theta)$. In the nucleon rest frame, this term describes that piece of the quark–scalar-diquark relative momentum correlation which possesses zero intrinsic quark–diquark orbital angular momentum, i.e. $L = 0$ before the propagator lines are reattached to form the Faddeev wave function. Referring to Fig. 1, $p = P/3 - p_q$ and $\cos\theta = p \cdot P/\sqrt{p^2 P^2}$. The amplitude is normalised such that its U_0 Chebyshev moment is unity at $|p| = 0$

work and a vector ⊗ vector contact interaction. It is our purpose here reviewing some of the most important outcomes and refer to the interested reader to the original works for details.

2 Baryon Structure

The existence of strong diquark correlations inside baryons is a dynamical prediction of Faddeev equation studies based on the observation that the attractive nature of quark-antiquark correlations in a colour-singlet meson is also attractive for $\bar{3}_c$ quark-quark correlations within a colour-singlet baryon [14].

In the quark+diquark picture, baryons are described by the Poincaré covariant Faddeev equation depicted in the left panel of Fig. 1. Two main contributions appear in the binding energy: (i) the formation of tight diquark correlations and (ii) the quark exchange depicted in the shaded area of the left panel of Fig. 1.[1] This exchange ensures that diquark correlations within the baryon are fully dynamical: no quark holds a special place because each one participates in all diquarks to the fullest extent allowed by its quantum numbers. Attending to the quantum numbers of the nucleon and Roper, scalar and pseudovector diquark correlations are dominant whereas only pseudovector ones are present inside the Δ-baryon.

The quark + diquark structure of the nucleon is elucidated in the right panel of Fig. 1, which depicts the leading component of its Faddeev amplitude: $S_1(|p|, \cos\theta)$ [11]. This function describes a piece of the quark + scalar-diquark relative momentum correlation. Notably, in this solution of a realistic Faddeev equation there is strong variation with respect to both arguments. Support is concentrated in the forward direction, $\cos\theta > 0$, so that alignment of p and P is favoured; and the amplitude peaks at $(|p| \simeq M_N/6, \cos\theta = 1)$, whereat $p_q \sim p_d \sim P/2$ and hence the natural relative momentum is zero. In the antiparallel direction, $\cos\theta < 0$, support is concentrated at $|p| = 0$, i.e. $p_q \sim P/3$, $p_d \sim 2P/3$.

The strong diquark correlations must be evident in many physical observables. We focus our attention on the flavour separated versions of the Dirac a Pauli form factors of the nucleon. Figure 2 displays the proton's flavour separated Dirac and Pauli form factors. The salient features of the data are: the d-quark contribution to F_1^p is far smaller than the u-quark contribution; $F_2^d/\kappa_d > F_2^u/\kappa_u$ on $x < 2$ but this ordering is reversed on $x > 2$; and in both cases the d-quark contribution falls dramatically on $x > 3$ whereas the u-quark contribution remains roughly constant.

It is natural to seek an explanation for the pattern of behaviour in Fig. 2. We have mentioned that the proton contains scalar and pseudovector diquark correlations. The dominant piece of its Faddeev wave function is $u[ud]$; namely, a u-quark in tandem with a $[ud]$ scalar correlation, which produces 62 % of the proton's normalisation. If this were the sole component, then photon—d-quark interactions within the proton would receive a $1/x$ suppression on $x > 1$, because the d-quark is sequestered in a soft correlation, whereas a spectator u-quark is always available to participate in a hard interaction. At large $x = Q^2/M_N^2$, therefore,

[1] Whilst an explicit three-body term might affect fine details of baryon structure, the dominant effect of non-Abelian multi-gluon vertices is expressed in the formation of diquark correlations [15].

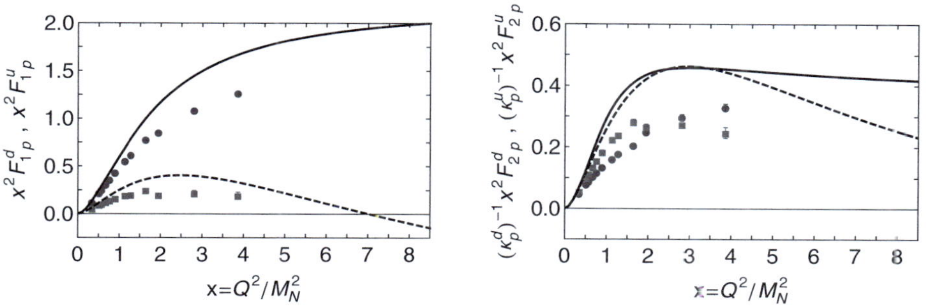

Fig. 2 *Left panel* Flavour separation of the proton's Dirac form factor as a function of $x = Q^2/M_N^2$. The results have been obtained using a framework built upon a Faddeev equation kernel and interaction vertices that possess QCD-like momentum dependence. The *solid-curve* is the u-quark contribution, and the *dashed-curve* is the d-quark contribution. Experimental data taken from Ref. [16] and references therein: *circles*—u-quark; and *squares*—d-quark. *Right panel* Same for Pauli form factor. We define $\kappa_u = F^u_{2p}(0)$ and $\kappa_d = F^d_{2p}(0)$ and they can be expressed in terms of the proton (κ_p) and neutron (κ_n) anomalous magnetic moments [12, 16]

scalar diquark dominance leads one to expect $F^d \sim F^u/x$. Available data are consistent with this prediction but measurements at $x > 4$ are necessary for confirmation.

3 The $\gamma^* N \to \Delta$ Transition

The electromagnetic $\gamma^* N \to \Delta$ transition is described by three Poincaré-invariant form factors [17]: magnetic-dipole, G^*_M, electric quadrupole, G^*_E, and Coulomb (longitudinal) quadrupole, G^*_C; that can be extracted in the Dyson–Schwinger approach by a sensible set of projection operators [18] The following ratios

$$R_{\rm EM} = -\frac{G^*_E}{G^*_M}, \qquad R_{\rm SM} = -\frac{|\mathbf{Q}|}{2m_\Delta}\frac{G^*_C}{G^*_M}, \tag{1}$$

are often considered because they can be read as measures of the deformation of the hadrons involved in the reaction and how such deformation influences the structure of the transition current.

The upper-left panel of Fig. 3 displays the magnetic transition form factor in the Jones–Scadron convention. Our prediction obtained with a QCD-based kernel agrees with the data on $x \gtrsim 0.4$, and a similar conclusion can be inferred from the contact interaction result. On the other hand, both curves disagree markedly with the data at infrared momenta. This is explained by the similarity between these predictions and the bare result determined using the Sato-Lee (SL) dynamical meson-exchange model [19]. The SL result supports a view that the discrepancy owes to omission of meson-cloud effects in the DSEs' computations. An exploratory study of the effect of pion-cloud contributions to the mass of the nucleon and the Δ-baryon has been performed within a DSEs' framework in Ref. [20].

Presentations of the experimental data associated with the magnetic transition form factor typically use the Ash convention. This comparison is depicted in the upper-right panel of Fig. 3. One can see that the difference between form factors obtained with the QCD-kindred and CI frameworks increases with the transfer momentum. Moreover, the normalized QCD-kindred curve is in fair agreement with the data, indicating that the Ash form factor falls unexpectedly rapidly mainly for two reasons. First: meson-cloud effects provide up-to 35 % of the form factor for $x \lesssim 2$; these contributions are very soft; and hence they disappear quickly. Second: the additional kinematic factor $\sim 1/\sqrt{Q^2}$ that appears between Ash and Jones–Scadron conventions and provides material damping for $x \gtrsim 2$.

Our predictions for the ratios in Eq. (1) are depicted in the lower panels of Fig. 3. The left panel displays the Coulomb quadrupole ratio. Both the prediction obtained with QCD-like propagators and vertices and the contact-interaction result are broadly consistent with available data. This shows that even a contact-interaction can produce correlations between dressed-quarks within Faddeev wave-functions and related features in the current that are comparable in size with those observed empirically. Moreover, suppressing the dressed-quark anomalous magnetic moment (DqAMM) in the transition current has little impact. These remarks highlight that R_{SM} is not particularly sensitive to details of the Faddeev kernel and transition current.

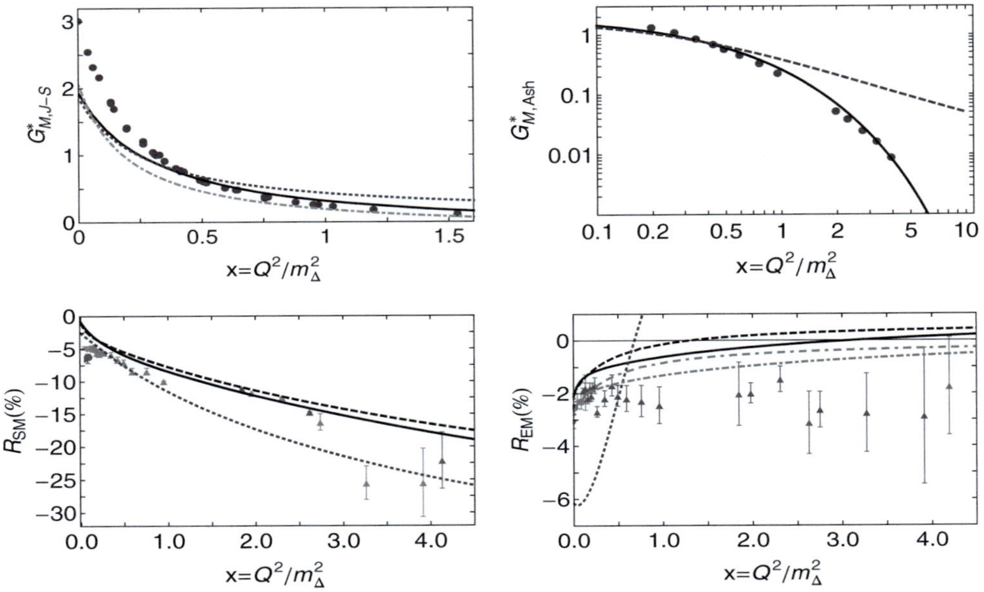

Fig. 3 *Upper-left panel*—$G^*_{M,J-S}$ result obtained with QCD-based interaction (*solid, black*) and with contact-interaction (CI) (*dotted, blue*); The *green dot-dashed curve* is the dressed-quark core contribution inferred using SL-model [19]. *Upper-right panel*—$G^*_{M,Ash}$ result obtained with QCD-based interaction (*solid, black*) and with CI (*dotted, blue*). *Lower-left panel* —R_{SM} prediction of QCD-based kernel including dressed-quark anomalous magnetic moment (DqAMM) (*black, solid*), nonincluding DqAMM (*black, dashed*), and CI result (*dotted, blue*). *Lower-right panel*—R_{EM} prediction obtained with QCD-kindred framework (*solid, black*); same input but without DqAMM (*dashed, black*); these results renormalised (by a factor of 1.34) to agree with experiment at $x = 0$ (*dot-dashed, red*—zero at $x \approx 14$; and *dot-dash-dashed, red*, zero at $x \approx 6$); and CI result (*dotted, blue*). The data in the *panels* are from references that can be found in [11] (colour figure online)

This is certainly not the case with R_{EM}. The differences between the curves displayed in the lower-right panel in Fig. 3 show that this ratio is a particularly sensitive measure of diquark and orbital angular momentum correlations. The contact-interaction result is inconsistent with data, possessing a zero that appears at a rather small value of x. On the other hand, predictions obtained with QCD-like propagators and vertices can be viable. We have presented four variants, which differ primarily in the location of the zero that is a feature of this ratio in all cases we have considered. The inclusion of a DqAMM shifts the zero to a larger value of x. Given the uniformly small value of this ratio and its sensitivity to the DqAMM, we judge that meson-cloud affects must play a large role on the entire domain that is currently accessible to experiment.

4 The $\gamma^* N \to$ *Roper* Transition

Jefferson Lab experiments [7,21–23] have yielded precise nucleon-Roper ($N \to R$) transition form factors and thereby exposed the first zero seen in any hadron form factor or transition amplitude. It has also attracted much theoretical attention; but Ref. [13] provides the first continuum treatment of this problem using the power of relativistic quantum field theory. That study begins with a computation of the mass and wave function of the proton and its first radial excitation. The masses are (in GeV): $M_{\text{nucleon (N)}} = 1.18$ and $M_{\text{nucleon−excited (R)}} = 1.73$. These values correspond to the locations of the two lowest-magnitude $J^P = 1/2^+$ poles in the three-quark scattering problem. The associated residues are the Faddeev wave functions, which depend upon $(p^2, p \cdot P)$, where p is the quark-diquark relative momentum. Figure 4 depicts the zeroth Chebyshev moment of all S-wave components in that wave function. The appearance of a single zero in S-wave components of the Faddeev wave function associated with the first excited state in the three dressed-quark scattering problem indicates that this state is a radial excitation.

The empirical values of the pole locations for the first two states in the nucleon channel are [24]: 0.939 and $1.36 - i\,0.091$ GeV, respectively. At first glance, these values appear unrelated to those obtained within the DSEs framework. However, deeper consideration reveals [25,26] that the kernel in the Faddeev equation omits all those resonant contributions which may be associated with the meson-baryon final-state interactions that are resummed in dynamical coupled channels models in order to transform a bare-baryon into the observed

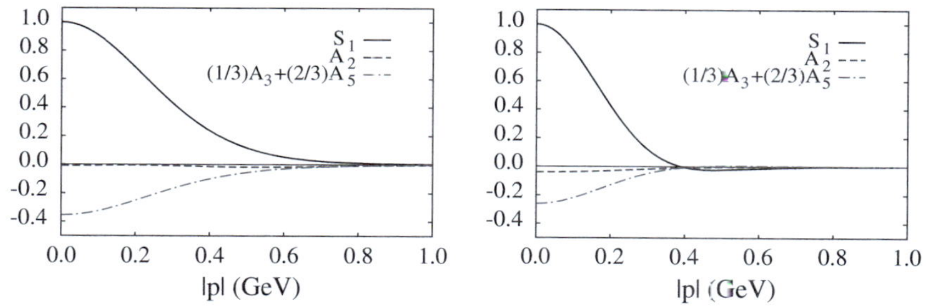

Fig. 4 *Left panel* Zeroth Chebyshev moment of all *S*-wave components in the nucleon's Faddeev wave function. *Right panel* Kindred functions for the first excited state. *Legend* S_1 is associated with the baryon's scalar diquark; the other two curves are associated with the pseudovector diquark; and the normalisation is chosen such that $S_1(0) = 1$

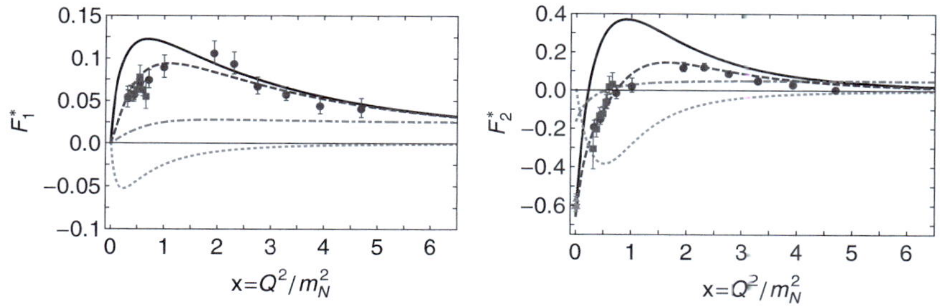

Fig. 5 *Left*—Dirac transition form factor, $F_1^*(x)$, $x = Q^2/m_N^2$. *Solid* (*black*) curve, QCD-kindred prediction; *dot-dashed* (*red*) curve, contact-interaction result; *dotted* (*green*) curve, inferred meson-cloud contribution; and *dashed* (*blue*) curve, anticipated complete result. *Right*—Pauli transition form factor, $F_2^*(x)$, with same legend. Data in both panels: *circles* (*blue*) [21]; *triangle* (*gold*) [23]; *squares* (*purple*) [7]; and *star* (*green*) [6] (colour figure online)

state [3,24]. This Faddeev equation should therefore be understood as producing the dressed-quark core of the bound-state, not the completely-dressed and hence observable object. Crucial, therefore, is a comparison between the quark-core mass, 1.73 GeV, and the value determined for the mass of the meson-undressed bare-Roper in Ref. [24] which is 1.76 GeV.

The transition form factors are displayed in Fig. 5. The results obtained using QCD-derived propagators and vertices agree with the data on $x \gtrsim 2$. The contact-interaction result simply disagree both quantitatively and qualitatively with the data. Therefore, experiment is evidently a sensitive tool with which to chart the nature of the quark-quark interaction and hence discriminate between competing theoretical hypotheses.

The mismatch between the DSE predictions and data on $x \lesssim 2$ is due to Meson-cloud contributions that are expected to be important on this domain. An inferred form of that contribution is provided by the dotted (green) curves in Fig. 5. These curves have fallen to just 20 % of their maximum value by $x = 2$ and vanish rapidly thereafter so that the DSE predictions alone remain as the explanation of the data. Importantly, the existence of a zero in F_2^* is not influenced by meson-cloud effects, although its precise location is.

5 Conclusions

We have described a unified study of Nucleon, Delta and Roper elastic and transition form factors that compares predictions made by a QCD-kindred framework with results obtained using a symmetry-preserving treatment of a vector ⊗ vector contact-interaction. The comparison emphasises that experiment is sensitive to the momentum dependence of the running coupling and masses in QCD. Amongst our results, the following are of particular interest: the presence of strong diquark correlations within the nucleon is sufficient to understand empirical extractions of the flavour-separated form factors. In connection with the $\gamma^* N \to \Delta$ transition, the momentum-dependence of the magnetic transition form factor, G_M^*, matches that of the proton's elastic form factor once the momentum transfer is high enough to pierce the meson-cloud; and the electric quadrupole ratio is a keen measure of diquark and orbital angular momentum correlations. Finally, the Roper resonance is at heart of the nucleon's first radial excitation, consisting of a dressed-quark core augmented by a meson cloud that reduces its mass by approximately 20 %. Our analysis shows that a meson-cloud obscures the dressed-quark core from long-wavelength probes, but that it is revealed to probes with $Q^2 \gtrsim 3m_N^2$.

Acknowledgements The material described in this contribution is drawn from work completed in collaboration with numerous excellent people, to all of whom I am greatly indebted. I would also like to thank V. Mokeev, R. Gothe, T.-S. H. Lee and G. Eichmann for insightful comments; and to express my gratitude to the organisers of the *Light Cone 2015*, whose support helped my participation. I acknowledge financial support from a postdoctoral IUFFyM contract at Universidad de Salamanca, Spain.

References

1. Aznauryan, I.G., et al.: Studies of nucleon resonance structure in exclusive meson electroproduction. Int. J. Mod. Phys. E **22**, 1330015 (2013)
2. Roberts, C.D.: Opportunities and challenges for theory in the N^* program. AIP Conf. Proc. **1432**, 19–25 (2011)
3. Kamano, H., et al.: Nucleon resonances within a dynamical coupled-channels model of πN and γN reactions. Phys. Rev. C **88**, 035209 (2013)
4. Cloët, I.C., et al.: Revealing dressed-quarks via the proton's charge distribution. Phys. Rev. Lett. **111**, 101803 (2013)
5. Chang, L., et al.: Pion electromagnetic form factor at spacelike momenta. Phys. Rev. Lett. **111**, 141802 (2013)
6. Olive, K.A., et al.: The review of particle physics. Chin. Phys. C **38**, 090001 (2014)
7. Mokeev, V.I., et al.: Experimental study of the $P_{11}(1440)$ and $D_{13}(1520)$ resonances from CLAS data on $ep \to e'\pi^+\pi^- p'$. Phys. Rev. C **86**, 035203 (2012)
8. Mokeev, V.I., et al.: Studies of N^* structure from the CLAS meson electroproduction data. Int. J. Mod. Phys. Conf. Ser. **26**, 1460080 (2013)
9. Segovia, J., et al.: Insights into the $\gamma^* N \to \Delta$ transition. Phys. Rev. C **88**, 032201 (2013)
10. Segovia, J., et al.: Elastic and transition form factors of the $\Delta(1232)$. Few Body Syst. **55**, 1–33 (2013)
11. Segovia, J., et al.: Nucleon and Δ elastic and transition form factors. Few Body Syst. **55**, 1185–1222 (2014)
12. Segovia, J., et al.: Understanding the nucleon as a Borromean bound-state. Phys. Lett. B **750**, 100–106 (2015)
13. Segovia, J., et al.: Completing the picture of the Roper resonance. Phys. Rev. Lett. **115**, 171801 (2015)
14. Cahill, R.T., et al.: Baryon structure and QCD. Aust. J. Phys. **42**, 129–145 (1989)
15. Eichmann, G., et al.: Nucleon mass from a covariant three-quark Faddeev equation. Phys. Rev. Lett. **104**, 201601 (2010)
16. Cates, G.D., et al.: Flavor decomposition of the elastic nucleon electromagnetic form factors. Phys. Rev. Lett. **106**, 252003 (2011)
17. Jones, H.F., et al.: Multipole $\gamma N \Delta$ form-factors and resonant photoproduction and electroproduction. Ann. Phys. **81**, 1–14 (1973)
18. Eichmann, G., et al.: Nucleon to Delta electromagnetic transition in the Dyson–Schwinger approach. Phys. Rev. D **85**, 093004 (2012)
19. Julia-Diaz, B., et al.: Extraction and interpretation of $\gamma N \to \Delta$ form factors within a dynamical model. Phys. Rev. C **75**, 015205 (2007)
20. Sanchis-Alepuz, H., et al.: Pion cloud effects on baryon masses. Phys. Lett. B **733**, 151–157 (2014)
21. Aznauryan, I.G., et al.: Electroexcitation of nucleon resonances from CLAS data on single pion electroproduction. Phys. Rev. C **80**, 055203 (2009)
22. Aznauryan, I.G., et al.: Electroexcitation of nucleon resonances. Prog. Part. Nucl. Phys. **67**, 1–54 (2012)
23. Dugger, M., et al.: π^+ photoproduction on the proton for photon energies from 0.725 to 2.875 GeV. Phys. Rev. C **79**, 065206 (2009)
24. Suzuki, N., et al.: Disentangling the dynamical origin of P_{11} nucleon resonances. Phys. Rev. Lett. **104**, 042302 (2010)
25. Eichmann, G., et al.: Perspective on rainbow-ladder truncation. Phys. Rev. C **77**, 042202 (2008)
26. Eichmann, G., et al.: Toward unifying the description of meson and baryon properties. Phys. Rev. C **79**, 012202 (2009)

DOI 10.1007/978-3-319-50699-9_17
Reprinted from *Few-Body System* Journal, DOI 10.1007/s00601-016-1113-1

M. T. Peña · Sofia Leitão · Elmar P. Biernat · Alfred Stadler · J. E. Ribeiro · Franz Gross

Covariant Spectator Theory and Hadron Structure

Received: 21 February 2016 / Accepted: 20 April 2016 / Published online: 11 May 2016

Abstract We present the first results of a study on meson spectroscopy using a covariant formalism based on the Covariant Spectator Theory. Our approach is derived directly in Minkowski space and it approximates the Bethe–Salpeter equation by taking effectively into account the contributions from both ladder and crossed ladder diagrams in the $q\bar{q}$ interaction kernel. A general Lorentz structure of the kernel is tested and chiral constraints on the kernel are discussed. Results for the pion form factor are also presented.

1 Introduction

Novel hadronic structures found at LHCb, and the forthcoming experimental programs as GlueX and PANDA at Jefferson Lab and FAIR/GSI, respectively, together with new measurements in the precision frontier for the search of physics beyond the Standard Model incite theoretical studies on hadron spectroscopy and structure. Advances in this challenging field are made through lattice QCD (LQCD) simulations [1,2]. Still, models are needed for a link between state-of-the-art LQCD data and experimental data. Continuum methods for implementation of models apply the Dyson–Schwinger/Bethe–Salpeter and mass gap equations [3–10], as well as the light-front formulation of quantum field theory [11,12]. In current applications, the Bethe–Salpeter/Dyson–Schwinger equations still leave out some features of QCD, and so far do not describe all decay properties of the meson spectrum, or need to re-adjust parameters obtained in the quarkonia sector when extended to the heavy-light sector.

M. T. Peña (✉) · S. Leitão · E. P. Biernat
Centro de Física Teórica de Partículas (CFTP), Instituto Superior Técnico (IST), Universidade de Lisboa, Av. Rovisco Pais, 1049-001 Lisbon, Portugal
E-mail: teresa.pena@tecnico.ulisboa.pt

A. Stadler
Departamento de Física, Universidade de Évora, 7000-671 Évora, Portugal

A. Stadler
Instituto Superior Técnico (IST), Universidade de Lisboa, Av. Rovisco Pais, 1049-001 Lisbon, Portugal

A. Stadler
Centro de Física Teórica de Partículas (CFTP), 1649-003 Lisbon, Portugal

J. E. Ribeiro
CeFEMA, Instituto Superior Técnico (IST), Universidade de Lisboa, Av. Rovisco Pais, 1049-001 Lisbon, Portugal

F. Gross
Thomas Jefferson National Accelerator Facility (JLab), Newport News, VA 23606, USA

In this talk we specialize on an alternative route to these approaches, the Covariant Spectator Theory (CST), and report on preliminary results for mesons. CST is a field-theory based approach established in *physical* Minkowski space. Therefore, our calculations are not restricted to bound state mass values and momentum regions which are free of propagator singularities—an advantage over Euclidean formulations but with a high cost in the numerical complexity to exactly handle those singularities—and form factors can be computed directly in the timelike region. Importantly also, the formalism does not destroy chiral symmetry requirements. This work is the beginning of a larger project to (1) calculate the meson spectrum, (2) include dynamical quark mass generation through the quark self-energy, calculated consistently from the quark–quark interaction kernel complying with spontaneous chiral symmetry breaking (SχSB), and (3) investigate the Lorentz structure of the confining term in the interaction kernel that reproduces the linear confining potential in the non-relativistic limit.

In this talk we show how confining forces with a large Lorentz scalar component, which were suggested by other works [13–15], are made consistent with SχSB, an essential ingredient to our understanding of the light meson spectrum. The dressed electromagnetic quark current can also be calculated consistently, and then many applications will be possible, such as calculation of hadron form factors and decay properties. The parameters of the interaction kernel will be determined by fits of the meson spectra, and we present here the first results for quarkonia and heavy-light mesons. They are encouraging and indicate that the CST yields very reasonable results.

2 Covariant Spectator Theory: Basic Concepts

2.1 The Dynamical Two-Body CST Equation

Momentum-space formulations are ideal for fully relativistic formalisms. The coupled set of CST-Bethe–Salpeter (CST-BS) equations is obtained directly in momentum space from the Bethe–Salpeter (BS) equation for the two-body vertex $\Gamma_{\mathrm{BS}}(p_1, p_2)$,

$$\Gamma_{\mathrm{BS}}(p_1, p_2) = \mathrm{i} \int \frac{\mathrm{d}^4 k}{(2\pi)^4} \mathcal{V}(p, k; P) S_1(k_1)\, \Gamma_{\mathrm{BS}}(k_1, k_2)\, S_2(k_2), \tag{1}$$

where P is the two-body total 4-momentum, $\mathcal{V}(p, k; P)$ is an irreducible interaction kernel depending on P (in general) and on the external and internal relative 4-momenta p and k, respectively, and $S_i(k_i)$ is the dressed one-body propagator depending on the individual 4-momentum k_i of quark i. Within CST, one only keeps the 4 pole contributions from the 2 intermediate state propagators in the energy loop integration. This assumes that the kernel pole contributions to the box diagrams are largely canceled by the higher order crossed-box diagrams, in which case the CST-BS equation takes effectively the contributions of both ladder and crossed-ladder diagrams into account. The obtained CST-BS set of coupled channel equations can therefore be viewed as a reorganization of the complete BS equation. Technical details can be found in [16,17].

Although a full four-channel CST (4CSE) formulation is needed for the vertex function of light mesons such as the pion, for heavy-light systems the positive-energy pole contribution from the heavy quark propagator in the energy integration dominates, and the full set of four-channel equations reduces to the one-channel CST equation (1CSE) represented in Fig. 1. This is a good approximation when the bound state mass μ is large and close to the sum of the quark masses. Moreover, in the limit of an infinite quark mass the 1CSE becomes identical to the Dirac equation, and in the non-relativistic limit it gives the Schrödinger equation. The existence of these limits confirms that the CST reorganization of the kernel series goes indeed beyond the BS ladder approximation.

At a practical level, the non-relativistic limit allows us to test, in calculations of heavy quarkonia with the Schrödinger equation (SE), our numerical methods involving subtraction techniques needed for the treatment

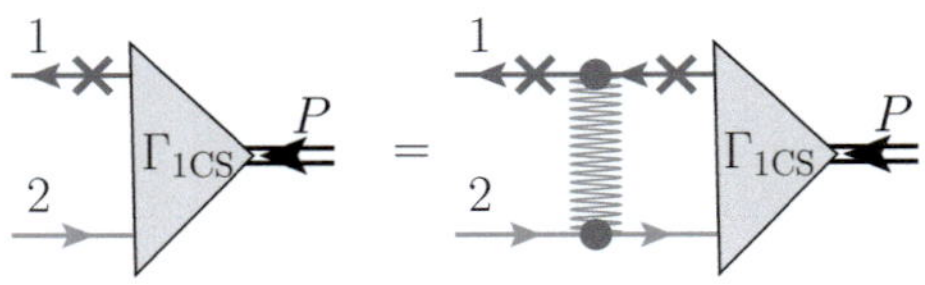

Fig. 1 Diagrammatic representation of the 1CSE. The cross indicates that particle 1 is on its mass-shell. P is the total momentum and $\Gamma_{1\mathrm{CS}}$ the vertex function of the bound-state

of the linear confining interaction in momentum space, since in S waves the SE solutions are known exactly. It turns out that the singularities arising from the kernel can be treated efficiently, yielding accurate and stable numerical solutions as described in Ref. [18]. More recently, the same methods were applied for the first time to solve the (relativistic) 1CSE, and the new results are reported in this talk.

The most general form for the kernel $\mathcal{V}$ is

$$\mathcal{V}(p-k) = V_L(p-k)\Big[\lambda_S(\mathbf{1}\otimes\mathbf{1}) + \lambda_P(\gamma^5\otimes\gamma^5) + \lambda_V(\gamma^\mu\otimes\gamma_\mu) + \lambda_A(\gamma^5\gamma^\mu\otimes\gamma^5\gamma_\mu) + \frac{\lambda_T}{2}(\sigma^{\mu\nu}\otimes\sigma_{\mu\nu})\Big] + V_C(p-k)\Big[\kappa_V(\gamma^\mu\otimes\gamma_\mu) + \kappa_A(\gamma^5\gamma^\mu\otimes\gamma^5\gamma_\mu)\Big], \tag{2}$$

where V_L and V_C are the Lorentz-invariant momentum-dependent parts of the linear-confining and remaining kernels, respectively. The corresponding weight parameters λ_i and κ_i [with $i = S$ (scalar), P (pseudoscalar), V (vector), A (axial-vector), and T (tensor)] are arbitrary constants. The form of V_L in Eq. (2) for the momentum dependence of the linear confining part satisfies the CST generalization of the non-relativistic condition $V_L(r=0)=0$,

$$\int \frac{d^3k}{E_k} V_L(p\pm\hat{k}) = 0, \tag{3}$$

where $E_k = \sqrt{m^2+\mathbf{k}^2}$, $\hat{k} = (E_k, \mathbf{k})$, and m is the dressed quark mass. In the next subsection we give more constrains, obtained from SχSB, on the general form (2) of the kernel.

2.2 Chiral Symmetry and the Axial-Vector Ward–Takahashi Identity

The condition (3) regularizes the singular confinement kernel and at the same time is crucial to prove the existence of a massless pion in the chiral limit as imposed by SχSB. This is ensured by imposing the axial-vector Ward–Takahashi identity (AV-WTI), which relates the dressed quark propagators and the dressed axial-vector and pseudoscalar vertices:

$$P_\mu\Gamma^{5\mu}(p',p) + 2m_0\Gamma^5(p',p) = S^{-1}(p')\gamma^5 + \gamma^5 S^{-1}(p), \tag{4}$$

where $\Gamma^{5\mu}(p',p)$ and $\Gamma^5(p',p)$ are the dressed axial-vector and pseudoscalar vertices, respectively, m_0 is the bare quark mass, p and p' are the incoming and outgoing quark momenta, respectively, and $S(p)$ is the dressed quark propagator. The LHS of Eq. (4) is called the dressed axial vertex $\Gamma^A(p',p)$. Notice that the AV-WTI reduces a vertex to two terms of a propagator line with a contact (γ_5) term, and that $\Gamma^A(p',p)$ is the solution of the inhomogeneous CST-BS equation,

$$\Gamma^A(p',p) = \gamma^A(p',p) + \mathrm{i}\int_{k0} \mathcal{V}(p-k)S(k')\Gamma^A(k',k)S(k), \tag{5}$$

where $\gamma^A(p',p)$ is the bare axial vertex and the notation "$k0$" stands for the CST prescription used in the k_0 contour integration. It follows (see Ref. [19]) from combining the AV-WTI (4) together with the CST-BS equation (5), and using the condition (3), that $S(p)$ is the solution of the one-body CST-Dyson equation

$$S^{-1}(p) = S_0^{-1}(p) - \mathrm{i}\int_{k0} \mathcal{V}(p-k)S(k), \tag{6}$$

where S_0 is the bare quark propagator (which satisfies a corresponding AV-WTI obtained by replacing the dressed vertex $\Gamma^A(p',p)$ with $\gamma^A(p',p)$ on the LHS of (4)), if and only if one has $\lambda_S = \lambda_P$ in V_L. Since in the chiral limit of vanishing bare quark mass, $m_0 \to 0$, and vanishing vertex momentum, $P \to 0$ (soft-pion limit), $\gamma^A(p',p)$ vanishes also, Eq. (5) becomes homogeneous and identical to the zero-mass pion CST equation for the pion vertex function Γ^π_χ in the chiral limit. Therefore one can write

$$\Gamma^A_\chi(p,p) \propto \Gamma^\pi_\chi(p,p). \tag{7}$$

Because (5) in the chiral limit becomes identical to the scalar part of (6), the existence of a zero-mass pion in the chiral limit—dynamical quark mass generation—is proven. The conclusion is that because of condition (3), and for equal weights of the scalar and pseudoscalar terms in the interaction kernel, in the chiral limit, the scalar, pseudoscalar and tensor structures of the linear-confinement terms of $\mathcal{V}_L$ entirely decouple from the two- and one-body equations and therefore do not violate chiral symmetry.

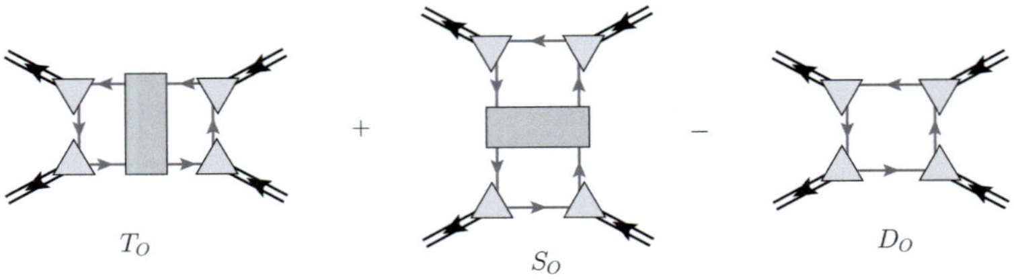

Fig. 2 The direct box contributions to π–π scattering. The *orange boxes* denote the quark–quark scattering amplitudes

2.3 The Adler Zero of π–π Scattering

Furthermore, consistency with chiral symmetry demands the vanishing of the π–π scattering amplitude in the chiral limit (the Adler zero). This amplitude is non-zero in impulse approximation and vanishes due to remarkable cancellations that can only be obtained by going beyond that approximation [20]. We have checked these cancellations for the CST scattering amplitude. The proof is rather lengthy, with all details given in Ref. [19], and it will only be sketched here. The first 2 diagrams in Fig. 2 show two of six possible insertions of the CST-BS ladder sum between two quark lines in the intermediate state. They correspond to a direct box topology whereas the other four insertions correspond to a crossed-box topology. The rest of the demonstration is based on the application of the AV-WTI in order to reduce the number of propagator lines and vertices from four to three. To this end one inserts an additional ladder sum at one pion vertex by using the spectral decomposition of the ladder sum. Then, by Eq. (7), Γ_χ^π is replaced by Γ_χ^A, allowing the application of the AV-WTI given by Eq. (4) between two ladder sums. The resulting terms either cancel against each other, or are proportional to the anticommutator $\{\mathcal{V}, \gamma^5\}$. Henceforth, all vector and axial-vector contributions vanish, while the scalar, pseudoscalar, and tensor structures from the confining potential integrate to zero because of the decoupling property in the chiral limit discussed in the previous subsection.

3 Results for the Meson Spectrum

For our test calculations with the 1CSE the kernel in Eq. (2) was made of the linear (L) confining potential piece,

$$\mathcal{V}_\mathrm{L}(p-k) = V_L\left[(1-y)(\mathbf{1}\otimes\mathbf{1}+\gamma^5\otimes\gamma^5) - y(\gamma^\mu\otimes\gamma_\mu)\right], \tag{8}$$

with V_L given by

$$V_L(p-k) = 8\sigma\pi\left[\frac{1}{q^4} - \frac{E_{p_1}}{m_1}(2\pi)^3\delta^3(\mathbf{q})\int\frac{d^3k'}{(2\pi)^3}\frac{m_1}{E_{k'_1}}\frac{1}{q'^4}\right], \tag{9}$$

together with a $\mathcal{V}_\mathrm{C}$ term, made of vector one-gluon exchange (OGE) and constant (C_0) contributions,

$$\mathcal{V}_\mathrm{OGE}(p-k) = -\tfrac{4\pi\alpha}{q^2}(\gamma^\mu\otimes\gamma_\mu), \qquad \mathcal{V}_{\mathrm{C}_0}(p-k) = (2\pi)^3\tfrac{E_{k_1}}{m_1}C\delta^3(\mathbf{q})(\gamma^\mu\otimes\gamma_\mu), \tag{10}$$

Here $q = p - k$ is the transferred four-momentum, and $E_{p_1} = (m_1^2 + \mathbf{p}^2)^{1/2}$ is the energy of the heavier quark with mass m_1. Notice first that the relativistic extension of the confinement potential given by (9) satisfies the important condition (3) for SχSB. The mixing parameter y dials continuously between a scalar-plus-pseudoscalar structure, also due to the chiral-symmetry constraint discussed in the previous sub-section, and a vector structure, while preserving the same nonrelativistic limit. By fitting the y parameter from the mesonic spectra further information can be gained.

The three coupling strengths σ, α and C, were treated as free parameters of a fit to the pseudoscalar meson spectra. Two disctinct models were tested: whereas model 2 has exactly the equally weighted scalar-pseudoscalar component, as required by chiral symmetry, mixed with a vector structure, model 1 omits the pseudoscalar part. The results are given in Fig. 3 and Table 1.

We found good fits with either model, and our predictions for the mesons with lighter quarks are very reasonable. The observed larger difference between the predictions and the data for the lightest systems is due to the omitted pole in the propagator of the lighter quark whose importance increases with decreasing bound-state mass. The increasing importance of relativistic effects when the bound state mass decreases is

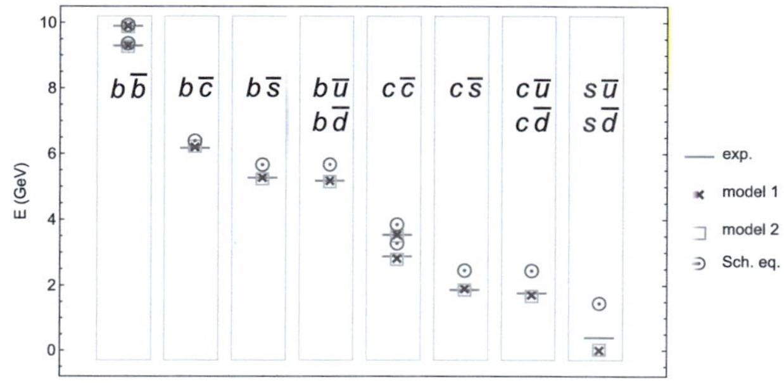

Fig. 3 Mass spectra obtained from the 1CSE for the $b\bar{b}$, $c\bar{c}$ quarkonia, and for the heavy-light $b\bar{s}$, $b\bar{u}/b\bar{d}$, $b\bar{c}$, $c\bar{u}/c\bar{d}$, $s\bar{u}/s\bar{d}$ pseudoscalar states. The *horizontal bars* represent the experimental $J^{PC} = 0^{--}$ states, the *centers of the crosses* and *open squares* indicate the calculated states for models 1 and 2, respectively. The states fitted to the data are those in the first 5 columns, whereas the remaining are predictions. The results from the SE (*purple circles*) are compared to the relativistic calculations.

Table 1 Fitted parameters for kernel models 1 and 2

	σ (GeV^2)	α	C (GeV)	y
Model 1	0.22	0.38	0.337	8.60×10^{-7}
Model 2	0.21	0.37	0.311	0.38×10^{-7}

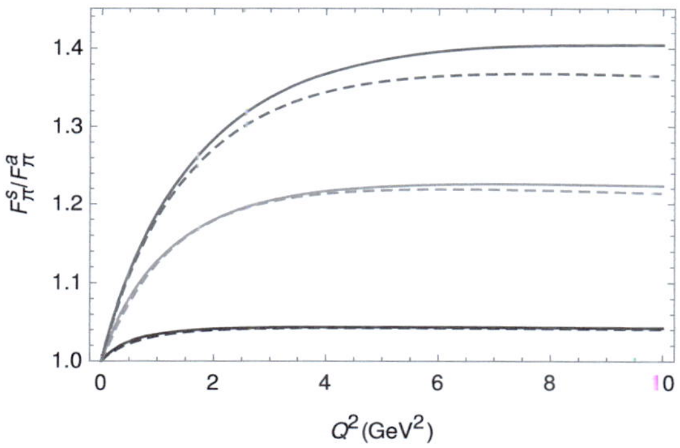

Fig. 4 The ratio F^s_π/F^a_π for fixed (*dashed lines*) and running (*solid lines*) quark masses, and different values of μ. The pairs of curves, from *top* to *bottom* are the results obtained with pion masses $\mu = 0.60$, 0.42 and 0.14 GeV

clearly seen in Fig. 3. Both models favor almost pure Lorentz scalar structures of the confining interaction and the pseudoscalar part in the confining interaction leads to a marginally better fit. It will be interesting to see if these features persist once the fits are extended to include also vector mesons, or are restricted to the light pseudo-scalar mesons only, where SχSB is decisive. Because the heavier mesons are almost nonrelativistic systems for which the pseudoscalar interaction is strongly suppressed, the difference between models 1 and 2 becomes noticeable only for the lighter states. An improvement will be to replace the 1CSE by a more complete version of the CST bound-state equations, namely the two-channel (2CSE) or—in the case of the pion—the complete 4CSE.

4 The Importance of the Active-Quark Poles to the Electromagnetic Pion form Factor

We have also computed the electromagnetic pion form factor, with an approximated pion vertex function near the chiral limit of the form $\Gamma^\pi(\hat{p}_1, p_2) \propto \gamma_5 h(p_2^2)$ [17], a quark current that satisfies the WTI and with a dressed quark mass function dynamically calculated from Eq. (6) with the parameters constrained through LQCD data in the spacelike region. The asymptotic behavior of the pion form factor was described correctly and scaling relations between results obtained with different pion masses were found. In Fig. 4 we show the importance of the positive- and negative-energy spectator-quark poles compared to the four active-quark poles [21]. The latter are the positive- and negative-energy poles of the quark that interacts with the photon, before and after this interaction. The inclusion of both, positive- and negative-energy poles is necessary in order to preserve charge-conjugation invariance in CST [16,17], hence we refer to the present calculation as the C-symmetric complete impulse approximation. For small pion mass μ, the active quark contributions are as important as the spectator contributions, over the whole range of Q^2. For large μ and large Q^2, the active pole contributions are smaller than the spectator contributions by about 30 %.

An extension of the calculations presented here, namely to include a description of the vector mesons spectrum, is underway.

Acknowledgments This work was supported by Fundação para a Ciência e a Tecnologia (FCT) under Grants Nos. PTDC/FIS/113940/2009, CFTP-FCT (UID/FIS/00777/2013), CERN/FP/123580/2011, and SFRH/BPD/ 100578/2014.

References

1. Edwards, R.G., Mathur, N., Richards, D.G., Wallace, S.J.: The flavor structure of the excited baryon spectra from lattice QCD. Phys. Rev. D **87**, 054506 (2013)
2. Guo, P., Dudek, J.J., Edwards, R.G., Szczepaniak, A.P.: Coupled-channel scattering on a torus. Phys. Rev. D **88**, 014501 (2013)
3. Bicudo, P.J., Ribeiro, J.E.F.T.: Current quark model in a p wave triplet condensed vacuum. 1. The dynamical breaking of chiral symmetry. Phys.Rev. D **42**, 1611–1162 (1990)
4. Bicudo, P.J., Ribeiro, J.E.F.T.: Current quark model in a p wave triplet condensed vacuum. 2. Salpeter equations: π, K, ρ, φ as qq^- bound states. Phys. Rev. D **42**, 1625–1634 (1990)
5. Bicudo, P.J., Ribeiro, J.E.F.T.: Current quark model in p wave triplet condensed vacuum. 3. Generalized r.G.m. equations: the ϕ and ρ resonances. Phys. Rev. D **42**, 1635–1650 (1990)
6. Alkofer, R., von Smekal, L.: The infrared behavior of QCD Green's functions: confinement dynamical symmetry breaking, and hadrons as relativistic bound states. Phys. Rep. **353**, 281 (2001)
7. Maris, P., Roberts, C.D.: Dyson–Schwinger equations: a tool for hadron physics. Int. J. Mod. Phys. E **12**, 297–365 (2003)
8. Nefediev, A.V., Ribeiro, J.E.F.T.: Mesonic states and vacuum replicas in potential quark models for QCD. Phys. Rev. D **70**, 094020 (2004)
9. Fischer, C.S.: Infrared properties of QCD from Dyson–Schwinger equations. J. Phys. G Nucl. Part. Phys. **32**, R253–R291 (2006)
10. Hilger, T., Gomez-Rocha, M., Krassnigg, A.: Masses of $J^{PC} = 1^{-+}$ exotic quarkonia in a Bethe–Salpeter-equation approach. Phys. Rev. D **91**, 114004 (2015)
11. Brodsky, S.J., Pauli, H.C., Pinsky, S.S.: Quantum chromodynamics and other field theories on the light cone. Phys. Rep. **301**, 299–486 (1998)
12. Sales, J., Frederico, T., Carlson, B., Sauer, P.: Light front Bethe–Salpeter equation. Phys. Rev. C **61**, 044003 (2000)
13. Allen, T.J., Olsson, M.G., Veseli, S.: From scalar to string confinement. Phys. Rev. D **62**, 094021 (2000)
14. Bali, G.S., Schilling, K., Wachter, A.: Complete O (v**2) corrections to the static interquark potential from SU(3) gauge theory. Phys. Rev. D **56**, 2566–2589 (1997)
15. Lucha, W., Schoberl, F.F., Gromes, D.: Bound states of quarks. Phys. Rep. **200**, 127–240 (1991)
16. Biernat, E.P., Gross, F., Peña, M.T., Stadler, A.: Confinement, quark mass functions, and spontaneous chiral symmetry breaking in Minkowski space. Phys. Rev. D **89**, 016005 (2014)
17. Biernat, E.P., Gross, F., Peña, M.T., Stadler, A.: Pion electromagnetic form factor in the covariant spectator theory. Phys. Rev. D **89**, 016006 (2014)
18. Leitão, S., Stadler, A., Peña, M.T., Biernat, E.P.: Linear confinement in momentum space: singularity-free bound-state equations. Phys. Rev. D **90**, 096003 (2014)
19. Biernat, E.P., Peña, M.T., Ribeiro, J.E., Stadler, A., Gross, F.: Chiral symmetry and π–π scattering in the covariant spectator theory. Phys. Rev. D **90**, 096008 (2014)
20. Bicudo, P., Cotanch, S., Llanes-Estrada, F., Maris, P., Ribeiro, J.E., Szczepaniak, A.: Phys. Rev. D **65**, 076008 (2002)
21. Biernat, E.P., Gross, F., Peña, M.T., Stadler, A.: Charge-conjugation symmetric complete impulse approximation for the pion electromagnetic form factor in the Covariant Spectator Theory. Phys. Rev. D **92**, 076011 (2015)

DOI 10.1007/978-3-319-50699-9_18
Reprinted from *Few-Body System* Journal, DOI 10.1007/s00601-016-1099-8

CrossMark

V. A. Karmanov

Non-perturbative Renormalization in Truncated Yukawa Model

Received: 29 January 2016 / Accepted: 26 March 2016 / Published online: 5 April 2016

Abstract An approach to non-perturbative calculations in the light-front quantum field theory and its new developments are briefly reviewed. We start with the decomposition of the state vector in Fock components. After truncation of this decomposition (main approximation in this approach), the eigenvalue equation for the light-front Hamiltonian generates, in Minkowski space, a finite system of integral equations for the Fock components. Solving this system numerically and performing the non-perturbative renormalization, we find the state vector of fermion in the quenched scalar Yukawa model, up to the four-body truncation (one fermion + three bosons), for rather large values of the coupling constant. With the state vector, found in this way, the fermion electromagnetic form factors are calculated. Comparing results obtained in the four-body truncation with those found in the previous, three-body truncation, we discover very good convergence relative to truncation, that indicates that we are close to the exact non-perturbative solution in this field-theoretical model. The approach can be extended to more realistic field theories and, after further development, it could constitute an alternative to the lattice calculations.

1 Introduction

After impressive achievements in the perturbative field theory (especially, QED) considerable efforts were devoted to find the non-perturbative solutions, needed in QCD. Main works in this field are performed in the framework of the lattice approach by large collaborations and using supercomputers.

In a series of papers [1–6] we developed another, quite different non-perturbative approach. It allows to find the state vector solving the eigenvector equation with a given field-theoretical Hamiltonian. Knowing the state vector, we then can find observables, e.g., electromagnetic form factors. The state vector is represented in the form of a decomposition in the Fock components. In principle, this decomposition contains infinite numbers of terms. It is truncated, i.e., it is replaced by a finite decomposition with N terms. Then the eigenvector equation is reduced to a finite system of N integral equations with the kernels determined by Hamiltonian. The solution of this system (analytical for $N = 2$ and numerical for $N \geq 3$) provides a set of Fock components, i.e., the state vector in a given truncation. It should be noted that there exist few forms of relativistic dynamics: the instant form, the front form and the point form [7]. We work in the front form (light-front dynamics—LFD) [8,9].

The truncation of the decomposition is the main approximation in this approach. The crucial point is the convergence of the results relative to the increase of the dimension of the Fock-space truncation. This convergence can be checked by increasing the number of terms in the decomposition and see, whether the observables converge relative to the truncation. In our last works, in scalar Yukawa model, we have obtained very promising results which show very fast convergence relative to truncation.

V. A. Karmanov (✉)
Lebedev Physical Institute, Leninskii Prospect 53, Moscow, Russia 119991
E-mail: karmanov@sci.lebedev.ru

As always, solving a field-theoretical problem, for instance, finding the state vector, one should carry out the renormalization procedure, i.e., express the bare constants (normally, the bare coupling constant and mass) in the initial Hamiltonian by the physical ones, similarly to what is done in the perturbation theory, though in different way technically.

In this contribution we give a brief review of this non-perturbative approach, including its last developments.

2 Truncations and the Sector Dependent Renormalization

As mentioned above, in quantum field theory, the state vector, satisfying the eigenvector equation, is an infinite superposition (Fock decomposition) of the states with different numbers of particles. After truncation by a finite number N of Fock components it reads:

$$|p\rangle = \sum_{n=1}^{N} \int \psi_n(k_1, \ldots, k_n, p)\, |n\rangle\, D_n. \tag{1}$$

ψ_n is the n-body wave function (Fock component) and D_n is an integration measure. The set of the Fock components ψ_n, $n = 1, \ldots, N$, determines the truncated state vector $|p\rangle$. Our aim is to find ψ_n.

In LFD one uses the variables $\mathbf{k}_{i\perp} = (k_{ix}, k_{iy})$, $k_{i+} = k_{i0} + k_{iz}$, $k_{i-} = k_{i0} - k_{iz}$ (with the on-mass shell condition $k_i^2 = m_i^2 \quad \rightarrow \quad k_{i-} = (\mathbf{k}_{i\perp}^2 + m_i^2)/k_{i+}^2$), so that the wave function depends on $\mathbf{k}_{i\perp}$, $x_i = k_{i+}/p_+$ and the decomposition (1) obtains the form:

$$|p\rangle = \sum_{n=1}^{N} \int \psi_n(\mathbf{k}_{1\perp}, x_1 \ldots, \mathbf{k}_{n\perp}, x_n)\, |n\rangle\, \delta^{(2)}(\mathbf{k}_{1\perp}+\cdots+\mathbf{k}_{n\perp})\delta(x_1+\cdots+x_n-1)2\frac{d^2k_{1\perp}dx_1}{2x_1}\cdots\frac{d^2k_{n\perp}dx_n}{2x_n} \tag{2}$$

In the explicitly covariant version of LFD [8], where the state vector is defined on the light-front plane $\omega\cdot x = 0$, $\omega = (\omega_0, \boldsymbol{\omega})$, $\omega^2 = 0$, one constructs the four-vectors $R_i = k_i - x_i p$, where $x_i = \omega\cdot k_i/\omega\cdot p$. By definition of R_i, $\omega \cdot R_i = 0$. Let $\mathbf{R}_{i\perp} \equiv \mathbf{k}_{i\perp}$ is the component of R_i, orthogonal to $\boldsymbol{\omega}$: $R_i = (R_{i0}, R_{i\|}, \mathbf{R}_{i\perp})$, $\boldsymbol{\omega}\cdot\mathbf{R}_{i\perp} = 0$. Then from $\omega \cdot R_i = \omega_0 R_{i0} - \omega_0 R_{i\|} = 0$ it follows: $R_{i0} = R_{i\|}$ and therefore $\mathbf{k}_{i\perp}^2 = -R_i^2$ and $\mathbf{k}_{i\perp}\cdot\mathbf{k}_{j\perp} = -R_i \cdot R_j$ are the Lorentz invariants.

To find the components ψ_n, we substitute the state vector $|p\rangle$ in the eigenvector equation:

$$H\,|p\rangle = M\,|p\rangle\,.$$

This equation was solved [5,6] in the Yukawa model (with a fermion, coupled to the scalar field), though in its quenched version (up to the truncation $N = 3$, that is, keeping the Fock components with one fermion f and up to two bosons $2b$) and also in the quenched scalar Yukawa model [10,11] (up to the truncation $N = 4$, that is keeping the Fock components with one spinless "fermion" f and up to three bosons $3b$).

Quenched model means that we do not take into account the fermion-antifermion pairs $f\bar{f}$ (contributing, for example, to the boson self-energy). We do not discuss here the validity of the quenched approximation. We need a theoretical polygon, where we can test our approach, develop the non-perturbative renormalization and study the convergence (if any) relative to the Fock-space truncation. Quenched theory, not overcomplicated by antifermions, is just a good candidate for such a polygon. In a quenched theory we still can make decomposition (1) and find the state vector $|p\rangle$ by solving Eq. (2) for a strong coupling constant in the Hamiltonian H (where the processes of the type $b \rightarrow f\bar{f}$ are omitted). In addition, the scalar Yukawa model has no stable vacuum (the ground state energy tends to $-\infty$) [12]. According to [13], this happens because of contributions of infinite number of antifermions. If the number of incorporated antifermions is finite (zero in the quenched theory), it is expected that the vacuum is stable. Some antiparticle contributions were taken into account in Refs. [5,6]. The full incorporation of the antiparticles and the study of their influence (in a given truncation) is, of course, a very interesting next step, to be done in future. In the quenched Yukawa model the Hamiltonian H generates two interaction vertices only. Namely, the vertex corresponding to creation of boson by fermion $f \rightarrow fb$ (and analogous vertex for absorption $fb \rightarrow f$); these vertices change the number of particles by ± 1; and the vertex with the fermion mass counter term δm on the fermion line; this vertex does not change the number of particles. The vertex $b \rightarrow f\bar{f}$ (and analogously, $f\bar{f} \rightarrow b$) is absent in a quenched theory. The latter also forbids the boson mass renormalization and, hence, the boson mass counter term $\delta\mu$.

The system of equations for the Fock components in the truncation $N = 2$ is shown in Fig. 1.

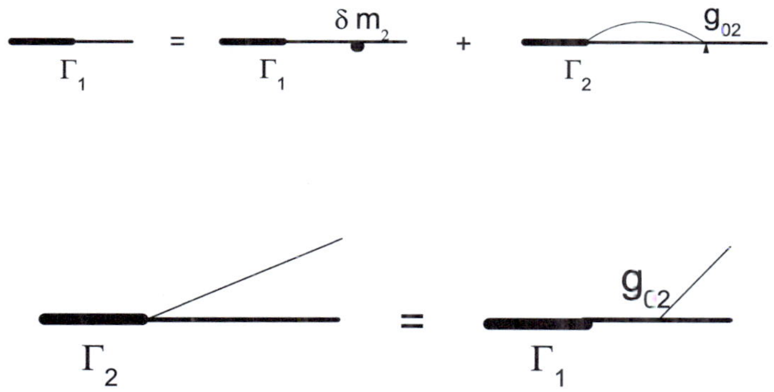

Fig. 1 System of equations for the vertex functions $\Gamma_{1,2}$ in the two-body truncation

The two vertices discussed above just appear in the graphs of Fig. 1, though the corresponding constants obtain subscripts referred to the order of truncation. The subscript "2" at the mass counter term δm_2 means the this counter term is found, using the renormalization condition, in the two-body truncation (in which the Fock components are determined by the system of equations shown in Fig. 1). The coupling constant g_{02} contains the subscripts "02". The subscript "0" means that it is bare coupling constant (to be expressed by the renormalization condition, discussed later, via the physical one), the subscript "2", like for δm_2, means that g_{02} is expressed via g in the two-body truncation. These notations will be extended to the higher truncations. As mentioned, the system defined in Fig. 1 has analytical solution. Dependence of the coupling constant and the mass counter term on the order of truncation results in the sector dependent renormalization scheme.

The essence of the sector dependent renormalization can be explained as follows. The non-truncated system of equations for the Fock components, being represented as an infinite series of perturbative graphs, contains all the possible intermediate states. Let us consider all the perturbative graphs in a given order g^n. This set of perturbative graphs contains the graphs with different numbers of intermediate states. Restricting the decomposition (1) by a finite number of terms N, we omit the perturbative graphs of the order g^n with the intermediate states containing more than N particles. This destroys cancellation of infinities after renormalization (for cancellation in a given order g^n all perturbative graphs are needed). The sector dependent renormalization scheme is constructed to ensure the cancellation of infinities even in a restricted set of perturbative graphs, which contribute in full set of graphs with a given number of intermediates states. The system of equations for the Fock components in the truncation $N = 3$ is shown in Fig. 2. First line of Fig. 2 contains true three-body states and therefore it contains the three-body constants δm_3, g_{03}, to be found from the renormalization conditions. However, other lines of this figure contain also two-body substates (which are under the extra line). Therefore the two-body constants δm_2, g_{02}, known from the $N = 2$ solution, appear in these substates. This procedure continues to higher truncations.

The system of equations for the Fock components in the truncation $N = 4$ is shown in Fig. 3. . This system of equations contains (already found) two-body constants δm_2, g_{02}, the three-body ones δm_3, g_{03} and (to be found from the renormalization conditions) the four-body constants δm_4, g_{04}.

The complexity of the problem increases because of the increase in the number of arguments of the many-body vertex function. One can always exclude from equations the highest Fock components (e.g., in the $N = 3$ truncation, using the last equation in Fig. 2 one can exclude Γ_3, i.e., to express it via Γ_2). Therefore in the three-body truncation the complexity is determined by the two-body vertex function depending (after using the momentum conservation) on two arguments: $\Gamma_2 = \Gamma_2(\mathbf{k}_\perp, x)$. In practice, to find Γ_2 in the truncation $N = 3$ is not a difficult numerical problem. For the four-body truncation, after excluding the four-body component, the system of equations contains the three-body vertex function depending (after using the momentum conservation) on two 2D vector and two scalar arguments: $\Gamma_3 = \Gamma_3(\mathbf{k}_{1\perp}, x_1; \mathbf{k}_{2\perp}, x_2)$, i.e., on five scalar arguments: $k_{1\perp}, k_{2\perp}, \mathbf{k}_{1\perp} \cdot \mathbf{k}_{2\perp}, x_1, x_2$. The numerical solution of this problem already requires high-power computers. Each extra particle adds three scalar arguments.

3 Results and Conclusion

The solution of the system of equations shown in Fig. 2 (the three-body truncation) was found in Refs. [5,6] for the quenched fermionic Yukawa model. As regulators, we introduced in the decomposition (1) the Pauli–Villars

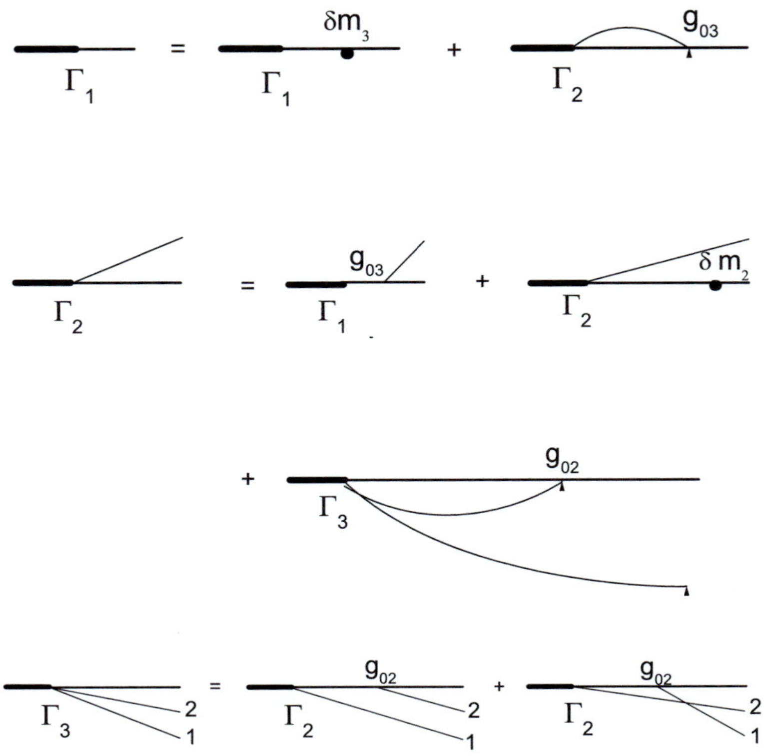

Fig. 2 System of equations for the vertex functions Γ_{1-3} in the truncation $N = 3$

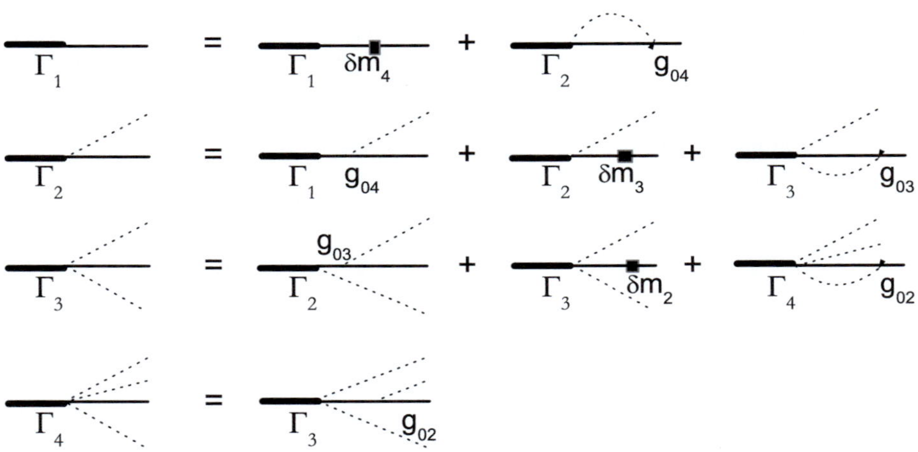

Fig. 3 System of equations for the vertex functions Γ_{1-4} in the truncation $N = 4$

particles and, at the end, took numerically their very large values and checked stability of our results relative this limit. By means of this solution for the Fock components the anomalous magnetic moment of fermion and electromagnetic form factors were calculated [5,6]. They turned out to be rather stable relative to relative to the increase of the Pauli–Villars regulator mass. This shows that the sector dependent non-perturbative renormalization scheme indeed cancels infinities, like it happens in perturbation theory. This is an important point.

In the scalar (with the spinless "fermion") quenched Yukawa model the solution of the system of equations with the four-body truncation (Fig. 3) was found in Refs. [10,11]. In Fig. 4 (taken from [10,11]) the left panel presents the two-body light-front wave function $\psi_2(k_\perp, x)$ for the coupling constant $\alpha = g^2/(16\pi m^2) = 1$ calculated in the four-body truncation. This wave function has typical behavior: for fixed x it decreases when $k_\perp \to \infty$ and for fixed $k_\perp$ it decreases when $x \to 0$ and $x \to 1$. However, it is not symmetric relative to

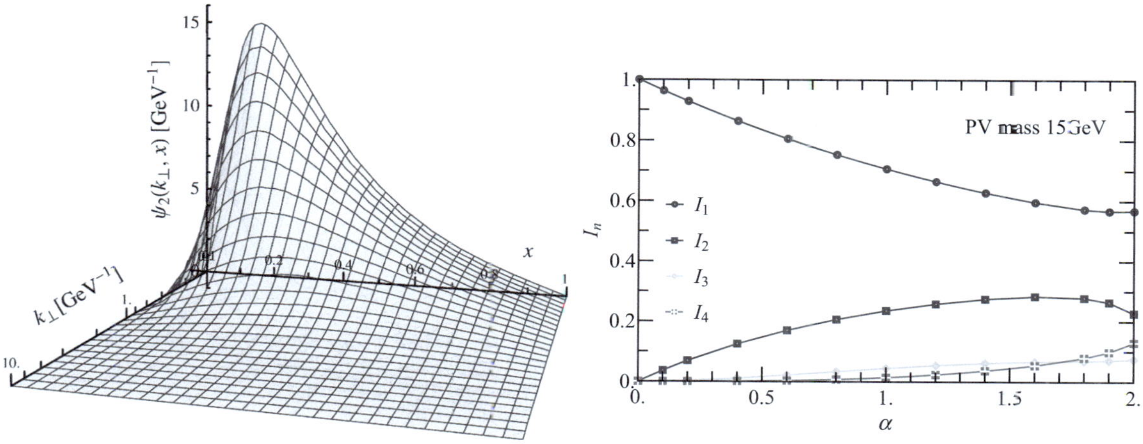

Fig. 4 *Left panel* The two-body light-front wave function $\psi_2(k_\perp, x)$ for $\alpha = 1.$ calculated in the four-body truncation. *Right panel* Contributions of the n-body Fock sectors I_n for $n = 1, 2, 3, 4$ to full norm ($= 1$) of the state vector found in the four-body truncation

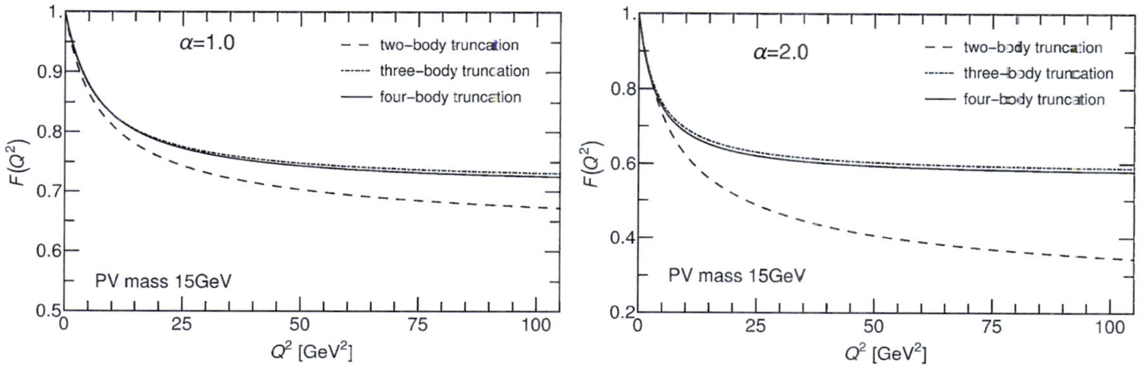

Fig. 5 Comparison of the form factors calculated in the two-, three and four-body truncations at $\alpha = 1.0$ (*left*) and $\alpha = 2.0$ (*right*)

$x = 1/2$, since the constituents (scalar fermion and bosons) have different masses m and μ. Therefore for $k_\perp = 0$ it has the maximum at $x = \mu/(\mu + m)$.

The right panel of Fig. 4 presents the contributions I_n to the normalization integral (equated to 1), of the one-, two-, three-, and four-body components also calculated in the four-body truncations for $\alpha = 1$ and for the mass of the Pauli–Villars (boson) regulator of 15 GeV (for which dependence on the regulator mass becomes negligible). The contributions I_n for $n = 1, 2, 3$ decrease when the number of particles n increases. The one-body component I_1 dominates.

Another very important point is the check of convergence relative to the increase in the dimension of the Fock-space truncation. Though we had some indirect indications of convergence [14], unambiguous and convincing conclusions can be done by comparing results with a higher truncation of the Fock-space. For the present, this study was performed in [10,11] in the scalar Yukawa model (still quenched, with spinless "fermion"), by comparison of electromagnetic form factor of two-, three- and four-body truncations. The electromagnetic form factor versus momentum transfer Q^2 (more precisely, the electric form factor; there is only one—electric—form factor in the model with spinless "fermion") calculated in the two-, three- and four-body truncations is shown in Fig. 5 for two values of the coupling constant $\alpha = 1.0$ and $\alpha = 2.0$. Though the two- and three-body truncation results considerably differ from each other, the three- and four-body ones show very good agreement even at the non-perturbative couplings, suggesting a convergence with respect to the Fock sector expansion.

This calculation demonstrates the validity and efficiency of non-perturbative approach to the quantum field theory presented above. It is constructed in Minkowski space and deals with most fundamental quantum-mechanical quantities: wave functions. Calculation of electromagnetic form factors in terms of the wave functions is direct and easy. After further development and adjusting to more realistic theories, it could constitute an alternative to lattice calculations.

References

1. Bernard, D., Cousin, T., Karmanov, V.A., Mathiot, J.-F.: Nonperturbative renormalization in a scalar model within light-front dynamics. Phys. Rev. D **65**, 025016 (2001)
2. Karmanov, V.A., Mathiot, J.-F., Smirnov, A.V.: Renormalized nonperturbative fermion model in covariant light front dynamics. Phys. Rev. D **69**, 045009 (2004)
3. Karmanov, V.A., Mathiot, J.-F., Smirnov, A.V.: Systematic renormalization scheme in light-front dynamics with Fock space truncation. Phys. Rev. D **77**, 085028 (2008)
4. Mathiot, J.-F., Smirnov, A.V., Tsirova, N.A., Karmanov, V.A.: Nonperturbative renormalization in light-front dynamics and applications. Few-Body Syst. **49**, 183 (2011)
5. Karmanov, V.A., Mathiot, J.-F., Smirnov, A.V.: Nonperturbative calculation of the anomalous magnetic moment in the Yukawa model within truncated Fock space. Phys. Rev. D **82**, 056010 (2010)
6. Karmanov, V.A., Mathiot, J.-F., Smirnov, A.V.: Ab initio nonperturbative calculation of physical observables in light-front dynamics. Application to the Yukawa model. Phys. Rev. D **86**, 085006 (2012)
7. Dirac, P.A.M.: Forms of relativistic dynamics. Rev. Mod. Phys. **21**, 392 (1949)
8. Carbonell, J., Desplanques, B., Karmanov, V.A., Mathiot, J.-F.: Explicitly covariant light-front dynamics and relativistic few-body systems. Phys. Rep. **300**, 215 (1998)
9. Brodsky, S.J., Pauli, H., Pinsky, S.S.: Quantum chromodynamics and other field theories on the light cone. Phys. Rep. **301**, 299 (1998)
10. Li, Y., Karmanov, V.A., Maris, P., Vary, J.P.: Non-Perturbative calculation of the scalar Yukawa theory in four-body truncation. Few-Body Syst. **56**, 495 (2015)
11. Li, Y., Karmanov, V.A., Maris, P., Vary, J.P.: Ab initio approach to the non-perturbative scalar Yukawa model. Phys. Lett. B **748**, 278 (2015)
12. Baym, G.: Inconsistency of cubic Boson–Boson interactions. Phys. Rev. **117**, 886 (1960)
13. Gross, F., Şavkli, Ç., Tjon, J.: Stability of the scalar $\chi^2\phi$ interaction. Phys. Rev. D **64**, 076008 (2001)
14. Hwang, D.S., Karmanov, V.A.: Many-body Fock sectors in Wick–Cutkosky model. Nucl. Phys. B **696**, 413 (2004)

DOI 10.1007/978-3-319-50699-9_19
Reprinted from *Few-Body System* Journal, DOI 10.1007/s00601-016-1081-5

W. N. Polyzou · Marc Herrmann

The Light-Front Vacuum

Received: 8 January 2016 / Accepted: 4 March 2016 / Published online: 15 March 2016

Abstract We discuss the relation between the trivial light-front vacuum and the non-trivial Heisenberg vacuum.

1 Introduction

The light-front representation [1] of quantum field theory [2–10] has the advantage that the vacuum of the interacting theory is identical to the Fock vacuum of the corresponding free field theory. This is in contrast to the canonical representation of field theory where the interacting vacuum has infinite norm in the free field Hilbert space. On the other hand the light-front and canonical representations of the same field theory should be equivalent. The purpose of this work is to understand the relation between the vacuum vectors in these two representations of the same field theory.

There is a large body of work devoted to the properties of the light front vacuum. Motivated by work by Fubini and Furlan [11] and Susskind [12] on current algebras in the infinite momentum limit, Weinberg pointed out the general suppression of vacuum diagrams in the infinite momentum frame [13]. A number of authors, including [14–23], have investigated properties of the vacuum and zero modes in light-front field theory and its role in chiral symmetry breaking. The discussion in this work is motivated by the algebraic aspects of this question discussed by Luetwyler Kaluder and Streit [3] and Schlieder and Seiler [7].

2 Triviality of the Light-Front Vacuum

The light front is a hyperplane tangent to the light cone defined by $x^+ = x^0 + \hat{\mathbf{n}} \cdot \mathbf{x} = 0$. Translations tangent to the light-front hyperplane are independent of interactions and are generated by the following components of the four momentum

$$P^+ := P^0 + \hat{\mathbf{n}} \cdot \mathbf{P} \geq 0 \qquad \mathbf{P}_\perp := \mathbf{P} - \hat{\mathbf{n}}(\hat{\mathbf{n}} \cdot \mathbf{P}) \tag{1}$$

Interactions, V, appear in the generator of translations normal to the light-front hyperplane

$$P^- = P_0^- + V. \tag{2}$$

Research Supported by the US Department of Energy under Grant Number No. DE-FG02-86ER40286 with the University of Iowa.

W. N. Polyzou (✉) · M. Herrmann
The University of Iowa, Iowa City, IA, USA
E-mail: polyzou@uiowa.edu

The translation generators P^+ and $\mathbf{P}_\perp$ are sums of single-particle operators

$$P^+ = \sum_n p_n^+; \qquad \mathbf{P}_\perp = \sum_n \mathbf{p}_{n\perp}, \tag{3}$$

where each $p_n^+ \geq 0$.

The origin of the triviality of the light-front vacuum follows from the observation that both P^- and P_0^- commute with P^+. This means that $V|0\rangle$, where $|0\rangle$ is the free Fock vacuum, is an eigenstate of P^+ with eigenvalue zero. If V is expressed in normal ordered form, the part of V with all creation operators can only increase the value of p_n^+

$$V = \int \mathcal{V}(\tilde{\mathbf{p}}_1, \ldots, \tilde{\mathbf{p}}_n) a^\dagger(\tilde{\mathbf{p}}_1) \ldots a^\dagger(\tilde{\mathbf{p}}_n) d\tilde{\mathbf{p}}_1 \ldots d\tilde{\mathbf{p}}_n + \ldots, \tag{4}$$

however the sum of all of the non-negative p_n^+ must vanish, so it follows that either $\mathcal{V}|0\rangle = 0$ or the pure creation operator coefficient is proportional to $\delta(p_n^+)$ or derivatives of $\delta(p_n^+)$. We refer to these contributions, if they appear, as zero modes. These contributions cannot appear if the kernel $\mathcal{V}$ is smooth. In this case the interaction does not change the Fock vacuum. This does not happen in the canonical case because although the interaction is translationally invariant, the single-particle momenta can sum to zero without being all zero.

3 Characterization of the Vacuum

The simplest way to understand the relation between the light-front and canonical vacua is to first consider the case of free fields, where the precise relations between the Heisenberg field, fields restricted to a fixed time surface and fields restricted to a light front are known. For a free scalar field the fixed time and light front representations of the field are given by the well-known expressions:

$$\phi(x) = \frac{1}{(2\pi)^{3/2}} \int \frac{d\mathbf{p}}{\sqrt{2\omega_m(\mathbf{p})}} (e^{ip\cdot x} a(\mathbf{p}) + e^{-ip\cdot x} a^\dagger(\mathbf{p})) \tag{5}$$

$$= \frac{1}{(2\pi)^{3/2}} \int \frac{d\tilde{\mathbf{p}}}{\sqrt{2p^+}} (e^{ip\cdot x} a(\tilde{\mathbf{p}}) + e^{-ip\cdot x} a^\dagger(\tilde{\mathbf{p}})), \tag{6}$$

where $\omega_m(\mathbf{p})$ is the energy of a particle of mass m and momentum $\mathbf{p}$ and $\tilde{\mathbf{p}} = (p^+, \mathbf{p}_\perp)$.

In either representation the vacuum is normally defined by the condition that it is annihilated by the annihilation operator. Identifying these two representations of the field and changing variables from the three momenta to the light-front components of the four momentum gives the relation between the annihilation operators

$$a_i(\mathbf{p}) = a_i(\tilde{\mathbf{p}}) \sqrt{\frac{p^+}{\omega_{m_i}(\mathbf{p})}}. \tag{7}$$

This suggest that the vacuum defined by the annihilation operator in these two representations is the same. When the field is restricted to the light front all of the mass dependent (dynamical) terms in (6) disappear. Furthermore, the field restricted to the light front is irreducible (i.e. it is possible to extract both the creation and annihilation operator directly from the field restricted to the light front - there is no need for derivatives off of the light front). It follows that all inner products and all expectation values of operators are identical, which means that free fields with different masses restricted to the light front are unitarily equivalent. This suggests that there is one vacuum vector, up to unitary equivalence, for free scalar fields of any mass. On the other hand, the creation and annihilation operators for two free scalar fields with different masses are related by the canonical transformation [24]

$$a_2(\mathbf{p}) = \cosh(\eta(\mathbf{p})) a_1(\mathbf{p}) + \sinh(\eta(\mathbf{p})) a_1^\dagger(\mathbf{p}) \tag{8}$$

where

$$\cosh(\eta(\mathbf{p})) = \frac{1}{2}\left(\sqrt{\frac{\omega_{m_2}}{\omega_{m_1}}} + \sqrt{\frac{\omega_{m_1}}{\omega_{m_2}}}\right). \tag{9}$$

The problem is that the generator of the associated unitary transformation

$$G = -i \int \frac{d\mathbf{p}\eta(\mathbf{p})}{2} \left(a_1(\mathbf{p})a_1(\mathbf{p}) - a_1^\dagger(\mathbf{p})a_1^\dagger(\mathbf{p}) \right) \tag{10}$$

has infinite norm when applied to the vacuum of particle 1, which means the canonical transformation (8) cannot be realized as a unitary transformation on the Hilbert space associated with particle 1.

These results have the appearance of being inconsistent - specifically identifying the vacuum of the canonical and light-front fields of the same mass, and the vacuum of the light-front fields with different masses, suggest that all of the vacua are identical or related by a unitary transformation. On the other hand a direct comparison on the vacua of canonical field theories with different masses show that they are not even in the same Hilbert space.

The resolution of this apparent inconsistency is that the annihilation operator does not completely characterize the vacuum [25]. Another characterization of the vacuum is as an invariant linear functional on an algebra of operators. The relevant algebras are the algebra of free fields integrated against test functions of four space-time variables, the algebra of free fields restricted to a light front, integrated against test functions in coordinates of the light-front hypersurface, and the algebra of free fields and their time derivatives restricted to a fixed time surface, integrated against test functions in three space variables. We call these algebras the local algebra, the light-front algebra, and the canonical algebra respectively. Each algebra is invariant with respect to a different symmetry group.

The formal expressions (5–6) of the fields in terms of creation and annihilation operators can still be used to construct elements of each of these algebras by making the appropriate restrictions and integrating against the appropriate test functions.

From the algebraic perspective it is clear that both the light-front and canonical algebras do not contain enough operators to formulate Lorentz invariance or locality. Schlieder and Seiler [7] give an example that illustrates the importance of the underlying algebra. They consider two free fields of different mass, but they restrict the algebra by limiting the test functions to functions with Fourier transforms that are related on the two different mass shells by

$$\frac{f(\sqrt{m_1^2+\mathbf{p}^2},\mathbf{p})}{(m_1^2+\mathbf{p}^2)^{1/4}} = \frac{f(\sqrt{m_2^2+\mathbf{p}^2},\mathbf{p})}{(m_2^2+\mathbf{p}^2)^{1/4}}. \tag{11}$$

For this class of test functions

$$\langle 0_1|\phi_1(f_1)\cdots\phi_1(f_n)|0_1\rangle = \langle 0_2|\phi_2(f_1)\cdots\phi_2(f_n)|0_2\rangle. \tag{12}$$

In this case the inner products are identical, and the two fields restricted to this algebra are related by a unitary transformation. In addition, because there is no restriction on the test functions on one mass shell, it follows that both of these algebras are irreducible. On the other hand, these properties are not preserved when the restricted class of test functions are enlarged to include the full set of test functions. This illustrates the importance of the role of the algebra.

In the case of the two representations of the free fields, the structure of the field (5–6) plays a key role in extending the algebra from an algebra of operators restricted to a hypersurface to the local algebra. In the case of the light-front representation the free Heisenberg field of mass m can be expressed as a linear operator acting on a field restricted to a light front

$$\phi(x) = \int F_m(x,\tilde{\mathbf{y}})\phi(\tilde{\mathbf{y}})d\tilde{\mathbf{y}} \tag{13}$$

where

$$F_m(x,\tilde{\mathbf{y}}) = \frac{1}{(2\pi)^2}\int \frac{d\tilde{\mathbf{p}}}{2} e^{-i\frac{p_\perp^2+m^2}{p^+}x^+ + i\tilde{\mathbf{p}}\cdot(\tilde{\mathbf{x}}-\tilde{\mathbf{y}})}. \tag{14}$$

Here the kernel $F_m(x,\tilde{\mathbf{y}})$ that defines this extension has the mass dependence, which is the dynamical information for a free field. We can also alternatively interpret this kernel as a map of test functions in four variables to test functions restricted to a light front. The Fourier transforms of these test functions are related by

$$\tilde{f}(\tilde{\mathbf{p}}) = \tilde{f}\left(\frac{\mathbf{p}_\perp^2+m^2}{p^+},\tilde{\mathbf{p}}\right). \tag{15}$$

The important observation is that because the Fourier transform of a Schwartz function in four variables is a Schwartz function, the function $\tilde{f}(\tilde{\mathbf{p}})$ vanishes faster than $(p^+)^n$ for any n at the origin. We will see that this observation has important implications when we discuss properties of the light-front Fock algebra.

We can interpret $F_m(x, \tilde{\mathbf{y}})$ as a mapping from the local Heisenberg algebra to a sub-algebra of the light-front Fock algebra that preserves all Wightman functions, and is hence unitary. Under this mapping the vacuum becomes the free field Fock vacuum, and the dynamics is contained in the mapping.

4 Zero Modes and Dynamics

In order to consider the potential role of zero modes it is useful to summarize the key properties of the light-front Fock algebra [3,7,21], that is that algebra generated by free fields restricted to a light front. Operators in this algebra have the form

$$\phi(\tilde{f}) = \int d\tilde{\mathbf{x}} \phi(\tilde{\mathbf{x}}, x^+ = 0) f(\tilde{\mathbf{x}}). \tag{16}$$

This algebra is closed under operator multiplication, which can be summarized by

$$e^{i\phi(\tilde{f})} e^{i\phi(\tilde{g})} = e^{i\phi(\tilde{f}+\tilde{g})} e^{-\frac{1}{2}((\tilde{f},\tilde{g})-(\tilde{g},\tilde{f}))} \tag{17}$$

where the light-front scalar product is given by

$$(\tilde{f}, \tilde{g}) = \int \frac{d\tilde{\mathbf{p}} \theta(p^+)}{p^+} f(-\tilde{\mathbf{p}}) g(\tilde{\mathbf{p}}). \tag{18}$$

One of the important properties of this algebra is that it is irreducible. The simplest way to understand this is to note that it is possible to extract the creation and annihilation parts of $\phi(\tilde{f})$ without extending the algebra. Specifically the creation and annihilation operators are related to the Fourier transform of the restricted field by

$$\phi_+(\tilde{\mathbf{p}}) := \theta(p^+)\phi(\tilde{\mathbf{p}}) = \frac{\theta(p^+)}{\sqrt{p^+}} a(\tilde{\mathbf{p}}) \tag{19}$$

$$\phi_-(\tilde{\mathbf{p}}) := \theta(-p^+)\phi(\tilde{\mathbf{p}}) = \frac{\theta(-p^+)}{\sqrt{-p^+}} a^\dagger(-\tilde{\mathbf{p}}) \tag{20}$$

$$\phi(\tilde{\mathbf{p}}) = \phi_+(\tilde{\mathbf{p}}) + \phi_-(\tilde{\mathbf{p}}). \tag{21}$$

One consequence of this is that there is a purely algebraic normal ordering, which can be summarized by

$$: e^{i\phi(\tilde{f})} := e^{i\phi_-(\tilde{f})} e^{i\phi_+(\tilde{f})}. \tag{22}$$

It follows, using the Campbell-Baker-Hausdorff formula that

$$e^{i\phi(\tilde{f})} =: e^{i\phi(\tilde{f})} : e^{\frac{1}{2}(\tilde{f},\tilde{f})}. \tag{23}$$

We note that independent of the choice of vacuum, if the test function vanishes for $p^+ = 0$, the light-front inner product in (23) is well behaved and the vacuum expectation value of the normal product must be one, since for $p^+ \neq 0$ the annihilation operator necessarily reduces the value of p^+, which means that it must annihilate the vacuum. In this case the vacuum is uniquely determined by this equation. Since the operator F_m maps four variable test functions into functions on the light front with Fourier transforms that vanish for $p^+ = 0$, the vacuum is always the free field Fock vacuum in this case and there can be no zero mode contributions.

It is well known that the equivalence of calculations of some observables based on light-front field theory and canonical field theory require 0-mode contributions in the light-front expressions. So far we have only considered free fields, so there is a question concerning whether zero modes are a consequence of a more complex dynamics or not.

To answer this first note that the classical Noether's theorem gives formal expressions for the Poincaré generators as local products of operators at the same point on the light front. These products are not elements of the light-front algebra. One can see the problem in the following expression for the two point Wightman function in light-front variables:

$$\langle 0|\phi(x)\phi(y)|0\rangle = \tag{24}$$

$$\frac{1}{(2\pi)^3}\int \frac{d\tilde{\mathbf{p}}}{p^+} e^{-i\frac{\mathbf{p}_\perp^2+m^2}{p^+}(x^+-y^+)+i\tilde{\mathbf{p}}(\tilde{\mathbf{x}}-\tilde{\mathbf{y}})}. \tag{25}$$

Although there is an apparent logarithmic singularity due to the p^+ integral, if $(x^+ - y^+)$ is not zero the $1/p^+$ singularity in the above expression is regulated by oscillations in the exponent This can be seen by considering the integral of the same form

$$\int_0^a \frac{e^{ic/p^+}}{p^+} dp^+ = \int_{c/a}^\infty \frac{e^{iu}}{u} du = \frac{\pi}{2} - (Ci(c/a) + i Si(c/a)). \tag{26}$$

If both fields are restricted to the light front, this mechanism that regulates the singularity at $p^+ = 0$ is turned off leading an infrared singularity. There will also be ultraviolet singularities associated with the other variables. In this case there is no reason for the mapped test functions to vanish at $p^+ = 0$. In this case if we try to make sense out of (25) when the test function does not vanish at $p^+ = 0$ two things happen. First the logarithmic singularity in the light front scalar product needs to be regularized. Typically this regularization breaks longitudinal boost invariance. A second thing that can happen is that because it is possible to have non-zero contributions from $p^+ = 0$, the algebraic normal product can be extended to include additional singular contributions that have support on $p^+ = 0$. These can have a non-trivial dependence on transverse momentum and may be needed to restore full rotational covariance and longitudinal boost invariance.

It follows that zero modes are a consequence of renormalizing local operator products. They are available tools that may be needed to make the theory finite and consistent with the corresponding canonical theory.

We note that Lorentz invariance of the S-matrix in a light-front quantum theory is equivalent to invariance with respect to a change of orientation of the light front [26]. Changing the orientation of the light front can exchange infrared and ultraviolet singularities. This suggest that the problem of zero modes and renormalization is more complicated in 3+1 dimensional theories than it is in 1+1 dimensional theories.

While the above analysis is limited to free fields, in an asymptotically complete field theory an interacting Heisenberg field can be expanded in normal ordered products of an irreducible set of asymptotic (in or out) fields [24,27]:

$$\phi(x) = \sum \int L(x; z_1, \ldots, z_n) : \phi_{in}(z_1) \ldots \phi_{in}(z_n) : dz_1 \ldots dz_n. \tag{27}$$

The asymptotic fields are free fields, and following the above analysis, can be expressed as products of free fields on the light front using the mapping (14):

$$\phi(x) = \sum \int \mathcal{L}(x; \tilde{\mathbf{y}}_1, \ldots, \tilde{\mathbf{y}}_n) : \phi_0(\tilde{\mathbf{y}}_1) \cdots \phi_0(\tilde{\mathbf{y}}_n) : d\tilde{\mathbf{y}}_1 \cdots d\tilde{\mathbf{y}}_n \tag{28}$$

$$\mathcal{L}(x; \tilde{\mathbf{y}}_1, \ldots, \tilde{\mathbf{y}}_n) = \int L(x; z_1, \ldots, z_n) \prod_i F_m(z_i, \tilde{\mathbf{y}}_i) d^4 z_1 \cdots d^4 z_n. \tag{29}$$

Furthermore, if the Heisenberg field $\phi(x)$ is an operator valued distribution we expect that the smeared coefficient will behave like (15):

$$\int f(x) \tilde{\mathcal{L}}(x; \tilde{\mathbf{p}}_1, \ldots, \tilde{\mathbf{p}}_n) d^4 x \to 0 \quad \text{any} \quad p_i^+ \to 0. \tag{30}$$

If this is the case then $\{\mathcal{L}(x; \tilde{\mathbf{y}}_1, \ldots, \tilde{\mathbf{y}}_n)\}$ is the kernel of a unitary mapping from the Heisenberg algebra of the local field field theory to a sub-algebra of the light-front Fock algebra, where the unitary transformation maps the Heisenberg vacuum to the trivial Fock vacuum:

$$\phi(x) = \sum \int \mathcal{L}(x; \tilde{\mathbf{y}}_1, \ldots, \tilde{\mathbf{y}}_n) : \phi_0(\tilde{\mathbf{y}}_1) \cdots \phi_0(\tilde{\mathbf{y}}_n) : d\tilde{\mathbf{y}}_1 \cdots d\tilde{\mathbf{y}}_n. \tag{31}$$

At this level zero modes cannot occur, however they can occur as intermediate steps in the construction of the coefficient functions $\{\mathcal{L}(x; \tilde{\mathbf{y}}_1, \ldots, \tilde{\mathbf{y}}_n)\}$.

The conclusion of this work is that there is a unitarily equivalent representation of the dynamics with the free-field vacuum. The Hilbert space is generated from this vacuum by a sub-algebra of the light-front Fock algebra. While the vacuum functional has no zero mode contributions on this sub algebra, zero modes can appear in intermediate steps in the construction of the operators that map the local Heisenberg algebra to this sub-algebra.

References

1. Dirac, P.A.M.: Forms of relativistic dynamics. Rev. Mod. Phys. **21**, 392 (1949)
2. Bardaki, K., Halpern, M.B.: Theories at infinite momentum. Phys. Rev. **176**, 1686 (1968)
3. Leutwyler, H., Klauder, J.R., Streit, L.: Quantum field theory on lightlike slabs Il. Nuovo Cimento A **66**, 536 (1970)
4. Kogut, J.B., Soper, D.E.: Quantum electrodynamics in the infinite-momentum frame. Phys. Rev. D **1**, 2901 (1970)
5. Bjorken, J.D., Kogut, J.B., Soper, D.E.: Quantum electrodynamics at infinite momentum: scattering from an external field. Phys. Rev. D **3**, 1382 (1970)
6. Neville, R.A., Rohrlich, F.: Quantum electrodynamics on null planes and Applications to lasers. Phys. Rev. D **3**, 1692 (1971)
7. Schlieder, S., Seiler, E.: Remarks on the null plane development of a relativistic quantum field theory. Commun. Math. Phys. **25**, 62 (1972)
8. Chang, S.J., Root, R.G., Yan, T.M.: Quantum field theories in the infinite momentum frame. 1. quantization of scalar and Dirac fields. Phys. Rev. D **7**, 1133 (1973)
9. Chang, S.J., Yan, T.M.: Quantum field theories in the infinite momentum frame. 2. scattering matrices of scalar and Dirac fields. Phys. Rev. D **7**, 1147 (1973)
10. Brodsky, S., Pauli, H.-C., Pinsky, S.: Quantum chromodynamics and other field theories on the light cone. Phys. Rep. **301**, 299 (1998)
11. Fubini, S., Furlan, G.: Renormalization effects for partially conserved currents. Physics **1**, 229 (1965)
12. Susskind, L.: Model of self-induced strong interactions. Phys. Rev. **165**, 1535 (1968)
13. Weinberg, S.: Dynamics at infinite momentum. Phys. Rev. **150**, 1313 (1966)
14. Maskawa, T., Yamawaki, K.: The problem of $p^+ = 0$ mode in the null plane field theory and Dirac's method of quantization. Prog. Theor. Phys. **56**, 270 (1976)
15. Harindranath, A., Vary, J.P.: Stability of the vacuum in scalar field models in $1+1$ dimensions. Phys. Rev. D **37**, 1067 (1988)
16. Heinzl, Th., Krusche, St., Werner, E.: Light cone quantization and manifestation of non-perturbative vacuum properties for scalar field theory. Z. Phys. A **334**, 443 (1989)
17. Lenz, F., Thies, M., Levit, S., Yazaki, K.: Hamiltonian formulation of two-dimensional gauge theories on the light cone. Ann. Phys. **208**, 1 (1991)
18. Hornbostel, K.: Nontrivial vacua from equal time to the light cone. Phys. Rev. D **45**, 3781 (1992)
19. McCartor, G., Robertson, D.G.: Bosonic zero modes in discretized light-cone field theory. Z. Phys. C **53**, 679 (1992)
20. Tsujimaru, Sho, Yamawaki, Koichi: Zero mode and symmetry breaking on the light front. Phys. Rev. D **57**, 4942 (1998)
21. Coester, F., Polyzou, W.N.: Vacuum Structures in Hamiltonian Light Front Dynamics. Found. Phys. **24**, 387 (1994)
22. Burkhardt, M.: Trivial vacua, high orders in perturbation theory, and nontrivial condensates. Phys. Rev. D **53**, 933 (1996)
23. Martinovic, Lubomir: Gauge symmetry and the light-front vacuum structure. Phys. Lett. B **509**, 355 (2001)
24. Haag, R.: On quantum field theories. Mat-Fys. Medd. K. Danske Vidensk. Selsk. **29**, 1 (1955)
25. Herrmann, M., Polyzou, W.N.: The light front vacuum. Phys. Rev. D **91**, 085043 (2015)
26. Polyzou, W.N.: Left Ccoset invariance and relativistic invariance. Few-Body Syst. **27**, 57 (1999)
27. Glaser, V., Lehmann, H., Zimmermann, W.: Field operators and retarded functions. Il Nuovo Cimento **6**, 1122 (1957)

DOI 10.1007/978-3-319-50699-9_20
Reprinted from *Few-Body System* Journal, DOI 10.1007/s00601-016-1044-x

CrossMark

Chandan Mondal · Dipankar Chakrabarti

A Comparative Study of Nucleon Structure in Light-Front Quark Models in AdS/QCD

Received: 6 January 2016 / Accepted: 18 January 2016 / Published online: 23 February 2016

Abstract We present a comparative study of nucleon structure such as electromagnetic form factors, transverse charge and magnetization densities in three different models within AdS/QCD framework.

1 Introduction

In recent years, AdS/QCD has emerged as one of the most promising techniques to unravel the structure of mesons and nucleons. AdS/CFT conjecture relates a strongly coupled gauge theory in d space-time dimensions by a dual weak coupling gravity theory in AdS_{d+1} space. To exploit this duality to address the problems in QCD, the conformal invariance needs to be broken. In the literature there are two methods to achieve this goal, one is called hard wall model where a sharp cut-off is put in the hologhaphic direction in the AdS space where the wave functions are made to vanish and the other is called the soft wall model in which a confining potential is introduced in the AdS space which breaks the conformal invariance and allows the QCD mass scale.

Electromagnetic form factors provide us insights into the structure of the nucleons and have been measured in many experiments. In the light-cone frame with $q^+ = q^0 + q^3 = 0$, the charge and anomalous magnetization densities in the transverse plane can be identified with the two-dimensional Fourier transform(FT) of the electromagnetic form factors. The contributions of individual quark to the nucleon charge and magnetization densities are obtained from the flavor decompositions of the transverse densities.

Here we present a detailed analysis of the nucleon form factors in AdS/QCD soft-wall models [1–6]. The flavor form factors are obtained by decomposing the Dirac and Pauli form factors for nucleons using the charge and isospin symmetry. We also present a comparative study of the nucleon as well as the flavor contributions to the nucleon charge and anomalous magnetization densities in the transverse plane [7].

2 Nucleon and Flavor Form Factors in ADS/QCD Models

Model I Model I refers to the AdS/QCD model for nucleon form factors proposed by Brodsky and Téramond [2]. The relevant AdS/QCD action for the fermion field is written as

C. Mondal (✉) · D. Chakrabarti
Department of Physics, Indian Institute of Technology Kanpur, Kanpur 208016, India
E-mail: mchandan@iitk.ac.in

D. Chakrabarti
E-mail: dipankar@iitk.ac.in

$$S = \int d^4x dz \sqrt{g} \Big(\frac{i}{2} \bar{\Psi} e^M_A \Gamma^A D_M \Psi - \frac{i}{2} (D_M \bar{\Psi}) e^M_A \Gamma^A \Psi - \mu \bar{\Psi} \Psi - V(z) \bar{\Psi} \Psi \Big), \tag{1}$$

where $e^M_A = (z/R)\delta^M_A$ is the inverse vielbein and $V(z)$ is the confining potential which breaks the conformal invariance and R is the AdS radius. In $d = 4$ dimensions, $\Gamma_A = \{\gamma_\mu, -i\gamma_5\}$. To map the Dirac equation in AdS space with the light front wave equation, one identifies $z \to \zeta$ (light front transverse impact variable) and substitutes $\Psi(x, \zeta) = e^{-iP\cdot x}\zeta^2 \psi(\zeta) u(P)$ and sets $| \mu R | = \nu + 1/2$ where $\nu = L + 1$ (more details can be found in [2]). For linear confining potential $U(\zeta) = (R/\zeta)V(\zeta) = \kappa^2 \zeta$, one gets the light front wave equation for the baryon which leads to the AdS solutions of nucleon wavefunctions $\psi_+(z)$ and $\psi_-(z)$ corresponding to different orbital angular momentum $L^z = 0$ and $L^z = +1$ [2]

$$\psi_+(\zeta) \sim \psi_+(z) = \frac{\sqrt{2}\kappa^2}{R^2} z^{7/2} e^{-\kappa^2 z^2/2}, \qquad \psi_-(\zeta) \sim \psi_-(z) = \frac{\kappa^3}{R^2} z^{9/2} e^{-\kappa^2 z^2/2}. \tag{2}$$

The Dirac form factors in this model are obtained by the SU(6) spin-flavor symmetry and given by

$$F_1^p(Q^2) = R^4 \int \frac{dz}{z^4} V(Q^2, z) \psi_+^2(z), \quad F_1^n(Q^2) = -\frac{1}{3} R^4 \int \frac{dz}{z^4} V(q^2, z)(\psi_+^2(z) - \psi_-^2(z)). \tag{3}$$

A precise mapping for the spin-flip nucleon form factor using the action in Eq. (1) is not possible. Thus, the Pauli form factors for the nucleons are modeled in this model as

$$F_2^{p/n}(Q^2) = \kappa_{p/n} R^4 \int \frac{dz}{z^3} \psi_+(z) V(Q^2, z) \psi_-(z). \tag{4}$$

The Pauli form factors are normalized to $F_2^{p/n}(0) = \kappa_{p/n}$ where $\kappa_{p/n}$ are the anomalous magnetic moment of proton/neutron. $V(Q^2, z)$ is the bulk-to-boundary propagator [2]. Here we use the value $\kappa = 0.4 GeV$ which is fixed by fitting the ratios of Pauli and Dirac form factors for proton with the experimental data [5,8].

Model II The other model of the nucleon form factors was formulated by Abidin and Carlson [1]. Since the action defined in Eq. (1) can not generate the Pauli form factors, they introduced an additional gauge invariant non-minimal coupling term

$$\int d^4x \, dz \, \sqrt{g} \, \bar{\Psi} \, e^A_M \, e^B_N \, [\Gamma_A, \Gamma_B] \, F^{MN} \psi. \tag{5}$$

This additional term also provides an anomalous contribution to the Dirac form factor. In this model the form factors are given by [1]

$$F_1^p(Q^2) = C_1(Q^2) + \eta_p C_2(Q^2), \qquad F_1^n(Q^2) = \eta_n C_2(Q^2), \tag{6}$$

$$F_2^p(Q^2) = \eta_p C_3(Q^2), \qquad F_2^n(Q^2) = \eta_n C_3(Q^2). \tag{7}$$

The functions $C_i(Q^2)$ are defined as $C_1(Q^2) = \frac{a+6}{(a+1)(a+2)(a+3)}$, $C_2(Q^2) = \frac{2a(2a-1)}{(a+1)(a+2)(a+3)(a+4)}$, and $C_3(Q^2) = \frac{48}{(a+1)(a+2)(a+3)}$, where $a = Q^2/(4\kappa^2)$. The value of $\kappa = 0.350 GeV$ is fixed by simultaneous fit to proton and rho meson masses. The other parameters are determined from the normalization conditions of the Pauli form factor at $Q^2 = 0$ and are given by $\eta_p = 0.224$ and $\eta_n = -0.239$ [1]. The Pauli form factors in Model I and Model II are identical, the main difference is in the Dirac form factor where we have an additional contribution in Model II coming the extra term added.

Quark-diquark model in AdS/QCD Here we consider a light front scalar quark-diquark model for nucleon [9] where the 2-particle wavefunction is modeled from the soft-wall AdS/QCD solution. In the light front overlap formalism, the electromagnetic form factors in this model are given by

$$F_1^q(Q^2) = \int_0^1 dx \int \frac{d^2\mathbf{k}_\perp}{16\pi^3} \Big[\psi_{+q}^{+*}(x, \mathbf{k}'_\perp) \psi_{+q}^{+}(x, \mathbf{k}_\perp) + \psi_{-q}^{+*}(x, \mathbf{k}'_\perp) \psi_{-q}^{+}(x, \mathbf{k}_\perp) \Big], \tag{8}$$

$$F_2^q(Q^2) = -\frac{2M_n}{q^1 - iq^2} \int_0^1 dx \int \frac{d^2\mathbf{k}_\perp}{16\pi^3} \Big[\psi_{+q}^{+*}(x, \mathbf{k}'_\perp) \psi_{+q}^{-}(x, \mathbf{k}_\perp) + \psi_{-q}^{+*}(x, \mathbf{k}'_\perp) \psi_{-q}^{-}(x, \mathbf{k}_\perp) \Big], \tag{9}$$

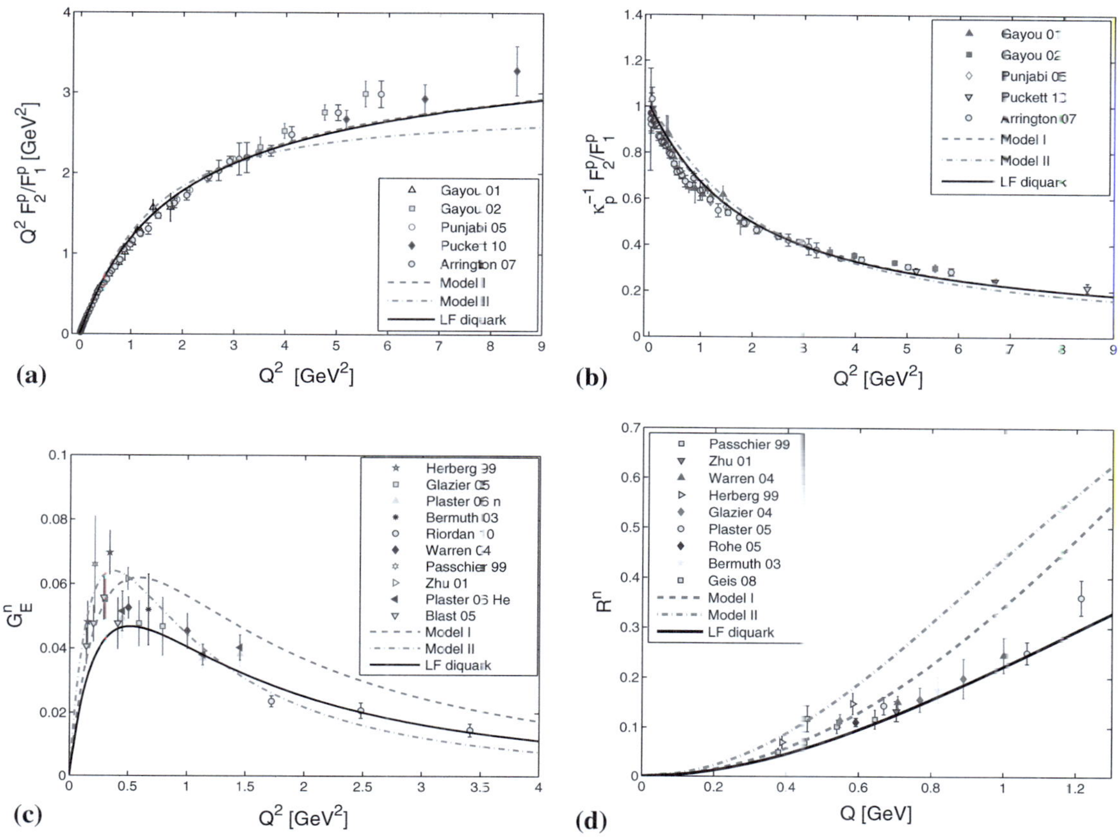

Fig. 1 Nucleon form factors **a** the ratio of Dirac and Pauli form factors for proton is multiplied by $Q^2 = -q^2 = -t$, **b** the ratio is divided by κ_p, **c** electric Sach form factor for neutron $G_E^n(Q^2)$, **d** the ratio $R^n = \frac{\mu_n G_E^n}{G_M^n}$. The *red dashed* and *pink dash dot lines* represent the Model I and Model II and the *solid black lines* represent the quark-diquark model [6]. The references of the experimental data can be found in Ref. [5]

where $\mathbf{k}'_\perp = \mathbf{k}_\perp + (1-x)\mathbf{q}_\perp$. $\psi^{\lambda_N}_{\lambda_q q}(x, \mathbf{k}_\perp)$ are the LFWFs with specific nucleon helicities $\lambda_N = \pm$ and for the struck quark $\lambda_q = \pm$, where plus and minus correspond to $+\frac{1}{2}$ and $-\frac{1}{2}$ respectively. The LFWFs are specified at an initial scale $\mu_0 = 313$ MeV [9]:

$$\begin{aligned}
\psi^{+}_{+q}(x, \mathbf{k}_\perp) &= \varphi_q^{(1)}(x, \mathbf{k}_\perp), & \psi^{+}_{-q}(x, \mathbf{k}_\perp) &= -\frac{k^1 + ik^2}{xM_n}\varphi_q^{(2)}(x, \mathbf{k}_\perp), \\
\psi^{-}_{+q}(x, \mathbf{k}_\perp) &= \frac{k^1 - ik^2}{xM_n}\varphi_q^{(2)}(x, \mathbf{k}_\perp), & \psi^{-}_{-q}(x, \mathbf{k}_\perp) &= \varphi_q^{(1)}(x, \mathbf{k}_\perp),
\end{aligned} \tag{10}$$

$$\varphi_q^{(i)}(x, \mathbf{k}_\perp) = N_q^{(i)}\frac{4\pi}{\kappa}\sqrt{\frac{\log(1/x)}{1-x}}x^{a_q^{(i)}}(1-x)^{b_q^{(i)}}\exp\left[-\frac{\mathbf{k}_\perp^2}{2\kappa^2}\frac{\log(1/x)}{(1-x)^2}\right]. \tag{11}$$

For $a_q^{(i)} = b_q^{(i)} = 0$, $\varphi_q^{(i)}(x, \mathbf{k}_\perp)$ reduces to the AdS/QCD prediction [2]. κ is the AdS/QCD scale parameter which is taken to be 0.4 GeV [5,8]. The parameters $a_q^{(i)}$ and $b_q^{(i)}$ with the constants $N_q^{(i)}$ are fixed by fitting the electromagnetic properties of the nucleons [10].

Flavor decompositions of the nucleon form factors Under the charge and isospin symmetry it is straightforward to write down the flavor decompositions of the nucleon form factors as [11]

$$F_i^p = e_u F_i^u + e_d F_i^d \quad \text{and} \quad F_i^n = e_d F_i^u + e_u F_i^d, \quad (i = 1, 2) \tag{12}$$

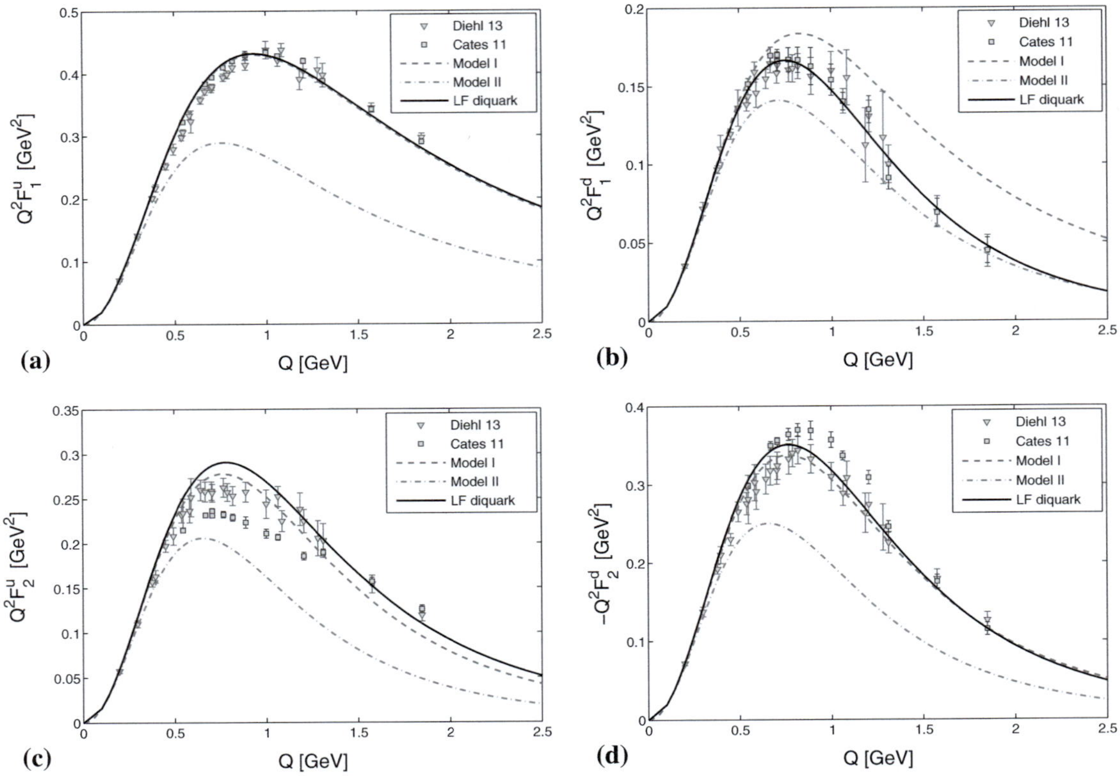

Fig. 2 Flavor dependent form factors for u and d quarks. The experimental data are taken from [11,12]. The lines are same as in Fig. 1

Table 1 Electromagnetic radii of the nucleons

Quantity	Model I	Model II	LF diquark	Measured data
r_E^p (fm)	0.810	0.980	0.786	0.877 ± 0.005
r_M^p (fm)	0.782	0.921	0.772	0.777 ± 0.016
$\langle r_E^2 \rangle^n$ (fm^2)	−0.088	−0.123	−0.085	-0.1161 ± 0.0022
r_M^n (fm)	0.796	0.937	0.7596	$0.862^{+0.009}_{-0.008}$

where e_u and e_d are charge of u and d quarks respectively. The normalizations of the flavor form factors are $F_1^u(0) = 2$, $F_2^u(0) = \kappa_u$ and $F_1^d(0) = 1$, $F_2^d(0) = \kappa_d$ where the anomalous magnetic moments for the up and down quarks are $\kappa_u = 2\kappa_p + \kappa_n = 1.673$ and $\kappa_d = \kappa_p + 2\kappa_n = -2.033$.

In Fig. 1 we compare the results for electromagnetic form factors of nucleons calculated in different AdS/QCD models. The flavor form factors are shown in Fig. 2. The figures show that the results of the Model I and the quark-diquark model are in good agreement with experimental data whereas the Model II deviates from the data. Only for F_1^d, Model I deviates at higher Q^2 from the data and also for $G_E^n(Q^2)$ Model II is better than Model I. The fitted results for the electromagnetic radii of the nucleons are listed in Table 1.

3 Transverse Charge and Magnetization Densities

The transverse charge density inside the nucleons is given by

$$\rho_{ch}(b) = \int \frac{d^2 q_\perp}{(2\pi)^2} F_1(q^2) e^{i q_\perp . b_\perp} = \int_0^\infty \frac{dQ}{2\pi} Q J_0(Qb) F_1(Q^2), \tag{13}$$

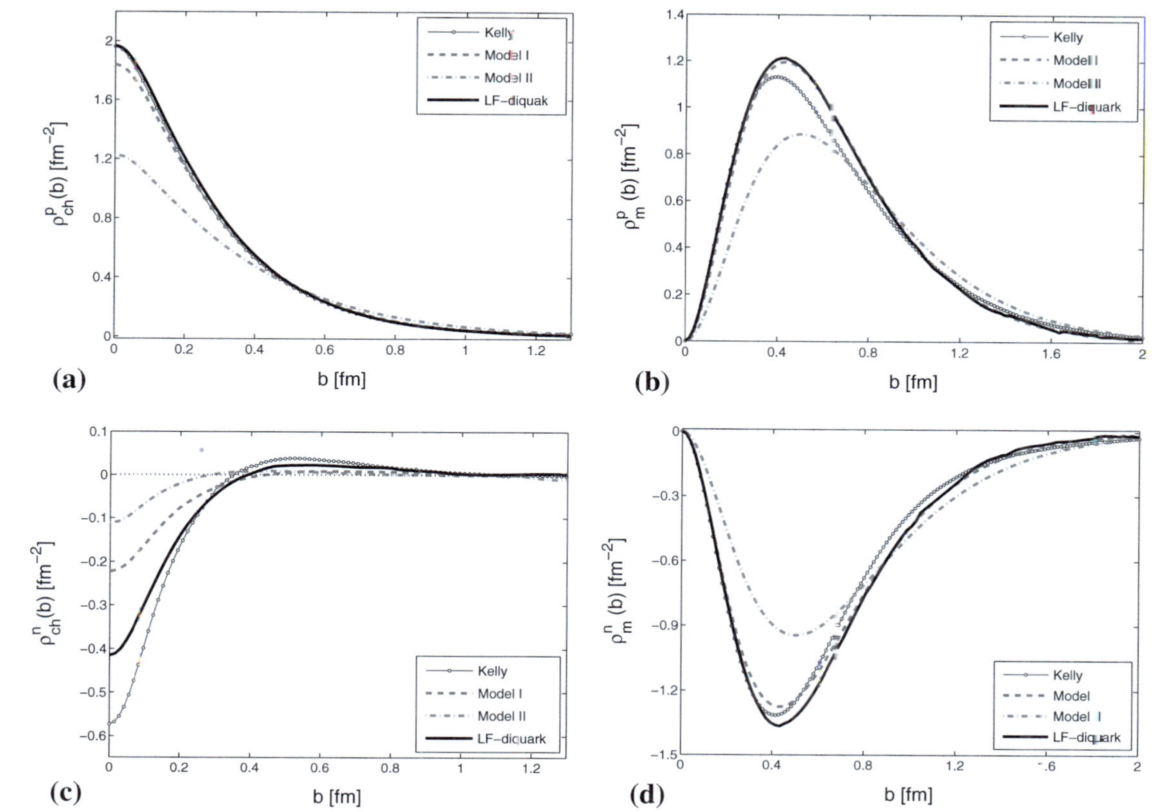

Fig. 3 Transverse charge and anomalous magnetization densities for nucleon. **a** and **b** represent ρ_{ch} and ρ_m for the proton. **c** and **d** same as proton but for neutron. *Lines with circle* represent the parametrization of Kelly [13]

where b represents the impact parameter and J_0 is the cylindrical Bessel function of order zero. One can define the magnetization density($\widetilde{\rho}_M(b)$) in the similar fashion with F_1 is replaced by F_2, whereas, $\rho_m(b) = -b(\partial\widetilde{\rho}_M(b)/\partial b)$ can be interpreted as anomalous magnetization density. We evaluate the quark contributions to the nucleon transverse densities using charge and isospin symmetry (see [6, 7]).

For transversely polarized nucleon, the charge density is given by

$$\rho_T(b) = \rho_{ch} - \sin(\phi_b - \phi_s)\frac{1}{2M_n b}\rho_m. \tag{14}$$

The transverse polarization of the nucleon is given by $S_\perp = (\cos\phi_s\hat{x} + \sin\phi_s\hat{y})$ and the transverse impact parameter $b_\perp = b(\cos\phi_b\hat{x} + \sin\phi_b\hat{y})$. Without loss of generality, the polarization of the nucleon is taken along x-axis ie., $\phi_s = 0$. The second term in Eq. (14), provides the deviation from circular symmetry of the unpolarized charge density. The nucleon charge and anomalous magnetization densities presented in Fig. 3 suggest that the quark-diquark model agrees with the phenomenological parametrizations [13] much better than the Model I and the Model II and Model I is better compare to Model II. The charge densities between unpolarized and transversely polarized nucleon in the quark-diquark model are compared in Fig. 4. The unpolarized densities are axially symmetric in transverse plane while for the transversely polarized nucleons they become distorted. For nucleon polarized along x direction, the densities get shifted towards negative y-direction. Due to large anomalous magnetization density, the distortion in neutron charge density is found to be stronger than that for proton.

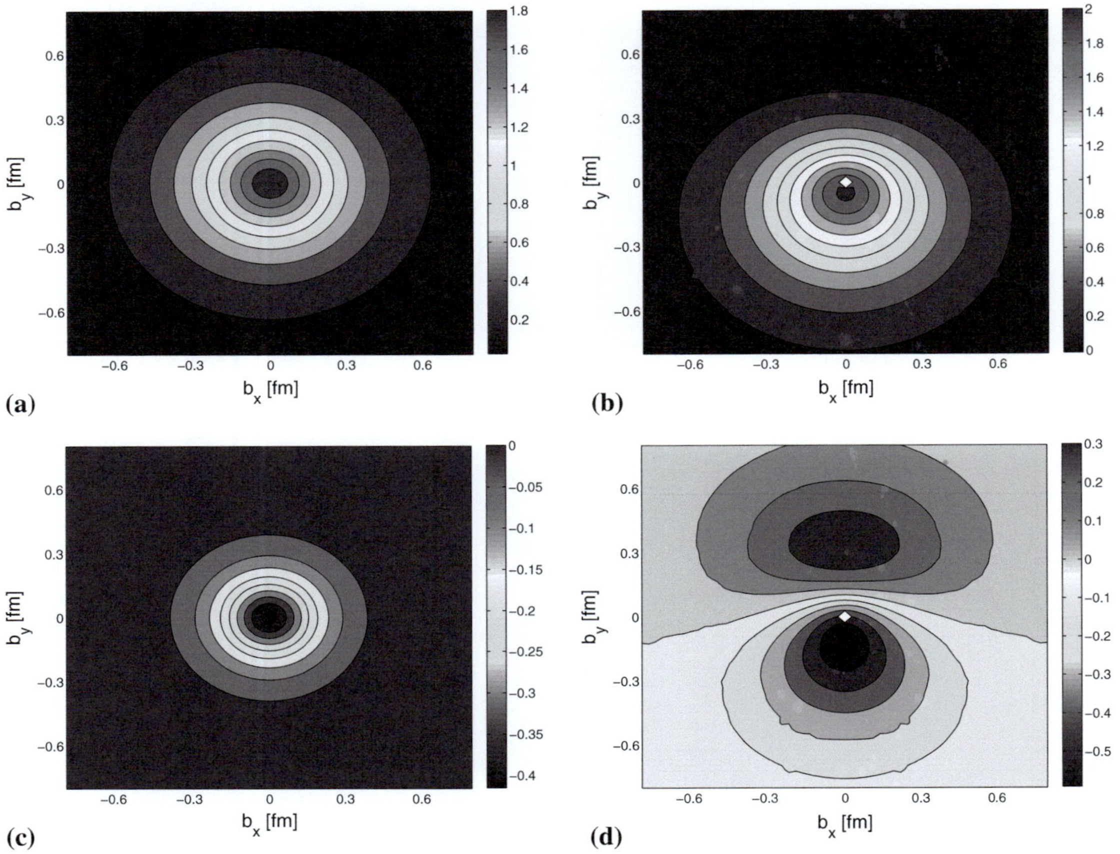

Fig. 4 Unpolarized and transversely polarized charge densities for proton(*upper*) and neutron(*lower*) in LF diquark model

Acknowledgments CM thanks the Sciencce and Engineering Research Board (SERB), Gorernment of India for supporting the travel grant under contract no. ITS/3444/2015-2016 to attend the conference LightCone2015 where the work was presented.

References

1. Abidin, Z., Carlson, C.E.: Nucleon electromagnetic and gravitational form factors from holography. Phys. Rev. D **79**, 115003 (2009)
2. de Téramond, G.F., Brodsky, S.J.: Hadronic Form Factor Models and Spectroscopy Within the Gauge/Gravity Correspondence. arXiv:1203.4025[hep-ph]
3. Vega, A., Schmidt, I., Gutsche, T., Lyubovitskij, V.E.: Generalized parton distributions in AdS/QCD. Phys. Rev. D **83**, 036001 (2011)
4. Gutsche, T., Lyubovitskij, V.E., Schmidt, I., Vega, A.: Nucleon structure including high Fock states in AdS/QCD. Phys. Rev. D **86**, 036007 (2012)
5. Chakrabarti, D., Mondal, C.: Nucleon and flavor form factors in a light front quark model in AdS/QCD. Eur. Phys. J. C **73**, 2671 (2013)
6. Mondal, C., Chakrabarti, D.: Generalized parton distributions and transverse densities in a light-front quark–diquark model for the nucleons. Eur. Phys. J. C **75**, 261 (2015)
7. Chakrabarti, D., Mondal, C.: Transverse charge and magnetization densities in holographic QCD. Eur. Phys. J. C **74**, 2962 (2014)
8. Chakrabarti, D., Mondal, C.: Generalized parton distributions for the proton in AdS/QCD. Phys. Rev. D **88**, 073006 (2013)
9. Gutsche, T., Valery, E. L., Schmidt, I., Vega, A.: Light-front quark model consistent with Drell-Yan-West duality and quark counting rules. Phy. Rev. D **89**, 054033 (2014)
10. Chakrabarti, D., Mondal, C.: Chiral-odd generalized parton distributions for proton in a light-front quark-diquark model. Phys. Rev. D **92**, 074012 (2015)
11. Cates, G.D., Jager, C.W.de , Riodian, S., Wojtsekhowski, B.: Flavor decomposition of the elastic nucleon electromagnetic form factors. Phys. Rev. Lett. **106**, 252003 (2011)
12. Diehl, M., Kroll, P.: Nucleon form factors, generalized parton distributions and quark angular momentum. Eur. Phys. J. C **73**, 2397 (2013)
13. Kelly, J.J.: Simple parametrization of nucleon form factors. Phys. Rev. C **70**, 068202 (2004)

DOI 10.1007/978-3-319-50699-9_21
Reprinted from *Few-Body System* Journal, DOI 10.1007/s00601-016-1045-9

CrossMark

L. L. Pappalardo

HERMES Results on the 3D Imaging of the Nucleon

Received: 11 December 2015 / Accepted: 19 January 2016 / Published online: 25 February 2016

Abstract It the last decades, a formalism of transverse momentum dependent parton distribution functions (TMDs) and of generalised parton distributions (GPDs) has been developed in the context of non-perturbative QCD, opening the way for a *tomographic* imaging of the nucleon structure. TMDs and GPDs provide complementary three-dimensional descriptions of the nucleon structure in terms of parton densities. They thus contribute, with different approaches, to the understanding of the full phase-space distribution of partons. A selection of HERMES results sensitive to TMDs is presented.

1 Introduction

Theoretical developments of the *parton model* in the framework of non-perturbative QCD, followed by decades of deep-inelastic scattering (DIS) experiments, have led to a relatively detailed description of the nucleon structure in terms of the three *collinear* PDFs $f_1(x)$, $g_1(x)$ and $h_1(x)$[1], where x denotes the longitudinal momentum fraction carried by the parton. In recent years, huge efforts are being focused towards a more comprehensive multi-dimensional description of the nucleon that includes transverse (i.e. *non-collinear*) degrees of freedom. The goal is to map the full quantum phase-space distribution of quarks and gluons inside the nucleon. This maximal knowledge on the partonic structure of nucleons is encoded by the *Wigner function* [1,2], not measurable due to Heisenberg's uncertainty principle. The main focus of present and future experiments in this field is thus concentrated on two independent classes of partially integrated non-perturbative objects, the transverse momentum dependent PDFs (TMDs) and the generalized parton distributions (GPDs). They allow for complementary descriptions of the nucleon in three dimensions spanned by the quark's longitudinal momentum and, respectively, by its transverse momentum components and transverse spatial coordinates. Furthermore, they provide complementary insights into the yet unmeasured quark orbital angular momentum. Experimentally, TMDs and GPDs can be accessed through the analysis of specific azimuthal asymmetries measured, respectively, in semi-inclusive DIS (SIDIS) and hard exclusive processes, such as hard lepto-production of real photons or mesons. The HERMES experiment has measured a variety of such asymmetries in scattering of a longitudinally polarised lepton (electron or positron) beam from HERA off unpolarised, longitudinally and transversely polarised internal gas targets.

[1] These are the only three leading-twist PDFs that survive integration over the parton momentum components that are transverse with respect to the direction of the hard probe.

For the Hermes Collaboration

L. L. Pappalardo (✉)
Dipartimento di Fisica e Scienze della Terra Polo Scientifico-Tecnologico - Edificio C, Università di Ferrara,
Via Saragat, 1, 44122 Ferrara, Italy
E-mail: pappalardo@fe.infn.it

2 Transverse Momentum Distributions

At leading-twist, eight TMDs, each with a specific probabilistic interpretation in terms of quark number densities, enter the SIDIS cross-section in conjunction with a fragmentation function (FF) [3]: the poorly known chiral-odd Collins function $H_1^{\perp}$, describing the correlation between the quark transverse spin and the transverse momentum of the produced hadron, or the relatively well known spin-independent chiral-even D_1 FF. Three of the eight TMDs are the transverse-momentum-dependent versions of the collinear PFDs f_1, g_1 and h_1, respectively known as the *momentum*, the *helicity* and the *transversity* distributions. The other five are the *Sivers* function $f_{1T}^{\perp}$, the *Boer-Mulders* function $h_1^{\perp}$, the *pretzelosity* function $h_{1T}^{\perp}$ and the so-called *worm-gear* functions $h_{1L}^{\perp}$ and $g_{1T}^{\perp}$.

The HERMES experiment at HERA has played a pioneering role in these studies, by measuring observables (azimuthal asymmetries) sensitive to these (and also some higher-twist) TMDs, in some cases providing the first evidence of non-zero effects. Most of the HERMES results are confirmed by other experiments (COMPASS, CLAS, etc.), though intriguing differences were observed in some cases. High-precision results from the next generation high-luminosity and large-acceptance experiments (at present or upgraded facilities and, possibly, at the future Electron-Ion Collider) will greatly improve our understanding of the complex dynamical structure of nucleons.

2.1 Transversity and Collins Functions

The first evidence for a non-zero transversity function was reported by the HERMES Collaboration in 2005 [4]. HERMES measured significant asymmetry amplitudes (*Collins amplitudes*) in the production of π^+ and π^- in SIDIS, using an unpolarised 27.5 GeV electron beam and a transversely polarised hydrogen target. These amplitudes are sensitive to the convolution of the transversity function and the Collins FF (the so-called *Collins effect*), which causes the observed azimuthal distribution of the final-state hadrons (pions in this case) generated in the fragmentation of transversely polarised quarks:

$$A_{UT}^{\sin(\phi+\phi_S)} \propto \sum_q e_q^2\, h_1^q(x, p_T^2) \otimes H_1^{\perp,q}(z, k_T^2). \tag{1}$$

Here, ϕ and ϕ_S are, respectively, the azimuthal angle of the detected hadron and of the target transverse polarisation with respect to the lepton scattering plane and around the virtual-photon direction, z denotes the fraction of the energy of the exchanged virtual photon carried by the produced hadron (in this case a pion), and p_T and k_T represent the transverse momentum of the struck quark and that of the fragmenting quark, respectively. The sum is over the quark flavours q.

The measured Collins amplitudes for π^+ and π^- have similar magnitude and opposite sign. This unexpected result can be interpreted assuming that the scattering process involves predominantly the u quarks (*u-quark dominance*) and that the unfavoured Collins function (describing the formation of a π^-) has approximately the same size and opposite sign of the favoured one (describing the formation of a π^+): $H_{1,unfav}^{u\to\pi^-} \approx -H_{1,fav}^{u\to\pi^+}$. This hypothesis is consistent with the recent measurements of azimuthal asymmetries sensitive to the Collins function performed at the BELLE [5] and BABAR [6] experiments. A global analysis of the Collins amplitudes for pions measured at HERMES, together with the analogous results from the COMPASS experiment obtained using a transversely polarised deuteron target [7] and the azimuthal asymmetries measured in e^+e^- collisions from BELLE [8] yielded the first extraction of the transversity function for u ad d quarks [9]. The final HERMES results [10], based on the full statistics available, confirm the main features of the early measurements for charged pions and include results for neutral pions and charged kaons. The amplitudes for K^+ are large and positive while those for π^0 and K^- are consistent with zero. The results are provided in 1-dimensional (1D) projections as a function of x, z or the final-state hadron transverse momentum $P_{h\perp}$. Recently, the data analysis was refined and the relevant asymmetry amplitudes were extracted in a 3-dimensional (3D) binning in x, z and $P_{h\perp}$. The 3D binning allows to map the kinematic dependence of the asymmetry amplitudes in a greatly detailed way. As an example, Fig. 1 shows the Collins amplitude for π^- as a function of x in simultaneous bins in z and $P_{h\perp}$. The rise in magnitude of the Collins amplitude for π^- as a function of x, as observed in the conventional 1D projection, reveals non-trivial kinematic dependencies in z and $P_{h\perp}$ when observed in the 3D representation, where most of the rise is clearly localised in the high-x and high-$P_{h\perp}$ region.

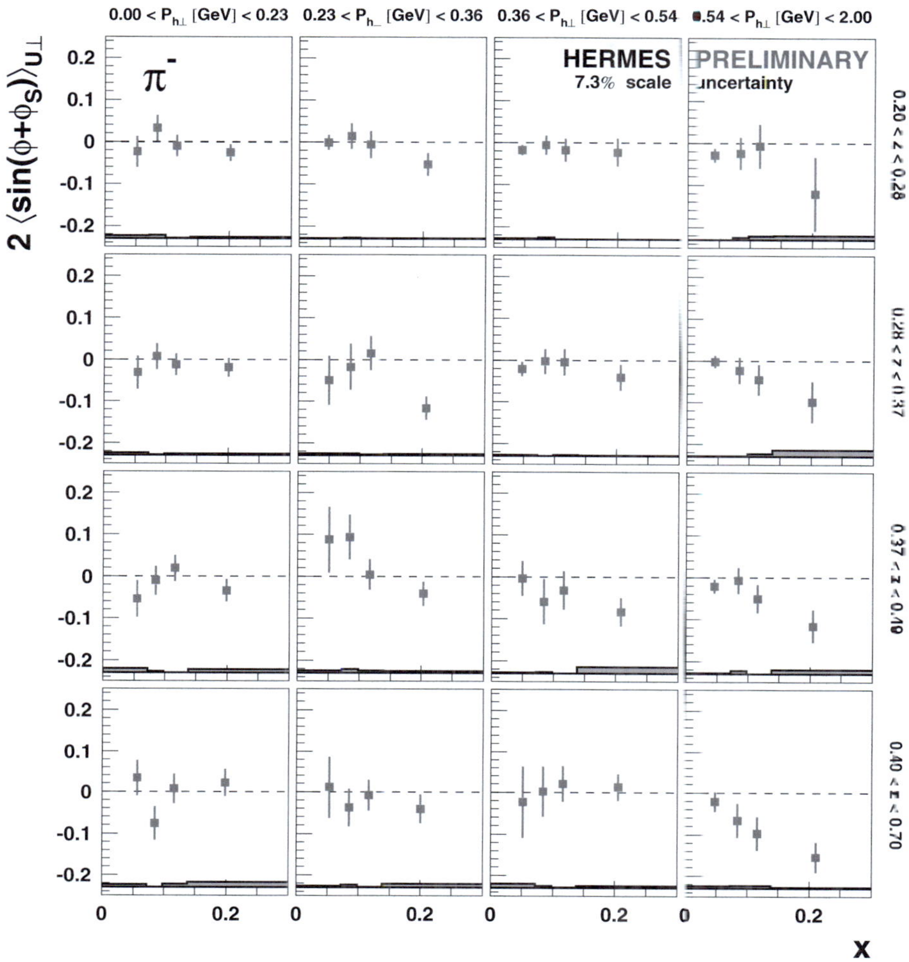

Fig. 1 Collins amplitudes for π^- in 3-dimensional binning in x, z and $P_{h\perp}$.

2.2 The Sivers Function

A particularly interesting TMD is the naive-T-odd Sivers function [11], describing the correlation between the quark transverse momentum and the transverse spin of the nucleon. The interest on the Sivers function suddenly increased after it was demonstrated [12] to be linked to the quark and gluon orbital angular momentum, the main still unmeasured contribution to the nucleon spin.

The first evidence for a non-zero Sivers function was reported by the HERMES Collaboration in 2005 [4], through the extraction of asymmetry amplitudes from data collected with an unpolarised 27.5 GeV electron beam and a transversely polarised hydrogen target. These amplitudes are sensitive to the convolution of the Sivers function $f_{1T}^{\perp}$ and the unpolarised fragmentation function D_1:

$$A_{UT}^{\sin(\phi-\phi_S)} \propto \sum_q e_q^2\, f_{1T}^{\perp,q}(x, p_T^2) \otimes D_1^q(z, k_T^2). \tag{2}$$

In particular, HERMES observed large positive amplitudes for π^+ and amplitudes consistent with zero for π^-. The vanishing π^- amplitudes require a d-quark Sivers function opposite in sign to the u-quark Sivers function. The final HERMES results [13] confirm the main features of the early measurements for charged pions and include results for neutral pions and charged kaons. The amplitudes for K^+ are large and positive while those for π^0 and K^- are slightly positive. Unexpectedly, the K^+ amplitudes are larger than the π^+ ones. This observation can be ascribed, for instance, to a significant role of the sea quarks or to possible higher-twist contributions in kaon production. Recently, the HERMES experiment has also measured single

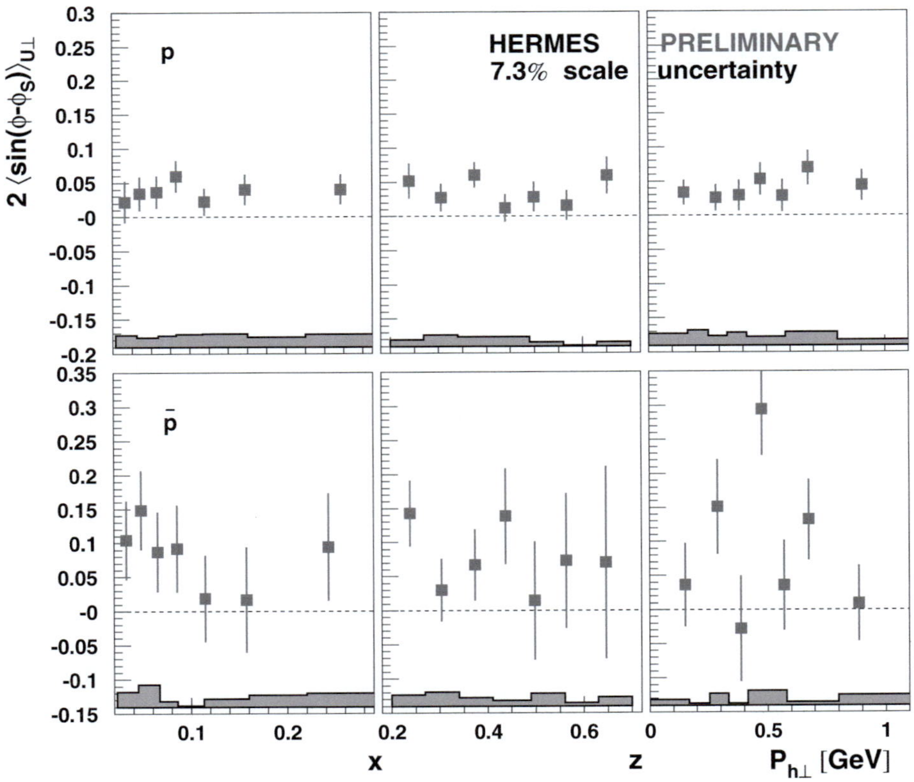

Fig. 2 Sivers amplitudes for protons and anti-protons in 1-dimensional binning in x, z or $P_{h\perp}$.

and double-spin asymmetries in SIDIS for protons and antiprotons. Figure 2 shows the Sivers amplitudes for protons and antiprotons in 1D bins in x, z or $P_{h\perp}$. While the antiproton amplitude is consistent with zero within the experimental uncertainties, the proton data exhibit a small but significantly positive amplitude. This constitutes, at present, the first measurement of a non-zero Sivers effect for protons in SIDIS. Figure 3 shows the Sivers amplitude for π^+ as a function of $P_{h\perp}$ in simultaneous bins in x and z. From the 3D representation it becomes clear that the rise of the Sivers amplitude for π^+ as a function of $P_{h\perp}$ is localised in the high-x and high-z region.

2.3 The $A_{LU}^{\sin(\phi)}$ Beam-Spin Asymmetries

At HERMES kinematics ($Q^2 \approx 2.5$ GeV^2) higher-twist effects can be sizeable. It is therefore extremely useful to study them, not only because they provide interesting pieces of information about the nucleon structure by themselves, but also because a precise knowledge of the higher-twist effects is necessary for a correct interpretation of the leading-twist results. Higher-twist structure functions, derived from the quark-gluon-quark correlator, are relatively complex objects as they depend on convolutions of leading-twist PDFs with higher-twist FFs and convolutions of higher-twist PDFs with leading-twist FFs. In particular, the $F_{LU}^{\sin(\phi)}$ structure function, accessible at HERMES using a longitudinally polarised lepton beam and an unpolarised hydrogen or deuterium target is sensitive to the leading-twist PDFs f_1 and $h_1^\perp$ (the *momentum* and *Boer-Mulders* TMDs), the twist-three PDFs e and $g^\perp$, the leading-twist FFs D_1, $H_1^\perp$ (the *unpolarised* and *Collins* FFs), and the twist-three FFs $\tilde{G}^\perp$ and $\tilde{E}$ [3]. The HERMES access to $F_{LU}^{\sin(\phi)}$ is achieved through the $\sin(\phi)$ Fourier component of the beam-spin asymmetry:

$$A_{LU} = \frac{1}{P_L}\frac{d\sigma^+(\phi) - d\sigma^-(\phi)}{d\sigma^+(\phi) - d\sigma^-(\phi)} \propto \frac{F_{LU}}{F_{UU}}. \tag{3}$$

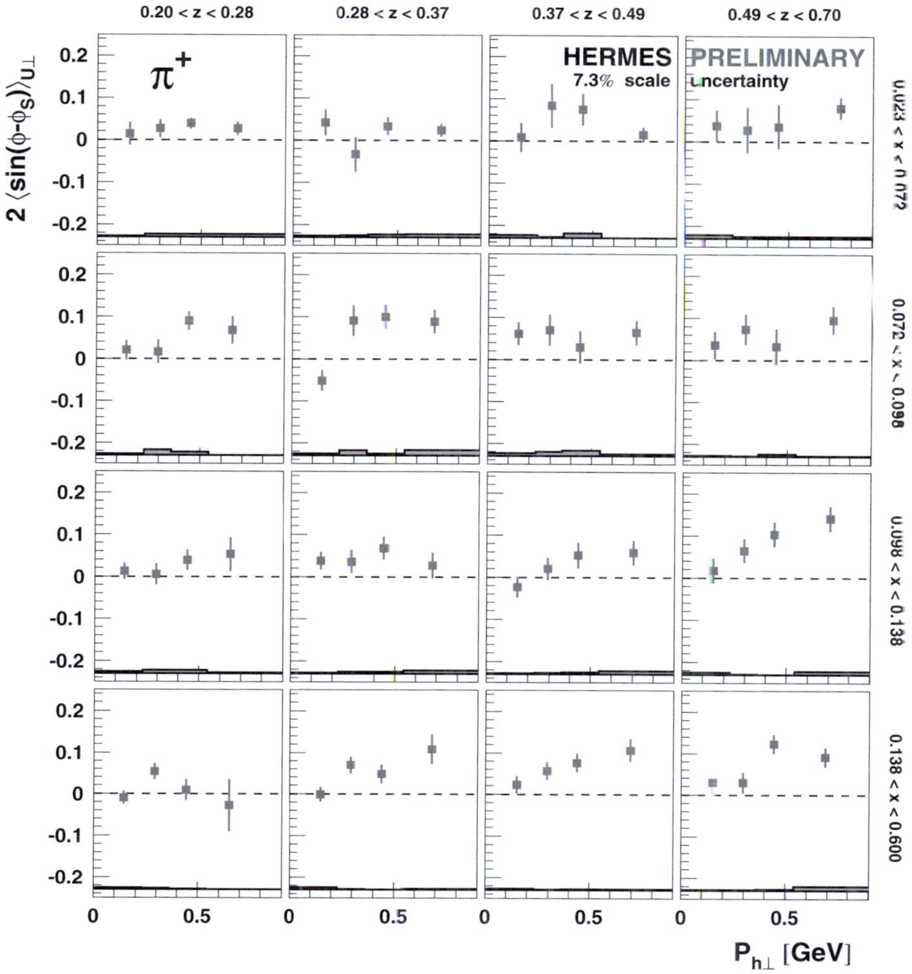

Fig. 3 Sivers amplitudes for π^+ in 3-dimensional binning in x, z and $P_{h\perp}$.

Figure 4 shows the $\sin(\phi)$ azimuthal amplitudes for charged mesons extracted in 1D bins in x, z or $P_{h\perp}$. The results obtained using hydrogen and deuterium targets are in general consistent with each other within the experimental uncertainties. Similar results were also obtained for protons/antiprotons (not shown). All amplitudes are consistent with zero except for the π^+ and the π^- ones, which are both positive. The π^+ 3D results, binned in x, z and $P_{h\perp}$ (not shown here), reveal that the largest contribution to the asymmetry amplitude comes from the high-x and high-z region.

3 Generalized Parton Distributions

The experimental access to GPDs is so far obtained from the measurements of hard exclusive lepton-nucleon scattering processes. These processes are described by four leading-twist, quark-chirality conserving GPDs (H, E, $\widetilde{H}$, and $\widetilde{E}$). The cleanest process to probe GPDs is deeply virtual compton scattering (DVCS), $\ell + N \to \ell + N' + \gamma$, where a real photon is emitted by the struck quark in the nucleon. During the last decade the DVCS process was extensively studied both theoretically and experimentally. The Bethe-Heitler (BH) process, where a real photon is radiated by the incoming or outgoing lepton, shares the same initial and final state with the DVCS process. These two processes thus interfere and the total cross-section of hard lepto-production of real photons can be decomposed as: $\sigma \propto |\mathcal{T}_{BH}|^2 + |\mathcal{T}_{DVCS}|^2 + I$. The individual terms of the cross-section, including the interference I, can be decomposed into Fourier harmonics in the relevant azimuthal angles. Different Fourier components depend on different Compton Form Factors, which in turn are convolutions of hard scattering amplitudes with the corresponding GPDs.

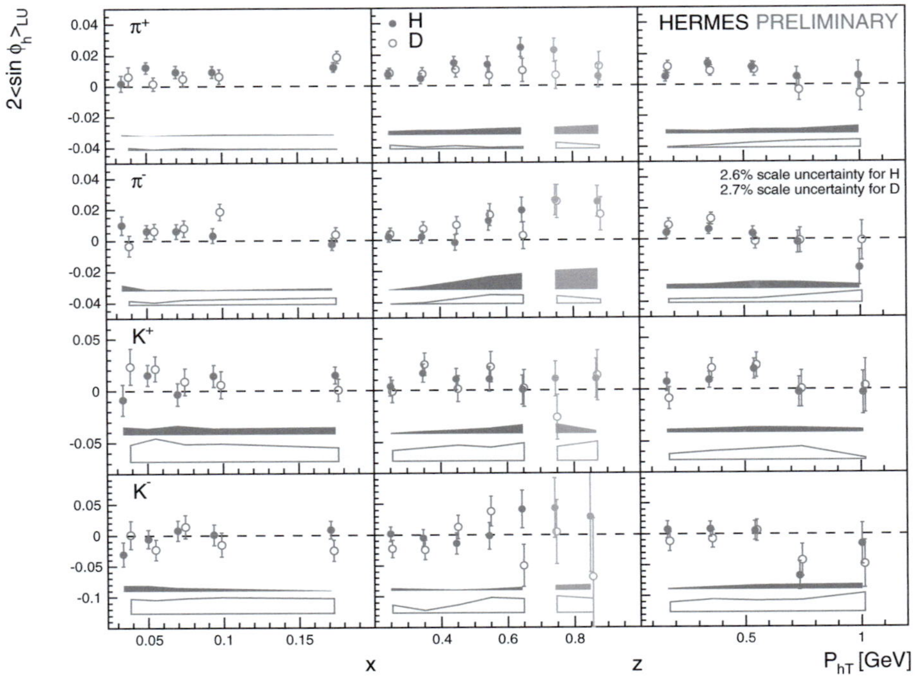

Fig. 4 sin(ϕ) asymmetry amplitudes for charged mesons in 1-dimensional binning in x, z or $P_{h\perp}$.

At HERMES, the DVCS process is accessed through measurements of cross-section asymmetries in the azimuthal distributions of the final-state photons. Using data collected with longitudinally polarised electron and positron beams and unpolarised, longitudinally and transversely polarised hydrogen and deuterium targets, it was possible to measure asymmetries with respect to beam charge, beam polarisation and target polarisation alone and also with respect to their different combinations. Full references for the HERMES DVCS results can be found in [14–20]. Complementary information on GPDs can be obtained from measurements of hard exclusive meson production processes $\ell + N \rightarrow \ell + N' + V$. Measurements of exclusive meson production processes with different final state mesons (ρ, ϕ, ω, etc.) provide a unique opportunity to study the flavour dependence of GPDs. Full references for the HERMES Spin Density Matrix Elements measurements can be found in [21–23].

4 Summary

The study of the 3D structure of nucleons is a young and fast evolving field. A rich phenomenology arises when the transverse degrees of freedom (spin, momentum, position) of quarks and gluons, for long time neglected, are not integrated out. HERMES, as a pioneering experiment, has played a key exploratory role in this field. New precision data from present and future experiments, operating in complementary kinematic domains, will push forward our present understanding of the nucleon structure and will allow to map the phase-space distributions of quarks and gluons with unprecedented precision.

References

1. Ji, X.D.: Viewing the proton through color filters. Phys. Rev. Lett. **91**, 062001 (2003)
2. Lorce, C., Pasquini, B.: Quark wigner distributions and orbital angular momentum. Phys. Rev. D **84**, 014015 (2011)
3. Bacchetta, A. et al.: Semi-inclusive deep inelastic scattering at small transverse momentum. JHEP **02**, 093 (2007)
4. Airapetian, A. et al.: (HERMES). Single-spin asymmetries in semi-inclusive deep-inelastic scattering on a transversely polarized hydrogen target. Phys. Rev. Lett. **94**, 012002 (2005)
5. Seidl, R. et al.: (BELLE). Erratum: Measurement of azimuthal asymmetries in inclusive production of hadron pairs in e^+e^- annihilation at $\sqrt{s} = 10.58$ GeV. Phys. Rev. D **86**, 039905 (2012)
6. Lees, J.P. et al.: (BABAR). Measurement of Collins asymmetries in inclusive production of charged pion pairs in e^+e^- annihilation at BABAR. Phys. Rev. D **90**, 052003 (2014)

7. Ageev, E.S. et al.: (COMPASS). A new measurement of the Collins and Sivers asymmetries on a transversely polarised deuteron target. Nucl. Phys. B **765**, 31–70 (2007)
8. Seidl, R. et al.: (BELLE). Measurement of azimuthal asymmetries in inclusive production of hadron pairs in e^+e^- annihilation at $\sqrt{s} = 10.58$ GeV. Phys. Rev. D **78**, 032011 (2008)
9. Anselmino, M. et al.: Transversity and Collins functions from SIDIS and e^+e^- data. Phys. Rev. D **75**, 054032 (2007)
10. Airapetian, A. et al.: (HERMES). Effects of transversity in deep-inelastic scattering by polarized protons. Phys. Lett. B **693**, 11–16 (2010)
11. Sivers, D.W.: Single-spin production asymmetries from the hard scattering of pointlike constituents. Phys. Rev. D **41**, 83–90 (1990)
12. Brodsky, S.J. et al.: Final-state interactions and single-spin asymmetries in semi-inclusive deep inelastic scattering. Phys. Lett. B **530**, 99–107 (2002)
13. Airapetian, A. et al.: (HERMES). Observation of the naive-T-odd sivers effect in deep-inelastic scattering. Phys. Rev. Lett. **103**, 152002 (2009)
14. Airapetian, A. et al.: (HERMES). Beam-helicity and beam-charge asymmetries associated with deeply virtual Compton scattering on the unpolarised proton. JHEP **07**, 032 (2012)
15. Airapetian, A. et al.: (HERMES). Exclusive leptoproduction of real photons on a longitudinally polarised hydrogen target. JHEP **06**, 019 (2010)
16. Airapetian, A. et al.: (HERMES). Measurement of azimuthal asymmetries with respect to both beam charge and transverse target polarization in exclusive electroproduction of real photons. JHEP **06**, 066 (2008)
17. Airapetian, A. et al.: (HERMES). Measurement of double-spin asymmetries associated with deeply virtual Compton scattering on a transversely polarized hydrogen target. Phys. Lett. **B 704**, 15–23 (2011)
18. Airapetian, A. et al.: (HERMES). Measurement of azimuthal asymmetries associated with deeply virtual Compton scattering on an unpolarized deuterium target. Nucl. Phys. **B 829**, 1–27 (2010)
19. Airapetian, A. et al.: (HERMES). Measurement of azimuthal asymmetries associated with deeply virtual Compton scattering on a longitudinally polarized deuterium target. Nucl. Phys. **B 842**, 265–298 (2011)
20. Airapetian, A. et al.: (HERMES). Beam-helicity asymmetry arising from deeply virtual Compton scattering measured with kinematically complete event reconstruction. JHEP **10**, 042 (2012)
21. Airapetian, A. et al.: (HERMES). Spin density matrix elements in exclusive ρ^0 electroduction on ${}^1H$ and ${}^2H$ targets at 27.6 GeV beam energy. Eur. Phys. J. C **62**, 659–694 (2009)
22. Airapetian, A. et al.: (HERMES). Spin density matrix elements in exclusive ω electroproduction on ${}^1H$ and ${}^2H$ targets at 27.6 GeV beam energy. Eur. Phys. J. C **74**, 3110 (2014)
23. Airapetian, A. et al.: (HERMES). Exclusive ρ^0 electroproduction on transversely polarized proton. Phys. Lett. B **679**, 100–105 (2009)

DOI 10.1007/978-3-319-50699-9_22
Reprinted from *Few-Body System* Journal, DOI 10.1007/s00601-016-1046-8

CrossMark

Gernot Eichmann

Towards a Microscopic Understanding of Nucleon Polarizabilities

Received: 15 January 2016 / Accepted: 27 January 2016 / Published online: 17 February 2016

Abstract We outline a microscopic framework to calculate nucleon Compton scattering from the level of quarks and gluons within the covariant Faddeev approach. We explain the connection with hadronic expansions of the Compton scattering amplitude and discuss the obstacles in maintaining electromagnetic gauge invariance. Finally we give preliminary results for the nucleon polarizabilities.

1 Introduction

There is much ongoing interest in the precision determination of the nucleon's polarizabilities; see [1] for a recent review. The electric polarizability α and magnetic polarizability β tell us how the nucleon responds to an external electromagnetic field, with current PDG values $\alpha = 11.2(4) \times 10^{-4}$ fm^3 and $\beta = 2.5(4) \times 10^{-4}$ fm^3 for the proton [2]. The polarizabilities are proportional to the volume and their smallness indicates that the proton is a rigid object due to the strong binding of its constituents. Whereas $\alpha + \beta$ is constrained by a sum rule, the small value for β is commonly believed to be due to a cancellation between the nucleon 'quark core' and the interaction with its pion cloud.

The polarizabilities are encoded in the nucleon Compton scattering (CS) amplitude $N\gamma^* \to N\gamma^*$ which has many applications also beyond polarizabilities. The integrated CS amplitude is relevant for two-photon corrections to nucleon form factors [3] and perhaps also for the proton radius puzzle [1]. So far, our knowledge of the CS amplitude is restricted to a few kinematic limits including the (generalized) polarizabilities in real and virtual CS [4], the nucleon structure functions in the forward limit, and deeply virtual CS (DVCS) from where generalized parton distributions are extracted [5]. In addition, the crossed process $p\bar{p} \to \gamma\gamma$ will be measured by PANDA.

The main theoretical tools to address CS are chiral perturbation theory which provides a systematic expansion of the CS amplitude at low energies [6], and dispersion relations with a direct link to experimental data [7,8]. On the other hand, handbag dominance in DVCS is well established and a key ingredient to factorization theorems. Is it then possible to connect these two facets by a common, underlying approach at the level of quarks and gluons that reproduces all established features, from hadronic poles to the handbag picture? In the following we will briefly outline such an approach and present first calculated results for the proton polarizabilities α and β.

This work is supported by the German Science Foundation DFG under project number DFG TR-16.

G. Eichmann (✉)
Institut für Theoretische Physik, Justus-Liebig-Universität Giessen, Heinrich-Buff-Ring 16, 35392 Giessen, Germany
E-mail: gernot.eichmann@theo.physik.uni-giessen.de

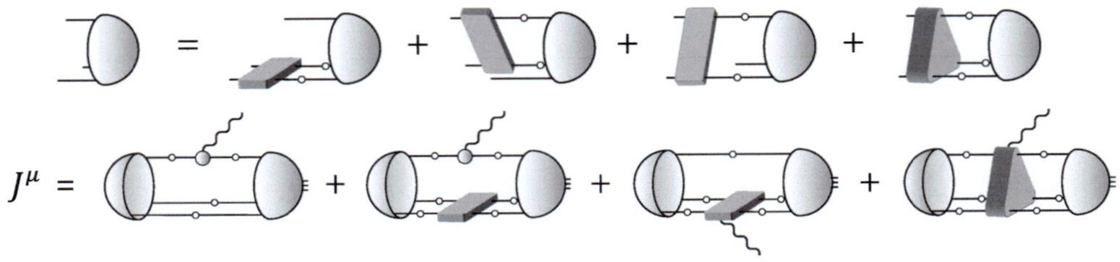

Fig. 1 Three-quark Faddeev equation (*top*) and electromagnetic current matrix element (*bottom*)

2 The Covariant Faddeev Approach

Our tool of choice is the covariant three-body Faddeev approach established in [9–11]. Its basic equations are illustrated in Fig. 1. The Faddeev equation determines the nucleon mass and bound-state amplitude (its 'wave function') by summing up all possible two- and three-body interactions between dressed quarks. The electromagnetic current matrix element couples the photon to all microscopic ingredients and thereby satisfies electromagnetic gauge invariance.

To solve the Faddeev equation one needs to specify its input. Three-body interactions have been neglected so far, and most studies have employed a rainbow-ladder truncation where the two-body kernel is given by a dressed gluon exchange. The dressed quark propagator is solved from its Dyson–Schwinger equation and the resulting quark mass function becomes momentum-dependent; it describes the transition from the input current-quark mass at large momenta to a nonperturbative, dressed 'constituent quark' mass of a few hundred MeV in the infrared. In general, any truncation must preserve chiral symmetry to ensure a massless pion in the chiral limit via the analogous Bethe–Salpeter equation; see e.g. [12] for recent advances in this area.

Whereas the applicability of rainbow-ladder in the light-meson sector is mainly limited to pseudoscalar and vector mesons, baryons fare much better: the approach reproduces the octet and decuplet ground state masses within 5–10 % [14]. Figure 2 shows results for the ρ−meson, nucleon and Δ masses as functions of m_π^2 (which is also calculated) compared to lattice data and experiment. The only input is the quark-gluon interaction for the two-body kernel whose model dependence is given by the bands. In particular, once the model scale is set to reproduce the pion decay constant, there are no further parameters or approximations and all subsequent results are predictions.

Apart from mass spectra, a range of form factors have been calculated as well within this setup. Among them are nucleon, Δ and hyperon electromagnetic form factors, the $N \to \Delta\gamma$ transition, and nucleon axial form factors [15,16]. All these cases exhibit good overall agreement with experimental data (where available) and also lattice results at larger pion masses, with discrepancies at low Q^2 where pion-cloud effects become important. While the three-body Faddeev approach does not depend on explicit diquark degrees of freedom, it is conceptually close to the quark–diquark framework which typically yields similar results and thereby establishes the presence of strong diquark correlations inside baryons [17–19]. An advantage is that the approach is not limited to two- and three-body systems: using the very same building blocks, it has been recently also applied to tetraquarks and the muon g-2 problem [20,21]. Given the body of results so far it is desirable to go a step further and ask: what can we learn about Compton scattering from such a microscopic perspective?

3 Compton Scattering

The nucleon CS amplitude depends on three independent momenta (see Fig. 3): the average nucleon momentum p, the average photon momentum $\Sigma = (Q + Q')/2$, and the momentum transfer $\Delta = Q - Q'$. The process is described by four Lorentz-invariant kinematic variables which we define as

$$\eta_+ = \frac{Q^2 + Q'^2}{2m^2}, \quad \eta_- = \frac{Q \cdot Q'}{m^2}, \quad \omega = \frac{Q^2 - Q'^2}{2m^2}, \quad \lambda = \frac{p \cdot \Sigma}{m^2}, \tag{1}$$

where m is the nucleon mass. The kinematic phase space in the variables $\{\eta_+, \eta_-, \omega\}$ is illustrated in Fig. 3. The spacelike region that is integrated over in nucleon-lepton scattering forms the interior of a cone around the η_+ axis. Its apex is where the static polarizabilities are defined, with momentum-dependent extensions to real

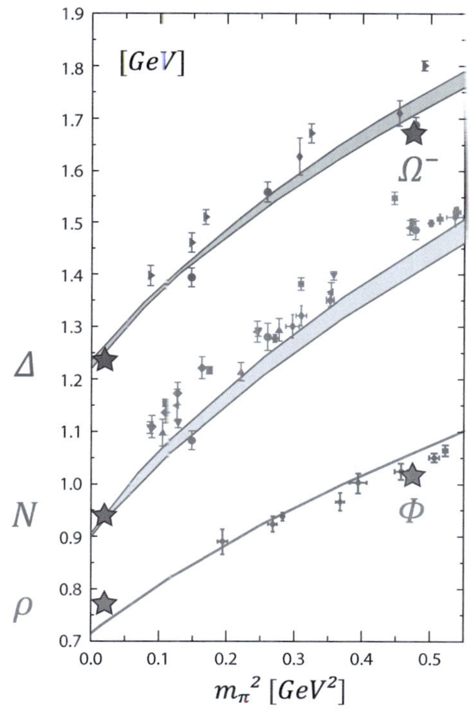

Fig. 2 ρ−meson [13], nucleon and Δ masses [9–11] calculated from their Bethe–Salpeter and Faddeev equations. *Stars* are PDG values and symbols with *error bars* are lattice data (see [9–11] for references)

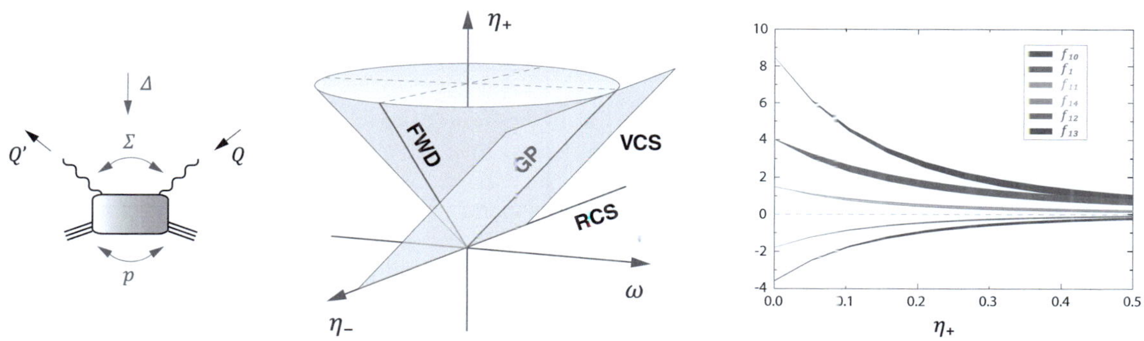

Fig. 3 *Left* kinematics and phase space in Compton scattering. *Right* dominant Compton form factors corresponding to the residue of the nucleon Born terms after removing the common pole factor. The bands contain the full kinematic dependence on all four variables inside the cone

CS ($\eta_+ = \omega = 0$), the doubly-virtual forward limit ($\eta_+ = \eta_-$, $\omega = 0$), and virtual CS ($\eta_+ = \omega$) including the generalized polarizabilities at $\eta_- = 0$.

Hadronic vs. microscopic decomposition At the hadronic level the CS amplitude is given by the sum of Born terms, which are determined by the nucleon form factors, and a one-particle-irreducible (1PI) structure part that carries the dynamics and encodes the polarizabilities, see Fig. 4. The latter contains $s/u-$channel nucleon resonances beyond the nucleon Born terms (including the Δ, Roper, etc.), $t-$channel meson exchanges (pion, scalar, axialvector, ...), and pion loops, with well-established low-energy expansions in chiral effective field theory. This is usually viewed as the 'correct' description at low energies, whereas the handbag picture is interpreted as the 'correct' one at large photon virtualities. Hence again the question: is there a common underlying description at the quark level that is valid in *all* kinematic regions and encompasses both approaches?

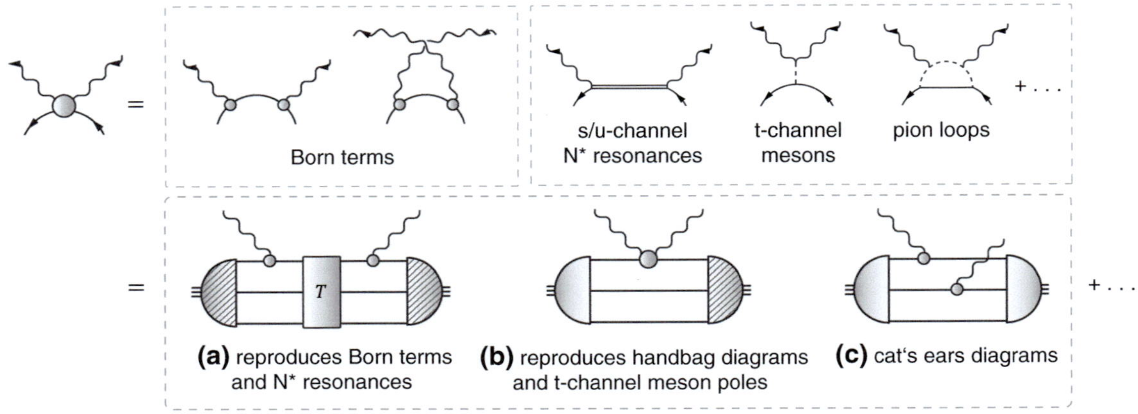

Fig. 4 Hadronic versus quark-level decomposition of the nucleon Compton scattering amplitude. The *first row* depicts the hadronic contributions as the sum of Born terms and a 1PI structure part. The latter encodes the polarizabilities and contains $s/u-$channel nucleon resonances, $t-$channel meson exchanges and pion loops. The *second row* shows the microscopic decomposition (in rainbow-ladder) featuring Faddeev amplitudes, quark propagators, quark-photon and quark Compton vertices, and the three-quark scattering matrix [23]

In analogy to the form factor diagrams in Fig. 1 one can derive a closed expression for the CS amplitude at the quark level [22,23]. The topologies that survive in a rainbow-ladder truncation (apart from permutations and symmetrizations) are collected in the second row of Fig. 4. Ambiguities stemming from intermediate offshell hadrons never arise here because hadronic degrees of freedom do not appear explicitly. Instead, the diagrams reproduce the onshell hadron pole contributions:

- Diagram (a) depends on the three-quark scattering matrix which contains all possible baryon poles, so it reproduces the nucleon Born terms as well as all $s/u-$channel resonances.
- Diagram (b) contains the quark two-photon (quark Compton) vertex, which has an analogous decomposition into quark Born terms and a 1PI part. The Born terms provide the handbag contributions. The 1PI part features a quark–antiquark scattering matrix that contains all possible $t-$channel meson poles and thereby reproduces the meson exchanges in the first row.

Neither the handbag nor the cat's-ears contributions from diagram (c) have a direct analogue in the hadronic expansion where they are rather absorbed into counterterms. Vice versa, the diagrams in the bottom do not contain the microscopic representation of pion loops because those only enter beyond rainbow-ladder. In any case, the sum of all graphs in the box satisfies electromagnetic gauge invariance so that the resulting CS amplitude is purely transverse; it is $s/u-$channel crossing symmetric; it reproduces all known hadronic poles; and it contains the handbag contributions which are perturbatively (and presumably also nonperturbatively) important.

Is it feasible to calculate all these microscopic diagrams in analogy to what has been achieved for form factors? The main obstacle is diagram (a): while there has been progress in the calculation of three- and four-point functions [24], the treatment of six-point functions is so far beyond reach. We therefore approximate this graph by summing up the leading hadronic diagrams in the form of nucleon resonances. Neglecting also diagram (c), we calculate graph (b) in rainbow-ladder but without further approximations: the quark propagator is obtained from its Dyson–Schwinger equation, the nucleon amplitude from the covariant Faddeev equation, and the quark Compton vertex including all (128) tensor structures from its inhomogeneous Bethe–Salpeter equation [23].

Gauge invariance The problem with this strategy is that only the sum of diagrams (a-c) is gauge invariant but not the individual graphs alone. Since transversality is connected with analyticity, a simple transverse projection does not suffice because an approximation that breaks electromagnetic gauge invariance can induce kinematic singularities that render its results meaningless. The problem can be illustrated with a textbook example, namely the photon vacuum polarization whose general form is $\Pi^{\mu\nu}(Q) = a\,\delta^{\mu\nu} + b\,Q^\mu Q^\nu$. The coefficients a and b are functions of Q^2 and must be analytic at $Q^2 = 0$; poles would correspond to intermediate massless particles but since the vacuum polarization is 1PI intermediate propagators are excluded by definition. Gauge invariance entails transversality, $Q^\mu \Pi^{\mu\nu} = 0$, which fixes $a = -b\,Q^2$. The vacuum polarization can then be written as the sum of a transverse part and a 'gauge part' (which is *not* longitudinal):

$$\Pi^{\mu\nu}(Q) = \Pi(Q^2)\, t^{\mu\nu}_{QQ} + \widetilde{\Pi}(Q^2)\, \delta^{\mu\nu}, \qquad t^{\mu\nu}_{AB} = A\cdot B\, \delta^{\mu\nu} - B^\mu A^\nu. \tag{2}$$

The transverse dressing function $\Pi(Q^2)$ is free of kinematic singularities and zeros. The gauge part $\delta^{\mu\nu}$ is the tensor that we eliminated in the first place, so $\widetilde{\Pi}(Q^2)$ must vanish due to gauge invariance. This is what happens in dimensional regularization, whereas a cutoff breaks gauge invariance and induces a quadratic divergence (only) in the gauge part. If we did not know about the decomposition (2) and performed a transverse projection, $\Pi(Q^2)$ would pick up a $1/Q^2$ pole from the gauge part which invalidates the extraction of $\Pi(Q^2=0)$. The transverse/gauge separation is also convenient if gauge invariance is broken by more than a cutoff, for instance by an incomplete calculation: ultimately the sum of all gauge parts must vanish, but the partial result for $\Pi(Q^2)$ is still free of kinematic problems and—ideally—not strongly affected by gauge artifacts.

This simple example also provides the template for the CS amplitude, where a complete decomposition into transverse and gauge parts free of kinematic singularities and zeros is necessary as well:

$$\mathcal{M}^{\mu\nu}(p,Q',Q) = \frac{1}{m}\,\bar{u}(p_f)\Bigg[\underbrace{\left(\frac{f_1}{m^4}\, t^{\mu\alpha}_{Q'p}\, t^{\alpha\nu}_{pQ} + \frac{f_2}{m^2}\, t^{\mu\nu}_{Q'Q} + \cdots\right)}_{\text{transverse part, 18 tensors}} + \underbrace{(\cdots)}_{\substack{\text{gauge part,}\\ \text{14 tensors}}}\Bigg]u(p_i), \tag{3}$$

with $t^{\mu\nu}_{AB}$ defined in Eq. (2). We employ the transverse tensor basis of Refs. [7,25] but insert factors of ω, λ and m where necessary, so that all Compton form factors (CFFs) $f_i(\eta_+, \eta_-, \omega, \lambda)$ are dimensionless and invariant under photon crossing and charge conjugation. The gauge part must be zero and vanish if all diagrams in Fig. 4 are included. However, even if one breaks gauge invariance by retaining only a subset of diagrams, the transverse CFFs still yield a well-defined prediction.

This can be understood already at the hadronic level. The definition of polarizabilities entails that both the Born and 1PI parts must be individually gauge invariant, so the expansion (3) must hold for both contributions alone. The Born terms are specified by the nucleon's electromagnetic current, but since the intermediate nucleon is offshell the half-offshell nucleon-photon vertex can have more tensor structures. It is well known that only an onshell Dirac current with Q^2-dependent Pauli and Dirac form factors ensures gauge invariance of the Born term.[1] The right panel in Fig. 3 shows the leading CFFs for this case after removing the common nucleon pole factor. Note that all CFFs are well-behaved and approach constant values for $\eta_+ \to 0$. As required, the gauge part is exactly zero. The implementation of offshell form factors destroys this property: the gauge part then no longer vanishes but within a reasonable range of model parametrizations the transverse CFFs remain almost unchanged.

Another remarkable feature is visible in Fig. 3: the bands contain the *full* kinematic dependence on all four variables η_+, η_-, ω and λ inside the cone, but effectively they only depend on η_+. The residues of the nucleon Born terms therefore scale with η_+, which reflects the symmetric makeup of the phase space. The hadronic poles form planes in the phase space that will generally counteract this symmetry property: the nucleon Born poles appear at $\eta_- = \lambda = 0$; the nucleon resonance poles form vertical planes at fixed $\eta_- < 0$, where the value of λ depends on the width of the resonance; and $t-$channel meson poles appear at fixed $\Delta^2 = -m_i^2 \Rightarrow \eta_- = \eta_+ + m_i^2/(2m^2)$.

Polarizabilities The nucleon polarizabilities α and β are related to the CFFs f_1 and f_2 in the limit where all kinematic variables are zero: $\{\alpha+\beta, \beta\} = \{f_1, f_2\} \times \alpha_{\text{QED}}/m^3$. In Fig. 5 we show preliminary results from the quark-level calculation extracted from the basis in Eq. 3. So far they are only ballpark estimates: the quark Compton vertex that enters in the calculation depends on 6 Lorentz invariants and 128 tensor structures and its transverse/gauge separation is extremely sensitive to the numerics. We extracted the momentum dependence of $\alpha+\beta$ from f_1 only whereas the standard definition [7] contains admixtures from higher CFFs at $\eta_+ > 0$, but for those our results are still too noisy.

The hatched bands in Fig. 5 are the outcome of diagram (b) inside the cone. For the total result we added the Δ resonance, the dominant hadronic contribution to diagram (a), using a parametrization for the experimental $N \to \Delta\gamma$ form factors. For comparison we plot the dispersion relation results for the generalized polarizabilities from Refs. [4,7]. The figure makes clear that the sum $\alpha+\beta$ is dominated by diagram (b) and, as it turns out, especially by the handbag contributions. The magnetic polarizability β is dominated by the Δ pole

[1] For arbitrary offshell form factors the Born terms must be combined with (non-diagrammatic) parts of the 1PI contribution to arrive again at a gauge-invariant expression, see [23] for a discussion.

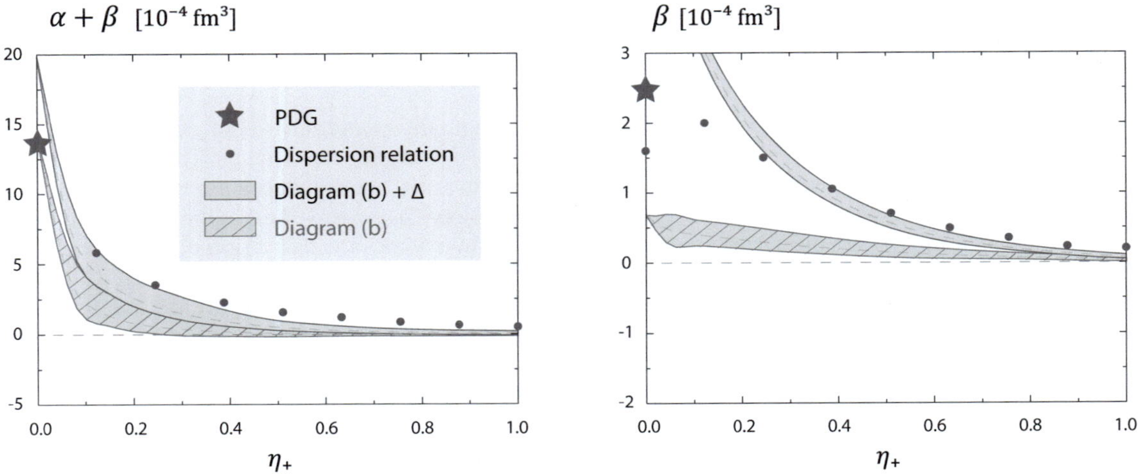

Fig. 5 Proton polarizabilities as functions of η_+. The bands were obtained from diagram (**b**) with and without the Δ contribution. The *dots* were extracted from Ref. [4] and the *stars* are the experimental values [2]

from diagram (a) whereas (b) contributes little due to cancellations. The discrepancy at low η_+ is presumably due to missing pion loops—β is subject to cancellations between the quark core (which then mainly comes from the Δ pole) and pion cloud effects.

To summarize, we demonstrated how to extract microscopic information on nucleon polarizabilities from the decomposition in Fig. 4. It will be further interesting to investigate spin polarizabilities and gather knowledge on the spacelike momentum dependence of the CS amplitude, which will improve our understanding of two-photon corrections to form factors as well as the proton radius puzzle. Finally, the same framework can be adapted to other processes such as pion electroproduction, which has contributed much to our knowledge of nucleon resonances and transition form factors. For their clean extraction one needs to know the non-resonant 'QCD background' beyond hadronic exchanges, which is information that a microscopic approach can provide.

Acknowledgments I am grateful to C. S. Fischer and G. Ramalho for valuable discussions, and I would like to thank the organizers of the *Light Cone 2015* conference for their support.

References

1. Hagelstein, F., Miskimen, R., Pascalutsa, V.: Nucleon Polarizabilities: from Compton scattering to hydrogen atom. arXiv:1512.03765 [nucl-th]
2. Olive, K.A. et al.: Review of particle physics. Chin. Phys. **C38**, 090001 (2014)
3. Arrington, J., Blunden, P., Melnitchouk, W.: Review of two-photon exchange in electron scattering. Prog. Part. Nucl. Phys. **66**, 782–833 (2011)
4. Downie, E.J., Fonvieille, H.: Real and virtual Compton scattering: the nucleon polarisabilities. Eur. Phys. J. Special Topics **198**, 287–306 (2011)
5. Guidal, M., Moutarde, H., Vanderhaeghen, M.: Generalized parton distributions in the valence region from deeply virtual Compton scattering. Rep. Prog. Phys. **76**, 066202 (2013)
6. Griesshammer, H.W. et al.: Using effective field theory to analyse low-energy Compton scattering data from protons and light nuclei. Prog. Part. Nucl. Phys. **67**, 841–897 (2012)
7. Drechsel, D., Pasquini, B., Vanderhaeghen, M.: Dispersion relations in real and virtual Compton scattering. Phys. Rep. **378**, 99–205 (2003)
8. Schumacher, M., Scadron, M.D.: Dispersion theory of nucleon Compton scattering and polarizabilities. Fortschr. Phys. **61**, 703–741 (2013)
9. Eichmann, G. et al.: Nucleon mass from a covariant three-quark Faddeev equation. Phys. Rev. Lett. **104**, 201601 (2010)
10. Sanchis-Alepuz, H. et al.: Delta and Omega masses in a three-quark covariant Faddeev approach. Phys. Rev. **D84**, 096003 (2011)
11. Eichmann, G.: Nucleon electromagnetic form factors from the covariant Faddeev equation. Phys. Rev. **D84**, 014014 (2011)
12. Williams, R., Fischer, C.S., Heupel, W.: Light mesons in QCD and unquenching effects from the 3PI effective action. arXiv:1512.00455 [hep-ph]
13. Maris, P., Tandy, P.C.: Bethe–Salpeter study of vector meson masses and decay constants. Phys. Rev. **C60**, 055214 (1999)
14. Sanchis-Alepuz, H., Fischer, C.S.: Octet and decuplet masses: a covariant three-body Faddeev calculation. Phys. Rev. **D90**(9), 096001 (2014)

15. Alkofer, R. et al.: Electromagnetic baryon form factors in the Poincaré-covariant Faddeev approach. Hyperfine Interact. **234**(1–3), 149–154 (2015)
16. Sanchis-Alepuz, H., Fischer, C.S.: Hyperon elastic electromagnetic form factors in the space-like momentum region. arXiv:1512.00833 [hep-ph]
17. Eichmann, G. et al.: Toward unifying the description of meson and baryon properties. Phys. Rev. **C79**, 012202 (2009)
18. Cloet, I.C. et al.: Survey of nucleon electromagnetic form factors. Few Body Syst. **46**, 1–36 (2009)
19. Segovia, J.: Elastic and transition form factors in DSEs. arXiv:1601.00973 [nucl-th]
20. Eichmann, G., Fischer, C.S., Heupel, W.: The light scalar mesons as tetraquarks. Phys. Lett. **B753**, 282–287 (2016)
21. Eichmann, G., Fischer, C.S., Heupel, W., Williams, R.: The muon g-2: Dyson–Schwinger status on hadronic light-by-light scattering. In: Confinement XI, (2014). arXiv:1411.7876 [hep-ph]
22. Eichmann, G., Fischer, C.S.: Unified description of hadron-photon and hadron-meson scattering in the Dyson–Schwinger approach. Phys. Rev. **D85**, 034015 (2012)
23. Eichmann, G., Fischer, C.S.: Nucleon Compton scattering in the Dyson–Schwinger approach. Phys.Rev. **D87**(3), 036006 (2013)
24. Eichmann, G., Fischer, C.S., Heupel, W.: Four-point functions and the permutation group S4. Phys. Rev. **D92**(5), 056006 (2015)
25. Tarrach, R.: Invariant amplitudes for virtual Compton scattering off polarized nucleons free from kinematical singularities, zeros and constraints. Nuovo Cimento **A28**, 409 (1975)

DOI 10.1007/978-3-319-50699-9_23
Reprinted from *Few-Body System* Journal, DOI 10.1007/s00601-016-1047-7

J. R. McKenney · Nobuo Sato · W. Melnitchouk · Chueng-Ryong Ji

SU(2) Flavor Asymmetry of the Proton Sea in Chiral Effective Theory

Received: 16 January 2016 / Accepted: 28 January 2016 / Published online: 17 February 2016

Abstract We refine the computation of the $\bar{d} - \bar{u}$ flavor asymmetry in the proton sea with a complementary effort to reveal the dynamics of pion exchange in high-energy processes. In particular, we discuss the efficacy of pion exchange models to simultaneously describe leading neutron electroproduction at HERA along with the $\bar{d} - \bar{u}$ flavor asymmetry in the proton. A detailed χ^2 analysis of the ZEUS and H1 data, when combined with constraints on the pion flux from Drell-Yan data, allows regions of applicability of one-pion exchange to be delineated. Based on the fit results, we also address a possible estimate for leading proton structure functions in upcoming tagged deep-inelastic scattering experiments at Jefferson Lab on the deuteron with forward protons.

1 Introduction

Last year in LC2014 at Raleigh, Wally Melnitchouk discussed the computation of the $\bar{d} - \bar{u}$ flavor asymmetry in the proton arising from pion loops using chiral effective field theory [1]. The calculation in his presentation included both nucleon and Δ degrees of freedom and used both the fully relativistic and heavy baryon frameworks. In analogy with low-energy observables such as masses and magnetic moments, the discussion included that moments of parton distribution functions (PDFs) could be systematically expanded in powers of m_π^2, with the coefficients of the leading nonanalytic (LNA) terms given in terms of model-independent constants [2–5]. This enabled an unambiguous connection to be established between chiral symmetry breaking in QCD and the existence of an SU(2) flavor asymmetry in the proton [2].

In this work, we discuss our refined computation of the $\bar{d} - \bar{u}$ flavor asymmetry in the proton sea with an extended analysis to reveal the dynamics of pion exchange in high-energy processes. This complementary effort has been the study of leading neutron production in semi-inclusive deep inelastic scattering (DIS) on the proton, in which a forward moving neutron is produced in coincidence with the scattered lepton in the high-energy reaction $ep \to enX$. Several dedicated experiments at the ep collider HERA [6–8] have collected high-precision data on the spectrum of leading neutrons carrying a large fraction of the proton's energy. Moreover, the upcoming tagged DIS (TDIS) experiment at Jefferson Lab [9] plans to take data on the production of leading protons from an effective neutron target in the reaction $en \to epX$, which, in analogy

Presented by Chueng-Ryong Ji at LIGHTCONE 2015, September 21-25, 2015, Frascati.

J. R. McKenney
University of North Carolina, Chapel Hill, NC 27599, USA

N. Sato · W. Melnitchouk
Jefferson Lab, Newport News, Virginia 23606, USA

C.-R. Ji (✉)
North Carolina State University, Raleigh, NC 27695, USA
E-mail: crji@ncsu.edu

with the HERA leading neutron leptoproduction, can be described through the exchange of a π^-. In the proposed experiment, the effective neutron target will be prepared by tagging spectator protons with momenta between 60 and 400 MeV at backward kinematics in the DIS of the electron from a deuteron nucleus, using the same technique that was developed for the measurement of the neutron structure function in the BONuS experiment at Jefferson Lab [10].

In particular, we have recently addressed [11] the question of whether one can reduce the model dependence of F_2^{π} extracted from the HERA data [6–8] by using additional constraints from the data on the SU(2) flavor asymmetry $\bar{d} - \bar{u}$, especially those from the E866 Drell-Yan experiment [12]. Because the E866 data are at relatively high x values compared with the HERA measurements, within the pion exchange framework they are sensitive to the pion PDFs at large x_π, where these are well determined from pion–nucleon Drell-Yan data [13–15]. The main variable in describing the $\bar{d} - \bar{u}$ asymmetry is therefore the pion distribution function in the nucleon. In our study [11], we use methodology adopted from global PDF analysis [16,17] to simultaneously fit both the HERA leading neutron and E866 $\bar{d}-\bar{u}$ asymmetry data. This study allows us to refine the analysis of the $\bar{d} - \bar{u}$ asymmetry and at the same time discuss a possible prediction of the future tagged DIS experiments in JLab.

2 Pion Exchange Models

In this section we briefly review the computation of the pion light-cone momentum distributions in the nucleon (sometimes also referred to as the pion splitting functions), for both πN and $\pi \Delta$ fluctuations of the proton.

For the fluctuation of a proton (with four momentum p) to a positively charged pion (momentum k) and a neutron $(p - k)$, illustrated by the "rainbow" diagram in Fig. 1a, the $p \to n\pi^+$ splitting function derived from chiral effective theory is expressed as a sum of on-shell and δ-function pieces [18,19],

$$f_{\pi^+ n}(y) = 2\Big[f_N^{(\text{on})}(y) + f_N^{(\delta)}(y)\Big], \tag{1}$$

where $y = k^+/p^+$ is the fraction of the protons light-cone momentum carried by the pion, and the "+" component of the four-vector is defined as $k^+ \equiv k^0 + k^z$. The on-shell contribution $f_N^{(\text{on})}$ corresponds to the region $y > 0$ and can be written as [20,21]

$$f_N^{(\text{on})}(y) = \frac{g_A^2 M^2}{(4\pi f_\pi)^2} \int dk_\perp^2 \, \frac{y\,(k_\perp^2 + y^2 M^2)}{(1-y)^2 D_{\pi N}^2}, \tag{2}$$

where M is the nucleon mass, $g_A = 1.267$ is the axial charge, $f_\pi = 93$ MeV is the pion decay constant, and

$$D_{\pi N} \equiv t - m_\pi^2 = -\frac{1}{1-y}\left[k_\perp^2 + y^2 M^2 + (1-y) m_\pi^2\right] \tag{3}$$

for an on-shell nucleon intermediate state, with the pion virtuality $t \equiv k^2 = -(k_\perp^2 + y^2 M^2)/(1 - y)$. The second term in Eq. (1), $f_N^{(\delta)}$, arises from off-shell nucleon contributions and is proportional to $\delta(y)$. The significance of this term has been discussed [22] with respect to the model-independent nonanalytic structure of the vertex renormalization constant as a function of the pion mass. One may regard this nonanalytic function of m_π^2 as the first principles constraint on the infrared behavior of the chiral effective theory consistent with the chiral symmetry of QCD. In scattering processes this term contributes only at $x = 0$, and is therefore relevant only for the lowest moment of the parton distribution. In this work we will be analyzing data at nonzero values of x, at which $f_N^{(\delta)}$ will play no direct role.

Note that the factor 2 in Eq. (1) is an isospin factor specific to the $p \to n\pi^+$ fluctuation; the distribution for the fluctuation $p \to p\pi^0$ is related to that in Eq. (1) by $f_{\pi^+ n}(y) = 2 f_{\pi^0 p}(y)$. In writing the coefficient in front of the integration in Eq. (2), we have assumed the GoldbergerTreiman relation, $g_A/f_\pi = g_{\pi NN}/M$, where $g_{\pi NN}^2/4\pi \approx 13.7$ gives the strength of the πNN coupling [23].

In addition to the nucleon intermediate states, contributions from Δ baryons in Fig. 1b are known to play an important role in hadron structure. Within the same chiral effective theory framework, using an effective $\pi N \Delta$ interaction [19], the $p \to \Delta^0 \pi^+$ splitting function can be written as a sum of three terms,

$$f_{\pi^+ \Delta^0}(y) = f_\Delta^{(\text{on})}(y) + f_\Delta^{(\delta)}(y) + f_\Delta^{(\text{end-pt})}(y). \tag{4}$$

Fig. 1 Contributions to the pion distributions in the proton from the rainbow diagrams involving **a** a nucleon (*solid lines*) and **b** a Δ isobar (*double solid line*) in the intermediate state. The external operators couple to the virtual pions (*dashed lines*)

The on-shell piece $f_{\Delta}^{(\text{on})}$, corresponding to the Δ pole, is given for $0 < y < 1$ by

$$f_{\Delta}^{(\text{on})}(y) = C_{\Delta} \int dk_{\perp}^2 \frac{y\,(\overline{M}^2 - m_{\pi}^2)}{(1-y) D_{\pi\Delta}^2} \Big[(\overline{M}^2 - m_{\pi}^2)(\Delta^2 - m_{\pi}^2) - [3(\Delta^2 - m_{\pi}^2) + 4 M M_{\Delta}] D_{\pi\Delta}\Big] \tag{5}$$

where

$$D_{\pi\Delta} \equiv t - m_{\pi}^2 = -\frac{1}{1-y}\big[k_{\perp}^2 - y(1-y)M^2 + yM_{\Delta}^2 + (1-y)m_{\pi}^2\big] \tag{6}$$

for an on-shell Δ intermediate state of mass M_{Δ}, with $\overline{M} \equiv M_{\Delta} + M$ and $\Delta \equiv M_{\Delta} - M$. The pion virtuality here is given by $t \equiv k^2 = -(k_{\perp}^2 - y(1-y)M^2 + yM_{\Delta}^2)/(1-y)$. The coefficient $C_{\Delta} = g_{\pi N\Delta}^2/[(4\pi)^2 18 M_{\Delta}^2]$ contains the $\pi N \Delta$ coupling constant, which is given from SU(6) symmetry by $g_{\pi N\Delta} = (3\sqrt{2}/5) g_A/f_{\pi} \approx 11.8\ \text{GeV}^{-1}$ [24]. For the other charge channels in the $p \to \Delta\pi$ dissociation, the splitting functions are related by $2 f_{\pi^- \Delta^{++}} = 3 f_{\pi^0 \Delta^+} = 6 f_{\pi^+ \Delta^0}$.

Note that the on-shell contribution in Eq. (5) differs from the "Sullivan" form often used in the literature [2,24–26], which is obtained by taking the Δ-pole contribution, $D_{\pi\Delta} \to M_{\Delta}^2$. In particular, it has a higher power of $k_{\perp}$ ($k_{\perp}^6$ compared with $k_{\perp}^2$ in Eq. (5)), which arises from the neglect of the end-point contributions in the Sullivan process.The other two terms in Eq. (4), $f_{\Delta}^{(\delta)}$ and $f_{\Delta}^{(\text{end-pt})}$, correspond to a δ-function contribution at $y = 0$ and an end-point contribution proportional to a δ-function at $y = 1$, respectively.

From the on-shell nucleon and Δ splitting functions in Eqs. (2) and (5), it is evident that integration over contributions from large $k_{\perp}$ will introduce logarithmic divergences in the point-like theory, which must be regularized in order to obtain finite results. Since the nucleon is not point-like, but has a finite spatial extent of $\mathcal{O}(1\ \text{fm})$, this introduces an additional scale into the effective theory, along with the chiral symmetry breaking scale [27]. The precise way that the finite range of the nucleon is implemented in order to regularize the ultraviolet divergences depends on the prescription adopted [27,28], although any prescription must correctly incorporate the infrared behavior of pion loops which is model independent. In practice, the model dependence amounts to a choice of form factor $F(y, k_{\perp}^2)$ multiplying the integrands of Eqs. (2) and (5) which suppresses the large-$k_{\perp}$ contributions. In our recent paper [11], we have extensively discussed the available regularization prescriptions and noted that the best results for the E866 data were obtained with the t-dependent exponential form factor given by

$$F = \exp\big[(t - m_{\pi}^2)/\Lambda^2\big] \qquad [t\text{-dependent exponential}], \tag{7}$$

where the best fit values of the ultraviolet cutoff parameter Λ were found as $\Lambda \approx 0.85(0.88)$ GeV with the SMRS [29] (ASV [30]) parametrization for the pion PDFs that we discuss in the next section.

3 SU(2) Flavor Asymmetry of the Sea

The E866 experiment at Fermilab measured the ratio σ^{pd}/σ^{pp} at high (projectile) proton momentum fractions x_1 and low target momentum fraction x_2, where at leading order in the strong coupling constant α_s it is approximately given by [12]

$$\frac{\sigma^{pd}}{2\sigma^{pp}} \approx \frac{1}{2}\left(1 + \frac{\bar{d}(x_2)}{\bar{u}(x_2)}\right), \qquad [x_1 \gg x_2]. \tag{8}$$

The cross sections were measured for x_2 between 0.015 and 0.35, at an average dimuon mass squared of $Q^2 = 54$ GeV^2, and the extracted $\bar{d}/\bar{u}$ ratio was found to exceed 1.5 for $x_2 \approx 0.1 - 0.2$.

We have examined [11] the constraints on the models of the pion cloud of the nucleon that can be inferred from a detailed analysis of the $\bar{d} - \bar{u}$ asymmetry in the proton. Within the effective chiral framework described in Sect. 2, the contributions to the $\bar{d} - \bar{u}$ difference from the pion loop diagrams in Fig. 1 can be written as [19]

$$\bar{d} - \bar{u} = \left(f_{\pi^+ n} - \frac{2}{3} f_{\pi^- \Delta^{++}} \right) \otimes \bar{q}_v^\pi , \tag{9}$$

where $\bar{q}_v^\pi \equiv \bar{d}^{\pi^+} - d^{\pi^+} = \bar{u}^{\pi^-} - u^{\pi^-}$ is the valence quark PDF in the pion, and the symbol "$\otimes$" denotes the convolution integral $f \otimes q = \int_0^1 dy \int_0^1 dz\, f(y)\, q(z)\, \delta(x - yz)$. Although there may be no *a priori* justification for ignoring the additional processes beyond this convolution [31], Eq. (9) appears justifiable for the pions under favorable kinematic conditions that we discuss in this work. As we discuss the efficacy of the pion exchange model [11] in Sect. 4, we focus on the kinematic region where it is natural to treat the peripheral modification of the nucleon in terms of pions [32]. The convolution in Eq. (9) follows from the crossing symmetry properties of the splitting functions $f(-y) = f(y)$ [33], and isospin symmetry relations have been assumed for the $\pi \Delta$ distributions. The contributions from neutral pions cancel in the asymmetry.

Performing a χ^2 fit to the E866 data, we have compared the results for the various regularization prescriptions with the best fit cutoff parameters and $\chi^2_{\rm dof}$ values [11]. For the valence antiquark distribution in the pion we used the SMRS parametrization [29] of the world's data from πN Drell-Yan and prompt photon production, evaluated at the E866 average Q^2 of 54 GeV^2. As indicated in Sect. 2, the fits with the lowest $\chi^2_{\rm dof}$ values were obtained with the t-dependent exponential regulator [Eq. (7)]. For much greater details including the analyses of other comparable regularizations, one should take a look at our recent paper [11]. We note that the biggest contributions to the χ^2 arise from the high-x data points, which have a steeper fall-off than can be accommodated in any of the models. We anticipate that the new SeaQuest experiment at Fermilab [34] will in the near future check the high-x behavior by measuring the $\bar{d}/\bar{u}$ ratio up to $x \approx 0.45$.

While the pion PDFs at small x values have never been directly measured, in the valence quark region the πN Drell-Yan data [13–15] provide strong constraints on the x dependence of $\bar{q}_v^\pi$ for $x \gtrsim 0.1$. Interestingly, the distributions at $x \to 1$ were observed [15] to be more consistent with a $\sim (1 - x)$ behavior [35–38] than with the $\sim (1 - x)^2$ expectation from perturbative QCD [39] or model calculations using the Dyson–Schwinger equations (see Ref. [40]). The large-x behavior in the SMRS parametrization [29] was consistent with the $\sim (1 - x)$ form indicated by the data.

Later, an analysis including next-to-leading order (NLO) corrections [41] found that the higher order effects soften the distributions, leading to a behavior that was intermediate between $(1-x)$ and $(1-x)^2$. More recently, Aicher et al. (ASV) [30] found that inclusion of threshold resummation at next-to-leading log (NLL) accuracy produces valence distributions that behave approximately as $(1 - x)^2$ at a low energy scale $Q_0 = 0.63$ GeV.

In order to assess the possible impact of the different $x \to 1$ behaviors of the valence pion PDF on the $\bar{d} - \bar{u}$ asymmetry, we repeated our analysis using the ASV parametrization [30], evolved from the low energy scale Q_0 to $Q^2 = 54$ GeV^2. The best fit results are compared in Fig. 2 for the t-dependent exponential form factor (7) with the result using the SMRS parametrization. As expected, the result with the ASV distribution leads to a softer asymmetry, with slightly better agreement at large x but marginally worse at $x \lesssim 0.1$. The overall $\chi^2_{\rm dof}$ values are slightly better for the ASV fit, mostly because the softer distribution allows a smaller asymmetry at $x \gtrsim 0.2$, as preferred by the E866 data, although the differences are not significant. The new results for the flavor asymmetry from the SeaQuest experiment [34] at large x may provide further insights into these comparisons.

4 Efficacy of the Pion Exchange Model in Leading Neutron Production at HERA

We have examined [11] the efficacy of the pion exchange model described in Sect. 3 in fitting the HERA leading neutron production data [6,7], and the compatibility of the results with the $\bar{d} - \bar{u}$ asymmetry extracted from the E866 Drell-Yan measurement [12]. While some dedicated analyses [42] have attempted to describe the HERA leading neutron spectra at all kinematics, our aim here was instead to maximize the sensitivity to the basic one-pion exchange contribution, which has the most direct connection to the chiral effective theory. This has been achieved by restricting the analysis to regions where one-pion exchange is expected to be the dominant process, and contributions from other backgrounds are minimal. In practice, since the calculation of the backgrounds is significantly more model dependent, the exact choice of kinematics may be somewhat

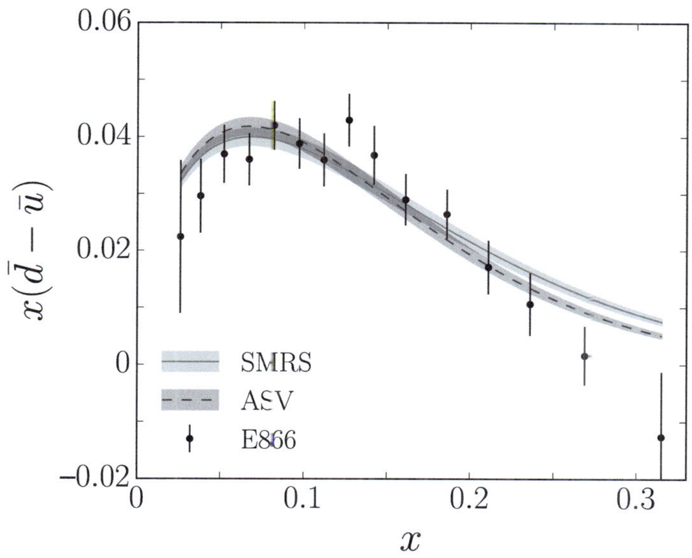

Fig. 2 Comparison of the pion model fits to the $\bar{d}-\bar{u}$ data from the E866 experiment [12] with the t-dependent exponential form factor (7) for the valence pion PDFs from the SMRS [29] and ASV [30] parametrizations

subjective. To determine in a more objective way the region of kinematics where the one-pion exchange is applicable, we performed a χ^2 analysis of the data as a function of the maximum value of y up to which the data were fitted, i.e. $y < y_{\rm cut}$. Although this reduced the total number of data points in the fit, the analysis of the more restrictive kinematic range allowed for a cleaner interpretation and extraction of the pion exchange parameters.

With the inclusion of the E866 $\bar{d}-\bar{u}$ asymmetry data in the fits together with the HERA leading neutron cross sections, the correlations between the pion flux and pion structure function parameters decreased dramatically for all cutoff models. For the t-dependent exponential model (7), for instance, the correlations between the Λ parameter and the normalization parameter of the pion structure function were reduced to between −8% to −16% over the range of cutoffs between $y_{\rm cut} = 0.1$ and 0.4. There was significantly greater discriminating power between the form factor models, with much stronger dependence of the fit results to the value of $y_{\rm cut}$. It is interesting to note that the behavior of the t variable at small y is given by $t \sim -k_\perp^2/(1-y)$ so that any undesirable suppression at small y such as the poor fit to the small-y HERA and E866 data does not occur for the t-dependent form factors. Through the convolution formula (9), one can see that less strength at small y also translates into suppression of the calculated PDFs at small x values. Again, for more details of this analysis, see Ref. [11].

5 Future Directions

Our analysis has sought to determine whether both the $\bar{d}-\bar{u}$ asymmetry in the proton extracted from the E866 Drell-Yan data [12] and the HERA leading neutron cross sections [6–8] can be obtained consistently within a common pion exchange framework. Rather than relying on assumptions about specific forms for the pion distributions in the nucleon, we have addressed the model dependence empirically, by performing the first comprehensive statistical analysis of the combined HERA leading neutron and E866 data sets, for a wide range of prescriptions adopted in the literature for regularizing the pion–nucleon amplitudes. Our findings suggest that we can indeed describe both HERA and E866 data within pion exchange models, if the cutoff parameters in the πNN form factors are fitted simultaneously with the pion structure function. The combined fits to both the HERA and E866 data are significantly more restrictive, with models with t-dependent form factors, such as exponential or monopole, giving the best descriptions of the combined data sets over the largest range of kinematics, up to $y_{\rm cut} \approx 0.3$ [24–26].

In the near future, the SeaQuest Drell-Yan experiment [34] at Fermilab will measure the $\bar{d}-\bar{u}$ difference up to larger values of x, $x \approx 0.45$, which should allow improved constraints on the models of the pion distribution function in the nucleon. Beyond that, the tagged DIS experiment [9] at Jefferson Lab will provide precise information on pion exchange in leading proton production from an effective neutron target at kinematics complementary to the range covered by the HERA and Drell-Yan measurements. This should reduce the uncertainty in F_2^π in the intermediate x_π region, $x_\pi \sim 0.1$.

One may also examine in more detail the $k_\perp$ dependence of leading neutron (or proton) cross sections, which was studied in some of the HERA measurements [8] and will be explored in the TDIS experiment. Comparison of the unintegrated pion flux with the empirical transverse momentum distributions could provide a more incisive test of the momentum dependence of the πNN form factor. In the longer term, a necessary goal would be to perform a global PDF fit, in terms of both sea and valence quark PDFs, to the πN Drell-Yan data at moderate and high values of x_π, together with HERA leading neutron data at small x_π, and the new TDIS data on leading proton production in the intermediate x_π region. We look forward to these endeavors revealing much more concisely and completely the partonic structure of the pion, and the role of the pion cloud in the structure of the nucleon.

Acknowledgments We are indebted to Y. Salamu and P. Wang for the earlier collaboration on this work. We thank T. J. Hobbs for helpful discussions and collaboration on portions of this work, and C. Keppel and B. Wojtsekhowski for discussions about the TDIS experiment. C.-R.Ji thanks the organizers of LC2015-Frascati for their invitation. This work was supported by the DOE contract No. DE-AC05-06OR23177, under which Jefferson Science Associates, LLC operates Jefferson Lab, and DOE Contract No. DE-FG02-03ER41260.

References

1. Salamu, Y., Ji, C.-R., Melnitchouk, W., Wang, P.: $\bar{d} - \bar{u}$ flavor asymmetry in the proton in chiral effective field theory. Few Body Syst. **56**(6-9), 355 (2015)
2. Thomas, A.W., Melnitchouk, W., Steffens, F.M.: Dynamical symmetry breaking in the sea of the nucleon. Phys. Rev. Lett. **85**, 2892 (2000)
3. Detmold, W., Melnitchouk, W., Negele, J.W., Renner, D.B., Thomas, A.W.: Chiral extrapolation of lattice moments of proton quark distributions. Phys. Rev. Lett. **87**, 172001 (2001)
4. Chen, J.-W., Ji, X.: Is the Sullivan process compatible with QCD chiral dynamics?. Phys. Lett. B **523**, 107 (2001)
5. Arndt, D., Savage, M.J.: Chiral corrections to matrix elements of twist-2 operators. Nucl. Phys. **A697**, 429 (2002)
6. Chekanov, S. et al.: Leading neutron production in e^+p collisions at HERA. Nucl. Phys. **B637**, 3 (2002)
7. Aaron, F.D. et al.: Measurement of leading neutron production in deep-inelastic scattering at HERA. Eur. Phys. J. C **68**, 381 (2010)
8. Chekanov, S. et al.: Leading neutron energy and pT distributions in deep inelastic scattering and photoproduction at HERA. Nucl. Phys. **B776**, 1 (2007)
9. Keppel, C., Wojtsekhowski, B., King, P., Dutta, D., Annand, J., as spokespersons: Measurement of tagged deep-inelastic scattering (TDIS). Jefferson Lab experiment PR12-15-006
10. Baillie, N. et al.: Measurement of the neutron F2 structure function via spectator tagging with CLAS. Phys. Rev. Lett. **108**, 142001 (2012)
11. McKenney, J.R., Sato, N., Melnitchouk, W., Ji, C.R.: Pion structure function from leading neutron electroproduction and SU(2) flavor asymmetry. Phys. Rev. D. arXiv:1512.04459 [hep-ph]
12. Towell, R.S. et al.: Improved measurement of the anti-d/anti-u asymmetry in the nucleon sea. Phys. Rev. D **64**, 052002 (2001)
13. Badier, J. et al.: Experimental determination of the π meson structure functions by the Drell-Yan mechanism. Z. Phys. C **18**, 281 (1983)
14. Betev, B. et al.: Observation of anomalous scaling violation in Muon Pair Production by 194-GeV/$c\pi^-$ Tungsten interactions. Z. Phys. C **28**, 15 (1985)
15. Conway, J.S. et al.: Experimental study of Muon pairs produced by 252-GeV pions on Tungsten. Phys. Rev. D **39**, 92 (1989)
16. Jimenez-Delgado, P., Melnitchouk, W., Owens, J.F.: Parton momentum and helicity distributions in the nucleon. J. Phys. G **40**, 093102 (2013)
17. Forte, S., Watt, G.: Progress in the determination of the partonic structure of the proton. Ann. Rev. Nucl. Part. Sci. **63**, 291 (2013)
18. Burkardt, M., Hendricks, K.S., Ji, C.-R., Melnitchouk, W., Thomas, A.W.: Pion momentum distributions in the nucleon in chiral effective theory. Phys. Rev. D **87**, 056009 (2013)
19. Salamu, Y., Ji, C.-R., Melnitchouk, W., Wang, P.: $\bar{d} - \bar{u}$ asymmetry in the proton in chiral effective theory. Phys. Rev. Lett. **114**, 122001 (2015)
20. Thomas, A.W.: A limit on the pionic component of the nucleon through SU(3) flavor breaking in the sea. Phys. Lett. B **126**, 97 (1983)
21. Drell, S.D., Levy, D.J., Yan, T.M.: A theory of deep inelastic lepton-nucleon scattering and lepton pair annihilation processes. 1. Phys. Rev. **187**, 2159 (1969)
22. Ji, C.-R., Melnitchouk, W., Thomas, A.W.: Anatomy of relativistic pion loop corrections to the electromagnetic nucleon coupling. Phys. Rev. D **88**, 076005 (2013)
23. Bugg, D.V.: The pion nucleon coupling constant. Eur. Phys. J. C **33**, 505 (2004)
24. Melnitchouk, W., Speth, J., Thomas, A.W.: Dynamics of light anti-quarks in the proton. Phys. Rev. D **59**, 014033 (1999)
25. Kumano, S.: Flavor asymmetry of anti-quark distributions in the nucleon. Phys. Rep. **303**, 183 (1998)
26. Speth, J., Thomas, A.W.: Mesonic contributions to the spin and flavor structure of the nucleon. Adv. Nucl. Phys. **24**, 83 (1998)

27. Donoghue, J.F., Holstein, B.R., Borasoy, B.: SU(3) baryon chiral perturbation theory and long distance regularization. Phys. Rev. D **59**, 036002 (1999)
28. Young, R.D., Leinweber, D.B., Thomas, A.W.: Leading quenching effects in the proton magnetic moment. Phys. Rev. D **71**, 014001 (2005)
29. Sutton, P.J., Martin, A.D., Roberts, R.G., Stirling, W.J.: Parton distributions for the pion extracted from Drell-Yan and prompt photon experiments. Phys. Rev. D **45**, 2349 (1992)
30. Aicher, M., Schafer, A., Vogelsang, W.: Soft-gluon resummation and the valence parton distribution function of the pion. Phys. Rev. Lett. **105**, 252003 (2010)
31. Jaffe, R.L.: Deep Inelastic Scattering with Application to Nuclear Targets, Lectures presented at the 1985 Los Alamos School on Quark Nuclear Physics, June 10-14 (1985)
32. Llewellyn Smith, C.H.: Nuclear effects in deep inealstic lepton scattering. Nucl. Phys **A434**, 35c–56c (1985)
33. Chen, J.-W., Ji, X.: Constructing parton convolution in effective field theory. Phys. Rev. Lett. **87**, 152002 (2001); *ibid* **88**, 249901(E) (2002)
34. Geesaman, D.F., Reimer, P.E., as spokespersons: Fermilab E906 experiment (SeaQuest), http://www.phy.anl.gov/mep/SeaQuest/
35. Melnitchouk, W.: Quark hadron duality in electron pion scattering. Eur. Phys. J. A **17**, 223 (2003)
36. Melnitchouk, W.: Higher twists in the pion structure function. Phys. Rev. D **67**, 077502 (2003)
37. Shigetani, T., Suzuki, K., Toki, H.: Pion structure function in the Nambu and Jona-Lasinio model. Phys. Lett. B **308**, 383 (1993)
38. Szczepaniak, A., Ji, C.-R., Cotanch, S.R.: Generalized relativistic meson wave function. Phys. Rev. D **49**, 3466 (1994)
39. Farrar, G.R., Jackson, D.R.: The pion form-factor. Phys. Rev. Lett. **43**, 246 (1979)
40. Holt, R.J., Roberts, C.D.: Distribution functions of the nucleon and pion in the valence region. Rev. Mod. Phys. **82**, 2991 (2010)
41. Wijesooriya, K., Reimer, P.E., Holt, R.J.: The pion parton distribution function in the valence region. Phys. Rev. C **72**, 065203 (2005)
42. Kopeliovich, B.Z., Potashnikova, I.K., Povh, B., Schmidt, I.: Pion structure function at small x from DIS data. Phys. Rev. D **85**, 114025 (2012)

DOI 10.1007/978-3-319-50699-9_24
Reprinted from *Few-Body System* Journal, DOI 10.1007/s00601-016-1051-y

CrossMark

Loredana Bellantuono

On $J^{PC} = 0^{--}$ exotic glueball

Received: 14 January 2016 / Accepted: 3 February 2016 / Published online: 25 February 2016

Abstract The mass spectrum of the gluonium with $J^{PC} = 0^{--}$ is examined in three bottom-up AdS/QCD models. The results are used to identify several production and decay modes useful for searching this state. Moreover, the properties of such glueball in a hot and dense quark medium are discussed.

1 Introduction

Glueballs, bound states of gluons arising from the non-Abelian nature of strong interactions, are an important testbed for non perturbative aspects of QCD. The main obstacle in their search is the mixing with quarkonia ($\bar{q}q$) with the same quantum numbers. A promising strategy for their identification is to focus on exotic states whose J^{PC} quantum numbers are unaccessible to quark-antiquark configurations. This is the case of several gluonia with negative C-parity, composed of an odd number of constituent gluons ("oddballs"), for which little theoretical information is available. In particular, for $J^{PC} = 0^{--}$, the mass predictions for the lightest state span the range from $m_{0^{--}} = 2.79\,\text{GeV}$ in the flux-tube model [1], to $m_{0^{--}} \approx 5.166\,\text{GeV}$ in lattice QCD simulations [2]. Two stable 0^{--} oddballs, with masses $m_{0^{--}} = 3.81 \pm 0.12\,\text{GeV}$ and $m_{0^{--}} = 4.33 \pm 0.13\,\text{GeV}$, have been predicted by QCD sum rules [3].

The mass spectrum of the $J^{PC} = 0^{--}$ oddball can be computed in a framework inspired by the AdS/CFT correspondence. This duality conjecture relates a strongly coupled gauge theory in a four dimensional ($4D$) Minkowski space to a semiclassical gravity theory defined in a five dimensional ($5D$) anti-de Sitter (AdS) geometry times a $5D$ sphere [4]. In Poincaré coordinates, the line element

$$ds^2 = \frac{R^2}{z^2}(dx_0^2 - d\mathbf{x}^2 - dz^2) \qquad z > 0\,, \tag{1}$$

with z the fifth holographic coordinate, describes the AdS bulk metric. The original formulation of gauge/gravity duality required the $4D$ theory to be conformal invariant. Holographic bottom-up models, constructed to reproduce QCD properties, break such a symmetry by introducing an infrared energy scale in the bulk; the $J^{PC} = 0^{--}$ oddball mass spectrum can then be determined either in vacuum or in a quark bath at finite temperature and density. Such an approach complements the top-down methods applied to analyze, e.g., the scalar 0^{++} gluonium [5,6].

2 $J^{PC} = 0^{--}$ Mass Spectrum in Three Holographic Models of QCD

The holographic correspondence is based on a dictionary, according to which a local gauge-invariant operator in the $4D$ field theory is dual to a field in $5D$ AdS [7,8]. In QCD, an interpolating current representing the

L. Bellantuono (✉)
Dipartimento di Fisica, Università di Bari, and INFN, Sezione di Bari, via Orabona 4, 70126 Bari, Italy
E-mail: loredana.bellantuono@ba.infn.it

glueball with quantum numbers $J^{PC} = 0^{--}$ can be written in terms of the gluon field strengths $G^a_{\mu\nu}(x)$ and $\widetilde{G}^a_{\mu\nu}(x) = \frac{1}{2}\epsilon_{\mu\nu\rho\sigma} G^a_{\rho\sigma}(x)$, namely

$$J_0(x) = g_s^3 d_{abc} [\eta^t_{\alpha\beta} \widetilde{G}^a_{\mu\nu}(x)][\partial_\alpha \partial_\beta G^b_{\nu\rho}(x)][G^c_{\rho\mu}(x)] , \tag{2}$$

with a, b, c color indices, d_{abc} the symmetric tensor defining the anticommutator of $SU(3)_c$ generators, and g_s the strong coupling constant. The transverse $\eta^t_{\alpha\beta}$ metric is defined as $\eta^t_{\alpha\beta} = \eta_{\alpha\beta} - \frac{\partial_\alpha \partial_\beta}{\partial^2}$, with α, β (as well as μ, ν) $4D$ Lorentz indices, and $\eta_{\alpha\beta}$ the Minkowski metric tensor [3,9]. The operator (2) has conformal dimension $\Delta = 8$, and its holographic dual field, $O_0(x, z)$, has mass obtained by the relation $M_5^2 R^2 = \Delta(\Delta-4)$ [7,8]. In the following, the AdS_5 radius is set to $R = 1$. The $5D$ action for $O_0(x, z)$ can be written as

$$S = \frac{1}{k} \int d^5x \sqrt{g} a(z) \left[g^{MN} \partial_M O_0 \, \partial_N O_0 - M_5^2 O_0^2 \right] , \tag{3}$$

where g_{MN} is the bulk metric, $g = |det(g_{MN})|$ and k is a parameter making the action dimensionless. To account for confinement in QCD, conformal invariance must be broken in the action (3). Three different models including such effect, either by a proper choice of the function $a(z)$ or by introducing a dynamical field in the metric, are discussed in the following. The mass spectrum can be determined solving the Euler-Lagrange equations for the field $O_0(x, z)$.

2.1 Hard-Wall Model

A simple way of modeling confinement in the holographic setup is by considering a slice of the AdS_5 space, with a sharp cutoff at a finite distance z_m along the fifth dimension [10]. The metric g_{MN} is given by (1), and the condition $z \le z_m$ is implemented in the action through $a(z) = \Theta\,(z_m - z)$, with Θ the Heaviside function. The cutoff z_m sets a mass scale. The choice $1/z_m = 346\,\text{MeV}$, obtained from analyses on axial and vector mesons [10], gives the results $m_0 = 2.80\,\text{GeV}$ and $m_1 = 4.14\,\text{GeV}$ for the lowest-lying and the first-excited 0^{--} oddball, respectively [11].

2.2 Soft-Wall Model

In this framework, the geometry is AdS_5 (1), and conformal invariance is smoothly broken by the function $a(z) = e^{-c^2 z^2}$ which introduces a mass scale c in the action [12]. The Regge-like mass spectrum is obtained [11]:

$$m_n^2 = 4c^2(n+4). \tag{4}$$

Setting $c = m_\rho/2 = 388\,\text{MeV}$ from the ρ meson mass computed in the model [12], one obtains $m_0 = 1.55\,\text{GeV}$ and $m_1 = 1.74\,\text{GeV}$ [11]. The mass spectrum (4) corresponds to the poles of the two-point correlation function of $J_0(x)$,

$$\Pi(p^2) = i \int d^4x \, e^{ipx} \langle 0|T[J_0(x) J_0^\dagger(0)]|0\rangle . \tag{5}$$

In the AdS/CFT dictionary, the QCD interpolating current $J_0(x)$ is interpreted as the source of the dual field $O_0(x, z)$. The $4D$ Fourier transforms of such operators, $\tilde{J}_0(p)$ and $\tilde{O}_0(p, z)$, are related by the bulk-to-boundary propagator $\tilde{K}(p, z)$ of the oddball field, through $\tilde{O}_0(p, z) = \tilde{K}(p, z)\tilde{J}_0(p)$. Holography prescribes the identification between the partition functions of the dual theories, and this allows to compute the two-point correlation function (5) as

$$\Pi\left(p^2\right) \approx \left. \frac{\delta^2 S_{os}}{\delta \tilde{J}_0 \delta \tilde{J}_0} \right|_{\tilde{J}_0 = 0} , \tag{6}$$

with S_{os} the $5D$ on-shell action. The poles of $\Pi\left(p^2\right)$ are given in (4).

Table 1 Production and decay modes of the $J^{PC} = 0^{--}$ oddball, for $m_{0^{--}} = 2.8\,\text{GeV}$

Radiative production	Hadronic production	Decay mode
$\chi_{c1}(3510) \to \gamma\, G(0^{--})$	$X(3872) \to \omega\, G(0^{--})$	$G(0^{--}) \to \gamma\, f_1(1285)$
$X(3872) \to \gamma\, G(0^{--})$	$h_c(3525) \to \pi\pi\ (I=0)\, G(0^{--})$	$G(0^{--}) \to \omega\, f_1(1285)$
$\chi_{c2}(3556) \to \gamma\, G(0^{--})$	$\chi_{b1}(10255) \to (\omega, \phi, J/\Psi)\, G(0^{--})$	$G(0^{--}) \to \rho\, a_1(1260)\ (I=0)$
$\chi_{c2}(3927) \to \gamma\, G(0^{--})$	$\Upsilon(nS) \to (f_1(1285), \chi_{c1}, X(3872))\, G(0^{--})$	
$\chi_{b1}(9892) \to \gamma\, G(0^{--})$	$h_b(9899) \to f_0(980)\, G(0^{--})$	$G(0^{--}) \to h_1(1270)\, f_0(980)$
$\chi_{b1}(10255) \to \gamma\, G(0^{--})$	$h_b(10260) \to f_0(980)\, G(0^{--})$	$G(0^{--}) \to \rho\pi\ (I=0)$
$\chi_{b2}(9912) \to \gamma\, G(0^{--})$	$h_b(9899) \to G(0^{++})\, G(0^{--})$	$G(0^{--}) \to K^* K\ (I=0)$
$\chi_{b2}(10269) \to \gamma\, G(0^{--})$	$h_b(10260) \to G(0^{++})\, G(0^{--})$	$G(0^{--}) \to (\eta, \eta')(\omega, \phi)$

2.3 Einstein-Dilaton Model

A third possibility is to consider a class of dynamical bottom-up models reproducing confinement through a distorsion of the bulk geometry [13,14]:

$$ds^2_{(ED)} = \frac{e^{2A_s(z)-\frac{4}{3}\Phi(z)}}{z^2}\left[dx_0^2 - d\mathbf{x}^2 - dz^2\right]. \tag{7}$$

$\Phi(z)$, a dilaton field, couples to the graviton, and its profile is determined solving the Einstein equations deduced from this metric. The function $A_s(z)$ can be chosen as $A_s(z) = \delta^2 z^2$, with the mass scale fixed to $\delta = 0.43\,\text{GeV}$ [13]. Once the profile of $\Phi(z)$ is obtained, the two lightest oddball states turn out to have mass $m_0 = 2.82\,\text{GeV}$ and $m_1 = 4.07\,\text{GeV}$ [11].

With these results, several production and decay modes can be identified, useful for the search of the 0^{--} gluonium. They are reported in Table 1 for the specific case $m_{0^{--}} = 2.8$ GeV [11].

3 Oddball in Thermalized and Dense Medium

The AdS/QCD duality allows to examine the properties of a $J^{PC} = 0^{--}$ oddball in a quark thermal bath, at finite temperature T and chemical potential μ. Stability of the gluon configuration against thermal and density fluctuations can be investigated in comparison with other quark and gluon bound states. In-medium effects can be incorporated in the holographic description using the action (3), with an appropriate bulk geometry. I explicitly discuss the soft-wall case. The $5D$ line element

$$ds^2 = \frac{1}{z^2}\left(f(z)dx_0^2 - d\mathbf{x}^2 - \frac{dz^2}{f(z)}\right) \tag{8}$$

is characterized by a metric with $f(z)$ which can be related to temperature and density of the boundary theory. The $5D$ Reissner-Nordström AdS geometry (AdS/RN), with

$$f(z) = 1 - \left(\frac{1}{z_h^4} + q^2 z_h^2\right) z^4 + q^2 z^6\,, \tag{9}$$

can be used as the holographic dual of a thermalized and dense medium in the deconfined phase [15,16]. This charged black hole bulk metric has an outer horizon $z = z_h$ and charge q which can be related to the temperature T and chemical potential μ of the dual boundary theory. Defining $Q = q z_h^3$ and imposing the condition $0 \le Q \le \sqrt{2}$, the black-hole temperature is

$$T = \frac{1}{4\pi}\left|\frac{df}{dz}\right|_{z=z_h} = \frac{1}{\pi z_h}\left(1 - \frac{Q^2}{2}\right). \tag{10}$$

In the QCD generating functional, μ multiplies the quark number operator $O_q(x) = q^\dagger(x)q(x)$. Thus, it can be interpreted, according to gauge/gravity correspondence, as the source of a bulk field dual to $O_q(x)$, the time component of a $U(1)$ gauge field $A_M(x,z)$. By rotational invariance, the spatial components A_i (with

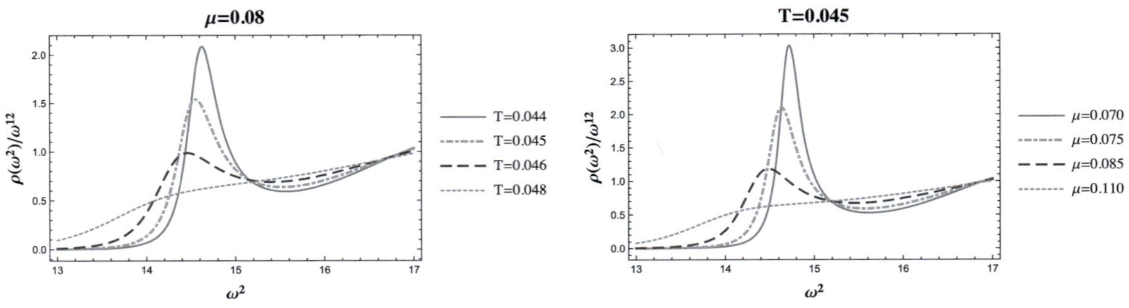

Fig. 1 Spectral function $\rho\left(\omega^2\right)/\left(\omega^2\right)^6$ computed in the soft-wall model with AdS/RN metric, at fixed μ (*left*) or T (*right*). The dimensionful quantities are in units of c, and the constant $10^{-8}k/\left(2z_h^8\right)$ has been factorized out in the spectral function [11]

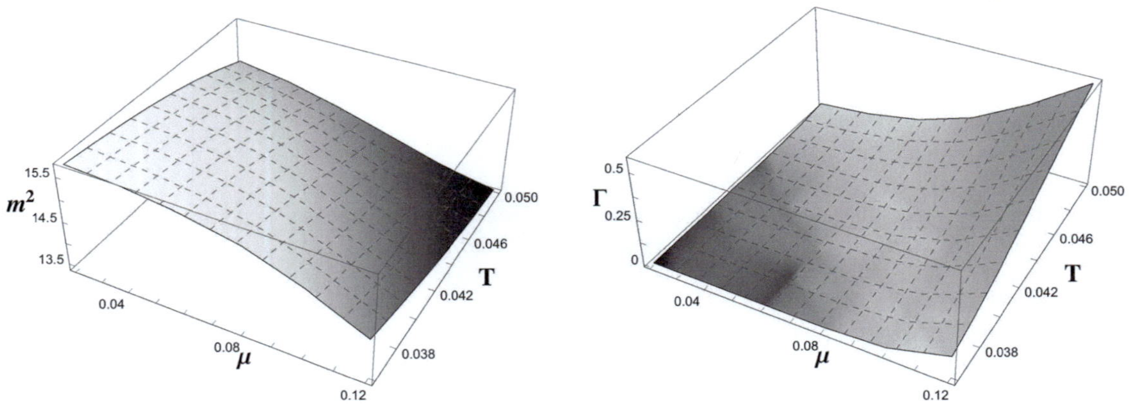

Fig. 2 Squared mass m^2 and width Γ of the lightest $J^{PC} = 0^{--}$ glueball, obtained from the soft-wall model with AdS/RN metric, in a range of temperature T and chemical potential μ, in units of c [11]

$i = 1, 2, 3, z$) vanish, while $A_0 \approx \mu - \kappa q z^2$ for small z. The condition $A_0(z_h) = 0$ provides a linear relation between μ and q:

$$\mu = \kappa \frac{Q}{z_h}, \tag{11}$$

with κ a dimensionless parameter that will be fixed to 1, a choice determining the quark chemical potential μ up to a numerical factor. Important information on the QCD bound states in a quark thermal bath can be inferred from the spectral function $\rho\left(\omega^2\right)$, the imaginary part of the retarded Green function. The spectral function of the 0^{--} oddball is represented in vacuum by an infinite number of delta functions centered at the eigenvalues of the mass spectrum (4). Figure 1 shows the changes of the in-medium $\rho\left(\omega^2\right)$ as temperature and chemical potential are switched on. At any finite and fixed value of μ, the peaks of the spectral function broaden and move towards lower values of ω^2 when T increases. An analogous behaviour is observed by keeping the temperature fixed, and increasing the chemical potential. The (T, μ)-dependence of the lightest oddball's squared mass and width is shown in Fig. 2. The values of the temperature and chemical potential at which the lowest-energy peak becomes indistinguishable in the profile of the spectral function, are smaller than those obtained for light vector [17,18], scalar $\bar{q}q$ mesons [19], the lightest scalar glueball [20], and hybrid mesons [21]. Hence, 0^{--} oddballs are found to suffer of larger in-medium instabilities with respect to other hadrons.

For points in the $T - \mu$ plane below the QCD deconfinement transition, the state described by AdS/RN metric is metastable [16]. The thermal-charged AdS metric (tcAdS), with line element (8) and deformation

$$f(z) = 1 + q^2 z^6 \tag{12}$$

is proposed as a dual of the confined phase of QCD at small temperature and finite chemical potential [16,22]. This geometry is related to the AdS/RN one by the Hawking-Page transition, which describes deconfinement in the holographic framework. The chemical potential, related to the finite density of the hadronic medium, grows linearly with q, while the temperature is implemented through a periodicity in the Euclidean time coordinate

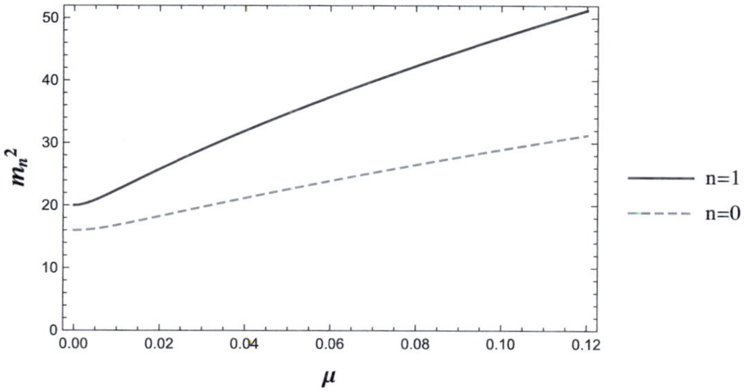

Fig. 3 Squared mass m^2 as a function of μ for the two lightest $J^{PC} = 0^{--}$ oddballs, in the tcAdS soft-wall model

$\tau = ix_0$ [22]. The results, depicted in Fig. 3, reveal that the masses of the two lightest oddballs increase with μ, a different behaviour with respect to what is found in the deconfined phase [11].

4 Conclusions

In AdS/QCD, the mass of the lowest-lying 0^{--} oddball is found to be lighter than as computed by other approaches. The effect of deconfined quark matter can be modeled by the AdS/RN geometry: oddballs are found to suffer from larger thermal and density instabilities with respect to other hadrons. On the other hand, the holographic description of a confined medium, obtained using the tcAdS metric, gives a mass which increases with the chemical potential.

Acknowledgments I thank P. Colangelo and F. Giannuzzi for collaboration, and F. De Fazio and S. Nicotri for discussions.

References

1. Isgur, N., Paton, J.E.: A flux tube model for hadrons in QCD. Phys. Rev. D **31**, 2910 (1985)
2. Gregory, E. et al.: Towards the glueball spectrum from unquenched lattice QCD. JHEP **1210**, 170 (2012)
3. Qiao, C.-F., Tang, L.: Finding the 0^{--} glueball. Phys. Rev. Lett. **113**, 221601 (2014)
4. Maldacena, J.M.: The large N limit of superconformal field theories and supergravity. Int. J. Theor. Phys. **38**, 1113 (1999)
5. Csaki, C., Ooguri, H., Oz, Y., Terning, J.: Glueball mass spectrum from supergravity. JHEP **9901**, 017 (1999)
6. Brunner, F., Rebhan, A.: Nonchiral enhancement of scalar glueball decay in the Witten–Sakai–Sugimoto model. Phys. Rev. Lett. **115**, 131601 (2015)
7. Witten, E.: Anti-de Sitter space and holography. Adv. Theor. Math. Phys. **2**, 253 (1998)
8. Gubser, S., Klebanov, I.R., Polyakov, A.M.: Gauge theory correlators from noncritical string theory. Phys. Lett. B **428**, 105 (1998)
9. Qiao, C.-F., Tang, L.: Mass spectra of 0^{+-}, 1^{-+}, and 2^{+-} exotic glueballs. arXiv:1509.00305 [hep-ph] (2015)
10. Erlich, J. et al.: QCD and a holographic model of hadrons. Phys. Rev. Lett. **95**, 261602 (2005)
11. Bellantuono, L., Colangelo, P., Giannuzzi, F.: Holographic oddballs. JHEP **1510**, 137 (2015)
12. Karch, A., Katz, E., Son, D.T., Stephanov, M.A.: Linear confinement and AdS/QCD. Phys. Rev. D **74**, 015005 (2006)
13. Li, D., He, S., Huang, M., Yan, Q.-S.: Thermodynamics of deformed AdS_5 model with a positive/negative quadratic correction in graviton–dilaton system. JHEP **1109**, 041 (2011)
14. Chen, Y., Huang, M.: Two-gluon and trigluon glueballs from dynamical holography QCD. arXiv:1511.07018 [hep-ph] (2015)
15. Colangelo, P., Giannuzzi, F., Nicotri, S.: In-medium hadronic spectral functions through the soft-wall holographic model of QCD. JHEP **1205**, 076 (2012)
16. Lee, B.-H., Park, C., Sin, S.-J.: A dual geometry of the hadron in dense matter. JHEP **0907**, 087 (2009)
17. Fujita, M., Fukushima, K., Misumi, T., Murata, M.: Finite-temperature spectral function of the vector mesons in an AdS/QCD model. Phys. Rev. D **80**, 035001 (2009)
18. Giannuzzi, F.: Holographic vector mesons in a hot and dense medium. Nucl. Phys. Proc. Suppl. **234**, 309 (2013)
19. Colangelo, P., De Fazio, F., Giannuzzi, F., Jugeau, F. et al.: Light scalar mesons in the soft-wall model of AdS/QCD. Phys. Rev. D **78**, 055009 (2008)

20. Colangelo, P., De Fazio, F., Jugeau, F., Nicotri, S.: On the light glueball spectrum in a holographic description of QCD. Phys. Lett. B **652**, 73 (2007)
21. Bellantuono, L., Colangelo, P., Giannuzzi, F.: Exotic $J^{PC} = 1^{-+}$ mesons in a holographic model of QCD. Eur. Phys. J. C **74**, 2830 (2014)
22. Park, C. et al.: The soft wall model in the hadronic medium. Phys. Rev. D **84**, 046007 (2011)

DOI 10.1007/978-3-319-50699-9_25
Reprinted from *Few-Body System* Journal, DOI 10.1007/s00601-016-1053-9

CrossMark

Ho-Meoyng Choi · Chueng-Ryong Ji

Light-Front Quark Model Analysis of Meson-Photon Transition Form Factor

Received: 11 January 2016 / Accepted: 4 February 2016 / Published online: 29 February 2016

Abstract We discuss $(\pi^0, \eta, \eta') \rightarrow \gamma^*\gamma$ transition form factors using the light-front quark model. Our discussion includes the analysis of the mixing angles for $\eta - \eta'$. Our results for $Q^2 F_{(\pi^0,\eta,\eta')\rightarrow\gamma^*\gamma}(Q^2)$ show scaling behavior for high Q^2 consistent with pQCD predictions.

1 Introduction

The pion-photon transition form factor (TFF) $F_{\pi\gamma}(Q^2)$ has been known to be the simplest exclusive process involving the strong interaction. It can be calculated asymptotically at leading twist as a convolution of a perturbative hard scattering amplitude and a gauge-invariant meson distribution amplitude (DA) which incorporates the nonperturbative dynamics of QCD bound state [1]. The prediction for $F_{\pi\gamma}(Q^2)$ at the asymptotic limit $Q^2 \rightarrow \infty$ is shown to satisfy the well-known asymptotic pQCD formula [1]: $Q^2 F_{\pi\gamma}(Q^2 \rightarrow \infty) = \sqrt{2} f_\pi$ GeV with $f_\pi \simeq 130$ MeV. However, the data for $Q^2 F_{\pi\gamma}(Q^2)$ measured from the $\gamma^*\gamma \rightarrow \pi^0$ process by the BaBar Collaboration in 2009 [2] have shown not only the serious violation of the asymptotic pQCD formula but also the rapid growth for $Q^2 > 15$ GeV^2. On the other hand, in 2012, the Belle Collaboration [3] has reported their measurement of the $\gamma^*\gamma \rightarrow \pi^0$ process and has shown that the measured values of $Q^2 F_{\pi\gamma}(Q^2)$ are consistent with the asymptotic limit of QCD for $Q^2 > 15\,\mathrm{GeV}^2$.

Hadronic DAs provide essential information on the QCD interaction of quarks, antiquarks and gluons inside the hadrons and play an essential role in applying QCD to hard exclusive processes. It has motivated many theoretical studies [4–7] using various forms of the pion DAs to understand the discrepancy of $Q^2 F_{\pi\gamma}(Q^2)$ data between the BaBar and Belle measurements. The general agreement on the analysis of the pion DA is that the broader the pion DA the steeper the slope of $Q^2 F_{\pi\gamma}(Q^2)$ as Q^2 is getting larger. For instance, the flat pion DA [5] $\phi(x) = 1$ shows the agreement with the BaBar data [2]. The subsequent BaBar data [8] for the $(\eta, \eta') \rightarrow \gamma^*\gamma$ TFFs, however, have shown that the use of flat DA for η and η' distributions strongly disagrees with the data. Both Belle data [3] for $Q^2 F_{\pi\gamma}(Q^2)$ and BaBar data [8] for $Q^2 F_{(\eta,\eta')\gamma}(Q^2)$ provided consistency with the perturbative QCD prediction and disfavored the flat DA $\phi(x) = 1$ which is far different from the lowest twist-2 DA $\phi(x) = 6x(1-x)$ predicted by the asymptotic QCD. Accordingly, careful analysis

This research work is supported by Kyungpook National University Bokhyeon Research Fund, 2015.

H.-M. Choi (✉)
Department of Physics, Teachers Colleges, Kyungpook National University, Daegu 702-701, Korea
E-mail: homyoung@knu.ac.kr

C.-R. Ji
Department of Physics, North Carolina State University, Raleigh, NC 27695-8202, USA
E-mail: crji@ncsu.edu

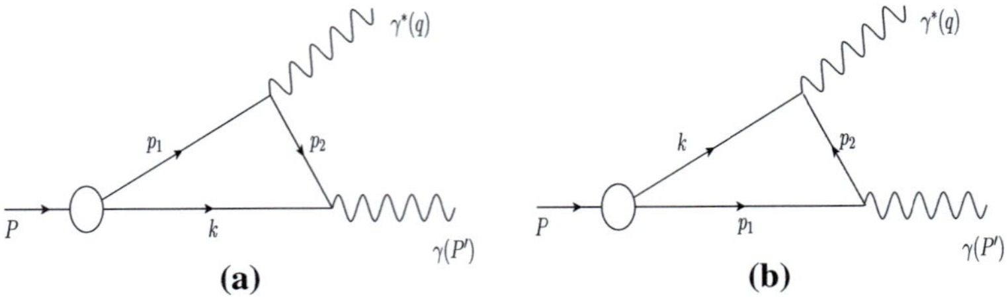

Fig. 1 One-loop Feynman diagrams that contribute to $P \to \gamma^*\gamma$

of $(\eta, \eta') \to \gamma^*\gamma$ transitions [8–10] appears particularly important in the ongoing discussion over the pion-photon TFF results.

The purpose of this work is to comprehensively investigate the $P \to \gamma^*\gamma (P = \pi^0, \eta, \eta')$ transitions using the light-front quark model (LFQM) based on the QCD motivated effective LF Hamiltonian [11]. The paper is organized as follows. In Sect. 2, we discuss the meson-photon TFFs in an exactly solvable model based on the covariant Bethe–Salpeter (BS) model of (3 + 1)-dimensional fermion field theory. Performing both manifestly covariant calculation and the LF calculation in the BS model, we show the equivalence between the two results and the absence of the zero-mode contribution to the TTF. The $\eta - \eta'$ mixing scheme for the calculations of the $(\eta, \eta') \to \gamma^*\gamma$ TFFs is also discussed. We then apply the manifestly covariant BS model to the standard LFQM using the Gaussian radial wave function. The self-consistent covariant descriptions of the meson TFFs in the standard LFQM are given in this section. In Sect. 3, we present our numerical results for the $(\pi^0, \eta, \eta') \to \gamma^*\gamma$ TFFs and compare them with the available experimental data [2,3,8,12,13]. Summary and discussion follow in Sect. 4.

2 Model Calculation

The transition form factor $F_{P\gamma}$ for the $P \to \gamma^*\gamma$ $(P = \pi^0, \eta$, and $\eta')$ transition is defined from the matrix element of electromagnetic current $\Gamma^\mu = \langle \gamma(P-q)|J^\mu|P(P)\rangle$ as follows: $\Gamma^\mu = ie^2 F_{P\gamma}(Q^2)\epsilon^{\mu\nu\rho\sigma} P_\nu \epsilon_\rho q_\sigma$, where P and q are the momenta of the incident pseudoscalar meson and virtual photon, respectively, and ϵ is the transverse polarization vector of the final (on-shell) photon. This process is illustrated by the Feynman diagram in Fig. 1, where Fig. 1a, b represents the amplitude $\Gamma^\mu_{(a)}$ $[\Gamma^\mu_{(b)}]$ of the virtual photon being attached to the quark [antiquark] line. The total amplitude is then given by $\Gamma^\mu_{\rm tot} = \Gamma^\mu_{(a)} + \Gamma^\mu_{(b)}$. In the exactly solvable manifestly covariant BS model, the amplitude $\Gamma^\mu_{(a)}$ is given by the following momentum integral

$$\Gamma^\mu_{(a)} = ie_Q e_{\bar{Q}} N_c \int \frac{d^4k}{(2\pi)^4} \frac{{\rm Tr}\left[\gamma_5 \left(\not{p}_1 + m_Q\right) \gamma^\mu \left(\not{p}_2 + m_Q\right) \not{\epsilon} \left(-\not{k} + m_Q\right)\right]}{(p_1^2 - m_Q^2 + i\varepsilon)(p_2^2 - m_Q^2 + i\varepsilon)(k^2 - m_Q^2 + i\varepsilon)} H_0, \tag{1}$$

where N_c is the number of colors and $e_{Q(\bar{Q})}$ is the quark (antiquark) electric charge of mass m_Q $(=m_{u(d)}, m_s)$. For the $\bar{q}q$ bound-state vertex function $H_0 = H_0(p_1^2, k^2)$ of the meson, we simply take the constant parameter g since the covariant loop is regularized with this constant vertex in this model calculation.

Performing both manifestly covariant calculation and the LF calculation of the amplitude $I^{m_Q}_{(a)}(q^2)$ obtained from $\Gamma^\mu_{(a)} = ie_Q e_{\bar{Q}} I^{m_Q}_{(a)}(q^2)\epsilon^{\mu\nu\rho\sigma} P_\nu \epsilon_\rho q_\sigma$ in Eq. (1), we explicitly show the equivalence between the two results $[I^{m_Q}_{(a)}]^{\rm Cov}$ and $[I^{m_Q}_{(a)}]^{\rm LF}$. Especially for the LF calculation, we take the reference frame where $P = (P^+, P^-, \mathbf{P}_\perp) = (P^+, M^2/P^+, 0)$ to investigate the LF zero-mode contribution. By the integration over k^- in Eq. (1) and using the plus component of the currents, we found that the LF zero-mode contribution is absent and only the on-shell propagator contributes in the valence region. The resulting LF amplitude $I^{m_Q}_{(a)}$ in this manifestly covariant model is given by

$$[I^{m_Q}_{(a)}]^{\rm LF} = \frac{N_c}{4\pi^3} \int_0^1 \frac{dx}{x(1-x)} \int d^2\mathbf{k}_\perp \frac{m_Q}{M_0'^2} \chi(x, \mathbf{k}_\perp), \tag{2}$$

where

$$\chi(x, \mathbf{k}_{\_}) = \frac{g}{x(M^2 - M_0^2)}, \tag{3}$$

and $M_0^{(\prime)2} = (\mathbf{k}^{(\prime)2}_{\perp} + m_Q^2)/x(1-x)$ with $\mathbf{k}'_{\perp} = \mathbf{k}_{\perp} + (1-x)\mathbf{q}_{\perp}$. Likewise, $[I^{m_Q}_{(b)}]^{\text{LF}}$ corresponding to the second amplitude $\Gamma^{+}_{(b)}$ is obtained as $[I^{m_Q}_{(b)}]^{\text{LF}}$=$[I^{m_Q}_{(a)}]^{\text{LF}}(x \to 1-x, \mathbf{q}_{\perp} \to -\mathbf{q}_{\perp})$ but the two results are found to give the same numerical values. Thus, we obtain the total LF result as $I^{m_Q}_{\text{tot}} = 2[I^{m_Q}_{(a)}]^{\text{LF}}$.

For $(\eta, \eta') \to \gamma^*\gamma$ transitions, we take into account the presence of two-nonstrange (u and d) and strange (s)-components in the η and η' mesons as well as their mixing. Making use of the $\eta - \eta'$ mixing scheme, the flavor assignment of η and η' mesons in the quark-flavor basis $\eta_q = (u\bar{u} + d\bar{d})/\sqrt{2}$ and $\eta_s = s\bar{s}$ is given by [14]

$$\begin{pmatrix} \eta \\ \eta' \end{pmatrix} = \begin{pmatrix} \cos\phi & -\sin\phi \\ \sin\phi & \cos\phi \end{pmatrix} \begin{pmatrix} \eta_q \\ \eta_s \end{pmatrix}. \tag{4}$$

Therefore, we obtain the transition form factors $F_{P\gamma}$ for $P \to \gamma^*\gamma$ $(P = \pi^0, \eta, \eta')$ transitions as follows [15]

$$\begin{aligned} F_{\pi\gamma}(q^2) &= \frac{(e_u^2 - e_d^2)}{\sqrt{2}} I^{m_{u(d)}}_{\text{tot}}, \\ F_{\eta\gamma}(q^2) &= \cos\phi \,\frac{(e_u^2 + e_d^2)}{\sqrt{2}} I^{m_{u(d)}}_{\text{tot}} - \sin\phi \, e_s^2 I^{m_s}_{\text{tot}}, \\ F_{\eta'\gamma}(q^2) &= \sin\phi \,\frac{(e_u^2 + e_d^2)}{\sqrt{2}} I^{m_{u(d)}}_{\text{tot}} + \cos\phi \, e_s^2 I^{m_s}_{\text{tot}}, \end{aligned} \tag{5}$$

where ϕ is related with the mixing angle θ in the flavor SU(3) octet-singlet basis via $\theta = \phi - \arctan\sqrt{2} \simeq \phi - 54.7°$.

In the standard LFQM [11,15], the wave function of a ground state pseudoscalar meson as a $q\bar{q}$ bound state is given by $\Psi_{\lambda\bar{\lambda}}(x, \mathbf{k}_{\perp}) = \phi_R(x, \mathbf{k}_{\perp})\mathcal{R}_{\lambda\bar{\lambda}}(x, \mathbf{k}_{\perp})$, where ϕ_R is the radial wave function and the spin-orbit wave function $\mathcal{R}_{\lambda\bar{\lambda}}$ with the helicity $\lambda(\bar{\lambda})$ of a quark (antiquark) is obtained by the interaction-independent Melosh transformation from the ordinary spin-orbit wave function assigned by the quantum numbers J^{PC}. The Gaussian wave function ϕ_R for $m_Q = m_{\bar{Q}}$ is given by

$$\phi_R(x, \mathbf{k}_{\perp}) = (4\pi^{3/4}/\beta^{3/2})\sqrt{\partial k_z/\partial x} e^{m_Q^2/2\beta^2} e^{-M_0^2/8\beta^2}, \tag{6}$$

where $\partial k_z/\partial x = M_0/4x(1-x)$ is the Jacobian of the variable transformation $\{x, \mathbf{k}_{\perp}\} \to \mathbf{k} = (\mathbf{k}_{\perp}, k_z)$ and β is the variational parameter fixed by the analysis of meson mass spectra [11]. In our previous analysis of the twist-2 and twist-3 DAs of pseudoscalar and vector mesons [16,17] and the pion electromagnetic form factor [17], we have shown that standard LF (SLF) results of the LFQM is obtained by the replacement of the LF vertex function χ in the BS model with the Gaussian wave function ϕ_R as follows (see, e.g., Eq. (35) in [17])

$$\sqrt{2N_c}\frac{\chi(x, \mathbf{k}_{\perp})}{1-x} \to \frac{\phi_R(x, \mathbf{k}_{\perp})}{\sqrt{\mathbf{k}_{\perp}^2 + m_Q^2}}, \quad M \to M_0, \tag{7}$$

where $M \to M_0$ implies that the physical mass M included in the integrand of BS amplitude has to be replaced with the invariant mass M_0 since the SLF results in the LFQM are obtained from the requirement of all constituents being on their respective mass-shell. The correspondence in Eq. (7) is valid again in this analysis of $P \to \gamma^*\gamma$ transition. Applying the correspondence given by Eq. (7) to Eq. (2), we obtain the corresponding SLF result $[I^{m_Q}_{(a)}]^{\text{SLF}}$ in the LFQM and the total result $[I^{m_Q}_{\text{tot}}]^{\text{LFQM}} = 2[I^{m_Q}_{(a)}]^{\text{SLF}}$ is given by

$$[I^{m_Q}_{\text{tot}}]^{\text{LFQM}} = \frac{\sqrt{2N_c}}{4\pi^3}\int_0^1 \frac{dx}{x(1-x)} \int d^2\mathbf{k}_{\perp} \frac{\phi_R(x, \mathbf{k}_{\perp})}{\sqrt{\mathbf{k}_{\perp}^2 + m_Q^2}} \frac{(1-x)m_Q}{M_0'^2}. \tag{8}$$

Our LFQM predictions of the TFFs for $P \to \gamma^*\gamma$ is then obtained by substituting $[I^{m_Q}_{\text{tot}}]^{\text{LFQM}}$ into Eq. (5).

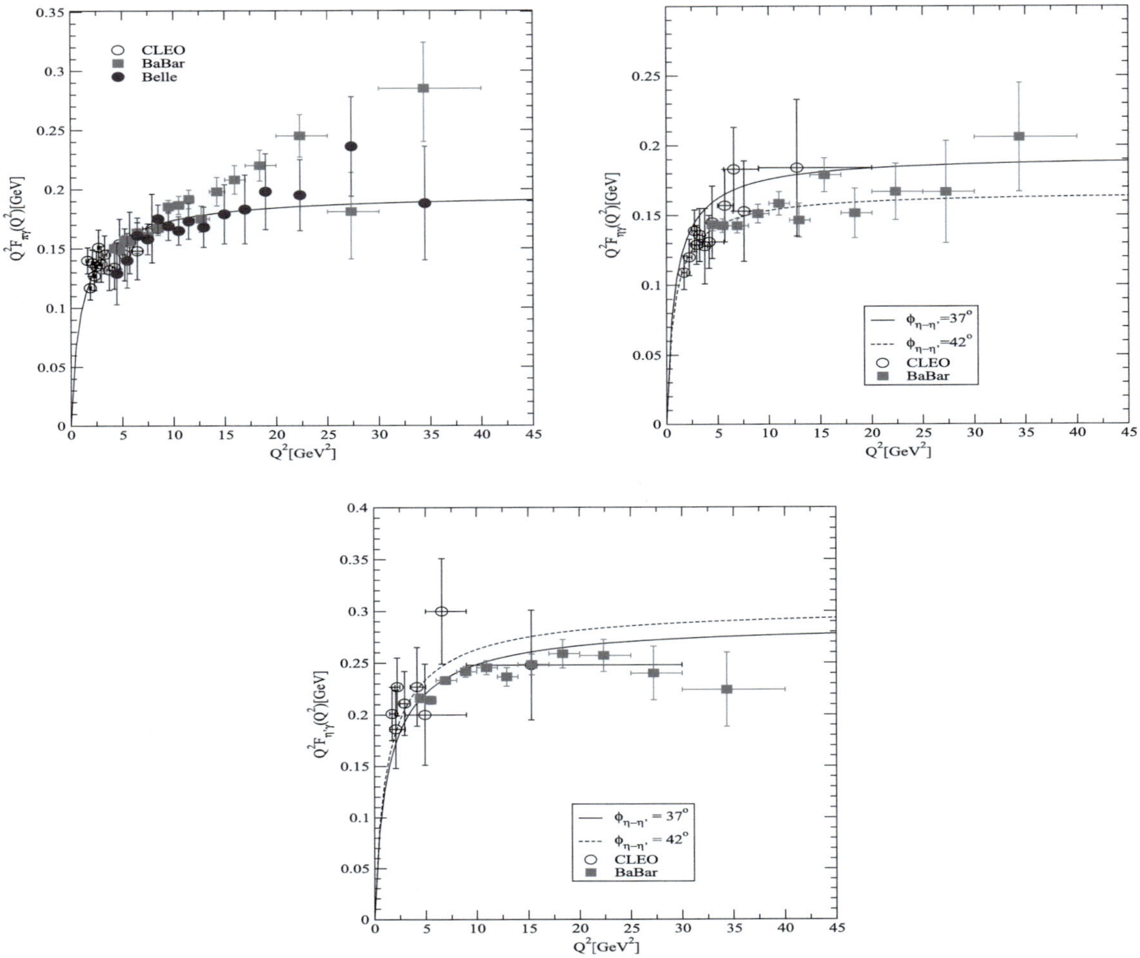

Fig. 2 The transition form factors $Q^2F_{\pi\gamma}(Q^2)$, $Q^2F_{\eta\gamma}(Q^2)$, and $Q^2F_{\eta'\gamma}(Q^2)$ up to $Q^2 = 45$ GeV^2. For $Q^2F_{(\eta,\eta')\gamma}(Q^2)$ case, the solid and dashed lines are results obtained from $\eta - \eta'$ mixing angles with $\phi_{\eta-\eta'} = 37°$ and 42°, respectively. The data are taken from [2,3,8,12,13]

3 Numerical Results

In our numerical calculations, we use the constituent quark masses $(m_q, m_s) = (0.22, 0.45)$ GeV and the gaussian parameters $(\beta_{q\bar{q}}, \beta_{s\bar{s}}) = (0.3695, 0.4128)$ GeV ($q = u$ and d), which were obtained from the calculation of meson mass spectra using the variational principle in our LFQM [11]. While the quadratic (linear) Gell–Mann–Okubo mass formula prefers $\theta \simeq -10°$, $\phi \simeq 44.7°$ ($\theta \simeq -23°$, $\phi \simeq 31.7°$), the KLOE Collaboration [18] extracted $\phi = (41.5 \pm 0.3_{\rm stat} \pm 0.7_{\rm syst} \pm 0.6_{\rm th})°$ by measuring the ratio $\mathrm{BR}(\phi \to \eta'\gamma)/\mathrm{BR}(\phi \to \eta\gamma)$ and RBC-UKQCD Collaboration [19] obtained $\phi = 40.6(2.8)°$. We thus use $\phi = 37° \sim 42°$ to check the sensitivity of our LFQM since the mixing angle for $\eta - \eta'$ is still not yet settled as a fixed value.

In Fig. 2, we show the transition form factors $Q^2F_{\pi\gamma}(Q^2)$, $Q^2F_{\eta\gamma}(Q^2)$, and $Q^2F_{\eta'\gamma}(Q^2)$ up to $Q^2 = 45$ GeV^2. For $Q^2F_{(\eta,\eta')\gamma}(Q^2)$ case, the solid and dashed lines are results obtained from $\eta - \eta'$ mixing angles with $\phi_{\eta-\eta'} = 37°$ and 42°, respectively. The data are taken from [2,3,8,12,13]. For $Q^2F_{\eta\gamma}(Q^2)$, we obtain the asymptotic result as $\lim_{Q\to\infty} Q^2F_{\pi\gamma}(Q^2) \simeq 0.195$ GeV, which is consistent with the asymptotic limit set by perturbative QCD: $Q^2F_{\pi\gamma}(Q^2) = \sqrt{2}f_\pi \simeq 0.185$ GeV [1]. We also note that our LFQM result for $Q^2F_{\pi\gamma}(Q^2)$ is in good agreement with the recent data from the Belle experiment [3] showing the asymptotic behavior for the region $10 \leq Q^2 \leq 45$ [GeV^2] but disagree with the BaBar data [2] showing the rapid growth for this Q^2 regime. For $Q^2F_{\eta\gamma}(Q^2)$ and $Q^2F_{\eta'\gamma}(Q^2)$ TFFs, our predictions $F_{\eta\gamma}(Q^2)$ using the mixing angle $\phi = 37°$ show slightly better agreement with the data compared to the results obtained from the mixing angle $\phi = 42°$. As in the case of $Q^2F_{\pi\gamma}(Q^2)$, the TFFs $Q^2F_{\eta\gamma}(Q^2)$ and $Q^2F_{\eta'\gamma}(Q^2)$ show asymptotic behavior for high Q^2 region.

The asymptotic values obtained in the spacelike region are obtained as follows: $\lim_{Q^2\to\infty} Q^2 F_{\eta\gamma}(Q^2) \simeq 0.192$ (0.167) GeV for $\phi = 37°$ (42°) and $\lim_{Q^2\to\infty} Q^2 F_{\eta'\gamma}(Q^2) \simeq 0.286$(0.302) GeV for $\phi = 37°(42°)$, respectively.

4 Summary and Discussion

In this work, we investigated $(\pi^0, \eta, \eta') \to \gamma^*\gamma$ TFFs using the standard LF (SLF) approach within the phenomenologically accessible, realistic LFQM [11]. Performing a LF calculation in the covariant BS model, we found that the TFF using the plus component of the currents is immune to the zero-mode. We then linked the covariant BS model to the standard LFQM following the same correspondence relation Eq. (7) between the two that we found in our previous analysis of two-point and three-point functions for pseudoscalar and vector meson [16, 17]. This link allows us to effectively substitute the LF vertex function in the covariant BS model with the more phenomenologically accessible Gaussian wave function provided by the LFQM analysis of meson mass [11].

For the $\pi \to \gamma^*\gamma$ transition, our numerical result of $Q^2 F_{\pi\gamma}(Q^2)$ does not show any steep rising behavior for high Q^2 region as measured from the BaBar Collaboration [2] but shows scaling behavior for high Q^2 consistent with the pQCD prediction. This may be ascribed to the fact that our twist-2 DA [17] is highly suppressed at the end points ($x = 0, 1$) unlike the flat DA [5] showing the enhancement at the end points. We should note that our results for the twist-2 pion DA and the pion-photon transition form factor are very similar to those obtained by the authors [20]. For the $(\eta, \eta') \to \gamma^*\gamma$ transitions, we use the $\eta - \eta'$ mixing angles $\phi = [37°, 42°]$ in the quark-flavor basis to check the sensitivity of our LFQM. Comparing the experimental data for $Q^2 F_{(\eta,\eta')\gamma}(Q^2)$, our optimum value of the $\eta - \eta'$ mixing angle seems to be $\phi \simeq 37°$. Our results of $Q^2 F_{(\eta,\eta')\gamma}(Q^2)$ show again scaling behavior for high Q^2 consistent with the pQCD prediction.

References

1. Lepage, G.P., Brodsky, S.J.: Exclusive processes in perturbative quantum chromodynamics. Phys. Rev. D **22**, 2157 (1980)
2. Aubert, B. et al.: (BABAR Collaboration), Measurement of the $\gamma\gamma* \to \pi^0$ transition form factor. Phys. Rev. D **80**, 052002 (2009)
3. Uehara, S. et al.: (Belle Collaboration), Measurement of $\gamma\gamma* \to \pi^0$ transition form factor at Belle. Phys. Rev. D **86**, 092007 (2012)
4. Mikhaillov, S.V., Stefanis, N.G.: Transition form factors of the pion in light-cone QCD sum rules with next-to-next-to-leading order contributions. Nucl. Phys. B **821**, 291 (2009)
5. Radyushkin, A.V.: Shape of pion distribution amplitude. Phys. Rev. D **80**, 094009 (2009)
6. Brodsky, S.J., Cao, F.-G., de Téramond Guy, F.: Evolved QCD predictions for the meson-photon transition form factors. Phys. Rev. D **84**, 033001 (2011)
7. de Melo, J.P.B.C., El-Bennich, B., Frederico, T.: The photon-pion transition form factor: incompatible data or incompatible models?. Few Body Syst. **55**, 373 (2014)
8. del Amo Sanchez, P. et al.: (BABAR Collaboration), Measurement of the $\gamma\gamma* \to \eta$ and $\gamma\gamma* \to \eta'$ transition form factors. Phys. Rev. D **84**, 052001 (2011)
9. Escribano, R., Masjuan, P., Sanchez-Puetas, P.: η and η' transition form factors from rational approximants. Phys. Rev. D **89**, 034014 (2014)
10. Kroll, P., Passek-Kumerički, K.: The η (η') gamma transition form factor and the gluon–gluon distribution amplitude. J. Phys. G **40**, 075005 (2013)
11. Choi, H.-M., Ji, C.-R.: Mixing angles and electromagnetic properties of ground state pseudoscalar and vector meson nonets in the light cone quark model. Phys. Rev. D **59**, 074015 (1999)
12. Behrend, H.-J., et al.: (CELLO Collaboration), A measurement of the π^0, η and η' electromagnetic form factors. Z. Phys. C **49**, 401 (1991)
13. Gronberg, J., et al.: (CLEO Collaboration), Measurements of the meson-photon transition form-factors of light pseudoscalar mesons at large momentum transfer. Phys. Rev. D **57**, 33 (1998)
14. Feldmann, T., Kroll, P., Stech, B.: Mixing and decay constants of pseudoscalar mesons. Phys. Rev. D **58**, 114006 (1998)
15. Jaus, W.: Relativistic constituent quark model of electroweak properties of light mesons. Phys. Rev. D **44**, 2851 (1991)
16. Choi, H.-M., Ji, C.-R.: Self-consistent covariant description of vector meson decay constants and chirality-even quark-antiquark distribution amplitudes up to twist-3 in the light-front quark model. Phys. Rev. D **89**, 033011 (2014)
17. Choi, H.-M., Ji, C.-R.: Consistency of the light-front quark model with chiral symmetry in the pseudoscalar meson analysis. Phys. Rev. D **91**, 014018 (2015)
18. Ambrosino, F., et al. Measurement of the pseudoscalar mixing angle and eta-prime gluonium content with KLOE detector. (KLOE Collaboration) Phys. Lett. B **648**, 267 (2007)
19. Christ N.H., et al.: The η and η' mesons from lattice QCD. Phys. Rev. Lett. **105**, 241601 (2010)
20. Stefanis, N.G., Pimikov, A.V.: Chimera distribution amplitudes for the pion and the longitudinally polarized ρ-meson. Nucl. Phys. A **945**, 248 (2014)

DOI 10.1007/978-3-319-50699-9_26
Reprinted from *Few-Body System* Journal, DOI 10.1007/s00601-016-1057-5

T. Frederico · J. Carbonell · V. Gigante · V. A. Karmanov

Inverting the Nakanishi Integral Relation for a Bound State Euclidean Bethe–Salpeter Amplitude

Received: 7 January 2016 / Accepted: 11 February 2016 / Published online: 18 March 2016

Abstract The extraction of the weight function g of the Nakanishi integral representation of the Bethe–Salpeter amplitude is investigated. We studied the numerical inversion of the discretized Nakanishi kernel in the case of an Euclidean bound state. The discretized kernel is regularized by adding the identity operator times a small regularisation parameter ε to avoid numerically unstabilities. We have found that the weight function g as well as the associated light-front valence wave function are unstable against variation of ε. These results suggest that the extraction of the Nakanishi weight function from an Euclidean amplitude, is an ill-defined problem. Without further assumptions on the solution or/and without developing more elaborate methods, the Nakanishi weight function, as well as the corresponding light-front valence wave function, cannot be safely determined.

1 Introduction

Ab initio calculations of hadron properties are nowadays performed by solving numerically the path-integral formulation of Quantum Chromodynamics (QCD) in a discretized Euclidean space-time lattice (LQCD) [1]. On another hand one serious issue is the calculation of Minkowski space observables, like e.g. the full x-dependence of hadronic amplitudes entering in the evaluation of parton distributions (PDFs) and generalized parton distributions (GPDs), which are associated with matrix elements of bi-local operators between Light-Front (LF) wavefunctions. So far, LQCD provided only the lowest moments of such distributions. Higher moments are increasingly difficult to compute as they involve higher derivatives of the matrix elements and are polluted by increasing errors. However the lowest moments are insufficient to reconstruct the full dependence in x of the PDFs and GPDs [2]. This limitation illustrates the difficult for obtaining LF nucleon wave function from LQCD.

A possible way to overcome this shortcoming was recently suggested in [3], where was shown that the parton physics on the light-cone can be obtained from the matrix elements of equal-time correlators in the

T. Frederico (✉) · V. Gigante
Instituto Tecnológico de Aeronáutica, DCTA, São José dos Campos 12.228-900, Brazil
E-mail: tobias@ita.br

V. Gigante
E-mail: vitorgigante@yahoo.com.br

J. Carbonell
Institut de Physique Nucléaire, Université Paris-Sud, IN2P3-CNRS, 91406 Orsay Cedex, France
E-mail: carbonell@ipno.in2p3.fr

V. A. Karmanov
Lebedev Physical Institute, Leninsky Prospekt 53, Moscow, Russia 119991
E-mail: karmanov@lebedev.ru

large momentum limit, allowing to compute these quantities on an Euclidean lattice [4]. This observation would allow to calculate parton physics—which involves correlations of quark and gluons fields separated along the light-cone in Minkowski space-time—in Euclidean LQCD. On the other hand such matrix elements of operators can be computed using hadron light-front wave functions defined in the Minkowski space [5].

A step forward is achieved by the direct computation of the hadron wave function by diagonalizing of the LF Hamiltonian in a basis function [6]. Applications for QED in the case of an electron in a transverse harmonic cavity has been done in [7], and to positronium in [8]. These efforts open the route to compute the hadron light-front wave function. However, it is not yet practical with the current computational facilities.

Another direction in the effort of calculating a realistic hadron LF wave function was given in [9]. These authors projected onto the LF the pion Bethe–Salpeter (BS) amplitude [10] computed by solving the QCD Dyson–Schwinger equations in Euclidean space. In addition, each scalar component of the pion BS amplitude was parametrized, expressed in terms of the Nakanishi representation [11–13] and the parameters were adjusted to reproduce the corresponding Euclidean functions.

Nakanishi representation was proposed in [14,15] to calculate the relativistic two-particle bound state in Minkowski space. In Refs. [16–19] this representation was used in connection with the LF projection to obtain manageable solution of BS equation, for bound and scattering states [20–22]. A key property underlying such approach is the uniqueness of the weight function, proved perturbatively in [13], but still lacking a proof in the non-perturbative case. It was found in [14,15] that the uniqueness holds non perturbatively by solving an integral equation for the weight function which relies on this property and provides the correct binding energies. Additional checks was done in Ref. [23].

We stress that all these computations were performed starting from Minkowski space. The real challenge would be to obtain the Nakanishi weight function from Euclidean results, which are much easier to compute. Because of the interest in obtaining a LF hadron wave function from an Euclidean BS amplitude, either computed in LQCD or from Dyson–Schwinger equations, our additional motivation is to check also the extraction of the Nakanishi weight function from an Euclidean amplitude in order to obtain the corresponding valence LF wave function as well.

We have udertaken an extensive study of the Nakanishi integral representation associated with the BS amplitude and the LF wave function of a relativistic two-particle bound state [24]. Our aim is to find an efficient method to extract the Nakanishi weigth function g from each of these quantities and use it as a natural link between the Minkowski and Euclidean space solutions, thus allowing the analytic continuation from one to other.

In this contribution we will limit ourselves to the following points: (i) the extraction of the Nakanishi weight function from the Euclidean BS amplitude using different numerical methods and (ii) the computation of the LF wave functions in terms of the Nakanishi weight function extracted from the Euclidean solutions. To this aim we have used two analytical models: a Gaussian form for the BS ampitude in Euclidean space and a Gaussian Nakanishi weight function.

2 Bethe–Salpeter Amplitude and LF Wave Function in the Nakanishi Representation

The Nakanishi integral representation of the BS amplitude for a s-wave state is given by [13]

$$\Phi(k,p) = \int_{-1}^{1} dz' \int_{0}^{\infty} d\gamma' \frac{g(\gamma',z')}{\left(\gamma' + \kappa^2 - k^2 - p\cdot k z' - i\epsilon\right)^3}, \tag{1}$$

where g is the Nakanishi weight function. The dependence on the bound state mass $M = \sqrt{p^2}$ is introduced throught the parameter $\kappa^2 = m^2 - \frac{M^2}{4} > 0$.

The corresponding Euclidean amplitude is obtained by changing the variable $k^0 \to i k_E^0$ in Eq. (1). In the rest frame—i.e. setting $p = (M, 0)$—one gets:

$$\Phi_E(k_E,p) = \int_{-1}^{1} dz' \int_{0}^{\infty} d\gamma' \frac{g(\gamma',z')}{\left(\gamma' + \kappa^2 + k_E^2 - iMk_E^0 z'\right)^3} \tag{2}$$

with $k_E^2 = \left(k_E^0\right)^2 + (\mathbf{k})^2$. The symmetry of the Nakanishi weight function $g(\gamma, z) = g(\gamma, -z)$ leads to:

$$\Phi_E(k_E, p) = \int_{-1}^{1} dz \int_{0}^{\infty} d\gamma \, \mathrm{Re}\left[\frac{1}{\left(\gamma + \kappa^2 + k_E^2 - iM\, k_E^0 z\right)^2}\right] g(\gamma, z)\,, \tag{3}$$

It is remarkable that the Nakanishi weight function g depend on dummy arguments which are—as g itself—independent on the metric. All the singularities and metric depending quantities of the BS amplitudes are collected in the denominator of Eqs. (1) and (2).

The projection onto the light-front of the BS amplitude gives the LF valence wave function. It can be obtained by using the procedure given in [16] and [20]. In terms of the variables (γ, z) defined as

$$k_E^0 = m \frac{z}{1 - z^2} \qquad |\,\mathbf{k}\,|^2 = \gamma \tag{4}$$

it reads

$$\Psi_{LF}(\gamma, z) = \frac{1 - z^2}{4} \int_{0}^{\infty} d\gamma' \frac{g(\gamma', z)}{\left[\gamma' + \gamma + z^2 m^2 + \left(1 - z^2\right)\kappa^2\right]^2} \tag{5}$$

Notice that the dependence of all these functions on the parameters m and M and κ is not depicted.

3 Extracting the Nakanishi Weight Function from Φ_E

The integral relation (2) has been discretized by using both the Gauss–Legendre quadrature points and the splines basis functions. In each case we end up with a generic linear system on the form

$$A\, g = b \tag{6}$$

with A a non symmetric matrix and $g_n \equiv g(\gamma_i, z_j)$ the values of the unknown weight function at the grid points. This linear system—or rather its normal form $A^\dagger A x = A^\dagger b$—turned to be ill-defined with a condition number, i.e. the ratio between the largest to the smallest eigenvalue, which can reach $\sim 10^{20}$ even with $N = 8$ Gauss points on each variable. This is a consequence of the large number of very small eigenvalues of the matrix A, despite that the continuous operator supposed to represent A is invertible. The numerical inversion of (6) is thus very unstable and was regularized as in [16] by shifting its diagonal with a term ϵI, where ϵ is a regulator parameter. Eq. (6) was thus replace by

$$(A - \epsilon I) g = b \tag{7}$$

In the first model case we considered, we have used as inhomogeneous term b, an Euclidean BS amplitude given by a Gaussian form:

$$\Phi_E(k_E, p) = e^{-k_E^2}, \tag{8}$$

and have fixed the mass parameters to the values $m = 1$ and $M = 1$. Despite the fact that g is metric independent, we have kept the index g_E in our results to remind that they have been obtained with Euclidean Nakanishi kernel (2) and inhomogeneous term.

The results of the numerical inversion of Eq. (2) are shown in Fig. 1 for the z-dependence of g and Fig. 2 for its γ-dependence. Calculation have been performed with $\varepsilon = 10^{-2}$ and 10^{-4}, using Gaussian-Legendre quadratures and splines basis in order to verify the numerical stability of our procedure. In addition, we varied the integration domain $0 < \gamma < \gamma_{max}$, with $\gamma_{max} = 10$ and $\gamma_{max} = \infty$.

Our first check consisted in comparing g_E—computed with the integration range $\gamma_{max} = 10$ and $\varepsilon = 10^{-2}$—provided by the two different numerical methods: Gaussian-Legendre with $N = 60 \times 60$ discretisation points and splines basis with $N = 20 \times 20$ functions. The results are displayed in the left panels of Fig. 1 (for the z-dependence) and Fig. 2 (for the γ-dependence): blue dots correspond to splines and solid lines to Gauss–Legrendre. They show a good agreement between both methods.

We would like to notice however that the regularisation procedure we use is not strictly equivalent in both discretisation methods and give rise to small differences when the value of ε is decreased. This is illustrated in

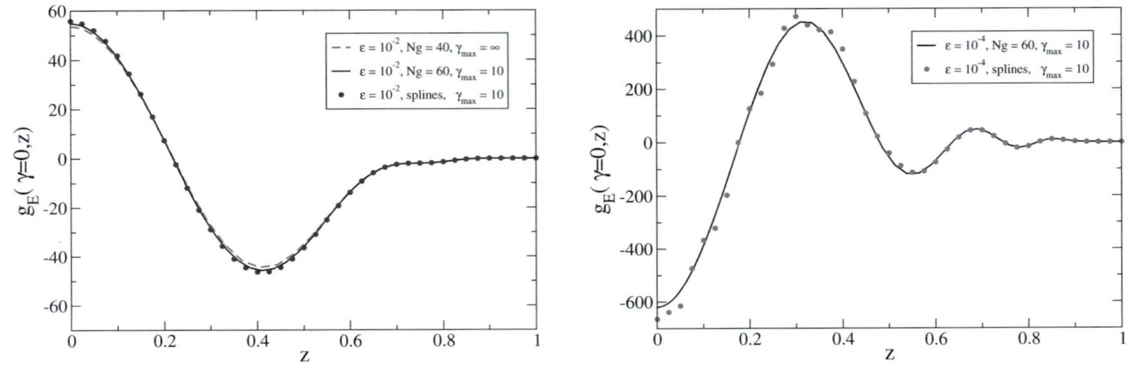

Fig. 1 Nakanishi weight function $g(0, z)$ extracted from the Euclidean BS amplitude (8) with $\varepsilon = 10^{-2}$ (*left panel*) and 10^{-4} (*right panel*). Results using Gauss–Legendre mesh points are indicated by *solid lines* for $\gamma_{max} = 10$ and with *dashed line* for $\gamma_{max} = \infty$. Spline results with $\gamma_{max} = 10$ are given by *full circles*

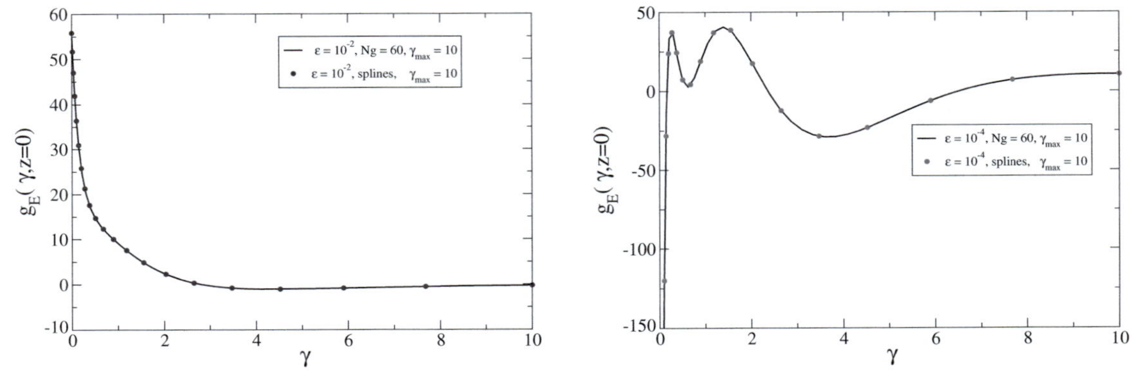

Fig. 2 Same results as Fig. 1 displaying the γ-dependence for a fixed value $z = 0$

Fig. 1 right panel corresponding to $\epsilon = 10^{-4}$, comparing spline results (red dots) with Gauss–Legendre ones (black solid line). Differences are due to distinct meaning of the identity operator I in both numerical schemes and can be decreased by using larger discretisation basis [16].

Next check was to study the stability of results when varying the integration domain. For this purpose we included in Fig. 1 (left panel, red dashed lines) results corresponding to $\gamma_{max} = \infty$ obtained by using a mapping in γ variable: $f : [0, 1] \to [0, \infty)$. As one can see there are no significant differences and we conclude that the use of a finite integration domain $\gamma_{max} = 10$ is enough to our purposes.

The evolution of g_E with ε is displayed in Fig. 1. Left panel corresponds to $\varepsilon = 10^{-2}$ and the right one to 10^{-4}. As one can see, the solution strongly dependents on ϵ. The same results are visible in the γ-dependence of the solution displayed in Fig. 2. Notice in particular that the value of $g_E(0, 0)$ exhibits a huge increase, when ε decreases from 10^{-2} to 10^{-4}, pointing to the singular character of the inversion procedure due to the presence of small eigenvalues in the kernel. Furthermore one can observe that the number of zeros of $g_E(\gamma, z)$ in both variables increases when $\varepsilon \to 0$. These features were also found to be independent of the numerical method we used.

In order to further study the stability in the extraction procedure of the Nakanishi weight function, we computed the light-front wave function Ψ by integrating numerically Eq. (5) with ε values of 10^{-2}, 10^{-4} and 10^{-6}. The results, dropping the kinematical prefactor $(1 - z^2)$, are shown in Fig. 3 with the normalisation condition $\Psi_{LF}(0, 0) = 1$ for the sake of comparison. The wiggles of the wave function with z (left panel) do not show any sign of convergence as a function of ε, a clear indication that the small eigenvalues of the discretized kernel are contributing considerably to g_E and to Ψ_{LF}.

This unstability in the LF wave function is in sharp contrast with the findings of Refs. [16,23], where the BS solution in Minkowski space was computed using the Nakanishi and LF projection. There, the stability of the LF wave-function versus ε was found to be very high, even if the Nakanishi weight function g was not stable.

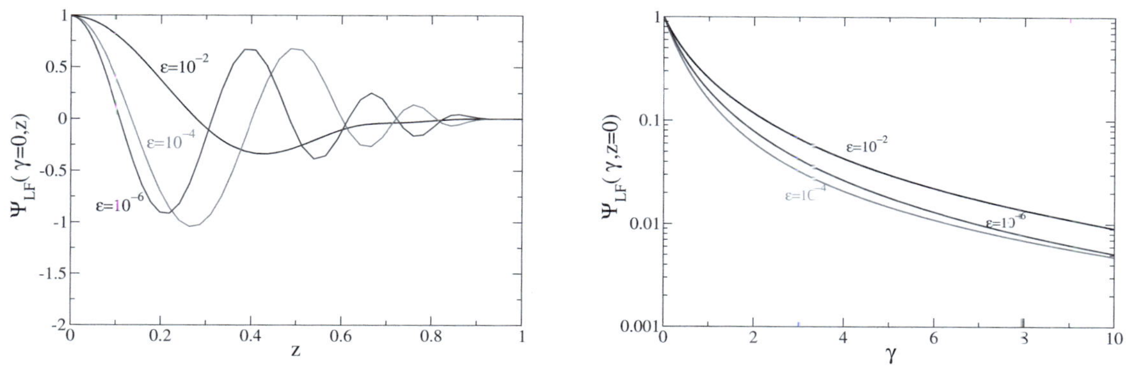

Fig. 3 ε dependence of the light-front wave function computed with the Nakanishi weight function g obtained from the Euclidean BS amplitude. The wave function is normalised to unity for $\gamma = z = 0$

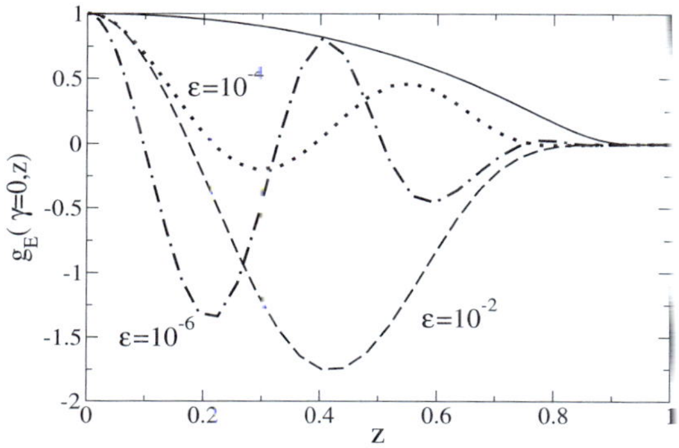

Fig. 4 ε-dependence of the Nakanishi weight function obtained from the Euclidean BS amplitude calculated with g_{test} given in (9) (*solid line*). The normalisation condition $\gamma = z = 0$ was used

Finally, we have investigated whether or not the failure in extracting the Nakanishi weight function from the numerical inversion of Eq. (2) could be attributed to the oversimplified Gaussian model of the BS amplitude (8), which does not have a realistic analytical structure in Minkowski space, i.e. as given by the cuts of the Nakanishi PTIR in Eq. (1). For that purpose, we performed a second test case, starting with the Nakanishi weight function model:

$$g_{test}(\gamma', z) = e^{-(\gamma+1)/(1-z^2)}, \tag{9}$$

having more realistic asymptotics. We have then obtained the corresponding Φ_E by numerically integrating the Nakanishi relation (2) with with $m = M = 1$ and proceed to compute again g_E, this time with the numerical inversion of (7).

Corresponding results—their z-dependence—are shown in Fig. 4 for different values of ε. As one can see, one finds a similar behaviour as for the above discussed model: an increasing number of nodes appear when decreasing ε, a clear indication that no convergence can be found in the limit $\varepsilon \to 0$. This again indicates that small eigenvalues of the kernel are present and not eliminated by our regularisation procedure when $\varepsilon \to 0$. This example shows that even with the correct analytic structure of the kernel no sign of stability is found. The situation is thus very striking: the solution is unique, is well known (at least in this model case) but we are badly unable to find it.

It is clear that more efficient methods—or further assumptions on $g_E(\gamma, z)$—are required to obtain a stable and unique solution from the inversion of (3) in the Euclidean space.

4 Summary

The possibility to extract the Nakanishi weight function from the Bethe–Salpeter amplitude in Euclidean space was investigated.

For this purpose we have considered the discretized version of the Nakanishi integral representation and performed its inversion using two different numerical methods: Gauss–Legendre and splines.

The resulting linear system turned to be ill-conditioned and the numerical results are uncontrolled, independently of the discretization scheme. It has been regularized by adding to the kernel the identity operator times a small regularisation parameter (ε). The integral equation for the Nakanishi weight function for a fixed value of ε is then safely solved and results provided by the two discretization methods are consistent to each other.

The inhomogeneous term is given by Euclidean Bethe–Salpeter amplitude. Two models have been considered: (i) Gaussian Φ_E and (ii) exponential weight function g with realistic boundary conditions.

Independently of the model considered, the computed Nakanishi weight function g is unstable against ε. The corresponding LF valence wave function, obtained by performing a one dimensional integral on g, displays the same instability.

Our conclusion, established only for a bound state of two scalars, show that the extraction of the Nakanishi weight function from an Euclidean amplitude, is an ill-defined problem, and unstable in a naive discretization scheme. The reason for such instability is the existence of a dense bulk of very small eigenvalues of the Nakanishi kernel.

More extensive investigations on the spectral properties of this kernel are in course to circunvent this problem and some stable methods—although with limited accuracy—have been found to compute the Nakaknishi weight function [24].

Alternatively, we cannot exclude that physically motivated assumptions on the weight function could help to safely extract it from Euclidean space. If this is so, we could dispose of a useful tool to analytically extend to Minkowski space the Euclidean amplitudes computed e.g. in LQCD.

Acknowledgments VG thanks support from CAPES, TF thanks partial support from FAPESP and CNPq.

References

1. Hägler, P.: Hadron structure from lattice quantum chromodynamics. Phys. Rep. **490**, 49 (2010)
2. Moutarde, H.: Contribution to the Special Issue "Light-Cone 2015", to be published in Few-Body Systems
3. Ji, Xiangdong: Parton physics on a Euclidean lattice. Phys. Rev. Lett. **110**, 262002 (2013)
4. Alexandrou, C.: Contribution to the Special Issue "Light-Cone 2015", to be published in Few-Body Systems
5. Brodsky, S.J., Pauli, H.C., Pinsky, S.S.: Quantum chromodynamics and other field theories on the light cone. Phys. Rep. **301**, 299 (1998)
6. Vary, J.P., Honkanen, H., Li, Jun, Maris, P., Brodsky, S.J. et al.: Hamiltonian light-front field theory in a basis function approach. Phys. Rev. C **81**, 035205 (2010)
7. Honkanen, H., Maris, P., Vary, J.P., Brodsky, S.J.: Electron in a transverse harmonic cavity. Phys. Rev .Lett. **106**, 061603 (2011)
8. Wiecki, P., Li, Y., Zhao, X., Maris, P., Vary, J.P.: Basis light-front quantization approach to positronium. Phys. Rev. D **91**, 105009(2015)
9. Chang, L., Cloet, I.C., Cobos-Martinez, J.J., Roberts, C.D. et al.: Imaging dynamical chiral-symmetry breaking: pion wave function on the light front. Phys. Rev. Lett. **110**, 132001 (2013)
10. Salpeter, E.E., Bethe, H.A.: A relativistic equation for bound-state problems. Phys. Rev. **84**, 1232 (1951)
11. Nakanishi, N.: Partial-wave Bethe-Salpeter equation. Phys. Rev. **130**, 1230 (1963)
12. Nakanishi, N.: A General Survey of the theory of the Bethe-Salpeter equation. Prog. Theor. Phys. Suppl. **43**, 1 (1969)
13. Nakanishi, N.: Graph theory and Feynman integrals. Gordon and Breach, New York (1971)
14. Kusaka, K., Williams, A.G.: Solving the Bethe-Salpeter equation for scalar theories in Minkowski space. Phys. Rev. D **51**, 7026 (1995)
15. Kusaka, K., Simpson, K., Williams, A.G.: Solving the Bethe-Salpeter equation for bound states of scalar theories in Minkowski space. Phys. Rev. D **56**, 5071 (1997)
16. Karmanov, V.A., Carbonell, J.: Solving Bethe-Salpeter equation in Minkowski space. Eur. Phys. J. A **27**, 1 (2006)
17. Carbonell, J., Karmanov, V.A.: Cross-ladder effects in Bethe-Salpeter and light-front equations. Eur. Phys. J. A **27**, 11 (2006)
18. Carbonell, J., Karmanov, V.A., Mangin-Brinet, M.: Electromagnetic form factor via Bethe-Salpeter amplitude in Minkowski space. Eur. Phys. J. A **39**, 53 (2009)
19. Carbonell, J., Karmanov, V.A.: Solving the Bethe-Salpeter equation for two fermions in Minkowski space. Eur. Phys. J. A **46**, 387 (2010)
20. Frederico, T., Salmè, G., Viviani, M.: Two-body scattering states in Minkowski space and the Nakanishi integral representation onto the null plane. Phys. Rev. D **85**, 036009 (2012)
21. Sales, J.H.O., Frederico, T., Carlson, B.V., Sauer, P.U.: Light-front Bethe-Salpeter equation. Phys. Rev. C **61**, 044003 (2000)
22. Frederico, T., Salmè, G.: Projecting the Betheâ Salpeter equation onto the light-front and back: A short review. Few-Body Syst. **49**, 163 (2011)

23. Frederico, T., Salmè, G., Viviani, M.: Quantitative studies of the homogeneous Bethe-Salpeter equation in Minkowski space. Phys. Rev. D **89**, 016010 (2014)
24. Frederico, T., Karmanov, V.A., Carbonell, J.: in preparation

DOI 10.1007/978-3-319-50699-9_27
Reprinted from *Few-Body System* Journal, DOI 10.1007/s00601-016-1066-4

L. Mantovani · A. Bacchetta · B. Pasquini

The Electron in Three-Dimensional Momentum Space

Received: 18 January 2016 / Accepted: 22 February 2016 / Published online: 18 March 2016

Abstract We study the electron as a system composed of an electron and a photon and derive the leading-twist transverse-momentum-dependent distribution functions for both the electron and photon in the dressed electron, thereby offering a three-dimensional description of the dressed electron in momentum space. To obtain the distribution functions, we apply both the formalism of light-front wave function overlap representation and the diagrammatic approach; we discuss the comparison of our results between light-cone gauge and Feynman gauge, discussing the role of the Wilson lines to obtain gauge-independent results. We provide examples of plots of the computed distributions.

1 Introduction

The electron is a structureless particle; however, thanks to its quantum fluctuations, we can think of it as an object surrounded by a composite virtual cloud of photons, electrons and positrons. The latter can be interpreted as partons contained in a "dressed" electron, in analogy to the partonic structure of hadrons, and therefore described through the formalism of parton correlation functions, which can be parametrized in terms of different types of distribution functions [1,2]. These functions not only allow us to view the dressed electron from a new perspective [3–5]; they can also help to shed light on several formal aspects of their QCD analogues, taking advantage of the perturbative nature of QED.

In this work we review the results obtained in Ref. [6] for the transverse-momentum dependent distribution functions (TMDs), which describe the distribution of QED partons in the dressed electron as a function of their longitudinal and transverse momentum, and therefore give access to the three-dimensional picture of the dressed electron in momentum space. In Sect. 2 we review the derivation of the leading-twist TMDs for the electron and the photon inside the electron and discuss the equivalence of the results in two different gauges, namely the Feynman and the light-cone gauge. We apply both the diagrammatic approach of Feynman diagrams and the representation in terms of overlap of light-front wave functions (LFWFs) [7]. In Sect. 3 we show some numerical results and plots of the TMDs, and we summarize our conclusions.

2 Dressed Electron TMDs

Let us consider a dressed electron of mass m, spin S and 4-momentum P, composed of a bare electron of momentum k and a photon of momentum $P - k$. We first focus on the distribution for the internal electron

L. Mantovani (✉) · A. Bacchetta · B. Pasquini
Dipartimento di Fisica, Università degli Studi di Pavia, and INFN, Sezione di Pavia, 27100 Pavia, Italy
E-mail: luca.mantovani@pv.infn.it

A. Bacchetta
E-mail: alessandro.bacchetta@unipv.it

B. Pasquini
E-mail: barbara.pasquini@unipv.it

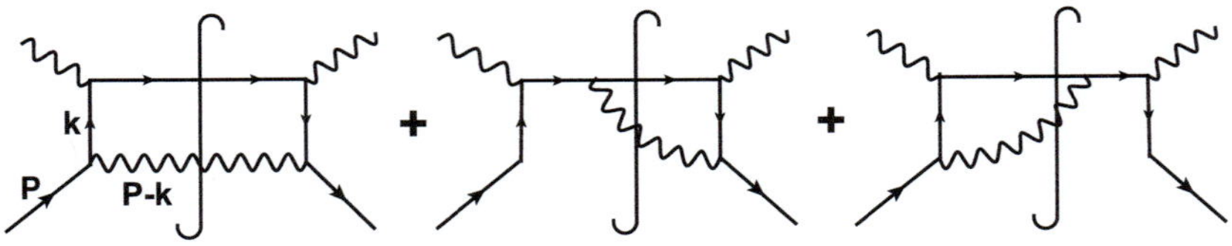

Fig. 1 Handbag diagrams contributing to the electron TMDs at order α

and we set its momentum, in light-cone coordinates, as $k \equiv (xP^+, k^-, \mathbf{k}_\perp)$, with the longitudinal momentum fraction x defined as $x := k^+/P^+$. If we probe the internal structure of the electron via the scattering of a virtual photon (see Fig. 1), the transverse-momentum dependent correlator function between initial and final electron states is given by [1,8]

$$\Phi(x, \mathbf{k}_\perp; P, S) = \int dk^- \, \Phi(k; P, S) = \int \frac{d\xi^- d^2\xi_\perp}{(2\pi)^3} e^{ik\cdot\xi} \left\langle P, S | \bar{\psi}(0) \mathcal{U}_{(0,\xi)} \psi(\xi) | P, S \right\rangle \Big|_{\xi^+=0}. \quad (1)$$

From a diagrammatic point of view, the correlator Φ corresponds to the lower part of the handbag diagrams in Fig. 1, i.e. the part obtained removing the external photons and the propagator of momentum $k + q$. The gauge link operator (or Wilson line) $\mathcal{U}_{(0,\xi)}$ in Eq. (1) is needed to make the correlator a gauge-invariant object; it is defined as [9]

$$\mathcal{U}_{(0,\xi)} := \exp\left[-ie \int_0^\xi d\eta^\mu A_\mu(\eta)\right], \quad (e > 0). \quad (2)$$

The integration path lies in the $\eta_\perp - \eta^-$ plane, since the plus component of the fields is fixed at 0 in Eq. (1); the path used in the TMD definition is shown in Fig. 2. It should be noticed that the path itself is gauge-dependent: in the Feynman gauge we have the boundary condition $\mathbf{A}_\perp(\infty) = 0$, therefore only the two lines along the light-like direction η^- are relevant; in the light-cone gauge $A^+ = 0$, instead, only the transverse component of the gauge link survives. As for the physical interpretation is concerned, the gauge link accounts for the (infinitely many, in principle) photons that can be absorbed or emitted by the incoming or outgoing particles. For this reason, the simple picture of the handbag diagram (a) in Fig. 1 is valid only if we take the $O(\mathbb{1})$ expansion of the Wilson line; if we go to higher order, we need to consider also different diagrams, such as (b) and (c) in Fig. 1. We remark that in the present work we neglect all the contributions that come from vertex corrections and self-energy diagrams (which show up only at the end-point $x = 1, \mathbf{k}_\perp = \mathbf{0}_\perp$), since they would require a regularization procedure, which is beyond the scope of the present work.

From now on, we consider a reference frame where $P = (P^+, P^-, \mathbf{0}_\perp)$, with $P^- = m^2/2P^+$ from the on-shell condition; moreover, we can set $S \equiv (S_z P^+/m, -S_z m/2P^+, \mathbf{S}_\perp)$. Introducing also the light-like vector $n_+ = (1, 0, \mathbf{0}_\perp)$ and the antisymmetric tensor $\epsilon_\perp^{12} = -\epsilon_\perp^{21} = 1$, the eight leading-twist electron TMDs are implicitly defined through the following decomposition of the correlator (1) in terms of the elements of the Dirac space basis $\mathcal{D} := \{\mathbb{1}, \gamma^\mu, \gamma^5, \gamma^5\gamma^\mu, i\sigma^{\mu\nu}\}$:

$$\begin{aligned}\Phi(x, \mathbf{k}_\perp; S) \approx & \frac{1}{2}\Bigg[f_1^e \not{n}_+ - \frac{\epsilon_\perp^{ij} k_\perp^i S_\perp^j}{m} f_{1T}^{\perp e} \not{n}_+ + S_z g_{1L}^e \gamma^5 \not{n}_+ + \frac{\mathbf{k}_\perp \cdot \mathbf{S}_\perp}{m} g_{1T}^e \gamma^5 \not{n}_+ + \frac{[\not{S}_\perp, \not{n}_+]}{2} \gamma^5 h_{1T}^e \\ & + S_z \frac{[\not{k}_\perp, \not{n}_+]}{2m} \gamma^5 h_{1L}^{\perp e} + \frac{\mathbf{k}_\perp \cdot \mathbf{S}_\perp [\not{k}_\perp, \not{n}_+]}{2m^2} \gamma^5 h_{1T}^{\perp e} + \frac{i[\not{k}_\perp, \not{n}_+]}{2m} h_1^{\perp e} \Bigg], \end{aligned} \quad (3)$$

where $i, j = 1, 2$ and $a_\perp^\mu = (0, 0, \mathbf{a}_\perp)$. From the above decomposition, one can recover the analytic expression of the TMDs by selecting proper projections of the correlator in Dirac space. Out of the eight TMDs in Eq. (3), the so-called Boer–Mulders function $h_1^{\perp e}$ and the Sivers function $f_{1T}^{\perp e}$ are T-odd, i.e. they change sign under naive time reversal (which is defined as usual time-reversal but without interchange of initial and final states), while the other six are T-even. If one excludes the contribution at the end-point $x = 1, \mathbf{k}_\perp = \mathbf{0}_\perp$, the T-odd TMDs $h_1^{\perp e}$ and $f_{1T}^{\perp e}$ are vanishing [6]. This is due to the fact that only diagrams containing a loop at one side of the cut can potentially give a non-zero contributions to T-odd TMDs [10], but the only possibility to have such a loop is either to consider diagrams of higher perturbative order, or take into account also the end-point.

The evaluation of the correlator, and consequently of the TMDs, can be performed in two different ways, which yield equivalent results: the LFWF overlap representation and the diagrammatic approach with Feynman diagrams. Let us describe the former first. In the light-cone gauge $A^+ = 0$, we introduce the state $|P, \Lambda\rangle$ for

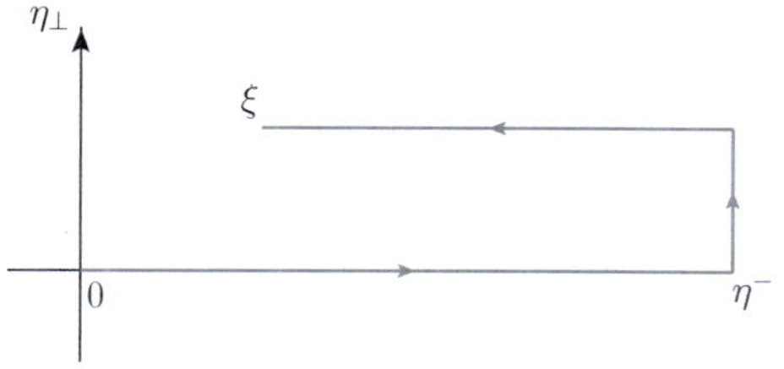

Fig. 2 Wilson line connecting two points at $0 = (0^-, \mathbf{0}_\perp)$ and $\xi = (\xi^-, \boldsymbol{\xi}_\perp)$. The *vertical line* parallel to the transverse axis is taken at $\eta^- = \infty$

the dressed electron with four-momentum P and light-front helicity Λ [11], and expand it in the Fock space of free quanta [12], i.e.

$$|P, \Lambda\rangle = |e\rangle + |e\gamma\rangle, \tag{4}$$

where[1]

$$|e\rangle = b^\dagger_\Lambda(\mathbf{P})|0\rangle, \quad |e\gamma\rangle = \int \frac{dx\, d^2\mathbf{k}_\perp}{2(2\pi)^3\sqrt{x(1-x)}} \sum_{\lambda,\lambda_\gamma} \Psi^\Lambda_{\lambda,\lambda_\gamma}(x, \mathbf{k}_\perp)|e\gamma;\ \mathbf{p}_e, \lambda;\ \mathbf{p}_\gamma, \lambda_\gamma\rangle. \tag{5}$$

The coefficients in the above expansion are the LFWFs $\Psi^\Lambda_{\lambda,\lambda_\gamma}(x, \mathbf{k}_\perp)$, which depend on Λ and the light-front helicities λ and λ_γ of the internal electron and photon, respectively; they can be calculated exactly in QED using time-ordered perturbation theory in the infinite momentum frame, as outlined e.g. in [6]. Inserting Eqs. (4) and (5) in the correlator function (1), one obtains

$$\Phi^{[\Gamma]}(x, \mathbf{k}_\perp; \Lambda) \equiv \frac{1}{2}\text{Tr}\,[\Phi\Gamma] = \frac{1}{4(2\pi)^3 x P^+} \sum_{\lambda,\lambda',\lambda_\gamma} \Psi^{\Lambda\,*}_{\lambda',\lambda_\gamma}(x, \mathbf{k}_\perp)\bar{u}_{\lambda'}(\mathbf{p}_e)\Gamma u_\lambda(\mathbf{p}_e)\Psi^\Lambda_{\lambda,\lambda_\gamma}(x, \mathbf{k}_\perp), \tag{6}$$

where $\Gamma \in \mathcal{D}$; consequently, the TMDs themselves can be expressed as overlap of LFWFs. Since the LFWFs are eigenstates of the total spin of the partons in the z direction, $J_z = \lambda/2 + \lambda_\gamma$, and of the total OAM $L_z = \Lambda/2 - J_z$, this representation allows one to disclose the different contributions to the TMDs from the spin and OAM configurations of the partons.

In the Feynman diagram approach, instead, one needs to evaluate the handbag diagram in Fig. 1a, by applying the usual QED Feynman rules, and then factorize out the lower part, corresponding to the correlator. In the Feynman gauge (FG), as one considers also the longitudinal gauge link, we need to take into account also diagrams (b) and (c) shown in Fig. 1, which however are vanishing in the light-cone gauge (LCG). Nonetheless, we find perfect agreement between the results in the two gauges: the mismatch coming from the different number of diagrams to be considered is compensated by the the fact that the photon propagator is proportional to the sum $d^{\mu\nu}(p) := \sum_\lambda \varepsilon^{\mu\,*}_\lambda(p)\varepsilon^\nu_\lambda(p)$, which in the two gauges reads

$$d^{\mu\nu}(p) = -g^{\mu\nu} \quad \text{(FG)}, \quad d^{\mu\nu}(p) = -g^{\mu\nu} + \frac{1}{p^+}\left(p^\nu n^\mu_- + p^\mu n^\nu_-\right) \quad \text{(LCG)}. \tag{7}$$

We stress that the in order to obtain the equivalence between the results for the two choices of the gauge, one needs to include the longitudinal gauge link when working in the Feynman gauge, while it is not necessary to take into account the transverse gauge link in the light cone-gauge. There is however a contribution to the TMDs that comes from the transverse gauge link [13] and that it is worth to analyze, even though it shows up only at the end-point $x = 1$. It comes from the consideration that one must regularize the denominator of the photon propagator in light-cone gauge, i.e. Eq. (7), since we integrate over the photon momentum. One still has, however, the freedom of choosing a certain prescription for the regularization, among the following possibilities: the retarded and advanced prescriptions, where $p^+ \longrightarrow p^+ \pm i\epsilon$, respectively, and the principal value (PV) prescription, where $1/p^+ \longrightarrow 1/2[1/(p^+ + i\epsilon) + 1/(p^+ - i\epsilon)]$. While the PV prescription yields

[1] One should distinguish between the $|e\gamma\rangle$ state, which describes the dressed electron as a 2-body system in Eq. (4), and the state of two free quanta $|e\gamma;\ \mathbf{p}_e, \lambda;\ \mathbf{p}_\gamma, \lambda_\gamma\rangle$.

a vanishing contribution to the TMDs, this is not the case for the other two prescriptions, where extra terms (proportional to $\delta(1-x)$) appear. The final result, however, should be prescription-independent, and coincide with the one found in the Feynman gauge, where the propagator (Eq. 7) does not need to be regularized; this is ensured by the contributions from the transverse gauge link since they exactly cancel the terms coming from the regularization of the propagator, as shown in Ref. [6].

We finally mention that an analysis similar to the one performed so far for the TMDs of the electron inside the electron can be extended also to the case of the internal photon. In this case, the active parton is the photon, and the relevant correlator reads

$$\Phi^{\gamma[ij]}(k; P, S) := \frac{1}{xP^+} \int \frac{d\xi^- d^2\boldsymbol{\xi}_\perp}{(2\pi)^3} e^{ik\cdot\xi} \left\langle P, S | F^{+j}(0) F^{+i}(\xi) | P, S \right\rangle \Big|_{\xi^+=0} . \tag{8}$$

The photon-field tensor $F^{\mu\nu}$ is by itself gauge-invariant; therefore, in this case, we do not need to insert a gauge link operator in the definition (8). This consideration allows us to state straightforwardly that the T-odd photon TMDs (namely $f_{1T}^{\perp\gamma}, h_{1L}^{\perp\gamma}, h_{1T}^{\gamma}$ and $h_{1T}^{\perp\gamma}$) are vanishing, at lest at order α and still neglecting the end-point $x = 1, \mathbf{k}_\perp = \mathbf{0}_\perp$. For the T-even TMDs, instead, one can use the same procedure applied to the case of the electron TMDs.

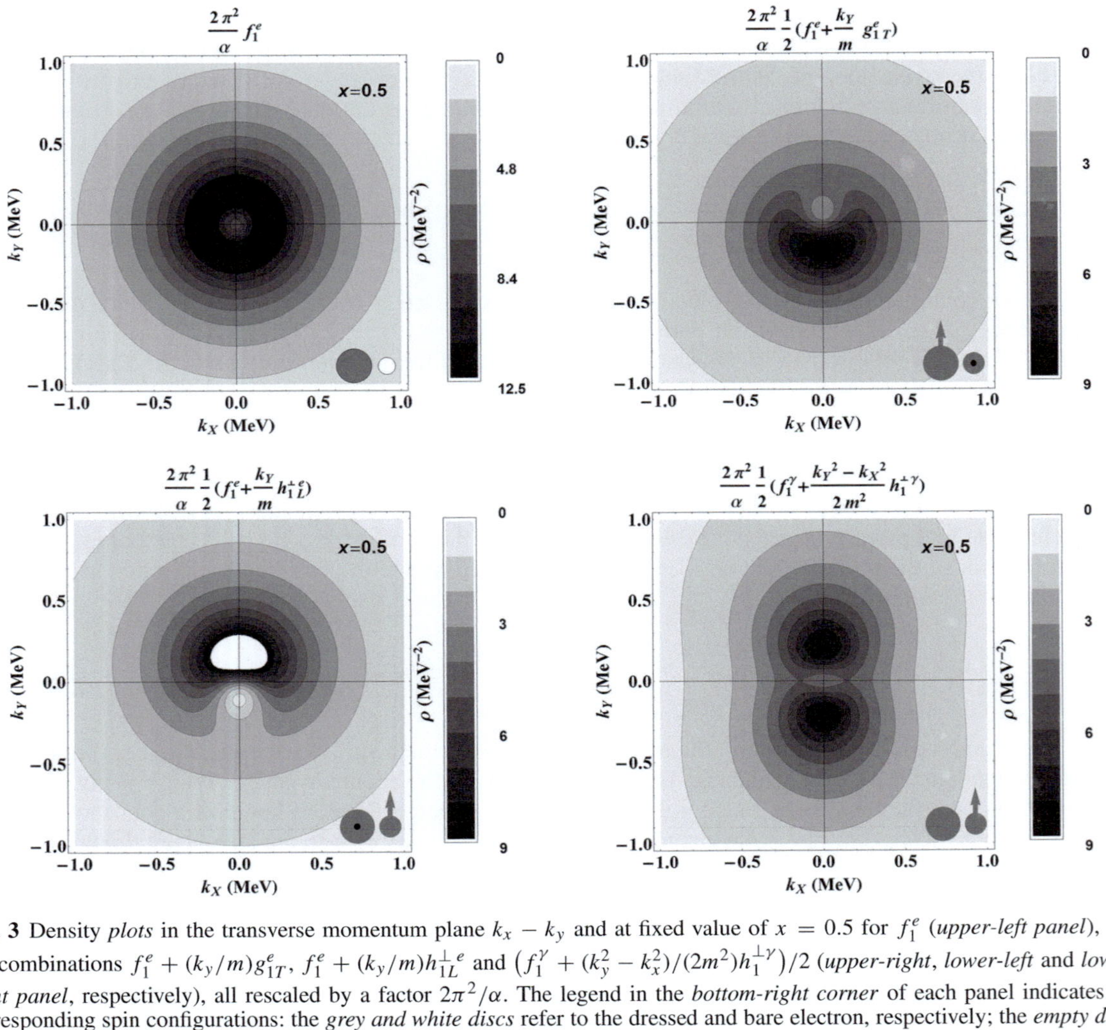

Fig. 3 Density *plots* in the transverse momentum plane $k_x - k_y$ and at fixed value of $x = 0.5$ for f_1^e (*upper-left panel*), and the combinations $f_1^e + (k_y/m)g_{1T}^e$, $f_1^e + (k_y/m)h_{1L}^{\perp e}$ and $\left(f_1^\gamma + (k_y^2 - k_x^2)/(2m^2)h_1^{\perp\gamma}\right)/2$ (*upper-right*, *lower-left* and *lower-right panel*, respectively), all rescaled by a factor $2\pi^2/\alpha$. The legend in the *bottom-right corner* of each panel indicates the corresponding spin configurations: the *grey and white discs* refer to the dressed and bare electron, respectively; the *empty discs* refer to the unpolarized case; the *circle* inside the discs stands for polarization along the longitudinal axes; the *arrows* indicate polarization along the y direction

3 Results and Conclusions

In this section we briefly review some results of the calculations of the electron and photon TMDs for $(x, \mathbf{k}_\perp) \neq (1, \mathbf{0})$ presented in [6]; we recall that these are perturbative-QED exact results at $\mathcal{O}(\alpha)$, hence they are model independent. We can form the following densities for electrons of definite longitudinal or transverse polarization:

$$\rho(x, k_x, k_y, (\lambda, s_\perp), (\Lambda, S_\perp)) = \frac{1}{2}\left[f_1^e + \lambda \Lambda\, g_{1L}^e + \lambda\, S_\perp^i k_\perp^i \frac{1}{m}\, g_{1T}^e + \Lambda\, s_\perp^i k_\perp^i \frac{1}{m}\, h_{1L}^{\perp e} + s_\perp^i\, S_\perp^i\, h_1^e \right], \quad (9)$$

where $s_\perp$ is the transverse spin of the internal electron. As an example, we show in Fig. 3 the densities in the transverse momentum plane, at fixed $x = 0.5$, for the electron TMD f_1^e and for the combinations $f_1^e + (k_y/m)g_{1T}^e$ and $f_1^e + (k_y/m)h_{1L}^{\perp e}$; they can all be interpreted as probability densities of finding the internal electron in a certain point in momentum space, in correspondence of a particular spin configuration. The TMD f_1^e describes the situation where both the internal and the dressed electron are unpolarized: it has a peak when $|\mathbf{k}_\perp|$ is around 0.2–0.3 keV and looks like a ring-shaped image with a radius of about 200 keV. The distributions $f_1^e + (k_y/m)g_{1T}^e$ and $f_1^e + (k_y/m)h_{1L}^{\perp e}$, instead, describe the situation for longitudinally polarized electrons in an electron target transversely polarized along the y direction and transversely polarized electrons in a longitudinally polarized target, respectively. We observe a dipole distortion along the direction of the transverse spin, which deforms the symmetric situation of the unpolarized case. Finally, we show the plot of the Boer–Mulders distribution $h_1^{\perp\gamma}$ for the photon, in the combination $\left(f_1^\gamma + (k_y^2 - k_x^2)/(2m^2)h^{\perp\gamma}\right)/2$, corresponding to the density of photons with linear polarization along the y direction in an unpolarized dressed electron. This distributions shows two peaks, shifted along the polarization direction by approximately ± 200 keV at $x = 0.5$.

This work can be extended further by including the contributions from the endpoints and the treatment of divergences; another possible direction of study would be to identify which observables are sensitive to the TMDs of a dressed electron.

References

1. Meissner, S., Metz, A., Schlegel, M.: Generalized parton correlation functions for a spin-1/2 hadron. JHEP **08**, 056 (2009)
2. Lorcé, C., Pasquini, B., Vanderhaeghen, M.: Unified framework for generalized and transverse-momentum dependent parton distributions within a 3Q light-cone picture of the nucleon. JHEP **05**, 041 (2011)
3. Hoyer, P., Kurki, S.: The transverse shape of the electron. Phys. Rev. D **81**, 013002 (2010)
4. Miller, G.A.: Electron structure: Shape, size, and generalized parton distributions in QED. Phys. Rev. D **90**, 113001 (2014)
5. Brodsky, S.J., Hwang, D.S., Ma, B., Schmidt, I.: Light cone representation of the spin and orbital angular momentum of relativistic composite systems. Nucl. Phys. B **593**, 311–335 (2001)
6. Bacchetta, A., Mantovani, L., Pasquini, B.: Electron in three-dimensional momentum space. Phys Rev. D **93**, 013005 (2016)
7. Pasquini, B., Cazzaniga, S., Boffi, S.: Transverse momentum dependent parton distributions in a light-cone quark model. Phys. Rev. D **78**, 034025 (2008)
8. Mulders, P.J., Tangerman, R.D.: The complete tree level result up to order 1/Q for polarized deep inelastic leptoproduction. Nucl. Phys. B **461**, 197–237 (1996)
9. Cherednikov, I.O., Mertens, T., Van der Veken, F.F.: Wilson Lines in Quantum Field Theory. De Gruyter, Berlin (2014)
10. Boer, D., Mulders, P.J., Pijlman, F.: Universality of T odd effects in single spin and azimuthal asymmetries. Nucl. Phys. B **667**, 201–241 (2003)
11. Lorcé, C., Pasquini, B.: On the origin of model relations among transverse-momentum dependent parton distributions. Phys. Rev. D **84**, 034039 (2011)
12. Brodsky, S.J., Pauli, H., Pinsky, S.S.: Quantum chromodynamics and other field theories on the light cone. Phys. Rept. **301**, 299–486 (1998)
13. Belitsky, A.V., Ji, X., Yuan, F.: Final state interactions and gauge invariant parton distributions. Nucl. Phys. B **656**, 165–198 (2003)

DOI 10.1007/978-3-319-50699-9_28
Reprinted from *Few-Body System* Journal, DOI 10.1007/s00601-016-1067-3

Emanuele Pace · Alessio Del Dotto · Leonid Kaptari · Matteo Rinaldi · Giovanni Salmé · Sergio Scopetta

Light-Front Dynamics and the ^3He Spectral Function

Received: 29 January 2016 / Accepted: 19 February 2016 / Published online: 10 March 2016

Abstract Two topics are presented. The first one is a novel approach for a Poincaré covariant description of nuclear dynamics based on light-front Hamiltonian dynamics. The key quantity is the light-front spectral function, where both normalization and momentum sum rule can be satisfied at the same time. Preliminary results are discussed for an initial analysis of the role of relativity in the EMC effect in ^3He. A second issue, very challenging, is considered in a non-relativistic framework, namely a distorted spin-dependent spectral function for ^3He in order to take care of the final state interaction between the observed pion and the remnant in semi-inclusive deep inelastic electron scattering off polarized ^3He. The generalization of the analysis within the light-front dynamics is outlined.

1 Introduction

The standard model of few-nucleon systems, where nucleon and pion degrees of freedom are taken into account, has achieved a very high degree of sophistication. Nonetheless, one should try to fulfill, as much as possible, the relativistic constraints, dictated by the covariance with respect to the Poincaré group, if processes involving nucleons with large 3-momentum are considered and a high precision is needed. This is the case if one studies, e.g., (1) the nucleon structure functions (unpolarized and polarized); (2) semi-inclusive deep inelastic (SIDIS) processes, (3) signatures of short-range correlations. At least, one should carefully deal with the boosts of the nuclear states, $|\Psi_{init}\rangle$ and $|\Psi_{fin}\rangle$!

In particular a relativistic treatment is important to accurately describe the JLab program @ 12 GeV for few-body systems (see, e.g., [1–3]).

E. Pace (✉)
Dipartimento di Fisica, Università di Roma Tor Vergata, Rome, Italy
Tel: +39-0672594565
Fax: +39-062023507
E-mail: pace@roma2.infn.it

E. Pace
Sezione di Roma Tor Vergata, INFN, Rome, Italy

A. Del Dotto · G. Salmé
Sezione di Roma, INFN, Rome, Italy

L. Kaptari
Bogoliubov Laboratory of Theoretical Physics, JINR, 141980 Dubna, Russia

M. Rinaldi · S. Scopetta
Dipartimento di Fisica e Geologia, Università di Perugia, Perugia, Italy

M. Rinaldi · S. Scopetta
Sezione di Perugia, INFN, Perugia, Italy

Our key tool for a relativistic description of few-body nuclei is a Poincaré covariant spectral function (SF) built up within the light-front Hamiltonian dynamics (LFHD).

Indeed, the Relativistic Hamiltonian Dynamics (RHD) of an interacting system with a fixed number of on-mass-shell constituents (see, e.g., [4]) plus the Bakamijan–Thomas construction of the Poincaré generators [5] allow one to give a fully Poincaré covariant description of DIS, SIDIS and deeply virtual Compton scattering (DVCS) off ^3He. The light-front (LF) form of RHD is adopted [4], which has a subgroup structure of the LF boosts (with a separation of the intrinsic motion from the global one: very important for us!) and a meaningful Fock expansion (see, e.g., [6]). Furthermore, within the LFHD one can take advantage of the whole successful non-relativistic (NR) phenomenology that has been developed for the nuclear interaction.

In Sect. 2 the LF spin-dependent (SD) SF obtained from the LF wave functions for the two- and the three-nucleon systems is described. In Sect. 3 the LF SF is applied to study the role of relativity for the EMC effect in ^3He and preliminary results are presented. In Sect. 4 the actual possibility to get information on the neutron structure from SIDIS experiments on ^3He is shown within a non-relativistic approach which include the final state interaction (FSI) through a distorted SF [7–9]. Preliminary results with the LF SF are reported. In Sect. 5 conclusions and perspectives are drawn.

2 Light-Front Dynamics and the Light-Front Spectral Function

An explicit construction of the 10 Poincaré generators that fulfills the proper commutation rules in presence of interactions was given by Bakamjian and Thomas [5] through the mass operator, M : (1) only M contains the interaction; (2) it generates the dependence upon the interaction of the three dynamical generators in LFHD, namely P^- and the transverse rotations $\mathbf{F}_\perp$. M is obtained adding to the free mass M_0 of the system an interaction V. There are two possibilities: $M^2 = M_0^2 + U$ (then for two particles one can easily embed the NR phenomenology) or $M = M_0 + V$. The interaction, U or V, must commute with all the kinematical generators, and with the non interacting spin. Then it has to be invariant for translations, as in the NR case, and the angular momentum is a conserved quantity.

The full theory must fulfill the macroscopic locality. This property could be implemented by using interaction-dependent, unitary operators: the packing operators (see, e.g., [4]). Their effects should be small, and therefore they will be neglected in what follows, but in principle should be investigated.

For the three-body case the mass operator is $M_{BT}(123) = M_0(123) + V^{BT}$, where $M_0(123) = \sqrt{m^2 + k_1^2} + \sqrt{m^2 + k_2^2} + \sqrt{m^2 + k_3^2}$ is the free mass operator, V^{BT} a Bakamjian-Thomas (BT) two-body and three-body force, and k_i $(i = 1, 2, 3)$ are intrinsic momenta with $\mathbf{k}_1 + \mathbf{k}_2 + \mathbf{k}_3 = 0$ [4].

The NR mass operator is written as $M^{NR} = 3m + \sum_{i=1,3} k_i^2/2m + V_{12}^{NR} + V_{23}^{NR} + V_{31}^{NR} + V_{123}^{NR}$ and must obey the commutation rules proper of the Galilean group, leading to translational and rotational invariance. Those properties are analogous to the ones in the BT construction. This allows us to consider the standard NR mass operator as a sensible BT mass operator, and embed it in a Poincaré covariant approach: $M_{BT}(123) \sim M^{NR}$.

Coupling angular momenta is accomplished within the Instant Form (IF) of RHD through the Clebsch-Gordan coefficients. To embed this machinery in the LFHD one needs the Melosh rotations that relate LF and IF spin wave functions. For a particle of LF momentum $\tilde{k} \equiv \{k^+, \mathbf{k}_\perp\}$ one has

$$|\tilde{\mathbf{k}}; s\sigma\rangle_{LF} = \sum_{\mu} D^{s}_{\mu\sigma}\left[\mathcal{R}^{\dagger}_{M}(\tilde{\mathbf{k}})\right] |\mathbf{k}; s\mu\rangle_{IF} \tag{1}$$

where $D^s_{\sigma,\sigma'}(R^{\dagger}_M(\tilde{\mathbf{k}}))$ is the standard Wigner function and $R_M(\tilde{\mathbf{k}})$ is the rotation between the rest frames of the particle reached through a LF boost or a canonical, rotationless boost.

Then for quantities, like the spin-dependent SF, taking care of the Melosh rotations, one obtains

$$O^{LF}_{\sigma',\sigma} = \sum_{\bar{\sigma}',\bar{\sigma}} D^{1/2}_{\sigma',\bar{\sigma}'}(R_M)\, O^{IF}_{\bar{\sigma}',\bar{\sigma}}\, D^{1/2}_{\bar{\sigma},\sigma}(R^{\dagger}_M)$$

The NR spin-dependent SF for a nucleus of mass number A is defined through its matrix elements

$$P^{\tau}_{\sigma,\sigma',\mathcal{M}}(\mathbf{p}, E) = \sum_{f_{(A-1)}} \langle \mathbf{p}, \sigma\tau; \psi_{f_{(A-1)}} | \psi_{\mathcal{J}\mathcal{M}}\rangle\, \langle \psi_{\mathcal{J}\mathcal{M}} | \psi_{f_{(A-1)}}; \mathbf{p}, \sigma'\tau\rangle\, \delta(E - E_{f_{(A-1)}} + E_A) \tag{2}$$

where $|\psi_{\mathcal{J}\mathcal{M}}\rangle$ is the ground state of the nucleus with energy E_A and polarized along $\mathbf{S}$, $|\psi_{f_{(A-1)}}\rangle$ an eigenstate of the (A-1)-nucleon system with energy $E_{f_{(A-1)}}$, interacting with the same interaction of the nucleus, $|\mathbf{p}, \sigma\tau\rangle$ the plane wave for the nucleon $\tau = \pm 1/2$, with momentum $\mathbf{p}$ in the rest frame and spin along the z-axis equal to σ [10,11].

To obtain within the LFHD a Poincaré-covariant spin-dependent SF for a three-particle system in the bound state $|\Psi_0; S, T_z\rangle$, polarized along $\mathbf{S}$, let us replace the NR overlaps $\langle \mathbf{p}, \sigma\tau; \psi_{f_{(A-1)}}|\psi_{\mathcal{J}\mathcal{M}}\rangle$ with their LF counterparts ${}_{LF}\langle \tau_S, T_S; \alpha, \epsilon; J_z J; \tau\sigma, \tilde{\kappa}|\Psi_0; S, T_z\rangle$, that depend upon the energy ϵ of the system of two fully interacting particles (say, particles 2 and 3) and upon the intrinsic momentum, $\tilde{\kappa}$, of the third particle (say, particle 1) in the intrinsic reference frame of the cluster (1,23). Then, the LF spin-dependent SF for the three-nucleon system (^3He or ^3H) is [12]

$$\mathcal{P}^{\tau}_{\sigma'\sigma}(\xi, \boldsymbol{\kappa}_\perp, \kappa^-, S) = \left|\frac{\partial \kappa^+}{\partial \xi}\right| \sum\!\!\!\!\!\!\int d\epsilon\; \rho(\epsilon)\; \delta\left(\kappa^- - M_3 + \frac{M_S^2 + |\boldsymbol{\kappa}_\perp|^2}{(1-\xi)M_3}\right) \times$$
$$\sum_{J J_z \alpha} \sum_{T_S \tau_S} {}_{LF}\langle \tau_S, T_S, \alpha, \epsilon; J J_z; \tau\sigma', \tilde{\kappa}|\Psi_0; S, T_z\rangle \langle S, T_z; \Psi_0|\tilde{\kappa}, \sigma\tau; J J_z; \epsilon, \alpha, T_S, \tau_S\rangle_{LF} \tag{3}$$

where $\tau = \pm 1/2$, M_3 is the nucleus mass, ϵ the intrinsic energy of the two-nucleon eigenstate, $\rho(\epsilon)$ the corresponding density of states ($\rho(\epsilon) = \sqrt{\epsilon\, m}\, m/2$ for the two-body continuum states and $\rho(\epsilon) = 1$ for the deuteron bound state), J the spin, T_S the isospin of the two-body state, α the set of quantum numbers needed to completely specify this eigenstate, and $M_S = 2\sqrt{m^2 + m\epsilon}$ its mass. From $\xi, M_S, \boldsymbol{\kappa}_\perp$ one can define $\kappa^+ = \xi \mathcal{M}_0(1, 23)$, where $\mathcal{M}_0(1, 23)$ is the free mass of the cluster (1,23)

$$\mathcal{M}_0^2(1, 23) = \frac{m^2 + |\boldsymbol{\kappa}_\perp|^2}{\xi} + \frac{M_S^2 + |\boldsymbol{\kappa}_\perp|^2}{(1-\xi)} \tag{4}$$

The overlap ${}_{LF}\langle \tau_S, T_S; \alpha, \epsilon; J_z J; \tau\sigma, \tilde{\kappa}|\Psi_0; S, T_z\rangle$ is defined as follows [12]

$${}_{LF}\langle \tau_S, T_S, \alpha, \epsilon; J_z J; \tau\sigma, \tilde{\kappa}|\Psi_0; S, T_z\rangle = \sum_{\sigma_1'} D^{\frac{1}{2}}[\mathcal{R}_M^{\dagger}(\tilde{\kappa})]_{\sigma\sigma_1'} \sum_{\tau_2,\tau_3} \int d\mathbf{k}_{23} \sum_{\sigma_2,\sigma_3} \sqrt{\frac{\kappa^+ E_{23}}{k^+ E_S}}$$
$$\times \sqrt{(2\pi)^3\, k^+ \frac{\partial k_z}{\partial k^+}}\; {}_{IF}\langle \tau_S, T_S, \alpha, \epsilon; J_z J|\mathbf{k}_{23}; \sigma_2, \sigma_3; \tau_2, \tau_3\rangle \langle \tau_3, \tau_2, \tau; \sigma_3, \sigma_2, \sigma_1'; \mathbf{k}, \mathbf{k}_{23}|\Psi_0; S, T_z\rangle_{IF} \tag{5}$$

where $\mathbf{k}_{23}$ is the intrinsic momentum of the (23) pair, $\mathbf{k}$ is the intrinsic nucleon momentum in the (123) system ($\mathbf{k}_\perp = \boldsymbol{\kappa}_\perp$, since we choose the ^3He transverse momentum $\mathbf{P}_\perp = 0$), $k^+ = \xi\, M_0(123) = \kappa^-\, M_0(123)/\mathcal{M}_0(1, 23)$, with $M_0(123)$ the free mass of the three-particle system

$$M_0^2(123) = \frac{m^2 + |\mathbf{k}_\perp|^2}{\xi} + \frac{M_{23}^2 + |\mathbf{k}_\perp|^2}{(1-\xi)} \tag{6}$$

and $M_{23}^2 = 4(m^2 + |\mathbf{k}_{23}|^2)$ the mass of the spectator pair without interaction ! In Eq. (5) one has $k_z = \frac{1}{2}\left[k^+ - (m^2 + |\boldsymbol{\kappa}_\perp|^2)/k^+\right]$, $E_{23} = \sqrt{M_{23}^2 + |\mathbf{k}|^2}$ and $E_S = \sqrt{M_S^2 + |\boldsymbol{\kappa}|^2}$.

In the actual calculations, we identified the IF overlaps of Eq. (5) with the NR wave functions for the two-nucleon and the three-nucleon [13] systems, corresponding to the NN interaction AV18 [14].

3 Light-Front Momentum Distribution and Preliminary Results for the EMC Effect

From the LF SF one can obtain the light-cone momentum distribution $f^A_{p(n)}(z)$

$$f^A_\tau(z) = \int_0^1 d\xi \int d\boldsymbol{\kappa}_\perp \int d\kappa^- \frac{1}{2(2\pi)^3\kappa^+}\, Tr\left[\mathcal{P}^\tau(\xi, \mathbf{k}_\perp, \kappa^-, S)\right]\, \delta\left(z - \frac{\xi M_A}{m}\right) \tag{7}$$

that fulfills the following relations, given the SF normalization

$$\int_0^{M_A/m} dz\, f_\tau^A(z) = 1 \qquad N_A = \frac{1}{A}\int_0^{M_A/m} dz \left[Z f_p^A(z) + (A-Z) f_n^A(z) \right] = 1 \tag{8}$$

The symmetry of the three-body bound state naturally entails the momentum sum rule (see [12])

$$MSR = \frac{1}{A}\int_0^{M_A/m} dz\, z \left[Z f_p^A(z) + (A-Z) f_n^A(z) \right] = \frac{1}{A} \tag{9}$$

We evaluated the nuclear structure function $F_2^A(x)$ ($x = Q^2/2m\nu$) as a convolution of the nuclear spectral function and of the structure function of the nucleon

$$F_2^A(x) = \int_0^1 d\xi \int d\boldsymbol{\kappa}_\perp \int \frac{d\kappa^-}{2(2\pi)^3\kappa^+} \left[Z Tr\, \mathcal{P}^p(\xi, \boldsymbol{\kappa}_\perp, \kappa^-) F_2^p\left(\frac{x}{\xi}\right) + N Tr\, \mathcal{P}^n(\xi, \boldsymbol{\kappa}_\perp, \kappa^-) F_2^n\left(\frac{x}{\xi}\right) \right] \tag{10}$$

Then, to investigate whether the LF SF can affect the EMC effect, the ratios

$$R_2^A(x) = \frac{F_2^A(x)}{Z\, F_2^p(x) + (A-Z)\, F_2^n(x)} \tag{11}$$

and $R_2^{He}(x)/R_2^D(x)$ were evaluated. The Pisa group wave function [13], corresponding to the NN interaction AV18 [14], was used. For the two-body channel an exact calculation was performed. In the three-body channel average values for k_{23} were inserted in Eq. (5), different for the proton and the neutron (113.53 MeV and 91.27 MeV, respectively), corresponding to the average kinetic energy of the intrinsic motion of the spectator pair in the continuum part of the spectrum. We checked in the two-body channel that using the average value $< k_{23} >= 136.37$ MeV, proper for this channel, a result similar to the exact calculation is obtained. Our preliminary results are shown in Fig. 1 and encourage us in performing the full LF calculation.

With the same approximation used for the three-body channel, we checked that $MSR_{calc} = 0.3331$.

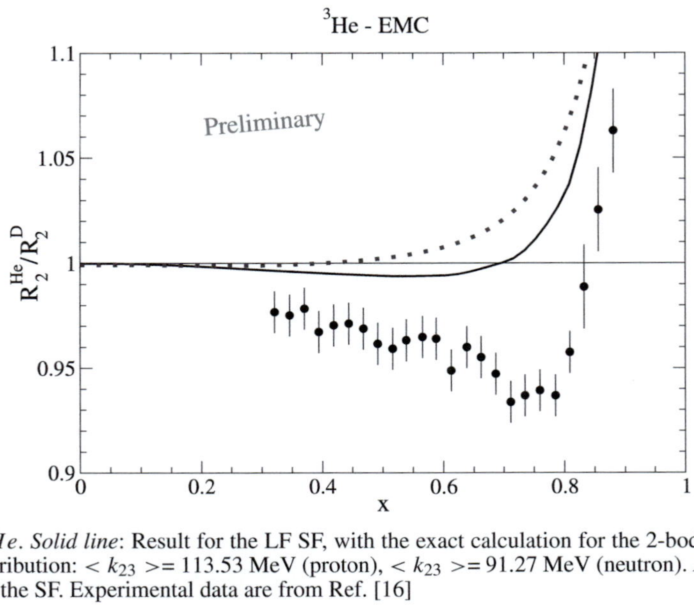

Fig. 1 EMC effect for ^3He. *Solid line*: Result for the LF SF, with the exact calculation for the 2-body channel, and an average energy in the 3-body contribution: $< k_{23} >= 113.53$ MeV (proton), $< k_{23} >= 91.27$ MeV (neutron). *Dotted line* Result with the approach of Ref. [15] for the SF. Experimental data are from Ref. [16]

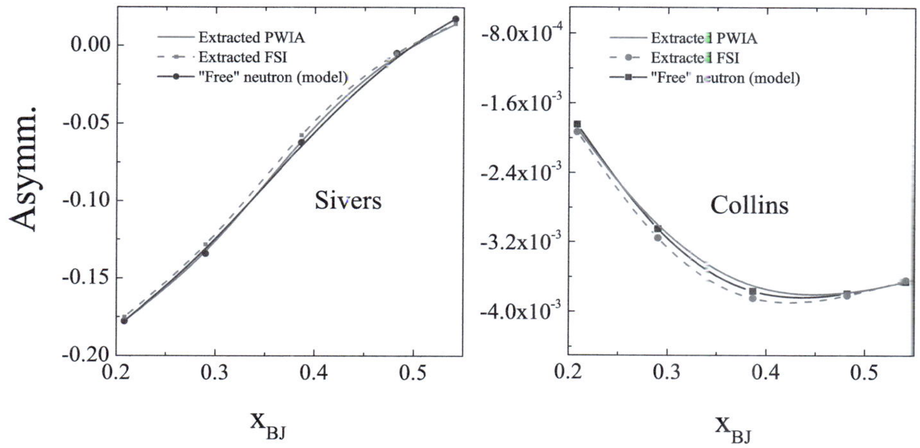

Fig. 2 Neutron asymmetries extracted through Eq. (12), from the Sivers (*left panel*) and Collins (*right panel*) asymmetries, with and without FSI, in the actual kinematics of JLab [2]. Preliminary results to appear in [9]

4 Extraction of Neutron Asymmetries from SIDIS Experiments Off ^3He

In Ref. [17] it was shown, within the plane wave impulse approximation (no interaction, after the electron scattering off one of the ^3He nucleons, between the measured fast π, the remnant debris and the interacting two-nucleon recoiling system) and using the NR SF of Ref. [10], that the formula [18]

$$A_n \simeq \frac{1}{p_n f_n} \left(A_3^{exp} - 2p_p f_p A_p^{exp}\right) , \tag{12}$$

already widely used to extract neutron asymmetries in DIS from experiments on ^3He, works also in SIDIS, both for the Collins and Sivers single spin asymmetries. Nuclear effects are hidden in the effective polarizations (EP) $p_p = -0.023$ and $p_n = 0.878$ and in the dilution factors, $f_{p(n)}$.

To investigate whether the formula (12) can be safely applied even in presence of the FSI, in Ref. [9] a generalized eikonal approximation (GEA) is used to take care of the FSI through a distorted spin-dependent SF in a NR approach. The relevant part of the (GEA-distorted) spin-dependent SF is

$$\mathcal{P}_{||}^{FSI} = \mathcal{O}_{\frac{1}{2}\frac{1}{2}}^{FSI} - \mathcal{O}_{-\frac{1}{2}-\frac{1}{2}}^{FSI} \qquad \mathcal{O}_{\lambda\lambda'}^{FSI}(p_N, E) =$$

$$= \sum\!\!\!\!\!\!\int_{\epsilon^*_{A-1}} \rho\left(\epsilon^*_{A-1}\right) \langle S_A, \mathbf{P_A}|(\hat{S}_{Gl})\{\Phi_{\epsilon^*_{A-1}}, \lambda', \mathbf{p}_N\}\rangle \langle(\hat{S}_{Gl})\{\Phi_{\epsilon^*_{A-1}}, \lambda, \mathbf{p}_N\}|S_A, \mathbf{P_A}\rangle \delta\left(E - B_A - \epsilon^*_{A-1}\right) \tag{13}$$

where $\hat{S}_{Gl}(\mathbf{r}_1, \mathbf{r}_2, \mathbf{r}_3) = \prod_{i=2,3}\left[1 - \theta(z_i - z_1)\Gamma(\mathbf{b}_1 - \mathbf{b}_i, z_1 - z_i)\right]$ is a Glauber operator [19] which takes care of hadronization and FSI. The model of Ref. [20] for the (generalized) profile function $\Gamma(\mathbf{b}, z)$, already successfully applied to $^2H(e, e'p)X$ [21], is adopted.

It occurs that effects of GEA-FSI in the dilution factors and in the EP compensate each other to a large extent: $p_p^{FSI} f_p^{FSI} \approx p_p f_p$, $p_n^{FSI} f_n^{FSI} \approx p_n f_n$ [8]. Then the usual extraction of Eq. (12) is safe, as shown at $E_i = 8.8$ GeV in Fig. 2.

Relativity effects seem weak, since LF longitudinal and transverse EPs change little from the ones obtained with the NR SF [22]. Then the usual extraction procedure should work well also within the LF approach. Concerning FSI, we plan to include in our LF description the FSI between the jet produced from the hadronizing quark and the two-nucleon spectator system through the GEA of [21].

5 Conclusions and Perspectives

A Poincaré covariant description of A = 3 nuclei, based on the LF Hamiltonian dynamics, has been proposed. The Bakamjian-Thomas construction of the Poincaré generators allows one to embed the successful NR phenomenology for few-nucleon systems in a Poincaré covariant framework. A LF SF can be defined that exactly fulfills both the normalization and the momentum sum rule. The nucleon SF for ^3He, has been evaluated by approximating the IF overlaps in Eq. (5) with their NR counterparts, calculated with the AV18 NN interaction, since fulfills rotational and translational symmetries.

A first test of our approach is the EMC effect for ^3He. The 2-body contribution to the nucleon SF has been calculated with the full expression, while for the 3-body contribution an average value for $\langle k_{23}^2 \rangle$ has been used. In the comparison with experimental data encouraging improvements clearly appear with respect to the non-relativistic result. The next step will be the full calculation of the EMC effect for ^3He, including the exact 3-body contribution.

An investigation of SIDIS processes off ^3He beyond the NR, impulse approximation (IA) approach is presently being carried out. A Generalized Eikonal Approximation has been used to deal with the FSI effects and a distorted spin-dependent spectral function, still non relativistic, has been defined. It has been shown that the formula (12) can be safely used to obtain both the Collins and Sivers neutron asymmetries from the measured Collins and Sivers asymmetries off ^3He.

Preliminary promising results (in IA) of the relativistic effects in SIDIS processes were obtained through an evaluation of the LF effective polarizations. The next step will be the introduction of the FSI through the GEA within the LFHD.

References

1. MARATHON Coll. E12-10-103 (Rating A): MeAsurement of the F_2^n/F_2^p, d/u Ratios and A=3 EMC Effect in Deep Inelastic Electron Scattering Off the Tritium and Helium MirrOr Nuclei, http://hallaweb.jlab.org/12GeV/
2. Gao, H. et al.: PR12-09-014 (Rating A), Target Single Spin Asymmetry in Semi-Inclusive Deep-Inelastic (e, e'$\pi^{\pm}$) Reaction on a Transversely Polarized ^3He Target, http://hallaweb.jlab.org/12GeV/; G. Cates et al., E12-09-018, JLAB approved experiment, http://hallaweb.jlab.org/collab/PAC/PAC38//E12-09-018-SIDIS.pdf; J.P. Chen et al., PR12-11-007 (Rating A): Asymmetries in Semi-Inclusive Deep-Inelastic (e, e'$\pi^{\pm}$) Reactions on a Longitudinally Polarized ^3He Target
3. Chen, J.-P., Jiang, X., Peng, J.-C.: E-06-010 Proposal to JLab-PAC29. In: Cisbani, E., Gao, H., Jiang, X. (eds.) E-06-011 Proposal to JLab-PAC29, Measurement of Single Target-Spin Asymmetry in Semi-Inclusive Pion Electroproduction on a Transversely Polarized ^3He Target
4. Keister, B.D., Polyzou, W.N.: Relativistic Hamiltonian dynamics in nuclear and particle physics. Adv. Nucl. Phys. **21**, 225–550 (1991)
5. Bakamjian, B., Thomas, L.H.: Relativistic particle dynamics. II*. Phys. Rev. **92**, 1300–1310 (1953)
6. Brodsky, S.J., Pauli, H.C., Pinsky, S.S.: Quantum chromodynamics and other field theories on the light cone. Phys. Rep. **301**, 299–486 (1998)
7. Kaptari, L., Del Dotto, A., Pace, E., Rinaldi, M., Salmé, G., Scopetta, S.: Distorted spin-dependent spectral function of an A=3 nucleus and semi-inclusive deep inelastic scattering processes. Phys. Rev. C **89**, 035206-1–035206-18 (2014)
8. Del Dotto, A., Kaptari, L., Pace, E., Salmé, G., Scopetta, S.: Towards an improved description of SiDIS by a polarized ^3He target. Few Body Syst. **55**, 877–880 (2014)
9. Kaptari, L., Del Dotto, A., Pace, E., Salmé G., Scopetta, S.: to be published
10. Ciofi degli Atti, C., Pace, E., Salmè, G.: Spin dependent spectral function of ^3He and the asymmetry in the process polarized ^3He (polarized e, e') X. Phys. Rev. C **46**, R1591–R1594 (1992)
11. Ciofi degli Atti, C., Pace, E., Salmè, G.: Investigation of the neutron form factors by inclusive quasielastic scattering of polarized electrons off polarized ^3He: a theoretical overview. Phys. Rev. C **51**, 1108–1119 (1995)
12. Del Dotto, A., Pace, E., Salmé, G., Scopetta, S.: Light-Front Spectral Function and Nucleon Momentum Distributions for a Three-Body System. To be published
13. Kievsky, A., Viviani, M., Rosati, S.: Study of bound and scattering states in three-nucleon systems. Nucl. Phys. A **577**, 511–527 (1994)
14. Wiringa, R.B., Stocks, V.G.J., Schiavilla, R.: Accurate nucleon–nucleon potential with charge-independence breaking. Phys. Rev. C **51**, 38–51 (1995)
15. Oelfke, U., Sauer, P., Coester, F.: Convolution models of deep inelastic scattering: the three-nucleon bound state as a test case. Nucl. Phys. A **518**, 593–616 (1990)
16. Seely, J. et al.: New measurements of the European muon collaboration effect in very light nuclei. Phys. Rev. Lett. **103**, 202301-1–202301-5 (2009)
17. Scopetta, S.: Neutron single spin asymmetries from semi-inclusive deep inelastic scattering off transversely polarized ^3He. Phys. Rev. D **75**, 054005-1–054005-10 (2007)
18. Ciofi degli Atti, C., Pace, E., Salmé, G., Scopetta, S.: Nuclear effects in deep inelastic scattering of polarized electrons off polarized ^3He and the neutron spin structure functions. Phys. Rev. C **48**, R968–R972 (1993)
19. Kopeliovich, B.Z., Nemchik, J., Predazzi, E., Hayashigaki, A.: Nuclear hadronization: within or without?. Nucl. Phys. A **740**, 211–245 (2004)
20. Ciofi degli Atti, C., Kopeliovich, B.Z.: Final state interaction in semiinclusive DIS off nuclei. Eur. Phys. J. **A17**, 133–144 (2003)
21. Ciofi degli Atti, C., Kaptari, L.: Semi-inclusive deep-inelastic scattering off few-nucleon systems: Tagging the EMC effect and hadronization mechanisms with detection of slow recoiling nuclei. Phys. Rev. C **83**, 044602-1–044602-8 (2011)
22. Pace, E., Del Dotto, A., Rinaldi, M., Salmé, G., Scopetta, S.: Neutron transverse-momentum distributions and polarized ^3He within light-front Hamiltonian dynamics. Few Body Syst. **54**, 1079–1082 (2013)

DOI 10.1007/978-3-319-50699-9_29
Reprinted from *Few-Body System* Journal, DOI 10.1007/s00601-016-1068-2

F. F. Van der Veken · N. G. Stefanis

Identifying the Universal Part of TMDs

Received: 17 January 2016 / Accepted: 22 February 2016 / Published online: 8 April 2016

Abstract We attempt to identify a path layout in the definition of transverse-momentum-dependent T-odd parton distribution functions (TMD)s which combines features of both, initial- and final-state interactions, so that it remains universal despite the fact that the Wilson lines entering such TMDs change their orientation. The generic structure of the quark correlator for this path layout is calculated.

1 Introduction

Collinear factorization is a fully grown and well understood framework to describe inclusive processes like deep inelastic scattering (DIS). The hadron is portrayed by a parton distribution function (PDF) which only depends on the longitudinal momentum fraction x shared by the struck quark, as the transversal momentum components are integrated out. Factorization is then achieved by convoluting the PDF with the hard part of the interaction and integrating over x, see [1] for a recent review.

The collinear framework is however less suitable to describe high-energy particle collisions that are not fully inclusive. This is, because in this case one has to take into account the transversal part of the quark momentum as well. This enlarged framework is based on $\boldsymbol{k}_\perp$-factorization and employs PDFs which depend on the transverse momentum—thus transverse-momentum-dependent PDFs or TMDs for short (see [2] for an exhaustive review). Exclusive processes are extremely hard to treat because one deals with the full nonperturbative quark-gluon dynamics which cannot be computed from first principles and has to be modeled. However, for some processes which are semi-inclusive and the set of the final states is not fully integrated out but is neither fully identified, factorization theorems have been proven. Among them, the most well known example is semi-inclusive deep inelastic scattering (SIDIS), which has been thoroughly treated in [3–5]. However, a few years later it was realized due to the non-vanishing Sivers effect [6] that the presence of the Wilson line inside the TMD definition should be taken into consideration [7]. The reason is that the Wilson line entails the existence of T-odd TMDs, breaking the universality of the TMD. Indeed, while in SIDIS the Wilson line is future-pointing, in initial-state processes like the Drell–Yan (DY), where a produced virtual photon decays into a muon pair, the Wilson line is past-pointing, leading to an overall sign change for T-odd TMDs [8–10]. This came as a surprise because

F. F. Van der Veken (✉)
Departement Fysica, Universiteit Antwerpen, 2020 Antwerp, Belgium
E-mail: frederikvanderveken@gmail.com

Present Address:
F. F. Van der Veken
CERN, Meyrin Site, Building 9, Room 1-012, 1211 Geneve 23, Switzerland

N. G. Stefanis
Institut für Theoretische Physik II, Ruhr-Universität Bochum, 44780 Bochum, Germany
E-mail: stefanis@tp2.ruhr-uni-bochum.de

the universality of the TMD was presumed taking recourse to the stringent validity of collinear factorization in DIS.

This important finding notwithstanding, it appears attractive to ask whether it is possible to create a path layout that combines features of both initial and final state interactions. Such a geometrical structure of Wilson lines could eventually be associated with a common and hence universal part of the SIDIS and the DY processes. This paper is devoted to the conceptual investigation of this issue. The treatment of the renormalization issues of the correlators, in particular questions related to the geometrical features of the Wilson lines—the appearance of cusps and their anomalous dimensions—is relegated to the literature (see, for instance, [2,11–13] and references cited therein). A more general assessment and a fuller treatment of the present approach will be given elsewhere. The work is organized as follows. In the next section, we recall the structure of the gauge-invariant quark correlator. The new path layout is discussed in Sec. 3, whereas the generic quark correlator employing the new path layout is considered in Sec. 4. Finally, Sec. 5 contains our main conclusions.

2 Gauge-Invariant Quark Correlator

The key quantities to describe the three-dimensional dynamics of partons within hadrons, e.g., a proton, in high-energy collisions is provided by TMD distribution and fragmentation functions. Typically, these quantities depend on a large scale, say, Q^2, which can be the invariant mass in a Drell–Yan process, and $q_\perp^2 \ll Q^2$, the transverse momentum of the lepton pair in the same process (with analogous definitions for SIDIS). The generic form of the TMD quark correlator is given by

$$\Phi_{ij}^{q[C]}(x, \mathbf{k}_\perp; n) \sim \int \frac{dz^- d^2\mathbf{z}_\perp}{(2\pi)^3} e^{-ik^+z^- + ik_\perp \cdot z_\perp} \langle p|\bar{\psi}_j(z)\mathcal{U}(0 \to z|C)\psi_i(0)|p\rangle_{z^+=0}, \tag{1}$$

where the Wilson line to ensure gauge invariance is a path-ordered exponential of the form

$$\mathcal{U}(a \to b|C) = \mathcal{P}\exp\left[ig\int_a^b dz \cdot A(z)\right] \tag{2}$$

and the path C has to be taken along the color flow from a to b for each considered process.

As already mentioned, for processes with final-state interactions (SIDIS) the path wraps around at $+\infty$ (left panel of Fig. 1), whereas for processes with initial-state interactions (DY process)—right panel of Fig. 1—the path wraps around $-\infty$. It can be shown [14–16] that changing the endpoint of a linear Wilson-line segment from $+\infty$ to $-\infty$ (or vice versa) amounts to the substitution $n \to -n$ (where n is the directional vector of the segment), i.e., $\mathcal{U}(a \to +\infty|n) = \mathcal{U}(a \to -\infty|-n)$. Because the Wilson line is the only factor with an n-dependence, this relation is repeated in the correlator:

$$\Phi_{ij}^{q[C]}(x, \mathbf{k}_\perp; n)\Big|_{\text{DY}} = \Phi_{ij}^{q[C]}(x, \mathbf{k}_\perp; -n)\Big|_{\text{SIDIS}}. \tag{3}$$

The fully unintegrated quark correlator can be parameterized by a set of scalar functions that are, after the k^--integration, related to their TMD counterparts [17]. The Lorentz structure of the correlator will only depend on contractions of the vectors p^μ, k^μ, S^μ and n^μ with themselves and with the Dirac basis. In the case of an unpolarized proton the fully unintegrated correlator can be cast in the form

$$\begin{aligned}\Phi(p,k,n) = {} & MA_1 + \gamma_\mu\left(p^\mu A_2 + k^\mu A_3 + \frac{M^2}{p\cdot n}n^\mu B_1\right) + \gamma_\mu\gamma_5\frac{1}{p\cdot n}\varepsilon^{\mu\nu\rho\sigma}n_\nu p_\rho k_\sigma B_4 \\ & + i\gamma_{\mu\nu}\left[\frac{1}{M}p^\mu k^\nu A_4 + \frac{M}{p\cdot n}\left(p^\mu B_2 + k^\mu B_3\right)n^\nu\right].\end{aligned} \tag{4}$$

The important thing to notice here is that the directional vector n_μ appears *scaleless*, i.e., only via factors of the form $\frac{n^\mu}{p\cdot n}$, being straightforward to show that this remains true in the polarized case as well. As the k^--integration does not influence n^μ, it is tempting to conclude that the TMD correlator is invariant under a sign change in n, and hence *universal* by virtue of Eq. (3). However, one should not forget that the scalar functions themselves can still depend on n^μ. Moreover, it has been demonstrated that those functions that are

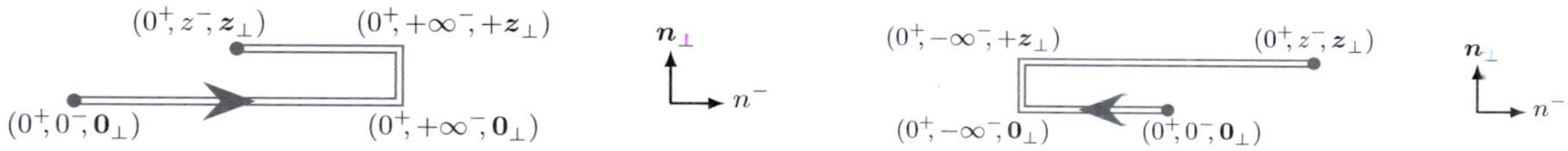

Fig. 1 Path layouts pertaining to SIDIS (*left*) and the Drell–Yan process (*right*)

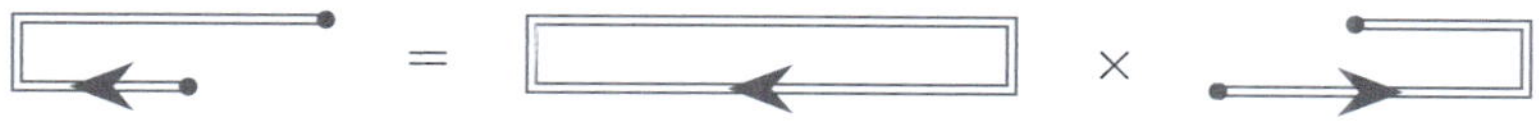

Fig. 2 Relation between the DY path layout (*left-hand side*) and that in SIDIS (*right-hand side*)

odd under time reversal flip sign when interchanging past-pointing with future pointing Wilson lines [7]. This can be summarized in terms of the following relations

$$f^{\rm DY}_{\text{T-even}} = f^{\rm SIDIS}_{\text{T-even}}, \qquad f^{\rm DY}_{\text{T-odd}} = -f^{\rm SIDIS}_{\text{T-odd}}. \tag{5}$$

3 New Path Layout

In what follows we will try to construct a path layout that encapsulates features of both initial- and final-state interactions in a single frame. First we note that due to the transitivity property of a Wilson line, the layouts of SIDIS and DY can be easily related to one another (see Fig. 2):

$$\mathcal{U}^{\rm DY} = \mathcal{U}^{\rm loop}\, \mathcal{U}^{\rm SIDIS}, \tag{6}$$

where $\mathcal{U}^{\rm loop}$ is a loop with infinite longitudinal length (in the light-cone minus direction), and with finite transversal width (equal to $\mathbf{z}_\perp$). The overlapping parts cancel out on the right-hand side of Eq. (6), resulting in the DY path layout. However, this relation is not reproduced at the level of the correlator $\Phi^{\rm DY} \neq \Phi^{\rm loop}\, \Phi^{\rm SIDIS}$ because of the intricacies related to the splitting of expectation values, so that in general

$$\langle \mathcal{U}^{\rm loop}\, \mathcal{U}^{\rm SIDIS} \rangle \neq \langle \mathcal{U}^{\rm loop} \rangle \langle \mathcal{U}^{\rm SIDIS} \rangle. \tag{7}$$

We could at most hope that this relation holds up to power corrections in α_s and/or $1/N_c$. Instead, we split the loop contour at transverse infinity into parts, each in turn assigned to one of the two considered path layouts so that one obtains the following configuration:

(8)

(9)

The emerging layout represents a new contour pattern which can be conceived of as being the "golden mean" between the SIDIS and the DY path layouts. Its structure is illustrated in Fig. 3: segments 1 and 2 correspond to the part on the left of the final-state cut in the DY layout, while segments 4 and 5 refer to the part on the right of the cut in the SIDIS layout. Segment 3 vanishes in Lorentz and axial gauges, but is needed for gauge invariance and in order to ensure that the full layout reduces to the regular PDF layout after performing the $\mathbf{k}_\perp$-integration (a simple straight line). Let us emphasize that we do not pretend that the correlator built from this layout should be considered as a physical quantity. It is merely a mathematical construct out of academic interest, trying to represent the core universal part of the regular quark correlator. If we need to introduce a final-state cut, we draw it somewhere through segment 3, the exact splitting point being irrelevant for practical calculations.

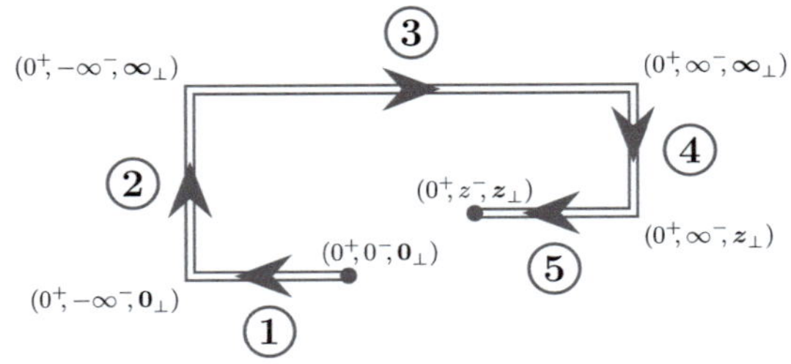

Fig. 3 The proposed new path layout for the Wilson line

Table 1 Values of the parameters n_1, r_2, r_4 and n_5 entering the various considered correlators and pertaining to two different gauges

	$\tilde{\Phi}$	Φ^{SIDIS}	Φ^{DY}
Feynman	$n_1 = -n^-$ $n_5 = -n^-$	$n_1 = +n^-$ $n_5 = -n^-$	$n_1 = -n^-$ $n_5 = +n^-$
Light-cone	$r_2 = -\infty^-$ $r_4 = +\infty^-$	$r_2 = +\infty^-$ $r_4 = +\infty^-$	$r_2 = -\infty^-$ $r_4 = -\infty^-$

The double lines denote Wilson lines

4 Generic Quark Correlator for the New Path Layout

The next step is to retrieve the quark correlators for SIDIS and DY from our new correlator. We denote the latter, i.e., the correlator constructed from our new path layout, as $\tilde{\Phi}$, so that we can write the original correlators as a convolution between a (process-dependent) correction factor and $\tilde{\Phi}$:

$$\Phi^{\text{SIDIS}} = C^{\text{SIDIS}} \otimes \tilde{\Phi}, \qquad \Phi^{\text{DY}} = C^{\text{DY}} \otimes \tilde{\Phi}, \tag{10}$$

where the C^i are expected to be perturbative quantities. In leading order, all Wilson lines reduce to unity and hence $\Phi^{\text{SIDIS}} = \Phi^{\text{DY}} = \tilde{\Phi}$. This implies that the correction factors at the leading-order level are just delta functions:

$$C_0^{\text{SIDIS}} = C_0^{\text{DY}} = \delta(1-x)\delta^{(2)}(\mathbf{k}_\perp). \tag{11}$$

A fruitful approach to continue to higher orders, is to focus on parameters which are peculiar to each of the three correlators, Φ^{SIDIS}, Φ^{DY}, $\tilde{\Phi}$. In Feynman gauge (or another Lorentz, i.e., covariant gauge), only segments 1 and 5 in Fig. 3 survive. What changes across different correlators is the orientation of these segments, as depicted graphically in Table 1. To maximize reproducibility, we assign different directional vectors n_1 and n_5 to these segments. This allows the retrieval of each correlator by a mere substitution of $n_{1,5}$. Similarly, in the light-cone gauge (or any other axial gauge), only segments 2 and 4 contribute, and the differing parameter is the transversal endpoint r^-. Hence, we assign different endpoints r_2 and r_4 to these segments.

In what follows, we perform our investigation of the parametric structure in Feynman gauge; the extension to the light-cone gauge is, however, trivial. Expanding Eq. (10) in terms of powers of α_s, it is easy to show that the first-order correction factors depend on the relevant substitution differences:

$$C_1^{\rm SIDIS} \sim \Phi_1\big|_{n_1\to -n_1} - \Phi_1, \qquad C_1^{\rm DY} \sim \Phi_1\big|_{n_5\to -n_5} - \Phi_1, \tag{12}$$

where Φ is the generic correlator, i.e., with both n_1 and n_5 left unspecified. Now to continue, we note that the generic correlator in leading order of α_s consists of three parts: a term not depending on the directional vector, and terms "even" or "odd" in n:[1]

$$\Phi = \Phi_0 + \alpha_s\left(A + \frac{n_1+n_5}{2}B + \frac{n_1-n_5}{2}C\right) + \mathcal{O}(\alpha_s^2), \tag{13}$$

where A, B, and C are the same for all three correlators. Comparing this generic form with Table 1, we observe that

$$\tilde{\Phi} = \Phi_0 + \alpha_s\left(A - n^- B\right) + \mathcal{O}(\alpha_s^2), \qquad C_1^{\rm SIDIS} = -\alpha_s n^-(B+C), \tag{14a}$$

$$\Phi^{\rm SIDIS,DY} = \Phi_0 + \alpha_s\left(A \pm n^- C\right) + \mathcal{O}(\alpha_s^2), \qquad C_1^{\rm DY} = -\alpha_s n^-(B-C). \tag{14b}$$

It is important to realize that A is the only term that is T-even, with both B and C being T-odd. If B would vanish, then $\tilde{\Phi}$ would be completely universal. A deep investigation of this term would help to shed light on the universality properties of $\tilde{\Phi}$ and TMDs in more general terms.

5 Conclusions

We have constructed a new path layout, which combines features of both initial- and final-state interactions entering the calculation of quark correlators in SIDIS and DY-type processes making use of Wilson lines to preserve gauge invariance. The ensuing generic correlator provides a theoretical tool for the investigation of the universality properties of the original correlators considered in the SIDIS and Drell–Yan processes in connection with particular process-dependent contours [17,18]. The proposed path layout reveals universal features and enhances the insight into the universality properties of the known correlators. Adopting an approach which employs two parameters, we were able to perform explicit calculations of these correlators in the Feynman and also in the light-cone gauge and demonstrate the usefulness of the proposed layout. A more general investigation of the universal features of TMDs, which includes the renormalization properties of the correlators, will be given elsewhere.

Acknowledgments F. F. Van der Veken gratefully acknowledges support from the Gary T. McCartor Fund Fellowship and thanks the Research Foundation-Flanders (FWO) for a travel grant. He is also thankful to the members of the Institute for Theoretical Physics II of Bochum University for their warm hospitality during a research stay when most part of this investigation was carried out.

References

1. Brambilla, N. et al.: QCD and strongly coupled gauge theories: challenges and perspectives. Eur. Phys. J. C **74**, 2981 (2014)
2. Boer, D., et al.: Gluons and the Quark Sea at High Energies: Distributions, Polarization, Tomography. arXiv:1108.1713 [nucl-th]
3. Mulders, P.J., Tangerman, R.D.: The complete tree level result up to order 1/Q for polarized deep inelastic leptoproduction. Nucl. Phys. B **461**, 197 (1996); Erratum: ibid. B **484**, 538 (1997)
4. Boer, D., Mulders, P.J.: Time reversal odd distribution functions in leptoproduction. Phys. Rev. D **57**, 5730 (1998)
5. Boer, D., Jakob, R., Mulders, P.J.: Angular dependences in electroweak semi-inclusive leptoproduction. Nucl. Phys. B **564**, 471 (2000)
6. Brodsky, S.J., Hwang, D.S., Schmidt, I.: Final state interactions and single spin asymmetries in semiinclusive deep inelastic scattering. Phys. Lett. B **530**, 99 (2002)
7. Collins, J.C.: Leading twist single transverse-spin asymmetries: Drell–Yan and deep inelastic scattering. Phys. Lett. B **536**, 43 (2002)
8. Ji, X.-d., Yuan, F.: Parton distributions in light cone gauge: Where are the final state interactions?. Phys. Lett. B **543**, 66 (2002)
9. Belitsky, A.V., Ji, X., Yuan, F.: Final state interactions and gauge invariant parton distributions. Nucl. Phys. **656**, 165 (2003)

[1] No other terms appear at the first-order level due to $(n_1)^2 = (n_5)^2 = n_1 \cdot n_5 = 0$. One can keep the "even/odd" distinction at higher orders using the *sign* of n_1 and n_5 instead of employing the vectors themselves.

10. Boer, D., Mulders, P.J., Pijlman, F.: Universality of T odd effects in single spin and azimuthal asymmetries. Nucl. Phys. B **667**, 201 (2003)
11. Collins, J.C.: Foundations of Perturbative QCD. Cambridge University Press, Cambridge (2011)
12. Cherednikov, I.O., Stefanis, N.G.: Wilson lines and transverse-momentum dependent parton distribution functions. A renormalization-group analysis. Nucl. Phys. **802**, 146 (2008)
13. Stefanis, N.G.: Worldline techniques and QCD observables. Acta Phys. Polon. Suppl. **6**, 71 (2013)
14. Van der Veken, F.F.: Working with Wilson lines. Int. J. Mod. Phys. Conf. Ser. (2015); in print
15. Van der Veken, F.F.: A new approach to piecewise linear Wilson lines. Int. J. Mod. Phys. Conf. Ser. **37**, 1560029 (2015)
16. Van der Veken, F.F.: Piecewise linear Wilson lines. EPJ Web Conf. **85**, 02004 (2015)
17. Goeke, K., Metz, A., Schlegel, M.: Parameterization of the quark–quark correlator of a spin-1/2 hadron. Phys. Lett. B **618**, 90 (2005)
18. Bacchetta, A., Mulders, P.J., Pijlman, F.: New observables in longitudinal single-spin asymmetries in semi-inclusive DIS. Phys. Lett. B **595**, 309 (2004)

DOI 10.1007/978-3-319-50699-9_30
Reprinted from *Few-Body System* Journal, DOI 10.1007/s00601-016-1069-1

Bernard L. G. Bakker · Chueng-Ryong Ji

Sensitivity of the DVCS Cross Section to the Compton form Factors in Scalar QED

Received: 18 January 2016 / Accepted: 23 February 2016 / Published online: 15 March 2016

Abstract Using deeply virtual Compton scattering as a tool to study the structure of hadrons in an exclusive process, one expresses the amplitudes in terms of invariant quantities: the Compton form factors. In this paper the sensitivity of the hadronic part of the cross section to the Compton form factors is determined.

1 Introduction

Deeply-virtual Compton scattering (DVCS)[1,2] has been proposed to determine the generalized-parton distributions (GPDs) of hadrons.

A hard photon, $q^2 = -Q^2$, with Q much larger than the characteristic hadronic scales, probes the quark content of the hadronic target. The detection of the outgoing real photon provides information not contained in deep-inelastic scattering (DIS). It is important to realize that GPDs are not invariant quantities. They are related asymptotically, i.e., for large Q, and for small Mandelstam t to Compton form factors (CFFs) [3,4]. Even if the experimental data are analysed in terms of these Lorentz-invariant quantities, it is not immediately clear what are the sensitivities of the data to the CFFs. In a simple case, namely DVCS on a scalar target where the minimal number of diagrams that are necessary to maintain EM current conservation are known, the corrections to the tree-level case can be calculated. We have found their scaling with Q^2. In practice, the DVCS amplitude interferes with the Bethe–Heitler amplitude. The latter one does only depend on the EM form factor(s) and thus give by itself no information about the CFFs. We shall concentrate on the DVCS part of the amplitude and even ignore the leptonic part, because it is independent of the CFFs.

2 Formalism

In virtual Compton scattering the physical amplitudes are written as contractions of a tensor with the polarization vectors of the photons:

$$A(h', h) = \epsilon^*(q'; h')_\mu T^{\mu\nu} \epsilon(q; h)_\nu \quad \text{with} \quad q'_\mu T^{\mu\nu} = 0, \quad T^{\mu\nu} q_\nu = 0 \quad (i.e.,\ T^{\mu\nu} \text{ transverse}). \qquad (1)$$

Presented by Ben Bakker at Light Cone 2015, 21–25 September, 2015, Frascati, Italy.

B. L. G. Bakker (✉)
Department of Physics and Astrophysics, Vrije Universiteit, Amsterdam, The Netherlands
E-mail: b.l.g.bakker@vu.nl

C.-R. Ji
Department of Physics, North Carolina State University, Raleigh, NC 27695-8202, USA
E-mail: crji@ncsu.edu

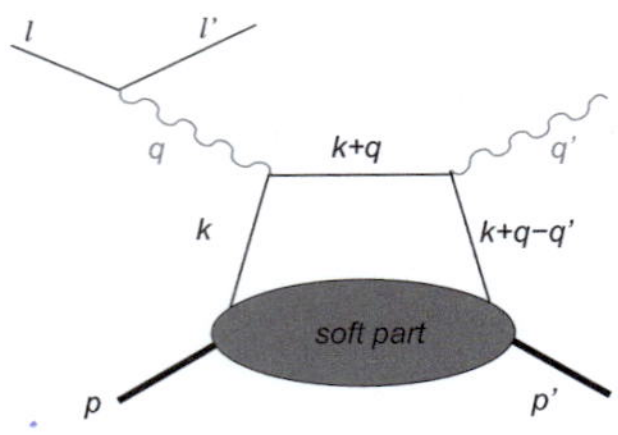

Fig. 1 Handbag diagram, including the leptonic part

The tensor is expressed in terms of CFFs and basis tensors. It is important to use the most general form of that tensor operator consistent with EM gauge invariance. We shall briefly discuss two proposals, one by Tarrach [1,2] and the other by Metz [3].

In the case we study, namely DVCS on a scalar hadron, the number of independent tensor structures is known, namely equal to five [1,2,5]. As these five independent tensor structures can be chosen in an infinite number of ways, we look for a synthetic way to construct the effective tensor.

Following Tarrach, one constructs the tensor $T^{\mu\nu}$ by applying a two-sided projector $\tilde{g}^{\mu\nu}(q, q')$ to the most general second rank tensor expressed in terms of our basis:

$$T^{\mu\nu} = \tilde{g}^{\mu m}\, t_{mn}\, \tilde{g}^{n\nu}, \quad t_{mn} = t_0\, g_{mn} + \sum_{i,j} t_{ij}\, k_{i\,m} k_{j\,n} \quad \text{with} \quad \tilde{g}^{\mu\nu}(q, q') = g^{\mu\nu} - \frac{q^\mu q'^\nu q \cdot q'}{q \cdot q'}. \tag{2}$$

We define the reduced momenta, $(k = \bar{P} = p + p',\ q',\ q)$:

$$\tilde{k}^\mu_{\mathrm{L}} = \tilde{g}^{\mu m} k_m, \quad \tilde{k}^\nu_{\mathrm{R}} = k_n\, \tilde{g}^{n\nu} \tag{3}$$

and find for unrestricted kinematics the following result for $T^{\mu\nu}$

$$T^{\mu\nu} = \mathcal{H}_0\, \tilde{g}^{\mu\nu} + \mathcal{H}_1\, \tilde{P}^\mu_{\mathrm{L}} \tilde{P}^\nu_{\mathrm{R}} + \mathcal{H}_2\, \tilde{P}^\mu_{\mathrm{L}} \tilde{q}^\nu_{\mathrm{R}} + \mathcal{H}_3\, \tilde{q}'{}^\mu{}_{\mathrm{L}} \tilde{P}^\nu_{\mathrm{R}} + \mathcal{H}_4\, \tilde{q}'{}^\mu{}_{\mathrm{L}} \tilde{q}^\nu_{\mathrm{R}}. \tag{4}$$

Contracting the tensor with $\epsilon^*_\mu(q')$ and $\epsilon_\nu(q)$ we find that all five pieces of the tensor contribute, if $q'^2 \neq 0$ and $q^2 \neq 0$. The number of independent tensor structures is equal to the number of independent physical matrix elements consistent with parity conservation: $A(-h', -h) = (-1)^{h'-h} A(h', h),\ h', h = \pm 1, 0,$

$$A(1,1),\ A(1,0),\ A(1,-1),\ A(0,1),\ \text{and}\ A(0,0). \tag{5}$$

We limit ourselves here to the case where one of the photons is real, say $q'^2 = 0$. Then the number of independent physical amplitudes reduces to three, say $A(1,1)$, $A(1,0)$, and $A(1,-1)$. In this case the CFFs $\mathcal{H}_3$ and $\mathcal{H}_4$ do not contribute.

The method using the projectors introduces a kinematical singularity at $q' \cdot q = 0$. In Tarrach's paper a method is described to remove these kinematic poles. Here we give the final result of that algorithm as obtained in the thesis of Metz. His CFFs in the scalar case are denoted as $\mathcal{B}_1$, $\mathcal{B}_2$, $\mathcal{B}_3$, $\mathcal{B}_4$, and $\mathcal{B}_{19}$. They are implicitly given in terms of the elementary tensor by the following equations:

$$\begin{aligned}
M^{\mu\nu} &= \mathcal{B}_1 M^{\mu\nu}_1 + \mathcal{B}_2 M^{\mu\nu}_2 + \mathcal{B}_3 M^{\mu\nu}_3 + \mathcal{B}_4 M^{\mu\nu}_4 + \mathcal{B}_{19} M^{\mu\nu}_{19}, \\
M^{\mu\nu}_1 &= -q' \cdot q\, g^{\mu\nu} + q^\mu q^\nu, \\
M^{\mu\nu}_2 &= -(\bar{P} \cdot q)^2\, g^{\mu\nu} - q' \cdot q\, \bar{P}^\mu \bar{P}^\nu + \bar{P} \cdot q\, (\bar{P}^\mu q' \nu + q^\mu \bar{P}^\nu), \\
M^{\mu\nu}_3 &= q^2 q^2\, g^{\mu\nu} + q' \cdot q\, q^\mu q^\nu - q^2\, q^\mu q^\nu - q^2\, q^\mu q^\nu, \\
M^{\mu\nu}_4 &= \bar{P} \cdot q\, (q^2 + q^2)\, g^{\mu\nu} - \bar{P} \cdot q\, (q^\mu q^\nu + q^\mu q^\nu) \\
&\quad - q^2\, \bar{P}^\mu q^\nu - q^2\, q^\mu \bar{P}^\nu + q' \cdot q\, (\bar{P}^\mu q^\nu + q^\mu \bar{P}^\nu), \\
M^{\mu\nu}_{19} &= (\bar{P} \cdot q)^2 q^\mu q^\nu + q^2 q^2\, \bar{P}^\mu \bar{P}^\nu - \bar{P} \cdot q\, q^2\, q^\mu \bar{P}^\nu - \bar{P} \cdot q\, q^2\, \bar{P}^\mu q^\nu.
\end{aligned}$$

The relations between the CFFs $\mathcal{H}_i$ and the CFFs $\mathcal{B}_j$ is found by identifying $M^{\mu\nu}$ and $T^{\mu\nu}$. (We shall not need them here.) If $q'^2 = 0$, which is the case we treat here, the independent CFFs are $\mathcal{B}_{1,2}$ and $\mathcal{B}_4$, which agrees with the number of independent amplitudes.

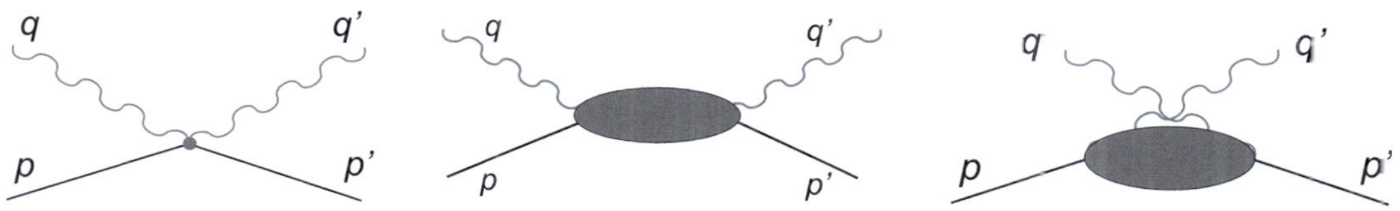

Fig. 2 Seagull, s- and u-channel amplitudes

The tree-level DVCS amplitude corresponding to Tarrach's basic formulation to the CFFs

$$\mathcal{H}_0^{\mathrm{tree}} = -2, \ \mathcal{H}_1^{\mathrm{tree}} = \frac{1}{s-M^2} + \frac{1}{u-M^2}, \ \mathcal{H}_3^{\mathrm{tree}} = 0, \tag{6}$$

and in Metz's formulation to

$$\mathcal{B}_1^{\mathrm{tree}} = \frac{1}{s-M^2} + \frac{1}{u-M^2}, \ \mathcal{B}_2^{\mathrm{tree}} = -\frac{2}{(s-M^2)(u-M^2)}, \ \mathcal{B}_4^{\mathrm{tree}} = 0, \tag{7}$$

$s = (p+q)^2, \ u = (p-q)'2$, M is the target hadron's mass.

3 Amplitudes

In a simple model where the internal structure of the target is included at one-loop level, it is known [6] that if all diagrams are included to guarantee EM gauge invariance to second order in the coupling constant g, one finds that the corrections at one-loop level scale as ($Q^2 = -q^2$)

$$\frac{1}{Q^n} \log(A + BQ^2), \tag{8}$$

where the functions A and B depend on the Mandelstam variables. The power n is 2 for $\mathcal{H}_0$ and 4 for $\mathcal{H}_{1,2}$. The imaginary parts scale as $1/Q^n$

We shall not consider this model in the present work. Instead we use the tree-level CFFs in Metz's formulation. In this formulation the whole interval $0 \leq \theta \leq \pi$, where θ is the photon scattering angle, can be explored when the sensitivities of the cross sections to variations of the CFFs are studied. As we do not include the Bethe-Heitler amplitudes, we shall consider the VCS amplitudes only and consider variations by $\pm 10\,\%$ for $\mathcal{B}_1$ and $\mathcal{B}_2$ to study the sensitivity of the cross section to them. Furthermore, we add a third CFF, namely $\mathcal{B}_4$ for which we take 0 (the tree-level value) and $\pm 0.1\ \mathcal{B}_2$.

We use the center-of mass (CMF) kinematics for the hadronic system and rotate the coordinate system such that the z-axis is along the three-momentum of the virtual photon.

To compute the amplitudes, we need the polarization vectors. We use circular polarisation in our calculation.

4 Numerical Calculations

We calculate the tree-level amplitudes for two realistic kinematical situations, described in a document written by Julie Roche and co-workers [7]. We take the kinematics for the largest value of the virtuality of the photon considered there, namely $Q^2 = 9\ \mathrm{GeV}^2$, because DVCS is mostly used to learn about GPDs [1,2] and the connection of these quantities to the invariant CFFs is most accurate for large Q^2. We chose a value $x_{\mathrm{Bj}} = 0.6$ because this value lies in the centre of the kinematical domain considered in Ref. [7]. The smallest hadronic target with spin zero is the α-particle, thus we concentrate our study to the case where $M = 3.7272\ \mathrm{GeV}/c^2$. Then we use the Metz tensor to calculate the amplitudes using the tree-level CFFs with variations of their numerical values in a range of $\pm 10\,\%$. The quantities we show in our results are *pseudo cross sections*, i.e., the sums of the squared amplitudes over the spin components.

We want to know in what kinematical region the greatest sensitivities of the *pseudo cross section*

$$\sum_{h'h} |A(h', h)|^2 \tag{9}$$

occurs. Therefore we need to explore for a given value of the Bjorken variable x_{Bj} and Q^2 the whole kinematical domain as parameterized by the photon scattering angle in the CMF $0 \leq \theta \leq \pi$. This means that $\theta = 0$ corresponds to the recoiled hadron moving in the backward direction in the CMF.

We stress that the parameterisation provided by Metz can be used in the whole kinematic range. The one given by Tarrach cannot, because there exists an angle where $q' \cdot q = 0$. Thus, the simplest construction will not be enough, unless one would limit oneself to a situation where $|t|$ is much smaller than Q^2, which limits the domain in θ.

First, we show the three independent amplitudes at tree level: $A(1,1)$, $A(1,0)$, and $A(1,-1)$. It is clear that for these values of Q^2 and x_{Bj} the amplitude involving a longitudinal virtual photon is dominant.

It is interesting to note that the largest magnitude of the longitudinal amplitude $|A(1,0)|$ occurs at the specific angle where the two transverse amplitudes coincide: $A(1,1) = A(1,-1)$. This angle can be computed rather easily from the consideration of the maximum production of the real photon in the target-rest-frame (TRF). When the virtual photon with the longitudinal polarization impinges on the scalar target at rest, the real photon is most abundantly produced at right angles with respect to the incoming virtual photon. Using the Lorentz transformation from the TRF to the CMF we find that the cosine of the scattering angle is given by

$$\cos\theta = -\frac{Q\sqrt{Q^2 + 4x_{\text{Bj}}^2 M^2}}{Q^2 + 2Mx_{\text{Bj}}}. \tag{10}$$

From the values of Q^2, M, and x_{Bj} given in Fig. 3 we can estimate that this angle is around 2.25, which is very close to $3\pi/4 \approx 2.36$. In the following figures we show the variations of the pseudo cross sections when the tree-level values of the CFFs are changed by $\pm 10\%$. Clearly, a variation of $\mathcal{B}_1$ with 10 % has hardly visible effects on the cross section, as seen in Fig. 4.

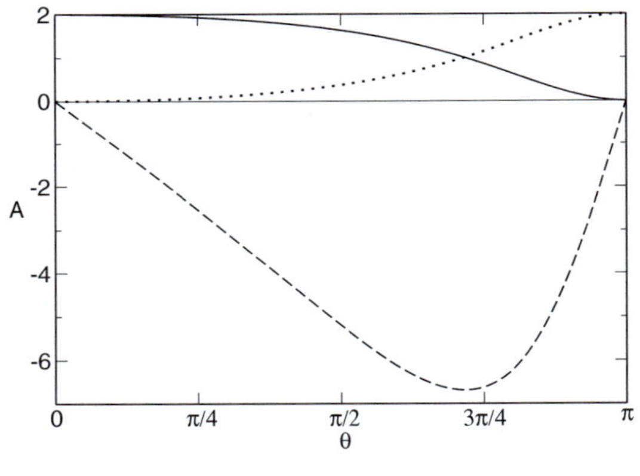

Fig. 3 The amplitudes $A(1,1)$ (*solid line*), $A(1,0)/i$ (*dashed line*), and $A(1,-1)$ (*dotted line*). $Q^2 = 9$ (GeV/c)^2, $M = 3.7273$ GeV/c^2, $x_{\text{Bj}} = 0.6$

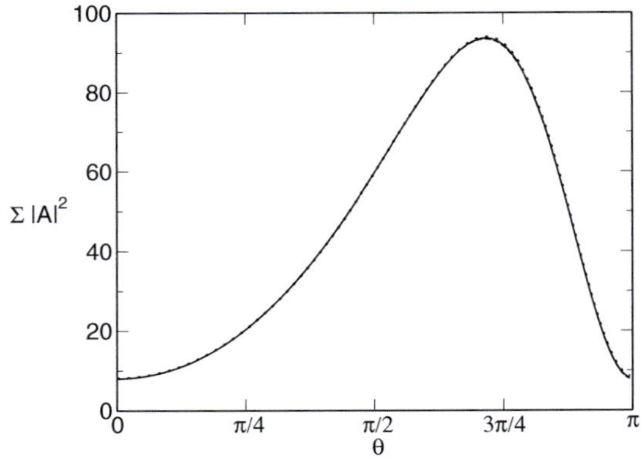

Fig. 4 Sum of squared amplitudes for $Q^2 = 9$ (GeV/c)^2, $M = 3.7273$ GeV/c^2, and $x_{\text{Bj}} = 0.6$. The CFF $\mathcal{B}_1$ is varied by ± 10 %

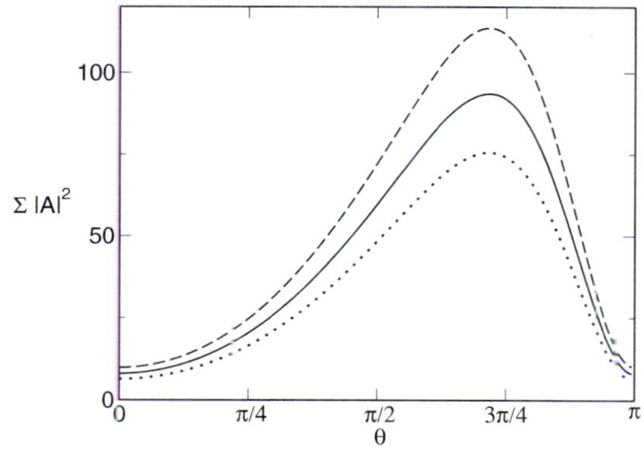

Fig. 5 Sum of squared amplitudes for $Q^2 = 9\ (\mathrm{GeV}/c)^2$, $M = 3.7273\ \mathrm{GeV}/c^2$, and $x_{\mathrm{Bj}} = 0.6$. The CFF $\mathcal{B}_2$ is varied by $\pm 10\,\%$

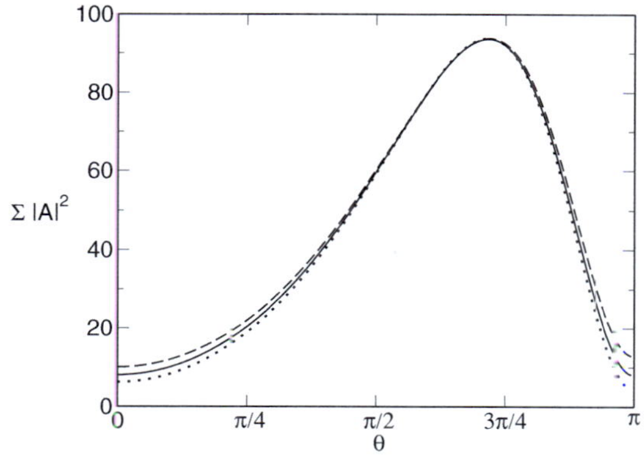

Fig. 6 Sum of squared amplitudes for $Q^2 = 9\ (\mathrm{GeV}/c)^2$, $M = 3.7273\ \mathrm{GeV}/c^2$, and $x_{\mathrm{Bj}} = 0.6$. The CFF $\mathcal{B}_4$ is varied by $\pm 10\,\%$ of $\mathcal{B}_2$

On the contrary, Fig. 5 shows that the same relative variation of $\mathcal{B}_2$ has a dramatic effect on the cross section. Replacing the tree-level CFF $\mathcal{B}_4$, which vanishes, to $\pm 0.1\mathcal{B}_2$, has only a significant effect on the cross section for extremely forward or backward angles, where the cross section itself is minimal.

This leads to the conclusion that the chance of determining all three CFFs is slim, even in this simplest possible case of a spinless target. The best one can do is to aim for the determination of $\mathcal{B}_4$ for angles in the extremely forward or backward kinematics.

Four more items should be considered before a final conclusion can be drawn. First, in the pseudo cross sections we have studied here, the kinematic factors connecting the differential cross section to the squares of the invariant amplitudes have not been included. Therefore, the kinematics where the cross section is most sensitive to the CFFs may differ somewhat from the ones indicated by our results.

Secondly, in an experiment, the complete amplitudes, hadronic plus Bethe–Heitler, must be used to calculate the cross section. Because the Bethe–Heitler amplitudes do not depend on the CFFS, but only on the well-measured EM form factor of the hadronic target, it would be surprising if the sensitivity of the cross section to the CFFs would be increased if the Bethe–Heitler amplitudes would be included.

Thirdly, in the full analysis, the interference term of the hadronic and the Bethe–Heitler amplitudes may be extracted for instance by measuring the beam-charge asymmetry, which may enhance the sensitivity to some of the CFFs in some kinematical domain.

Finally, because the cross sections are strongly varying with the scattering angle θ, one may expect that the CFFs may only be accurately determined from the data in a limited domain in θ, which corresponds for any fixed Bjorken variable x_{Bj} and Q^2 to a limited domain in the Mandelstam invariant t. Because the GPD description of DVCS implicitly assumes that Q^2 is much larger than any hadronic mass scale and that $|t|$ is much smaller than Q^2, the limitations on the domain where the CFFs can be reliably determined may not have sufficient overlap with the domain where the GPD description makes sense.

5 Summary and Conclusions

- In the comparatively simple case of DVCS on a scalar target, it is not clear that one can disentangle the differential cross section to determine all three CFFs.
- Only the interference with the Bethe–Heitler may give additional information on the CFFs.
- The amplitudes $A(\pm 1, 0)$ are dominant. In the hemisphere where the recoiled target moves forward, these amplitudes are most sensitive to $\mathcal{B}_2$.
- Using a non-singular basis like Metz's is essential to the analysis of the differential cross section data in the whole kinematical domain.

References

1. Belitsky, A.V., Müller, D.: Refined analysis of photon leptoproduction off a spinless target. Phys. Rev. D **79**, 014017 (2009)
2. Kumerički, K., Müller, D.: Deeply virtual compton scattering at small x_B and the access to the GPD H. Nucl. Phys. B **841**, 1 (2010)
3. Tarrach, R.: Invariant amplitudes for virtual compton scattering off polarized nucleons free from kinematic singularities, zeros and constraints. Nuovo Cim. **28A**, 409 (1975)
4. Metz, M.: Virtuelle comptonstreuung und die polarisierbarkeiten des nukleons, Ph.D. thesis, Universität Mainz (1997) (**in German**)
5. Ji, C.-R., Bakker, B.L.G.: Extraction of compton form factors in scalar QED. Few-Body Syst. **56**, 275 (2015)
6. Ji, C.-R., Bakker, B.L.G.: Conceptual issues concerning generalized parton distributions. Int. J. Mod. Phys. E **22**, 1330002 (2013)
7. Roche, J., et al.: Measurements of the electron-helicity dependent cross section of deeply virtual compton scattering with CEBAF at 12 GeV, arXiv:nucl-ex/0609015v1 (2006)

DOI 10.1007/978-3-319-50699-9_31
Reprinted from *Few-Body System* Journal, DOI 10.1007/s00601-016-1070-8

Stanley J. Brodsky

Light-Front Holography, Color Confinement, and Supersymmetric Features of QCD

Received: 14 January 2016 / Accepted: 24 February 2016 / Published online: 28 March 2016

Abstract Light-Front Quantization—Dirac's "Front Form"—provides a physical, frame-independent formalism for hadron dynamics and structure. Observables such as structure functions, transverse momentum distributions, and distribution amplitudes are defined from the hadronic light-front wavefunctions. One obtains new insights into the hadronic spectrum, light-front wavefunctions, and the functional form of the QCD running coupling in the nonperturbative domain using light-front holography—the duality between the front form and AdS_5, the space of isometries of the conformal group. In addition, superconformal algebra leads to remarkable supersymmetric relations between mesons and baryons of the same parity. The mass scale κ underlying confinement and hadron masses can be connected to the parameter $\Lambda_{\overline{MS}}$ in the QCD running coupling by matching the nonperturbative dynamics, as described by the effective conformal theory mapped to the light-front and its embedding in AdS space, to the perturbative QCD regime. The result is an effective coupling defined at all momenta. This matching of the high and low momentum transfer regimes determines a scale Q_0 which sets the interface between perturbative and nonperturbative hadron dynamics. The use of Q_0 to resolve the factorization scale uncertainty for structure functions and distribution amplitudes, in combination with the principle of maximal conformality for setting the renormalization scales, can greatly improve the precision of perturbative QCD predictions for collider phenomenology. The absence of vacuum excitations of the causal, frame-independent front form vacuum has important consequences for the cosmological constant. I also discuss evidence that the antishadowing of nuclear structure functions is non-universal; i.e., flavor dependent, and why shadowing and antishadowing phenomena may be incompatible with the momentum and other sum rules for nuclear parton distribution functions.

1 Introduction

Light-front quantization provides a physical, frame-independent formalism for hadron dynamics and structure. When one makes a measurement of a hadron, such as in deep inelastic lepton–proton scattering $\ell p \to \ell' X$, the hadron is observed along a light-front (LF)—in analogy to a flash photograph—not at a fixed time t. In effect, the LF time variable is $\tau = x^+ = t + z/c$, the time tangent to the light-cone. This is the underlying principle of the "front form" postulated by Dirac [1].

The front form has the maximum number of kinematic generators of the Lorentz group, and most remarkably, the formalism is boost invariant. The LF time evolution operator $P^- \equiv P^0 - P^z = i\frac{d}{d\tau}$ and the corresponding LF Hamiltonian $H_{LF} = P^+P^- - \mathbf{P}_\perp^2$, where $P^+ = P^0 + P^z$ and $\mathbf{P}_\perp$ are kinematical, can be derived directly from the Lagrangian. In the case of QCD, the eigenvalues of the LF invariant Hamiltonian are the squares of the hadron masses M_H^2: $H_{LF}|\Psi_H> = M_H^2|\Psi_H>$ [3], and the corresponding eigensolutions provide the n-particle hadronic LF Fock state wavefunctions (LFWFs) $< n|\Psi_H> = \psi_n^H(x_i, \mathbf{k}_{\perp i}, \lambda_i)$, where

S. J. Brodsky (✉)
SLAC National Accelerator Laboratory, Stanford University, Stanford, CA, USA
E-mail: sjbth@slac.stanford.edu

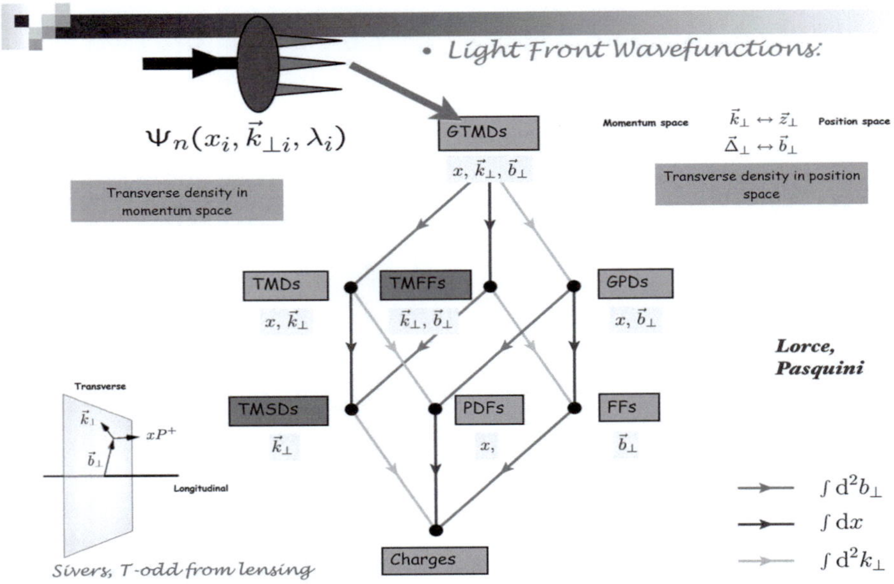

Fig. 1 Measures such as the generalized transverse momentum distribution (GTMD) which are directly defined from overlaps of hadronic light-front wavefunctions. In some cases, such as the leading twist Sivers pseudo-T-odd spin correlation in polarized deep inelastic lepton–proton scattering [11] or diffractive deep inelastic scattering $\ell p \to \ell' p' X$ [12], one must include final-state "lensing" interactions. Adopted from a figure designed by Pasquini and Lorce [13]

$|n>$ projects the eigenstate on the free Fock basis. The constituents' physical momenta are $p_i^+ = x_i P^+$, and $\mathbf{p}_{\perp i} = x_i \mathbf{P}_\perp + \mathbf{k}_{\perp i}$, and the λ_i label the spin projections S_i^z. One can avoid ghosts and longitudinal gluonic degrees of freedom by choosing to work in the light-cone gauge $A^+ = 0$.

The LFWFs are thus the Fock state projections of the eigenstates of the QCD LF Hamiltonian. The LFWFs are boost invariant; i.e., independent of the hadron's longitudinal $P^+ = P^0 + P^z$ and transverse momentum $\mathbf{P}_\perp$. Observables such as hadron structure functions, form factors, transverse momentum distributions, weak-decay amplitudes and distribution amplitudes are defined directly from the hadronic LFWFs. An example of a calculation of deeply virtual Compton scattering is given in Ref. [2].

The LF analyses are causal and frame independent. Operators appearing in commutators are automatically normal-ordered since there are no quantum fluctuations created from the LF vacuum. The LF wavefunctions thus play the role of Schrödinger wavefunctions in atomic physics, but they are relativistic and frame independent, see Fig. 1. Since they are independent of $P^+ = P^0 + P^z$ and $\mathbf{P}_\perp$, the LFWFs are the same whether measured in the hadron rest frame, as in the SLAC experiments or at a hadron collider, such as at an electron-proton collider. There is thus no concept of length contraction of a moving hadron or nucleus in the front form since observations of the collisions of the composite hadrons are not made at fixed instant time t. The absence of length contraction of a photographed object was first noted by Terrell [4], Penrose [5], and Weisskopf [6]. LF quantization is thus the natural formalism for particle physics.

There is an effort in the lattice gauge theory community to boost LGTh results to infinite momentum in order to simulate the causal light-front results [7]. An alternative lattice method—the " transverse lattice" introduces light-front coordinates in an effective 1 + 1 gauge theory connected to a lattice in the remaining transverse dimensions [8–10].

A review of the light-front formalism is given in Ref. [3].

A key result for hadron physics is the Drell–Yan–West formula [14–16] for electromagnetic form factors measured in elastic lepton hadron scattering. If one chooses the special frame $q^+ = \ell^+ - \ell'^+ = 0$, then the current matrix element $\langle p + q | j^+(0) | p \rangle$, which defines form factors, is a simple overlap integral of the initial $\psi_H(x_i\ \mathbf{k}_{\perp i}, \lambda_i)$ and final-state LFWFs $\psi_H(x_i, \mathbf{k}'_{\perp i}, \lambda_i)$, summed over Fock states n. This yields the exact results for the electromagnetic, weak, and gravitational form factors. Here $\mathbf{k}'_{\perp i} = \mathbf{k}'_{\perp i} + (1 - x_i)\mathbf{q}_\perp$ for the struck quark and $\mathbf{k}'_{\perp i} = \mathbf{k}'_{\perp i} - x_i \mathbf{q}_\perp$ for the spectators. For example, one can easy show that the anomalous gravitomagnetic moment $B(q^2 = 0)$ vanishes identically for any LF Fock state [17], in agreement with a theorem by Okun, Kobzarev, and Teryaev [18–20] which follows from the equivalence theorem of gravity.

The LF predictions for current matrix elements are equivalent to the corresponding Bethe–Salpeter formulae for form factors, but one must sum over the currents of an infinite number of irreducible kernels. The LFWFs are related to the Bethe Salpeter amplitudes if one integrates over k^-.

The LFWFs are boost invariant. This contrasts with the wavefunctions defined at a fixed time t—the Lorentz boost of an instant-form wavefunction is a dynamical problem, much more complicated than a simple Melosh transform [21]—even the number of Fock constituents changes under a boost. Thus there is no practical formula for calculating form factors using ordinary instant time analogous to the Drell–Yan West formula since the boost of the wavefunction of a composite state is unknown except for weakly bound systems; and worse, one must include the contributions of an infinite number of connected currents which arise from the vacuum.

PQCD factorization theorems and the DGLAP [22–26] and ERBL [27–31] evolution equations can also be derived using the light-front Hamiltonian formalism [28]. In the case of an electron-ion collider, one can represent the cross section for $e - p$ colisions as a convolution of the hadron and virtual photon structure functions times the subprocess cross-section in analogy to hadron-hadron colisions. This nonstandard description of $\gamma^* p \to X$ reactions gives new insights into electroproduction physics—physics not apparent in the usual infinite-momentum frame description, such as the dynamics of heavy quark-pair production. Intrinsic heavy quarks at high x also play an important role [32].

The LF Heisenberg equation can in principle be solved numerically by matrix diagonalization using the "Discretized Light-Cone Quantization" (DLCQ) [33] method. Anti-periodic boundary conditions in x^- render the k^+ momenta discrete as well as limiting the size of the Fock basis. In fact, one can easily solve $1+1$ quantum field theories such as QCD$(1+1)$ [34] for any number of colors, flavors and quark masses using DLCQ. Unlike lattice gauge theory, the nonpertubative DLCQ analysis is in Minkowski space, is frame-independent and is free of fermion-doubling problems.

As I shall discuss, AdS/QCD, together with light-front holography, is now providing a color-confining approximation to H^{QCD}_{LF}, for QCD(3+1), thus giving a first approximation to the meson and baryon spectra and their hadronic LFWFs. A new method for solving nonperturbative QCD "Basis Light-Front Quantization" (BLFQ) [35], uses the eigensolutions of a color-confining approximation to QCD (such as LF holography) as the basis functions, rather than the plane-wave basis used in DLCQ. LFWFs can also be determined from the covariant Bethe–Salpeter wavefunction by integrating over k^- [36].

2 Calculations Using LF-Time-Ordered Perturbation Theory

There are a number of advantages if one uses LF Hamiltonian methods for perturbative QCD calculations.

Propagating particles are on their respective mass shells: $k_\mu k^\mu = m^2$, and intermediate states are off-shell in invariant mass; *i.e.*, $P^- \neq \sum k^-_i$. Unlike instant form, where one must sum $n!$ frame-dependent amplitudes, only τ-ordered diagrams where every line has positive $k^+ = k^0 + k^z$ contribute. An excellent example of LF-time-ordered perturbation theory is the computation of multi-gluon scattering amplitudes by Cruz-Santiago and Stasto [37]. The number of nonzero amplitudes is also greatly reduced by noting that the total angular momentum projection $J^z = \sum_i^{n-1} L^z_i + \sum_i^n S^z_i$ and the total P^+ are conserved at each vertex. In addition, in a renormalizable theory the change in orbital angular momentum is limited to $\Delta L^z = 0, \pm 1$ at each vertex.

A remarkable advantage of LF time-ordered perturbation theory (LFPth) is that the calculation of a subgraph of any order in pQCD only needs to be done once; the result can be stored in a "history" file. This is due to the fact that in LFPth the numerator algebra is independent of the process; the denominator changes, but only by a simple shift of the initial P^-. Another simplification is that loop integrations are three dimensional: $\int d^2\mathbf{k}_\perp \int_0^1 dx$. Unitarity and explicit renormalization can be implemented using the "alternate denominator" method which defines the required subtraction counterterms [38].

3 The Light-Front Vacuum

It is important to distinguish the LF vacuum from the conventional instant-form vacuum.

The eigenstates of the instant-form Hamiltonian describe a state defined at a single instant of time t over all space, and they are thus acausal as well as frame-dependent. The instant-form vacuum is defined as the lowest energy eigenstate of the instant-form Hamiltonian. Quantum loops in the instant-form vacuum typically vanish as $1/P^2$.

In contrast, the vacuum in LF Hamlitonian theory is defined as the eigenstate of H_{LF} with lowest invariant mass. It is defined at fixed LF time τ within the causal horizon and is frame-independent. The loop diagrams which occur in the usual instant-form vacuum do not appear in the front-form vacuum since the $+$ momenta are positive: $k_i^+ = k_i^0 + k_i^z \geq 0$, and the sum of $+$ momenta is conserved at every vertex. Thus the creation of particles cannot arise from the LF vacuum since $\sum_i k^{+i} \neq P^+_{vacuum} = 0$. Since propagation with negative k^+ does not appear, the LF vacuum is trivial up to possible $k^+ = 0$ "zero" modes. The usual quark and gluon QCD vacuum condensates of the instant form are replaced by physical effects contained within the hadronic LFWFs in the hadronic domain. This is referred to as "in-hadron" condensates [39–41]. In the case of the Higgs theory, the traditional Higgs vacuum expectation value (VEV) is replaced by a "zero mode", analogous to a classical Stark or Zeeman field [42]. This again contrasts with the traditional view of the vacuum based on the instant form.

The cosmological constant is of order 10^{120} times larger than what is observed if one computes the effects of quantum loops from QED using the instant form vacuum. QCD instantons and condensates in the same vacuum give a contribution of order 10^{42}. The contribution of the Higgs VEV computed in the instant form vacuum is 10^{54} times too large [43]. Conventional wisdom [44] suggests that this problem can be solved by assuming the existence of 10^{500} possible universes, fine-tuning to 120 decimal places, together with the anthropic principle.

However, the universe is observed within the causal horizon, not at a single instant of time. The causal, frame-independent light-front vacuum can thus provide a viable match to the empty visible universe [41]. The cosmological constant problem thus does not appear if one notes that the causal, frame-independent light-front vacuum has no quantum fluctuations—in dramatic contrast to to the acausal, frame-dependent instant-form vacuum; the cosmological constant vanishes if one uses the front form.

In the case of electroweak theory, the Higgs LF zero mode [42] has no energy-momentum density, so it also gives zero contribution to the cosmological constant. However, it is possible that if one solves electroweak theory in a curved universe, the Higgs LF zero mode will be replaced with a field of nonzero curvature which could give a nonzero contribution.

4 Color Confinement and Supersymmetry in Hadron Physics from LF Holography

A key problem in hadron physics is to obtain a first color-confining approximation to QCD which can predict both the hadron spectrum and the hadronic LFWFs. If one neglects the Higgs couplings of quarks, then no mass parameter appears in the QCD Lagrangian, and the theory is conformal at the classical level. Nevertheless, hadrons have a finite mass. De Téramond, Dosch, and I [45] have shown that a mass gap and a fundamental color confinement scale can be derived from a conformal action when one extends the formalism of de Alfaro, Fubini and Furlan (dAFF) [46] to light-front Hamiltonian theory. Remarkably, the resulting light-front potential has a unique form of a harmonic oscillator $\kappa^4\zeta^2$ in the light-front invariant impact variable ζ where $\zeta^2 \hat{E} = b_\perp^2 x(1-x)$. The result is a single-variable frame-independent relativistic equation of motion for $q\bar{q}$ bound states, a "Light-Front Schrödinger Equation" [47], analogous to the nonrelativistic radial Schrödinger equation in quantum mechanics. The Light-Front Schrödinger Equation incorporates color confinement and other essential spectroscopic and dynamical features of hadron physics, including a massless pion for zero quark mass and linear Regge trajectories with the same slope in the radial quantum number n and internal orbital angular momentum L. The same light-front equation for mesons of arbitrary spin J can be derived [48] from the holographic mapping of the "soft-wall model" modification of AdS_5 space with the specific dilaton profile $e^{+\kappa^2 z^2}$, where one identifies the fifth dimension coordinate z with the light-front coordinate ζ. The five-dimensional AdS_5 space provides a geometrical representation of the conformal group. It is holographically dual to 3+1 spacetime at fixed light-front time $\tau = t + z/c$. The derivation of the confining LF Schrodinger Equation is outlined in Fig. 2. The reduction to an effective Hamiltonian acting on the valence Fock state of hadrons in QCD is analogous to the reduction used in precision analyses in QED for atomic physics.

The combination of light-front dynamics, its holographic mapping to AdS_5 space, and the dAFF procedure provides new insight into the physics underlying color confinement, the nonperturbative QCD coupling, and the QCD mass scale. A comprehensive review is given in Ref. [50]. The $q\bar{q}$ mesons and their valence LF wavefunctions are the eigensolutions of the frame-independent relativistic bound state LF Schrödinger equation. The mesonic $q\bar{q}$ bound-state eigenvalues for massless quarks are $M^2(n, L, S) = 4\kappa^2(n+L+S/2)$. The equation predicts that the pion eigenstate $n = L = S = 0$ is massless at zero quark mass. The Regge spectra of the pseudoscalar $S = 0$ and vector $S = 1$ mesons are predicted correctly, with equal slope in the principal quantum number n and the internal orbital angular momentum L. The comparison with experiment is shown in Fig. 3.

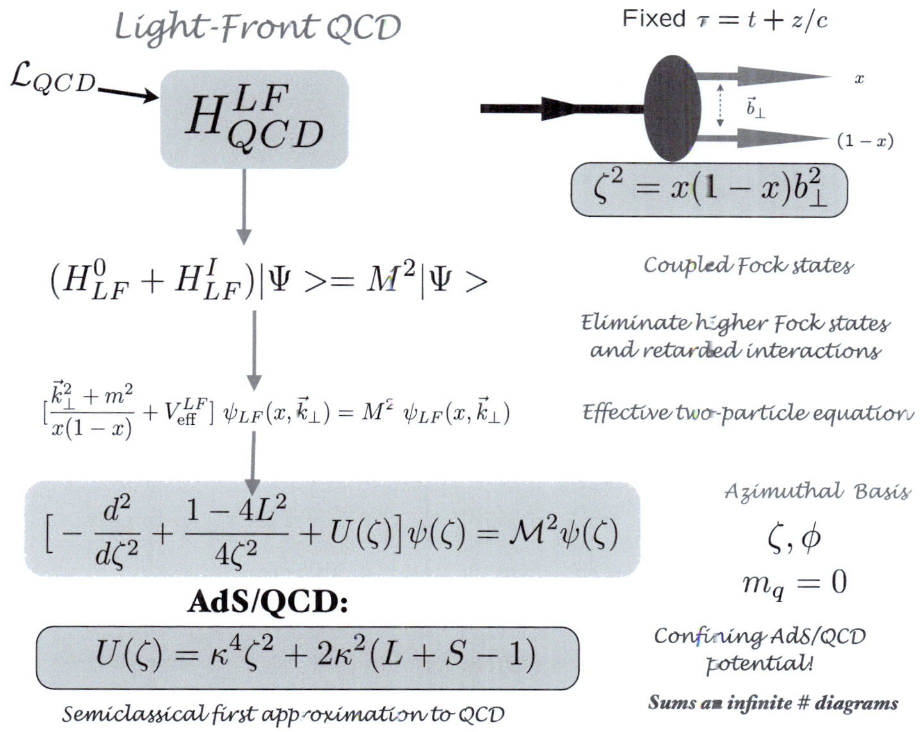

Fig. 2 Derivation of the Effective Light-Front Schrödinger Equation from QCD. As in QED, one reduces the LF Heisenberg equation $H_{LF}|\Psi >= M^2|\Psi >$ to an effective two-body eigenvalue equation for $q\bar{q}$ mesons by systematically eliminating higher Fock states. One utilizes the LF radial variable ζ, where $\zeta^2 = x(1-x)b^2_\perp$ is conjugate to the $q\bar{q}$ LF kinetic energy $\frac{k^2_\perp}{x(1-x)}$ for $m_q = 0$. This allows the reduction of the dynamics to a single-variable bound state equation acting on the valence $q\bar{q}$ Fock state. The confining potential $U(\zeta)$, including its spin-J dependence, is derived from the soft-wall AdS/QCD model with the dilaton $e^{+\kappa^2 z^2}$, where z is the fifth coordinate of AdS_5 holographically dual to ζ. See Ref. [45]. The resulting light-front harmonic oscillator confinement potential $\kappa^4\zeta^2$ for light quarks is equivalent to a linear confining potential for heavy quarks in the instant form [49]

The AdS/QCD light-front holographic eigenfunction for the ρ meson LFWF $\psi_\rho(x, \mathbf{k}_\perp)$ gives excellent predictions for the observed features of diffractive ρ electroproduction $\gamma^* p \to \rho p'$, as shown by Forshaw and Sandapen [51]. Note that the prediction for the LFWF is a function of the LF kinetic energy $\mathbf{k}^2_\perp/x(1-x)$ – the conjugate of the LF radial variable $\zeta^2 = b^2_\perp x(1-x)$ – times a function of $x(1-x)$. It does not factorize as a function of $\mathbf{k}^2_\perp$ times a function of x. The resulting nonperturbative pion distribution amplitude $\phi_\pi(x) = \int d^2\mathbf{k}_\perp \psi_\pi(x, \mathbf{k}_\perp) = (4/\sqrt{3}\pi) f_\pi \sqrt{x(1-x)}$, see Fig. 4, is consistent with the Belle data for the photon-to-pion transition form factor [52].

5 Supersymmetric Aspects of Hadron Physics

These results can be extended [53–55] to effective QCD light-front equations for both mesons and baryons by using the generalized supercharges of superconformal algebra [56]. In effect the baryons are color-singlet bound-states of color-triplet quarks and $\bar{3}_C$ $[qq]$ diquarks.

The supercharges connect the baryon and meson spectra and their Regge trajectories to each other in a remarkable manner: each meson has internal angular momentum one unit higher than its superpartner baryon $L_M = L_B + 1$. See Fig. 5a. Only one mass parameter κ again appears; it sets the confinement and the hadron mass scale in the chiral limit, as well as the length scale which underlies hadron structure. "Light-Front Holography" not only predicts meson and baryon spectroscopy successfully, but also hadron dynamics, including vector meson electroproduction, hadronic light-front wavefunctions, distribution amplitudes, form factors, and valence structure functions.

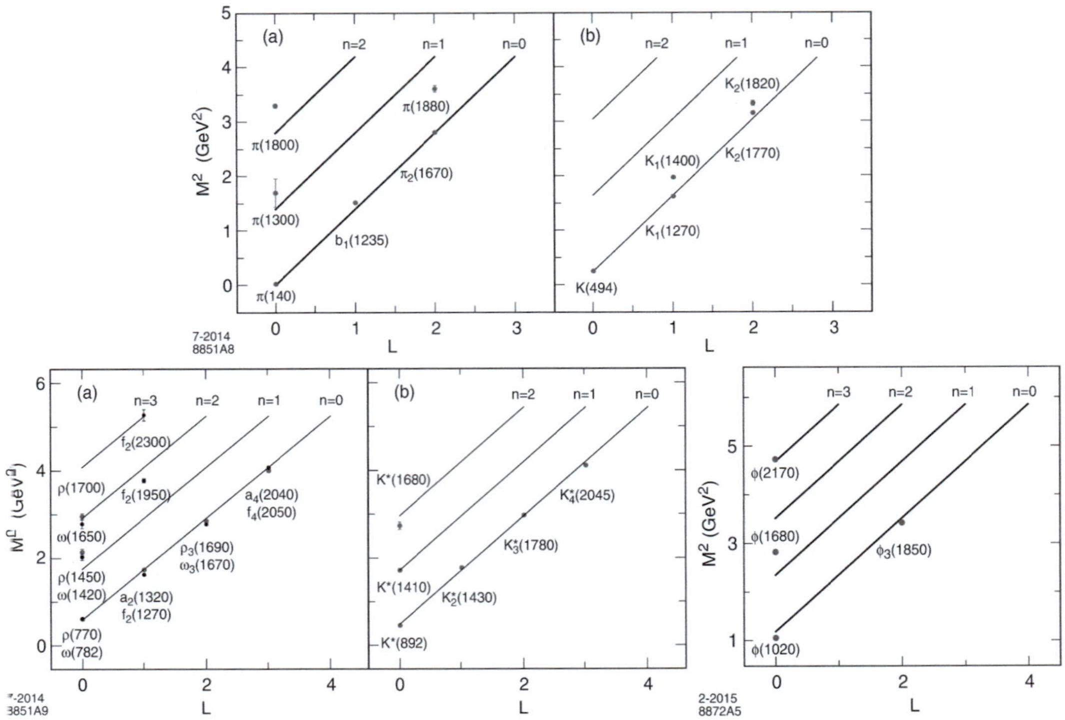

Fig. 3 Comparison of the AdS/QCD prediction $M^2(n, L, S) = 4\kappa^2(n + L + S/2)$ for the orbital L and radial n excitations of the meson spectrum with experiment. The pion is predicted to be massless for zero quark mass. The u, d, s quark masses can be taken into account by perturbing in $\langle m_q^2/x \rangle$. The fitted value of $\kappa = 0.59$ MeV for pseudoscalar mesons, and $\kappa = 0.54$ MeV for vector mesons

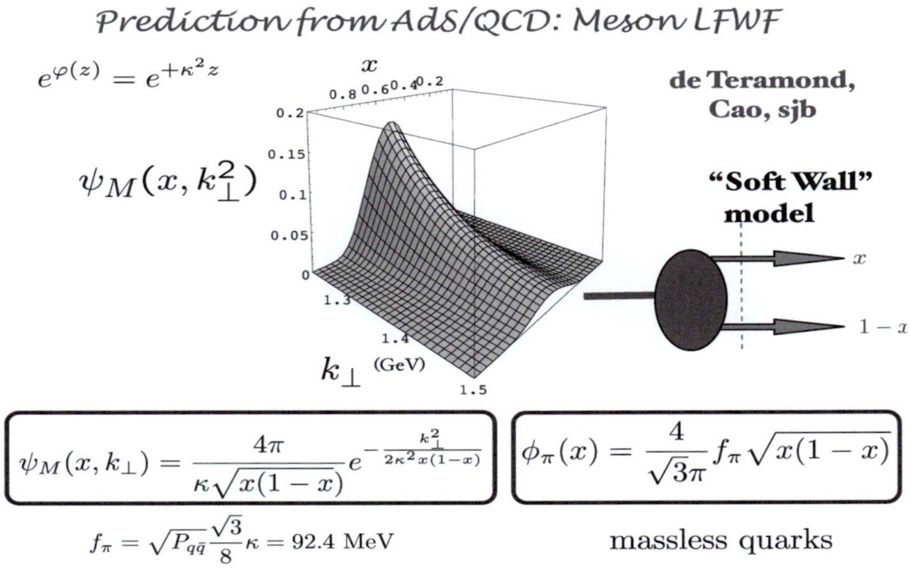

Fig. 4 Prediction from AdS/QCD and Light-Front Holography for meson LFWFs $\psi_M(x, \mathbf{k}_\perp)$ and the pion distribution amplitude

The LF Schrödinger Equations for baryons and mesons derived from superconformal algebra are shown in Fig. 5. In effect the baryons on the proton (Delta) trajectory are bound states of a quark with color 3_C and scalar (vector) diquark with color $\bar{3}_C$ The proton eigenstate labeled ψ^+ (parallel quark and baryon spins) and

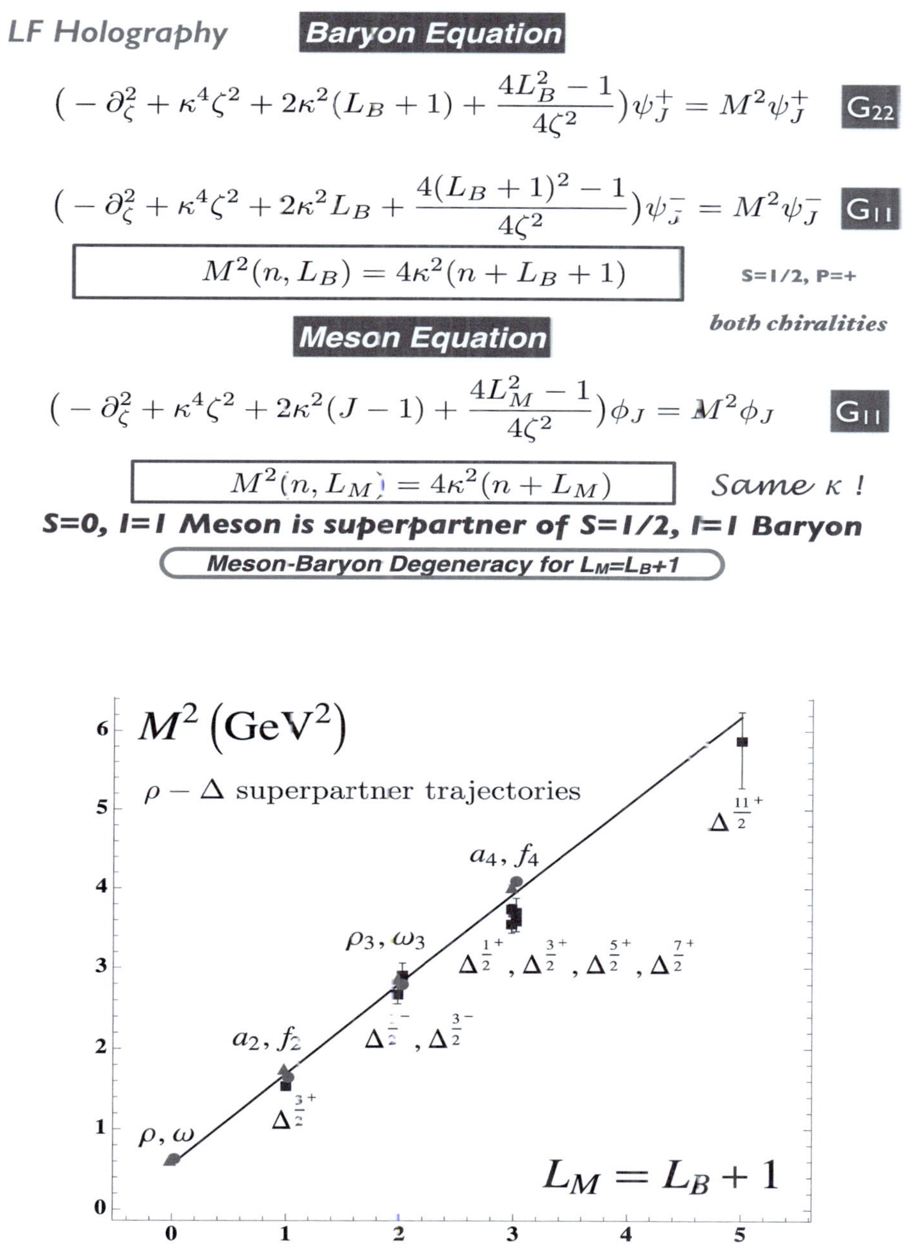

Fig. 5 **a** The LF Schrödinger equations for baryons and mesons for zero quark mass derived from the Pauli 2×2 matrix representation of superconformal algebra. The $\psi^{\pm}$ are the baryon quark-diquark LFWFs where the quark spin $S_q^z = \pm 1/2$ is parallel or antiparallel to the baryon spin $J^z = \pm 1/2$. The meson and baryon equations are identical if one identifies a meson with internal orbital angular momentum L_M with its superpartner baryon with $L_B = L_M - 1$. See Refs. [53–55]. **b** Comparison of the ρ/ω meson Regge trajectory with the $J = 3/2$ Δ baryon trajectory. Superconformal algebra predicts the degeneracy of the meson and baryon trajectories if one identifies a meson with internal orbital angular momentum L_M with its superpartner baryon with $L_M = L_B + 1$. See Refs. [53,54]

ψ^- (anti parallel quark and baryon spins) have equal Fock state probability—a remarkable feature of chirality invariance. The static properties of the nucleons is discussed in Ref. [58]

The comparison between the meson and baryon masses of the ρ/ω Regge trajectory with the spin-3/2 Δ trajectory is shown in Fig. 5b. Superconformal algebra predicts that the bosonic meson and fermionic baryon masses are equal if one identifies each meson with internal orbital angular momentum L_M with its superpartner baryon with $L_B = L_M - 1$. Since $|L_B - L_M| = 1$, the meson and baryon superpartners thus have have the same parity as well as the same twist.

Notice that the twist $2+L_M = 3+L_B$ of the interpolating operators for the meson and baryon superpartners are the same. Superconformal algebra also predicts that the LFWFs of the superpartners are identical, and thus they have identical LFWFs, and the corresponding elastic and transition form factors are equal. The predicted identity of meson and baryon timelike form factors can be tested in $e^+e^- \to H\bar{H}'$ reactions.

In the case of $ep \to e'X$, one can consider the collisions of the confining QCD flux tube appearing between the q and $\bar{q}$ of the virtual photon with the flux tube between the quark and diquark of the proton. Since the $q\bar{q}$ plane is aligned with the scattered electron's plane, the resulting "ridge" of hadronic multiplicity produced from the $\gamma^* p$ collision will also tend to be aligned with the scattering plane of the scattered electron. The virtual photon's flux tube will also depend on the photon virtuality Q^2, as well as the flavor of the produced pair arising from $\gamma^* \to q\bar{q}$. In the case of high energy $\gamma^*\gamma^*$ collisions, one can control the produced hadron multiplicity and ridge geometry using the scattered electrons' planes. The resulting dynamics [57] is a natural extension of the flux-tube collision description of the ridge produced in $p - p$ collisions [59].

6 The QCD Coupling at all Scales

The QCD running coupling $\alpha_s(Q^2)$ sets the strength of the interactions of quarks and gluons as a function of the momentum transfer Q. The dependence of the coupling Q^2 is needed to describe hadronic interactions at both long and short distances.

The QCD running coupling can be defined [60,61] at all momentum scales from a perturbatively calculable observable, such as the coupling $\alpha^s_{g_1}(Q^2)$, which is defined from measurements of the Bjorken sum rule. At high momentum transfer, such "effective charges" satisfy asymptotic freedom, obey the usual pQCD renormalization group equations, and can be related to each other without scale ambiguity by commensurate scale relations [62].

The dilaton $e^{+\kappa^2 z^2}$ soft-wall modification of the AdS_5 metric, together with LF holography, predicts the functional behavior of the running coupling in the small Q^2 domain [63]: $\alpha^s_{g_1}(Q^2) = \pi e^{-Q^2/4\kappa^2}$. Measurements of $\alpha^s_{g_1}(Q^2)$ are remarkably consistent [64,65] with this predicted Gaussian form; the best fit gives $\kappa = 0.513 \pm 0.007$ GeV (see Fig. 6a).

Deur, de Teramond, and I [63,66,67] have shown how the parameter κ, which determines the mass scale of hadrons in the zero quark mass limit, can be connected to the mass scale Λ_s controlling the evolution of the perturbative QCD coupling. The high momentum transfer dependence of the coupling $\alpha_{g1}(Q^2)$ is specified by pQCD and its renormalization group equation. The matching of the high and low momentum transfer regimes of $\alpha_{g1}(Q^2)$—both its value and its slope—then determines the scale Q_0 setting the interface between perturbative and nonperturbative hadron dynamics. This connection can be done for any choice of renormalization scheme, such as the $\overline{MS}$ scheme, as seen in Fig. 6b. The result of this perturbative/nonperturbative matching is an effective QCD coupling defined at all momenta.

The predicted value of $\Lambda_{\overline{MS}} = 0.341 \pm 0.024$ GeV from this analysis agrees well the measured value [68] $\Lambda_{\overline{MS}} = 0.339 \pm 0.016$ GeV. These results, combined with the AdS/QCD superconformal predictions for hadron spectroscopy, allow us to compute hadron masses in terms of $\Lambda_{\overline{MS}}$: $m_p = \sqrt{2}\kappa = 3.21\ \Lambda_{\overline{MS}}$, $m_\rho = \kappa = 2.2\ \Lambda_{\overline{MS}}$, and $m_p = \sqrt{2} m_\rho$, meeting a challenge proposed by Zee [69]. The pion is predicted to be massless for $m_q = 0$ consistent with chiral theory.

The value of Q_0 can be used to set the factorization scale for DGLAP evolution of hadronic structure functions and the ERBL evolution of distribution amplitudes. Deur, de Téramond, and I [70], have also computed the dependence of Q_0 on the choice of the effective charge used to define the running coupling and the renormalization scheme used to compute its behavior in the perturbative regime.

7 Hadronization at the Amplitude Level and Other New Directions

- The new insights into color confinement given by AdS/QCD suggest that one could compute hadronization at amplitude level [71] using LF time-ordered perturbation theory, but including the confinement interaction. For example, if one computes $e^+e^- \to q\bar{q} \to q\bar{q}g \ldots$, the quarks and gluons only appear in intermediate states, and only hadrons can be produced. LF perturbation theory provides a remarkably efficient method for the calculation of multi-gluon amplitudes [72].

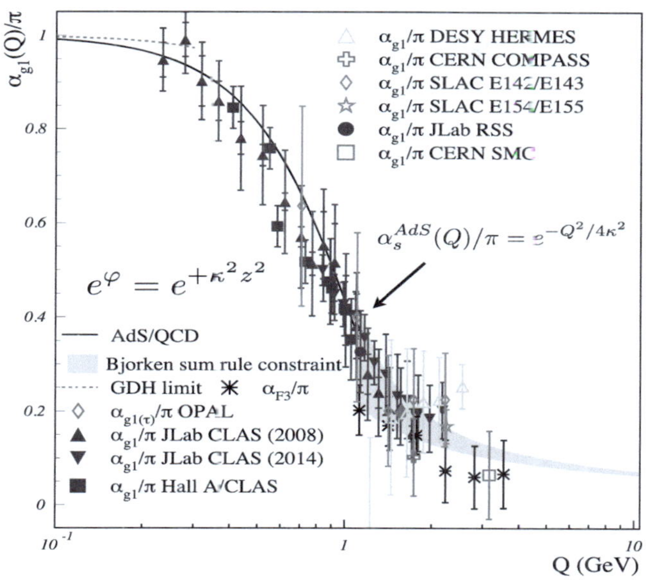

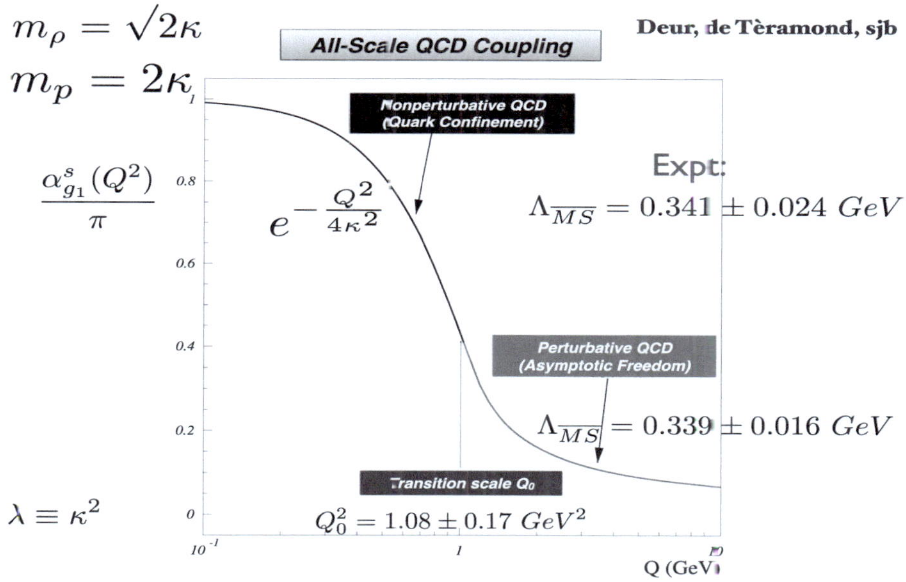

Fig. 6 **a** Comparison of the predicted nonpertubative coupling, based on the dilaton exp $(+\kappa^2 z^2)$ modification of the AdS_5 metric, with measurements of the effective charge $\alpha^s_{g_1}(Q^2)$ defined from the Bjorken sum rule. **b** Prediction from LF Holography and pQCD for the QCD running coupling $\alpha^s_{g_1}(Q^2)$ at all scales The magnitude and derivative of the perturbative and nonperturbative coupling are matched at the scale Q_0. This matching connects the perturbative scale $\Lambda_{\overline{MS}}$ to the nonpertubative scale κ which underlies the hadron mass scale. See Ref. [67]

- The eigensolutions of the AdS/QCD LF Hamiltonian can used to form an orthonormal basis for diagonalizing the complete QCD LF Hamiltonian. This method, "basis light-front quantization" [73] is expected to be more efficient than the DLCQ method [33] for obtaining QCD 3+1 solutions.
- All of the hadron physics predictions discussed in this report are independent of the value of κ; only dimensionless ratios are predicted, such as $m_p = \sqrt{2} m_\rho$ and the ratio $\Lambda_{\overline{MS}}/m_\rho$. The ratio can be obtained in any renormalization scheme. One thus retains dilatation invariance $\kappa \to \gamma\kappa$ of the prediction.
- The $\kappa^4\zeta^2$ confinement interaction between a q and $\bar{q}$ will induce a κ^4/s^2 correction to $R_{e^+e^-}$, replacing the $1/s^2$ signal usually attributed to a vacuum gluon condensate.

- The kinematic condition that all $k^+ = k^0 + k^3$ are positive and conserved precludes QCD condensate contributions to the $P^+ = 0$ LF vacuum state, which by definition is the causal, frame-independent lowest invariant mass eigenstate of the LF Hamiltonian [40,74].
- It is interesting to note that the contribution of the *'H'* diagram to the heavy quark potential is IR divergent as the transverse separation between the Q and the $\bar{Q}$ increases [75]. This is a signal that pQCD is inconsistent without color confinement. The sum of such diagrams could sum to the confinement potential $\kappa^4 \zeta^2$ dictated by the dAFF principle that the action remains conformally invariant despite the mass scale in the Hamiltonian.

8 Elimination of Renormalization and Factorization Scale Ambiguities

The "Principle of Maximum Conformality", (PMC) [76] systematically eliminates the renormalization scale ambiguity in perturbative QCD calculations, order-by-order. The PMC predictions are also insensitive to the choice of the initial renormalization scale μ_0. The PMC sums all of the non-conformal terms associated with the QCD β function into the scales of the coupling at each order in pQCD, systematically extending the BLM procedure [77] to all orders. The resulting conformal series is free of renormalon resummation problems. The number of active flavors n_f in the QCD β function is also correctly determined at each order.

The R_δ scheme—a generalization of t'Hooft's dimensional regularization systematically identifies the nonconformal β contributions to any perturbative QCD series, thus allowing the automatic implementation of the PMC procedure [78]. The resulting scale-fixed predictions for physical observables using the PMC are *independent of the choice of renormalization scheme*—a key requirement of renormalization group invariance. A related approach is given in Refs. [79–81].

The elimination of renormalization scale ambiguities greatly increases the precision, convergence, and reliability of pQCD predictions. For example, PMC scale-setting has been applied to the pQCD prediction for $t\bar{t}$ pair production at the LHC, where subtle aspects of the renormalization scale of the three-gluon vertex and multi-gluon amplitudes, as well as large radiative corrections to heavy quarks at threshold play a crucial role. The large discrepancy of pQCD predictions with the $t\bar{t}$ forward-backward asymmetry measured at the Tevatron is significantly reduced from 3 σ to approximately 1 σ [82,83,85,86].

The use of the scale Q_0 discussed in the previous section to resolve the factorization scale uncertainty in structure functions and fragmentation functions, in combination with the PMC for setting the renormalization scales, can greatly improve the precision of pQCD predictions for collider phenomenology.

9 Is the Momentum Sum Rule Valid for Nuclear Structure Functions?

Sum rules for deep inelastic scattering are usually analyzed using the operator product expansion of the forward virtual Compton amplitude, assuming it depends in the limit $Q^2 \to \infty$ on matrix elements of local operators such as the energy-momentum tensor. The moments of structure functions and other distributions can then be evaluated as overlaps of the target hadron's light-front wavefunction, as in the Drell–Yan–West formulae for hadronic form factors [16,87–89]. The real phase of the resulting DIS amplitude and its OPE matrix elements reflects the real phase of the stable target hadron's wavefunction.

The "handbag" approximation to deeply virtual Compton scattering also defines the "static" contribution [90,91] to the measured parton distribution functions (PDF), transverse momentum distributions, etc. The resulting momentum, spin and other sum rules reflect the properties of the hadron's light-front wavefunction. However, final-state interactions which occur *after* the lepton scatters on the quark, can give non-trivial contributions to deep inelastic scattering processes at leading twist and thus survive at high Q^2 and high $W^2 = (q + p)^2$. For example, the pseudo-T-odd Sivers effect [11] is directly sensitive to the rescattering of the struck quark. Similarly, diffractive deep inelastic scattering (DDIS) involves the exchange of a gluon after the quark has been struck by the lepton [12]. In each case the corresponding DVCS amplitude is not given by the handbag diagram since interactions between the two currents are essential. These "lensing" corrections survive when both W^2 and Q^2 are large since the vector gluon couplings grow with energy. Part of the final state phase can be associated with a Wilson line as an augmented LFWF [92] which does not affect the moments.

The Glauber propagation of the vector system V produced by the DDIS interaction on the nuclear front face and its subsequent inelastic interaction with the nucleons in the nuclear interior $V + N_b \to X$ occurs after the lepton interacts with the struck quark. Because of the rescattering dynamics, the DDIS amplitude acquires a complex phase from Pomeron and Regge exchange; thus final-state rescattering corrections lead to nontrivial "dynamical" contributions to the measured PDFs; i.e., they involve physical aspects of the scattering process itself [93]. The $I = 1$ Reggeon contribution to DDIS on the front-face nucleon leads to flavor-dependent antishadowing [94,95]. This could explain why the NuTeV charged current measurement $\mu A \to \nu X$ scattering does not appear to show antishadowing in contrast to deep inelastic electron nucleus scattering as discussed in Ref. [96]. Again the corresponding DVCS amplitude is not given by the handbag diagram since interactions between the two currents are essential.

Diffractive deep inelastic scattering is leading-twist and is the essential component of the two-step amplitude which causes shadowing and antishadowing of the nuclear PDF. It is important to analyze whether the momentum and other sum rules derived from the OPE expansion in terms of local operators remain valid when these dynamical rescattering corrections to the nuclear PDF are included. The OPE is derived assuming that the LF time separation between the virtual photons in the forward virtual Compton amplitude $\gamma^* A \to \gamma^* A$ scales as $1/Q^2$. However, the propagation of the vector system V produced by the DDIS interaction on the front face and its inelastic interaction with the nucleons in the nuclear interior $V + N_b \to X$ are characterized by a longer LF time which scales as $1/W^2$. Thus the leading-twist multi-nucleon processes that produce shadowing and antishadowing in a nucleus are evidently not present in the $Q^2 \to \infty$ OPE analysis.

It should be emphasized that shadowing in deep inelastic lepton scattering on a nucleus involves nucleons at or near the front surface; i.e, the nucleons facing the incoming lepton beam. This geometrical orientation is not built into the frame-independent nuclear LFWFs used to evaluate the matrix elements of local currents. Thus the dynamical phenomena of leading-twist shadowing and antishadowing appear to invalidate the sum rules for nuclear PDFs. The same complications occur in the leading-twist analysis of deeply virtual Compton scattering $\gamma^* A \to \gamma^* A$ on a nuclear target.

Acknowledgments Presented at Light-Cone 2015, *Theory and Experiment for Hadrons on the Light-Front*, September 21–25, 2015, INFN, Frascati, Italy. The results presented here are based on collaborations and discussions with James Bjorken, Kelly Chiu, Alexandre Deur, Guy de Téramond, Guenter Dosch, Susan Gardner, Fred Goldhaber, Paul Hoyer, Dae Sung Hwang, Rich Lebed, Simonetta Liuti, Cedric Lorce, Matin Mojaza, Michael Peskin, Craig Roberts, Robert Shrock, Ivan Schmidt, Peter Tandy, and Xing-Gang Wu. This research was supported by the Department of Energy, contract DE-AC02-76SF00515. SLAC-PUB-16453.

References

1. Dirac, P.A.M.: Forms of relativistic dynamics. Rev. Mod. Phys. **21**, 392 (1949)
2. Brodsky, S.J., Diehl, M. and Hwang, D.S.: Light cone wave function representation of deeply virtual Compton scattering. Nucl. Phys. B. **596**,99 (2001) [arXiv:hep-ph/0009254]
3. Brodsky, S.J., Pauli, H.C. and Pinsky, S.S.: Quantum chromodynamics and other field theories on the light cone. Phys. Rept. **301**,299 (1998) [arXiv:hep-ph/9705477]
4. Terrell, J.: Invisibility of the lorentz contraction. Phys. Rev. **116**, 1041 (1959)
5. Penrose, R.: The Apparent shape of a relativistically moving sphere. Proc. Cambridge Phil. Soc. **55**, 137 (1959)
6. Weisskopf, V.: Physics Today. **13**, 24 (1960)
7. Ma, Y.Q. and Qiu, J.W.: Extracting parton distribution functions from lattice QCD calculations. arXiv:1404.6860 [hep-ph].
8. Bardeen, W.A.: Transverse lattice theory of quantum chromodynamics. FERMILAB-CONF-78-023-THY, FERMILAB-CONF-78-023-T
9. Bardeen, W.A., Pearson, R.B., Rabinovici, E.: Hadron masses in quantum chromodynamics on the transverse lattice. Phys. Rev. D **21**, 1037 (1980)
10. Burkardt, M. and Dalley, S.: The relativistic bound state problem in QCD: transverse lattice methods. Prog. Part. Nucl. Phys. **48**,317 (2002) [arXiv:hep-ph/0112007]
11. Brodsky, S.J., Hwang, D.S. and Schmidt, I.: Final state interactions and single spin asymmetries in semiinclusive deep inelastic scattering. Phys. Lett. B **530**,99 (2002) [arXiv:hep-ph/0201296]
12. Brodsky, S.J., Hoyer, P., Marchal, N., Peigne, S. and Sannino, F.: Structure functions are not parton probabilities. Phys. Rev. D **65**, 114025 (2002) [arXiv:hep-ph/0104291]
13. Pasquini, B., Lorc, C.: Wigner distributions in light-front quark models. Few Body Sys. **55**, 287 (2014)
14. Drell, S.D., Yan, T.M.: Connection of elastic electromagnetic nucleon form-factors at large Q**2 and deep inelastic structure functions near threshold. Phys. Rev. Lett. **24**, 181 (1970)
15. West, G.B.: Phenomenological model for the electromagnetic structure of the proton. Phys. Rev. Lett. **24**, 1206 (1970)
16. Brodsky, S.J., Drell, S.D.: The anomalous magnetic moment and limits on fermion substructure. Phys. Rev. D **22**, 2236 (1980)
17. Brodsky, S.J., Hwang, D.S., Ma, B.Q. and Schmidt, I.: Light cone representation of the spin and orbital angular momentum of relativistic composite systems. Nucl. Phys. B **593**, 311 (2001) [arXiv:hep-th/0003082]
18. Kobzarev, I.Y., Okun, L.B.: Gravitational interaction of fermions. Zh. Eksp. Teor. Fiz. **43**, 1904 (1962)
19. Kobzarev, I.Y., Okun, L.B.: Gravitational interaction of fermions. Sov. Phys. JETP. **16**, 1343 (1963)
20. Teryaev, O.V.: Spin structure of nucleon and equivalence principle. arXiv:hep-ph/9904376
21. Brodsky, S.J., Primack, J.R.: The electromagnetic interactions of composite systems. Annals Phys. **52**, 315 (1969)
22. Gribov, V.N., Lipatov, L.N.: Deep inelastic e p scattering in perturbation theory. Sov. J. Nucl. Phys. **15**, 438 (1972)
23. Gribov, V.N., Lipatov, L.N.: Deep inelastic e p scattering in perturbation theory. Yad. Fiz. **15**, 781 (1972)

24. Altarelli, G., Parisi, G.: Asymptotic freedom in parton language. Nucl. Phys. B. **126**, 298 (1977)
25. Dokshitzer, Y.L.: Calculation of the structure functions for deep inelastic scattering and e+ e- annihilation by perturbation theory in quantum chromodynamics. Sov. Phys. JETP **46**, 641 (1977)
26. Dokshitzer, Y.L.: Calculation of the structure functions for deep inelastic scattering and e+ e- annihilation by perturbation theory in quantum chromodynamics. Zh. Eksp. Teor. Fiz. **73**, 1216 (1977)
27. Lepage, G.P., Brodsky, S.J.: Exclusive processes in quantum chromodynamics: evolution equations for hadronic wave functions and the form-factors of mesons. Phys. Lett. B **87**, 359 (1979)
28. Lepage, G.P., Brodsky, S.J.: Exclusive processes in perturbative quantum chromodynamics. Phys. Rev. D **22**, 2157 (1980)
29. Efremov, A.V., Radyushkin, A.V.: Factorization and asymptotical behavior of pion form-factor in QCD. Phys. Lett. B **94**, 245 (1980)
30. Efremov, A.V., Radyushkin, A.V.: Asymptotical behavior of pion electromagnetic form-factor in QCD. Theor. Math. Phys. **42**, 97 (1980)
31. Efremov, A.V., Radyushkin, A.V.: Asymptotical behavior of pion electromagnetic form-factor in QCD. Teor. Mat. Fiz. **42**, 147 (1980)
32. Brodsky, S.J. and Gardner, S.: Comment on “New Limits on Intrinsic Charm in the Nucleon from Global Analysis of Parton Distributions”. arXiv:1504.00969 [hep-ph]
33. Pauli, H.C., Brodsky, S.J.: Solving field theory in one space one time dimension. Phys. Rev. D **32**, 1993 (1985)
34. Hornbostel, K., Brodsky, S.J., Pauli, H.C.: Light cone quantized QCD in (1+1)-dimensions. Phys. Rev. D **41**, 3814 (1990)
35. Vary, J.P., Zhao, X., Ilderton, A., Honkanen, H., Maris, P. and Brodsky, S.J.: Applications of basis light-front quantization to QED. Nucl. Phys. Proc. Suppl. **251-252**, 10 (2014) [arXiv:1406.1838 [nucl-th]]
36. Brodsky, S.J., et al.: QCD and hadron physics. arXiv:1502.05728 [hep-ph]
37. Cruz-Santiago, C., Kotko, P., Stasto, A.M.: Scattering amplitudes in the light-front formalism. Prog. Part. Nucl. Phys. **85**, 82 (2015)
38. Brodsky, S.J., Roskies, R., Suaya, R.: Quantum electrodynamics and renormalization theory in the infinite momentum frame. Phys. Rev. D. **8**, 4574 (1973)
39. Casher, A., Susskind, L.: Chiral magnetism (or magnetohadrochironics). Phys. Rev. D **9**, 436 (1974)
40. Brodsky, S.J. and Shrock, R.: Condensates in quantum chromodynamics and the cosmological constant. Proc. Nat. Acad. Sci. **108**,45 (2011) [arXiv:0905.1151 [hep-th]]
41. Brodsky, S.J., Roberts, C.D., Shrock, R. and Tandy, P.C.: Essence of the vacuum quark condensate. Phys. Rev. C **82**, 022201 (2010) [arXiv:1005.4610 [nucl-th]]
42. Srivastava, P.P. and Brodsky, S.J.: A unitary and renormalizable theory of the standard model in ghost free light cone gauge. Phys. Rev. D **66**, 045019 (2002) [arXiv:hep-ph/0202141]
43. Zee, A.: Dark energy and the cosmological constant paradox. Mod. Phys. Lett. A. **23**, 1336 (2008)
44. D. H. Bailey and J. M Borwein, “Data vs Theory: The Mathematical Battle for the Soul of Physics”, Huffington Post, December 30, (2016)
45. Brodsky, S.J., de Téramond, G.F. and Dosch, H.G.: Threefold Complementary Approach to Holographic QCD. Phys. Lett. B **729**, 3 (2014) [arXiv:1302.4105 [hep-th]]
46. Alfaro, V.de , Fubini, S., Furlan, G.: Conformal invariance in quantum mechanics. Nuovo Cim. A **34**, 569 (1976)
47. de Téramond, G.F. and Brodsky, S.J.: Light-front holography: a first approximation to QCD. Phys. Rev. Lett. **102**,081601 (2009) [arXiv:0809.4899 [hep-ph]]
48. de Téramond, G.F., Dosch, H.G. and Brodsky, S.J.: Kinematical and dynamical aspects of higher-spin bound-state equations in holographic QCD. Phys. Rev. D **87**, 075005 (2013) [arXiv:1301.1651 [hep-ph]]
49. Trawinski, A.P., Glazek, S.D., Brodsky, S.J., de Téramond, G.F. and Dosch, H.G.: Effective confining potentials for QCD. Phys. Rev. D **90**, 074017 (2014) [arXiv:1403.5651 [hep-ph]]
50. Brodsky, S.J., de Téramond, G.F., Dosch, H.G. and Erlich, J.: Light-front holographic QCD and emerging confinement. Phys. Rept. **584**, 1 (2015) [arXiv:1407.8131 [hep-ph]]
51. Forshaw, J.R. and Sandapen, R.: An AdS/QCD holographic wavefunction for the rho meson and diffractive rho meson electroproduction. Phys. Rev. Lett. **109**, 081601 (2012) [arXiv:1203.6088 [hep-ph]]
52. Brodsky, S.J., Cao, F.G. and de Téramond, G.F.: Meson transition form factors in light-front holographic QCD. Phys. Rev. D **84**, 075012 (2011) [arXiv:1105.3999 [hep-ph]]
53. de Téramond, G.F., Dosch, H.G. and Brodsky, S.J.: Baryon spectrum from superconformal quantum mechanics and its light-front holographic embedding. Phys. Rev. D **91**, 045040 (2015) [arXiv:1411.5243 [hep-ph]]
54. Dosch, H.G., de Téramond G.F. and Brodsky, S.J.: Superconformal baryon-meson symmetry and light-front holographic QCD. Phys. Rev. D **91**, 085016 (2015) [arXiv:1501.00959 [hep-th]]
55. Dosch, H.G., de Téramond, G.F. and Brodsky, S.J.: Supersymmetry across the light and heavy-light hadronic spectrum. Phys. Rev. D **92**, 074010 (2015) [arXiv:1504.05112 [hep-ph]]
56. Fubini, S., Rabinovici, E.: Superconformal quantum mechanics. Nucl. Phys. B. **245**, 17 (1984)
57. Brodsky, S.J.: Novel QCD Phenomena at the LHC: The Ridge, Digluon-Initiated Subprocesses, Direct Reactions, Non-Universal Antishadowing, and Forward Higgs Production. Nucl. Part. Phys. Proc. **258-259**, 23–30 (2015) [arXiv:1410.0404 [hep-ph]]
58. Liu, T. and Ma, B.Q.: Baryon properties from light-front holographic QCD. Phys. Rev. D **92**, no. 9, 096003 (2015) doi:10.1103/PhysRevD.92.096003 [arXiv:1510.07783 [hep-ph]]
59. Bjorken, J.D., Brodsky, S.J. and Goldhaber, A. Scharff: Possible multiparticle ridge-like correlations in very high multiplicity proton-proton collisions. Phys. Lett. B **726**, 344 (2013) [arXiv:1308.1435 [hep-ph]]
60. Grunberg, G.: Renormalization group improved perturbative QCD. Phys. Lett. B. **95**, 70 (1980)
61. Grunberg, G.: Renormalization group improved perturbative QCD. Phys. Lett. B. **110**, 501 (1982)
62. Brodsky, S.J. and Lu, H.J.: Commensurate scale relations in quantum chromodynamics. Phys. Rev. D **51**, 3652 (1995) [arXiv:hep-ph/9405218]

63. Brodsky, S.J., de Téramond G.F. and Deur, A.: Nonperturbative QCD coupling and its β-function from light-front holography. Phys. Rev. D **81**, 096010 (2010) [arXiv:1002.3948 [hep-ph]]
64. Deur, A., Burkert, V., Chen, J.P., Korsch, W.: Experimental determination of the effective strong coupling constant. Phys. Lett. B **650**, 244 (2007)
65. Deur, A., Burkert, V., Chen, J.P., Korsch, W.: Experimental determination of the effective strong coupling constant. Phys. Lett. B **665**, 349 (2008)
66. Deur, A., Brodsky, S.J. and de T'eramond, G.F.: Connecting the hadron mass scale to the fundamental mass scale of quantum chromodynamics. Phys. Lett. B **750**, 528 (2015) [arXiv:1409.5488 [hep-ph]]
67. Brodsky, S.J., de Téramond, G.F., Deur, A., Dosch, H.G.: The light-front schršdinger equation and the determination of the perturbative QCD scale from color confinement: a first approximation to QCD. Few Body Syst. **56**, 621 (2015) [arXiv:1410.0425 [hep-ph]]
68. Olive, K.A. et al.: [Particle Data Group Collaboration], Review of particle physics. Chin. Phys. C. **38**, 090001 (2014)
69. Zee, A.: Quantum field theory in a nutshell. Princeton University Press, Princeton (2010)
70. Deur, A., Brodsky, S.J., de Téramond, G.F.: On the interface between perturbative and nonperturbative QCD (2016) [arXiv:1601.06568 [hep-ph]]
71. Brodsky, S.J. and de Teramond, G.F.: Light-front holography and QCD hadronization at the amplitude level. arXiv:0901.0770 [hep-ph]
72. Cruz-Santiago, C., Kotko, P. and Stasto, A.: Recursion relations for multi-gluon off-shell amplitudes on the light-front and Wilson lines. Nucl. Phys. B **895**, 132 (2015) [arXiv:1503.02066 [hep-ph]]
73. Vary, J.P. et al.: Hamiltonian light-front field theory in a basis function approach. Phys. Rev. C **81**, 035205 (2010) [arXiv:0905.1411 [nucl-th]]
74. Brodsky, S.J., Roberts, C.D., Shrock, R. and Tandy, P.C.: Confinement contains condensates. Phys. Rev. C **85**, 065202 (2012) [arXiv:1202.2376 [nucl-th]]
75. Smirnov, A.V., Smirnov, V.A. and Steinhauser, M.: Three-loop static potential. Phys. Rev. Lett. **104**, 112002 (2010) [arXiv:0911.4742 [hep-ph]]
76. Wu, X.G., Brodsky, S.J. and Mojaza, M.: The renormalization scale-setting problem in QCD. Prog. Part. Nucl. Phys. **72**, 44 (2013) [arXiv:1302.0599 [hep-ph]]
77. Brodsky, S.J., Lepage, G.P., Mackenzie, P.B.: On the elimination of scale ambiguities in perturbative quantum chromodynamics. Phys. Rev. D **28**, 228 (1983)
78. Mojaza, M., Brodsky, S.J. and Wu, X.G.: Systematic all-orders method to eliminate renormalization-scale and scheme ambiguities in perturbative QCD. Phys. Rev. Lett. **110**, 192001 (2013) [arXiv:1212.0049 [hep-ph]]
79. Kataev, A.L. and Mikhailov, S.V.: β-expansion in QCD, its conformal symmetry limit: theory + applications. Nucl. Part. Phys. Proc. **258-259**, 45 (2015) [arXiv:1410.0554 [hep-ph]]
80. Kataev, A.L. and Mikhailov, S.V.: Generalization of the Brodsky-Lepage-Mackenzie optimization within the β-expansion and the principle of maximal conformality. Phys. Rev. D **91**(1),014007 (2015) [arXiv:1408.0122 [hep-ph]]
81. Kataev, A.L.: The generalized BLM approach to fix scale-dependence in QCD: the current status of investigations. J. Phys. Conf. Ser. **608**(1),012078 (2015) [arXiv:1411.2257 [hep-ph]]
82. Brodsky, S.J. and Wu, X.G.: Eliminating the renormalization scale ambiguity for top-pair production using the principle of maximum conformality. Phys. Rev. Lett. **109**, 042002 (2012) [arXiv:1203.5312 [hep-ph]]
83. Brodsky, S.J., Wu, X.G.: Application of the principle of maximum conformality to top-pair production Phys. Rev. D. **86**, 014021 (2012)
84. Brodsky, S.J. and Wu, X.G. (2013) Application of the principle of maximum conformality to top-pair production. Phys. Rev. D. **87**, 099902 [arXiv:1204.1405 [hep-ph]]
85. Wu, X.G., Wang, S.Q. and Brodsky, S.J.: The importance of proper renormalization scale-setting for testing QCD at colliders. Physics. **11**, 111201 (2016) [arXiv:1508.02332 [hep-ph]]
86. Wang, S.Q., Wu, X.G., Si, Z.G. and Brodsky, S.J.: Predictions for the top-quark forward-backward asymmetry at high invariant pair mass using the principle of maximum conformality. arXiv:1508.03739 [hep-ph]
87. Liuti, S. Rajan, A., Courtoy, A., Goldstein, G.R. and Hernandez, J.O.Gonzalez: Partonic picture of GTMDs. Int. J. Mod. Phys. Conf. Ser. **25**, 1460009 (2014) [arXiv:1309.7029 [hep-ph]]
88. Mondal, C. and Chakrabarti, D.: Generalized parton distributions and transverse densities in a light-front quark-diquark model for the nucleons. Eur. Phys. J. C **75**, no. 6, 261 (2015) [arXiv:1501.05489 [hep-ph]]
89. Lorce, C., Pasquini, B. and Vanderhaeghen, M.: Unified framework for generalized and transverse-momentum dependent parton distributions within a 3Q light-cone picture of the nucleon. JHEP. **1105**, 041 (2011) [arXiv:1102.4704 [hep-ph]]
90. Brodsky, S.J.: Dynamic versus static structure functions and novel diffractive effects in QCD. AIP Conf. Proc. **1105**, 315 (2009) [arXiv:0811.0875 [hep-ph]]
91. Brodsky, S.J.: Dynamic versus static hadronic structure functions. Nucl. Phys. A **827**, 327C (2009) [arXiv:0901.0781 [hep-ph]]
92. Brodsky, S.J., Pasquini, B., Xiao, B.W. and Yuan, F.: Phases of augmented hadronic light-front wave functions. Phys. Lett. B **687**, 327 (2010) [arXiv:1001.1163 [hep-ph]]
93. Brodsky, S.J., Hwang, D.S., Kovchegov, Y.V., Schmidt, I. and Sievert, M.D.: Single-spin asymmetries in semi-inclusive deep inelastic scattering and drell-yan processes. Phys. Rev. D **88**, 014032 (2013) [arXiv:1304.5237 [hep-ph]]
94. Brodsky, S.J., Lu, H.J.: Shadowing and antishadowing of nuclear structure functions. Phys. Rev. Lett. **64**, 1342 (1990)
95. Brodsky, S.J., Schmidt, I. and Yang, J.J.: Nuclear antishadowing in neutrino deep inelastic scattering. Phys. Rev. D **70**, 116003 (2004) [arXiv:hep-ph/0409279]
96. Schienbein, I., Yu, J.Y., Keppel, C., Morfin, J.G., Olness, F. and Owens, J.F.: Nuclear parton distribution functions from neutrino deep inelastic scattering. Phys. Rev. D **77**, 054013 (2008) [arXiv:0710.4897 [hep-ph]]

DOI 10.1007/978-3-319-50699-9_32
Reprinted from *Few-Body System* Journal, DOI 10.1007/s00601-016-1073-5

Constantia Alexandrou

Parton Distribution Functions from Lattice QCD

Received: 30 January 2016 / Accepted: 29 February 2016 / Published online: 18 March 2016

Abstract We present recent results on the first moments of parton distributions using gauge configurations generated with two degenerate flavors of light twisted mass quarks with pion mass fixed approximately to its physical value. We also present a first study of the vector parton distribution function using a twisted mass ensemble at pion mass of 373 MeV.

1 Introduction

Although nucleon structure has been an area of intensive activity within lattice QCD for a number of years, it is only recently that simulations with near physical parameters both in terms of the value of the pion mass (referred to as physical point), as well as, with respect to the lattice volume and lattice spacing have become available [1–9]. Generalized parton distributions (GPDs) encode information related to nucleon structure that complements the information extracted from form factors [10–12]. Their forward limit coincides with the usual parton distribution functions (PDFs) and, using Ji's sum rule [13], allows one to determine the contribution of a specific parton to the nucleon spin. In the context of the "proton spin puzzle", which refers to the unexpectedly small fraction of the total spin of the nucleon carried by quarks, this has triggered intense experimental activity [14–18]. Calculations using gauge configurations produced with physical values of the light quark mass, pose new challenges associated with the slower convergence of the inversion of the Dirac operator and the increase of gauge noise requiring very large computational resources. Therefore, techniques to speedup the inversions are critical in order to obtain results at the physical point. In Fig. 1 we show the speedup achieved when deflation of lowest eigen-modes is applied. The best improvement is achieved with exact deflation of the lowest about 1800 eigen-modes, which yields a speed-up of a factor of 20 as compared to the conjugate gradient method after about 1000 right hand sides (rhs) that is the typical number of rhs in our production runs.

2 Evaluation of Matrix Elements in Lattice QCD

In Fig. 1 we show the contributions entering the evaluation of a hadron matrix element. One needs to compute the three-point function given by

$$G^{\mu\nu}(\Gamma, \mathbf{q}, t_s, t_{\rm ins}) = \sum_{\mathbf{x}_s, \mathbf{x}_{\rm ins}} e^{i\mathbf{x}_{\rm ins}\cdot\mathbf{q}}\, \Gamma_{\beta\alpha}\, \langle J_\alpha(\mathbf{x}_s, t_s)\mathcal{O}_\Gamma^{\mu\nu}(\mathbf{x}_{\rm ins}, t_{\rm ins})\overline{J}_\beta(\mathbf{x}_0, t_0)\rangle \tag{1}$$

C. Alexandrou (✉)
Department Physics, University of Cyprus, P.O. Box 20537, 1678 Nicosia, Cyprus
Tel.: +357-22-892829
E-mail: alexand@ucy.ac.cy

C. Alexandrou
Computation-based Science and Technology Research Research, The Cyprus Institute, 20 Kavafi Str., 2121 Nicosia, Cyprus

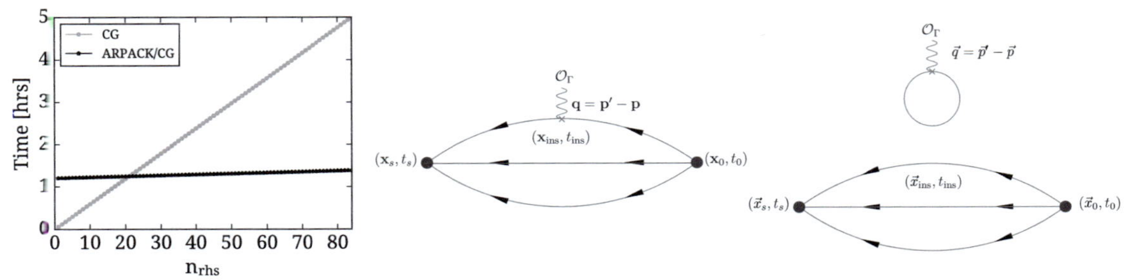

Fig. 1 Speedup achieved with exact deflation as a function of the number of rhs $n_{\rm rhs}$ (*left*). Connected (*middle*) and disconnected (*right*) contributions to a three-point function

and then form a ratio by dividing the three-point function by an appropriate combination of two-point functions:

$$R(t_s, t_{\rm ins}, t_0) \overset{(t_s - t_{\rm ins})\Delta \gg 1, (t_{\rm ins} - t_0)\Delta \gg 1}{\longrightarrow \longrightarrow \longrightarrow} \mathcal{M}\left[1 + \cdots e^{-\Delta(\mathbf{p})(t_{\rm ins} - t_0)} + \cdots e^{-\Delta(\mathbf{p}')(t_s - t_{\rm ins})}\right]. \tag{2}$$

In Eq. 2 $\mathcal{M}$ is the desired matrix element, $t_s, t_{\rm ins}, t_0$ the sink, insertion and source time-slices and $\Delta(\mathbf{p})$ the energy gap with the first excited state. Summing over the current insertion time $t_{\rm ins}$ we obtain

$$\sum_{t_{\rm ins}=t_0}^{t_s} R(t_s, t_{\rm ins}, t_0) = \text{Const.} + \mathcal{M}[(t_s - t_0) + \mathcal{O}(e^{-\Delta(\mathbf{p})(t_s - t_0)}) + \mathcal{O}(e^{-\Delta(\mathbf{p}')(t_s - t_0)})]. \tag{3}$$

In the summed ratio, excited state contributions are suppressed by exponentials decaying with $t_s - t_0$, rather than $t_s - t_{\rm ins}$ and/or $t_{\rm ins} - t_0$. However, one needs to fit the slope rather than to a constant, which increases the statistical uncertainty in the extracted value of the desired matrix element.

In order to extract the physical value of the matrix element from the lattice results one needs to carry out the renormalization of the lattice matrix element by evaluating the renormalization function $Z(\mu, a)$. The renormalization functions are typically computed non-perturbatively. In this work we performed a perturbative subtraction of $\mathcal{O}(a^2)$ lattice artifacts, which in general leads to a better determination of the renormalization functions [19].

3 Generalized Parton Distributions

The standard procedure to study generalized parton distributions in lattice QCD is to use factorization of the light cone matrix element in terms of local operators. There are three types of GPDs involving the following operators: i) the vector operator $\mathcal{O}_{V^a}^{\mu_1\cdots\mu_n} = \bar\psi(x)\gamma^{\{\mu_1} i \overleftrightarrow{D}^{\mu_2} \dots i \overleftrightarrow{D}^{\mu_n\}} \frac{\tau^a}{2}\psi(x)$, ii) the axial-vector operator $\mathcal{O}_{A^a}^{\mu_1\cdots\mu_n} = \bar\psi(x)\gamma^{\{\mu_1} i \overleftrightarrow{D}^{\mu_2} \dots i \overleftrightarrow{D}^{\mu_n\}}\gamma_5 \frac{\tau^a}{2}\psi(x)$ and iii) the tensor operator $\mathcal{O}_{T^a}^{\mu_1\cdots\mu_n} = \bar\psi(x)\sigma^{\{\mu_1,\mu_2} i \overleftrightarrow{D}^{\mu_3} \dots i \overleftrightarrow{D}^{\mu_n\}} \frac{\tau^a}{2}\psi(x)$. In the special case when no derivatives are present then one obtains the nucleon form factors, while for $Q^2 = 0$ one has the parton distribution functions (PDFs). In this work, we limit ourselves to one-derivative operators determining the first moments of these PDFs.

Generalized form factor decomposition of the vector one-derivative operator yields

$$\begin{aligned} &\langle N(p', s')|\mathcal{O}_{V^3}^{\mu\nu}|N(p, s)\rangle \\ &= \bar u_N(p', s')\left[A_{20}(q^2)\gamma^{\{\mu}P^{\nu\}} + B_{20}(q^2)\frac{i\sigma^{\{\mu\alpha}q_\alpha P^{\nu\}}}{2m} + C_{20}(q^2)\frac{q^{\{\mu}q^{\nu\}}}{m}\right]\frac{1}{2}u_N(p, s) \end{aligned} \tag{4}$$

The nucleon spin due to a quark q is then determined by $J^q = \frac{1}{2}\left[A_{20}^q(0) + B_{20}^q(0)\right]$, while the momentum fraction is given by $\langle x\rangle_q = A_{20}^q(0)$.

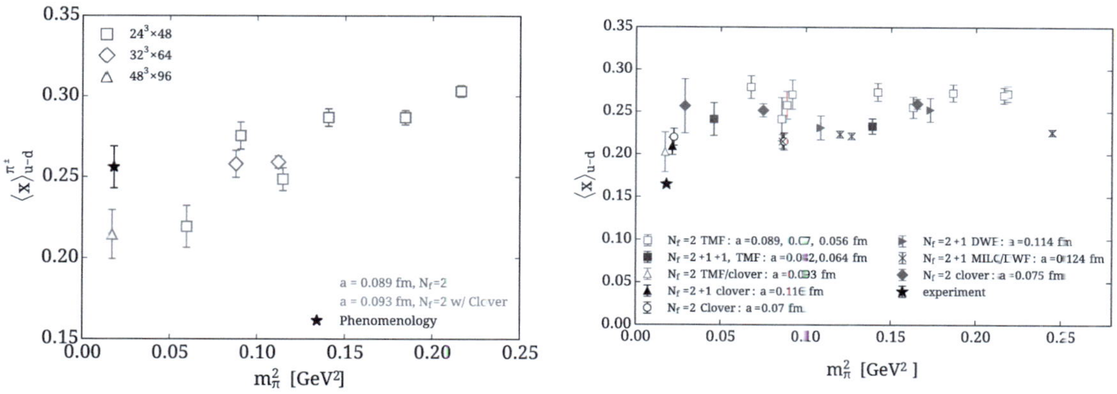

Fig. 2 $\langle x\rangle_{u-d}$ in the $\overline{MS}$ scheme at $\mu = 2\,\text{GeV}$ for the pion (*left*) and for the nucleon (*right*). The experimental value for $\langle x\rangle_{u-d}$ shown here is from Ref. [20]

3.1 Momentum Fraction of the Pion and the Nucleon

In Fig. 2, we show results on the isovector $\langle x\rangle_{u-d}$ for which disconnected contributions are zero in the isospin limit for the pion and the nucleon in the $\overline{MS}$ scheme at $\mu = 2$ GeV.

In the case of the pion we show results obtained using $N_f = 2$ twisted mass fermions. There is agreement between the clover-improved and non-clover improved ensembles [1] while no volume effects are observed within our statistical errors. The phenomenological value is extracted from a next-to-leading order analysis from Fermilab E-615 pionic Drell–Yan data [21]. The nucleon momentum fraction has been studied by a number of collaborations. In particular, near the physical point, in addition to $N_f = 2$ twisted mass plus clover-improved fermions from ETMC [1,22,23,23,24], there are results using $N_f = 2+1$ clover fermions with 2-HEX smearing from LHPC [9] and $N_f = 2$ clover fermions [25,26] from QCDSF/UKQCD. As can be seen for the latest lattice results close to the physical pion mass, $\langle x\rangle_{u-d}$ approaches the physical value. Furthermore, these studies show that for bigger sink-source time separations the value decreases towards to the physical one. We note that, as the sink-source time separation increases, one needs increasingly larger statistics to attain statistical errors that can yield a meaningful result and this analysis is on-going.

3.2 Nucleon Gluon Moment

We consider the nucleon matrix element of the gluon operator $O_{\mu\nu} = -\text{Tr}[G_{\mu\rho}G_{\nu\rho}]$, where we take $\langle N\ O_{44} - \frac{1}{3}O_{jj}|N\rangle$ at zero momentum, which yields directly $\langle x\rangle_g$. We use HYP-smearing to reduce noise and, for this operator, apply perturbative renormalization. We analyze two ensembles: one $N_f = 2+1+1$ twisted mass ensemble with $a = 0.082$ fm, $m_\pi = 373$ MeV, using 34,470 statistics and one $N_f = 2$ twisted mass plus clover ensemble, with $a = 0.093$ fm, $m_\pi = 132$ MeV and ~155,800 statistics. For the latter ensemble, we find $\langle x\rangle_g = 0.282(39)$ in $\overline{\text{MS}}$ scheme at $\mu = 2$ GeV, which gives for the first time a determination of this important quantity at the physical value of the pion mass.

3.3 Nucleon Spin

Having computed the generalized form factors A_{20} and B_{20} we can determine the nucleon spin carried by the quarks via the relation $J^q = \frac{1}{2}\left(A_{20}^q + B_{20}^q\right)$. The spin sum given by $\frac{1}{2} = \sum_q J^q = \left(\frac{1}{2}\Delta\Sigma^q - L^q\right) + J^G$ includes a gluon contribution J^G. The intrinsic quark spin is given by $\Delta\Sigma^q = g_A^q$, which has been computed within lattice QCD. Knowing J^q and g_A^q one can extract the angular momentum L^q.

In order to extract the individual quark contribution one needs to compute both the connected and disconnected parts shown schematically in Fig. 1. Connected contributions have been computed using the well established sequential inversion through the sink method. Disconnected contributions have been computed with new methods developed for graphics cards (GPUs). This was carried out for ensembles using $\mathcal{O}(150,000)$ statistics for the $N_f = 2+1+1$ ensemble of twisted mass fermions at $m_\pi = 373$ MeV [27] as well as using an

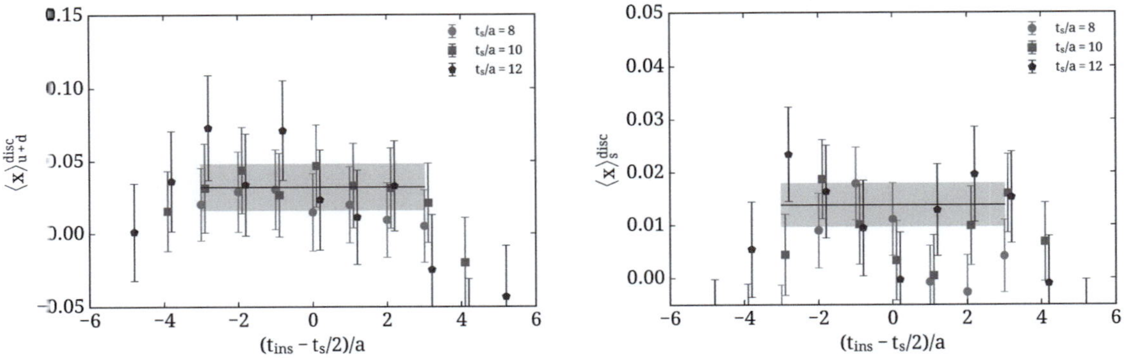

Fig. 3 Disconnected contributions to the nucleon momentum fraction at the physical point for the light (*left*) using currently 39000 measurements and for the strange (*right*) using about 185000 measurements

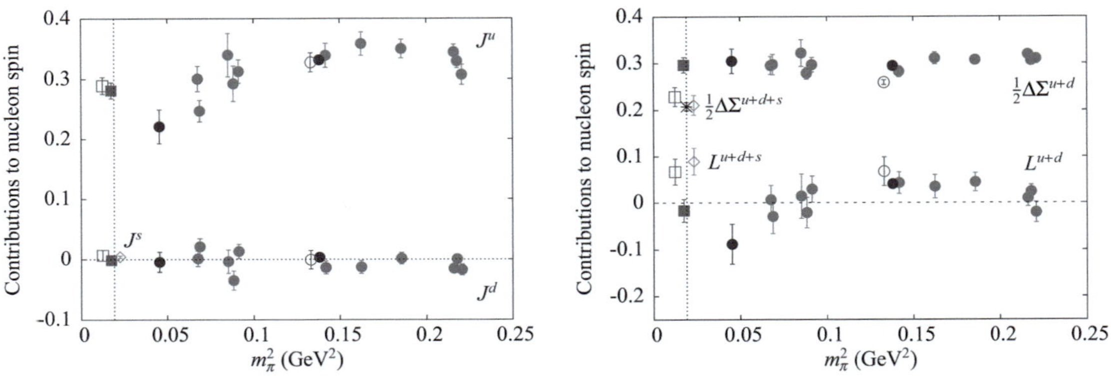

Fig. 4 The total quark spin (*left*) carried by the light and the strange quarks and $\Delta\Sigma^{u+d}$ and L^{u+d} (*right*) in the $\overline{\text{MS}}$ scheme at 2 GeV using $N_f = 2$ and $N_f = 2+1+1$ twisted mass fermions. *Open circles* and *open squares* include disconnected contributions from the u and d quarks, while the *open diamonds* also include the strange quark contribution. The *asterisk* is the experimental value of the intrinsic spin

$N_f = 2$ ensemble of twisted mass fermions with a clover term at $m_\pi = 132$ MeV [28], which is still on-going. We show in Fig. 3 the momentum fraction $\langle x \rangle_q = A^q_{20}(0)$ arising from the disconnected part. As can be seen, it is non-zero for both the light and the strange quarks. Whereas $A^q_{20}(0)$ is accessible directly at $Q^2 = 0$, $B_{20}(0)$ is not, requiring extrapolation to $Q^2 = 0$. Although we have computed the disconnected contributions to B_{20} for the few lowest Q^2-values the results are consistent with zero and we will ignore them in what follows. The disconnected intrinsic spin $\Delta\Sigma^q$ is also found to be non-zero. We show in Fig. 4 the total spin J^q as well as $\Delta\Sigma^{u+d}$ at the physical point as well as $\Delta\Sigma^{u+d+s}$. As can be seen, including the disconnected contributions brings agreement with the experimental value in the case of $\Delta\Sigma$. Using $L^q = J^q - \frac{1}{2}\Delta\Sigma^q$ we obtain results on the angular momentum. The major outcome is that disconnected contributions produce a non-zero angular momentum at the physical point. At the physical point, our preliminary values are $J^{u+d} = 0.296(17)$ and $L^{u+d} = 0.067(28)$, while $\frac{1}{2}\Delta\Sigma^{u+d} = 0.229(20)$ and $\frac{1}{2}\Delta\Sigma^{u+d+s} = 0.211(21)$, where for the first time, disconnected contributions are included.

3.4 Direct Evaluation of Parton Distribution Functions: An Exploratory Study

We consider the matrix element: $\tilde{q}(x, \Lambda, P_3) = \int_{-\infty}^{+\infty} \frac{dz}{4\pi} e^{-izxP_3} \langle P|\bar{\psi}(z,0)\,\gamma_3\, W(z)\psi(0,0)|P\rangle_{h(P_3,z)}$ where $\tilde{q}(x)$ is the quasi-distribution defined in Ref. [29], which can be computed in lattice QCD. First results are obtained for $N_f = 2+1+1$ clover fermions on HISQ sea [30] and for an $N_f = 2+1+1$ TMF ensemble with $m_\pi = 373$ MeV [31] for which we show results in Fig. 5 on the isovector distribution $q^{u-d}(x)$ for five steps of HYP smearing. The matching to the PDF $q(x)$ is done using

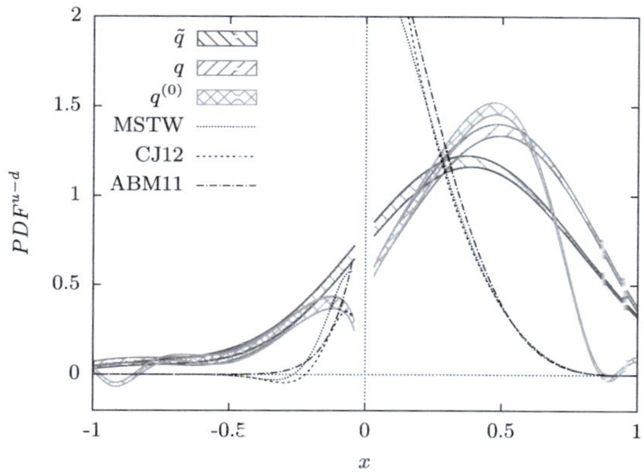

Fig. 5 Results on the unrenormalized $q(x)$ for 5-HYP steps, $P_3 = 4\pi/L$ from Ref. [31]

$$q(x,\mu) = \tilde{q}(x,\Lambda,P_3) - \frac{\alpha_s}{2\pi}\tilde{q}(x,\Lambda,P_3)\delta Z_F^{(1)}\left(\frac{\mu}{P_3},\frac{\Lambda}{P_3}\right) - \frac{\alpha_s}{2\pi}\int_{-1}^{1}\frac{dy}{y}Z^{(1)}\left(\frac{x}{y},\frac{\mu}{P_3},\frac{\Lambda}{P_3}\right)\tilde{q}(y,\Lambda,P_3) - \mathcal{O}(\alpha_s^2) \tag{5}$$

We note that: (1) The calculation of the leading UV divergences in $\tilde{q}$ perturbatively is done keeping P_3 fixed while taking $\Lambda \to \infty$ (in contrast to first taking $P_3 \to \infty$ for the renormalization of q); (2) The renormalization procedure is still under study and thus here we identify the UV regulator as μ for $q(x)$ and as Λ for the case of the quasi-distribution $\tilde{q}(x)$. The dependence on the UV regulator Λ will be translated, in the end, into a renormalization scale μ after proper renormalization; (3) Single pole terms cancel when combining the vertex and wave function corrections, and double poles are reduced to a single pole that are taken care via the principal value prescription; (4) A divergent term remains in $\delta Z^{(1)}$ that depends on the cut-off x_c

4 Conclusions

Simulations at near physical parameters of QCD are yielding important results on the structure of hadrons. In this work we have shown results on nucleon observables taking into account for the first time disconnected contributions at the physical point, which are shown to be crucial to obtain agreement with experiment for the intrinsic spin of the nucleon. Exploration of new techniques to compute hadron PDFs, charge radii and electric dipole moments is on-going, as well as, the development of techniques for resonances and for ab initio Nuclear Physics [32]. This thus represents a very rich program for zero-temperature hadron and nuclear physics and we expect rapid progress in many of these areas in the near future.

Acknowledgments I would like to thank all members of the ETM Collaboration for a most enjoyable collaboration and in particular A. Abdel-Rehim, M. Constantinou, K. Jansen, K. Hadjiyiannakou, Ch. Kallidonis, G. Koutsou, B. Kostrzewa, F. Steffens, C. Urbach, C. Wiese and A. Vaquero for their invaluable contributions to the the results presented here. This work was supported by a grant from the Swiss National Supercomputing Centre (CSCS) under project ID s540 and in addition used computational resources from the John von Neumann-Institute for Computing on the Juropa system and the BlueGene/Q system Juqueen at the research center in Jülich, resources from the Gauss Centre for Supercopmuting on HazelHen (HLRS) and PRACE project access to the Tier-0 computing resources Curie (CEA), Fermi (CINECA) and SuperMUC (LRZ).

References

1. Abdel-Rehim, A. et al.: Nucleon and pion structure with lattice QCD simulations at physical value of the pion mass. Phys. Rev. D **92**(11), 114513 (2015)
2. Abdel-Rehim, A., et al.: Simulating QCD at the physical point with $N_f = 2$ Wilson twisted mass fermions at maximal twist. arXiV:1507.05068 (2015)
3. Bali, G., Collins, S., Gockeler, M., Rödl, R., Schäfer, A., Sternbeck, A.: Nucleon generalized form factors from lattice QCD with nearly physical quark masses. arXiv:1601.04818 (2016)

4. Bhattacharya, T., Cirigliano, V., Cohen, S., Gupta, R., Joseph, A., Lin, H.-W., Yoon, B.: Iso-vector and Iso-scalar tensor charges of the nucleon from lattice QCD. Phys. Rev. D **92**(9), 094511 (2015)
5. Durr, S., et al.: Lattice computation of the nucleon scalar quark contents at the physical point. arXiv:1510.08013 (2015)
6. Ohta, S.: Systematics analyses on nucleon isovector observables in 2+1-flavor dynamical domain-wall lattice QCD near physical mass. In: PoS, LATTICE2014, p. 149 (2014)
7. Green, J.R., Negele, J.W., Pochinsky, A.V., Syritsyn, S.N., Engelhardt, M., Krieg, S.: Nucleon electromagnetic form factors from lattice QCD using a nearly physical pion mass. Phys. Rev. D **90**, 074507 (2014)
8. Horsley, R., Nakamura, Y., Nobile, A., Rakow, P.E.L., Schierholz, G., et al.: Nucleon axial charge and pion decay constant from two-flavor lattice QCD. arXiV:1302.2233 (2013)
9. Green, J.R., Engelhardt, M., Krieg, S., Negele, J.W., Pochinsky, A.V., et al.: Nucleon structure from lattice QCD using a nearly physical pion mass. arXiv:1209.1687, (2012)
10. Mueller, D., Robaschik, D., Geyer, B., Dittes, F.M., Horejsi, J.: Wave functions, evolution equations and evolution kernels from light ray operators of QCD. Fortsch. Phys. **42**, 101 (1994)
11. Ji, X.-D.: Deeply virtual Compton scattering. Phys. Rev. D **55**, 7114–7125 (1997)
12. Radyushkin, A.V.: Nonforward parton distributions. Phys. Rev. D **56**, 5524–5557 (1997)
13. Ji, X.-D.: Gauge invariant decomposition of nucleon spin and its spin-off. Phys. Rev. Lett. **78**, 610–613 (1997)
14. Airapetian, A. et al.: Separation of contributions from deeply virtual Compton scattering and its interference with the Bethe–Heitler process in measurements on a hydrogen target. JHEP **0911**, 083 (2009)
15. Chekanov, S. et al.: A measurement of the Q**2, W and t dependences of deeply virtual Compton scattering at HERA. JHEP **0905**, 108 (2009)
16. Aaron, F.D. et al.: Measurement of deeply virtual Compton scattering and its t-dependence at HERA. Phys. Lett. B **659**, 796–806 (2008)
17. Munoz Camacho, C. et al.: Scaling tests of the cross-section for deeply virtual compton scattering. Phys. Rev. Lett. **97**, 262002 (2006)
18. Stepanyan, S. et al.: First observation of exclusive deeply virtual Compton scattering in polarized electron beam asymmetry measurements. Phys. Rev. Lett. **87**, 182002 (2001)
19. Alexandrou, C., Constantinou, M., Korzec, T., Panagopoulos, H., Stylianou, F.: Renormalization constants of local operators for Wilson type improved fermions. Phys. Rev. D **86**, 014505 (2012)
20. Alekhin, S., Blumlein, J., Moch, S.: Parton distribution functions and benchmark cross sections at NNLO. Phys. Rev. D **86**, 054009 (2012)
21. Wijesooriya, K., Reimer, P.E., Holt, R.J.: The pion parton distribution function in the valence region. Phys. Rev. C **72**, 065203 (2005)
22. Abdel-Rehim, A., Alexandrou, C., Dimopoulos, P., Frezzotti, R., Jansen, K., et al.: Progress in simulations with twisted mass fermions at the physical point. In: PoS, LATTICE2014, p. 119 (2014)
23. Abdel-Rehim, A., Boucaud, Ph, Carrasco, N., Deuzeman, A., Dimopoulos, P., et al.: A first look at maximally twisted mass lattice QCD calculations at the physical point. In: PoS, LATTICE2013, p. 264 (2013)
24. Abdel-Rehim, A., Burger, F., Deuzeman, A., Jansen, K., Kostrzewa, B., Scorzato, L., Urbach, C.: Recent developments in the tmLQCD software suite. PoS, LATTICE2013, 414 (2014)
25. Bali, G.S., Collins, S., Glle, B., Gckeler, M., Najjar, J., Rdl, R.H., Schfer, A., Schiel, R.W., Sternbeck, A., Sldner, W.: The moment $\langle x \rangle_{u-d}$ of the nucleon from $N_f = 2$ lattice QCD down to nearly physical quark masses. Phys. Rev. D **90**(7), 074510 (2014)
26. Pleiter D., et al.: Nucleon form factors and structure functions from N(f)=2 Clover fermions. In: PoS, LATTICE2010, p. 153 (2010)
27. Abdel-Rehim, A., Alexandrou, C., Constantinou, M., Drach, V., Hadjiyiannakou, K., Jansen, K., Koutsou, G., Vaquero, A.: Disconnected quark loop contributions to nucleon observables in lattice QCD. Phys. Rev. D **89**(3), 034501 (2014)
28. Abdel-Rehim, A., Alexandrou, C., Constantinou, M., Hadjiyiannakou, K., Jansen, K., Kallidonis, C., Koutsou, G., Avils-Casco, A.V.: Disconnected quark loop contributions to nucleon observables using $N_f = 2$ twisted clover fermions at the physical value of the light quark mass. In: Proceedings, 33rd International Symposium on Lattice Field Theory (Lattice 2015) (2015)
29. Ji, X.: Parton physics on a euclidean lattice. Phys. Rev. Lett. **110**, 262002 (2013)
30. Lin, H.-W., Chen, J.-W., Cohen, S.D., Ji, X.: Flavor structure of the nucleon sea from lattice QCD. Phys. Rev. D **91**, 054510 (2015)
31. Alexandrou, C., Cichy, K., Drach, V., Garcia-Ramos, E., Hadjiyiannakou, K., Jansen, K., Steffens, F., Wiese, C.: Lattice calculation of parton distributions. Phys. Rev. D **92**, 014502 (2015)
32. Savage, M.J.: Nuclear physics from lattice quantum chromodynamics. In: 12th Conference on the Intersections of Particle and Nuclear Physics (CIPANP 2015) Vail, Colorado, USA, 19–24 May 2015 (2015)

DOI 10.1007/978-3-319-50699-9_33
Reprinted from *Few-Body System* Journal, DOI 10.1007/s00601-016-1075-3

Henry Lamm · **Richard F. Lebed**

Nonperturbative True Muonium on the Light Front with TMSWIFT

Received: 21 January 2016 / Accepted: 2 March 2016 / Published online: 28 March 2016

Abstract The true muonium ($\mu\bar{\mu}$) bound state presents an interesting test of light-cone quantization techniques. In addition to exhibiting the standard problems of handling non-perturbative calculations, true muonium requires correct treatment of $e\bar{e}$ Fock-state contributions. Having previously produced a crude model of true muonium using the method of iterated resolvents, our current work has focused on the inclusion of the box diagrams to improve the cutoff-dependent issues of the model. Further, a parallel computer code, TMSWIFT, allowing for smaller numerical uncertainties, has been developed. This work focuses on the current state of these efforts to develop a model of true muonium that is testable at near-term experiments.

1 Introduction

The current state of flavor physics has a "muon problem". Several muon observables, of which the most prominent are the $(g-2)_\mu$ [5] and the $r_P(\mu H)$ [1], show disagreement with Standard Model calculations. A strong candidate for providing clues to resolve this problem is the bound state ($\mu\bar{\mu}$), dubbed "true muonium" [17]. Traditional bound states have limited new physics reach due to small reduced masses $\mu \approx m_e$ or nuclear-structure uncertainties. In contrast, true muonium's $\mu = m_\mu/2$ and leptonic nature make it an ideal probe through the Lamb shift or hyperfine splitting [23–25,37].

True muonium has not been observed due to difficulties in producing associated low-energy muon pairs and its short lifetime ($\tau \approx 1$ ps). Many proposed methods of production channels exist [2,3,6,11,14,16,21,30,31]. The Heavy Photon Search (HPS) experiment plans to search for true muonium in 2016 [3,8]. Additionally, the DImeson Relativistic Atom Complex (DIRAC) could observe it in an upgraded run [4,12]. Given enough statistics, DIRAC could measure the Lamb shift using methods developed for ($\pi\bar{\pi}$) [32]. These experiments produce relativistic true muonium. Unfortunately, instant-form (conventional fixed time) wave functions are not boost-invariant; thus production and decay rates are modified. To reduce this uncertainty, we are producing boost-invariant wave functions through light-front techniques [26].

Quantizing at light-front time $x^+ \equiv t + z$ (called *front form*) [13] allows one to develop a Hamiltonian formalism [7]. In this formalism, a analogue of the Schrödinger equation exists as an infinite but denumerable set of coupled integral equations. One may then truncate the equations by limiting the set of Fock states

Work supported by the National Science Foundation under Grant Nos. PHY-1068286 and PHY-1403891 and by the International Light Cone Advisory Committee under the McCartor Fellowship program.

H. Lamm (✉) · R. F. Lebed
Department of Physics, Arizona State University, P.O. Box 871504, Tempe, AZ 85287-1504, USA
E-mail: hlammiv@asu.edu

R. F. Lebed
E-mail: Richard.Lebed@asu.edu

included and discretizing momenta. This finite problem can then be solved by a computer. This technique is called Discretized Light-Cone Quantization (DLCQ) [34].

Our work is a direct continuation of the DLCQ methods developed in Refs. [18,22,35,36] to true muonium. The explicit $|\gamma\rangle$ component allows mixing between $|e\bar{e}\rangle$, $|\mu\bar{\mu}\rangle$, and $|\tau\bar{\tau}\rangle$ Fock states. We also implemented a counterterm subtraction to the interaction amplitudes that improves the ultraviolet behavior. Following Ref. [36], we adopted the larger value $\alpha = 0.3$. In this regime, QED perturbative calculations potentially become unreliable. We use this strong coupling value of α to study flavor-mixing and regularization dependence. In this work, we also investigate the approach to the physical coupling constant of QED.

2 True Muonium model

In front-form, the eigenvalue equation for a bound state is given by:

$$\left(M^2 - \sum_i \frac{m_i^2 + \boldsymbol{k}_{\perp i}^2}{x_i}\right)\psi(x_i, \boldsymbol{k}_{\perp i}; h_i) = \sum_{h_j} \int_D \mathrm{d}x'_j \mathrm{d}^2\boldsymbol{k}'_{\perp j} \langle x_i, \boldsymbol{k}_{\perp i}; h_i\, |V|\, x'_j, \boldsymbol{k}'_{\perp j}; h_j\rangle \psi(x'_j, \boldsymbol{k}'_{\perp, j}; h_j), \tag{1}$$

where M is the invariant mass of the state, m indicates a mass term, i, j are particle indices, x and $\boldsymbol{k}_\perp$ are the conventional light-front coordinates, h is shorthand for all intrinsic quantum numbers of a state, and V are interaction terms given by the light-front Hamiltonian. The domain D in Eq. (1) is defined by introducing a cutoff Λ, and we choose [27]

$$\frac{m^2 + \boldsymbol{k}_\perp^2}{x(1-x)} \leq \Lambda^2 + 4m^2. \tag{2}$$

In our model, we are considering the charge-zero, lepton family number-zero states in the truncated Fock space of

$$|\Psi\rangle = \psi_{\mu\bar{\mu}}|\mu\bar{\mu}\rangle + \psi_{e\bar{e}}|e\bar{e}\rangle + \psi_{\tau\bar{\tau}}|\tau\bar{\tau}\rangle + \psi_\gamma|\gamma\rangle + \psi_{\mu\bar{\mu}\gamma}|\mu\bar{\mu}\gamma\rangle + \psi_{e\bar{e}\gamma}|e\bar{e}\gamma\rangle + \psi_{\tau\bar{\tau}\gamma}|\tau\bar{\tau}\gamma\rangle \tag{3}$$

Solving for the eigenstates of H_{LC} with this limited Fock space nonetheless gives the bound states of positronium ($e\bar{e}$), true muonium ($\mu\bar{\mu}$), and true tauonium ($\tau\bar{\tau}$), as well as associated continuum states (up to effects from neglected higher-order Fock states). The wave functions are in the form of Eq. (3) with helicity states for only $|\mu\bar{\mu}\rangle$, $|e\bar{e}\rangle$, and $|\tau\bar{\tau}\rangle$ components. The $|\gamma\rangle$ and $|\ell\bar{\ell}\gamma\rangle$ components are folded into V_{eff} by means of the method of iterated resolvents [33,36].

Since we have restricted ourselves to QED, the inclusion of quark flavors is qualitatively the same as additional leptons and TMSWIFT would allow this. We consider the benchmark model used here of true muonium with a single particle flavor lighter (e) and a single heavier flavor (τ) to encapsulate all the numerical complications that will need to be resolved, and therefore have neglected all hadronic effects (with the understanding that these will be critical for precision physics).

It has been shown [22,26,36] that strong dependence in ${}^1S_0$ states on Λ arises from the matrix element between antiparallel-helicity states (called G_2 in App. F.3 of Ref. [36]). This element approaches a constant as $k_\perp \equiv |\boldsymbol{k}_\perp|$ or $k'_\perp \equiv |\boldsymbol{k}'_\perp| \to \infty$. The result is a δ function-like behavior in configuration space. Ref. [22] chose to regularize this singularity by deleting the entire term. We instead subtract only its limit as $k_\perp$ or $k'_\perp \to \infty$, retaining part of the term (including x and x' dependence). This scheme removes the strong Λ dependence of ${}^1S_0$ states in both QED [26,38] and QCD [28] models. At the cost of unknown regularization dependence, the model allows for taking the $\Lambda \to \infty$ limit. We found that the 1^1S_0 state eigenvalue can be fit by

$$M^2(N, \Lambda) = M_\infty^2(1 - be^{-cN})(1 - de^{-f\Lambda}). \tag{4}$$

where N is the number of discretization points utilized in each of the two dimensions that are used.

3 Results from Mesonix and TMSWIFT

Our initial results [26] were for the system of electrons and muons. Using a modified version of the Mesonix code [36], we produced the entire spectrum of positronium and true muonium. The largest N we explored with this serial code was $N_\mu = N_e = 37$. This required a full day of CPU time on a 3.0 GHz core to produce

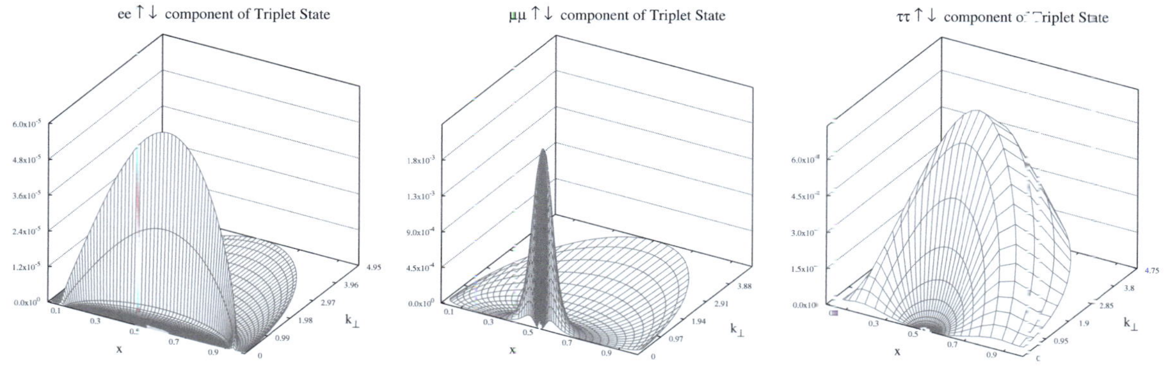

Fig. 1 The $1^3S_1^0$ probability density of (*left*) $\uparrow\downarrow e\bar{e}$, (*center*) $\uparrow\downarrow \mu\bar{\mu}$, and (*right*) $\uparrow\downarrow \tau\bar{\tau}$ components of true muonium with $J_z = 0$, as functions of x and $k_\perp$, for $\alpha = 0.3$, $m_e = \frac{1}{2}m_\mu$, $m_\tau = 2m_\mu$ $\Lambda_i = 10\alpha m_i/2$, and $N_\mu = N_\tau = 37, N_e = 71$

Table 1 M^2 of the true muonium ground state 1^1S_0 for a range of α in units of m_μ

α	$M^2(1^1S_0)_{LF}$	$M^2(1^1S_0)_{IF}$
0.3	3.8944(3)	3.9053
0.1	3.98988(2)	3.98994
0.07	3.99507088(8)	3.99508575
0.05	3.99749246(8)	3.99749629
0.02	3.9996087(2)	3.99959990
0.01	3.999900017(5)	3.999899994

The first column is from TMSWIFT and the second column presents $\mathcal{O}(\alpha^4)$ instant-form calculations. The light-front results were obtained by fitting a set of calculations with $N_\mu = (13, 71)$ and $\Lambda_\mu = (1, 30)$ in units of $\alpha m_\mu/2$ to Eq. (4). The reported errors are only from fitting

a single J_z spectrum. These results agreed with the instant-form corrections to the annihilation channel from electron loops, albeit with a large uncertainty due to difficulty in properly sampling continuum electron states.

We have written a new parallel code, TMSWIFT (True Muonium Solver With Front-form Techniques), to overcome the numerical limitations. This code is more flexible, as well as using the parallel eigenvalue solver package SLEPc [15]. TMSWIFT allows an arbitrary number of flavors each specified by a mass, Λ, and N. Different discretization schemes are available for exploration of numerical errors and efficiency. TMSWIFT allows easy implementation of new effective interactions (e.g., from $|\gamma\gamma\rangle$ states). We have performed three-flavor calculations with $N_{\mu,\tau} = 37$ and $N_e = 71$ for $J_z = 0$ in 2 h using 512 cores of the Stampede supercomputer. In Fig. 1, we present anti-parallel helicity wave-function components in the 1^3S_0 state of true muonium.

Calculations with lower α were also performed. The numerical stability becomes worse at decreasing α for fixed N. The lowest achieved value is $\alpha = 0.01 \approx 1.4\alpha_{\text{QED}}$. Fitting multiple values of N and Λ, we were able to extract values of $M^2_\infty(1^1S_0)$ and compare them to instant-form calculations at $\mathcal{O}(\alpha^4)$ (Table 1). Currently the source of the small discrepancy is undetermined, but two sources to investigate are regularization dependence and the mismatch of higher-order corrections between the $\alpha-$perturbative instant-form and the nonperturbative Fock-state front-form calculations.

4 The Two-Photon Fock State

We are currently working to include the $|\gamma\gamma\rangle$ state, which dominates the decay of singlet states of true muonium (and in particular should have a pronounced effect on $^1S_0$ wave functions). To do this, we have computed the sum of 19 time-ordered light-front diagrams (Fig. 2 and permutations of the vertices, plus diagrams where each internal particle is instantaneous), which include some $|\ell\bar{\ell}\ell\bar{\ell}\rangle$ and $|\ell\bar{\ell}\ell'\bar{\ell}'\rangle$ diagrams to preserve gauge invariance. The diagrams result in two integrals for the effective interactions, for the $|\gamma\gamma\rangle$ sector we have

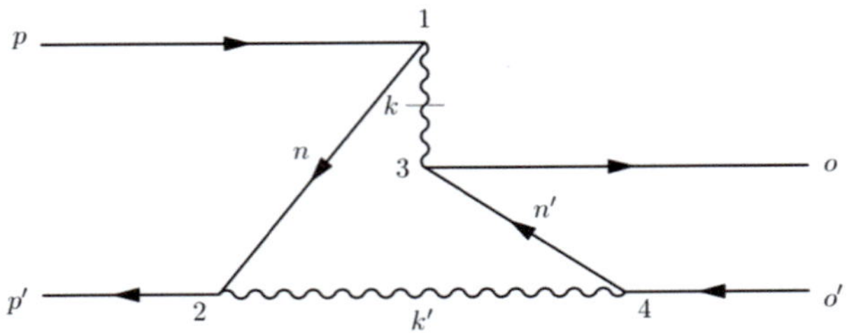

Fig. 2 Example of a $|\gamma\gamma\rangle$ intermediate-state diagram that must be computed

$$H_{\gamma\gamma} = -\frac{\alpha^2}{(2\pi)^4}\int d^3k \left(\frac{G_{\ell^+\ell^-\gamma}G_{\gamma\gamma}G'_{\ell^+\ell^-\gamma}}{\sqrt{p^+o^+(o')^+(p')^+}}\right)\frac{\bar{u}_1\gamma^\mu\left(\not{l}+m+\frac{\gamma^+}{2G_{\ell^+\ell^-\gamma}}\right)\gamma^\sigma v_2\bar{v}_4\gamma^\rho\left(\not{l}'+m+\frac{\gamma^+}{2G'_{\ell^+\ell^-\gamma}}\right)\gamma^\nu u_3}{|k^+||P^+-k^+||p^+-k^+||o^+-k^+|}$$
$$\times\left(d_{\mu\nu}d_{\sigma\rho}\Theta_k\Theta_{-k}+\frac{\eta_{\mu\nu}d_{\sigma\rho}}{G_{\gamma\gamma}|k^+|}\Theta_k\Theta_{-p}\Theta_{-o}-\frac{d_{\mu\nu}\eta_{\sigma\rho}}{G_{\gamma\gamma}|(P-k)^+|}\Theta_{-k}\Theta_p\Theta_o\right), \quad (5)$$

and a similar integral occurs for the $|\ell\bar{\ell}\ell\bar{\ell}\rangle$ sector. In this integral, l^μ and $(l')^\mu$ are not free momenta but fixed functions of the external momenta and k^μ. We further used the notation

$$\Theta_k = \theta(k^+), \qquad \Theta_{-k} = \theta((P-k)^+), \qquad \Theta_{\pm i} = \theta(\pm(i-k)^+), \quad (6)$$

where P^+ is the total + component at any time, i indicates an external momentum, and k is the k^μ photon's + component. We introduce $\eta_{\mu\nu}$, which is zero except for the $\eta_{++}=1$ component, and $d^{\mu\nu}(k)$ is the standard light-front photon propagator (Eq.(A.23) in Ref. [7]).

The non-perturbative propagators G_i need to be defined such that the instantaneous diagrams are canceled correctly in the vein of Ref. [33]. While in the $|\ell\bar{\ell}\gamma\rangle$ sector only one G_i occurs, the mixing between $|\gamma\gamma\rangle$ and $|\ell\bar{\ell}\ell\bar{\ell}\rangle$ has presented unique problems in ensuring the proper cancellations. We are looking at different forms of this function. Additionally, since we include a pure photon state, IR features of the interaction must be dealt with. The simplest method to would be to include a photon mass to the interaction and study the limit as $m_\gamma \to 0$. From these results, we can compute helicity matrix elements and then implement them in TMSWIFT.

5 Summary and Conclusion

Here, we have computed light-front wave functions for true muonium. This was done in a three-flavor model of QED with only 7 explicit Fock states. Our new, parallelized code is capable of simulating such larger Fock spaces. We have shown that extrapolations to $\Lambda, N \to \infty$ are possible. Finally, TMSWIFT also allows for the study of lower values of α. We are working to include the $|\gamma\gamma\rangle$ state which requires including $|\ell\bar{\ell}\ell\bar{\ell}\rangle$ and $|\ell\bar{\ell}\ell'\bar{\ell}'\rangle$ for gauge invariance. These corrections are crucial for precision predictions and are a necessary step for QCD. A next step would be to investigate observables like decay constants and production cross-sections. Having shown that numerical simulations at physical α_{QED} are possible, renormalization remains the only hurdle to making experimental predictions. The initial studies here indicate that the numerical error is small in the energy levels, but how renormalization effects this picture is unclear. Limitations on the Fock space have been greatly decreased via TMSWIFT. This allows for explicit Fock-state renormalization methods like Pauli–Villars regulators [9,29] and sector-dependent counterterms [19,20]. Using the exchange properties of leptons could reduce the number of basis states, similar to Ref. [10].

Acknowledgments This work was supported by the National Science Foundation under Grant Nos. PHY-1068286 and PHY-1403891 and through the McCartor Fellowship program from the International Light-Cone Advisory Committee. This work used the Extreme Science and Engineering Discovery Environment (XSEDE), which is supported by National Science Foundation grant number ACI-1053575. The author would also like to thank the to the organizers of the Light Cone 2015, whose support helped my participation.

References

1. Antognini, A., Nez, F., Schuhmann, K., Amaro, F.D., Biraben, F. et al.: Proton structure from the measurement of $2S-2P$ transition frequencies of muonic hydrogen. Science **339**, 417–420 (2013)
2. Arteaga-Romero, N., Carimalo, C., Serbo, V.: Production of bound triplet $\mu^+\mu^-$ system in collisions of electrons with atoms. Phys. Rev. A **62**, 032501 (2000)
3. Banburski, A., Schuster, P.: The production and discovery of true muonium in fixed-target experiments. Phys. Rev. D **86**, 093007 (2012)
4. Benelli, A.: DIRAC experiment at CERN. EPJ Web Conf. **37**, 01011 (2012)
5. Bennett, G. et al.: Final report of the muon E821 anomalous magnetic moment measurement at BNL. Phys. Rev. D **73**, 072003 (2006)
6. Brodsky, S.J., Lebed, R.F.: Production of the smallest QED atom: true muonium ($\mu^+\mu^-$). Phys. Rev. Lett. **102**, 213401 (2009)
7. Brodsky, S.J., Pauli, H.C., Pinsky, S.S.: Quantum chromodynamics and other field theories on the light cone. Phys. Rep. **301**, 299–486 (1998)
8. Celentano, A.: The heavy photon search experiment at Jefferson laboratory. J. Phys. Conf. Ser. **556**(1), 012064 (2014)
9. Chabysheva, S.S., Hiller, J.R.: Nonperturbative Pauli-Villars regularization of vacuum polarization in light-front QED. Phys. Rev. D **82**, 034004 (2010)
10. Chabysheva, S.S., Hiller, J.R.: Basis of symmetric polynomials for many-boson light-front wave functions. Phys. Rev. E **90**(6), 063,310 (2014)
11. Chen, Y., Zhuang, P.: Dimuonium ($\mu^+\mu^-$) Production in a Quark-Gluon Plasma. arXiv:1204.4389. (2012)
12. Chliapnikov, P.: Numbers of $\mu^\pm\pi^\mp$, $\mu^+\mu^-$ and K^+K atoms and Coulomb pairs in the DIRAC experiment. DIRAC-NOTE-2014-05
13. Dirac, P.A.M.: Forms of Relativistic Dynamics. Rev. Mod. Phys. **21**, 392–399 (1949)
14. Ginzburg, I., Jentschura, U., Karshenboim, S.G., Krauss, F., Serbo, V. et al.: Production of bound $\mu^+\mu^-$ systems in relativistic heavy ion collisions. Phys. Rev. C **58**, 3565–3573 (1998)
15. Hernandez, V., Roman, J.E., Vidal, V.: SLEPc: a scalable and flexible toolkit for the solution of eigenvalue problems. ACM Trans. Math. Softw. **31**(3), 351–362 (2005)
16. Holvik, E., Olsen, H.A.: Creation of relativistic fermionium in collisions of electrons with atoms. Phys. Rev. D **35**, 2124 (1987)
17. Hughes, V., Maglic, B.: True Muonium. Bull. Am. Phys. Soc. **16**, 65 (1971)
18. Kaluza, M., Pauli, H.C.: Discretized light cone quantization: $e^+e^-(\gamma)$ model for positronium. Phys. Rev. D **45**, 2968–2981 (1992)
19. Karmanov, V.A., Mathiot, J.F., Smirnov, A.V.: Systematic renormalization scheme in light-front dynamics with Fock space truncation. Phys. Rev. D **77**, 085028 (2008)
20. Karmanov, V.A., Mathiot, J.F., Smirnov, A.V.: Ab initio nonperturbative calculation of physical observables in light-front dynamics Application to the Yukawa model. Phys. Rev. D **86**, 085006 (2012)
21. Kozlov, G.: On the problem of production of relativistics lepton bound states in the decays of light mesons. Sov. J. Nucl. Phys. **48**, 167–171 (1988)
22. Krautgartner, M., Pauli, H.C., Wolz, F.: Positronium and heavy quarkonia as testing case for discretized light cone quantization. Phys. Rev. D **45**, 3755–3774 (1992)
23. Lamm, H.: Can Galileons solve the muon problem? Phys. Rev. D **92**(5), 055007 (2015a)
24. Lamm, H.: Electroweak corrections to the true muonium hyperfine splitting. Phys. Rev. D **91**, 073008 (2015b)
25. Lamm, H.: True muonium: the atom that has it all. In: CIPANP 2015 Vail, Colorado, USA, 19–24 May 2015. arXiv:1509.09306
26. Lamm, H., Lebed, R.F.: True muonium ($\mu^+\mu^-$) on the light front. J. Phys. G **41**(12), 125003 (2014)
27. Lepage, G.P., Brodsky, S.J.: Exclusive processes in perturbative quantum chromodynamics. Phys. Rev. D **22**, 2157 (1980)
28. Li, Y., Maris, P., Zhao, X., Vary, J.P.: Heavy Quarkonium in a Holographic Basis. arXiv:1509.07212 (2015)
29. Malyshev, M.Y., Paston, S.A., Prokhvatilov, E.V., Zubov, R.A.: Renormalized light front Hamiltonian in the Pauli-Villars regularization. Int. J. Theor. Phys. **54**(1), 169–184 (2015)
30. Moffat, J.: Does a heavy positronium atom exist?. Phys. Rev. Lett. **35**, 1605 (1975)
31. Nemenov, L.: Atomic decays of elementary particles. Yad Fiz **15**, 1047–1050 (1972)
32. Nemenov, L.L., Ovsyannikov, V.D.: DC field control of decay processes in a ($\pi^+\pi^-$) atom: A possibility to measure the energy splitting between 2s- and 2p-states. Phys. Lett. B **514**, 247–252 (2001). doi:10.1016/S0370-2693(01)00802-4
33. Pauli, H.C.: Solving gauge field theory by discretized light cone quantization. hep-th/9706036. (1997)
34. Pauli, H.C., Brodsky, S.J.: Discretized light cone quantization: solution to a field theory in one space one time dimensions. Phys. Rev. D **32**, 2001 (1985)
35. Tang, A.C., Brodsky, S.J., Pauli, H.C.: Discretized light cone quantization: formalism for quantum electrodynamics. Phys. Rev. D **44**, 1842–1865 (1991)
36. Trittmann, U., Pauli, H.C.: Quantum electrodynamics at strong couplings. hep-th/9704215. (1997)
37. Tucker-Smith, D., Yavin, I.: Muonic hydrogen and MeV forces. Phys. Rev. D **83**, 101702 (2011)
38. Wiecki, P., Li, Y., Zhao, X., Maris, P., Vary, J.P.: Basis light-front quantization approach to positronium. Phys. Rev. D **91**(10), 105009 (2015)

DOI 10.1007/978-3-319-50699-9_34
Reprinted from *Few-Body System* Journal, DOI 10.1007/s00601-016-1076-2

Daya Shankar Kulshreshtha · Usha Kulshreshtha · James P. Vary

Light-Front Quantization of the Restricted Gauge Theory of QCD_2

Received: 24 January 2016 / Accepted: 2 March 2016 / Published online: 15 March 2016

Abstract In this talk we study the light-front quantization of the restricted gauge theory of QCD_2 *à* la Cho et al.

In this talk, we study the light-front (LF) quantization (LFQ) of the restricted gauge theory of QCD_2 *à* la Cho et al. [1–7] on the hyperplanes defined by the equal light-cone (LC) time ($\tau = x^+ = \frac{1}{\sqrt{2}}(x^0 + x^1)) =$ constant [8–11], using the Hamiltonian, path integral and BRST [9–13] quantization procedures under the appropriate LC gauge-fixing conditions (GFC's). The theory makes use of the so-called "Cho-decomposition", which is, in fact, the gauge independent decomposition of the non-Abelian potential into the restricted potential and the valence potential and it helps in the clarification of the topological structure of the non-Abelian gauge theory, and it also takes care of the topological characters in the dynamics [2–6]. An important consequence of the decomposition is that it allows one to view *QCD* as the restricted gauge theory (made of the restricted potential) which is coupled to a gauge-covariant colored vector field (the valence potential). The restricted potential is defined in such a way that it allows a co-variantly constant unit iso-vector $\hat{n}$ everywhere in space-time, which enables one to define the gauge-independent color direction everywhere in space-time and at the same time allows one to define the magnetic potential of the non-Abelian monopoles. Furthermore it has the full SU(2) gauge degrees of freedom, in spite of the fact that it is restricted. Consequently, the restricted *QCD* made of the restricted potential describes a very interesting dual dynamics of its own, and plays a crucial role in the understanding of *QCD*. On the other hand, the restricted *QCD* is a constrained system, due to the presence of the topological field $\hat{n}$ which is constrained to have the unit norm. A natural way to accommodate

Presented by Daya Shankar Kulshreshtha at LIGHTCONE2015, 21–25 September, 2015, LMF, INFN, Frascati, Italy.

D. S. Kulshreshtha (✉) · U. Kulshreshtha · J. P. Vary
Department of Physics and Astronomy, Iowa State University, Ames, IA 50011, USA
E-mail: dskulsh@gmail.com

U. Kulshreshtha
E-mail: ushakulsh@gmail.com

J. P. Vary
E-mail: jvary@iastate.edu

D. S. Kulshreshtha
Department of Physics and Astrophysics, University of Delhi, Delhi 110007, India

U. Kulshreshtha
Department of Physics, Kirori Mal College, University of Delhi, Delhi 110007, India

the topological degrees into the theory is to introduce a topological field $\hat{n}$ of unit norm, and to decompose the connection into the Abelian projection part which leaves $\hat{n}$ a covariant constant and the remaining part which forms a covariant vector field [2–6]:

$$\mathbf{A}_\mu = A_\mu \hat{n} - \frac{1}{g}\hat{n} \times \partial_\mu \hat{n} + \mathbf{X}_\mu = \hat{A}_\mu + \mathbf{X}_\mu, \quad A_\mu = \hat{n} \cdot \mathbf{A}_\mu, \quad \hat{n}^2 = 1 \tag{1}$$

where A^μ is the "electric" potential and the Abelian projection A^μ is precisely the connection which leaves $\hat{n}$ invariant under the parallel transport and makes $\hat{n}$ a covariant constant: $\hat{D}_\mu \hat{n} = \partial_\mu \hat{n} + g\hat{A}_\mu \times \hat{n} = 0$ Also, under the infinitesimal gauge transformation:

$$\delta\hat{n} = -\vec{\alpha} \times \hat{n}, \quad \delta\mathbf{A}_\mu = \frac{1}{g} D_\mu \vec{\alpha}, \quad \delta A_\mu = \frac{1}{g}\hat{n} \cdot \partial_\mu \vec{\alpha}, \quad \delta\hat{A}_\mu = \frac{1}{g}\hat{D}_\mu \vec{\alpha}, \quad \delta\mathbf{X}_\mu = -\vec{\alpha} \times \mathbf{X}_\mu \tag{2}$$

This shows that A^μ by itself describes an $SU(2)$ connection which enjoys the full $SU(2)$ gauge degrees of freedom. The restricted potential A^μ is defined by the Abelian projection and the connection space (the space of all gauge potentials) forms an affine space. Indeed the affine nature of the connection space guarantees that one can describe an arbitrary potential simply by adding a gauge-covariant piece X_μ to the restricted potential.

Above decomposition is known as Cho-decomposition or the Cho-Faddeev-Niemi decomposition (introduced in an attempt to demonstrate the monopole condensation in QCD), and the importance of this decomposition in clarifying the non-Abelian dynamics has been studied by many authors [1–7]. The restricted potential A^μ actually has a dual structure and the field strength made of the restricted potential is decomposed as:

$$\hat{F}_{\mu\nu} = (F_{\mu\nu} + H_{\mu\nu})\,\hat{n}, \quad F_{\mu\nu} = (\partial_\mu A_\nu - \partial_\nu A_\mu), \quad H_{\mu\nu} = -\frac{1}{g}\hat{n} \cdot (\partial_\mu \hat{n} \times \partial_\nu \hat{n}) = \left(\partial_\mu \tilde{C}_\nu - \partial_\nu \tilde{C}_\mu\right)$$

where C_μ is the "magnetic" potential (it is possible to introduce the magnetic potential (at least locally section-wise), since one has the identity: $\partial_\mu \tilde{H}_{\mu\nu} = 0$, $\tilde{H}_{\mu\nu} = \frac{1}{2}\epsilon_{\mu\nu\rho\sigma} H_{\rho\sigma}$. This allows one to identify the non-Abelian monopole potential by: $\mathbf{C}_\mu = -\frac{1}{g}\hat{n} \times \partial_\mu \hat{n}$, in terms of which the magnetic field is expressed as: $\mathbf{H}_{\mu\nu} = (\partial_\mu \mathbf{C}_\nu - \partial_\nu \mathbf{C}_\mu + g\mathbf{C}_\mu \times \mathbf{C}_\nu) = -\frac{1}{g}(\partial_\mu \hat{n} \times \partial_\nu \hat{n}) = H_{\mu\nu}\hat{n}$ With the above connection (albeit decomposition) one has:

$$\mathbf{F}_{\mu\nu} = \left[(F_{\mu\nu} + H_{\mu\nu})\hat{n} + \hat{D}_\mu \mathbf{X}_\nu - \hat{D}_\nu \mathbf{X}_\mu + g\mathbf{X}_\mu \times \mathbf{X}_\nu\right], \quad \hat{n} \cdot \mathbf{X}_\mu = 0, \quad \hat{n} \cdot \hat{D}_\mu \mathbf{X}_\nu = 0 \tag{3}$$

and for the Yang-Mills Lagrangian density one has: [2–7]:

$$\begin{aligned}\tilde{\mathcal{L}} = \Bigg[&-\frac{1}{4}\hat{F}_{\mu\nu}\hat{F}^{\mu\nu} - \frac{1}{4}\left(\hat{D}_\mu \mathbf{X}_\nu - \hat{D}_\nu \mathbf{X}_\mu\right)^2 - \frac{g}{2}\hat{F}_{\mu\nu} \cdot (\mathbf{X}_\mu \times \mathbf{X}_\nu) \\ &- \frac{g^2}{4}(\mathbf{X}_\mu \times \mathbf{X}_\nu)^2 + \lambda\,(\hat{n}^2 - 1) + \lambda_\mu \hat{n} \cdot \mathbf{X}_\mu\Bigg]\end{aligned} \tag{4}$$

where λ and λ_μ are the Lagrange multiplier fields and $g^{\mu\nu} := g_{\mu\nu} = diag(+1, -1)$. The Lagrangian density of the so-called restricted gauge theory made of the Abelian projection without $\mathbf{X}_\mu$ is therefore defined by [2–7]:

$$\mathcal{L} = \left[-\frac{1}{4}\hat{F}_{\mu\nu}\hat{F}^{\mu\nu} + \lambda\,(\hat{n}^2 - 1)\right] \tag{5}$$

Theory defined by the above Lagrangian density has a full $SU(2)$ gauge invariance and it describes the dual dynamics of QCD with the dynamical degrees of the maximal Abelian subgroup U(1) as the electric component and the topological degrees of SU(2) as the magnetic component. We now study the Hamiltonian and path integral quantization of the above restricted gauge theory of QCD_2 (made of the Abelian projection without $\mathbf{X}_\mu$) defined by above Lagrangian density re-expressed as [2–7]:

$$\begin{aligned}\mathcal{L} &= \left[-\frac{1}{4}\hat{F}_{\mu\nu}\hat{F}^{\mu\nu} + \lambda\,(\hat{n}^2 - 1)\right] = \left[-\frac{1}{4}(F_{\mu\nu} + H_{\mu\nu})(F^{\mu\nu} + H^{\mu\nu}) + \lambda\,(\hat{n}^2 - 1)\right] \\ &= \left[-\frac{1}{4}F_{\mu\nu}F^{\mu\nu} - \frac{1}{2}F_{\mu\nu}H^{\mu\nu} - \frac{1}{4}H_{\mu\nu}H^{\mu\nu} + \lambda\,(\hat{n}^2 - 1)\right]\end{aligned} \tag{6}$$

In the light-front quantization (LFQ), the above Lagrangian density reads:

$$\mathcal{L} = \left[\frac{1}{2}\left(\partial_+ A^+ - \partial_- A^-\right)^2 + \frac{1}{g}\left(\partial_+ A^+ - \partial_- A^-\right)\Phi + \frac{1}{2g^2}\Phi^2 + \lambda\left(\hat{n}^2 - 1\right)\right] \tag{7a}$$

$$\Phi = \left[\left(\partial_+ \hat{n}\right)\cdot\left(\hat{n}\times\partial_-\hat{n}\right)\right] \tag{7b}$$

Here Φ is another topological scalar field constructed out of the topological field $\hat{n}$ (of unit norm) and its space derivative $(\partial_- \hat{n})$ as well as its time derivative $(\partial_+ \hat{n})$, as defined above. The canonical momenta obtained for the above theory are:

$$\Pi_\lambda = \frac{\partial \mathcal{L}}{\partial\left(\partial_+ \lambda\right)} = 0\ ,\quad \Pi^+ = \frac{\partial \mathcal{L}}{\partial(\partial_+ A^-)} = 0 \tag{8a}$$

$$E(:= \Pi^-) = \frac{\partial \mathcal{L}}{\partial(\partial_+ A^+)} = (\partial_+ A^+ - \partial_- A^-) + \frac{1}{g}\Phi \tag{8b}$$

$$\hat{\Pi}_{\hat{n}} = \frac{\partial \mathcal{L}}{\partial(\partial_+ \hat{n}}) = \frac{1}{g}\left(\partial_+ A^+ - \partial_- A^-\right)\left(\hat{n}\times\partial_-\hat{n}\right) + \frac{1}{g^2}\Phi(n\times\partial_- n) \tag{8c}$$

Here Π_λ, Π^+, Π^-, and $\hat{\Pi}_{\hat{n}}$ are the momenta conjugate canonically to λ, A^-, A^+, and $\hat{n}$ respectively. Above equations imply the existence of three primary constraints: $\chi_1 = \Pi_\lambda \approx 0$, $\Omega_1 = \Pi^+ \approx 0$, $\psi_1 = \hat{n}\cdot\hat{\Pi}_{\hat{n}} \approx 0$ After including the primary constraint χ_1 , Ω_1, and ψ_1 in the canonical Hamiltonian density $\mathcal{H}_c$ with the help of Lagrange multiplier field $u(x,t)$, $v(x,t)$ and $w(x,t)$ which are treated as dynamical, the total Hamiltonian density of the theory $\mathcal{H}_T$ could be written as:

$$\mathcal{H}_T^N = \left[\frac{1}{2}(\Pi^-)^2 + \Pi^-\partial_- A^- - \lambda(\hat{n}^2 - 1) + \chi_1 u + \Omega_1 v + \psi_1 w\right] \tag{9}$$

Preservation of primary constraints χ_1 and Ω_1 with time leads to the secondary Gauss-law constraints χ_2 and Ω_2 respectively: $\chi_2 = (\hat{n}^2 - 1) \approx 0$, $\Omega_2 = \partial_- \Pi^- \approx 0$. Preservation of ψ_1 with time does not lead to any secondary constraint. Preservation of χ_2 and Ω_2 also does not lead to any new constraints The theory is therefore seen to possess a set of five constraints: $\rho_1 = \Omega_1 = \Pi^+ \approx 0$, $\rho_2 = \Omega_2 = \partial_- \Pi^- \approx 0$, $\rho_3 = \chi_1 = \Pi_\lambda \approx 0$, $\rho_4 = \chi_2 = (\hat{n}^2 - 1) \approx 0$ and $\rho_5 = \psi_1 = \hat{n}\cdot\hat{\Pi}_{\hat{n}} \approx 0$. These constraints form a set of first-class constraints and the theory is seen to be invariant under the local vector gauge transformations:

$$\delta A_\mu = \frac{1}{g}\hat{n}\cdot\partial_\mu\vec{\alpha} = \frac{1}{g}\partial_\mu\alpha\ ,\ \delta\lambda = \frac{1}{g}\partial_+\alpha\ ,\ \delta\hat{n} = \hat{n}\times\vec{\alpha} = 0 \tag{10a}$$

$$\delta\Pi^+ = \delta\Pi^- = \delta\Pi_\lambda = \delta\hat{\Pi}_{\hat{n}} = \delta\Pi_u = \delta\Pi_v = \delta\Pi_w = 0 \tag{10b}$$

where the gauge parameter $\alpha(x,t)$ is an arbitrary function of its arguments. Dirac quantization of this theory under the gauge-fixing conditions (GFC's): $\zeta_1 = \lambda \approx 0$, $\zeta_2 = A^+ \approx 0$ and $\zeta_3 = A^- \approx 0$ leads to the non-vanishing equal-time Dirac brackets (DB's) of the theory as:

$$\{\Pi_{\hat{n}}^a(x^-,\tau)\ ,\ \Pi_{\hat{n}}^b(y^-,\tau)\}_{DB} = \frac{1}{\hat{n}^2}\left(n^a\Pi_{\hat{n}}^b - \Pi_{\hat{n}}^a n^b\right)\delta(x^- - y^-) \tag{11a}$$

$$\{n^a(x^-,\tau)\ ,\ \Pi_{\hat{n}}^b(y^-,\tau)\}_{DB} = \frac{1}{\hat{n}^2}\left(n^a\Pi_{\hat{n}}^b - \Pi_{\hat{n}}^a n^b\right)\delta(x^- - y^-) \tag{11b}$$

$$\{\Pi_{\hat{n}}^a(x^-,\tau)\ ,\ n^b(y^-,\tau)\}_{DB} = \frac{-1}{\hat{n}^2}\left(n^a\Pi_{\hat{n}}^b - \Pi_{\hat{n}}^a n^b\right)\delta(x^- - y^-) \tag{11c}$$

Here one needs to remember that while making a transition from equal-time Dirac brackets (ETDB's) to the equal-time commutation relations (ETCR's) using the Dirac quantization rule one needs to take in to account the problem of operator ordering (which occurs here because the results of the ETCR's involve the product of the operators). Also, the roman ind-ices a and b here, are the color ind-ices of the gauge theory of QCD_2. While considering the path integral formulation, the transition to quantum theory is made again by writing the vacuum to vacuum transition amplitude for the theory, called the generating functional $Z[J_k]$ of the theory, following again the Senjanovic procedure for a theory possessing a set of second-class constraints, appropriate for our present theory, considered under the gauge-fixing conditions η_i, in the presence of the external sources: J_k as follows [10,11]:

$$Z[J_k] = \int [d\mu] \exp\left[i \int dx^+ dx^- \left[J_k \Phi^k + \Pi_\lambda \partial_+ \lambda + \Pi^+ \partial_+ A^- + \Pi^- \partial_+ A^+ \right.\right.$$
$$\left.\left. + \hat{\Pi}_{\hat{n}} \cdot \partial_+ \hat{n} + \Pi_u \partial_+ u + \Pi_v \partial_+ v + \Pi_w \partial_+ w - \mathcal{H}_T \right]\right] \tag{12}$$

where the phase space variables of the theory are: $\Phi^k \equiv (\lambda, A^-, A^+, \hat{n}, u, v, w)$ with the corresponding respective canonical conjugate momenta: $\Pi_k \equiv (\Pi_\lambda, \Pi^+, \Pi^-, \hat{\Pi}_{\hat{n}}, \Pi_u, \Pi_v, \Pi_w)$. The functional measure $[d\mu]$ where the generating functional $Z[J_k]$ is:

$$[d\mu] = \left[[2\hat{n}^2 (\delta(x^- - y^-))^3 \partial_1 \delta(x^- - y^-)][d\lambda][dA^-][dA^+][d\hat{n}][du][dv][dw][d\Pi_\lambda] \right.$$
$$\times [d\Pi^+][d\Pi^-][d\hat{\Pi}_{\hat{n}}][dp_u][dp_v][dp_w] \delta[\Pi^+ \approx 0] \delta[\partial_- \Pi^- \approx 0] \delta[\Pi_\lambda \approx 0]$$
$$\left. \times \delta[(\hat{n}^2 - 1) \approx 0] \delta[\hat{n} \cdot \hat{\Pi}_{\hat{n}} \approx 0] \delta[\lambda \approx 0] \delta[A^+ \approx 0] \delta[A^- \approx 0] \right] \tag{13}$$

In the following, we study the BRST formulation of the theory. For the BRST formulation of the model, we rewrite the theory as a quantum system that possesses the generalized gauge invariance called BRST symmetry. For this, we first enlarge the Hilbert space of our gauge-invariant theory and replace the notion of gauge-transformation, which shifts operators by c-number functions, by a BRST transformation, which mixes operators with Bose and Fermi statistics. We then introduce new anti-commuting variables c and $\bar{c}$ (Grassman numbers on the classical level and operators in the quantized theory) and a commuting variable b such that [10,11,13]:

$$\hat{\delta}\lambda = \frac{1}{g} \partial_+ c \,,\ \hat{\delta} A^- = \frac{1}{g} \partial_+ c \,,\ \hat{\delta} A^+ = \frac{1}{g} \partial_- c \,,\ \hat{\delta} u = \frac{1}{g} \partial_+ \partial_+ c \tag{14a}$$
$$\hat{\delta} \hat{\Pi}_{\hat{n}} = \hat{\delta} \Pi^+ = \hat{\delta} \Pi^- = \hat{\delta} \Pi_\lambda = \hat{\delta} \Pi_u = \hat{\delta} \Pi_v = \hat{\delta} \Pi_w = 0 \tag{14b}$$
$$\hat{\delta} \hat{n} = \hat{\delta} v = \hat{\delta} w = 0 \,,\ \hat{\delta} c = 0 \,,\ \hat{\delta} \bar{c} = b \,,\ \hat{\delta} b = 0 \tag{14c}$$

with the property $\hat{\delta}^2 = 0$. The BRST gauge-fixed quantum Lagrangian density $\mathcal{L}_{BRST}$ could be obtained by adding to the first-order Lagrangian density $\mathcal{L}_{I0}$, a trivial BRST-invariant function:

$$\mathcal{L}_{BRST} = \left[\frac{1}{2} \left(\partial_+ A^+ - \partial_- A^- \right)^2 - \frac{1}{2g^2} \Phi^2 + \lambda \left(\hat{n}^2 - 1 \right) \right.$$
$$\left. + \hat{\delta} \left[\bar{c} \left(g \partial_+ A^- + g \partial_- A^- - g \partial_+ A^+ + \frac{1}{2} b \right) \right] \right] \tag{15}$$

Last term in the above equation is the BRST-invariant gauge-fixing term. Euler-Lagrange equation for b reads: $-b = (g\partial_+ A^- + g\partial_- A^- - g\partial_+ A^+)$ which in turn (with the requirement $\hat{\delta} b = 0$) then implies: $\partial_+ \partial_+ c = 0$. The bosonic momenta is: $\Pi^+ := gb$ and the fermionic momenta with directional derivatives are defined as: $\Pi_c := \partial_+ \bar{c}$ and $\Pi_{\bar{c}} := \partial_+ c$. implying that the variable canonically conjugate to c is ($\partial_+ \bar{c}$) and the variable conjugate to $\bar{c}$ is ($\partial_+ c$). The Hamiltonian density is obtained from the Lagrangian density in the usual manner:

$$\mathcal{H}_{BRST} = \left[\Pi_u \partial_+ u + \Pi_v \partial_+ v + \hat{n} \cdot \hat{\Pi}_{\hat{n}} w + \frac{1}{2} \left(\Pi^- \right)^2 + \Pi^- \partial_- A^- - \lambda \left(\hat{n}^2 - 1 \right) \right.$$
$$\left. + \frac{1}{2} g^2 b^2 + \Pi_c \Pi_{\bar{c}} - gb \left(\partial_+ A^- - \partial_+ A^+ + \partial_- A^- \right) \right] \tag{16}$$

$$Q = \int dx^- \left[ic \left(\partial_- \Pi^- + \hat{n}^2 - 1 \right) - i \partial_+ c \left(\Pi^+ + \Pi_\lambda + \hat{n} \cdot \hat{\Pi}_{\hat{n}} \right) \right] \tag{17}$$

Now because $Q|\psi>=0$, the set of states annihilated by Q contains not only the set for which the constraints of the theory hold but also additional states for which the constraints of the theory do not hold. The Hamiltonian is also invariant under the anti-BRST transformations given by:

$$\bar{\hat{\delta}}\lambda=-\frac{1}{g}\partial_+\bar{c},\quad \bar{\hat{\delta}}A^-=-\frac{1}{g}\partial_+\bar{c},\quad \bar{\hat{\delta}}A^+=-\frac{1}{g}\partial_-\bar{c},\quad \bar{\hat{\delta}}u=-\frac{1}{g}\partial_+\partial_+\bar{c} \tag{18a}$$

$$\bar{\hat{\delta}}\hat{\Pi}_{\hat{n}}=\bar{\hat{\delta}}\Pi^+=\bar{\hat{\delta}}\Pi^-=\bar{\hat{\delta}}\Pi_\lambda=\bar{\hat{\delta}}\Pi_u=\bar{\hat{\delta}}\Pi_v=\bar{\hat{\delta}}\Pi_w=0 \tag{18b}$$

$$\bar{\hat{\delta}}\hat{n}=\bar{\hat{\delta}}v=\bar{\hat{\delta}}w=0\quad \bar{\hat{\delta}}\bar{c}=0,\quad \bar{\hat{\delta}}c=-b,\quad \bar{\hat{\delta}}b=0 \tag{18c}$$

with generator or anti-BRST charge

$$\bar{Q}=\int dx^-\left[-i\bar{c}\left(\partial_-\Pi^-+\hat{n}^2-1\right)+i\partial_+\bar{c}\left(\Pi^++\Pi_\lambda-\hat{n}\cdot\hat{\Pi}_{\hat{n}}\right)\right] \tag{19}$$

We also have

$$\partial_+Q=[Q,H_{BRST}]=0,\quad \partial_+\bar{Q}=\left[\bar{Q},H_{BRST}\right]=0,\quad H_{BRST}=\int dx^-\mathcal{H}_{BRST} \tag{20}$$

We now impose the dual condition that both Q and $\bar{Q}$ annihilate physical states, implying that: $Q|\psi>=0$ and $\bar{Q}|\psi>=0$. The states for which the constraints of the theory hold, satisfy both of these conditions and are in fact, the only states satisfying both of these conditions. Thus the only states satisfying $Q|\psi>=0$ and $\bar{Q}|\psi>=0$ are those that satisfy the constraints of the theory and they belong to the set of BRST-invariant as well as to the set of anti-BRST-invariant states. This completes the LF Hamiltonian, path integral and BRST quantization of the theory.

Acknowledgments One of the authors (DSK) thanks University of Delhi not only for granting him Sabbatical Leave for May–November, 2015 during which this work was carried out but also for the award of a Full Travel Grant to him for attending the International Conference on Light-Cone Physics, LC2015, held at LMF, INFN, Frascati, Italy, during September 15–21, 2015, where this work was presented by him as an Invited Contributed Talk. The authors thank Giovani Salmè, Barbara Pasquini and other organizers of the above Conference (LC2015), held at Frascati, Italy, for their great hospitality during the conference. This work was supported in part by the US Department of Energy under Grant No. DE-FG02-87ER40371 and by the US National Science Foundation under Grant No. PHY-0904782.

References

1. Hooft, G.'t: Topology of the gauge condition and new confinement phases in non-Abelian gauge theories. Nucl. Phys. B **190**, 455 (1981)
2. Cho, Y.M., Hong, S.-T., Kim, J.H., Park, Y.-J.: Dirac quantization of restricted QCD. Mod. Phys. Lett. A **22**(37) 2799–2813 (2007)
3. Faddeev, L., Niemi, A.: Stable knot-like structures in classical field-theory. Nature **387**, 58 (1997)
4. Faddeev, L., Niemi, A.: Partially dual variables in SU (2) Yang–Mills theory. Phys. Rev. Lett. **82**, 1624 (1999)
5. Faddeev, L., Niemi, A.: Partial duality in SU (N) Yang–Mills theory. Phys. Lett. B **449**, 214 (1999)
6. Langman, E., Niemi, A.: Towards a string representation of infrared SU (2) Yang–Mills theory. Phys. Lett. B **463**, 252 (1999)
7. Kulshreshtha, D.S.: Quantization of the restricted Gauge theory of QCD_2. Few Body Syst. **56**(6–9), 565–569 (2015)
8. Dirac, P.A.M.: Forms of relativistic dynamics. Rev. Mod. Phys. **21**, 392 (1949)
9. Brodsky, S.J., Pauli, H.C., Pinsky, S.S.: Quantum chromodynamics and other field theories on the light cone. Phys. Rep. **301**, 299 (1998)
10. Kulshreshtha, U., Kulshreshtha, D.S., Vary, J.P.: Hamiltonian, path integral and BRST formulations of large N scalar QCD_2 on the light-front and spontaneous symmetry breaking. Eur. J. Phys. C **75**, 174 (2015)
11. Kulshreshtha, Usha, Kulshreshtha, D.S., Vary, James: Hamiltonian, path integral and BRST formulations of vector schwinger model with a photon mass term with faddeevian regularization. Int. J. Theor. Phys. **55**(1) 338–360 (2015)
12. Dirac, P.A.M.: Generalized Hamiltonian dynamics. Can. J. Math. **2**, 129 (1950)
13. Nemeschansky, D., Preitschopf, C., Weinstein, M.: A BRST primer. Ann. Phys. N. Y. **183**, 226 (1988)

DOI 10.1007/978-3-319-50699-9_35
Reprinted from *Few-Body System* Journal, DOI 10.1007/s00601-016-1077-1

Jian-Hui Zhang · Xiangdong Ji · Yong Zhao

Proton Spin Structure from Large Momentum Effective Theory

Received: 29 January 2016 / Accepted: 2 March 2016 / Published online: 15 March 2016

Abstract Recently a large momentum effective theory approach has been proposed to compute parton observables. We discuss its application to the spin structure of the proton. In particular, we explain how the partonic contribution to the proton spin in the Jaffe–Manohar sum rule can be computed within the large momentum effective theory framework.

1 Introduction

Understanding the spin structure of the proton has been an important goal in hadron physics since the European Muon Collaboration discovered that the quark spin contributes only a small fraction to the proton spin. In the past few decades, two well-known spin sum rules have been proposed to understand the spin decomposition of the proton. One of them is the Jaffe–Manohar sum rule [1] motivated from a free-field expression of the QCD angular momentum boosted to the infinite momentum frame (IMF) of the proton. The other is the frame-independent and manifestly gauge-invariant sum rule proposed by Ji [2]. In the latter case, the angular momentum can be related to generalized parton distributions (GPDs), and therefore are experimentally measurable [2–4]; but it does not have a natural partonic interpretation. On the other hand, the Jaffe–Manohar sum rule defined in the light-cone gauge $A^+ = 0$ has a clear parton picture

$$\mathbf{J} = \int d^3x\ \psi^\dagger \frac{\mathbf{\Sigma}}{2}\psi + \int d^3x\ \psi^\dagger \mathbf{x} \times (-i\nabla)\psi + \int d^3x\ \mathbf{E}_a \times \mathbf{A}^a + \int d^3x\ E_a^i\ \mathbf{x} \times \nabla A^{i,a}, \tag{1}$$

where all the angular momentum operators are free-field ones, $E^i = F^{i+}$, a and i are the color and spatial indices. However, except for the quark spin, all the other terms are gauge dependent, and it is not clear how they are connected to experimental measurements.

In the past two decades, there has been a long list of literatures attempting to justify the Jaffe–Manohar sum rule as physical (see e.g., [5–13]). There are strong motivations behind this: First, ΔG as defined in the

J.-H. Zhang (✉)
Institut für Theoretische Physik, Universität Regensburg, 93040 Regensburg, Germany
E-mail: jianhui.zhang@physik.uniregensburg.de

X. Ji
INPAC, Department of Physics and Astronomy, Shanghai Jiao Tong University, Shanghai 200240, People's Republic of China

X. Ji · Y. Zhao
Maryland Center for Fundamental Physics, Department of Physics, University of Maryland, College Park, MD 20742, USA

light-cone gauge is measurable in high-energy experiments, although this appears to be a theoretical puzzle by itself—while ΔG is easy to define from the Feynman parton picture, there is no natural gauge-invariant notion for the spin of gauge particles [14]. Second, when the proton is probed in IMF, some of its physical properties can be understood from simple addition of those of free quarks and gluons. For example, the longitudinal momentum of the proton is the sum of that of the quark and gluon partons:

$$1 = \int dx\, x \left(\sum_q q(x) + g(x) \right), \tag{2}$$

where $q(x)$ and $g(x)$ are the unpolarized quark and gluon momentum distribution functions. This simple parton picture may work for the proton spin as well.

It was first proposed that although the free-field angular momentum operators are gauge dependent, their physical matrix elements are gauge invariant [6,7]. A similar claim was also made recently [8]. However, this is invalidated by a one-loop calculation by the present authors [10], where the matrix element of the gluon spin operator was shown to be different in the Coulomb and light-cone gauges [15] (for more general discussions see Ref. [9]). Actually, as argued in Refs. [10,11], for the bound-state proton, there is no physically meaningful notion of gluon spin or orbital angular momentum due to the existence of longitudinal gluons. Only when the proton is boosted to IMF, the longitudinal component of gluons is suppressed by the infinite boost and the gluons can be regarded as free radiation. This is the well-known Weizsäcker–Williams (WW) approximation [16]. The gluon spin then acquires a clear physical meaning and can be represented by $\mathbf{E} \times \mathbf{A}$, but is subject to a class of "physical" gauge conditions that leave the transverse polarizations of the gluon field intact [12]. Similar arguments also apply to the quark and gluon orbital angular momentum. Therefore, we can regard the free-field form in the Jaffe–Manohar sum rule as physical if we work in IMF with a "physical" gauge condition. This is equivalent to using the light-cone coordinates and gauge [1], and the reason is simple: All the "physical" gauges will flow into the light-cone gauge in the IMF limit [12].

Nevertheless, the Jaffe–Manohar sum rule still poses difficulty for a practical lattice calculation of its individual contributions, due to the explicit usage of light-cone coordinates and gauge. This difficulty can, however, be circumvented in the framework of *large-momentum effective field theory* (LaMET) [17] proposed by one of the authors. Suppose one is to calculate some light-cone or parton observable $\mathcal{O}$. Instead of computing it directly, one defines, in the LaMET framework, a quasi-observable $\tilde{\mathcal{O}}$ that depends on a large hadron momentum P^z. In general, both $\mathcal{O}$ and $\tilde{\mathcal{O}}$ suffer from ultraviolet (UV) divergences. If $P^z \to \infty$ is taken prior to a UV regularization, the quasi-observable $\tilde{\mathcal{O}}$ becomes the parton observable $\mathcal{O}$ by construction. However, what one can calculate in practice is the quasi-observable $\tilde{\mathcal{O}}$ at large but finite P^z with UV regularization imposed first. This is the case in lattice computations. The difference between $\mathcal{O}$ and $\tilde{\mathcal{O}}$ is just the order of limits. This is similar to an effective field theory set-up. The difference is that here the role of heavy degree of freedom is played by the large momentum of the external state, hence it cannot be arranged into a Lagrangian formalism. Nevertheless, one can bridge the quasi- and parton observables through

$$\tilde{\mathcal{O}}(P^z/\Lambda) = Z\left(P^z/\Lambda, \mu/\Lambda\right) \mathcal{O}(\mu) + \frac{c_2}{(P^z)^2} + \frac{c_4}{(P^z)^4} + \cdots, \tag{3}$$

where Λ is a UV cutoff imposed on the quasi-observable, and c_i's denote contributions suppressed by powers of P^z. That is, the quasi-observable $\tilde{\mathcal{O}}(P^z/\Lambda)$ can be factorized into the parton observable $\mathcal{O}(\mu)$ and a matching coefficient Z, up to power suppressed corrections. Taking the $P^z \to \infty$ limit does not change the infrared (IR) behavior of the observable, but only affects its UV behavior. Therefore $\mathcal{O}(\mu)$ captures all the IR physics in $\tilde{\mathcal{O}}(P^z/\Lambda)$, and the matching coefficient Z is completely perturbative.

Some explicit examples of Eq. (3) have been studied in Refs. [18–22] for the case of quark parton distributions and GPDs, where the matrix elements of equal-time quark bilinear operators were used as quasi observables. As shown there, the factorization formula for the quark parton distributions or GPDs has a convoluted form since they are non-local correlators, and the perturbative matching coefficients transform the P^z dependence of the quasi parton distributions into the renormalization scale dependence of the parton distributions. Using these results, the first direct lattice calculation of the isovector sea-quark parton distributions has been available [23], the result of which was later confirmed in Ref. [24]. A similar factorization formula was also proposed in Ref. [25] to extract PDFs from lattice QCD calculations based on QCD factorization of lattice "cross sections".

2 Partonic Contribution to Proton Spin Within LaMET

Within the LaMET framework, we are able to calculate the proton spin content from suitably chosen quasi-observables. According to our discussions above, these quasi-observables can be defined as the free-field QCD angular momentum operators in a "physical" gauge condition that has the correct WW approximation in the IMF limit [12]. A possible choice of the "physical" gauge condition is the expression in terms of nonlocal operators introduced by Chen et al. [26,27]:

$$\mathbf{J}_{\rm QCD} = \int d^3x\, \psi^\dagger \frac{\mathbf{\Sigma}}{2}\psi + \int d^3x\, \psi^\dagger \mathbf{x}\times(-i\nabla - g\mathbf{A}_\parallel)\psi \\ + \int d^3x\, \mathbf{E}_a \times \mathbf{A}_\perp^a + \int d^3x\, E_a^i\, (\mathbf{x}\times\nabla) A_\perp^{i,a}, \tag{4}$$

where $\mathbf{x}$ are the spatial coordinates, and $\mathbf{A}$ is decomposed into a pure-gauge part $\mathbf{A}_\parallel$ and a physical part $\mathbf{A}_\perp$ which satisfy (see also Ref. [28])

$$\partial^i A_\parallel^{j,a} - \partial^j A_\parallel^{i,a} - g f^{abc} A_\parallel^{i,b} A_\parallel^{j,c} = 0, \\ \partial^i A_\perp^i - ig[A^i, A_\perp^i] = 0, \tag{5}$$

so that each term in Eq. (4) is gauge invariant. From Eq. (5), one can show that in the Coulomb gauge $\nabla\cdot\mathbf{A} = 0$, $\mathbf{A}_\perp$ equals $\mathbf{A}$ order by order in perturbation theory. Therefore, Eq. (4) corresponds to choosing the Coulomb gauge as the "physical" gauge.

It has been shown in Ref. [10] that $\mathbf{E}_a \times \mathbf{A}_\perp^a$ in Eq. (4) is equivalent to the total gluon spin operator in the IMF limit. It is easy to see that the other nonlocal terms in Eq. (4) also have the correct WW approximation as the parton orbital angular momentum. Therefore, we can choose the nonlocal operators in Eq. (4) as the quasi-observables for parton angular momentum in the LaMET approach.

The advantage of the expression in Eq. (4) is that it is time independent and thus allows for a direct calculation in lattice QCD. Suppose we evaluate the matrix elements of these quasi-observables with finite momentum P^z, we should have

$$\frac{1}{2} = \frac{1}{2}\Delta\widetilde{\Sigma}(\mu, P^z) + \Delta\widetilde{G}(\mu, P^z) + \Delta\widetilde{L}_q(\mu, P^z) + \Delta\widetilde{L}_g(\mu, P^z), \tag{6}$$

where μ is the renormalization scale, the P^z-dependence in the above equation is expected since Eq. (4) is a frame-dependent expression [10]. Following the effective theory argument above, we can relate these quasi-observables to the corresponding parton observables through the following factorization formula:

$$\Delta\widetilde{\Sigma}(\mu, P^z) = \Delta\Sigma(\mu), \\ \Delta\widetilde{G}(\mu, P^z) = z_{qg}\Delta\Sigma(\mu) + z_{gg}\Delta G(\mu) + O\left(\frac{M^2}{(P^z)^2}\right), \\ \Delta\widetilde{L}_q(\mu, P^z) = P_{qq}\Delta L_q(\mu) + P_{gq}\Delta L_g(\mu) + p_{qq}\Delta\Sigma(\mu) + p_{gq}\Delta G(\mu) + O\left(\frac{M^2}{(P^z)^2}\right), \\ \Delta\widetilde{L}_g(\mu, P^z) = P_{qg}\Delta L_q(\mu) + P_{gg}\Delta L_g(\mu) + p_{qg}\Delta\Sigma(\mu) + p_{gg}\Delta G(\mu) + O\left(\frac{M^2}{(P^z)^2}\right), \tag{7}$$

where M is the proton mass, and all the matrix elements are renormalized in dimensional regularization and $\overline{\rm MS}$ scheme. $\Delta\widetilde{\Sigma}(\mu, P^z)$ is the same as $\Delta\Sigma(\mu)$ because the quark spin operator is gauge invariant and the same as that in Eq. (1). The z_{ij}, P_{ij} and p_{ij}'s are the matching coefficients to be calculated in perturbative QCD.

To obtain the matching coefficients in Eq. (7), we need to calculate the matrix elements of the quasi angular momentum operators both at finite momentum and in the IMF limit (before UV regularization). As discussed above, in the IMF limit these quasi angular momentum operators become those in the Jaffe–Manohar sum rule, and therefore have the same matrix elements as the latter. To illustrate the procedure, let us take the one-loop calculation of z_{qg} and z_{gg} as an example, for which we need to calculate the matrix elements of $\mathbf{E}_a \times \mathbf{A}_\perp^a$ at finite P^z and in the IMF limit (before UV regularization). We will choose the external quark and gluon states to be on-shell and massless, and use dimensional regularization with spacetime dimension $d = 4 - 2\epsilon$ to regularize both UV, IR and collinear divergences.

Since the angular momentum operators we consider are all gauge invariant, we can work in any gauge, and for simplicity we choose the Coulomb gauge. As mentioned before, the Coulomb gauge condition is equivalent to the condition for $\mathbf{A}_\perp$ in QCD (see Eq. (5)).

At tree level, $\Delta\widetilde{G}^{\text{tree}} = \Delta G^{\text{tree}}$. At one-loop level, the matrix element of $\mathbf{E}_a \times \mathbf{A}_\perp^a$ is

$$\Delta\widetilde{G}^{(1)} = \frac{\alpha_S C_F}{4\pi}\left(\frac{5}{3}\frac{1}{\epsilon'_{UV}} + \frac{4}{3}\ln\frac{(P^z)^2}{\mu^2} - \frac{3}{\epsilon'_{IR}} + R_1\right)\Delta\Sigma^{\text{tree}}$$
$$+\frac{\alpha_S}{4\pi}\left[\frac{4C_A - 2n_f}{3}\frac{1}{\epsilon'_{UV}} - \frac{11C_A - 2n_f}{3}\frac{1}{\epsilon'_{IR}} + C_A\left(\frac{7}{3}\ln\frac{(P^z)^2}{\mu^2} + R_2\right)\right]\Delta G^{\text{tree}}, \tag{8}$$

where n_f is the number of flavors of active quarks, and

$$\frac{1}{\epsilon'} = \frac{1}{\epsilon} - \gamma_E + \ln 4\pi, \quad C_F = \frac{N_c^2 - 1}{2N_c}, \quad C_A = N_c, \tag{9}$$

with N_c being the number of colors. R_1 and R_2 are finite constants

$$R_1 = \frac{8}{3}\ln 2 - \frac{64}{9}, \quad R_2 = \frac{14}{3}\ln 2 - \frac{121}{9}. \tag{10}$$

The corresponding IMF (or light-cone) matrix elements are [15]

$$\Delta G^{(1)} = \frac{\alpha_S C_F}{4\pi}\left(\frac{3}{\epsilon'_{UV}} - \frac{3}{\epsilon'_{IR}}\right)\Delta\Sigma^{\text{tree}} + \frac{\alpha_S}{4\pi}\left[\frac{11C_A - 2n_f}{3}\left(\frac{1}{\epsilon'_{UV}} - \frac{1}{\epsilon'_{IR}}\right)\right]\Delta G^{\text{tree}}. \tag{11}$$

Apparently the anomalous dimensions (coefficients of $1/\epsilon'_{UV}$'s) are different between $\Delta\widetilde{G}^{(1)}$ and $\Delta G^{(1)}$, but the IR divergence (coefficients of $1/\epsilon'_{IR}$'s) is the same for both. After renormalization, we substitute the $1/\epsilon'_{IR}$ terms in $\Delta\widetilde{G}$ with ΔG, and obtain at $O(\alpha_S)$

$$\Delta\widetilde{G} = \frac{\alpha_S C_F}{4\pi}\left(\frac{4}{3}\ln\frac{(P^z)^2}{\mu^2} + R_1\right)\Delta\Sigma + \left[1 + \frac{\alpha_S C_A}{4\pi}\left(\frac{7}{3}\ln\frac{(P^z)^2}{\mu^2} + R_2\right)\right]\Delta G. \tag{12}$$

Therefore, the matching coefficients for $\Delta\widetilde{G}$ can be read off as

$$z_{qg}(\mu/P^z) = \frac{\alpha_S C_F}{4\pi}\left(\frac{4}{3}\ln\frac{(P^z)^2}{\mu^2} + R_1\right),$$
$$z_{gg}(\mu/P^z) = 1 + \frac{\alpha_S C_A}{4\pi}\left(\frac{7}{3}\ln\frac{(P^z)^2}{\mu^2} + R_2\right). \tag{13}$$

Following the same procedure, we can obtain all the other matching coefficients in Eq. (7) at one-loop order. The complete results are as follows:

$$P_{qq} = 1 + \frac{\alpha_S C_F}{4\pi}\left(-2\ln\frac{(P^z)^2}{\mu^2} + R_3\right), \quad P_{gq} = 0,$$
$$P_{qg} = \frac{\alpha_S C_F}{4\pi}\left(2\ln\frac{(P^z)^2}{\mu^2} - R_3\right), \quad P_{gg} = 1,$$
$$p_{qq} = \frac{\alpha_S C_F}{4\pi}\left(-\frac{1}{3}\ln\frac{(P^z)^2}{\mu^2} + R_4\right), \quad p_{gq} = 0,$$
$$p_{qg} = \frac{\alpha_S C_F}{4\pi}\left(-\ln\frac{(P^z)^2}{\mu^2} - R_1 - R_4\right),$$
$$p_{gg} = \frac{\alpha_S C_A}{4\pi}\left(-\frac{7}{3}\ln\frac{(P^z)^2}{\mu^2} - R_2\right), \tag{14}$$

where

$$R_3 = -4\ln 2 + \frac{28}{3}, \quad R_4 = -\frac{2}{3}\ln 2 + \frac{13}{9}. \tag{15}$$

With the above results, we are able to convert the quasi-observables in Eq. (4) evaluated at a large finite momentum to the parton spin and orbital angular momentum in IMF. This can be done by a simple inversion of Eq. (7). Although for realistic lattice computations the above matching coefficients have to be recalculated using lattice perturbation theory with lattice action [29], the leading logarithmic term of the matching coefficients is independent of the regularization scheme, and therefore is the same in dimensional and lattice regularization. The above one-loop matching coefficients can thus be used for a lattice computation of parton angular momentum to leading logarithmic accuracy.

3 Conclusions

The LaMET approach provides a practical possibility to compute parton observables using lattice QCD. In the case of parton distributions, it enables us to evaluate the full distribution instead of only the first few moments. In this talk, we explained the main idea of this approach, and applied it to the Jaffe–Manohar spin sum rule for the proton. We showed that the computation of partonic contributions in this sum rule can be divided into two steps which are separately computable: the matrix elements of time-independent operators at a finite nucleon momentum can be computed on the lattice, and the matching condition is perturbative.

References

1. Jaffe, R.L., Manohar, A.V.: The G(1) problem: fact and fantasy on the spin of the proton. Nucl. Phys. B **337**, 509–546 (1990)
2. Ji, X.-D.: Gauge-invariant decomposition of nucleon spin. Phys. Rev. Lett. **78**, 610–613 (1997). [arXiv:hep-ph/9603249]
3. Ji, X.D.: Deeply virtual compton scattering. Phys. Rev. D **55**, 7114–7125 (1997). [arXiv:hep-ph/9609381]
4. Hoodbhoy, P., Ji, X.D., Lu, W.: Quark orbital–angular–momentum distribution in the nucleon. Phys. Rev. D **59**, 014013 (1999). [arXiv:hep-ph/9804337]
5. Jaffe, R.L.: Gluon spin in the nucleon. Phys. Lett. B **365**, 359–366 (1996). [arXiv:hep-ph/9509279]
6. Chen, X.-S., Wang, F.: Gauge invariance and hadron structure. arXiv:hep-ph/9802346
7. Wang, F., Sun, W.-M., Lu, X.-F.: Quantum version of gauge invariance and nucleon internal structure. arXiv:hep-ph/0510285
8. Leader, E.: On the controversy concerning the definition of quark and gluon angular momentum. Phys. Rev. D **83**, 096012 (2011). [Erratum-ibid. D **85**, 039905 (2012)] [arXiv:1101.5956 [hep-ph]]
9. Hoodbhoy, P., Ji, X.D.: Does the gluon spin contribute in a gauge invariant way to nucleon spin? Phys. Rev. D **60**, 114042 (1999). [arXiv:hep-ph/9908275]
10. Ji, X., Zhang, J.H., Zhao, Y.: Physics of the gluon-helicity contribution to proton spin. Phys. Rev. Lett. **111**, 112002 (2013). [arXiv:1304.6708 [hep-ph]]
11. Ji, X., Xu, Y., Zhao, Y.: Gluon spin, canonical momentum, and gauge symmetry. JHEP **1208**, 082 (2012)
12. Hatta, Y., Ji, X., Zhao, Y.: Gluon helicity Delta G from a universality class of operators on a lattice. Phys. Rev. D **89**, 085030 (2014). [arXiv:1310.4263 [hep-ph]]
13. Ji, X., Zhang, J.H., Zhao, Y.: Justifying the naive partonic sum rule for proton spin. Phys. Lett. B **743**, 180–183 (2015). [arXiv:1409.6329 [hep-ph]]
14. Berestetskii, V.B., Pitaevskii, L.P., Lifshitz, E.M.: Quantum electrodynamics, 2nd edn, Vol. 4 in Courses in Theoretical Physics, Butterworth-Heineman (1982)
15. Ji, X.-D., Tang, J., Hoodbhoy, P.: The spin structure of the nucleon in the asymptotic limit. Phys. Rev. Lett. **76**, 740–743 (1996). [arXiv:hep-ph/9510304]
16. Jackson, J.D.: Classical Electrodynamics, 3rd edn. Wiley, New York (1999)
17. Ji, X.: Parton physics from large-momentum effective field theory. Sci. China Phys. Mech. Astron. **57**(7), 1407 (2014). [arXiv:1404.6680 [hep-ph]]
18. Ji, X.: Parton physics on a euclidean lattice. Phys. Rev. Lett. **110**, 262002 (2013). [arXiv:1305.1539 [hep-ph]]
19. Xiong, X., Ji, X., Zhang, J.H., Zhao, Y.: One-loop matching for parton distributions: non-singlet case. Phys. Rev. D **90**, 014051 (2014). [arXiv:1310.7471 [hep-ph]]
20. Ji, X., Zhang, J.H.: Renormalization of quasiparton distribution. Phys. Rev. D **92**, 034006 (2015). [arXiv:1505.07699 [hep-ph]]
21. Ji, X., Schfer, A., Xiong, X., Zhang, J.H.: One-loop matching for generalized parton distributions. Phys. Rev. D **92**, 014039 (2015). [arXiv:1506.00248 [hep-ph]]
22. Xiong, X., Zhang, J.H.: One-loop matching for transversity generalized parton distribution. Phys. Rev. D **92**(5), 054037 (2015). [arXiv:1509.08016 [hep-ph]]

23. Lin, H.W., Chen, J.W., Cohen, S.D., Ji, X.: Flavor structure of the nucleon sea from lattice QCD. Phys. Rev. D **91**, 054510 (2015). [arXiv:1402.1462 [hep-ph]]
24. Alexandrou, C., Cichy, K., Drach, V., Garcia-Ramos, E., Hadjiyiannakou, K., Jansen, K., Steffens, F., Wiese, C.: Lattice calculation of parton distributions. Phys. Rev. D **92**, 014502 (2015). [arXiv:1504.07455 [hep-lat]]
25. Ma, Y.Q., Qiu, J.W.: Extracting parton distribution functions from lattice QCD calculations. arXiv:1404.6860 [hep-ph]
26. Chen, X.-S., Lü, X.-F., Sun, W.-M., Wang, F., Goldman, T.: Spin and orbital angular momentum in gauge theories: proton spin structure and multipole radiation revisited. Phys. Rev. Lett. **100**, 232002 (2008)
27. Chen, X.-S., Sun, W.-M., Lü, X.-F., Wang, F., Goldman, T.: Do gluons carry half of the nucleon momentum? Phys. Rev. Lett. **103**, 062001 (2009)
28. Treat, R.P.: Gauge invariant decompositon of Yang-Mills potentials. J. Math. Phys. **13**, 1704–1714 (1972). [Erratum-ibid. **14**, 1296 (1973)]
29. Capitani, S.: Lattice perturbation theory. Phys. Rept. **382**, 113–302 (2003). [arXiv:hep-lat/0211036]

DOI 10.1007/978-3-319-50699-9_36
Reprinted from *Few-Body System* Journal, DOI 10.1007/s00601-016-1078-0

María Gómez-Rocha · Stanisław D. Głazek

Asymptocic Freedom of Gluons in Hamiltonian Dynamics

Received: 28 January 2016 / Accepted: 15 February 2016 / Published online: 18 March 2016

Abstract We derive asymptotic freedom of gluons in terms of the renormalized $SU(3)$ Yang–Mills Hamiltonian in the Fock space. Namely, we use the renormalization group procedure for effective particles to calculate the three-gluon interaction term in the front-form Yang–Mills Hamiltonian using a perturbative expansion in powers of g up to third order. The resulting three-gluon vertex is a function of the scale parameter s that has an interpretation of the size of effective gluons. The corresponding Hamiltonian running coupling constant exhibits asymptotic freedom, and the corresponding Hamiltonian β-function coincides with the one obtained in an earlier calculation using a different generator.

1 Introduction

The renormalization group procedure for effective particles (RGPEP) has been developed during the last years [1–3] as a non-perturbative tool for constructing bound-states in quantum chromodynamics (QCD) [4]. It introduces the concept of effective particles, which differ from the bare or canonical ones, by having size s, corresponding to the momentum scale $\lambda = 1/s$. Creation and annihilation of effective particles in the Fock space are described by the action of effective particle operators, $a_s^\dagger$ and a_s, on states built from the vacuum state $|0\rangle$ using $a_s^\dagger$; bare particle operators, $a_0^\dagger$ and a_0, appearing in the canonical Hamiltonian, create and annihilate pointlike particles (with size $s = 0$).

We are interested in calculating the evolution of quark and gluon quantum states, describing their dynamics and studying their binding. In a single formulation, the sought effective Hamiltonian must provide a means for the constituent-like behavior of quarks and gluons in hadrons with the measured quantum numbers, and also an explanation for the short-distance phenomena of weakly interacting pointlike partons.

In this work, we apply the RGPEP to interacting gluons in the absence of quarks. We will demonstrate that the RGPEP passes the test of describing asymptotic freedom, which is a precondition for any approach aiming at using QCD, especially for tackling nonperturbative issues, such as the ones that emerge when one allows effective gluons to have masses [4].

We start from the regularized canonical Hamiltonian for quantum Yang–Mills field in the Fock space, obtained from the corresponding Lagrangian density. We use the RGPEP to introduce effective particles and calculate a family of effective Hamiltonians characterized by a scale or size parameter s. These Fock-space

Presented by María Gómez-Rocha at Light Cone 2015, 21–25 September, 2015, INFN Frascati National Laboratories.

M. Gómez-Rocha (✉)
ECT*, Strada delle Tabarelle, 286, 38123 Villazzano, TN, Italy
E-mail: mariagomezrocha@gmail.com

S. D. Głazek
University of Warsaw, Pasteura 5, 02-093 Warsaw, Poland

Hamiltonians depend on the effective-particle size parameter in an asymptotically free way: the coupling constant in a three-gluon interaction term vanishes with the inverse of $\ln(1/s)$.

In the following, we summarize the procedure, which is general and can be applied to any other quantum field theory. For a more extended and detailed explanation, we refer the reader to Ref. [3].

2 Renormalization Group Procedure for Effective Particles

2.1 Initial Hamiltonian

We derive the canonical Hamiltonian for Yang–Mills theories from the Lagrangian density:

$$\mathcal{L} = -\frac{1}{2}\mathrm{tr}\, F^{\mu\nu} F_{\mu\nu}, \tag{1}$$

where $F^{\mu\nu} = \partial^\mu A^\nu - \partial^\nu A^\mu + ig[A^\mu, A^\nu]$, $A^\mu = A^{a\mu} t^a$, $[t^a, t^b] = i f^{abc} t^c$, which leads to the energy–momentum tensor,

$$\mathcal{T}^{\mu\nu} = -F^{a\mu\alpha}\partial^\nu A^a_\alpha + g^{\mu\nu} F^{a\alpha\beta} F^a_{\alpha\beta}/4. \tag{2}$$

We choose the front-form (FF) of dynamics [5] which consists of setting the quantization surface on the hyperplane $x^+ = x^0 + x^3 = 0$. Using the gauge $A^+ = 0$, the Lagrange equations lead to the condition

$$A^- = \frac{1}{\partial^+} 2\,\partial^\perp A^\perp - \frac{2}{\partial^{+2}}\, ig\,[\partial^+ A^\perp, A^\perp]\,, \tag{3}$$

so that the only degrees of freedom are the fields $A^\perp$.

Integration of $\mathcal{T}^{+-}$ over the front $x^+ = 0$ leads to the FF energy of the constrained gluon field:

$$P^- = \frac{1}{2}\int dx^- d^2x^\perp \mathcal{T}^{+-}\,|_{x^+=0} \quad . \tag{4}$$

This operator contains a series of products of 2nd, 3rd or 4th powers of the field A^μ or their derivatives. The energy momentum tensor $\mathcal{T}^{+-}$ can be written as

$$\mathcal{T}^{+-} = \mathcal{H}_{A^2} + \mathcal{H}_{A^3} + \mathcal{H}_{A^4} + \mathcal{H}_{[\partial AA]^2}, \tag{5}$$

where [6,7]

$$\mathcal{H}_{A^2} = -\frac{1}{2} A^{\perp a} (\partial^\perp)^2 A^{\perp a}, \tag{6}$$

$$\mathcal{H}_{A^3} = g\, i\partial_\alpha A^a_\beta [A^\alpha, A^\beta]^a, \tag{7}$$

$$\mathcal{H}_{A^4} = -\frac{1}{4} g^2\, [A_\alpha, A_\beta]^a [A^\alpha, A^\beta]^a, \tag{8}$$

$$\mathcal{H}_{[\partial AA]^2} = \frac{1}{2} g^2\, [i\partial^+ A^\perp, A^\perp]^a \frac{1}{(i\partial^+)^2} [i\partial^+ A^\perp, A^\perp]^a. \tag{9}$$

The quantum canonical Hamiltonian is obtained by replacing the field A^μ by the quantum field operator

$$\hat{A}^\mu = \sum_{\sigma c} \int [k] \left[t^c \varepsilon^\mu_{k\sigma} a_{k\sigma c} e^{-ikx} + t^c \varepsilon^{\mu *}_{k\sigma} a^\dagger_{k\sigma c} e^{ikx} \right]_{x^+=0}, \tag{10}$$

where $[k] = \theta(k^+) dk^+ d^2k^\perp/(16\pi^3 k^+)$, the polarization four-vector is defined as $\varepsilon^\mu_{k\sigma} = (\varepsilon^+_{k\sigma} = 0, \varepsilon^-_{k\sigma} = 2k^\perp \varepsilon^\perp_\sigma / k^+, \varepsilon^\perp_\sigma)$ and the indices σ and c denote spin and color quantum numbers, respectively. The creation and annihilation operators satisfy the commutation relations

$$\left[a_{k\sigma c}, a^\dagger_{k'\sigma' c'}\right] = k^+ \tilde{\delta}(k - k')\, \delta^{\sigma\sigma'}\, \delta^{cc'}, \quad [a_{k\sigma c}, a_{k'\sigma' c'}] = \left[a^\dagger_{k\sigma c}, a^\dagger_{k'\sigma' c'}\right] = 0, \tag{11}$$

with $\tilde{\delta}(p) = 16\pi^3 \delta(p^+)\delta(p^1)\delta(p^2)$.

The canonical Hamiltonian is divergent and needs regularization. At every interaction term, every creation and annihilation operator in the canonical Hamiltonian is multiplied by a regulating factor[1]

$$r_{\Delta\delta}(\kappa^{\perp}, x) = \exp(-\kappa^{\perp}/\Delta)\, x^{\delta}\theta(x-\epsilon), \tag{12}$$

where x is the relative momentum fraction $x_{p/P} = p^{+}/P^{+}$, κ is the relative transverse momentum, $\kappa_{p/P} = p^{\perp} - xP^{\perp}$, and P is the total momentum in the term one considers. The regulating function prevents the interaction terms from acting if the change of transverse momentum between gluons were to exceed Δ, or if the change of longitudinal momentum fraction x were to be smaller than δ.

2.2 Derivation of the Effective Hamiltonian

The RGPEP transforms bare, or point-like creation and annihilation operators into effective ones [1]. Effective particle operators of size $s = t^{1/4}$ are related to bare ones by certain unitary transformation

$$a_t = \mathcal{U}_t\, a_0\, \mathcal{U}_t^{\dagger}. \tag{13}$$

The fact that the Hamiltonian operator cannot be affected by this change requires,

$$\mathcal{H}_t(a_t) = \mathcal{H}_0(a_0), \tag{14}$$

which is equivalent to writing:

$$\mathcal{H}_t(a_0) = \mathcal{U}_t^{\dagger}\mathcal{H}_0(a_0)\mathcal{U}_t. \tag{15}$$

Differentiating both sides of (15) leads to the RGPEP equation:

$$\mathcal{H}_t'(a_0) = [\mathcal{G}_t(a_0), \mathcal{H}_t(a_0)], \tag{16}$$

where $\mathcal{G}_t = -\mathcal{U}_t^{\dagger}\mathcal{U}_t'$, and therefore, $\mathcal{U}_t = T\exp\left(-\int_0^t d\tau\, \mathcal{G}_\tau\right)$. T denotes ordering in τ. The RGPEP equation (16) is the engine of this procedure. It governs the evolution of effective particles with the scale parameter t. It encodes the relation between pointlike quantum gluons appearing in the canonical Hamiltonian and the effective, or constituent ones referred to by effective phenomenological models describing bound states.

We choose the generator to be the commutator $\mathcal{G}_t = [\mathcal{H}_f, \mathcal{H}_{Pt}]$,[2] where $\mathcal{H}_f$ is the non-interacting term of the Hamiltonian and $\mathcal{H}_{Pt}$ is defined in terms of $\mathcal{H}_t$.

$\mathcal{H}_t$ is a series of normal-ordered products of creation and annihilation operators,

$$\mathcal{H}_t(a_0) = \sum_{n=2}^{\infty} \sum_{i_1, i_2, \ldots, i_n} c_t(i_1, \ldots, i_n)\, a_{0i_1}^{\dagger} \cdots a_{0i_n}. \tag{17}$$

$\mathcal{H}_{Pt}$ differs from $\mathcal{H}_t$ by the vertex total +-momentum factor,

$$\mathcal{H}_{Pt}(a_0) = \sum_{n=2}^{\infty} \sum_{i_1, i_2, \ldots, i_n} c_t(i_1, \ldots, i_n) \left(\frac{1}{2}\sum_{k=1}^{n} p_{i_k}^{+}\right)^2 a_{0i_1}^{\dagger} \cdots a_{0i_n}. \tag{18}$$

The initial condition for the differential equation (16) is given by the regularized canonical Hamiltonian given in Sect. 2.1 plus counterterms. More precisely, the initial condition is given by the physical fact that at very small distances or very high energies, the regularized canonical Hamiltonian must be recovered and any regularization dependence must be removed.

We solve the RGPEP equation (16) for the effective Hamiltonians using an expansion in powers of the coupling constant g up to third order and we focus our studies on the structure of the three-gluon term [1,3].

[1] Other regulating functions are available [3]. Finite dependence of the effective Hamiltonian on the small-x regularization may be thought to be related to the vacuum state problem, the phenomena of symmetry breaking and confinement [4].

[2] Other generators are also allowed but may lead to more complicated expressions [8]. Our choice is similar to Wegner's [9].

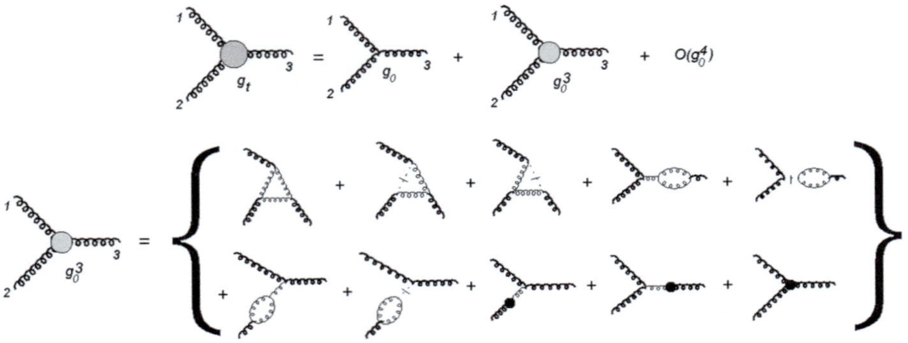

Fig. 1 Graphical representation of terms contributing to the effective three-gluon vertex (third-order expansion) [3]. *Thin internal lines* correspond to intermediate bare gluons and *thick external lines* correspond to the creation and annihilation operators that appear in the three-gluon FF Hamiltonian interaction term for effective gluons of size s. *Dashed lines* with transverse *bars* represent the combined contributions of terms (8) and (9). The *black dots* indicate counterterms

3 The Three-Gluon Vertex

The third-order effective Hamiltonian expansion have the following structure:

$$H_t = H_{11,0,t} + H_{11,g^2,t} + H_{21,g,t} + H_{12,g,t} + H_{31,g^2,t} + H_{13,g^2,t} + H_{22,g^2,t} + H_{21,g^3,t} + H_{12,g^3,t}. \tag{19}$$

The first and second subscripts indicate the number of creation and annihilation operators, respectively. The third subscript labels the order in powers of g. Finally, the last label indicates the dependence on the scale parameter t.

The initial condition at $t = 0$ has the form:

$$H_0 = H_{11,0,0} + H_{11,g^2,0} + H_{21,g,0} + H_{12,g,0} + H_{31,g^2,0} + H_{13,g^2,0} + H_{22,g^2,0} + H_{21,g^3,0} + H_{12,g^3,0}, \tag{20}$$

and consists of the regularized canonical Hamiltonian plus counterterms. The latter are calculated in such a way that H_t remains finite when $\Delta \to \infty$. It is not possible to remove the small-x cutoff δ at these point. However, this dependence will be of interest in higher-order calculations, since the small-x phenomena are thought to be related to the vacuum-state behavior. The last step in the RGPEP is to replace bare creation and annihilation operators by effective ones.

The three-gluon vertex and the running of the Hamiltonian coupling are encoded in the sum of first- and third-order terms:

$$H_{(1+3),t} = (H_{21,g,t} + H_{12,g,t}) + (H_{21,g^3,t} + H_{12,g^3,t}). \tag{21}$$

The third-order solution requires the knowledge of the first and second-order solutions. The sum of all these contributions has the form[3] [3] (see Fig. 1):

$$H_{(1+3),t} = \sum_{123} \int [123]\, \tilde{\delta}(k_1 + k_2 - k_3)\, f_{12t} \left[\tilde{Y}_{21\,t}(x_1, \kappa_{12}^{\perp}, \sigma) a_{1t}^{\dagger} a_{2t}^{\dagger} a_{3t} + \tilde{Y}_{12\,t}(x_1, \kappa_{12}^{\perp}, \sigma)\, a_{3t}^{\dagger} a_{2t} a_{1t} \right] \tag{22}$$

where $f_{12t} = e^{-(k_1+k_2)^4 t}$ is a form factor and $\tilde{Y}_{21\,t}(x_1, \kappa_{12}^{\perp}, \sigma)$ is the object of our study. We define the Hamiltonian coupling constant g_t as the coefficient in front of the canonical color, spin and momentum dependent factor $Y_{123}(x_1, \kappa_{12}^{\perp}, \sigma) = i f^{c_1 c_2 c_3} [\varepsilon_1^* \varepsilon_2^* \cdot \varepsilon_3 \kappa_{12}^{\perp} - \varepsilon_1^* \varepsilon_3 \cdot \varepsilon_2^* \kappa_{12}^{\perp} \frac{1}{x_{2/3}} - \varepsilon_2^* \varepsilon_3 \cdot \varepsilon_1^* \kappa_{12}^{\perp} \frac{1}{x_{1/3}}]$ in the limit $\kappa_{12}^{\perp} \to 0$, for some value of x_1 denoted by x_0. So,

$$\lim_{\kappa_{12}^{\perp} \to 0} \tilde{Y}_t(x_1, \kappa_{12}^{\perp}, \sigma) = \lim_{\kappa_{12}^{\perp} \to 0} \left[c_t(x_1, \kappa_{12}^{\perp}) Y_{123}(x_1, \kappa_{12}^{\perp}, \sigma) + g^3 \tilde{T}_{3\,\mathrm{finite}}(x_1, \kappa_{12}^{\perp}, \sigma) \right] \tag{23}$$

[3] The subscripts 1,2,3 refer to the gluon lines indicated in Fig. 1. So, e.g. $\kappa_{12}^{\perp} = x_{2/3}\kappa_{1/3}^{\perp} - x_{1/3}\kappa_{2/3}^{\perp}$.

where $\tilde{T}_{3\,\mathrm{finite}}(x_1, \kappa_{12}^\perp, \sigma)$ is a finite part contained in the counterterm and does not contribute to the running coupling,

$$\lim_{\kappa_{12}^\perp \to 0} c_t(x_1, \kappa_{12}^\perp) = g + g^3 \lim_{\kappa_{12}^\perp \to 0} \left[c_{3t}(x_1, \kappa_{12}^\perp) - c_{3t_0}(x_1, \kappa_{12}^\perp) \right] . \tag{24}$$

And assuming some value for g_0 at some small t_0, $g_{t_0} = g_0$,

$$g_t \equiv c_t(x_1) \;=\; g_0 + g_0^3 \left[c_{3t}(x_1) - c_{3t_0}(x_1) \right] . \tag{25}$$

We introduce now the momentum scale parameter $\lambda = t^{-1/4}$. This yields,

$$g_\lambda = g_0 - \frac{g_0^3}{48\pi^2} N_c \, 11 \, \ln \frac{\lambda}{\lambda_0} . \tag{26}$$

Differentiation of the latter with respect to λ leads to

$$\lambda \frac{d}{d\lambda} g_\lambda = \beta_0 g_\lambda^3 , \quad \text{with} \quad \beta_0 \;=\; -\frac{11 N_c}{48\pi^2} . \tag{27}$$

This result equals the asymptotic freedom result in Refs. [10,11], when one identifies λ with the momentum scale of external gluon lines in Feynman diagrams. Our result also coincides with the expression obtained in [8], where an analogous calculation were performed using a different generator.

4 Summary and Conclusion

We have applied the RGPEP to the quantum $SU(3)$ Yang-Mills theory and extracted the running coupling from the three-gluon-vertex term in the third-order effective Hamiltonian. The result turns out to be independent of the choice of the generator, as it coincides with the one obtained in an analogous calculation performed in [8], using a different generator. The present generator, however, leads to simpler equations than the older one, which is desired and needed for our forthcoming forth-order calculations, required for any attempt at description of physical systems using QCD [4]. The obtained running coupling is of the form that is familiar from other formalism and renormalization schemes and passes the test of producing asymptotic freedom, which any method aiming at solving QCD must past.

Acknowledgments Part of this work was supported by the Austrian Science Fund (FWF) under Project No. P25121-N27. Figure 1 was produced with JaxoDraw [12].

References

1. Glazek, S.D.: Perturbative formulae for relativistic interactions of effective particles. Acta Phys. Polon. **B43**, 1843–1862 (2012)
2. Glazek, S.D., Trawiski, A.P.: Model of the AdS/QFT duality. Phys. Rev. **D88**(10), 105025 (2013)
3. Gomez-Rocha, M., Glazek, S.D.: Asymptotic freedom in the front-form Hamiltonian for quantum chromodynamics of gluons. Phys. Rev. **D92**(6), 065005 (2015)
4. Wilson, K.G., Walhout, T.S., Harindranath, A., Zhang, W.-M., Perry, R.J., Glazek, S.D.: Nonperturbative QCD: a weak coupling treatment on the light front. Phys. Rev. **D49**, 6720–6766 (1994)
5. Dirac, P.A.M.: Forms of relativistic dynamics. Rev. Mod. Phys. **21**, 392–399 (1949)
6. Casher, A.: Gauge fields on the null plane. Phys. Rev. **D14**, 452 (1976)
7. Thorn, C.B.: Asymptotic freedom in the infinite momentum frame. Phys. Rev. **D20**, 1934 (1979)
8. Glazek, S.D.: Dynamics of effective gluons. Phys. Rev. **D63**, 116006 (2001)
9. Wegner, F.: Flow-equations for Hamiltonians. Ann. Phys. **506**, 77 (1994)
10. Gross, D.J., Wilczek, F.: Ultraviolet behavior of nonabelian gauge theories. Phys. Rev. Lett. **30**, 1343–1346 (1973)
11. Politzer, H.D.: Reliable perturbative results for strong interactions?. Phys. Rev. Lett. **30**, 1346–1349 (1973)
12. Binosi, D., Theussl, L.: JaxoDraw: a graphical user interface for drawing Feynman diagrams. Comput. Phys. Commun. **161**, 76–86 (2004)

DOI 10.1007/978-3-319-50699-9_37
Reprinted from *Few-Body System* Journal, DOI 10.1007/s00601-016-1079-z

Valery E. Lyubovitskij · Thomas Gutsche · Ivan Schmidt · Alfredo Vega

Deuteron Electromagnetic Form Factors in AdS/QCD

Received: 3 January 2016 / Accepted: 3 March 2016 / Published online: 18 March 2016

Abstract We extend a soft-wall AdS/QCD approach to a description of deuteron properties. Our framework is based an effective action formulated in terms of AdS fields, which are holographically equivalent to the deuteron and photon fields. This action produces the equation of motion for the deuteron wave function and the Q^2-dependent electromagnetic current, which are then used to calculate the deuteron electromagnetic form factors and structure functions in the Euclidean region. We show that the predicted deuteron quantities are expressed through a universal function, which is defined by a single scale parameter κ and which has the correct $1/Q^{10}$ power scaling at large Q^2.

Form factors are the main observables to study the structure of the deuteron (for the current status of experimental and theoretical progress see the recent review [1]). A variety of theoretical approaches have been used to calculate the deuteron form factors. For example, potential models [2,3] work with the nonrelativistic impulse approximation in order to factorize the deuteron form factors in terms of the isoscalar combinations of the elementary nucleon form factors. Unfortunately, potential approaches can only be applied to the description of data up to momentum transfer squared of about 0.5 GeV^2. The predictions for higher Q^2 are not consistent

V. E. Lyubovitskij (✉) · T. Gutsche
Institut für Theoretische Physik, Kepler Center for Astro and Particle Physics, Universität Tübingen,
Auf der Morgenstelle 14, 72076 Tübingen, Germany
E-mail: valeri.lyubovitskij@uni-tuebingen.de

T. Gutsche
E-mail: thomas.gutsche@uni-tuebingen.de

V. E. Lyubovitskij
Department of Physics, Tomsk State University, Tomsk, Russia 634050

V. E. Lyubovitskij
Mathematical Physics Department, Tomsk Polytechnic University, Lenin Avenue 30, Tomsk, Russia 634050

I. Schmidt
Departamento de Física y Centro Científico Tecnológico de Valparaíso (CCTVal),
Universidad Técnica Federico Santa María, Casilla 110-V, Valparaiso, Chile
E-mail: ivan.schmidt@usm.cl

A. Vega
Instituto de Física y Astronomía, Universidad de Valparaíso, Avenida Gran Bretaña 1111, Valparaiso, Chile

A. Vega
Centro de Astrofísica de Valparaíso, Avenida Gran Bretaña 1111, Valparaiso, Chile
E-mail: alfredo.vega@uv.cl

with data and contradict quark counting rules. Relativistic effects have been implemented by various methods. One possibility consists of taking into account relativistic corrections in a v/c expansion of the nonrelativistic current (leading to two-body interaction current diagrams) [4–6]. This extension allows for a validity in the description of data up to Q^2 of about 1–2 GeV^2. Another way to include relativistic corrections is based on the relativistic Hamiltonian constraint dynamics, which uses specific phenomenological potentials and the point, instant or front form in the quantization procedure [7,8]. Large progress in the study of deuteron form factors has been achieved using field-theoretical approaches. These methods deal with quark-gluon or hadronic degrees of freedom and include solutions with a quasipotential [9] or of the Bethe-Salpeter [10] equation. They can be based on field theories quantized on the light cone [11,12], on phenomenological Lagrangian approaches [13], on effective field theories treating the long-range dynamics explicitly and parametrizing the short-distance effects by contact interactions [14], on nonrelativistic quark models [15,16] or in pQCD [17]. In particular, the last method of pQCD [17] resulted in a model-independent prediction for the asymptotic large Q^2 behavior of the deuteron form factors consistent with the quark counting rules. A class of theories based on gauge/gravity duality can reproduce the scaling of hadronic form factors [18–20] as predicted by pQCD, but was not applied to the deuteron yet.

In present note we discuss the calculation of deuteron form factors in a soft-wall AdS/QCD approach [21–29], which is based on gauge/gravity duality and is constructed as a holographic dual to pQCD. The main advantage of our framework is that it gives a description of the deuteron electromagnetic form factors in terms of an universal form factor depending on a single dimensional parameter κ with the correct power scaling $1/Q^{10}$ at $Q^2 \to \infty$ in agreement with pQCD. The approach gives a good starting point for studying more complicated many-body nuclear systems, like three- and four-nucleon systems, etc.

The formalism is based on an effective action given in terms of the AdS fields $d^M(x,z)$ and $V^M(x,z)$. These fields are dual to the Fock component contributing to the deuteron with twist $\tau = 6$ and the electromagnetic field, respectively. The action is given by [30]

$$S = \int d^4x dz\, e^{-\varphi(z)} \biggl[-\frac{1}{4} F_{MN} F^{MN} - D^M d^\dagger_N D_M d^N - i c_2 F^{MN} d^\dagger_M d_N + \frac{c_3}{4M_d^2}\, e^{2A(z)}\, \partial^M F^{NK} \Bigl(i D_K d^\dagger_M d_N - d^\dagger_M i D_K d_N + \text{H.c.} \Bigr) + d^\dagger_M \Bigl(\mu^2 + U(z) \Bigr) d^M \biggr], \tag{1}$$

where $A(z) = \log(R/z)$, F^{MN} is the stress tensor of V^M, D^M is the covariant derivative, $\mu^2 R^2 = (\Delta - 1)(\Delta - 3)$ is the five-dimensional mass; R is the AdS radius, $\varphi(z) = \kappa^2 z^2$ is the background dilaton field; $\Delta = \tau + L$ is the dimension of the $d^M(x,z)$ field; L is the orbital angular momentum, and M_d is the deuteron mass. $U(z) = U_0 \varphi(z)/R^2)$ is the confinement potential, where the constant U_0 is fixed by the value of the deuteron mass. We work in the axial gauge for both vector fields $d^z(x,z) = 0$ and $V^z(x,z) = 0$. The parameters c_2 and c_3 are constrained by the normalization of the deuteron electromagnetic form factors (see discussion below). Therefore, we only have one free parameter κ.

Our calculation of deuteron form factors includes several steps. First, we perform a Kaluza-Klein (KK) decomposition for the vector AdS field dual to the deuteron $d^\mu(x,z) = e^{\frac{\varphi(z)-A(z)}{2}} \sum_n d_n^\mu(x) \Phi_n(z)$, where $d_n^\mu(x)$ is the tower of the KK fields dual to the deuteron fields with radial quantum number n and twist-dimension $\tau = 6$, and $\Phi_n(z)$ are their bulk profiles in the fifth direction of the AdS space.

Second, we derive the Schrödinger-type equation of motion (EOM) for the profile $\Phi_n(z)$

$$\left[-\frac{d^2}{dz^2} + \frac{4(L+4)^2 - 1}{4z^2} + \kappa^4 z^2 + \kappa^2 U_0 \right] \Phi_n(z) = M_{d,n}^2 \Phi_n(z), \tag{2}$$

which is solved analytically

$$\Phi_n(z) = \sqrt{\frac{2n!}{(n+L+4)!}}\, \kappa^{L+5}\, z^{L+9/2}\, e^{-\kappa^2 z^2/2}\, L_n^{L+4}(\kappa^2 z^2), \quad M_{d,n}^2 = 4\kappa^2 \left[n + \frac{L+5}{2} + \frac{U_0}{4} \right], \tag{3}$$

where $L_n^m(x)$ are the generalized Laguerre polynomials. Restricting our considerations to the ground state ($n = 0$, $L = 0$) we get $M_d = 2\kappa\sqrt{\frac{5}{2} + \frac{U_0}{4}}$. Using the central experimental value for the deuteron mass $M_d = 1.875613$ GeV and setting the parameter $\kappa = 190$ MeV (in a fit to the data of the electromagnetic deuteron form factors), we fix $U_0 = 87.4494$. Note that the deuteron scale parameter is two times smaller than the analogous parameter $\kappa_N \simeq 380$ GeV entering in nucleon properties—mass and electromagnetic form factors [24–29]. The difference between the nucleon and deuteron scale parameters can be related to the change of size of the hadronic systems—the deuteron as a two-nucleon bound state is 2 times larger than the nucleon.

Third, we perform a Fourier transformation of the vector field $V(x, z)$ with respect to the Minkowski coordinate

$$V_\mu(x, z) = \int \frac{d^4q}{(2\pi)^4} e^{-iqx} V_\mu(q) V(q, z) \tag{4}$$

and derive a EOM for the vector bulk-to-boundary propagator $V(q, z)$ dual to the q^2-dependent electromagnetic current

$$\partial_z \left(\frac{e^{-\varphi(z)}}{z} \partial_z V(q, z) \right) + q^2 \frac{e^{-\varphi(z)}}{z} V(q, z) = 0. \tag{5}$$

The solution of this equation in terms of the gamma $\Gamma(n)$ and Tricomi $U(a, b, z)$ functions reads

$$V(q, z) = \Gamma\Big(1 - \frac{q^2}{4\kappa^2}\Big) U\Big(-\frac{q^2}{4\kappa^2}, 0, \kappa^2 z^2\Big). \tag{6}$$

In the Euclidean region it is convenient to use the integral representation for $V(Q, z)$ [33]

$$V(Q, z) = \kappa^2 z^2 \int_0^1 \frac{dx}{(1-x)^2} x^a e^{-\kappa^2 z^2 \frac{x}{1-x}}, \tag{7}$$

where x is the light-cone momentum fraction and $a = Q^2/(4\kappa^2)$.

The gauge-invariant matrix element describing the interaction of the deuteron with the external vector field reads

$$\begin{aligned} M^\mu_{\rm inv}(p, p') = &-\Big(G_1(Q^2)\epsilon^*(p') \cdot \epsilon(p) - \frac{G_3(Q^2)}{2M_d^2} \epsilon^*(p') \cdot q\, \epsilon(p) \cdot q\Big) (p + p')^\mu \\ &- G_2(Q^2) \Big(\epsilon^\mu(p)\, \epsilon^*(p') \cdot q - \epsilon^{*\mu}(p')\, \epsilon(p) \cdot q\Big) \end{aligned} \tag{8}$$

where $\epsilon(\epsilon^*)$ and $p(p')$ are the polarization and four-momentum of the initial (final) deuteron, and $q = p' - p$ is the momentum transfer. The three EM form factors $G_{1,2,3}$ of the deuteron are related to the charge G_C, quadrupole G_Q and magnetic G_M form factors by

$$G_C = G_1 + \frac{2}{3}\tau_d G_Q, \quad G_M = G_2, \quad G_Q = G_1 - G_2 + (1 + \tau_d) G_3, \quad \tau_d = \frac{Q^2}{4M_d^2}. \tag{9}$$

The form factors are normalized at $Q^2 = 0$ as

$$G_C(0) = 1, \quad G_Q(0) = M_d^2 Q_d = 25.83, \quad G_M(0) = \frac{M_d}{M_N} \mu_d = 1.714, \tag{10}$$

where M_N is the nucleon mass, $Q_d = 7.3424$ GeV^{-2} and $\mu_d = 0.8574$ are the quadrupole and magnetic moments of the deuteron.

With the present set-up the deuteron form factors are expressed in terms of the universal form factor $F(Q^2)$:$G_1(Q^2) = F(Q^2)$, $\;G_i(Q^2) = c_i F(Q^2)$, $\;i = 2, 3$. The couplings c_2 and c_3 are fixed by normalization of the deuteron form factors with $c_2 = G_M(0) = 1.714$, $c_3 = G_M(0) + G_Q(0) - 1 = 26.544$. $F(Q^2)$ is the twist-6 form factor which is given by the overlap of the square of the bulk profile dual to the deuteron wave function (twist-6 hadronic wave function) and the confined electromagnetic current

$$F(Q^2) = \int_0^\infty dz\, \Phi_0^2(z)\, V(Q, z) = \frac{\Gamma(6)\, \Gamma(a+1)}{\Gamma(a+6)}. \tag{11}$$

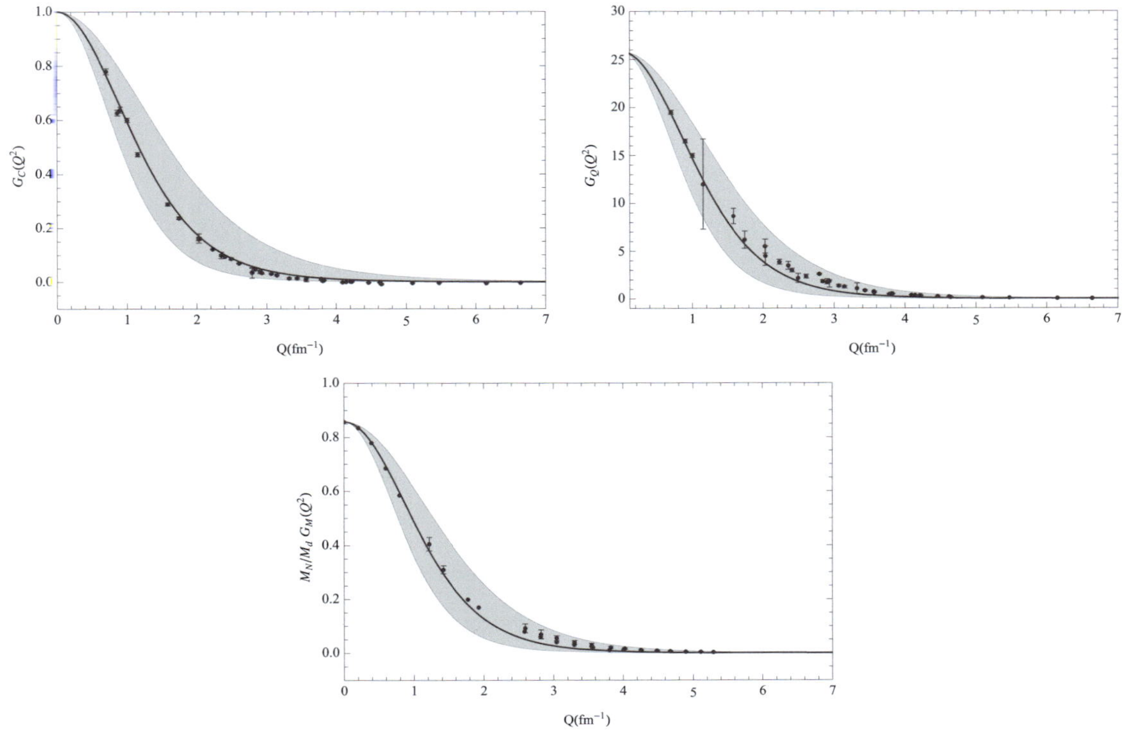

Fig. 1 Charge $G_C(Q^2)$, quadrupole $G_Q(Q^2)$ and magnetic $G_M(Q^2)$ deuteron form factor

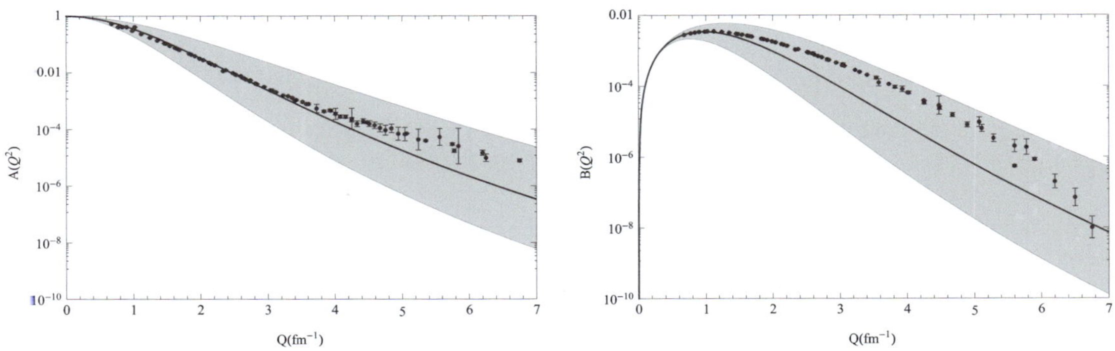

Fig. 2 Deuteron structure functions $A(Q^2)$ and $B(Q^2)$

This formula follows from the general formula for the hadronic form factor with twist τ obtained in Ref. [23]. Note that $F(Q^2)$ has the correct power-scaling $F(Q^2) \sim 1/(Q^2)^5$ at large $Q^2 \to \infty$. Our results for the charge $G_C(Q^2)$, quadrupole $G_Q(Q^2)$ and magnetic $G_M(Q^2)$ form factors are shown in Fig. 1. Data points are taken from Refs. [31,32]. The shaded band corresponds to values of the scale parameter κ in the range of 150–250 MeV. An increase of the parameter κ leads to an enhancement of the form factors. The best description of the data on the deuteron form factors is obtained for $\kappa = 190$ MeV and is shown by the solid line. We also determine the deuteron charge $r_C = (-6dG_C(Q^2)/dQ^2|_{Q^2=0})^{1/2} = \sqrt{137/(40\kappa^2) - Q_d}$ and magnetic $r_C = (-6dG_M(Q^2)/dQ^2|_{Q^2=0}/G_M(0))^{1/2} = \sqrt{137/(40\kappa^2)}$ radii, which for $\kappa = 190$ MeV are equal to 1.85 fm and 1.92 fm, respectively. These values compare well with the data of $r_C = 2.13 \pm 0.01$ fm and $r_M = 1.90 \pm 0.14$ fm. Finally, in Fig. 2 we present our predictions for the structure functions $A(Q^2)$ and $B(Q^2)$ defined as

$$A(Q^2) = G_C^2\left(Q^2\right) + \frac{2}{3}\tau_d G_M^2\left(Q^2\right) + \frac{8}{9}\tau_d^2 G_Q^2\left(Q^2\right), \quad B\left(Q^2\right) = \frac{4}{3}\tau_d(1+\tau_d)G_M^2\left(Q^2\right). \tag{12}$$

and compare them with data [31,32].

Acknowledgments This work was supported by the German Bundesministerium für Bildung und Forschung (BMBF) under Project 05P2015—ALICE at High Rate (BMBF-FSP 202): "Jet- and fragmentation processes at ALICE and the parton structure of nuclei and structure of heavy hadrons", by CONICYT (Chile) Research Project No. 80140097, by CONICYT (Chile) under Grant No. 7912010025, by FONDECYT (Chile) under Grants Nos. 1140390 and 1141280, by Tomsk State University Competitiveness Improvement Program and the Russian Federation program "Nauka" (Contract No. 0.1526.2015, 3854).

References

1. Marcucci, L.E., et al.: Electromagnetic structure of few-nucleon ground states. J. Phys. G Nucl. Part. Phys. **43**, 023002 (2016)
2. De Forest Jr., T., Walecka, J.D.: Electron scattering and nuclear structure. Adv. Phys. **15**, 1 (1966)
3. Donnelly, T.W., Walecka, J.D.: Electron scattering and nuclear structure. Ann. Rev. Nucl. Part. Sci. **25**, 329 (1975)
4. Friar, J.L.: Relativistic effects on the wave function of a moving system. Phys. Rev. C **12**, 695 (1975)
5. Mathiot, J.F.: Electromagnetic meson exchange currents at the nucleon mass scale. Phys. Rep. **173**, 63 (1989)
6. Schiavilla, R., Riska, D.O.: Threshold electrodisintegration and electromagnetic form-factors of the deuteron. Phys. Rev. C **43**, 437 (1991)
7. Lev, F.M., Pace, E., Salmè, G.: Deuteron magnetic and quadrupole moments with a Poincare covariant current operator in the front form dynamics. Phys. Rev. Lett. **83**, 5250 (1999)
8. Allen, T.W., Payne, G.L., Polyzou, W.N.: Comparison of relativistic nucleon–nucleon interactions. Phys. Rev. C **62**, 054002 (2000)
9. Blankenbecler, R., Sugar, R.: Linear integral equations for relativistic multichannel scattering. Phys. Rev. **142**, 1051 (1966)
10. Buck, W.W., Gross, F.: A family of relativistic deuteron wave functions. Phys. Rev. D **20**, 2361 (1979)
11. Carbonell, J., Desplanques, B., Karmanov, V.A., Mathiot, J.F.: Explicitly covariant light front dynamics and relativistic few body systems. Phys. Rep. **300**, 215 (1998)
12. Karmanov, V.A., Smirnov, A.V.: Electromagnetic form-factors in the light front dynamics. Nucl. Phys. A **546**, 691 (1992)
13. Dong, Y.B., Faessler, A., Gutsche, T., Lyubovitskij, V.E.: Phenomenological Lagrangian approach to the electromagnetic deuteron form factors. Phys. Rev. C **78**, 035205 (2008)
14. Epelbaum, E., Gasparyan, A.M., Gegelia, J., Schindler, M.R.: Deuteron electromagnetic form factors in a renormalizable formulation of chiral effective field theory. Eur. Phys. J. A **50**, 51 (2014)
15. Buchmann, A., Yamauchi, Y., Faessler, A.: The one gluon exchange current in the quark cluster model of the deuteron. Phys. Lett. B **225**, 301 (1989)
16. Ito, H., Kisslinger, L.S.: Tensor polarization of the elastic e^2H scattering and the scaling behavior of six quark dynamics. Phys. Rev. C **40**, 887 (1989)
17. Brodsky, S.J., Ji, C.R., Lepage, G.P.: Quantum chromodynamic predictions for the deuteron form-factor. Phys. Rev. Lett. **51**, 83 (1983)
18. Polchinski, J., Strassler, M.J.: Hard scattering and gauge/string duality. Phys. Rev. Lett **88**, 031601 (2002)
19. Maldacena, J.M.: The large N limit of superconformal field theories and supergravity. Adv. Theor. Math. Phys. **2**, 231 (1998)
20. Maldacena, J.M.: The large N limit of superconformal field theories and supergravity. Int. J. Theor. Phys. **38**, 1113 (1999)
21. Karch, A., Katz, E., Son, D.T., Stephanov, M.A.: Linear confinement and AdS/QCD. Phys. Rev. D **74**, 015005 (2006)
22. Andreev, O.: $1/q^2$ corrections and gauge/string duality. Phys. Rev. D **73**, 107901 (2006)
23. Brodsky, S.J., de Teramond, G.F.: Light-front dynamics and AdS/QCD correspondence: the pion form factor in the space- and time-like regions. Phys. Rev. D **77**, 056007 (2008)
24. Gutsche, T., Lyubovitskij, V.E., Schmidt, I., Vega, A.: Dilaton in a soft-wall holographic approach to mesons and baryons. Phys. Rev. D **85**, 076003 (2012)
25. Gutsche, T., Lyubovitskij, V.E., Schmidt, I., Vega, A.: Nucleon structure including high Fock states in AdS/QCD. Phys. Rev. D **86**, 036007 (2012)
26. Gutsche, T., Lyubovitskij, V.E., Schmidt, I., Vega, A : Nucleon resonances in AdS/QCD. Phys. Rev. D **87**, 016017 (2013)
27. Gutsche, T., Lyubovitskij, V.E., Schmidt, I., Vega, A.: Chiral symmetry breaking and meson wave functions in soft-wall AdS/QCD. Phys. Rev. D **87**, 056001 (2013)
28. Vega, A., Schmidt, I., Branz, T., Gutsche, T., Lyubovitskij, V.E.: Meson wave function from holographic models. Phys. Rev. D **80**, 055014 (2009)
29. Branz, T., Gutsche, T., Lyubovitskij, V.E., Schmidt, I., Vega, A.: Light and heavy mesons in a soft-wall holographic approach. Phys. Rev. D **82**, 074022 (2010)
30. Gutsche, T., Lyubovitskij, V.E., Schmidt, I., Vega, A.: Nuclear physics in soft-wall AdS/QCD: deuteron electromagnetic form factors. Phys. Rev. D **91**, 114001 (2015)
31. Abbott, D., et al., (JLAB t20 Collaboration): Phenomenology of the deuteron electromagnetic form-factors. Eur. Phys. J. A **7**, 421 (2000)
32. Holt, R.J., Gilman, R.: Transition between nuclear and quark-gluon descriptions of hadrons and light nuclei. Rep. Prog. Phys. **75**, 086301 (2012)
33. Grigoryan, H.R., Radyushkin, A.V.: Structure of vector mesons in holographic model with linear confinement. Phys. Rev. D **76**, 095007 (2007)

DOI 10.1007/978-3-319-50699-9_38
Reprinted from *Few-Body System* Journal, DOI 10.1007/s00601-016-1083-3

Jerzy A. Przeszowski · Elżbieta Dzimida-Chmielewska · Jan Żochowski

Light-Front Perturbation Without Spurious Singularities

Received: 12 January 2016 / Accepted: 8 March 2016 / Published online: 28 March 2016

Abstract A new form of the light front Feynman propagators is proposed. It contains no energy denominators. Instead the dependence on the longitudinal subinterval $x_L^2 = 2x^+x^-$ is explicit and a new formalism for doing the perturbative calculations is invented. These novel propagators are implemented for the one-loop effective potential and various 1-loop 2-point functions for a massive scalar field. The consistency with results for the standard covariant Feynman diagrams is obtained and no spurious singularities are encountered at all. Some remarks on the calculations with fermion and gauge fields in QED and QCD are added.

1 Introduction

Wightman function for a free massive scalar field $\langle 0|\phi(x)\phi(0)|0\rangle = W_2(x)$ has its LF momentum representation

$$W_2(x^+, x^-, \boldsymbol{x}_\perp) = \int_{\mathbb{R}^2} \frac{d^2\boldsymbol{k}_\perp}{(2\pi)^2} e^{-i\boldsymbol{k}_\perp\cdot\boldsymbol{x}_\perp} \int_0^\infty \frac{dk^+}{4\pi k^+} e^{-ik^+x^-} e^{-i\frac{m^2+\boldsymbol{k}_\perp^2}{2k^+}x^+}. \tag{1}$$

The LF propagator $\Delta_{LF}(x)$ is defined by the chronological (in x^+) ordering

$$\begin{aligned}\Delta_{LF}(x) = \langle 0|T^+\phi(x)\phi(0)|0\rangle &:= \Theta(x^+)\langle 0|\phi(x)\phi(0)|0\rangle + \Theta(-x^+)\langle 0|\phi(0)\phi(x)|0\rangle \\ &= \Theta(x^+)W_2(x^+, x^-, \boldsymbol{x}_\perp) + \Theta(-x^+)W_2(-x^+, -x^-, -\boldsymbol{x}_\perp).\end{aligned} \tag{2}$$

Within the standard LF approach [1] (for review see [2]) one introduces the Fourier representation for the Heaviside step function

$$\Theta(x^+) = \int_{\mathbb{R}} \frac{dk^-}{2\pi} e^{-ik^-x^+} \frac{i}{k^- + i0}, \tag{3}$$

J. A. Przeszowski (✉) · E. Dzimida-Chmielewska · J. Żochowski
Faculty of Physics, University of Białystok, ul. Ciołkowskiego 4a, 15-424 Białystok, Poland
E-mail: j.przeszowski@uwb.edu.pl

E. Dzimida-Chmielewska
E-mail: elka@alpha.uwb.edu.pl

J. Żochowski
E-mail: j.zochowski@uwb.edu.pl

then changes the order of integrations and finally one shifts the integration variable which gives

$$\int_{\mathbb{R}} \frac{dk^-}{2\pi} \frac{i\, e^{-ik^- x^+}}{k^- + i0} \int_0^\infty \frac{dk^+}{k^+} e^{-ik^+ x^-} e^{-ix^+(m^2+k_\perp^2)/(2k^+)} = \int_0^\infty \frac{dk^+}{k^+} e^{-ik^+ x^-} \int_{\mathbb{R}} \frac{dk^-}{2\pi} e^{-ik^- x^+} \frac{i}{k^- - \frac{m^2+k_\perp^2}{2k^+} + i0}. \tag{4}$$

For $k^+ \to 0$, the pole in k^- moves to infinity, so the naive implementation of the residua theorem can lead to false results. This problem overlaps with the usual LF singularity due to $1/k^+$ pole. The very trick, presented in (4), is analogous to the equal-time propagators, where one makes the ordering in x^0 temporal variable, so one introduces

$$\Theta(x^0) = \int_{\mathbb{R}} \frac{dk_0}{2\pi} e^{-ik_0 x^0} \frac{i}{k_0 + i0}, \tag{5}$$

and then one proceeds as follows

$$\int_{\mathbb{R}} \frac{dk_0}{2\pi} e^{-ik_0 x^0} \frac{i}{k_0 + i0} \frac{1}{2\omega_k} e^{-ix^0 \omega_k} = \int_{\mathbb{R}} \frac{dk_0}{2\pi} \frac{e^{-ik_0 x^0}}{2\omega_k} \frac{i}{k_0 - \omega_k + i0}. \tag{6}$$

In this calculation the shifting of k_0 is reliable since the limit $\omega_k \to \infty$ is suppressed by the inverse power of ω_k in the invariant measure factor, on contrary in the Eq. (4). Therefore it is not strange that the LF propagator with the pole structure as in (4) may lead to various artificial singularities of Feynman diagrams, which are absent in the analogous equal-time calculation.

2 Novel LF Representation of Propagator and Convolutions of Propagators

We observe that we may make the following changes of integration variables: for $x^+ > 0$ we take $k^+ = 2\lambda x^+$

$$W_2(x) \simeq \int_0^\infty \frac{dk^+}{4\pi k^+} e^{-ik^+ x^-} e^{-i\frac{m^2+k_\perp^2}{2k^+} x^+} = \int_0^\infty \frac{d\lambda}{4\pi\lambda} e^{-i\lambda \boldsymbol{x}_L^2} e^{-i(m^2+\boldsymbol{k}_\perp^2)/(4\lambda)}, \tag{7}$$

while for $x^+ < 0$ we take $k^+ = -2\lambda x^+$

$$W_2(-x) \simeq \int_0^\infty \frac{dk^+}{4\pi k^+} e^{ik^+ x^-} e^{i\frac{m^2+k_\perp^2}{2k^+} x^+} = \int_0^\infty \frac{d\lambda}{4\pi\lambda} e^{-i\lambda \boldsymbol{x}_L^2} e^{-i(m^2+\boldsymbol{k}_\perp^2)/(4\lambda)} \tag{8}$$

This leads to the LF propagator in the form, which we call λ-representation,

$$\langle 0|T^+\phi(x)\phi(0)|0\rangle = \int_{\mathbb{R}^2} \frac{d^2\boldsymbol{k}_\perp}{(2\pi)^2} e^{-i\boldsymbol{k}_\perp \cdot \boldsymbol{x}_\perp} \int_0^\infty \frac{d\lambda}{4\pi\lambda} e^{-i\lambda \boldsymbol{x}_L^2} e^{-iM^2/(4\lambda)} = \Delta_{LF}(x), \tag{9}$$

where $\boldsymbol{x}_L^2 = 2x^+x^-$ and $M^2 = m^2 + \boldsymbol{k}_\perp^2$. If one wishes to compare this new representation with the covariant formula in the 4-momentum space, then one takes the Fourier transform in the 2-dimensional longitudinal subspace (31) and then integrates over λ, as defined in the sense of distributions in (30). This gives the equivalence between λ-representation and covariant Feynman propagators

$$\Delta_{LF}(x) = \int_{\mathbb{R}^2} \frac{d^2\boldsymbol{k}_\perp}{(2\pi)^2} \int_{\mathbb{R}^2} \frac{d^2\boldsymbol{k}_L}{(4\pi)^2} e^{-i\boldsymbol{k}_\perp \cdot \boldsymbol{x}_\perp} e^{-i\boldsymbol{k}_L \cdot \boldsymbol{x}_L} \frac{4i}{k_L^2 - k_\perp^2 + m^2 + i0} = \Delta_F(x). \tag{10}$$

Evidently λ-representation of $\Delta_{LF}(x)$ is singular at $x^+ = 0$, but one may check its consistency with the help of its Volterra equation, which is quite similar to the Volterra equation for the Wightman function of a free massive scalar field [3]. Next we consider the convolutions of LF propagators and we begin with the convolution of two LF propagators defined as follows

$$[\Delta_{LF} * \Delta_{LF}](x-z) = \Delta_{LF}^2(x-z) := \int_{\mathbb{R}^4} d^4y \Delta_{LF}(x-y)\Delta_{LF}(y-z). \tag{11}$$

By inserting λ-representation, with $(\lambda_1, \boldsymbol{p}_{1\perp})$ and $(\lambda_2, \boldsymbol{p}_{2\perp})$ for respective propagators, one may directly integrate over the transverse and longitudinal coordinates using (32a) and (32c)

$$\Delta_{LF}^2(x-z) = \int_{\mathbb{R}^2} \frac{d^2\boldsymbol{p}_{1\perp}}{(2\pi)^2} \int_{\mathbb{R}^2} d^2\boldsymbol{p}_{1\perp}\delta^2(\boldsymbol{p}_{1\perp}-\boldsymbol{p}_{2\perp})e^{-i\boldsymbol{p}_{1\perp}\cdot(\boldsymbol{x}_\perp-\boldsymbol{z}_\perp)} \int_0^\infty \frac{d\lambda_1}{4\pi\lambda_1} \int_0^\infty \frac{d\lambda_2}{4\pi\lambda_2}$$
$$\times \frac{\pi}{\lambda_1+\lambda_2} e^{-i\frac{2\lambda_1\lambda_2}{\lambda_1+\lambda_2}(x^+-z^+)(x^--z^-)} e^{-iM_1^2/(4\lambda_1)} e^{-iM_1^2/(4\lambda_2)}. \tag{12}$$

where $M_1^2 = m^2 + \boldsymbol{p}_{1\perp}^2$ and $M_2^2 = m^2 + \boldsymbol{p}_{2\perp}^2$. While the integration over $\boldsymbol{p}_{2\perp}$ is immediate, then for evaluating the integrals over $\lambda_{1,2}$ one needs to parameterize them as $\lambda_1 = \lambda/\xi$, $\lambda_2 = \lambda/(1-\xi)$, with new parameters $\lambda \in (0,\infty)$ and $\xi \in (0,1)$. This allows to evaluate the integration over ξ explicitly, and finally we obtain λ-representation for the convolution of two LF propagators, (where we put $\boldsymbol{p}_{1\perp} = \boldsymbol{p}_\perp$)

$$\Delta_{LF}^2(x-z) = \frac{1}{4}\int_{\mathbb{R}^2} \frac{d^2\boldsymbol{p}_\perp}{(2\pi)^2} e^{-i\boldsymbol{p}_\perp\cdot(\boldsymbol{x}_\perp-\boldsymbol{z}_\perp)} \int_0^\infty \frac{d\lambda}{4\pi\lambda^2} e^{-i\lambda(\boldsymbol{x}_L-\boldsymbol{z}_L)^2} e^{-i\frac{m^2+p_\perp^2}{4\lambda}}. \tag{13}$$

By induction one finds the convolution of n propagators

$$\Delta_{LF}^n(x-z) = \frac{1}{4^{n-1}}\frac{1}{(n-1)!}\int_{\mathbb{R}^2} \frac{d^2\boldsymbol{p}_\perp}{(2\pi)^2} e^{-i\boldsymbol{p}_\perp\cdot(\boldsymbol{x}_\perp-\boldsymbol{z}_\perp)} \int_0^\infty \frac{d\lambda}{4\pi\lambda^n} e^{-i\lambda(\boldsymbol{x}_L-\boldsymbol{z}_L)^2} e^{-i\frac{m^2+p_\perp^2}{4\lambda}}. \tag{14}$$

The 1-loop effective potential (for the $g/(4!)\phi^4$ theory) is given by [4]

$$V_{eff}^{(1)}[\phi_c] = \frac{i}{2}\sum_{n=1} \frac{(-i)^n}{n}\left(\frac{g}{2}\phi_c^2\right)^n \Delta_{LF}^n(0), \tag{15}$$

thus we need to evaluate $\Delta_{LF}^n(0)$, where using the general formula (30) we find

$$\Delta_{LF}^n(0) = \frac{1}{4^{n-1}}\frac{1}{(n-1)!}\int_{\mathbb{R}^2} \frac{d^2\boldsymbol{p}_\perp}{(2\pi)^2} \int_0^\infty \frac{d\lambda}{4\pi\lambda^n} e^{-i\frac{m^2+p_\perp^2}{4\lambda}} = \frac{i^{n-1}}{4\pi(n-1)}\int_{\mathbb{R}^2} \frac{d^2\boldsymbol{p}_\perp}{(2\pi)^2}\frac{1}{(m^2+\boldsymbol{p}_\perp^2)^{n-1}}. \tag{16}$$

This expression gives nontrivial contribution to the effective potential, which can be compared with the result obtained within the standard LF formulation [4]

$$\bar{\Delta}_{LF}^n(0) = \int_{\mathbb{R}^2} \frac{d^2\boldsymbol{p}_\perp}{(2\pi)^2} \int_{\mathbb{R}} \frac{dp^+}{4\pi(p^+)^n} \int_{\mathbb{R}} dp^- \frac{1}{\left[p^- - \frac{m^2+p_\perp^2}{2p^+} + i\,\mathrm{sgn}(p^+)0\right]^n}, \tag{17}$$

which for $n > 1$, vanishes by residua, unless the contribution from the arc is properly taken into account.

3 One Loop 2-Point Diagrams

Now we will consider one loop 2-point diagrams with a flow of non-zero external 4-momentum q^μ, starting with the simplest scalar self-energy diagram

$$\Sigma(q) = \int_{\mathbb{R}^4} d^4x e^{+iq\cdot x} \Delta_F(x)\Delta_F(x) = \int_{\mathbb{R}^4} \frac{d^4k}{(2\pi)^4} \frac{i}{k^2 - m^2 + i0} \frac{i}{(k+q)^2 - m^2 + i0}, \tag{18}$$

where the standard covariant Feynman propagators are inserted and further after the Wick rotation one may evaluate the Euclidean momentum integrals. Our aim is to calculate $\Sigma(q)$ with λ-representation for $\Delta_{LF}(x)$

$$\Sigma(q) = \int_{\mathbb{R}^4} d^4x\, e^{+iq\cdot x} \Delta_{LF}(x)\Delta_{LF}(x). \tag{19}$$

We denote $(\lambda_1, \boldsymbol{p}_{1\perp})$ and $(\lambda_2, \boldsymbol{p}_{2\perp})$ for $\Delta_{LF}(x)$ respectively and start with the integration over the space-time coordinates, using (32b),

$$\Sigma(q) = \int_{\mathbb{R}^2} \frac{d^2\boldsymbol{p}_{1\perp}}{(2\pi)^2} \int_{\mathbb{R}^2} d^2\boldsymbol{p}_{2\perp} \delta^2(\boldsymbol{p}_{1\perp} + \boldsymbol{p}_{2\perp} - \boldsymbol{q}_\perp) \int_0^\infty \frac{d\lambda_1}{4\pi\lambda_1} \int_0^\infty \frac{d\lambda_2}{4\pi\lambda_2} e^{-i(\lambda_1+\lambda_2)(\boldsymbol{x}_L)^2} e^{-iM_1^2/(4\lambda_1)} e^{i\frac{\boldsymbol{q}_L^2}{4(\lambda_1+\lambda_2)}}, \tag{20}$$

where $M_1^2 = m^2 + \boldsymbol{p}_{1\perp}^2$ and $M_2^2 = m^2 + \boldsymbol{p}_{2\perp}^2$. Then we parameterize $\lambda_1 = \lambda\xi$, $\lambda_2 = \lambda(1-\xi)$ with the Jacobian $\mathscr{J} = \lambda$ and the transverse momenta as $\boldsymbol{p}_{1\perp} = \boldsymbol{p}_\perp\xi + \boldsymbol{k}_\perp$, $\boldsymbol{p}_{2\perp} = \boldsymbol{p}_\perp(1-\xi) - \boldsymbol{k}_\perp$ with the Jacobian $\mathscr{J} = 1$, which leads to

$$\Sigma(q) = \int_{\mathbb{R}^2} \frac{d^2\boldsymbol{p}_\perp}{(2\pi)^2} \int_{\mathbb{R}^2} d^2\boldsymbol{k}_\perp \delta^2(\boldsymbol{p}_\perp - \boldsymbol{q}_\perp) \int_0^\infty \frac{d\lambda}{16\pi\lambda^2} \int_0^1 \frac{d\xi}{\xi(1-\xi)} e^{i\frac{\boldsymbol{q}_L^2}{4\lambda}} e^{-iM_1^2/(4\xi\lambda)} e^{-iM_2^2/(4(1-\xi)\lambda)}. \tag{21}$$

The integration over $\boldsymbol{p}_\perp$ is simple and due to the property (33), one finds

$$\Sigma(q) = \int_{\mathbb{R}^2} \frac{d^2\boldsymbol{k}_\perp}{(2\pi)^2} \int_0^\infty \frac{d\lambda}{16\pi\lambda^2} \int_0^1 \frac{d\xi}{\xi(1-\xi)} e^{i(\boldsymbol{q}_L^2 - \boldsymbol{q}_\perp^2)/(4\lambda)} e^{-i(m^2+\boldsymbol{k}_\perp^2)/(4\xi(1-\xi)\lambda)}. \tag{22}$$

The integral over λ can be performed explicitly according to (30) for $n = 1$, so

$$\Sigma(q) = \frac{i}{4\pi} \int_0^1 d\xi \int_{\mathbb{R}^2} \frac{d^2\boldsymbol{k}_\perp}{(2\pi)^2} \frac{1}{q^2\xi(1-\xi) - m^2 - \boldsymbol{k}_\perp^2 + i0}, \tag{23}$$

which coincides with the result of calculation with the covariant Feynman propagators. Then we wish to consider another 2-point function

$$\Sigma_\mu(q) = i \int_{\mathbb{R}^4} d^4x e^{iq\cdot x} \partial_\mu \Delta_F(x)\Delta_F(x) = \int_{\mathbb{R}^4} \frac{d^4k}{(2\pi)^4} \frac{k_\mu}{k^2 - m^2 + i0} \frac{i^2}{(k+q)^2 - m^2 + i0}. \tag{24}$$

We now insert the propagators in λ-representation and take steps analogous to those in the calculation of $\Sigma(q)$. There is a slight difference,because $i\partial_\mu \Delta_{LF}(x)$ appears instead of $\Delta_{LF}(x)$. First, for the longitudinal partial derivatives (32d) one obtains the extra factor $\lambda_1 q_\pm/(\lambda_1 + \lambda_2)$, which after the re-parametrization of $\lambda_{1,2}$ boils down to the extra factor $q_\mu\xi$. Second, for the transverse partial derivatives one obtains another extra factor

$\boldsymbol{p}_{1\perp} = \boldsymbol{p}_\perp \xi + \boldsymbol{k}_\perp$. Third, the term linear in $k\_$, due to its antisymmetry, will vanish during the integration over $\boldsymbol{k}_\perp$. Ultimately one obtains the covariant expression

$$\Sigma_\mu(q) = \frac{i\,q_\mu}{4\pi} \int_0^1 d\xi \int_{\mathbb{R}^2} \frac{d^2\boldsymbol{k}_\perp}{(2\pi)^2} \frac{\xi}{q^2\xi(1-\xi) - m^2 - \boldsymbol{k}_\perp^2 + i0}. \tag{25}$$

Finally we may consider the 2-point function, which has been used by Melikhov and Simula in [5] for their discussion of spurious end-point singularities

$$\Sigma_\mu^{ms}(q) = i \int_{\mathbb{R}^4} d^4x e^{iq\cdot x} \partial_\mu \Delta_F(x) \Delta_F^2(x) = \int_{\mathbb{R}^4} \frac{d^4k}{(2\pi)^4} \frac{k_\mu}{k^2 - m^2 + i0} \frac{i^3}{[(k+q)^2 - m^2 + i0]^2}. \tag{26}$$

From λ-representations for $\Delta_F(x)$ in (9) and the convolution $\Delta_F^2(x)$ in (13), we see that $\Sigma_\mu^{ms}(q)$ contains extra term $1/(4\lambda_2) = [4\lambda(1-\xi)]^{-1}$ in comparison with $\Sigma_\mu(q)$. Thus following all steps preceding (25) we find

$$\Sigma_\mu^{ms}(q) = q_\mu \int_{\mathbb{R}^2} \frac{d^2\boldsymbol{k}_\perp}{(2\pi)^2} \int_0^\infty \frac{d\lambda}{16\pi\lambda^3} \int_0^1 \frac{d\xi}{\xi(1-\xi)} \frac{\xi}{1-\xi} e^{i(\boldsymbol{q}_L^2 - \boldsymbol{q}_\perp^2)/(4\lambda)} e^{-i(m^2+\boldsymbol{k}_\perp^2)/(4\xi(1-\xi)\lambda)}. \tag{27}$$

The integral over λ can be performed explicitly according to (30) for $n = 1$, which leads to the desired form

$$\Sigma_\mu^{ms}(q) = \frac{iq_\mu}{4\pi} \int_0^1 d\xi \int_{\mathbb{R}^2} \frac{d^2\boldsymbol{k}_\perp}{(2\pi)^2} \frac{\xi^2}{[q^2\xi(1-\xi) - m^2 - \boldsymbol{k}_\perp^2 + i0]^2}, \tag{28}$$

with no sign of any end-point singularities.

4 Conclusions and Prospects

We conclude that the novel λ-representation for the scalar field LF propagator, which appears naturally within the LF quantization, is a very useful tool for evaluation 1-loop diagrams. For 2-point functions, we have derived the desired form of covariant expression with no spurious end-point singularities. For the effective potential we arrived at the nontrivial integrals over the transverse momenta, which are consistent with the equal-time results. Evidently, the next step should be to evaluate 3-point functions and to compare our method with the existing LF alternative approach proposed by Heinzl [6].

Also we would like to mention what happens for fields with non-zero spin. For the gauge field in the LC gauge $A^+ = 0$ one obtains for the chronological ordering in x^+

$$\langle 0|T^+ A_+(x) A_+(0)|0\rangle = 2\partial_- \int_0^{x^-} d\tau\, D_{LF}(x^+, \tau, \boldsymbol{x}_\perp) - i\delta(x^+)\delta^2(\boldsymbol{x}_\perp)|x^-|, \tag{29}$$

with the LF massless Feynman propagator function $D_{LF}(x) = \lim_{m\to 0} \Delta_{LF}$. For the first term, which has the form proposed by Bassetto [7], one may introduce λ-representation consistently, but the second term, which is local in time x^+ needs to be subtracted. This agrees with the standard LF quantization, where the LF propagator also has this additional local term. Analogous situation occurs for the fermion field propagator, where one obtains an additional local term for the "bad components" ψ_-, $\psi_-^\dagger$ of the fermion field. However in the LF Hamiltonian there are local interaction terms which exactly cancel the contribution from local terms in the LF propagators, thus effectively one obtains Feynman diagrams for the perturbative QED and QCD without local term contributions.

We hope that our novel representation for propagators may shed a new light on the equivalence problem between the LF and equal-time perturbative calculations [8]. At last, it will be quite interesting to apply λ-regularization for the LF Bethe-Salpeter equations as in [9].

Appendix: Definitions and Useful Formulas

The LF longitudinal coordinates are defined as $x^{\pm} = (x^0 \pm x^3)/\sqrt{2}$ and the partial derivatives are denoted as $\partial_{\pm} = \partial/\partial x^{\pm}$. The Minkowski space-time metric tensor has non-vanishing components $g_{+-} = g_{-+} = 1, \quad g_{ij} = -\delta_{ij}$.

We have the formula for $n \in \mathbb{N} - \{1\}$

$$\int_0^{\infty} \frac{d\lambda}{\lambda^{n+1}} e^{iA/\lambda} = \frac{i^n (n-1)!}{(A+i0)^n}. \tag{30}$$

The Fourier transform in the longitudinal coordinates

$$e^{-i\lambda \boldsymbol{x}_L^2} = e^{-i2x^+x^-\lambda} = \int_{\mathbb{R}^2} \frac{dk^+dk^-}{4\pi\lambda} e^{-i(k^+x^- + k^-x^-)} e^{ik^+k^-/(2\lambda)} = \int_{\mathbb{R}^2} \frac{d^2\boldsymbol{k}_L}{4\pi\lambda} e^{-i\boldsymbol{k}_L\cdot\boldsymbol{x}_L} e^{i\boldsymbol{k}_L^2/(4\lambda)}, \tag{31}$$

The integrations over the transverse and longitudinal coordinates give respectively

$$\int_{\mathbb{R}^2} d^2\boldsymbol{y}_\perp e^{-i\boldsymbol{p}_{1\perp}\cdot(\boldsymbol{x}_\perp - \boldsymbol{y}_\perp)} e^{-i\boldsymbol{p}_{2\perp}\cdot(\boldsymbol{y}_\perp - \boldsymbol{z}_\perp)} = (2\pi)^2\delta^2(\boldsymbol{p}_{1\perp} - \boldsymbol{p}_{2\perp}) e^{-i\boldsymbol{p}_{1\perp}\cdot\boldsymbol{x}_\perp} e^{i\boldsymbol{p}_{2\perp}\cdot\boldsymbol{z}_\perp}, \tag{32a}$$

$$\int_{\mathbb{R}^2} d^2\boldsymbol{x}_\perp e^{-i\boldsymbol{x}_\perp\cdot\boldsymbol{p}_{1\perp}} e^{-i\boldsymbol{x}_\perp\cdot\boldsymbol{p}_{2\perp}} e^{i\boldsymbol{q}_\perp\cdot\boldsymbol{x}_\perp} = (2\pi)^2\delta^2(\boldsymbol{p}_{1\perp} + \boldsymbol{p}_{2\perp} - \boldsymbol{q}_\perp), \tag{32b}$$

$$\int_{\mathbb{R}^2} d^2\boldsymbol{y}_L e^{-i\lambda_1(\boldsymbol{x}_L - \boldsymbol{y}_L)^2} e^{-i2\lambda_2(\boldsymbol{y}_L - \boldsymbol{z}_L)^2} = \frac{\pi}{\lambda_1 + \lambda_2} \exp -i\left(\frac{\lambda_1\lambda_2}{\lambda_1+\lambda_2}(\boldsymbol{x}_L - \boldsymbol{z}_L)^2\right), \tag{32c}$$

$$\int_{\mathbb{R}^2} d^2\boldsymbol{x}_L \left[e^{-i\boldsymbol{x}_L^2\lambda_1}, \left(i\partial_{\pm} e^{-i\boldsymbol{x}_L^2\lambda_1}\right)\right] e^{-i\boldsymbol{x}_L^2\lambda_2} e^{i\boldsymbol{q}_L\cdot\boldsymbol{x}_L} = \frac{\pi}{\lambda_1+\lambda_2}\left[1, \frac{\lambda_1 q_{\pm}}{\lambda_1+\lambda_2}\right] \exp\frac{i\,\boldsymbol{q}_L^2}{4(\lambda_1+\lambda_2)}. \tag{32d}$$

The parameterization of transverse momenta leads to

$$\frac{M_1^2}{\xi} + \frac{M_2^2}{1-\xi} = \frac{m^2 + (\boldsymbol{p}_\perp\xi + \boldsymbol{k}_\perp)^2}{\xi} + \frac{m^2 + (\boldsymbol{p}_\perp(1-\xi) - \boldsymbol{k}_\perp)^2}{1-\xi} = \frac{m^2 + \boldsymbol{k}_\perp^2}{\xi(1-\xi)} + \boldsymbol{p}_\perp^2. \tag{33}$$

Acknowledgments We would like to thank the organizers of the LC2015 Workshop for all their efforts and hospitality.

References

1. Chang, S.-J., Yan, T.-M.: Quantum field theories in the infinite-momentum frame. II. Scattering matrices of scalar and Dirac fields. Phys. Rev. D **7**, 1147–1161 (1973)
2. Brodsky, S.J., Pauli, H.-C., Pinsky, S.: Quantum chromodynamics and other field theories on the light cone. Phys. Lett. C (Phys. Rep.) **301**, 299–486 (1998)
3. Przeszowski, J.A.: Lorentz symmetry for the light-front Wightman functions. Acta Phys. Pol. Proc. Suppl. B **6**, 327–333 (2013)
4. Convery, M.E., Taylor, C.C., Jun, J.W.: Vacuum structure, zero modes, and the effective potential in light-cone quantization. Phys. Rev. D **51**, 4445–4450 (1995)
5. Melikhov, D., Simula, S.: End-point singularities of Feynman graphs on the light cone. Phys. Lett. B **556**, 135–141 (2003)
6. Heinzl, T.: Alternative approach to light-front perturbation theory. Phys. Rev. D **75**, 025013 (2007)
7. Bassetto, A.: Free vector propagator in the light-cone gauge and the Mandelstam-Leibbrandt prescription. Phys. Rev. D **46**, 3676–3677 (1992)
8. Bakker, B.L.G., DeWitt, M.A., Ji, C.-R., Mischenko, Y.: Restoring the equivalence between the light-front and manifestly covariant formalisms. Phys. Rev. D **72**, 076005 (2005)
9. Sales, J.H.O., Frederico, T., Carlson, B.V., Sauer, P.U.: Light-front Bethe-Salpeter equation. Phys. Rev. C **61**, 044003 (2000)

DOI 10.1007/978-3-319-50699-9_39
Reprinted from *Few-Body System* Journal, DOI 10.1007/s00601-016-1084-2

Usha Kulshreshtha · Daya Shankar Kulshreshtha · James P. Vary

Light-Front Quantization of the Vector Schwinger Model with a Photon Mass Term in Faddeevian Regularization

Received: 24 January 2016 / Accepted: 1 March 2016 / Published online: 18 March 2016

Abstract In this talk, we study the light-front quantization of the vector Schwinger model with photon mass term in Faddeevian Regularization, describing two-dimensional electrodynamics with mass-less fermions but with a mass term for the $U(1)$ gauge field. This theory is gauge-non-invariant (GNI). We construct a gauge-invariant (GI) theory using Stueckelberg mechanism and then recover the physical content of the original GNI theory from the newly constructed GI theory under some special gauge-fixing conditions (GFC's). We then study LFQ of this new GI theory.

In this work, we study the light-front (LF) quantization (LFQ) of the vector Schwinger model (VSM) [1–15] with a photon mass term (PMT) [6–15] with Faddeevian Regularization (FR) [9–15] describing two-dimensional (2D) electrodynamics with mass-less fermions [1–15]. In this theory, the left-handed and right-handed fermions are coupled to the U(1) gauge field with equal couplings. This theory has been studied by several authors in various contexts [1–15]. This model is known for its exact solubility, ensured by a remarkable property of one-dimensional fermion systems, namely that they could be described in terms of canonical one-dimensional boson fields. This fermion-boson equivalence has led to the discovery of many interesting properties of 2D field theories [1–15]. They are often used as toy models to investigate many interesting theoretical techniques in field theories. These are some of the motivations for this study.

This model has been studied in the literature more often without a PMT (which resulted by demanding the regularization for the VSM to be gauge-invariant (GI) [1–5]). The theory with a mass term for the $U(1)$ gauge field [6–15] represents a new class of models in the 2D quantum electrodynamics with mass-less fermions but with a PMT [6–15]. We have studied this model with a mass term for the $U(1)$ gauge field in a series of papers [6–8] in the standard regularization (SR) [6–8]. This model could also be studied in the so-called Faddeevian regularization (FR) [9–15]. For a proper understanding of different regularizations like the SR and FR, their

Presented by Usha Kulshreshtha at LIGHTCONE2015, 21–25 September, 2015, LMF, INFN, Frascati, Italy.

U. Kulshreshtha (✉) · D. S. Kulshreshtha · J. P. Vary
Department of Physics and Astronomy, Iowa State University, Ames, IA 50011, USA
E-mail: ushakulsh@gmail.com

D. S. Kulshreshtha
E-mail: dskulsh@gmail.com

J. P. Vary
E-mail: jvary@iastate.edu

U. Kulshreshtha
Department of Physics, Kirori Mal College, University of Delhi, Delhi 110007, India

D. S. Kulshreshtha
Department of Physics and Astrophysics, University of Delhi, Delhi 110007, India

physical significance and importance in field theories, we refer to the work of Refs. [9–15]. The mass term for the $U(1)$ gauge field in the FR has been derived explicitly by Mukhopadhyay and Mitra in Ref. [13] using Pauli–Villers method of regularization and reviewed by us in details in Ref. [10] and for more details on the SR and FR we refer to the work of Refs. [10–13].

We have studied the instant-form (IF) quantization (IFQ) (on the hyperplanes defined by the IF time $t = x^0 =$ constant) and light-front (LF) quantization (LFQ) (on the hyperplanes defined by the equal light-cone (LC) time ($\tau = x^+ = \frac{1}{\sqrt{2}}(x^0 + x^1)) =$ constant) [16–19] of this theory with mass-less fermions but with a PMT in the SR [6–8] in a series of papers [6–8]. More recently, we have studied this theory in the FR [9, 10], however, only in the IFQ. In the present work, we present its LFQ [17–19] using the Hamiltonian, path integral and BRST quantization [16–19].

The bosonized VSM without a PMT is defined by the Lagrangian density [1–15]:

$$\mathcal{L}_1 = \left[\frac{1}{2}\partial_\mu\phi\partial^\mu\phi - e\epsilon^{\mu\nu}\partial_\mu\phi A_\nu - \frac{1}{4}F_{\mu\nu}F^{\mu\nu}\right], \quad \epsilon^{\mu\nu} = \epsilon^{-\nu\mu} = \begin{pmatrix} 0 & +1 \\ -1 & 0 \end{pmatrix}$$
$$g^{\mu\nu} = g_{\mu\nu} = \begin{pmatrix} +1 & 0 \\ 0 & -1 \end{pmatrix} \ (IFQ), \quad g^{\mu\nu} = g_{\mu\nu} = \begin{pmatrix} 0 & +1 \\ -1 & 0 \end{pmatrix} \ (LFQ)$$
$$F^{\mu\nu} = (\partial^\mu A^\nu - \partial^\nu A^\mu) \tag{1}$$

This theory is a GI theory possessing a set of two first-class constraints. We now modify the above theory by including a mass term for the $U(1)$ gauge-field A^μ: ($\mathcal{L}_m^{SR} = \frac{1}{2}ae^2 A_\mu A^\mu$), where a is the regularization parameter in the so-called standard regularization (SR). The modified resulting theory then describes the Schwinger model with a PMT defined by the Lagrangian density [6–15]:

$$\mathcal{L}_2 := \mathcal{L}_1 + \mathcal{L}_m^{SR} = \left[\frac{1}{2}\partial_\mu\phi\partial^\mu\phi - e\epsilon^{\mu\nu}\partial_\mu\phi A_\nu - \frac{1}{4}F_{\mu\nu}F^{\mu\nu} + \frac{1}{2}ae^2 A_\mu A^\mu\right] \tag{2}$$

We next consider this theory with a different regularization called the Faddeevian regularization (FR). A new regularization called FR was introduced by Mitra in Refs. [11–13] and was explicitly derived by Mukhopadhyay and Mitra [13] using Pauli–Villars method of regularization, in the context of chiral Schwinger model (CSM) [13]. It however remains valid for a large class of 2D field theories in general. In this regularization, the mass term for the $U(1)$ gauge field A^μ in LFQ ($with\ \mu, \nu = +, -$) reads [11–15]:

$$\mathcal{L}_m^{FR} = \frac{1}{2}e^2\left[A_\mu M^{\mu\nu} A_\nu\right] = \left[2e^2 A^+ A^- - e^2 A^- A^-\right], \quad M^{\mu\nu} = \begin{pmatrix} +1 & -1 \\ -1 & -3 \end{pmatrix} \tag{3}$$

Therefore the Lagrangian density of VSM-PMT in the FR could be defined as [11–15]:

$$\mathcal{L}^N := \mathcal{L}_1 + \mathcal{L}_m^{FR} = \left[\frac{1}{2}\partial_\mu\phi\partial^\mu\phi - e\epsilon^{\mu\nu}\partial_\mu\phi A_\nu - \frac{1}{4}F_{\mu\nu}F^{\mu\nu} + \frac{1}{2}e^2 A_\mu M^{\mu\nu} A_\nu\right] \tag{4}$$

Lagrangian density of the theory in FR in the LFQ becomes:

$$\mathcal{L}^N = \left[\frac{1}{2}(\partial_+ A^+ - \partial_- A^-)^2 + 2e^2 A^+ A^- - e^2 A^- A^- + \partial_+\phi\, \partial_-\phi + eA^+\partial_+\phi - eA^-\partial_-\phi\right] \tag{5}$$

Canonical momenta obtained from the above action are:

$$\Pi^+ = \frac{\partial\mathcal{L}}{\partial(\partial_+ A^-)} = 0, \ \Pi^- = \frac{\partial\mathcal{L}}{\partial(\partial_+ A^+)} = \partial_+ A^+ - \partial_- A^+, \ \pi = \frac{\partial\mathcal{L}}{\partial(\partial_+\phi)} = \partial_-\phi + eA^+ \tag{6}$$

Here Π^+, Π^- and π are the momenta canonically conjugate respectively to A^-, A^+ and ϕ respectively. Above equations imply two primary constraints: $\psi_1 = \Pi^+ \approx 0$; $\psi_2 = (\pi - \partial_-\phi - eA^+) \approx 0$. The symbol $\approx$ here denotes a weak equality in the sense of Dirac [16]. After including the primary constraints ψ_1 and ψ_2 in the canonical Hamiltonian density $\mathcal{H}_c$ with the help of Lagrange multiplier fields $u(x, t)$ and $v(x, t)$ (which are treated as dynamical), total Hamiltonian density of the theory $\mathcal{H}_T$ becomes:

$$\mathcal{H}_T = \left[\frac{1}{2}(\Pi^-)^2 + \Pi^-\partial_- A^- + eA^-\partial_-\phi - 2e^2 A^+ A^- + e^2 A^- A^- + u\Pi^+ + v(\pi - \partial_-\phi - eA^+)\right] \tag{7}$$

The theory also possesses a secondary Gauss-law constraint: $\psi_3 = [\partial_-\Pi^- - e\partial_-\phi + 2e^2(A^- - A^-)] \approx 0$. For constructing a GI theory corresponding to $\mathcal{L}^N$, we enlarge the Hilbert space of the theory and introduce a Stueckelberg field θ through the following redefinition of fields ϕ and A^μ: $\phi \to \Phi = \phi - \theta,\ A^\mu \to A^\mu + \partial^\mu\theta$. We thus obtain the modified Lagrangian density $\mathcal{L}^I = (\mathcal{L}^N + \mathcal{L}^S)$ with

$$\mathcal{L}^S = \Big[(e-1)\partial_+\phi\,\partial_-\theta - (e+1)\partial_+\theta\,\partial_-\phi - (2e^2-1)\partial_+\theta\,\partial_-\theta \\ - e^2\partial_+\theta\,\partial_+\theta - 2e^2A^-\partial_+\theta - e(2e+1)(A^+\partial_+\theta - A^-\partial_-\theta)\Big] \tag{8}$$

Where $\mathcal{L}^S$ is the Stueckelberg term which when added to $\mathcal{L}^N$ gives a new theory defined by $\mathcal{L}^I$. The canonical momenta for the GI theory described by $\mathcal{L}^I$ are:

$$\Pi^+ = \frac{\partial\mathcal{L}^I}{\partial(\partial_+A^-)} = 0;\quad \Pi^- = \frac{\partial\mathcal{L}^I}{\partial(\partial_+A^+)} = (\partial_+A^+ - \partial_-A^-) \tag{9a}$$

$$\pi = \frac{\partial\mathcal{L}^I}{\partial(\partial_+\phi)} = [\partial_-\phi + eA^+ + (e-1)\partial_-\theta] \tag{9b}$$

$$\Pi_\theta = \frac{\partial\mathcal{L}^I}{\partial(\partial_+\theta)} = \Big[-e^2(A^+ + 2A^- + \partial_-\theta + 2\partial_+\theta) - (e+1)\pi\Big] \tag{9c}$$

Here Π^+, Π^-, π and Π_θ are the momenta canonically conjugate respectively to A^-, A^+, ϕ and θ respectively. Above equations imply two primary constraints: $\xi_1 = \Pi^+ \approx 0,\ \xi_2 = [\pi - \partial_-\phi + eA^+ + (e-1)\partial_-\theta] \approx 0$. After including the primary constraints ξ_1 and ξ_2 in the canonical Hamiltonian density with the help of Lagrange multiplier fields $u(x,t)$ and $v(x,t)$ (which are dynamical), the total Hamiltonian density of the theory $\mathcal{H}_T$ becomes:

$$\mathcal{H}_T^N = \Big[\frac{1}{2}\pi^2 + \Pi^-\partial_-A^- + eA^-\partial_-\phi - 2e^2A^+A^- + e^2A^-A^- - e(2e+1))A^-\partial_-\theta \\ + \Pi^+u + [\pi - \partial_-\phi - eA^+ - (e-1)\partial_-\theta]\,v \\ - \frac{1}{4e^2}\Big(\Pi_\theta + (e+1)\pi + e^2\partial_-\theta + e^2A^+ + 2e^2A^-A^-\Big)^2\Big] \tag{10}$$

Demanding that the primary constraint ξ_1 be preserved in the course of time leads to a secondary Gauss-law constraint: $\xi_3 = [\Pi_\theta + e^2(A^+ + 2A^- + \partial_-\theta + 2\partial_+\theta) + (e+1)\pi] \approx 0$. Also the matrix of Poission brackets of the constraints ξ_i is singular implying that the constraints ξ_i form a set of first-class constraints, and that the theory described by $\mathcal{L}^I$ is GI. The theory is in fact, seen to be invariant under the local vector gauge transformations:

$$\delta A^+ = \partial_-\beta,\ \delta A^- = \partial_+\beta,\ \delta\phi = \delta\theta = -\beta,\ \delta u = \partial_+\partial_+\beta \\ \delta\Pi^+ = \delta\Pi^- = \delta\Pi_\theta = \delta\pi = \delta\Pi_u = \delta\Pi_v = 0,\ \delta v = -\partial_+\beta \tag{11}$$

where the gauge parameter $\beta(x^-,\tau)$ is an arbitrary function of its arguments. We now quantize this theory under the GFC's:

$$\zeta_1 = [-(\partial_-\theta]) \approx 0,\quad \zeta_2 = [\Pi_\theta + (e+1)\partial_-\phi + (2e^2+e)A^+ - 2e^2A^-] \approx 0 \tag{12}$$

The above GFC's reproduce precisely, the physical content of the GNI theory described by $\mathcal{L}^N$ from that of the GI theory described by $\mathcal{L}^I$. Thus the addition of $\mathcal{L}^S$ to $\mathcal{L}^N$ enlarges only the unphysical part of the full Hilbert space of the theory $\mathcal{L}^N$, without modifying the physical content of the GNI theory $\mathcal{L}^N$. In path integral formulation [18,19], the transition to quantum theory is made by writing the vacuum to vacuum transition amplitude for the theory, called the generating functional $Z[J_k]$ of the GI theory in the presence of the external sources: J_k as:

$$Z[J_k] = \int [d\mu]\exp\Big[i\int d^2x(J_k\Phi^k + \pi\partial_+\phi + \Pi^+\partial_+A^- + \Pi^-\partial_+A^+ + \Pi_u\partial_+u + \Pi_v\partial_+v - \mathcal{H}_T)\Big]$$

where the phase space variables of the theory are: $\Phi^k \equiv (\phi, A^-, A^+, u, v)$ with the corresponding respective canonical conjugate momenta: $\Pi_k \equiv (\pi, \Pi^+, \Pi^-, \Pi_u, \Pi_v)$. The functional measure $[d\mu]$ of the generating functional $Z[J_k]$ under the above LC gauges is:

$$\begin{aligned}[d\mu] = {} & (e^2\delta(x^- - y^-))(2\partial_-\delta(x^- - y^-))^{3/2}\, [d\phi]\, [dA^+]\, [dA^-]\, [du]\, [dv]\, [d\pi] \\ & [d\Pi^+]\, [d\Pi^-]\, [d\Pi_u]\, [d\Pi_v]\, \delta[\Pi^+ \approx 0]\, \delta[(\pi - \partial_-\phi - eA^+) \approx 0] \\ & \delta[(\partial_-\Pi^- - e\partial_-\phi + 2e^2(A^+ - A^-)) \approx 0]\, \delta[(-\partial_-\theta) \approx 0] \\ & \delta[(\Pi_\theta + (e+1)\partial_-\phi + (2e^2 + e)A^+ - 2e^2A^-) \approx 0] \end{aligned} \tag{13}$$

For BRST formulation, we rewrite the theory as a quantum system that possesses the generalized gauge invariance called BRST symmetry. We enlarge the Hilbert space of GI theory and replace the notion of gauge-transformation (which shifts operators by c-number functions) by a BRST transformation, which mixes operators with Bose and Fermi statistics and introduce anti-commuting variables c and $\bar{c}$ (Grassman numbers at classical level and operators in quantized theory) and a commuting variable b such that:

$$\hat{\delta}\phi = \hat{\delta}\theta = -c\ \hat{\delta}A^- = \partial_+ c,\ \hat{\delta}A^+ = \partial_- c,\ \hat{\delta}u = \partial_+\partial_+ c,\ \hat{\delta}v = -\partial_+ c \tag{14a}$$

$$\hat{\delta}\pi = \hat{\delta}\Pi^+ = \hat{\delta}\Pi^- = \hat{\delta}\Pi_\theta = \hat{\delta}\Pi_u = \hat{\delta}\Pi_v = 0,\ \hat{\delta}c = 0,\ \hat{\delta}\bar{c} = b,\ \hat{\delta}b = 0 \tag{14b}$$

with the property $\hat{\delta}^2 = 0$. Now the BRST gauge-fixed quantum Lagrangian density $\mathcal{L}_{BRST}$ for the theory could be obtained by adding to the first-order Lagrangian density $\mathcal{L}_{I0}$, a trivial BRST-invariant function:

$$\begin{aligned}\mathcal{L}_{BRST} = & \Bigg[\Pi^-\partial_+A^+ + \Pi_\theta\partial_+\theta - \frac{1}{2}\Pi^-\Pi^- - \Pi^-\partial_-A^- - eA^-\partial_-\phi + 2e^2A^+A^- - e^2A_-A^- \\ & + [\partial_+\phi + eA^+ + (e-1)\partial_-\theta](\partial_+\phi) + e(2e+1)A^-\partial_-\theta + \Pi_u\partial_+u + \Pi_v\partial_+v \\ & + \frac{1}{4e^2}\left(\Pi_\theta + (e+1)\pi + e^2\partial_-\theta + e^2A^+ + 2e^2A^-\right)^2 + \hat{\delta}[\bar{c}(\partial_+A^- - \phi - \frac{b}{2})]\Bigg] \end{aligned} \tag{15}$$

The last term in the above equation is the BRST-invariant gauge-fixing term. The Euler-Lagrange equation for b reads: $-b = (g\partial_+A^- + g\partial_-A^- - g\partial_+A^+)$ which in turn (with the requirement $\hat{\delta}b = 0$) then implies: $\partial_+\partial_+c = 0$. The bosonic momenta is: $\Pi^+ := gb$ and the fermionic momenta with directional derivatives are defined as: $\Pi_c := \partial_+\bar{c}$ and $\Pi_{\bar{c}} := \partial_+c$, implying that the variable canonically conjugate to c is ($\partial_+\bar{c}$) and the variable conjugate to $\bar{c}$ is (∂_+c). The Hamiltonian density is:

$$\begin{aligned}\mathcal{H}_{BRST} = & \Bigg[\frac{1}{2}\Pi^-\Pi^- + \Pi^-\partial_-A^- + eA^-\partial_-\phi - 2e^2A^+A^- - e(2e+1)A^-\partial_-\theta \\ & + e^2A_-A^- + \frac{1}{2}(\Pi^+)^2 + \Pi_c\Pi_{\bar{c}} + \bar{c}c - \Pi^+(\partial_+A^- - \phi) \\ & - \frac{1}{4e^2}\left(\Pi_\theta + (e+1)\pi + e^2\partial_-\theta + e^2A^+ + 2e^2A^-\right)^2\Bigg] \end{aligned} \tag{16}$$

BRST charge operator Q is the generator of BRST transformations. It is nil-potent and mixes operators with Bose and Fermi statistics. It is given by:

$$\begin{aligned}Q = \int dx^- \Bigg[& ic\left(\Pi_\theta + \partial_-\Pi^- + (2e+1)\pi + 2e^2A^+ - 2e\partial_-\phi + 2e(e+1)\partial_-\theta\right) \\ & - i\partial_+c\left(\Pi^+ + \pi - \partial_-\phi - eA^+ - (e-1)\partial_-\theta\right)\Bigg] \end{aligned} \tag{17}$$

Now because $Q|\psi >= 0$, the set of states annihilated by Q contains not only the set for which the constraints of the theory hold but also additional states for which the constraints of the theory do not hold. The Hamiltonian is also invariant under the anti-BRST transformations given by:

$$\bar{\hat{\delta}}\theta = \bar{\hat{\delta}}\phi = \bar{c},\ \bar{\hat{\delta}}A^- = -\partial_+\bar{c},\ \bar{\hat{\delta}}A^+ = -\partial_-\bar{c},\ \bar{\hat{\delta}}u = -\partial_+\partial_+\bar{c},\ \bar{\hat{\delta}}v = \partial_+\bar{c} \tag{18a}$$

$$\bar{\hat{\delta}}\Pi_\theta = \bar{\hat{\delta}}\pi = \bar{\hat{\delta}}\Pi^+ = \bar{\hat{\delta}}\Pi^- = \bar{\hat{\delta}}\Pi_u = \bar{\hat{\delta}}\Pi_v = 0\ \bar{\hat{\delta}}\bar{c} = 0,\ \bar{\hat{\delta}}c = -b,\ \bar{\hat{\delta}}b = 0 \tag{18b}$$

with generator or anti-BRST charge

$$\bar{Q} = \int dx^- \Big[-i\bar{c}\Big(\Pi_\theta + \partial_- \Pi^- + (2e+1)\pi + 2e^2 A^+ - 2e\partial_- \varphi + 2e(e+1)\partial_- \theta \Big)$$

$$+ i\partial_+ \bar{c}\Big(\Pi^+ + \pi - \partial_- \phi - eA^+ - (e-1)\partial_- \theta \Big) \Big] \quad (19)$$

We also have $\partial_+ Q = [Q, H_{BRST}] = 0$, $\partial_+ \bar{Q} = [\bar{Q}, H_{BRST}] = 0$, $H_{BRST} = \int dx^- \mathcal{H}_{BRST}$ We now impose the dual condition that both Q and $\bar{Q}$ annihilate physical states, implying that: $Q|\psi>=0$ and $\bar{Q}|\psi>=0$. The states for which the constraints of the theory hold, satisfy both of these conditions and are in fact, the only states satisfying both of these conditions. Thus the only states satisfying $Q|\psi>=0$ and $\bar{Q}|\psi>=0$ are those that satisfy the constraints of the theory and they belong to the set of BRST-invariant as well as to the set of anti-BRST-invariant states. This completes the LF Hamiltonian, path integral and BRST quantization of the theory.

The authors thank Giovani Salmè and Barbara Pasquini and other organizers of the International Conference on Light-Cone Physics: LC2015, held at LMF, INFN, Frascati, Italy, during September 21–15, 2015, for their warm hospitality during the conference. The Award of the partial Travel Grants to UK by the CSIR and INSA, New Delhi, India, which enabled her to attend this Conference where this work was presented by her as an Invited Contributed Talk are gratefully acknowledged. This work was supported in part by the US Department of Energy under Grant No. DE-FG02-87ER40371 and by the US National Science Foundation under Grant No. PHY-0904782.

References

1. Schwinger, J.: Gauge invariance and mass. Phys. Rev **125**, 397–398 (1962)
2. Schwinger, J.: Gauge invariance and mass 2. Phys. Rev. **128**, 2425–2429 (1962)
3. Kulshreshtha, U., Kulshreshtha, D.S.: Generalized Schwinger model in the light-front frame. Contributed paper at the 8th International Workshop on Light-Cone QCD and Non-perturbative Hadronic Physics, held at Lutsen, University of Minnesota at Duluth, USA, August, pp. 11–22 (1997)
4. Kulshreshtha, U., Kulshreshtha, D.S., Mueller-Kirsten, H.J.W.: Hamiltonian and BRST formulations of the Schwinger model. Helv. Phys. Acta **66**, 743–757 (1993)
5. Kulshreshtha, U., Kulshreshtha, D.S., Mueller-Kirsten, H.J.W.: Front-form Hamiltonian and BRST formulations of the Schwinger model. Int. J. Theor. Phys. **37**, 2539 (1998)
6. Kulshreshtha, U.: Vector Schwinger model with a photon mass term: gauge-invariant reformulation, operator solutions and Hamiltonian and path integral formulations. Mod. Phys. Lett. A **22**, 2993–3001 (2007)
7. Kulshreshtha, U., Kulshreshtha, D.S.: Gauge-invariant reformulation of the vector Schwinger model with a photon mass term and its Hamiltonian, path integral and BRST formulations. Int. J. Mod. Phys. A **22**, 6183–6201 (2007)
8. Kulshreshtha, U.: Vector Schwinger model with a Photon Mass term on the light-front. **PoS** LC2008, 008 (2008)
9. Kulshreshtha, Usha, Kulshreshtha, D.S., Vary, James: Vector Schwinger model with a photon mass term with Faddeevian regularization. Few Body Syst. **56**(Issue Nos. 6–9), 559–563 (2015)
10. Kulshreshtha, Usha, Kulshreshtha, D.S., Vary, James: Hamiltonian, path integral and BRST formulations of vector Schwinger model with a photon mass term with Faddeevian regularization. Int. J. Theor. Phys. **55**(Issue No. 1), 338–360 (2016)
11. Mitra, P.: Gauge Non-invariant Regularization and non-confining solutions of the Schwinger Model. Saha Inst. Preprint, Report no. SINP/TNP/92-09, (9 pp.) (1992)
12. Mitra, P.: Chiral Schwinger model with a Faddeevian regularization. Phys. Lett. B **284**, 23–26 (1992)
13. Mukhopadhyay, Subir, Mitra, P.: Chiral Schwinger model with new regularization. Z. Phys. **C67**, 525–530 (1995)
14. Bhattacharya, G., Ghosh, A., Mitra, P.: A generalized gauge invariant regularization of the Schwinger model. Phys. Rev. D **50**, 4183–4188 (1994)
15. Mitra, P., Rahaman, A.: The non-confining Schwinger model. Ann. Phys. **249**, 34–43 (1996)
16. Dirac, P.A.M.: Generalized Hamiltonian dynamics. Can. J. Math **2**, 129 (1950)
17. Dirac, P.A.M.: Forms of relativistic dynamics. Rev. Mod. Phys. **21**, 392–399 (1949)
18. Brodsky, S.J., Pauli, H.C., Pinsky, S.S.: Quantum chromodynamics and other field theories on the light-cone. Phys. Rep. **301**, 299–486 (1998)
19. Kulshreshtha, U., Kulshreshtha, D.S., Vary, J.P.: Hamiltonian, path Integral and BRST formulations of large N Scalar QCD_2 on the light-front and spontaneous symmetry breaking. Eur. J. Phys. **C75**, 174 (12 pages), (2015)

DOI 10.1007/978-3-319-50699-9_40
Reprinted from *Few-Body System* Journal, DOI 10.1007/s00601-016-1088-y

CrossMark

Miguel G. Echevarria

T-Odd Gluon TMDs Inside a Transversely Polarized Hadron

Received: 31 January 2016 / Accepted: 13 March 2016 / Published online: 31 March 2016

Abstract We consider the relevant gluon transverse momentum dependent distributions (TMDs) in the spin asymmetries generated by the scattering off transversely polarized hadrons. At large transverse momentum they can be expressed in terms of different collinear distributions, via perturbatively calculable Wilson coefficients. We calculate these coefficients at next-to-leading order, and show that when the small-x limit is taken only one independent function remains for dipole-type gluon TMDs: the so-called spin-dependent odderon. This universal origin for the spin asymmetries is of importance to better understand hadron substructure.

1 Introduction

Hadron substructure can be investigated in high energy collisions. And as the energy of the scattering process grows, gluons play an increasingly relevant role. This limit corresponds to the case where the gluons carry an small fraction x of the longitudinal momentum of the parent hadron. In this scenario, the description of the scattering processes gets simplified.

However recently it has been realized that the transverse momentum distributions of gluons inside the hadron are not unique quantities, but depend on the process, which can probe different distributions. This is due to the fact that TMDs are sensitive to the color flow in the process [1]. There have been studies in this direction that dealt with the unpolarized gluon distribution at small x [2–4]. In [3, 4] both the small-x treatment and the TMD formalism where combined, but no spin effects where considered. This is our aim, pursued in [5].

We study the T-odd gluon TMDs inside a transversely polarized hadron, and consider the limit of high gluon density or small x. We find that even if there are several independent TMDs at the beginning, once the small-x limit is taken only one distribution survives, highly reducing the degree of nonuniversality [6]. This result is compatible with the picture that arises in a small-x treatment, and actually the found distribution is linked to what is known as the spin-dependent odderon.

The universality of gluon effects, although likely smaller than the ones driven by quarks, is of high experimental relevance, as it can be studied at RHIC and at the proposed Electron-Ion Collider (EIC) or a polarized fixed-target experiment at LHC (AFTER@LHC).

Gluon TMDs for a transversely polarized hadron, with spin vector S_T, are given in terms of the following bi-local correlator,

$$\Gamma^{\mu\nu[U,U']} = \frac{1}{xP^+}\int \frac{dy^- d^2y_T}{(2\pi)^3} e^{ik\cdot y}\langle P, S_T|2\mathrm{Tr}\left[F_T^{+\nu}(0)U F_T^{+\mu}(y)U'\right]|P, S_T\rangle\big|_{y^+=0}, \tag{1}$$

M. G. Echevarria (✉)
Departament d'Estructura i Constituents de la Matèria, Universitat de Barcelona, Martí i Franquès 1, 08028 Barcelona, Spain
E-mail: mgechevarria@ecm.ub.edu

where U and U' stand for process dependent gauge links in the fundamental representation (see e.g. Dominguez et al. [4]). At leading twist, this correlator is parametrized in terms of six independent tensor structures [7],

$$\Gamma^{\mu\nu} = \delta_T^{\mu\nu} f_1^g - \left(\frac{2k_T^\mu k_T^\nu}{k_T^2} + \delta_T^{\mu\nu}\right) h_1^{\perp g} - \delta_T^{\mu\nu} \frac{\epsilon_{T\alpha\beta} k_T^\alpha S_T^\beta}{M} f_{1T}^{\perp g}$$
$$- i\epsilon_T^{\mu\nu} \frac{k_T \cdot S_T}{M} g_{1T}^g - \frac{\tilde{k}_T^{\{\mu} S_T^{\nu\}} + \tilde{S}_T^{\{\mu} k_T^{\nu\}}}{2M} h_{1T}^g + \frac{\tilde{k}_T^{\{\mu} k_T^{\nu\}}}{k_T^2} \frac{k_T \cdot S_T}{M} h_{1T}^{\perp g}, \tag{2}$$

where all six distributions are functions of x and k_T^2. $\epsilon_T^{\mu\nu} = \epsilon^{\rho\sigma\mu\nu} n_\rho p_\sigma$, with $\epsilon_T^{12} = 1$ and $\delta_T^{\mu\nu} = -g^{\mu\nu} + p^{\{\mu} n^{\mu\}}/p \cdot n$. We also use the short-hand notation $\tilde{k}_T^\nu = \epsilon_T^{\mu\nu} k_{T\mu}$. As it is well-known by now, the correlator above contains spurious rapidity divergences that prevent it from being a well-defined hadronic quantity [8–10]. To cure this problem one needs to add to it a soft factor that cancels these divergences and renders the new TMDs properly defined. However for the sake of simplicity we omit such soft factor in the discussion below, as it does not affect the results at all.

The first two gluon TMDs, f_1^g and $h_1^{\perp g}$, stand for the unpolarized and linearly polarized gluon distributions. Among the rest, we have three T-odd gluon TMDs: $f_{1T}^{\perp g}$, $h_{1T}^{\perp g}$ and h_{1T}^g, which are relevant for spin asymmetry studies. As mentioned, these TMDs are not universal and should have a label $[U, U']$ to denote their process-dependence. However we restrict ourselves to the two most important cases, which involve only single future or past pointing staple-like gauge links, denoted by $+$ and $-$, respectively. In the notation of [11], there are two T-odd combinations labeled with (f) and (d), namely,

$$\Gamma_{(f)}^{(\text{T-odd})} = \frac{1}{2}\left(\Gamma^{[+,+\dagger]} - \Gamma^{[-,-\dagger]}\right),$$
$$\Gamma_{(d)}^{(\text{T-odd})} = \frac{1}{2}\left(\Gamma^{[+,-\dagger]} - \Gamma^{[-,+\dagger]}\right). \tag{3}$$

For the unpolarized gluon distribution in the small-x formalism, the first type is referred to as the Weizsäcker–Williams (WW) distribution, while the second one is known as the dipole gluon distribution [3,4].

2 NLO Calculation

The T-odd quark TMDs can be perturbatively calculated at large transverse momentum in terms of twist-3 collinear distributions [12–14]. In the same way, below we perform an analogous calculation for the T-odd gluon TMDs, both (f) and (d) type ones, which are C-even and C-odd, respectively. The T-odd collinear twist-3 functions that appear at large k_T are the (chiral-even) quark-gluon and tri-gluon Qiu–Sterman functions [15–17], with corresponding C-parity. Leaving the gauge-link structure aside, the two twist-3 collinear correlators are:

$$T_{F,q}(x,x) = \frac{M}{P^+} \int \frac{dz^- d\eta}{2\pi} e^{ik\cdot z} \frac{(\mathbf{P} \times \mathbf{S}_T)^j}{2M} \langle P, S|\, \bar{\psi}(0)\gamma^+ F^{+j}(\eta z)\psi(z)\, |P, S\rangle\, |_{z^+=|\mathbf{z}_\perp|=\mathbf{0}} \tag{4}$$

and

$$T_G^{\pm}(x,x) = -\frac{2M\delta_T^{lm}}{x(P^+)^2} \int \frac{dz^- d\eta}{2\pi} e^{ik\cdot z} \frac{(\mathbf{P} \times \mathbf{S}_T)^j}{2M} \langle P, S|\, C_\pm^{abc} F_a^{+l}(0) F_b^{+j}(\eta z) F_c^{+m}(z)\, |P, S\rangle\, |_{z^+=|\mathbf{z}_\perp|=\mathbf{0}}, \tag{5}$$

where $C_+^{abc} = if^{abc}$ and $C_-^{abc} = d^{abc}$.

First we compute the large k_T tails of the T-odd gluon TMDs, represented in Fig. 1 for the flavor-singlet case, and then we take the small-x limit. Details of the calculation can be found in [5], but in summary we obtain for the f-type functions

$$f_{1T}^{\perp g\,(f)}(x \to 0, k_\perp^2) \approx h_{1T}^{g\,(f)}(x \to 0, k_\perp^2) \approx h_{1T}^{\perp g\,(f)}(x \to 0, k_\perp^2) \tag{6}$$
$$\approx 0 \tag{7}$$

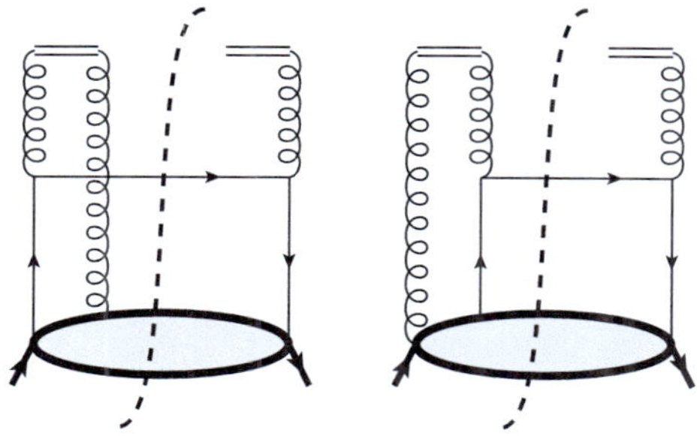

Fig. 1 Diagrams contributing to T-odd gluon TMDs at large transverse momentum in the flavor-singlet case. **a** soft-gluon pole contribution. **b** hard-gluon pole contribution. (Mirror diagrams are not shown)

and for d-type ones

$$f_{1T}^{\perp g\,(d)}(x \to 0, k_\perp^2) \approx h_{1T}^{g\,(d)}(x \to 0, k_\perp^2) \approx h_{1T}^{\perp g\,(d)}(x \to 0, k_\perp^2) \tag{8}$$

$$\approx C_2 \frac{M}{k_T^4} \frac{4}{x} \int dz \Big\{ \sum_q T_{F,q}(z,z) + T_G^{(-)}(z,z) \Big\}, \tag{9}$$

with $C_2 = \frac{N_c^2-4}{2N_c}\frac{\alpha_s}{2\pi^2}$. As can be seen, the (f)-type distributions vanish at small x, while the (d)-type ones are enhanced. Moreover, the latter turn out to be identical, gaining a higher degree of universality.

3 Universal Origin: The Spin-Dependent Odderon

Let us start from the T-odd part of the dipole gluon TMD correlator

$$\begin{aligned} \Gamma^{\mu\nu}_{\text{T-odd}}(x, k_T; S_T) = &\frac{1}{xP^+} \int \frac{dy^- d^2 y_T}{(2\pi)^3} e^{ik\cdot y} \\ &\times \Big\{ \langle P, S_T | \mathrm{Tr} \Big[F_T^{+\nu}(0) U^{[+]} F_T^{+\mu}(y) U^{[-]\dagger} \\ &- F_T^{+\nu}(0) U^{[-]} F_T^{+\mu}(y) U^{[+]\dagger} \Big] | P, S_T \rangle \Big\} \Big|_{y^+=0}, \end{aligned} \tag{10}$$

where we omit the label (d). Next we take the small-x limit by approximating the exponential $e^{ik^+ y^-}$ by 1, which in [3,4] is argued to be a good approximation as long as $k^+ = xP^+$ is very small. Then, Eq. (10) can be written as

$$\begin{aligned} \Gamma^{\mu\nu}_{T-odd}(x, k_T; S_T) = &\frac{k_T^\mu k_T^\nu}{g^2 V x P^+} \int \frac{d^2 y_T}{(2\pi)^3} e^{ik_T \cdot y_T} \\ &\times \langle P, S_T | \mathrm{Tr} \Big[U^{[\square]}(0_T, y_T) - U^{[\square]\dagger}(0_T, y_T) \Big] | P, S_T \rangle, \end{aligned} \tag{11}$$

where $V = \int dy^-$ and $U^{[\square]}$ stands for a rectangular Wilson loop with light-like Wilson lines at transverse separation y_T. This result is obtained by using translational invariance and the following relation,

$$\partial_T^\mu U(y_T) = -ig \int_{-\infty}^{+\infty} dy^-\, U[-\infty^-, y^-; y_T] \partial_T^\mu A_+(y^-, y_T) U[y^-, \infty^-; y_T], \tag{12}$$

where $\partial_T^\mu A_+(y^-, y_T)$ belong to the gluon field strength $F_T^{+\mu}$. The $\partial_+ A_T^\mu(y^-, y_T)$ part corresponds to the transverse gauge link at lightcone infinity, which can be neglected in a calculation performed in a covariant gauge [1]. Finally, the remaining part is power suppressed.

One can see that $\mathrm{Tr}\big[U^{[\square]}(0_T, y_T) - U^{[\square]\dagger}(0_T, y_T)\big]$ corresponds to the dipole odderon operator [18]. The spin-dependent odderon was considered in [19], and also in other studies not related to TMDs, e.g. [20–22].

The matrix element of the odderon operator can only have one S_T dependence. For a transversely polarized nucleon, the T-odd part of $\Gamma^{\mu\nu}$ can be parametrized only by one leading-twist tensor structure [19]:

$$\Gamma^{\mu\nu}_{T-odd}(x, k_T; S_T) = \frac{k_T^\mu k_T^\nu N_c}{2\pi^2 \alpha_s x} \frac{\epsilon_T^{\alpha\beta} S_{T\alpha} k_{T\beta}}{M} O_{1T}^{\perp}(x, k_T^2), \tag{13}$$

where $O_{1T}^{\perp}(x, k_T^2)$ is identified as the spin-dependent odderon in [19]. Then this leads to the following relation:

$$\begin{aligned}\frac{k_T^\mu k_T^\nu N_c}{2\pi^2 \alpha_s x} \frac{\epsilon_T^{\alpha\beta} S_{T\alpha} k_{T\beta}}{M} O_{1T}^{\perp}(x, k_T^2) &= -\delta_T^{\mu\nu} \frac{\epsilon_{T\alpha\beta} k_T^\alpha S_T^\beta}{M} f_{1T}^{\perp g} - \frac{\tilde{k}_T^{\{\mu} S_T^{\nu\}} + \tilde{S}_T^{\{\mu} k_T^{\nu\}}}{2M} h_{1T}^g \\ &\quad + \frac{\tilde{k}_T^{\{\mu} k_T^{\nu\}}}{k_T^2} \frac{k_T \cdot S_T}{M} h_{1T}^{\perp g}. \end{aligned} \tag{14}$$

Finally, given that the dipole type T-odd gluon TMDs are equal at small x, one has

$$x f_{1T}^{\perp g} = x h_{1T}^g = x h_{1T}^{\perp g} = \frac{-k_T^2 N_c}{4\pi^2 \alpha_s} O_{1T}^{\perp}(x, k_T^2), \tag{15}$$

which is the main result of Boer et al. [5]. This universal function should govern the single transverse spin asymmetries in the small-x regime.

In order to test these expectations in experiments, one could study single spin asymmetries in different processes that in the small-x regime probe the dipole distributions, see e.g. Dominguez et al. [4]. In $p^\uparrow A$ collisions these are backward hadron production (with C-odd final states, because the odderon is C-odd, like $h^\pm X$), γ^* production and back-to-back γ^* jet production. BRAHMS data on single spin asymmetries in backward charged hadron production allow for a 10 % level asymmetries [28].

4 Conclusions

In summary, we have performed a perturbative calculation of the three leading-twist T-odd gluon TMDs inside a transversely polarized hadron at large k_T and in the small-x limit. We found that the dipole type ones are enhanced at small-x, whereas the WW type ones are suppressed. Moreover, the three dipole type T-odd gluon TMDs become equal in the small-x limit, and are given by the spin-dependent odderon. This leads to a high universality and thus a simplification of the picture of transversely polarized hadrons at small x, which can be tested in current and future experimental facilities like RHIC, EIC or AFTER@LHC.

Acknowledgments I would like to thank Daniel Boer, Piet Mulders and Jian Zhou for fruitful discussions and collaborations.

References

1. Collins, J.C.: Leading twist single transverse-spin asymmetries: Drell–Yan and deep inelastic scattering. Phys. Lett. B **536**, 43 (2002). doi:10.1016/S0370-2693(02)01819-1. arxiv:hep-ph/0204004
2. Kharzeev, D., Kovchegov, Y.V., Tuchin, K.: Nuclear modification factor in d+Au collisions: onset of suppression in the color glass condensate. Phys. Lett. B **599**, 23 (2004). doi:10.1016/j.physletb.2004.08.034. arXiv:hep-ph/0405045
3. Dominguez, F., Xiao, B.W., Yuan, F.: k_t-factorization for hard processes in nuclei. Phys. Rev. Lett. **106**, 022301 (2011). doi:10.1103/PhysRevLett.106.022301. arXiv:1009.2141 [hep-ph]
4. Dominguez, F., Marquet, C., Xiao, B.W., Yuan, F.: Universality of unintegrated gluon distributions at small x. Phys. Rev. D **83**, 105005 (2011). doi:10.1103/PhysRevD.83.105005. arXiv:1101.0715 [hep-ph]
5. Boer, D., Echevarria, M.G., Mulders, P., Zhou, J.: Single spin asymmetries from a single Wilson loop. Phys. Rev. Lett. **116**(12), 122001 (2016). doi:10.1103/PhysRevLett.116.122001. arXiv:1511.03485 [hep-ph]
6. Buffing, M.G.A., Mukherjee, A., Mulders, P.J.: Generalized universality of definite rank gluon transverse momentum dependent correlators. Phys. Rev. D **88**, 054027 (2013). doi:10.1103/PhysRevD.88.054027. arXiv:1306.5897 [hep-ph]
7. Mulders, P.J., Rodrigues, J.: Transverse momentum dependence in gluon distribution and fragmentation functions. Phys. Rev. D **63**, 094021 (2001). doi:10.1103/PhysRevD.63.094021. arXiv:hep-ph/0009343
8. Echevarria, M.G., Idilbi, A., Scimemi, I.: Factorization theorem for Drell-Yan at low q_T and transverse momentum distributions on-the-light-cone. JHEP **1207**, 002 (2012). arXiv:1111.4996 [hep-ph]
9. Echevarria, M.G., Kasemets, T., Mulders, P.J., Pisano, C.: QCD evolution of (un)polarized gluon TMDPDFs and the Higgs q_T-distribution. JHEP **1507**, 158 (2015). doi:10.1007/JHEP07(2015)158. arXiv:1502.05354 [hep-ph]

10. Collins, J.: Foundations of perturbative QCD (Cambridge monographs on particle physics, nuclear physics and cosmology. 32, 2011)
11. Bomhof, C.J., Mulders, P.J.: Non-universality of transverse momentum dependent parton distribution functions. Nucl. Phys. B **795**, 409 (2008). doi:10.1016/j.nuclphysb.2007.11.024. arXiv:0709.1390 [hep-ph]
12. Ji, X., Qiu, J.W., Vogelsang, W., Yuan, F.: A Unified picture for single transverse-spin asymmetries in hard processes. Phys. Rev. Lett. **97**, 082002 (2006). doi:10.1103/PhysRevLett.97.082002. arXiv:hep-ph/0602239
13. Zhou, J., Yuan, F., Liang, Z.T.: Hyperon polarization in unpolarized scattering processes. Phys. Rev. D **78**, 114008 (2008). doi:10.1103/PhysRevD.78.114008. arXiv:0808.3629 [hep-ph]
14. Dai, L.Y., Kang, Z.B., Prokudin, A., Vitev, I.: Next-to-leading order transverse momentum-weighted Sivers asymmetry in semi-inclusive deep inelastic scattering: the role of the three-gluon correlator. Phys. Rev. D **92**(11), 114024 (2015). doi:10.1103/PhysRevD.92.114024. arXiv:1409.5851
15. Qiu, J.W., Sterman, G.F.: Single transverse spin asymmetries. Phys. Rev. Lett. **67**, 2264 (1991). doi:10.1103/PhysRevLett.67.2264
16. Ji, X.D.: Gluon correlations in the transversely polarized nucleon. Phys. Lett. B **289**, 137 (1992). doi:10.1016/0370-2693(92)91375-J
17. Braun, V. M., Manashov, A. N., Pirnay, B.: Scale dependence of twist-three contributions to single spin asymmetries. Phys. Rev. D **80**, 114002 (2009) (Erratum: Phys. Rev. D **86**, 119902 (2012)). doi:10.1103/PhysRevD.80.114002. doi:10.1103/PhysRevD.86.119902. arXiv:0909.3410 [hep-ph]
18. Hatta, Y., Iancu, E., Itakura, K., McLerran, L.: Odderon in the color glass condensate. Nucl. Phys. A **760**, 172 (2005). doi:10.1016/j.nuclphysa.2005.05.163. arXiv:hep-ph/0501171
19. Zhou, J.: Transverse single spin asymmetries at small x and the anomalous magnetic moment. Phys. Rev. D **89**(7), 074050 (2014). doi:10.1103/PhysRevD.89.074050. arXiv:1308.5912
20. Ryskin, M.G.: Odderon and polarization phenomena in QCD. Sov. J. Nucl. Phys. **46**, 337 (1987) [Yad. Fiz. **46**, 611 (1987)]
21. Buttimore, N.H., Kopeliovich, B.Z., Leader, E., Soffer, J., Trueman, T.L.: The spin dependence of high-energy proton scattering. Phys. Rev. D **59**, 114010 (1999). doi:10.1103/PhysRevD.59.114010. arXiv:hep-ph/9901339
22. Leader, E., Trueman, T.L.: The odderon and spin dependence of high-energy proton proton scattering. Phys. Rev. D **61**, 077504 (2000). doi:10.1103/PhysRevD.61.077504. arXiv:hep-ph/9908221
23. McLerran, L.D., Venugopalan, R.: Computing quark and gluon distribution functions for very large nuclei. Phys. Rev. D **49**, 2233 (1994). arXiv:hep-ph/9309289
24. Jeon, S., Venugopalan, R.: A classical odderon in QCD at high energies. Phys. Rev. D **71**, 125003 (2005). arXiv:hep-ph/0503219
25. Bartels, J., Lipatov, L.N., Vacca, G.P.: A new odderon solution in perturbative QCD. Phys. Lett. B **477**, 178 (2000). arXiv:hep-ph/9912423
26. Bartels, J.: High-energy behavior in a nonabelian gauge theory. 2. First corrections to T(n—>m) beyond the leading LNS approximation. Nucl. Phys. B **175**, 365 (1980)
27. Kovchegov, Y.V., Szymanowski, L., Wallon, S.: Perturbative odderon in the dipole model. Phys. Lett. B **586**, 267 (2004). arXiv:hep-ph/0309281
28. Arsene, I. et al.: [BRAHMS Collaboration], Single transverse spin asymmetries of identified charged hadrons in polarized p+p collisions at s**(1/2) = 62.4-GeV. Phys. Rev. Lett. **101**, 042001 (2008). doi:10.1103/PhysRevLett.101.042001. arXiv:0801.1078 [nucl-ex]

DOI 10.1007/978-3-319-50699-9_41
Reprinted from *Few-Body System* Journal, DOI 10.1007/s00601-016-1093-1

Franco Bradamante

Overview of the COMPASS Experimental Programme on TMDs and GPDs

Received: 28 January 2016 / Accepted: 16 March 2016 / Published online: 31 March 2016

Abstract The investigation of the recently introduced transverse momentum dependent (TMD) parton distribution functions (PDFs) and generalized parton distributions (GPDs) is an important part of the programme of the COMPASS experiment at CERN. A variety of Semi-inclusive deeply inelastic scattering (SIDIS) measurements on transversely polarized proton and deuteron targets from the 2002–2010 data have allowed to obtain a wealth of results on the transversity distribution and on the TMD PDFs. Further SIDIS measurements are planned in 2016 and 2017 on a pure hydrogen target. In a complementary approach, the first ever Drell–Yan measurement on a transversely polarized proton target is the objective of the 2015 run, still ongoing at the time of this conference, to address the fundamental issue of the change of sign of the Sivers function when measured in the Drell–Yan process as compared to that measured in SIDIS. GPDs are also being investigated. Measurements of the exclusive muon-production of vector mesons, in particular the ρ^0 and the ω, on transversely polarized proton targets have provided information on the GPD E. Preliminary results from a test run on deeply virtual compton scattering of 160 GeV muons on a liquid hydrogen target are also given, as well as plans for a two-year long run in 2016 and 2017 with a suitably modified spectrometer.

1 Introduction

Twenty years ago the COMPASS spectrometer was conceived as "COmmon Muon and Proton apparatus for structure and spectroscopy", capable of addressing a large variety of open problems in both hadron structure and spectroscopy. By now, an impressive list of results has been published. In this review, I will mention only the results which are relevant to the field of TMD PDFs and GPDs.

The description of the nucleon structure in terms of collinear parton distributions, i. e. distributions depending only on Bjorken x and the squared four-momentum transfer Q^2, has recently been generalized to take into account the parton transverse momentum $\mathbf{k}_T$, thus leading to a total of eight TMD PDFs for a complete picture of the nucleon at leading order. After integration over the transverse momentum three of them reduce to the well-known "collinear" PDFs, namely the number density, the helicity and the transversity distributions. The other five TMD PDFs are sensitive to the direction of the quark transverse momentum, they vanish when integrating over $\mathbf{k}_T$, and their measurement provides important information on the dynamics of the partons in the transverse plane in momentum space. In particular the T-odd Sivers and the Boer-Mulders functions are related to the correlations between the transverse momentum of quarks and the nucleon transverse spin and between the transverse momentum and the transverse spin of quarks in unpolarized nucleons, respectively. All these

On behalf of the COMPASS Collaboration.

F. Bradamante (✉)
INFN, Trieste Section, Via Valerio 2, Trieste, Italy
E-mail: Franco.Bradamante@ts.infn.it

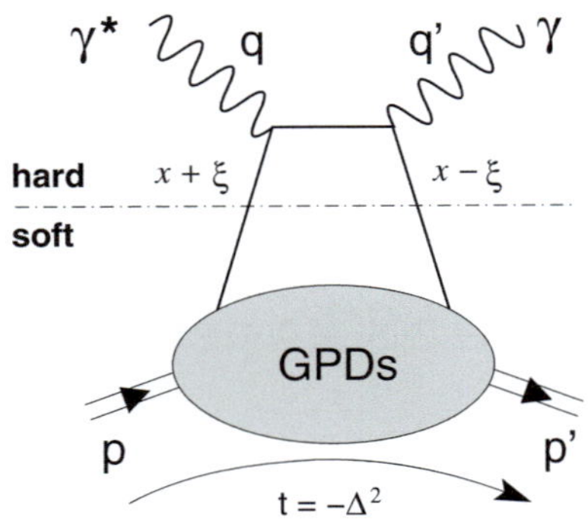

Fig. 1 Handbag diagram for the DVCS at leading twist

PDFs have been investigated by HERMES and COMPASS through measurements of the azimuthal asymmetries of hadrons produced in SIDIS processes [1,2], and are presently being investigated in the Drell–Yan process $\pi p \to \mu^+\mu^- X$ on transversely polarized protons [3].

In a different approach, PDFs have been generalized to GPDs introducing a transverse spatial dependence. The GPDs constitute a new, three dimensional parametrization of the nucleon structure [4–6]. They correlate the longitudinal momentum distribution of the partons inside the nucleon, parametrized by the PDFs, to the transverse spatial distribution of those partons, parametrized by the form factors. Most important, these objects grant access to the orbital angular momentum of the quarks [7,8]. GPDs can be accessed via exclusive processes such as deeply virtual compton scattering (DVCS, $lp \to l' p\gamma$) or hard exclusive meson production (HEMP) off the nucleon. In such reactions, the virtual photon (with a high virtuality Q^2) emitted by the lepton selects a quark in the nucleon with a longitudinal momentum fraction ξ^+, which re-emits a photon (in the case of DVCS) or produces a hadron (in the case of meson production), and is reabsorbed in the proton with a longitudinal momentum fraction ξ^-. During the reaction, a four-momentum t is transferred to the nucleon, as schematically shown in Fig. 1.

2 Measurements of TMDs in SIDIS

Impinging 160 GeV muons from the M2 beam line at the CERN SPS on either proton (NH_3) or deuteron (^{6}LiD) transversely polarized targets, the COMPASS collaboration has produced a wealth of results, the most important being the assessment, together with the HERMES Collaboration, that the so-called Collins and Sivers asymmetries on the proton are definitely non-zero and sizable. The fact that these effects are non-vanishing implies that both the transversity and the Sivers PDFs are non-zero and several extractions of these PDFs have already been performed. In particular a non-zero Sivers PDF for a given quark implies non-zero orbital angular momentum for that quark. Results have already been published, both for unidentified and identified hadrons, as function of x, z (the fraction of the available energy taken by the hadron) and p_T (the hadron transverse momentum). As an example, Fig. 2 shows the recently published COMPASS results for the Sivers asymmetries on the proton for π^+ and K^+ [9]: the asymmetries are definitely non-zero and positive, even down to the sea-quark region. A multi-dimensional analysis of the SIDIS data collected, presently in progress, should provide a deeper sensitivity on the kinematic. An example of the binning presently chosen in the $x - Q^2$ domain is shown in Fig. 3. Preliminary results for the x-dependence of the Sivers asymmetries in different Q^2 bins are shown in Fig. 4. Corresponding results have been produced for all the other transverse spin asymmetries. Clear non-zero signals over the whole Q^2 range are also observed in the case of the Collins asymmetry.

Further investigations of the Collins asymmetry as measured on transversely polarized protons in 2010 have led to interesting conclusions [10] about the properties of the Collins fragmentation function (FF) and of the di-hadron FF. In a nutshell, the amplitudes of the Collins-like (CL) asymmetries in a sample of SIDIS events in which at least a h^+h^- pair is detected are well compatible to those of the usual Collins asymmetries of h^+ and h^- in the "standard" SIDIS event sample (i.e. a sample containing at least one charged hadron), i.e. they have the same modulus and opposite sign, as shown in Fig. 5, where they are plotted as a function of $\Delta\phi$, the difference of the azimuthal angles of the two hadrons. Moreover, an analytical calculation allows to compute the di-hadron asymmetry $A_{CL,2h}$ as a function of $\Delta\phi$ from the CL asymmetries, giving convincing evidence that the Collins and the di-hadron asymmetries are driven by a common mechanism, as suggested

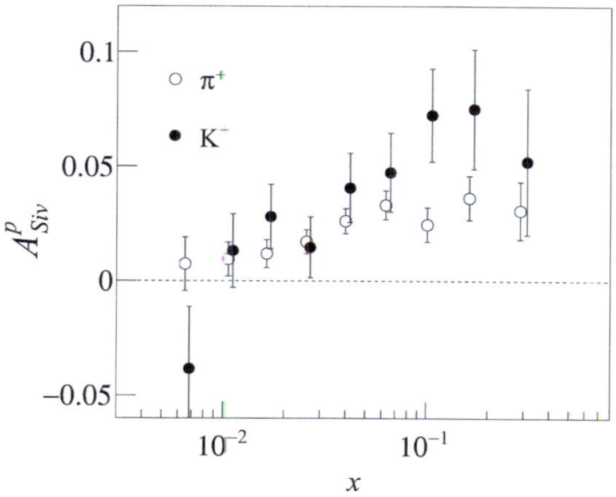

Fig. 2 The Sivers asymmetries for positive pions and kaons, as a function of x (COMPASS results)

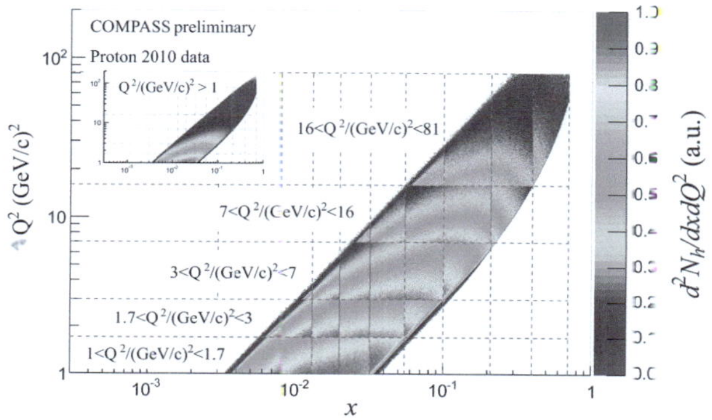

Fig. 3 The $x - Q^2$ grid which has been chosen to evaluate the transverse spin asymmetries. The region with $Q^2 > 16\ (\mathrm{GeV}/c)^2$ is the so-called safe region for the Drell–Yan measurement

by the $^3P_0$ recursive string fragmentation model [11]. Further studies of SIDIS will be performed in 2016 and 2017, in parallel to the GPD measurement, on a liquid hydrogen target, mainly to extract identified hadron multiplicities, p_T distributions and the azimuthal modulations of the cross-section [3]. These measurements, if precise enough, will allow to extract information on the k_T distribution and the Boer-Mulders PDFs.

3 Measurements of TMDs in Drell–Yan

After the long CERN shutdown of 2013, starting in October 2014 COMPASS had a 2-months test run with a 190 GeV/c momentum π^- beam to commission a modified spectrometer aiming at a first ever measurement of the Drell–Yan process $\pi^- p \to \mu^+\mu^- X$ on a transversely polarized proton target. In order to keep to a tolerable degree the radiation level in the spectrometer, a sophisticated absorber was inserted between the target and the first trackers to stop most of the beam and the hadrons created in the target. The cross-section for the process can be written as the sum of several azimuthal modulations, whose amplitudes are given by convolutions of different TMD PDFs. Of particular interest is the term $\sin\phi_{CS}$, where ϕ_{CS} is the azimuthal angle between the hadron plane and the lepton plane in the Collins-Sopper frame, whose amplitude depends on the Sivers function of the u quark of the target proton ($f^{\perp}_{1T/p}$) and the number density of the $\bar{u}$ quark in the incident π^- ($f_{1/\pi}$). Knowing $f_{1/\pi}$ it is thus possible to extract the Sivers function of the u quark in the transversely polarized proton, and compare it with the corresponding one measured in SIDIS. Being T-odd, the Sivers function is expected to change sign going from SIDIS to Drell–Yan. Such a prediction is deeply rooted in the present formulation of the theory [12], in particular in the way gauge invariance is guaranteed, and its experimental verification is the main motivation of the COMPASS Drell–Yan programme. After the successful test run of 2014 the measurements have been resumed this year, at the end of April, and data will

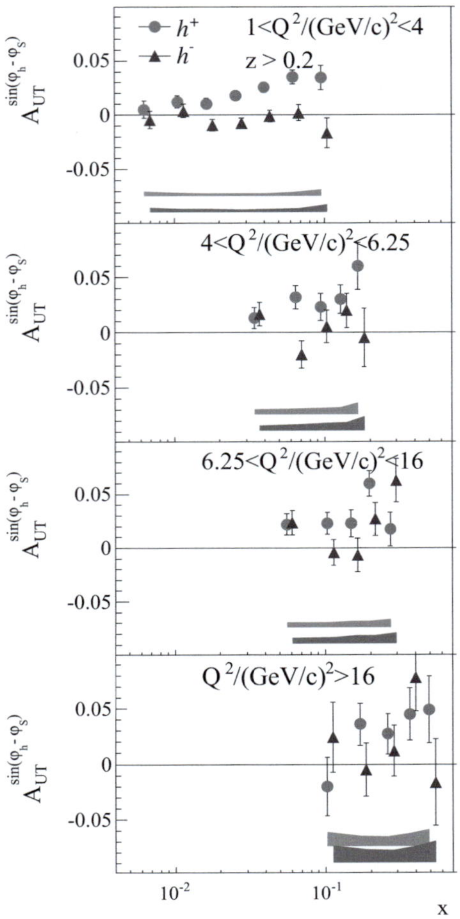

Fig. 4 The proton Sivers asymmetries as a function of x in 5 different Q^2 bins for positive and negative hadrons (COMPASS 2010 data, *preliminary*)

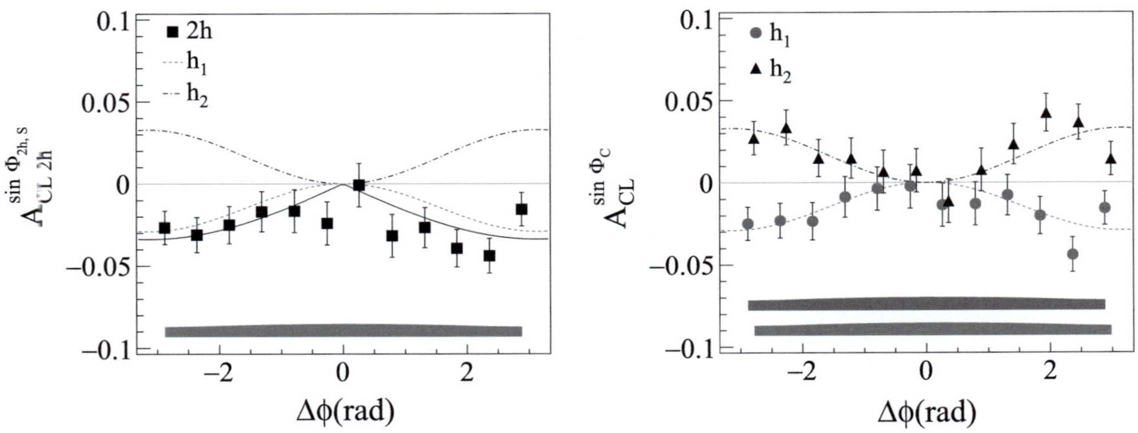

Fig. 5 COMPASS results. *Left* Dihadron asymmetry. *Right* Collins-like asymmetry in h^+h^- sample; h_1 indicates the first hadron of the pair, which is always the positive hadron. As expected from the mirror symmetry of the h^+ and h^- A_{CL}, all the three asymmetries can be fitted with simple functions which depend on only one parameter (see Ref. [10] for details)

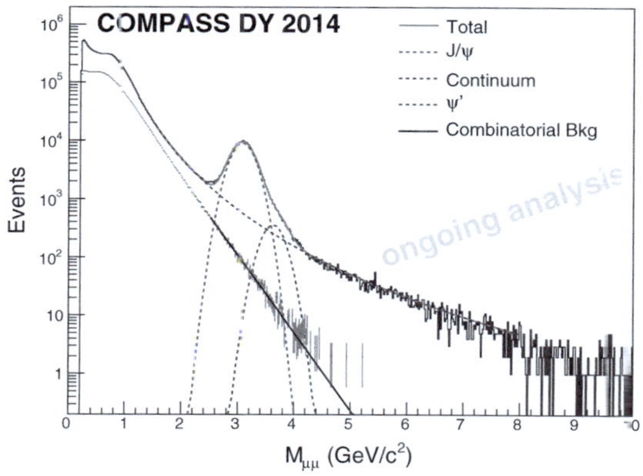

Fig. 6 $\mu^+\mu^-$ pair invariant mass spectrum from the 2014 Drell–Yan test run together with the measured combinatorial background from uncorrelated muons

be taken until middle of November. The $\mu^+\mu^-$ pair invariant mass spectrum measured in 2014 is given in Fig. 6. As shown in Fig. 4, in the "safe" Drell–Yan range $Q^2 > 16\,(\mathrm{GeV}/c)^2$, the Sivers asymmetry measured in SIDIS is clearly different from zero, therefore we should definitely be able to assess the expected change of sign in the Drell–Yan process.

4 Measurements of GPDs: HEMP

The measurements of DVCS and exclusive channels on a polarized proton target give access to the GPD E, which allows for the nucleon spin flip. The exclusive production of ρ^0 mesons on a transversely polarized proton target has been performed with muon data recorded at COMPASS in 2007 and 2010. Since no recoil detector could be used around the polarized target, exclusivity was ensured imposing energy and momentum conservation between the incoming muon and the outgoing muon and the vector meson. Eight target spin asymmetries, depending on ϕ (the azimuthal angle of the vector meson) and φ_S (the angle between the muon scattering plane and proton polarization) have been extracted [13] and successfully interpreted in terms of GPDs [14]. From the data collected on a transversely polarized proton target in 2010 it has also been possible to study the exclusive production of the ω meson, with $\omega \rightarrow \pi^+\pi^-\pi^0$ and $\pi^0 \rightarrow \gamma\gamma$ (the two photons were detected in the electromagnetic calorimeters of the COMPASS spectrometer). Due to the different quark combinations in the flavor dependent wave functions of the mesons, certain asymmetries are expected to be larger for ω production than the corresponding ones for ρ^0. Also, certain azimuthal asymmetries for ω production are sensitive to the pion pole contribution and hence allow in principle to determine the sign of the $\omega\pi$ transition form factor. Figure 7 shows the preliminary results on the single transverse spin asymmetries in bins of Q^2, x and p_T^2. The dashed red and dotted blue curves represent the predictions [15] with the positive and negative contributions of the pion pole, respectively, while the solid black curve represents the predictions without the pion pole contribution.

5 Measurements of GPDs: DVCS

In order to ensure exclusivity of DVCS (and also HEMP) events, a new recoil detector has been built, to surround a 2.5 m long liquid hydrogen target [3]. The DVCS cross section will be determined as a function of both the momentum transfer t between initial and final nucleons and the fraction of the longitudinal nucleon momentum carried by the struck parton. A new electromagnetic calorimeter (ECAL0) will provide coverage of substantially higher values of this fraction as compared to the existing calorimeters ECAL1 and ECAL2. With such an apparatus, COMPASS can measure the DVCS process on a wide x range (from 0.005 to 0.3), in a Q^2 range from 1 to 20 $(\mathrm{GeV}/c)^2$ at the expected integrated luminosity. In 2012 a short test run was performed with an almost complete apparatus (the new ECAL0 was only partially available), and data taking is planned in 2016 and 2017. A competing process to DVCS is the Bethe-Heitler (BH) process of elastic lepton-nucleon scattering with a hard Bremsstrahlung photon emitted by either the incoming or the outgoing lepton. It produces the same final state as DVCS so that the two processes interfere at the level of amplitudes.

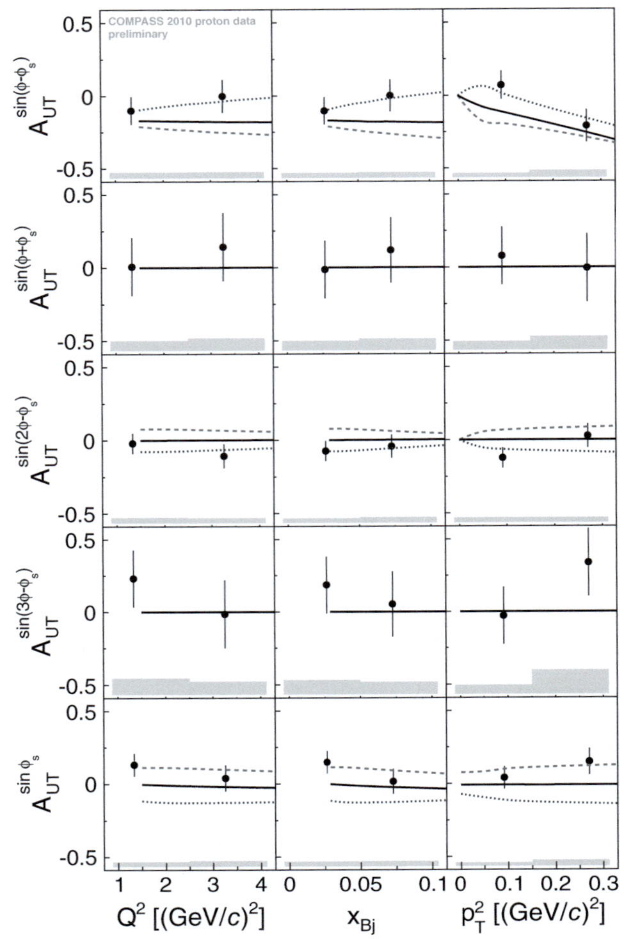

Fig. 7 Preliminary results on the transverse spin asymmetries for ω production in bins of Q^2, x and p_T^2

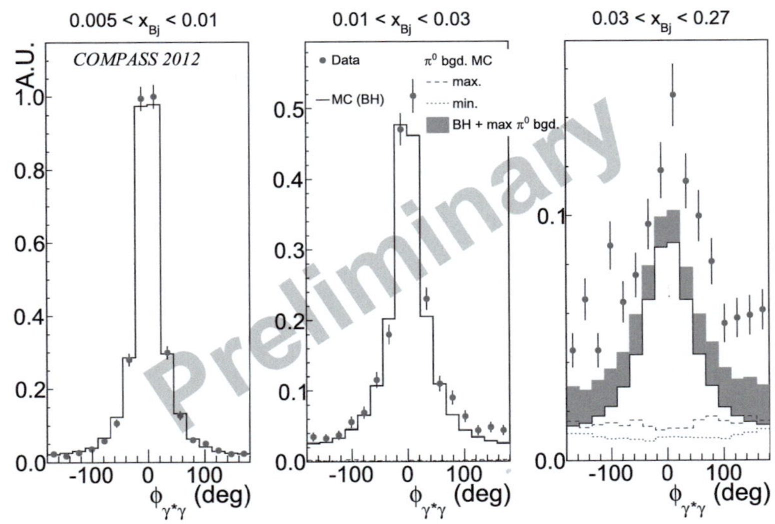

Fig. 8 Preliminary results from the 2012 test run and Monte Carlo (MC) predictions

In the COMPASS experiment, kinematic domains are accessible where either BH or DVCS dominate. At high x almost pure DVCS events will be collected, which will also allow for a measurement of the t dependence of the cross section, related to the tomographic imaging of the nucleon. At low x almost pure BH events are recorded which can be used as an excellent reference yield. In the intermediate domain the DVCS contribution will be enhanced by the BH process through their interference. Some preliminary results from the 2012 test run are shown in Fig. 8, together with Monte Carlo predictions. At large x a clear DVCS signal is visible, and the BH contribution can easily be subtracted. The analysis of these data is still ongoing to improve the estimate of the π^0 background. All COMPASS DVCS measurements will exploit the availability of positive and negative muon beams with opposite polarization. This allows to measure with the same apparatus the beam-charge and beam-spin cross-section sum and difference. From these measurements the real and imaginary part of the Compton form factors can be extracted allowing a detailed study of the GPD H.

6 Outlook

In 2018 COMPASS will take one more year of data. The most likely options are either a second year of Drell–Yan or the continuation of the GPD/SIDIS data taking: both measurements are statistically limited, and the decision will be taken on the basis of the significance of the data collected in 2015, 2016 and 2017. Starting from 2019 all CERN accelerators will stop for a second long shut-down, but our Collaboration has already made plans for a new phase of measurements [16], which include GPD measurements on a transversely polarized proton target and SIDIS measurements on a transversely polarized deuteron target. These plans are presently being revisited.

References

1. Aidala, C.A., Bass, S.D., Hasch, D., Mallot, G.K.: The spin structure of the nucleon. Rev. Mod. Phys. **85**, 655 (2013)
2. Barone, V., Bradamante, F., Martin, A.: Transverse-spin and transverse-momentum effects in high-energy processes. Prog. Part. Nucl. Phys. **65**, 267 (2010)
3. COMPASS Collaboration (2010) The COMPASS-II Proposal. SPSC-2010-014/P-340
4. Müller, D., et al.: Wave functions, evolution equations and evolution kernels from light ray operators of QCD. Fortschr. Phys. **42**, 101 (1994)
5. Radyushkin, A.V.: Asymmetric gluon distributions and hard diffractive electroproduction. Phys. Lett. B **385**, 333 (1996)
6. Radyushkin, A.V.: Nonforward parton distributions. Phys. Rev. D **56**, 5524 (1997)
7. Ji, X.D.: Deeply virtual Compton scattering. Phys. Rev. D **55**, 7114 (1997)
8. Ji, X.D.: Gauge-invariant decomposition of nucleon spin. Phys. Rev. Lett. **78**, 610 (1997)
9. Adolph, C., [COMPASS Collaboration], et al.: Collins and Sivers asymmetries in muonproduction of pions and kaons off transversely polarised protons. Phys. Lett. B **744**, 250 (2015)
10. Adolph, C., [COMPASS Collaboration], et al.: Interplay among transversity induced asymmetries in hadron leptoproduction. Phys. Lett. B **753**, 406 (2016)
11. Artru, X., Czyzewski, J., Yabuki, H.: Single spin asymmetry in inclusive pion production, Collins effect and the string model. Z. Phys. C **73**, 527 (1997)
12. Collins, J.C.: Leading twist single transverse-spin asymmetries: Drell–Yan and deep inelastic scattering. Phys. Lett. B **536**, 43 (2002)
13. Adolph, C., [COMPASS Collaboration], et al.: Transverse target spin asymmetries in exclusive ρ^0 muoproduction. Phys. Lett. B **731**, 19 (2014)
14. Goloskokov, S.V., Kroll, P.: Transversity in exclusive vector-meson leptoproduction. Eur. Phys. J. C **74**, 2725 (2014a)
15. Goloskokov, S.V., Kroll, P.: The pion pole in hard exclusive vector-meson leptoproduction. Eur. Phys. J. A **50**(9), 146 (2014b)
16. COMMPASS Collaboration (2012) COMPASS planned measurements in the next 5 years and longer term perspectives on the study of the nucleon structure. Report prepared for the European Strategy Preparatory Group, July 30, 2012

DOI 10.1007/978-3-319-50699-9_42
Reprinted from *Few-Body System* Journal, DOI 10.1007/s00601-016-1095-z

Enrique Ruiz Arriola · Wojciech Broniowski

Proton–Proton On Shell Optical Potential at High Energies and the Hollowness Effect

Received: 18 March 2016 / Accepted: 22 March 2016 / Published online: 8 April 2016

Abstract We analyze the usefulness of the optical potential as suggested by the double spectral Mandelstam representation at very high energies, such as in the proton-proton scattering at ISR and the LHC. Its particular meaning regarding the interpretation of the scattering data up to the maximum available measured energies is discussed. Our analysis reconstructs 3D dynamics from the effective transverse 2D impact parameter representation and suggests that besides the onset of gray nucleons at the LHC there appears an inelasticity depletion (hollowness) which precludes convolution models at the attometer scale.

1 Introduction

The history of proton–proton scattering at high energies has been marked by continuous surprises: extrapolations have often been contradicted by actual measurements such as at ISR (see, e.g., [1–4] for comprehensive accounts and references therein). The first run of the CERN-LHC on pp collisions has unveiled new features at CM energies about 7 TeV, measured by the TOTEM collaboration [5]. Contrary to naive expectations, probing strong interactions at such high energies, corresponding to a de Broglie wavelength of about half an attometer, becomes more intricate and may still be far from the asymptotics [6]. Regge phenomenology [7] motivated Barger and Phillips [8] to propose a parameterization which, when suitably extended, describes all ISR and the LHC data simultaneously [9, 10]. Dynamical calculations display the intricacy of non perturbative phenomena at these high energies from a fundamental viewpoint (see e.g. [11, 12]) In the present talk we return to a phenomenological level and unveil interesting features of pp scattering via the so-called *on-shell optical potential model*.

The optical potential was first suggested to deal with inelastic neutron-nucleus scattering above the compound nucleus regime [13]. There, the concept of the black disk limit was first proposed and tested along with the observed Fraunhofer diffraction pattern [14], which also applies to the eikonal approximation [15].

Talk presented by ERA at the Light Cone 2015 Conference, Frascati, Italy. 21-25 September 2015.
Supported by Spanish DGI and Feder funds (grant FIS2014-59386-P), by Junta de Andalucía (grant FQM225), and by Polish National Science Center (grant DEC-2012/06/A/ST2/00390).

E. R. Arriola (✉)
Departamento de Física Atómica, Molecular y Nuclear and Instituto Carlos I de Física Teórica y Computacional, Universidad de Granada, 18071 Granada, Spain
E-mail: earriola@ugr.es

W. Broniowski
The H. Niewodniczański Institute of Nuclear Physics, Polish Academy of Sciences, 31342 Kraków, Poland

W. Broniowski
Jan Kochanowski University, 25406 Kielce, Poland
E-mail: Wojciech.Broniowski@ifj.edu.pl

The Serber model [16] was an incipient extension of the optical eikonal formalism to high energy particle physics. Based on the Mandelstam double spectral representation of the scattering amplitude, Cornwall and Ruderman [17] delineated a definition of the optical potential directly rooted in field theory. Field-theoretic discussions using the multichannel Bethe–Salpeter equation shed some further light [18,19] (see, e.g., an early review on optical models [20]). The off-shell versus on-shell interplay was analyzed and an on-shell-type equation was proposed by Namyslowski [21]. The high energy grayness of the nucleon has been a matter of discussion since the 70s [22].

2 Amplitudes and Parameterizations

The invariant proton–proton elastic scattering differential cross section is given by

$$\frac{d\sigma_{\rm el}}{dt} = \frac{\pi}{p^2}\frac{d\sigma_{\rm el}}{d\Omega} = \frac{\pi}{p^2}|f(s,t)|^2 , \tag{1}$$

where $f(s,t)$ is the scattering amplitude having both the partial wave and the impact parameter expansions [23]

$$f(s,t) = \sum_{l=0}^{\infty}(2l+1)f_l(p)P_l(\cos\theta) = \frac{p^2}{\pi}\int d^2b\, h(\mathbf{b},s)\, e^{i\mathbf{q}\cdot\mathbf{b}} = 2p^2\int_0^{\infty} bdbJ_0(bq)h(b,s) , \tag{2}$$

where $s = 4(p^2+M^2)$, p is the CM momentum, and $t = -\mathbf{q}^2$ with $q = 2p\sin(\theta/2)$ denoting the momentum transfer. In the eikonal approximation one has $bp = l + 1/2 + \mathcal{O}(s^{-1})$, hence $h(b,s) = f_l(p) + \mathcal{O}(s^{-1})$ and $P_l(\cos\theta) \to J_0(qb)$. The total, elastic, and total inelastic cross sections read, respectively [23]

$$\sigma_T = \frac{4\pi}{p}{\rm Im}f(\theta=0) = 4p\int d^2b{\rm Im}h(\mathbf{b},s) , \tag{3}$$

$$\sigma_{\rm el} = \int d\Omega|f(s,t)|^2 = 4p^2\int d^2b|h(\mathbf{b},s)|^2 , \tag{4}$$

$$\sigma_{\rm in} \equiv \sigma_T - \sigma_{\rm el} = \int d^2bn_{\rm in}(b) . \tag{5}$$

Here, the transverse probability inelastic profile fulfills $n_{\rm in}(b) \le 1$ and is given by

$$n_{\rm in}(b) = 4p\left[{\rm Im}h(b,s) - p|h(b,s)|^2\right] . \tag{6}$$

For our purposes we just need a working parameterization of the scattering amplitude. Here, and for definiteness we use the work in Ref. [9], and more specifically, their MBP2 form, which have been fitted separately for all known differential cross sections for $\sqrt{s}$ = 23.4, 30.5, 44.6, 52.8, 62.0, and 7000 GeV[1] and read

$$f(s,t) = \sum_n c_n F_n(t)s^{\alpha_n(t)} , \tag{7}$$

where $F_n(t)$ are form factors, $\alpha_n(t) = \alpha_n(0) + \alpha_n'(0)t$ and c_n are complex numbers which at variance with Regge theory [7], are assumed to be energy dependent. The fits produce χ^2/d.o.f. $\sim 1.2 - 1.7$ [9,10].

3 The On-Shell Optical Potential

A general field theoretic approach requires solving a coupled channel Bethe–Salpeter equation involving all open channels, a most impractical a procedure, since their number becomes huge for the large energies at ISR or the LHC. Of course, a viable approach would be to determine the kernel, operating as a phenomenological optical potential, from the available NN scattering data. In the geometric picture, the diffraction pattern is manifest as a shadow of the inelastic scattering, such that the diffraction peak in the forward direction is due to a coherent interference. From a Quantum Mechanics point of view, the inelastic process can be interpreted as

[1] A compilation of high energy scattering data can be found at http://www.theo.phys.ulg.ac.be/alldata-v2.zip.

a leakage in the probability current. We propose a scheme below where a local and energy dependent potential can be directly computed from the data, unveiling the structure of the *inelasticity hole*.

A standard tool for handling the two-body relativistic scattering is the Bethe–Salpeter equation (we use conventions of Ref. [24]), which in the operator form reads

$$T = V + VG_0T \,. \tag{8}$$

For the $2 \to 2$ sector with the kinematics $(P/2+k, P/2-k) \to (P/2+p, P/2-p)$ it can be written as

$$T_P(p,k) = V_P(p,k) + \mathrm{i}\int \frac{d^4q}{(2\pi)^4} T_P(q,k)S(q_+)S(q_-)V_P(p,q)\,, \tag{9}$$

where $q_\pm = (P/2 \pm q)$, P is the total momentum, $T_P(p,k)$ is the total scattering amplitude, and $S(q_\pm)$ denotes the nucleon propagator. The kernel V represents the irreducible four-point Green's function and it is generically referred to as the *potential*. Equation (9) is a linear four-dimensional equation. It requires the off-shell behavior of the potential $V_P(k',k)$ and generally depends on the choice of the interpolating fields, although one expects the scattering amplitude for the on-shell particles to be independent of the field choice.[2]

An approach which is manifestly independent of the off-shell ambiguities deduces an on-shell equation by separating explicitly those states which are on-shell and elastic from the rest [21] (for a similar and related ideas see, e.g., Ref. [24]). In the operator form, the final result can be written as (see also Ref. [17])

$$T_{\rm el} = W + T_{\rm el}G_0T_{\rm el}^\dagger\,, \tag{10}$$

where W is the *on-shell optical potential* and $T_{\rm el}$ is the elastic on-shell scattering amplitude,

$$T_{P,{\rm el}}(p,k) = T_P(p,k)|_{p^2=k^2=s/4-M^2} = -8\pi\sqrt{s}f(s,t)\,. \tag{11}$$

When written out explicitly, the equation becomes

$$T_{P,{\rm el}}(p,k) = W_P(p,k) + \mathrm{i}\int \frac{d^4q}{(2\pi)^4} T_{P,{\rm el}}(q,k)S(q_+)S(q_-)T_{P,{\rm el}}(q,p)^*\,. \tag{12}$$

At the level of partial waves, we have a simplified form, which using $p(s) = (s/4 - M_N^2)^{\frac{1}{2}}$ reads

$$f_l(s) = w_l(s) + \frac{1}{\pi}\int_{s_0}^\infty ds' \frac{f_l(s')p(s')f_l(s')^\dagger}{s'-s-i0^+}\,, \tag{13}$$

where $s_0 = 4M_N^2$. Note that only the on-shell amplitude enters here, whereas the equation is non-linear. As usual, the scattering amplitudes are defined as boundary values of analytic complex functions, such that $f_l(s) \equiv f_l(s+i0^+)$. Note that due to the Schwartz reflection principle $f_l(s)^\dagger \equiv f_l(s-i0^+)$, and thus the unitarity condition corresponds to a right-hand discontinuity cut $2i{\rm Im}\, f_l(s) = f_l(s+i0^+) - f_l(s-i0)$, which reads ${\rm Im}\, f_l(s) - p(s)|f_l(s)|^2 = {\rm Im}\, w_l(s)$ at $s > 4M_N^2$ and yields two contributions. The exchange, written as a left cut condition becomes ${\rm Im}\, f_l(s) = {\rm Im}\, w_l(s)$ for $s < 0$, whereas the causality implies a dispersion relation in energy along the left and right cuts. Invoking the eikonal approximation, which works phenomenologically for $\sqrt{s} \geq 23.5$GeV, and using Eq. (6) we get

$$w_l(s)|_{l-1/2=bp} = n_{\rm in}(b)/4p + \mathcal{O}(s^{-1}) \tag{14}$$

Solving the BS equation becomes very complicated and for a phenomenological kernel is not truly essential. Instead, we use a *minimal relativistic approach* [25] based on the squared mass operator [26] defined as $\mathcal{M}^2 = P^\mu P_\mu + \mathcal{V}$, where $\mathcal{V}$ represents the (invariant) interaction determined in the CM frame by matching to the non-relativistic limit with a local and energy-dependent phenomenological optical potential, $V(\mathbf{r},s) = {\rm Re}V(\mathbf{r},s) + i{\rm Im}V(\mathbf{r},s)$. This yields $\mathcal{V} = 8M_N V(\mathbf{r},s)$; it could be obtained by fitting elastic scattering data [25]. After quantization we have $\hat{\mathcal{M}}^2 = 4(\hat{p}^2 + M_N^2) + 8M_N V$, with $\hat{p} = -i\nabla$, such that the relativistic equation can be written as $\hat{\mathcal{M}}^2\Psi = 4(k^2 - M_N^2)\Psi$, i.e., as a non-relativistic Schrödinger equation

$$(-\nabla^2 + M_N V)\Psi = \left(s/4 - M_N^2\right)\Psi\,. \tag{15}$$

[2] This point is not made very clear in the literature; see, e.g., Ref. [24] for an explicit demonstration in the particular case of $\pi\pi$ scattering that the field transformations preserving the on-shell potential modify the on-shell T-matrix.

This equation incorporates the necessary physical ingredients which were also present in the BS equation: relativity and inelasticity. The optical potential V *does not* yet correspond to the on-shell one defined by Eq. (10). The argument given in Ref. [17] uses the optical theorem from the continuity equation, yielding

$$\sigma_T - \sigma_{\rm el} \equiv \sigma_{\rm in} = -\frac{M_N}{p} \int d^3x {\rm Im}\, W(\mathbf{x}, s)\,, \tag{16}$$

where the on-shell optical potential is defined by its imaginary part ${\rm Im}\, W(\mathbf{x}, s) \equiv {\rm Im}\, V(\mathbf{x}, s)|\Psi(\mathbf{x})|^2$ and can be interpreted as the local density of inelasticity at a given CM energy $\sqrt{s}$; it also becomes $W = V + \ldots$ perturbatively. Using $\sum_{l=0}^{\infty}(2l+1)\,[j_l(pr)]^2 = 1$, Eq. (16) becomes consistent with Eq. (14). We may further rewrite this in the impact parameter space by taking $\mathbf{x} = (\mathbf{b}, z)$ and integrating over the longitudinal component. As a result we get Eq. (5), where the transverse probability profile function is given by

$$n_{\rm in}(b) = -\frac{M_N}{p} \int_{-\infty}^{\infty} dz\, {\rm Im}\, W(\mathbf{x}, s) = -\frac{M_N}{p} \int_{b}^{\infty} \frac{2r dr}{\sqrt{r^2 - b^2}} {\rm Im}\, W(r, s)\,. \tag{17}$$

In the last step the spherical symmetry has been exploited. To determine $W(r, s)$ we recognize this formula as an integral equation of the Abel type, hence it can be inverted using the standard method (see e.g. Ref. [27]) to give the on-shell optical potential *directly* from the inelasticity profile and hence from data,

$${\rm Im} W(r, s) = \frac{2p}{\pi M_N} \int_{r}^{\infty} db \frac{n'_{\rm in}(b)}{\sqrt{b^2 - r^2}}\,. \tag{18}$$

This new formula is remarkable as it reconstructs the 3D on-shell dynamics from the effective transverse 2D impact parameter representation where the longitudinal physics has been integrated out.

4 Numerical Results and Discussion

For the MBP2 parameterization of Ref. [9] we obtain the inelastic profile function and its derivative for the measured and fitted energies $\sqrt{s} = 23.4$, 30.5, 44.6, 52.8, 62.0, and 7000 GeV. The result for 14000 GeV is an extrapolation proposed in Ref. [9]. The amplitudes of the on-shell potentials depend strongly on the CM energy $\sqrt{s}$, with a power-like behavior. Thus, in Fig. 1 we show the ratio normalized to the value at the origin, ${\rm Im} W(r, s)/{\rm Im} W(0, s)$. As can be vividly seen, the lower energy values have a maximum at the origin, whereas the LHC pp data develop a dip in the origin, which suggests that the inelasticity becomes maximal at a finite value, around $r = 1$fm. The fact that the optical potential has its maximum away from the origin is most remarkable; a feature shared by the profile function $n_{\rm in}(b)$, which shows that in an inelastic collision most damage is not necessarily produced by central collisions. The "hollowness" effect is less evident in 2D as the 3D hole is integrated over the longitudinal variables which effectively fill the 3D-hole. This goes beyond the idea that the protons become "gray" above 13TeV, as recently suggested by Dremin [28] (see also [29,30]), rather than a black disk. Actually, this shows in particular that the hollowness effect cannot be reproduced by an intuitive folding structure. Indeed, for small r we get

$$W(r) = \int d^3y \rho(\mathbf{y} + \mathbf{r}/2)\rho(\mathbf{y} - \mathbf{r}/2) = \int d^3y \rho(\mathbf{y})^2 - \frac{1}{4}\int d^3y [\mathbf{r} \cdot \nabla\rho(\mathbf{y})]^2 + \cdots \tag{19}$$

showing that $W(0)$ is a local maximum, in contrast to the phenomenological result, see Fig. 1. This conclusion also holds if the folding is made between wave functions with no extra weight. Finally, we note that the mean squared radius of ${\rm Im} W(r)$; $\langle r^2 \rangle = \frac{3}{2}\langle b^2 \rangle$ displays a logarithmic growth with the energy. These surprising new high energy features as well as the fluctuations in $n_{\rm in}(b)$ [31] become relevant in heavy ions collisions and will be addressed in more detail elsewhere. The energy interpolation of [9] suggests that the 3D depletion already happens at $0.5 - 1$TeV, below LHC energies, generating a flattening of the 2D impact parameter dependence.

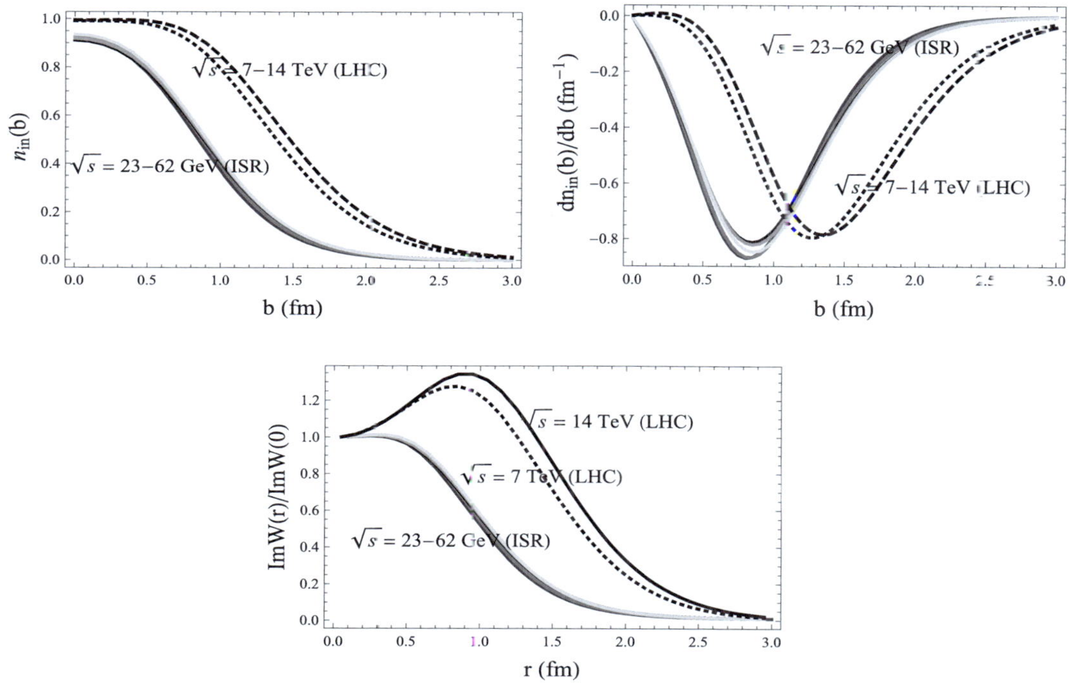

Fig. 1 Inelasticity properties of the proton-proton scattering at the CM energies $\sqrt{s}$ = 23.4, 30.5, 44.6, 52.8, 62.0, 7000, and 14000 GeV. *Top*: The inelastic profile (*left*) and its derivative (*right*) as a function of the impact parameter. *Bottom*: The imaginary part of the on-shell optical potential normalized to the value at the origin, plotted as a function of the radial distance

5 Conclusions

The on-shell optical potential is a meaningful concept under the most common and general assumption of the Mandelstam double spectral representation. We have shown that it is also a useful quantity when interpreting the proton-proton scattering data at very high energies; the shape of the inelasticity hole changes dramatically when going from ISR to the LHC. The shoulder-like form of the imaginary part of the on-shell potential resembles very much the traditional pattern found in the absorptive part of optical potential in the neutron-nucleus reactions beyond the compound model regime [32]. For a heavy nucleus, the surface is much smaller than the volume and the shoulder merely shows that most inelastic processes occur at the surface. This can pictorially be imagined as derivatives of a Fermi-type distribution. In the case of the proton-proton scattering, extremely high energies seem necessary to resolve between the surface from the volume effects at the attometer scales. The puzzling hollowness effect sets in at $0.5 - 1$TeV and awaits a dynamical explanation.

References

1. Amaldi, U., Klaus, R.: Impact parameter interpretation of proton–proton scattering from a critical review of all ISR data. Nucl. Phys. B **166**, 301 (1980)
2. Cheng, H., Wu, T.T.: Expanding Protons: Scattering at High Energies. Mit Press, Cambridge (1987)
3. Matthiae, G.: Proton and anti-proton cross-sections at high-energies. Rep. Prog. Phys. **57**, 743–790 (1994)
4. Barone, V., Predazzi, E.: High-Energy Particle Diffraction. Springer, Berlin (2002)
5. Antchev, G., et al.: Luminosity-independent measurement of proton–proton elastic scattering and total cross-section at $\sqrt{s} = 7$ TeV. Europhys. Lett. **101**, 21002 (2013)
6. Fagundes, D.A., Menon, M.J., Silva, P.V.R.G.: Exploring central opacity and asymptotic scenarios in elastic hadron scattering. Nucl. Phys. A **946**, 194–226 (2016)
7. Collins, P.D.B.: An Introduction to Regge Theory and High-Energy Physics. Cambridge University Press, Cambridge (1977)
8. Phillips, R.J.N., Barger, V., Zeppenfeld, D.: Model independent analysis of the structure in pp scattering. Phys. Lett. B **46**, 412–414 (1973)
9. Fagundes, D.A., Grau, A., Pacetti, S., Pancheri, G., Srivastava, Y.N.: Elastic pp scattering from the optical point to past the dip: an empirical parametrization from ISR to the LHC. Phys. Rev. D **88**(9), 094019 (2013)
10. Selyugin, O.V.: Nucleon structure and the high energy interactions. Phys. Rev. D **91**(11), 113003 (2015) [Erratum: Phys. Rev. D92, no.9, 099901 (2015)]

11. Khoze, V.A., Martin, A.D., Ryskin, M.G.: Elastic scattering and diffractive dissociation in the light of LHC data. Int. J. Mod. Phys. A **30**(08), 1542004 (2015)
12. Gotsman, E., Levin, E., Maor, U.: A comprehensive model of soft interactions in the LHC era. Int. J. Mod. Phys. A **30**(08), 1542005 (2015)
13. Fernbach, S., Serber, R., Taylor, T.B.: The scattering of high energy neutrons by nuclei. Phys. Rev. **75**, 1352–1355 (1949)
14. Blatt, J., Weisskopf, V.: Theoretical Nuclear Physics. Springer, Berlin (2012)
15. Glauber, R.J.: High Energy Collision Theory. Lectures in Theoretical Physics. Interscience, NewYork (1959)
16. Serber, R.: High-energy proton–proton scattering. Rev. Mod. Phys. **36**(2), 649 (1964)
17. Cornwall, J.M., Ruderman, M.A.: Mandelstam representation and Regge poles with absorptive energy-dependent potentials. Phys. Rev. **128**(3), 1474 (1962)
18. Torgerson, R.: Field-theoretic formulation of the optical model at high energies. Phys. Rev. **143**(4), 1194 (1966)
19. Arnold, R.C.: Optical potential for high-energy physics: theory and applications. Phys. Rev. **153**(5), 1523 (1967)
20. Islam, M.M.: The optical model at high energies. Phys. Today **25**, 23 (1972)
21. Namyslowski, J.M.: Relativistic, 3-dimensional, 2-body integral equations. On-shell and off-shell formalisms. Phys. Rev. **160**(5), 1522 (1967)
22. Kaidalov, A.B.: Diffractive production mechanisms. Phys. Rep. **50**, 157 (1979)
23. Blankenbecler, R., Goldberger, M.L.: Behavior of scattering amplitudes at high energies, bound states, and resonances. Phys. Rev. **126**, 766–786 (1962)
24. Nieves, J., Ruiz Arriola, E.: Bethe–Salpeter approach for unitarized chiral perturbation theory. Nucl. Phys. A **679**, 57–117 (2000)
25. Ruiz Arriola, E., Broniowski, W.: Low energy nuclear structure from ultra-relativistic heavy-light ion collisions. J. Phys. Conf. Ser. **630**(1), 012060 (2015)
26. Allen, T.W., Payne, G.L., Polyzou, W.N.: Comparison of relativistic nucleon–nucleon interactions. Phys. Rev. C **62**, 054002 (2000)
27. Buck, U.: Inversion of molecular scattering data. Rev. Mod. Phys. **46**(2), 369 (1974)
28. Dremin, I.M.: Will protons become gray at 13 TeV and 100 TeV? arXiv:1511.03212, hep-hp, (2015)
29. Dremin, I.M.: Elastic scattering of hadrons. Phys. Uspekhi **56**, 3–28 (2013). [Usp. Fiz. Nauk183,3(2013)]
30. Troshin, S.M., Tyurin, N.E.: Are there any clues for hadron asymptotics in the LHC data? arXiv:1601.00483 [hep-ph]
31. Rybczyński, M., Włodarczyk, Z.: The nucleon–nucleon collision profile and cross section fluctuations. J. Phys. **G41**, 015106 (2013)
32. Hodgson, P.E.: Nuclear Reactions and Nuclear Structure. Clarendon Press, Oxford (1971)

DOI 10.1007/978-3-319-50699-9_43
Reprinted from *Few-Body System* Journal, DOI 10.1007/s00601-016-1096-y

Xiaonu Xiong

Hadron Structures from Large Momentum Effective Field Theory

Received: 15 January 2016 / Accepted: 23 March 2016 / Published online: 19 April 2016

Abstract The large momentum effective field theory (LaMET) is aiming to calculate parton distributions directly on lattice. In this approach a class of so-called quasi distributions are defined as pure spatial correlations in a finite momentum frame and they are directly calculated on lattice. The light-cone distributions are extracted from the quasi distributions by perturbative matching conditions. We present the application of LaMET on parton distribution functions and generalized parton distributions. Besides, heavy meson's quasi and light-cone distribution amplitudes are studied under nonrelativistic QCD (NRQCD) factorization framework

1 Introduction

Parton distribution functions are essential for revealing the inner structure of hadron. There are several ways to determine parton distribution functions: higher energy scattering experiments, lattice QCD calculation, Ads/CFT theory and quark models. Among those methods, lattice QCD is currently the only reliable first principal calculation. However, parton distribution functions are defined as non-local light-cone correlation, e.g.

$$q(x) = \int \frac{d\xi^-}{4\pi} e^{-ixp^+\xi^-} \left\langle PS \left| \bar{\psi}\left(\frac{\xi^-}{2}\right) \gamma^+ \mathcal{L}\left[\frac{\xi^-}{2}; -\frac{\xi^-}{2}\right] \psi\left(-\frac{\xi^-}{2}\right) \right| PS \right\rangle, \tag{1}$$

where $\mathcal{L}$ is the gauge link ensures gauge invariance of PDF. The field separation ξ^- becomes $-i\xi^0 - \xi^z$ which is complex on lattice. Therefore it can't be calculated directly on lattice. So far, lattice calculations are focusing on calculating PDF's Mellin moments, but high moments' calculation is hard and expensive computationally because higher moments will mix with lower moments and require a fine lattice spacing to simulate higher derivatives.

The recently proposed concept of quasi PDF is defined as pure spatial correlation (Euclidian) [1,2]

$$\tilde{q}(x) = \int \frac{dz}{4\pi} e^{ixp^z z} \left\langle PS \left| \bar{\psi}\left(\frac{z}{2}\right) \gamma^z \mathcal{L}\left[\frac{z}{2}; -\frac{z}{2}\right] \psi\left(-\frac{z}{2}\right) \right| PS \right\rangle, \tag{2}$$

where the hadron momentum $P^\mu = \left(P^0, \mathbf{0}^\perp, P^z\right)$, $x = k^z/P^z$ is the longitudinal momentum fraction carried by the quark and $\mathcal{L}$ is the gauge link along the z-direction. As this definition doesn't depend on time, it can be directly simulated on lattice.

The LaMET approach works with the following procedure:

X. Xiong (✉)
Istituto Nazionale di Fisica Nucleare, Sezione di Pavia, 27100 Pavia, Italy
E-mail: xxionglbl@gmail.com
E-mail: xiaonu.xiong@pv.infn.it

Table 1 Comparison between quasi lattice calculations and high energy scattering processes

	High-energy scattering	Quasi lattice calculation
"Observable"	Cross section	Quasi quantities
Scale	Large momentum transfer Q	Hadron momentum P
Factorization	$\sigma = \sigma_H\left(x, Q^2\right) \otimes f\left(x, Q^2\right) + \mathcal{O}\left(Q^{-n}\right)$	$\tilde{f}(P) = \sigma_Z(P, \mu) \otimes f(\mu) + \mathcal{O}\left(P^{-n}\right)$

1. First, we define a quasi operator $\tilde{O}$ which can be directly calculated on the lattice, that means it should be Euclidean. The quasi quantity is defined as matrix element on a state with finite but large momentum,

$$\tilde{f} = \langle P \left| \tilde{O} \right| P \rangle. \tag{3}$$

2. The light-cone quantity is defined as the matrix element in infinite momentum frame (IMF)

$$f = \langle P_\infty \left| O \right| P_\infty \rangle. \tag{4}$$

3. As the difference between the above quasi quantity and light-cone quantity is the fact that they are defined in finite momentum frame and infinite momentum frame respectively, the above two quantities share the same infrared (IR) behavior and the difference is controlled by ultraviolet (UV) behavior. The two quantities could be related by the following matching condition (which can also be viewed as a factorization formula),

$$\tilde{f} = Z \otimes f + \mathcal{O}\left(\frac{M^n}{P^n}, \frac{\Lambda^n_{\text{QCD}}}{P^n}\right), \tag{5}$$

where Z is the matching factor, $\otimes$ denotes a convolution (will be explained in the next section) and $\mathcal{O}\left(M^n/P^n, \Lambda^n_{\text{QCD}}/P^n\right)$ denotes the target mass corrections and higher twist corrections. These corrections must be taken into consideration because that hadron's momentum is close to the hadron mass in the current quasi PDF calculations.

The matching between quasi quantities and light-cone quantities is very similar to the factorization of high energy scattering processes. The comparison between the matching and factorization of higher energy scattering process is shown in Table 1. At the present stage, the renormalization of quasi quantities is not settled yet and we will not discuss this issue in this paper.

2 Application of LaMET

2.1 Example 1: PDF and GPD

The matching condition between quasi PDF and light-cone PDF is [3,4]

$$\tilde{q}(x) = \int_{-1}^{1} \frac{dy}{|y|} Z\left(\frac{x}{y}\right) q(y), \tag{6}$$

where the higher order corrections in Eq. (5) have been omitted. We only concentrate on the non-singlet quark distribution, therefore there is no contribution from gluon mixing. As mentioned before, the matching factor between quasi PDF and light-cone PDF is perturbatively calculable, thus we can perform a perturbative expansion of quark-in-quark PDF and matching factor

$$q(x) = q^{(1)}(x) + \delta Z_F^{(1)} \delta(1-x) \quad \text{and} \quad \tilde{q}(x) = \tilde{q}^{(1)}(x) + \delta \tilde{Z}_F^{(1)} \delta(1-x), \tag{7}$$

where $q^{(1)}$ and $\tilde{q}^{(1)}$ denote contribution from the gluon exchange and $\delta Z_F^{(1)}$ and $\delta \tilde{Z}_F^{(1)}$ denote the contribution from quark's wave function renormalization (shown in Fig. 1). $\delta(1-x)$ is the tree-level PDF (both light-cone and quasi). Substituting the above expansion into the matching condition Eq. (6) leads to

$$Z^{(0)}\left(\xi, \frac{P^z}{\mu}\right) = \delta(1-\xi) \quad \text{and} \quad Z^{(1)}\left(\xi, \frac{P^z}{\mu}\right) = \tilde{q}^{(1)}(\xi, P^z) - q^{(1)}(\xi, \mu) + \left[\delta \tilde{Z}_F(P^z) - \delta Z_F(\mu)\right] \delta(1-\xi), \tag{8}$$

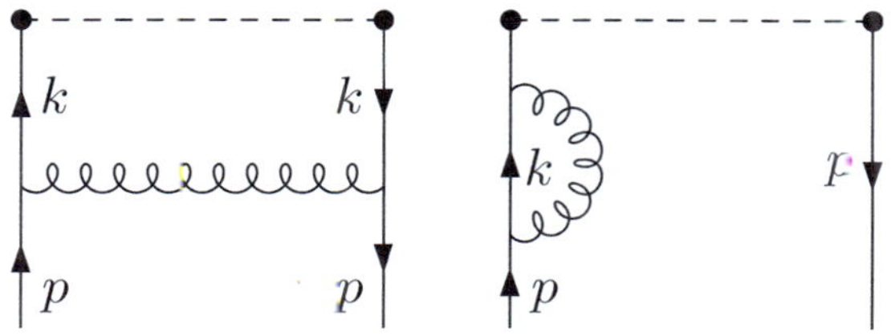

Fig. 1 One-loop corrections to quasi quark distribution

where $\xi = x/y$, the superscript (0) and (1) refer to the tree level and one-loop level results. The perturbative matching factor at one-loop level is the difference between quasi and light-cone PDF.

The nucleon's momentum is set as $P^{\mu} = \left(P^0, \mathbf{0}^{\perp}, P^z\right)$. We choose the gauge $n \cdot A = 0$ where $n^{\mu} = (1, 0, 0, -1)/\sqrt{2}$ for light-cone distribution and $n^{\mu} = (0, 0, 0, -1)$ for quasi distribution, therefore the gauge link in Eqs. (1, 2) reduces to the unit operator. The gluon propagator in $n \cdot A = 0$ gauge is

$$D_n^{\mu\nu}(q) = \frac{-i}{q^2}\left(g^{\mu\nu} - \frac{q^{\mu}n^{\nu} + n^{\mu}q^{\nu}}{n \cdot q} + n^2\frac{q^{\mu}q^{\nu}}{n \cdot q^2}\right). \tag{9}$$

The related Feynman diagrams in $n \cdot A = 0$ gauge are shown in Fig. 1.

In the one-loop calculation, we use a transverse cut-off $\int d^2\mathbf{k}_{\perp} = \int_0^{\Lambda} dk_{\perp}\, k_{\perp}$ to regularize UV divergence, quark mass m as collinear regulator and we keep the gluon massless. We choose transverse cut-off regularization in order to mimic lattice calculation, because the inverse of lattice spacing (a^{-1}) acts as a UV cut-off regulator in lattice calculation.

The resulting matching factor for unpolarized non-singlet quark distribution is [3,4]

$$Z^{(1)}\left(\xi, \frac{P^z}{\Lambda}\right) = \frac{\alpha_S C_F}{2\pi}\begin{cases}\left(\frac{1+\xi^2}{1-\xi}\right)\ln\frac{\xi}{\xi-1} + 1 + \frac{1}{(1-\xi)^2}\frac{\Lambda}{P^z}, & \xi > 1,\\ \left(\frac{1+\xi^2}{1-\xi}\right)\ln\frac{(P^z)^2}{\Lambda^2} + \left(\frac{1+\xi^2}{1-\xi}\right)\ln\left[4\xi(1-\xi)\right] - \frac{2\xi}{1-\xi} + 1 + \frac{1}{(1-\xi)^2}\frac{\Lambda}{P^z}, & 0 < \xi < 1,\\ \left(\frac{1+\xi^2}{1-\xi}\right)\ln\frac{\xi-1}{\xi} - 1 + \frac{1}{(1-\xi)^2}\frac{\Lambda}{P^z}, & \xi < 0.\end{cases}$$

$$+\,\delta(1-\xi)\frac{\alpha_S C_F}{2\pi}\int dy\begin{cases}-\frac{1+y^2}{1-y}\ln\frac{y}{y-1} - 1 - \frac{\Lambda}{(1-y)^2P^z} & y > 1\\ -\frac{1+y^2}{1-y}\frac{\ln(P_z^2)}{\Lambda^2} - \frac{1+y^2}{1-y}\ln\left[4y(1-y)\right] + \frac{4y^2-2y}{1-y} + 1 - \frac{\Lambda}{(1-y)^2P^z} & 0 < y < 1\\ -\frac{1+y^2}{1-y}\ln\frac{y-1}{y} + 1 - \frac{\Lambda}{(1-y)^2P^z} & y < 0,\end{cases} \tag{10}$$

In the above result, we have taken the limit $\Lambda \gg P^z \gg m$ because the nucleon momentum in quasi PDFs should be large but finite, while the quark mass m is a collinear regulator. The limit $P^z \gg \Lambda \gg m$ of quasi PDFs equal to the light-cone PDFs calculated with light-cone coordinates. The quasi PDF doesn't vanish in the region $x < 0$ or $x > 1$, corresponding to the process where a mother parton splits into a daughter parton carrying larger momentum than the mother. This process is suppressed in the infinite momentum frame. However, quasi PDFs are defined in finite momentum frame where this process is allowed.

It can be checked that the matching factor forms a + function and is free of collinear divergence $\ln m$ or soft divergence at $\xi = 1$. This is because the quasi and light-cone PDF share the same IR behavior, which is canceled in the matching. As a consequence, the matching factor is free of IR divergences and transfers the momentum dependence into the UV scale dependence.

The lattice results of non-singlet quasi distribution together with one-loop matching and target mass correction can be found in Ref. [5,6].

The definitions of light-cone [7–10] and quasi GPDs are [1]

$$2P^+ \int \frac{dz}{2\pi} e^{ixP^+z^-} \left\langle P+\tfrac{\Delta}{2}, S \middle| \bar{\psi}\left(-\tfrac{z^-}{2}\right)\gamma^+ \mathcal{L}\left[-\tfrac{z^-}{2};\tfrac{z^-}{2}\right]\psi\left(\tfrac{z^-}{2}\right) \middle| P-\tfrac{\Delta}{2}, S\right\rangle$$
$$= H(x,\xi,\Delta^2)\bar{U}\left(P+\tfrac{\Delta}{2}\right)\gamma^+ U\left(P-\tfrac{\Delta}{2}\right) + E(x,\xi,\Delta^2)\bar{U}\left(P+\tfrac{\Delta}{2}\right)\frac{i\sigma^{+\rho}\Delta_\rho}{2m}U\left(P-\tfrac{\Delta}{2}\right) \quad (11a)$$

$$2P^z \int \frac{dz}{2\pi} e^{-ixP^z z} \langle P+\tfrac{\Delta}{2}, S | \bar{\psi}\left(-\tfrac{z}{2}\right)\gamma^z \mathcal{L}\left[-\tfrac{z}{2};\tfrac{z}{2}\right]\psi\left(\tfrac{z}{2}\right) | P-\tfrac{\Delta}{2}, S\rangle$$
$$= \mathcal{H}(x,\xi,\Delta^2)\bar{U}\left(P+\tfrac{\Delta}{2}\right)\gamma^z U\left(P-\tfrac{\Delta}{2}\right) + \mathcal{E}(x,\xi,\Delta^2)\bar{U}\left(P+\tfrac{\Delta}{2}\right)\frac{i\sigma^{z\rho}\Delta_\rho}{2m}U\left(P-\tfrac{\Delta}{2}\right) \quad (11b)$$

We calculated the one-loop light-cone and quasi GPD(quark-in-quark) [11,12]. The quasi and light-cone GPDs also share the same IR behavior and the soft gluon poles are regularized to + functions. We also checked that the forward limit of GPD reduce to PDF.

2.2 Heavy Meson Distribution Amplitudes: A Perturbative Test of LaMET

The light-cone and quasi DA (longitudinally polarized vector meson as an example) are defined as

$$\Phi_V^\parallel(x,\mu) = -if_V^\parallel \varepsilon_\parallel^{*+} M_V \phi_V^\parallel(x,\mu)$$
$$= \int \frac{d\xi^-}{2\pi} e^{-i\left(x-\frac{1}{2}\right)P^+\xi^-} \left\langle V(P,\varepsilon_\parallel) \middle| \bar{\psi}\left(\tfrac{\xi^-}{2}\right)\gamma^+ \mathcal{L}\left[\tfrac{\xi^-}{2};\tfrac{-\xi^-}{2}\right]\psi\left(\tfrac{-\xi^-}{2}\right) \middle| 0\right\rangle, \quad (12a)$$

$$\tilde{\Phi}_V^\parallel(x,\mu) = -if_V^\parallel \varepsilon_\parallel^{*z} M_V \tilde{\phi}_V^\parallel(x,\mu)$$
$$= \int \frac{dz}{2\pi} e^{i\left(x-\frac{1}{2}\right)P^z z} \langle V(P,\varepsilon_\parallel) | \bar{\psi}\left(\tfrac{z}{2}\right)\gamma^z \mathcal{L}\left[\tfrac{z}{2};\tfrac{-z}{2}\right]\psi\left(\tfrac{-z}{2}\right) | 0\rangle, \quad (12b)$$

where ε is the polarization vector, $f_V^\parallel$ is the corresponding decay constant. M_V is the mass of the quarkonium state

Heavy quarkonium can be viewed as a non-relativistic system and one can apply NRQCD factorization. The NRQCD factorization [13] factorizes DA into product of long-range NRQCD matrix elements and short distance coefficient functions [14–16]. The decay and production of meson occur in the distance scale $\sim \mathcal{O}\left(m^{-1}\right)$, where m is the mass of the heavy quark. In the meson rest frame, the quark anti-quark pair are non-relativistic system characterized by the relative velocity v between the quark and anti-quark. The NRQCD factorization leads to

$$\Phi_H(x,\mu) \sim \sum_n \langle H | \mathcal{O}_n^{\text{NRQCD}} | 0\rangle \; \phi_H^n(x,\mu), \quad (13)$$

where $\langle H|\mathcal{O}_n^{\text{NRQCD}}|0\rangle$ denotes the long-range NRQCD matrix element encoding the non-perturbative information, $n = 0, 1, 2, \ldots$ denotes the s,p,d... wave in the series of v^n. $\phi_H^n(x)$ are the short distance coefficient functions which are perturbatively calculable.

Because the NRQCD factorization is a consequence of nonrelativistic property of heavy quarkonia that can be considered as a IR behavior, NRQCD factorization should also be valid for quasi DA since it has the same IR behavior as light-cone DA. Therefore, the light-cone and quasi DA are both perturbatively calculable, we can compare the two DAs and determine the magnitude of P^z needed in order to recover the light-cone DA from quasi DA through a boost.

We performed an one-loop calculation on the s-wave DA of η_c, longitudinally and transversely polarized J/ψ meson. The analytical results [17] of DAs show that all the IR regulator are absent and the soft poles at $x = 1/2$ are organized to + and ++ functions. These are required by NRQCD factorization which absorbs all IR information into the long-range NRQCD matrix element.

The comparison between the light-cone and quasi DA under different P^z is shown in Fig. 2. We define the *degree of resemblance* as

$$\Delta_H(P^z) = \frac{\int_0^{\frac{1}{2}} dx\,(1-2x)^4\left[\phi_H^{(1)}(x;\Lambda,m) - \tilde{\phi}_H^{(1)}(x,P^z;\Lambda,m)\right]^2}{\int_0^{\frac{1}{2}} dx\,(1-2x)^4\left[\phi_H^{(1)}(x;\Lambda,m)\right]^2}. \quad (14)$$

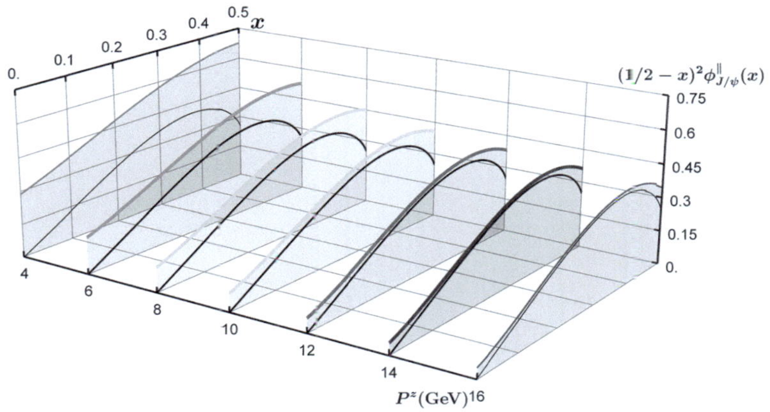

Fig. 2 Numerical result of $J/\psi^{\parallel}$'s quasi and light-cone DA. The *black curve* is the light-cone DA. The *colored curves* are the quasi DA with differen meson momentum

to measure the "distance" between light-cone DA and quasi DA quantitatively. Numerical results [17] of Δ_H from J/ψ and η_c meson show that when $P^z > 1.66\Lambda + 3.86\text{GeV}$, Δ_H is within 5%. Particularly for $\Lambda = 3\text{GeV} \approx M_{\eta_c} \approx M_{J/\psi}$ which is a characteristic scale, $P^z \approx 3M_{\eta_c} \approx 3M_{J/\psi}$ is large enough to bring quasi DA close to light-cone DA. This result is also consistent with the comparison between the first-inverse moments of quasi and light-cone DAs.

Acknowledgments This work was supported partially through the McCartor Fund Fellowship program under the auspices of International Light-Cone Advisory Committee (ILCAC).

References

1. Ji, X.: Parton Physics on Euclidean lattice. Phys. Rev. Lett. **110**, 262002 (2013). [arXiv:1305.1539 [hep-ph]]
2. Ji, X.: Sci. China Phys. Mech. Astron. **57**, 1407 (2014). doi:10.1007/s11433-014-5492-3. [arXiv:1404.6680 [hep-ph]]
3. Xiong, X., Ji, X., Zhang, J.H., Zhao, Y.: One-loop matching for parton distributions: nonsinglet case. Phys. Rev. D **90**(1), 014051 (2014). [arXiv:1310.7471 [hep-ph]]
4. Ma, Y.-Q., Qiu, J.-W.: Extracting Parton Distribution Functions from Lattice QCD Calculations, arXiv:1404.6860 [hep-ph]
5. Lin, H.-W., Chen, J.-W., Cohen, S.D., Ji, X.: Flavor Structure of the Nucleon Sea from Lattice QCD, arXiv:1402.1462 [hep-ph]
6. Alexandrou, C., Cichy, K., Drach, V., Garcia-Ramos, E., Hadjiyiannakou, K., Jansen, K., Steffens, F., Wiese, C.: Lattice calculation of parton distributions. Phys. Rev. D **92**(1), 014502 (2015). [arXiv:1504.07455 [hep-lat]]
7. Ji, X.-D.: Gauge-invariant decomposition of nucleon spin. Phys. Rev. Lett. **78**, 610 (1997). arXiv:hep-ph/9603249
8. Radyushkin, A.V.: Scaling limit of deeply virtual Compton scattering. Phys. Lett. B **380**, 417 (1996). arXiv:hep-ph/9604317
9. Ji, X.-D.: Deeply virtual Compton scattering. Phys. Rev. D **55**, 7114 (1997). arXiv:hep-ph/9609381
10. Radyushkin, A.V.: Asymmetric gluon distributions and hard diffractive electroproduction. Phys. Lett. B **385**, 333 (1996). arXiv:hep-ph/9605431
11. Ji, X., Schäfer, A., Xiong, X., Zhang, J .H.: One-loop matching for generalized parton distributions. Phys. Rev. D **92**(1), 014039 (2015). [arXiv:1506.00248 [hep-ph]]
12. Xiong, X., Zhang, J .H.: One-loop matching for transversity generalized parton distribution. Phys. Rev. D **92**(5), 054037 (2015). doi:10.1103/PhysRevD.92.054037. [arXiv:1509.08016 [hep-ph]]
13. Bodwin, G.T., Braaten, E., Lepage, G.P.: Rigorous QCD analysis of inclusive annihilation and production of heavy quarkonium. Phys. Rev. D **51**, 1125 (1995) [Phys. Rev. D **55**, 5853 (1997)] arXiv:hep-ph/9407339
14. Ma, J.P., Si, Z.G.: NRQCD factorization for twist-2 light-cone wave-functions of charmonia. Phys. Lett. B **647**, 419 (2007). arXiv:hep-ph/0608221
15. Bell, G., Feldmann, T.: Modelling light-cone distribution amplitudes from non-relativistic bound states. JHEP **0804**, 061 (2008). [arXiv:0802.2221 [hep-ph]]
16. Wang, X.P., Yang, D.: The leading twist light-cone distribution amplitudes for the S-wave and P-wave quarkonia and their applications in single quarkonium exclusive productions. JHEP **1406**, 121 (2014). [arXiv:1401.0122 [hep-ph]]
17. Jia, Y., Xiong, X.: arXiv:1511.04430 [hep-ph]

DOI 10.1007/978-3-319-50699-9_44
Reprinted from *Few-Body System* Journal, DOI 10.1007/s00601-016-1097-x

R. Pimentel · W. de Paula

Excited States of the Wick–Cutkosky Model with the Nakanishi Representation in the Light-Front Framework

Received: 18 February 2016 / Accepted: 24 March 2016 / Published online: 16 April 2016

Abstract The Wick–Cutkosky model is investigated using the framework of the Nakanishi Perturbative Integral Representation projected in the Light-Front hyperplane for an s-wave amplitude. We developed a new technique based on sucessive integration by parts of the Nakanishi Representation which enabled us to transform one of the integrations into a sum. With a set of boundary conditions it was possible to guess the format of the solution, which was in fact a distribution. The eigenequations obtained were the same as the originals of Cutkosky (1954). Finally, a numerical check confirmed that the final equation reproduce the same eigenvalues as the initial Bethe–Salpeter equation.

1 Introduction

The Bethe–Salpeter (BS) equation [2] was proposed to treat the problem of the bound state in a relativistic framework. However, since it's kernel had poles, the numerical solution could not be obtained using standard numerical algorithms for integral equations. In order to solve this difficulty, in 1954 Wick [3] and Cutkosky [1] published articles proposing new methods. In particular, the Wick rotation avoids the poles and, since then, this has been the canonical way to solve the BS equation. Nevertheless, Wick also proposed the strategy of using a particular type of integral representation for the amplitude of the BS equation with a massless interaction, and using a Dirac delta Ansatz for his weight function he was able to obtain a simpler equation for the massless case. Moreover, Cutkosky extended Wick's work to the full spectrum of the massless ladder interaction case. This model has been known as the Wick–Cutkosky model, specially important as a toy model because of it's simplicity.

In the seventies, Nakanishi explored the idea of an integral representation, very similar to the type used by Wick, for Feynman Diagrams and scattering amplitudes [4–6]

$$\phi(k) = \int_0^\infty d\gamma' \int_0^1 \prod_h dz'_h \delta\left(\sum_h z'_h - 1\right) \frac{g^{(n)}(\gamma', z')}{(\gamma' - \sum_h z'_h s_h - i\epsilon)^n}, \tag{1}$$

where s_h are scalar products of the external momenta. He analyzed this representation and demonstrated important properties such as analicity and uniqueness. Interestingly, although Nakanishi studied this formula for perturbative purposes, the initial ideia of Wick was to use it to solve the nonperturbative BS equation.

In 1995 Kusaka and Williams [7,8] observed that the Nakanishi representation was useful to solve the massive interaction BS equation in Minkowski space, but the method was still not very practical for numerical purposes. Thus, in 2006 Karmanov and Carbonell proposed a new technique based on the application of the Nakanishi [9–12] representation to the BS equation and then projecting it in the Light-Front hyperplane, and

R. Pimentel · W. de Paula (✉)
DCTA, Instituto Tecnológico de Aeronáutica, 12.228-900 São José dos Campos, Brazil
E-mail: wayne@ita.br

this resulted in an generalized integral equation with smooth kernels. Then Frederico, Salmè and Viviani extended the Light Front projected BS equation for scattering states and also developed a new method based on the uniqueness of the Nakanishi representation [13–15].

Following the recent success of the Light Front projected BS equation technique, we decided to look again to the massless interaction case, which was the first one described by an integral representation. We found that the solutions that before were thought of as an Ansatz could be obtained by using integration by parts and suitable boundary conditions, even for the excited states. Moreover we checked numerically the eigenvalues of the obtained equation and the original one.

2 Wick–Cutkosky Model with the Nakanishi Representation

The scalar BS equation with a ladder massless interaction is given by

$$\phi(k,p) = \frac{1}{m^2 - \left(\frac{p}{2}-k\right)^2}\frac{1}{m^2 - \left(\frac{p}{2}+k\right)^2} i g^2 \int \frac{d^4k'}{(2\pi)^4}\frac{\phi(k',p)}{(k-k')^2 + i\epsilon}, \tag{2}$$

where g is the coupling constant, p is the total momentum of the bound state and k is the relative momentum.

The BS amplitude $\phi(k,p)$ can be written as a three leg Nakanishi representation with one momentum on-shell (p) and two off-shell ($\frac{p}{2}-k$ and $\frac{p}{2}+k$),

$$\phi(k,p) = \frac{-i}{4\pi}\int_0^\infty d\gamma' \int_{-1}^{1} dz' \frac{g(\gamma',z')}{(\gamma' + \kappa^2 - k^2 - p\cdot k z' - i\epsilon)^3}. \tag{3}$$

where $\kappa^2 = m^2 - \frac{p^2}{4}$. Note that the symmetry $\phi(k_0,\mathbf{k},p) = \phi(-k_0,\mathbf{k},p)$, such as for an s-wave, is translated into $g(\gamma',z') = g(\gamma',-z')$ for the weight function in this representation. Substituting this representation in the previous BS equation and projecting the equation in the Light Front by integrating in k^- we obtain [9]

$$\int_0^\infty d\gamma' \frac{g(\gamma',z)}{\left(\gamma' + \gamma + z^2m^2 + (1-z^2)\kappa^2\right)^2} = \int_0^\infty d\gamma' \int_{-1}^{1} dz'\, V(\gamma,z;\gamma',z') g(\gamma',z'). \tag{4}$$

The Kernel of the integral equation is

$$\begin{aligned} V(\gamma,z;\gamma',z') = {}& \frac{\alpha m^2}{2\pi}\frac{1}{\left(\gamma + z^2m^2 + (1-z^2)\kappa^2\right)}\frac{1}{\left(\gamma' + z'^2m^2 + (1-z'^2)\kappa^2\right)} \\ & \times\left(\frac{\theta(z-z')}{\left(\gamma + \gamma'\frac{1-z}{1-z'} + z^2m^2 + (1-z^2)\kappa^2\right)}\frac{1-z}{1-z'}\right. \\ & \left. + \frac{\theta(z'-z)}{\left(\gamma + \gamma'\frac{1+z}{1+z'} + z^2m^2 + (1-z^2)\kappa^2\right)}\frac{1+z}{1+z'}\right), \end{aligned} \tag{5}$$

where we defined $\alpha = g^2/4\pi$. The BS equation can be rewritten in a suitable way as

$$\begin{aligned} & \int_0^\infty d\gamma' \frac{g(\gamma',z)}{\left(\gamma' + \gamma + z^2m^2 + (1-z^2)\kappa^2\right)^2} \\ & = \frac{\alpha m^2}{2\pi}\frac{1}{d_0(z)}\int_{-1}^{1} dz' \left[\omega(z,z')\theta(z-z') + \omega(-z,-z')\theta(z'-z)\right], \end{aligned} \tag{6}$$

where we introduced the auxiliary functions

$$\begin{aligned} \omega(z,z') &= \int_0^\infty d\gamma' \frac{g(\gamma',z')}{(\gamma' + a(z'))(\gamma' + c(\gamma,z,z'))}; \quad d_0(\gamma,z) = \gamma + z^2m^2 + (1-z^2)\kappa^2; \\ a(z') &= z'^2m^2 + (1-z'^2)\kappa^2; \quad c(\gamma,z,z') = \frac{1-z'}{1-z} d_0(\gamma,z). \end{aligned} \tag{7}$$

Now, we intend to transform the integration in γ' into a sum, using integration by parts. To do that, we first define the sequence $G^{(n)}\left(\gamma', z\right)$ as

$$G^{(0)}(\gamma', z) = g(\gamma', z); \quad G^{(n+1)}(\gamma', z) = -\int_{\gamma'}^{\infty} d\gamma''\, G^{(n)}(\gamma'', z); \quad G^{(n)}(\gamma', z) = \frac{\partial}{\partial \gamma'} G^{(n+1)}(\gamma', z). \quad (8)$$

First, we perform an integration by parts in γ' in the LHS of Eq. (4)

$$\begin{aligned} \int_0^\infty d\gamma' \frac{g(\gamma', z)}{\left(\gamma' + \gamma + z^2 m^2 + (1 - z^2)\kappa^2\right)^2} &= \int_0^\infty d\gamma' \frac{g(\gamma', z)}{(\gamma' + d_0(\gamma, z))^2} \\ &= \int_0^\infty d\gamma' \left(\frac{\partial}{\partial \gamma'} G^{(1)}(\gamma', z)\right) \frac{1}{(\gamma' + d_0(\gamma, z))^2} \\ &= G^{(1)}(0, z) \frac{1}{d_0(\gamma, z)^2} \\ &\quad - \int_0^\infty d\gamma'\, G^{(1)}(\gamma', z) \frac{\partial}{\partial \gamma'} \left(\frac{1}{(\gamma' + d_0(\gamma, z))^2}\right). \end{aligned} \quad (9)$$

Each integration by parts generates a new term for the sum. In general, the relationship that generates the n-th term of the sum is

$$\begin{aligned} &\int_0^\infty d\gamma'\, G^{(n)}(\gamma', z) \frac{\partial^{(n)}}{\partial \gamma'^{(n)}} \left(\frac{1}{(\gamma' + d_0(\gamma, z))^2}\right) \\ &= \int_0^\infty d\gamma' \left(\frac{\partial}{\partial \gamma'} G^{(n+1)}(\gamma', z)\right) \frac{\partial^{(n)}}{\partial \gamma'^{(n)}} \frac{1}{(\gamma' + d_0(\gamma, z))^2} \\ &= \left(G^{(n+1)}(\gamma', z) \frac{\partial^{(n)}}{\partial \gamma'^{(n)}} \left(\frac{1}{(\gamma' + d_0(\gamma, z))^2}\right)\right)\Bigg|_0^\infty \\ &\quad - \int_0^\infty d\gamma'\, G^{(n+1)}(\gamma', z) \frac{\partial^{(n+1)}}{\partial \gamma'^{(n+1)}} \left(\frac{1}{(\gamma' + d_0(\gamma, z))^2}\right). \end{aligned} \quad (10)$$

Thus, after performing n integrations by parts we have

$$\begin{aligned} &\int_0^\infty d\gamma' \frac{g(\gamma', z)}{\left(\gamma' + \gamma + z^2 m^2 + (1 - z^2)\kappa^2\right)^2} \\ &= \sum_{i=1}^{n} (-1)^{i-1} \left(G^{(i)}(\gamma', z) \frac{\partial^{(i-1)}}{\partial \gamma'^{(i-1)}} \left(\frac{1}{(\gamma' + d_0(\gamma, z))^2}\right)\right)\Bigg|_0^\infty \\ &\quad + (-1)^n \int_0^\infty d\gamma'\, G^{(n)}(\gamma', z) \frac{\partial^{(n)}}{\partial \gamma'^{(n)}} \left(\frac{1}{(\gamma' + d_0(\gamma, z))^2}\right). \end{aligned} \quad (11)$$

The term in the limit of large n vanishes and each term of the sum is proportional to $G^{(i)}(0, z)$. So, we define this term as a sequence of functions in the variable z

$$b_i(z) = G^{(i)}(0, z) \quad (12)$$

Finally, in the limit of $n \to \infty$

$$\int_0^\infty d\gamma' \frac{g(\gamma', z)}{\left(\gamma' + \gamma + z^2 m^2 + (1 - z^2)\kappa^2\right)^2} = \sum_{i=1}^{\infty} i! \frac{b_i(z)}{d_0(\gamma, z)^{i+1}}. \quad (13)$$

Note that we successfully rewrote the integration in γ' as a sum. We now intend to do the same for the RHS of Eq. (4). In the following we give some details of the derivation.

$$\begin{aligned}
&\int_0^\infty d\gamma' \frac{g(\gamma',z')}{(\gamma'+a(z'))(\gamma'+c(\gamma,z,z'))} \\
&= \sum_{j=1}^{\infty}(-1)^{j-1}\left(G^{(j)}(\gamma',z')\frac{\partial^{(j-1)}}{\partial\gamma'^{(j-1)}}\left(\frac{1}{(\gamma'+a(z'))\,(\gamma'+c(\gamma,z,z'))}\right)\right)\Bigg|_0^\infty \\
&= \sum_{j=1}^{\infty}(-1)^{j-1}b_j(z')\sum_{n=0}^{j-1}\binom{j-1}{n}\frac{\partial^{(j-n-1)}}{\partial\gamma'^{(j-n-1)}}\left(\frac{1}{\gamma'+a(z')}\right)\frac{\partial^{(n)}}{\partial\gamma'^{(n)}}\left(\frac{1}{\gamma'+c(\gamma,z,z')}\right) \\
&= \sum_{j=1}^{\infty}b_j(z')\sum_{n=0}^{j-1}\frac{(j-1)!}{n!(j-n-1)!}\frac{(j-n-1)!}{a(z')^{j-n}}\frac{n!}{c(\gamma,z,z')^{n+1}} \\
&= \sum_{j=1}^{\infty}b_j(z')\sum_{n=0}^{j-1}\frac{(j-1)!}{a(z')^{j-n}c(\gamma,z,z')^{n+1}} \\
&= \sum_{i=1}^{\infty}\frac{1}{c(\gamma,z,z')^{i}}\sum_{j=i}^{\infty}\frac{(j-1)!b_j(z')}{a(z')^{j-i+1}} \\
&= \sum_{i=1}^{\infty}\left(\frac{1-z}{(1-z')d_0(z)}\right)^{i}\sum_{j=i}^{\infty}\frac{(j-1)!b_j(z')}{a(z')^{j-i+1}}
\end{aligned} \tag{14}$$

Collecting the terms, the BS equation for the Wick–Cutkosky model can be expressed as

$$\begin{aligned}
\sum_{i=1}^{\infty} i!\frac{b_i(z)}{d_0(\gamma,z)^{i+1}} &= \frac{\alpha m^2}{2\pi}\frac{1}{d_0(z)}\int_{-1}^{1}dz'\sum_{i=1}^{\infty}\left(\frac{1-z}{(1-z')d_0(z)}\right)^{i}\sum_{j=i}^{\infty}\frac{(j-1)!b_j(z')}{a(z')^{j-i+1}}\theta(z-z') \\
&\quad + z\to -z \text{ and } z'\to -z'
\end{aligned} \tag{15}$$

Note that for $\gamma\to\infty \Rightarrow d_0\to\infty$ and therefore the small i terms dominates the series. Then we can match the $\frac{1}{d_0^i}$ terms of the series and we have, since $a(z)$ and $b_i(z)$ are even:

$$b_i(z) = \frac{\alpha m^2}{2\pi}\int_{-1}^{1}dz'\frac{1}{i!}\left(\left(\frac{1-z}{1-z'}\right)^{i}\theta(z-z') + \left(\frac{1+z}{1+z'}\right)^{i}\theta(z'-z)\right)\sum_{j=i}^{\infty}\frac{(j-1)!b_j(z')}{a(z')^{j-i+1}}. \tag{16}$$

To solve the eigenequation we set $b_i(z)=0$ if $i>i_{max}$ for some i_{max}. Note that this is consistent because all the equations for $i>i_{max}$ becomes trivially solved, since they are homogeneous and each b_i depends only on b_j with $j>i$. Thus, it is only necessary to solve the equations with $i\leqslant i_{max}$. Since the resulting matrix is triangular by blocks, the eigenvalues are solely determined by the $i=j=i_{max}$ equations:

$$b_i(z) = \frac{\alpha m^2}{2\pi}\int_{-1}^{1}dz'\frac{1}{i}\left(\left(\frac{1-z}{1-z'}\right)^{i}\theta(z-z') + \left(\frac{1+z}{1+z'}\right)^{i}\theta(z'-z)\right)\frac{b_i(z')}{a(z')}, \tag{17}$$

which is exactly the Wick–Cutkosky equation originally obtained. It is known that the support for the Nakanishi weight function for the Wick–Cutkosky model is only $\gamma'=0$. Previously it was used the Ansatz $g(\gamma',z')=\delta(\gamma')f(z')$ which is equivalent to set $b_i(z')=0$ for $i>1$ in the present formalism. To fully reconstruct $g(\gamma',z')$ from the $b_i(z')$ we propose the following Ansatz:

$$g(\gamma',z') = \sum_{i=1}^{i_{max}} b_i(z')\delta^{(i-1)}(\gamma'). \tag{18}$$

This distribution has support only in $\gamma' = 0$ and clearly reproduces Eqs. (13) and (14). This happens because the derivative of delta acts on the kernel in the same way as when one performs the integration by parts , thus this is the solution for the Nakanishi weight function. At this point we notice that the integration by parts is ill-defined since the weight function is a distribution and not a function. To be mathematically precise, first one should set a small mass μ for the interaction particle and then take the limit of $\mu \to 0$ in the end of the calculation. To reconstruct the BS amplitude we can use the Nakanishi Perturbative Integral Representation (3). Using the weight function (18) in the representation (3) we have

$$\phi(k, p) = \frac{-i}{4\pi} \sum_{n=1}^{i_{max}} \frac{(n+1)!}{3} \int_{-1}^{1} dz' \frac{b_n(z')}{(\kappa^2 - k^2 - p \cdot kz' - i\epsilon)^{n+2}}. \tag{19}$$

From (19) we can also obtain the light-front wave function using the relation

$$\psi(\mathbf{k}_\perp, z) = \frac{(\omega.k_1)(\omega.k_2)}{\pi(\omega.p)} \int_{-\infty}^{\infty} \phi(k + \beta\omega, p) d\beta, \tag{20}$$

where $\omega = (1, 0, 0, -1)$. Finally, we have

$$\psi(\mathbf{k}_\perp, z) = \frac{1 - z^2}{8\sqrt{\pi}} \sum_{i=1}^{i_{max}} i! \frac{b_i(z)}{\left(\mathbf{k}_\perp^2 + m^2 - (1 - z^2)\frac{M^2}{4}\right)^{i+1}}. \tag{21}$$

3 Numerical Results

In this section we compare the numerical values of the first four eigenvalues by solving the Bethe–Salpeter equation for zero mass exchange, Eq. (4), and the ones obtained from Eq. (17), which is obtained using the method described in the last section.

We solve the BS equation, Eq. (4), using a Laguerre basis in γ and γ', and a Legendre basis in z and z'. Moreover, with five Laguerre basis functions and five Legendre polynomials, we obtain the eigenvalues of the ground state ($\alpha_1^{(\gamma',z')}$) and the first excited state ($\alpha_2^{(\gamma',z')}$) for various bounding energies B/m. Also, we solve Eq. (17) using a Legendre basis in z and z', obtaining the eigenvalues of the ground state ($\alpha_1^{(z')}$) and the first excited state ($\alpha_2^{(z')}$). The results are in Table 1.

For the ground state and the first excited state the numerical results has a relative error of about 1 %, which support the method presented in this work.

4 Summary

The Wick–Cutkosky model was reexamined using the new light front projected Nakanishi method.

In this case, the expansion in basis functions is not efficient, because the solution is a distribution. Thus, it was necessary to find an alternative way to deal with it. Our strategy was to transform the integration in γ' into a sum, using integration by parts and a suitable boundary condition for the solution. This enabled us not only to reproduce the equations originally obtained by Cutkosky, but also to show that the weight-function for

Table 1 Values of $\alpha = g^2/4\pi$ as a function of binding energy B/m for a massless scalar exchange for the ground state and the first excited state

B/m	$\alpha_1^{(\gamma',z')}$	$\alpha_1^{(z')}$	$\alpha_2^{(\gamma',z')}$	$\alpha_2^{(z')}$
0.1	1.11	1.12	2.90	2.93
0.2	1.78	1.79	4.90	4.85
0.3	2.34	2.35	6.58	6.53
0.4	2.84	2.84	8.09	8.05
0.5	3.29	3.30	9.44	9.42

The first and third column shows the results for Eq. (4). The second and fourth column shows the results for Eq. (17)

the amplitude reduces to a distribution in the massless limit. So, we were able to use these weight-functions to write the final formulas for the Bethe–Salpeter amplitude and the light-front wave function.

Moreover, we checked the final and initial equations numerically, expanding the solution in Laguerre and Legendre basis, and we obtained the same eigenvalues for the ground state and for the first excited states in both equations.

The integration-by-part method may also be useful in the calculation of the BS equation with a small interacting mass, specially if used with the uniqueness equation. This is importante because the basis expansion performs better when the interacting mass is bigger and thus the weight function has a more distributed support in γ'.

Following this procedure, the next step is to obtain the eigenvalues of higher excited states and their corresponding wave functions, which is a work in progress.

Acknowledgments We acknowledge partial support from the Brazilian agencies CAPES and CNPq.

References

1. Cutkosky, R.E.: Solutions of a Bethe–Salpeter equation. Phys. Rev. **96**, 1135 (1954)
2. Salpeter, E.E., Bethe, H.A.: A relativistic equation for bound-state problems. Phys. Rev. **84**, 1232 (1951)
3. Wick, G.C.: Properties of Bethe–Salpeter wave functions. Phys. Rev. **96**, 1124 (1954)
4. Nakanishi, N.: Partial-wave Bethe–Salpeter equation. Phys. Rev. **130**, 1230 (1963)
5. Nakanishi, N.: A general survey of the theory of the Bethe–Salpeter equation. Prog. Theor. Phys. Suppl. **43**, 1 (1969)
6. Nakanishi, N.: Graph Theory and Feynman Integrals. Gordon and Breach, New York (1971)
7. Kusaka, K., Williams, A.G.: Solving the Bethe–Salpeter equation for scalar theories in Minkowski space. Phys. Rev. D **51**, 7026 (1995)
8. Kusaka, K., Simpson, K., Williams, A.G.: Solving the Bethe–Salpeter equation for bound states of scalar theories in Minkowski space. Phys. Rev. D **56**, 5071 (1997)
9. Karmanov, V.A., Carbonell, J.: Solving Bethe–Salpeter equation in Minkowski space. Eur. Phys. J. A **27**, 1 (2006)
10. Carbonell, J., Karmanov, V.A.: Cross-ladder effects in Bethe–Salpeter and light-front equations. Eur. Phys. J. A **27**, 11 (2006)
11. Carbonell, J., Karmanov, V.A., Mangin-Brinet, M.: Electromagnetic form factor via Bethe–Salpeter amplitude in Minkowski space. Eur. Phys. J. A **39**, 53 (2009)
12. Carbonell, J., Karmanov, V.A.: Solving Bethe–Salpeter equation for two fermions in Minkowski space. Eur. Phys. J. A **46**, 387 (2010)
13. Frederico, T., Salmè, G., Viviani, M.: Two-body scattering states in Minkowski space and the Nakanishi integral representation onto the null plane. Phys. Rev. D **85**, 036009 (2012)
14. Frederico, T., Salmè, G.: Projecting the Bethe–Salpeter equation onto the light-front and back: a short review. Few Body Syst. **49**, 163 (2011)
15. Frederico, T., Salmè, G., Viviani, M.: Quantitative studies of the homogeneous Bethe–Salpeter equation. Phys. Rev. D **89**, 016010 (2014)

DOI 10.1007/978-3-319-50699-9_45
Reprinted from *Few-Body System* Journal, DOI 10.1007/s00601-016-1098-9

J. R. Hiller

Pauli–Villars Regularization of Non-Abelian Gauge Theories

Received: 29 December 2015 / Accepted: 26 March 2016 / Published online: 5 April 2016

Abstract As an extension of earlier work on QED, we construct a BRST-invariant Lagrangian for SU(N) Yang–Mills theory with fundamental matter, regulated by the inclusion of massive Pauli–Villars (PV) gluons and PV quarks. The underlying gauge symmetry for massless PV gluons is generalized to accommodate the PV-index-changing currents that are required by the regularization. Auxiliary adjoint scalars are used, in a mechanism due to Stueckelberg, to attribute mass to the PV gluons and the PV quarks. The addition of Faddeev–Popov ghosts then establishes a residual BRST symmetry. Although there are drawbacks to the approach, in particular the computational load of a large number of PV fields and a nonlocal interaction of the ghost fields, this formulation could provide a foundation for renormalizable nonperturbative solutions of light-front QCD in an arbitrary covariant gauge.

1 Introduction

Nonperturbative Hamiltonian approaches to quantum field theories [1–8] in more than two dimensions require regularization. The standard approach to regularization of non-Abelian gauge theories is dimensional regularization [9–11], but this is inherently perturbative. For Abelian theories, Pauli–Villars (PV) regularization has proven useful [12–14]. However, the ordinary PV regularization of non-Abelian theories fails. Gauge invariance is violated, blocking any hope of BRST invariance, which confounds proofs of renormalizability [15–18]. To implement a PV regularization [19],[1] we repair gauge invariance and then break it properly with a gauge-fixing term. Ghost fields [21] are added to build a BRST invariance.

The causes of gauge non-invariance in ordinary PV regularization are the PV gluon mass, the non-degenerate PV quark masses, and the PV-index changing currents for null interactions. A mass is acceptable for PV photons in an Abelian theory [22–24], but the approach used does not generalize to non-Abelian theories. Null interactions (meaning interactions that involve only null combinations of positive and negative metric fields) are necessary to provide the subtractions that regulate theory. So, how can one restore gauge invariance?

The Higgs mechanism is, of course, one possible approach, but investigations of this invariably found only null combinations of gluon fields to be massless, rather than a massless physical gluon. There is, however, another mechanism for generating masses [25–28], by coupling the PV gluons and PV quarks to a PV scalar, with all removed from the spectrum in the infinite-mass limit. The mixing of currents can be overcome by extension of the gauge transformation to include mixing PV fields; the original gauge transformation is obtained when the fields are removed from the spectrum. The construction is done is such a way as to facilitate

[1] An alternative proposal for PV regularization has been made with the purpose of equivalence with covariant perturbation theory [20]; however, the use of additional cutoffs and limiting procedures make nonperturbative applications difficult beyond the simplest Fock representation [12].

J. R. Hiller (✉)
Department of Physics and Astronomy, University of Minnesota-Duluth, Duluth, MN 55812, USA
E-mail: jhiller@d.umn.edu

quantization in an arbitrary covariant gauge, as has been done previously for QED [14]. The remaining gauge invariance can be tested by studying the dependence of physical quantities on the gauge-fixing parameter.

In the remainder of this paper, we present the PV-regulated Lagrangian term by term, along with the extended gauge transformation and the BRST transformations. We then discuss light-front quantization in an arbitrary covariant gauge and the light-front coupled-cluster method [29] for solution of the Hamiltonian eigenvalue problem in Fock space.

2 A Pauli–Villars Regulated Lagrangian

For the purposes of discussion, we divide the PV-regulated Lagrangian into four terms

$$\mathscr{L} = \mathscr{L}_{\text{massless}} + \mathscr{L}_{\text{gluon}} + \mathscr{L}_{\text{quark}} + \mathscr{L}_{\text{ghost}}. \tag{1}$$

Here $\mathscr{L}_{\text{massless}}$ is a gauge-invariant Lagrangian for massless gluons and quarks, with null interactions; $\mathscr{L}_{\text{gluon}}$ is the mass and gauge-fixing term for gluons and auxiliary scalars; $\mathscr{L}_{\text{quark}}$ is the mass term for quarks; and $\mathscr{L}_{\text{ghost}}$ is the Faddeev–Popov ghost term. The first is given by

$$\mathscr{L}_{\text{massless}} = -\frac{1}{4}\sum_k r_k F_{ak}^{\mu\nu} F_{ak\mu\nu} + \sum_i s_i \bar{\psi}_i i\gamma^\mu \partial_\mu \psi_i + g\sum_{ijk} \beta_i\beta_j\xi_k \bar{\psi}_i \gamma^\mu T_a A_{ak\mu}\psi_j, \tag{2}$$

where the field tensor is

$$F_{ak}^{\mu\nu} = \partial^\mu A_{ak}^\nu - \partial^\nu A_{ak}^\mu - r_k\xi_k g f_{abc}\sum_{lm}\xi_l\xi_m A_{bl}^\mu A_{cm}^\nu. \tag{3}$$

The indices are k for (PV) gluons and i, j for (PV) quarks, with values of zero for the physical fields. The gauge transformations of the fields are

$$A_{ak}^\mu \longrightarrow A_{ak}^\mu + \partial^\mu \Lambda_{ak} + r_k\xi_k g f_{abc}\Lambda_b A_c^\mu, \tag{4}$$

$$\psi_i \longrightarrow \psi_i + igs_i\beta_i T_a \Lambda_a \psi, \tag{5}$$

with $\Lambda_a \equiv \sum_k \xi_k \Lambda_{ak}$ and $[T_a, T_b] = i f_{abc} T_c$.

The regularization is provided by the null field combinations:

$$A_a^\mu \equiv \sum_k \xi_k A_{ak}^\mu, \quad \psi \equiv \sum_i \beta_i \psi_i, \tag{6}$$

with $\sum_k r_k\xi_k^2 = 0$ and $\sum_i s_i\beta_i^2 = 0$. Here r_k and s_i equal ± 1 [e.g. $(-1)^k$ and $(-1)^i$], A_a^μ is Abelian, and ψ is gauge invariant. With these definitions, this term of the Lagrangian can be written as

$$\mathscr{L}_{\text{massless}} = -\frac{1}{4}\sum_k r_k \left(\partial^\mu A_{ak}^\nu - \partial^\nu A_{ak}^\mu\right)^2 + g f_{abc}\partial^\mu A_a^\nu A_{b\mu} A_{c\nu} + \sum_i s_i\bar{\psi}_i i\gamma^\mu\partial_\mu\psi_i + g\bar{\psi}\gamma^\mu T_a A_{a\mu}\psi. \tag{7}$$

The free terms are those for fields with metrics r_k and s_i, and the interaction terms involve only null fields. The four-gluon interaction is implicit in the infinite-PV-mass limit through a contraction of two three-gluon interactions.

The null field combinations provide the necessary two PV subtractions to regulate any loop. For example, consider a loop with vertices from interactions of the form $g\bar{\psi}\gamma^\mu T_a A_{a\mu}\psi$. Assume that there is one gluon contraction and one quark contraction to form the loop. The contraction of the kth gluon field yields the metric signature r_k and contraction of the ith quark field, s_i. The coupling coefficients are selected as ξ_k, ξ_k', β_i, and β_i'. The loop contribution then contains the factors $r_k\xi_k\xi_k'$ and $s_i\beta_i\beta_i'$. These factors provide the two subtractions, because on their own they sum to zero.

The gluon mass and gauge-fixing term is

$$\mathscr{L}_{\text{gluon}} = \frac{1}{2}\sum_k r_k\left(\mu_k A_{ak}^\mu - \partial^\mu\phi_{ak}\right)^2 - \frac{\zeta}{2}\sum_k r_k\left(\partial_\mu A_{ak}^\mu + \frac{\mu_k}{\zeta}\phi_{ak}\right)^2. \tag{8}$$

The first term is gauge-invariant, and the second, gauge-fixing. The scalar fields ϕ_{ak} obey the gauge transformation

$$\phi_{ak} \longrightarrow \phi_{ak} + \mu_k \Lambda_{ak} + \mu_k r_k \xi_k g f_{abc} \int^{x} dx'_\mu \Lambda_b(x') A_c^\mu(x'). \tag{9}$$

The line integral allows the derivative to transform as

$$\partial^\mu \phi_{ak} \longrightarrow \partial^\mu \phi_{ak} + \mu_k \partial^\mu \Lambda_{ak} + \mu_k r_k \xi_k g f_{abc} \Lambda_b A_c^\mu. \tag{10}$$

This part of the Lagrangian can be reduced to the form

$$\mathscr{L}_{\text{gluon}} = \frac{1}{2}\sum_k r_k \mu_k^2 \left(A_{ak}^\mu\right)^2 - \frac{\zeta}{2}\sum_k r_k \left(\partial_\mu A_{ak}^\mu\right)^2 + \frac{1}{2}\sum_k r_k \left[\left(\partial_\mu \phi_{ak}\right)^2 - \frac{\mu_k^2}{\zeta}\phi_{ak}^2\right], \tag{11}$$

where the gluon mass μ_k is explicit and the scalar is given a mass $\mu_k/\sqrt{\zeta}$ and inherits the metric r_k. This is a non-Abelian extension of a Stueckelberg mechanism [25,26,28].

The quark mass term is

$$\mathscr{L}_{\text{quark}} = \sum_i s_i m_i \left(\bar{\psi}_i + ig\frac{s_i\beta_i}{\mu_{\text{PV}}}\widetilde{\phi}_a \bar{\psi} T_a\right)\left(\psi_i - ig\frac{s_i\beta_i}{\mu_{\text{PV}}}\widetilde{\phi}_a T_a \psi\right), \tag{12}$$

with the combination

$$\widetilde{\phi}_a \equiv \sum_k \xi_k \frac{\mu_{\text{PV}}}{\mu_k}\phi_{ak} \tag{13}$$

made null by the additional constraint $\sum_k r_k \frac{\xi_k^2}{\mu_k^2} = 0$ and with $\mu_{\text{PV}} \equiv \max_k \mu_k$. The gauge transformation of the combination is Abelian:

$$\widetilde{\phi}_a \longrightarrow \widetilde{\phi}_a + \mu_{\text{PV}}\Lambda_a. \tag{14}$$

To make all couplings null, we define

$$\widetilde{\psi} = \sum_i \beta_i \frac{m_i}{m_{\text{PV}}}\psi_i, \tag{15}$$

with $m_{\text{PV}} \equiv \max_i m_i$, and impose the constraints $\sum_i s_i m_i^2 \beta_i^2 = 0$ and $\sum_i s_i m_i \beta_i^2 = 0$. The second constraint makes $\widetilde{\psi}$ and ψ mutually null, to allow couplings between these combinations, and cancels the quartic coupling term. The quark mass term then simplifies to

$$\mathscr{L}_{\text{quark}} = -\sum_i s_i m_i \bar{\psi}_i \psi_i - ig\frac{m_{\text{PV}}}{\mu_{\text{PV}}}\left[\bar{\psi} T_a \widetilde{\phi}_a \widetilde{\psi} - \bar{\widetilde{\psi}} T_a \widetilde{\phi}_a \psi\right] \tag{16}$$

A standard construction [30] yields the ghost term [21]

$$\mathscr{L}_{\text{ghost}} = \sum_k r_k \partial_\mu \bar{c}_{ak} \partial^\mu c_{ak} - \sum_k r_k \frac{\mu_k^2}{\zeta}\bar{c}_{ak} c_{ak} + g f_{abc}\left[\partial_\mu \bar{c}_a c_b A_c^\mu - \frac{\mu_{\text{PV}}^2}{\zeta}\bar{\bar{c}}_a \int^{x} dx'_\mu c_b(x') A_c^\mu(x')\right], \tag{17}$$

for ghosts c_{ak} and anti-ghosts $\bar{c}_{ak}$, with null combinations defined as

$$c_a \equiv \sum_k \xi_k c_{ak}, \quad \bar{c}_a \equiv \sum_k \xi_k \bar{c}_{ak}, \quad \bar{\bar{c}}_a \equiv \sum_k \xi_k \frac{\mu_k^2}{\mu_{\text{PV}}^2}\bar{c}_{ak}. \tag{18}$$

For these to be (mutually) null, we require $\sum_k r_k \mu_k^2 \xi_k^2 = 0$ and $\sum_k r_k \mu_k^4 \xi_k^2 = 0$

As a summary of the various constraints, the following need to be satisfied for the adjoint fields:

$$\sum_k r_k \xi_k^2 = 0, \quad \sum_k r_k \frac{\xi_k^2}{\mu_k^2} = 0, \quad \sum_k r_k \mu_k^2 \xi_k^2 = 0, \quad \sum_k r_k \mu_k^4 \xi_k^2 = 0, \tag{19}$$

and for the quark fields:

$$\sum_i s_i \beta_i^2 = 0, \quad \sum_i s_i m_i^2 \beta_i^2 = 0, \quad \sum_i s_i m_i \beta_i^2 = 0. \tag{20}$$

For the PV masses to be chosen independently, these constraints require four PV gluons, four PV ghosts and antighosts, five PV scalars, and three PV quarks. For pure Yang–Mills theory, the number of PV scalars is four, with the $k = 0$ field dropped and $\mu_0 = 0$; the fields ϕ_{a0} are used only in splitting the masses of the PV quarks. In either case, the number of PV fields translates to a large, but necessary computational load.

The BRST transformations are

$$\delta A^\mu_{ak} = \varepsilon \partial^\mu c_{ak} + \varepsilon r_k \xi_k g f_{abc} c_b A^\mu_c, \tag{21}$$

$$\delta \psi_i = i\varepsilon g s_i \beta_i T_a c_a \psi, \quad \delta \bar{\psi}_i = -i\varepsilon g s_i \beta_i \bar{\psi} T_a c_a, \tag{22}$$

$$\delta \phi_{ak} = \varepsilon \mu_k c_{ak} + \varepsilon r_k \xi_k \mu_k g f_{abc} \int^x dx'_\mu c_b(x') A^\mu_c(x'), \tag{23}$$

$$\delta \partial^\mu \phi_{ak} = \varepsilon \mu_k \partial^\mu c_{ak} + \varepsilon r_k \xi_k \mu_k g f_{abc} c_b A^\mu_c, \tag{24}$$

$$\delta \bar{c}_{ak} = -\zeta \varepsilon \left(\partial_\mu A^\mu_{ak} + \frac{\mu_k}{\zeta} \phi_{ak} \right), \quad \delta c_{ak} = \frac{1}{2} \varepsilon r_k \xi_k g f_{abc} c_b c_c, \tag{25}$$

with ε a real Grassmann constant, for which $\varepsilon^2 = 0$. For the various null combinations, we then find

$$\delta A^\mu_a = \varepsilon \partial^\mu c_a, \quad \delta \widetilde{\phi}_a = \varepsilon \mu_{\rm PV} c_a, \quad \delta c_a = 0, \quad \delta \psi = 0, \quad \delta \widetilde{\psi} = 0. \tag{26}$$

The full Lagrangian is invariant with respect to these transformations.

3 Massive Vector Quantization

The construction of the Lagrangian makes no reference to a choice of quantization coordinates; however, the construction was done in such a way as to keep open the possibility of light-front quantization in an arbitrary covariant gauge. To see how this quantization can take place, consider the case of QED [14]. The Lagrangian is $\mathscr{L} = -\frac{1}{4}F^2 + \frac{1}{2}\mu A^2 - \frac{1}{2}\zeta(\partial \cdot A)^2$, which yields the field equation: $(\Box + \mu^2)A_\mu - (1-\zeta)\partial_\mu(\partial \cdot A) = 0$. The light-front Hamiltonian density is

$$\mathscr{H} = \mathscr{H}|_{\zeta=1} + \frac{1}{2}(1-\zeta)(\partial \cdot A)\left(\partial \cdot A - 2\partial_- A^+ - 2\partial_\perp \cdot \mathbf{A}_\perp\right), \tag{27}$$

with

$$\mathscr{H}|_{\zeta=1} = \frac{1}{2} \sum_{\mu=0}^{3} \varepsilon^\mu \left[(\partial_\perp A^\mu)^2 + \mu^2 (A^\mu)^2 \right] \tag{28}$$

and $\varepsilon^\mu = (-1, 1, 1, 1)$. From this we obtain the light-front Hamiltonian

$$\mathscr{P}^- = \int d\underline{x} \mathscr{H}|_{x^+=0} = \int d\underline{k} \sum_\lambda \varepsilon^\lambda \frac{k_\perp^2 + \mu^{(\lambda)2}}{k^+} a^{(\lambda)\dagger}(\underline{k}) a^{(\lambda)}(\underline{k}) \tag{29}$$

with $\underline{k} = (k^+, \mathbf{k}_\perp)$, $\mu^{(\lambda)} = \mu$ for $\lambda = 1, 2, 3$, but $\mu^{(0)} = \tilde{\mu} \equiv \mu/\sqrt{\zeta}$. The nonzero commutator is

$$[a^{(\lambda)}(\underline{k}), a^{(\lambda')\dagger}(\underline{k}')] = \varepsilon^\lambda \delta_{\lambda\lambda'} \delta(\underline{k} - \underline{k}'). \tag{30}$$

The normal-mode expansion for the field is

$$A_\mu(x) = \int \frac{d\underline{k}}{\sqrt{16\pi^2 k^+}} \left\{ \sum_{\lambda=1}^{3} e_\mu^{(\lambda)}(\underline{k}) \left[a^{(\lambda)}(\underline{k}) e^{-ik\cdot x} + a^{(\lambda)\dagger}(\underline{k}) e^{ik\cdot x} \right] \right.$$
$$\left. + e_\mu^{(0)}(\underline{k}) \left[a^{(0)}(\underline{k}) e^{-i\tilde{k}\cdot x} + a^{(0)\dagger}(\underline{k}) e^{i\tilde{k}\cdot x} \right] \right\}, \quad (31)$$

with polarization vectors

$$e^{(1,2)}(\underline{k}) = (0, 2\hat{e}_{1,2}\cdot \mathbf{k}_\perp/k^+, \hat{e}_{1,2}), \quad e^{(3)}(\underline{k}) = \frac{1}{\mu}((k_\perp^2 - \mu^2)/k^+, k^+, \mathbf{k}_\perp), \quad e^{(0)}(\underline{k}) = \tilde{k}/\mu. \quad (32)$$

Here $\underline{\tilde{k}} = \underline{k}$, $\tilde{k}^- = (k_\perp^2 + \tilde{\mu}^2)/k^+$, and $\hat{e}_{1,2}$ are transverse unit vectors. These vectors satisfy: $k \cdot e^{(\lambda)} = 0$ and $e^{(\lambda)} \cdot e^{(\lambda')} = -\delta_{\lambda\lambda'}$ for $\lambda, \lambda' = 1, 2, 3$. The first term in A_μ satisfies $(\Box + \mu^2)A_\mu = 0$ and $\partial \cdot A = 0$. The $\lambda = 0$ term violates each, but the field equation is satisfied [25,31].

The nondynamical components of the fermion fields satisfy the constraints

$$is_i\partial_-\psi_{i-} + eA_-\beta_i \sum_j \psi_{j-} = (i\gamma^0\gamma^\perp)\left[s_i\partial_\perp\psi_{i+} - ieA_\perp\beta_i \sum_j \psi_{j+} \right] - s_i m_i \gamma^0 \psi_{i-}. \quad (33)$$

When multiplied by $s_i\beta_i$ and summed over i, this becomes

$$i\partial_-\psi_- = (i\gamma^0\gamma^\perp)\partial_\perp\psi_+ - \gamma^0 \sum_i \beta_i m_i \psi_{i+}, \quad (34)$$

which is the same constraint as for a free fermion field, in any gauge. Because only null combinations are coupled to the gauge fields, the individual fermion fields can be quantized as though free, and the constrained null field ψ_- constructed from them [12]. Quantization can then proceed without need of light-cone gauge.

4 Light-Front Coupled-Cluster Method

Given a field-theoretic light-front Hamiltonian $\mathcal{P}^-$, we wish to solve the fundamental eigenvalue problem $\mathcal{P}^-|\psi\rangle = \frac{M^2+P_\perp^2}{P^+}|\psi\rangle$. The LFCC method [29] is designed to do so in terms of a Fock-state expansion but without the usual truncation of Fock space. Instead, write the eigenstate as $|\psi\rangle = \sqrt{Z}e^T|\phi\rangle$. Here Z controls the normalization, which is fixed as $\langle\psi'|\psi\rangle = \delta(\underline{P}' - \underline{P})$. The ket $|\phi\rangle$ is the valence state, with norm $\langle\phi'|\phi\rangle = \delta(\underline{P}' - \underline{P})$. The operator T contains terms that only increase particle number; it does, however, conserve the quantum numbers of the full state, such as J_z, light-front momentum $\underline{P}$, and charge. Because p^+ is always positive, T must include annihilation operators, and powers of T include contractions. With construction of the effective Hamiltonian $\overline{\mathcal{P}^-} = e^{-T}\mathcal{P}^- e^T$ and the definition of a projection P_v onto the valence Fock sector, we have the coupled system:

$$P_v\overline{\mathcal{P}^-}|\phi\rangle = \frac{M^2 + P_\perp^2}{P^+}|\phi\rangle, \quad (1 - P_v)\overline{\mathcal{P}^-}|\phi\rangle = 0 \quad (35)$$

This is still an infinite system of equations. The LFCC approximation is to truncate T at a fixed increase in particle count, but not truncate the exponential of T. The projection $(1 - P_v)$ is truncated to a consistent set of Fock sectors above the valence sector, enough to have a finite, but sufficient system of (nonlinear) equations for the functions in T and in the valence state. The leading approximation of T for QCD would include quark → quark + gluon, gluon → quark + antiquark, and gluon → gluon + gluon transitions.

The LFCC approach requires no Fock-space truncation and no sector dependence or spectator dependence of self-energies or bare parameters. It is systematically improvable by adding terms to T and sectors to $(1 - P_v)$ with more constituents. Various applications have been carried out, including a heavy-fermion model [29], ϕ^4 theory [32], and QED [33].

5 Summary

With these tools in place, the time has come for a concentrated assault on QCD. The procedure would be to invoke PV regularization and light-front quantization in an arbitrary covariant gauge with variable gauge parameter. Any (numerical) approximation [6–8,34–37] would then be applied to a finite theory where the continuum limit can be taken independent of any chosen renormalization scheme. Fock-space truncation and the resulting uncanceled divergences can be avoided by using the LFCC method. The best places to start would be simpler systems in quenched approximation, such as heavy-quark mesons and glueballs. The computational load and effort are likely comparable to lattice gauge theory; the large number of PV modes and the discretizations being the analogs of lattice size and spacing.

There also remains the open question of a proof of renormalizability. The BRST invariance of the Lagrangian is intended as the key ingredient of such a proof.

Acknowledgments This work was done in collaboration with S. S. Chabysheva. Travel to the conference was supported in part by the University of Minnesota Global Programs and Strategy Alliance.

References

1. Brodsky, S.J., Pauli, H.-C., Pinsky, S.S.: Quantum chromodynamics and other field theories on the light cone. Phys. Rep. **301**, 299–486 (1998)
2. Carbonell, J., Desplanques, B., Karmanov, V.A., Mathiot, J.F.: Explicitly covariant light front dynamics and relativistic few body systems. Phys. Rep. **300**, 215–347 (1998)
3. Miller, G.A.: Light front quantization: a technique for relativistic and realistic nuclear physics. Prog. Part. Nucl. Phys. **45**, 83–155 (2000)
4. Burkardt, M.: Light front quantization. Adv. Nucl. Phys. **23**, 1–74 (2002)
5. Bakker, B.L.G., et al.: Light-front quantum chromodynamics: a framework for the analysis of hadron physics. Nucl. Phys. Proc. Suppl. **251–252**, 165–174 (2014)
6. Vary, J.P., et al.: Hamiltonian light-front field theory in a basis function approach. Phys. Rev. C **81**, 035205 (2010)
7. Honkanen, H., Maris, P., Vary, J.P., Brodsky, S.J.: Electron in a transverse harmonic cavity. Phys. Rev. Lett. **106**, 061603 (2011)
8. Zhao, X., Honkanen, H., Maris, P., Vary, J.P., Brodsky, S.J.: Electron $g-2$ in light-front quantization. Phys. Lett. B **737**, 65–69 (2014)
9. 't Hooft, G., Veltman, M.: Regularization and renormalization of gauge fields. Nucl. Phys. B **44**, 189–213 (1972)
10. Bollini, C.G., Giambiagi, J.J.: Lowest order divergent graphs in ν-dimensional space. Phys. Lett. B **40**, 566–568 (1972)
11. Ashmore, J.F.: A method of gauge invariant regularization. Nuovo Cim. Lett. **4**, 289–290 (1972)
12. Brodsky, S.J., Franke, V.A., Hiller, J.R., McCartor, G., Paston, S.A., Prokhvatilov, E.V.: A nonperturbative calculation of the electron's magnetic moment. Nucl. Phys. B **703**, 333–362 (2004)
13. Chabysheva, S.S., Hiller, J.R.: A nonperturbative calculation of the electron's magnetic moment with truncation extended to two photons. Phys. Rev. D **81**, 074030 (2010)
14. Chabysheva, S.S., Hiller, J.R.: A first nonperturbative calculation in light-front QED for an arbitrary covariant gauge. Phys. Rev. D **84**, 034001 (2011)
15. Becchi, C., Rouet, A., Stora, R.: The abelian Higgs–Kibble model, unitarity of the S-operator. Phys. Lett. B **52**, 344–346 (1974)
16. Becchi, C., Rouet, A., Stora, R.: Renormalization of the abelian Higgs–Kibble model. Commun. Math. Phys. **42**, 127–162 (1975)
17. Becchi, C., Rouet, A., Stora, R.: Renormalization of gauge theories. Ann. Phys. **98**, 287–321 (1976)
18. Tyutin, I.V.: Gauge invariance in field theory and in statistical physics in the operator formulation. Lebedev preprint FIAN No. 39, unpublished (1975)
19. Chabysheva, S.S., Hiller, J.R.: BRST-invariant Pauli–Villars regularization of QCD. arXiv:1506.05429 (2015)
20. Malyshev, M.Y., Paston, S.A., Prokhvatilov, E.V., Zubov, R.A., Franke, V.A.: Pauli–Villars regularization and light front Hamiltonian in (2 + 1)-dimensional Yang–Mills theory. Teor. Mat. Fiz. **184**, 487–498 (2015). [Theor. Math. Phys. 184: 1314–1323]
21. Faddeev, L.D., Popov, V.N.: Feynman diagrams for the Yang–Mills field. Phys. Lett. B **25**, 29–30 (1967)
22. Matthews, P.T.: The S-matrix for meson–nucleon interactions. Phys. Rev. **76**, 1254 (1949)
23. Salam, A.: An equivalence theorem for partially gauge invariant vector meson interactions. Nucl. Phys. **18**, 681–690 (1960)
24. Kamefuchi, S.: On Salam's equivalence theorem in vector meson theory. Nucl. Phys. **18**, 691–696 (1960)
25. Stueckelberg, E.C.G.: The interaction forces in electrodynamics and in the field theory of nuclear forces (I). Helv. Phys. Acta **11**, 225–244 (1938)
26. Stueckelberg, E.C.G.: The interaction forces in electrodynamics and in the field theory of nuclear forces (II). Helv. Phys. Acta **11**, 299–311 (1938)
27. Stueckelberg, E.C.G.: The interaction forces in electrodynamics and in the field theory of nuclear forces (III). Helv. Phys. Acta **11**, 312–328 (1938)
28. Marnelius, R., Sogami, I.S.: Generalized BRST quantization and massive vector fields. Int. J. Mod. Phys. A **13**, 3101–3120 (1998)

29. Chabysheva, S.S., Hiller, J.R.: A light-front coupled-cluster method for the nonperturbative solution of quantum field theories. Phys. Lett. B **711**, 417–422 (2012)
30. Bailin, D., Love, A.: Introduction to Gauge Field Theory. Institute of Physics, Bristol (1993)
31. Coester, F.: Quantum electrodynamics with nonvanishing photon mass. Phys. Rev. **83**, 798–800 (1951)
32. Elliott, B., Chabysheva, S.S., Hiller, J.R.: Application of the light-front coupled-cluster method to ϕ^4 theory in two dimensions. Phys. Rev. D **90**, 056003 (2014)
33. Chabysheva, S.S., Hiller, J.R.: An application of the light-front coupled-cluster method to the nonperturbative solution of QED. arXiv:1203.0250 (2012)
34. Glazek, S.D.: Renormalization group procedure for effective particles: elementary example of exact solution with finite mass corrections and no involvement of vacuum. Phys. Rev. D **85**, 125018 (2012)
35. Gomez-Rocha, M., Glazek, S.D.: Asymptotic freedom in the front-form Hamiltonian for quantum chromodynamics of gluons. Phys. Rev. D **92**, 065005 (2015)
36. Karmanov, V.A., Mathiot, J.-F., Smirnov, A.V.: Systematic renormalization scheme in light-front dynamics with Fock space truncation. Phys. Rev. D **77**, 085028 (2008)
37. Karmanov, V.A., Mathiot, J.-F., Smirnov, A.V.: Nonperturbative calculation of the anomalous magnetic moment in the Yukawa model within truncated Fock space. Phys. Rev. D **82**, 056010 (2010)

DOI 10.1007/978-3-319-50699-9_46
Reprinted from *Few-Body System* Journal, DOI 10.1007/s00601-016-1101-5

Patrizia Rossi

Jefferson Lab: Present and Future

Received: 4 February 2016 / Accepted: 1 April 2016 / Published online: 19 April 2016

Abstract The continuous electron beam accelerator facility and associated experimental equipment at Jefferson Lab comprise a unique facility for nuclear physics research whose upgrade is presently underway, with completion expected in 2017. The upgraded facility will accelerate electron beams to 11 GeV for experiments in the existing Halls A, B and C. In addition, a 12 GeV beam can be provided to a new experimental hall, Hall D, to generate a 9 GeV tagged photon beam. This upgrade will enable a new experimental program with substantial discovery potential to address important topics in hadronic, nuclear, and electroweak physics. Further in the future, it is envisioned that the Laboratory will evolve into an electron-ion colliding beam facility.

1 Introduction

Just as Quantum Mechanics and Coulomb's Law explained atoms as bound states of electrons and nuclei, and molecules as bound states of atoms, it is CEBAF's scientific goal to use the theory of Quantum Chromodynamics (QCD) to describe nucleons as bound states of quarks and gluons and nuclei as bound states of nucleons. This machine and its capabilities will enable us to fill in what are some of the most glaring gaps in our understanding of nature: "Why do we have nucleons instead of a quark-gluon soup?", "Why do we always find quarks bound together in two and three, but never alone?", "What is the quark structure of nuclei?"

2 CEBAF and the 12 GeV Upgrade

The core of Jefferson Lab is the continuous electron beam accelerator facility (CEBAF) based on superconducting radiofrequency technology and designed to circulate a near continuous-wave electron beam with one to five passes through the accelerator (Fig. 1). Jefferson Lab started physics operations in 1995, and provided up to 6-GeV electron beams to three experimental halls, Halls A, B and C, simultaneously. In May 2012, the 6-GeV beam operations were stopped, with Jefferson Lab upgrading its facility to expand opportunity for discovery. Construction for doubling the maximum energy of the accelerator from 6 to 12 GeV already had begun in 2008. The so-called 12-GeV Upgrade is a $338 million project that, in addition to the accelerator scope also includes the construction of a new, fourth, experimental hall (Hall D) and upgrades to beam line and equipment in the existing halls. The base equipment in the four Halls is shown in Fig. 2. Hall A will retain the present capability with 2 high resolution spectrometers (HRS). CLAS12 in Hall B is a large acceptance spectrometer, designed to enable luminosity operation with electron beams, up to 10^{35} cm^{-2} s^{-1}. In Hall C the existing high momentum spectrometer (HMS) will be complemented with a new super high momentum spectrometer (SHMS). The SHMS is capable of a maximum central momentum of 11 GeV, a solid angle

P. Rossi (✉)
12000 Jefferson Avenue, Newport News, VA 23606, USA
E-mail: rossi@jlab.org

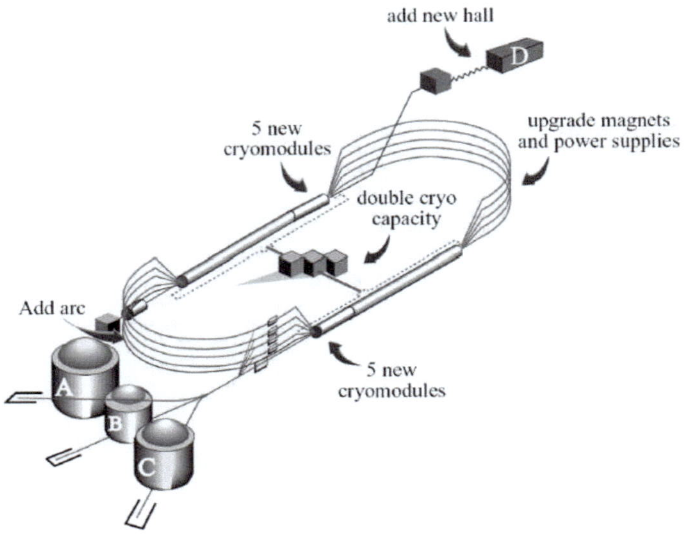

Fig. 1 Jefferson Lab 12 GeV upgrade concept

acceptance of 5 msr and a momentum resolution better than 10^{-3}. Hall C can run experiments with luminosity up to 10^{38} cm^{-2} s^{-1}. Hall D houses a large hermitic detector that is based on a solenoid magnet, and has tracking capabilities and photon detection over nearly the full 4π solid angle. There are additional pieces of equipment proposed that are outside the scope of the 12 GeV project and require separate funding. In Hall A a super big bite spectrometer (SBS) with large solid angle is under construction. This experimental hall will also be the site of a future spectrometer (Moller) to measure parity violation in electron–electron scattering, and of a solenoid detector for parity violation experiments in DIS and for SIDIS experiments (SOLID). There are also efforts underway to incorporate a large acceptance RICH detector into CLAS12 for improved charged particle identification capabilities, and various ancillary detectors for neutron detection and quasi-real photon tagging.

As Jefferson Lab performs beam development activities to bring the accelerator up to its new, full capabilities, the project continues to work toward the start up of the remaining two halls, B and C in late 2016—beginning of 2017, and finally the completion of the project, which is scheduled for September 2017. At the time of this conference the 12 GeV Upgrade was 95 % complete, and both experimental Hall A and Hall D received beam: the very first data of the 12 GeV era were recorded!

3 The 12 GeV Physics Program

The high intensity—up to 90 μA, and high polarization—up to 90 %, of the CEBAF electron beam will enable unprecedented study of the inner structure of protons and atomic nuclei and promises to solve several outstanding mysteries. An upgraded CEBAF will also allow physicists to study the limits of the Standard Model at the low-energy precision frontier, searching for a potential discovery of new physics complementing information from the high energy frontier at the Large Hadron Collider. For a complete overview of the JLab 12 GeV physics program see [1].

The physics program to be addressed with the Jefferson Lab 12 GeV upgrade has been developed in collaboration with the user community and with the guidance of the Program Advisory Commitee (PAC). There are presently 70 approved experiments. These proposals fall into four general categories:

- the physical origin of quark confinement (including meson and baryon spectroscopy),
- the three-dimensional structure of the nucleon,
- the quark structure of nuclei,
- the discovery of new physics through high precision tests of the Standard Model.

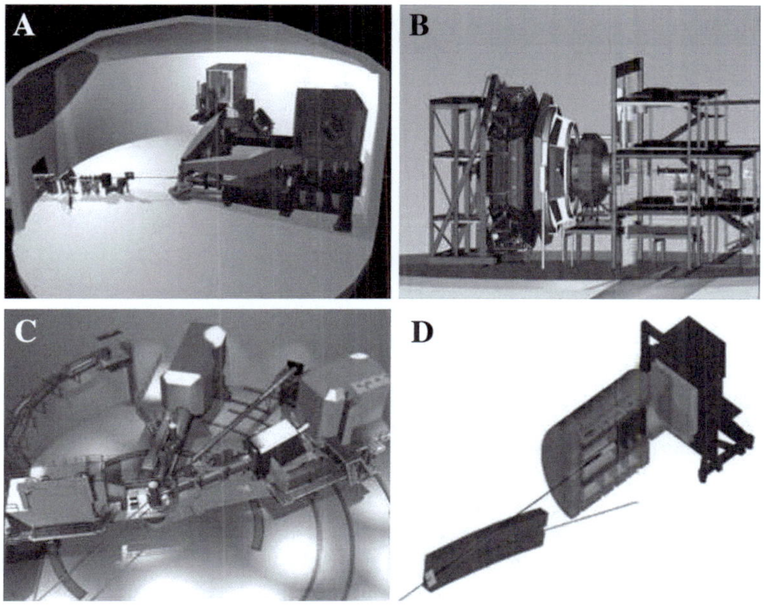

Fig. 2 The base equipment in the four experimental halls: 2 high resolution spectrometers (HRS) in Hall A, the large acceptance spectrometer CLAS12 in Hall B, the super high momentum spectrometer (SHMS) and high momentum spectrometer (HMS) in Hall C, and the 4π hermetic GlueX detector with the 9 GeV tagged polarized photons in Hall D

3.1 The Mystery of Quark Confinement

The new Hall D facility has been designed as a photon-only hall. It hosts the GlueX experiment. The primary physics goal of GlueX is to offer insight into the role of gluon self-interactions and the nature of quark confinement, through the discover and study of the properties of hybrid mesons-particles where the gluonic field contributes directly to the J^{PC} quantum numbers of the mesons. Lattice QCD calculations indicate that several of the nonets of these hybrid mesons have exotic quantum numbers, forbidden J^{PC} for a simple fermion-antifermion system, with expected masses for the lightest hybrids, well-matching the energy and kinematics accessible to the GlueX experiment. The GlueX experiment will makes use of a dedicated spectrometer and will use linearly polarized 9 GeV photons with expected fluxes to reach 10^8/s (produced by coherent bremsstrahlung from a 12 GeV electron beam in the Tagger Facility) incident on a liquid hydrogen target. The photon polarization facilitates a partial wave analysis of any new observed states to determine whether they have allowed or exotic quantum numbers. The solenoid-based hermetic detector will collect data on meson resonance production through decays to both charged and neutral final-state particles. Commissioning of the GlueX experiment started in late 2014 and continued through the spring of 2015. During this period, nearly two billion triggered events were recorded under a variety of trigger, detector, and magnetic field configurations. These data have been used to align, calibrate, and assess the overall performance of the detector. All systems are within 30 % of design specifications, and once sufficient data are obtained to complete calibrations, all are expected to reach design specifications. The data have been also used to extract the first physics result. For vector meson photoproduction using linearly polarized photons, one of the observables that can be measured is the beam asymmetry, Σ, which can be interpreted as the percentage transfer of the linear polarization of the photon to the vector meson. Using a 5.5 GeV electron beam and a 50 μm thick diamond radiator, linearly polarized photons were produced in the 2.5 to 3.0 GeV energy range. From the all the collected events, exclusive $\gamma p \rightarrow \pi^+\pi^- p$ events were selected. The resulting $\pi^+\pi^-$ invariant mass distribution is shown in the left-hand side of Fig. 3. For determining the polarization transfer to the $\rho\pi^+ \pi^-$ candidates have been selected with $0.60 < M_{\pi^+\pi^-} < 0.95$ GeV/c^2 and beam photons with $2.5 < E_\gamma < 3.0$ GeV, where the coherent bremsstrahlung is enhanced. The measured decay distribution for both unpolarized and linearly polarized photons is shown in the right-hand side of Fig. 3. For the case of the aluminum radiator (unpolarized photons), the angular distribution is uniform, while for the diamond radiator (linearly polarized photons), there is a clear $\cos 2\psi$ dependence (where ψ is the azimuthal angle between the photon polarization vector and the decay plane of the vector meson). The results from a few hours of running with polarized photons of the measurement

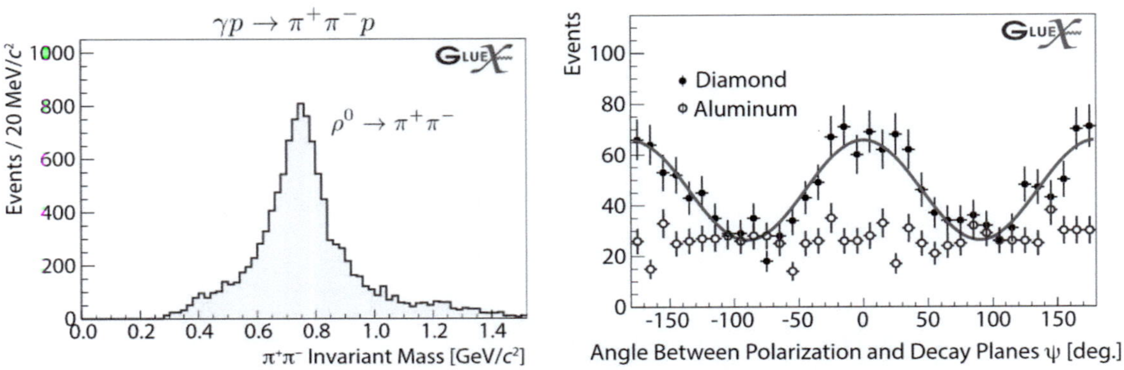

Fig. 3 The GlueX reconstructed $\pi^+\pi^-$ invariant mass for events from the exclusive reaction $\gamma p \to \pi^+\pi^- p$ (*left*). The decay angular distribution for the ρ meson for linearly polarized photons (*red curve on solid black markers*) and for unpolarized photons (*blue open markers*) (*right*) (color figure online)

of the transfer of the photons polarization to photo-produced mesons are consistent with old measurements and comparable in statistics to the older data. The first physics quality beam in Hall D is expected in 2016.

3.2 The Structure of the Nucleon

Over the last decade or so a multidimensional description of nucleon structure is emerging that is providing profound new insights. This new description should help us to solve the many puzzles that the protons inner structure still contains. One outstanding mystery is what exactly makes up the spin-1/2 of a proton. In 1987 it was discovered that quark spins generate only a small part (even compatible to zero) of it, which subsequent experiments have shown to be 30 %. Recently it has been shown that gluons do not generate the remainder. The quark orbital angular momentum is the natural candidate to explain this deficiency. To address this fundamental question, namely how the protons spin is apportioned between the spin of its constituents and their orbital angular momentum, we need a new representation of the protons inner structure. Form factors tell us about the distribution of charge and magnetization but contain no direct dynamical information. Parton distribution functions (PDFs) allow us to access information on the underlying quarks and their longitudinal momentum but tell us nothing about spatial locations. It has now been established that both form factors and PDFs are special cases of a more general class of distribution functions that merge spatial and dynamic information. A three-dimensional description of the nucleon can now be obtained by the generalized parton distributions (GPDs) and the transverse momentum-dependent parton distributions (TMDs). GPDs can be viewed as spatial densities at different values of the longitudinal momentum of the quark. Due to the space–momentum correlation information encoded in the GPDs, it is possible through the so-called Ji sum rule to link a certain combination of GPDs to a partons angular momentum. The TMDs are functions of both the longitudinal and transverse momentum of the partons, and they offer a momentum tomography of the nucleon complementary to the spatial tomography of GPDs. The tomography of the nucleon is one of the flagship science programs of Jefferson Lab that is fully enabled by the combination of the foreseen equipment in Halls A, B and C.

In Hall A, the very first experiment on the floor in the 12-GeV era [2] will allow a verification of our understanding of the reaction mechanism of the so-called deeply virtual compton scattering (DVCS) process, the simplest and cleanest reaction to access GPDs. A successful commissioning of this experiment was carried out in the Fall 2014 and Spring 2015, and the analysis of the acquired data continues.

The transverse momentum-dependent parton distributions functions will be accessed through semi-inclusive deep inelastic scattering experiments (SIDIS). The JLab PAC already approved several proposals by Hall A, B and C aiming at the extraction of SIDIS related observables. The joint efforts from the three halls involved, where the high-precision, high-statistics measurements in Hall A and C will be combined with the wide kinematics ones to be performed in Hall B, by using different targets and several target/beam polarization combinations, will allow a thorough exploration of the 3D structure of the nucleon in the momentum space and of its flavor structure.

With the 12 GeV beam, measurements of the unpolarized and polarized parton distribution function at large x (the fraction of the longitudinal momentum of the proton carried by the quark), and measurements of the proton an neutron electromagnetic form factors will be performed as well. The measurement of the proton elastic cross section to a statistical precision of better than 1 % will allow the precise extraction of the proton magnetic form factor G_M^p at high Q^2. This experiment [3] is now on the floor and it has been commissioned concurrently with the DVCS experiment.

JLab experimental program includes also the proton charge radius (r_p) measurement, one of the fundamental quantities in physics. New results from the recent muonic hydrogen experiments seriously questioned our knowledge of the r_p. The new value, with its unprecedented less than sub-percent precision, is currently up to eight standard deviation smaller than the average value from all previous experiments. The PRad experiment [4] using a novel, magnetic-spectrometer-free *ep* scattering experiment, has been approved to run in Hall B for a new independent r_p measurement. This experiment will critically improve all systematic uncertainties by: (i) normalizing the *ep* cross sections to the well known Moller scattering; (ii) achieving very forward scattering angles while keeping the Q^2 range $[10^{-4} - 10^{-2}\,(\mathrm{GeV/c})^2]$ large enough for the practically model independent extraction of r_p; (iii) using a windowless hydrogen gas flow target to sufficiently reduce the background. PRad, with its sub-percent precision, is expected to have a direct impact on the proton radius puzzle. The experiment is currently in preparation and will take data in Spring 2016 before the start of the CLAS12 commissioning.

3.3 Beyond the Standard Model

The discoveries of dark matter, dark energy, and the flavor oscillations of neutrinos (associated with their small but finite masses) are all indications that the standard model of the strong and electroweak interactions requires modification. The existence of heavy photons, called *A's*, new hypothesized massive vector bosons that have a small coupling to electrically charged matter, including electrons, could explain several intriguing dark matter-related anomalies as well as the discrepancy between the measured and observed anomalous magnetic moment of the muon. New electron fixed-target experiments proposed at Jefferson Lab, with its high-quality and high-luminosity electron beams, present a unique and powerful probe for *A's*. These experiments include: the *A'* Experiment (APEX) [5], the heavy photon search (HPS) [6], and detecting a resonance kinematically with electrons incident on a gaseous hydrogen target (dark light) [7].

The HPS experiment successfully installed, commissioned, and ran in Hall B during CEBAF's 2015 Spring Run. A magnet chicane directed the CEBAF electron beam onto a tungsten foil, producing heavy photons. Their decay to e^+e^- pairs were measured by the silicon vertex tracker inside an analyzing magnet while a $PbWO_4$ electromagnetic calorimeter provided a fast trigger. All aspects of the experiment worked very well and in the end roughly 2 PAC days of 1 GeV data were taken. The collaboration anticipates to release the first physics results in Spring 2016.

4 Electron Ion Collider (EIC)

The 2015 US Nuclear Science Long-Range Planning effort [8] included a future electron-ion collider (EIC) [9] as a recommendation for future construction. Two options are under study at Jefferson Lab and at Brookhaven National Lab. The basic scientific requirements for such a facility are:

- Highly polarized (~0 %) electron and proton beams,
- Ion beams from deuteron to the heaviest nuclei (Uranium or Lead),
- Variable center of mass energies from ~20–~100 GeV, upgradable to ~150 GeV,
- High collision luminosity $\sim 10^{33-34}\,\mathrm{cm}^{-2}\,\mathrm{s}^{-1}$.

The EIC will be unique in colliding polarised electrons off polarised protons and light nuclei, providing the spin degrees of freedom essential to pursue its physics program driven by spin structure, multi-dimensional tomographic images of protons and nuclei, and discovery of the role of collective effects of gluons in nuclei. Such a facility would capitalize on the powerful new experimental techniques for exploring nucleon structure that are being developed for the 12 GeV JLab program, and apply them to the low x region where the dynamics is dominated by the gluons. Addressing this kinematic regime with high luminosity and fully polarized beams is necessary to complete our understanding of the basic partonic structure of the nucleon. The Accelerator and Physics Divisions at Jefferson Lab have been working for several years on a novel design for an EIC that would utilize the 12 GeV CEBAF as an injector to a collider facility [10]. The storage rings would be in a figure 8 layout to mitigate the effects of depolarizing resonances and facilitate high beam polarization. A medium energy version, MEIC, is envisioned that would collide 12 GeV electrons with 100 GeV protons. This could be upgraded to the full EIC facility with 12 GeV electrons colliding with 250 GeV protons.

5 Conclusions

The CEBAF accelerator upgrade is now a reality: the first beam of the 12 GeV era has been delivered to Hall A, B, D. In Hall D successful commissioning runs were carried out in Fall 2014 and Spring 2015 with signals observed in several exclusive reactions. In Hall A, equipment is on the floor to execute together both a 3rd generation DVCS experiment, and a precision measurement of elastic *ep* cross section. First physics beam to Hall B has been delivered for the HPS experiment engineering run. Also PRad will be ready to use opportunistic running while 12 GeV construction continues in Hall B. The CLAS12 detector in Hall B and SHMS in Hall C are completed at ~87 %. Commissioning of the Hall C/B are foreseen in late 2016/beginning of 2017. Further in the future, it is envisioned that Jefferson Lab will evolve into an electron-ion colliding beam facility (EIC).

Acknowledgments This material is based upon work supported by the U.S. Department of Energy, Office of Science, Office of Nuclear Physics under contract DE-AC05-06OR23177.

References

1. Dudek, J., et al.: Physics opportunities with the 12 GeV upgrade at Jefferson Lab. Eur. Phys. J. A **48**, 187 (2012)
2. Roche, J., et al.: E12-06-114: Measurements of the Electron-Helicity Dependent Cross Sections of Deeply Virtual Compton Scattering with CEBAF at 12 GeV. http://hallaweb.jlab.org/collab/PAC/PAC30/PR12-06-114-DVCS (2006)
3. Moffit, B., et al.: E12-07-108: Precision Measurement of the Proton Elastic Cross Section at High Q^2. http://hallaweb.jlab.org/collab/PAC/PAC32/PR12-07-108-GMP (2007)
4. Gasparian, A., et al.: E12-11-106: High Precision Measurement of the Proton Charge Radius. https://www.jlab.org/exp_prog/proposals/11/PR12-11-106 (2011)
5. Wojtsekhowski, B., et al.: E12-10-009: The A' Experiment: Search for a New Vector Boson A' Decaying to e^+e^-. http://hallaweb.jlab.org/collab/PAC/PAC37/E12-10-009-APEX (2010)
6. Stepanyan, S., et al.: E12-11-006: HPS Heavy Photon Search. https://www.jlab.org/exp_prog/proposals/11/PR12-11-006 (2011)
7. Milner, R., et al.: E12-11-008: A Proposal for the DarkLight Experiment at the Jefferson Laboratory Free Electron Laser. https://www.jlab.org/exp_prog/proposals/12/C12-11-008 (2011)
8. Geesaman, D., et al.: Reaching for the Horizon. http://inspirehep.net/record/1398831 (2015)
9. Accardi, A., et al.: Electron Ion Collider: The Next QCD Frontier—Understanding the glue that binds us all. arXiv:1212.1701 (2012)
10. Abeyratne, S., et al.: Science Requirements and Conceptual Design for a Polarized Medium Energy Electron-Ion Collider at Jefferson Lab. arXiv:1209.0757 (2012)

DOI 10.1007/978-3-319-50699-9_47
Reprinted from *Few-Body System* Journal, DOI 10.1007/s00601-016-1102-4

Andrea Signori

Gluon TMDs in Quarkonium Production

Received: 31 January 2016 / Accepted: 4 April 2016 / Published online: 22 April 2016

Abstract I report on our investigations into the impact of (un)polarized transverse momentum dependent parton distribution functions (TMD PDFs or TMDs) for gluons at hadron colliders, especially at A Fixed Target Experiment at the LHC (AFTER@LHC). In the context of high energy proton-proton collisions, we look at final states with low mass (e.g. η_b) in order to investigate the nonperturbative part of TMD PDFs. We study the factorization theorem for the q_T spectrum of η_b produced in proton-proton collisions relying on the effective field theory approach, defining the tools to perform phenomenological investigations at next-to-next-to-leading log and next-to-leading order accuracy in the perturbation theory. We provide predictions for the unpolarized cross section and comment on the possibility of extracting nonperturbative information about the gluon content of the proton once data at low transverse momentum are available.

1 Gluon TMD PDFs

TMD PDFs describe the probability of finding a parton inside a hadron in 3D momentum space, taking into account the possible polarization states of both the parton and the hadron. They encode all the possible spin–spin and spin–orbit interaction terms between a hadron and its constituents. For this reason TMDs play a key role in understanding the spin structure of hadrons. For a gluon inside a proton we can introduce the following correlators [1,2]:

$$G_g^{\mu\nu\,[U]} = -\frac{g_\perp^{\mu\nu}}{2} f_1^g(x, |\boldsymbol{k}_{n\perp}|) + \frac{1}{2}\left(2\frac{k_{n\perp}^\mu k_{n\perp}^\nu}{|\boldsymbol{k}_{n\perp}|^2} + g_\perp^{\mu\nu}\right) h_1^{\perp g}(x, |\boldsymbol{k}_{n\perp}|), \tag{1}$$

$$G_g^{\mu\nu\,[L]} = i\lambda \frac{\epsilon_\perp^{\mu\nu}}{2} g_{1L}^g(x, |\boldsymbol{k}_{n\perp}|) - \frac{\epsilon_\perp^{k_{n\perp}\{\mu} k_{n\perp}^{\nu\}}}{2|\boldsymbol{k}_{n\perp}|^2} \lambda h_{1L}^{\perp g}(x, |\boldsymbol{k}_{n\perp}|), \tag{2}$$

$$\begin{aligned} G_g^{\mu\nu\,[T]} = {} & g_\perp^{\mu\nu} \frac{\epsilon_\perp^{k_{n\perp} S_\perp}}{|\boldsymbol{k}_{n\perp}|} f_{1T}^{\perp g}(x, |\boldsymbol{k}_{n\perp}|) + i\epsilon_\perp^{\mu\nu} \frac{\boldsymbol{k}_{n\perp}\cdot \boldsymbol{S}_\perp}{|\boldsymbol{k}_{n\perp}|} g_{1T}^g(x, |\boldsymbol{k}_{n\perp}|) \\ & - \frac{\epsilon_\perp^{k_{n\perp}\{\mu} k_{n\perp}^{\nu\}}}{2|\boldsymbol{k}_{n\perp}|^2} \frac{\boldsymbol{k}_{n\perp}\cdot \boldsymbol{S}_\perp}{|\boldsymbol{k}_{n\perp}|} h_{1T}^{\perp g}(x, |\boldsymbol{k}_{n\perp}|) - \frac{\epsilon_\perp^{k_{n\perp}\{\mu} S_\perp^{\nu\}} + \epsilon_\perp^{S_\perp\{\mu} k_{n\perp}^{\nu\}}}{4|\boldsymbol{k}_{n\perp}|} h_{1T}^g(x, |\boldsymbol{k}_{n\perp}|). \end{aligned} \tag{3}$$

The subscripts U, L, T refer to the polarization state of the proton (unpolarized, longitudinally and transversely polarized) and the functions are the gluon TMD PDFs for the proton.

A. Signori (✉)
Nikhef and Department of Physics and Astronomy, VU University Amsterdam, Science Park 105, 1098 XG Amsterdam, The Netherlands
E-mail: asignori@nikhef.nl

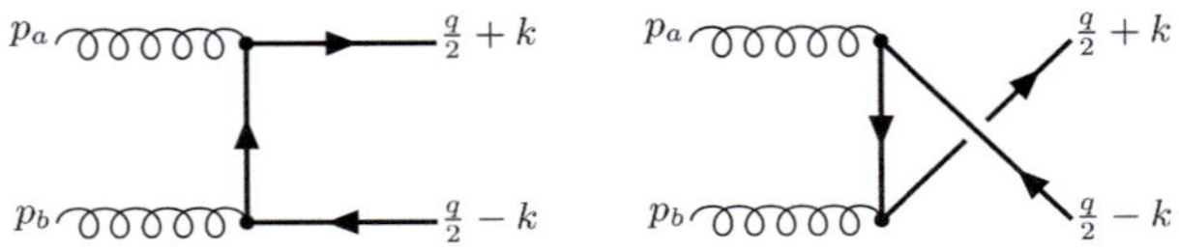

Fig. 1 Feynman diagrams for the process $gg \to Q\bar{Q}$ at leading order (LO) in pQCD

The evolution of TMD PDFs is multiplicative in $\boldsymbol{b}_T$-space ($\boldsymbol{b}_T$ being the variable conjugate to transverse momenta). For this reason, we introduce the Fourier transform of the correlator:

$$\tilde{G}_g^{\mu\nu[\text{pol}]}(x, \boldsymbol{b}_T) \stackrel{\text{def}}{=} \int d^2\boldsymbol{k}_{n\perp}\, e^{i\boldsymbol{k}_{n\perp}\cdot \boldsymbol{b}_T}\, \tilde{G}_g^{\mu\nu[\text{pol}]}(x, \boldsymbol{k}_{n\perp}). \tag{4}$$

TMD PDFs in $\boldsymbol{b}_T$-space are defined as the coefficients in the parametrization of (4) with the same Lorentz structures as in (1), (2), (3), with $\boldsymbol{b}_T$ replacing $\boldsymbol{k}_{n\perp}$. The dimension of TMDs changes according to (4).

The "observability" of TMDs in a particular process strictly depends on the possibility of factorizing hard contributions from the soft ones [3]. TMD PDFs can be further factorized onto collinear PDFs at large parton transverse momentum by means of an operator product expansion (OPE). This allows one to distinguish the emission of a parton with a high transverse momentum (calculable in perturbation theory) from a low transverse momentum emission, for which a nonperturbative model is needed.

2 TMD Factorization

Let us consider the process

$$p(P_A) + p(P_B, S_B) \to \{Q\bar{Q}\}[^{2S+1}L_J^{(1)}](q) + X, \tag{5}$$

where the colliding protons have four-momenta P_A and P_B, the first proton is unpolarized and the second one is in a polarized state described by the spin vector S_B, with $S_B^2 = -1$ and $S_B \cdot P_B = 0$. We assume that a colorless heavy quark-antiquark pair $Q\bar{Q}[^{2S+1}L_J^{(1)}]$ with four-momentum q is produced and forms a bound state described by a nonrelativistic wave function with spin S, orbital angular momentum L and total angular momentum J. The S, L, J quantum numbers are indicated in the spectroscopic notation, while the color assignment of the pair is specified by the singlet or octet superscript, (1) or (8). Following the color-singlet model, we assume that the two quarks are produced in a color singlet state. The squared invariant mass of the resonance is $M_h^2 = q^2$ and M_h is twice the heavy quark mass, up to relativistic corrections (which are usually neglected).

To lowest order in perturbative QCD (pQCD), we have only the gluon fusion process

$$g(p_a) + g(p_b) \to \{Q\bar{Q}\}[^{2S+1}L_J^{(1)}](q), \tag{6}$$

described by the Feynman diagrams in Fig. 1.

The production of the heavy quark and antiquark in a color singlet state will be described by soft-collinear effective theory (SCET), while the transition of the quark-antiquark pair into the color singlet quarkonium will be described by non relativistic QCD (NRQCD).

TMD factorization can be viewed as a multi-step matching procedure:

$$\text{QCD} \to \text{NRQCD} \oplus \text{SCET}_{q_T} \to \text{NRQCD} \oplus \text{SCET}_{\Lambda_{\text{QCD}}}. \tag{7}$$

In the first step, the hard scale M_h associated with the process is integrated out and we perform the matching of full QCD onto a combination of SCET_{q_T} and NRQCD operators. This step already factorizes the cross section in terms of TMDs (describing the initial state), a NRQCD matrix element (describing the transition into the quarkonium) and a spin-independent matching coefficient. In the second step, valid when $\Lambda_{\text{QCD}} \ll q_T \ll M$, the TMDs are further factorized in terms of the collinear PDFs. This matching is performed by means of spin-dependent Wilson coefficients.

The first step has already been investigated at NLO in [4], but with TMDs defined off the light-cone and with rapidity divergences. Here we investigate TMD factorization on the light-cone [5]. Another recent study can be found in [6]. In order to check if TMD factorization holds (or, following the SCET terminology, in order to "establish" TMD factorization) at NLO, we need to check that the cross section expressed in terms of TMDs has the same infrared behavior of the cross section evaluated at $\mathcal{O}(\alpha_S)$ in pQCD.

Let us introduce an effective operator to describe the Feynman amplitudes in Fig. 1:

$$\mathcal{O}(\xi) = C_H(-q^2; \mu^2) \{\chi^\dagger \, \Gamma^{(h)}_{\mu\nu} \, \psi \, \mathcal{B}^{\mu,a}_{n\perp}(\xi) \, (\mathcal{S}^\dagger_n \mathcal{S}_{\bar{n}})^{ab} \, \mathcal{B}^{\nu,b}_{\bar{n}\perp}(\xi)\}, \tag{8}$$

where C_H is the spin-independent matching coefficient used to integrate out the hard scale of the process, χ and ψ are the fermion fields, $\mathcal{B}_{n\perp}$ is the SCET gluon field including collinear Wilson lines, $\mathcal{S}_n$ is the SCET soft Wilson line (for the definition of SCET quantities see [2,7]), a, b are the gauge group indexes, Γ is a Lorentz matrix and its role will be specified later. Using (8), we can write the cross section for the process in (5):

$$d\sigma = \frac{1}{2s} \frac{d^3q}{(2\pi)^3 2q^0} \int d^4y \, e^{-iqy} \sum_X \langle P_A, P_B S_B | \mathcal{O}(y) | X + h \rangle \langle X + h | \mathcal{O}(0) | P_A, P_B S_B \rangle. \tag{9}$$

Introducing the operator definitions [2,7] for the correlators in (1), (2), (3), we cast the cross section in (9) in a factorized form:

$$\begin{aligned} \frac{d\sigma_{U[\text{pol}]}}{dy d^2\boldsymbol{q}_T} &= \frac{\pi}{8sM^2} \, \mathcal{O}^{Q\bar{Q}}(h) \, |C_H(-q^2; \mu^2)|^2 \, \Gamma^\dagger_{\mu\alpha} \Gamma_{\nu\beta} \\ &\times \int \frac{d^2\boldsymbol{b}_T}{(2\pi)^2} \left[\tilde{G}^{\mu\nu\,[U]}_{g/A}(x_a, \boldsymbol{b}_T; \mu, \zeta_a) \, \tilde{G}^{\alpha\beta\,[\text{pol}]}_{g/B}(x_b, \boldsymbol{b}_T, S_B; \mu, \zeta_b) \right] \\ &+ Y(q_T; M) + \mathcal{O}(\Lambda_{\text{QCD}}/M), \end{aligned} \tag{10}$$

where h is the produced resonance, $|C_H|^2$ is the hard function and $\mathcal{O}^{Q\bar{Q}}(h)$ refers to the NRQCD matrix element:

$$\mathcal{O}^{Q\bar{Q}}(h) = |\langle 0 | \chi^\dagger \psi(y) | h \rangle|^2 = \frac{N_c}{2\pi} |R_{nl}(0)|^2 [1 + \mathcal{O}(v^4)]. \tag{11}$$

In the last equation, N_c is the number of colors, R is the radial wave function of the hadron h and v is the relative velocity of Q and $\bar{Q}$. In (10), Y represents corrections for large q_T. In order for this effective description to be valid, we enforce it to reproduce the leading order QCD result for production of pseudoscalar quarkonium [8] by fixing Γ. Its expression is:

$$\Gamma_{\mu\nu} = \frac{\alpha_S \pi}{3\sqrt{M}} \frac{2\sqrt{2}\epsilon_{\perp\mu\nu}}{\sqrt{(d-2)(d-3)}} \sqrt{N_c^2 - 1}, \tag{12}$$

where d is the dimension of the space and $\epsilon_\perp^{\mu\nu} = \epsilon^{n\bar{n}\mu\nu}$.

Now we investigate how legitimate (10) is beyond the LO of QCD, namely if it reproduces the structure of infrared poles of the QCD calculation at NLO. Diagrams in Fig. 1 plus the emission of a real gluon do not suffer of infrared divergences because the transverse momentum of the emitted gluon is fixed and finite. For this reason, we focus only on virtual diagrams:

$$\sigma^{(1)}_{\text{virt}} \overset{\text{IR}}{\longleftrightarrow} [\tilde{f}_1^{g/A} \tilde{f}_1^{g/B}]^{(1)}_{\text{virt}}. \tag{13}$$

If the IR poles of the NLO calculation for the virtual part of the cross section (LHS of (13)) are the same as the ones generated by the two TMD PDFs (RHS of (13)), TMD factorization is established at $\mathcal{O}(\alpha_S)$, namely the factorized form based on SCET and NRQCD reproduces the physical (QCD) result, up to a finite matching coefficient that can be calculated subtracting the RHS from the LHS of (13). Comparing the results for the cross section in [9,10] and for the TMD PDFs in [2], we check that their IR poles are the same (more details in [7]). The finite matching coefficient (hard part) is:

$$\begin{aligned} \mathcal{H} = |C_H|^2 &= \sigma^{(1)}_{\text{virt}} - [\tilde{f}_1^{g/A} \tilde{f}_1^{g/B}]^{(1)}_{\text{virt}} \\ &= 1 + \frac{\alpha_S}{2\pi} \left[-C_A \ln^2 \frac{\mu^2}{M^2} + 2C_A \left(1 + \frac{\pi^2}{3}\right) + 2C_F \left(-5 + \frac{\pi^2}{4}\right) \right]. \end{aligned} \tag{14}$$

This is a byproduct of the factorization theorem and it is fundamental for phenomenology.

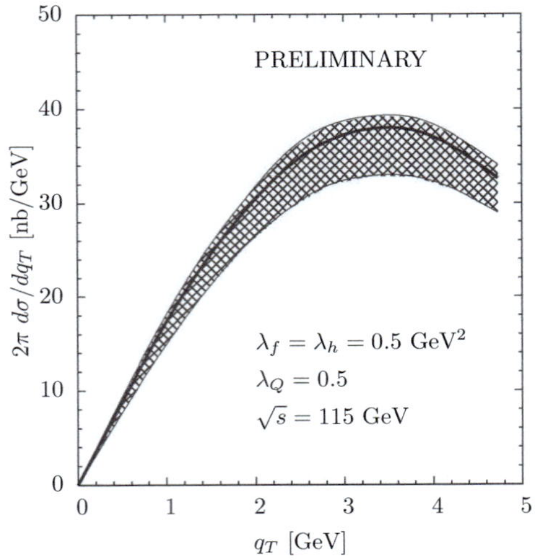

Fig. 2 η_b production from unpolarized proton collisions at AFTER@LHC. The graph shows the cross section in the TMD factorization regime, namely where the transverse momentum is small compared to the hard scale (we restricted $q_T \leq M_h/2$). Nonperturbative parameters are fixed: $\lambda_Q = 0.5$, $\lambda_{f,h} = 0.5$ GeV^2. The thick line represents the choice described in Sect. 3 for the factorization scale μ_f and the rapidity scale ζ_f. The band comes from variations of μ_f and ζ_f by a factor of 2

The second matching step in (7) consists in expanding the TMD PDFs onto a basis of collinear PDFs with perturbative coefficients:

$$\tilde{T}_g(x, b_T; \mu, \zeta) = \left\{ \sum_{j=q,\bar{q},g} \tilde{C}^T_{g/j}(x, b_T; \mu, \zeta) \otimes t_j(x; \mu) \right\} \tilde{T}^{\rm NP}_g(x, b_T, Q; \{\boldsymbol{\lambda}\}), \tag{15}$$

where the summation runs over quarks, antiquarks and gluons, $\tilde{T}_g$ is a generic gluon TMD PDF in b_T-space, t is its collinear counterpart and $\tilde{C}^T_{g/j}$ are the calculable Wilson coefficients which match the TMD $\tilde{T}_g$ onto PDFs. The expansion is only valid at low values of b_T, corresponding to high values of partonic transverse momentum. At low transverse momentum (high b_T), due to the divergence of the coupling constant, a nonperturbative factor $\tilde{T}^{\rm NP}_g$ is needed: it depends on the kinematic variables and on a set of parameters $\{\boldsymbol{\lambda}\}$ to be fixed on experimental data.

In the following we focus on collisions of unpolarized protons, involving TMD PDFs for unpolarized gluons (f_1^g) and linearly polarized gluons ($h_1^{\perp g}$). Their Wilson coefficients are available in [2].

3 Phenomenology

Knowing the hard part of the process (14) and the Wilson coefficients for the TMDs in unpolarized protons, we can predict the q_T-spectrum of σ_{UU} (10) for η_b ($^1S_0^{(1)}$) production at AFTER@LHC [11]. Substituting (1) in both the correlators in (10), we get:

$$\frac{d\sigma_{UU}}{dy\, d^2\boldsymbol{q}_T} \sim \mathcal{H} \int \frac{d^2\boldsymbol{b}_T}{(2\pi)^2} \left[f_1^{g/A}(x_a, b_T; \mu, \zeta_a) f_1^{g/B}(x_b, b_T; \mu, \zeta_b) - h_1^{\perp g/A}(x_a, b_T; \mu, \zeta_a) h_1^{\perp g/B}(x_b, b_T; \mu, \zeta_b) \right]. \tag{16}$$

We implement (15) for f_1^g and $h_1^{\perp g}$ at NNLL+NLO, consistently with (14), and we fix the value of the radial wavefunction from [12]. We choose a Gaussian model to describe the high b_T behavior of f_1^g:

$$f_1^{g{\rm NP}}(b_T, Q; \{\boldsymbol{\lambda}\}) = \exp[-b_T^2(\lambda_f + \lambda_Q \ln M^2)]. \tag{17}$$

We use the same model for $h_1^{\perp g{\rm NP}}$, with λ_f replaced by λ_h. $4\lambda_{f/h}$ represent the average square intrinsic transverse momenta, whereas λ_Q accounts for emission of soft gluons. The values of $\lambda_{f/h}$ and λ_Q are not well known yet. Experimental data at low q_T are needed to better constrain them.

Because of the medium value of its mass (9.39 GeV), η_b production is an ideal process to extract information about the nonperturbative part of twist-2 gluon TMDs. The latter plays here a relatively clean role, because the energy is high enough to safely neglect higher twist and factorization breaking effects and at the same time, it is low enough to avoid that the perturbative effects dominate the nonperturbative (which happens, e.g., in Higgs production). We choose the initial scales $\zeta_i = \mu_i^2 = \mu_{\hat{b}}^2$, where $\mu_{\hat{b}}$ is defined through the b^* prescription [3]. The final scales are $\zeta_f = \mu_f^2 = M^2$. The cross section sketched in (16) (without Y term) is displayed in Fig. 2.

4 Conclusions

In this work we discussed TMD factorization at NLO for the q_T-spectrum of color singlet quarkonium production in terms of gluon TMDs, using for the first time the effective field theory (SCET and NRQCD) approach. With the tools available from the factorization theorem, we can make accurate predictions for (un)polarized cross sections at AFTER@LHC. Once experimental data are available at low q_T, this formalism will allow the extraction of the nonperturbative part of the involved gluon TMD PDFs. The distribution of linearly polarized gluons in unpolarized protons ($h_1^{\perp g}$) will be especially relevant in forthcoming studies at hadron colliders [13].

Acknowledgments This report is based on ongoing work in collaboration with M. Garcia Echevarria, T. Kasemets, J.P. Lansberg, C. Pisano. The work of AS is part of the program of the Stichting voor Fundamenteel Onderzoek der Materie (FOM), which is financially supported by the Nederlandse Organisatie voor Wetenschappelijk Onderzoek (NWO).

References

1. Mulders, P.J., Rodrigues, J.: Transverse momentum dependence in gluon distribution and fragmentation functions. Phys. Rev. D **63**(094), 021 (2001). arXiv:hep-ph/0009343
2. Echevarria, M.G., Kasemets, T., Mulders, P.J., Pisano, C.: QCD evolution of (un)polarized gluon TMDPDFs and the Higgs q_T-distribution. JHEP **07**(158) (2015). arXiv:1502.05354
3. Collins, J.: Foundations of Perturbative QCD. Cambridge University Press. (2013). http://www.cambridge.org/de/knowledge/isbn/item5756723
4. Ma, J.P., Wang, J.X., Zhao, S.: Transverse momentum dependent factorization for quarkonium production at low transverse momentum. Phys. Rev. D **88**(1), 014,027 (2013). arXiv:1211.7144
5. Echevarria, M.G., Idilbi, A., Scimemi, I.: Factorization theorem for Drell–Yan at low q_T and transverse-momentum distributions on-the-light-cone. JHEP **07**, 002 (2012). doi:10.1007/JHEP07(2012)002. arXiv:1111.4996
6. Ma, J.P., Wang, C.: QCD factorization for quarkonium production in hadron collions at low transverse momentum. (2015). arXiv:1509.04421
7. Echevarria, M.G., et al.: Gluon TMDs and quarkonium production in hadronic collisions (2016) (in preparation)
8. Boer, D., Pisano, C.: Polarized gluon studies with charmonium and bottomonium at LHCb and AFTER. Phys. Rev. D **86**(094), 007 (2012). arXiv:1208.3642
9. Kuhn, J.H., Mirkes, E.: QCD corrections to toponium production at hadron colliders. Phys. Rev. D **48**, 179–189 (1993). arXiv:hep-ph/9301204
10. Petrelli, A., Cacciari, M., Greco, M., Maltoni, F., Mangano, M.L.: NLO production and decay of quarkonium. Nucl. Phys. B **514**, 245–309 (1998). arXiv:hep-ph/9707223
11. Brodsky, S.J., Fleuret, F., Hadjidakis, C., Lansberg, J.P.: Physics opportunities of a fixed-target experiment using the LHC beams. Phys. Rep. **522**, 239–255 (2013). arXiv:1202.6585
12. Maltoni, F., Polosa, A.D.: Observation potential for eta(b) at the tevatron. Phys. Rev. D **70**(054), 014 (2004). arXiv:hep-ph/0405082
13. Angeles-Martinez, R., et al.: Transverse momentum dependent (TMD) parton distribution functions: status and prospects. Acta Phys. Polon. B **46**(12), 2501–2534 (2015). arXiv:1507.05267

DOI 10.1007/978-3-319-50699-9_48
Reprinted from *Few-Body System* Journal, DOI 10.1007/s00601-016-1104-2

O. V. Teryaev

Radon Transform and Light-Cone Distributions

Received: 31 January 2016 / Accepted: 5 April 2016 / Published online: 3 May 2016

Abstract The relevance of Radon transform for generalized and transverse momentum dependent parton distributions is discussed. The new application for conditional (fracture) parton distributions and dihadron fragmentation functions is suggested.

1 Introduction

Radon transform (RT) will soon celebrate its centennial [1]. Finding a lot of applications since that time, it was also applied to generalized Parton Distributions [2]. The different slopes of straight lines over which the RT integration is performed correspond to the way of analytic continuation which involves Generalized Distribution Amplitudes [3]. RT leads to the holographic sum rule for GPDs (GDAs), when information about the 2 dimensional x, ξ domain relevant for LO amplitudes is contained in one-dimensional section $x = \pm\xi$.

In the framework of TMDs, the holography is manifested in the appearance of the variable $x + p_T^2/(M^2x)$ combining longitudinal and transverse degrees of freedom. To relate GPDs and TMDs one may use their representations in terms of AdS/QCD light-cone wave functions.

Here we suggest the new application to Fracture Functions, resulting from the Lorentz invariance and polynomiality. Originally, Fracture Functions were suggested [4,5] to describe the diffractive region at high energies. At the same time, it is possible to apply them [6] at relatively low momentum transfers in SIDIS due to early scaling in Q^2 and "handbag dominance".

One point on terminology: to be more strict, one should distinguish structure functions and parton distributions, so that for Fracture Functions one should introduce respective Conditional Parton Distributions (CPDs).[1]

The cross-related objects implied by the different slopes of RT integration lines are in this case the dihadron fragmentation functions (DFFs), so that symbolically

$$\frac{\textit{Generalized Parton Distributions}}{\textit{Generalized Distribution Amplitudes}} \leftrightarrow \frac{\textit{Conditional Parton Distributions}}{\textit{Dihadron Fragmentation Functions}}. \tag{1}$$

[1] Term is suggested by L. Frankfurt.

O. V. Teryaev (✉)
BLTP, JINR and International University, Dubna, Russia 141980
E-mail: teryaev@theor.jinr.ru

2 GPDs and Radon Transform

The representation on terms of Radon transform of Double Distributions (DDs)

$$H(z,\xi,t) = \int_{-1}^{1} dx \int_{|x|-1}^{1-|x|} dy(F(x,y,t) + \xi G(x,y,t))\delta(z - x - \xi y). \tag{2}$$

allows one to express the latter (more exactly, the effective DD, which will be discussed later) by the inverse Radon transform [2] which was elaborated in the case of photon GPDs [7]. The recovery requires the notion of both GPDs ($|\xi| < 1$) and GDAs ($|\xi| > 1$). By introducing the "unintegrated" ξ-dependent DDs one may compare the role of GPDs and GDAs channels [8].

The RT implies [9,10] the "holographic" sum rule for LO Compton formfactors

$$\mathcal{H}_i(\xi,t) = \int_{-1}^{1} dx H_i(x,\xi,t)\left[\frac{1}{x-\xi+i\epsilon} \pm \frac{1}{x+\xi-i\epsilon}\right], \tag{3}$$

where index i describes the type of GPD, defining also the choice of $\pm$ sign. For the time being we drop the dependence on the momentum transfer t. The sum rule states that

$$\Delta\mathcal{H}(t) = \int_{-1}^{1} dx \frac{H(x,x,t) - H(x,\xi,t)}{x-\xi+i\epsilon}, \tag{4}$$

where the subtraction term is

$$\Delta\mathcal{H}(t) = \int_{-1}^{1} dx \int_{|x|-1}^{1-|x|} dy \frac{G(x,y,t)}{1-y}. \tag{5}$$

As a result, the information on 2-dimensional x, ξ region sufficient to recover LO Compton formfactor is stored in 1-dimensional section.

3 TMDs and Abel Transform

The holographic property of TMDs is manifested in the existence of the variable [11,12]

$$z = x + \frac{p_T^2}{xM^2}$$

implying relations between longitudinal and transverse momentum dependence. Note that its appearance in the original framework of "Zavada model" is due to the integrals (see e.g. eq. C4,C5 in [11,12])

$$\int d^3p\, \delta\left(x - \frac{p_0 - p_1}{M}\right)\ldots$$

which may be considered as a 3-dimensional Abel transform (which is particular case of RT), and as usual at the odd numbers of dimensions the inverse transform has a simple local form. From the other side, this variable is rooted in the Lorentz [13] and, in particular, rotational [14] symmetries.

4 GPDs and TMDs

It would be interesting to look for common origins of GPDs and TMDs holographic properties [15]. Probably they may correspond to the respective properties of Wigner function [16] being their "common ancestor".

Note that the RT of Wigner function in the (x, p) plane is the positive object widely used in optics and quantum theory (see e.g. [17] and Ref. therein). One may use this experience to define the RT (more exactly, the product of RT's in the two transverse directions) in the $(\mathbf{p}_T, \mathbf{b}_T)$ plane to define the object interpolating between TMDs and GPDs. As GPDs are, generally speaking, non-positive, this interpolating objects also do not have positivity property.

At the same time, the "horizontal" relations are certainly of interest. It may be supported by the light-cone models [16, 18], when GTMDs correspond to the overlap of wave functions of the type

$$\bar{\Psi}(p_T + q_T/2)\Psi(p_T - q_T/2)$$

where $q_T^2 = -t$ and reflection symmetry implies the symmetry w.r.t. interchange $q_T \leftrightarrow p_T$. Recently [19] this possibility was explored in the framework of quark/diquark light-front AdS/QCD wave functions and the (approximate) relation between GPDs and Tranverse Momentum Dependent Distributions (TMDs) was obtained:

$$\frac{\partial}{\partial |t|}[\ln(\text{GPD})] = \frac{(1-x)^2}{4}\frac{\partial}{\partial p_\perp^2}[\ln(\text{TMD})]. \tag{6}$$

The t and $p_\perp^2$ dependence of the GPDs and TMDs comes from the AdS/QCD wave function. The GPDs obtained from the model satisfy the Regge behavior with the Regge slope $\alpha' = 1/(4\kappa^2)$. The average value of Gaussian transverse momentum distribution is then obtained as

$$\langle p_\perp^2 \rangle = \frac{(1-x)^2}{4\alpha' \ln 1/x} = \frac{\kappa^2(1-x)^2}{\ln 1/x}, \tag{7}$$

which takes the maximum value of $\langle p_\perp^2 \rangle_{max} \sim (250\,\text{MeV})^2$ for $x \sim 1/3$.

5 Radon Transform for Conditional Parton Distributions

The general factorization framework [20] implies that for description of initial and final hadron in SIDIS by the separate non-perturbative objects (parton distributions and fragmentation functions) the magnitude of the momentum transfer squared between the initial and final hadron should exceed some hadronic scale

$$|t| > \mu^2.$$

Neglecting hadron masses one get [6]

$$Q^2 \frac{z}{x} > \mu^2.$$

It is convenient to perform the account for mass corrections (which should be especially important when Λ hyperon production [21] is considered) in the target rest frame

$$E_f > \frac{M_i^2 + m_f^2 + \mu^2}{2M_i}.$$

This criterion was used in the analysis of SIDIS data [22]. One may also account for transverse momentum dependence of CPDs and consider them together with ordinary fragmentation functions for asymmetric hadron pair production [23].

Let us define the simplest spin-averaged quark (with flavour k) CPD as

$$F^k(x,\xi,t) = (f_f^k)^2 \sum_X \int \frac{d\lambda}{4\pi} < p_i|\bar{\psi}_\alpha^k(0)|p_f, X > \hat{n}_{\alpha\beta} < p_f, X|\psi_\beta^k(\lambda n)|p_i >, \tag{8}$$

where n is the light-cone vector ($P \cdot n = \frac{1}{2}(p_i - p_f)\cdot n = 1$) and axial gauge ($A \cdot n = 0$) is assumed. Here f_f^k (which is just f_π^k in the case of final pion) is the dimensionful coupling of given quark to the final hadron compensating the dimension of matrix element. The variables ξ, t are convenient to define in complete analogy with GPDs, to which CPDs are related [24] in the $\xi \to 1$ limit. The semi-exclusive nature of the process leads to the constraint

$$\xi \geq \frac{x_B}{2 - x_B},$$

in the collinear limit, while the equality in the case of GPDs holds.

Taking the moments of (8) in x and using the Lorentz invariance, one get the polynomials in ξ in complete analogy to GPDs case. Following the analogy with GPDs case further, this property can be satisfied if the representation in terms of double CPDs (DCPDs) holds.

$$F(z,\xi,t)=\int\int dxdy(F_{CPD}(x,y)+\xi G_{CPD}(x,y))\delta(z-x-\xi y). \tag{9}$$

The values $|\xi|>1$ which are required to perform the inverse RT [2,7] for the case of CPDs correspond to dihadron fragmentation functions which were introduced [25] and explored [26] in the chiral- and T-odd case related to transversity measurements. Note, that the simpler spin-averaged DFFs appearing here are the necessary ingredients for calculation of respective asymmetries where they contribute to the denominators.

Note that the analog of Compton formfactor defining the SIDIS cross-section is

$$\mathcal{F}(\xi,t)_k=\int_{-1}^{1}dxF_k(x,\xi,t)\left[\frac{1}{x-x_B+i\epsilon}\pm\frac{1}{x+x_B-i\epsilon}\right]. \tag{10}$$

Due to the appearance of different variables in the numerator (ξ) and denominator (x_B) there is no straightforward analog of the holographic sum rule.

At the same time, the notion of effective DCPD,

$$N_{CPD}(x,y)=\frac{\partial F_{CPD}(x,y)}{\partial x}+\frac{\partial G_{CPD}(x,y)}{\partial y}, \tag{11}$$

analogs with gauge invariance in 2-dimensional magnetostatics

$$\begin{aligned}F_{CPD}(x,y)&\to F_{CPD}(x,y)+\frac{\partial\alpha(x,y)}{\partial y},\\ G_{CPD}(x,y)&\to G_{CPD}(x,y)-\frac{\partial\alpha(x,y)}{\partial x},\\ \alpha(x,y)&=-\alpha(x,-y),\end{aligned} \tag{12}$$

and D-term corresponding to

$$\begin{aligned}\alpha_{PW}(x,y)&=\frac{1}{2}\Big(\int dtG_{CPD}(t,y)-\int G_{CPD}(t,y)-sign(x)D_{CPD}(y)\Big)\\ &=\theta(x<0)\int dtG_{CPD}(t,y)-\theta(x>0)\int dtG_{CPD}(t,y),\end{aligned} \tag{13}$$

may be directly inherited from GPDs case [2].

6 Conclusions

The Radon transform naturally appears in the domain of light-cone distributions. In all the explored examples it corresponds to the Lorentz invariance or some of its subgroups. The new field of applicability is provided by Conditional Parton Distributions describing the target fragmentation region in SIDIS. The use of handbag dominance and early scaling allows their application at JLab kinematic region. The Lorentz invariance implies for CPDs the polynomiality property, similar to the case of Generalized Parton Distributions. It can be naturally fulfilled by introducing the Double Conditional Parton Distributions.

Acknowledgments I am indebted to Organizers of LC-2015 for warm hospitality in Frascati and to participants for many interesting discussions. This work was partially supported through the RFBR grant No 14-01-00647.

References

1. Radon, J.: On determination of functions by their integral values along certain multiplicities. Ber. der Sachische Akademie der Wissenschaften Leipzig, Germany, 69, 262–277 (1917)
2. Teryaev, O.V.: Crossing and Radon tomography for generalized parton distributions. Phys. Lett. B **510**, 125 (2001). doi:10.1016/S0370-2693(01)00564-0. arXiv:hep-th/0102303
3. Diehl, M., Gousset, T., Pire, B., Teryaev, O.: Probing partonic structure in $\gamma^*\gamma \longrightarrow \pi\pi$ near threshold. Phys. Rev. Lett. **81**, 1782 (1998). doi:10.1103/PhysRevLett.81.1782. arXiv:hep-th/9805380
4. Trentadue, L., Veneziano, G.: Fracture functions: an improved description of inclusive hard processes in QCD. Phys. Lett. B **323**, 201 (1994). doi:10.1016/0370-2693(94)90292-5
5. Grazzini, M., Trentadue, L., Veneziano, G.: Fracture functions from cut vertices. Nucl. Phys. B **519**, 394 (1998). doi:10.1016/S0550-3213(97)00840-7. arXiv:hep-th/9709452
6. Teryaev, O.V.: Spin dependent, interference and T odd fragmentation and fracture functions. Acta Phys. Polon. B **33**, 3749 (2002). arXiv:hep-ph/0211027
7. Gabdrakhmanov, I.R., Teryaev, O.V.: Analyticity and sum rules for photon GPDs. Phys. Lett. B **716**, 417 (2012). doi:10.1016/j.physletb.2012.08.041. arXiv:1204.6471 [hep-ph]
8. Gabdrakhmannov, I., Müller, D., Teryaev, O.V.: Radon transform for separate channels. (**in preparation**)
9. Teryaev, O.V.: Analytic properties of hard exclusive amplitudes. arXiv:hep-th/0510031
10. Anikin, I.V., Teryaev, O.V.: Dispersion relations and subtractions in hard exclusive processes. Phys. Rev. D **76**, 056007 (2007). doi:10.1103/PhysRevD.76.056007. arXiv:0704.2185 [hep-ph]
11. Efremov, A.V., Schweitzer, P., Teryaev, O.V., Zavada, P.: Transverse momentum dependent distribution functions in a covariant parton model approach with quark orbital motion. Phys. Rev. D **80**, 014021 (2009). arXiv:0903.3490 [hep-ph]
12. Efremov, A.V., Schweitzer, P., Teryaev, O.V., Zavada, P.: The relation between TMDs and PDFs in the covariant parton model approach. Phys. Rev. D **83**, 054025 (2011). arXiv:1012.5296 [hep-ph]
13. D'Alesio, U., Leader, E., Murgia, F.: On the importance of Lorentz structure in the parton model: target mass corrections, transverse momentum dependence, positivity bounds. Phys. Rev. D **81**, 036010 (2010). arXiv:0909.5650 [hep-ph]
14. Lorce, C., Pasquini, B.: On the origin of model relations among transverse-momentum dependent parton distributions. Phys. Rev. D **84**, 034039 (2011). arXiv:1104.5651 [hep-ph]
15. Teryaev, O.: Transverse spin effects and holography. EPJ Web Conf. **85**, 02008 (2015) doi:10.1051/epjconf/20158502008
16. Pasquini, B., Lorce, C.: Wigner distributions in light-front quark models. Few Body Syst. **55**, 287 (2014)
17. Klimov, A.B., Manko, O.V., Manko, V.I., Smirnov, Y.F., Tolstoy, V.N.: Tomographic representation of spin and quark states. J. Phys. A **35**, 6101 (2002). doi:10.1088/0305-4470/35/29/312
18. Muller, D., Hwang, D.S.: The concept of phenomenological light-front wave functions–Regge improved diquark model predictions. arXiv:1407.1655 [hep-ph]
19. Maji, T., Mondal, C., Chakrabarti, D., Teryaev, O.V.: Relating transverse structure of various parton distributions. JHEP **1601**, 165 (2016). doi:10.1007/JHEP01(2016)165. arXiv:1506.04560 [hep-ph]
20. Efremov, A., Radyushkin, A.: On perturbative QCD of hard and soft processes. Mod. Phys. Lett. A **24**, 2803 (2009). doi:10.1142/S0217732309001029. arXiv:0911.1195 [hep-ph]
21. Ceccopieri, F.A.: Lambda production in the DIS target fragmentation region. arXiv:1508.07459 [hep-ph]
22. Solovtsova, O., Sidorov, A.: Polarized nonsinglet $Delta q_3$ and nonsinglet fragmentation function $D_{u_v}^{pi+}$ in the analytic approach to QCD. PoS BaldinISHEPPXXII (2014) 019 arXiv:1411.6975 [hep-ph]
23. Anselmino, M., Barone, V., Kotzinian, A.: Double hadron lepto-production in the current and target fragmentation region. Phys. Lett. B **706**, 46 (2011). doi:10.1016/j.physletb.2011.10.064. arXiv:1109.1132 [hep-ph]
24. Hebecker, A., Teubner, T.: Skewed parton distributions and F_2^D at $\longrightarrow$ 1. Phys. Lett. B **498**, 16 (2001). doi:10.1016/S0370-2693(00)01372-1. arXiv:hep-th/0010273
25. Jaffe, R.L., Jin, X.m, Tang, J.: Interference fragmentation functions and the nucleon's transversity. Phys. Rev. Lett. **80**, 1166 (1998). doi:10.1103/PhysRevLett.80.1166. arXiv:hep-th/9709322
26. Bacchetta, A., Courtoy, A., Radici, M.: First glances at the transversity parton distribution through dihadron fragmentation functions. Phys. Rev. Lett. **107**, 012001 (2011). doi:10.1103/PhysRevLett.107.012001. arXiv:1104.3855 [hep-ph]

DOI 10.1007/978-3-319-50699-9_49
Reprinted from *Few-Body System* Journal, DOI 10.1007/s00601-016-1105-1

J. Carbonell · V. A. Karmanov

Scattering Solutions of Bethe–Salpeter Equation in Minkowski and Euclidean Spaces

Received: 27 January 2016 / Accepted: 6 April 2016 / Published online: 2 June 2016

Abstract We shortly review different methods to obtain the scattering solutions of the Bethe–Salpeter equation in Minkowski space. We emphasize the possibility to obtain the zero energy observables in terms of the Euclidean scattering amplitude.

1 Introduction

The Bethe–Salpeter (BS) equation [1] deals with a—preexisting—Quantum Field Theory (QFT) object defined as a vacuum expectation value of the T-ordered product of Heisenberg field operators [2]. For the two-body case in a scalar theory (ϕ), this object has the expression

$$\Phi(x_1, x_2; p) = \langle 0 \mid T\{\phi(x_1)\phi(x_2)\} \mid p\rangle, \tag{1}$$

where $\mid p\rangle$ is the state vector with total momentum p.

After removing some trivial dependence related to translational invariance,

$$\Phi(x_1, x_2; p) = \frac{1}{(2\pi)^{3/2}}\,\Phi(x; p)\,e^{-ip\cdot(x_1+x_2)/2},$$

its Fourier transform

$$\Phi(x; p) = \int d^4x\;\Phi(k; p)\,e^{-ik\cdot x} \qquad x = x_1 - x_2$$

defines the momentum space BS amplitude. Bethe and Salpeter wrote a covariant 4D integral equation for the amplitude $\Phi(k; p)$, which for the two-body bound state reads

$$\Phi(k; p) = S_1(k, p)\,S_2(k, p)\int \frac{d^4k'}{(2\pi)^4}\,iK(k, k'; P)\,\Phi(k'; p) \tag{2}$$

where

$$S_1(k, p) = \frac{i}{\left(\frac{p}{2}+k\right)^2 - m^2 + i\epsilon} \qquad S_2(k, p) = \frac{i}{\left(\frac{p}{2}-k\right)^2 - m^2 + i\epsilon} \tag{3}$$

J. Carbonell (✉)
Institut de Physique Nucléaire, IN2P3-CNRS, Université Paris-Sud, 91406 Orsay Cedex, France
E-mail: carbonell@ipno.in2p3.fr

V. A. Karmanov
Lebedev Physical Institute, Leninsky Prospekt 53, Moscow, Russia 119991
E-mail: karmanov@sci.lebedev.ru

are the free constituent propagators and iK the interaction kernel. If the propagators were dressed and the kernel was a sum of all the irreducible Feynman graphs of the theory, solving (2) would be equivalent to solve the full QFT. This is however a wishful thinking. In practice only very poor representations of the interaction are used: most of nuclear physics calculations use ladder approximation and free propagators, though dressed propagators are systematically used in the Schwinger-Dyson approach.

There could be some misunderstanding on what BS amplitude denotes: either the quantity (1) computed whit the full QFT solution or the quantity Φ obtained by solving (2) with a—necessarily—truncated interaction kernel. These quantities can strongly differ from each other, but we live with it.

Besides the bound states equation (2), we have also considered the corresponding scattering equation for the process $k_{1s} + k_{2s} \to k_1 + k_2$. The two-body scattering amplitude $F(k; p)$ satisfies an inhomogeneous BS equation which, for spinless particles in Minkowski space, reads:

$$F(k;p) = K(k,k_s) - i\int \frac{d^4k'}{(2\pi)^4} \frac{K(k,k')F(k';p)}{\left[\left(\frac{p}{2}+k'\right)^2 - m^2 + i\epsilon\right]\left[\left(\frac{p}{2}-k'\right)^2 - m^2 + i\epsilon\right]} \tag{4}$$

with $2k = k_1 - k_2$ and $p = k_1 + k_2$.

This equation is plagued with singularities: (1) in the two-body propagators, (2) in most of the interaction kernels even the simplest ones, (3) in the (inhomogeneous) Born term and (4) in the amplitude F itself, being difficult to represent it in terms of smooth functions.

To avoid these troubles—and get some bound state solutions—Wick proposed [3] to introduce a new variable $k_0 = ik_4$ in terms of which the 4D Minkowski metric is changed into an Euclidean one:

$$k^2 = k_0^2 - \mathbf{k}^2 = -(k_4^2 + \mathbf{k}^2) \equiv -k_E^2$$

This change of metric leads to a smooth integral equation for the Euclidean amplitude, related to the Minkowski one by

$$F_E(k_4, \mathbf{k}) = F_M(ik_4, \mathbf{k}) \tag{5}$$

which is easily solvable by standard methods. Until very recently most of the existing solutions were found in this way. The are however some problems in performing the so called Wick rotation:

1. It is not a simple change of variable but underlies an application of the Cauchy theorem with two different integration contours. The equivalence of both theories—Minkowski and Euclidean—requires a careful analysis of the singularities crossed when moving from one to other. For the bound state case, the singularities come only in the interaction kernel and the validity has been proved only for simple cases. For the scattering case, they come also in the propagators and the corresponding residues must be taken into account.
2. The total mass M of the system is invariant under such transformation but the amplitude not. In practice it is very difficult—if possible at all—to recover Φ_M from Φ_E (see contribution [4])
3. Other observable—scattering amplitudes, form factors—are defined and computable in terms of Minkowski amplitudes. It is worth noticing, however, that the distinction between Euclidean and Minkowski amplitudes is meaningful only when keeping for both of them real arguments $(k_0, \mathbf{k})$ and $(k_4, \mathbf{k})$ and the later statement must be understood in this conventional sense. Euclidean amplitudes with imaginary arguments are equivalent to Minkowski ones and can lead to correct results.

All these reasons motivated several authors to solve the BS equation in Minkowski space, first for the bound state [5–11] and later for the scattering problem [12–15]. Obtaining these numerical solution is indeed quite difficult and was not tried until very recently.

2 Minkowski Space Solutions: Light-Front Projection and Nakanishi Representation

Some years ago we proposed [7,8] a method based on an integral transform of the original BS equation which make all things regular. It was inspired by the relation between the Light-Front (LF) wave function Ψ and the BS amplitude Φ

$$\Psi(k_1, k_2, p, \omega) = \frac{(\omega\cdot k_1)(\omega\cdot k_2)}{\pi(\omega\cdot p)} \int_{-\infty}^{+\infty} d\beta\ \Phi(k + \beta\omega)$$

This integral transform, denoted symbollicaly by $\Psi = L\Phi$, maps the singular BS amplitude into a smooth wave function Ψ and was applied to the full BS equation (2) according to

$$\Phi = G_0 K \Phi \quad \Longrightarrow \quad L\Phi = L G_0 K \Phi \quad \Longrightarrow \quad \Psi = L G_0 K \Phi \tag{6}$$

In order to use the same unknown function on both sides of (6) and avoid the inversion of L, we have used the Nakanishi representation [16] of the BS amplitude

$$\Phi(k,p) = \int_{-1}^{+1} dz \int_0^{\infty} d\gamma \, \frac{1}{[\gamma + m^2 - \frac{M^2}{4} - k^2 - p\cdot kz - i\epsilon]^3} \, g(\gamma, z) \tag{7}$$

and obtained an integral equation for the Nakanishi weight function g, on the form:

$$LNg = (LG_0KN)g \tag{8}$$

The details of this derivation and the analytic expressions of the kernels can be found in [7,8].

The function g is metric-independent and finite although it can display—at least for zero energy scattering—Heaviside-like discontinuities [15]. All the Minkowski space poles of Φ are contained in the denominator of (7). Once g is known, the BS amplitude—either in Minkowski or in Euclidean space—can be computed by means of (7). A similar, one-dimensional, expression exists for the Light-Front wave function Ψ in terms of g [7,8]

$$\Psi(\gamma, z) = \frac{1-z^2}{4} \int_0^{\infty} d\gamma' \, \frac{g(\gamma', z)}{\left[\gamma + m^2 - \frac{M^2}{4}(1-z^2) + \gamma'\right]^2} \tag{9}$$

These two possibilites make the Nakanishi weight function g especially well adapted for establishing a relation between Euclidean and Minkowski BS solutions. Indeed, if g could be determined either from the LF wave function—inverting (9)—or from an Euclidean BS solution—inverting (7)—it could be used to computed the Minkowski amplitude and related observables. Care must be taken however [4] with the ill-defined character in the numerical inversion of (7) and (9).

Further applications of this LF/Nakanishi approach have been developed in the recent years.

- The solutions of (8) were used to compute the electromagnetic form factors [17,18] in terms of g.
- Equation (8) was generalized to the two-fermion system. The usual one-boson exchange kernels have been derived and the bound state solution for the $J^{\pi} = 0^{+}$ states with the Yukawa coupling have been obtained in [10].
- In [11], equation (8) was written in the form $g = K' g$. This non trivial analytic inversion of the left-hand-side kernel of (8), allows to compute unambiguously the function g and study its analytic properties. The bound state solutions found by solving this equation were in perfect agreement with the preceding results. The equation was extended to the scattering states and the scattering length could be accurately computed [15].

3 Minkowski Space Solutions: Direct Approach

A second method was developed in [12,13] aiming to directly solve the original equation without any transform and taking into account all the singularities of equation (4). This direct approach has proved its efficiency both for the bound and, especially, for the scattering states. It provided for the first time the full off-shell scattering amplitude at the price of a careful analytical work to transform all the singular terms, from the kernels as well as from the propagators, into a smooth integrands.

For instance, the BS equation for the S-wave can be written in the following form

$$
\begin{aligned}
F_0(k_0,k) &= F_0^B(k_0,k)\\
&+\frac{i\pi^2 k_s}{8\varepsilon_{k_s}} W_0^S(k_0,k,0,k_s)F_0(0,k_s)\\
&+\frac{\pi}{2M}\int_0^\infty \frac{dk'}{\varepsilon_{k'}(2\varepsilon_{k'}-M)}\left[k'^2 W_0^S(k_0,k,a_-,k')F_0(|a_-|,k') - \frac{2{k_s}^2\varepsilon_{k'}}{\varepsilon_{k'}+\varepsilon_{k_s}} W_0^S(k_0,k,0,k_s)F_0(0,k_s)\right]\\
&-\frac{\pi}{2M}\int_0^\infty \frac{k'^2 dk'}{\varepsilon_{k'}(2\varepsilon_{k'}+M)} W_0^S(k_0,k,a_+,k')F_0(a_+,k')\\
&+\frac{i}{2M}\int_0^\infty \frac{k'^2 dk'}{\varepsilon_{k'}}\int_0^\infty dk'_0\left[\frac{W_0^S(k_0,k,k'_0,k')F_0(k'_0,k') - W_0^S(k_0,k,a_-,k')F_0(|a_-|,k')}{k'^2_0 - a_-^2}\right]\\
&-\frac{i}{2M}\int_0^\infty \frac{k'^2 dk'}{\varepsilon_{k'}}\int_0^\infty dk'_0\left[\frac{W_0^S(k_0,k,k'_0,k')F_0(k'_0,k') - W_0^S(k_0,k,a_+,k')F_0(a_+,k')}{k'^2_0 - a_+^2}\right]
\end{aligned}
\tag{10}
$$

where W_0^S is the S-wave kernels and F_0^B the corresponding Born term. Notice, apart from the inhomogeneous term, the existence of zero-, one- and two-dimensional integral terms. The details of the derivation and some illustrative results can be found in [13].

Having computed the scattering solutions, the corresponding "bound → scattering state" transition electromagnetic form factor was found in [19].

4 Deriving an Euclidean Equation

In contrast to the bound state case, the scattering BS equation cannot be in general transformed into an equation for the Euclidean amplitude F_E alone. Indeed, let us consider equation (4) and assume we want to obtain from it an equation for the Euclidean amplitude F_E defined in (5). In order to properly apply the Wick rotation, we have to take into account the pole singularities associated with the propagators in (4) which are given by

$$
\begin{aligned}
k'^{(1)}_0(k,k_s) &= \ \ \varepsilon_{k_s}+\varepsilon_{k'}-i\epsilon = +a_+ - i\epsilon\\
k'^{(2)}_0(k,k_s) &= \ \ \varepsilon_{k_s}-\varepsilon_{k'}+i\epsilon = -a_- + i\epsilon\\
k'^{(3)}_0(k,k_s) &= -\varepsilon_{k_s}+\varepsilon_{k'}-i\epsilon = +a_- - i\epsilon\\
k'^{(4)}_0(k,k_s) &= -\varepsilon_{k_s}-\varepsilon_{k'}+i\epsilon = -a_+ + i\epsilon
\end{aligned}
\tag{11}
$$

with $a_\pm(k',k_s)=\varepsilon_{k'}\pm\varepsilon_{k_s}$

In the case $k'<k_s$, their positions in the complex k_0-plane are shown in Fig. 1. When the integration contour is rotated, the singularities $k'^{(2)}_0$ and $k'^{(3)}_0$ are crossed and the corresponding residues of the integrand at these poles will contribute to the integral term.

Equation (4) is then transformed into:

$$
F^E(k_4,\mathbf{k};\mathbf{k}_s) = V^B(k_4,\mathbf{k};\mathbf{k}_s) + \int \frac{d^4k'}{(2\pi)^4}\frac{V(k_4,\mathbf{k}';k'_4,\mathbf{k}')F^E(k'_4,\mathbf{k}';\mathbf{k}_s)}{({k'_4}^2+a_-^2)({k'_4}^2+a_+^2)} + S(k_4,k,k_s) \tag{12}
$$

where V and V^B are respectively the Wick-rotated interaction kernel and Born term.

We emphasize that the obtained equation does not simply follow from the original one after the variable change $k_0\to ik_4$. The additional term S is the contribution of the two poles $k'^{(2)}_0$ and $k'^{(3)}_0$ displayed in Fig. 1. It can be shown [13,20] that S depends on the Minkowski amplitude at the particular value of its k_0 argument

$$
F_M(k_0=\varepsilon_{k_s}-\varepsilon_k,k)
$$

Consequently, when rotating the initial BS scattering equation, the Euclidean and Minkowski amplitudes are in general coupled and one cannot obtain a scattering equation involving F_E alone.

A second integral equation is required and after some algebraic manipulations [13] one ends with a system of two coupled integral equations involving both F_M and F_M and denoted symbolically

$$
F_E(k_4,\mathbf{k}) = \mathcal{I}_1[F_E(k'_4,\mathbf{k}'),F_M(\varepsilon_{k_s}-\varepsilon_{k'},\mathbf{k}')]
$$

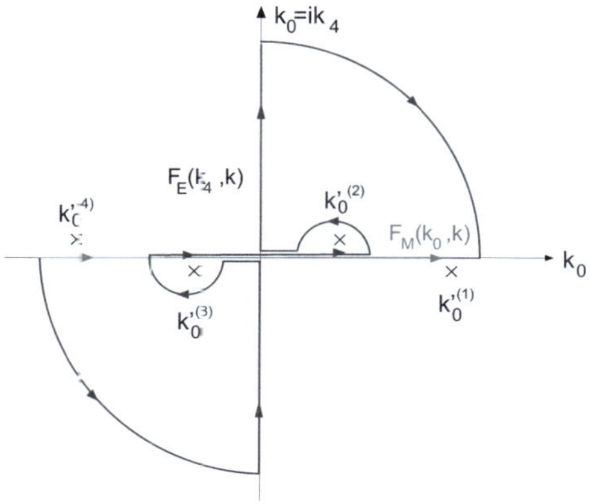

Fig. 1 Singularities of propagators for scattering state and integration contour in complex k_0 plane ($k' < k_s$)

$$F_M(\varepsilon_{k_s} - \varepsilon_k, \mathbf{k}) = \mathcal{I}_2[F_E(k_4', \mathbf{k}'), F_M(\varepsilon_{k_s} - \varepsilon_{k'}, \mathbf{k}')] \tag{13}$$

This system of equations—whose Minkowski part remains singular—was first derived in [21,22], re-derived and solved in [13] to check our direct Minkowski solution F_M at the particular value $k_0 = \varepsilon_{k_s} - \varepsilon_k$. The solution of (13) provides F_E in the full domain but F_M is limited to a restricted manifold and does not provide the full solution of the scattering problem. However, on mass-shell $k_0 = k_4 = 0$ and since $F_M(0, k) = F_E(0, k)$, the Euclidean amplitude computed this way can be also used to obtain the phase shifts. The results provided by our direct method (10) are in agreement with the solution of (13) and with those of [21,22].

4.1 The Case of Zero Energy Scattering

Of particular interest is the zero energy scattering. It can be shown [20] that in the limit $k_s \to 0$ the term S in Eq. (12) vanishes and the four-dimensional BS equation for the Euclidean amplitude in the zero energy limit takes the form

$$F^E(k_4, \mathbf{k}; 0) = V^B(k_4, \mathbf{k}; 0) + \int \frac{d^4k'}{(2\pi)^4} \frac{V(k_4, \mathbf{k}'; k_4', \mathbf{k}')}{(k_4'^2 + a_-^2)(k_4'^2 + a_+^2)} F^E(k_4', \mathbf{k}'; 0) \tag{14}$$

This is a self-consistent integral equation for determining $F^E(k_4, \mathbf{k}; \mathbf{k}_s = 0)$ without any coupling to the Minkowski amplitude F. This decoupling is valid only in the limit $k_s \to 0$.

In the S-wave case, Eq. (14) reduces to the following two-dimensional integral equation:

$$F_0(k_4, k) = V_0^B(k_4, k) + \int_0^\infty k'^2 dk' \int_0^\infty dk_4' \frac{V_0(k_4, k; k_4', k')}{(k_4'^2 + a_-'^2)(k_4'^2 + a_+'^2)} F_0(k_4', k') \tag{15}$$

with $a_\pm' = \sqrt{m^2 + k'^2} \pm m$ and where we have omitted the superscript on F^E. This equation determines the S-wave zero energy half-mass shell scattering amplitude F_0. The scattering length is given by

$$a_0 = -\frac{1}{m} F_0(k_4 = 0, k = 0) \tag{16}$$

A more detailed derivation can be found in [20].

The interest in deriving Eq. (15) is not only a significant simplification in computing the BS scattering length but in the fact that this fundamental quantity can be obtained from a purely Euclidean solution. Although considered here in the approximate framework of BS equation, the "exact" Euclidean amplitude (1)—solution of the full QFT problem—is nowadays accessible in the Lattice approach. It is the basic ingredient of the HAL-QCD collaboration to obtain the *ab initio* Nucleon-Nucleon potential from QCD [23] but it has never been used to extract the scattering observables.

The possibility to compute a scattering amplitude from Euclidean correlators in infinite space is forbidden by the Maiani-Testa theorem [24]. This result is in agreement with the impossibility discussed above to obtain

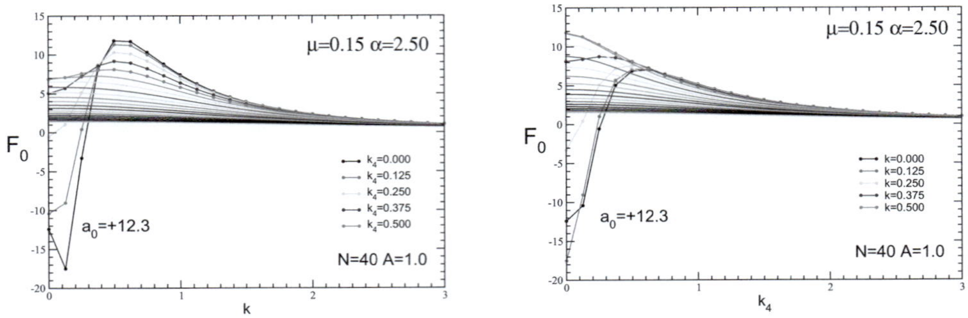

Fig. 2 Euclidean scattering amplitude $F_0(k_4, k)$ as a function of k for different values of k_4 (left) and as a function of k_4 for different values of k (right). It corresponds to $m = 1$, $\mu = 0.15$ and $\alpha = 2.5$. The scattering length value is given by $a_0 = -F(0, 0) = 12.3$

BS scattering observable without a coupling to the Minkowski amplitudes, but does not apply to zero scattering energy where the phase shifts vanish.

Luscher et al. [25–28] circumvented this problem and proposed a method to compute the phaseshifts based on the V-dependence of the 2-particle energies confined in a box with periodic boundary conditions. This approach was successfully applied to several hadronic systems in *ab initio* QCD [29].

The use of Eq. (16) constitutes an alternative method to compute the scattering length in the Lattice calculations, directly from the Fourier transform of the Euclidean BS amplitude defined in (1). In the framework of BS equation, such a possibility is justified by the existence of Eq. (15) allowing to compute the Euclidean BS amplitude in a purely Euclidean formalism as in the Lattice approach.

The numerical solutions of Eq. (15) with one-boson exchange kernel were obtained in [20]. The scattering length a_0 was extracted from the Euclidean amplitude (16) at the origin. The value is in full agreement with the results of our previous work [12] and from those of Ref. [15], obtained by independent methods.

Figure 2 displays the amplitude $F_0(k_4, k)$ as a function of k at fixed values of k_4 (left panel) and as a function of k_4 at fixed values of k (right panel) for the parameters $m = 1$, $\mu = 0.15$ and $\alpha = 2.5$. The scattering length $a_0 = 12.3$ is directly readable in both panels. For these parameters the two-body system has two bound states, which are responsible for the structure in the amplitude.

5 Conclusion

We reviewed different methods to obtain the scattering solutions of the Bethe–Salpeter equation in Minkowski space.

We showed that the Bethe–Salpeter scattering amplitude at zero energy obeys a purely Euclidean equation, as it is the case for the bound states. This decoupling between Euclidean and Minkowski BS amplitudes is only possible for zero energy scattering and allows determining the scattering length from the Euclidean BS amplitude.

Such a possibility suggests to extract the scattering length in Lattice calculations from a direct computation of the Euclidean Bethe–Salpeter amplitude (1) in momentum space and provides an alternative to the Luscher method.

References

1. Salpeter, E.E., Bethe, H.: A relativistic equation for bound-state problems. Phys. Rev. **84**, 1232 (1951)
2. Gell-Mann, M., Low, F.: Bound states in quantum field theory. Phys. Rev. **84**, 350 (1951)
3. Wick, G.C.: Properties of Bethe–Salpeter wave functions. Phys. Rev. **96**, 1124 (1954)
4. Frederico T., Carbonell J., Gigante V., Karmanov, V.A.: Inverting the Nakanishi integral relation for a bound state Euclidean Bethe-Salpeter amplitude. Few-Body Syst. (2016). doi:10.1007/s00601-016-1057-5
5. Kusaka, K., Williams, A.G.: Solving the Bethe–Salpeter equation for scalar theories in Minkowski space. Phys. Rev. D **51**, 7026 (1995)
6. Kusaka, K., Simpson, K., Williams, A.G.: Solving the Bethe-Salpeter equation for bound states of scalar theories in Minkowski space. Phys. Rev. D **56**, 5071 (1997)
7. Karmanov, V.A., Carbonell, J.: Solving Bethe–Salpeter equation in Minkowski space. Eur. Phys. J. A **27**, 1 (2006)

8. Carbonell, J., Karmanov, V.A.: Cross-ladder effects in Bethe–Salpeter and Light–Front equations. Eur. Phys. J. A **27**, 11 (2006)
9. Sauli, V.: Solving the Bethe–Salpeter equation for a pseudoscalar meson in Minkowski space. J. Phys. G **35**, 035005 (2008)
10. Carbonell, J., Karmanov, V.A.: Solving Bethe–Salpeter equation for two fermions in Minkowski space. Eur. Phys. J. A **46**, 387–397 (2010)
11. Frederico, T., Salmè, G., Viviani, M.: Two-body scattering states in Minkowski space and the Nakanishi integral representation onto the null plane. Phys. Rev. D **85**, 036009 (2012)
12. Carbonell, J., Karmanov, V.A.: Bethe–Salpeter scattering amplitude in Minkowski space. Phys. Lett. B **727**, 319 (2013)
13. Carbonell, J., Karmanov, V.A.: Solving Bethe–Salpeter scattering state equation in Minkowski space. Phys. Rev. D **90**, 056002 (2014)
14. Frederico, T., Salmé, G., Viviani, M.: Quantitative studies of the homogeneous Bethe-Salpeter equation in Minkowski space. Phys. Rev. D **89**, 016010 (2014)
15. Frederico, T., Salmè, G., Viviani, M.: Solving the inhomogeneous Bethe-Salpeter equation in Minkowski space: the zero-energy limit. Eur. Phys. J. C **75**, 398 (2015)
16. Nakanishi, N.: Partial-wave Bethe–Salpeter equation. Phys. Rev. **130**, 1230 (1963)
17. Carbonell, J., Karmanov, V.A., Mangin-Brinet, M.: Electromagnetic form factor via Bethe–Salpeter amplitude in Minkowski space. Eur. Phys. J. A **39**, 53 (2009)
18. Carbonell, J., Karmanov, V.A.: Solutions of the Bethe-Salpeter equation in Minkowski space and applications to electromagnetic form factors. Few-Body Syst. **49**, 205 (2011)
19. Carbonell, J., Karmanov, V.A.: Transition electromagnetic form factor and current conservation in the Bethe–Salpeter approach. Phys. Rev. D **91**, 076010 (2015)
20. Carbonell, J., Karmanov, V.A.: Zero energy scattering calculation in Euclidean space. Phys. Lett. B **754**, 270 (2016)
21. Levine, M.J., Tjon, J.A., Wright, J.: Nonsingular Bethe–Salpeter equation. Phys. Rev. Lett. **16**, 962 (1966)
22. Levine, M.J., Wright, J., Tjon, J.A.: Solution of the Bethe–Salpeter equation in the inelastic region. Phys. Rev. **154**, 1433 (1967)
23. Ishii, N., Aoki, S., Hatsuda, T.: The nuclear force from lattice QCD. Phys. Rev. Lett. **99**, 022001 (2007)
24. Maiani, L., Testa, M.: Final state interactions from euclidean correlation functions. Phys. Lett. B **245**, 585 (1990)
25. Luscher, M.: Volume dependence of the energy spectrum in massive quantum field theories. Commun. Math. Phys. **104**, 177 (1986)
26. Luscher, M.: Volume dependence of the energy spectrum in massive quantum field theories. Commun. Math. Phys. **105**, 153 (1986)
27. Luscher, M.: Two-particle states on a torus and their relation to the scattering matrix. Nucl. Phys. B **354**, 531 (1991)
28. Luscher, M.: Signatures of unstable particles in finite volume. Nucl. Phys. B **364**, 237 (1991)
29. Detmold, W.: Nuclear physics from lattice QCD. Lect. Notes Phys. **889**, 153 (2015)

DOI 10.1007/978-3-319-50699-9_50
Reprinted from *Few-Body System* Journal, DOI 10.1007/s00601-016-1106-0

S. S. Chabysheva

Light-Front ϕ^4_{1+1} Theory Using a Many-Boson Symmetric-Polynomial Basis

Received: 29 December 2015 / Accepted: 7 April 2016 / Published online: 26 April 2016

Abstract We extend earlier work on fully symmetric polynomials for three-boson wave functions to arbitrarily many bosons and apply these to a light-front analysis of the low-mass eigenstates of ϕ^4 theory in 1+1 dimensions. The basis-function approach allows the resolution in each Fock sector to be independently optimized, which can be more efficient than the preset discrete Fock states in DLCQ. We obtain an estimate of the critical coupling for symmetry breaking in the positive mass-squared case.

1 Introduction

Our objective is to use newly developed multivariate polynomials [1,2] as a basis set for the computation of the odd and even-parity massive eigenstates of light-front ϕ^4_{1+1} theory [3–8]. The eigenvalue problem is solved in the form of a truncated Fock-state expansion, with each Fock wave function expanded in the polynomials. This allows separate tuning of the resolution in each Fock sector, which should be more efficient than a discrete light-cone quantization (DLCQ) [9–11] calculation [12]. We can then study convergence with respect to the Fock-space truncation and compare results with those obtained with the light-front coupled-cluster (LFCC) method [13,14]. We also estimate the value of the critical coupling for symmetry breaking.

The Lagrangian for two-dimensional ϕ^4 theory is $\mathcal{L} = \frac{1}{2}(\partial_\mu\phi)^2 - \frac{1}{2}\mu^2\phi^2 - \frac{\lambda}{4!}\phi^4$, where μ is the mass of the boson and λ is the coupling constant. The light-front Hamiltonian density is $\mathcal{H} = \frac{1}{2}\mu^2\phi^2 + \frac{\lambda}{4!}\phi^4$. The mode expansion for the field at zero light-front time is

$$\phi = \int \frac{dp^+}{\sqrt{4\pi p^+}}\left\{a(p^+)e^{-ip^+x^-/2} + a^\dagger(p^+)e^{ip^+x^-/2}\right\}, \tag{1}$$

with the modes quantized such that $[a(p^+), a^\dagger(p'^+)] = \delta(p^+ - p'^+)$. The light-front Hamiltonian is $\mathcal{P}^- = \mathcal{P}^-_{11} + \mathcal{P}^-_{13} + \mathcal{P}^-_{31} + \mathcal{P}^-_{22}$, with

$$\mathcal{P}^-_{11} = \int dp^+ \frac{\mu^2}{p^+} a^\dagger(p^+)a(p^+), \tag{2}$$

$$\mathcal{P}^-_{13} = \frac{\lambda}{6}\int \frac{dp_1^+ dp_2^+ dp_3^+}{4\pi\sqrt{p_1^+ p_2^+ p_3^+(p_1^+ + p_2^+ + p_3^+)}} a^\dagger(p_1^+ + p_2^+ + p_3^+)a(p_1^+)a(p_2^+)a(p_3^+), \tag{3}$$

S. S. Chabysheva (✉)
Department of Physics and Astronomy, University of Minnesota-Duluth, Duluth, MN 55812, USA
E-mail: schabysh@d.umn.edu

$$\mathcal{P}_{31}^{-} = \frac{\lambda}{6}\int \frac{dp_1^+dp_2^+dp_3^+}{4\pi\sqrt{p_1^+p_2^+p_3^+(p_1^++p_2^++p_3^+)}} a^\dagger(p_1^+)a^\dagger(p_2^+)a^\dagger(p_3^+)a(p_1^++p_2^++p_3^+), \tag{4}$$

$$\mathcal{P}_{22}^{-} = \frac{\lambda}{4}\int \frac{dp_1^+dp_2^+}{4\pi\sqrt{p_1^+p_2^+}}\int \frac{dp_1'^+dp_2'^+}{\sqrt{p_1'^+p_2'^+}}\delta(p_1^++p_2^+-p_1'^+-p_2'^+)a^\dagger(p_1^+)a^\dagger(p_2^+)a(p_1'^+)a(p_2'^+). \tag{5}$$

The eigenstate with momentum P^+ is expanded as

$$|\psi(P^+)\rangle = \sum_m (P^+)^{\frac{m-1}{2}}\int \prod_i^m dy_i\delta(1-\sum_i^m y_i)\psi_m(y_i)\frac{1}{\sqrt{m!}}\prod_{i=1}^m a^\dagger(y_iP^+)|0\rangle. \tag{6}$$

The sum over m is restricted to odd or even numbers. The Hamiltonian does not mix the two cases, and we solve for the lowest eigenstate in each case.

With use of the Fock-state expansion, the light-front Hamiltonian eigenvalue problem $\mathcal{P}^-|\psi(P)\rangle = \frac{M^2}{P}|\psi(P)\rangle$ becomes

$$\begin{aligned}
&\frac{m}{y_1}\psi_m(y_i) + \frac{g}{4}\frac{m(m-1)}{\sqrt{y_1y_2}}\int\frac{dx_1dx_2}{\sqrt{x_1x_2}}\delta(y_1+y_2-x_1-x_2)\psi_m(x_1,x_2,y_3,\dots,y_m)\\
&+\tfrac{g}{6}m\sqrt{(m+2)(m+1)}\int\tfrac{dx_1dx_2dx_3}{\sqrt{y_1x_1x_2x_3}}\delta(y_1-x_1-x_2-x_3)\psi_{m+2}(x_1,x_2,x_3,y_2,\dots,y_m)\\
&\quad+\tfrac{g}{6}\tfrac{(m-2)\sqrt{m(m-1)}}{\sqrt{y_1y_2y_3(y_1+y_2+y_3)}}\psi_{m-2}(y_1+y_2+y_3,y_4,\dots,y_m) = \tfrac{M^2}{\mu^2}\psi_m(y_i).
\end{aligned} \tag{7}$$

Here $g=\lambda/4\pi\mu^2$ is a dimensionless coupling. It is to this system of equations that we apply our basis-function expansion for each Fock-state wave function, to convert the coupled integral equations to a matrix eigenvalue problem.

2 Solution of the Integral Equations

We solve this coupled system by first truncating the Fock-state expansion at some maximum number of constituents and then expanding each wave function in a basis of symmetric multivariate polynomials $P_{ni}^{(m)}$

$$\psi_m(y_i) = \sqrt{\prod_i y_i}\sum_{ni} c_{ni}^{(m)}P_{ni}^{(m)}(y_1,\dots,y_m). \tag{8}$$

The sum is truncated in order to have a finite-sized numerical calculation. This truncation can be different in each Fock sector, which allows the resolution to be tuned separately in each sector. Thus, more polynomials can be added in those Fock sectors where they are needed.

The $P_{ni}^{(m)}$ are of order n and fully symmetric with respect to interchange of momenta [1,2]. The subscript i differentiates the various possibilities at a given order n. For $m=2$ constituents there is only one possibility at each order, but for $m>2$ there can be more than one. For example, for three constituents there are two sixth-order polynomials, $P_{61}^{(3)}=(y_1y_2y_3)^2$ and $P_{62}^{(3)}=(y_1y_2+y_1y_3+y_2y_3)^3$.

The number of linearly independent polynomials of a given order is restricted by the constraint of momentum conservation, $\sum_i y_i=1$. For example, $P_2^{(3)}=y_1y_2+y_1y_3+y_2y_3$ is equivalent to $y_1^2+y_2^2+y_3^2$, up to a constant, when y_3 is replaced by $1-y_1-y_2$.

The linearly independent symmetric polynomials can be written as products of powers of simpler polynomials, in the form $P_{ni}^{(N)}=C_2^{n_2}C_3^{n_3}\cdots C_N^{n_N}$, with the powers restricted by $n=\sum_j jn_j$. Each different way of decomposing n into a sum of integers greater than 1 yields a different polynomial. The C_m are sums of simple monomials $\prod_j^N y_j^{m_j}$, where m_j is 0 or 1 and $\sum_j^N m_j=m$. The sum over the monomials ranges over all possible choices for the m_j, making each C_m fully symmetric.

As examples of the C_m, consider the general case of N longitudinal momentum variables. Then C_2 is just $\sum_j^N\left(y_j\sum_{k>j}^N y_k\right)$; C_{N-1} is $\sum_j^N\prod_{k\neq j}y_k$; and C_N is $y_1y_2\cdots y_N$. In particular, for N=3, $C_2=y_1y_2+y_1y_3+$

$y_2 y_3$ and $C_3 = y_1 y_2 y_3$. The first-order polynomial $C_1 = \sum_j y_j$ does not appear because the momentum constraint reduces it to a constant.

Projection of the coupled system onto the basis functions yields the matrix equations

$$\sum_{n'i'} \left[T^{(m)}_{ni,n'i'} + g V^{(m,m)}_{ni,n'i'}\right] c^{(m)}_{n'i'} + g \sum_{n'i'} V^{(m,m+2)}_{ni,n'i'} c^{(m+2)}_{n'i'} + g \sum_{n'i'} V^{(m,m-2)}_{ni,n'i'} c^{(m-2)}_{n'i'} = \frac{M^2}{\mu^2} \sum_{n'i'} B^{(m)}_{ni,n'i'} c^{(m)}_{n'i'}, \tag{9}$$

with $T^{(m)}$ the kinetic-energy matrix in the mth Fock sector

$$T^{(m)}_{ni,n'i'} = m \int \left(\prod_j dy_j\right) \delta(1 - \sum_j y_j) \left(\prod_{j=2}^m y_j\right) P^{(m)}_{ni}(y_j) P^{(m)}_{n'i'}(y_j), \tag{10}$$

$B^{(m)}$ the basis-function overlap matrix

$$B^{(m)}_{ni,n'i'} = \int \left(\prod_j dy_j\right) \delta(1 - \sum_j y_j) \left(\prod_j^m y_j\right) P^{(m)}_{ni}(y_j) P^{(m)}_{n'i'}(y_j), \tag{11}$$

and $V^{(m,m')}$ the potential-energy matrices

$$V^{(m,m)}_{ni,n'i'} = \frac{g}{4} m(m-1) \int \left(\prod_j dy_j\right) \delta(1 - \sum_j y_j \tag{12}$$

$$\times \int dx_1 dx_2 \delta(y_1 + y_2 - x_1 - x_2) \left(\prod_{j=3}^m y_j\right) P^{(m)}_{ni}(y_j) P^{(m)}_{n'i'}(x_1, x_2, y_3, \ldots, y_m),$$

$$V^{(m,m+2)}_{ni,n'i'} = \frac{g}{6} m \sqrt{(m+2)(m+1)} \int \left(\prod_j dy_j\right) \delta(1 - \sum_j y_j) \tag{13}$$

$$\times \int dx_1 dx_2 dx_3 \delta(y_1 - x_1 - x_2 - x_3) \left(\prod_{j=2}^m y_j\right) P^{(m)}_{ni}(y_j) P^{(m+2)}_{n'i'}(x_1, x_2, x_3, y_2, \ldots, y_m),$$

$$V^{(m,m-2)}_{ni,n'i'} = \frac{g}{6} (m-2) \sqrt{m(m-1)} \int \left(\prod_j dy_j\right) \delta(1 - \sum_j y_j) \tag{14}$$

$$\times \left(\prod_{j=4}^m y_j\right) P^{(m)}_{ni}(y_j) P^{(m-2)}_{n'i'}(y_1 + y_2 + y_3, y_4, \ldots, y_m).$$

All of the integrals can be done analytically in terms of a generalized beta function.

We now have a generalized eigenvalue problem of the form $H\mathbf{c} = (M^2/\mu^2) B\mathbf{c}$. A standard approach would be to factorize B and convert the original problem to an ordinary eigenvalue problem. However, factorization can fail in practice due to round-off errors in the implicit orthogonalization of the basis. Round-off errors also plague an explicit orthogonalization. A reliable factorization is a singular-value decomposition [15] $B = UDU^T$, where the columns of the matrix U are the eigenvectors of B and D is a diagonal matrix of the eigenvalues of B. We then solve $H'\mathbf{c}' = (M^2/\mu^2)\mathbf{c}'$, with $H' = D^{-1/2} U^T H U D^{-1/2}$ and $\mathbf{c}' = D^{1/2} U^T \mathbf{c}$. Results for particular truncations of the polynomial basis are extrapolated to an infinite basis size in each Fock sector.

3 Results

The results for the odd and even cases are shown in Fig. 1, where M^2 is plotted in units of the bare mass squared μ^2 for a range of dimensionless coupling strengths g. The convergence with respect to Fock sector truncation is easily seen to be rapid, with the last two truncations yielding identical results to within errors in each case. We also compare the lowest order LFCC results [14] for the odd case and find that these are quite consistent with the converged Fock-space calculation, even though the LFCC calculation involves only a three-body function.

For both the odd and even cases, the mass crosses zero at a finite value of the coupling, with the massive eigenstate becoming degenerate with the Fock vacuum. We interpret this as the appearance of symmetry breaking and extract a value of the critical coupling, with use of the results as plotted in Fig. 2. The odd and even cases cross zero at nearly the same value. As a check, we plot points for four times the M^2 values in the odd case; in an exact calculation these should coincide with the even values, and here they are close, consistent with the errors in each. From the figure we estimate the critical coupling to be $g_c = 2.1 \pm 0.05$.

Comparison with other calculations of the critical coupling, as summarized in Table 1, can be readily made. However, the values compiled by Rychkov and Vitale [16] are normalized in a slightly different manner; the translation from our g to theirs is $\bar{g} = \frac{\pi}{6}g$. Clearly there is a systematic difference between equal-time and light-front values. However, this is consistent with the expectation that the renormalization of the mass μ is different in the two quantizations [17]. Calculations to evaluate this difference quantitatively are underway. We should also mention that the equal-time calculations [16] include a separate calculation of the vacuum state, an unnecessary step for light-front calculations where the Fock vacuum is trivial.

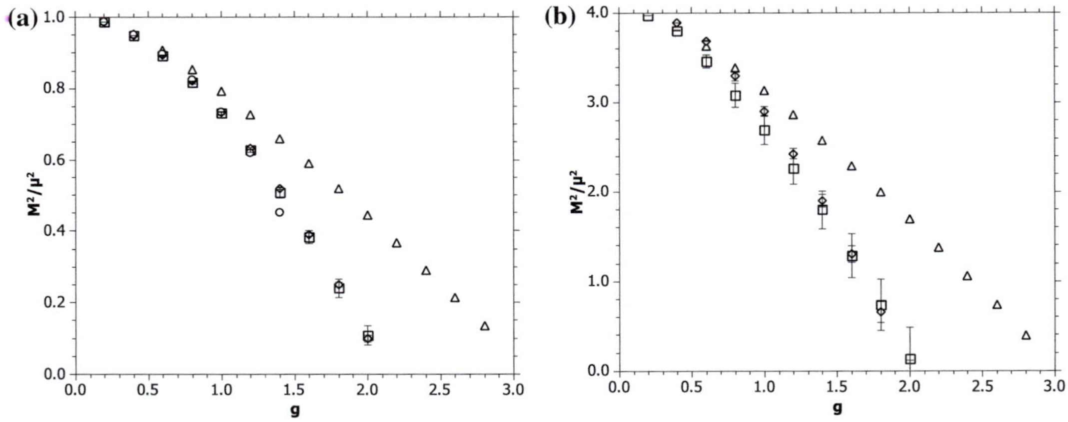

Fig. 1 Mass squared vs coupling strength for an **a** odd and **b** even number of constituents. The different Fock-space truncations in **a** are the three-body (*triangles*), five-body (*squares*), and seven-body (*diamonds*) Fock sectors. Results for the LFCC method (*circles*) are also included. In **b** the different truncations are the four-body (*triangles*), six-body (*squares*), and eight-body (*diamonds*) Fock sectors. *Error bars* are determined by the fits to extrapolation in the polynomial basis size

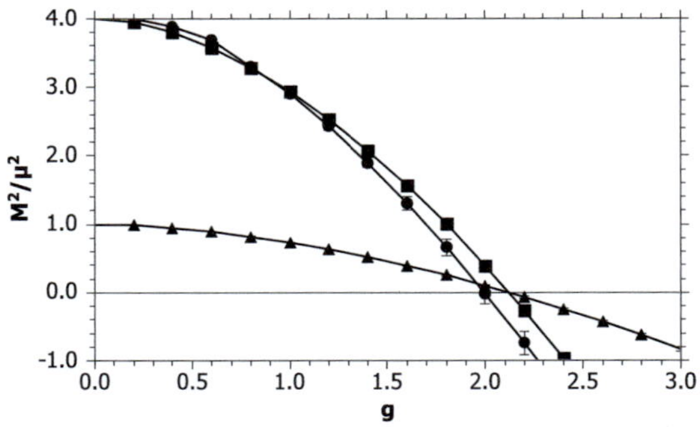

Fig. 2 Mass squared vs coupling strength, as used to estimate the critical coupling. The odd (*triangles*) and even (*circles*) cases are represented by the seven-body and eight-body truncations, respectively. Points at twice the mass of the odd case (*squares*) are also plotted

Table 1 Comparison of values for the critical coupling, with $\bar{g} = \frac{\pi}{6}g$, as compiled in Ref. [16]. The first two use light-front quantization; the remainder use equal-time quantization

Method	$\bar{g}_c$	Reported by
LF symmetric polynomials	1.1 ± 0.03	This talk
DLCQ	1.38	Harindranath and Vary [12]
Quasi-sparse eigenvector	2.5	Lee and Salwen [18]
Density matrix renormalization group	2.4954 (4)	Sugihara [19]
Lattice Monte Carlo	$2.70\left\{\begin{matrix}+0.025\\-0.013\end{matrix}\right.$	Schaich & Loinaz [20]
Uniform matrix product	2.766 (5)	Milsted et al. [21]
Renormalized Hamiltonian truncation	2.97 (14)	Rychkov and Vitale [16]

4 Summary

We have developed a high-order method for (1+1)-dimensional light-front theories that is distinct from DLCQ. The method employs function expansions in terms of fully symmetric multivariate polynomials that respect the constraint of momentum conservation. It allows separate tuning of resolutions in each Fock sector, and could be combined with transverse discretization or basis functions for applications to (3+1)-dimensional theories.

The method has been applied to ϕ^4_{1+1} theory, to compute the lowest mass eigenvalues and to extract an estimate of critical coupling for the positive μ^2 case. We have identified a systematic difference with equal-time quantization which can be associated with the difference in mass renormalizations of the two quantizations [17]. We have also compared these converged high-order Fock space truncations with the lowest-order LFCC calculation [14] and found good agreement, which implies that the LFCC method shows promise for rapid convergence.

Acknowledgments This work was done in collaboration with J.R. Hiller and was supported in part by the Minnesota Supercomputing Institute of the University of Minnesota with grants of computing resources. We thank M. Burkardt and L. Martinovic for insightful comments.

References

1. Chabysheva, S.S., Hiller, J.R.: Basis of symmetric polynomials for many-boson light-front wave functions. Phys. Rev. E **90**, 063310 (2014)
2. Chabysheva, S.S., Elliott, B., Hiller, J.R.: Symmetric multivariate polynomials as a basis for three-boson light-front wave functions. Phys. Rev. E **88**, 063307 (2013)
3. Rozowsky, J.S., Thorn, C.B.: Spontaneous symmetry breaking at infinite momentum without P^+ zero modes. Phys. Rev. Lett **85**, 1614–1617 (2000)
4. Kim, V.T., Pivovarov, G.B., Vary, J.P.: Phase transition in light-front ϕ^4_{1+1}. Phys. Rev. D **69**, 085008 (2004)
5. Chakrabarti, D., Harindranath, A., Martinovic, L., Vary, J.P.: Kinks in discrete light cone quantization. Phys. Lett. B **582**, 196–202 (2004)
6. Chakrabarti, D., Harindranath, A., Martinovic, L., Pivovarov, G.B., Vary, J.P.: Ab initio results for the broken phase of scalar light front field theory. Phys. Lett. B **617**, 92–98 (2005)
7. Chakrabarti, D., Harindranath, A., Vary, J.P.: Transition in the spectrum of the topological sector of ϕ^4_2 theory at strong coupling. Phys. Rev. D **71**, 125012 (2005)
8. Martinovic, L.: Spontaneous symmetry breaking in light front field theory. Phys. Rev. D **78**, 105009 (2008)
9. Pauli, H.-C., Brodsky, S.J.: Solving field theory in one space and one time dimension. Phys. Rev. D **32**, 1993–2000 (1985)
10. Pauli, H.-C., Brodsky, S.J.: Discretized light-cone quantization: solution to a field theory in one space and one time dimension. Phys. Rev. D **32**, 2001–2013 (1985)
11. Brodsky, S.J., Pauli, H.-C., Pinsky, S.S.: Quantum chromodynamics and other field theories on the light cone. Phys. Rep. **301**, 299–486 (1998)
12. Harindranath, A., Vary, J.P.: Solving two-dimensional ϕ^4 theory by discretized light-front quantization. Phys. Rev D **36**, 1141–1147 (1987)
13. Chabysheva, S.S., Hiller, J.R.: A light-front coupled-cluster method for the nonperturbative solution of quantum field theories. Phys. Lett. B **711**, 417–422 (2012)
14. Elliott, B., Chabysheva, S.S., Hiller, J.R.: Application of the light-front coupled-cluster method to ϕ^4 theory in two dimensions. Phys. Rev. D **90**, 056003 (2014)
15. Press, W.H., Teukoksky, S.A., Vetterling, W.T., Flannery, B.P.: Numerical Recipes: The Art of Scientific Computing, 3rd edn. Cambridge University Press, Cambridge (2007)
16. Rychkov, S., Vitale, L.G.: Hamiltonian truncation study of the ϕ^4 theory in two dimensions. Phys. Rev. D **91**, 085011 (2015)
17. Burkardt, M.: Light-front quantization of the sine-Gordon model. Phys. Rev. D **47**, 4628–4633 (1993)

18. Lee, D., Salwen, N.: The diagonalization of quantum field hamiltonians. Phys. Lett. B **503**, 223 (2001)
19. Sugihara, T.: Density matrix renormalization group in a two-dimensional lambda ϕ^4 Hamiltonian lattice model. J. High Energy Phys. **05**(2004), 007 (2004)
20. Schaich, D., Loinaz, W.: An improved lattice measurement of the critical coupling in ϕ_2^4 theory. Phys. Rev. D **79**, 056008 (2009)
21. Milsted, A., Haegeman, J., Osborne, T.J.: Matrix product states and variational methods applied to critical quantum field theory. Phys. Rev. D **88**, 085030 (2013)

DOI 10.1007/978-3-319-50699-9_51
Reprinted from *Few-Body System* Journal, DOI 10.1007/s00601-016-1108-y

Alfredo Vega · Ivan Schmidt · Thomas Gutsche · Valery E. Lyubovitskij

Light Front Wave Function for Hadrons with Arbitrary Twist

Received: 1 February 2016 / Accepted: 9 April 2016 / Published online: 6 May 2016

Abstract We present a phenomenological light-front wave function for hadrons with arbitrary twist dimension (mesons, baryons and multiquark states), which gives the correct scaling behavior of structure functions and form factors. Some examples of his uses are presented.

1 Introduction

In this talk we presented a phenomenological light-front wave function (LFWF) for hadrons with arbitrary twist dimension (mesons, baryons, and multiquark states) introduced previously in [1]. At an initial scale, this LFWF is constrained by the soft-wall AdS/QCD model, and at higher scales it gives the correct scaling behavior of parton distribution functions (PDFs) and form factors, both for pions and nucleons. The explicit form of the wave function at large scales is extracted from the hard evolution of PDFs and generalized parton distributions (GPDs). The wave function produces form factors consistent with quark counting rules and also gives predictions for the corresponding parton distributions. This model is by construction consistent with the Drell–Yan–West (DYW) relation [2] between the large-Q^2 behavior of nucleon electromagnetic form factors and the large-x behavior of the structure functions and with quark counting rules [3].

As examples of uses of this LFWF, we present some results for PDFs in Pions and nucleons, form factors in a quark-diquark model and calculation of $s - \bar{s}$ asymmetry in Proton by using the Brodsky–Ma model with LFWF proposed [1,4].

A. Vega (✉)
Instituto de Física y Astronomía y Centro de Astrofísica de Valparaíso, Universidad de Valparaíso, A. Gran Bretaña 1111, Valparaiso, Chile
E-mail: alfredo.vega@uv.cl

I. Schmidt
Departamento de Física y Centro Científico Tecnológico de Valparaíso (CCTVal), Universidad Técnica Federico Santa María, Casilla 110-V, Valparaiso, Chile

T. Gutsche · V. E. Lyubovitskij
Institut für Theoretische Physik, Kepler Center for Astro and Particle Physics, Universität Tübingen, Auf der Morgenstelle 14, 72076 Tübingen, Germany

V. E. Lyubovitskij
Department of Physics, Tomsk State University, Tomsk, Russia 634050

V. E. Lyubovitskij
Mathematical Physics Department, Tomsk Polytechnic University, Lenin Avenue 30, Tomsk, Russia 634050

2 Light Front Wave Function for Hadron with Arbitrary Twist

In [1] we propose a phenomenological LFWF $\psi_\tau(x, k_\perp, \mu)$ for hadrons with arbitrary twist dimension τ (number of constituents) without referring to the flavor structure and with scale dependence μ, with fraction of momentum x and transverse momentum $k_\perp$. The main idea is that at some initial scale, these functions are derived from the matching of the electromagnetic form factors of hadrons with arbitrary spin in the soft-wall AdS/QCD approach and in light-front QCD. Then we propose that these wave functions have a scale dependence, which is also encoded in the corresponding PDFs and GPDs. Next, doing the hard evolution of PDFs and GPDs, we find the evolution of the LFWFs, which fulfill the following constraints: (1) at large scales $\mu \to \infty$ and for large $x \to 1$, this function must produce the scaling of PDFs as $(1-x)^\tau$; (2) at large Q^2, the form factor scales as $1/(Q^2)^{\tau-1}$.

The LFWF at initial scale μ_0 is

$$\psi_\tau(x, k_\perp) = \sqrt{\tau - 1}\frac{4\pi}{\kappa}\sqrt{log(1/x)}(1-x)^{\frac{\tau-4}{2}} exp\left[-\frac{k_\perp^2}{2\kappa^2}\frac{log(1/x)}{(1-x)^2}\right]. \quad (1)$$

The PDFs $q_\tau(x)$ and GPDs at zero skewness $H_\tau(x, Q^2)$ in terms of the LFWFs at the initial scale are defined as

$$q_\tau(x) = \int \frac{d^2k_\perp}{16\pi^3}|\psi_\tau(x, k_\perp)|^2, \quad (2)$$

$$H_\tau(x, Q^2) = \int \frac{d^2k_\perp}{16\pi^3}\psi_\tau^\dagger(x, k'_\perp)\psi_\tau(x, k_\perp). \quad (3)$$

A straightforward calculation gives

$$q_\tau(x) = (\tau - 1)(1-x)^{\tau-2} \quad (4)$$

$$H_\tau(x, Q^2) = q_\tau(x) exp\left[-\frac{Q^2}{4\kappa^2}log(1/x)\right]. \quad (5)$$

In [1] interested reader can find details of an extension of (1), that include an scale that produce expressions for PDFs and GPDs scale dependent.

3 Examples of Uses of this LFWF

3.1 Pion PDFs

Following Ref. [5] in evolving the pion PDF to $\mu = 4$ GeV, we are able reproduce the analysis of the Drell–Yan data from the E615 experiment [6]. Evolving the pion PDF to very large scales $\mu \to \infty$, we reproduce the power scaling $(1-x)^2$ for large $x \to 1$. In Fig. 1, we compare our results, which exactly coincide with the prediction of Ref. [5], to the ones of the E615 experiment [6] for $\mu = 4$ GeV. At this scale, the theoretical result for the pionic PDF is very well approximated in our approach to the form suggested by analysis of the E615 experiment [6].

3.2 PDFs in Nucleons

In Fig. 2, we plot the evolution of the valence u quark PDF in the nucleon and perform a comparison to the global fit of Martin, Stirling, Thorne, and Watt at leading order (MSTWLO) [7]. For x values in the range $0.2 = x = 1$, we get qualitative agreement with the valence u quark PDF extracted from data analysis [7].

3.3 Light Front Quark–Diquark Model for the Nucleon

In [1] we considered the application of the phenomenological LFWF in setting up a light-front quark-diquark model for the nucleon. Here, as an example we show only one Fig. 3 for Dirac proton form factor multiplied by Q^4, but additional properties calculated in the quark-diquark model considered can be found in [1].

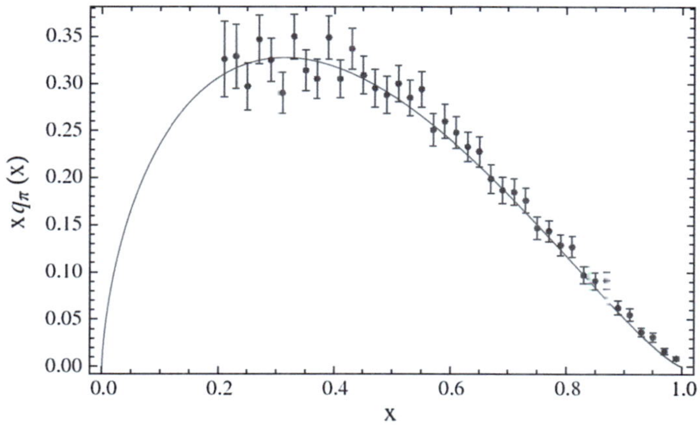

Fig. 1 Comparison of the evolved pion PDF at the scale $\mu = 4$ GeV in Ref. [5] and in our approach to the analysis of the E615 experiment [6]

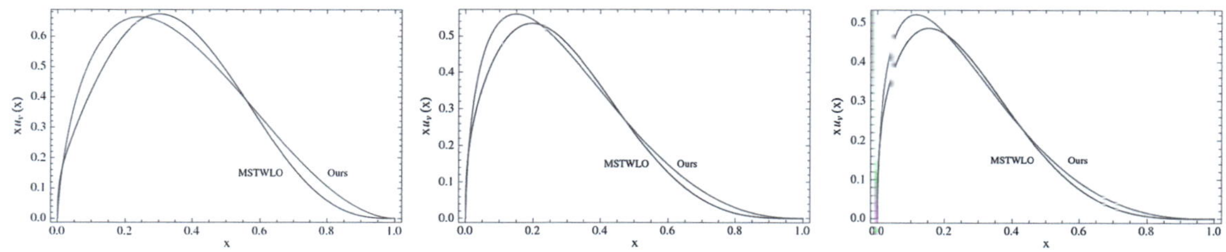

Fig. 2 PDF $x\ u_v(x,\mu)$ at $\mu = 1$ GeV (*left*), $\mu = 10$ GeV (*center*) and $\mu = 100$ GeV (*right*)

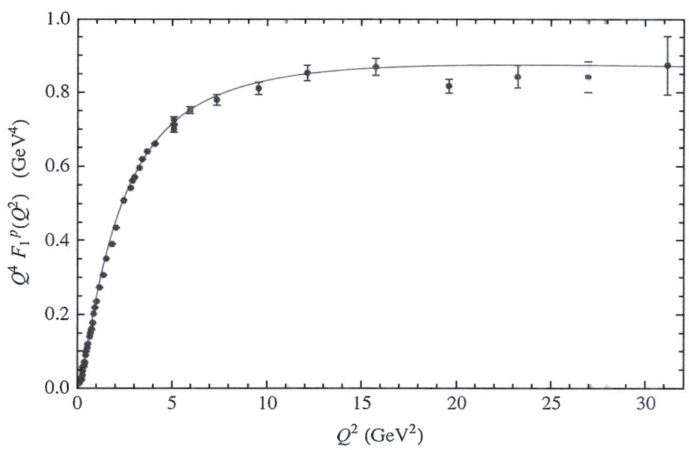

Fig. 3 Dirac proton form factor multiplied by Q^4

3.4 $s(x) - \bar{s}(x)$ Asymmetry in Protons

Recently, in [4] we used the LFWF for hadrons with arbitrary number of constituents to calculate $s - \bar{s}$ asymmetry. In this paper we use three different kinds of the LFWFs: (i) a typical Gaussian LFWF [8] (ii) a so-called holographic LFWFs obtained using light-front holography ideas [9–14] at large x, that we called variant I and (iii) a further generalization which is extracted from matching at any value of x and that further allows to describe hadrons with an arbitrary number of constituents [1] (variant II). In Fig. 4 we show our calculation for $s - \bar{s}$ asymmetry

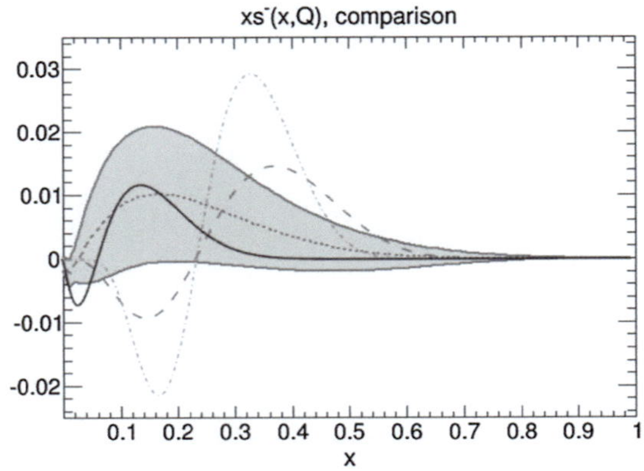

Fig. 4 $xS^- = x(s(x) - \bar{s}(x))$. *Green region* and *small dashed line* correspond to MMHT [15]) that it was generated with APFEL [16]. Other *lines* correspond to Gaussian LFWF (*large dashed line* $\kappa = 330$ MeV, holographic LFWF (variant I), (*dot dashed line* $\kappa = 350$ MeV) and holographic LFWF (variant II) (*continuous line* $\kappa = 350$ MeV) (color figure online)

4 Conclusion

We present a light-front quark model (or wave function) consistent with model independent scaling lawsthe DYW duality [2] and quark counting rules [3]and which is explicitly dependent on the scale resolution. The proposed LFWF it is versatil (e.g. [1,4,17]), and can be used in several hadron model as an alternative to the traditional gaussian wave function, and it can also be generalized to sea quarks, antiquarks and gluons because it explicitly depends on twist dimension (number of constituents). Such an extension work is planned to be done.

In this talk we show some examples where this LFWF has been used successfully.

Acknowledgments This work was supported by the German Bundesministerium für Bildung und Forschung (BMBF) under Project 05P2015—ALICE at High Rate (BMBF-FSP 202): "Jet- and fragmentation processes at ALICE and the parton structure of nuclei and structure of heavy hadrons", Tomsk State University Competitiveness Improvement Program and the Russian Federation program "Nauka" (Contract No. 0.1526.2015, 3854), by FONDECYT (Chile) under Grant Nos. 1100287 and 1141280 and by CONICYT (Chile) Research Project No. 80140097, and under Grant No. 7912010025.

References

1. Gutsche, T., Lyubovitskij, V.E., Schmidt, I., Vega, A.: Light-front quark model consistent with Drell-Yan-West duality and quark counting rules. Phys. Rev. D **89**(5), 054033 (2014) [Phys. Rev. D **92**, no. 1, 019902 (2015)] arXiv:1306.0366 [hep-ph]
2. Drell, S.D., Yan, T.M.: Connection of elastic electromagnetic nucleon form-factors at large Q**2 and deep inelastic structure functions near threshold. Phys. Rev. Lett. **24**, 181 (1970)
3. Brodsky, S.J., Farrar, G.R.: Scaling laws at large transverse momentum. Phys. Rev. Lett. **31**, 1153 (1973)
4. Vega, A., Schmidt, I., Gutsche, T., Lyubovitskij, V.E.: Nonperturbative contribution to the strange-antistrange asymmetry of the nucleon sea. Phys. Rev. D **93**(5), 056001 (2016). arXiv:1511.06476 [hep-ph]
5. Broniowski, W., Ruiz Arriola, E.: Generalized parton distributions of the pion in chiral quark models and their QCD evolution. Phys. Rev. D **77**, 034023 (2008). arXiv:0712.1012 [hep-ph]
6. Conway, J.S., et al.: Experimental study of muon pairs produced by 252-GeV pions on tungsten. Phys. Rev. D **39**, 92 (1989)
7. Martin, A.D., Stirling, W.J., Thorne, R.S., Watt, G.: Parton distributions for the LHC. Eur. Phys. J. C **63**, 189 (2009). arXiv:0901.0002 [hep-ph]
8. Brodsky, S.J., Ma, B.-Q.: The quark-antiquark asymmetry of the nucleon sea. Phys. Lett. B **381**, 317 (1996)
9. Brodsky, S.J., de Teramond, G.F.: Hadronic spectra and light-front wavefunctions in holographic QCD. Phys. Rev. Lett. **96**, 201601 (2006)
10. Brodsky, S.J., de Teramond, G.F.: Light-front dynamics and AdS/QCD correspondence: the pion form factor in the space-and time-like regions. Phys. Rev. D **77**, 056007 (2008)
11. Brodsky, S.J., de Teramond, G.F., Dosch, H.G., Erlich, J.: Light-front holographic QCD and emerging confinement. Phys. Rep. **584**, 1 (2015)
12. Vega, A., Schmidt, I., Branz, T., Gutsche, T., Lyubovitskij, V.E.: Meson wave function from holographic models. Phys. Rev. D **80**, 055014 (2009)

13. Branz, T., Gutsche, T., Lyubovitskij, V.E., Schmidt, I., Vega, A.: Light and heavy mesons in a soft-wall holographic approach. Phys. Rev. D **82**, 074022 (2010)
14. Gutsche, T., Lyubovitskij, V.E., Schmidt, I., Vega, A.: Dilaton in a soft-wall holographic approach to mesons and baryons. Phys. Rev. D **85**, 076003 (2012)
15. Harland-Lang, L.A., Martin, A.D., Motylinski, P., Thorne, R.S.: Parton distributions in the LHC era: MMHT 2014 PDFs. Eur. Phys. J. C **75**(5), 204 (2015)
16. Carrazza, S., Ferrara, A., Palazzo, D., Rojo, J.: APFEL Web: a web-based application for the graphical visualization of parton distribution functions. J. Phys. G **42**(5), 057001 (2015)
17. Gutsche, T., Lyubovitskij, V.E., Schmidt, I., Vega, A.: Pion light-front wave function, parton distribution and the electromagnetic form factor. J. Phys. G **42**(9), 095005 (2015) arXiv:1410.6424 [hep-ph]

DOI 10.1007/978-3-319-50699-9_52
Reprinted from *Few-Body System* Journal, DOI 10.1007/s00601-016-1109-x

L'ubomír Martinovič · Pierre Grangé

Two-Dimensional Massless Light Front Fields and Solvable Models

Received: 8 February 2016 / Accepted: 9 April 2016 / Published online: 9 May 2016

Abstract Quantum field theory formulated in terms of light front (LF) variables has a few attractive as well as some puzzling features. The latter hindered a wider acceptance of LF methods. In two space–time dimensions, it has been a long-standing puzzle how to correctly quantize massless fields, in particular fermions. Here we show that two-dimensional massless LF fields (scalar and fermion) can be recovered in a simple way as limits of the corresponding massive fields and thereby quantized without any loss of physical information. Bosonization of the fermion field then follows in a straightforward manner and the solvable models can be studied directly in the LF theory. We sketch the LF operator solution of the Thirring-Wess model and also point out the closeness of the massless LF fields to those of conformal field theory.

1 Introduction

Quantum field theory (QFT) formulated in terms of the Dirac's light-front (LF) space–time and field variables [1,2] differs in a few important aspects from the conventional (instant or space-like—SL) form of the theory. It has only three dynamical Poincaré generators and all three boost operators are kinematical—they do not contain interaction. The vacuum structure is drastically simplified (the Fock vacuum is very often an eigenstate of the full Hamiltonian) and permits one the probabilistic interpretation of the bound-state wave functions. However, it also raises serious questions concerning the description of the symmetry breaking and vacuum condensation in the LF theory. These and additional features (enhanced role of field zero modes, proliferation of constrained variables, etc.) seemed to many physicists so confusing that the LF field theory, despite its great potential [3–9], did not achieve so far a status comparable to the conventional theory.

A particularly disturbing feature was a lack of description of two-dimensional (2D) massless fields. The free massless scalar and fermion fields are the essential ingredients of the exact solutions of the 2D interacting models (Thirring, Thirring-Wess, Federbush, Schwinger…), that have played a major role in understanding the QFT non-perturbatively. If its LF version is to be considered as a more efficient and transparent formulation, it should provide a simple and consistent operator solution of this class of models. This however was not the case. A partial exception is the Schwinger model [10,11]. Unfortunately, even with an extension of the canonical

L. Martinovič (✉)
Institute of Physics, Slovak Academy of Sciences, Dúbravská cesta 9, 845 11 Bratislava, Slovakia
E-mail: fyziluma@savba.sk

Present Address
L. Martinovič
Bogoliubov Laboratory of Theoretical Physics, JINR, Dubna, Russia 141980

P. Grangé
LUPM, Université Montpellier II, Pl. E. Bataillon, 34095 Montpellier Cedex 05, France

LF framework (by introducing additional fermion degrees of freedom with negative LF momenta) the results were not comparable with the generally accepted physical picture of this prototype gauge theory.

At the previous LC meeting [12], we have shown how the 2D massless fermions can be consistently described in the LF theory. In the present contribution, we extend this solution to the LF massless scalar field (including the LF version of bosonization), demonstrating thereby that the problem with 2D massless LF fields is only fictitious. The key observation is that the two-point functions of the 2D massive scalar and fermion fields have non-vanishing massless limit, indicating the presence of corresponding massless fields. Their components can be obtained directly as limits of massive fields, or in some cases after a change of variables. Let us note that quantum fields should be rigorously treated as operator-valued distributions [13]. Such a formulation of the studied problem can be found in [14]. In the present paper, we shall be less formal.

The consistent quantum representation of the 2D massless LF fields enables one to address the solvable models in a genuine LF approach. We give some details of the LF operator solution of the Thirring-Wess model. A few remarks related to the LF version of conformal field theory are also included.

We use the following LF notation. The LF coordinate is $x^\mu = (x^+, x^-) = (x^0 + x^1, x^0 - x^1)$, while $k^\mu = (k^+, k^-)$ (or p^μ) denotes the momentum two-vector. Further symbols are $\partial_\pm = \frac{\partial}{\partial x^\pm}$, $\hat{k}^- = \frac{\mu^2}{k^+}$, $\hat{k}\cdot x = \frac{1}{2}k^+x^- + \frac{1}{2}\hat{k}^-x^+$. k^+ is the LF momentum and $\hat{k}^-$ the on-shell LF energy. It is evident that there is no sign ambiguity analogous to $E(k^1) = \pm\sqrt{(k^1)^2 + \mu^2}$ of the SL theory and both k^+, k^- can be taken positive.

2 Massless LF Scalar Field

The covariant Lagrangian density $\mathcal{L}$ and the corresponding field equation describing a free massive scalar field, $\mathcal{L} = \frac{1}{2}\partial_\mu\phi\partial^\mu\phi - \frac{1}{2}\mu^2\phi^2$, $(\partial_\mu\partial^\mu + \mu^2)\,\phi(x) = 0$, takes, in terms of the LF variables, the form

$$\mathcal{L} = 2\partial_+\phi\partial_-\phi - \frac{1}{2}\mu^2\phi^2, \quad (4\partial_+\partial_- + \mu^2)\,\phi(x) = 0. \tag{1}$$

The field time derivative and the conjugate momentum are $\theta(x) = 2\partial_+\phi(x) = -\frac{1}{2}\mu^2\partial_-^{-1}\phi(x)$, $\pi(x) = 2\partial_-\phi(x)$, where ∂_-^{-1} is the inverse derivative (unambiguously defined after prescribing boundary conditions).

The components of the energy-momentum tensor corresponding to the Lagrangian (1) are

$$T^{++}(x) = 4\partial_-\phi(x)\partial_-\phi(x), \quad T^{--}(x) = 4\partial_+\phi(x)\partial_+\phi(x), \quad T^{+-}(x) = \mu^2\phi^2(x) = T^{-+}(x). \tag{2}$$

The quantum solution of the field equation (1) takes the form [15]

$$\phi(x) = \int_0^{+\infty} \frac{dk^+}{\sqrt{4\pi k^+}}\left[a(k^+)e^{-\frac{i}{2}\hat{k}\cdot x} + a^\dagger(k^+)e^{\frac{i}{2}\hat{k}\cdot x}\right] f(k^+, \frac{\mu^2}{k^+}), \ \left[a(k^+), a^\dagger(l^+)\right] = \delta(k^+ - l^+), \tag{3}$$

with a test function $f(k^+, \frac{\mu^2}{k^+}) \in \mathcal{S}$, the space of rapidly decreasing functions [13]. Here we adopt the most simple form $\exp\left[-(k^+\delta^- + \frac{\mu^2}{k^+}\delta^+)\right]$ for f. Even if not displayed (like e.g. in Eq. (5) below), it is implicitly present to give mathematically well defined expressions. The field commutation relation at equal LF time is

$$\left[\varphi(x^+, x^-), \phi(x^+, y^-)\right] = \frac{1}{4i}\epsilon(z^-) = \int_0^{\infty} \frac{dk^+}{4\pi}\left[e^{-\frac{i}{2}k^+(z^- - i\delta^-)} - e^{\frac{i}{2}k^+(z^- + i\delta^-)}\right], \ z^- = x^- - y^-, \tag{4}$$

where $\epsilon(z^-)$ is the sign function. From (3) we directly find

$$\theta(x) = \int_0^{+\infty} \frac{dk^+}{i\sqrt{4\pi k^+}}\frac{\mu^2}{k^+}\left[a(k^+)e^{-\frac{i}{2}\hat{k}\cdot x} - a^\dagger(k^+)e^{\frac{i}{2}\hat{k}\cdot x}\right], \ \pi(x) = \int_0^{+\infty} \frac{dk^+}{i\sqrt{4\pi k^+}}k^+\left[a(k^+)e^{-\frac{i}{2}\hat{k}\cdot x} - H.c.\right]. \tag{5}$$

The LF Hamiltonian and the momentum operator are $P^\mu = \frac{1}{2}\int_{-\infty}^{+\infty} dx^- T^{+\mu}(x) = \int_0^{+\infty} dk^+ \hat{k}^\mu a^\dagger(k^+)a(k^+)$.

One can form a few two-point correlation functions from the three field operators above:

$$D_0^{(+)}(z) = \langle 0|\phi(x)\phi(y)|0\rangle, \quad D_1^{(+)}(z) = \langle 0|\phi(x)\pi(y)|0\rangle, \quad D_2^{(+)}(z) = \langle 0|\phi(x)\theta(y)|0\rangle, \tag{6}$$

$$D_i^{(+)}(z) = \int\limits_0^\infty \frac{dk^+}{4\pi} g_i(k^+) e^{-\frac{i}{2}k^+(z^- - i\delta^-) - \frac{i}{2}\frac{\mu^2}{k^+}(z^+ - i\delta^+)}, \quad g_0 = \frac{1}{k^+}, \quad g_1 = 1, \quad g_2 = \frac{\mu^2}{(k^-)^2}. \tag{7}$$

The above integrals are easily evaluated in terms of the (modified) Bessel functions $J_\nu(z)$, $N_\nu(z)$, $K_\nu(z)$, $\nu = 0, 1$. As in the SL theory, the first one logarithmically diverges for vanishing mass μ. The second is given by

$$D_1^{(+)}(z) = -\theta\left(-z^2\right)\operatorname{sgn}(z^-)\frac{\mu}{4\pi}\sqrt{-\frac{z^+}{z^-}}K_1\left(\mu\sqrt{-z^2}\right)$$
$$-\theta\left(z^2\right)\frac{i\mu}{4}\sqrt{\frac{z^+}{z^-}}\left[J_1\left(\mu\sqrt{z^2}\right) - i\operatorname{sgn}(z^+)N_1\left(\mu\sqrt{z^2}\right)\right]. \tag{8}$$

The important property is that both $D_1^{(+)}$ and $D_2^{(+)} = D_1^{(+)}(z^+ \leftrightarrow z^-)$ have a non-vanishing massless limit,

$$D_1^{(+)}(x-y;\mu^2=0) = \frac{1}{2\pi}\frac{1}{(x^- - y^- - i\delta^-)}, \quad D_2^{(+)}(x-y;\mu^2=0) = \frac{1}{2\pi}\frac{1}{(x^+ - y^+ - i\delta^+)}. \tag{9}$$

Technically, this is due to the behaviour of the Bessel function $K_1(z) \sim \frac{1}{z}$ for small value of z (the same is true for $N_1(z)$ in the timelike region), so that $\mu\sqrt{-\frac{z^\mp}{z^\pm}}K_1\left(-\mu\sqrt{z^+z^-}\right)$ has the finite massless limit [16]. The results (9) suggest that there must exist massless analogs of the fields $\phi(x)$, $\pi(x)$, $\theta(x)$, with correlation functions reproducing the above massless limits of the massive correlators. The massless Klein–Gordon equation $\partial_+\partial_-\tilde{\phi}(x) = 0$ tells us that one should expect a general solution of the form

$$\tilde{\phi}(x) = \tilde{\phi}(x^+) + \tilde{\phi}(x^-). \tag{10}$$

The massless limit of the massive solution (3) yields $\tilde{\phi}(x^-)$ in a straightforward way:

$$\tilde{\phi}(x^-) = \int\limits_0^\infty \frac{dk^+}{\sqrt{4\pi k^+}}\left[a(k^+)e^{-\frac{i}{2}k^+x^-} + a^\dagger(k^+)e^{\frac{i}{2}k^+x^-}\right], \; [a(k^+), a^\dagger(l^+)] = \delta(k^+ - l^+)]. \tag{11}$$

Remarkably, the second piece $\tilde{\phi}(x^+)$ is also contained in (3). A simple change of variables $k^+ \mapsto \frac{\mu^2}{k^-}$ yields $\phi(x) = \int_0^{+\infty}\frac{dk^-}{\sqrt{4\pi k^-}}\left[\frac{\mu}{k^-}a\left(\frac{\mu^2}{k^-}\right)e^{-\frac{i}{2}\frac{\mu^2}{k^-}x^- - \frac{i}{2}k^-x^+} + H.c.\right]$. The Fock commutator changes to

$$\left[\frac{\mu}{k^-}a\left(\frac{\mu^2}{k^-}\right), \frac{\mu}{l^-}a^\dagger\left(\frac{\mu^2}{l^-}\right)\right] = \frac{\mu^2}{k^-l^-}\delta\left(\frac{\mu^2}{k^-} - \frac{\mu^2}{l^-}\right) = \delta(k^- - l^-). \tag{12}$$

The rhs is mass independ and hence survives the massless limit. This should then be true for the lhs as well:

$$\lim_{\mu\to 0}\frac{\mu}{k^-}a\left(\frac{\mu^2}{k^-}\right) \equiv \tilde{a}(k^-) \neq 0, \; \left[\tilde{a}(k^-), \tilde{a}^\dagger(l^-)\right] = \delta(k^- - l^-), \; \left[a(k^+), \tilde{a}^\dagger(l^-)\right] = 0. \tag{13}$$

Performing the massless limit in the expression for $\phi(x)$ above Eq. (12) and in $\pi(x)$, $\theta(x)$ (5), we find[1]

$$\tilde{\phi}(x^+) = \int\limits_0^{+\infty}\frac{dk^-}{\sqrt{4\pi k^-}}\left[\tilde{a}(k^-)e^{-\frac{i}{2}k^-x^+} + \tilde{a}^\dagger(k^-)e^{\frac{i}{2}k^-x^+}\right], \tag{14}$$

$$\tilde{\theta}(x^+) = \int\limits_0^{+\infty}\frac{dk^-k^-}{i\sqrt{4\pi k^-}}\left[\tilde{a}(k^-)e^{-\frac{i}{2}k^-x^+} - H.c.\right], \quad \tilde{\pi}(x^-) = \int\limits_0^{+\infty}\frac{dk^+k^+}{i\sqrt{4\pi k^+}}\left[a(k^+)e^{-\frac{i}{2}k^+x^-} - H.c.\right]. \tag{15}$$

[1] Strictly speaking, this derivation should be done on a classical level with field amplitudes and then taken over to the quantum field operators. Also, the above change of variables has to be done for θ to make it compatible with the massless field equation.

The field commutators, following from the Fock algebra in (11) and from (13) are

$$\left[\tilde{\phi}(x^-), \tilde{\phi}(y^-)\right] = -\frac{i}{4}\epsilon(x^- - y^-), \quad \left[\tilde{\phi}(x^+), \tilde{\phi}(y^+)\right] = -\frac{i}{4}\epsilon(x^+ - y^+). \tag{16}$$

In this way, the second half of the solution of the field equation has been recovered from the massive solution. Note also that the variables k^+ and k^- in fact coincide. This is analogous to the SL case where $k^0 = |k^1|$.

As a consistency check, the two-point functions calculated from the massless fields should coincide with the massless limit of the massive functions. This is indeed the case for $D_1^{(+)}(z)$ and $D_2^{(+)}(z)$. The massive $D_0^{(+)}(z)$ can be evaluated for small mass μ, it diverges as $\ln\mu$. On the other hand, $D_0^{(+)}(z)$ calculated from the massless solution is ill defined with μ carelessly set to zero from the start of the calculation. Heuristically it can be seen that if we introduce the infrared cutoffs $\lambda^+ = \lambda^- \equiv \lambda$ in the corresponding integrals,

$$\tilde{D}_0^{(+)}(z) = \int_\lambda^\infty \frac{dk^-}{4\pi k^-} e^{-\frac{i}{2}k^-(z^+ - i\delta^+)} + \int_\lambda^\infty \frac{dk^+}{4\pi k^+} e^{-\frac{i}{2}k^+(z^- - i\delta^-)}, \tag{17}$$

We find the same (regularized) $\ln\lambda$ divergent behaviour (see below).

It is instructive to study the massless limit of the LF momentum and Hamiltonian. One expects that these operators survive the limit since for a consistent quantum theory we have to find the Heisenberg equations $2i\partial_+\phi(x) = -[P^-, \phi(x)]$, $2i\partial_-\phi(x) = -[P^+, \phi(x)]$. The massless limit of the momentum operator P^+ is straightforward, while the above change of variables is necessary for the Hamiltonian. The resultant operators

$$P^+ = \int_0^{+\infty} dk^+ k^+ a^\dagger(k^+) a(k^+), \quad P^- = \int_0^{+\infty} dk^+ k^- a^\dagger(k^-) a(k^-) \tag{18}$$

generate, by virtue of (13) and due to $[P^+, \phi(x^+)] = [P^-, \phi(x^-)] = 0$, the correct Heisenberg equations

$$2i\partial_+\phi(x^+) = -[P^-, \phi(x^+)], \quad 2i\partial_-\phi(x^-) = -[P^+, \phi(x^-)]. \tag{19}$$

However, the massless limit of the LF Hamiltonian density $T^{+-}(x)$ itself vanishes, in agreement with the conformal symmetry of the massless theory, which requires Tr $T^{\mu\nu} = T^\mu_\mu = T^{+-} = 0$. It is also interesting to note that the correlation functions of the scalar conformal field [17] (and of its energy-momentum tensor) are directly reproduced from Eq. (9) by going over to the Euclidean time $t \to -i\tau$ and by defining the complex variables $\zeta = \tau + ix^1$, $\bar{\zeta} = \tau - ix^1$. This link to CFT deserves a further study.

3 Massless Light Front Fermion Field and Bosonization

The construction of the massless LF fermion fields proceeds in a completely analogous manner. The derivation is in a sense simpler since there are no infrared divergencies in this case. The only difference is that one has to carefully treat the constraint for the non-dynamical fermion field component.

The massless LF Dirac equation requires that [12] $\psi_2(x) = \psi_2(x^-)$, $\psi_1(x) = \psi_1(x^+)$. The massless limit of the solution of the massive Dirac equation then indeed yields

$$\tilde{\psi}_2(x^-) = \int_0^{+\infty} \frac{dp^+}{\sqrt{4\pi}} \left[b(p^+)e^{-\frac{i}{2}p^+x^-} + d^\dagger(p^+)e^{\frac{i}{2}p^+x^-}\right], \quad \{b(p^+), b^\dagger(q^+)\} = \{d(p^+), d^\dagger(q^+)\} = \delta(p^+ - q^+), \tag{20}$$

$$\tilde{\psi}_1(x^+) = \int_0^{+\infty} \frac{dp^-}{\sqrt{4\pi}} \left[\tilde{b}(p^-)e^{-\frac{i}{2}p^-x^+} - \tilde{d}^\dagger(p^-)e^{\frac{i}{2}p^-x^+}\right], \quad \{\tilde{b}(p^-), \tilde{b}^\dagger(q^-)\} = \{\tilde{d}(p^-), \tilde{d}^\dagger(q^-)\} = \delta(p^- - q^-), \tag{21}$$

with $\tilde{b}(p^-)$ and $\tilde{d}(p^-)$ given analogously to the scalar case (13). The field anticommutators acquire the form

$$\left\{\tilde{\psi}_1(x^+), \tilde{\psi}_1^\dagger(y^+)\right\} = \delta(x^+ - y^+), \quad \left\{\tilde{\psi}_2(x^-), \tilde{\psi}_2^\dagger(y^-)\right\} = \delta(x^- - y^-). \tag{22}$$

All this shows that the massless fields can be recovered from the original massive solution. No new fermionic variables have to be introduced artificially. This is confirmed by the fact that while the mixed two-point function $S_{12}^{(+)}$ vanishes for $m = 0$, the two-point functions calculated from the massless representation (20, 21)

$$S_{22}^{(+)}(x-y; m=0) = \frac{1}{2i\pi}\frac{1}{(x^- - y^- - i\delta^-)}, \quad S_{11}^{(+)}(x-y; m=0) = \frac{1}{2i\pi}\frac{1}{(x^+ - y^+ - i\delta^+)} \tag{23}$$

coincide with the massless limits of the massive two-point functions [12].

Since the solvable models are based on free Heisenberg fields, the above derivation of the 2D massless LF fermion fields opens the prospect for the genuine light-front solution of this class of models [12,14]. Following the consistent quantization of the massless LF scalar and fermion fields, it is natural to analyze the celebrated bosonization property [18] in a genuine LF form. Since the massless $\phi(x)$ and $\psi(x)$ fields decompose as $\phi(x) = \phi(x^+) + \phi(x^-)$, $\psi^T(x) = (\psi_1(x^+), \psi_2(x^-))$ (we omit tilde for the massless fields henceforth), the bosonization is straightforward. One simply observes [14] that the normal-ordered exponentials

$$\hat{\varphi}_1(x^-) = C : e^{i\alpha\phi(x^+)} := Ce^{i\alpha\phi^{(-)}(x^+)}e^{i\alpha\phi^{(+)}(x^+)}, \quad \hat{\varphi}_2(x^-) = C : e^{i\alpha\phi(x^-)} := Ce^{i\alpha\phi^{(-)}(x^-)}e^{i\alpha\phi^{(+)}(x^-)}, \tag{24}$$

with $C = \sqrt{\frac{\lambda e^{\gamma_E}}{4\pi}}$, $\alpha = 2\sqrt{\pi}$ (γ_E is the Euler's constant), satisfy the fermionic properties

$$\hat{\varphi}_1(x^+)\hat{\varphi}_1(y^+) = -\hat{\varphi}_1(y^+)\hat{\varphi}_1(x^+), \quad \hat{\varphi}_2(x^-)\hat{\varphi}_1(y^-) = -\hat{\varphi}_2(y^-)\hat{\varphi}_1(x^-), \tag{25}$$

$$\left\{\hat{\varphi}_1(x^+), \hat{\varphi}_1^\dagger(y^+)\right\} = \delta(x^- - y^+), \quad \left\{\hat{\varphi}_2(x^-), \hat{\varphi}_2^\dagger(y^-)\right\} = \delta(x^- - y^-). \tag{26}$$

The above construction allows one to study the bosonized LF Thirring model as well as the LF version of the sine-Gordon—massive Thirring model correspondence [19,20]. Here we demonstrate the simplicity of the LF treatment with the example of the exactly solvable Thirring-Wess model.

4 The LF Thirring-Wess Model

The dynamics of the model is characterized by the covariant-form Lagrangian

$$\mathcal{L} = \frac{i}{2}\overline{\Psi}\gamma^\mu \overleftrightarrow{\partial_\mu} \Psi - \frac{1}{4}\tilde{G}_{\mu\nu}\tilde{G}^{\mu\nu} + \frac{1}{2}\mu_0^2\tilde{B}_\mu\tilde{B}^\mu - e\tilde{B}_\mu J^\mu, \quad \tilde{G}_{\mu\nu} = \partial_\mu\tilde{B}_\nu - \partial_\nu\tilde{B}_\mu, \tag{27}$$

where $\Psi(x)$ ($\tilde{B}^\mu(x)$) is the interacting massless fermion (massive vector) field. The solvability of the theory implies that one can find an operator solution of the coupled system of the Dirac and Proca equations:

$$i\gamma^\mu\partial_\mu\Psi = e\gamma_\mu\tilde{B}^\mu\Psi, \quad \partial_\mu\tilde{G}^{\mu\nu} + \mu_0^2\tilde{B}^\nu = eJ^\nu, \quad J^\mu(x) = \overline{\Psi}(x)\gamma^\mu\Psi(x). \tag{28}$$

For the conserved current $J^\mu(x)$ the vector field satisfies $\partial_\mu\tilde{B}^\mu(x) = 0$. The LF form of the field equations is

$$2i\partial_+\Psi_2 = e\tilde{B}^-\Psi_2, \quad 2i\partial_-\Psi_1 = e\tilde{B}^+\Psi_1, \quad 4\partial_+\partial_-\tilde{B}^+ + \mu_0^2\tilde{B}^+ = eJ^+, \quad 4\partial_+\partial_-\tilde{B}^- + \mu_0^2\tilde{B}^- = eJ^- \tag{29}$$

The classical solution of the Dirac equation involves the free massless LF fermion field components:

$$\Psi_1(x) = \exp\left\{-\frac{ie}{2}\int_{-\infty}^{+\infty} dy^- \frac{1}{2}\epsilon(x^- - y^-)\tilde{B}^+(x^+, y^-)\right\}\psi_1(x^+),$$

$$\Psi_2(x) = \exp\left\{+\frac{ie}{2}\int_{-\infty}^{+\infty} dy^- \frac{1}{2}\epsilon(x^- - y^-)\tilde{B}^+(x^+, y^-)\right\}\psi_2(x^-). \tag{30}$$

The interacting current appearing in the Proca equation can be determined from the solution (30) (or rather its quantum version, see below), using the point-split regularized definition

$$J^+(x) = \lim_{\epsilon\to 0}\left[\Psi_2^\dagger\left(x+\frac{\epsilon}{2}\right)\Psi_2\left(x-\frac{\epsilon}{2}\right) + H.c.\right], \quad J^-(x) = \lim_{\epsilon\to 0}\left[\Psi_1^\dagger\left(x+\frac{\epsilon}{2}\right)\Psi_1\left(x-\frac{\epsilon}{2}\right) + H.c\right]. \tag{31}$$

The interacting current is hence equal to the (normal-ordered) free current plus a quantum correction,

$$J^+(x) = 2 : \psi_2^\dagger(x^-)\psi_2(x^-) : -\frac{e}{2\pi}\tilde{B}^+(x^+,x^-),\ J^-(x) = 2 : \psi_1^\dagger(x^+)\psi_1(x^+) : -\frac{e}{2\pi}\tilde{B}^-(x^+,x^-). \tag{32}$$

Nevertheless, the interacting quantum current is conserved due to the condition $\partial_\mu \tilde{B}^\mu = 0$. On the other hand, the axial-vector current $J^\mu(x) = \overline{\Psi}(x)\gamma^\mu\gamma^5\Psi(x)$ is "anomalous", $\partial_\mu J_5^\mu = -\frac{e}{2\pi}\partial_+\tilde{B}^+ = \frac{e}{4\pi}\epsilon_{\mu\nu}\tilde{G}^{\mu\nu}$. When (32) is inserted to the Proca equation, the quantum correction renormalizes the bare mass μ_0 as $\mu_0^2 + \frac{e^2}{2\pi} \equiv \mu^2$:

$$4\partial_+\partial_-\tilde{B}^\nu + \mu^2\tilde{B}^\nu = ej^\nu \Rightarrow \tilde{B}^\nu = B^\nu + \frac{e}{\mu^2}j^\nu, \quad \text{where } (4\partial_+\partial_- + \mu^2)B^\nu = 0. \tag{33}$$

The interacting vector-meson field is thus given entirely in terms of the free fields. This permits us to regularize the solutions (30) on the quantum level by normal ordering:

$$\Psi_1(x) = e^{-\frac{ie}{2}F^{(-)}(x^+,x^-)}e^{-\frac{ie}{2}F^{(+)}(x^+,x^-)}\psi_1(x^+),\ \Psi_2(x) = e^{\frac{ie}{2}F^{(-)}(x^+,x^-)}e^{\frac{ie}{2}F^{(+)}(x^+,x^-)}\psi_2(x^-), \tag{34}$$

where the fields were decomposed into the positive and negative frequency parts,

$$F^{(\pm)}(x^+,x^-) = \int\limits_{-\infty}^{+\infty} dy^- \frac{1}{2}\epsilon(x^- - y^-)\left[B^{+(\pm)}(x^+,y^-) + \frac{e}{\mu^2}j^{+(\pm)}(y^-)\right]. \tag{35}$$

The free current $j^\mu(x)$ built from the components (20, 21) can be bosonized by a Fourier transformation:

$$j^+(x^-) = -\frac{i}{\sqrt{\pi}}\int\limits_0^\infty \frac{dk^+\, k^+}{\sqrt{4\pi k^+}}\left[c(k^+)e^{-\frac{i}{2}k^+x^-} - H.c.\right] = j^{+(+)} + j^{+(-)},\quad \left[c(k^+), c^\dagger(l^+)\right] = \delta(k^+ - l^+),$$

$$j^-(x^+) = -\frac{i}{\sqrt{\pi}}\int\limits_0^\infty \frac{dk^-\, k^-}{\sqrt{4\pi k^-}}\left[c(k^-)e^{-\frac{i}{2}k^-x^+} - H.c.\right],\quad \left[c(k^-), c^\dagger(l^-)\right] = \delta(k^- - l^-). \tag{36}$$

The composite boson operators $c(k^+), c(k^-)$ are bilinear in the fermion Fock operators present in (20, 21) [12]. The free LF vector-meson field, quantized by $[a(k^+), a^\dagger(l^+)] = \delta(k^+ - l^+)$, is expanded as

$$B^+(x) = \int\limits_0^{+\infty} \frac{dk^+}{\sqrt{4\pi k^+}}\frac{k^+}{\mu}\left[a(k^+)e^{-i\hat{k}\hat{x}} + a^\dagger(k^+)e^{i\hat{k}\hat{x}}\right],\quad B^-(x) = -\int\limits_0^{+\infty} \frac{dk^+}{\sqrt{4\pi k^+}}\frac{\hat{k}^-}{\mu}\left[a(k^+)e^{-i\hat{k}\hat{x}} + H.c.\right],$$

$$\left[B^+(x^-), \Pi_{B^+}(y^-)\right] = i\delta(x^- - y^-),\quad \Pi_{B^+} = \partial_+B^+ - \partial_-B^- = 2\partial_+B^+. \tag{37}$$

The field expansions (37, 36) can be used to obtain the Fock form of the corresponding LF Hamiltonian

$$P^- = \frac{1}{2}\int\limits_{-\infty}^{+\infty} dx^-\left[-\mu_0^2B^+B^- - \mu_0^2\frac{e}{\mu^2}\left(B^+j^- + B^-j^+\right) - \mu_0^2\frac{e^2}{\mu^4}j^+j^-\right]. \tag{38}$$

This, along with the nonperturbative computation of the correlation functions of the model, will be published separately [14]. Finally, as indicated at the end of Sect. 2, there is a direct link [14] between the quantum theory of two-dimensional massless LF fields, developed in the present work, and the conformal field theory [17].

Acknowledgments L.M. thanks the Grant VEGA 2/0072/2013 and the Slovak CERN Committee for support.

References

1. Dirac, P.A.M.: Forms od relativistic dynamics. Rev. Mod. Phys. **21**, 392 (1949)
2. Dirac, P.A.M.: The mathematical foundations of quantum theory. In: Marlow, A. (ed.) Mathematical Foundation of Quantum Theory. Academic Press, New York (1978). ISBN 978-0-12-473250-6
3. Weinberg, S.: Dynamics at infinite momentum. Phys. Rev. **150**, 1313 (1966)
4. Leutwyler, H.: Current algebra and lightlike charges. Springer Tracts Mod. Phys. **50**, 29 (1969)
5. Neville, R.A., Rohrlich, F.: Quantum field theory off null planes. Nuovo Cim. A **1**, 625 (1971)
6. Kogut, J.B., Soper, D.E.: Quantum electrodynamics in the infinite momentum frame. Phys. Rev. D **1**, 2901 (1971)
7. Chang, S.-J., Root, R.G., Yan, T.-M.: Quantum field theories in the infinite momentum frame. 1. Quantization of scalar and Dirac fields. Phys. Rev. D **7**, 1133 (1973)
8. Leutwyler, H., Klauder, J.R., Streit, L.: Quantum field theory on light like slabs. Nuovo Cim. A **66**, 536 (1970)
9. Pauli, H.C., Brodsky, S.J.: Solving field theory in one space one time dimension. Phys. Rev. **32**, 1993 (1985)
10. McCartor, G.: Schwinger model in the light cone representation. Z. Phys. C **64**, 349 (1994)
11. McCartor, G., Pinsky, S.S., Robertson, D.G.: Vacuum structure of two-dimensional gauge theories on the light front. Phys. Rev. D **56**, 1035 (1997)
12. Martinovič, L., Grangé, P.: Solvable models with massless light front fermions. Few Body Syst. **56**, 607 (2015)
13. Schwartz, L.: Théorie des distributions. Hermann, Paris (1966)
14. Martinovič, L., Grange, P.: to be published
15. Grangé, P., Werner, E.: The Taylor–Lagrange scheme as a template for symmetry preserving renormalization procedures. J. Phys. A Math. Theor. **44**, 385402 (2011)
16. Bergknoff, T.: Physical particles of the massive Schwinger model. Nucl. Phys. B **122**, 215 (1977)
17. Di Francesco, P., Mathieu, P., Sénéchal, D.: Conformal Field Theory, Graduate Texts in Contemporary Physics. Springer, New York (1997)
18. Stone, M.: Bosonization. World Scientific, Singapore (1995)
19. Coleman, S.: The quantum sine-Gordon equation as the massive Thirring model. Phys. Rev. D **11**, 2088 (1975)
20. Mandelstam, S.: Soliton operators for the quantized sine-Gordon equation. Phys. Rev. D **11**, 3026 (1975)

DOI 10.1007/978-3-319-50699-9_53
Reprinted from *Few-Body System* Journal, DOI 10.1007/s00601-016-1117-x

James P. Vary · Lekha Adhikari · Guangyao Chen · Yang Li · Pieter Maris · Xingbo Zhao

Basis Light-Front Quantization: Recent Progress and Future Prospects

Received: 31 January 2016 / Accepted: 2 May 2016 / Published online: 20 May 2016

Abstract Light-front Hamiltonian field theory has advanced to the stage of becoming a viable non-perturbative method for solving forefront problems in strong interaction physics. Physics drivers include hadron mass spectroscopy, generalized parton distribution functions, spin structures of the hadrons, inelastic structure functions, hadronization, particle production by strong external time-dependent fields in relativistic heavy ion collisions, and many more. We review selected recent results and future prospects with basis light-front quantization that include fermion-antifermion bound states in QCD, fermion motion in a strong time-dependent external field and a novel non-perturbative renormalization scheme.

1 Introduction

Light-front (LF) Hamiltonian theory is a natural ab initio framework to investigate the structures and reactions of hadrons and nuclei (see, e.g., Ref. [1] and references therein). Basis light-front quantization (BLFQ) [2–4] and its time-dependent generalization tBLFQ [5,6] represent practical non-perturbative frameworks for these investigations.

The goal of the light-front Hamiltonian approach is to solve the Hamiltonian eigenvalue problem,

$$\hat{P}^{+}\hat{P}^{-}|\psi_h\rangle = M_h^2|\psi_h\rangle. \tag{1}$$

Here $\hat{P}^{\pm} = \hat{P}^0 \pm \hat{P}^3$ are the longitudinal momentum (+) and the light-front quantized Hamiltonian operator (−). The eigenvalues directly produce the invariant-mass spectrum. The eigenvectors, or light-front wavefunctions (LFWFs), play a pivotal role in the study of the hadron structures in deep inelastic scattering (DIS) [7], deeply virtual Compton scattering (DVCS) [8] and diffractive hadron production in high energy nuclear collisions [9,10].

The hadron state vector $|\psi_h\rangle$ is then represented in the Fock space and the problem becomes a quantum many-body problem. By adopting a suitable single-particle basis, BLFQ preserves all the kinematical symmetries of the full Hamiltonian. Such a choice is in parallel with the no-core shell model (NCSM) used successfully in nuclear many-body theory [11–14] but retains a fully covariant framework. It has been applied to a range of non-perturbative problems, including the electron anomalous magnetic moment [3,15], non-linear Compton scattering [5,6] and the positronium spectrum [16]. In this paper, we briefly summarize the recent BLFQ applications to heavy quarkonium [17] in Sect. 2, the extension to tBLFQ [5,6,18] in Sect. 3, and a

J. P. Vary (✉) · L. Adhikari · G. Chen · Y. Li · P. Maris · X. Zhao
Department of Physics and Astronomy, Iowa State University, Ames, IA 50011, USA
E-mail: jvary@iastate.edu

X. Zhao
Institute of Modern Physics, Chinese Academy of Sciences, Lanzhou 730000, China

promising approach to non-perturbative renormalization [19] in Sect. 4, each of which points pathways to future developments and applications.

2 Heavy Quarkonium in BLFQ

We formulate an initial effective Hamiltonian based on the holographic QCD Hamiltonian and one-gluon exchange. We select the light-front AdS/QCD soft-wall Hamiltonian for the first part [20], which provides a reasonable first description of the hadron spectrum [21], so that

$$H_{\text{eff}} \equiv P^+\hat{P}^-_{\text{eff}} - \boldsymbol{P}_\perp^2 = \frac{\boldsymbol{k}_\perp^2}{x(1-x)} + \kappa^4 x(1-x)\boldsymbol{r}_\perp^2 + V_{\text{g}} \tag{2}$$

where, $x = p_q^+/P^+$ is the longitudinal momentum fraction of the quark, $\boldsymbol{k}_\perp = \boldsymbol{p}_{q\perp} - x\boldsymbol{P}_\perp$ is the intrinsic transverse momentum, $\boldsymbol{r}_\perp = \boldsymbol{r}_{q\perp} - \boldsymbol{r}_{\bar{q}\perp}$ is the transverse separation of the partons. κ is the strength of the confining potential. V_{g} is the one-gluon exchange interaction. Note that the "light-cone Hamiltonian" has mass squared dimension producing squared invariant masses as eigenvalues.

Following Brodsky and de Téramond [20], it is convenient to introduce the holographic coordinate $\boldsymbol{\zeta}_\perp = \sqrt{x(1-x)}\boldsymbol{r}_\perp$, and its conjugate $\boldsymbol{q}_\perp = \boldsymbol{k}_\perp/\sqrt{x(1-x)} \equiv -i\nabla_{\zeta_\perp}$. In light-front holography, $\zeta_\perp$ is mapped to the fifth coordinate z of the AdS space. In these coordinates, H_{SW} is a harmonic oscillator (HO),

$$H_{\text{SW}} = \boldsymbol{q}_\perp^2 + \kappa^4\boldsymbol{\zeta}_\perp^2. \tag{3}$$

Its eigenvalues follow the Regge trajectory $M^2 = 2\kappa^2(2n + |m| + 1)$. Its eigenfunctions are 2D HO functions in the holographic variables,

$$\phi_{nm}(\boldsymbol{q}_\perp) = e^{im\theta}\left(\frac{q_\perp}{\kappa}\right)^{|m|} e^{-q_\perp^2/(2\kappa^2)} L_n^{|m|}(q_\perp^2/\kappa^2). \tag{4}$$

Here $q_\perp = |\boldsymbol{q}_\perp|$, $\theta = \arg \boldsymbol{q}_\perp$, and $L_n^m(x)$ is the associated Laguerre polynomial. We adopt these functions as our basis. This basis has the advantage that in the many-body sector, it allows the exact factorization of the center-of-mass motion in the single-particle coordinates. This is a very valuable property, because the boson/fermion symmetrization/anti-symmetrization in the relative coordinates quickly becomes intractable, as the number of identical particles increases [22,23]. We work here with relative coordinates but results would be identical in single-particle coordinates due to our basis selection leading to factorization of relative and center-of-mass motions in BLFQ, as noted.

The soft-wall Hamiltonian Eq. (3) is designed for massless quarks, and it is inherently 2-dimensional. For the heavy quarkonium systems and other applications we also incorporate longitudinal dynamics,

$$H_{\text{SW}} \to H_{\text{CON}} = \frac{\boldsymbol{k}_\perp^2 + m_q^2}{x} + \frac{\boldsymbol{k}_\perp^2 + m_{\bar{q}}^2}{1-x} + \kappa^4 x(1-x)\boldsymbol{r}_\perp^2 + V_L(x) \tag{5}$$

Here V_L is a longitudinal confining potential. Several longitudinal confining potentials have been proposed [24–26]. Here we propose a new longitudinal confinement with some similarity to those,

$$V_L(x) = -\frac{\kappa^4}{(m_q + m_{\bar{q}})^2}\partial_x\big(x(1-x)\partial_x\big). \tag{6}$$

This term combined with the mass term from the kinetic energy forms a Sturm–Liouville problem,

$$-\partial_x\big(x(1-x)\partial_x\chi_\ell(x)\big) + \frac{1}{4}\Big(\frac{\alpha^2}{1-x} + \frac{\beta^2}{x}\Big)\chi_\ell(x) = \lambda_\ell^{(\alpha,\beta)}\chi_\ell(x), \tag{7}$$

where $\alpha = 2m_{\bar{q}}(m_q + m_{\bar{q}})/\kappa^2$, $\beta = 2m_q(m_q + m_{\bar{q}})/\kappa^2$, and $\lambda_\ell^{(\alpha,\beta)} = (\ell + \frac{1}{2}(\alpha+\beta))(\ell + \frac{1}{2}(\alpha+\beta) + 1)$, $(\ell = 0, 1, 2, \ldots)$ is the eigenvalue. The solutions are analytically known,

$$\chi_\ell(x) = x^{\frac{1}{2}\alpha}(1-x)^{\frac{1}{2}\beta}P_\ell^{(\alpha,\beta)}(2x-1) \tag{8}$$

Here $P_\ell^{(a,b)}(z)$ is the Jacobi polynomial. We adopt these as our longitudinal basis functions.

Our proposed longitudinal confinement has several appealing features. For example, the basis functions resemble the known asymptotic parton distribution $\sim x^{\alpha}(1-x)^{\beta}$. The basis function is also analytic providing numerical convenience. In addition, in the non-relativistic limit $m_f \gg \kappa$, our longitudinal confinement combines with transverse confinement to form a 3D HO potential, $V_{\text{con}} = \frac{m_q m_{\bar{q}}}{(m_q+m_{\bar{q}})^2}\kappa^4 \boldsymbol{r}^2$, and rotational symmetry is preserved. This non-relativistic reduction also provides us an order-of-magnitude estimate of the model parameters. On the other hand, in the massless limit $m_f \ll \kappa$, the longitudinal mode stays in the ground state and the longitudinal wavefunction $\chi_0(x) = \text{const}$. Thus we restore the massless model of Brodsky and de Téramond [21].

Our primary results for heavy quarkonium are presented in Table 1 and in Figs. 1, 2 and 3. The results depend on the basis space cutoffs that restrict the dimension of the Hamiltonian matrix. These cutoffs are defined with N_{max} as the maximum $2n + |m| + 1$ and L_{max} as the maximum l.

Since total angular momentum is dynamical in light-front quantized field theory our truncation to the $q\bar{q}$ sector leads to a weak dependence on the total angular momentum projection m_J. While our fits to the heavy quarkonia states below open flavor thresholds are done with $m_J = 0$, we illustrate the weak dependence on m_J in Table 1 and in Fig. 1. Figure 2 presents the calculated decay constants for the scalar and vector states and compares with other approaches and with experiment where available.

In view of the limited fitting, the results in Table 1 and in Figs. 1 and 2 appear reasonable, even competitive with other approaches. Future investigations will examine the role of running coupling and present results in increasing basis spaces so that accurate extrapolations may be performed.

Table 1 Summary of the model parameters

	α_s	μ_g (GeV)	κ (GeV)	m_q (GeV)	$\delta M_{m_J=0}$ (MeV)	$\delta \overline{M}$ (MeV)	$N_{\max} = L_{\max}$
$c\bar{c}$	0.3595	0.02	0.950	1.510	64 (8 states)	52 (8 states)	16
$b\bar{b}$	0.2500	0.02	1.491	4.761	56 (14 states)	51 (14 states)	16
$c\bar{c}$	0.3595	0.02	0.938	1.522	65 (8 states)	52 (8 states)	24
$b\bar{b}$	0.2500	0.02	1.490	4.763	54 (14 states)	50 (14 states)	24

The coupling α_s and the gluon mass μ_g are fixed. We choose the bottomonium $\alpha_s(9.5\,GeV) = 0.25$ and then obtain the charmonium coupling from pQCD. The confining strength κ and the quark mass m_q are fitted with $m_J = 0$ using the experimental data below the $D\overline{D}$ or $B\overline{B}$ threshold. The r.m.s. deviations for the $m_J = 0$ spectrum, $\delta M_{m_J=0}$, and the r.m.s. average-m_J spectrum, $\delta\overline{M}$, are computed for states below the threshold

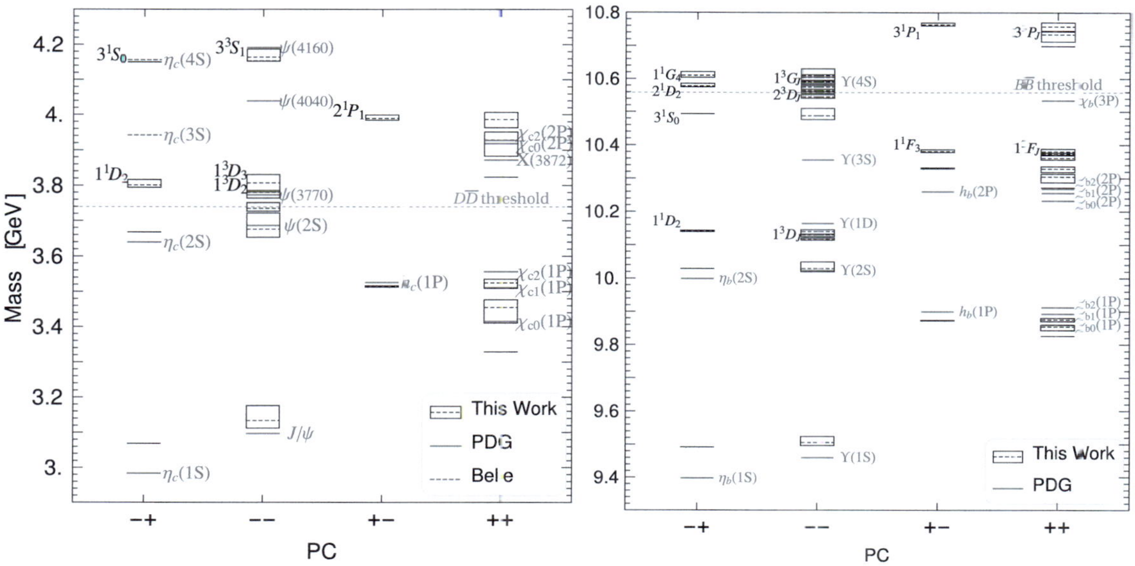

Fig. 1 (Color online) *Left(Right)* Charmonium (bottomonium) spectrum in GeV/c^2; known states are labeled by their PDG symbols [27] while others are labeled by non-relativistic symbols. A *box* indicates the spread of M_{m_J} and *dashed lines* signify $\overline{M}$. Parameters are given in Table 1 ($N_{\max} = L_{\max} = 24$)

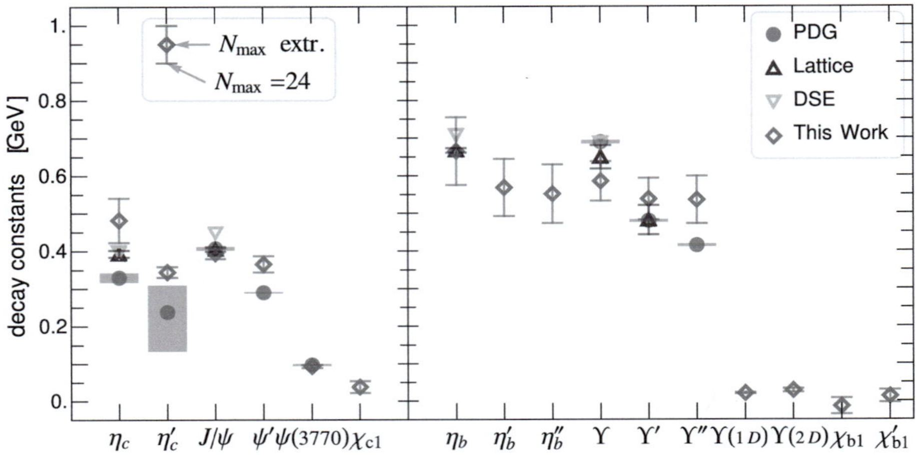

Fig. 2 (Color online) The decay constants of pseudo-scalar and (pseudo-)vector quarkonia compared with PDG data [27], with Lattice QCD (Lattice) [28–31] and with Dyson–Schwinger (DSE) [32] approaches. Note Υ(1D) (Υ(2D)) refer to $\Upsilon(1^3D_1)$ ($\Upsilon(2^3D_1)$). Simple polynomial extrapolations (extr.) in $1/N_{\max}$ produce the results indicated by the *open square symbols* and the uncertainty is the defined as the difference between the extrapolated result and the $N_{\max} = L_{\max} = 24$ result

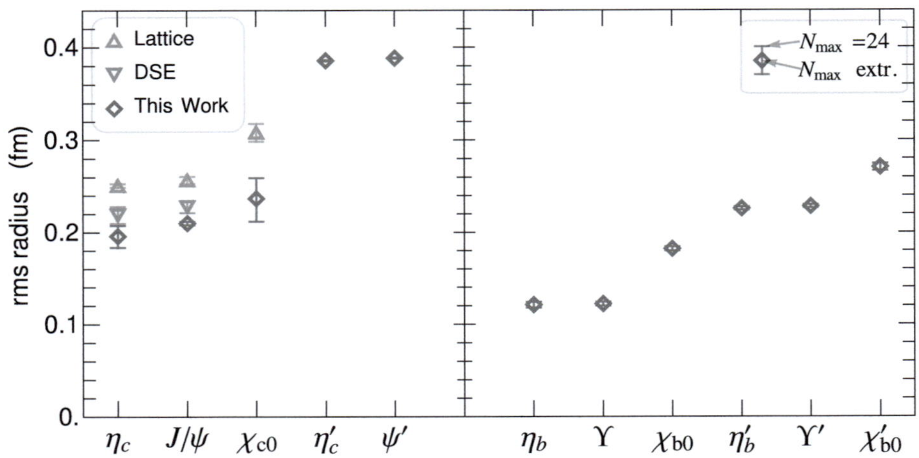

Fig. 3 (Color online) The rms radii of pseudo-scalar and (pseudo-)vector quarkonia compared with Lattice QCD (Lattice) [28–31] and with Dyson–Schwinger (DSE) [32] approaches. See the caption to Fig. 2

Before closing this section, we present an example of additional observables that can be accessed with the BLFQ results for the LFWFs. Here, we choose the positronium system at strong coupling where the spectroscopy, LFWFs and form factors are available [16]. For brevity, we present only the helicity-non-flip GPDs $H(x, \xi = 0, t = -\Delta_\perp^2)$ for a selection of four low-lying bound states in Fig. 4. The helicity-non-flip GPDs in the region can be written, for the case where the photon couples only to the electron in this toy problem, as overlap integrals between LFWFs [8,34–36]

$$H(x, \xi = 0, t = -\mathbf{\Delta}_\perp^2) = \sum_{\lambda_e, \lambda_{\bar{e}}} \int d^2\mathbf{k}_\perp\, \psi^*(\mathbf{k}'_\perp, x, \lambda_e, \lambda_{\bar{e}})\psi(\mathbf{k}_\perp, x, \lambda_e, \lambda_{\bar{e}}), \tag{9}$$

Here, x is the longitudinal momentum fraction of the electron, $\mathbf{k}_\perp$ and $\mathbf{k}'_\perp$ are the respective relative momenta of the electron before and after being struck by the virtual photon, Δ is the momentum transfer (we choose the Drell–Yan frame $\Delta^+ = 0$, $t \equiv \Delta^2 = -\mathbf{\Delta}_\perp^2$, ξ is the skewness parameter and in the Drell–Yan frame, $\xi = 0$, and $\lambda_e(\lambda_{\bar{e}})$ is the spin of the electron (positron). This is a demonstration case from Ref. [33] illustrating the potential for hadronic GPDs in BLFQ, another of our future projects.

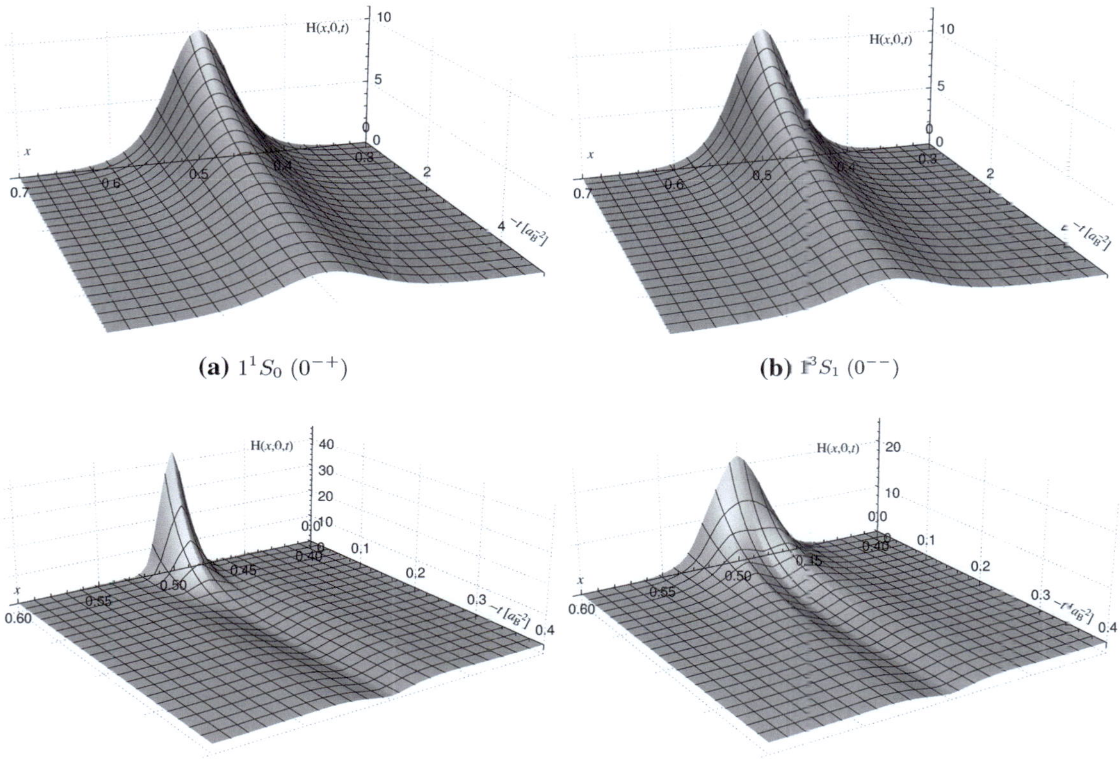

(a) 1^1S_0 (0^{-+}) **(b)** 1^3S_1 (0^{--})

(c) 2^1S_0 (0^{-+}) **(d)** 2^3P_0 (0^{++})

Fig. 4 3D plot of helicity-non-flip GPDs $H(x, \xi = 0, t = -\Delta_\perp^2)$ (Eq. 9) for the few lowest bound states of positronium with $N_{\max} = 31$, $K = 61$ (K is the total longitudinal momentum in dimensionless units for plane waves in a discretized basis [2,7]), angular momentum projection $m_J = 0$, coupling constant $\alpha = 0.3$, and photon mass $\mu = 0.02m_e$. Note $a_B = 1/(\alpha m_e)$ is the Bohr radius

3 Electron Motion in External Field with tBLFQ

We adopt the tBLFQ framework [5,6] to investigate the effects of electromagnetic (EM) fields generated by ultra-relativistic heavy ions on charged particles [18]. This study is motivated by the fact that strong EM fields are generated during heavy ion collisions [37], and a quantitative study of their effects on charged particles is essential for extracting properties of the quark gluon plasma (QGP). For this initial project, we study the real-time evolution of the quantized electron field under influence of such fields. When the coupling between the electron and the fields is small, e.g , for EM fields generated by an ultra-relativistic proton with coupling $\alpha = \alpha_{\text{em}} \approx 1/137$, the transition rate between two kinetic energy eigenstates calculated within tBLFQ approach shows agreement with light-front perturbation theory (LFPT) calculation. On the other hand, for EM fields generated by an ultra-relativistic gold nucleus, the coupling between an electron and the fields is $\alpha = Z_{\text{Au}}\alpha_{\text{em}} \approx 79/137$. The transition rate between the same two kinetic energy eigenstates calculated within tBLFQ approach deviates from LFPT calculations [both leading-order (LO) and next-to-leading order (NLO)], and the differences among these calculations is a clear indication that higher order effects are significant. See Fig. 5 for an illustration.[1]

The EM fields generated by an ultra-relativistic heavy ions have sizable influence on physical observables. We demonstrate such influence with the real-time evolution of the transverse and longitudinal momentum distribution and helicity configuration of an electron evolving in the electromagnetic fields generated by a gold nucleus moving along positive z-axis with rapidity $y = 5.3$. Figure 6 shows a snap shot of the transverse and longitudinal momentum distribution of the electron after evolving 50 GeV^{-1} (roughly 10 fm/c) inside such strong EM fields. We observe about 30 % reduction in the peak of the transverse momentum distribution

[1] The kinetic energy eigenstates are obtained through diagonalization of kinetic energy matrix in the BLFQ basis. We then evolve the electron amplitude in light-front time using the external field in the interaction picture. Figure 5 shows transition rates between two states with energy 0.0648 and 1.886 GeV.

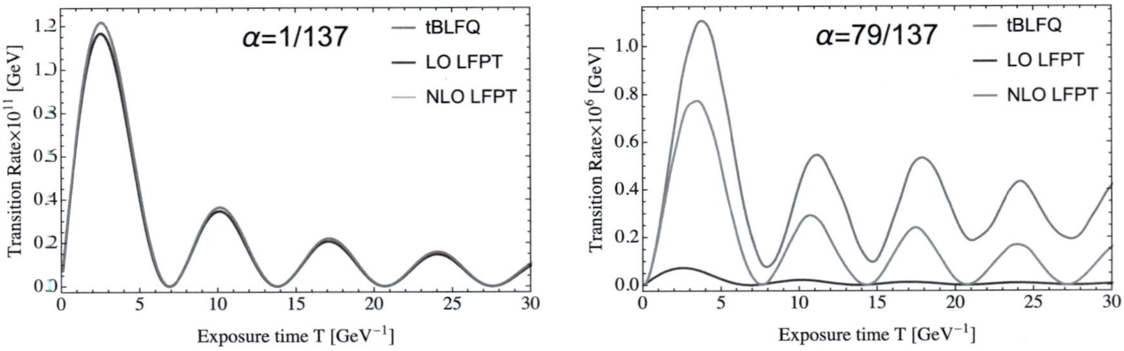

Fig. 5 (Color online) The transition rate between two kinetic energy eigenstates of an electron inside EM fields generated by a proton (*left*) and a gold nucleus (*right*) moving along positive z-axis with rapidity $y = 5.3$. The *horizontal axis* represents the total time the electron been exposed to the fields. The tBLFQ calculation is performed using $N_{max} = 30$, $K = 30$, electron localization scale $L = 10\ \text{GeV}^{-1}$ and HO basis scale $b = 1000\ m_e$

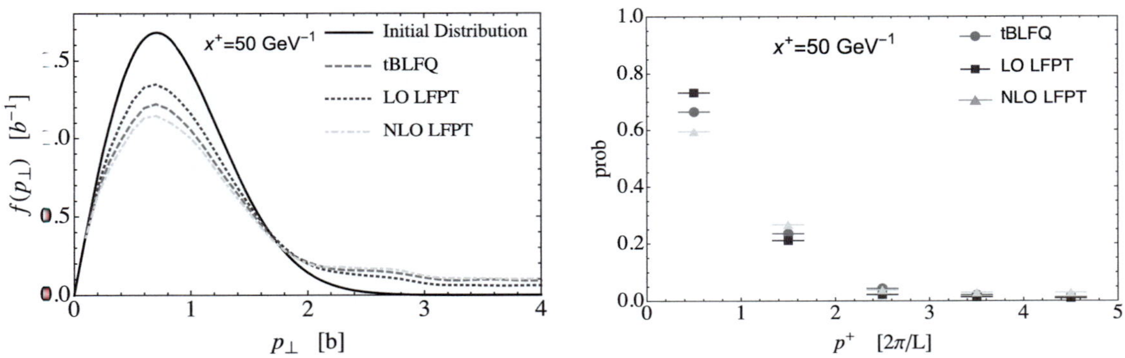

Fig. 6 (Color online) The transverse (*left*) and longitudingnal (*right*) momentum distribution of an electron after evolving in the EM fields generated by a gold nucleus moving along positive z-axis with rapidity $y = 5.3$ for $50\ \text{GeV}^{-1}$. The initial state of the electron is a BLFQ eigenstate with $p^+ = \frac{\pi}{L}$ GeV, $n = 0$, $m = 0$ and helicity $\lambda = 1/2$. The calculation is performed using $N_{max} = 30$, $K = 30$, electron localization scale $L = 10\ \text{GeV}^{-1}$ and HO basis scale $b = 1000 m_e$. *Bars* in the plot of the longitudinal momentum distribution indicated the adoption of discretized plane waves in the longitudinal direction

with up to an order-of-magnitude increase in the high transverse momentum region. The probability that the electron has been excited to higher longitudinal momentum is about 35 % according to the tBLFQ simulation. Calculations in Fig. 6 confirm our speculation that EM fields generated by ultra-relativistic heavy ions have significant influences on charged particles.

The prospects for applying the tBLFQ formalism to heavy ion collisions and electron ion collisions are bright and efforts are underway to achieve this goal. For instance, we will adopt the classical description of gluon fields in high energy nuclear collisions from the color glass condensate (CGC) effective theory [38], and study the real-time evolution of colored objects interacting with these classical gluon fields. Within the tBLFQ framework, we can study the effects of initial gluon fields generated in the initial stage of relativistic heavy-ion collisions on heavy quarks and jets. Furthermore, we can study the diffractive processes in electron ion collisions using the dipole picture. The advantages of tBLFQ framework are distinctive: the calculation is at the amplitude level thereby incorporating quantum interference effects, and we can naturally extend our calculation to higher Fock sectors and go beyond Eikonal approximation.

4 Renormalization via Embedding Process

Non-perturbative renormalization within BLFQ represents a significant challenge with several recent efforts reporting promising methods [15,19,39]. Most recently, an embedding process has been introduced [19] whereby each constituent field, an electron and a positron in the positronium test case, is treated independently in BLFQ with coupling constant and mass renormalization at a sequence of cutoffs.

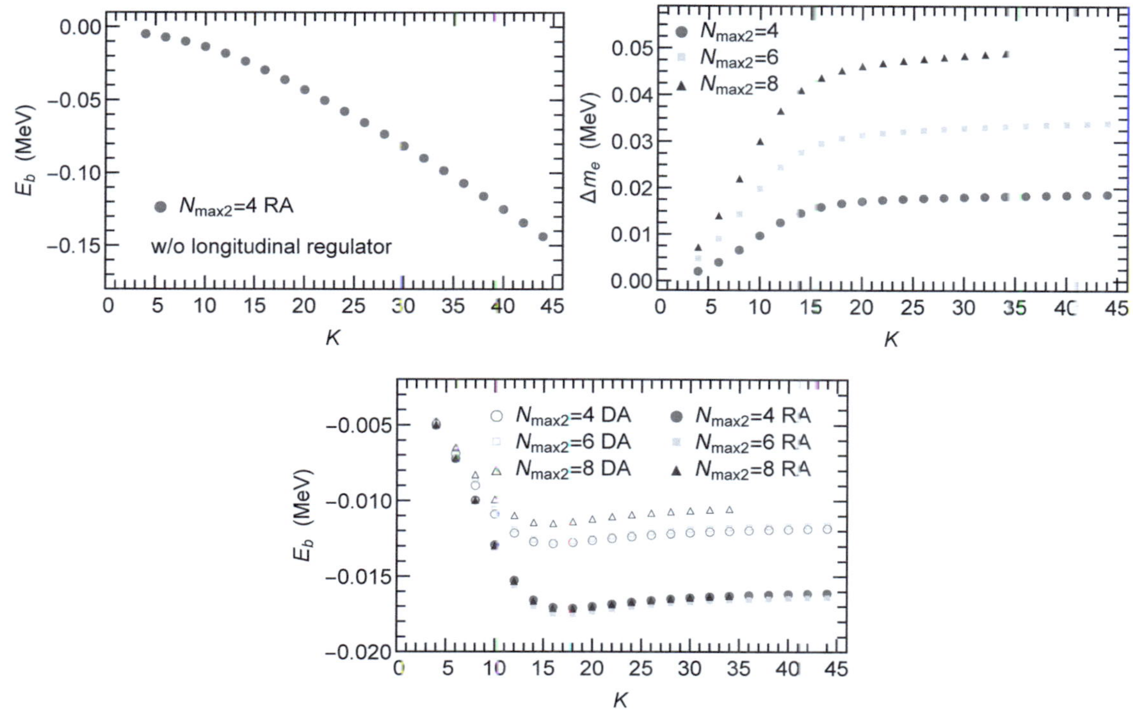

Fig. 7 (Color online) *Top left* Ground state binding energy E_b of positronium as a function of the total longitudinal momentum K *before* longitudinal regulators are applied; *Top right* K dependence of the mass counterterm Δm_e for basis states in the $|e\bar{e}\rangle$ sector with $n = m = 0$ for both e and $\bar{e}$ and with an equal longitudinal momentum partition ($K/2$) between them *after* longitudinal regulators are applied. *Bottom* The ground state binding energy E_b of the positronium as a function of K. Open symbols: E_b evaluated with the light-front amplitudes from direct diagonalization (DA). *Solid symbols* E_b evaluated with the rescaled light-front amplitudes (RA). These results are adapted from Ref. [19]

The renormalized constituent fields define single-particle basis states that are then woven together for the basis of the positronium system, the chosen demonstration case, in such a way that each two-particle state in the positronium basis dictates the cutoffs and associated renormalizations of the constituents. This weaving, or embedding, process leads to considerable computational complexity, as may be anticipated for a non-perturbative renormalization scheme. Initial results for positronium are encouraging and are displayed as a sequence of panels in Fig. 7 [19].

The BLFQ applications summarized here potentially fruitful research paths. Beyond those enumerated above, we would like to mention the extension of BLFQ to the other meson sectors (mixed flavor and light mesons), to the baryons and to multi-baryon systems.

Acknowledgments This work was supported by the Department of Energy under Grant Nos. DE-FG02-87ER40371 and DESC0008485 (SciDAC-3/NUCLEI). Computational resources were provided by NERSC, which is supported by the Office of Science of the U.S. DOE under Contract No. DE-AC02-05CH11231 and by an INCITE award, "Nuclear Structure and Nuclear Reactions", from the US DOE Office of Advanced Scientific Computing. This research also used resources of the Oak Ridge Leadership Computing Facility, which is supported by the US DOE Office of Science under Contract DE-AC05-00OR22725. X. Zhao is supported by new faculty startup funding by the Institute of Modern Physics, Chinese Academy of Sciences.

References

1. Bakker, B.L.G., et al.: Light-front quantum chromodynamics: a framework for the analysis of hadron physics. Nucl. Phys. Proc. Suppl. **251–252**, 165 (2014)
2. Vary, J.P., Honkanen, H., Li, J., Maris, P., Brodsky, S.J., Harindranath, A., de Teramond, G.F., Sternberg, P., et al.: Hamiltonian light-front field theory in a basis function approach. Phys. Rev. C **81**, 035205 (2010)
3. Honkanen, H., Maris, P., Vary, J.P., Brodsky, S.J.: Electron in a transverse harmonic cavity. Phys. Rev. Lett. **106**, 061603 (2011)
4. Vary, J.P.: Hamiltonian light-front field theory: recent progress and tantalizing prospects. Few Body Syst. **52**, 331 (2012)

5. Zhao, X., Ilderton, A., Maris, P., Vary, J.P.: Non-perturbative quantum time evolution on the light-front. Phys. Lett. B **726**, 856 (2013)
6. Zhao, X., Ilderton, A., Maris, P., Vary, J.P.: Scattering in time-dependent basis light-front quantization. Phys. Rev. D **88**, 065014 (2013)
7. Brodsky, S.J., Pauli, H.-C., Pinsky, S.: Quantum chromodynamics and other field theories on the light cone. Phys. Rep. **301**, 299 (1998)
8. Brodsky, S.J., Diehl, M., Hwang, D.S.: Light-cone wavefunction representation of deeply virtual compton scattering. Nucl. Phys. B **596**, 99–124 (2001)
9. Afanasiev, S., et al.: Photoproduction of J/psi and of high mass e+e- in ultra-peripheral Au+Au collisions at $s^{**}(1/2) = 200$-GeV. (PHENIX Collaboration). Phys. Lett. B **679**, 321 (2009)
10. Abbas, E., et al.: Charmonium and e^+e^- pair photoproduction at mid-rapidity in ultra-peripheral Pb-Pb collisions at $\sqrt{s_{NN}}$=2.76 TeV. (ALICE Collaboration). Eur. Phys. J. C **73**(11), 2617 (2013)
11. Navratil, P., Vary, J.P., Barrett, B.R.: Properties of C-12 in the ab initio nuclear shell model. Phys. Rev. Lett. **84**, 5728 (2000)
12. Navratil, P., Vary, J.P., Barrett, B.R.: Large basis ab initio no-core shell model and its application to C-12. Phys. Rev. C **62**, 054311 (2000)
13. Maris, P., Vary, J.P., Shirokov, A.M.: Ab initio no-core full configuration calculations of light nuclei. Phys. Rev. C **79**, 014308 (2009)
14. Barrett, B.R., Navratil, P., Vary, J.P.: Ab initio no core shell model. Prog. Part. Nucl. Phys. **69**, 131 (2013)
15. Zhao, X., Honkanen, H., Maris, P., Vary, J.P., Brodsky, S.J.: Electron g-2 in light-front quantization. Phys. Lett. B **737**, 65 (2014)
16. Wiecki, P., Li, Y., Zhao, X., Maris, P., Vary, J.P.: Basis light-front quantization approach to positronium. Phys. Rev. D **91**(10), 105009 (2015)
17. Li, Y., Maris, P., Zhao, X., Vary, J.P.: Heavy quarkonium in a holographic basis. Phys. Letts. B **758**, 118 (2016) arXiv:1509.07212 [hep-ph]
18. Chen, G. et al.: Light front time evolution in intense fields, in preparation
19. Zhao, X.: Advances in basis light-front quantization. Few Body Syst. **56**(6–9), 257 (2015)
20. de Teramond, G.F., Brodsky, S.J.: Light-front holography: a first approximation to QCD. Phys. Rev. Lett. **102**, 081601 (2009)
21. Brodsky, S.J., de Teramond, G.F., Dosch, H.G., Erlich, J.: Light-front holographic QCD and emerging confinement. Phys. Rep. **584**, 1 (2015)
22. Yang, L.I., Wiecki, P.W., Zhao, X., Maris, P., Vary, J.P.: Introduction to basis light-front quantization approach to QCD bound. In: Schirokov, A.M., Mazur, A.I. (eds.) Proceedings of International Conference on Nuclear Theory Suppercomputing Era (NTSE-2013), Ames, IA, USA, May 1–17, 2013, p. 136. Pacific National University, Khabarovsk (2013)
23. Maris, P., Wiecki, P., Li, Y., Zhao, X., Vary, J.P.: Bound state calculations in QED and QCD using basis light-front quantization. Acta Phys. Polon. Supp. **6**, 321 (2013)
24. Głazek, S.D.: Similarity flow of a neutral scalar coupled to a fixed source. Acta Phys. Polon. B **42**, 1933 (2011)
25. Trawiński, A.P., Głazek, S.D., Brodsky, S.J., de Téramond, G.F., Dosch, H.G.: Effective confining potentials for QCD. Phys. Rev. D **90**, 074017 (2014)
26. Chabysheva, S.S., Hiller, J.R.: Dynamical model for longitudinal wave functions in light-front holographic QCD. Ann. Phys. **337**, 143–152 (2013)
27. Olive, K.A., et al.: (Particle Data Group). Review of Particle Physics. Chin. Phys. **C38**, 090001 (2014). http://pdg.lbl.gov
28. Davies, C.T.H., McNeile, C., Follana, E., Lepage, G.P., Na, H., Shigemitsu, J.: Update: Precision Ds decay constant from full lattice QCD using very fine lattices. (HPQCD Collaboration). Phys. Rev. D **82**, 114504 (2010)
29. McNeile, C., Davies, C.T.H., Follana, E., Hornbostel, K., Lepage, G.P.: Heavy meson masses and decay constants from relativistic heavy quarks in full lattice QCD. (HPQCD Collaboration). Phys. Rev. D **86**, 074503 (2012)
30. Donald, G.C., Davies, C.T.H., Dowdall, R.J., Follana, E., Hornbostel, K., Koponen, J., Lepage, G.P., McNeile, C.: Precision tests of the J/ψ from full lattice QCD: mass, leptonic width and radiative decay rate to η_c. (HPQCD Collaboration). Phys. Rev. D **86**, 094501 (2012)
31. Colquhoun, B., Dowdall, R.J., Davies, C.T.H., Hornbostel, K., Lepage, G.P.: Υ and Υ' Leptonic Widths, a_μ^b and m_b from full lattice QCD. (HPQCD Collaboration). Phys. Rev. D **91**, 074514 (2015)
32. Blank, M., Krassnigg, A.: Bottomonium in a Bethe-Salpeter-equation study. Phys. Rev. D **84**, 096014 (2011)
33. Adhikari, L., Li, Y., Zhao, X., Maris, P., Vary, J.P., El-Hady, A.A.: Form factors and generalized parton distributions in basis light-front quantization. Phys. Rev. C (to appear). arXiv:1602.06027 [nucl-th]
34. Frederico, T., Pace, E., Pasquini, B., Salmè, G.: Pion generalized parton distributions with covariant and light-front constituent quark models. Phys. Rev. D **80**, 054021 (2009)
35. Diehl, M.: Generalized parton distributions. Phys. Rep. **388**, 41 (2003)
36. Brodsky, S.J., de Teramond, G.F.: Light-front dynamics and AdS/QCD correspondence: the pion form factor in the space- and time-like regions. Phys. Rev. D **77**, 056007 (2008)
37. Tuchin, K.: Particle production in strong electromagnetic fields in relativistic heavy-ion collisions. Adv. High Energy Phys. **2013**, 490495 (2013)
38. McLerran, L.D., Venugopalan, R.: Computing quark and gluon distribution functions for very large nuclei. Phys. Rev. D **49**, 2233 (1994)
39. Li, Y., Karmanov, V.A., Maris, P., Vary, J.P.: Ab initio approach to the non-perturbative scalar yukawa model. Phys. Lett. B **748**, 278 (2015)

DOI 10.1007/978-3-319-50699-9_54
Reprinted from *Few-Body System* Journal, DOI 10.1007/s00601-016-1118-9

Harut Avakian

Studies of the 3D Structure of the Nucleon at JLab

Received: 15 February 2016 / Accepted: 1 May 2016 / Published online: 27 May 2016

Abstract Studies of the 3D structure of the nucleon encoded in transverse momentum dependent distribution and fragmentation functions of partons and generalized parton distributions are among the key objectives of the JLab 12 GeV upgrade and the electron ion collider. Main challenges in extracting 3D partonic distributions from precision measurements of hard scattering processes include clear understanding of leading twist QCD fundamentals, higher twist effects, and also correlations of hadron production in target and current fragmentation regions. In this contribution we discuss some ongoing studies and future measurements of spin-orbit correlations at Jefferson Lab.

1 Introduction

The quark–gluon dynamics manifests itself in a set of non-perturbative functions describing all possible spin-spin and spin-orbit correlations, in particular, describing transverse momentum dependent distributions of partons (TMDs), transverse space distributions of partons or generalized parton distributions (GPDs), and fracture functions (FFs), defined as conditional probabilities to produce a hadron in the target fragmentation for a given flavor of a struck quark (see Table 1). In recent years, measurements of azimuthal moments of polarized hadronic cross sections in hard processes have emerged as a powerful tool to probe nucleon structure in general and access the TMDs and GPDs, in particular. Many experiments worldwide are currently trying to pin down various effects related to the nucleon structure through semi-inclusive deep-inelastic scattering (HERMES at DESY [1–4], COMPASS at CERN [5–9], Jefferson Lab [10–15,25]) polarized proton–proton collisions (PHENIX, STAR and BRAHMS at RHIC) [16–18], and electron–positron annihilation (Belle and BaBar) [19–21].

Azimuthal distributions of final state particles in semi-inclusive deep inelastic scattering, in particular, are sensitive to the orbital motion of quarks and play an important role in the study of TMDs in the nucleon. The first results on transverse single spin asymmetries (SSAs) in azimuthal distributions of final state hadrons have been reported by the HERMES and COMPASS [3,5–9,22] collaborations, indicating significant azimuthal moments generated both by the Collins [23] and the Sivers [24] effects, describing spin orbit correlations in the final hadronization and initial distribution processes, respectively.

Measurements of SSAs at JLab, performed with longitudinally polarized NH_3 [13] and transversely polarized ^{3}He [14,15,25,26], indicate that spin orbit correlations may be significant for certain combinations of spins of quarks and nucleons and transverse momentum of scattered quarks. Single spin asymmetries measured by CLAS with longitudinally polarized protons (NH_3 target) related to the TMD $h_{1L}^{\perp}$, describing transversely polarized quarks in the longitudinally polarized nucleon, and the one measured by Hall-A with the transversely polarized neutron (polarized ^{3}He target) related to the TMD $g_{1T}^{\perp}$, describing longitudinally polarized quarks

H. Avakian (✉)
Jefferson Lab, Newport News, VA, USA
E-mail: avakian@jlab.org

Table 1 Leading twist TMD distribution functions (left), GPDs (middle) and fracture functions (right)

N/q	U	L	T		U	L	T		U	L	T
U	$\mathbf{f_1}$		$h_1^{\perp}$	U	$\mathcal{H}$		$\mathcal{E}_T$	U	$\hat{u}_1$	$\hat{l}_1^{\perp h}$	$\hat{t}_1^h, \hat{t}_1^{\perp}$
L		$\mathbf{g_{1L}}$	$h_{1L}^{\perp}$	L		$\widetilde{\mathcal{H}}$	$\widetilde{\mathcal{E}}_T$	L	$\hat{u}_{1L}^{\perp h}$	$\hat{l}_{1L}$	$\hat{t}_{1L}^h, \hat{t}_{1L}^{\perp}$
T	$f_{1T}^{\perp}$	$g_{1T}^{\perp}$	h_1 , $h_{1T}^{\perp}$	T	$\mathcal{E}$	$\widetilde{\mathcal{E}}$	$\mathcal{H}_T, \widetilde{\mathcal{H}}_T$	T	$\hat{u}_{1T}^h, \hat{u}_{1T}^{\perp}$	$\hat{l}_{1T}^h, \hat{l}_{1T}^{\perp}$	$\hat{t}_{1T}, \hat{t}_{1T}^{hh}, \hat{t}_{1T}^{\perp h}, \hat{t}_{1T}^{\perp\perp}$

The U,L,T correspond to unpolarized, longitudinally polarized, and transversely polarized nucleons (rows) and quarks (columns)

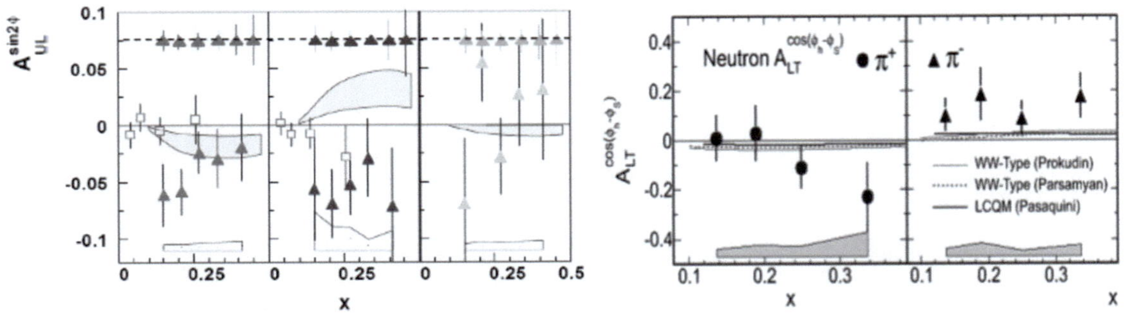

Fig. 1 The Kotzinian–Mulders asymmetry from CLAS (*left plot*). The *open squares* and *triangles* show the existing measurement from HERMES and the preliminary results from CLAS 5.7 GeV CLAS data sets [13], respectively. The curves are calculated using Ref. [27]. The *left panel* shows the SSA from Hall-A [26]

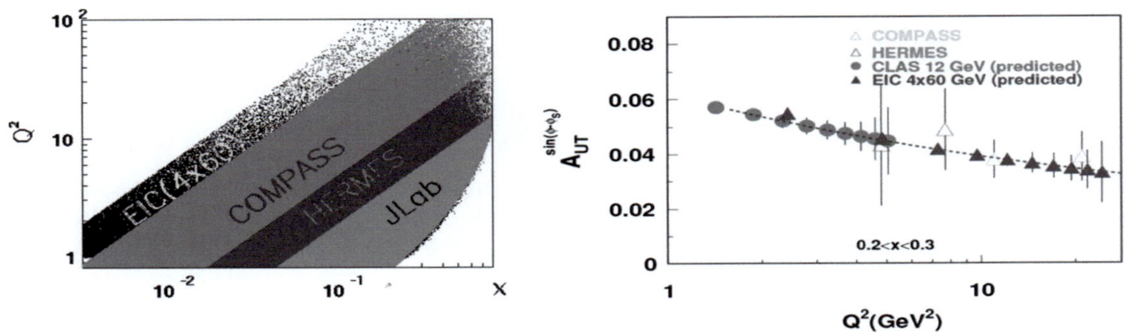

Fig. 2 Comparing CLAS12 and EIC [39] projected data with HERMES [46] and COMPASS [8] results on the Sivers asymmetry

in the transversely polarized nucleon (see Fig. 1), are consistent with the light-cone quark model predictions [28], as well as lattice calculation [29] for corresponding TMDs, for the expected relative sign of SSAs $A_{UL}^{\sin 2\phi}$ in $e\pi^+X$ and $A_{LT}^{\cos(\phi-\phi_S)}$ in $e\pi^-X$.

Significant azimuthal moments in leptoproduction ($A_{UU}^{\cos\phi}$), which have been measured in SIDIS already by EMC collaboration [30,31], were reproduced by latest measurements at CERN, HERMES and JLab [11,12,32–34]. The first unambiguously measured single spin phenomena in SIDIS, which triggered important theoretical developments, were the sizable longitudinal target ($A_{UL}^{\sin\phi}$) and beam ($A_{LU}^{\sin\phi}$) spin asymmetries observed at HERMES and JLab [1,2,4,10]. They could be interpreted in terms of higher twist distribution functions.

Recent measurements of multiplicities and double spin asymmetries as a function of the final transverse momentum of pions in SIDIS at JLab and HERMES [11,13,47] suggest that transverse momentum distributions may depend on the polarization of quarks and possibly also on their flavor. Calculations of transverse momentum dependence of TMDs in different models [28,35–37] and on lattice [29,38] indicate that the dependence of the transverse momentum distributions on the quark polarization and flavor may be very significant. Combination of measurements in a large Q^2 range from HERMES, COMPASS and JLab would allow studies of evolution effects and control possible higher twist contributions in the measurements of TMD observables in general, and of the Sivers asymmetry in particular. Comparison of JLab12 data with HERMES, COMPASS, and future electron ion collider (EIC) (Fig. 2) will pin down the non-trivial Q^2 evolution of Sivers asymmetry.

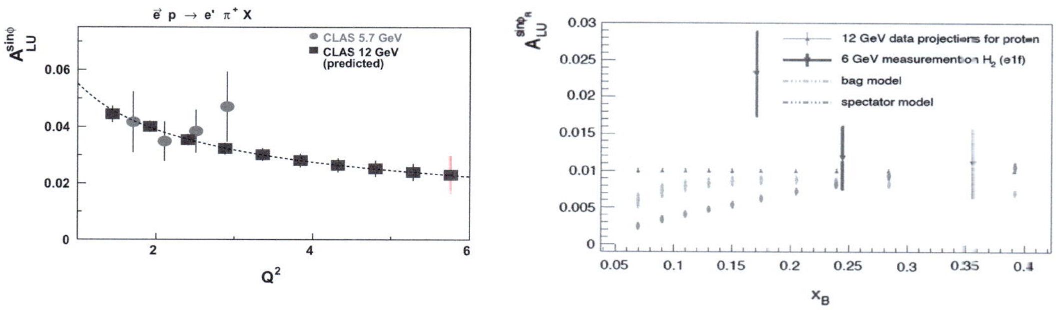

Fig. 3 Projections for higher-twist lepton spin asymmetry $A_{LU}^{\sin\phi}$ for positive pion production as a function of Q^2 (*right*) compared to published data from CLAS [10] and projected CLAS12 [40] in one x, z bin ($0.2 < x < 0.3, 0.5 < z < 0.55$). The *right panel* shows projections for x-dependence of the SSA measurements with CLAS12 using two hadron production [41]. *Lines* correspond to model predictions based on bag [58] and spectator [42] models

Much higher Q^2 range accessible at JLab12 would allow for studies of Q^2-dependence of different higher twist spin-azimuthal asymmetries (Fig. 3), which, apart from providing important information on quark–gluon correlations are needed for understanding of possible corrections from higher twists to leading twist observables. An important process, which can provide independent information on twist-3 TMDs, is the dihadron production in SIDIS [49]. The projections for corresponding CLAS12 measurements [41] of the SSA in dihadron production as a function of Bjorken-x are shown in Fig. 3.

2 Target Fragmentation and Back-to-Back Hadron Production

Although the target fragmentation region (TFR) of DIS, when the hadrons are created from the target remnant, carries important information about the spin and flavor structure of the nucleon, it has not been studied systematically in experiments due to lack of theory fundamentals. The main physical question in the TFR is how the diquark-like remnant system after the DIS process dresses itself up to become a full-fledged hadron, i.e., by which mechanism the quark-antiquark pairs restoring color neutrality are produced, and how this process is correlated with the spin of the target or/and the produced particles. Recently the leading twist formalism for spin and transverse-momentum dependent fracture functions (FFs) has been developed [48]. The structure functions defining the cross section for hadron production in TFR for different polarization states involve unpolarized, longitudinally and transversely polarized quark FFs (see Table 1). With hadrons detected in the TFR, the beam SSAs appear already at leading order. With two hadrons detected in the final state, structure functions may depend also on the relative azimuthal angle of the two hadrons, generating a long range correlation between hadrons produced in CFR and TFR. The kinematic plane for back-to-back hadron production is shown on Fig. 4.

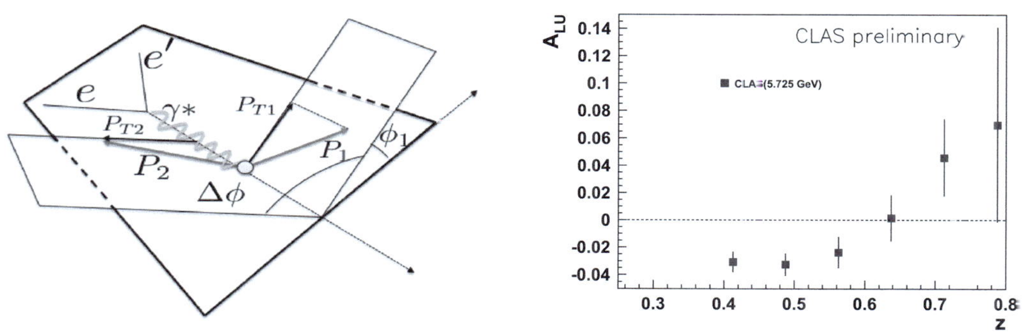

Fig. 4 Kinematic plane for back-to-back (b2b) hadron production in SIDIS (*left*) and measured SSA in back-to-back proton and π^+ production as a function of the product of their transverse momenta $P_{T1}P_{T2}$.

Choosing as independent azimuthal angles $\Delta\phi = \phi_2 - \phi_1$ and ϕ_2, the beam spin asymmetry could be defined as

$$A_{LU}(x, z, P_{T1}^2, P_{T2}^2, \Delta\phi, \dots) = \frac{\int \mathrm{d}\phi_2\, \sigma_{LU}}{\int \mathrm{d}\phi_2\, \sigma_{UU}} = \frac{-\frac{P_{T1}P_{T2}}{m_2 m_N} F_{k1}^{\hat{l}\perp h \cdot D_1} \sin(\Delta\phi)}{F_0^{\hat{u}\cdot D_1}}, \tag{1}$$

where z is the fraction of the virtual photon energy carried by the forward pion, P_{T1}, P_{T2} are transverse momenta of the final state hadrons, and structure functions $F_0^{\hat{u}\cdot D_1}$ and $F_{k1}^{\hat{l}\perp h \cdot D_1}$ are convolutions of unpolarized and polarized FFs with the standard unpolarized fragmentation function D_1. First measurements of single spin asymmetry defined by Eq. (1) in semi-inclusive production of protons and charged pions in coincidence with the scattered electron in hard scattering kinematics ($Q^2 > 1\,\mathrm{GeV}^2$, $W^2 > 4\,\mathrm{GeV}^2$) have been performed by the CLAS collaboration using 5.5 and 5.7 GeV longitudinally polarized electron beams scattering off a 5-cm-long liquid-hydrogen target (CLAS e1f and e16 experiments). The beam spin asymmetry, $A_{LU}^{\sin\Delta\phi}$, has been calculated as a sinusoidal modulation of the difference of azimuthal angles of proton and π^+ with respect to the lepton scattering plane, for different electron helicity states. The modulation was extracted for different bins in x, z of the pion and the product of transverse momenta of final state proton and pion with respect to the virtual photon in the CM frame. Target and current fragmentation regions were selected by cuts on the x_F variables of protons ($x_F < 0$) and pions ($x_F > 0$).

Significant asymmetries in b2b production of protons and pions have been observed for all 3 pions in the final state. The z-dependence of the SSA shows a trend for asymmetry (see Fig. 4) to change the sign at large z, indicating opposite sign in the exclusive limit. With a cut on the missing mass of the $ep\pi^+X$ system, $M_X > 1.4$ we have almost no data at $z > 0.7$.

The x-dependence of the the $A_{LU}^{\sin\phi}$ is consistent with asymmetry being large in the large-x region, were the valence quark presence is very significant. Similar behavior has been observed also for events with neutral and negative pions in the final state, detected in coincidence with scattered electron and proton in the target fragmentation region.

3 Hard Exclusive Processes and GPDs

Measurements of various observables in hard exclusive production versus the momentum transfer to the nucleon, t, on other hand, provide the information necessary for transverse nucleon imaging [50]. The different nucleon spin components of the GPDs [43,44] describing unpolarized and longitudinally polarized quarks (Table 1) can be extracted by measuring spin asymmetries in deeply virtual Compton scattering (DVCS) with polarized targets. Information about the flavor decomposition requires measurements with both protons and neutrons. The latest measurements of DVCS cross sections from unpolarized proton target from JLab Halls A&B [52,59] are consistent with theory predictions for a rapid change of the cross section with azimuthal angle, due to Bethe–Heitler propagators, pushing the DVCS to smaller bins in the azimuthal angle ϕ for separation of different azimuthal moments. Further separation of contributions from unpolarized GPDs, and in particular polarized GPDs, require measurements with longitudinally and transversely polarized targets. Latest measurements of target spin asymmetries in DVCS from CLAS collaboration (see Fig. 5) suggests that t-dependence of polarized quark GPDs may be significantly different than for the unpolarized GPDs, indicating that transverse space distributions of quarks may depend on their spin.

Additional information about the spin/flavor separation can come from meson production data. In particular pseudoscalar meson electroproduction provides access to transversely polarized quark GPDs, H_T^q, E_T^q, $\tilde{H}_T^q$, $\tilde{E}_T^q$, corresponding to parton helicity-flip (chiral-odd) processes [53], not accessible in DVCS studies [56]. For example, the chiral-odd GPD H_T encode unique information on the transverse spin structure of the proton and on the tensor charge since it is defined as a generalization of the quarks transversity distribution, $h_1(x)$. Exclusive pseudoscalar meson electroproduction provides a unique possibility to access the "transversity" or chiral-odd GPDs through polarized scattering measurements [45]. Other important observables accessible in hard exclusive production of pseudoscalar mesons include the GPD $\bar{E}_T$ describing the transverse space distributions of transversely polarized quarks in the unpolarized nucleon and the longitudinal-transverse (LT) correlation provided by $\tilde{E}_T$, which may be relevant for interpretation of orbital angular momentum [51].

Measurements of different final state mesons, π^o, η and K will provide important complementary information for flavor decomposition of the various GPDs [54,55], and in particular, help to further improve the

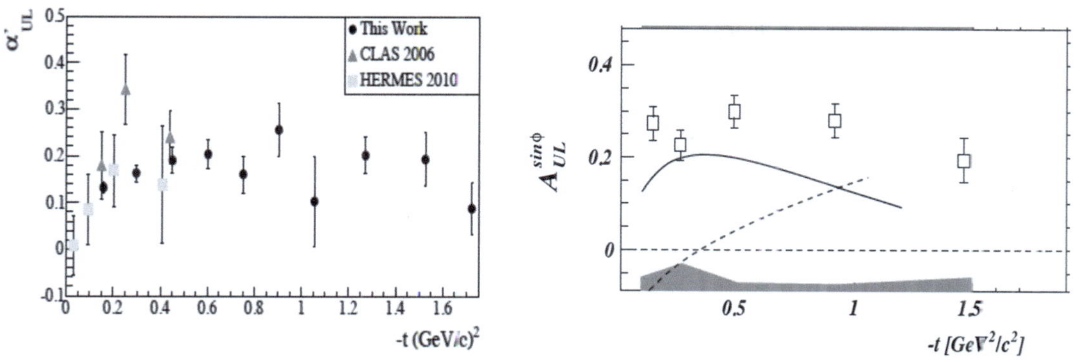

Fig. 5 Spin asymmetries using the Hall-B polarized NH_3 target for exclusive photons [61] and neutral pions [60] for $\langle Q^2 \rangle = 1.94$ $(\mathrm{GeV/c})^2$, $\langle x \rangle = 0.25$

precision of the tensor charge. The measurements for π^o and η SSAs over a large kinematic range are challenging, since their cross sections are small (compared to charged pions). In addition the clean detection of π^os and ηs requires the measurement of their two decay photons. Since the precision attainable in charge particle detection is typically much higher, charged kaons provide a very attractive alternative to neutral mesons channels in studies of transversity GPDs. It has been argued that the H_T effects are essential in the $K^+\Lambda$ channel while in the $K^+\Sigma^0$ leptoproduction the $\bar{E}_T$ contribution is mostly important [57]. Predictions for the cross section of the $K^+\Lambda$ and $K^+\Sigma^0$ leptoproduction [57] has been used to justify future measurements with CLAS12 [62].

The future measurements with the 12 GeV upgrade of Jefferson Lab will be essential for the study of the transverse momentum and transverse space dependent distributions in the valence region. At JLab there are several closely-related proposals approved to measure spin and azimuthal asymmetries in all three Halls, providing complementary studies of different aspects of the complex structure of the nucleon in terms of flavor, momentum and spin.

References

1. Airapetian, A., et al.: Observation of a single-spin azimuthal asymmetry in semi-inclusive pion electro-production. Phys. Rev. Lett. **84**, 4047–4051 (2000). arXiv:hep-ex/9910062
2. Airapetian, A., et al.: Single-spin azimuthal asymmetries in electroproduction of neutral pions in semi-inclusive deep-inelastic scattering. Phys. Rev. D **64**, 097101 (2001). arXiv:hep-ex/0104005
3. Airapetian, A., et al.: Single-spin asymmetries in semi-inclusive deep-inelastic scattering on a transversely polarized hydrogen target. Phys. Rev. Lett. **94**, 012002 (2005). arXiv:hep-ex/0408013
4. Airapetian, A., et al.: Beam-spin asymmetries in the azimuthal distribution of pion electroproduction. Phys. Lett. B **648**, 164–170 (2007). arXiv:hep-ex/0612059
5. Alexakhin, V.Y., et al.: First measurement of the transverse spin asymmetries of the deuteron in semi-inclusive deep inelastic scattering. Phys. Rev. Lett. **94**, 202002 (2005). arXiv:hep-ex/0503002
6. Ageev, E.S., et al.: A new measurement of the collins and sivers asymmetries on a transversely polarised deuteron target. Nucl. Phys. **B765**, 31–70 (2007). arXiv:hep-ex/0610068
7. Alekseev, M.G., et al.: Azimuthal asymmetries of charged hadrons produced by high energy muons scattered off longitudinally polarised deuterons. Eur. Phys. J. **C70**, 39–49 (2010a). doi:10.1140/epjc/s10052-010-1461-9. arXiv:1007.1562
8. Alekseev, M.G., et al.: Measurement of the Collins and Sivers asymmetries on transversely polarised protons. Phys. Lett. B **692**, 240–246 (2010). doi:10.1016/j.physletb.2010.08.001. arXiv:1005.5609
9. Adolph, C., et al.: Transverse spin effects in hadron-pair production from semi-inclusive deep inelastic scattering. Phys. Lett. **B713**, 10–16 (2012). doi:10.1016/j.physletb.2012.05.015. arXiv:1202.6150
10. Avakian, H., et al.: Measurement of beam-spin asymmetries for deep inelastic pi+ electroproduction. Phys. Rev. D **69**, 112004 (2004). arXiv:hep-ex/0301005
11. Mkrtchyan, H., et al.: Transverse momentum dependence of semi-inclusive pion production. Phys. Lett. B **665**, 20–25 (2008). arXiv:0709.3020 [hep-ph]
12. Osipenko, M., et al.: Measurement of unpolarized semi-inclusive pi+ electroproduction off the proton. Phys. Rev. D **80**, 032004 (2009). arXiv:0809.1153 [hep-ex]
13. Avakian, H., et al.: Measurement of single and double spin asymmetries in deep inelastic pion electroproduction with a longitudinally polarized target. Phys. Rev. Lett. **105**, 262,002 (2010). doi:10.1103/PhysRevLett.105.262002. arXiv:1003.4549 [hep-ex]
14. Zhang, Y., et al.: Measurement of pretzelosity asymmetry of charged pion production in semi-inclusive deep inelastic scattering on a polarized ^{3}He target. Phys. Rev. C **90**(5), 055209 (2014). doi:10.1103/PhysRevC.90.055209. arXiv:1312.3047

15. Zhao, Y.X., et al.: Single spin asymmetries in charged kaon production from semi-inclusive deep inelastic scattering on a transversely polarized ^{3}He target. Phys. Rev. C **90**(5), 055–201 (2014). doi:10.1103/PhysRevC.90.055201. arXiv:1404.7204
16. Adams, J., et al.: Cross sections and transverse single-spin asymmetries in forward neutral pion production from proton collisions at s**(1/2) = 200-GeV. Phys. Rev. Lett. **92**, 171,801 (2004). arXiv:hep-ex/0310058
17. Chiu, M.: Single spin transverse asymmetries of neutral pions at forward rapidities in s**(1/2) = 62.4-GeV polarized proton collisions in PHENIX. AIP Conf. Proc. **915**, 539–542 (2007). arXiv:nucl-ex/0701031
18. Arsene, I., et al.: Single transverse spin asymmetries of identified charged hadrons in polarized p+p collisions at $\sqrt{s}$ = 62.4 GeV. Phys. Rev. Lett. **101**, 042,001 (2008). arXiv:0801.1078 [nucl-ex]
19. Abe, K., et al.: Measurement of azimuthal asymmetries in inclusive production of hadron pairs in e+ e- annihilation at Belle. Phys. Rev. Lett. **96**, 232002 (2006). arXiv:hep-ex/0507063
20. Vossen, A., et al.: Observation of the interference fragmentation function for charged pion pairs in e^+e^- annihilation near $\sqrt{s} = 10.58$ GeV. Phys. Rev. Lett. (2011). arXiv:1104.2425
21. Lees, J.P., et al.: Measurement of Collins asymmetries in inclusive production of charged pion pairs in e^+e^- annihilation at BABAR. Phys. Rev. D **90**(5), 052003 (2014). doi:10.1103/PhysRevD.90.052003. arXiv:1309.5278
22. Airapetian, A., et al.: Effects of transversity in deep-inelastic scattering by polarized protons. Phys. Lett. B **693**, 11–16 (2010). doi:10.1016/j.physletb.2010.08.012. arXiv:1006.4221
23. Collins, J.C.: Fragmentation of transversely polarized quarks probed in transverse momentum distributions. Nucl. Phys. **B396**, 161–182 (1993). arXiv:hep-ph/9208213
24. Sivers, D.W.: Hard scattering scaling laws for single spin production asymmetries. Phys. Rev. D **43**, 261–263 (1991)
25. Qian, X., et al.: Single spin asymmetries in charged pion production from semi-inclusive deep inelastic scattering on a transversely polarized ^{3}He target. Phys. Rev. Lett. **107**(072), 003 (2011). doi:10.1103/PhysRevLett.107.072003. arXiv:1106.0363
26. Huang, J., et al.: Beam-target double spin asymmetry A_LT in charged pion production from deep inelastic scattering on a transversely polarized he-3 target at $1.4 < q^2 > 2.7gev^2$. Phys. Rev. Lett. **108**(052), 001 (2012). doi:10.1103/PhysRevLett.108.052001. arXiv:1108.0489
27. Efremov, A.V., Goeke, K., Schweitzer, P.: Transversity distribution in spin asymmetries in semi- inclusive dis and in the Drell–Yan process. Czech. J. Phys. **55**, A189–A208 (2005). arXiv:hep-ph/0412420
28. Pasquini, B., Cazzaniga, S., Boffi, S.: Transverse momentum dependent parton distributions in a light-cone quark model. Phys. Rev. D **78**(034), 025 (2008). arXiv:0806.2298 [hep-ph]
29. Hagler, P., Musch, B.U., Negele, J.W., Schafer, A.: Intrinsic quark transverse momentum in the nucleon from lattice QCD. Europhys. Lett. **88**(61), 001 (2009). arXiv:0908.1283 [hep-lat]
30. Aubert, J.J., et al.: Measurement of hadronic azimuthal distributions in deep inelastic muon proton scattering. Phys. Lett. B **130**, 118 (1983)
31. Arneodo, M., et al.: Measurement of hadron azimuthal distributions in deep inelastic muon proton scattering. Z. Phys. C **34**, 277 (1987)
32. Kafer, W.:Measurements of Unpolarized Azimuthal Asymmetries at COMPASS. Transversity 2008 Proceedings (2008). arXiv:0808.0114 [hep-ex]
33. Giordano, F., Lamb, R.: Measurement of azimuthal asymmetries of the unpolarized cross section at HERMES. AIP Conf. Proc. **1149**, 423–426 (2009). arXiv:0901.2438 [hep-ex]
34. Gohn, W., Avakian, H., Joo, K., Ungaro, M.: Beam spin asymmetries from semi-inclusive pion electroproduction in deep inelastic scattering. AIP Conf. Proc. **1149**, 461–464 (2009). doi:10.1063/1.3215689
35. Lu, Z., Ma, B.Q.: Sivers function in light-cone quark model and azimuthal spin asymmetries in pion electroproduction. Nucl. Phys. A **741**, 200–214 (2004). doi:10.1016/j.nuclphysa.2004.06.006. arXiv:hep-ph/0406171
36. Anselmino, M., Efremov, A., Kotzinian, A., Parsamyan, B.: Transverse momentum dependence of the quark helicity distributions and the Cahn effect in double-spin asymmetry a(ll) in semi inclusive dis. Phys. Rev. D **74**(074), 015 (2006). arXiv:hep-ph/0608048
37. Bourrely, C., Buccella, F., Soffer, J.: Semiinclusive DIS cross sections and spin asymmetries in the quantum statistical parton distributions approach. Phys. Rev. **D83**(074), 008 (2011). doi:10.1103/PhysRevD.83.074008. arXiv:1008.5322
38. Musch, B.U., Hagler, P., Negele, J.W., Schafer, A.: Exploring quark transverse momentum distributions with lattice QCD. Phys. Rev. **D83**(094), 507 (2011). doi:10.1103/PhysRevD.83.094507. arXiv:1011.1213
39. Accardi, A., Albacete, J., Anselmino, M., Armesto, N., Aschenauer, E., et al.: Electron Ion Collider: The Next QCD Frontier—Understanding the Glue that Binds Us All (2012). arXiv:1212.1701
40. Avakian, H., et al.: Transverse spin effects in sidis at 12 GeV with transversely polarized target. JLab. Exp. **E12**, 06–015 (2008)
41. Pisano, S., et al.: Higher-twist collinear structure of the nucleon through di-hadron sidis on unpolarized hydrogen and deuterium. JLab. Exp. E12-06-112B/E12-09-008B (2014). http://www.jlab.org/exp_prog/proposals/14/E12-06-112B_E12-09-008B.pdf
42. Jakob, R., Mulders, P.J., Rodrigues, J.: Modelling quark distribution and fragmentation functions. Nucl. Phys. A **626**, 937–965 (1997). arXiv:hep-ph/9704335
43. Ji, X.D.: Deeply-virtual compton scattering. Phys. Rev. **D55**, 7114–7125 (1997). arXiv:hep-ph/9609381
44. Radyushkin, A.V.: Scaling limit of deeply virtual compton scattering. Phys. Lett. B **380**, 417–425 (1996). arXiv:hep-ph/9604317
45. Ahmad, S., Goldstein, G.R., Liuti, S.: Nucleon tensor charge from exclusive π^o electroproduction. Phys. Rev. D **79**(054), 014 (2009). doi:10.1103/PhysRevD.79.054014. arXiv:0805.3568
46. Airapetian, A., et al.: Observation of the naive-T-odd Sivers effect in deep inelastic scattering. Phys. Rev. Lett. **103**(152), 002 (2009). doi:10.1103/PhysRevLett.103.152002. arXiv:0906.3918
47. Airapetian, A., et al.: Multiplicities of charged pions and kaons from semi-inclusive deep-inelastic scattering by the proton and the deuteron. Phys. Rev. **D87**(074), 029 (2013). doi:10.1103/PhysRevD.87.074029. arXiv:1212.5407
48. Anselmino, M., Barone, V., Kotzinian, A.: SIDIS in the target fragmentation region: polarized and transverse momentum dependent fracture functions. Phys. Lett. B **699**, 108–118 (2011). doi:10.1016/j.physletb.2011.03.067. arXiv:1102.4214

49. Bacchetta, A., Radici, M.: Two-hadron semi-inclusive production including subleading twist. Phys. Rev. D **69**(074), 026 (2004). doi:10.1103/PhysRevD.69.074026. arXiv:hep-ph/0311173
50. Burkardt, M.: Impact parameter space interpretation for generalized parton distributions. Int. J. Mod. Phys. A **18** 173–208 (2003). doi:10.1142/S0217751X03012370. arXiv:hep-ph/0207047
51. Courtoy, A., Goldstein, G.R., Hernandez, J.O.G., Liuti, S., Rajan, A.: On the observability of the Quark orbital angular momentum distribution. Phys. Lett. B **731**, 141–147 (2014). doi:10.1016/j.physletb.2014.02.017. arXiv:1310.5157
52. Defurne, M., et al.: E00–110 experiment at Jefferson Lab Hall A: deeply virtual compton scattering off the proton at 6 GeV. Phys. Rev. C **92**(5), 055202 (2015). doi:10.1103/PhysRevC.92.055202. arXiv:1504.05453
53. Diehl, M.: Generalized parton distributions. Phys. Rep. **388**, 41–277 (2003). doi:10.1016/j.physrep.2003.08.002. arXiv:hep-ph/0307382
54. Frankfurt, L., Pobylitsa, P., Polyakov, M.V., Strikman, M.: Hard exclusive pseudoscalar meson electroproduction and spin structure of a nucleon. Phys. Rev. **D60**(014), 010 (1999). doi:10.1103/PhysRevD.60.014010. arXiv:hep-ph/9901429
55. Goeke, K., Polyakov, M.V., Vanderhaeghen, M.: Hard exclusive reactions and the structure of hadrons. Prog. Part Nucl. Phys. **47**, 401–515 (2001). doi:10.1016/S0146-6410(01)00158-2
56. Goloskokov, S.V., Kroll, P.: An attempt to understand exclusive pi+ electroproduction. Eur. Phys. J. **C65**, 137–151 (2010). doi:10.1140/epjc/s10052-009-1178-9. arXiv:0906.0460
57. Goloskokov, S.V., Kroll, P.: Transversity in hard exclusive electroproduction of pseudoscalar mesons. Eur. Phys J. A **47**, 112 (2011). doi:10.1140/epja/i2011-11112-6. arXiv:1106.4897
58. Jaffe, R.L., Ji, X.: Chiral odd parton distributions and Drell–Yan processes. Nucl. Phys. **B375**, 527–560 (1992)
59. Jo, H.S., et al.: Cross sections for the exclusive photon electroproduction on the proton and generalized parton distributions. Phys. Rev. Lett. **115**(21), 212003 (2015). doi:10.1103/PhysRevLett.115.212003. arXiv:1504.02009
60. Kim, A., et al.: Target and Double Spin Asymmetries of Deeply Virtual π^0 Production with a Longitudinally Polarized Proton Target and CLAS. Submitted to PRL (2015). arXiv:1511.03338
61. Pisano, S., et al.: Single and double spin asymmetries for deeply virtual Compton scattering measured with CLAS and a longitudinally polarized proton target. Phys. Rev. D **91**(5), 052014 (2015). doi:10.1103/PhysRevD.91.052014. arXiv:1501.07052
62. Kim A, et al.: Studies of chiral-odd gpds using the clas12 detector. LOI-12-14-003 (2014)

DOI 10.1007/978-3-319-50699-9_55
Reprinted from *Few-Body System* Journal, DOI 10.1007/s00601-016-1120-2

Silvia Pisano · for the CLAS Collaboration

Precise Measurements of DVCS at JLab and Quark Orbital Angular Momentum

Received: 24 February 2016 / Accepted: 5 May 2016 / Published online: 6 June 2016

Abstract Deeply-virtual Compton scattering provides the cleanest access to the 3D imaging of the nucleon structure encoded in the generalized parton distributions, that correlate the fraction of the total nucleon momentum carried by a constituent to its position in the transverse plane. Besides the information on the spatial imaging of the nucleon, GPDs provide an access, through the Ji relation, to the contribution of the angular momentum of quarks to proton spin. An accurate estimate of such a contribution will lead to a better understanding of the origin of the proton spin. Jefferson Lab has been an ideal environment for the study of exclusive processes, thanks to the combination of the high-intensity and high-polarization electron beam provided by the CEBAF, with the complementary equipments of the three experimental halls. This has allowed high-precision measurements of the DVCS observables in a wide kinematic region, with focus on those observable s that provide access to the GPDs entering the Ji relation. These studies will be further widened by the projected data from the 12-GeV era, which will improve the existing measurements both in terms of precision and phase-space coverage. The important results on the proton DVCS obtained during the 6-GeV era will be discussed, together with the upcoming experiments approved for the 12-GeV upgrade, that foresees measurements with both proton and quasi-free neutron targets and that, when combined, will lead to the extraction of the Compton Form Factors for separate quark flavors.

1 Introduction

In the last decades, Deeply-Virtual Compton Scattering (DVCS) proved to be an essential tool to study the nucleon structure in terms of the QCD degrees of freedom [1]. It allows access to the generalized parton distributions [2–13], structure functions that relate the parton position in the transverse plane to the momentum fraction they carry. GPDs encode, in this way, the complementary information underlying the electromagnetic form factors and the 1D parton distribution functions (PDF). Beyond shedding light on the hadron structure, GPDs also play a role in exploring the nucleon spin content. One of the open puzzles of hadronic physics is a complete understanding of how the nucleon spin is generated from the relevant degrees of freedom. Since the contribution coming from the quark spin turned out to be less than expected [14, 15] and the gluon one—while recently found to be non zero [16]—is still not sufficient to account for the whole nucleon spin budget a novel interest arose in the last years on the contribution from the quark orbital angular momentum.

Quark total angular momentum can be extracted through the so-called Ji's relation [3]. It relates the angular momentum of a parton of flavor q to the integral of the GPDs H and E over the momentum fraction x, extrapolated to the limit of a zero transverse momentum transferred. By combining this value with the contribution from the quark spin measured through the inclusive deep-inelastic scattering, the quark orbital momentum can be accessed. Different DVCS observable s provide access to combinations of Compton Form

S. Pisano (✉)
Laboratori Nazionali di Frascati, Istituto Nazionale di Fisica Nucleare, Via Enrico Fermi, 40, 00044 Frascati, Italy
E-mail: silvia.pisano@lnf.infn.it

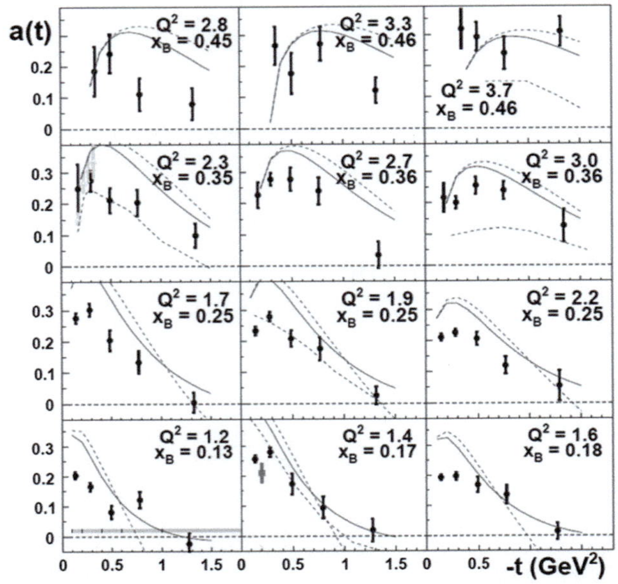

Fig. 1 Dependence of the beam-spin asymmetry at $\phi = 90$ deg on the transferred four-momentum $-t$ in different (x_B, Q^2) bins. The *grey band* in the *lower left plot* shows systematic uncertainties. *Red squares* are from [17] and *green triangles* from [18,19]. Different *curves* are described in Ref. [20] and references therein

Factors, each of the latter being related to different GPDs. In order to exploit the Ji's relation, a set of DVCS observables have to be measured, combining different polarization degrees of freedom (e.g., beam or target polarization) and different targets (proton, deuterium). In fact, the combination of proton and neutron data allows to perform a flavor decomposition, eventually leading to the extraction of the GPDs of the u and d flavors. Depending on the target involved, different observables will be sensitive to different GPDs. Such a sensitivity is dictated by the electromagnetic form factors and the kinematical terms that appear in the expressions relating asymmetries (or cross-sections) to Compton Form Factors.

2 DVCS at Jefferson Lab

In ref. [3], the ideal facility to access DVCS observables, whose cross-section is down by an order of α_s with respect to the inclusive case, is described as being capable of collecting high-statistics data, so to provide statistically-significant measurements. Wide kinematic coverage is also important, in order to map the dependence s of the observables on the relevant variables in a kinematic region as extended as possible. The `Continuous Electron Beam Accelerator Facility` (`CEBAF`) at Jefferson Lab provided a high-intensity, longitudinally-polarized electron beam to three experimental halls (Hall-A, B, C), allowing to perform a vast set of high-precision measurements, ranging from spectroscopy to nucleon structure. It is presently undergoing an upgrade [21] that will double the energy of the electron beam. In the same time, experimental equipments in the halls are being upgraded, in order to optimize their performances in the kinematic region accessible with the increased beam energy. Hall-B is equipped with a large-acceptance spectrometer, `CLAS` [22]. It is undergoing an upgrade to a new large-acceptance spectrometer `CLAS12` [23], that will be able to operate at a luminosity one order of magnitude larger than `CLAS`. In a similar way, Hall-A and Hall-C are upgrading their equipments, in order to adapt their performances to the new precision frontier that the 12-GeV operations will challenge. During the 6-GeV era, many important measurements have been performed both in Hall-A and Hall-B. They mapped, with a high precision, the kinematical dependence of several DVCS observables in an extended kinematical region. Starting from a pioneering measurement in Hall-B, that led to the first observation of the interference among the DVCS and the Bethe-Heitler processes [17], DVCS scaling properties have been tested in Hall-A [18,19]. A deeper investigation of the DVCS scaling property are expected from both Hall-A [24] and Hall-C [25] in the next 12-GeV era, when high-statistics measurements will allow to test the DVCS scaling behaviour on an extended Q^2 range. As to the Orbital Angular Momentum (`OAM`), main ingredients to exploit the Ji's relation are the GPDs H and E. The latter have to be measured both on proton

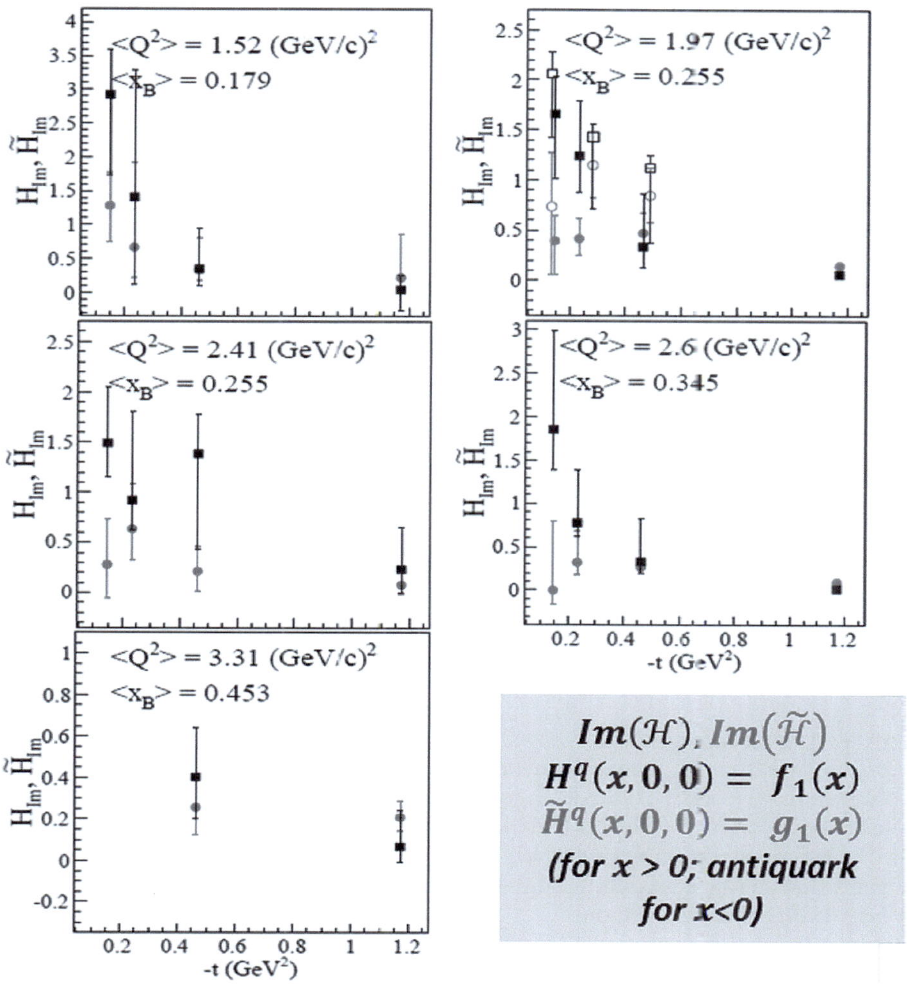

Fig. 2 *t*-dependence of $H(Im)$ (*black squares*) and $\tilde{H}(Im)$ (*red circles*) for different (Q^2, x_B) bins. See Ref. [27] for details

and neutron targets, in order to access the single-flavor GPDs by implementing a flavor-separation procedure. A high-statistics mapping of the beam-spin asymmetry A_{LU} on an unpolarized proton target, mainly sensitive to the imaginary part of H, has been performed in Hall-B [20]. A_{LU} is defined as the difference among the cross-sections corresponding to a positive and a negative beam helicity, divided by their sum. It is shown in Fig. 1, where the dependence of the amplitude of the dominant $\sin\phi$ modulation is shown as a function of the transferred four-momentum $-t$ for different (x_B, Q^2) bins. Comparison to different theoretical models shows the dominance, already at CLAS kinematics, of the leading-twist contribution. Extraction of A_{LU} has been recently extended to a longitudinally-polarized NH_3 target [27], that led to the measurement of both single (beam and target [28,35]) and double-spin asymmetries. This last measurement allowed to compare the distribution inside the nucleon of electric and axial charges (see Fig. 2), respectively related to the GPDs H and $\tilde{H}$.

Beyond asymmetries, also DVCS cross-section, sensitive to both the real and imaginary part of H, has been recently measured by Hall-B [26], extending the kinematic coverage of the afore-mentioned measurements by Hall-A [18,19]. The dependences of $Im(H)$ and $Re(H)$ on the transferred four-momentum $-t$ are reported in Fig. 3. These dependences are fitted with the function Ae^{bt}, where b is related to the transverse size of the nucleon. The decreasing of b for increasing x_B seems to suggest that, going to the large-x_B region (*valence quark*), the nucleon is more compact than in the case where also low-x_B, *sea quarks* are considered. A possible interpretation is that the valence quarks are confined in the inner core of the nucleon volume, while sea quarks extend to its periphery.

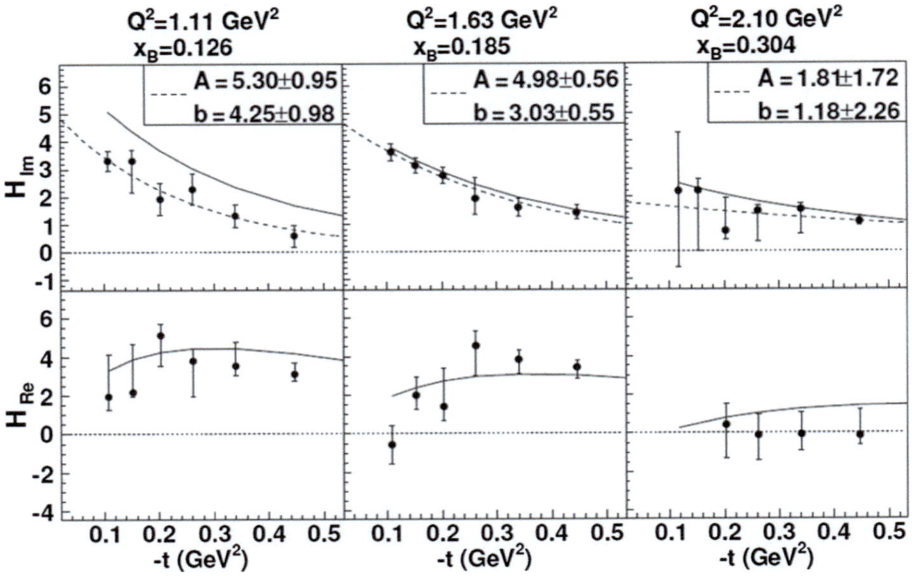

Fig. 3 Dependences on $-t$ of the Compton Form Factors $H(\text{Im})$ (*upper panel*) and $H(\text{Re})$ (*lower panel*) for three different (x_B, Q^2) bins. The *blue solid curves* are predictions from the VGG model. The *black dashed curves* show the fit to the function Ae^{bt}. More details in [26]

While A_{LU} on a proton target is mainly sensitive to the GPD H_p, the access to E_p requests transverse degrees of freedom. In particular, a measurement of the cross-section differences for two different transverse target polarization is needed. To this end, part of the physics program of the CLAS12 activity will be devoted to measurements on a transversely-polarized proton target. The conditionally-approved experiment `PR12-12-010` [29] aims at extracting single-spin asymmetries on a transversely polarized HD-Ice target. The azimuthal moments that can be accessed through these polarization degree of freedom, indeed, will provide access to the elusive GPD E for the proton.

As anticipated in the previous sections, to get at the level of parton GPDs measurements on proton and neutron targets have to be combined. A first measurement on neutron DVCS has been performed in Hall-A [30]. By exploiting the interference between DVCS and BH, the imaginary part of DVCS amplitudes on the neutron and on the deuteron was extracted from the helicity-dependent $D(\mathbf{e}, e'\gamma)X$ cross section, at the kinematical point $Q^2 = 1.9\,\text{GeV}^2$ and $x_B = 0.36$. More data are expected during the 12-GeV era. `CLAS12` approved, for its physics program, different measurements on the neutron. In particular, experiment `E12-11-003` will measure the longitudinal beam-spin asymmetry on the neutron in a wide kinematical range [31], that is mainly sensitive to the GPD E. Projections are shown in Fig. 4, where the high precision of the CLAS12 measurement can be appreciated. In order to complete the set of GPDs entering the Ji's relation, also the GPD H_n is needed. Since no measurement are currently available, a new proposal [32] that foresees the usage of CLAS12 with a longitudinally polarized deuterium target (ND_3) was recently submitted. Its goal is to measure target-spin asymmetries and double-spin asymmetries, similarly to the CLAS experiment described in Refs. [27,28]. These asymmetries will be combined with the A_{LU} described above, and will allow a first model-independent extraction of neutron CFFs.

Recently, an alternative path to access the quark OAM is being explored [33,34]. It identified the relevant observables in the higher-twist target-spin asymmetries accessible through DVCS on a longitudinally-polarized target, raising a new interest in higher-twist modulations.

3 Conclusions

In the last years, GPDs started to play a fundamental role in understanding nucleon structure. They allow to relate the macroscopic nucleon properties to the dynamics of its constituents, shedding light on how the hadronic degrees of freedom depend and are related to the partonic ones. A major question in this field is how the nucleon spin is formed in terms of its constituent polarizations. However, despite years of both theoretical and experimental quest, the mechanism of proton spin generation is far from being fully understood. Also by

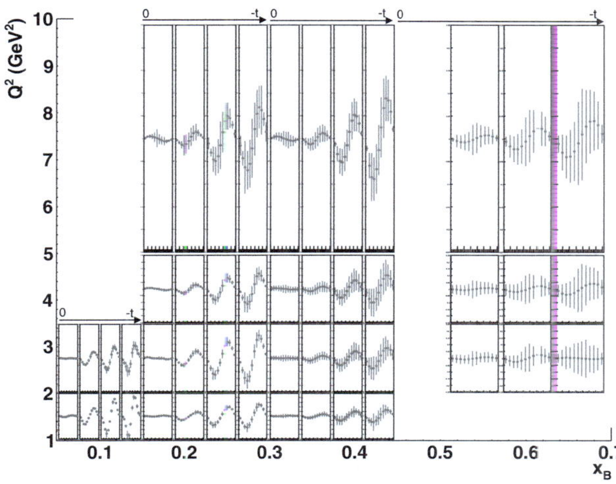

Fig. 4 Projected A_{LU} from the CLAS12 experiment in Ref. [31]. The y-scale range is (0 25:0.25)

combining the latest extractions of the quark and gluon spin contributions, the nucleon spin budget cannot be fully described. As a consequence, the investigation focused on the quark orbital angular momentum, that can significantly contribute to the proton spin. In order to get an estimate of this quantity, information from inclusive DIS, that gives access to the quark-spin contribution, can be combined with the total angular momentum J^q of a given flavor, that appears in the so-called Ji's relation. In the latter J^q is expressed in terms of the GPDs H and E, that appear then as strategic quantities toward a better understanding of the quark orbital angular momentum contribution. Extraction of H and E on both proton and neutron target is mandatory in order to perform a flavor separation and access the GPDs of the u and d constituents. Jefferson Lab provided important measurements during its 6-GeV operations, when several DVCS longitudinal observables were measured. Several kinematical dependences have been extracted based on JLab data, that allowed to map the GPD H in a wide kinematical region. More data are expected to come during the 12-GeV operations [24,25 36], that will now extend also to transverse observables [29] and to neutron targets [31,32]. The combination of present and future data, and the following flavor separation of the Compton Form Factors, will represent a major step toward a better understanding of the role that quark angular momentum play in the making up of the nucleon spin.

References

1. Jaffe, R.L.: Spin, twist and hadron structure in deep inelastic processes. Lect. Notes Phy. **496**, 178 (1997)
2. Müller, D., Robaschik, D., Geyer, B., Dittes, F.-M., Horejsi, J.: Wave functions, evolution equations and evolution kernels from light ray operators of QCD. Fortschr. Phys. **42**, 101 (1994)
3. Ji, X.: Gauge-invariant decomposition of nucleon spin. Phys. Rev. Lett. **78**, 610 (1997)
4. Ji, X.: Deeply virtual Compton scattering. Phys. Rev. D **55**, 7114 (1997)
5. Radyushkin, A.V.: Scaling limit of deeply virtual compton scattering. Phys. Lett. B **380**, 417 (1996)
6. Radyushkin, A.V.: Nonforward parton distributions. Phys. Rev. D **56**, 5524 (1997)
7. Collins, J.C., Frankfurt, L., Strikman, M.: Factorization for hard exclusive electroproduction of mesons in QCD. Phys. Rev. D **56**, 2982 (1997)
8. Goeke, K., Polyakov, M.V., Vanderhaeghen, M.: Hard exclusive reactions and the structure of hadrons. Prog. Part. Nucl. Phys. **47**, 401 (2001)
9. Diehl, M.: Generalized parton distributions. Phys. Rept. **388**, 41 (2003)
10. Belitsky, A.V., Radyushkin, A.V.: Unraveling hadron structure with generalized parton distributions. Phys. Rept. **418**, 1 (2005)
11. Burkardt, Matthias: Impact parameter dependent parton distributions and offforward parton distributions for $\zeta > 0$. Phys. Rev. D **62**, 071503 (2000)
12. Burkardt, Matthias: Impact parameter space interpretation for generalized parton distributions. Int. J. Mod. Phys. A **18**, 173–208 (2003)
13. Burkardt, Matthias: Generalized parton distributions for large x. Phys. Lett. B **595**, 245–249 (2004)
14. Ashman, J., et al.: (European Muon): A measurement of the spin asymmetry and determination of the structure function g1 in deep inelastic muon-proton scattering. Phys. Lett. B **206**, 364 (1988)
15. Ashman, J., et al.: (European Muon): An investigation of the spin structure of the proton in deep inelastic scattering of polarised muons on polarised protons. Nucl. Phys. B **328**, 1 (1989)

16. de Florian, et al.: Evidence for polarization of gluons in the proton. Phys. Rev. Lett. **113**, 012001
17. Stepanyan, S., et al.: (CLAS Collaboration): Observation of exclusive deeply virtual compton scattering in polarized electron beam asymmetry measurements. Phys. Rev. Lett. **87**, 182002 (2001)
18. Muñoz Camacho, C., et al.: (Hall-A Collaboration): Scaling tests of the cross section for deeply virtual compton scattering. Phys. Rev. Lett. **97**, 262002 (2006)
19. Defurne, M., et al.: (Hall-A Collaboration): E00110 experiment at Jefferson Lab hall a: deeply virtual Compton scattering off the proton at 6 GeV. Phys. Rev. C **92**, 055202 (2015)
20. Girod, F.-X., et al.: (CLAS Collaboration): Measurement of deeply virtual Compton scattering beam-spin asymmetries. Phys. Rev. Lett. **100**, 162002 (2008)
21. Burkert, V.D.: The JLab 12GeV upgrade and the initial science program. (2012). arXiv:1203.2373 [nucl-ex]
22. Mecking, B.A., et al.: (CLAS Collaboration): Nucl. Inst. Meth. A **503**, 513 (2003)
23. CLAS12 Technical Design Report, version 5.1, 208 (2008). https://www.jlab.org/Hall-B/clas12_tdr
24. Hall-A proposal: Measurements of the electron-helicity dependent cross sections of the DVCS with CEBAF at 12 GeV. E12-06-114
25. Hall-C proposal: Exclusive deeply virtual compton and neutral pion cross-section measurements in Hall C. PR12-13-010
26. Jo, H.S., et al. (CLAS Collaboration): Cross sections for the exclusive photon electroproduction on the proton and generalized Parton distributions. Phys. Rev. Lett. **115** (2015) 21, 212003. arXiv:1504.02009 [hep-ex]
27. Pisano, S., et al. (CLAS Collaboration): Single and double spin asymmetries for deeply virtual Compton scattering measured with CLAS and a longitudinally polarized proton target. Phys. Rev. D **91**, 052014 (2015). arXiv:1501.07052 [hep-ex]
28. Seder, E., et al. (CLAS Collaboration): Longitudinal targetspinasymmetries for deeply virtual Compton scattering. Phys. Rev. Lett. **114**, 032001 (2015). arXiv:1410.6615 [hep-ex]
29. Avakian, H., et al.: Deeply virtual compton scattering at 11 GeV with transversely polarized target using the CLAS12 Detector. 12-12-010
30. Mazouz, M., et al.: (Hall-A Collaboration): Deeply virtual compton scattering off the neutron. Phys. Rev. Lett. **99**, 242501 (2007)
31. Niccolai, S., et al.: (CLAS12 proposal): Deeply virtual compton scattering on the neutron with CLAS12 at 11 GeV. 12-11-003
32. Niccolai, S., Biselli, A., Keith, C., Pisano, S., Sokhan, D., et al. (CLAS12 proposal): Deeply virtual compton scattering on the neutron with a longitudinally polarized deuteron target. JLab experiment C12-15-004 (conditionally approved)
33. Courtoy, A., Goldstein, G.R., Hernandez, J.O.G., Liuti, S., Rajan, A.: On the observability of the quark orbital angular momentum distribution. Phys. Lett. B **731**, 141 (2014). doi:10.1016/j.physletb.2014.02.017. arXiv:1310.5157 [hep-ph]
34. Rajan, A., Courtoy, A., Engelhardt, M., Liuti, S.: Parton transverse momentum and orbital angular momentum distributions. (2015). arXiv:1601.06117 [hep-ph]
35. Chen, S., et al.: (CLAS Collaboration): Measurement of deeply virtual compton scattering with a polarized-proton target. Phys. Rev. Lett. **97**, 072002 (2006)
36. Biselli, A., et al.: (CLAS12 proposal): Deeply-virtual compton scattering with CLAS12 at 11 GeV. 12-06-119

MIX
Papier aus verantwortungsvollen Quellen
Paper from responsible sources
FSC® C105338

If you have any concerns about our products,
you can contact us on
ProductSafety@springernature.com

In case Publisher is established outside the EU,
the EU authorized representative is:
Springer Nature Customer Service Center GmbH
Europaplatz 3, 69115 Heidelberg, Germany

Printed by Libri Plureos GmbH
in Hamburg, Germany